Infinite Science
Publishing

Student Conference Proceedings 2024

13th Student Conference on Medical Engineering Science
9th Student Conference on Medical Informatics
7th Student Conference on Biomedical Engineering
6th Student Conference on Auditory Technology
4rd Student Conference on Biophysics
4rd Student Conference on Robotics and Autonomous Systems
2nd Student Conference on Medical Microtechnology
1st Student Conference on Psychology - Cognitive Systems

Lübeck, March 6-8, 2024

Editors in Chief

T. M. Buzug, H. Handels, C. Hübner, A. Mertins, S. Müller, P. Rostalski, N. Bunzeck

Associate Editors

K. Gräfe, J.-H. Wrage, K. Ehlers, R. Pallenberg, Y.-H. Song, S. Venker, M. Haller

Editors

H. Abbas, A. Aboud, M. Ahlborg, E. Barth, M. Berekoic, R. Birngruber, m. Brandenburger, R. Brinkmann, N. Bunzeck, T. M. Buzug, C. Damiani, J. Ehrhardt, F. Ernst, S. Fudickar, J. Ghofrani, M. Gräser, M. Grzegorzek, H. Handels, H. Hamann, M. Heinrich, M. Heldmann, K. Hiltawsky, D. Hollfelder, R. Huber, C. Hübner, H. Husstedt, G. Hüttmann, J. Ingenerf, T. Jürgens, T. Käster, M. Koch, A. Korff, U. Krämer, F.-L. Lau, M. Leucker, N. Linz, K. Lüdtke-Buzug, A. Madany Mamlouk, E. Maehle, L. Marshall, A. Mertins, S. Müller, C. Nehls, N. T. Nguyen, S. Otte, H. Paulsen, C. Peifer, U. Pilz, M. Rafecas, M. Rahlves, R. Rahmanzadeh, P. Rostalski, G. Schildbach, A. Schweikard, Y.-H. Song, F. Spitzenberger, M. Stille, M. Urban, R. Wendlandt, I. Wohlers

Student Conference Proceedings 2024

Preface and Acknowledgements

After the great success of the previous meetings from 2012 to 2023, the Student Conference 2024 shows continuing growth in quality of scientific contributions. In this year, the 13th Student Conference on Medical Engineering Science is held together with the 9th Student Conference on Medical Informatics, the 7th Student Conference on Biomedical Engineering, the 6th Student Conference on Auditory Technology, the 4rd Student Conference on Biophysics, the 4rd Student Conference on Robotics and Autonomous Systems, the 2nd Student Conference on Medical Microtechnology and the 1st Student Conference on Psychology - Cognitive Systems.

The organization team has worked to provide an excellent conference, where master students of the campus present their recent research results to a broad public of academics.

The contributions show how new approaches and methods in medical engineering and medical informatics can advance medicine, health, and health care. Students from the Life Sciences programs present their results from projects carried out at the laboratories, clinics, and institutes of Lübeck's Universities, in international research facilities, or research-oriented industrial companies. The conference focus has been placed on topics ranging from medical engineering to medical informatics. The interdisciplinary field of medical engineering has been established at the Lübeck University of Applied Sciences for decades, and Medical Engineering Science (Medizinische Ingenieurwissenschaft – MIW) is an important bachelor and master program at the Universität zu Lübeck as well. Both universities jointly offer the international master degree programs Biomedical Engineering (BME), Medical microtechnology and Auditory Technology (Hörakustik und Audiologische Technik – HAT). Furthermore, in the master program Medical Informatics (Medizinische Informatik – MI) the Student Conference on Medical Informatics is integrated as an important element where project results in the emerging field of digital medicine are presented by the students. These subject areas are widened by the master's programs Robotics and Autonomous Systems (Robotik und Autonome Systeme – RAS) and Biophysics (Biophysik – BP). In 2024, master students of Psychology - Cognitive Systems, a degree program offered by the Universität zu Lübeck will participate in the conference for the first time. and complement the conference with content from scientific psychological research and its links with programming, data analysis and computer simulation. The conference is thus complemented by content from scientific-psychological research and its links to programming, data analysis and computer simulation.

As Conference Chairs, we want to thank everybody who worked with enthusiasm and dedication to make the conference a successful event. We want to thank Infinite Science for producing these proceedings. Personally, and on behalf of all colleagues of the Student Conference Committee, we especially want to thank Ksenija Gräfe, Jan-Hinrich Wrage, Silke Venker, René Pallenberg, Kristian Ehlers, Young-Hwa Song and Michaela Haller. Their in-depth overview of all details of this event is the key to the success of the Student Conference 2024.

.

Lübeck, March 6-8, 2024

Prof. Dr. Thorsten M. Buzug
Chair of the 12th Student Conference on Medical Engineering Science

Prof. Dr. Heinz Handels
Chair of the 8th Student Conference on Medical Informatics

Prof. Dr.-Ing. Stefan Müller
Chair of the 6th Student Conference on Biomedical Engineering

Prof. Dr.-Ing. Alfred Mertins
Chair of the 5th Student Conference on Auditory Technology

Prof. Dr. Christian Hübner
Chair of the 3rd Student Conference on Biophysics

Prof. Dr.-Ing. Philipp Rostalski
Chair of the 3rd Student Conference on Robotics and Autonomous Systems

Prof. Dr. Nico Bunzeck,
Chair of the 1st Student Conference on Psychology - Cognitive Systems

Contents

Medical Electronics

Biomedical Engineering

Robotics

Sensor Data Analysis

Control technology

Safety and Quality

Machine Learning

Biochemical Physics

E-Health

Artificial Intelligence

Psychology

Medical imaging

Biomedical optics

Auditory technology

Image Processing

1

Medical Electronics

Analyzing the Impact of Thermal Cycles on the Performance of Solid-State SMD Batteries (TDK)

Abdulkader Chukr [1], Stefan Müller [2], and Sebastian Jungbauer [3]

[1] Biomedical Engineering, TH Lübeck University, AbdulkaderChukr@gmail.com
[2] Biomedical Engineering Department, TH Lübeck University, Stefan.Mueller@th-luebeck.de
[3] R&D Electronics, Olympus Surgical Technologies Europe, Sebastian.Jungbauer@olympus.com

Abstract

This study delves into the performance of TDK CeraCharge Surface-Mount Device (SMD) batteries following exposure to Thermal Cycle (TC) conditions, specifically concentrating on their Charging/Discharging (C/D) capabilities. The TC protocol incorporated temperature fluctuations ranging between 10°C and 140°C in each cycle mimicking a standard autoclave cycle, with the experimentation employing a Constant Current batteries testing setup. Initially, Baseline C/D tests were conducted on six battery cells at room temperature. Subsequent assessments involved additional C/D procedures after 200 TCs, followed by a concluding set of tests after an additional 200 TCs, resulting in a cumulative total of 400 cycles for each cell. The data collected at each stage underwent meticulous analysis utilizing Python scripts. The findings offer valuable insights into the degradation of capacity stemming from thermal cycling, thereby providing a nuanced comprehension of the performance of CeraCharge cells across diverse environmental conditions.

1 Introduction

The burgeoning demand for reliable energy storage systems necessitates a comprehensive understanding of battery behavior under diverse conditions. This study focuses on TDK CeraCharge solid-state SMD batteries, investigating the impact of TCs on C/D capacities. Addressing the need for robust and temperature-resistant energy storage solutions that can withstand Autoclave Cycles, this research employs a USB Potentiostate/Galvanostate setup[1] that utilizes a constant current method to measure the C/D capacities of the tested batteries. The exploration involves subjecting cells to thermal cycling (between 10°C and 140°C)(Fig.3) and analyzing resultant performance shifts. By elucidating the relationship between thermal stress and battery performance, our work contributes to advancing the field of solid-state SMD battery technology.

2 Material and Methods

2.1 Preparation of TDK CeraCharge Cells

In the preparation of CeraCharge cells [2], which are SMD components characterized by compact dimensions, a mounting mechanism was imperative for seamless integration into our testing apparatus. A rectangular PCB[1] copper plate measuring (3x2) centimeters served as the foundation, featuring a 2mm milled gap on one side to facilitate the segregation of cell leads. Employing the solder re-flow method[2], the SMD battery was soldered onto the milled section, and subsequent verification of proper connections was conducted using a Multimeter. To safeguard the cells from external humidity, a 0.5mm layer of Silicon/Risen compound was applied, adhering strictly to manufacturer specifications [2] (Fig.1). This meticulous process was replicated for all six cells in the study.

Figure 1: CeraCharge cell, soldered Cell on the mounting plate (left) and Test-Coupon (right).

2.2 Hardware Testing Setup

The testing apparatus featured a Galvanostate/Potentiostate circuit[1] governed by a PIC16F1459 microprocessor. This circuit utilized four-leads configuration that interfaces with the cell under examination, which not only provides Voltage and Current measurement points (as close as possible to the cell) but also facilitate diverse battery testing modes. Controlled by the microprocessor, a mode switch facilitated the selection of desired mode, including Constant-Current C/D, VC[2], Test Rating, and OCV[3] testing[1]. Long, high-temperature-resistant wires were soldered to spring-loaded copper pins, which were in turn affixed to two parallel plates made of PEEK[4] connected with bolts and nuts. Two electrodes were secured on each acrylic plate, creating a configuration where the tested cell was positioned between the electrodes (Fig.1). This setup ensured secure cell placement and proper electrical connection during testing. To power up the circuit and establish communication, a USB connection was established between the circuit and a Raspberry Pi, streamlining the testing process.

2.3 Software Environment

The Galvanostate/Potentiostate circuit interfaced with a Raspberry Pi 4B. Equipped with the latest Raspbian Operating System (OS)(Kernel version 6.1), this choice was logical as Python seamlessly integrates with Linux OS. Python version 3.10 was installed, alongside essential libraries such as *NumPy*[3] and *pandas*[4]. The Raspberry Pi configuration included enabling VNC[5] and SSH[6], facilitating remote communication, test control, and data transfer over the shared WiFi network.

The Galvanostate/Potentiostate script "*tdstatv3.py*"[1](Fig. 2) was downloaded onto the Raspberry Pi and rigorously tested. Upon the completion of each test, generated two txt files. One contained metadata, voltage and current curve values against timestamps, while the other documented the number of test cycles and corresponding C/D capacities. To enhance efficiency, additional Python scripts were developed:

"*Name_Generator.py*"A script to generate a naming scheme for each test, facilitating documentation by integrating "cell make", "module number", "cell number", "test number", "test temperature", and "remarks". This scheme, including time and date stamps, ensured traceability.

"*File_Grouping.py*" A file grouping script that ensures the systematic organization of test data, preventing confusion, especially in cases of multiple tests stored in a single directory.

"*VI-Plotter.py*". A visualization script with a Graphical User Interface (GUI) interface for importing test data (the two txt files generated per test) and plotting voltage-current curves and C/D capacities as a Bar-chart.

This feature allows users to quickly analyze test results, identify outliers, assess C/D cycles, and verify the integrity of the test. These Python scripts were designed for the use on the Raspberry Pi. Post-test completion, the VNC file transfer feature facilitated the wireless transfer of test files to a connected laptop or PC. Subsequently, all test results were compiled into a single file, serving as a quasi-database. Finally, Data Processing was performed using Python scripts to extract information for data analysis and plotting of combined test results.

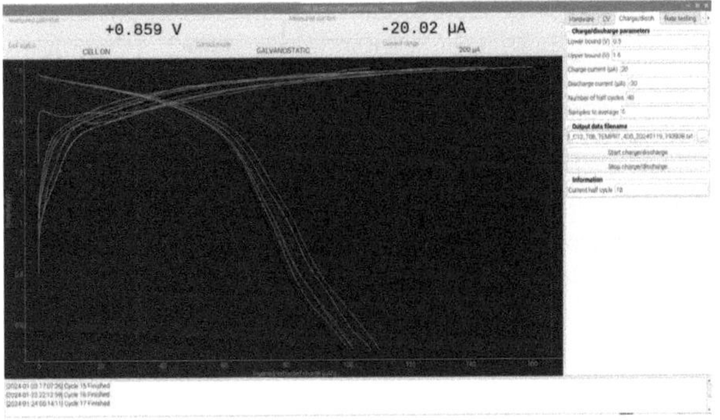

Figure 2: Python script used for C/D testing "*tdstatv3.py*".

2.4 Test Methodology

An initial set of 20 C/D testing cycles was conducted on all six cells at Room Temperature (RT) to establish a *Baseline (Stage 1)*. These cycles captured standard C/D capacities, along with typical voltage curves.

Following this baseline phase, the cells were enclosed within a plastic container made out off PEEK and exposed to 200 TC in a thermal chamber (10°C to 140°C)(Fig.3).

Subsequent to this thermal stress, at RT, the cells underwent an additional 20 C/D testing cycles *(Stage 2)* to observe any impact on voltage curves and capacity Values.

After the second set of testing cycles, the cells were once again placed in the same container and exposed to another 200 TCs, reaching a cumulative total of 400 TC.

Following this extended cycling, at RT, the cells underwent a final set of 20 C/D testing cycles *(Stage 3)* for evaluation. A comparison of the baseline data with results obtained after 200 TC and 400 TC revealed discernible differences in curve shapes and C/D capacities, providing insights into the impact of thermal cycling on the cells' performance.

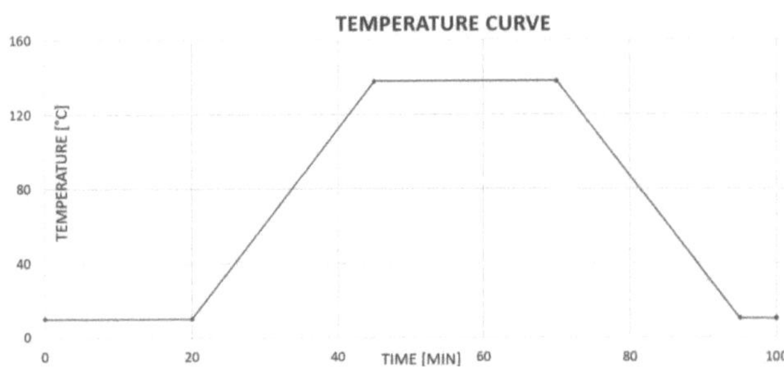

Figure 3: Thermal Chamber's thermal cycle mimicking a standard Autoclave cycle, (100 mins, 10°C - 140°C).

2.5 Data Processing

The data acquisition process encompassed the collection of relevant information subsequent to each testing iteration for every cell, which was systematically stored in files conforming to the defined naming scheme. Each test yielded two distinct txt files, documenting the C/D voltage curves, as well as the corresponding capacity values for every cycle. Subsequently, a Python script was employed to extract the amassed data into CSV file format, thereby facilitating streamlined data processing and visualization. A second Python script was implemented to join the data of each cell into a single CSV file, aligning the entries based on their respective *timestamps* and adhering to the devised file naming scheme ("cell number" and "test number").

A third Python script was utilized to generate visual representations of the combined test data for each cell, focusing on the C/D curves(Fig.5)(Fig.6). This visual inspection played a pivotal role in the identification and subsequent exclusion of outliers and failed cycles. Only cycles exhibiting healthy curves underwent further consideration(Fig.4), emphasizing the reliability of the dataset. Following the meticulous curation of viable data, the scripting process transitioned to statistical analysis. The average values for each cell at every stage were calculated and meticulously organized into tables for enhanced legibility, utilizing an Excel spreadsheet. These average values , derived from every cell at each stage, were subsequently employed to compute the overall *average*(Mean) and *standard deviation* for each stage. A bar chart, depicting the overall average values at each stage along with the corresponding standard deviation as an Error Bar (Fig.7), was instrumental in discerning the extent to which thermal cycling affected the C/D capacities of the TDK CeraCharge SMD battery cells.

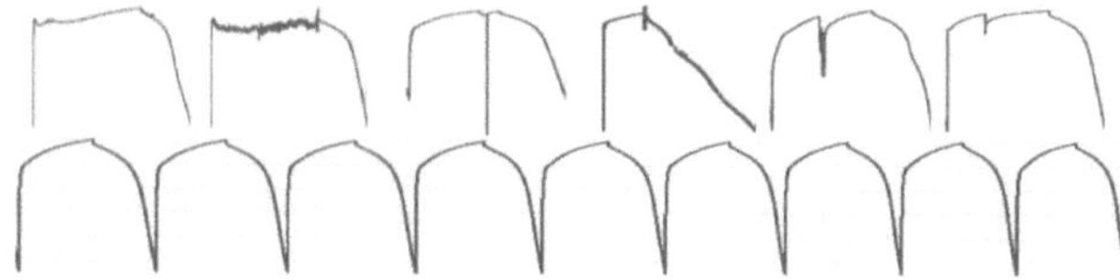

Figure 4: Healthy C/D Voltage curves (bottom) and curves which are considered as Outliers (top)

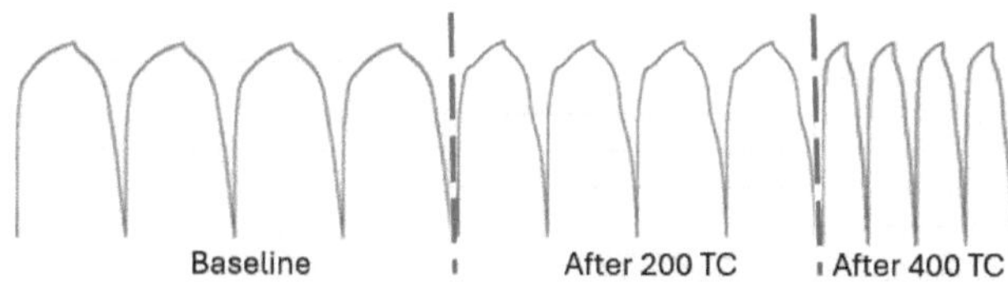

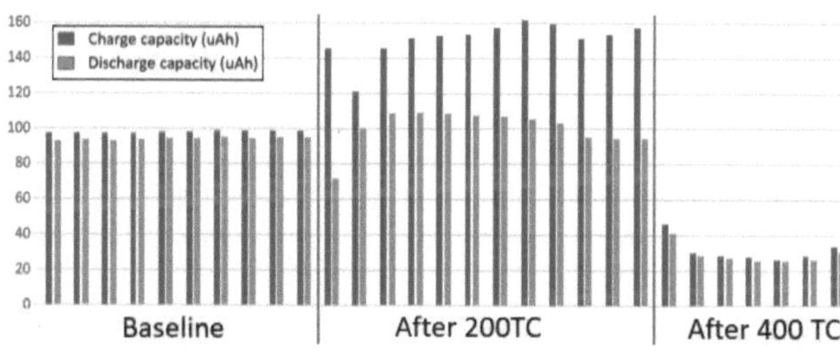

Figure 5: C/D Voltage curves at Baseline (left), after 200 TCs (middle) and after 400 TCs (right).

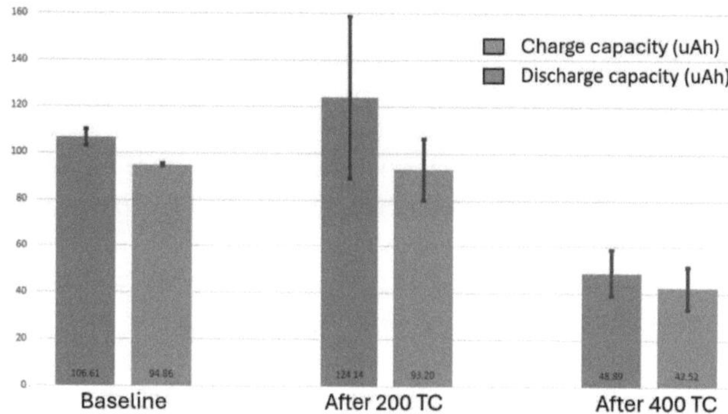

Figure 6: C/D capacity values (in μAh) at Baseline (left), after 200 TCs (middle) and after 400 TCs (right).

3   Results and Discussion

The investigation involved the examination of combined average values of C/D capacity values at three distinct stages: Baseline, *Stage 2* (after 200 TCs), and *Stage 3* (after 400 TCs) for TDK CeraCharge cells. A bar chart representation was employed to visualize the impact of thermal cycling on C/D capacities, with error bars representing standard deviations(Fig.7). The baseline measurements established a foundation for comparison, revealing an average charging capacity of 106.61μAh with a standard deviation of 3.38, and an average discharging capacity of 94.86μAh with a standard deviation of 0.66. which aligns perfectly with the expected values stated in the cells' data-sheet[2].

In terms of charging capacity, after 200 TCs, a significant increase of 16.22% from the baseline was observed, resulting in an average charging capacity of 124.14μAh with a standard deviation of 34.76. Conversely, after 400 TCs, a substantial decrease of 54.16% from the baseline was noted, with an average charging capacity of 48.89μAh and a standard deviation of 9.86.

Regarding discharging capacity, after 200 TCs, a marginal decrease of 1.81% from the baseline was identified, with an average discharging capacity of 93.20μAh and a standard deviation of 13.17. After 400 TCs, a notable decrease of 55.11% from the baseline was observed, resulting in an average discharging capacity of 42.52μAh with a standard deviation of 9.17.

Figure 7: Overall Averaged C/D throughout the different testing stages, with error bars representing the Standard deviation.

Comparing the Baseline with the results after thermal cycling raises intriguing considerations for both battery management and future utilization scenarios. Notably, the observation that a significantly higher amount of charging is required compared to discharging capacities suggests a potential imbalance in the charging and discharging capabilities of the batteries post-thermal cycling. This asymmetry could have implications for battery management circuits, as it may necessitate a more intricate approach to maintain optimal performance. One plausible avenue for further investigation is the analysis of internal resistances through frequency analysis[5]. Exploring the frequency-dependent behavior of the batteries could unveil insights into changes in internal resistances, aiding in the development of more nuanced battery management systems.

The substantial increase in standard deviation post-thermal cycling is a noteworthy finding. This heightened variability in the C/D capacities demands careful consideration, especially in the context of future developments in battery management electronics. The increased standard deviation signifies a greater degree of variability among individual cells, highlighting potential inconsistencies in their performance. Another significant observation is the noteworthy increase in the C/D cycle duration after doing the thermal cycling. The extended C/D duration, coupled with a lower C/D current (10μA compared to the baseline value of 40μA), prompts a crucial question regarding the underlying factors influencing these changes.

3.1 Names, Abbreviations and Acronyms

PCB[1](Printed Circuit Board). VC[2] (Voltage Cyclometry)
[1]. OCV[3] (Open Circuit Voltage)[1].
PEEK[4] (PolyetherEtherKetone);"it is a high-performance,
temperature-resistant semicrystalline polymer which is considered as one of the most excellent thermoplastic" [6].
VNC[5] (Virtual Network Computing) and SSH[6] (Home Subscriber Server) are networking tool that is used to access
systems in the network remotely and facilitate file transfer.

4 Conclusion

This study delved into the impact of thermal cycling on the
performance of TDK CeraCharge SMD batteries, employing a comprehensive experimental approach. The batteries
underwent 400 TCs, ranging from 10°C to 140°C, with testing facilitated by a Constant Current batteries setup, providing insights into the C/D capabilities post thermal cycling.
The outcomes underscore the sensitivity of CeraCharge
cells to thermal stress, revealing dynamic shifts in performance metrics. For instance, charging capacity witnessed
a substantial initial increase of 16.22% after the first 200
TCs, only to undergo a subsequent significant decrease of
54.16% after the cumulative 400 TCs. Similarly, discharging capacity experienced a marginal reduction of 1.81% after 200 TCs, followed by a notable decline of 55.11% after
the entire 400 TCs. These findings illuminate the intricate
interplay between thermal cycling and battery performance,
emphasizing the need for further exploration into the underlying mechanisms governing these shifts.
Looking ahead, avenues for extended inquiry include
an examination of internal resistances through frequency
analysis[5]. This approach holds promise in uncovering insights into alterations in internal resistances, thereby contributing to the refinement of sophisticated battery management systems. Additionally, a microscopic cross-sectional
examination of the batteries before and after thermal cycling is proposed to gain a deeper understanding of changes
in capacities at a microscopic level.
To enhance the robustness and statistical accuracy of future studies, employing a larger test sample from different batch numbers is recommended. This broader dataset
would provide a clearer understanding of the overall performance trends. Furthermore, considering the time-intensive
nature of the tests, exploring ways to optimize the testing
process, could expedite large-scale studies without compromising accuracy.

Acknowledgement

The work has been carried out at Olympus Surgical Technologies Europe and supervised by the Department of Applied Sciences at TH Lübeck University.

Authors' Statement

Conflict of interest: Authors state no conflict of interest.

5 References

[1] T. Dobbelaere, P. M. Vereecken, and C. Detavernier,
"A USB-controlled potentiostat/galvanostat for thin-film battery characterization," HardwareX, vol. 2, pp.
34–49, 2017. doi:10.1016/j.ohx.2017.08.001

[2] "*TDK CearaCharge Data-Sheet and Application-note*", TDK Electronics AG, https://www.tdk-electronics.tdk.com/en/ceracharge (accessed Jan.
16, 2024).

[3] *NumPy*, https://numpy.org/ (accessed Feb. 4, 2024).

[4] *Pandas*, pandas, https://pandas.pydata.org/ (accessed
Feb. 4, 2024).

[5] R. Di Fonso, P. Bharadwaj, R. Teodorescu and C. Cecati, "*Internal Resistance Estimation of Li-ion Batteries
using Wavelet Analysis*", 2022 IEEE 13th International
Symposium on Power Electronics for Distributed Generation Systems (PEDG), Kiel, Germany, 2022, pp. 1-5,
doi: 10.1109/PEDG54999.2022.9923094.

[6] C. Yang et al., "Influence of thermal processing conditions in 3D printing on the crystallinity and mechanical properties of Peek Material," Journal of Materials
Processing Technology, vol. 248, pp. 1–7, Oct. 2017.
doi:10.1016/j.jmatprotec.2017.04.027

Orthopedic sockets and methods for adjustments
– An approach with shape-memory-alloy actuators –

Benjamin Förster [1], Benjamin Finke [2], Jennifer Vibert [3], and Stefan Müller [4]

[1,3] Biomedical Engineering, Luebeck University of Applied Sciences, {benjamin.foerster,jennifer.vibert}@stud.th-luebeck.de

[2] Global Research Department, Duderstadt Ottobock, benjamin.finke@ottobock.de

[4] Department of Applied Natural Sciences, Medical Sensors and Devices Lab, Luebeck University of Applied Sciences, stefan.mueller@th-luebeck.de

Abstract

Novel shape-memory-alloy actuators (SMAA) represent a serious alternative to commonly used drive technologies like stepper or linear motors. To adapt the fit of a prosthetic socket to different situational conditions and user behaviors, an implementation of SMAA into a locking unit can adjust the rigidity of the socket brim. This is necessary to ensure a perfect fit and force distribution on the residual limb during different gait situations. Since an even socket surface and a low noise generation is essential for good usability, conventional actuators are quickly reaching their limits. SMAA could provide a remedy, but their high energy consumption requires a bistable locking unit. Therefore, electrical considerations are carried out to use the actuators efficiently. As a result, a current-flow pattern in combination with a bistable mechanism was conceptualized. Depending on the patients' daily walking behavior, a temperature controlled system will be required to deflect the actuators safely.

1   Introduction

Prostheses, especially sockets are highly individual components that are customized for every patient. During and after the rehabilitation, the residual limb undergoes major changes due to a decrease of muscle mass as well as vascular and osseous changes [1]. To regain the ability to walk requires not only a resilient limb without any surface pain, but also a precisely fitted socket for use during varying walking situations [1]. Nevertheless, a change of the tissue volume due to weight loss, a maladaptation of the socket or bad stability due to perspiration can cause severe injuries of tissue structure as well as pressure marks. By conceptualising an auto-adaptive socket, Ottobock carries out research to vary first the available socket volume and second tighten specific areas to change the load locus. Fig. 1 shows the expected load distribution during a variety of walking situations.

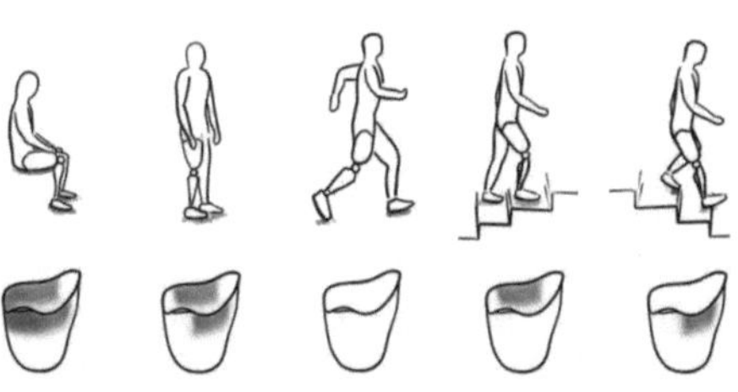

Figure 1: Different walking situations that lead to a varying load distribution along the socket brim

The adaptive socket will ensure a better wearing comfort and improve the quality of rehabilitation and quality of life. The research goal is to develop a locking mechanism demonstrator that is based on SMAA, which can potentially be used to adjust the brim of the socket depending on the patient's needs with a transfemoral amputation. This requires a consideration of a sensible mechanical layout as well as a suitable control circuit to use the actuators safely. By opening and closing the mechanism, two bistable states should be selectable to keep the energy consumption as low as possible. The actual adjustment relies on the rigidity of stiffening elements which are either additional components or embedded into the socket wall [2]. Due to their low overall height, complexity and silent actuation, SMAA are expected to be a convincing alternative to stepper or linear motors. Particularly in the area of the socket, a low material application and noise generation is essential for increasing the acceptability of prostheses by patients. The purpose of the article is to provide a brief introduction to nitinol transformation, the operating principle of the actuators and the electromechanical setup that was used to first use the actuators safely and second make them applicable as a locking mechanism on the socket wall.

2   Material and Methods

Thermal shape memory alloys (SMA) are a specific type of material combination made from nickel and titanium known as nitinol. SMAs have the ability to recover from plastic deformation caused by a heating input. Both an external heat source and an electrical current (Joule heating) would lead to a deformation of the wire due to its phase transition from martensite to austenite, which returns them to their original shape [3]. The used actuators consist of wires that are connected at both ends. If a certain heating current is applied the contraction between the stationary points will bend the actuator, creating a tensile force. The counter force of a spring steel, on which the actuator is mounted to, will restore the initial position after the heating input is turned off [4]. The phase transition temperature relies on the alloy ratio of the two materials and can be freely varied. In this way a reversible actuation is possible, which utilizes the movement of the nitinol wires for various applications.

2.1 SMA-Actuator

The utilized Curve® actuators from CompActive GmbH, can be seen as bending elements that can create a deflection force of 3-20N at a transition temperature between 60-70°C and will reach their maximum deflection at 100-110°C. In accordance to the inner resistance of the serial wires, the heating current and time can be set [5]. Both values are coherent and decisive for the subsequent strength of the supply voltage and its duration. Fig. 2 shows the coherence between temperature, deflection and heating current. It shows that the heating current and deflection are proportional.

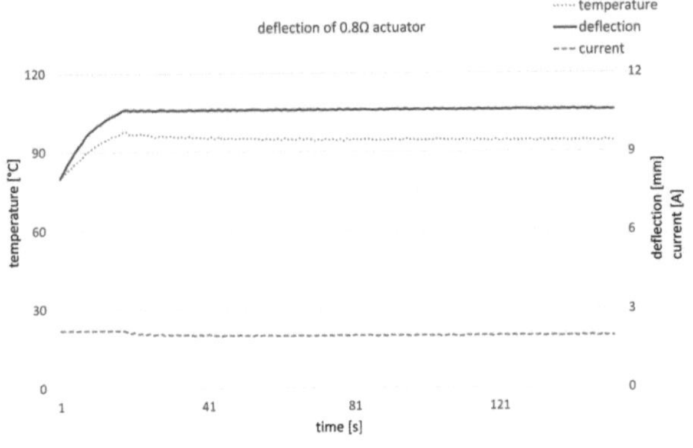

Figure 2: Relation between temperature, heating current and deflection of a 0.2 Ω actuator [5]

When the maximum possible activation time is reached, either the heating current can be turned off or lowered to 1.7 - 2.2 A, which hold the actuator in a deflected position [5]. This is based on the fact that the heating input balances out with the confection to the environment ($\cong$ 23°C). However, the so-called 'holding current' is strongly affected by the ambient conditions [5]. Table 1 shows the fundamental values to energize the actuators. The activation time should not be exceeded under any circumstances due to overheating of the nitinol wires.

Since the electrical heating input is converted into thermal energy and released to the environment, a slow heating process leads to higher energy consumption. A fast activation

Table 1: Coherence between activation time and heating current [5]

activation time	heating current
∞	1.7 - 2.2A
15s	3A
4s	4A
1.5s	5A
0.5s	7A
0.35s	8A

will reduce the heat loss during the heating interval. Another drawback is the slow cooling rate of 10 - 15s back to the initial state [5]. A sequential activation without a sufficient cooling time or lowered heating current will lead to an accumulation of the remaining temperature in the actuator and inevitably lead to overheating and fatal damage of the actuator. The frequency of activation strongly depends on the design of the mechanism as well as the number of changes to the walking situation of the patient. For visualization Fig. 3 shows a illustration of an SMAA in a flat and deflected state.

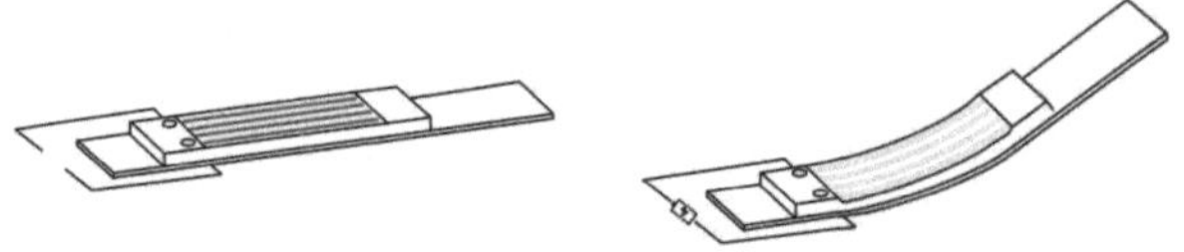

Figure 3: Left: flat actuator (no current); Right: deflected actuator due to heating current

2.2 Electrical Consideration

To control the actuators in a proper way, a control circuit was built. As a main component, an Arduino Nano was used to set the energizing time and to control the heating current with pulse width modulation (PWM). Fig. 4 shows the schematic of the used circuit.

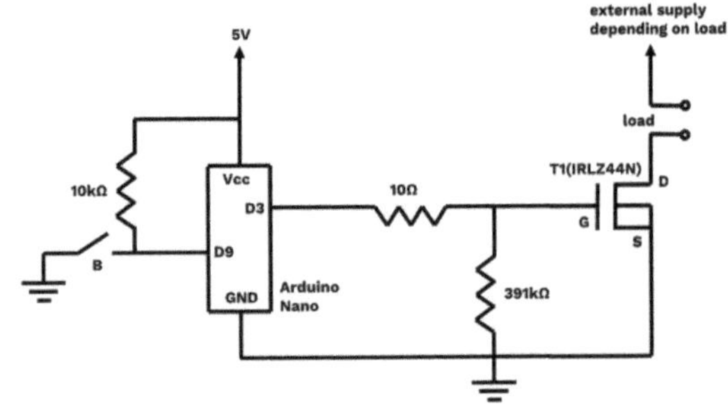

Figure 4: Circuit schematic

The Arduino is supplied with 5V at 0.1A, while the actuator has an external supply to individually adjust the input voltage for different actuator sizes or activation times. The corresponding code allows to set not only an activation time, but also a delay until a second activation is possible to prevent overheating. A fast-switching N-Channel

MOSFET transistor (IRLZ44N) by Infineon Technologies AG was used in the circuit. This transistor was selected because it can withhold a drain current up to 47 A, which is high enough to consider a generous safety factor. The PWM can be used to lower the heating current artificially.

To keep the energy consumption as low as possible, a current-flow pattern for a bistable system is set up. This requires a relatively high initiation current to lower the ascertained heating loss. In conjunction with a mechanism, the current will be high enough to move a locking unit from the first stable position into the second one. Followed by a second activation that leads to an unlocking from the second position back to the initial position. Fig. 5 shows the schematic of the flow pattern.

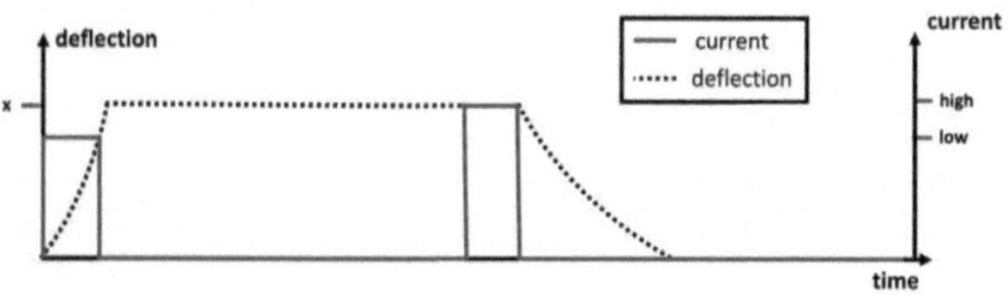

Figure 5: Current flow pattern that corresponds to the mechanical properties of a bistable system

It can be seen that the current pulses have the same time duration, but not the same height. This corresponds to the mechanism that was constructed.

2.3　Mechanical Consideration

Two main mechanisms were developed based on the electronic consideration in conjunction with the requirements of a closure system applicable on a socket brim. Both of them are based on the same locking concept, but using different techniques with regard to how they change the rigidity of the socket wall. Fig. 6 shows the technical drawing of the first concept, which consists of an actuator with an inner resistance of 0.8 Ω, neodymium magnets, a guiding shaft and a small lever.

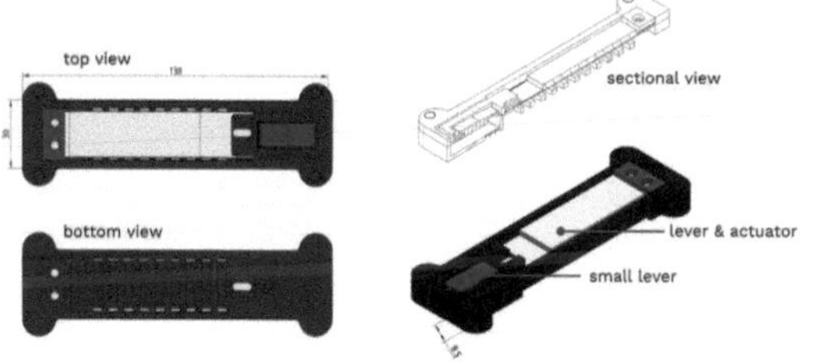

Figure 6: Mechanism that uses an actuator with a inner resistance of 0.8 Ω and a bistable lever mechanism

The actuator is mounted on a thin (0.5 mm) 3D-printed structure with a flexure bearing, that allows a unhindered deflection. During the actuation the small lever, which is connected to a small pathway, will move in an upright position. By turning off the heating current the deflected actuator will rest on the short lever and be held in this posi-

tion. To get a closer impression on how the bistable locking-mechanism works, Fig. 7 shows a detailed view of the section.

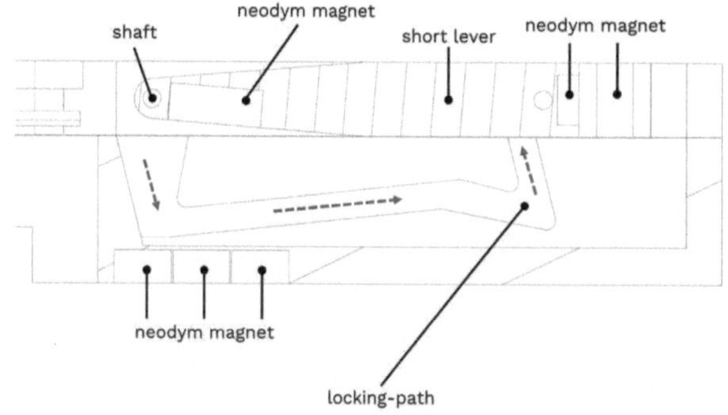

Figure 7: Path mechanism that uses a short lever to keep the actuator in a deflected state without a heating current

The path shape allows the short lever to be pulled up during the deflection of the actuator until it gets stuck at the upper apex. After the heating current is turned off, the shaft falls to the bottom right corner of the path, which leaves the actuator in this position. A second, stronger heating current will pull the shaft out of the path while one pair of magnets causes the short lever to move back into its original position. The mechanism is unlocked and goes back into its initial position. A second pair of neodymium magnets pull the short lever back and another activation cycle can follow. Since the movement of the used SMAA is very limited, the bistable mechanism needs to be transferred into a suitable locking unit, shown in Fig. 8. Therefore, a second mechanism was developed which is based on the path-locking unit.

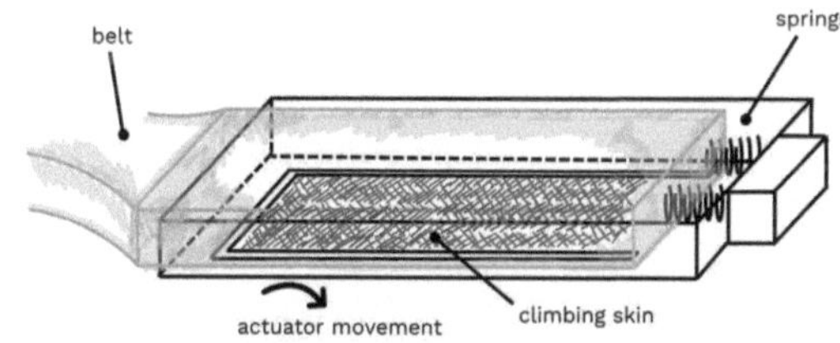

Figure 8: Climbing skin mechanism that is based on the former bistable lever mechanism

The concept uses a climbing skin that is sewed on top of a belt, which will be also implemented on top of the big lever of the bistable mechanism. Climbing skin is a material that only allows a movement in one direction if the two layers are placed on top of each other. The small hairs on the surface jam the belt in one pulling direction, while allowing free movement to the other direction. If the actuator is deflected the belt can move freely in both directions. A retraction of the belt will be realized by a spring element to support the flexible element on the socket. Due to the iterative development of the locking unit, a commercial Markforged Mark Two printer has been used to manufacture the prototypes. Fig. 9 shows an image of the 3D-printed bistable lever mechanism. On the left side the first state is shown in which the mechanism is closed in a flat position. The right image shows the opened, deflected position. Both states can be held without applying a constant heating current.

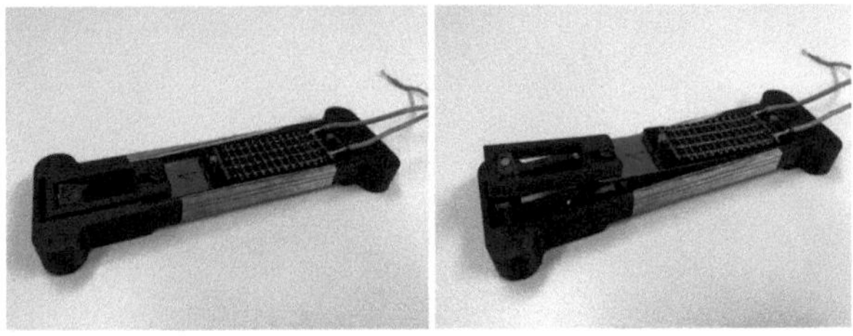

Figure 9: 3D-printed lever mechanism. left: stable state (flat); right: stable state (deflected)

3 Results and Discussion

To implement an SMAA actuator into a fully functional mechanism that can be used to change the rigidity of the socket brim requires further development. The conceptualized demonstrator is merely an approach to show that SMAA in context of a locking unit can be utilized. With regard to a fully functional system a battery supply is indispensable. This would require further investigations on how often a mechanism needs to open and close during everyday situations. In conjunction with the current flow pattern, this would give a direct indication about the required battery capacity. If the user's behaviour requires high switching rates between the two bistable states, the heating current should likely be controlled via a temperature sensor. For this purpose, the actuator could be used as a self-sensing element to adjust the heating current in respect to the remaining heating input of the actuator. This will prevent the nitinol wires from uncontrollable overheating during a high number of actuation intervals. However, a fully deflected actuator requires a cooling interval of 10-15s, which makes a rapidly successive opening and closing untenable. By lowering the deflection that is need to activate the mechanism, the cooling rate can be decreased due to the reduced heating input. Another conceivable solution would be an application for less frequent situational changes like the change between sitting and standing. Since the actuators are installed in a small space, additional cooling seems ineligible. A major disadvantage of the mechanical concept is the low resilience and actuation force. By applying a direct force, either vertical or horizontal, the deflection is not strong enough to unlock the mechanism, especially in case of a pre-loaded socket brim. Due to this, only partial loads should act on the mechanism. In addition, manufacturing limitations that come with 3D printed parts restrict the accuracy of smaller structures. Particularly the pathway of the locking unit is hard to produce in order to create a reliable mechanism. A subsequent step should involve to manufacture several parts from aluminum to increase the precision. This will increase the durability and improve operation characteristics. Regarding the locking unit, further development and testing is required. Especially when it comes to different gait situations, the constantly changing load conditions and alignment of the mechanism on the socket wall can influence the functionality of the concept.

4 Conclusion

The presented design proves that a shape memory alloy actuator could be used as mechatronic element in a locking unit. Therefore, it represents a sensible alternative to conventional actuators and could be suitable for different applications in the field of prosthetics. If it is manageable to use the nitinol wires as self-sensing temperature sensors in combination with a reliable bistable locking mechanism, it can be expected that they could be implemented into socket systems. This would provide patients an individual wear comfort that is situationally adjustable, which would increase the rehabilitation success and acceptance of a prosthetic device.

Acknowledgement

The work has been carried out at the Ottobock headquaters, Global Research Department. We are acknowledging the support of the Electronics and Global Research Department that provided insights, expertise and greatly assisted the research.
We would also like to thank Dr. Moritz Hübler and Nils Neblung from CompActive GmbH for their expertise and insights on shape memory alloy actuators.

Authors' Statement

Conflict of interest: Authors state no conflict of interest.

5 References

[1] M. Sandschulte, *Auto Adaptive Sockets*. Master-Thesis, HAWK Goettingen, 2021.

[2] *Orthopaedic Sockets and Method for Adjustment thereof,* by M. Koppe, S. Reinelt, B. L. Finke, L. Vier, J. Siegel. (October 6,2022). WO2022207635 [online]

[3] M.-S. Kim, J.-K. Heo, H. Rodrigue, H.-T. Lee, S. Pané, M.-W. Han et. al., *Shape Memory Alloy (SMA) Actuators: The Role of Material, Form, and Scaling Effect*. In: Advanced Material Volume 35, Issue 33, 2023. https://doi.org/10.1002/adma.202208517

[4] Q. Liu, S. Ghodrat, G. Huisman, K. Jansen, *Shape memory alloy actuators for haptic wearables: A review*. In: ScienceDirect, Material and Design Volume 233, 2023. https://doi.org/10.1016/j.matdes.2023.112264

[5] M. Huebler, D. Vogelsanger, N. Hummel, N. Neblung, *White Paper - how to power Curve® bending actuators based on shape memory alloy (SMA) wires*, 2023 [publisher, CompActive GmbH]

Indoor Air Quality Monitoring and Guiding Device to Decrease the Risk of Coronavirus and Similar Airborne Diseases

Akin Caglayan [1], Max Urban [2]

[1] Biomedical Engineering, Luebeck University of Applied Sciences, akin.caglayan@stud.th-luebeck.de

[2] Medical Sensors and Devices Laboratory,Luebeck University of Applied Sciences, max.urban@th-luebeck.de

Abstract

This study addresses a gap in existing Indoor Air Quality (IAQ) measurement devices by introducing a comprehensive monitoring and guiding system. The system, incorporating sensors like a micro-controller unit, Total Volatile Organic Compound (TVOC) - Carbon Dioxide (CO_2) sensor, temperature-humidity sensor, and dust sensor, aims to measure indoor air quality. Two different rooms were selected for air quality measurements. Two distinct conditions were established for each room: ventilated and unventilated. The air quality within the rooms was assessed. The results aligned with U.S. EPA standards and Kaiterra's TVOC index, underscore the pivotal role of indoor air quality management in preventing airborne diseases like Covid-19.

1 Introduction

Humankind encounters a new virus variant almost every day. It shows that new diseases that will affect the whole world may occur again in the coming years. Devices that measure indoor air quality are needed, especially to prevent respiratory diseases such as Covid-19 from spreading rapidly in society. It was noticed that spending a lot of time indoors means that air quality decreases [1]. Studies show that the spreading of respiratory diseases are highly dependent on indoor environments [2]. Covid-19 and similar airborne diseases spread rapidly in indoor environments which have bad air conditions. Indoor environments should be regularly ventilated to reduce the risk of contamination [2]. In this study, a device was developed to guide and monitor the indoor air quality. With the help of guiding system, it was aimed to reduce the risk of coronavirus and other airborne diseases.

1.1 Air Quality

Air quality depends on the concentration of pollutants in the air. Therefore, it can vary from day to day, from hour to hour. Indoor Air Quality (IAQ) dependent on the concentration of pollutants in indoor environments such as smoking, perfumes, personal products, mold, viruses, and, bacteria. In addition to these, indoor air could be more polluted than outdoor, if in-house ventilation systems are not built properly. Indoor pollution that releases gases or particulate matters (PM10) into the air are primary cause of indoor air quality problems for human health [3]. PM10 is an inhalable particle with diameter 10 micrometers or smaller.

1.2 VOCs-TVOC and CO_2

The World Health Organization (WHO) states that a variety of indoor environmental pollutants, including humidity, temperature, volatile organic compounds (VOCs),CO_2, formaldehyde (CH_2O), and particulate matter, have a significant negative impact on human health [4, 5, 6]. These factors may interact with viruses and cause airborne contamination. VOCs encompass a category of compounds characterized by their elevated vapor pressure and limited solubility in water. These substances exhibit a tendency to vaporize easily (volatile) and do not dissolve in water (organic) [7]. Common everyday items like construction materials, personal products, glues, paints, and cleaning products constitute the main components of VOCs which are released into the air as gases. That is why VOCs are harmful to human health and can be combined with viruses. Furthermore monitoring all the numerous VOCs continuously is not possible. Consequently, a metric called Total Volatile Organic Compounds (TVOC) was introduced to determine the overall concentration of VOCs in a specific area [8].

Table 1: Index of Air Quality [9]

Air Condition	TVOC(ppb)	CO_2(ppm)	Description
Good	0-50	<=600	Fresh air
Modarate	51-100	1000	Acceptable
Upper Limit	100-150	1500	Not Fresh
High	151-200	1501-2000	Cough
Very High	201-300	>2000	Unhealthy

Table 1 shows the classification of the indoor air quality. Air quality measurements were compared with the data in this table.

2 Material and Methods

The electronic components intended for use in the project were identified for making measurements. The Arduino Nano Every was used as the micro-controller unit for this project. It was integrated with the ATMega4809 processor, which has a 20 MHz clock speed, 256 bytes of memory, and supports the Inter-Integrated Circuit (I2C) serial communication bus. This allows it to handle larger programs for the Arduino Nano Every. Additionally, this micro-controller unit works with a 5V operating voltage.

In Fig. 1, the first digital sensor was the Air Quality Sensor CJMCU-811-CCS811, which has a 1-second response time. It operates on a voltage range of 1.8-3.6V and supports the I2C communication protocol. The CO_2 sensing range is from 400 to 29,206 parts per million (ppm), and it also measures TVOC sensing from 0 to 32,768 parts per billion (ppb). For efficient sensor operation, the room temperature should be within the range of $(-40) - (+85)$ Celsius.

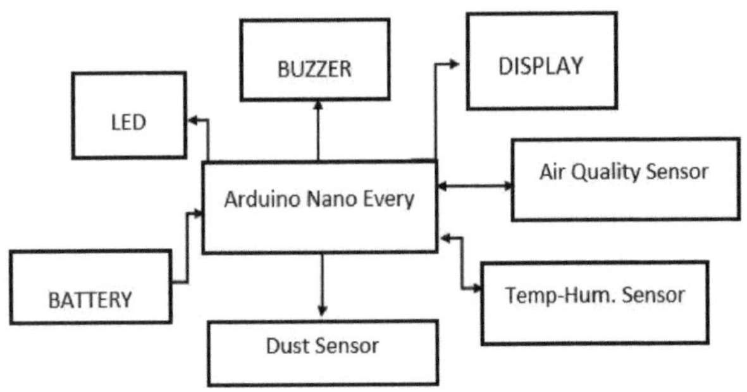

Figure 1: The block diagram of the air quality monitoring and guiding device.

Secondly, the Si7021 temperature-humidity sensor was used to measure indoor temperature and humidity levels. This sensor operates with a voltage range of 1.9-3.6V and has a 1-millisecond response time. The Si7021 also supports the I2C communication protocol, and its temperature measurement range is between $(-10) - (+85)$ Celsius. Humidity measurement of Si7021 0-80% relative humidity. This sensor can be affected by air flow. It may not take stable measurements in very hot and very cold conditions.

Figure 2: Electronic circuit of the indoor air quality monitoring and guiding device.

The last sensor used in this project was the Sharp Dust Sensor - GP2Y1010AU0F, which has a 1-second response time. It operates on 5V. This analog sensor is equipped with IR LED and lenses, enabling it to detect PM10 particles. To facilitate the pulse drive of the LED in the sensor, a 150-ohm resistor and a 220 microfarad capacitor are required. In addition to these, the micro-controller was programmed by using the Arduino IDE version 1.8.13. In Fig. 2 all circuit elements were assembled on the breadboard.

2.1 Experimental Setup

A thorough examination of air quality ensued in various locations, including a bedroom, cellar, and two additional rooms. For this study, results from only 2 rooms were shared. Prior to testing, one person slept in Room 1, and two people slept in Room 2 throughout the night, resulting in slightly reduced air quality in both rooms.

Room 1 was isolated from other occupants within the household. Both the window and door of Room 1 remained closed throughout the night to ensure controlled environmental conditions. Furthermore, the test subject in Room 1 experienced uninterrupted sleep during the designated period. Testing commenced subsequent to the test subject vacating Room 1 in the morning. The identical experimental procedure was applied to Room 2, maintaining consistent environmental parameters. Similarly, the test subjects in Room 2 also experienced uninterrupted sleep overnight.

The air conditions of the rooms were categorized into two groups: ventilated and unventilated rooms. This division aimed to measure and examine air pollutant changes in both scenarios. After ventilating the rooms, fresh air measurements were taken to allow for stabilization of air flow in the indoor environment. Tests were performed on different days.

3 Results and Discussion

Air quality measurements in the rooms were conducted at one-minute intervals over an hour. It was observed that there were fluctuations in the data in measurements at intervals of 10 and 20 seconds. It was decided that a 60-second interval for measurements compared to 10 and 20-second measurements. Additionally, the rooms' locations within the house, relative to the street, proved crucial. During the tests, factors such as opening the door, cleaning, weather condition and air pressure differences caused inconsistencies in the measurements in the data. Tests were repeated taking these factors into consideration.

3.1 Unventilated Room 1

It was observed that the average CO_2 level was close to 1500 ppm and the average TVOC level was close to 160 ppb. In other words, according to Table 1, Index of Air Quality, it was clarified that the indoor conditions was not fresh for sensitive. It was thought that Covid-19 and other respiratory diseases may spread more easily in such polluted air conditions. Change of TVOC and CO_2 in indoor environment within one hour was shown in Fig. 3.

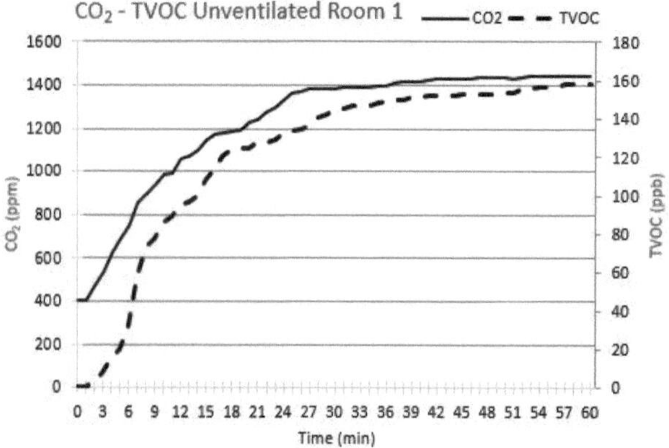

Figure 3: IAQ Variations within unventilated Room 1

3.1.1 Ventilated in Room 1

In this part of the experiment, Room 1 was ventilated lasting approximately 30 minutes. Subsequently, a 30-minute interval was observed to allow for stabilization of air circulation within the room before conducting the test. Analysis revealed that the average concentration of carbon dioxide (CO_2) approached 600 ppm, while the average concentration of TVOC neared 40 ppb. Based on these measurements, the indoor air quality conditions were assessed as meeting the criteria for fresh air with respect to 1. This indicates a reduction in TVOC and CO_2 levels in the ventilated room compared to the unventilated room, thus exerting a positive influence on indoor air quality.

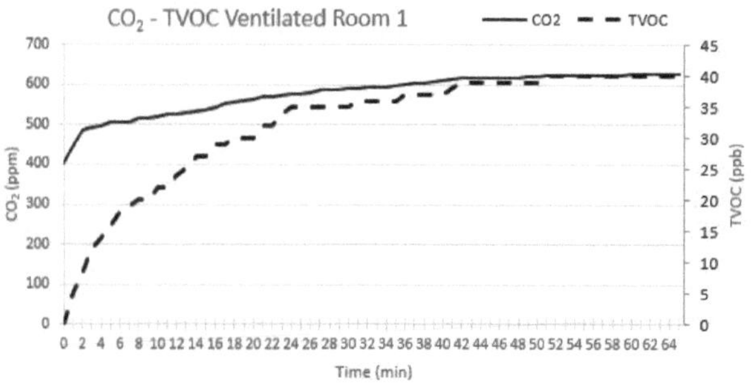

Figure 4: IAQ Variations within ventilated Room 1

Fig. 4 showed that variations in air quality were noted based on the levels of both CO_2 and TVOC. After a thirty-minute period of ventilation, it was observed that the concentration of total volatile organic compounds (TVOC) decreased by a factor of four, while the concentration of carbon dioxide (CO_2) decreased by approximately half.

3.2 Unventilated Room 2

When the data was examined an hour later, it was noticed that the air condition of Room 2 was worse than that of Room 1. It was observed that the average CO_2 level was close to 1600 ppm and the average TVOC level was more than 200 ppb. The indoor air quality conditions was classified as unhealthy and people may cough. It has been determined that indoor air quality can increase the transmission rate of airborne diseases.

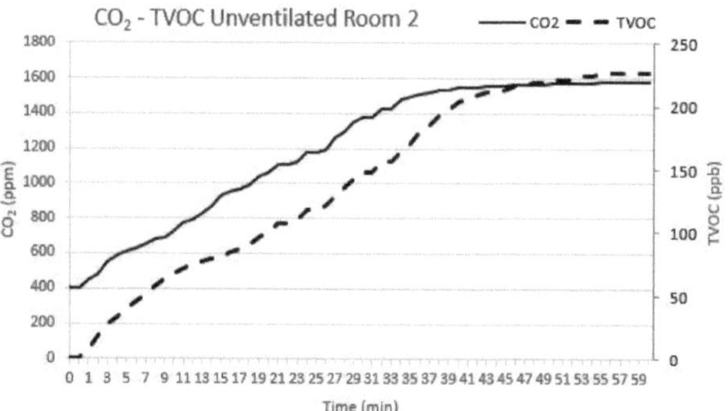

Figure 5: IAQ Variations within unventilated Room 2

In Fig. 5, the change in indoor air quality throughout the night depending on the number of people was measured. Comparing Fig. 3 and 5 revealed that indoor air quality was notably influenced by the number of people present in the room. This highlights the significant impact of human occupancy on indoor air quality dynamics, necessitating effective monitoring and management strategies. These findings underscore the importance of considering occupancy levels as a crucial factor in regulating indoor air quality across different environments.

3.2.1 Ventilated Room 2

In this part of the experiment, Room 2 was ventilated for approximately 30 minutes. To prevent air circulation from occurring in the room, another 30 minutes were waited, and then the test was performed. It was determined that the average CO_2 level was around 720 ppm and the average TVOC level was around 70 ppb. The indoor air quality conditions was classified as moderate and good condition.

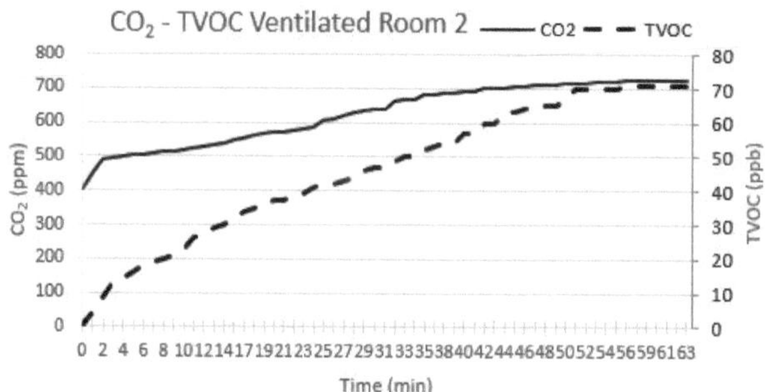

Figure 6: IAQ Variations within ventilated Room 2

In addition, Fig. 4 and Fig. 6 showed that depending on the number of people staying in a room, it may be beneficial to keep the room ventilated for a little longer to reduce the air pollutants.

It was understood that routine ventilation of indoor environment decreases the concentration of airborne pollutants. Consequently, inhabiting adequately ventilated environments is hypothesized to be able diminish the likelihood of respiratory diseases. Other studies have mostly been done on only the air quality measurement and built quite complex system. To avoid the complexity, a prototype was designed as compact and convenient as possible in this study.

4 Conclusion

In recent years, there has been a growing recognition of the necessity for indoor air quality measurement devices, particularly in mitigating the rapid spread of respiratory diseases such as Covid-19 or similar airborne diseases. Advancements in technology offer promise for more consistent measurements through the utilization of precision sensors and techniques. Furthermore, the integration of new software algorithms, such as Artificial Intelligence, can hold potential for detecting airborne contaminants and accurately measuring organic compounds to reduce the risk.

This study investigated changes in indoor TVOC and CO_2 levels, alongside real-time monitoring of temperature-humidity and PM10 data. Several challenges were encountered with the sensors employed in this research, particularly concerning the sensitivity of the air quality sensor to airflow and specific chemicals.

Additionally, the Si7021 digital temperature-humidity sensor demonstrated versatility in temperature and humidity studies, while the Arduino Nano Every proved effective as a micro-controller unit, offering superior processing speed and storage capacity compared to the Arduino Nano. However, compatibility issues between the Arduino Nano Every and sensors may arise, necessitating the availability of updated libraries.

Acknowledgement

The work has been carried out at Luebeck University of Applied Sciences and supervised by Prof. Max Urban. I would like to thank Prof. Max Urban and dear Silke Venker for their help.

Authors' Statement

Conflict of interest: Authors state no conflict of interest.

5 References

[1] S. M. Saad, A. Y. M. Shakaff, *Development of indoor environmental index: Air quality index and thermal comfort index*. American Institute of Physics, pp. 1-11, 2017.

[2] U.S. Department of Health and Human Services. Available: https://www.cdc.gov/coronavirus/2019-ncov/prevent-getting-sick/prevention.html. [last accessed on 2024-01-21].

[3] EPA, *United States Environmental Protection Agency*. Available: https://www.epa.gov/pm-pollution/particulate-matter-pm-basics [last accessed on 2024-01-21]

[4] J. M. Daisey, W. J. Angell, M. G. Apte, *Indoor air quality, ventilation and health symptoms in schools: an analysis of existing information*. In University of Minnesota, pp. 53-64, 2002.

[5] World Health Organization. *WHO Global Air Quality Guidelines: Particulate Matter (PM2.5 and PM10), Ozone, Nitrogen Dioxide, Sulfur Dioxide and Carbon Monoxide*. World Health Organization: Geneva, Switzerland, 2021.

[6] A. Ferreira, N. Barros. *COVID-19 and Lockdown: The Potential Impact of Residential Indoor Air Quality on the Health of Teleworkers*. International Journal of Environmental Research and Public Health, MDPI, Basel, Switerland, 17-05-2022.

[7] Smith, Dianna, *Understanding TVOC: What You Need To Know About Volatile Organic Compounds*. Available: https://learn.kaiterra.com/en/air-academy/understanding-tvoc-volatile-organic-compounds [last accessed on 2024-01-21]

[8] C. Meyer, *Overview of TVOC and Indoor Air Quality*. Renesas Electronics Corporation, pp. 2-5, 10-05-2021.

[9] Pietrucha, Tomasz, *Ability to Determine the Quality of Indoor Air in Classrooms without Sensors*, E3S Web of Conferences, Poland, 2017.

[10] Ian Buckey, *Maker Board Monday: Arduino Nano Every*, Electromaker, Available: https://www.electromaker.io/blog/article/arduino-nano-every-specs-and-more [last accessed on 2024-01-21]

Improvement of SNR and Sensitivity in MRI receiving array coils for neonatal patients by preamplifier decoupling

Abdelhalim Elashram[1], Martin A. Koch[2], Luca Belloi[3]
[1] Biomedical Engineering, Lübeck University of Applied Science, abdelhalim.elashram@stud.th-luebeck.de
[2] Institute of Medical Engineering, Universität zu Lübeck, martin.koch@uni-luebeck.de
[3] LMT Medical Systems, Lübeck, Germany, luca.belloi@lmt-medicalsystems.com

Abstract

A high signal-to-noise ratio (SNR) in magnetic resonance imaging is vital for ensuring an accurate diagnosis and treatment. The received signals from the patient are relatively weak and require amplification. In an array coil with more than two channels, the preamplifier decoupling technique is used to minimize the effects of mutual inductance between the nearest coils. In this study, a four-channel MRI receiver coil array and matching circuit have been developed to improve the SNR of MRI images and to minimize coupling between channels by utilizing preamplifier decoupling. As the result illustrates, the higher the preamplifier decoupling, the better the SNR. It is noticed that the signal does not change much, but the noise has decreased.

1   Introduction

Magnetic resonance imaging (MRI) is a medical imaging method that uses strong magnetic fields and radio frequency pulses to generate signals from the body's tissues [1]. RF coils in MRI operate on the principle of an LC (inductance-capacitance) circuit, operating to receive radio frequency signals. In this study, a 1.5 T MRI scanner was employed, utilizing RF coil arrays commonly known as phased array coils. Phased arrays consist of multiple coil channels organized in a geometric pattern to cover specific regions of the patient's body without any gaps. These coils operate at 63.6 MHz. The purpose of these coil arrays is to enhance the signal-to-noise ratio [2]-[4]. When two coils are positioned close to each other, they have different types of connection. When there is a weak coupling, changes of the current behavior in one coil do not affect the other coil much. But in case of strong coupling, changes in one coil significantly impact the other. In cases involving both weakly and strongly coupled coils, the resonance frequency curve shows a split, causing a drop in the middle value. This middle point means a decrease in the current (Fig. 1). When two coils overlap geometrically, the minimization of the coupling between neighboring coils occurs based on the principle of electromagnetic induction. In the first coil, the current flows in a specific direction, creating an outward-extending magnetic field from the coil. In the second coil, the current circulates in the opposite direction, generating an inward magnetic field within the coil. Opposing currents in the two coils create magnetic fields that can partially or completely cancel each other, leading to a reduction in their mutual induction. Mutual induction describes that a chang-

ing current in one coil can produce an Electromotive Force (EMF) in another. When handling more than two coils, a preamplifier decoupling is preferred because the difficulty in achieving geometric overlap stems from the challenge of controlling or deciding the direction of the current flow. Moreover, it is not possible to entirely eliminate mutual inductance between circuits [3].

How the preamplifier decoupling works is given by

$$V_{out} = V_{signal} + (R_1 + i(\omega L_1 - \frac{1}{\omega C_1}))I_1 + i\omega M_{12}I_2, \quad (1)$$

where V_{out} represents the output voltage of the circuit, V_{signal} is the signal voltage, i is the imaginary unit, ω is the angular frequency, R_1, L_1, C_1, and I_1 are resistance, inductance, capacitance and current for the first coil, respectively. M_{12} is the mutual inductance between coil 1 and coil 2, I_2 is the current of the second coil. The term associated with I_1 on the right-hand side denotes the noise associated with coil 1. The term related to I_2 represents the noise generated by the coupling between coil 1 and coil 2. The tuning and matching coil 1 at its resonance removes the imaginary component (L_1, C_1) from coil 1, leaving behind only its resistance(R_1). To eliminate the third term($i\omega M_{12}I_2$), two scenarios are possible: $M_{12} = 0$ or $I_2 = 0$. The first case, when M_{12} is zero, it means that coil 1 and coil 2 overlap in a way that cancels out their mutual inductance, is not considered in this study that involves more than two coils. Consequently, the term in question is eliminated by setting $I_2 = 0$. If using a preamplifier with low input impedance (Fig. 2) L_2 and C_2 form a parallel resonant circuit at the target frequency, resulting in a high impedance that tends towards infinity. This prevents any current circulating in coil

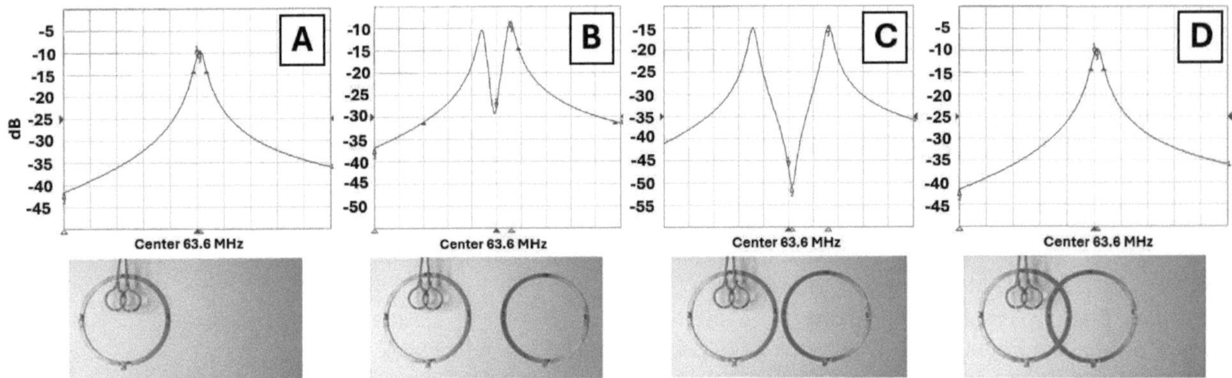

Figure 1:   The tuning curve at 63.6 MHz undergoes changes in its shape based on the position of the coils. (A) single coil, (B) weak coupling, (C) strong coupling, (D) geometrical overlap.

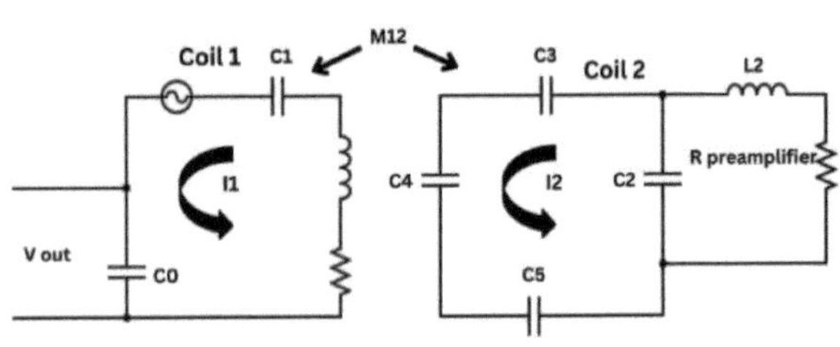

Figure 2: Equivalent circuit of two coils, each being an LC circuit, coupled through the nonzero mutual inductance.

2 (I_2=0). A decrease in the preamplifier's input impedance results in stronger decoupling [3]-[5].

2 Material and Methods

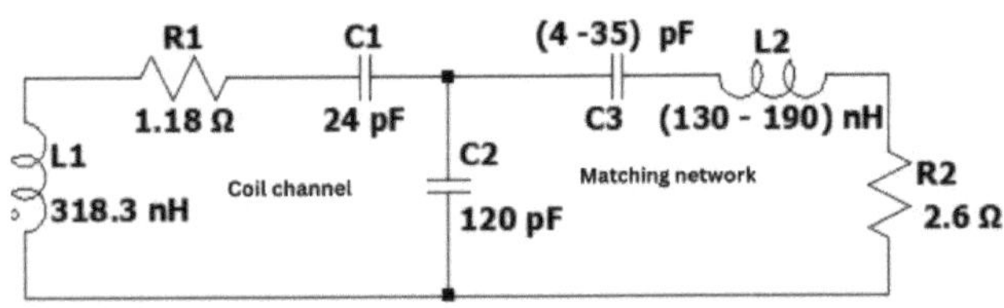

Figure 3: Circuit of one coil channel connected to matching network.

In the initial phase, LTspice XVII, a high-performance simulation software allowing users to simulate and analyze electronic circuit behavior (Analog Devices, MA, USA), was utilized to target tuning at 63.6 MHz. However, upon measurement, the observed value of tuning deviated to 67.6 MHz. Subsequently, modifications were applied to either the inductance or capacitance values based on

$$f = \frac{1}{2\pi\sqrt{LC}}, \qquad (2)$$

where f is the frequency, L is the inductance and C is the capacitance [1].

When applying the circuit (Fig. 3) and studying its characteristics with a network analyzer (part of the Series Network Analyzer ENA by Agilent Technologies, Santa Clara, California, USA). The observed tuning frequency did not agree with the expected 63.6 MHz. Following mathematical analysis of the circuit and subsequent adjustments to the components, we achieved two tuning curves: one centered at 63.6 MHz without utilizing a preamplifier and another while employing a preamplifier (Fig. 1). Initially, we set up a circuit comprising capacitors and resistors, using capacitance values of C_1= 24 pF and C_2= 120 pF by using (2) the determined inductance value equated to 318.3 nH. Analyzing this on a network analyzer allowed us to tune the circuit precisely to 63.6 MHz. We identified the Q factor= 120.44, enabling us to calculate the value of resistance ($R1$) which equals 1.18 Ω (Fig. 3) using

$$Q = \frac{1}{R}\sqrt{\frac{L}{C}}, \qquad (3)$$

when checking the tuning curve without the preamplifier, the matching circuit operates as an open circuit. Therefore, there is no flow of current through it. When adding a preamplifier, current flow increased. To counteract the increase in current, we introduced an inductor L_2 into the circuit. This helped to minimize the current by using the higher impedance in parallel connections. A trimmer capacitor C_3 was added in series with L_2. When the result of the trimmer inductor L_2 and trimmer capacitor C_3 acts as an inductance, it usually leads to high impedance. For instance, C_3= 35 pF has an impedance value Z_3= $-71.49i$, L_2= 190 nH has an impedance value Z_4= 75.88i, the combined impedance results in 4.38i, it generally indicates an inductive behavior in the circuit. When the result of L_2 and C_3 behave as a capacitance, it leads to low impedance. For example, when C_3= 4 pF has an impedance value Z_3=$-625.88i$, L_2 has an impedance value Z_2= 75.88i, the combined impedance results in $-549.72i$, it generally indicates a capacitive behavior in the circuit [3].

2.1 Prototyping

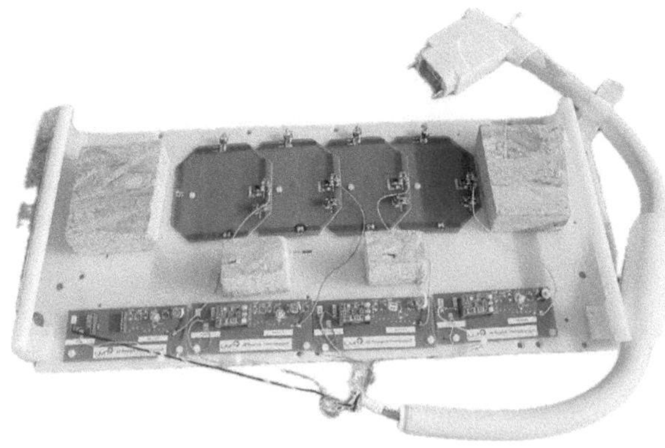

Figure 4: Prototype of phased array coil.

The phased array has four coil channels linked by coaxial cables to four matching networks. Each channel has the same parts like capacitance and inductance. They are firmly fixed to the board using plastic screws to prevent any movement. The matching networks with preamplifiers located at the bottom of the board connect the coil channels and help to reduce noise. For stability when carrying the phantom, a wooden block is positioned behind the board for testing purposes. The board has a coil plug that is used for connecting the prototype to the MRI scanner. Before real tests, the circuit is first tested virtually using LTspice XVII. Then, its behavior is studied further using a network analyzer to understand how it works at different frequencies. After these simulations, the circuit mounted on the board is tested experimentally with an MRI system.

2.2 Experiments

The following steps outline the essential procedures and conditions to ensure optimal performance of the coil, a series of crucial procedures and conditions are followed, initially the establishment of the setup and connection occurs by connecting the coil plug to the test bench, which used to test coil functionality. This connection facilitates control over each channel through the network analyzer. The assessment of the scattering parameter S21 allows the observation of amplitude and phase variations during the transmission of signals from Port 1 to Port 2. Following this, an evaluation of tuning is conducted at 63.6 MHz across all channels. This evaluation verifies the alignment of the coil's resonance at this specific frequency. Additionally, the detuning analysis involves an examination of both active and passive detuning components, also important for safety measures. Verification of preamplifier decoupling for each channel is performed to ensure that the recorded values exceed (-25 dB) in comparison with the reference, this step is critical in minimizing undesirable noise and guaranteeing high SNR. The assessment of channel gain utilizes S21 in conjunction with a single-pick probe to gauge the gain of each channel. This measurement represents sensitivity, with a requirement that the gain exceeds 2 dB. Upon fulfillment of all specified conditions, the coil was used within

the MRI scanner. The measurements involved adjusting the trimmer capacitor in the matching network circuit to four different current values (0 dB, 10 dB, 20 dB , and 30 dB). These dB values correspond to 5, 12, 20, 21 and 28 pF, respectively. These values were chosen to evaluate the impact on the SNR.

3 Results and Discussion

As discussed above, the four channel-phased array configuration involves four coil channels and matching networks. Each channel (Fig. 4) integrates a 47 pF capacitor in parallel with a trimmer capacitor (ranging from 4 to 35 pF), this capacitor plays a significant role to set the tuning curve at 63.6 MHz. The trimmer capacitor in conjunction with a 120 pF capacitor in parallel. A 15 cm coaxial cable connects the channels and the matching network. The matching network consists of the low noise preamplifier and a trimmer capacitor (ranging from 4 to 35 pF). This capacitor is employed to measure precise current values in decibels (dB) on a scale, along with an adjustable inductor ranging from 131 to 190 nH. These values were determined through calculations and confirmed by simulating the circuit using LTspice.

3.1 Network analyzer tests

The final verification step before usage with the MRI scanner involves ensuring that the prototype meets all specified conditions. Table 1 presents the measured values for essential parameters, including tuning, detuning, preamplifier decoupling, and gain.

Table 1: Network analyzer results

Number of channels	Tuning	Detuning	Preamp decoupling	Gain
1	-8.91 dB	-45.74 dB	-43.0 dB	12.4 dB
2	-7.53 dB	-38.5 dB	-44.0 dB	11.4 dB
3	-8.07 dB	-40.1 dB	-42.4 dB	10.6 dB
4	-8.50 dB	-40.9 dB	-43.2 dB	10.37 dB

3.2 MRI scanner tests

Table 2: MRI scanner test values

dB	Signal	Undesired signal	SNR
0	286 a.u	1.9 a.u	150.52
10	230 a.u	1.4 a.u	164.28
20	296.5 a.u	1.6 a.u	185.31
30	251 a.u	1.2 a.u	209.16

Table 2 displays signals originating from a phantom and measured in arbitrary unit. The effectiveness of preamplifier decoupling and its impact on the SNR are quantitatively assessed using decibels (dB). Undesired signal represents the noise. Observations on these values reveal that the dB

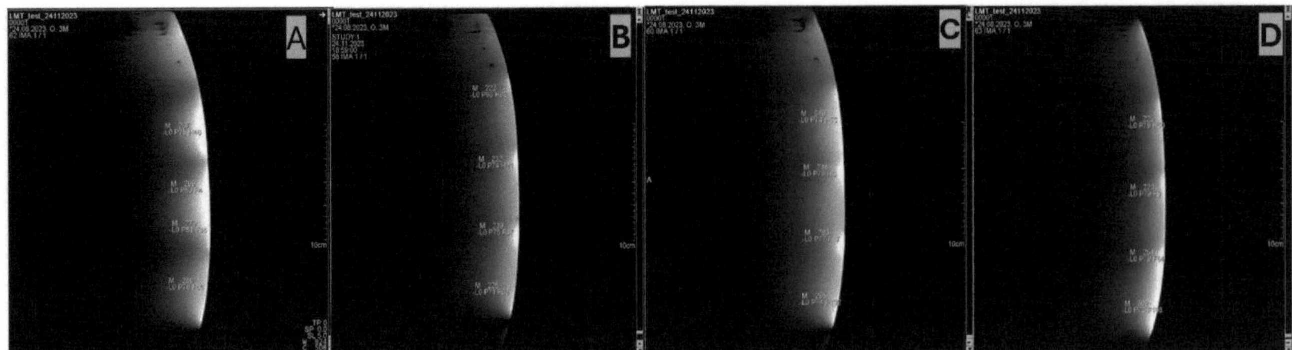

Figure 5: MRI images of a water phantom acquired with the prototype of four channels. (A) 0 dB, (B) 10 dB, (C) 20 dB, (D) 30 dB.

levels rise, the signal strength remains relatively constant without displaying a clear trend of increase or decrease. In contrast, the noise levels are decreasing as the dB levels increase. Consequently, SNR tends to increase with higher dB levels.

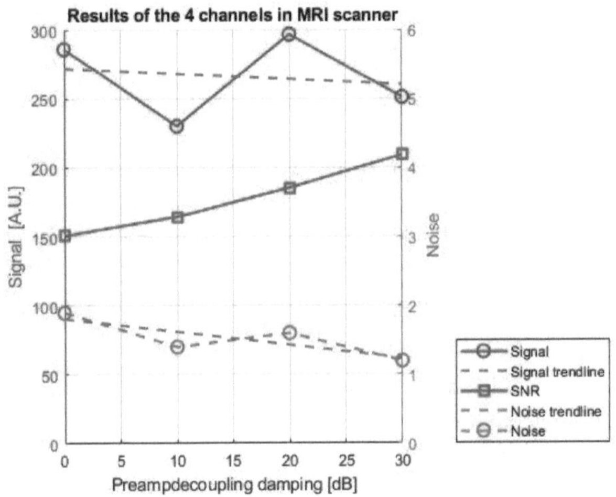

Figure 6: Relation between signal and preamplifier decoupling based on measurements.

the problem with noise is not solely caused by the preamplifier decoupling. It also originates from thermal noise (Johnson-Nyquist noise), a type of electronic noise .

4   Conclusion

The study focused on integrating RF phased array coils, which consist of four channels connected to four matching circuits with coaxial cables, working at 63.6 MHz. The main goal was to enhance SNR and reduce the interference between coil channels by using preamplifier decoupling. The study employed LTspice XVII and a network analyzer for virtual and practical assessments. The relation between signal and preamplifier decoupling was studied, revealing a constant signal throughout adjustments while observing an increase in SNR due to decreased noise resulting from increased preamplifier decoupling. In the future, exploring advanced signal processing algorithms or machine learning techniques could aid in extracting diagnostic information from enhanced SNR images.

Acknowledgement

The work was carried out at LMT Medical Systems in Lübeck and supervised by Prof. M. Koch, Institute of Medical Engineering, Universität zu Lübeck.

Authors' Statement

Conflict of interest: Authors state no conflict of interest.

5   References

[1] R. H. Hashemi, W. G. Bradley and C. J. Lisanti, *MRI: The Basics.* Wolters Kluwer Health, Philadelphia, 2017.

[2] B. Gruber, M. Froeling, T. Leiner and D. W. J. Klomp, *RF coils: A Practical Guide for Nonphysicists,* Journal of Magnetic Resonance Imaging, vol. 48, pp. 590–604, 2018.

[3] H. Fujita, T. Zheng, X. Yang, M. J. Finnerty and S. Handa, *RF Surface Receive Array Coils: The Art of an LC Circuit.* Journal of Magnetic Resonance Imaging, vol. 38, pp. 12–25, 2013.

[4] P. B. Roemer, W. A. Edelstein, C. E. Hayes, S. P. Souza and O. M. Mueller, *The NMR Phased Array.* Magnetic Resonance in Medicine, vol. 16, pp. 192–225, 1990.

[5] K. L. Moody, N. A. Hollingsworth, F. Zhao, J. Nielsen, D. C. Noll, S. M. Wright and M. P. McDougall, *An eight-channel T/R head coil.* Journal of Magnetic Resonance, vol. 246, pp. 62–68, 2014.

Test Standard for Controlling the Gas Temperature for Heated Breathing Hoses by Using Different Temperature Sensors

Sophia Merrath [1], Ludger Tappehorn [2], and Philipp Rostalski [3]

[1] Medical Engineering Science, Universität zu Lübeck, sophia.merrath@student.uni-luebeck.de
[2] Drägerwerk AG & Co. KGaA, Lübeck, ludger.tappehorn@draeger.com
[3] Institute for Electrical Engineering in Medicine, Universität zu Lübeck, philipp.rostalski@uni-luebeck.de

Abstract

During mechanical ventilation, the respiratory gas supplied to the patient needs to be humidified and warmed to ensure the maintenance of the lungs' cleansing function. However, since the gas must have the correct temperature, regulation shall be done by using temperature sensors inserted externally into the gas stream. Commonly, Pt1000 sensors are used, which have led to hygiene and application issues in the past. The objective of this article is to evaluate an alternative method of temperature measurement using an integrated Negative Temperature Coefficient (NTC) sensor and to compare temperature readings of both sensors. Therefore, an electrical circuit for implementing the temperature control was designed and controlled with a microcontroller. The results indicate that the temperature control of the respiratory gas can be optimally realised by a developed electrical control of the heating wires of the breathing hoses. The NTC proves to be a suitable alternative for temperature measurement, as it exhibits high accuracy in temperature readings and in comparison to the Pt1000, demonstrates a more dynamic behaviour.

1 Introduction

Respiration is essential for supplying oxygen to the human organism. The primary function of the upper airways is to contribute to the warming and moistening of the inhaled air. Insufficient humidity within the airways causes the ciliated epithelium and mucous membranes to stop working. Those are largely responsible for the self-cleaning of the respiratory tract by producing mucus in the inside of the airways. This mucus effectively transports dust and foreign particles to the trachea, preventing infections and other damages of the lungs [1].

Mechanical ventilation bypasses the upper respiratory tract through the ventilation tubing system and the endotracheal tube, thereby limiting its natural function of warming, humidifying, and cleansing the air (respiratory gas conditioning). If these functions are restricted for an extended period, it is imperative to artificially replace the warming and humidifying of the inspired air. This is necessary since the gases from the central gas supply system have a humidity lower than 0.5 mg/L [1]. For humidifying the respiratory gas in clinical practice, active 'Heated Humidifiers' (Pass-over humidifier) can be used. The field of active humidification with a heated breathing tube system is gaining more significance over the last years. In a Pass-over humidifier, the breathing air passes through a water chamber, where it undergoes simultaneous heating and moistening as the water chamber is heated by a heating plate. Afterwards, the warm and humid air flows into the patient by guiding through a heated inspiratory hose. Heated breathing hoses are used to ensure the ventilation performance and prevent the formation of condensate caused by the temperature difference from inside and outside the hose on the way to the patient. To achieve an optimal gas temperature in case of mechanical ventilation, the air should have a temperature of 37 °C (100 % relative humidity) when leaving the water chamber and 39 °C shortly before reaching the patient [1]. Only the temperature is measured and regulated, but not the humidification. In order to ensure that the control circuit of the humidifier is working correctly and the gas temperature is not too high when reaching the patient's lung, temperature measurements on both the chamber side and the patient side are carried out. Currently, this is mostly done by Pt1000 Positive Temperature Coefficient (PTC) sensors, which are inserted directly into the breathing tube and are reusable [1]. This may lead to a risk of cross contamination, as those sensors can only be disinfected and not sterilized due to their sensitive electronics. Moreover, an incorrect insertion of the sensor might lead to additional errors in the temperature measurements. To overcome those hygiene and application issues, this work proposes an alternative temperature measurement method using an permanently integrated Negative Temperature Coefficient (NTC) for single use. The aim is to investigate whether the NTC can achieve temperature measurements as precise as the currently used Pt1000. To accomplish this, electrical circuits were designed that are specific to the sensors (Pt1000 and NTC) and control the heating wires of the heated breathing hoses.

1.1 Temperature Sensors

There are different types of temperature sensors on the market, which are either inserted or integrated into the tubing system. A thermistor is a variable electrical heat-sensitive resistor whose resistance changes with the temperature. Thermistors are classified into PTC and NTC sensors based on their temperature behaviour (R-T-curve) [2].

The Pt1000 is a PTC sensor and consists of a platinum resistance thermometer. It has an approximately linear resistance-temperature characteristic, providing information on the correlated values of temperature and electrical resistance. At 0 °C, the Pt1000 has a specific resistance of 1 kΩ, which increases with increasing temperature. It is a plug-in sensor enclosed in a steel cap that protrudes into the flow channel and is in direct contact with the breathing gas. NTC thermistors have an inverse resistance-temperature characteristic. The resistance decreases non-linearly with increasing temperature. Compared to Pt1000 sensors, they have a higher sensitivity. In this work, a permanently integrated NTC sensor with a nominal resistance of 100 kΩ at 25 °C was used [3].

2 Material and Methods

2.1 Basic Structure of the System

The basic structure of the test setup is shown in Fig. 1. To obtain the correct pressure and flow rate, a pressure regulator followed by a flow meter (4000 Series, TSI, Aachen, Germany) were connected. The incoming air is then heated and humidified by the humidifier (HC100, Fisher & Paykel, Auckland, New Zealand) and passed through the heated breathing hose. For controlling the temperature on the patients side, different temperature sensors were placed 1.8 cm apart. First the integrated NTC, the reference temperature sensor (Ref 1), followed by the Pt1000 and the second reference sensor (Ref 2). It is assumed that the temperature decreases linearly with increasing distance from the heating wires, so the temperature of the Pt1000 should correspond to the mean value of Ref 1 and Ref 2. The values of both Ref 1 and Ref 2 were recorded by an Almemo datalogger at 1 s intervals. The NTC and Pt1000 were connected to a specific electrical circuit that implements the data readout of each sensor with a microcontroller (Arduino Uno), which allows for a serial communication with the computer. Several test cases were conducted for different flows rates (10 - 60 L/min) and temperatures of the hose in the range of 30 - 42 °C with a chamber temperature range of 35 - 37 °C.

2.2 Electrical Circuit of the Temperature Control of the Breathing Hoses

In general, the connections of heated breathing systems are initially designed for use with a humidifier, which can supply a maximum of 30 W to the tube. However, if higher power is required for measurements, it is necessary to control the breathing tube heating wires separately. Consequently, two electrical circuits have been designed for the Pt1000 and the NTC sensor. With those circuits and the use of an Arduino Uno, it is possible to control the temperature of the breathing hose according to the user's specifications. Since the Arduino can only measure analogue voltages, a resistance-to-voltage conversion is required. The circuit designed for the Pt1000 is depicted in Fig. 2. The Pt1000 is connected in series with a resistor R3 to form a voltage divider. Changes of the Pt1000 resistance will modulate the voltage across it. The second voltage divider (R1 & R2) generates a constant reference voltage. In this study, accurate temperature measurements of the sensor in the range of 20 - 50 °C, corresponding to the operating range of a heating tube, are crucial. As the resistance-temperature characteristic shows, the resistance of the Pt1000 changes only slightly over the operating range. Therefore, the Arduino's ADC, with a resolution of 10 bits ($2^{10} = 1023$ values), can not accurately detect the resulting small voltage differences. To amplify the voltage differences, an inverting amplifier is implemented using R2, R4 and the operational amplifier (AD8552). The amplifier compares the reference voltage with the actual voltage. This voltage difference is measured by the Arduino. The resistor values have been chosen to best match the voltage range of the Arduino for accurate temperature readings. A software program was written to convert the voltage into temperature values. The MOSFET (IRLZ334N) receives a digital signal from the Arduino to control the heating wires by switching the solid-state-relay. The heater requires a higher supply voltage than the 5 V provided by the Arduino and is powered by an external power supply.

For measurements with the NTC sensor, a voltage divider is implemented with R1, having an equivalent resistance value of 100 kΩ (see Fig. 3). Given that the NTC resistance undergoes considerably larger variations in the temperature range compared to the Pt1000 sensor, no voltage amplification is required. The control of the heating wires follows a scheme analogous to that depicted in Fig. 2.

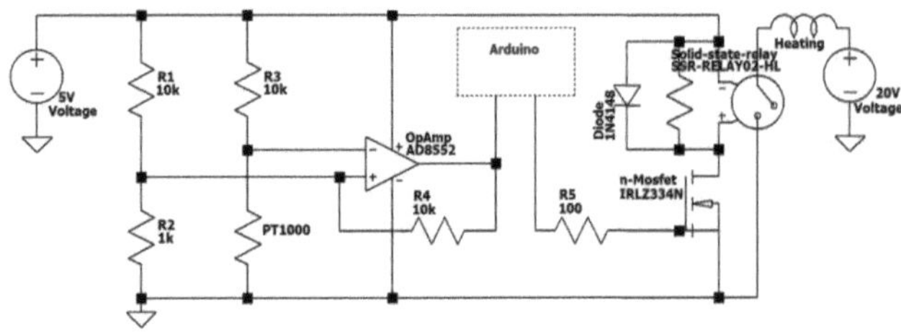

Figure 2: Electrical circuit with the Pt1000 sensor for controlling the temperature of breathing hoses.

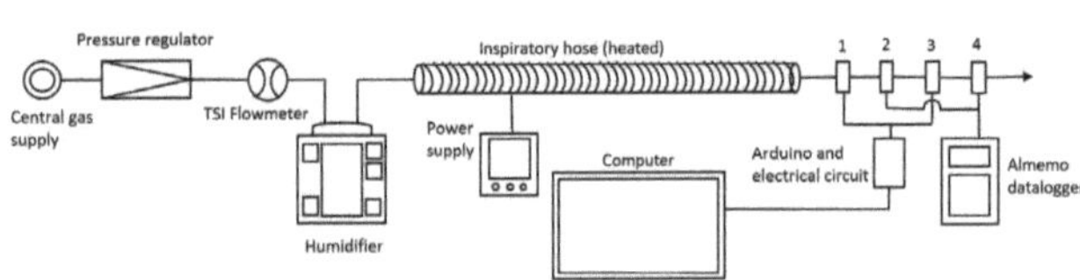

Figure 1: A scheme of the test setup with the different sensors: NTC (1), Ref 1 (2), PT1000 (3), Ref 2 (4).

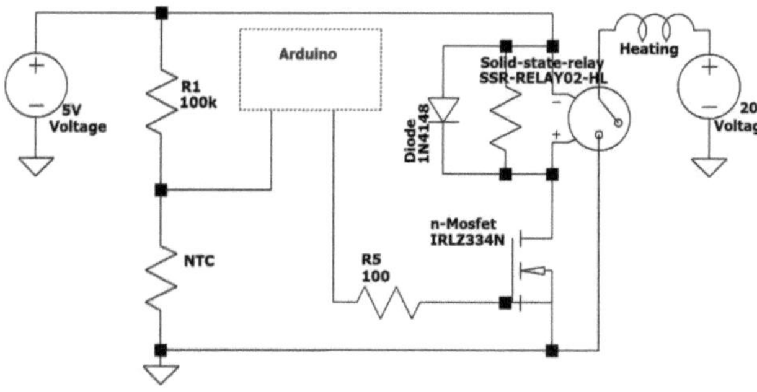

Figure 3: Electrical circuit with the NTC sensor for controlling the temperature of breathing hoses.

2.3 Arduino Software Design

The temperature was regulated by a two-point controller. This sets the output to "on" or "off" depending on the switching criterion of the input signal. In control theory, regulation is the process of bringing the actual value closer to the setpoint and then maintaining it [4].

Fig. 4 shows the structure of the Arduino code. The Arduino outputs an ADC value that corresponds to the input voltage from the electrical circuit, which is then used in the C++ code to calculate the corresponding resistance value. The resistance values were converted into temperature values using the sensor-specific R-T characteristic. To achieve this, a function was determined that approximates each of the Pt1000- and NTC-specific characteristics. This function is most accurate within the range considered in this study, which is 20 - 50 °C. The temperature measurement is used as feedback for temperature control. Initially, the user inputs the desired setpoint temperature, which is then compared to the actual temperature. If the measured temperature is lower than the setpoint, the MOSFET turns on, closing the relay switch and completing the circuit. The external power source supplies current to the tube's heating wire, causing the temperature of the respiratory gas to increase. The circuit remains closed until the actual temperature exceeds the desired temperature. When this happens, the circuit will be interrupted, and the temperature of the breathing gas will decrease. Temperature values will be determined and compared at 1 s intervals.

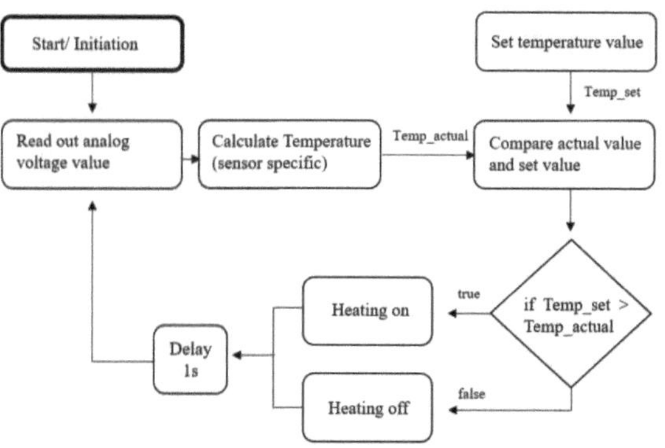

Figure 4: Flowchart of the Arduino software.

2.4 Temperature Control - Requirements

In order to ensure the functionality of the temperature control of a heated breathing hose connected to the humidifier, the requirements are summarized in Table 1. The requirements are derived from the particular standard for requirements for basic safety and essential performance of respiratory humidifying equipment (ISO 80601-2-74) [5].

	Requirements
1.	The measured gas temperature shall have a rated range of 25 °C to 45 °C [5].
2.	The measured gas temperature should not differ by more than $\pm$ 2 °C from the set temperature during normal use [5].
3.	The specific enthalpy should not exceed 197 kJ/m^3 during the 240 s period, nor the averaged specific enthalpy over 120 s [5].

Table 1: Requirements for the performance of a humidifier.

3 Results and Discussion

Various test cases were conducted at different flow rates and temperature controls of the breathing hose wires to investigate the accuracy of the comparative temperature sensors (Pt1000, NTC) (see Sect. 2.1). An example of a measurement of the gas temperature control is shown in Fig. 5. The measurement was taken with a 14 Ω tubing system at a constant flow rate of 12 L/min, an ambient temperature of 22.3 °C, a power of 25 W and a desired temperature of 39 °C. It can be seen that all four sensors have a similar temperature behaviour. All temperature sensors measure an initial temperature of 23.9 °C. After approximately 2.5 min, all sensors reach a local maximum and the temperature drops by approximately 1 °C before rising again. After a settling period of 6 min, the NTC reaches the set temperature of 39 °C and the existing two-point control takes effect. It can be seen that as soon as the NTC reaches 39 °C, the heating is switched off and the temperature drops slightly, causing the controller to readjust. It remains to be seen whether the observed spikes are due to measurement noise or represent a real phenomenon, since the NTC has a more dynamic behaviour than the other sensors. After the local maximum, the Pt1000 drops by 1 °C in line with the NTC, but remains at 33.5 °C for another 3 min before starting to rise again. After approximately 10 min, the Pt1000 reaches a temperature of 36.7 °C and remains constant at this value. The reference sensor Ref 1 reaches a temperature of 38.4 °C after 6 min, slightly lower than the NTC. The reference sensor Ref 2, at the end of the test setup, shows a lower temperature compared to the NTC and stabilises at a temperature of 36.7 °C, which is 2.3 °C lower and in agreement with the expected linear temperature drop. The results fulfil the first two requirements, as of both the temperatures are between 25 - 45 °C and temperatures of $\pm$ 2 °C from the set value is achieved.

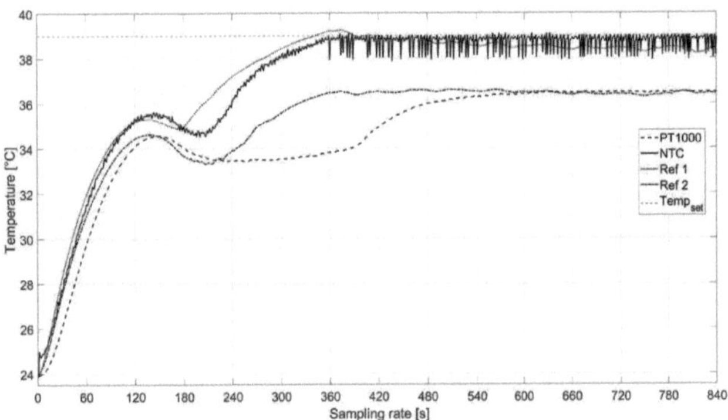

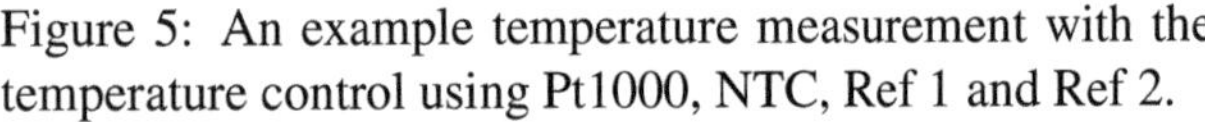

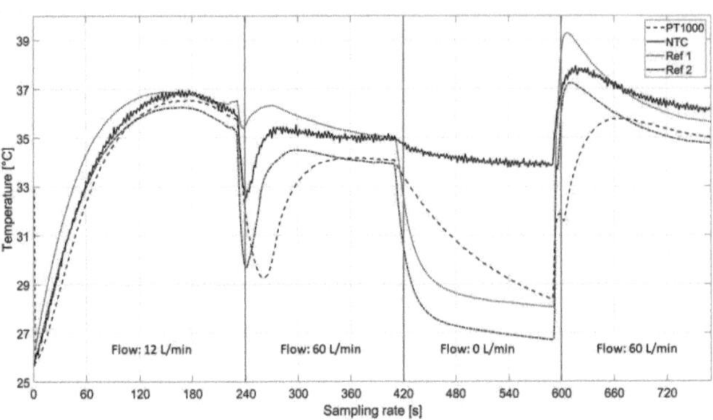

Figure 5: An example temperature measurement with the temperature control using Pt1000, NTC, Ref 1 and Ref 2.

Figure 6: An example temperature measurement at different flow rates using Pt1000, NTC, Ref 1 and Ref 2.

In general, the results suggest that the Pt1000 has a greater inertia than the other sensors. This is affirmed by the temperature curve of the Pt1000 after 10 min, which is approximately 1 °C lower than expected. The higher inertia may be explained by the metal cap on the Pt1000, which provides a greater mass. In addition, the sensor connector could form a thermal bridge to the outside and thus emit more heat into the environment.

The temperature control was validated by conducting tests in a temperature-controlled water bath for different temperatures (20 °C, 35 °C, 39 °C, 42 °C) for 5 min each using the Pt1000, NTC, Ref 1 and Ref 2. The results show that both the Pt1000 and NTC measure accurately, as for 25 °C the Pt1000 has a $\Delta T = 0.121$ °C and NTC a $\Delta T = 0.12$ °C.

Another example of a temperature measurement according to the requirements of a thermal hazard test case (based on the ISO 80601-2-74) is shown in Fig. 6. The measurement was made under the same test conditions but with varying flow rates. After 4 min with a flow of 12 L/min, the flow is increased to 60 L/min for 3 more minutes. It can be seen that the temperature of all the sensors drops briefly and then rises again. The Pt1000 responds with a 20 s delay and rises more slowly than the NTC. After 7 min, there is no flow for 3 min, during which time the temperatures briefly drop exponentially as no heated air is being transported through the hose. The temperature of the NTC decreases only slightly, which could be due to the NTC being closest to the end of the heating wires. The Pt1000 again shows thermally inert behaviour with a delayed temperature drop. After 10 min, the flow rate is suddenly increased to 60 L/min and the temperature of all the temperature sensors rises sharply. Again, the thermal inertia of the Pt1000 is evident. Nevertheless, the specific enthalpy of 197 kJ/m^3 is not exceeded and the third requirement is fulfilled. It is worth noting that the chamber temperature in this experiment was 35 °C, which is 2 °C lower than the standard. This is because a higher chamber temperature of 37 °C would result in more condensation on the patient side, where the temperature sensors under investigation are located, leading to errors in temperature measurement. In order to make a conclusive statement about the functionality of the control, further test cases of the thermal hazard test should be carried out.

4 Conclusion

In this work it was investigated, whether temperature measurements with an integrated NTC represents a suitable alternative to temperature measurements with a conventional plug-in Pt1000 during active humidification. For this purpose, a two-point temperature control system was developed and carried out with an Arduino. Overall, the results indicate that the NTC sensor is a viable alternative and as it shows a more dynamic thermal behaviour compared to the Pt1000. It was observed, that the NTC adjusts more quickly to temperature changes and has a higher temperature accuracy. Thus, with an integrated NTC sensor, hygiene and application-related issues can be circumvented. Next steps could address the occurring but unexplained drop in temperature of the sensors after a few minutes of heating.

Acknowledgement

The work has been carried out at Drägerwerk AG Co. KGaA, Lübeck and supervised at the Institute of Electrical Engineering in Medicine, Universität zu Lübeck.

Authors' Statement

Conflict of interest: Authors state no conflict of interest.

5 References

[1] H. Lang, *Beatmung für Einsteiger*. Springer-Verlag, vol.3, 2019.

[2] F. Henning, H. Moser, *Temperaturmessung*. Springer-Verlag, 1977.

[3] TE Connectivity Sensors, *GA100K6D619 NTC Temperature Sensor*, 2022 (internal).

[4] J. Lunze, *Regelungstechnik 1*. Springer-Verlag, 2020.

[5] ISO 80601-2-74, *Medical electrical equipment - Particular requirements for basic safety and essential performance of respiratory humidifying equipment*, 2021.

Testing for Service-Oriented Device Connectivity Application

Ahmad Alsalhani [1]

[1] Medical Engineering Science, Universität zu Lübeck, Ahmad.Alsalhani@student.uni-luebeck.de

Abstract

Widespread, reliable and cross-manufacturer communication systems in the hospital help saving costs in healthcare and simplify clinical workflows. The German OR.NET association has developed and implemented a concept for open interoperability between medical devices, and it has standardized the system according to the ISO/IEEE standards. The communication system, known as Service-Oriented Device Connectivity (SDC), enables secure and dynamic data exchange between medical devices in the operating room. The SDC system is also connected to the IEEE 11073 standard family. In order to achieve complete interoperability, the communication scenarios, a test environment and test tools must be developed. This paper presents how a device service provider and consumer communicate with each other according to the IEEE 11073 SDC standard family. An evaluation strategy for a test tool is developed to assess the conformity of the device under Test (DUT) with the standard.This builds the basis for an extensive analysis of the test tool.

1 Introduction

Interoperative networking is crucial for the exchange of data between different medical devices and for linking with the information system in the hospital. In addition, such communication plays a pivotal role in patient safety and helps to save costs in the operating room [1]. Networking can help to prevent false alarm, so that doctors avoid the risk of "alarm fatigue". To reduce the workload, a situation where different medical devices in the hospital communicate with each other based on a generalised native communication system is beneficial. The OR.NET Association has developed a communication protocol connected through the ISO/IEEE 11073 Service Oriented Devices Connectivity (SDC) system family of standards [2]. This system facilitates secure, dynamic data exchange, and remote control between systems of different medical devices from various manufacturers. Despite the significance of this communication system, a considerable number of medical devices are still neither SDC-enabled nor compatible with the ISO/IEEE 11073 system.To ensure secure connectivity, only approved medical devices that are guaranteed to be compatible with the standard may be used. The approval process for these devices involves extensive testing. it is an unresolved problem how to design a detailed testsuite that thoroughly examines all protocol aspects, ensuring that a wide range of potential sources of errors can be ruled out.

1.1 State of the Art

Numerous studies have been conducted to evaluate the implementation (SDC) with latency measurements diverse metrics, including Latency Measurement. This paper describes experiments conducted in an isolated Testbed environment comprising a single service provider and consumer. Moreover, a control PC initiates the measurement process and gathers timestamp data upon completion of the data exchange [3]. Another work involves the simulation of Network Medical Devices. The paper employs test tool integrated in the form Hardware-in-the-loop (HIL) simulation [4].Through this test tool, issues can be detected proactively before they manifest in real-world usage.

1.2 Service-Oriented Device Connectivity (SDC)

Service-Oriented Architecture (SOA) serves as the foundation for IEEE 11073 SDC. SOA defines an approach to achieving interoperability between software components and interfaces. This interoperability finds applications in diverse systems, such as the medical environment, facilitating interaction between devices or enabling data exchange between these devices, as well as hospital information systems. SOA plays two crucial roles: Service Provider (referred to as Device) and Service Consumer (referred to as Client). The service provider extends its services to all network participants, while service consumers utilize these services as needed, without contributing services or information to the network [5]. In literature and language usage, the term "device" is often used for the service provider, and "client" for the service consumer. In everyday medical communication, a physical device may implement both services (consumer and provider) in a single device. The most common implementation of SOA is the Web Service(WS), defined by the World-Wide-Web-Consortium (W3C) as a software system acting as an interface for machine-to-machine communication. These software systems interact with the web service using the Simple Object Access Protocol (SOAP). SOAP processes messages, storing them as Extensible Markup Language (XML). Communication often occurs through Hypertext Transfer Protocol (http) based on Transmission Control Protocol (TCP) and Universal Description Protocol (UDP). the http networking is utilized for exchange, while Universal Description Discovery and Integration (UDDI) is employed for service registration.

SOA components are modeled and further developed in devices, leading to the creation of the Service-Oriented Device Architecture (SODA). For medical applications, a Service-Oriented Medical Device Architecture (SOMDA) is developed to adapt to medical regulations and meet networking requirements. The practical implementation of SODA is the Device Profile for Web Service (DPWS). Based on DPWS,the MDPWS extension specializes in medicine, requiring essential extensions such as dual-channel support, safety context, streaming, and compression.

1.3   IEEE 11073 SDC Core Standards

1.3.1   Medical DPWS IEEE 11073-20702

The ISO/IEEE 11073-20702:2016 "Health informatics - Point-of-care medical device communication-Part 20702" realises the fundamental interoperability. The development of Medical Device Profile for Webservice (MDWPS) was based on the existing DPWS standard. The DPWS implements the SODA, which is specialised for devices and embedded systems. The DPWS is implemented by the Organisation for the Advancement of structured Information standards(OASIS) WS-Discovery. This involves identifying mechanisms for the dynamic search of services at runtime, for example through dynamic discovery. These mechanisms utilise two discovery methods: explicit and implicit discovery. With the explicit method, the service consumer actively searches for providers in the network. In contrast, with implicit discovery, the provider presents itself in the network without being requested. Thanks to the dynamic search, coupled with the ability to exchange meta information and self-descriptions of interfaces, DPWS offers a solid basis for the requirements of networking between medical devices. The MDPWS application protocol uses the same communication principle as DPWS, i.e. data transmission is independent of the actual network. Simple Object Access Protocol (SOAP), Hypertext Transfer Protocol (http), Transmission Control Protocol (TCP) and User Datagram Protocol (UDP) are used to transmit messages. The http is the central protocol of the World Wide Web (3W). Documents that are displayed in the web browser are transferred from the web server to the browser using http. The documents available on the 3W contain hyperlinks, also known as links, which can refer to other hypertext documents. These documents can in turn contain further links, creating a network (web) of interlinked documents, and https is used for secure data exchange. The TCP is used for point-to-point connection between data with confirmation of arrival. In medical environments such as operating rooms or intensive care units, Ethernet is usually used. However,

wireless technologies such as WLAN (IEEE 802.11 family) can also be expected to be used, taking into account appropriate safety and reliability mechanisms. For example mobile monitors that are connected wirelessly, especially during patient transport inside hospitals. The DPWS and the Organisation for the Advacement of structured Information Standard (OASIS) were set up for the development of MD-PWS. There are extensions in the standard from DPWS to Medical DPWS and these are:

1. Reliable Data Transmission: Reliable data transmission is essential in medical networking. The extension with two-channel transmission allows the first error-proof data transmission via a single physical transport medium[6].

2. Safety Context: With the safety context, it is possible to integrate additional, context-related information into the header of a message. This mechanism is primarily used in the context of remote control commands.

3.   MDPWS Mechanisms for Transporting Data Streams: An important application in data transmission in medical technology are curves and waves, such as in electrodiagrams (ECG) or electroencephalograms (EEG). Here you need a large number of individual measured values that are transmitted as data streams. It is recommended to use the core protocol http for this.

4. MDPWS Mechanisms for compressed Data Transmission: MDPWS uses the XML data format for data transmission, but specifies a compact UTF-8 encoded representation of the XML information set. As XML has a comparatively high overhead, the use of Efficient XML Interchange (EXI) is recommended for efficient data transfer. EXI offers compact, loss-free compression of the XML information set and optimises data transfer through efficient binary coding and the use of string tables. The "schema-informed" mode further improves efficiency by excluding components defined in the schema from the EXI stream. In the case of XML representations with extensions these are stored in the string tables.

5. Secure channel: MDPWS uses the essential security methods used in DPWS. "Domain Information and Service Model for Service-Oriented Point-of-Care Medical Device Communication". The non-standardised name of the 11073-10207 standard is (Basic Integrated Clinical Environment Protocol Specification) BICEP and this name is used more frequently. This standard defines both the structure of the data to be transmitted and the possibility of interacting with the information and functions provided. The standard is designed to fulfil the requirements of both current and future medical devices, regardless of the manufacturer or area of application.

1.3.2   Domain Information and Service Model IEEE 11073-10207

Participant model: defines the self-description of the network participants.

Communication Model: describes the services and message model.

Discovery Model: shows the entry and exit of participants in the network (implicit discovery) and that participants are actively searched for.

Extension Model: allows participant model and message model to be introduced in many places via standard.

The participant model consists of two parts - MdDescription and Mdstate. MdDescription is the self-description of the medical device while MdState describes the current state. These two modules are summarised in a medical device in the so-called medical data information base (MDIB). The MDIB and a containment tree are labelled from root node to leaf node. Medical Device System (MDS), Virtual Medical Device (VMD), Channel and Metric. The metrics located in the child nodes in the containment tree. Each element in this tree has a description. The elements in the description tree are called descriptors. The descriptor is shown as a handle and is used to identify the element. Each descriptor has a version counter. The counter is constantly incremented with every change of the descriptor. **Metrics:** The change in measured values or calculation is modulated as metrics, defined by a type and a unit. There are different classes of metric description to be used, including numeric metrics, string-based metrics, string-based metrics with fixed predefined values and datastream-based metrics. **MDDescription:** Each metric is described semantically on the basis of its type. This description is made using coded values that belong to specific nomenclatures. A coded value is made up of a coding system (the nomenclature or a reference to it) and the code (identifier of the facts to be described). The normative nomenclature for the IEEE 11073 SDC family is the IEEE 11073-1010X series, such as IEEE 11073-10101. For a pulse oxymeter e.g. the code 149.530 is used. **MDState:** The device status (MdState) represents the second sub-area of the MDIB. It represents the current status of the medical device, including current metric values. Each element of the device description (MdDescription) has a corresponding state in the entirety of the device state, whereby the assignment is made via the reference to the handle of the descriptor. In contrast to the description and its bijective mapping concept, which allows multistates for context states, the state is dynamic and must be communicated frequently. By separating description and state data economy is achieved without compromising semantic information and the processing effort for devices is reduced. A tree structure in the MDState is avoided by mapping states directly to the elements in the description tree. **Communication Model:** The communications model is divided into two models: Service Model and Message Model. **Message Model:** defines the messages between consumer and provider.

2 Implementation of an SDC model

For the implementation of an SDC model, e.g. a file with the name "Tutorial" is created in the project. After importing all required packages a Universally Unique Identifier (UUID) is created for the two services. In order to help the consumer finding the provider, a definition is created to create an ensemble context.

2.1 Service Provider

Here, the provider should match the consumer in the network. The provider constantly sends out metrics with different values. First, the local ensemble context must be determined in advance. The entire current associated locations should also be changed to *disassociated*. The main function in the provider starts with discovery model (MDPS) to look for the one that is currently running with the name adaptor on *Ethernet*. We set a *local context* to allow easy detection and set model information for the detection. In addition, an information model with the model name, number and URL is defined for the discovery model. Creating a provider class helps to complete the SDC process. The class assumes the role of the provider and supports the ensemble context. It is started with the local device and shows itself as "Discovered" in the network. The local ensemble context based on ensemble ID should be set for easy discovery. The location for the device must be determined. To process the metrics, a local numeric metric must first be generated and then all metrics are brought from MDIB or MdDescription. The metrics brought in are all changed in one transaction. The MdState also reads and calculates all metrics using a specific metric. Afterwards, values from the metrics are memorised and set. These processes are repeated every few seconds.

2.2 Service Consumer

To implement the consumer, the consumer first searches for a provider with the UUID. Two definitions are then created: With the callback function *on-metrics-update* we will receive all changes in the handle or metrics from the provider as parameters. The definition *set-ensemble-context* has the task of calling the operations of the remote devices. It will receive the container for the element in the MDIB. Then the context of the provider (client) is retrieved. An empty handle should be started. The call is repeated continuously until the searched providers are found and a match is made. The main function starts the search for a known provider. Meanwhile, the connection keeps on running and constantly consumes metrics. After the correct provider is found in the network, a response is received with data from MDPWS mechanisms. The response confirms that this is the right provider we are looking for. We perform an infinite loop to keep the consumers running and to notify changes in the metrics.

3 Method for a Test Tool "Compatibility testing"

Presentation of a possible strategy for the development of a test tool to test the compatibility between SDC and the standard. The test tool has the role of the service consumer to connect to the Device Under Test (DUT) and integrate with the DUT. It is necessary for the test to create a one-to-one connection between the devices during the test run. This means that only the test tool and the DUT must be in the network during the test. All offered reports and streams are subscribed to, and all outgoing messages are stored in a database. The test request is divided into two parts, direct test and invariant test. The direct tests also establish a connection to the service provider DUT. The direct test covers the required interaction with the device. Conformity with the associated requirement is assessed by checking the exchange of messages. The invariant test is performed after the connection is terminated. The stored messages from the database are used as the test base. This test covers the requirement to ensure that a particular operation is always performed during operation. To ensure that the invariant tests have sufficient data, they can declare preconditions that must be met before the DUT is disconnected. This test method is performed in four test phases: 1. The test tool attempts to perform a remote test procedure from each of the DUTs. If none occur and the Service Provider behaves correctly, a success message is displayed at the end of the first phase. 2. The direct tests are performed. 3. A check is made to ensure that all the prerequisites for the invariant tests have been met. If not, the appropriate messages are triggered or manipulations are performed before the connection is terminated. 4. The invariant tests are performed.

4   Evaluation

The development of a test tool requires a long period of preparation. The standard to which the software will be developed has to be worked out, evaluated and planned for in advance. The areas to which the standard will apply have to be defined. Then the structure and design have to be analysed. The programming of this specific test tool has not yet been implemented. Therefore, no results can be presented yet. The strategy has been successfully developed and lays the foundation for the programming and implementation of the testing tool. One problem with the strategy is currently the division of the test sections. While the direct test covers the required interaction with the device, the invariant test of the tool tests a specific operation, for example the suitable ventilation pressure for a specific ventilation setting. Since the two tests are not independent of each other, in case of a failure in the first test, the second test is not carried out. This ensures a suitable communication between the devices at the necessary performance required for medical device software. The vulnerabilities in the system prevent our communication between the devices. If the function is tested and the device responds perfectly, the weak points in the system are not discovered. By integrating a fault into the system and monitoring the systems reactions the system's response could be evaluated. The next testing of errors for each operation in the communication protocol could show this.

5   Conclusion

In summary, this paper shows how the communication between the SDC standard works. The IEEE 11073-SDC standard family and the sub-standards are presented. Finally, a test strategy is developed using test methods to create a test tool. This test tool allows the conformance of the device to the standard to be evaluated. In the next step we will try to implement the test tool and use it on our software. The concept will also be tested with an already developed test method. The test results will be compared and the final results will be evaluated. Missing information about the existing software on the manufacturer's devices makes it difficult to collect the required data for the tests. This could lead to problems in programming and errors in the results.

Acknowledgement

The work has been carried out at the Institute for Software Engineering and Programming Languages, Universität zu Lübeck.

Author's Statement

Conflict of interest: Author states no conflict of interest.

6   References

[1] Westhealth Institute, "The value of medical device interoperability," 2013. [Online]. Available: https://www.westhealth.org/wp-content/uploads/2015/02/The-Value-of-Medical-Device-Interoperability.pdf

[2] M. Kasparick, M. Schmitz, B. Andersen, M. Rockstroh, S. Franke, S. Schlichting, F. Golatowski, and D. Timmermann, "Or.net: a service-oriented architecture for safe and dynamic medical device interoperability," *Biomedical Engineering / Biomedizinische Technik*, vol. 63, no. 1, pp. 11–30, 2018.

[3] M. Kasparick, B. Beichler, B. Konieczek, A. Besting, M. Rethfeldt, F. Golatowski, and D. Timmermann, "Measuring latencies of IEEE 11073 compliant service-oriented medical device stacks," in *IECON 2017 - 43rd Annual Conference of the IEEE Industrial Electronics Society, Beijing, China, 2017*. IEEE, 2017, pp. 8640–8647.

[4] T. Hottenbacher and J. Haase, "Functional testing and performance evaluation of networked medical devices with hardware-in-the-loop simulation," in *49th Annual Conference of the IEEE Industrial Electronics Society, IECON 2023, Singapore, October 16-19, 2023*. IEEE, 2023, pp. 1–7.

[5] I. Melzer, *Service-orientierte Architekturen mit Web Services - Konzepte, Standards, Praxis (2. Aufl.)*. Spektrum Akademischer Verlag, 2007.

[6] S. Pöhlsen, W. Schöch, and S. Schlichting, "A protocol for dual channel transmission in service-oriented medical device architectures based on web services," Jan. 2011.

2

Biomedical Engineering

Development of an automated resistance measurement device

Frederic Stichler [1,2], Dennis Kusnik [2], and Stefan Müller [3]

[1] Biomedical Engineering, Luebeck University of Applied Sciences, frederic.pablo.stichler@stud.th-luebeck.de
[2] Dr. Langer Medical, Waldkirch, kussnik@medical-langer.de
[3] Medical Sensors and Devices Laboratory, Luebeck University of Applied Sciences, stefan.mueller@th-luebeck.de

Abstract

This study aimed to develop an integrated, automated cable testing system for Dr. Langer Medical, focusing on multi-wire cable quality, specifically breakage and insulation resistance. A product vision was established, followed by the evaluation and implementation of a measurement method into a prototype for testing resistance values in devices. A printed circuit board (PCB) was subsequently developed, achieving the targeted measurement accuracy of 1 Ω within a 1-100 Ω range and extending to 5 MΩ, with an average error of 0.35 Ω in the low ohm range. This precision enables testing cables with up to 50 wires. The system features a touch display for manual and automatic measurements, including calibration and external multimeter integration, ensuring comprehensive cable quality assurance.

1 Introduction

Intraoperative neuromonitoring (IONM) is the continuous monitoring of nerves during surgery, e.g. monitoring of the recurrent laryngeal nerve during thyroid surgery [1], with over 60,000 operations per year, thyroid surgery is one of the most common in Germany [2]. Failure of such a device can have serious consequences. This project therefore undertook the development of an automated cable tester to determine its ability to accurately and reliably assess the functionality of both internal and external cables manufactured by Dr. Langer Medical, a leading manufacturer of IONM devices. Currently, testing is done manually with a multimeter, which is time-consuming, complicated, error-prone and difficult to document. The idea was to automate this process, which would allow for fast, reliable and well-documented testing. The requirements for the device to be developed were derived from current test methods and the desired effects. Thus, the cables must be tested for cable breakage and insulation breakage (for multi-wire cables). It is important to achieve an accuracy of 1 Ω in the low-impedance range (cables resistance), whereas a measuring range of up to several MΩ is most important for insulation determination. As previous described, automatic measurement is the main requirement. In addition, cables with up to 50 wires are to be tested, which means that it must be possible to test 50 pins on both sides (between related pins (same wire) and between different pins (two wires)). It must also be possible to connect an external multimeter to perform a second test. In addition, the device must be operable via a touch screen as a standalone device.

This project was carried out as part of an internship and involved the development of a product vision (by defining System Specification Requirements and System Architec-ture Requirements according to the V-model). A modular approach was implemented and the final product was divided into five modules (Power supply unit, User interface unit, microcontroller unit, network interface unit and resistance measurement unit). This research paper describes the development and testing of the resistance measurement module.

2 Materials and Methods

In this section, the development of a prototype of the resistance measuring unit is described in more detail. For this purpose, first the chosen design is described in closer detail and then the development steps: proof-of-concept, PCB production and validation.

Resistance measurement is a commonly used tool, as some physical units can often be derived from a resistance measurement, such as strain gauges, where the desired quantity to measure, compression and strain, can be related to a changing resistance [3]. Therefore, their are several circuits that can be used to measure resistance, for example, bridge circuits for accurate determination in there reference region [4], [5].

Several measuring circuit configurations were evaluated where the main challenge was to obtain an accurate measurement in the low resistance range while maintaining a wide measuring range (up to 5 MΩ).

The simplest and most direct solution is to use a constant current source and measure the voltage that drops across the resistor under the applied current. However, in order to achieve the desired measurement range and at the same time stay within the low voltage range of the chosen analog-to-digital converter (ADC), high gain factors are required, which would lead to high noise sensitivity. Another possibility would be to use a constant voltage source and measure the flowing current. However, since an ideal cable has very a low resistance, this would lead to high currents. To avoid this, a voltage divider configuration can also be chosen, where a second resistor with a known value is connected in series with the cable to be measured and the voltage drop across the comparison resistor is measured. With the help of the voltage divider formula, the resistance of the cable can now be determined. In order to obtain a large measuring range, a voltage divider with two selectable comparison resistors was chosen. A comparison resistor of 68 Ω for the low resistance range and 100 kΩ for the high resistance range was chosen. These values result from the required accuracy, the voltage range of the ADC, the accuracy of the ADC and the applied constant voltage. Thus the measured voltage can be calculated using 1:

$$U_{measured} = U_{constant} * \frac{R_{comparison}}{R_{cable} + R_{comparison}} \quad (1)$$

To calculate the cable resistance from the measured value, the formula is rearranged:

$$R_{cable} = R_{comparison} * (\frac{U_{constat}}{U_{measured}} - 1) \quad (2)$$

It's important to acknowledge that the transfer functions seen in 1 and 2 are non linear, although the measurement points of the ADC a equidistant, but the related measurement points in the resistance space are not. This leads to the change of the measurement range between two bits and the accuracy can only be expressed as an averaged accuracy. As stated in the requirements, each of the measurement methods shall be applicable on all 50 pins on both connector sides. This requires a switching option of the pins and the method. The correct connection (which can be programmed) between the method and the outgoing pins is ensured by a three level multiplex system, consisting of two 4:1 and one 1:16 multiplex. This enables 50 pins for the cables testing plus the remaining 14 pins for calibration etc. The first multiplex selects the method (resistance measurement with the developed unit or with an external multimeter), which is implemented using a 4:1 multiplexers, and the next two multiplexers route the signal to the correct pin in a 4:1 and 1:16 configuration. This configuration is implemented twice for the two pathways, in total the multiplex systems consist out of four 4:1 multiplex and eight 1:16 multiplexers.

Since the multiplexers have a non-negligible resistance (also called R_{on}), the equations (1) and (2) change to:

$$U_{measured} = U_{const} * \frac{R_{comparison}}{R_{cable} + R_{comparison} + R_{on}} \quad (3)$$

respectively

$$R_{cable} = R_{comparison} * (\frac{U_{const}}{U_{measured}} - 1) - R_{on} \quad (4)$$

It's clear that the correct determination of the R_{on} value is important for the accuracy of the calculation.

To test the developed measurement method, a proof of concept was created, for which an STM32F429I evaluation board was used as a basis. Since it has an already integrated display, as well as ADC and enough General Purpose Input/Output (GPIO) pins, it is perfectly suited to promote rapid development. To stay within the acceptable voltage range of the evaluation board, the constant measurement voltage was set to 3.3V, which was provided by a low-dropout regulator (LDO). As previously described, the voltage was then applied to the desired pin of the cable via multiplexers and fed back to the measuring circuit from a desired pin (again via multiplexers) (therefore it is possible to measure directly, i.e. from one end of a wire to the other end, or to measure crosswise, i.e. between two wires). The measuring circuit consists of two resistors connected in parallel, which in turn are connected in series with an NMOS-FET. Thus, by driving the NMOSFETs, it can be determined which comparison resistor is connected in series with the cable to be measured. The voltage is measured across the comparison resistor (as described above), first using a buffer to ensure as far as possible that the measurement does not influence the voltage. Then a passive RC low-pass filter was used to suppress possible high-frequency noise. The final voltage measurement is made with a 12-bit ADC, which is already part of the evaluation board.

The entire circuit can be controlled via an integrated touch display. A firmware was written in C++ that allows the user to select which pins of the cable are to be measured and in which range. The firmware also implements the conversion of the measured voltage into the measured resistance. A digital low pass filter with a deviation of more than 20 values was also implemented to filter out further noise.

To verify the results of the proof-of-concept circuit, the circuit was additionally simulated in LT Spice. Based on the results of both approaches, the circuit was extended by another measurement module (with the same structure: buffer, RC filter and ADC), which monitors the output of the LDO. Based on this circuit, a printed circuit board was designed and assembled. The evaluation board can be easily plugged onto the PCB board. To measure the performance of this prototype, a test board was created. Various resistors in the range of 1 to 50 Ω were placed on the board, and this board can also simply be plugged onto the actual measurement prototype (Fig. 1 shows the complete setup). It should be noted here that the final power supply is still via a laboratory power supply unit. For the final prototype, additional functionality was added to the firmware. Firstly, an automatic measurement function was added to allow a predefined series of measurements to be performed automatically. Secondly, a calibration function was added that attempts to determine the R_{on} resistances of the multiplexers as accurately as possible using known resistances on the PCB.

A total of five different measurement series were recorded in order to test and validate all the possibilities of the system. The first measurement series was performed in manual measurement mode with a fixed R_{on} value (10 Ω). This value was determined from the typical values of the multiplexers given in the data sheet. The second and third series

Figure 1: The complete system, with the STM32F429 on the right with a touch screen to control the device and the test device with various resistors on the left. Both devices are mounted on the developed PCB below.

of measurements are directly related, since here the ability to test with an external multimeter was checked. For this purpose, the measurement method was changed from manual measurement to multimeter measurement and the various test resistors were also measured. In test series two, the raw measurement data is provided, while in test series three, a fixed value for R_{on} is subtracted. However, this value was determined by the automatic calibration function of the instrument. Measurement series four is the normal manual resistance measurement, but with the R_{on} value determined automatically as described above. Measurement series five is the fully automatic measurement, where the user only starts the test case and then the results are displayed.

3 Results and Discussion

The different measurement results for the 5 measurement series are shown in Fig. 2, the respective values in Table 1 (for the low ohm region, in all measurements a range up to 5 MΩ was achieved.). It can be seen that the measurement with fixed R_{on}, based on the typical R_{on} values, is the least accurate. The graph shows that there is a systematic error due to the poor estimation of R_{on} in the series of measurements (mean error -1.49 Ω). When the calibration function of the device is used, the accuracy improves by about 1 Ω, which is the difference between the fixed and the calibrated R_{on}. This gives an average error of -0.37 Ω, but a closer look at the measurements still shows a constant offset. This indicates that even with automatic calibration the R_{on} value is set too high. The same effect occurs when measuring with an external multimeter. Subtracting the R_{on} value obtained by calibration from the multimeter value gives an average error of 0.42 Ω. However, this time

a positive offset voltage is observed, indicating that the R_{on} value is underestimated this time, which could be caused by contact resistances at the multimeter terminals. The automatic measurement achieves the best accuracy with a mean error of -0.35 Ω, but the difference to the manual measurement is very minimal that it cannot be considered a significant improvement over the manual measurement. There are other sources of error to consider other than miscalculating the R_{on} value. For example, the contact resistance between the pins and the wires, but also the contacts of the various integrated circuits (IC) and the PCB. A common approach to avoid contact resistance is to use a four probe measurement, but this is not applicable in this case as the measurement method is multiplexed to the desired pin. In addition, other sources of noise such as multiplexer and operational amplifier leakage current have been neglected.

Test seris	mean error [Ω]	SD
manual, fix R_{on}	-1,49	0.18
multi	9.42	0.16
multi, fix R_{on}	0.42	0.16
manual, cal R_{on}	-0.37	0.18
auto, cal R_{on}	-0.35	0.21

Table 1: Measurement results for the measurement in the low ohm region (SD: standard deviation).

4 Conclusion

As described above, a resistance measurement circuit was successfully developed to achieve the desired accuracy of 1 Ω in the range 1-100 Ω with a measurement range of 5 MΩ for cable measurement. It was shown that a resistance

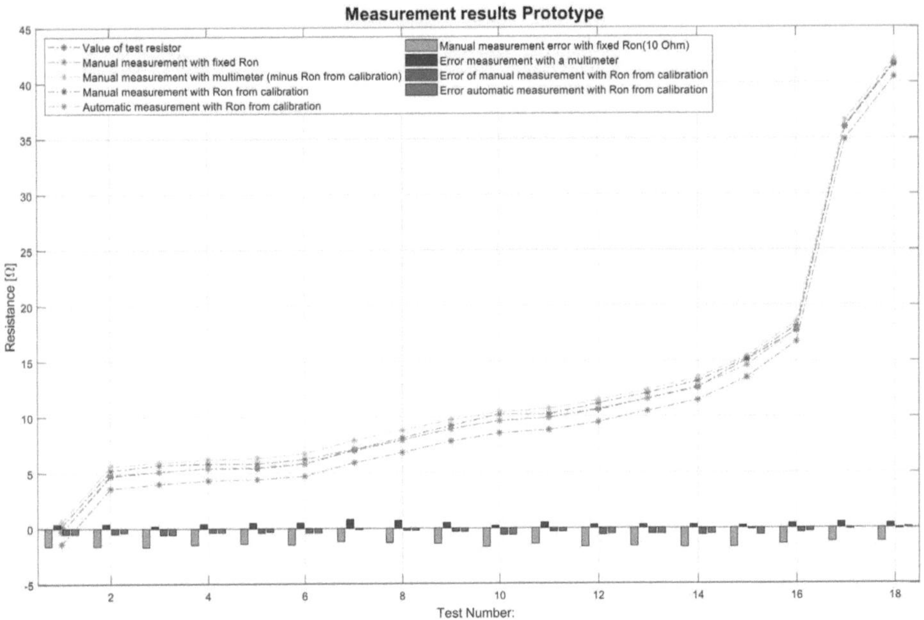

Figure 2: Results of the different measurement series conducted with the PCB board.

divider with selectable comparison resistors was a suitable solution to meet the requirements, and even measure significantly more accurately. In an iterative process, the circuit was improved and finally a prototype with a printed circuit board was produced for testing.

Although the solution meets all the requirements, some improvements could easily be made by using three LDOs, one as a power source for the ICs, one as a reference voltage for the ADC and one for the measurement circuit, thus reducing the load on the LDO that powers the measurement circuit and improving stability and accuracy. A further improvement would be to develop a more sophisticated calibration circuit and calculation that assigns an individual R_{on} value to each multiplexer. And for each measurement, the different R_{on} of the measurement path could be summed and used.

The next step will be to integrate the measurement module into the larger concept of an overall device, for which concepts have already been developed.

Acknowledgement

The work has been carried out at Dr. Langer Medical, Waldkirch and supervised by Steffan Müller, Medical Sensors and Devices Laboratory, Luebeck University of Applied Sciences.

Authors' Statement

Conflict of interest: Authors state no conflict of interest.

5   References

[1] Y. Liu, C. Chen, Z. Fu, Y. Zhang, Y. Han, B. Chen et al. "Effectiveness of the recurrent laryngeal nerve monitoring during endoscopic thyroid surgery: systematic review and meta-analysis." International journal of surgery, vol. 109,no. 7, pp. 2070–2081, 2023.

[2] Statistisches Bundesamt *Fallpauschalenbezogene Krankenhausstatistik (DRG-Statistik) Operationen und Prozeduren der vollstationären Patientinnen und Patienten in Krankenhäusern (4-Steller) 2021* 2022, Wiesbaden. Available: https://shorturl.at/xyH38. [last accessed on 2023-09-26].

[3] D. D. L. Chung *A critical review of piezoresistivity and its application in electrical-resistance-based strain sensing.* Journal of Materials Science, vol. 55, no. 32, pp. 15367–15396, 2020.

[4] A. Agarwal, J. Lang *Foundations of Analog and Digital Electronic Circuits.* Elsevier Morgan Kaufmann, Boston, 2005.

[5] E. Schruefer, L. Reindl and B. Zagar *Messung von ohmschen Widerständen; Widerstandsaufnehmer.* In: Elektrische Messtechnik, Carl Hanser Verlag, München, pp. 195–247, 2012.

Prototyping of a Standalone WiFi Sensing System using Single-Antenna Devices

Robert Schumann [1], Frédéric Li [2] and Marcin Grzegorzek [2]

[1] Medical Informatics, Universität zu Lübeck, robert.schumann@student.uni-luebeck.de
[2] Institute of Medical Informatics, Universität zu Lübeck, {fr.li, marcin.grzegorzek}@uni.luebeck.de

Abstract

Contactless methods of human sensing have seen a rise in popularity in the last years. As they require no device on the tracked object and can cover larger areas, they are of particular interest for health related use-cases. Sensing via reflected WiFi signals, known as WiFi Sensing, is very promising due to the ubiquity of WiFi networks as infrastructure. However research around WiFi sensing is still hindered by the limitations of the hardware needed to collect signal data, consequently limiting data gathering and testing in realistic setups. This work aimed to develop a novel standalone WiFi Sensing system along particular requirements for health-related use-cases. We were able to successfully develop a prototype system enabling quick and easy gathering of channel state information. Additionally the system is able to scale to multiple devices and can be configured to realistically emulate a variety of setups.

1 Introduction

Technologies using reflected WiFi signals to observe the environment of a transceiver pair have become increasingly popular over the recent years. As WiFi signals propagate through the environment away from the transmitter (Tx), they are reflected by static objects like furniture and walls, as well as dynamic objects like humans (see Fig. 1). The effects of such objects can be observed in the signal at the receiver (Rx) expressed in the channel state information (CSI) as the change in amplitude and phase. This effect allows for tracking without any device on a target object. These technologies, broadly gathered under the term *WiFi sensing*, offer valuable characteristics when compared to wearable sensors or camera-based systems, as they require no physical contact or optical line of sight with the tracked object. Furthermore their range is mostly restricted by the reach of the radio signal and they are able to recognize even fine-grained movements [1]. When compared to camera-based solutions, WiFi sensing systems are privacy-preserving and often low-cost with a mass market for commercial-off-the-shelf (COTS) WiFi devices. As WiFi has become a ubiquitous technology WiFi sensing has a high potential for adoption, promising the (re-)use of existing infrastructure.

A wide array of WiFi sensing applications have been proposed in the fields of localization, activity recognition, gesture recognition, crowd counting and many more. This work specifically focuses on its application in the field of ambient assisted living (AAL). AAL encompasses methods, concepts, services and technologies targeted at assisting people in care in their everyday life. Those primarily include people with disabilities and the elderly,

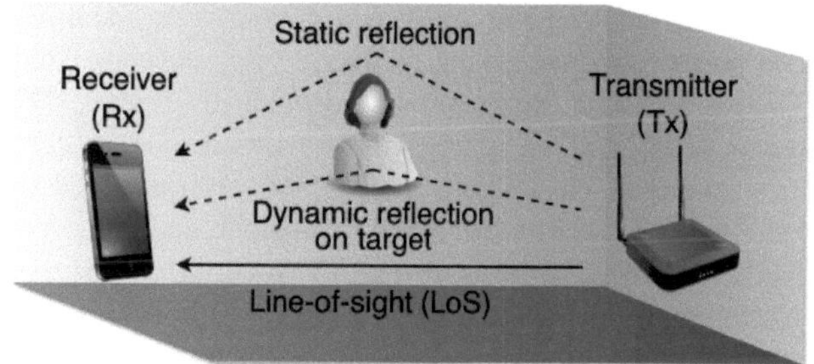

Figure 1: Indoor signal reflection paths

who are potentially in need of medical support and monitoring. Explored applications of WiFi sensing relevant for AAL include sleep monitoring, fall detection [1], respiration monitoring [2] and physical rehabilitation tracking [3]. While many WiFi sensing applications have been proven, general re-use of WiFi infrastructure has not. The accuracy of systems is still strongly dependent on the used Tx-Rx-hardware. The majority of the published systems deploy powerful multi-antenna Rx-devices, while few use devices with a single antenna [4]. Considering the transfer to a real world scenario and the re-use of existing WiFi devices in an AAL setting, the amount and location of devices available for sensing is limited and dictated by the primary non-sensing purpose of each device. As multi-antenna devices access points (APs) with external antennas and (Mini-) PCs equipped with specific network interface cards (NICs), are most commonly used for WiFi Sensing. The majority of WiFi devices use multiple antennas for multiple-input and multiple-output (MIMO) data transmissions to increase bandwidth. As this is not needed for most WiFi devices it is likely only few of these type of devices are placed in a single residential apartment (e.g. routers). Additionally they are potentially bound to outlets for ethernet and power. In contrast, many existing WiFi-enabled devices are single-antenna and that serve a wide variety of functions. Those include smartphones, smart speakers, environmental sensors and smart plugs. A multitude of those devices might already exist in AAL environments, possibly scattered or re-positioned periodically (e.g. smartphones). The antennas of such devices are often internal and more susceptible to noise. Based on this imbalance in WiFi sensing research and already existing WiFi infrastructure the hypothesis is formed, that a sensing system using multiple receivers single-antenna devices offers advantages over traditional systems with a single multi-antenna receiver. Further an easy-to-use standalone system can be developed allowing for quick capture of signal data for WiFi sensing overcoming existing hindrances and allowing adaptability to AAL scenarios.

In support of this hypothesis our previous work surveyed existing single-antenna WiFi sensing systems and their application for AAL [4]. It found a lack in systems leveraging multiple devices and mobile receivers. Specialized sensing devices that enable flexible capture of signal data for WiFi sensing are a possible solution. In this work requirements of such a system were identified and single-antenna sensing hardware was conceptualized. This specialized hardware was developed using COTS components and integrated into a standalone system. This improves on existing sensing software proposed by Hernandez et. al. by offering a standalone and integrated hardware and software system rather, generalized CSI data collection software for single devices. This holistic approach allowed to address existing problems and uniquely enabled the use of multiple devices. It lowers the burden for further research, while also providing a high degree of flexibility. This should help close the gap in WiFi sensing research for single-antenna devices.

2 Material and Methods

From the identified shortcomings of existing systems and the targeted application in an AAL setting the following requirements for the development of a WiFi sensing device were deduced. Firstly, the sensing system needs to be realistically transferable to existing WiFi infrastructure in care facilities and in residential settings. Secondly, it needs to be flexible in its deployment, making it adaptable to a wide range of testing environments. This includes the ability of mobile deployment, allowing for configurations with varying number of devices and also enabling adaptation of signal characteristics with adjustable antennas. The last requirement is that the system should be standalone, not requiring any other secondary devices for data capture. Using these requirements, additional practical considerations were made. The sensing setup needs to be small, cheap, use off-the-shelf components and be easy-to-use.

2.1 Channel State Information (CSI)

For WiFi sensing CSI data is collected to observe the environment. In order to sense, the transmitter broadcasts WiFi packages over a preset channel in the 2.4 GHz band. The receiver then calculates the CSI Tensor H based on the Long Training Field (LTFs), which estimate the transmitted signal x, the received signal y and an estimation of the noise n such that $y = Hx + n$. According to the Orthogonal Frequency-Division Multiplexing (OFDM) used by WiFi, each channel gets divided into sub-carriers. Hence the calculation is performed for each sub-carrier, resulting in a 4-dimensional tensor of complex numbers:

$$H \in \mathbb{C} \underbrace{N \times M}_{\text{Spatial}} \underbrace{\times K}_{\text{\# Sub-Carriers}} \underbrace{\times T}_{\text{Time}}$$

The spatial dimensions N and M refer to the number of sending and receiving antennas. Because we focus on single-antenna devices, this results in a 2D CSI tensor for the recorded data. Each complex number can be differentiated into amplitude and phase of the signal. The phase is often disregarded, as analysis requires very precise clock timings between transmitter and receiver, only achieved by a physical connection defeating the purpose of wireless communication.

2.2 Hardware

As a pre-requisite of WiFi sensing, the collection of CSI data from devices has been a limiting factor in past research, since CSI does not get reported by default on WiFi chipsets [1]. Data collection subsequently mostly relied on the modification of firmware for specific chipsets to enable collection of CSI on devices using this chipset. This limits the number of devices that can be used for WiFi sensing.

Table 1: Devices available for WiFi sensing

Software Tool	Device	Single-Antenna	Cost
Linux 802.11n CSI Tool (Intel Wireless Link 5300)	Mini-PC/Laptop	O	>100
	Hummingbird board	O	330
Atheros CSI Tool	Various Routers	O	>100
	COMPEX WPJ558	O	100
ESP32 CSI Toolkit Wi-ESP	ESP32	●	3
Nexmon	Nexus smartphones	●	300
	Raspberry Pi since 3A+	●	25-80
	Router RT-AC86U	O	150
AX-CSI	Router RT-AX86U	O	250

● := yes, O := no

Table 1 shows an overview of all currently available software tools enabling CSI data collection and their associated devices. With the promise of leveraging existing infrastructure and providing low-cost solutions, this results in a limited choice for the implementation of a WiFi sensing system. As single-antenna devices, the ESP32 family of controller from Espressif, the Raspberry Pi family of mono board computers and the Nexus 5, 6 and 6P smartphones are available for research. With the exception of the smartphones, they are all low-cost, thus lowering the financial barrier for the introduction of new devices. Unfortunately the Nexus line of smartphones compatible with the Nexmon has reached its end-of-life in 2018, meaning current smartphones are not available for research. Between the Raspberry Pi and the ESP32, a choice was made for the embedded tcontroller as it offers very low-power consumption, high flexibility and a wide market for off-the-shelf development board. Consequently the hardware of sender and receiver are built around an ESP32-S3 tcontroller board, specifically the TTGO T-QT Pro by LillyGo. The ESP32 tcontroller family is a variety WiFi and Bluetooth enabled System-on-a-Chip (SoC) microchips, popular for Internet-of-Things (IoT) devices [5]. The ESP32-S3 variant is the flagship chip, with a dual-core processor running at 240MHz. It is integrated on the T-QT Pro board with 4 MB of Flash RAM and 2 MB of PSRAM. The board further allows for the usage of either a directional on-board antenna or usage of an external antenna via a uFL connector. Programming of the ESP32-S3 is done via the Espressif IoT Development Framework (ESP-IDF) and code written in C. The ESP-IDF integrates FreeRTOS as an open source real-time operating system (RTOS) kernel.

2.3　Multi-Device Sensing

Single-antenna sensing systems could overcome their qualitative disadvantages to multi-antenna systems, by being able to leverage a greater number of devices for sensing. The simultaneous use of multiple devices for sensing, also known as multi-link sensing, has shown to increase sensing performance [4]. Generally devices can be organized in two different configurations for sensing:

1. *Tx-Rx-Pairs:* Each receiving device has a unique sending device transmitting only to it. This is the common sensing setup for WiFi sensing. As every Tx-Rx-pair acts independently, they are not benefiting from proximity and each pair introduces additional noise for others.

2. *Central Rx/Tx:* Multiple receiving devices are broadcast to by a single sender or vise-versa. For existing WiFi infrastructure, this is very common as central routers will handle communication to all devices. Sensing coverage over a specific area can be improved by multiple devices nearby. For this, the positioning of the central device becomes critical. Central receivers split their maximum sampling frequency over all transmitters, resulting in a upper limit of possible links.

Currently very few single-antenna sensing systems are using multiple receivers, which is connected to the problem of accurately synchronizing data between the devices in these networked configurations [5]. With a normal sampling frequency between 100 to 1000Hz matching samples over multiple hours requires a high clock accuracy. For example the highest precision clocks of ESP32 tcontrollers have an accuracy of 10 parts-per-million (ppm) and will drift $(60 * 60) * 1e^{-5} = 0.036$ seconds or 36 milliseconds every hour. For accurate synchronization we use an external DS3231 real-time clock (RTC) module, also suggested by Hernandez et al. [5], to accurately (2 ppm or

7ms/h) keep time during the sensing. This module also permits maintaining synchronization over power cycles, using a rechargeable LIR2032 lithium battery as backup. In order to synchronize to other data (e.g. cameras) via a global time, a single NTP synchronization is performed at the start of each recording, if internet access is available.

3　Results and Discussion

For the development of a suitable sensing system, an iterative prototyping process was followed, the final result of which can be seen in Fig. 2. The DS3231 high-precision clock and SD-card reader are connected to via soldered connections. All components are housed in a custom 3D-printed enclosure and can be powered either via the USB-C port of the T-QT Pro board or the custom external Lithium-Polymer battery pack magnetically attached on the back. The battery packs are charged slowly when the devices are powered through the USB-C port. They additionally can be charged faster using a custom charging station also indicating when they are at full capacity. While the receiver requires the SD-card reader and precision clock, it is not needed for the sender. In an effort to make the system as easy to use as possible, senders and receivers are made identical in hardware and only the final software configuration differentiates both.

Figure 2: Minimal set of sensing devices: sender, receiver, charging station and LiPo battery pack (left to right)

The software deployed onto the devices is tasked with coordinating with all the other devices, maintaining time through the precision clock, saving data to the SD-card. As the T-QT Pro board also has a 1.5 cm x 1.5 cm TFT display and two push-buttons, a GUI is also used to indicate the status of the devices, provide information (e.g. system time, MAC-address, disk space) and trigger the start of a recording via the push-buttons. All software was developed from scratch only relying on internal ESP-IDF components and the LVGL Graphics Library. Using the prototype system, we were able to record experimental data, a sample of which can be seen in Fig. 3. It was recorded with a single human present between a single-link, performing large-scale movement (waving arms). General movement activity can be observed in the signal by high amplitudes. This only constitutes an initial test and further evaluation

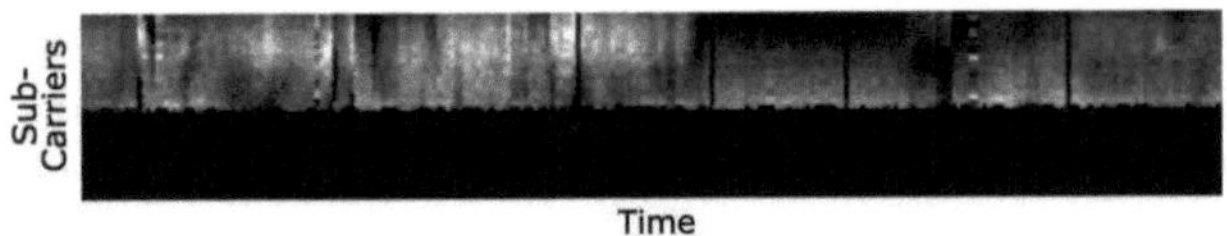

Figure 3: Sample CSI amplitude data, with movement observable in the signal. Y-axis represents the 56 sub-carriers and the x-axis the received packets received over time ($\approx$ 5 seconds at 100Hz). Higher intensity indicates higher values

is needed. Generally speaking, complex movements are not normally identifiable in the signal data and complex machine-learning methods are needed to interpret the multipath reflections all taking effect on the signal data.

For the evaluation of the system along the requirements small test were conducted to verify the capabilities. Firstly the change of signal characteristic by switching between the internal and a variety of external antennas. Secondly a runtime test of the battery pack and data collection over approximately 4 hours. In all tests no additional hardware was necessary for data capture making the system standalone. The cost for a device including battery pack is $\approx$ 25 , with a theoretical limit of 20 devices for the prototypes. The dimensions of the devices are $\approx$ 80x59x29 mm including the battery pack, but excluding the optional antenna. All components used are COTS. Configurations with multiple receivers and ease-of-use were not evaluated. With the increased flexibility in the sensing hardware, it is possible to now imitate a variety of devices that where not adequately covered by research. For example smartphones can be imitated by using external antennas that mimic the characteristics of the specific model. With a fully mobile sensing device, it is also possible to record data while the device is moving. So far, this aspect has not been explored widely in research [4, 5]. Another aspect of scaling the sensing up to a multitude of devices can also now be more easily addressed. While this does alleviate some hardware limitations, hindering leverage of multiple links, challenges still remain regarding software development and data analysis. While this work only focused on the data capture, the existing approaches of federated on-the-edge learning[6] should benefit as multi-link data can be generated more easily.

4 Conclusion

This work introduced WiFi Sensing with a focused application of single-antenna sensing system to health-related uses-cases. Building on previous work, shortcomings of existing sensing systems were identified together with requirements for testing in AAL scenarios, to conceptualize a novel standalone and scalable WiFi sensing system. We were able to successfully develop a low-cost prototype system and gather experimental data using it. This novel standalone sensing system was evaluated along preset requirements and enables easier and more flexible capture of CSI data compared with existing systems. This could help close the research gap between multi-antenna and single-antenna WiFi sensing. Further work will focus on using the prototype of the sensing system in a study focusing on room-scale device-free presence detection, localization and activity recognition. This should also help in validating the system and inform further iteration on the hardware and software. Eventually an open-source publication of all hardware and software, as well as build instructions is planned to enable easy replication.

Acknowledgement

This work is partially funded by the German *Bundesministerium für Bildung und Forschung (BMBF)* under the grant number 16KISA074 and was carried out and supervised by the Institute of Medical Informatics, Universität zu Lübeck.

Authors' Statement

Conflict of interest: Authors state no conflict of interest.

5 References

[1] Y. Ma, G. Zhou, and S. Wang, "Wifi sensing with channel state information: A survey," *ACM Comput. Surv.*, vol. 52, no. 3, Jun. 2019. [Online]. Available: https://doi.org/10.1145/3310194

[2] F. Li, M. Valero, H. Shahriar, R. A. Khan, and S. I. Ahamed, "Wi-COVID: A COVID-19 symptom detection and patient monitoring framework using WiFi," *Smart Health*, vol. 19, p. 100147, 3 2021. [Online]. Available: https://www.sciencedirect.com/science/article/pii/S2352648320300398

[3] S. M. Hernandez, M. Touhiduzzaman, P. E. Pidcoe, and E. Bulut, "Wi-PT: Wireless Sensing based Low-cost Physical Rehabilitation Tracking," in *2022 IEEE International Conference on E-health Networking, Application & Services (IEEE Healthcom 22)*, Genoa, Italy, Oct. 2022.

[4] R. Schumann, F. Li, and M. Grzegorzek, "Wifi sensing with single-antenna devices for ambient assisted living," in *Proceedings of the 8th International Workshop on Sensor-Based Activity Recognition and Artificial Intelligence*, ser. iWOAR '23. New York, NY, USA: Association for Computing Machinery, 2023. [Online]. Available: https://doi.org/10.1145/3615834.3615841

[5] S. M. Hernandez, "Wifi sensing at the edge towards scalable on-device wireless sensing systems," 2023.

[6] S. M. Hernandez and E. Bulut, "Wifederated: Scalable wifi sensing using edge-based federated learning," *IEEE Internet of Things Journal*, vol. 9, no. 14, p. 1262812640, Jul. 2022.

Developing a portable data logger for medical devices using the Arduino Portenta H7

Jennifer Vibert[1], Niklas Hainz[2], Pascal Krauß[3], Michael Nolte[4], and Stefan Müller[5]

[1] Biomedical Engineering, Luebeck University of Applied Sciences, jennifer.vibert@stud.th-luebeck.de
[2,3,4] Ottobock SE & Co. KGaA, Duderstadt, {niklas.hainz, pascal.krauss, michael.nolte}@ottobock.de
[5] Luebeck University of Applied Sciences, Luebeck, stefan.mueller@th-luebeck.de

Abstract

Data logging is used by Ottobock to record and store readable data regarding forces on and space alignment of prosthetic and other orthopaedic devices. This data can then be used to optimize products by analyzing patterns and differences in data for individual prosthesis users. The current data logger used by Ottobock is too large for the desired applications due to the dimensions of the Arduino Breakout board. By carefully selecting components and desired functionalities, a new breakout board was designed with smaller dimensions than the old board. An inertial measurement unit device was included directly on the board to ensure more accurate data readings for future uses. Printed circuit board manufacturing and testing need to be completed before the device is ready for use.

1 Introduction

Data logging in technical orthopaedics research and development is important because each device is tailored and fitted specifically for individual users [1]. Accordingly, data is collected about how different patients will use a device in order to optimize the end product both generally and on a case-by-case basis. The use of a data logger in prosthetics research has lead to a better understanding of how users interact with their device on a daily basis [2].

A data logger is used by Ottobock to record physical alignment in space, and force data on prosthetic devices. A possible use for the collected data is to analyze a prosthetic user's gait and track changes in their gait over time, or to use pattern recognition to customize the prosthesis' movement [2]. Some Ottobock employees have said that the data loggers used in the past at the company tend to break and lose connection while collecting data, and it is relatively simple to build and customize a data logger with desired features using an Arduino (Sommerville, MA, USA) microcontroller [3], [4].

Currently, the Arduino Portenta H7 is used in unison with the Arduino Portenta Breakout as a data logger at Ottobock. The H7 is a high functioning microcontroller with a dual core processor, allowing it to run tasks in real time in parallel [5]. It attaches to the Portenta Breakout via high density joint test action group (JTAG) connectors on both boards so that the pins are accessible to the user. The issue with the current data logger set up is that the H7 has dimensions 25.40 mm x 66.04 mm, [5] while the breakout board has dimensions 72.00 mm x 164.00 mm [6]. The dimensions of the breakout board make it difficult to use the data logger as a portable device. The idea for this project was to reduce the size of the breakout board by creating a new board with similar - and some additional - functionalities on a smaller scale. The portable data logger will be small enough to place close to the devices from which data is extracted in order to monitor and evaluate data. The logger will be able to extract data from devices using multiple different communication interfaces, and will output the data in a readable format.

2 Material and Methods

2.1 Board Functionalities

The first task in designing the new breakout board was to decide which of the board functionalities would need to be included in the final design. It was initially determined that the data logger would need to definitely include connection points to a display screen and buttons to control it, and various analog connections to different medical device components or sensors. The display screen would show real-time data in a readable format. The buttons had already been programmed in a previous iteration of the data logger, and their purposes when scrolling through a list include "Up", "Down", "Back", and "Ok", allowing the user to indicate which device the logger is extracting data from. The functionality of the buttons is of course changeable through programming, and an extra button will be included on the new board in the event that an additional function is required in the future.

Furthermore, the device had to be compact and portable, meaning the new breakout board needed to have the same dimensions as the H7 board, and the whole device was to be powered by an external battery. Arduino boards are easily powered by batteries, thus this issue was irrelevant for the data logger development. It was also deemed necessary that data storage and transfer via a microSD card should be possible, so the data logger needed to be equipped with an on-board microSD connector. The microSD card would be used to increase the storage volume on the data logger [3] and allow for easy transfer of stored data from the data logger to a binary file, which could then be converted to a readable text file.

To attach to the H7 board, the new breakout board would need to have the same high density connectors as the Portenta Breakout. To make the pins accessible, a connection would need to be made from the pin on the JTAG connector to a solder pad or other component on the new board. The necessary pins for this application included all of the power and ground pins, all of the analog to digital converter (ADC) pins, all of the Serial Peripheral Interface pins, all of the secure digital (SD) pins, and a handful of the general purpose input/output (GPIO) and pulse width modulation (PWM) pins. More pins are available on the H7 board, and could be made accessible for future use cases.

In terms of devices from which data could be extracted, it was determined that an inertial measurement unit (IMU) device would be implemented directly on the board in order to make the data logger more compact. In the past, an IMU has been an external component to the data logger, but including it right on the board would allow for more accurate measurements being logged. There also needed to be an analog connection with a built in filter between the Arduino board and some connector pads where an Ottobock-made instrumentation device for internal testing could be connected. The internal testing device can be simply described as a force sensor for this purpose, however the device is much more complex in reality.

2.2 Component Selection

2.2.1 High Density Connectors

The exact connector part on the H7 board is DF40C-80DP-0.4V(51) from Hirose Electric Co. Ltd. (Amsterdam, Netherlands). The mating receptacle component for this part, as can be found on the Portenta Breakout, is DF40HC(3.5)-80DS-0.4V(51), and this is the component that was used on the newly developed board. The receptacle comes in different stacking heights so that the boards that are being connected can be separated at different lengths. The 4.00 mm high connector was chosen for the application at hand so that components can fit on either side of the final printed circuit board (PCB) while still leaving some room between the stacked boards.

2.2.2 MicroSD Card

A standard microSD card connector part was chosen from the Alps Alpine (Unterschleissheim, Germany) SCHA Series. The microSD connector component needed to be pow-

ered by 3.3 V, which could theoretically be powered by the Arduino, however, an ultra-low dropout voltage regulator was used in order to ensure that the 3.3 V supply would be constant. The Texas Instruments (Dallas, TX, USA) LP2985-N voltage regulator was used for this purpose. The recommended basic application circuit from the data sheet was used to wire the component. A very low equivalent series resistance (ESR) output capacitor was necessary to keep the dropout voltage low, and thus the output voltage stable [7]. For this application, a ceramic output capacitor of 4.7 µF was used. The output capacitance was calculated using (1), where C is capacitance, Q is charge, and V is voltage. A ceramic capacitor was chosen as opposed to a tantalum capacitor due to their low ESR and low cost [8]. A disadvantage of ceramic capacitors is that they are susceptible to mechanical stress and vibrations called piezonoise at high frequencies [8], but this was not an issue for this application.

$$C = \Delta Q / \Delta V \qquad (1)$$

Furthermore, the voltage regulator required an input capacitor greater than or equal to 1 µF, which was included between the input and the ground. The input capacitor was used to minimize input voltage ripples. Including a bypass capacitor of 10 µF on the output side of the regulator reduces output noise to a typical level of about 30 µV [7]. The SD data pins all connected to their respective pins on the high density connector, with pull up resistors of 50 kΩ each to the regulated voltage on the lines. The pull up resistors were used to pull the data lines to a high logic level for operation. A pull up resistor value of 50 kΩ ensures that the data lines are at a high logic level while maintaining a low current as to not cause any current leakage [9].

2.2.3 Buttons

The buttons used on the circuit were standard surface mount technology (SMT) buttons connected to different PWM pins on the Arduino. Each button connected to the Arduino's 3.3 V power supply via a 100 kΩ pull up resistor. These pull up resistors provide the same function to the buttons as to the data pins on the SD card. In parallel with each of the buttons, there was a varistor, which was used in this application as an electrostatic discharge (ESD) suppressor to protect the circuit.

2.2.4 Data Collection Devices

The IMU device was connected via SPI to the Arduino. Between each pin on the IMU and the H7, a 0 Ω resistor was placed. These resistors act as placeholders for other potential resistors or connections. The value of the resistance can be changed out in the final device, and the placeholder where a resistor should be allows for the configuration to be changed easily. Using this method, it is also possible to leave an open circuit in place of the resistor by simply not soldering anything in that place. Using 0 Ω resistors reduces the need to change PCB designs once things have been finalized in the event of an error, because different connections can be made easily, even after the boards have been manufactured.

The IMU is rated to be powered between 2.8 V and 4.5 V, with the optimal power voltage being 3.6 V. Since the H7 power outputs are 3.3 V or 5 V, it was considered that a buck converter from 5 V to 3.6 V could be used in order to power the IMU at the proper rating, however, it was decided that this would take up too much space on the already space-limited board. Although the system voltage (Vsys) pins on the H7 board are reserved, these pins were measured to output 4.2 V when the board was connected to power. For testing purposes, the IMU was connected to the Vsys pin via a 0 Ω resistor, and was also connected to a 3.3 V pin via a 0 Ω resistor. This means that the IMU could theoretically be powered at either 4.2 V or 3.3 V, depending on which line is connected during assembly.

The force sensor connected to the board through a large filter combination. The sensor has an eight-pin analog connection, and the typical output signal from these pins is between 0 V and 5 V. From the force sensor, there is an immediate ESD protection MicroClamp connected to each pin. The signal through each pin then travels through a low-pass filtering network which includes a ferrite bead that filters out high frequency noise. The signal then passes through a voltage divider which reduces the voltage to a level that the Arduino can take as an input. Without this filtering and voltage reduction, the signal would be too noisy and too high for the Arduino to be usable.

2.3 Board Schematic and Layout

The board schematic was laid out in PADS VX2.4. PADS is a circuit design program from Siemens (Plano, TX, USA) and MentorGraphics (Wilsonville, OR, USA). Three different programs are required to create a complete PCB with the PADS software. PADS Designer is the program where the circuit schematic is laid out, including component selection and connection. PADS Layout is used to create a PCB, while PADS Router is used to route all of the components on the PCB. Users can switch between all three programs to ensure changes in the circuit schematic follow through to the PCB.

Each pin on the high density connector component was wired to the appropriate components in the PADS Designer program. It was also at this point where different nets were defined, so all pins that belonged to the same net were automatically connected. It was especially important to define nets on the ground and power pins, so as to not accidentally short circuit any of the pins.

In the PADS Layout program, the design rules for the board were defined, including trace width and clearance rules. A four-layered PCB was chosen for the application due to the simplicity of the circuit. The top and bottom layers were flooded with ground to reduce cross talk between traces, and to minimize ground bounce by providing a low-impedance return path for signals. The second layer was defined as an internal ground layer, and the third layer was defined as a power layer connected to 3.3 V, since this was the most common voltage needed for the on-board components.

The components were placed on the PCB in groups defined by function. Groups were placed in a way that minimized trace lengths as much as possible, although some longer traces were unavoidable due to the board dimensions and the placement of the pin connectors on either end of the board. Components that need to be accessible, such as the buttons and the soldering pads for analog device connection, were placed on the bottom side of the board. The top side housed the high density connectors, the IMU, and a few smaller components.

All components were placed such that current would flow through the components in the intended order. Each component was translated and rotated until the perfect orientation was found, so that similar pins, such as ground pins for example, were located closely together. Capacitors were placed as closely as possible to other components without breaking any design rules to enhance signal integrity, and to decouple the intended signal from noise before continuing through the functional group.

3 Results and Discussion

Using the methods described above, a breakout board PCB for the Arduino Portenta H7 was designed to be used with the H7 as a data logger for prosthetic devices. The final board had dimensions of 25.40 mm x 66.04 mm, which match the dimensions of the H7. The dimensions of the board allow for the H7 board to be stacked on top, while ensuring that the data logger is compact and portable. A data logger of this size could be placed in a much closer proximity to the device that data is being logged from, making measurements more accurate due to shorter connections.

The breakout board schematic has undergone a schematic review in the electronics department at the Ottobock headquarters in Duderstadt. The expected PCB delivery date from the manufacturing company caused a time constraint that did not allow for the PCB to be properly tested by this papers submission date. The PCB design, however, has undergone a similar review to the schematic by the lead PCB designer in Duderstadt. With these revisions, it is expected that PCB testing will be successful.

The top view of the final PCB is shown below (Fig. 1). The hatched lines indicate keep-out areas, which are defined areas where components or copper cannot intersect. This means that traces cannot travel beneath components and vias cannot be placed within a keep-out area. The area under the IMU and the microSD connector were mostly keep-out areas. Since these are the two largest components on the board, this caused a significant decrease in useable space on the board, and was one of the greatest challenges in placing and routing components.

Using the new breakout board in unison with the Portenta H7 and the previously implemented code, the data logger will be able to track acceleration data, angular velocity and acceleration through the IMU, and force data in different directions when the force sensor is connected via the ana-

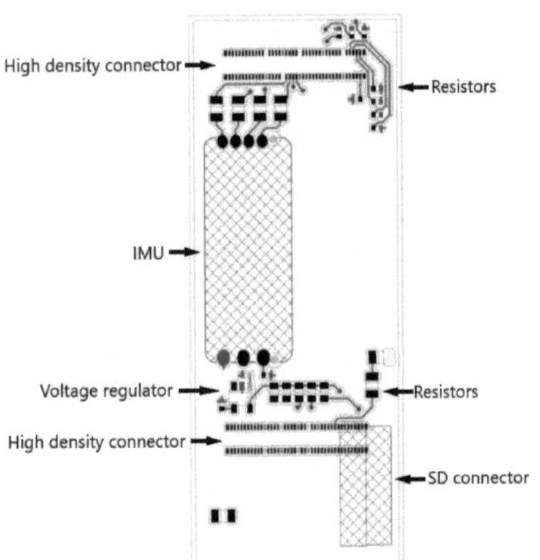

Figure 1: Top view of PCB

log pins. The force sensor could, in theory, be replaced by any analog device, and the data logger would be able to track the incoming signals from said device with a few minor changes in the code.

Furthermore, for the current purposes of the data logger, an SPI bus and the analog pins on the Arduino are supported both by the hardware and software. The new hardware outlined in this paper also supports inter-integrated circuit (I^2C) communication. This is currently not implemented in the software, as there is no current use for it. I^2C, however, is a powerful and common communication protocol [9], thus has been implemented for future possibilities.

4   Conclusion

The final PCB created for this project has the same dimensions as the H7 board and includes all of the necessary components and functionalities for the data logger. The other boards in the Portenta Family include shields that have functionalities like photo and video transfer, as well as GPS tracking. These uses could be implemented in future data logger applications by stacking the shields with the data logger.

At the time of the submission of this paper, the breakout board had not yet been tested. Remaining work on the data logger includes soldering components and testing each of the board functionalities to ensure everything works as expected. Due to some added testing features, including $0\,\Omega$ resistors implemented on connecting wires, it is possible to switch or rewire components in the event that functions do not work as expected during testing.

Upon completing the data logger circuit, the next step would be to design a 3D-printed casing for the device that would help to improve portability. The casing would ideally house all of the components and boards, including all of the optional Portenta shields, while also leaving an open frame for access to the buttons and the display.

The data logger will be used to track the physical alignment in space and the forces acting upon prosthetic devices. The new data logger design will allow for more accurate measurements due to the on-board IMU, and the portability of the new data logger.

Acknowledgement

The work has been carried out at Ottobock SE & Co. KGaA, Duderstadt, and supervised by Prof. Dr.-Ing. Dipl.-Ing. Stefan Müller, Faculty of Applied Natural Sciences, Luebeck University of Applied Sciences, and Michael Nolte, Electronics and Software Department, Ottobock.

Authors' Statement

Conflict of interest: Authors state no conflict of interest. Informed consent: Informed consent has been granted by all listed co-authors.

5   References

[1] S. Laing, P. V. S. Lee, J. C. H. Goh, *Engineering a trans-tibial prosthetic socket for the lower limb amputee.* Annals Academy of Medicine, Singapore, vol. 40, no. 5, p. 252-259, 2011.

[2] L. Osborn et. al, *Monitoring at-home prosthesis control improvements through real-time data logging.* Journal of Neural Engineering, vol. 19, no. 3, p. 036021, 2022. doi: 10.1088/1741-2552/ac6d7b.

[3] M. Gandra, R. Seabra, and F. P. Lima, *A low-cost, versatile data logging system for ecological applications.* Limnology and Oceanography: Methods, vol. 13, no. 3, p. 103-155, 2015. doi: 10.1002/lom3.10012.

[4] A. D. Wickert, C. T. Sandell, B. Schulz and G. H. C. Ng, *Open-source Arduino-compatible data loggers designed for field research.* Hydrology and Earth System Science, vol. 23, no. 4, p. 2065-2076, 2019. doi: 10.5194/hess-23-2065-2019.

[5] Arduino, *Arduino Portenta H7 Collective Datasheet,* January 2020 [Last Modified: December 2023].

[6] Arduino, *Arduino Portenta Breakout Board,* March 2021 [Last Modified: January 2024].

[7] Texas Instruments, *LP2985-N 150-mA, low-noise, low-power, ultra-low-dropout regulator in a SOT-23 package,* March 2000 [Last Modified July 2023]

[8] J. Pelcak, B. Vrana, and T. Zednicek, *Benchmark of Tantalum versus Ceramic Capacitors.* Proceeings of CARTS USA 2005, vol. 1, p. 47-56, 2005.

[9] F. Leens, *An introduction to I2C and SPI protocols.* IEEE Instrumentation & Measurement Magazine, vol. 12, no. 2, p. 8-13, 2009. doi: 10.1109/MIM.2009.4762946.

Airtightness Testing Procedure for Electronic Anaesthesia Vaporizers
– Ensuring Leakage-Free Operation –

Maron Alexander Vagt [1], Max Urban [2]

[1] Biomedical Engineering, Lübeck University of Applied Sciences, maron.alexander.vagt@stud.th-luebeck.de
[2] Lübeck University of Applied Sciences, max.urban@th-luebeck.de

Abstract

This paper presents the development of an automated test device for prototype electronic anesthetic gas vaporizers, prioritizing airtightness. Eight tests, enforcing a specified leakage limit, are conducted using a Dräger Atlan A350 chassis [1]. The device, featuring a vaporizer holder, pneumatic system, and integration with internal software using precise flow sensors, subjects all electronic components and sensors to thorough testing. The methodology delves into airtightness significance, identifies leakage causes, and discusses appropriate test techniques. The rig undergoes real-world tests, improvements, and validation, efficiently checking newly built vaporizers for functionality and airtightness on the prototype production line. The paper concludes with the successful integration of the testing device into an assembling process for over 200 prototypes, prioritizing quality and safety. Due to the sensitive nature of the ongoing development, detailed technical information cannot be disclosed.

1 Introduction

The first anaesthetic device has been a crucial part of anaesthesia equipment ever since the first publicly performed anaesthetic surgery at Massachusetts General Hospital in 1846 [2]. Back then these early gadgets were simple inhalers that relied on the anaesthetic chemical evaporating. A ventilator was added to the gas machines starting in the 1930s, and by the 1950s, it was a standard part of the anaesthesia apparatus [3]. The development of vaporizers has been affected by the introduction of strong inhalational anaesthetics with distinctive features. Modern anaesthetic vaporizers have been created to deliver precise dosages of anaesthetic gas while reducing the effects of temperature and barometric pressure on the evaporation process. As a result, anaesthesiologists can work more precise and safely. The idea is to bring a new device to market which, instead of a mechanical evaporation process, can be controlled electronically. In the current developmental stage, the recently designed vaporizers exist in a conceptual state. Despite this, the sophistication of these concepts allows for the initiation of comprehensive evaluation procedures and tests. Initial models were constructed individually; however, the subsequent phase involves the assembly of electronic vaporizers in a prepared limited production run. The objective is to assess the functionality and proper assembly of these newly manufactured vaporizers using a test device. Conducting tests directly on the production line yields triple advantages: prompt corrective action by the mechanic for any failed devices, and the assurance of safety for individuals engaged in subsequent work or tests involving the vaporizers. Last but not least, further technical and design issues can be addressed, which could otherwise arise to even bigger concerns further down the development process. For each anaesthetic used, up to eight tests are required to verify airtightness within the vaporizer compartments, with a maximum permissible leakage of only a fraction of Milliliters per minute. There are several possibilities that could lead to leakages here, such as broken seals and O-rings from improper manufacturing processes, accidental usage of wrong torques and thereby fitment issues of components [4]. A forgotten screw, even though highly unlikely, could lead to improper assembly of the vaporizer and thereby lead to leakages.

2 Requirements

In the course of this development phase, the objective is to create a testing apparatus specifically designed for assessing the integrity and overall functionality of these vaporizers. The test rig must meet stringent criteria, ensuring the delivery of dependable and reproducible results that yield meaningful insights into the behavior and performance of the Vaporizers. Emphasis is placed on automating the testing process as much as possible. Additionally, it is crucial that the test rig is configured to be placed in the Test-Assembly factory. Similar test rigs are already being used inside the production lines of current anaesthesia vaporizers. For this internship a cost effective but trustworthy test rig should be developed, that could not only find leakages but in the best case can also point the mechanic directly to the leakage area for direct, onside repairs.

2.1 Internal Airtightness Reqirements

The performance of a vaporizer can be highly affected by leakages. If the leak reaches a certain size it could result in an ineffective anaesthesia, since not enough narcotic agent reaches the chamber where it is being evaporated. Also, the needed internal pressures cannot be achieved which is directly connected with the narcotic agents transformation from a liquid into a gaseous state. Additionally, it can directly impact the accuracy of the vaporizer. Leaks can lead to an improper dosage of anaesthetic agent to the patient which can have serious impacts on the state of the patient during operation. A vaporizer that fails to meet its requirements will have severe legal consequences which can be an economical disaster for the manufacturer.

A general guideline for the allowed maximum leakage of anaesthetic agent is the DIN EN ISO 5360 (ISO 5360:2016) which states that a maximum of 0,5 ml of pure agent, evaporated or liquid, is allowed to leak [5]. This however refers to the (re-)filling process of an anaesthetic vaporizer, but has been taken into consideration for finding the leakage limits for this development task. In total up to eight specific tests need to be addressed, to completely check the device for its Airtightness. The number of eight tests is due to the complex internal structure and components of the evaporating device, so that each path and chamber of the Vaporizer is being tested thoroughly.

Test 1: Leakage at Entrance Valve
Investigation of airtightness at the entrance valve internally and externally from the tank and the pneumatic connection point.

Test 2: Leakage of tank
Testing the tank pressure sensor by comparing it to the test equipment sensor (reference sensor) at ambient pressure and at test pressure. Maximum deviation of tank sensor from the reference sensor is +/- 10 mbar. Checking the tightness of the valve on the tank outlet side externally and internally. Checking the tightness of the tank and the filling device with external lid.

Test 3: Deflation time
Test 3 must be carried out immediately after test 2 by decreasing the test pressure at the entrance valve down to ambient pressure (simulation of tank pressure relief during operation). The tank pressure must have fallen to ambient pressure in a maximum of 3 seconds. The values measured by the tank pressure sensor must be recorded.

Test 4: Leakage at tank outlet
Checking the tank outlet side for external and internal leaks.

Test 5: Leakage of fill port (Desflurane only)
Test 5 can only be carried out with the electronic vaporizer for desflurane. Testing the Desflurane filling device or filling valve for external leaks without a lid or closure cap.

Test 6: Main internal Leakages
Checking valve on the tank outlet side for external and internal leaks. Testing the pneumatic interface and all the chambers within, as well as the pneumatic lines between the chambers and paths which could lead to external leaks.

Test 7: Leakage at Safety valve
Checking Safety valve from the direction of the mixing chamber for external and internal leaks.

Test 8: Leakage at Injector side
Checking the Injector from the direction of the internal chambers for external and internal leaks.

3 Material and Methods

This chapter describes how the test stand is set up and operates. In the course of constructing this test stand, all elements are applied and connected in various ways.

3.1 Leak detection methods

In the pursuit of efficient leak detection for electronic vaporizers, a combination of scientific techniques are employed. *Visual inspection*, supported by a leak detection spray, unveils leaks through bubble formation, yet is impractical for series production due to manual processes and limited access to internal components. *Pressure decay testing* reveals leaks by monitoring pressure changes under positive pressure as can be seen in equation (1) for volumetric flow rate in a system. The relationship indicates that the volumetric flow rate (Q_V) is influenced by factors such as the change in pressure (dP), volume (V), time (dt), and pressure gradient (P_G).

$$Q_V = \frac{dP * 60 * V}{dt * P_G} \tag{1}$$

This method however is not applicable in series production and constrained by the vaporizer's volume and extended stabilization times. *Volume Flow Measuring* for assessing leaks in components with larger volumes and minimal rates, mass flow sensors prove pertinent. These sensors utilize laminar flows, temperature differentials, and thermal energy transfer for precise leak rate measurements. The Bronkhorst "EL-Flow" mass flow meter is a frequent choice in Dräger laboratories due to its reliability and versatility [6]. This method provides a comprehensive approach to detecting air leakages, considering factors such as volume, rate, and practicality in various production scenarios.

3.2 Test Setup

This subchapter outlines the design process for pneumatic components, valves, and sensors in the versatile test rig. While not exclusively for leak testing, the rig comprehensively examines various vaporizer components, including temperature sensors, heat-up times, locking sensors, internal pressure sensors, and fill-level sensors. After passing preliminary tests, the vaporizer is filled with anesthetic fluid to verify fill-level sensors efficiently using a tilting mechanism. Operational testing follows, assessing mixing accuracy in flows and anesthesia/air percentages. The subsequent detailed discussion primarily focuses on leak tests.

The design of the test setup involves connecting the electronic vapor's intake and exhaust sides to a new pneumatic system, that is shown in Fig. 1. The intake-pressure is regulated to about 2.8 bar (relative) using a pressure regulator, with a Bronkhorst Flow Meter bypass, ensuring timely refill. This bypass is controlled through the valve *(B)*. Various valves control the airflow within the vaporizer, and a pressure sensor checks the values against the internal pressure sensor. Depending on the particular test scenario (Test 1 to 8) specific valves determine the direction of airflow for both the "air" intake side valve *(C)* and the "Gas Mixer" exhaust side valve *(D)*. Internal vapor components, valves, and sensors are integral but are not detailed here due to confidentiality. An additional Bronkhorst Flow Controller for dosing tests and a Piezocon Gas Concentration Sensor further enhance the test scheme's capabilities. These valves, including the pressure sensor marked as *(3.1)* are controlled via an Arduino Mega, which is implemented into the rest of the test software, communicating directly with other viable components. The code which brings all this together is written in Pearl, however it would be too much to go into further details here.

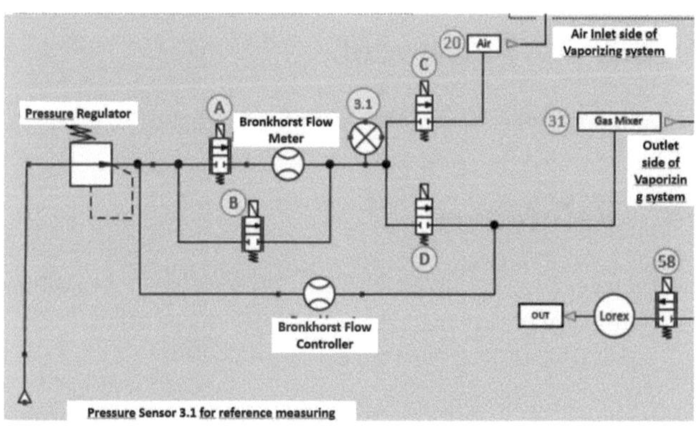

Figure 1: Pneumatic Plan

To minimize air leakage risks by reducing connections, a valve block was designed to integrate the components seen in Fig. 1, except for the two Bronkhorst sensors. Crafted from a solid block of aluminum, the design prioritized short tubing distances to minimize volume and expedite settling times under pressure. Fig. 2 showcases the valve block, featuring four spaces for the valves, a pressure sensor connection in the top left corner, and a pressure reducer connection in the bottom left. The block includes air outlets for the Bronkhorst sensors, as well as the Air intake and Mixer outlet. The block itself was tested for leakages which accumulated for 0,016 to 0,02 ml/min and need to be taken into account for the general leakage test of the vaporizer.

The assembled valve block including tubing and aforementioned components is shown in Fig. 3. These parts are mounted on a wooden back plate, which itself is fixed onto the Atlan Chassis' backside.

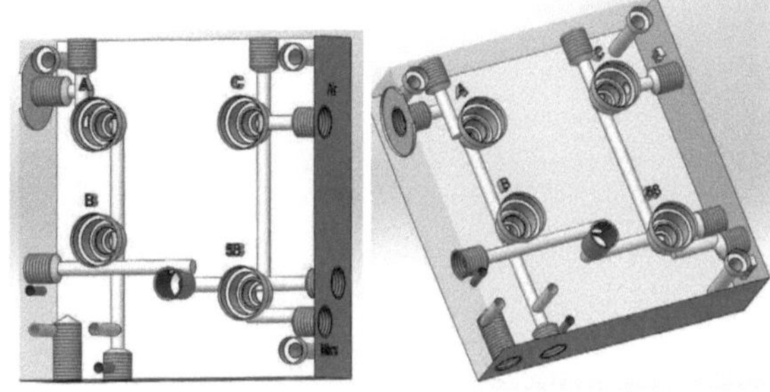

Figure 2: Valve Block

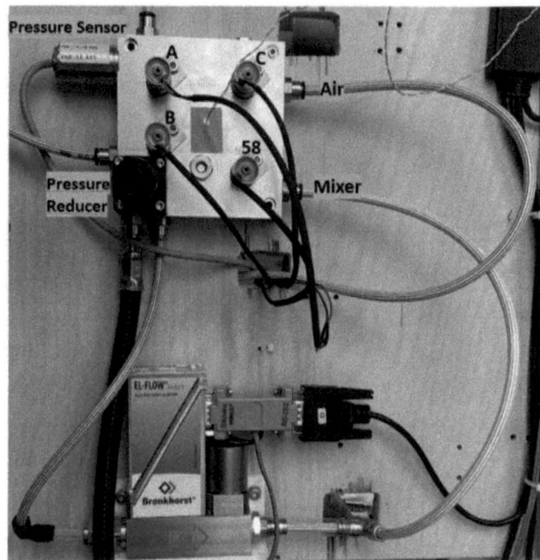

Figure 3: Improved pneumatic structure with valve block

The front view of the complete set-up (not shown here) consists of a laptop computer, which is needed to activate the automated test program and for saving the acquired data. From the front the pneumatic parts and sensors are not directly noticeable and out of reach for misuse. The main holder for the vaporizer is to the right of the Atlan chassis. Here the vaporizer can be attached and once the program has started the vaporizer will be recognized. Further to the right is the tilting mechanism, in which the vaporizer needs to be put to control the function of the fill level sensors.

4 Results and Discussion

To ensure device performance and accuracy, a systematic series of diverse tests was conducted, rigorously examining each construction step for potential air leakages. Cross-referencing automated program-generated data with manual testing methods provided comprehensive validation. The electronic Vapors, when placed in the holder, offer individualized communication capabilities, allowing precise control over specific compartments, including valve adjustments. This capability facilitates reintroduction of flow meters or enables comparative analyses between flow meter measurements and those obtained from pressure decay tests over defined intervals. Figure 4 illustrates a test Log-File, showcasing all steps and results, with particular focus on the leakage results of the eight tests.

By integrating these accumulated datasets into MATLAB,

detection techniques, and the chosen mass flow measurement method were detailed. The testing device setup, featuring pneumatic components, valves, and sensors, included a functional prototype with Arduino Mega control. Verification and validation highlighted the significance of test files and experimental results. Additional tests beyond air leakages, focusing on temperature control, fill-level sensors, and dosing accuracy, were discussed. Recommendations for the future include enhancing the user interface, real-time data visualization through a Graphical User Interface, and implementing advanced data analysis tools. Remote monitoring and continuous improvement processes were suggested. In conclusion, the testing device signifies a significant advancement in ensuring the quality of electronic anaesthetic gas vaporizers, with ongoing improvements crucial for prototype production and quality control.

```
EVa assembly serial                                                      ASSC-0045
Pressure source check                                                    2788.58 [mbar]
Pressure sensor comparison                                               1.80, -3.17 [mbar] (Ref. sensor, deviation)
Pressure sensor comparison                                               2797.09, 5.51 [mbar] (Ref. sensor, deviation)
Test 1: Leakage at 59                                                    0.154696666666666 [mL/min]
Test 2: Leakage of tank / V11                                            0.03233[mL/min] at +2807.2[mbar]
Test 3: Deflation Time                                                   3300 [ms] to go from +2807.2 to +13.3 [mbar]
Test 4: Leakage at V12                                                   0.0147066666666667[mL/min] at +2808.2[mbar]
Test 6: Leakage at 17, 19, 18, 13 and V66                                0.00288[mL/min] at +2805.4[mbar]
Test 7: Leakage at V11                                                   0[mL/min] at 2803.65[mbar]
Test 8: Leakage at V12                                                   0.0363066666666666[mL/min] at 2806.725[mbar]
Static Temp. sensortest [Ifa,Vapa,Vapb,Mixa,Mixb,Ifb]                    -1.10 -1.00 -1.30 -1.50 -1.30 -1.20  at 25.1 [degC]
Temps after heating vap. chamber for 10 [s] [Ifa,Vapa,Vapb,Mixa,Mixb,Ifb] -1.10 11.60 15.50 -1.50 -1.30 -1.30  at 25.1 [degC]
Vap. chamb heat up time and sensor match                                 74900 [ms] to heat vap. chamb to  +61.8 [degC]
Vap. chamb heat up time and sensor match                                 -1.20 [degC] difference at  +60.7 [degC]
Temps after heating mix. chamber for 10 [s] [Ifa,Vapa,Vapb,Mixa,Mixb,Ifb] -0.00 35.90 36.90 6.90 8.10 -1.00  at 25 [degC]
Mix. chamb heat up time and sensor match                                 55300 [ms] to heat vap. chamb to  +60.6 [degC]
Mix. chamb heat up time and sensor match                                 -0.40 [degC] difference at  +62.1 [degC]
Check fill level indicator voltages (empty)                              3074.95, 2489.01,[mV], (Out, Ref) at Indicator number: 0
Check fill level indicator voltages (empty)                              3084.72, 2489.01,[mV], (Out, Ref) at Indicator number: 1
Check fill level indicator voltages (empty)                              3070.07, 2489.01,[mV], (Out, Ref) at Indicator number: 2
Check fill level indicator voltages (empty)                              3050.54, 2489.01,[mV], (Out, Ref) at Indicator number: 3
Check fill level indicator voltages (empty)                              3071.29, 2489.01,[mV], (Out, Ref) at Indicator number: 4
Check fill level indicator voltages (empty)                              3074.95, 2489.01,[mV], (Out, Ref) at Indicator number: 5
Check fill level indicator voltages (filled)                             3825.68, 2495.12,[mV], (Out, Ref) at Indicator number: 5
Check fill level indicator voltages (filled)                             3789.06, 2493.90,[mV], (Out, Ref) at Indicator number: 4
Check fill level indicator voltages (filled)                             3734.13, 2493.90,[mV], (Out, Ref) at Indicator number: 3
Check fill level indicator voltages (filled)                             3778.08, 2493.90,[mV], (Out, Ref) at Indicator number: 2
Check fill level indicator voltages (filled)                             3820.80, 2495.12,[mV], (Out, Ref) at Indicator number: 1
Check fill level indicator voltages (filled)                             3813.48, 2493.90,[mV], (Out, Ref) at Indicator number: 0
Dosagetest
Dosagetest
Dosagetest                                                               Sensitive information
Dosagetest
Dosagetest
*** Test terminated. Overall verdict is PASSED ***
```

Figure 4: Logfile of accumulated test data

results across various tests become easily comparable. An excerpt of the whole plot is shown in Fig. 5. The top diagram will show the leakage tests. Since the tested Vaporizer data of Fig. 4 had very little leakages, only Test 1 can be seen clearly, with about 0,15 ml/min. The second row depicts heat-up times of individual internal components and the last plot highlights the deflation time of the pressurized system. Subplots showing fill-level indications, pressure sensor comparisons, dosage tests and also dosage test temperatures are not shown here.

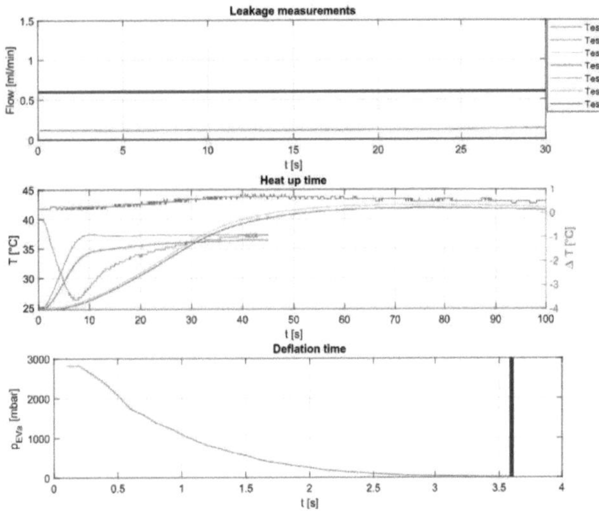

Figure 5: Acquired data during a complete test cycle

5   Conclusion

This internship provided a thorough understanding of addressing air leakages in medical devices, particularly electronic anaesthetic gas vaporizers. Practical knowledge covered causes, detection techniques, and mitigation strategies, enhancing problem-solving skills and commitment to high manufacturing standards. The development and implementation of this testing device underscored the importance of addressing air leakages for patient safety. Various causes,

Acknowledgement

The work has been carried out at Drägerwerk AG & Co. KGaA and supervised by Prof. Dr. sc. nat. Max Urban, Lübeck University of Applied Sciences for which I have to thank for. A special consideration goes to the whole team of the anaesthetic development / Components & Sensors for their constant support.

Authors' Statement

Conflict of interest: Authors state no conflict of interest.

6   References

[1] *Dräger Atlan® A350.* Available: https://www.draeger.com/de_de/Products/Atlan-A300-A300-XL [last accessed on 2023-07-25].

[2] Daniel H. Robinson, Alexander H. Toledo. *Historical development of modern anesthesia.* s.l. : Journal of Investigative Surgery, 2012.

[3] Pablo Romero-Ávila, Carlos Márquez-Espinós, Juan R. Cabrera Afonso. *Historical development of the anesthetic machine: from Morton to the integration of the mechanical ventilator.* Published online 2021 Feb 10 [last accessed on 2023-02-02: https://www.ncbi.nlm.nih.gov/pmc/articles/PMC9373687/]

[4] Marco Rubber and Plastics. [last accessed on 2023-07-08.] https://www.marcorubber.com/o-ring-failure.htm.

[5] Deutsches Institut für Nromungen e.V. DIN EN ISO 5360. *Anästhesiemittelverdampfer - Substanzspezifische Füllsysteme (ISO 5360-2016).* Berlin : Beuth Berlag GmbH, 2016.

[6] Bronkhorst.com. [Online] F-110C, EL-FLOW Select gas flow meter - 0,014...9 mln/min | Bronkhorst.

Review of CPU-Requirements in Medical Nanonetworks

Nils Stumpf [1], Florian-Lennert Lau [2] Stefan Fischer [3]

[1] Medical Informatics, Universität zu Lübeck, nils.stumpf@student.uni-luebeck.de
[2] Institute of Telematics, Universität zu Lübeck, f.lau@uni-luebeck.de
[3] Institute of Telematics, Universität zu Lübeck, fischer@itm.uni-luebeck.de

Abstract

With the gaining popularity of nanonetworks and its research, many new ideas and proposals have been published. From continuous data collection, and precise tumor treatment to quality analysis of veins. Many proposals by the nanonetwork research community are solving problems, that are already solved in macro applications. This paper aims to explain why many problems have to be solved again in a different way, condense the solutions into basic operations and assess whether the solutions are feasable. From analyzing the proposed algorithms, we gain an understanding of what nanodevices have to be capable of. This leads us to a list of instructions, which are necessary to enable the proposed algorithms for nanonetworks. Finally, we discuss the feasibility of implementing these instructions in nanodevices.

1 Introduction

Since 2008, the development of nanonetworks has been a focal point of research. Nanonetworks constitute densely populated networks characterized by nodes possessing low computational complexity and operating on minimal power. These nodes have little memory capacity and potentially support a singular application. Within these networks it is often suggested, that nodes possess the capacity for unrestricted movement throughout the body, leading to random and dynamic positioning within the system. In this work, we focus on in-body nanonetworks, where nanosensors are placed in the human body to monitor and improve health. Lau et al. [7] prove several operations as possible at the nanoscale, but more complex operations are often required by macroscale applications. Meanwhile, many proposals for nanonetworks or applications in nanonetworks are already solved in macro applications. By analyzing proposed algorithms and reducing them into their basic operations, we get a list of operations, which either are already possible or have to be solved. Though it is not yet clear, how computation will look at nanoscale, we assume, that it is desirable to reuse existing algorithms and programs. A source-to-source compiler could translate existing programs into a representation suitable for nanodevices. As a promising approach to nanocomputation, we can use DNA tiles [8]. We then reduce the unsolved operations to the corresponding CPU instructions or analyze their implementations in different systems to assess, whether they are already possible and what are the computational and memory requirements. This gives us an understanding of what a nanomachine should be capable of. A compiler can not only help with simulation and testing but also create opportunities for reusing programs that are built for macro applications. Therefore we want to assess proposals to learn the requirements of

such a compiler.

2 Communication Paradigms

The term network refers to a collection of interconnected devices. In the context of medical nanonetworks, the devices are nanomachines in blood or other liquids, which are often referred to as nanonodes. These nanonodes communicate by sharing messages, data, resources or services. The effectiveness of a network is often measured in bytes per second, which means the effectiveness lies in the speed of communication and the size of the data exchanged. In this paper, we focus on ideas surrounding communication in nanonetworks. Within the nano research community, discourse primarily revolves around four communication paradigms: molecular, electromagnetic, acoustic, and quantum communication. However, this paper focuses exclusively on molecular and electromagnetic communication due to their prominent status as the most promising and extensively investigated methods in recent years. This section will provide a brief overview of these two paradigms.

2.1 Electromagnetic Communication

In the nanonetwork research community, electromagnetic communication is an often discussed potential and appealing method of communication. While retaining the capability to transport a substantial amount of data, high-frequency electromagnetic waves are of 'nano-length'.
Electromagnetic communication involves the transmission of electromagnetic waves of varying wavelengths categorized into frequency bands. These transmissions can be directed or undirected, across different communication mediums like air or water. Further electromagnetic communica-

tion is already widely used at the macroscale where many protocols, standards and technologies already exist to learn and derive from. However, a primary challenge associated with immensely high bandwidths lies in their limited reach of merely 2mm. This is attributed to the pronounced absorption of electromagnetic waves by water molecules.

2.2 Molecular Communication

The innate ability of the body's hormonal communication system to swiftly influence widespread bodily functions motivates the emulation of such mechanisms in artificial systems. An adrenalin rush can impact a human's glucose level, heart frequency, and the cell's oxygen supply. Pheromones extend this communication paradigm to inter-body interactions, akin to hormones but traversing different communication channels. The communication itself is quite similar to hormones, though the channel is different.

As mentioned in the introduction, DNA tiles are capable of building a molecular nanonetwork. DNA tiles are building blocks made from DNA and are used to create two- and three-dimensional structures. They have connective DNA strands, which allow the tiles to connect with other specific tiles [8]. Hence, these tiles can be used as logical building blocks, by allowing or denying specific connections by coding the connective strands. This makes them suitable for creating binary numbers, messages, logical machines or running algorithms. Problems with this type of communication can include the long lifetime of DNA messages and the security issues coming with it. Another challenge is the aimed transportation of these messages.

3 Problems and Operations

In this section, we divide nanonetworks into a selection of components and we assess algorithms presented to solve a problem in the given section. The assessment is done by either condensing the algorithm or the equations into their fundamental operations and then sectioning them into already discussed or yet to be solved. Further, we want to find an instruction set, which can be seen as sufficiently powerful, but not too big to be implemented in nanodevices. To gain an understanding of what is necessary to solve the problems, we look at the CPU instructions and what they do to solve the problem. Another way of gaining information is to look at the implementation of less powerful devices like *Microcontroller Units* (MCU) and assess their implementation. Often they lack precision in comparison to normal CPUs but are less computational or memory expensive.

3.1 Logical Operations

Code is often written in a high-level language like C/C++ and then compiled into machine code, which contains instructions. These instructions can be logical operations, mathematical functions, memory operations, or control flow instructions. In this section, we will focus on logical operations, e.g. $+, *, \wedge, \vee$.

In [7], the authors listed several logical operations that are needed to fulfill the wishes of the nano research community to realize nanonetworks and communication at nanoscale. We aim to add to that list of operations, further, we state operations, that are not yet solved but are needed to solve problems in nanonetworks.

For the electromagnetic paradigm, logic gates are often of nanosize, which makes them applicable to nanonetworks.

3.2 Energy

Especially in the electrocommunication paradigm, energy is a crucial factor. The energy storage of nanomachines is very limited, therefore we assess energy-related proposals. A potential solution to energy supply was presented by Asghari [3], where semi-permanent wired nano clusters build themselves. The authors discussed the potential existence of different nanomachines, e.g. nanobatteries as nodes. They can connect to nanomachines in need of energy and either recharge them or deliver power for their operation. To recharge the battery, another cluster can form, where energy harvesting nodes can connect with empty batteries. In their paper, the authors designed an algorithm using $\gg, +, -, /, \geq, *$, and $\sqcup$.

A different approach to solving the energy problem was presented by Arrabal et al. [2], where the authors proposed an energy-efficient alternative for network density estimation. In their proposed algorithm, the authors only used integer operations, which are proven memory and energy-efficient. The instructions in use are $>, \geq, +, *, {}^x$, and $\binom{n}{k}$. All of those are already listed as wished for by Lau et al. [7] or can be reduced to operations that the authors stated.

3.3 Communication

The realm of network communications contains multiple different tasks, the network has to fulfill. In this section, we look at synchronization, location, and identification as a selection of tasks. For each task, we then assess a proposed algorithm and discuss the needed operations.

3.3.1 Synchronisation

The importance of synchronization in nanonetworks lies in the consistency and integrity of information, the correct ordering of events, the shared access to the shared gateway and the performance of the very dense network.

As a potential algorithm for network synchronization, Khattak [6] presents a nature-inspired approach, calling it the firefly algorithm. The authors minimize the complexity of the algorithm such that it only needs a few operations. Operations like $+, -, *, /, {}^2$, and $<$ are listed as wished for by Lau et al. [7], while $\sqrt{\ }$ and ln is usually solved by the CPU. To calculate the natural logarithm, the nanomachine would be required to store the constant e, which is in x86 CPU stored as a 66-bit number [4]. The IF statement used in the code requires five different instructions, three of which are already discussed by the authors, yet JMP and MOV would have to be solved.

Another approach would be the decentralized Lamport algorithm for clock synchronization in nanonetworks. There, $+, =, ==$, and $<$ is as in the firefly algorithm possible and the IF statement has the exact problems with JMP and MOV.

3.3.2 Location

To send data more efficiently or to deliver drugs precisely to e.g. a tumor, the location of a network node is important. In this section, we discuss different approaches to gaining information on location.

Although Torres et al. [10] aim for use in ultrasonic nanonetworks, the idea could also hold up in electromagnetic nanonetworks. The authors use a combination of *Age of Information* (AoI) and a *Markov Decision Process* (MDP) to predict the location of a device in the nanonetwork. Because of the ultrasonic communication, the authors have to consider the Doppler effect, which introduces the cosine function into the algorithm. This has to be solved either by a CPU instruction comparable to the FCOS [4] instruction, an implementation like CORDIC [11], or by using look-up tables. The original CORDIC implementation for calculating sine, cosine and tangent was designed in 1959 and was then used to locate aircraft. Only $+$ and BIT-SHIFTING are necessary to calculate the cosine function, which makes this algorithm applicable. As an alternative, we consider look-up tables. They have the advantage of being computationally cheap and fast, yet require more memory space, which might be very limited. Other than the cosine function, handling the Doppler effect requires division, which is already listed by Lau et al. [7]. To fulfill the remaining algorithm, we need the instructions $+, -, *$, an IF, a $\sqrt{}$, as well as the constants π and e.

An alternative location method was presented by Stelzner [9], where multiple gateways in the network and triangulation were used to gain location information. The algorithms operations are completely covered by Lau et al. [7], where only IF, , $>$, and $+$ are needed.

3.3.3 Identification

In computer communication, some kind of identification is needed, to send the information to the right device. While in many protocols, addresses in the form of IDs are assigned to devices, in nano communication it is very likely, that addressing has to be solved differently. In electromagnetic communication with e.g. the internet protocol, every device needs at least one unique ID. With the constraint, that every node in a nanonetwork needs an ID, there have to be at least millions of IDs, to serve every node. A major problem arises when millions of nodes would need unique IDs. On the one hand, this requires more memory space and leads to longer messages, on the other hand, it is likely, that there are few types of nodes in large amounts. In wireless sensor networks, the identification of nodes can be solved by location and function.

Fahim et al. [5] uses the firefly algorithm to cluster nodes to reduce the identification complexity. Besides the instructions mentioned in Section 3.3.1, the authors introduced a Mamdani fuzzy logic system, which relies on $\land$ and $\lor$, as well as a lookup table. The added operations are also covered by Lau et al. [7], but the lookup table used for defuzzification might be not applicable due to the low memory of nanodevices.

3.4 Sensors

Nanonetworks are often described in the context of *Internet of Bio-Nano Things* (IoBNT) Networks which are derived from *Internet of Things* (IoT) Networks. These IoT Networks are especially good at collecting and processing sensory data. Almost all sensory data is collected as analog data and has to be transformed into digital data representation. Therefore many MCUs have *Analog to Digital Converters* (ADC). Alongside concretizing the data, ADCs correct the values depending on the MCUs power. This step only depends on the operations $+$ and $-$.

3.5 Movement

As we consider intra-body nanonetworks, there is possibly a lot of natural movement coming from the environment. Nanodevices can be in the blood, where the bloodstream moves the devices constantly. Further Brownian motion can appear as well. Therefore nanodevices are likely to be in constant motion. These types of movements are called passive movement, while the counterpart, active movement, is more interesting to us.

Generally, we want to move along a vector or rotate around an axis. While moving will involve vector addition, rotating around an axis will likely involve sin and cos computations. So computing the forward motion can be solved with $+, -$, and $*$ which is listed by Lau et al. [7] and rotating can be solved using CORDIC-implementations of sin and cos.

3.6 Memory

In nanonetworks, memory is a scarce resource. Yet, many algorithms require memory to store data, for calculations, or need to store constants. In this section, we briefly discuss different possible memory requirements.

As seen in Section 3.3.2, we can solve mathematically complex problems by using less precise approximations of the functions. This also comes with the cost of needing more memory. Also, the precision of irrational numbers and functions including them is memory-bound, as more dedicated memory leads to more precise results.

We cannot say, how powerful a nanomachine can be in the future, but we can assume, that it can compute simple operations. Therefore the precision of operations is likely to be bounded by the memory of the nanomachine.

4 Summary and Conclusion

This paper reviewed some recent conceptual works and discussed the solutions proposed by the authors. By taking [7] as a baseline instruction set, we assess further operations by

Instructions	Explanation
JMP	A CPU instruction to jump over the next instruction.
MOV	A CPU instruction for moving the content from one register to another.
FSQRT	A CPU instruction to calculate the square root of a positive number.
FLDLN2	A CPU instruction to calculate the natural logarithm.
CORDIC SINCOS	A CPU instruction to calculate the sine or cosine function using the CORDIC method.

Table 1: CPU instruction to add to [7]

taking CPU instructions and other implementations as a reference. We were able to add sin, cos, and tan to the possible operations in nanonetworks and listed CPU-Instructions that are desirable in Table 4. With an implementation as by Volder [1], we would also be able to calculate the $\sqrt{}$ and the log of a number. This could make all the shown proposals at least computationally possible, whereas the memory issue still is in place. Because some operations, especially more complex ones, require constants, mostly irrational numbers like π or e. The precision of the calculations with these numbers is dependent on the precision of the floating point value of the irrational numbers, which makes the computational precision bound to the memory.

Finally, we can say that there are more complex computations as proposed by Lau [7] which need to be implemented in nanodevices. Yet, looking at implementations of MCUs or techniques like CORDIC, we might be able to solve them in nanonetworks.

Acknowledgement

The work has been carried out at the Institute of Telematics, Universität zu Lübeck and supervised by Dr. Florian-Lennert Adrian Lau.

Authors' Statement

There is no conflict of interest regarding the publication of this paper.

5   References

[1] ADVANCED MICRO DEVICES, I. CORDIC SQRT Vivado Design Suite Reference Guide: Model-Based DSP Design Using System Generator (UG958) Reader AMD Adaptive Computing Documentation Portal, 2020.

[2] ARRABAL, T., DHOUTAUT, D., AND DEDU, E. Efficient Density Estimation Algorithm for Ultra Dense Wireless Networks. In *2018 27th International Conference on Computer Communication and Networks (ICCCN)* (Hangzhou, July 2018), IEEE, pp. 1–9.

[3] ASGHARI, M. Integration of semi-permanent wired clusters into intrabody wireless perpetual nanonetworks. *Telecommunication Systems 84*, 3 (Nov. 2023), 285–301.

[4] CORPORATION, I. Intel 64 and IA-32 Architectures Software Developers Manual, Combined Volumes: 1, 2A, 2B, 2C, 2D, 3A, 3B, 3C, 3D, and 4.

[5] FAHIM, H., LI, W., JAVAID, S., SADIQ FAREED, M. M., AHMED, G., AND KHATTAK, M. K. Fuzzy Logic and Bio-Inspired Firefly Algorithm Based Routing Scheme in Intrabody Nanonetworks. *Sensors 19*, 24 (Jan. 2019), 5526. Number: 24 Publisher: Multidisciplinary Digital Publishing Institute.

[6] KHATTAK, M. Fuzzy Logic and Bio-Inspired Firefly Algorithm Based Routing Scheme in Intrabody Nanonetworks. *Sensors 19* (Dec. 2019).

[7] LAU, F.-L. A., BÜTHER, F., AND GERLACH, B. Computational requirements for nano-machines: there is limited space at the bottom. In *Proceedings of the 4th ACM International Conference on Nanoscale Computing and Communication* (Washington D.C., Sept. 2017), ACM, pp. 1–6.

[8] LAU, F.-L. A., BÜTHER, F., GEYER, R., AND FISCHER, S. Computation of decision problems within messages in DNA-tile-based molecular nanonetworks. *Nano Communication Networks 21* (sep 2019), 100245.

[9] STELZNER, M., AND TRAUPE, I. FCNN: Location Awareness Based on a Lightweight Hop Count Routing Body Coordinates Concept. In *Proceedings of the Sixth Annual ACM International Conference on Nanoscale Computing and Communication* (Dublin Ireland, Sept. 2019), ACM, pp. 1–2.

[10] TORRES GÓMEZ, J., ANGJO, J., AND DRESSLER, F. Age-of-Information-Based Performance of Ultrasonic Communication Channels for Nanosensor-to-Gateway Communication. *IEEE Transactions on Molecular, Biological and Multi-Scale Communications 9*, 2 (June 2023), 112–123.

[11] VOLDER, J. The CORDIC computing technique. In *Papers presented at the the March 3-5, 1959, western joint computer conference on XX - IRE-AIEE-ACM '59 (Western)* (San Francisco, California, 1959), ACM Press, pp. 257–261.

Optimizing Anaesthesia Devices: Temperature Sensor Integration for Enhanced Safety

Jonathan Ahnfeldt [1] and Stefan Müller [2]

[1] Biomedical Engineering, University of Applied Sciences Lübeck, Jonathan.Ahnfeldt@stud.th-luebeck.de
[2] Medical Sensors and Devices Lab, University of Applied Sciences Lübeck, Stefan.Mueller@th-luebeck.de

Abstract

Without full comprehension about the temperature inside the dosing unit of an anaesthesia device, reliable application of anaesthetic gas is not guaranteed. For a specific development project of Dräger, the temperature of the dosage module holder shall be monitored by the adjacent gas sensor unit, which has temperature sensors inbuilt. Investigations about temperature accuracy were carried out in a reference, regular and error case supervised by a reference temperature sensor. In these three trials, a temperature step response was induced by a 1 W heat source and measured in six points. The results affirm the measuring capability of the gas sensor unit, even though its firm attachment is indispensable.

1 Introduction

In order to perform surgical operations in hospitals, patients are relieved of consciousness, pain and muscle reflexes with the help of inhalation anaesthesia machines. The precise administration of the required medication in modern devices is carried out electronically and serves to ensure the safety of the patient, reduce costs and protect the environment [1]. In an anaesthesia device from Dräger, the fresh gas is enriched with anaesthetic gas by the dosage module [2], which is plugged into the dosage module holder. To monitor the anaesthetic gas concentration, the so-called sensor AGAS (Anaesthetic GAs Sensor) is attached to the holder with two screws and accesses the enriched gas, as shown in Fig. 1.

Full control about the anaesthetic gas in any complex environmental condition requires continuous measurement of the temperature T also in the dosage module holder as another instance to prevent dosage deviations. In combination with the pressure p in the given volume V, the total number of gas molecules N can be obtained, linked by the Boltzmann constant k_B [3] within the ideal gas equation (1):

$$p \cdot V = N \cdot k_\mathrm{B} \cdot T. \tag{1}$$

Here, p in Dalton's law (2) contains the partial pressures of any present gases, where the gas concentrations can be derived from:

$$p = \sum p_{gas,i}. \tag{2}$$

Since temperature and pressure sensors are present in the AGAS but not the dosage module holder, the question is: *Can the temperature of the dosage module holder be represented by the temperature sensor in the bottom housing of the AGAS both in a regular and error case?*

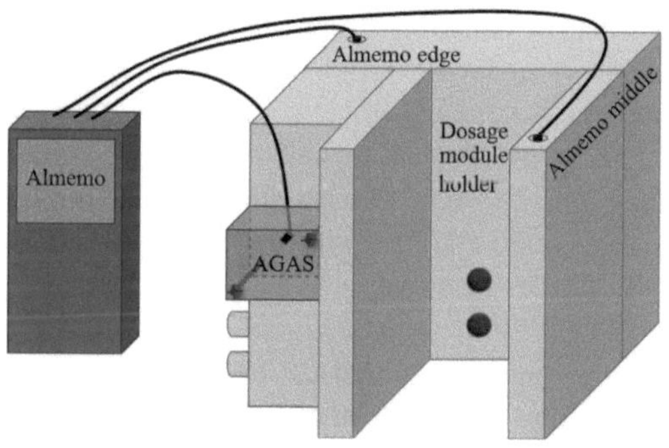

Figure 1: Set-up of AGAS and dosage module holder without the dosage module. The Almemo is used for reference temperature measurements, with two locations in the dosage module holder and two within the AGAS.

2 Material and Methods

Here, components and measuring devices are outlined, followed by their integration into the experimental process.

2.1 AGAS

The AGAS is an anaesthetic gas concentration sensor based on infrared absorption. The two-part housing contains a cuvette with gas inlet and outlet, which is surrounded by a PCB, where infrared radiator and detector are mounted for measuring the gas concentration. The cuvette can be heated by a transistor and is thermally detached from the bottom housing. Therefore NTC temperature sensors exist both in cuvette and bottom housing, as seen in Fig. 2, which have a lower resistance with increasing temperature [4].

2.2 Dosage module holder

The dosage module holder is the port for the dosage module, which enriches the fresh gas with anaesthetic gas. The AGAS sensor is attached onto the dosage module holder with its bottom housing in downstream direction.

2.3 Ahlborn ALMEMO temperature measuring instrument 2590A

For reference temperature measurements the Almemo temperature measuring instrument is used. It is a four-channel data logger using Pt100 and Pt1000 sensors with a system accuracy of 0.03% of the measuring value ± 3 digits at 2.5 measuring operations per second [5].

2.4 Experimental setup

The temperature in the dosage module holder needs to be supervised. For this case, it will be proved, whether the temperature sensor in the bottom housing of the AGAS called *AGAS Temp2 bottom* is sufficient for that. If not, another sensor has to be built directly into the dosage module holder. The accuracy of the measurement depends on the thermal conductivity between both elements [6] and must not exceed ± 1 K in an unheated mounted case [7].

The final objective is to compare the actual temperature in the dosage module holder with the measured temperature from the AGAS to prove its suitability for the measuring purpose.

First, the accuracy of the AGAS' two temperature measuring points must be validated by two reference sensors of the Almemo, as shown in Fig. 2. Then, two reference sensors of the Almemo are placed in the dosage module holder edge and middle, as shown in Fig. 1. In total this gives six measuring points, summarised in Table 1.

The Almemo sensors in the dosage module holder are placed within 1.5 cm deep holes, which are filled with heat sink compound, and fixed with tape to assure a high thermal coupling. In the AGAS, the Almemo sensors are placed as close as possible to the AGAS temperature sensors, as it can be seen in Fig. 2. The Almemo sensor in the cuvette is also placed in a 1.5 cm deep hole filled with heat sink compound, whereas the bottom housing sensor is covered with hear sink compound and stuck onto the surface with tape.

Table 1: Temperature sensor setup in AGAS and dosage module holder (DMH)

Position	AGAS Naming	Almemo Naming
AGAS cuvette	AGAS Temp1	Almemo Temp1
AGAS bottom	AGAS Temp2	Almemo Temp2
DMH edge	n/a	Almemo edge
DMH middle	n/a	Almemo middle

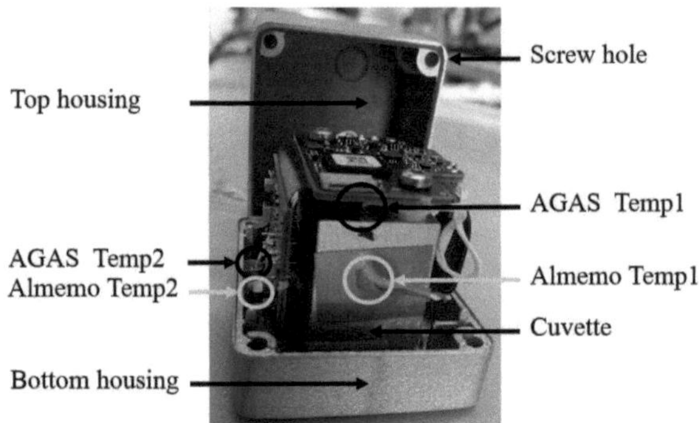

Figure 2: Temperature sensor set-up in AGAS with internal sensors (black) and external reference sensors (grey).

2.5 Experimental procedure

The best way to measure the temperature impact from one to another physical body is to trigger a step response by heating one of the bodies, measuring the temperature in both bodies and comparing these. Here, the heating body is the AGAS, since it has a transistor heater built in at the cuvette. Temperature sensors are used as described.

All trials were executed in room temperature conditions of 22 °C. Before starting, AGAS and dosage module holder are unsupplied and at room temperature. For the first ten minutes, the sensor will be switched on, which takes the PCB's supply power of 0.65 W. After that, the AGAS will be screwed onto the dosage module holder (except in the reference case) and waited for another ten minutes to evaluate the thermal coupling in an unheated case. Lastly, the 1 W powered heater element is switched on for the next 60 mins to assess the thermal coupling in a heated case. This procedure is summed up in Table 2.

Table 2: Experimental procedure

Time	Sensor state
Default	Sensor off
Min 0	Supply power 0.65 W + no heater power
Min 10	AGAS connected to dosage module holder
Min 20	Supply power 0.65 W + heater power 1 W
Min 80	Sensor off

To face the introduced question, the temperature behavior will be investigated in three different cases:

1. Reference measurement: AGAS is not attached to the dosage module holder, providing temperature behavior information solely during its heating up at 1 W, as there is no heat transfer onto the dosage module holder.

2. Regular case measurement: AGAS is firmly attached with two screws, enabling maximal heat energy transfer to the dosage module holder.

3. Error case measurement: AGAS is loosely attached with a single screw, simulating improperly mounting or potential loss of contact with the dosage module holder over time due to vibrations.

3 Results and Discussion

The step responses to 1 W heating are shown in Fig. 3 − 5. As a detailed analysis, all relevant temperature values and differences of interest are displayed in Table 3, after extracting and calculating them manually from the raw-data.
Finally a comprehensive discussion of the results follows, where the evaluation criteria are further explained.

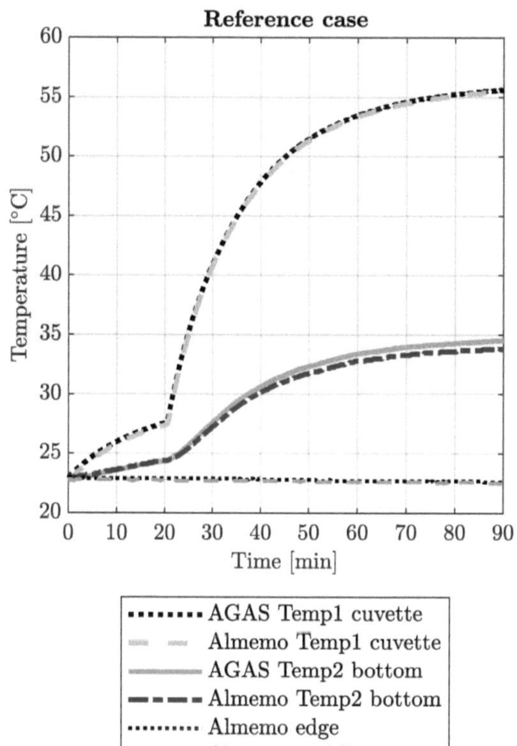

Figure 3: Step response to 1 W heating - Reference case

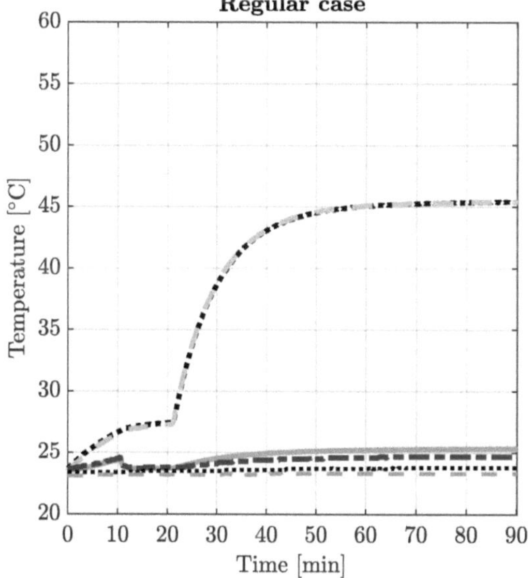

Figure 4: Step response to 1 W heating - Regular case

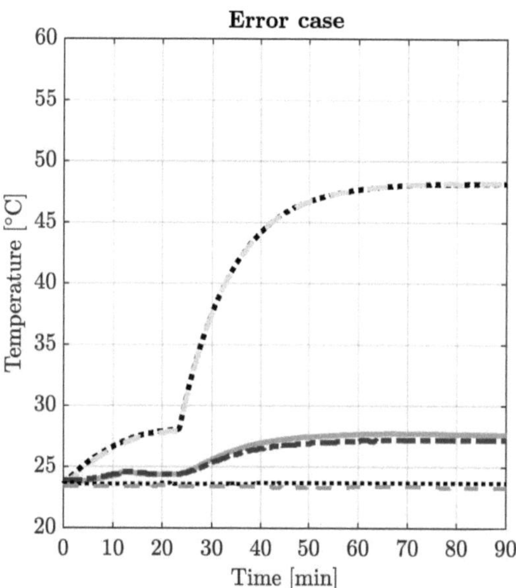

Figure 5: Step response to 1 W heating - Error case

Table 3: Results of AGAS and Almemo temperature measurements after 1h 1W heating (if not explicitly stated)

Criteria	Regular case	Error case	Ref. case
ΔT (AGAS bottom − Almemo bottom)	0.63 K	0.57 K	0.64 K
ΔT (AGAS bottom − Almemo edge) (unheated mounting)	0.25 K	0.69 K	n/a
ΔT (AGAS bottom − Almemo edge)	1.53 K	4.06 K	11.64 K
ΔT (AGAS bottom − Almemo middle)	2.03 K	4.36 K	11.64 K
AGAS Temp1 cuvette	45.40°C	48.20°C	55.33°C

Now the results are comprehensively discussed in ascending order according to the criteria outlined in Table 3:

The *measuring ability of AGAS Temp2 bottom* is evaluated with respect to the reference sensor. In all three cases the Almemo bottom temperature is reproductively 0.60±0.04 K lower than AGAS Temp2 bottom after one hour of heating up with 1 W, despite identical initial temperatures. In comparison to that, there is no significant temperature difference (< 0.1 K) between the other measuring point AGAS Temp1 cuvette and its reference Almemo cuvette (not listed in the table). The reason for the temperature difference is rather the superficial tape-fixed and heat sink compound surrounded attachment of the Almemo sensor at the bottom housing, leading to degraded thermal coupling.
However, the retaining temperature difference between them was reproduced in the three trials, even though the thermal coupling was different. Therefore, the measuring

capability of AGAS Temp2 bottom is rated as sufficient.

The *thermal coupling between unheated AGAS and dosage module holder* is rated as the following:

When the unheated AGAS is mounted to the dosage module holder in regular and error cases, the temperatures of AGAS Temp2 bottom drop with remaining differences towards Almemo edge of 0.25 K and 0.69 K respectively. In the regular case, the temperature drops immediately (T_{90} = 1 min), while in the error case, the temperature drops slower (T_{90} = 9 min).

These behaviours show, that the heat transfer from the AGAS to the dosage module holder is better in the regular, well-mounted case.

Furthermore, the *thermal coupling between heated AGAS and dosage module holder after 1 h of heating with 1 W* states a temperature difference of AGAS Temp2 bottom towards Almemo edge and Almemo middle of 1.53 K and 2.03 K, respectively for the regular case, while 4.06 K and 4.36 K for the error case respectively, and also 11.64 K and 11.64 K for the unmounted reference case, respectively.

This demonstrates, that the heat energy transfer from the AGAS towards the dosage module holder is higher if the AGAS is mounted properly, leading to a lower temperature difference between them.

Finally, the *maximum temperature of the AGAS cuvette after 1 h of heating with 1 W* is obviously the highest in the case of worst thermal coupling. The better the AGAS is mounted to the dosage module holder, the lower the maximum temperature of the cuvette, because the thermal coupling between both elements is higher, so their temperature is more similar.

In the reference case, measurements stop before reaching maximum temperatures, as it has been shown that the maximum temperature is significantly higher than in regular and error cases. Further experimentation would offer no additional information, since the temperature increase follows a stable step response of a PT2 element.

4   Conclusion

Initial tests confirmed AGAS' temperature values realism. For the specific purpose of the AGAS as a safety instance to monitor the temperature of the dosage module holder it was identified, that regardless the source of heat introduction, AGAS Temp2 is sufficient for temperature measuring under normal circumstances. However, it may fail in extreme circumstances, where thermal coupling obviously is reduced due to improperly mounting. Besides the fact, that leakages would occur, a misinterpretation of the dosage module holders temperature could lead to undesirable behaviour of the system including deviating concentrations.

In these normal circumstances, temperature changes would arise from non-local outer impacts, e.g. by change of room temperature or by the gas flowing through, affecting the dosage module and AGAS simultaneously. This is shown, when the AGAS is mounted to the dosage module holder without heating. Here, the temperature difference between AGAS temp1 bottom and Almemo edge is 0.25 K for the regular case and 0.69 K for the error case. Both meet the requirements of ± 1 K.

More extreme circumstances were simulated by introducing a local heat source to create a temperature step response to be measured by the AGAS. Here, the maximum temperature differences were 2.03 K and 4.36 K, which do not meet the requirements. Nevertheless, applying 1 W additional heat leading to cuvette temperatures over 50 °C does not occur in normal operation of the AGAS.

Technically it could be possible to integrate a routine to detect improperly mounting by executing the proposed or a simplified temperature test and comparing the results with those from a regular case performed during production.

During development, CFD simulation can be an alternative to the physical tests for gaining insights into the temperature behaviour. The biggest challenge here is to model the mechanical coupling affecting the thermal coupling.

The overall knowledge gain from this experiment is that temperature measurements in solid state volumes like the dosage module holder can be done by externally mounted sensors, as long as they are thermally well attached.

Acknowledgement

This work was carried out at Drägerwerk AG & Co. KGaA, Dr. Robert Jahns, and supervised by the Medical Sensors and Devices Lab, University of Applied Sciences Lübeck.

Author's Statement

Authors state no conflict of interest.

5   References

[1] C. Engeln, A. Herden, K. Fritz, E. Siegel, 2017, *Anästhesie-Arbeitsplatz*. In: Medizintechnik: Verfahren - Systeme - Informationsverarbeitung, Springer Berlin Heidelberg, pp. 445-465.

[2] G. Briggs, J. Maycock, 2016. *The anaesthetic machine.* Anaesthesia&Intensive Care Medicine 17, pp. 115-119.

[3] K. M. Tenny, J. S. Cooper, 2022, *Ideal Gas Behavior.* StatPearls Publishing, Treasure Island, Florida

[4] B. Padilla, 2020. *Temperature sensing with thermistors.* Texas Instruments: Dallas, Texas

[5] Ahlborn Mess- und Regelungstechnik GmbH 2023 Holzkirchen, *General technical specifications of the ALMEMO® Measuring Instruments.* Available: [last accessed on 2024-01-16].

[6] C. Uher, 2004. *Thermal conductivity of metals.* In: Thermal conductivity: theory, properties, and applications. Boston, MA: Springer US. pp. 21-91.

[7] Drägerwerk AG & Co. KGaA, 2022, Internal Project Requirements.

Reduction of interference noises on the adjustable-pressure-limiting valve of an anaesthesia machine

Annika Quack [1], Jens Lembke [2], Hauke Paulsen [3]
[1] Medical Engineering Science, Universität zu Lübeck, annika.quack@student.uni-luebeck.de
[2] Program PoC, Drägerwerk AG & Co. KGaA, Jens.Lembke@draeger.com
[3] Institut für Physik, Universität zu Lübeck, paulsen@physik.uni-luebeck.de

Abstract

This study investigates the noise generated by pressure-limiting valves in anesthesia machines that can interfere with patient examination, particularly chest listening, in surgical facilities. It is suspected that the hum is caused by the vibration of the valve piston through a mass-spring system. Various prototypes of valve plates, breathing systems and craters are being developed and tested, which differ in shape, weight and manufacturing process. These are tested both directly on the anesthesia machine and on a test stand with eight valve systems. An eccentric crater, which was initially assumed to be an improvement, shows no advantage over the standard design. No significant noise reduction was also observed with three different valve disk designs. However, a noise reduction was observed with two valve models in which the piston is provided with a notch on one side and a weight on the other side, which is supported on the outer wall. A third valve breaks off when the piston is replaced, suggesting that a more robust design is required for future prototypes. This summary presents the initial results and ongoing efforts to improve the adjustable pressure relief valve design and reduce noise in surgical environments.

1 Introduction

On the anaesthesia machine, the adjustable pressure-limiting valve (APL-valve) is an adjustable pressure relief valve that is used during spontaneous and manual ventilation of the patient. The valve serves a dual purpose: it restricts the maximum pressure during manual ventilation and also expels excess gas into the transfer system during both manual ventilation by hand and spontaneous breathing. It consists of a rotary knob for setting the pressure, a spring that transmits the pressure and a valve plunger [1]. Recently, there have been isolated cases in production and in the field where the valve has started to hum during the function test and during manual ventilation. The resulting pneumatic vibrations spread into the breathing system, to which the APL valve is attached. This noise impairs listening to the patient with a stethoscope. It is assumed that the valve disk is the trigger due to the plate flow phenomenon, as this was also present in the previously used titanium component and was modified [2]. The change to a plastic part has resulted in a significant difference in weight and a change in geometry [3]. It is therefore suspected that one or more of these changes are causing the noise. Several attempts have already been made to eliminate the disturbing noise, for example by changing the geometry of the crater of the breathing system on which the valve disk rests. The idea was to create an eccentric crater so that the air flow lifts the disk more strongly on one side, creating an irregular air flow that prevents the tappet from swinging [4]. Inserting

a vibration damper and increasing of the mass, was so far without success [3]. The new idea is to first check whether the oscillation is triggered at the APL valve, or whether the air flow in the respiratory system could be responsible. Several prototypes are being created for this purpose.

2 Material and Methods

To create the prototypes of the breathing system and the valve piston, the models and technical drawings are created in the CAD program SOLIDWORKS. As the new and old APL valves have different threads for screwing onto the breathing system, the thread has to be modified in order to fit the old valve onto the new breathing system. To do this, a push-on ring with a suitable thread for the old valve is first modelled in the program and the ventilation system is modified. The push-on ring is affixed to the breathing system and sealed with glue to ensure an airtight connection. The push-on ring is constructed from 1.4301 grade steel. To adapt the breathing system, an existing component is ground down so that the ring can be attached. The modified breathing system is attached to an anaesthesia machine to test the influence of the air flow on the hum. The parameters at which the noise occurs are used to test whether the old valve can also be stimulated to vibrate. Three different models (Fig. 1, Fig. 2, Fig. 3) are created to test changes to the valve disk. Two of the prototypes are 3D-printed from PA12 (polyamide 12), the other is made from 1.4301 steel

Figure 1: 3D-printed valve plunger prototype in drop shape

Figure 2: 3D-printed valve plunger prototype with notch on the side to increase the air flow on one side

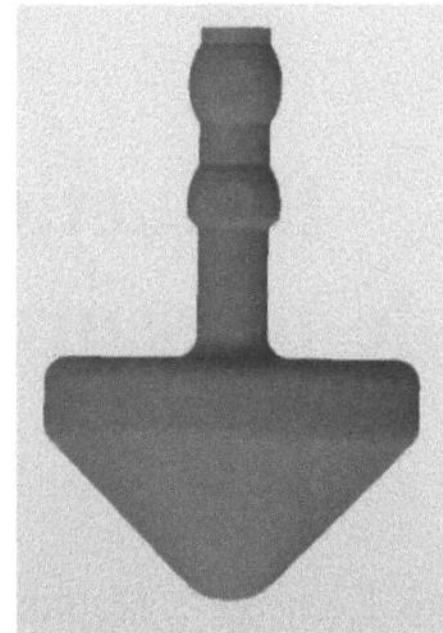

Figure 3: 3D-printed valve disk prototype "Stopfen" with 45° angle that fits directly into the crater and seals at the angle

to achieve a higher weight. The shape of the model in Fig. 1 is designed to optimize airflow dynamics. The current flat configuration of the plate lies planar on the crater, subjecting it to the full force of incoming air. In contrast, the drop-like shape of the plate in this configuration induces a shallower airflow, resulting in a gentler lift. In Fig. 2, the introduced indentation serves the purpose of directing airflow past the plate on one side, enhancing lift on that specific side. This unilateral lifting mechanism minimizes plate fluttering, consequently reducing the associated humming noise. In the shape shown in Fig. 3, the plunger exhibits a pitch set at a 45° angle. The corresponding edge of the crater, on which the plunger rests, also maintains a 45° angle. This meticulous alignment ensures a precise fit of the plate into the crater, promoting a smooth and seamless contact. Additionally, the rounded bottom of the plunger contributes to a more streamlined airflow. Another way to reduce the humming noise is to create air turbulence in the crater so that the airflow hits the piston differently. The crater insert is a 1.3cm plate with five holes for the air flow. In the first experiment, the cardboard plate is punched out and glued into the crater. The APL valve is then tested with the standard plunger on the anaesthesia machine. An eccentric crater has already been produced as a prototype from 1.4301 steel and is being tested as part of this work.

Eight noticeably strong humming valves with the numbers 3,4,5,7,9,10,12,13 from production are used to test the valve tappets. Each of the three different prototypes is inserted into each of the conspicuous valves and tested with the same parameters. The valve tappets are tested both on a test stand, which only tests the APL valve without exposing it to the influence of the breathing system, and on the anaesthesia

device. For this purpose, a compressed air cylinder unit is connected to a test stand specially built for the APL valve in order to supply the valve with air flow. The software flowplotter is used to create a flow ramp to simulate different flow velocities. The software is developed by a Dräger employee to control a volume flow controller. The pause between the individual flow rates is set to 100ms. The flow rate is set in steps of 0.25 L/min from 0 to 20 L/min. The test rig consists of a hollow metal block onto which the APL valve is screwed. A compressed air connection leads to the block, in front of which a flow sensor is fitted to measure the flow rate used. On the opposite side of the block, there is a connection point where various hoses or additional accessories can be attached. With this set-up, the conspicuous valves can be stimulated more reliably and thus tested for noise. The 3D-printed parts are also tested with additional weights. To do this, washers are pushed onto the valve discs from above and glued on with simple adhesive so that they can be removed again for any further tests. The valves are stimulated in two distinct manners: firstly, by the O_2 flush, where O_2 continuously flows through the system. The second method is manual bagging, where airflow is manually introduced into the system by compressing a ventilation bag [5]. The tests were carried out in the following order, and the results in Part 3 are listed in chronological order. First, the eccentric crater and the cardboard plate are tested on the test stand. Then the same craters are tested on the anaesthesia device. After that, the modified breathing system is tested on the device. The next tests are carried out on the test stand with the different valve plates. The same valve plates are then tested on the anaesthesia device. The results of the tests are entered into an Excel spreadsheet and compared with the reference data (Fig. 4). A distinction is made between the different types of excitation. A 0 is entered in the table if no oscillation occurs. A 1 is entered if the valve can be excited with external help, i.e. by pushing up or pulling up and releasing the valve head. An example of the table is shown in Table 1.

Table 1: Result inscription for the frequency and type of humming with the normal crater. The different shades of gray represent different types of air flow. The results highlighted in white are obtained with O_2 flush, the light grey ones are manually bagged and with an additional ventilation tube of one meter and the dark grey ones are also manually bagged without additional tube. The zero represents no excitation, the one is an excitation with an external pulse and the two represents an independent excitation.

valve number	pressure position, static (mbar)					
	5	10	20	30	40	70
3	0	0	1	1	1	0
4	0	0	0	0	0	0
5	0	0	0	0	0	0
7	0	0	0	2	2	0
9	0	0	0	2	1	0
10	0	0	0	2	2	0
12	0	0	0	2	2	1
13	0	0	0	2	2	0
3	0	0	1	1	1	0
4	0	0	1	1	1	0
5	0	0	2	0	1	0
7	0	0	1	2	2	0
9	0	0	1	1	1	0
10	0	0	1	1	1	0
12	0	0	0	1	1	0
13	0	0	2	2	2	0
3	0	0	0	0	0	0
4	0	0	0	0	0	0
5	0	1	2	1	0	0
7	0	0	1	1	0	0
9	0	0	1	0	0	0
10	0	0	0	0	0	0
12	0	0	0	0	0	0
13	0	2	1	1	0	0

3   Results and Discussion

The conspicuous valves with the normal flat piston are tested on the test bench and on the anaesthesia machine as a reference. All eight different valves can be stimulated on the test stand with a flow rate of at least eight litres per minute. The valve position specified in mbar is also important. With all valves, the humming occurs from 20 mbar at the latest, Fig. 4.Depending on the size and age of an adult, between 15 and 30 mbar is set in the field. The first test is carried out with the eccentric crater on the test bench. The eight conspicuous valves are all tested on the crater. As a control, two identical prototypes are tested and the results compared. For both eccentric craters, there are no significant changes in the vibration development at different flow rates and mbar positions of the rotatable part of the valve in contrast to the normal crater. The eccentric crater is tested directly on the anaesthesia device using the same procedure,

the results can be seen in Fig. 5. The test directly on the device also shows no improvements. It can be seen that the number of independently excited oscillations in the standard operating range of 30 mbar has increased, which could even indicate a deterioration. The cardboard panels have no influence on the noise, so it does not make sense to have 3D-printed prototypes produced and carry out further tests with them.

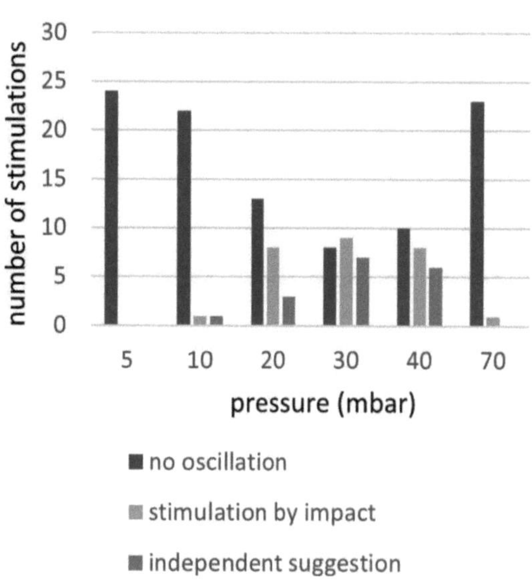

Figure 4: Number of excitations on the x-axis at the different pressures set on the y-axis with the normal crater on the device as a reference. The different bars represent the different types of excitations of the valves. The number of excitations is averaged over all valves.

To test the new breathing system with the old APL valve, various old APL valves are screwed onto the new breathing system and tested. None of the valves can be stimulated to hum, which indicates that the noise is not caused by the breathing system, but comes directly from the APL valve. The tests with the different valve disks are first carried out on the test setup. Both the drop-shaped steel disk and the two 3D-printed plungers can be excited to vibrate. The drop-shaped disk begins to vibrate at a pressure of 5 mbar with a volume flow of 1 l/min (litre per minute). This could be due to the increased weight of 9.3 grams and the resulting increased pressure on the crater. The plunger vibrates continuously at 10, 20, 30, 40 mbar and starts to hum at 70 mbar with a flow rate of 4 l/min. With the 3D-printed piston "Stopfen", the oscillation behaviour is similar; from 5 mbar to 40 mbar, the valve begins to oscillate continuously at a flow rate of 1 l/min. At 70 mbar, it hums from 4 l/min. The third valve disk with the notch behaves slightly better than the other two. It starts to oscillate at 5 mbar with a flow rate of 6 l/min, at 10 mbar with 5 l/min, at 20 and 30 mbar with 9 l/min. It does not oscillate at 70 mbar, which is probably due to the fact that at 70 mbar more pressure is exerted on the plate and it can therefore be lifted more stably by the air flow on one side. These values are an improvement on

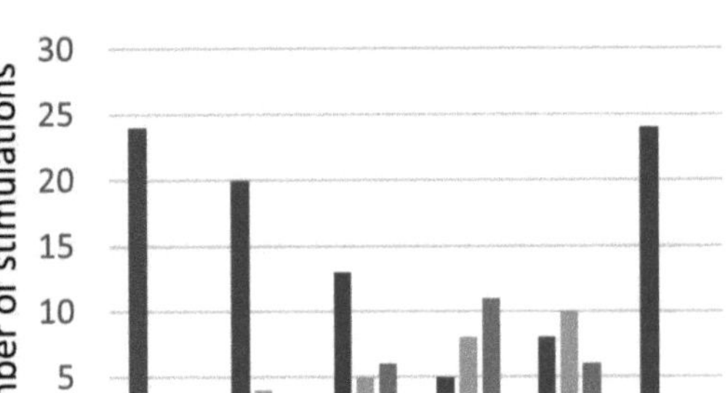

Figure 5: Number of sound excitations on the x-axis at the different pressures set on the y-axis with the eccentric crater on the breathing system. The different bars represent the different types of excitations of the valves. The number of excitations is averaged over all valves.

the other two plates, but not an improvement on the plate currently used in the valve (comparison with Fig. 4). As there is a difference between the behaviour of the plungers on the test setup and directly on the device, these results must also be taken into account. As the results of the tests do not yet represent the desired solution, the weighted 3D-printed plungers are being tested. The plunger "Stopfen" is weighted over its entire surface so that the weight is distributed evenly from above. The additional force makes the plate more stable and makes it more difficult to vibrate. The results show that the additional weight has no influence on the hum. A humming noise can be heard at all mbar settings, as with the plunger without additional weight. The piston with a notch is weighted on the notched side to compensate for the increased air flow on this side. For a second test, it is weighted on the opposite side to support the increased airflow on the notched side. The washer on the opposite side protrudes slightly. The experiment is carried out with two valves. When the plunger is moved to the next valve, the pin of the plate broke off, so new prototypes have to be ordered. No humming occurred with the two valves tested so far. On closer inspection, it can be seen that the washer is so far above the edge of the plunger that it is minimally supported. It is possible that the vibration and thus the humming is suppressed by the support. To test this, another valve disk is modelled and ordered in SOLID-WORKS, which has two devices for support at the edge. This will then be tested in a further experiment.

Due to the rare occurrences in the field and the unclear origin of the noise, it is difficult to determine a clear approach to the problem. As every user can dismantle and adjust the valve themselves, the humming can occur under a wide variety of circumstances. As the user assumes that the device works without problems, they does not pass on the parameters under which the noise occurs, but only that it occurs. The difficulty therefore lies in reproducing the conditions and therefore the noise under which it occurs in the field.

4   Conclusion

After numerous tests, the following conclusion can be drawn. The shapes and weights of the valve discs tested so far have not resulted in any clear improvement. The most promising variant so far is the valve disk with the notch and the additional weight and the resulting support. Due to the damage to the plunger and the resulting delay, new valve plungers have to be manufactured in order to continue the tests. If the tests with the supported plunger are successful, further tests must be carried out to investigate the influence of the supports on the force of the spring and the resulting pressures of the APL valve.

Acknowledgement

The work has been carried out at Drägerwerk AG & Co. KGaA, Moislinger Allee 53-55, 23558 Lübeck and supervised by PD Dr. Hauke Paulsen, Institute of Physik, Universität zu Lübeck.

Authors' Statement

Annika Quack and Jens Lembke are working for Drägerwerk AG & Co. KGaA. PD Dr. Hauke Paulsen has no conflict of interest.

5   References

[1] J. Thomas, M. Weiss, A. R. Schmidt, P. K. Buehler, *Performance of adjustable pressure-limiting (APL) valves in two different modern anaesthesia machines.* Anaesthesia , vol. 72, no. 1, pp. 28–34, 2017.

[2] Spektrum Akademischer Verlag 1998 Heidelberg, *Plattenumströmung.* Available: https://www.spektrum.de/lexikon/physik/plattenumstroemung/11390 [last accessed on 2024-01-18]

[3] J. Zindel, *Fehleranalyse des APL Ventils.* Drägerwerk AG & Co. KGaA, Lübeck, 2023.

[4] M. Rein, *Einführung in die Stömungsmechanik.* Universität Göttingen, 2020.

[5] T. Prien, H. Bürkle, M. Hölzl, C. Hönemann, J. Grensemann, T. Muders, R. Sattler, D. Schädler, T. Krauß, *Funktionsprüfung des Narkosegerätes zur Gewährleistung der Patientensicherheit.* Anästh Instensivmed , vol. 60, pp. 75–83, 2019.

Dimensionality Reduction for Innovative Cell-View Imaging Supported Flow Cytometry Data

Kim Paulke [1,4], Jochen Behrends [2], Thomas Scholzen [2], Linda Zemke [3], Norbert Reiling [3] and Inken Wohlers [4,5]

[1] Medical Engineering Science, Universität zu Lübeck, kim.paulke@student.uni-luebeck.de
[2] Core Facility Fluorescence Cytometry, Research Center Borstel, 23845 Borstel,{jbehrends, tscholzen}@fz-borstel.de
[3] Microbial Interface Biology, Research Center Borstel, 23845 Borstel, {lzemke, nreiling}@fz-borstel.de
[4] Biomolecular Data Science in Pneumology, Research Center Borstel, 23845 Borstel, iwohlers@fz-borstel.de
[5] Universität zu Lübeck, 23562 Lübeck, Germany, inken.wohlers@uni-luebeck.de

Abstract

Flow cytometry is a powerful technique for characterizing cells based on various morpholological and molecular parameters recorded simultaneously. The very recent integration of cell imaging into flow cytometry has led to an exponential increase in the complexity and dimensionality of the related datasets. In this work, we investigate dimensionality reduction approaches to provide deeper insight into image-enhanced flow cytometry data. Specifically, we apply PCA (Principal Component Analysis), t-SNE (t-Distributed Stochastic Neighbor Embedding) and UMAP (Uniform Manifold Approximation and Projection) to analyse cells according to two molecular features marked by fluorescence using spectral flow cytometric parameters as well as innovative image parameters provided by newest flow cytometry technology. The results show that by considering different dimensionality reduction approaches as well as novel image-deduced parameters, a comprehensive cell characterization both on molecular and morphological level can be achieved.

1 Introduction

The ongoing development of technologies in the field of cell analysis has led to groundbreaking advances in biomedical research. The flow cytometer, crucial in this context, operates by hydrodynamically focusing suspended cells, causing them to separate from each other in a liquid stream. The detected forward scatter light (FSC) is proportional to size, while side scatter light (SSC) is proportional to the granularity of the cells, i.e. the internal complexity such as the structure of their nucleus [1]. Modern flow cytometers are limited to conventional or spectral parameters in combination with the usage of anitbody-coupled fluorochromes.

A very recent technological advancement is the computational generation of images from a special laser technique within the flow cytometer, see Fig. 1 (a) for an example. Such imaging enables a more precise characterization of cell morphology as well as a spatial resolution of fluorescent markers. With this, entirely novel investigations, e.g. addressing colocalization of molecular markers, are possible, which have the potential to significantly deepen knowledge in the field of cell biology. This prominent technology is BD Biosciences' CellView™ Image Technology. This method not only enables the quantitative recording of cellular parameters, but also provides an unique insight into the morphological properties of individual cells and increases the number of parameters (combined spectral and images) that can be measured per cell to over 400. This diversity of information is accompanied by the challenge of high dimensionality, which not only complicates the analysis, but also the visualization and interpretation of the data. Therefore, dimensionality reduction is a crucial transformation, aiming to preserve data variability and similarity in fewer dimensions. This facilitates interpretable visualizations, offering insights into cellular relationships.

2 Material and Methods

2.1 FCS file format

Flow cytometer data is stored in a standardized file format known as FCS. This standardization enhances interoperability among diverse flow cytometry instrument manufacturers and compatibility with various analysis platforms. It not only encourages data exchange but also simplifies the development of new analysis software and algorithms.

FCS, a binary file format, includes a text segment with metadata (keyword/value pairs), a data segment creating an expression values matrix (list mode format), and a rarely used analysis segment. The International Society for Advancement of Cytometry (ISAC) traditionally develops and maintains FCS specifications. The current version, FCS3.2 [2], has made significant progress to address modern flow cytometry challenges. Upgrades to 64-bit storage in list mode with floating-point data type, parameter, and event numbers, along with new data field integration, enhance representation of complex experimental conditions.

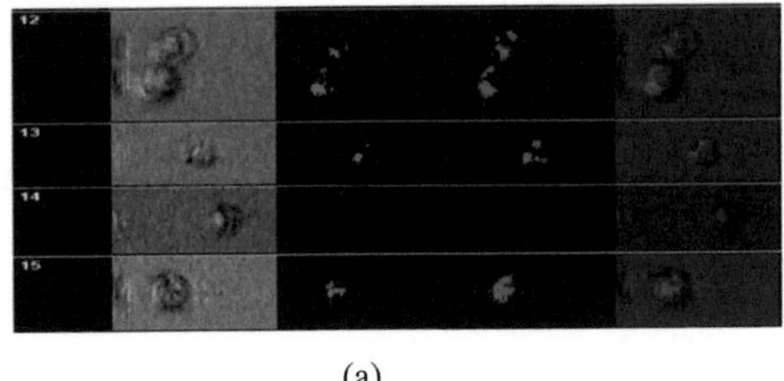

(a)

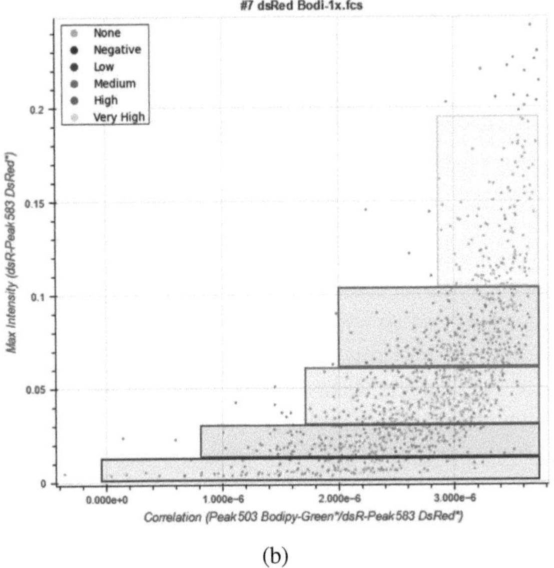

(b)

Figure 1: **(a)** Example of cell view images from *in vitro* generated mouse macrophages. In gray the cell morphology (left). In green and red the two fluorescence markers (center). On the right an overlay image of these three images. **(b)** Scatter plot of correlation of red and green fluorescence markers (x axis) versus the maximum intensity of the red fluorescence marker (y axis). Five gates were manually defined at level four of the gating strategy and are denoted in different colors. They refer to negative red fluorescence as well as to low, medium, high and very high correlation between red and green fluorescent markers.

2.2   BD FACSDiscover™ S8 data

The BD FACSDiscover™ S8 Cell Sorter has been launched in May 2023 and represents the most innovative flow cytometry device to date. It features two novel technologies: BD SpectralFX™ technology and BD CellView™ image technology. It expands the possibilities of cell analysis and sorting by combining flow cytometry data with spatial and morphological information through image features with high speed (10.000 events per second). For this purpose, the BD FACSDiscover™ S8 has a total of 86 separate optical detectors (78 spectral detection channels, three fluorescence imaging channels, three scatter imaging channels, two scatter channels). Images are generated from digitized signals on a per-event basis and include light loss, forward scatter (FSC), side scatter (SSC) images, and different fluorescent channels. Each photodetector produces a pulse with high-frequency modulations encoding the image (waveform). Fourier analysis is performed to reconstruct the image from the modulated pulse [3]. The parameters derived from images and available for downstream analyses are: center of mass X, center of mass Y, correlation, delta center of mass, diffusivity, eccentricity, maximum intensity, moment (long), moment (short), radial moment, size and total intensity.

2.3   Dimensionality reduction

2.3.1   PCA

Principal component analysis (PCA) is a multivariate statistical method that aims to identify the principal components by calculating the eigenvalues and eigenvectors of the covariance matrix. The eigenvectors represent the directions of maximum variance along which the data vary. The application of PCA thus allows the dimensionality of the data set to be reduced, while retaining most of the original variability.

2.3.2   tSNE

The t-distributed stochastic neighborhood embedding (t-SNE) is a widely used global non-linear method that allows local structures of high-dimensional data to be determined and global structures to be revealed at the same time. It converts Euclidean distances of the initial data points into conditional probabilities representing similarities between points. In this way, complex high-dimensional data is projected into low-dimensional spaces and the local structures are preserved [4].

2.3.3   UMAP

Similar to t-SNE, uniform manifold approximation and projection (UMAP) offers a powerful way to transform complex high-dimensional data into a low-dimensional space while preserving the local structure. UMAP is based on the idea of approximating a manifold. It strives to represent the data in a new space in which the geometric relationships between the points are preserved as accurately as possible [5].

2.4   Gating strategy

A crucial step in the analysis of flow cytometric data is the identification of multidimensional regions that contain functionally and morphologically homogeneous cell groups for further analysis, so-called gating. This entails defining threshold values for parameters like fluorescence signal intensity, typically done manually by examining two-dimensional projections, e.g. as shown in Fig. 1 (b). Choosing the right gating strategy and parameters is crucial to obtain precise and reproducible results. By setting thresholds for certain fluorescence parameters, specific cell types or markers labeled by fluorescent dyes or antibodies can be identified. Multiple hierarchical gating steps are then performed to isolate cell populations sequentially.

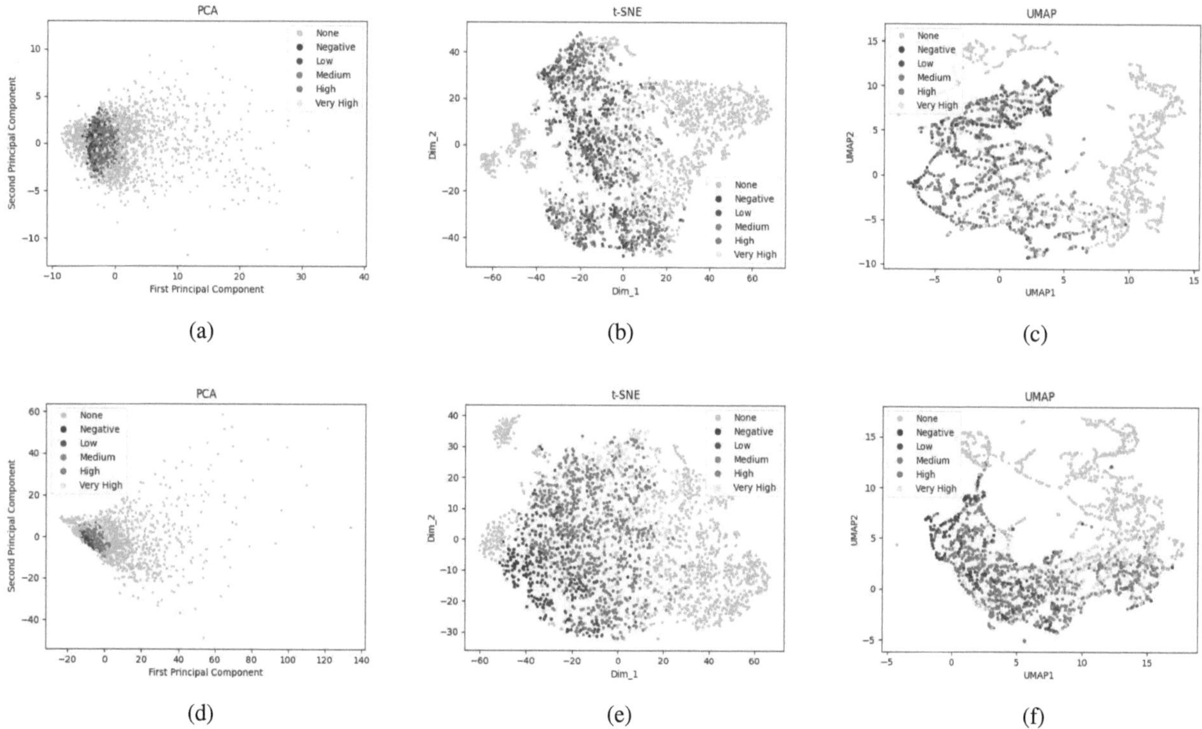

Figure 2: Comparison of the different dimensionality reduction algorithms. The events are colored by their gate membership (negative, low, medium, high, very high). **(a-c)** shows the results without fluorescence-related parameters. Also only morphological information captures roughly the differences characterized via the flurorescent markers. **(d-f)** shows the results with fluorescence and spatial information.

2.5 Data acquisition and analysis

In vitro generated mouse macrophages were infected with Discosoma Red fluorescent protein (red) expressing intracellular bacteria. After immobilization of the bacteria, samples were stained with Bodipy™ 493/503 NHS Ester (green), washed, and measured on the BD FACSDiscover™ S8 using the BD FACSChorus™ Software. In total 2793 events were recorded and exported in FCS file format version FCS3.2.

This file was imported into the Python environment with the FlowKit [6] package, which is one of the few open source software in bioinformatics to support this file format version as of January 2024. FlowKit is designed to facilitate the accurate pre-processing of FCS data, including the correct interpretation of event data using the metadata and the convenient extraction of processed data for use in external libraries. The dimension reduction approaches PCA, t-SNE and UMAP were implemented using the Python library scikit-learn [7] (version 1.4.0).

Parameters of the dimensionality reduction algorithms were set as follows: For PCA, the first two principal components were extracted, for t-SNE the perplexity, a smooth measure of the effective number of neighbors, was set to $perplexity = 30$, for UMAP the size of the locally considered neighborhood was limited to $n\_neighbors = 4$, while the default values were used for the remaining parameters.

3 Results and Discussion

3.1 Defining cell populations

The manual gating (Table 1), which was performed by an experienced flow cytometry expert, played a crucial role in defining cell populations, e.g. artefacts like debris and doublets are excluded. The first level gating was based on the parameters Lightloss-A and SSC-A. In the second step, the gating was refined by analyzing the parameters SSC-H and SSC-W. The third level gating was carried out using the Lightloss H and Lightloss W parameters. In the last step of the manual gating process, the correlation between green and red fluorescence versus the maximum intensity of the red fluorescent marker was considered, see Fig. 1 (b). This allows distinguishing cell populations with respect to presence and co-occurrence of the two fluorescence markers (red and green) and ensures that cells with similar marker expression are grouped together within respective level 4 gates (see Table 1). The five populations defined are used as labels in the visualizations after dimensionality reduction. A detailed overview of the hierarchical gating applied and its results is presented in Table 1.

3.2 Cell characterization

To determine how the dimensionality reduction approaches handle different types of derived image parameters, we tested the capabilities of PCA, t-SNE and UMAP based on

Table 1: Shown is the hierarchical gating strategy that was manually devised by an expert and used to identify and characterize precise cell populations. The table summarizes the remaining cell numbers and percentages at each gating level.

Gate	Count	Absolute	Relative	Level
Scatter	2256	80.77	80.77	1
SSC Singlets	1679	60.11	74.42	2
FSC Singlets	1464	52.42	87.19	3
Negative	123	4.40	8.40	4
Low	189	6.77	12.91	4
Medium	424	15.18	28.96	4
High	458	16.4	31.28	4
Very High	235	8.41	16.05	4

parameters without fluorescence but considering spatial information (Fig. 2 (a-c)) compared to all parameters, i.e. with spatial as well as fluorescent information (Fig. 2 (d-f)). The results show how the integration of different information can lead to an improved characterization of cell populations. Without fluorescence but with spatial information, cell populations tend to organize themselves based on morphological features and spatial patterns. However, no recognizable clusters are generated in this homogeneous data set. Instead, a smooth transition between the groups can be seen, similar to manual gating. With fluorescence information, more subtle differences in the cell populations become visible, indicating different expression levels of the fluorescent markers and their correlation. A clear differentiation of the five gates is possible. Although different cell populations according to manual gating were not always split into completely different clusters by t-SNE or UMAP, manually defined cell populations remain similarly identifiable in UMAP as in t-SNE, with both techniques outperforming PCA. However, all three approaches were able to clearly separate the outliers that were removed by manual gating. These results underline the benefit of integrating fluorescence information into image analysis. Nonetheless, also the sole consideration of spatial information enables a coarse classification of cells.

4   Conclusion

This work presented several dimensionality reduction approaches to characterize in detail the unprecedented amount of multidimensional data generated by the integration of morphological and molecular parameters at the single cell level. The results show that this not only helps to improve data visualization, but can also highlight subtle patterns and groups that could easily be overlooked when manually examining parameter pairs in two-dimensional projections, as is still common in flow cytometric approaches. This could be achieved by considering both morphological and fluorescence information, with the latter approach showing better results in this homogeneous data set. Therefore, it should be noted that the choice of dimensionality reduction technique and the choice of parameters depends significantly on the specific data, the research question and the specific biological context. Future research will focus on improving the selection of combinations of dimensionality reduction approaches and morphological, fluorescence and image parameters to improve the characterization of cell populations for specific applications.

Acknowledgements

This work has been carried out at the Research Center Borstel and utilized the center's BD FACSDiscover™ S8 device administered by the Core Facility Fluorescence Cytometry and the center's computing infrastructure administered by the Data Science group.

Authors' Statement

Conflict of interest: Authors state no conflict of interest.

5   References

[1] Aysun Adan & Nalbant, A. Flow cytometry: basic principles and applications. *Critical Reviews In Biotechnology.* **37**, 163-176 (2017)

[2] Spidlen, J., Moore, W., Parks, D., Goldberg, M., Blenman, K., Cavenaugh, J., Force, I. & Brinkman, R. Data file standard for flow cytometry, Version FCS 3.2. *Cytometry Part A.* **99**, 100-102 (2021)

[3] Schraivogel, D., Kuhn, T., Rauscher, B., Rodríguez-Martínez, M., Paulsen, M., Owsley, K., Middlebrook, A., Tischer, C., Ramasz, B., Ordoñez-Rueda, D., Dees, M., Cuylen-Haering, S., Diebold, E. & Lars M. Steinmetz High-speed fluorescence image–enabled cell sorting. *Science.* **375**, 315-320 (2022)

[4] Maaten, L. & Hinton, G. Visualizing data using t-SNE. *Journal Of Machine Learning Research.* **9** (2008)

[5] Becht, E., McInnes, L., Healy, J., Dutertre, C., Kwok, I., Ng, L., Ginhoux, F. & Newell, E. Dimensionality reduction for visualizing single-cell data using UMAP. *Nature Biotechnology.* **37**, 38-44 (2019,1)

[6] White, S., Quinn, J., Enzor, J., Staats, J., Mosier, S., Almarode, J., Denny, T., Weinhold, K., Ferrari, G. & Chan, C. FlowKit: A Python Toolkit for Integrated Manual and Automated Cytometry Analysis Workflows. *Frontiers In Immunology.* **12** (2021)

[7] Pedregosa, F., Varoquaux, G., Gramfort, A., Michel, V., Thirion, B., Grisel, O., Blondel, M., Prettenhofer, P., Weiss, R., Dubourg, V., Vanderplas, J., Passos, A., Cournapeau, D., Brucher, M., Perrot, M. & Duchesnay, E. Scikit-learn: Machine Learning in Python. *Journal Of Machine Learning Research.* **12** pp. 2825-2830 (2011)

Developmental Acquisition, and Processing of Immunoassay Data through an Automated Laboratory Analyzer – Expanding and Optimizing a Dashboard Visualization

Marike Kallweit [1]

[1] Medical Engineering Science, Universität zu Lübeck, marike.kallweit@student.uni-luebeck.de

Abstract

The automation of antigen and antibody tests is substantial in improving diagnostic efficiency and was therefore investigated as part of the "Accentis" project. A series of verification and validation tests were carried out solving hardware and software problems that arose during the development phase of the Accentis analyzer. Naturally, data analysis must be preceded by data acquisition, followed by processing and visualization of the results. The first approach was to reprogram and update the *Device-Viewer* visualization tool. Due to the deprecated application code, the integrated dashboard had been deprived of its functionality. Following a thorough analysis, multiple scripts were restructured and reprogrammed. The tool could be adjusted to restore all functions. Several new features were added to improve the user experience, and development issues were resolved while working on the prototype devices. As a result, data analysis remains more efficient and productive for the rest of the development phase.

1 Introduction

EUROIMMUN Labordiagnostika AG is the worldwide leader in automatic diagnostic application development and manufacturing around the world. Therefore, test automation is particularly important for the company. Especially antigen and antibody test automation using chemiluminescence immunoassays to improve the efficiency and productivity of blood sample testing is a valued development field for EUROIMMUN as an institution. Automation of these tests ensure an efficient and quicker diagnostic than regular manually executed detection tests.

Based on these principles, the project "Accentis" focuses on the development of a random-access device as a laboratory analyzer. The execution of the immunoassays functions autonomously. Resulting datasets are logged and saved locally by the Accentis Software. The data collection and analysis of logged assay data is based on a collection of development scripts, labeled *Accentis-Data-Science*. The visualization of the data is carried out internally using the *Accentis-Device-Viewer* application. The application serves as a user interface showing status and settings for all Accentis devices. In an overview, it displays the device state and error messages for every device in real time. The respective device card shows the following states: Processing, offline, under maintenance, error, or system stop.

In addition to the overview of all devices, the *Device-Viewer* has two further components. On the one hand, the analyzed data is displayed in the integrated live dashboard with different graphs and lists based on the test results.

Also, the current embedded configurations, software settings, teachings, calibrations and operation counters are listed for each device. These configurations as well as the graphs, can be viewed retrospectively and in real time for each assay. Due to essential outdated Dash and Flask packages, changed hosts and ports, as well as outdated scripts within the *Device-Viewer* and *Data-Science*, the integrated dashboard had not been functioning for months. The data could no longer be retrieved via the interface and the configurations were no longer updated. Since the data was not available, the *Device-Viewer* was displayed essentially empty in the browser. Naturally, the main task was to fully restore the functionality of the *Device-Viewer*. Also, due to regular software updates, it was essential to maintain and monitor the alterations in the logged data to ensure accuracy when evaluating the test results. The application performance was additionally optimized and supplemental features were integrated and tested on the device. Ultimately, new prototypes and corresponding systems were integrated into the *Device-Viewer* to ensure data collection, analysis and visualization of assays on the newer prototype devices.

Disclaimer

The project was carried out during the development phase of laboratory analyzer prototypes. The description of the main research and the actual results are therefore based on internal and confidential information of EUROIMMUN Medizinische Labordiagnostika AG. For this reason, laboratory work will not be covered in this article. Consequently, only the most fundamental information and basic programming assignments during the project are presented.

2 Material and Methods

The analyzer is programmed to run a two-step assay specifically designed for antigen and antibody testing in blood samples. This depends on the experimental design. The two-step assay is derived from the chemiluminescence immunoassay (ChLIA) developed by EUROIMMUN.

2.1 Antibody detection with ChLIA

Fig 1. roughly illustrates the antibody detection. The antigen-coated magnetic particles are incubated with diluted patient samples. If the patient sample contains specific antibodies against the antigen, these bind to the antigen-coated magnetic particles. In a further step, an antibody labeled with acridinium, functioning as conjugate, is added, which binds to the specific antibodies. Trigger solutions are then added to the reaction mixture to trigger a chemiluminescence reaction. The resulting light signal is proportional to the antibody concentration in the patient sample [1].

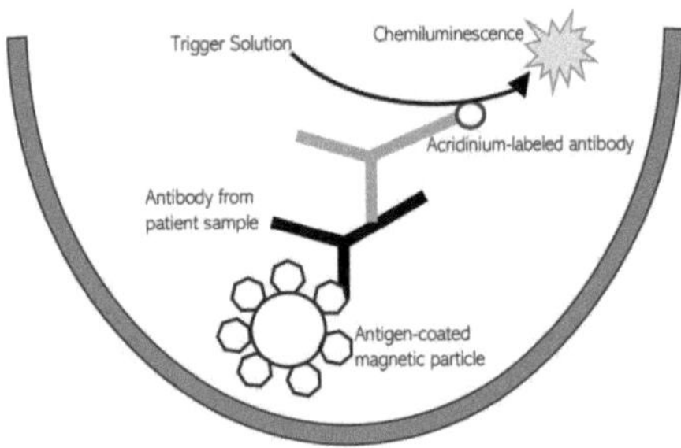

Figure 1: Antibody detection schematic

2.2 Antigen detection with ChLIA

As displayed in Fig. 2 the antibody-coated magnetic particles are incubated with the patient sample and an antigen-specific biotinylated antibody. During incubation, the patients antigen is bound by both the antibody coupled to the magnetic particles and the biotinylated antibody. In a further step, acridinium-labeled ExtrAvidin as the conjugate is added, which binds the biotinylated antibody. Trigger solutions are added to the reaction mixture to initiate a chemiluminescence reaction. The resulting light signal is automatically converted into the proportional antigen concentration by the device [1].

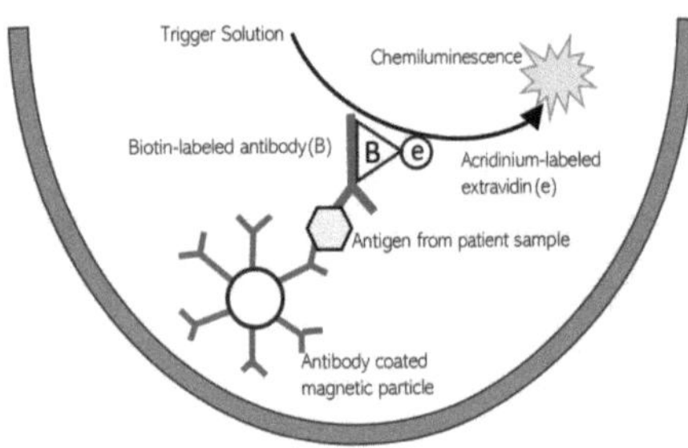

Figure 2: Antigen detection schematic

2.3 Procedure for immunoassay device

Preparation for any assay involves device preparation, including stocking and pipetting samples into 5ml test tubes. Due to the development status, system buffer is predominantly used as the sample in virtue of mimicking a biological solution. Additionally, the conjugate is also replaced with system buffer on the condition that the assays are not of biological nature. These "Dummy" assays are designed for stress tests on the device to imitate a real run without a biological sample. The focus here lies on the device data and module calibration accuracy rather than the actual sample results. The internal device process can be roughly described as follows: Magnetic beads are located individually in small capillaries within a cartridge. During a capillary rinsing process, the beads are each rinsed into a cuvette and transported through the device. The rinsing liquid is then rinsed out of the cuvette. The beads are washed to remove residues of the rinsing liquid and diluted with a desired sample dilution. The diluted sample is aspirated from the previously pipetted test tubes and dispensed into the cuvette. Incubation takes place while the diluted sample vibrates for several minutes. In addition, a conjugate is pipetted into the cuvette and a further incubation follows, which is concluded with the addition of a solution to trigger the light emission.

Countless immunoassays using the prototypes to collect data for processing and visualization were performed during the project.

2.4 Data Processing and Visualization

The application itself is divided into two main components. The existing *Device-Viewer* is configured with a Plotly Dash framework. The Dash ecosystem is not only described for data exploration, the framework can be used in all phases of data processing [2]. As for the back-end, Dash uses Flask for creating a dynamic dashboard application and architecture through callbacks [2]. As a data analysis and visualization tool for graphical statistics and analytics, Plotly is utilized [3]. Component handling is covered by React, which uses rendering to create a single-page React application [2]. The existing scripts were written with Python v3.11.4 as programming language, respectively standard HTML and JavaScript. In addition, Vue v9.8.1 is used as the JavaScript framework for front-end development and Node v20.3.1 as the runtime environment for the server [3]. In addition, the main framework packages, that were upgraded in order to restore functionality of the application are the following: Dash v2.11.1, Dash-daq v0.5.0, Flask-caching v2.0.2, Flask-HTTPAuth v4.8.0, Pandas v2.0.3, NumPy v12.25.0, Plotly v5.15.0, Natsort v8.4.0, Gunicorn v20.1.0, Werkzeug v2.2.3.

The Accentis software generates a new log folder each time the system is restarted. Within, all device data, module functionalities and test results are saved as individual log files. The *Data-Science* as the second component includes python scripts for processing the data inside these log folders.

Depending on the mode, the log files generated by the Accentis software are parsed at a certain interval. When a test run is visualized in the *Device-Viewer* in real time, data is analyzed every minute and updated accordingly in the interface. Otherwise, there is a task schedule for the automatic execution of the scripts at predetermined times. To parse acquired data, generated log files are localized by the script through Pandas to compare log file patterns and filter out required data for parsing. A regular expression is used to create ReGex patterns of log files to obtain specific data points from each step of the assay. ReGex or regular expression is a formula function for a query-based filter, such as a text, timestamp or marker that matches a pattern based on a regular expression [4]. This query-based search can be used to examine and filter particularly large amounts of data or rows of data for matches [4]. The filtering is carried out using various operators to determine the pattern [4]. Based on the ReGex pattern, the required data is filtered out of the respective logs in the script for calculations and plots.

The application was built using Docker for container management. The container as a standardized component combines the application source code with the operating system libraries and dependencies, to enable the capability of running the code in any environment [5]. Subsequently, the Docker Build runs via a Nginx server in the provided web browser, thus every developer with user privileges has access to the *Device-Viewer*.

2.5 Application optimization

In addition to the updates of the framework packages, all scripts of the *Device-Viewer* were revised. Among other things, deprecated and obsolete code was deleted or replaced, countless paths were changed accordingly, as well as ports and hosts. Following thorough examination and comparison of the logs at the current time and before the dashboard lost functionality, 35 data mining patterns were continuously changed as these files are constantly modified by software updates. In addition, each time the log folder was expanded, new patterns were written in association with the folder version in order to avoid losing data by reanalyzing or overwriting data. Furthermore, for each new data mining pattern, new error flags were created for each data extension and additional information was extracted for the hover text.

After data aggregation, the interactive diagrams are generated with the help of callbacks to create interactivity within the dashboard. Due to the expanding amount of data, performance is suffering. In order to counteract this for user-friendliness, lazy loading was initially integrated and further scripts were split within the callbacks during the project in order to load the desired data only after request changes in the web browser. Therefore, datasets can be dynamically visualized, in which case interactions with specific datapoints and statistics are possible, without affecting performance. To further improve the dashboard visualization, new graphs were integrated using the extended log files.

Furthermore, graph dropdown menus were added to show supplemental data. In particular calculated deviations, different units or calibrated limits. In addition to the existing devices within the interface, further prototypes were added to the system during the project. For this purpose, the existing python installation file was revised and integrated into the new device system. In order to collect states and device configurations, the device was integrated in the *Device-Viewer* and the data analysis was tested manually to verify the correct visualization in the application.

3 Results and Discussion

Due to the fact that the actual results such as screenshots as part of the visualization are of classified nature, the *Device-Viewer* and dashboard are presented as schematic architectures for illustration. First, an overview and subsequently a small part of the serviced and updated dashboard is presented to illustrate the designed intent of the statistical tool. Afterwards, parts of the added features are presented. This includes a discussion about current progress and ideas for further improving the *Device-Viewer*.

3.1 Results

Fig. 3 shows the schematic application as seen by the user after the functionality has been restored. This overview includes every device state and runtime in real time. The new integrated devices are shown as well. Previously, this interactive page was blank. The user is therefore once again able to monitor the device states and interact with the *Device-Viewer* in real time.

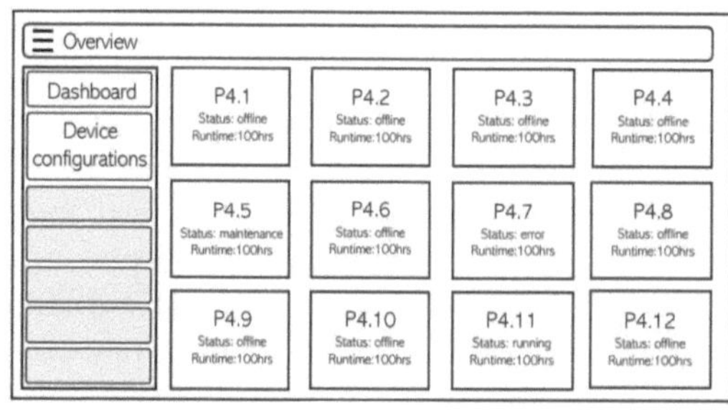

Figure 3: Schematic overview of the *Device-Viewer* interface as seen by the user in the web browser

Fig. 4 shows the rough structure of the functioning dashboard and some of its components as seen in the web browser. The left-hand side is equipped with an interactive menu to manage the devices and view the respective configurations for each device at any desired time during the last three years. In the upper part of the dashboard, the type of data display can be set in each case. This includes the time view, job view for each individual sample run, statistical view for all evaluated results and transitions view for information about the current processing status of the sample. Displayed interactive diagrams include measurement results, capillary rinsing and washing pressure, the aspiration monitoring, and temperatures of individual modules among other data depending on the request. Individual data points and information of each dataset can be displayed thanks to the interactivity and hover text, that appears while hovering over each graph. This feature adds supplemental data for the user such as the specific numerical readings, the exact sample number and name and the exact location of this specific sample inside the cartridge. The log file search for these specific readings is therefore obsolete.

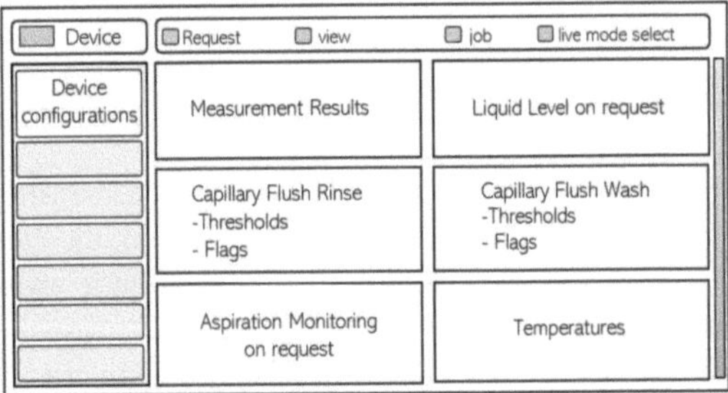

Figure 4: Schematic interactive dashboard interface. Displays module data and test results as graphs

Not shown in the schematic diagrams are, among other things, the additional error flags of measurement results outside the calibrated limits. These are displayed with icons at each data point and are also displayed in a list of all device steps with the corresponding error message. Any resulting system stops are immediately reported to the user. Moreover, due to new data mining patterns, the constantly changing set pressure limits of the pipettes within the limits are automatically displayed for each device.

3.2 Discussion

Based on the current status of the *Device-Viewer*, the functionality is fully restored. Developers can monitor and troubleshoot their respective assays once more and are up to date with every step of their trials. The display of additional data and enhanced visualization enable a more efficient evaluation without searching through raw data logs and manual data aggregation. Further additional functions in the *Device-Viewer* are planned on an ongoing basis to increase efficiency during development. These include the implementation of system qualification, whereby manual evaluation of raw data is no longer necessary and qualification reports can be viewed for each device in the web browser immediately after every performed system qualification. In addition, log data from each assay is parsed every night depending on the script schedule, thus visualization is currently only possible on the following day, or as long as the software is still running and the real-time mode in the viewer is switched on. As external data is further relevant for development, implementing a feature that allows raw data to be fed into the system and moreover automatically be analyzed and visualized, could be essential. A few scripts have already been written for this characteristic, which are still being tested for functionality.

4 Conclusion

The *Device-Viewer* has been successfully reprogrammed and updated and is currently enabled for all functions that originally worked in the tool. This means that the functionality for the entire department, which was the focus of the project, has been restored. Furthermore, deprecated and obsolete code was manually reprogrammed. Additionally, all log folders were successfully maintained and the *Data-Science* scripts were updated accordingly to allow for new callbacks and additional data acquisition along with parsing. The dashboard was further enhanced to support different file patterns and more dynamic data visualization through advanced detail processing. Experimental studies benefit from a great amount of stored data, which leads to performance degradation when loading certain data in the web browser. Therefore, an improvement had to be made by programming lazy loading and split codes within the application. The integrated dashboard has ultimately a visual standard and documentation which allows for easier implementation of more devices and addition of more functionalities or even new dynamics, without the loss of functionality.

Acknowledgement

The work has been carried out at EUROIMMUN, Medizinische Labordiagnostika AG in Dassow and supervised by Dr. Matthias Brandenburger, Fraunhofer-IMTE.

Authors' Statement

Conflict of interest: Author states no conflict of interest.

5 References

[1] ChLIA, *EUROIMMUN ChLIA*.
Available:https://www.euroimmun.de/de/produkte/ nachweismethoden/chlia/ [last accessed on 2024-02-05].

[2] E. Dabbas, *Interactive Dashboards and Data Apps with Plotly and Dash.* Packt Publishing Ldt, Birmingham, 2021.

[3] K. Dale, *Data Visualization with Python and JavaScript.* University Science, Mill Valley, 1989.

[4] ReGex, *Search for string using the regular expression pattern.*
Available:https://www.ibm.com/docs/de/tcamfma/6.3.0 ?topic=cf-regex-scan-string-based-regular-expression-pattern [last accessed on 2024-01-01].

[5] Docker, *Docker as an Open-Source-Platform.*
Available:https://www.ibm.com/de-de/topics/docker [last accessed on 2024-01-03].

Finite Element Analysis of Fracture Plates for Periprosthetic Humerus Fractures: An Evaluation and Comparison of Non-linear Effects and Design Improvements

Christopher Carlisle [1], Joerg Schroeter [2], and Robert Wendlandt [2]

[1] Medical Microtechnology, Technical University of Luebeck, christopher.carlisle@stud.th-luebeck.de

[2] Biomechanics Laboratory, University Medical Center of Schleswig-Holstein, {joerg.schroeter,robert.wendlandt}@uksh.de

Abstract

A new design for the treatment of periprosthetic humerus fractures has been proposed by Innovations Medical GmbH, which is thought to be less invasive and more versatile than current designs. The aim of this work was to perform a nonlinear finite element analysis (FEM) of this new design and to evaluate previous linear analysis data performed by another author in this regard. It was found that applying nonlinear simulations had a significant impact on the calculated stress concentrations, with stresses differing by over 50 %. Furthermore, based on the results obtained from a shifted plate placement and consultation with medical staff, it was determined that the previous linear analysis used a suboptimal placement of the plate, which led to artificially high stresses.

1. Introduction

Total joint replacement has become a common procedure to be performed in cases of damaged joints in order to alleviate pain. For a shoulder joint replacement, Figure 1, the humeral head is removed and a metal implant made of some alloy, commonly Titanium-Aluminium (Ti-6Al-4V) or Cobalt-Chromium (CoCrMo), is placed into the humerus.

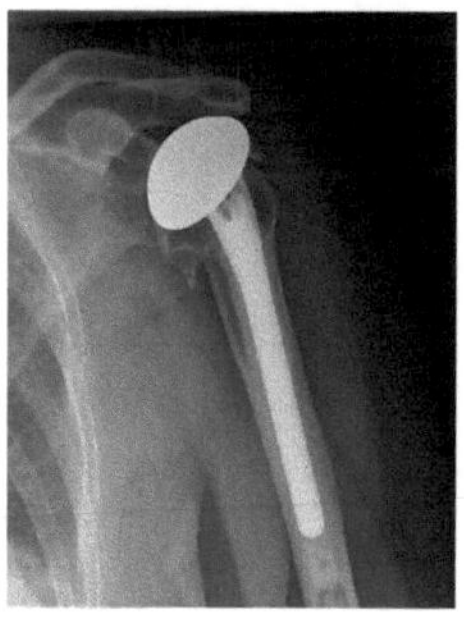

Figure 1: X-Ray of a shoulder joint replacement [1].

While such an operation can improve a patient's quality of life, it can also increase the risk of a periprosthetic fracture within the humerus, due to vastly different stiffnesses of the implant and natural bone. Currently, using traditional plates to treat such fractures requires the removal of muscles such as the pectoralis and deltoid, in order to properly fixate it to the humerus. Furthermore, inserting fasteners can be problematic, as these plates were originally designed for use in humeri without a metal implant inserted. To address these issues, a new design for the treatment of such fractures has

been proposed by Innovations Medical GmbH, which wraps around the humerus. By using such a design, Figure 2(b), the amount of tissue which will be damaged can be minimized, and the fastener placements have more possible variations.

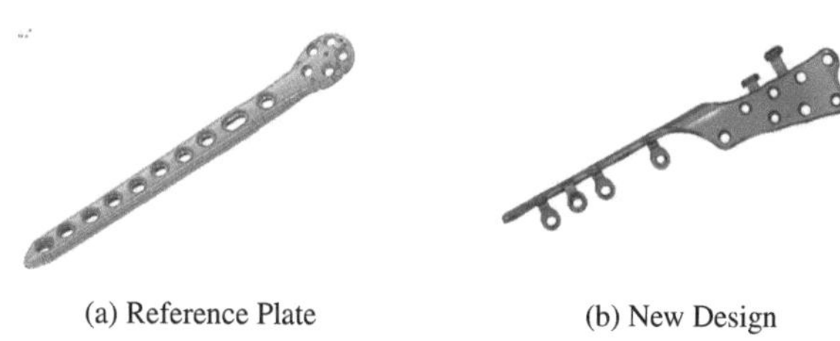

(a) Reference Plate   (b) New Design

Figure 2: Periproshtetic humerus fracture plates.

The purpose of this work was to perform a nonlinear finite element analysis (FEM) of this new design and to evaluate previous linear analysis data performed by Christopher Riches, a former student research assistant of the University Medical Center of Schleswig-Holstein (UKSH). This was thought to be necessary as the previous data showed high stress concentrations, far exceeding the yield strength of Grade II Titanium (Ti II), which the implant is made of. In his work, Riches notes that a further nonlinear analysis is necessary to support this conclusion [2]. A Grade I Titanium reference plate was tested for comparison.

2. Materials and Methods

All simulations were undertaken within Autodesk Fusion 360 (v. 2.0.17721) and Autodesk Inventor Nastran In-CAD (2024 Build 153). To this end, 3D Geometries were prepared using the open source softwares NetGen (v. 6.2.2304) and 3D Slicer (v. 5.2.2), while post-processing and material generation were done using R Studio (v. 2023.06.2).

2.1 Material Data Generation

First, the material databases were created for use in Fusion 360 and Nastran, since Grade I, Grade II, and Grade IV Titanium are not included in the native library. As the exact material properties of the titanium which will be used for the implant were not known at the time of writing, and a sample of material for testing was not available, the necessary quantities were taken from online databases and suppliers [3]. If a range of strengths was provided, the lower limit was always used as a conservative estimate. Table 1 summarizes the material properties used in this work.

Table 1: Material properties of Titanium Grades I, II, and IV

	Ti I	Ti II	Ti IV
Young's Modulus	103 GPa	103GPa	105GPa
Poisson's Ratio	0.37	0.37	0.37
Shear Modulus	45 GPa	45 GPa	40 GPa
Density	4.51g/cc	4.51g/cc	4.51g/cc
Yield Strength	170MPa	275 MPa	483MPa
Ultimate Strength	240MPa	344 MPa	552MPa
Ultimate Strain	0.24	0.2	0,15

Having this data available, theoretical stress-strain curves could then be produced without the need for a physical material to test. This was accomplished through the use of a R script using the modified Ramberg-Osgood equation [4].

As proposed by Hill in 1944, this equation can accurately estimate the plastic strain of ductile materials through the following relationship:

$$\varepsilon = \frac{\sigma}{E} + 0.002(\frac{\sigma}{\sigma_y})^n \qquad (1)$$

where ε is strain in the plastic region, σ is stress, σ_y is yield stress, n is the inverse of the strain hardening coefficient, and E is the modulus of elasticity (Young's Modulus).

Once the theoretical plastic strain was obtained, the entire stress-strain curve for all titanium grades were created by combining the plastic data, remembering to convert engineering to true stress/strains for use in Fusion 360, with data points generated using the modulus of elasticity and yield point for the elastic region. Databases for PMMA bone cement, cortical bone, and the joint replacement implant were taken over from Riches' work, as these showed stresses far below their yield point, and so were deemed unnecessary to simulate nonlinearly - in doing so the computational cost and time of the simulation was decreased.

2.2 Assembly

After the material databases had been produced and verified to be working, the model assembly was created. To this end, CT scan data in the form of the DICOM file format was segmented into cortical, cancellous (spongiosa), and other components of the bone using 3D Slicer. The final and most accurate model was provided by UKSH and uploaded into Fusion 360.

The models of the reference plate and the newly designed plate were provided by Innovations Medical and imported into Fusion 360 as well. However, during importation, inconsistencies in the model geometry occurred, which caused overlapping and negative volume regions. To remedy this, both models were imported into NetGen and healed using the IGES/STEP Topology Explorer/Doctor/Heal Geometry. Afterwards, any areas that were ill-fitted, based on a visual inspection, were manually patched using surface tools within Fusion 360.

Additional components necessary for the completion of the assembly included the bone cement, joint replacement implant, and fasteners. The joint replacement implant was modelled as a steel cylinder with a diameter of 10 mm, to reflect the same rod that would be used in a physical compression test by UKSH. The PMMA bone cement was modelled as a 2 mm thick cylinder around this steel implant, both having a length of 115 mm from the top of the Greater Tuberosity.

The fastener dimensions were based on *Tifix* locking screws, in keeping with Riches' work. These screws were simplified as cylinders with a diameter of 2.5 mm, and a head of 5 mm. As these screws cut into the implant during operation, the heads were later slightly extended as needed to allow the screws to "cut" into the plate for direct bonded contacts.

A fracture gap was created with a length of 10 mm and all components were dragged into place. Fasteners were then Boolean subtracted from any components they came into contact with to create their mounting holes. Based on consultation with an experienced surgeon and Innovations Medical staff, approximately six fastener connections above the fracture and approximately four below the fracture are expected in the new design, with some variation allowable based on the condition of the patient's bone. To this end, seven unicortical connections above and five connections below the fracture were placed, with one extra screw included at the fracture location to ensure a more even distribution of the load.

For all simulations undertaken within Autodesk Nastran, the assemblies from Fusion 360 were simply imported into Inventor. Any material library properties were reassigned within the Nastran environment, ensuring to make any necessary adjustments for values which were not imported directly - notably the plastic region of the stress-strain curves had to be reinserted.

Figure 3 shows a comparison of the original simplified model by Riches and the newly updated model used in this work. Note how the bottom of the bone in (b) was removed, and the point of rotation was extended via a virtual node at the top and bottom of the system, in order to decrease overall simulation time by reducing the number of nodes.

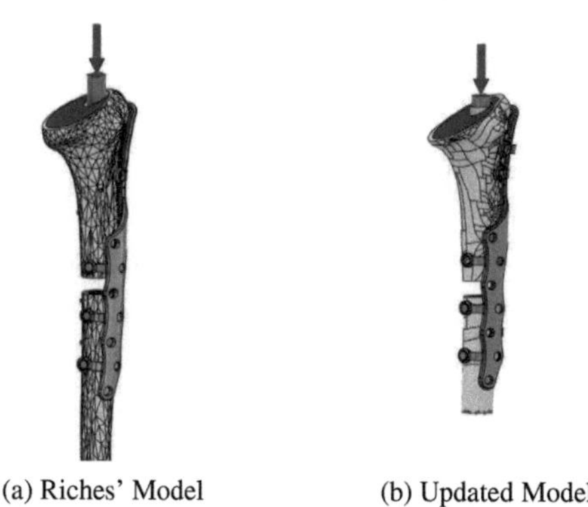

(a) Riches' Model   (b) Updated Model

Figure 3: Old model vs updated model.

After further consultation with the UKSH staff, it was determined that the original placement of the plates (as described above and used in the linear work) would be sub-optimal in this case, as there is more bone material on the dorsal side of the *SAWBONES* humerus, which was used to create the humerus scan. Keeping the plates in the original position could run the risk of splintering the cortex at its thinnest points.

Additionally, the original linear simulations assumed an implant shaft that was shorter than one which would be used in an actual joint replacement procedure today, so an updated length was recommended. The length of the implant shaft was updated to be 150 mm from the top of the Greater Tuberosity, which was used in a subsequent simulation of the reference plate. The reference plate fasteners for both the original and shifted locations were picked by UKSH staff, with four bicortical connections above, and three bicortical connections below the fracture.

2.3 Simulation Setup

After all materials were assigned, a load of 650 N was prescribed at the top of the implant, assuming that all forces would be transmitted through this in a real world scenario. Bonded contacts were applied to all surfaces that would be held by a fastener. Rotational degrees of freedom were given free at the top and bottom of the system, in order to reflect a gimbal system. By constraining surfaces around a virtual point of rotation, the number of nodes and therefore the complexity of the simulation were greatly reduced, with negligible effects on accuracy. The new design simulation used 34,863 nodes and 19,204 elements, and the reference 98,220 nodes and 60,919 elements, when considering only the plates (excluding components such as the bone, which had a less fine mesh). With the reference plate, the plate is free to collide with the outer surface of the bone, as a fastener can not

be placed here due to the joint replacement implant obstructing the locations where a fastener hole would be placed. As such, separation contacts with true surface-to-surface friction were applied in Nastran. The friction coefficient of these was set to 0.5, as typical values range from 0.5-0.7 based on various production parameters and surface treatments [5]. Finally, during troubleshooting within Fusion 360, it was determined that contact stiffnesses needed to be updated to aid the convergence of the system, which was set to 0.01, as recommended in Nastran's Handbook.

Once all components and their relations were defined, the mesh was generated. Within both softwares, parabolic and curved elements were used, as these tend to more accurately represent curved elements at a lower density. Within Nastran, small features were suppressed at 2 %.

After the simulations were completed, a script written in R was used to take the output files from Nastran, and filter the data into a usable form to be plotted in Plotly using R. A script provided by UKSH was then used to find the best fit function for the elastic region, using calculated data based on the relative displacement of two points in the fracture zone. From this, the yield point was calculated based on the assumption of a 0.2 % yield point offset [6].

3. Results

Initial simulations within Fusion 360 showed a maximum stress of 1588 MPa using Riches' simulation setup, and 890 MPa using the newly updated nonlinear model, both under the same loading of 650 N, revealing that the nonlinear simulation experienced a peak stress of nearly 56 % that of the linear simulation. This trend continued throughout the model, where the majority of stresses, based on a visual inspection, were approximately 50 % of the linear model. This result is as expected, and shown in the Nastran handbook as well - explained by the continuously updated stiffness of the material. However, in both cases, the yield point of Ti II, 275 MPa, was exceeded.

To ensure that the results were reproducible and not limited to Fusion 360, the same setup was run in Nastran. Stress concentrations at the same locations were obtained within an estimated value (from a visual inspection using several probes), of ± 25 MPa.

Measuring the deformation of the fracture zone and using the 0.002 offset of the linear region for the yield point, it was concluded that the initial plastification of the plate actually begins to occur much earlier than 650 N. First starting at 359 N, this force is far away from the application where yielding was expected and cause for further considerations to be made. The reference plate showed similar results, with permanent deformation occurring at 355N.

Note that the reference plate was not initially in contact with the bone, but did become so after the applied force increased. This was assumed to have a negligible effect on loading capabilities based on the fact that cerclage wire application, which is meant to keep fracture treatment plates closer to the bone, have had no significant impact on biomechanical loading in other studies [7].

While the strength of the reference plate and new plate seem to be nearly identical in the simulation, both yield far earlier than expected. As previously mentioned, an updated length and position for the implant was suggested by UKSH staff as more material and better mounting would be possible. This new shifted position of the plate was simulated as well, with results showing a significant impact on the loading of the system. A new yield point of 499.2 N was identified - an increase of almost 41 % over the previous nonlinear simulation.

Figure 4 shows the above discussed Nastran results of the new design in (a), the initial reference plate placement in (b), and the updated position in (c).

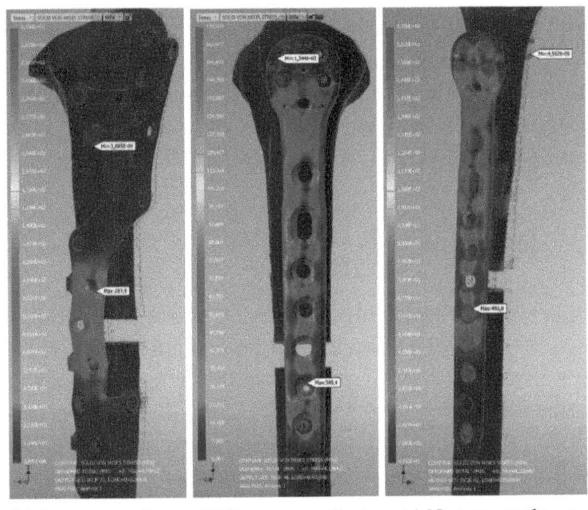

(a) Nastran results at 652N   (b) Nastran results at 403N   (c) Nastran results at 650N

Figure 4: Results of calculated stress distributions after shifted placement of the plate.

Note that in (b), the loading only reached 403N before failure, while in (a) and (c) the loading reached the full 650N.

4. Discussion and Conclusion

While all simulations failed to withstand the applied loading condition, several important observations can be made from these results. Firstly, running a nonlinear simulation is necessary to accurately capture the yielding of the metallic implant material, with changes of calculated stress of approximately 50 %. Secondly, shifting the location of the plate and updating the fracture zone/implant to reflect expected conditions increases the loading capability of the reference plate by an additional 41 %. As such, it is recommended to reconsider the initial placement of the new periprosthetic plate design. Finally, the reference and new design seem to be nearly identical in terms of their loading capabilities, with yield points of 359 N and 355 N respectively.

While these results do indicate comparable strength when wholly supported by the *Tifix* locking screws, it is nevertheless recommended to repeat a nonlinear simulation for the new design. A new position similar to the updated reference plate position should be used, along with actual stress-strain data obtained from material testing for each grade of titanium, in order to achieve the most accurate results. Finally, a validation of the system should involve compression testing to obtain actual experimental data.

Acknowledgements

This work has been carried out at the Biomechanics Laboratory for Trauma Surgery of the University Medical Center of Schleswig-Holstein, supervised by the Institute of Natural Sciences, University of Lübeck.

Authors' Statement

Conflict of interest: Authors state no conflict of interest.

5. References

[1] L. Monfils, 2011, "Shoulderprotesis", From: `https://en.wikipedia.org/wiki/Shoulder_replacement#/media/File:Shoulderprotesis.jpg`, Accessed: 04.12.2023.

[2] C. P., Riches, "Finite Element Analysis For The Evalutation Of Biomechanical Strength of Angular Stable Osteosynthesis Plates In Periprosthetic Humerus Fractures", Nat. Sci., TH Lübeck, SH, 2023.

[3] MatWeb LLC., 2024, "MatWeb, Your Source for Materials Information", From: `https://www.matweb.com/`, Accessed 16.01.2024

[4] R. Wendlandt, 2023, "Synthesize Test Data", From: `https://github.com/soylentOrange/sdarr/blob/HEAD/R/synthesize_test_data.R`, Accessed 04.12.2023.

[5] J. E., Biemond, et al., 2011, "Frictional and bone ingrowth properties of engineered surface topographies produced by electron beam technology", Archives of Orthopaedic and Trauma Surgery, From: `https://www.ncbi.nlm.nih.gov/pmc/articles/PMC3078515/`, Accessed 04.12.2023.

[6] R. Wendlandt, 2023, "sdarr: Slope Determination by Analysis of Residuals" From: `https://github.com/soylentOrange/sdarr` Accessed 04.12.2023

[7] R. Gupta, et al., 2022, "The addition of cerclage wiring does not improve proximal bicortical fixation of locking plates for Type C periprosthetic fractures in synthetic humeri", From: `https://doi.org/10.1016/j.clinbiomech.2022.105709`

Construction of a Test Rig to Quantify the Cutting Characteristics of Additively Manufactured Phantoms

Lea Peuckert [1] and Thomas Friedrich [2]

[1] Medical Engineering Science, Universität zu Lübeck, lea.peuckert@student.uni-luebeck.de

[2] Fraunhofer Research Institution for Individualized and Cell-Based Medical Engineering IMTE, thomas.friedrich@imte.fraunhofer.de

Abstract

To date, mainly subjective assessments from surgeons [1] or testing of mechanical properties have been used to evaluate additively manufactured phantoms. This research is about constructing a test device to analyse the quantitative cutting properties for the simulation of surgical procedures. To achieve this, both a scalpel holder and a sample holder which can be clamped into the testing machine were designed and manufactured. The resulting testing device allows to record the cutting force along the cut. This gives a new insight into the understanding of the cutting process in the individual sample, and is crucial for the comparison and development of surgical phantoms.

1   Introduction

In the field of additive manufacturing, there are currently numerous innovative projects for creating organ phantoms or synthetic tissue using 3D printing. Among other things, these phantoms can be used for the realistic training of surgeons [2].

In order to make cutting with scalpels into printed tissue as realistic as possible, the mechanical cutting properties of biological tissue must be measured quantitatively and transferred to additively manufactured phantoms.

While 3D printed phantoms have been evaluated in their biomechanical behaviour through tensile testing or measuring of the puncture force [3], their behaviour during surgical cutting has hardly been evaluated.

The concept of combining a testing machine with a custom-build test station to analyse high-speed forces during the cutting of food was shown in [4]. This device was however used for blocks of material where a part of the material was completely cut off and it was not a replication of a surgical cutting process.

The aim of this practical project was to expand a state of the art universal testing machine with a device for clamping a scalpel and a tissue or phantom sample. This makes it possible to measure the force progression via a load cell during the cutting process or when piercing the tissue. Therefore the construction has to be designed in such a way, that it can be easily installed and removed from the testing machine without compromising the reproducibility of the measurements.

2   Material and Methods

2.1   Universal testing machine

The Galdabini Quasar 100 universal testing machine was used to carry out the corresponding measurements. It can measure forces up to 100 kN. The loadcell that was used has a measurement range up to 50 N. The testing machine has a maximum testing velocity of 500 mm/min. It is used with the LabTest (Galdabini) software which records data such as the force, elongation or deformation over the time or the traverse (movement of the mechanical beam). It enables the user to adjust the testing velocity and the displacement of the traverse as well as various other settings.

2.2   Sample material and scalpel

Patches made of Stratasys agilus clear were used as sample material. Agilus is a rubber-like material with a Shore A hardness value of 30. The patches were 3D printed with a Stratasys J850 printer. They have a tickness of 1 mm and are (60 x 60) mm^2 in size.

Gouda cheese slices with a thickness of about 2 mm were used as another sample material to enable a comparison.

The scalpel to be clamped is a disposable scalpel with a steel blade from Teqler. It has the blade shape No. 11.

2.3   Concept and construction of the mounting brackets

Two corresponding devices to clamp both a scalpel and a sample into the testing machine were designed using the CAD Software SOLIDWORKS 2023 and then additively manufactured using a Stratasys F370 FDM printer with ABS M30 filament.

A fixture for the scalpel that can enable cutting with the testing machine was designed for the upper clamping head that is moving during the measurement. Another mount was constructed to be installed in the lower clamping head to ensure the sample material that is to be cut remains in a fixed position without putting stress on the material.

The geometric properties of the scalpel were measured and transferred to the construction design accordingly. Fig.1 shows the finished construction of the scalpel holder with a matching cover. The scalpel fits into the inlay without scope for movement but can still be easily taken out and put back in. The design allows for cutting the sample at a 90-degree angle as is realistic for surgical procedures especially with blade No. 11 [5].

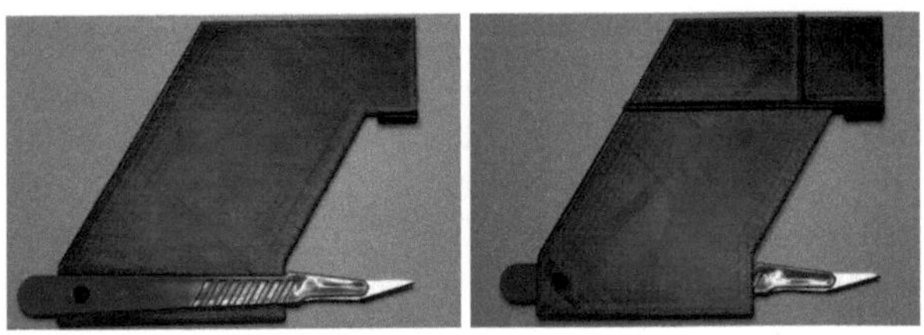

Figure 1: 3D printed scalpel holder without cover (left) and with cover (right)

The varying thickness of the scalpel handle has been compensated for by a moulded bevel in the design to ensure the scalpel is positioned straight so it can perform a straight cut. The holder has boundaries fitted to the size of the clamping head to ensure it will always be clamped at the same location to allow for reproducible measurements.

The construction was designed with two contact surfaces between the holder and the cover so that they cannot move against each other and the cover remains aligned with the holder. The scalpel was successfully prevented from being dislodged during cutting. When clamped into the machine the holder allows for a straight movement of the scalpel centered to the clamping heads so the influence of a torque on the measurements can be avoided. A preliminary stability test showed that it could easily withstand the maximum force required for the selected load cell.

Requirements for the design of the sample holder were for it to make the clamping of flat sample materials into the testing machine possible. It also allows this to always be done in the same way.

The finished design is shown in Fig. 2. The holder was designed with two parts that are almost identical to fit the sample between them. A contact surface was added to ensure they are aligned to each other.

The parts of the holder that are to be clamped into the machine are designed with boundaries to fit precisely between the clamping heads. The finished sample holder allows for flat samples with a thickness of about 1 - 2 mm to be clamped into the testing machine. The thickness is limited by the sample holder as well as by the length of the blade. With this thickness it is to be assumed that the behaviour

Figure 2: CAD rendering of the sample holder with a cutting area of (40 x 40) mm^2

of cutting through the first layer of tissue can be approximated.

For this project this is sufficient since it would not be expected for a scalpel to cut all the way through a thick material at once.

The sections of the holder that are clamping the sample have a total area of (60 x 60) mm^2. It has a cut out area in the middle which measures (40 x 40) mm^2, where the scalpel can cut through the sample material. On the upper side of the area the cut out is extended in the middle of the holder with a width of 12 mm and a height of 5 mm to lengthen the possible cutting path of the scalpel. As seen in Fig. 2 the part that is holding the sample stands on an 18-degree angle toward the part that is to be clamped into the machine. This is to ensure that the blade length can penetrate almost completely into the material when cutting instead of just using the tip of the blade. Extra material between the two parts is added to achieve more stability.

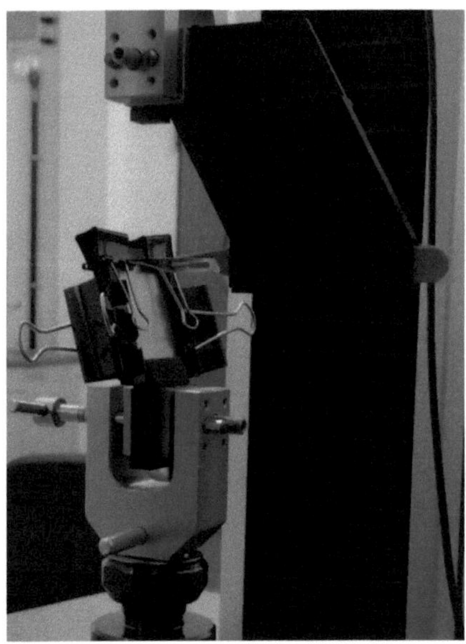

Figure 3: Scalpel and sample material clamped into the testing machine using the mounting brackets

In Fig. 3 the finished overall construction clamped into the testing machine is shown. The mounting brackets are aligned with each other so that the beginning of the blade can be positioned at the start of the open area of the sample material. The scalpel and an agilus patch are clamped into the mounting brackets. Foldback clips are used to keep the sample material in place.

2.4 Measurement process

The measurements were carried out with a fixed starting position. The sample holder was clamped in a position just before the scalpel would contact the sample material except for the first measurement where it was moved further away from the blade so the beginning of the cutting process was more visible. The maximum traverse was set to 36 mm so the scalpel would only cut through the sample material and not strike the edge of the holder. The cutting velocity was set to 500 mm/min based on what would be used in standardised tensile testing of elastomers [6].

For the first measurement the process of cutting through the agilus patch was filmed with a frame rate of 50 fps. Afterwards the video was split into individual frames and the frames were matched to the corresponding values of the recorded force-time diagram.

To evaluate whether a comparison between different materials can be achieved, agilus patches as well as cheese slices were cut. For each material the force-traverse diagrams of two samples of the same material were obtained.

3 Results and Discussion

3.1 Analysing the cutting behaviour of 3D printed material

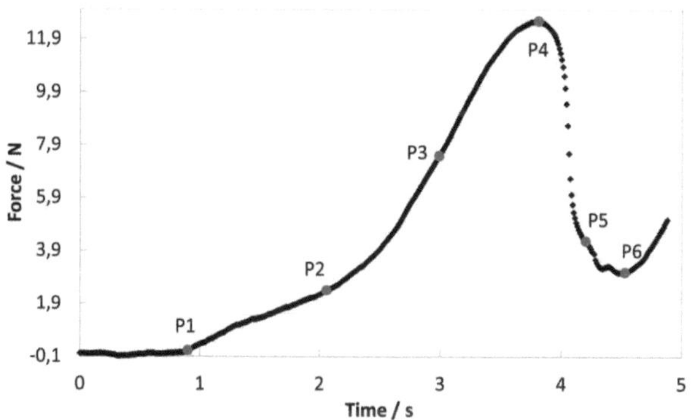

Figure 4: Force-time diagram of a cut through an agilus patch with a thickness of 1 mm

In Fig. 4 the force-traverse diagram resulting from cutting through an agilus patch is shown. Fig. 5 (a) - (f) shows the movement of the scalpel blade during the cutting process. In Fig. 5 (a) the blade is starting to contact the material. This is marked as the first point P1 in Fig. 4 where the force starts to build up. In Fig. 5 (b) the blade is starting to pull on the material at point P2 in Fig. 4. Afterwards the gradient increases. It has not yet pierced through the tissue, which

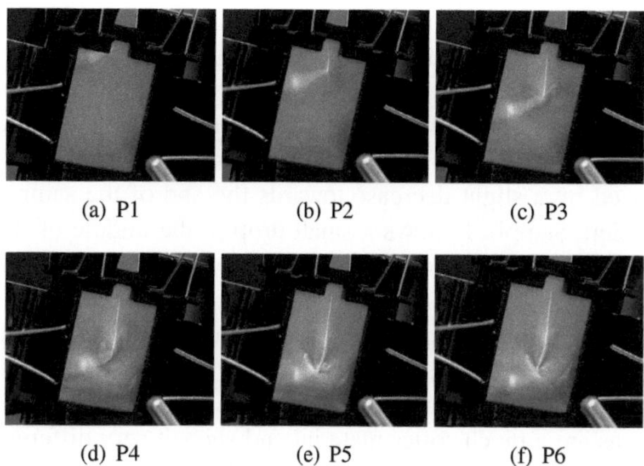

(a) P1   (b) P2   (c) P3

(d) P4   (e) P5   (f) P6

Figure 5: Pictures of the scalpel movement at different points in time throughout the cutting process

is happening at point P3 in the middle of the increase in force and can be seen in the matching picture in Fig. 5 (c). Fig. 5 (d) shows point P4 where the force reaches its maximum. It is visible how the material has been dragged down by the blade and is bulging in front of it. Shortly thereafter the blade has cut through the bulged part as can be seen in Fig. 5 (e) A steep decrease in force can be seen between P4 and the corresponding point P5. At point P6 the blade has finished cutting through the dent as can be seen in Fig. 5 (f) as the force reached a local minimum before starting to increase again.

As expected the process of cutting through the material leads to an increase in force. The process of the blade starting to pull on the material before cutting it seems to result in an increase of the slope. Contrary to this, the moment the blade is cutting through the material does not seem to have an impact on the force progression at first but it starts a bulging of the material in front of the blade. After the point of the maximum force the blade manages to cut trough the bulge. This sudden separation of the accumulated material leads to a sudden decrease in force before it is starting to increase again due to the continuation of the cutting process towards the end.

3.2 Comparing the cutting behaviour of different materials

The force-traverse diagrams for cutting through the materials can be seen in Fig. 6 for two agilus patches and in Fig. 7 for two slices of gouda cheese.

The measurement curves of the cheese slices were smoothed using the 5-point moving average, as they exhibit a high level of noise due to the low forces compared to the maximum force of the load cell.

The force curve for the scalpel cutting through the agilus material has a distinctive progression as is discussed in 3.1. Towards the end of the curve the force is even higher than at the first maximum. This seems to be caused by an accumulation of the material that has been dragged down by the blade towards the end of the sample area.

When repeating the measurement with gouda cheese slices the force that is required to cut through the material is significantly lower. Both graphs exhibit an entirely different force curve comparing to the agilus samples. An increase in force throughout almost the whole measurement is followed by a slight decrease towards the end of the sample holder. Sample 1 shows a small drop in the middle of the measurement. This difference in the force progression can be explained as a result of the high level of noise and varying degrees of bulging of the slices during the cutting and clamping process.

When comparing both materials it is evident that the cheese slices are a much softer material and show a very different cutting behaviour. In order to reduce the noise of the signal of very soft material, the testing machine would need to be used with a different load cell for smaller forces.

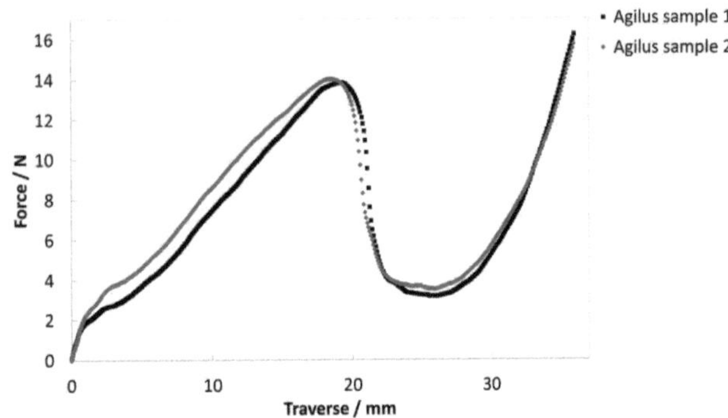

Figure 6: Force-traverse diagram of cutting through agilus patches

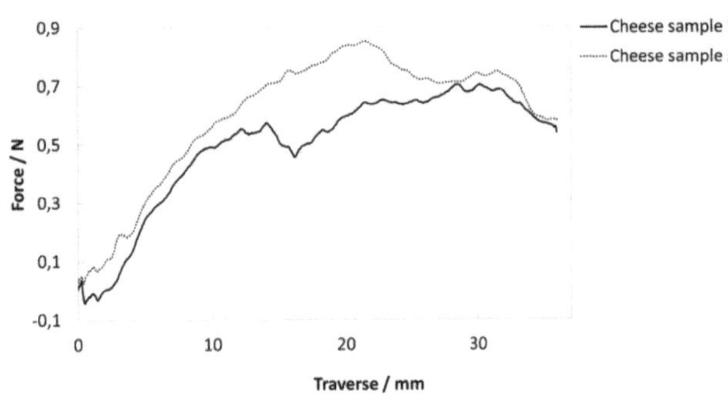

Figure 7: Force-traverse diagram of cutting through slices of gouda cheese

4 Conclusion

A test rig to quantify the cutting characteristics of different materials was successfully constructed. It allows the comparison of additively manufactured flat phantoms with other biological material.

To examine whether a 3D printed sample has realistic cutting properties it can be compared to the biological sample it means to simulate by comparing their force-traverse diagrams obtained using the test rig. When both materials have a force in the order of magnitude of the load cell, differences in gradient or the maximum force can be used for an evaluation. Even materials with forces in different orders of magnitude can be compared through their varied force progressions or can be measured with a different load cell to avoid a high level of noise. For further experiments the test rig can be expanded to also enable testing with different cutting angles.

Acknowledgement

The work has been carried out at the Fraunhofer Research Institution for Individualized and Cell-Based Medical Engineering IMTE and was supervised by Prof. Dr. rer. nat. Thorsten M. Buzug, Institute of Medical Engineering, Universität zu Lübeck.

Authors' Statement

Conflict of interest: Authors state no conflict of interest.

5 References

[1] P. Weinstock, R. Rehder, S.P. Prabhu, P.W. Forbes, C.J. Roussin and A.R. Cohen, *Creation of a novel simulator for minimally invasive neurosurgery: fusion of 3D printing and special effects.* In: Journal of neurosurgery. Pediatrics, 20(1),pp. 1–9, 2017

[2] H. Riedle, *Haptische, generische Modelle weicher anatomischer Strukturen für die chirurgische Simulation*, FAU University Press, 2020

[3] J.S. Steger, *Development of a novel colon anastomosis procedure and an endoscopic, single-use platform by using additive manufacturing*,TUM School of Engineering and Design, 2023

[4] S. Schuldt, T. Witt, C. Schmidt, Y. Schneider, T. Nündel, J.-P. Majschak and H. Rohm *High-speed cutting of foods: Development of a special testing device* In: Journal of Food Engineering, Volume 216, pp. 36–41,2018

[5] J. Vujevich, A. Kimyai-Asadi and L. Goldberg, *The four angles of cutting.* In: Dermatologic surgery : official publication for American Society for Dermatologic Surgery,pp. 1082-4, 2008

[6] DIN 53504:2017-03, *Prüfung von Kautschuk und Elastomeren - Bestimmung von Reißfestigkeit, Zugfestigkeit, Reißdehnung und Spannungswerten im Zugversuch*

Accurate Estimation of Yield Stress from Bending Test and FEM Analysis

Finjas Künnemann [1] and Robert Wendlandt [2]

[1] Medical Engineering Science, Universität zu Lübeck, finjas.kuennemann@student.uni-luebeck.de
[2] Clinic for Orthopaedics and Trauma Surgery, UKSH Campus Lübeck, robert.wendlandt@uni-luebeck.de

Abstract

The four-point bending test is a standard procedure in the quality testing of osteosynthesis plates, in which a material sample is bent between four contact points under load. The data obtained is informative, but not sufficient as a basis for non-linear finite element analysis and tests that provide the required data are not worthwhile for smaller companies and departments. Therefore, in this paper a procedure was designed to derive further material parameters with less error than before possible from the results of a four-point bending test. The implemented combination of mathematical formulas for estimating uniaxial stress-strain curves from bending data and iterative FEM simulations proved to achieve a better determination of the yield stress of an unknown material. A more extensive implementation of this method could provide an excellent alternative to complex material testing methods such as the tensile test.

1 Introduction

Osteosynthetic treatment of a bone fracture using metal plates is one of the most common types of surgical fracture treatment. The plates are screwed directly to the bone and enable good and rapid bone regeneration. As a medical device that is implanted in the body and has to withstand not only everyday stresses, but also stress peaks caused by minor falls, for example, careful quality testing is essential. If the plate breaks or becomes deformed, it could possibly cause additional injury to the bone or tissue and makes a new operation necessary. For these reasons, or to enable possible optimisation of the implant, various quality testing methods are used.

The standard procedure for medical implants is the four-point bending test (FPBT). In this easy-to-realise material test, an available material sample is placed between four contact points and bent under load P. However, this test only provides simple material parameters such as the elastic (Young's) modulus E. For a comprehensive analysis of a material using non-linear finite element simulations, however, further parameters such as yield strength and ultimate strength are required. Although these can be derived from the result of a four-point bending test, they cannot be determined with sufficient accuracy due to occurring non-linear effects.

One method of quality testing that provides accurate values is the tensile test, in which a standardised material sample is subjected to tensile stress and sensors measure the elongation and change in cross-section. This test provides all the necessary mechanical material parameters, but requires highly specialised measuring devices and equipment that only a few facilities can provide.

The aim of this work was to develop an algorithm by combining mathematical equations for the estimation of stress-strain curves and Finite Element Method (FEM) simulations, by means of which all important material parameters can be derived from the results of a simple four-point bending test, with higher accuracy than previously possible, without the need for a complex tensile test.

2 Material and Methods

The core approach of the methods described in the following chapter was to take data from the FPBT of a known material and assume that these data originate from the FBPT of an unknown material whose material parameters are to be determined as accurately as possible. Any information beyond the FPBT data was used only for the final assessment of the results.

In a first step, the model of a FPBT was generated in a FEM simulation program. As a template, the *ASTM* (American Society for Testing and Materials) F382 standard, which is required for CE certification of a medical device, was used.

2.1 Four-Point-Bending-Test

As shown schematically in Fig.1, the FPBT consists of a test sample resting on two support rollers and two load application rollers. The load application rollers are displaced at a constant speed in the z direction during the test. The distance between the two load applicators is a while the distance between a load point and support point is h.

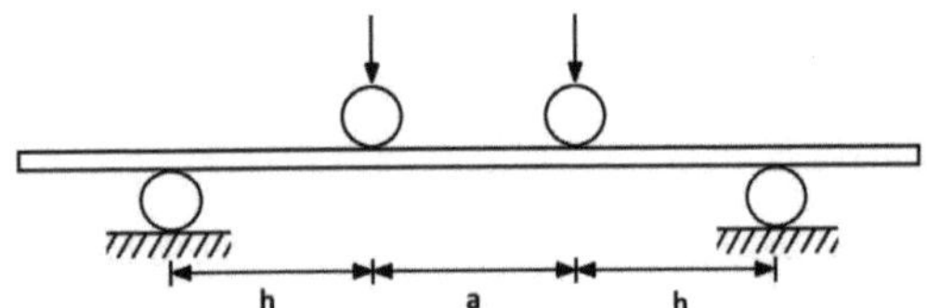

Figure 1: Schematic representation of a four-point bending test with two load application bodies on the upper side and two support bodies on the lower side.

From a physical point of view, the shear force induced by the bending load triggers a bending moment via the acting lever arm. Due to the two load application points of the four-point bending test, the bending moment is maximum over the entire distance between the two points and decreases linearly towards the support points. Due to the bending, the the convex curvature is subjected to tensile stress and the concave curvature is subjected to compressive stress. The maximum stresses occurring at the edges of the specimen are referred to as bending stresses σ. The displacement of the load application rollers is referred to as displacement w.

2.2 FEM Simulation of the FPBT

Simulation of the four-point bending test (FPBT) was carried out using the FEM programme *ANSYS 2023R2*. As shown in Fig. 2, the simulated test consists of a sample of height $d = 3$ mm, width $b = 15$ mm and length $l = 150$ mm. To simplify the simulation, the function of the two support rollers and two load application rollers was approximated by half cylinders with a radius of $r = 5$ mm. During the simulation, the load application bodies moved at a constant speed by -10 mm in the z-direction, while the support bodies were fixed in all spatial directions.

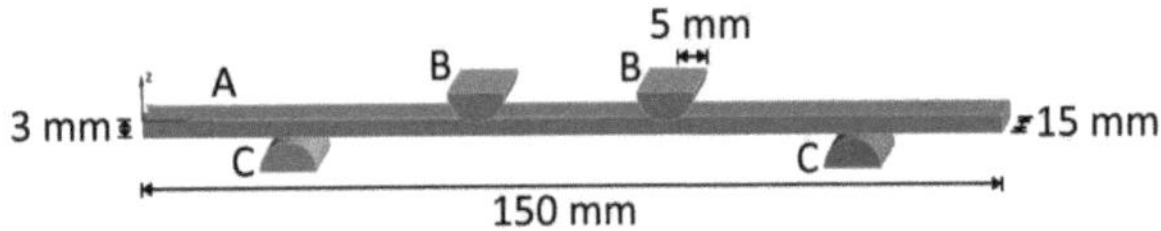

Figure 2: Simulated FPBT in *ANSYS 2023R2* consisting of a test piece (A), two force application bodies (B) which are moved in the z-direction and two support bodies (C).

Prior to the simulation, non-linear material parameters with defined multi-linear isotropic hardening were created for an aluminium alloy of the type Aluminium 6060 T66 [1]. Stainless steel from the *ANSYS* internal material library was used for the material of the force application and support bodies. Furthermore, a friction coefficient of $\mu = 0.19$ [2] was entered for all contacts and large deformations were enabled.
The result of the FEM simulation are 200 data points of the force reaction in the z-direction over the displacement of the load application bodies. Fig. 3 shows the corresponding

graph. In the following calculations, this simulated force-displacement (FD) curve is regarded as the measurement result of the four-point bending test of an unknown material.

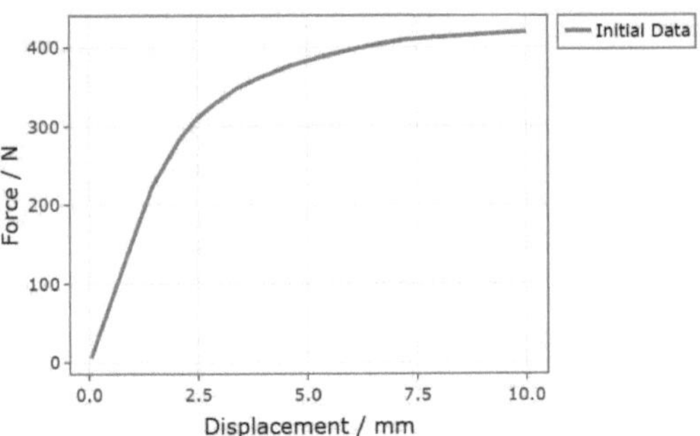

Figure 3: Force-displacement data of the simulated FPBT of aluminium 6060 T66. Used as initial data for algorithm development.

2.3 Calculation of Material Properties from FPBT

The calculation of the material properties from our unknown material sample was implemented using *R (version 4.3.2)*.
Firstly, the bending strain ϵ was calculated using (1) from values of the displacement w. As well as the simple conversion of the force P into stress σ_{lin} using (2) [3]. However, it should be noted here that (2) is a linear conversion that is only valid in the ideal elastic range of the deformation and is too imprecise beyond that. Using (3), the Young's modulus E_{lin} was calculated from the slope of the linear stress-strain data.

$$\epsilon = \frac{wd}{(2h+a)h} \cdot \frac{3(2h+a)}{3(2h+a)-4h} \tag{1}$$

$$\sigma_{lin} = \frac{3Fh}{bd^2} \tag{2}$$

$$E = \frac{3(2h+a)^2h}{4bd^3}\left(1 - \frac{4h^2}{3(2h+a)^2}\right)\frac{\triangle F}{\triangle w} \tag{3}$$

In order to improve the calculation of the stress σ, the next step utilised Kato's equations (4), (5) and (6) to calculate the bending stress from the values of the force P [4]. The equations achieve an improved approximation of the stress curve by including the derivative of the bending curve.

$$\frac{d\epsilon_1}{dP} = C_1 \quad and \quad \frac{d\epsilon_2}{dP} = C_2 \tag{4}$$

$$\sigma_1 = \frac{d}{bh^2C_1}\left(P(C_1+C_2) + \frac{\varepsilon_1+\varepsilon_2}{2}\right) \tag{5}$$

$$\sigma_2 = \frac{d}{bh^2C_2}\left(P(C_1+C_2) + \frac{\varepsilon_1+\varepsilon_2}{2}\right) \tag{6}$$

Assuming that tension strain ϵ_1 and compression strain ϵ_2 are identical, the equations were simplified to

$$\sigma = \frac{d}{bh^2C} \left(2PC + \varepsilon\right). \qquad (7)$$

The resulting data was idealised by first shifting the y-intercept and then replacing the first part of the measured data with an ideal elastic straight line. The slope for this, which is equivalent to Young's modulus E, was determined using the *SDAR* algorithm [5]. A generalised additive model (GAM) was fitted to the resulting data for smoothing (Fig. 4). The slope in the initial region of this stress-strain data corresponds to the Young's modulus E of the sample.

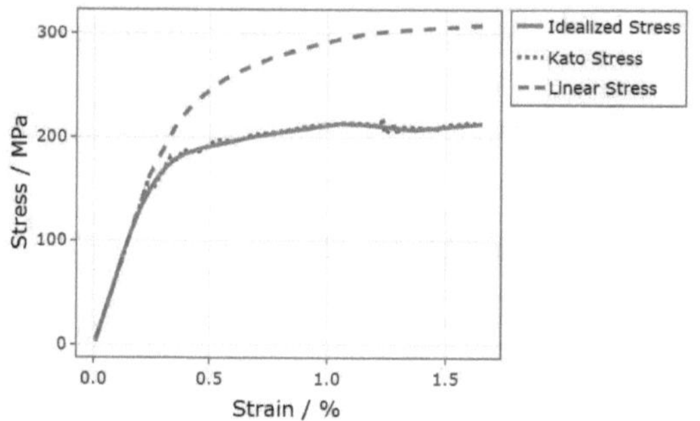

Figure 4: Linear stress calculated from the initial data as well as raw and idealized stress calculated according to Kato's equation.

In the next step, the yield point of the sample material was determined. As shown in Fig. 5, the approximation of the elastic region of the idealised and smoothed stress-strain data is used to create an ideal-elastic straight line that was shifted by 0.2% of the strain. The intersection of the graphs corresponds to the calculated yield stress σ_y.

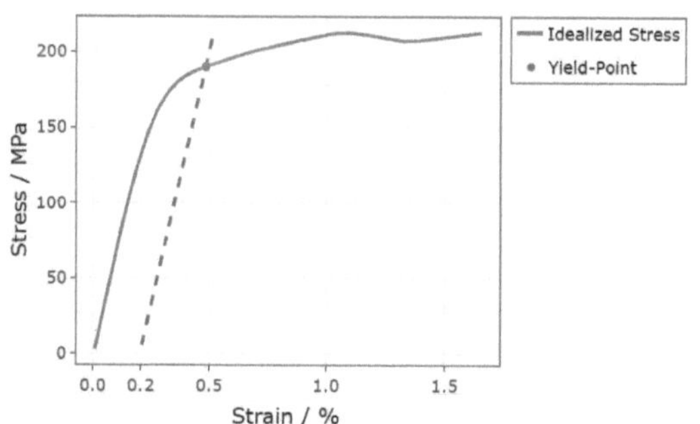

Figure 5: Determination of the yield point from the intersection of idealized stress according to Kato's equation and an ideal elastic gradient shifted by 0.2 %.

The Ramberg-Osgood equation (8) allows the stress-strain curve of a tensile test to be reconstructed using only the Young's modulus and two secant yield stresses [6].

$$\epsilon = \frac{\sigma}{E} + K \cdot \left(\frac{\sigma}{E}\right)^n \qquad (8)$$

The constant K was calculated using (9) from the strain ϵ_y and the stress σ_y at the yield point and Young's modulus E [6].

$$K = \epsilon_y \cdot \left(\frac{E}{\sigma_y}\right)^n \qquad (9)$$

The exponent n was given by (10), where σ_1 and σ_2 correspond to the intersection points of the previous stress-strain curve and two straight lines $E \cdot m_1$ and $E \cdot m_2$ intersecting the coordinates of the origin [6].

$$n = 1 + \frac{\ln\left(\frac{m_2}{m_1} \cdot \frac{1-m_1}{1-m_2}\right)}{\ln\left(\frac{\sigma_1}{\sigma_2}\right)} \qquad (10)$$

Using the previously calculated values of Young's Modulus E and yield-stress σ_y and the exponent n as a variable, a stress-strain curve was determined by the fit of the Ramberg-Osgood equation (8) to the calculated idealised stress-strain data. The result of this fit is shown in Fig. 6.

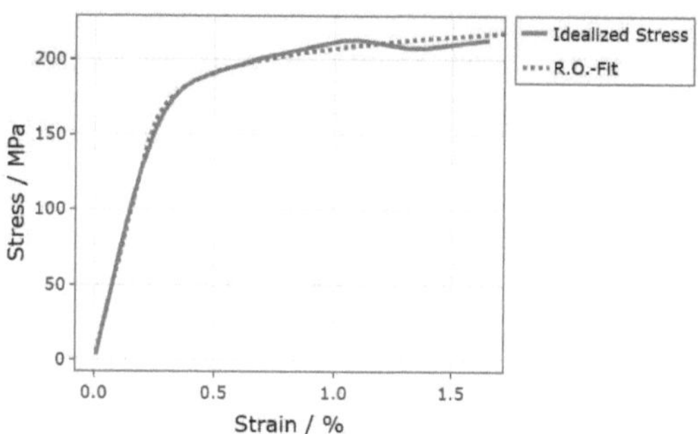

Figure 6: Fit of the Ramberg-Osgood equation to idealised data.

In the final step, an optimisation algorithm was implemented. The aim of this algorithm was to reduce the deviations of the material properties calculated up to this point. For this purpose, the Ramberg-Osgood equation was used to create stress-strain curves from variations of -25% to +10% of the calculated yield stress σ_y and the Young's Modulus E. From these variations in material properties, force displacement data was generated again using a simulated four-point bending test in *ANSYS*. The algorithm then calculated an error value as an integral of the deviations of the new force-displacement curves compared to the original in the area before the yield point (see Fig. 7). Finally, the errors were plotted over the variations of the respective material parameters and the curve approximated. The optimum material parameters correspond to the value at the zero point of the error [3].

3 Results and Discussion

The material parameters used for the simulation of the force-displacement curve of our material considered as unknown corresponded to Aluminium 6060 T66 with a true Young's modulus $E = 68\ GPa$, a yield stress $\sigma_y = 170\ MPa$ and an exponent $n = 22$. In the following section, the results of the different calculation methods for E and σ_y are listed and the deviations from the correct values are quantified.

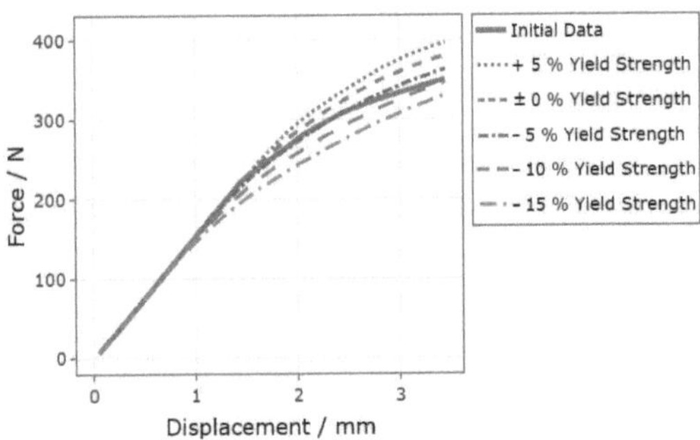

Figure 7: FD data of the initial FPBT as well as a selection of FPBTs simulated from calculated material parameters with varying yield strengths σ_y.

3.1 Young's Modulus

The calculation of Young's modulus E using (3) from the linear stress-strain data resulted in $E_{Lin} = 69.32\ GPa$, which corresponds to a deviation from the actual value of +1.94 %. For the Young's modulus, calculated from the slope of the stress-strain values determined using Kato's equations (7), on the other hand, was $E_{Kato} = 67.38\ GPa$, which corresponds to a deviation from the actual value of only +0.92 %.

3.2 Yield Stress

Yield stress σ_y, derived from the stress value at a strain of 0.2 % from the linear stress-strain data gave $\sigma_y = 256.94\ MPa$, which corresponds to a deviation from the actual value of +51.14 %.

The same calculation from the stress-strain values determined using the Kato equation, on the other hand, resulted in $\sigma_y = 190.18\ MPa$, which corresponds to a deviation from the actual value of +11.87 %.

The determination of the error integrals in the garaged area between 0 mm displacement and the yield point determined via Kato's approach, generates the plot of the error values over the simulated yield stress shown in Fig. 8. A yield stress of $\sigma_y = 179.88\ MPa$ can be read from the approximate zero crossing point of the error, which corresponds to a deviation from the actual value of only +5.81 %.

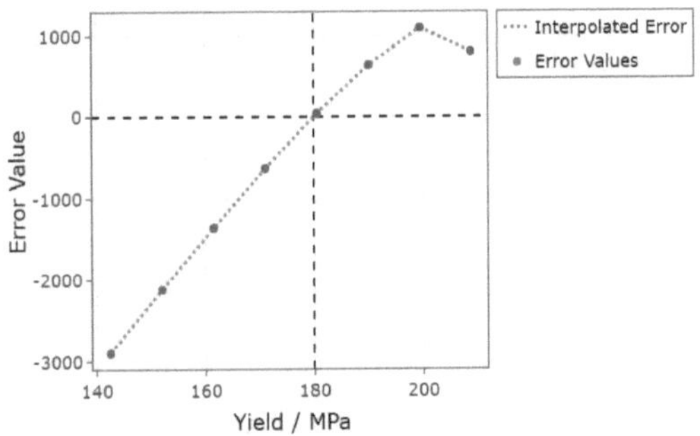

Figure 8: Error between original and varied force displacement data over yield-stress σ_y with zero crossing.

4 Conclusion

The results show that already after one iteration the developed algorithm has a higher accuracy in the determination of the yield stress than non-iterative methods and it can be assumed that further iterations would provide even better results.

The algorithm seems to be a promising approach for the determination of further material parameters from FPBT and should be validated for other materials and real FPBT in the future.

The exponent n of the Ramberg-Osgood equation, which describes the behaviour of stress-strain curves beyond the yield point, was not considered in this work. This value is also important for non-linear simulations of materials, but could also be determined in a similar way in future work.

Acknowledgement

The work has been carried out in the Laboratory for Biomechanics, Clinic for Orthopaedics and Trauma Surgery, UKSH Campus Lübeck.

I would like to thank the entire team at the Laboratory for Biomechanics, Clinic for Orthopaedics and Trauma Surgery, UKSH Campus Lübeck for their support.

Authors' Statement

Conflict of interest: Authors state no conflict of interest.

5 References

[1] Iron Boar Labs Ltd., *6060-T66 Aluminum*. Available: https://www.makeitfrom.com/material-properties /6060-T66-Aluminum [last accessed on 2024-01-12].

[2] A. Meissen, *Festkörperreibung: Reibungszahlen verschiedener Werkstoffe*. Schweizer Ingenieur und Architekt, vol. 111, pp. 28, 1993.

[3] M. Gaszow, *Bestimmung von Materialeigenschaften für die numerische Testung von Osteosyntheseplatten*, B.Sc. Thesis at University of Lübeck, pp. 16–26, 2020.

[4] H. Kato, Y. Tottori and K. Sasaki, *Four-point bending test of determining stress-strain curves asymmetric between tension and compression*. In: Experimental Mechanics, vol. 54, no. 3, pp. 489-492, 2014.

[5] R. Wendlandt, *sdar: Slope Determination by Analysis of Residuals*. Available: https://github.com/soylentOrange/sdarr [last accessed on 2024-01-14].

[6] W. Ramberg and W.R. Osgood, *DESCRIPTION OF STRESS-STRAIN CURVES BY THREE PARAMETERS*. In: National Advisory Commitee for Aeronautics, vol. 902, 1943.

Advancing Prostate Transurethral Resectoscope Testing: Enhancing Hydrogel-Based Tissue Mimics and Ensuring Storage Stability

Sevda Golkar [1]
[1] Biomedical Engineering. Luebeck University of Applied Science. sevdagolkar@stud.th-luebeck.de

Abstract

This research aims to enhance training materials for transurethral resectoscope devices (TUR) used in prostate treatment. Traditionally, porcine tissue has been used to mimic the prostate. Initial attempts for making an artificial mimic with alginate-based materials proved to be promising but lacked the necessary toughness and temperature stability. This study focuses on enhancing hydrogel-based material and introducing color for improved usability. Through systematic experimentation, novel composites were formulated by combining 3% kappa carrageenan with 3% alginate solved in 0.9% sodium cloride(NaCl), and crosslinked by 5Mol calcium chloride (CaCl2), resulting in a significant improvement in material toughness. Food color was introduced to enhance visibility. Additionally, investigations into storage viability were conducted.

1 Introduction

The most common sign of benign prostatic hyperplasia (BPH) is the growth of the prostate gland, which can result in a variety of symptoms, including kidney issues and incontinence. Transurethral resection of the prostate (TURP), the conventional treatment, is an endoscopic operation that uses bipolar technology to remove obstructive tissue. The understanding of heat transport during this process is still lacking, though. An in vitro model that resembles the lower urinary tract has been created to address this. Initially, porcine tissue was employed, but there were issues with its unpredictability. For reliable thermal experiments, a more constant tissue substitute is therefore needed [1]. The aim of this work is to improve alginate and kappa-carrageenan-based hydrogels as viable alternatives for prostate tissue simulation. Alginate, known for its gelling and biocompatibility properties, and kappa-carrageenan, known for its gelling and rheological properties, offer potential approaches to more closely mimic tissue mechanical properties such as toughness [2], [3]. Furthermore, the investigation focuses on refining the storage stability of these hydrogel-based materials to ensure sustained material properties under varying conditions. By highlighting the importance of maintaining high storage stability for real-world applications in medical device testing and prostate treatment technology, this study underscores the importance of advancing hydrogel-based materials in clinical settings [4], [5].

2 Material and Methods

2.1 Material

A comprehensive list of the materials used in the experimental investigations is given in Table 1.

Table 1: Materials and Chemicals used in this work.

Experiment	Material	Manufacturer
Alginate Samples	Alginate,	Roth GmbH
Production	calcium chloride,	Roth GmbH
	0.9% sodium chloride	
	solution,	Roth GmbH
Kappa Carrageenan	Potassium chloride,	
Samples Preparation	Kappa carrageenan	Roth GmbH
Enzyme-linked	Collagen, Gelatine,	Alpha Foods
sample production	Agar powder,	Sigma-Aldrich
	Meat glue	Roth GmbH
Pigmentation Experiment	Red food colorant	Food Colours Perczak
Storage Test Holder	1) 0.5% NaCl	
Solutions	2) 0.9% NaCl	
	3) 1.3% NaCl	
	4) 0.9% NaCl + 2.5 mol CaCl2	
	5) 0.9% NaCl + 5 mol CaCl2	
	6) 0.9% NaCl + 10 mol CaCl2	

2.2   Methods

Alginate samples were first prepared by dissolving 1g of alginate powder in 100ml of 0.9% sodium chloride (NaCl) solution. The solution was then stirred at 300 rpm and heated to 80°C to accelerate dissolution. The next step was to prepare a 5 mol calcium chloride (CaCl2) solution. Equation (1) was used to determine the amount of CaCl2 required:

$$W = C.V.M \tag{1}$$

If C is the molar mass (here C = 5 mol), V is the volume of the solution (here V = 100 ml = 0.1 L), M is the molarity of the material (here 147.02 g/mol for CaCl2), and W is the weight (g). Then from (1) it can be calculated that:

$$W = 5Mol/L.0.1L.147.02g/Mol = 73.51g \tag{2}$$

This was followed by dissolving 73.51 g of CaCl2 in 100 ml of water. Filter sheets were immersed in a 5mol CaCl2 solution for fifteen minutes. After complete dissolution of the alginate, the solution was cooled, filtered into PLA molds layered with filter paper and refrigerated at 5 Celsius degree for 24 hours as shown in Fig. 1. The same procedure was used to prepare other alginate samples, such as 2% and 3%. Similarly, kappa-carrageenan samples were prepared using

Figure 1: Alginate production procedure.

exactly the same process as alginate, except that the cross-linking agent - potassium chloride - was used instead of calcium chloride. The amount of potassium chloride chosen was exactly the same as the amount of kappa-carrageenan used. The same procedure was used for the combined alginate and kappa-carrageenan samples as for the alginate samples, and 2g of food coloring was added to the solution during the stirring phase to color the combined samples. For the preparation of enzyme cross-linked samples, the following three mixed solutions were prepared: 1) 20 ml gelatine 10% + 20 ml alginate 1% + 8 ml meat glue, 2) 20 ml gelatine 10% + 20 ml collagen 10% + 8 ml meat glue, and 3) 20 ml gelatine 10% + 20 ml carrageenan 1% + 8 ml meat glue. To dissolve the substances, they were placed in a water bath for one hour. After pouring these solutions into molds, they were kept refrigerated for 24 hours and their mass, length, width and height were measured. Samples were then placed in six different solutions listed in Table 1 at room temperature and normal room humidity (range between 40-60%), covered with paraffine paper, and the storage stability was evaluated using these solutions, with daily mass measurements.

After measuring the dimensions of each sample carefully, a container was filled with 0.9% NaCl saline solution, and a

thermal bath thermostat was set at the desired temperature (37 °C). The samples were then placed in the thermal bath for 15 minutes to check their stability at this temperature. After the heat resistance test, The mechanical test procedure was started with the aim of measuring Young's modulus for each sample. A scale was placed on a lift, and afterwards, the lift table and scale were placed between two metal structures. A cylindrical indenter fixated to a metal bar was placed on top of the sample as shown in Fig. 2. The lifting table was moved upward until the sample barely touched the indenter. The weight shown in the scale should not exceed 2 g. The scale was then tared and the data adquisiton software was started. After 40 s, the two metal sheets on either side were removed rapidly by pulling them out by hand. As a result, the indenter was pressed into the sample causing a deformation of 2 mm. The sample's resistance to deformation was them measured gravimetrically using the scale readings. The resistance serves as an indicator of the sample's stiffness. This test was repeated three times for each sample.

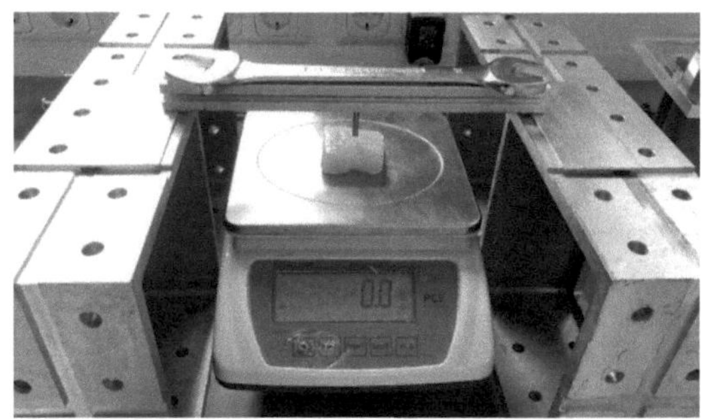

Figure 2: Mechanical test setup.

For the beginning of the electrical test, the probes and the oscilloscope were connected, as shown in Fig. 2. The next step was to prepare the tissue in saline and heat it to 37°C before starting the sectioning process. Each sample was cut five times, and the current and voltage over time from each cut were carefully stored. The data was then transferred to a PC, and a python script was used to calculate the RMS values of current, voltage, and power for each of the measurements.

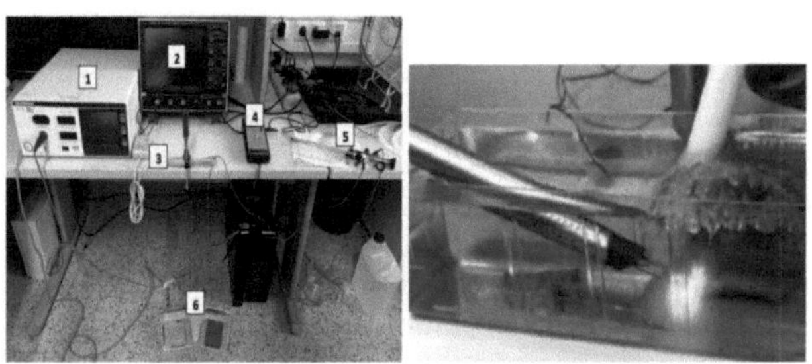

Figure 3: Electrical test procedure.

3 Results and Discussion

3.1 Result

Certain samples were rejected in the first step, including all six samples prepared using enzyme-linked techniques and four samples prepared using 2% kappa-carrageenan cross-linked with 2% potassium chloride from four samples prepared in the general, because they were not gelled at all. In addition, two samples prepared with 1% kappa-carrageenan melted at 37 degrees Celsius, failing the heat resistance test and rendering them unsuitable for practical use. Only the 3% kappa-carrageenan showed proper gelation and stability at 37 degrees Celsius.The four samples of 4% kappa-carrageenan cross-linked with 4% potassium chloride were initially heat resistant, but two of them broke in the mechanical test and the other two were difficult to use in the cutting process as they were very brittle. Better gelled samples, which were also stable in a heat resistance test, were obtained after using 5Mol CaCl2 to crosslink kappa-carrageen. It is worth noting that all ten samples prepared with 3% kappa-carrageen and 3% alginate cross-linked with 5Mol CaCl2 showed excellent gelation and stability.

The results obtained so far led to the decision that the combination of 3% alginate and 3% kappa-carrageenan was preferable to 3% alginate only, which had shown better results in previous research. Both combinations showed good gela-tion, shape and size and satisfactory results in heat resis-tance; mechanical tests were carried out on both. In the mechanical tests of 3% alginate alone, significant variations in the calculated Young's modulus (E) were observed. For pure alginate, the maximum Young's modulus (E max) was found to be 81.6Pa, with a subsequent decrease to 13.4Pa after 120 seconds of suppressed loading (E after 120s), re-sulting in a significant difference (E diff) of 68.39Pa. Con-versely, when a 5 mol CaCl2 solution was introduced, the maximum Young's modulus increased to 86.6Pa, with a higher residual value of 31 after 120 seconds and a reduced difference (E diff) of 55.9Pa. In addition, the results of the electrical test can be seen in Fig. 4, and Fig. 5, where the current is plotted against the sample's number, illustrating how the conductivity varies between different samples. Samples with 3% alginate have a mean value of 0.43A, while 3% alginate + 3% kappa carrageenan samples have a mean value (the average conductivity of the samples) of 0.76A. The results of the storage test showed varying degrees of mass loss over the course of one month as can be seen in Fig. 6. Solution 1, containing 0.5% NaCl, showed a mass loss of 14.8%. In contrast, solutions 2 (0.9% NaCl) and 3 (1.3% NaCl) showed lower mass losses of 12.7% and 12.3%, respectively. In particular, the last three solutions, namely solution 4 (0.9% NaCl + 2.5 mol CaCl2), solution 5 (0.9% NaCl + 5 mol CaCl2), and solution 6 (0.9% NaCl + 10 mol CaCl2), showed higher mass losses of 27.7%, 44.06%, and 70.05%, respectively.

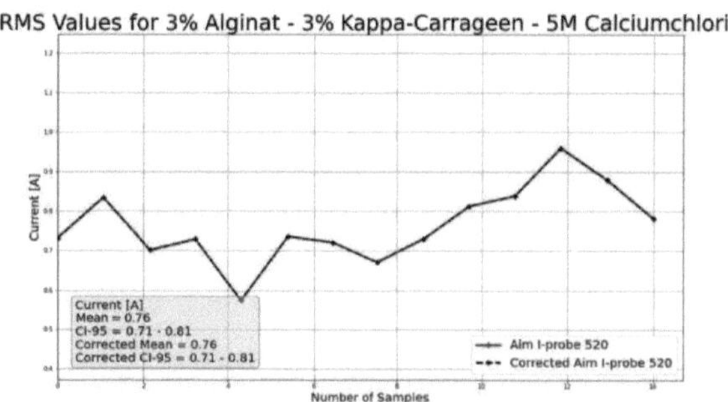

Figure 4: RMS values for Alginate 3% - Kappa- Carrageen 5 Mol Calcium chloride.

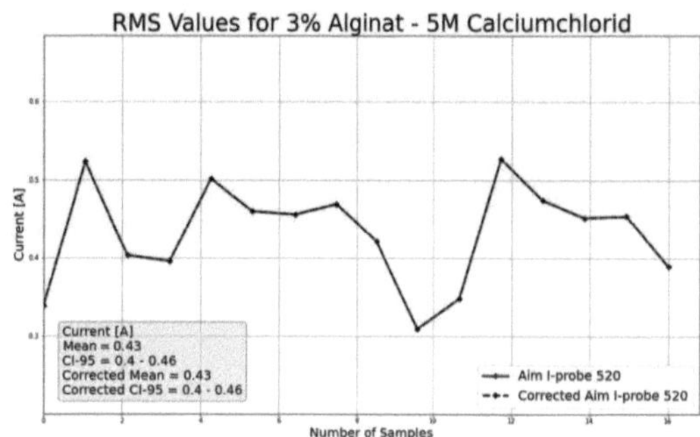

Figure 5: RMS values for Alginate 3% - 5Mol Calcium chloride.

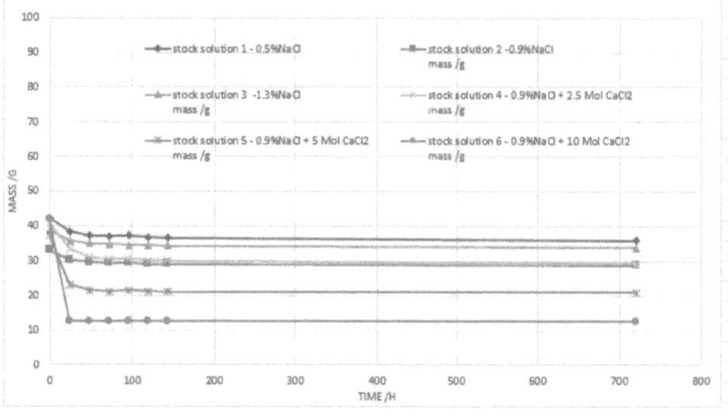

Figure 6: Mass (g) / time (h) for six different solutions.

3.2 Discussions

The decrease in Young's modulus for pure alginate indicates that the material has deformed over time, indicating strong viscoelastic effect. When a 5Mol CaCl2 solution was introduced to the combined 3% alginate and 3% kappa carrageenan, the maximum Young's modulus increased to 86.6Pa, indicating greater stiffness. In addition, the higher residual value of 31Pa after 120 seconds and a reduced difference (E diff) of 55.9Pa indicate improved material stability under the applied load. This suggests that the combination of 3% alginate and 3% kappa carrageenan has contributed to improved mechanical properties, potentially strengthening the structural integrity of the material.

In electrical tests, 3% alginate had a mean value of 0.43A, while the combination with 3% kappa-carrageenan increased this to 0.76A. Comparing these mean values with the mean value for pork, which is 0.3A, the pure alginate 3% is positioned as a closer option to pork.

The inclusion of CaCl2 affected their middle term stability in solution: the more CaCl2 in the samples the solution contained, the more the samples lost mass in the first two days. In particular, solution 2 (0.9% NaCl) and solution 3 (1.3% NaCl) without any CaCl2 had the lowest losses, making them favourable for long-term storage and improving the usability of the material.

4   Conclusion

The combination of alginate and kappa-carrageenan shows promise in achieving improved mechanical properties, while solutions 2 and 3 (0.9% NaCl and 1.3%) emerge as the optimal choices for storage stability considering mass loss.For further scope, it would be nice to work on more storage solutions, taking into account other factors such as temperature and PH value, and to test the samples mechanically and electrically during storage.

Acknowledgement

The work has been supervised by Dr Damiani, at the MSGT Lab of the University of Applied Science Lübeck.

Authors' Statement

Conflict of interest: Author states no conflict of interest.

5   References

[1] M. Ramien, P. Nana, N. Radtke, C. Knopf, S. Klein, and C. Damiani, *Artificial Prostate Tissue for the Simulation of the Transurethral Resection of the Prostate* . In: Current Directions in Biomedical Engineering 2022, vol. 8, no. 2, 2022, pp. 237-240.

[2] Zhai L, Madden J, Foo WC, Mouraviev V, Polascik TJ, Palmeri ML, Nightingale KR, *Characterizing stiffness of human prostates using acoustic radiation force.* In: Ultrason Imaging. 2010 Oct;32(4):201-13. doi: 10.1177/016173461003200401. PMID: 21213566; PMCID: PMC3413332.

[3] Krouskop TA, Wheeler TM, Kallel F, Garra BS, Hall T , *Elastic moduli of breast and prostate tissues under compression.* In: Ultrason Imaging. 1998 Oct;20(4):260-74. doi: 10.1177/016173469802000403. PMID: 10197347.

[4] Mohamadnia Z, Zohuriaan-Mehr MJ, Kabiri K, Jamshidi A, Mobedi H, *Ionically cross-linked carrageenan-alginate hydrogel beads.* In: J Biomater Sci Polym Ed. 2008;19(1):47-59. doi: 10.1163/156856208783227640. PMID: 18177553.

[5] Weikang Hu, Zijian Wang, Yu Xiao, Shengmin Zhanga and Jianglin Wang, *Advances in crosslinking strategies of biomedical hydrogels*. In: Biomaterials Science. Issue 3, 2019.

Probability Filtering for AI Monocular Video Motion Analysis

George Makhoulchahin [1] and Robert Wendlandt [2]

[1] Biomedical Engineering, Technische Hoschschule Lübeck, george.makhoulchahin@stud.th-luebeck.de
[2] Clinic for Orthopedics and Trauma Surgery, Universitätsklinikum Schleswig - Holstein, robert.wendlandt@uksh.de

Abstract

This research explores the possibility of using Google Research's MediaPipe Pose as a cost-effective alternative to traditional methods of gait analysis. Mediapipe Pose was used to estimate the joint coordinates from monocular videos of 10 subjects walking on a treadmill. While it provides real-time pose estimation, inaccuracies in distance estimation pose challenges. To address this, a probability filtering strategy was introduced to enhance the precision of joint coordinate data. The study compares results obtained from MediaPipe Pose after filtering with reference data from GaitLab, employing evaluation metrics such as the number of steps detected, step length, and joint angles. The findings indicate that while filtering improves step detection significantly, it proves insufficient for accurate distance estimation. This implies that, presently, MediaPipe Pose is still not recommended as a substitute for established gait analysis systems in clinical assessments.

1 Introduction

Walking is a fundamental form of human movement. It is the outcome of a complex process that involves the brain, spinal cord, muscles, nerves, bones and joints [1]. Identifying and assessing variations in this process is essential in both scientific and clinical settings. In some cases, a basic physical examination may fall short in accurately diagnosing or identifying specific irregularities in an individual's walking patterns [2]. Gait analysis offers a more comprehensive and precise assessment of a person's gait, proving valuable in treating neuromuscular disorders like cerebral palsy [2]. Additionally, it plays a crucial role in sports medicine by identifying inefficient movement patterns, which allows physicians to tailor training strategies and prevent chronic lower extremity injuries [3]. However, the widespread implementation of gait analysis in clinical routine is hindered by the substantial installation and operational costs associated with traditional instrumentation [4]. In this context, the integration of Artificial Intelligence (AI) presents a promising avenue to overcome these barriers, potentially transforming a ubiquitous device like a mobile phone into the sole equipment required for 3D gait analysis. Google Research's MediaPipe Pose stands as a Machine Learning (ML) solution designed for tracking the human body from monocular red, green and blue (RGB) video frames, detecting 3D landmarks in the process [5]. While its in-plane detection capabilities remain stable, the one-shot detection nature of the AI algorithm introduces instability in depth estimation, resulting in elevated noise levels and unreliable spatial joint coordinates [5]. This scientific paper introduces a probability filtering strategy designed to enhance the precision of joint coordinate data extracted from MediaPipe Pose and examines the potential of this strategy to improve the accuracy and practical utility of the data, providing insights into its practical efficacy for gait analysis applications.

2 Methods

2.1 Preliminary work

Previously recorded videos capture the walking patterns of 10 healthy subjects on a treadmill with a camera fixed on a tripod positioned in front of the subjects. MediaPipe Pose, a component of the ML package MediaPipe developed by Google, is designed for the real-time estimation of an individual's pose through the utilization of BlazePose GHUM 3D, a Convolutional Neural Network (CNN) architecture [5]. Each test subject underwent two recordings: one in standing position lasting approximately 10 seconds, and another while walking on the treadmill for at least 90 seconds [5]. The former served as a reference for movement deflections. The recorded data was processed using a Python script extracting and saving the body points coordinates (x, y, z) estimated by MediaPipe Pose for each frame as a CSV file [5]. The origin of the coordinate system established by Mediapipe Pose is the midpoint between the two hip points. The x-coordinate represents the lateral distance from the center point, with points on the left from the camera view considered negative. The y-coordinate represents the vertical distance, with points above the hip carrying a negative sign. The z-coordinate denotes the front-to-back distance, with points between the camera and the origin bearing a negative sign. These coordinates estimated by MediaPipe Pose show inaccuracies, especially the z-coordinates, and are subjected to subsequent filtering. For comparison, these coordinates are juxtaposed with reference data obtained from GaitLab. GaitLab, or AS-200 system, by the company LaiTronic, captures the positions of 16 infrared markers on the subject's back [5]. These markers' three spatial coordinates are recorded by three infrared cameras, providing high-quality raw data for medical gait analysis [5].

2.2 Setting up data in R

The raw data acquired from both MediaPipe and GaitLab were imported into R, version 4.3.1. R is a versatile and widely used programming language for statistical computing and data analysis [6]. The initial step involved aligning the coordinate systems of the data obtained from both MediaPipe and GaitLab. MediaPipe's world coordinate system, centered at the midpoint of the hips, as depicted in Fig. 2, was chosen as the reference [5]. In GaitLab, the camera served as the center of the coordinate system, placed at a known distance from the subjects. To convert the GaitLab data into this coordinate system, the following adjustments were necessary. The center between the left and right hip was calculated and used as a reference point, allowing for the translation of the coordinate system from the camera to the center of the hips and the x and z coordinates were interchanged.

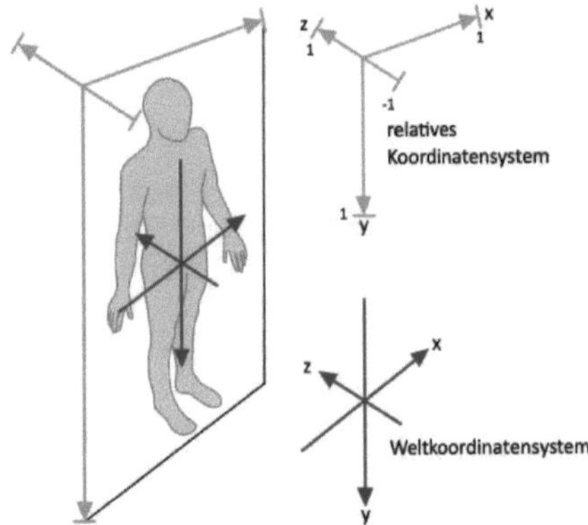

Figure 1: Direction of the axes of the Mediapipe coordinate system.

2.3 Probability Filtering Strategy

Given the placement of the camera in front of the subjects and since the measurements are only obtained from one camera angle, the primary challenge arises in the accurate determination of the z-coordinate. While the x and y coordinates of the joints can be readily obtained from the video, the z-coordinate shows inaccuracies due to the camera's inability to measure depth. MediaPipe Pose cannot predict the actual distance between the origin and the joints resulting in an unpredictable flipping of the z-coordinates between positive and negative values. The filtering strategy implemented focuses on the elimination of these flipping instances to ensure the consistent and reliable progression of the detected data between positive and negative values. One

way of identifying these irregularities is by studying the accelerations. Any sudden large change in the coordinate values is associated with a jump in the joint acceleration. The main idea of the filtering strategy is therefore to calculate the joint accelerations and store them in a dataset, identify the outliers in the dataset, eliminate the coordinate values corresponding to these outliers then use interpolation methods to replace them with more coherent values. Since the center of the coordinate system is the center of the hips, the hip coordinates were filtered first. The time, left hip coordinates and right hip coordinates were extracted into a separate dataset. The z-coordinates of the hips were then derived twice per respect to time to obtain the accelerations. To detect outliers, the Z-scores of each of the accelerations were calculated. The Z-score is a statistical measure that calculates the number of standard deviations by which a data point is above or below the average or mean of the dataset [7]. In other words, a Z-score close to 0 indicated that a data point is close to the average and a Z-score above 2 or below -2 indicates an outlier [7]. Equation (1) is used to calculate the Z-scores where x represents the data point, μ the mean and σ the standard deviation [7].

$$z = (x - \mu)/\sigma. \tag{1}$$

In R, the Z-scores can be easily calculated using the *scale* function. A threshold of 2 was defined and the Z-scores were evaluated for outlier detection. A new boolean column was created with true values for outliers and false values for other data points. The z-coordinates that correspond to the true values were omitted then interpolated using spline interpolation. The process was repeated until no outliers were detected anymore. The vectors between the hips and knees were then calculated and the same process was repeated to filter the knee coordinates, then the ankles and the feet. Furthermore, the length of the thighs and shins were calculated using the coordinates of the hips, knees and ankles to ensure that these remain constant for each subject. The algorithm was implemented for all 10 subjects.

2.4 Angle Calculation

To better visualize the movement of the different joints, some joint angles were calculated from the MediaPipe data and the GaitLab data and compared. These angles include the movement of the hips in the frontal and transverse planes, the flexion/extension of the thighs, which occur in the sagittal plane, the flexion/extension of the knee, which also occur in the sagittal plane, the varus/valgus of the knee, which occur in the frontal plane and the adduction/abduction of the hip joint, which also occur in the frontal plane. To calculate these angles, 3 functions were created in R to project vectors on the sagittal, frontal and transverse planes. The vectors between the left and right hips, between the hips and knees and between the knees and the ankles were then calculated. These vectors were then projected on the respective planes to get the different angles. For example, to calculate the flexion/extension of the left thigh, the vector between the left hip and the left knee was calculated and projected on the sagittal plane, then the angle between the projected vector and the reference vector (0,1,0) was calculated. The cross product between these vectors was also calculated and used to account for the orientation or sign of the angles. The gait data from the treadmill were corrected by calculating the same angles at the neutral-zero position and substracting them from the actual angles. Finally, the angles were smoothed with a moving average to eliminate noise and fluctuations. After the angles were calculated, they were plotted and compared to the angles calculated with the GaitLab data.

3 Results and Discussion

3.1 Evaluation code

A code was developed by Dr. Robert Wendlandt to evaluate the results of the filtering. The main purpose of the code is to perform gait phase detection to identify different gait cycles. The gait cycle is defined as the movement that occurs from the initial contact of the heel with the ground of one foot to the next heel strike of the same foot [2]. It is comprised of 2 main phases: the stance phase, where the foot is still in contact with the ground, and the swing phase, where the foot is moving forward before the next heel strike [2]. The code first reads the knee and ankle data of the subjects, calculates the shin vector as a directional vector using these coordinates, projects the vector on the sagittal plane to compute the flexion/extension of the knee angle then calculates the corresponding angular velocity. The angular velocity of the shin was used to calculate the initial contact of the foot and identify the beginning of the gait cycle as well as the next contact to indicate the end of the cycle. The algorithm was used on all subjects for the GaitLab data, which was used as reference, the unfiltered MediaPipe data and the filtered MediaPipe data. The code successfully computed a few parameters, which were used as a basis for comparing the unfiltered and filtered MediaPipe data to the reference data and evaluate whether the filtering strategy was successful at enhancing the z-coordinates obtained by MediaPipe. Some of these parameters include: the number of detected steps, the step length, which represents the anterior/posterior distance between one heel strike and the next, the step width, which represents the distance between the ankle joints of the 2 feet and the step time, which is the time from initial contact to toe off.

3.2 Results of the evaluation code

The parameters extracted from the evaluation code for all subjects were plotted into boxplots. Figure 3 shows the number of steps detected from the GaitLab data, unfiltered MediaPipe data and filtered MediaPipe data. In Figure 3, the plot on the left represents the GaitLab data, the one in the middle the unfiltered MediaPipe data and the one on the right the filtered MediaPipe data. As shown on the graph, the number of steps detected in the filtered data significantly outperforms the number of steps detected in the unfiltered

data when both are compared to the reference data obtained from GaitLab. The number of steps detected for the filtered MediaPipe data is around 160 steps on average, which is similar to the GaitLab data. The number of steps detected for the unfiltered data however, is around 110 on average. The filtering algorithm therefore enhances step detection.

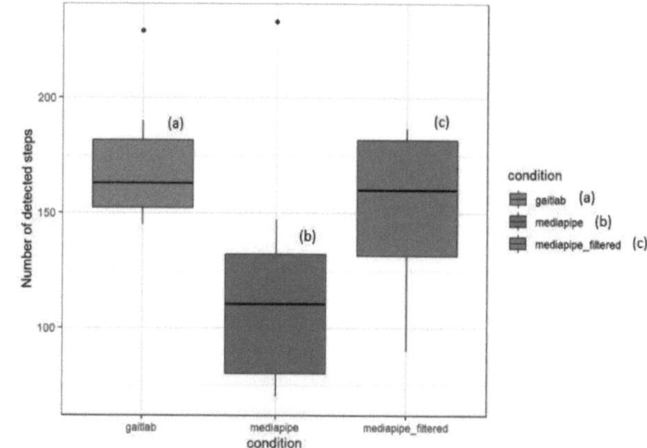

Figure 2: Number of steps detected for GaitLab (a), unfiltered MediaPipe (b) and filtered MediaPipe data (c).

Fig. 4 shows the step length in cm obtained from all 3 datasets. As seen from the graph, the average step length obtained from both the filtered and unfiltered MediaPipe data is around 16 cm. This is much lower than the step length obtained from the GaitLab data, which is 58 cm on average. The step length is measured in the faulty direction of MediaPipe, or the z direction. This result shows that even though the filtering of MediaPipe data proves successful for a better and more accurate step detection, it remains faulty in providing an accurate distance estimation in the z direction.

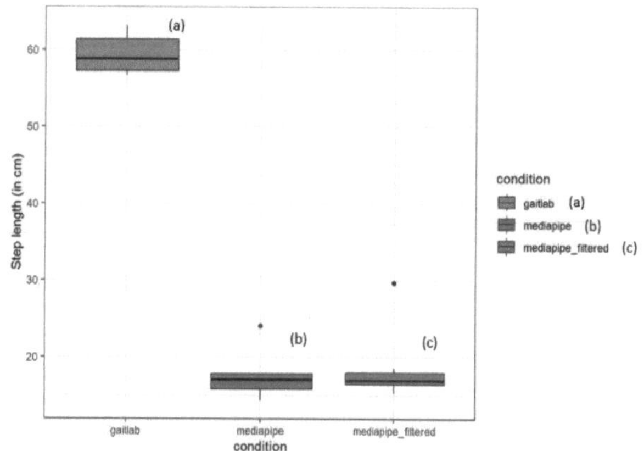

Figure 3: Step length in cm for GaitLab (a), unfiltered MediaPipe (b) and filtered MediaPipe data (c).

3.3 Results of the angle calculation

Fig. 5 shows the plotted flexion/extension of the knee angles. These angles are measured from the axis of the thigh to the shin vector. The x-axis of the graphs represents the time in seconds and the y-axis the angles in degrees. The angles obtained from the original and filtered data are plotted in the first and second graph respectively. The third graph shows the same filtered angles plot but after further smoothing with a moving average. The last graph illustrates the angles obtained from GaitLab, which are to be used as reference. To better visualize the graphs, only a part of the measurement was plotted, from the start to 8 seconds, covering 7 gait cycles.

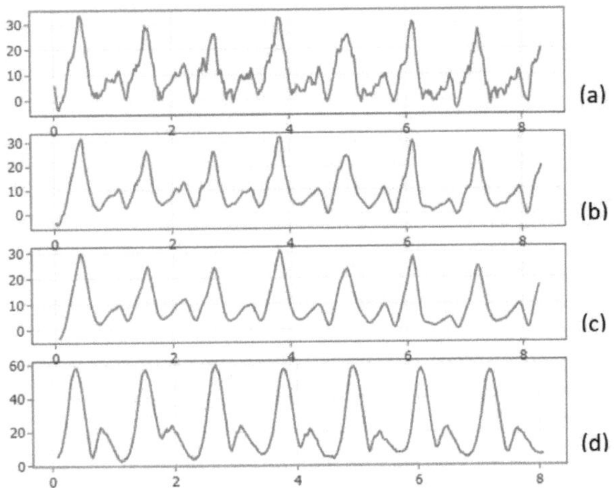

Figure 4: The original (a), filtered (b), smoothed (c) and GaitLab (d) flexion/extension of the knee angles in degrees with respect to time in seconds.

Comparing (c) and (d) in Fig. 4, it is clear that they exhibit a similar behavior having one big peak followed by a smaller peak. These peaks are also aligned and occuring at around the same time. Every peak represents the start of a new gait cycle. This shows that the step detection is accurate. However, the values of the angles differ greatly. The big peak happens at 30 degrees and the small peak at 10 degrees for MediaPipe while for GaitLab the peaks happen at 60 degrees and 20 degrees respectively. The factor of 2 appears to be random in this measurement. This further proves that a simple filtering of the MediaPipe coordinates is not enough to estimate values of the distance in the z direction.

4 Conclusion

The aim of this research paper was to assess the reliability of filtering MediaPipe Pose coordinates for gait analysis. In the x and y directions, this poses no problems at all. However, when it comes to the z direction, the MediaPipe estimated coordinates show great differences when compared to reference data. This is mainly due to the camera's inability to measure depth correctly from one angle. The filtering strategy introduced in this paper has a positive effect on step detection and the overall smoothness of the data, however it proves insufficient to accurately estimate z coordinates. Inconsistencies are still seen between the filtered data and reference GaitLab data, especially with

parameters that involve the depth measurement. To be able to use this estimated data as a reliable source for gait analysis, further post processing techniques need to be developed to correct the distance estimation inaccuracies produced by MediaPipe Pose.

Acknowledgement

The work has been carried out at Technische Hochschule Lübeck, Labor für Biomechanik und Biomechatronik and supervised by Universitätsklinikum Schleswig - Holstein, Klinik für Orthopädie und Unfallchirurgie. The raw data utilized in this study was taken from the previous work of Tabea Volpert, a Biomedical Engineering student at Technische Hochschule Lübeck.

Authors' Statement

Conflict of interest: Authors state no conflict of interest. Informed consent: Informed consent has been obtained from all individuals included in this study.

5 References

[1] M.W. Whittle, *Gait Analysis: An Introduction*. Oxford: Butterworth-Heinemann, 1991.

[2] J.R. Gage, P.A. Deluca and T.S. Renshaw, *Gait Analysis: Principles and Applications*, The Journal of Bone & Joint Surgery, 77(10):p 1607-1623, October 1995.

[3] A.F. DeJong and J. Hertel, *Gait-training devices in the treatment of lower extremity injuries in sports medicine: current status and future prospects*, Expert Review of Medical Devices, 15:12, 891-909, December 2018, doi: 10.1080/17434440.2018.1551130.

[4] A.A. Hulleck, D.M. Mohan, N. Abdallah, M. El Rich and K. Khalaf, *Present and future of gait assessment in clinical practice: Towards the application of novel trends and technologies*, Front Med Technol., December 2022, doi: 10.3389/fmedt.2022.901331.

[5] T. Volpert, *Development and evaluation of a gait analysis system based on AI-assisted evaluation of monocular video recordings*, Technische Hochschule Lübeck, Department of Applied Sciences Institute of Biomechanics and Biomechatronics, May 2023.

[6] The R Foundation. *What is R?*. Accessed: January 2024 [Online]. Available: https://www.r-project.org/about.html .

[7] Khan Academy. *Z-scores Review*. Accessed: January 2024 [Online]. Available: https://www.khanacademy.org/math/statistics-probability/modeling-distributions-of-data/z-scores/a/z-scores-review .

Development and validation of a process for the reprocessing of components of a washer-disinfector for endoscopes

Marcel Zulim [1]
[1] Medical Engineering Science, Universität zu Lübeck, marcel.zulim@student.uni-luebeck.de

Abstract

This paper deals with the development and validation of a method to reprocess components of a washer-disinfector for endoscopes (WD). The reprocessing of flexible, thermolabile endoscopes with WD necessitates pressure and flow meters to ensure standard-compliant disinfection. Defective components pose a risk of contamination, requiring hygienic preparation before returning to the third-party manufacturer for analysis, and thus depending on a validated reprocessing procedure. The knowledge gained from examination contributes to reducing sources of error, which is not only economically beneficial due to increased reliability and possibly reusability but will potentially further increase patient safety. The newly developed process described in this article allows analysis by utilizing the Matachana 130LF low-temperature gas and formaldehyde sterilizer. Hygiene tests, conducted in collaboration with a hygiene institute and Olympus OSTE, confirmed the reprocessing device's efficacy in killing defined microorganisms, ensuring safe handling by the manufacturer.

1   Introduction

Olympus Surgical Technologies Europe (OSTE) is specialized in the development and production of a wide variety of medical devices – including rigid and flexible endoscopes – which are of great importance in modern clinical examination. The reprocessing of flexible endoscopes is carried out using endoscope washer-disinfectors (WD) called Endo-Thermo-Disinfectors (ETD. They are developed by the Infection-Prevention-Systems (IPS) department at the Hamburg-Jenfeld site. The ETD ensures standard-compliant, chemothermic disinfection after each medical procedure by utilizing specialized chemicals, program sequences as well as pressure- and flow sensors. [1]

The Flow Control 2 (FC2) is a pressure- and flowmeter, which is installed inside each modern ETD. Defective FC2 covered by the warranty must be sent to the third-party manufacturer for inspection. As contamination of the component cannot be ruled out due to the area of application, occupational safety is not guaranteed. In addition, the third-party manufacturer only has one test- and production line. Contamination of this line would result in a whole series of problems.

The procedure of reprocessing the FC2 does not yet exist at Olympus, which is why a suitable operation has to be identified, tested and validated in order to guarantee the third-party manufacturer safe handling of the device.

2   Material and Methods

Selecting a suitable machine to use for validation and reprocessing requires complying with current hygiene regulations as defined by the DIN EN ISO 15883-4. These state, that reprocessing of a flexible endoscope must at least use chemothermic reprocessing techniques [2]. Following this statement, any other components within the cycle of reprocessing – including the FC2 – need to follow these hygienic regulations. Furthermore, it is to be tested if electronical components, especially pressure meters inside the FC2, are unharmed by fluids, chemicals, pressures or other substances of a selected reprocessing machine, thus functioning as intended after reprocessing. After these steps, validation requires ensuring that the FC2 can actually be reprocessed in accordance with the specified hygiene standard of the selected machine.

Compliance with hygiene regulations

The machine, which shall be tested and used for validation, is called Matachana 130LF low-temperature gas and formaldehyde sterilizer [3]. Sterilization is defined as the probability of pathogenic organisms being present on a given surface of less or equal than 10^{-6} to make it sterile, thus complying to DIN EN ISO 15883-4 [4].

Electronic Integrity

To ensure that gas-sterilization is not harmful to pressure meters, the setup seen in Fig. 1 is used to measure water pressure in each channel before and after sterilization respectively. Two FC2 are used, referred to as test unit 1 and test unit 2. A target pressure of 1.2 bar – which is a typical value during regular reprocessing – is provided by a pump underneath the table (cf. Fig. 1). It circulates water with a 3/2-way-valve into the test units and through six water outlets back to the pump.

Each measurement is recorded over a period of at least 60 seconds. The difference in pressure ΔP is calculated for each channel to derive, if the pressure meters are affected by sterilization. This is indicated by large ΔP values. To reduce fluctuations in pressure, ΔP is calculated after the target pressure levels out.

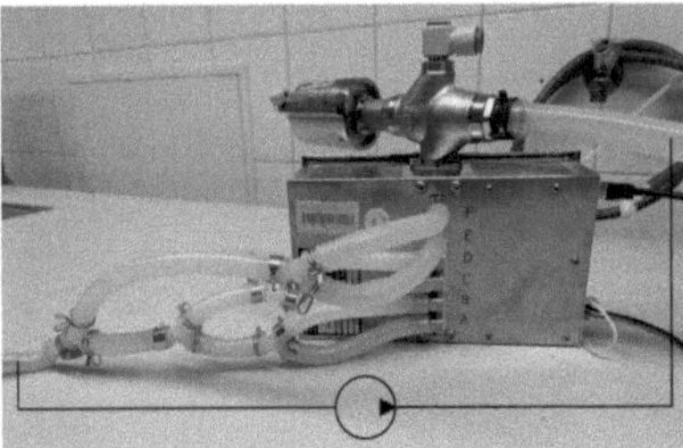

Figure 1: Setup to measure water pressure of pressure meters inside all channels. Target pressure is 1.2 bar. The pressure is measured for at least 60 seconds.

Fig. 2 and Fig. 3 depict pressures for each channel before and after sterilization and the according ΔP. Both diagrams show, that ΔP_{max} lies at 1.29 %, whereas ΔP_{mean} is at 0.53 % for test unit 1 and 0.52 % for test unit 2. As these values pose no significant fluctuations, it can be assumed that sterilization neither damages nor influences the electronics.

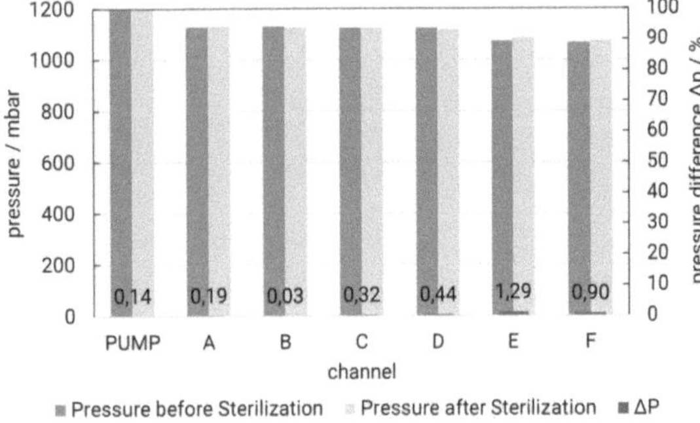

Figure 2: Test unit 1: Pressure P inside each FC2 channel, target pump pressure and difference in pressure ΔP before and after sterilization.

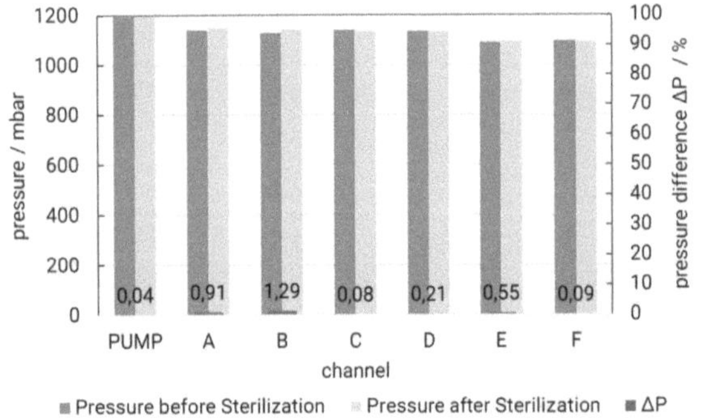

Figure 3: Test unit 2: Pressure P inside each FC2 channel, target pump pressure and difference in pressure ΔP before and after sterilization.

Validation of sterilization efficacy

Validating the FC2 to be sterilized by the Matachana 130LF is carried out by introducing a pre-selected microorganism (also referred to as test soil) with a defined amount of Colony-Forming-Units (CFU) into each channel of the FC2. After this the device is sterilized. For the validation to be successful, the total amount of CFU being detected after sterilization shall be < 1. For safety reasons, a hygiene institute is responsible for attaching the test soil to the component and for the subsequent examination after sterilization. As the microorganism used is genetically modified, it must be handled in an S2 laboratory, for which special safety precautions apply. Due to this, Test Specifications for Laboratories (LABS) are designed, serving as instructions for the hygiene institute. The LABS follow the structure illustrated in Fig. 4. They specify the handling of the FC2, the microorganism to be used as well as its insertion into each channel, its packaging and shipping to Olympus – and back. Lastly, the procedure for cultivation and elution, which is "the process of extracting a substance that is adsorbed to another by washing it with a solvent." [5], to determine CFU, is carried out.

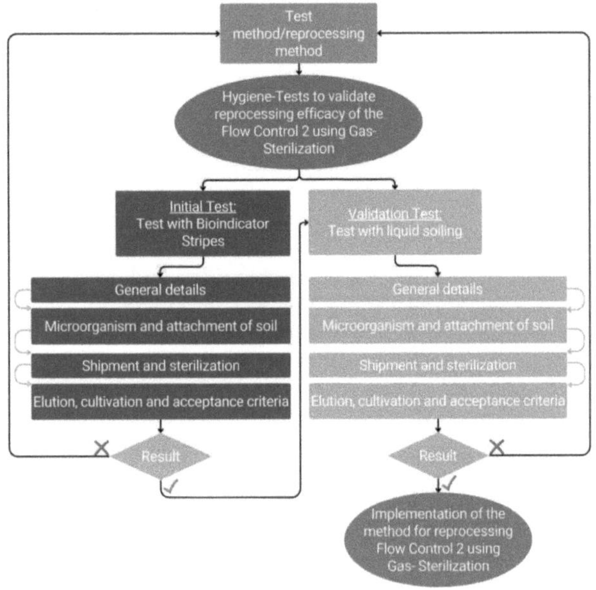

Figure 4: Flowchart of the hygiene test procedure.

Initial Test

The *initial test* examines efficacy of sterilization using Bioindicator Stripes (BI-Stripes) inserted in each channel of one *test unit*. The BI-Stripes are a validated method for testing the sterilization performance of gas sterilizers. Thus, the result can be used to draw conclusions about the general effectiveness of the sterilization. However, this test serves as an initial investigation, since the test soil within the *validation test* – as in the case of theoretical contamination during regular use – shall be dissolved in liquid.

The microorganism used is called *Geobazillus Stearothermophilus* (ATCC 7953). It is a spore-forming, thermophilic bacterium, which is attached onto BI-Stripes by manufacturer Simicon. The average population is $\geq 10^5$ CFU [6].

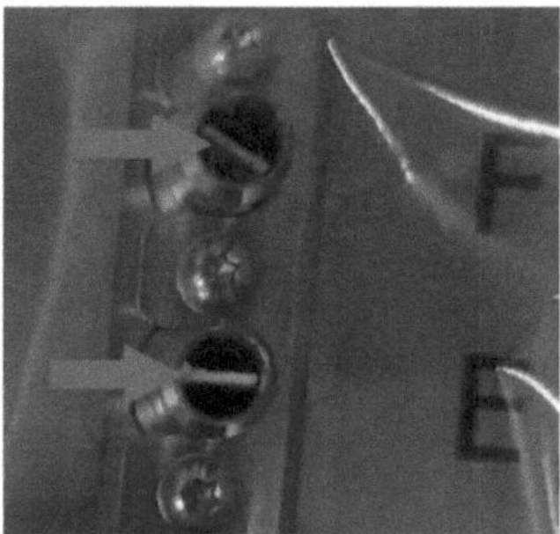

Figure 5: BI-Stripes inside channels E and F of the *test unit*.

BI-Stripes shall be fully inserted into each channel as seen in Fig. 5. One *transport control* and three *positive controls* are used besides the *test unit*. Each of the three constituents are wrapped in sterile, gas- and formaldehyde-permeable foil for transportation from the hygiene institute to Olympus as well as for sterilization. This is to reduce the risk of contamination. Further protection is achieved by using another layer of foil after sterilization and for transportation back to the hygiene institute.

For the *test unit* and *transport control*, the elution- and cultivation procedure is followed by the instructions given by manufacturer Simicon as in [6]. Each *positive control* is immersed in a container with 5 ml sterile water for 10 minutes and shaken manually, detaching the bacteria from the filter paper. The solution is then treated in an ultrasonic bath for 15 minutes and vortexed for 1 minute. After the incubation time of 48 hours at 50 °C - 60 °C, CFU is determined. Due to the high density of bacteria, the colonies are diluted several times. The counted quantity is then multiplied by the dilution factor [7].
The test is successful, if the *test unit* shows no specific growth of the microorganism. Both *transport-* and *positive control* shall show specific growth.

Validation Test

For the *validation test*, test soil consisting of carrier liquid and the microorganism utilized during the *initial test* is used in order to make the validation as realistic as possible. According to Table 1, the test consists of four *test units* as well as one *positive-*, *negative-* and *investigative control*.
The test soil is filled in the corresponding devices, using 30 ml/device. The initial concentration is specified with $2.1 * 10^8$ CFU/ml, but is reduced to $4.3 * 10^7$ CFU/ml by the hygiene institute as a result of dilution, incubation and counting. [8] This results in a more precise determination of CFU. Test soil shall remain in each device for one hour, after which it is poured out. This leads to only the residual moisture – carrying the test soil – remaining in the devices. It would otherwise leak in the sterilizer.
The CFU of the *positive control* is determined at the point of the *test units* being sterilized at Olympus. This is to check, if the CFU in the devices are decreasing over time. The *investigative control* is used to determine the CFU of the residual moisture immediately after the test soil

Constituent	Amount	Number	Soiling	Sterilization
Investigative Control	1	1	Yes	No
Test Unit	4	2 3 4 5	Yes	Yes
Positive Control	1	6	Yes	No
Negativ Control	1	7	No	Yes

Table 1: Overview of the constituents of the validation test.

was poured out. The channels of the *negative control* are eluted in a non-soiled state after sterilization, as its result represents an untreated device. Packaging is done in the same way as in the *initial test*, but each channel is plugged to prevent the residual moisture leaking out. Before sterilization, the plugs need to be detached carefully through the foil.

Elution- and cultivation procedure is carried out by filling each channel with an eluting solution. After shaking for one hour, the solution is extracted and cultivated on the culture medium Tryptone-Soy-Agar (TSA). Incubation time is 48 hours for the *positive control* and seven days for the *test units* and *negative control*. Temperature is at 50 °C - 60 °C. The residual bacteria count is determined by counting the CFU on the TSA-plates, whereby the dilution factor must be taken into account.
For the validation to be successful, the *positive control* and/or the *investigative control* must show specific growth of the test bacterium. The sterilized *test units* must not show any specific growth of the test bacterium [8].

3　Results and Discussion

The outcome of both tests is available as draft due to the limited period of the internship. The results shown are correct and form the basis to the implementation of the process aimed at by Olympus.

Results Initial Test

Table 2 shows that BI-Stripes of the *test unit* showed no specific growth, whereas BI-Stripes of the *transport control* showed specific growth. The *positive controls* were not evaluable under the given conditions [7]. It can be concluded from the results that the *initial test* has been passed despite the fact that the *positive controls* could not be evaluated. This is because the requirements for both *test unit* and *transport control* were met. It is shown, that the Matachana 130LF has neutralized all bacteria and, that they were not killed during transportation, passing the *initial test*.

Constituent	Amount	Channel	Growth
Test Unit	1	1 A - F	-
Transport Control	1	/	+
Positive Control	3	/	N/A N/A N/A

Table 2: Results of the initial test. Specific growth: "+" / No specific growth: "-".

Results Validation Test

The *investigative control* is an indicator of the amount of bacterial colonies solved in the residual moisture, which amounts to $7.8 * 10^6$ CFU/device as in Table 3. Neither the *test units* nor the *negative control* showed specific growth. The *positive control* could not be used due to the volume on the TSA-plates being too high to count. It is to be assumed, that a large amount of CFU were present at the time of sterilization. This can be explained by the fact, that the TSA-plates were overgrown, and because of the many bacterial colonies within the residual moisture [8]. As a result, the *validation test* is passed.

Constituent	Amount	Number	Growth
Investigative Control	1	1	$7.8 * 10^6$ CFU/device
Test Unit	4	2	< 1 CFU/device
		3	< 1 CFU/device
		4	< 1 CFU/device
		5	< 1 CFU/device
Positive Control	1	6	N/A
Negative Control	1	7	< 1 CFU/device

Table 3: Results of the validation test.

4 Conclusion

The reprocessing of a FC2 could be validated successfully. This was achieved by identifying and testing the sterilizer Matachana 130LF, which complies with hygiene regulations, electronic integrity and adequate sterilization performance. Test reports written by the hygiene institute prove the sterilizers' efficacy in eleminating the defined microorganism. From this follows the implementation of a new process at Olympus for reprocessing as well as for the third-party manufacturer for safe handling. Ultimately, this makes an analysis of the FC2 possible, contributing to reducing sources of errors, which brings many benefits. At the time of writing, this process is incorporated into the existing processes at Olympus and the third-party manufacturer.

Acknowledgement

The work has been carried out at Olympus Surgical Technologies Europe, Hamburg, Germany and supervised by Prof. Dr. P. Rostalski, Institute for Electrical Engineering in Medicine, Universität zu Lübeck.
Furthermore, I would like to thank Oliver Hopf and the IPS-Department.

Authors' Statement

Conflict of interest: Authors state no conflict of interest.

5 References

[1] Olympus OSTE. *Produkte: Reinigungs -und Desinfektionsgeräte*. Available: https://www.olympus.de/medical/de/Produkte-und-Lösungen/Lösungen-für-den-medizinischen-Bereich/Hygiene-und-Aufbereitung/Automated-Reprocessing.html [last accessed on 2024-01-11]

[2] DIN EN ISO 15883-4:2018. *Reinigungs-Desinfektionsgeräte - Teil 4*. Beuth, Berlin, 2018.

[3] Matachana. *NTDF-Sterilisator: DER LF - 130LF*. Available: https://www.matachana.com/de/2-ntdf-lf-serie/ [last accessed on 2023-08-22].

[4] Thieme via medici. *Reinigung, Desinfektion, Sterilisation*. Available: https://viamedici.thieme.de/lernmodul/5098089/4915496/reinigung+desinfektion+sterilisation [last accessed on 2024-01-11].

[5] The American Heritage Science Dictionary. *Elution*. Available: https://www.dictionary.com/browse/elution [last accessed at 2024-01-16], 2011.

[6] Simicon GmbH. *Bio-Indikator SIMICON FA für die Formaldehyd-Gassterilisation*. Available: https://www.simicon.de/fileadmin/produkte_pdf/bi_fa_4401.pdf [last accessed on 2024-01-14].

[7] Eurofins BioPharma Product Testing. *Report: Draft Version 1*. Internal Literature: Eurofins Munich Study No.: STUGC23AA1705-2. Munich, 2023.

[8] Eurofins BioPharma Product Testing. *Report: Draft Version 1*. Internal Literature: Eurofins Munich Study No.: STUGC23AA1967-3. Munich, 2023.

Concept Study for the Development of a Motionless Skin Lead-through for the Energy Supply of an Implanted Ventricular Assist Device

Lea Antonia Rossa [1], Michael Scharfschwerdt [2], Stephan Ensminger [2], and Anas Aboud [2]

[1] Medical Engineering Science, Universität zu Lübeck, lea.rossa@student.uni-luebeck.de

[2] Department of Cardiac and Thoracic Vascular Surgery, University Hospital of Schleswig-Holstein, Lübeck, Germany, {michael.scharfschwerdt, stephan.ensminger, anas.aboud}@uksh.de

Abstract

A battery, which is located outside the body, supplies an implanted ventricular assist device with energy via a cable that goes into the body. Because the cable cannot be fixed properly, the wound cannot heal and infections may occur. In order to reduce the infection rate, a concept for a motionsless skin lead-through was compiled. For this purpose, possible solutions were found based on the idea of using two cables with a connection point that can be fixed on the skin instead of using one cable. These were evaluated with the help of a risk evaluation and an utility analysis. Based on these analyses, the best solution for the motionless skin lead-through was extracted from the various approaches.

1 Introduction

In Germany, around four million people are affected by heart failure, which limits their quality of life [1]. Heart failure is a disease in which the heart's pumping capacity is significantly reduced. The heart muscle is no longer able to maintain blood flow in such a way that enough blood is pumped through the body to supply the organs with sufficient oxygen and nutrients [2].

In the early stages, heart failure can be treated with medication. If the disease is more advanced, it may be necessary to implant a defibrillator [1]. In the last stage of heart failure, which develop up to 10% of all patients with heart failure, the heart is so severely damaged that it is no longer able to pump enough oxygen-rich blood through the body, even at rest [3]. The gold standard for the treatment of terminal heart failure is a heart transplant. However, there are not enough donor organs available. Worldwide, the need for donor hearts is estimated at hundreds of thousands per year. However, only 3,500 organs are available [4].

Ventricular assist devices (VAD) are used either as an alternative to heart transplants or to bridge the waiting time until a donor organ is received. However, such systems are also implanted in patients who can no longer receive a donor organ [5]. VAD support the work of the ventricles and thus improve blood flow and oxygen supply to the body [6]. Around 5,000 VAD are implanted worldwide every year. The survival rates of patients who have received a VAD are comparable to those with a heart transplant. Ten years ago, only every second patient was still alive after two years of implantation. Because of the further development of VAD, survival rates of over 80% are now achieved in the same period [4].

Left ventricular assist devices (LVAD) are implanted most frequently. Such a system is implanted directly into the left ventricle and pumps the blood from there into the aorta, which transports oxygen-rich blood from the heart to the organs. A LVAD consists of a pump, an electric control unit and batteries [5]. The batteries are attached outside the body, for example to a belt. The batteries are connected to the pump via a power cable, the so-called driveline, which reaches the outside through a hole in the skin of the abdominal wall [5]. However, this can be lead to complications such as infections. These infections occur where the driveline passes through the skin [4]. As the cable cannot be fixed at this point, it is always in motion, which means that the wound cannot heal. This allows bacteria to enter the body, which triggers the infections [5]. These infections are found in around 60% of VAD patients and are the most common adverse event both in the first three months and in the follow-up period [7].

In order to reduce this infection rate, the aim of this study was to develop a concept for a motionless skin lead-through for the energy supply of an implanted VAD.

2 Material and Methods

The first step was to develop the idea for the skin lead-through. A list of requirements was then drawn up based on this idea. Also a risk list was drawn up to identify possible sources of errors or risks for which solutions had to be found during concept development.

2.1   Morphological Box

To find a solution for the skin lead-through, a morphological box was used. First, the problem is broken down into its elementary components. Corresponding possible solutions are sought for each element. By combining the respective solutions, an overall solution to the problem is then obtained [8].

2.2   Risk Evaluation

Then the resulting possible solutions should be subjected to a risk evaluation. This involves evaluating the probability of each risk (column) and the potential impact it could have on the skin lead-through or the energy supply to the pump (row). A risk matrix is used for this purpose [9].

Scaling tables for the extent of damage (EOD) and the probability of occurrence (POO) must be defined for a largely unified evaluation. These are shown in Table 1 and Table 2.

Table 1: Scaling table for the extent of damage (EOD) in the risk matrix.

Rating	Description
very low	It is unlikely that the error will have any noticeable effect.
low	The error is insignificant. The skin lead-through is only slightly affected.
medium	The error is moderatly serious, but the functionality is given.
high	Functionality is restricted and there is a risk of injury.
very high	The fault causes an interruption in the energy supply and therefore loss of function of the pump.

Table 2: Scaling table for the probability of occurrence (POO) in the risk matrix.

Rating	Description
impossible	The error will not occur.
unlikely	It is unlikely that the error will occur.
possible	The error will occur from time to time.
probable	The error will occur frequently.
very likely	It is almost certain that the error will occur.

Based on the risk matrices, the risk value (RV) is calculated according to Equation (1).

$$RV = POO \cdot EOD \qquad (1)$$

The ratings "very low" to "very high" for the extend of damage and "impossible" to "very probable" for the probability of occurrence take the values one to five. The calculated risk values per risk can then be added together to form an overall risk value [10]. The solution variant with the lowest value is considered to be the best solution.

2.3   Utility Analysis

In addition to the risk evaluation, an utility analysis was carried out, which also serves to evaluate different alternatives. The first step was to define evaluation criteria on the basis of which the proposed solutions were evaluated. These evaluation criteria were then weighted, with a higher weighting given to criteria that are very important for the skin lead-through or the function of the pump. The proposed solutions were then evaluated by rating points between 1 (poor) and 10 (very good) for each solution approach and each criterion. In the end, the corresponding points were multiplied by the respective weighting and added up. The result is the utility value. The approach with the highest utility value is selected as the best solution [11].

3   Results and Discussion

Until now, one cable has been used for the energy supply of the implanted VAD. The idea of motionless skin lead-through is based on the principle of using two cables that are connected at one point, as shown in Fig. 1. This connection point can be fixed on or under the abdominal wall. It is intended to reduce movement at the skin passage so that the wound can heal and thus reduce the risk of infection.

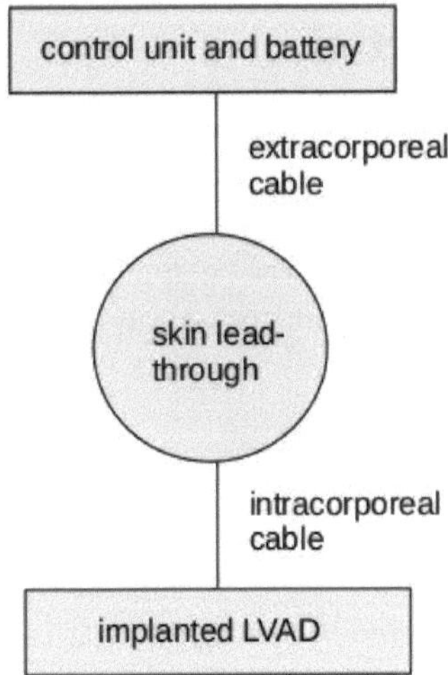

Figure 1: The idea of motionless skin lead-through is based on using two cables instead of one. So there is a connection point that can be fixed on the skin.

With reference to this scheme, the individual components have different tasks to fulfill. The most important requirement is that the energy transmission of the cables must be ensured so that the pump retains its functionality. For this purpose, the connection point of the two cables must be electrically conductive. In addition, it must be possible to fix the skin lead-through securely so that the extracorporeal cable's freedom of movement is restricted. To ensure that the patient is not further restricted in their daily life, the skin lead-through should be as light, flat and small as possible. In addition, it may not break in the event of a fall or other major forces. This would otherwise lead to an interruption in the energy supply and thus to a loss of function of the LVAD. A short circuit would also lead to an interruption in the energy supply. Therefore, it must be ensured that no water enters the sockets of either the extracorporeal or intracorporeal cable. The cables should also have strain relief on both sides to prevent the risk of the cable unintentionally becoming detached from the lead-through.

These requirements were used to identify potential sources of errors and risks. These include, for example, the unintentional detachment of the cables from the skin lead-through. The created risk list is shown in Table 3.

Table 3: Identified risks of the skin lead-through. The numbers of the risks are used as references for the risks in the following risk matrices.

Number	Risk
1	The intracorporeal cable detaches from the skin lead-through.
2	The extracorporeal cable detaches from the skin lead-through.
3	The extracorporeal cable is damaged by kinking.
4	The cables corrode.
5	The skin lead-through breaks.
6	Water enters the sockets.
7	The lead-through does not hold.
8	The solder joint breaks.

3.1 Morphological Box and Risk Evaluation

In order to find a solution for the skin passage, the problem was divided into several smaller parts. Different solutions were found for each component. These were combined, as can be seen in Fig. 2, resulting in two possible solutions.

Parameters	Characteristics		
Fixation of the skin feed-through	On the skin	Under the skin	
Type of fixation	glue	Sew on	Threads with barbs
Reinforcement of the abdominal wall	Prolene net	Goretex net	3D-printed net
Cables may not become loose – strain relief	Cable clamp	Cable gland	Two-hole plates
Connection of the two cables	Soldering	Connectors	

Figure 2: Morphological box with various solutions to different problems which are combined into two solutions indicated by the two lines with dots.

The first solution is based on a holding device that is glued onto the skin. A connector is soldered into the holder to connect the two cables and link them together. A cable gland is attached to protect against water entering and to provide strain relief. The abdominal wall is reinforced with a Prolene net during surgery. There is still room for improvement in terms of the adhesive material or the adhesive technique used for fixation of the lead-through, as it has to be able to tighten it again when the dressing is changed. The risk matrix for this approach is shown in Fig. 3. The risk value is calculated as 80.

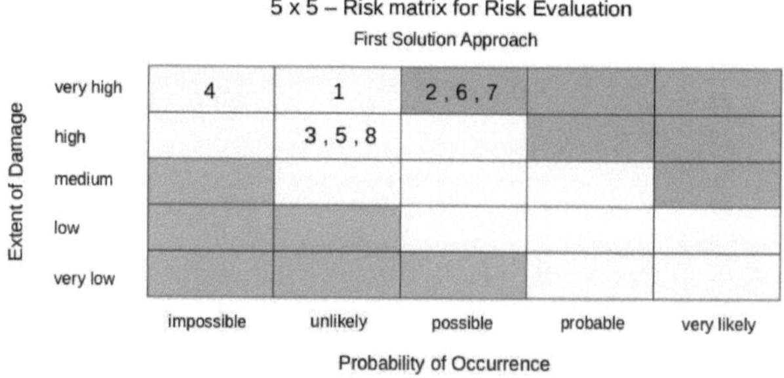

Figure 3: Risk matrix for the first solution approach. The risk value for this approach is 80.

The second approach is based on a construction that combines connectors and Prolene net and is implanted in the abdominal wall. The Prolene net is again used to reinforce the abdominal wall. In this solution, however, the mesh is integrated directly into the skin. The holding can be sewn in until the net has grown into the tissue. Cable glands again can be used here as strain relief. However, it must be investigated whether the cables can be easily inserted into the sockets, as there could be too little resistance from the abdominal wall in this approach. The risk matrix for this approach is shown in Fig. 4. The risk value is calculated as 73 which would make this approach more suitable as a skin lead-through than the first approach.

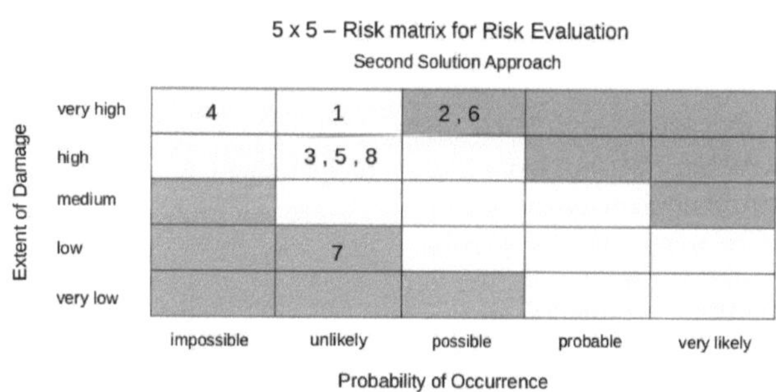

Figure 4: Risk matrix for the second solution approach. The risk value for this approach is 73 and thus lower than the risk value of the first solution approach, making this the better solution.

However, there is a third approach which is not based on the morphological box, because it is a modification of the cable. The original cable is round and contains an outer insulation and an inner insulation in which there are six further insulated cables. The idea is to fan out the outer cable sheating to create a flat rectangle in which the individual insulated cables are arranged in waves. This provides strain relief. The rectangle can be sewn onto the abdominal muscles so that the cable is fixed well in place. After fixation, the cable is merged in the same round shape and structure as before and is then passed through the skin to the battery. The risk matrix for this approach is shown in Fig. 5. The risk value is calculated as 29 and therefore this is the best solution of these three solution approaches.

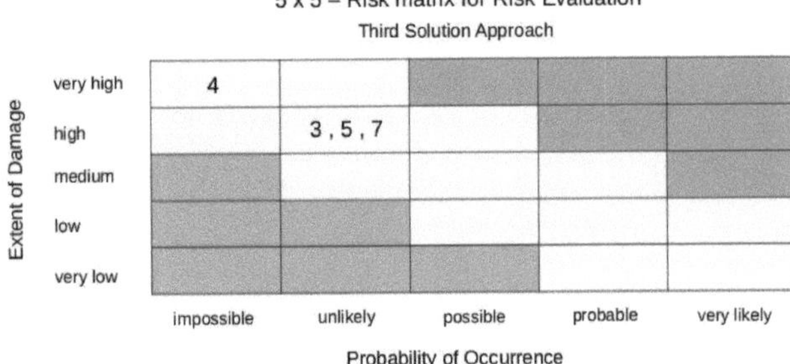

Figure 5: Risk matrix for the third solution approach. The risk value for this approach is 29. Therefore, this solution approach is the best of the three solutions.

One explanation for the low risk value may be that this solution approach contains fewer components and therefore fewer values that are included in the risk assessment. Nevertheless, fewer components also mean a lower number of components that can be damaged.

3.2 Utility Analysis

To validate the results of the risk analysis, an utility analysis was carried out. Four evaluation criteria were defined. The first criterion is that the energy supply to the pump must be guaranteed, which is included in the analysis with a weighting of 40%. The second criterion is also weighted at 40% and stands for the fact that the skin lead-through can be fixed well. The third criterion is that the skin lead-through is as small, flat and light as possible and the last criterion is that the risk of water damage can be excluded as far as possible. The last two criteria are each given a weighting of 10% in the analysis.

Table 4: Evaluation of the solution approaches (SA) with the utility analysis. The third solution approach has the highest utility value and is selected as the best solution.

Criterion	SA 1	SA 2	SA 3
1	9	9	10
2	7	8	10
3	9	10	10
4	8	9	9
Utility value	8.1	8.7	9.9

As Table 4 shows, the third solution approach has the highest utility value. Both in the risk evaluation and in the utility analysis, the third solution is assumed to be the best. Therfore, this solution is selected for the motionless skin lead-through for the energy supply of an implanted VAD.

4   Conclusion

This paper presents the development of a concept for a motionless skin lead-through for the energy supply of an implanted VAD. A solution was developed that has a high probability of achieving the desired goal and thus reducing the infection rate. This study should be seen as the basis for product development. The next step would be to produce a prototype and carry out tests to determine whether this solution is feasible and can fulfill the desired requirements.

Acknowledgement

The work was carried out in the research laboratory of the Department of Cardiac and Thoracic Vascular Surgery of the University Hospital of Schleswig-Holstein, Lübeck Campus.

Authors' Statement

Conflict of interest: Authors state no conflict of interest.

5   References

[1] K. Eulenburg, *Herzschwäche*, Deutsches Zentrum für Herz-Kreislauf-Forschung E.V., Berlin, 2023.

[2] D. Hansmann, *Herzinsuffizienz (Herzleistungsschwäche)*, Mediclin, Offenburg, 2021.

[3] C. Schulze and M. Barten and U. Boeken and G. Färber and C. Hagl and C. Jung et al., *Implantation mechanischer Unterstützungssysteme und Herztransplantation bei Patienten mit terminaler Herzinsuffizienz*, Springer Medizin Verlag GmbH, pp. 1-12, 2022.

[4] U. Steinseifer, *VDE-Expertenbericht - Biomedizinische Technik*, Verband der Elektrotechnik (VDE), pp. 64-65, 2015.

[5] A. Diegeler, *Herzunterstützungssysteme*, Rhön-Klinikum Campus Bad Neustadt, pp. 3-5 , 2020.

[6] J. Schulze, *Herzunterstützungs-Systeme*, Kreuzlingen, 2023.

[7] T. Schlöglhofer and P. Michalovics and J. Riebrandt and P. Angleitner and M. Stoiber and G. Laufer et al., *Left ventricular assist device driveline infections in three contemporary devices*, Artificial Organs, vol. 45, pp. 464-472, 2021.

[8] C. Schawel and F. Billing, *Morphologischer Kasten*, In: Top 100 Management Tools, Gabler Verlag, Wiesbaden, 2012.

[9] S. Sandhaus, *Risikomatrix: So messen und bewerten Sie Risiken*, HubSpot, Cambridge, USA, 2022.

[10] C. Niklas, *Risikowert*, projektmagazin, Berleb Media GmbH, 2018.

[11] J. Kühnapfel, *Das Vorgehen bei der Nutzwertanalyse*, In: Nutzwertanalysen in Marketing und Vertrieb, Springer Gabler Wiesbaden, pp. 5-14, 2014.

Dilute Injection of Anesthetic Gases to Reduce an Unintended Bolus

Corinna Warnecke [1], Björn Bargstädt [2] and Andreas Kater [2]

[1] Medical Engineering Science, Universität zu Lübeck, corinna.warnecke@student.uni-luebeck.de
[2] R&D Perioperativ Care, Drägerwerk AG & Co. KGaA, {bjoern.bargstaedt, andreas.kater}@draeger.com

Abstract

Target Controlled Anesthesia (TCA) is used for automated dosage of volatile anesthetics and fresh gas, so the user can focus on the patient better and the risk for patient harm is lowered. But in case of decreasing the desired concentration of expiratory anesthetic gas, the fresh gas flow is increased by the device and a so-called bolus occurs, that can irritate the user and tempt to incorrect readjustments by the user that may harm the patient. Aim of this study is to determine a time during the breathing cycle, where triggering a bolus leads to its lowest magnitude. For this purpose measurements with different fresh gas flow levels are carried out over a changing delay time to fully cover several breathing cycles. We see that the time for the bolus to reach its maximum and minimum depends on the level of fresh gas flow. Therefore, we eliminate this influence by calculating its dead time. Based on the results, we propose that both the highest and lowest bolus are reached during the first half of the expiration. At the point of the lowest bolus maximum, it would be the best time for increasing the fresh gas flow.

1 Introduction

In Germany the number of surgical interventions has risen by over 30 % in the last 17 years [1]. But not only the number but also the complexity of these interventions has increased. For this reason the requirements for anesthesia devices are becoming higher. Modern devices shall not only ventilate the patient and apply a given amount of volatile anesthetics, but also relieve the anesthetist. One approach for this is the Target Controlled Anesthesia (TCA) introduced in the anesthesia device *Zeus*. The inspiratory oxygen concentration and the expiratory anesthetic gas concentration are specified as target values for the TCA control circuit. Based on this, the controlled variables are the fresh gas flows of air, oxygen, nitrous oxide and anesthetic gases, which the TCA adjusts using a Direct Injection of Volatile Anesthetics (DIVA) and an electronic fresh gas mixer [2, 3]. This efficient use of these gases leads to a reduction in consumption and therefore also to a reduction in costs [2]. If, on the other hand, the desired anesthetic gas concentration in the patient gas decreases due to a change in the target value, the fresh gas flow through the device is increased. This may create a bolus, which can irritate the user and lead to readjustments made incorrectly, possibly harming the patient. In order to reduce or even avoid this, this work will investigate at which point in the breathing cycle the bolus reaches its maximum and minimum.

2 Material and Methods

To simulate the situation of a bolus formation, an anesthesia device and its patient gas module are deployed using the rapid prototyping device dSpace MicrolabBox. This allows to integrate state flows that manipulate the fresh gas flow $\dot{V}$ without external dependencies. The patient gas module measures the concentration of anesthetics agents c with an accuracy of $\pm(0.2\,\text{Vol.}\% + 15\,\%\,\text{rel.})$ [4]. The state flow as seen in Fig. 1 consists of three main phases: dosage of anesthetic gas, bolus triggering and flush. In the first phase, the anesthetic gas (agas) is dosed over a determined time. Next, the model waits for the inspiration to start and the set delay time Δt_{delay} to pass. Then the fresh gas flow $\dot{V}$ is increased and the anesthetic dosage concentration is lowered, so the bolus appears. After a limited time, the fresh gas flow is risen again, while no anesthetics are dosed until the lower limit of anesthetic gas concentration is measured by the patient gas module. After each cycle the delay time is incremented, so the whole breathing cycle is covered. For simulating an adult ventilation, the parameters recommended by [5] are used. These are a respiratory rate $RR = 12\,\frac{1}{\text{min}}$, a tidal volume $V_t = 500\,\text{mL}$ and an inspiration to expiration ratio of 1:2 [5]. To evaluate the extent of a bolus, its height, duration and dead time between fresh gas increase and bolus maximum are analyzed.

2.1 Determination of Dead Time

The fresh gas flow requires a certain amount of time to reach the breathing system. This is the so-called delay

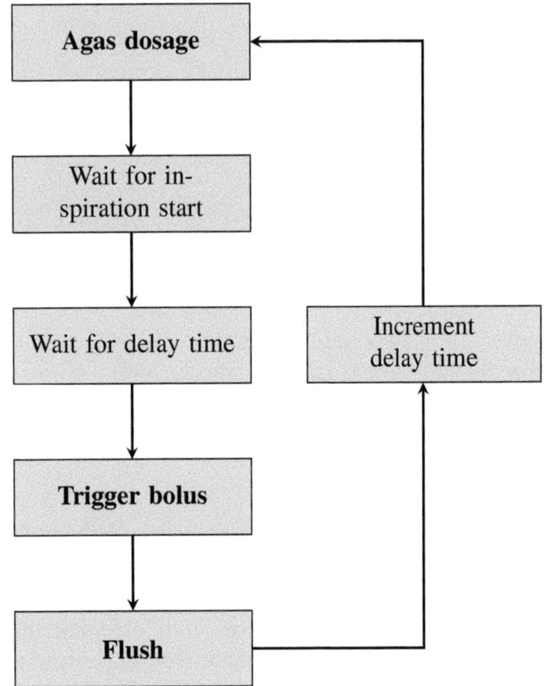

Figure 1: Simplified state flow to manipulate the fresh gas level.

time Δt_{total}. To determine this time, the state flow in Fig. 1 is adjusted to an increasing of the fresh gas level by $\dot{V} = 0.1 \frac{L}{min}$ in each cycle and no longer the delay time. In addition, the patient gas module is directly attached to the breathing system of the anesthesia device.

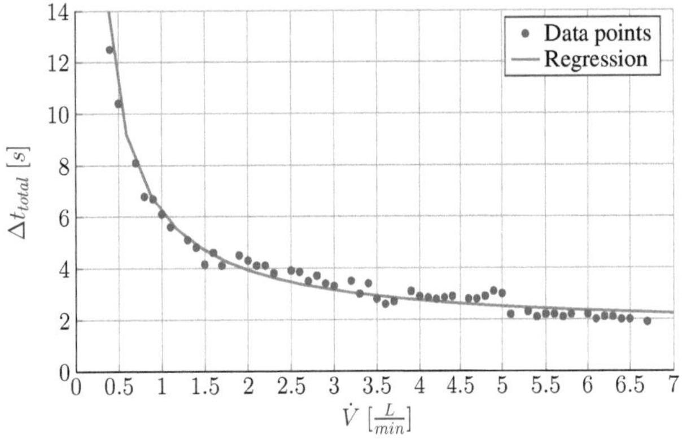

Figure 2: Dead time depending on the fresh gas flow level. The line displays the regression.

As shown in Fig. 2 the measurement starts with a fresh gas flow $\dot{V} = 0.2 \frac{L}{min}$ and is executed until a fresh gas flow $\dot{V} = 6.8 \frac{L}{min}$ is reached. In Fig. 2 a cycle is called interval and the level of fresh gas is displayed on the x-axis. The measurement is repeated three times and the dead time mean values are calculated and displayed on the y-axis. It can be seen that the average dead time decreases as the fresh gas level increases and is fitted best with a power function with two terms. The calculated equation of the power function

$$\Delta t_{total} = \frac{\Delta t_{delay}}{1000} + 4.705\dot{V}^{-0.9448} + 1.4844 \quad (1)$$

is used for correcting the bolus dead time as seen in the next section.

3 Results and Discussion

The following measurements were performed using the in Fig. 1 proposed software model.

3.1 Bolus Formation

As described in the previous section, the software model consists of three main phases, that can be also seen in Fig. 3. It displays the anesthetic gas concentration c in the patient gas plotted over time t in seconds. The first bolus in this data snipped starts at $t = 1,040$ s. At this point, the fresh gas flow is increased from $\dot{V} = 0.2 \frac{L}{min}$ to $\dot{V} = 2 \frac{L}{min}$. After a time of ten seconds, the flush state is reached and the system is flooded with $c = 100$ Vol.% air, until the measured anesthetic gas concentration is smaller than the limit of $c = 0.5$ Vol.%. Now the cycle starts again with a longer delay time Δt and the system starts dosing anesthetic gas over a defined time period. In the following the bolus maximum is defined as the difference between the anesthesia gas concentration before the fresh gas flow increment and the maximum anesthesia gas concentration in this cycle.

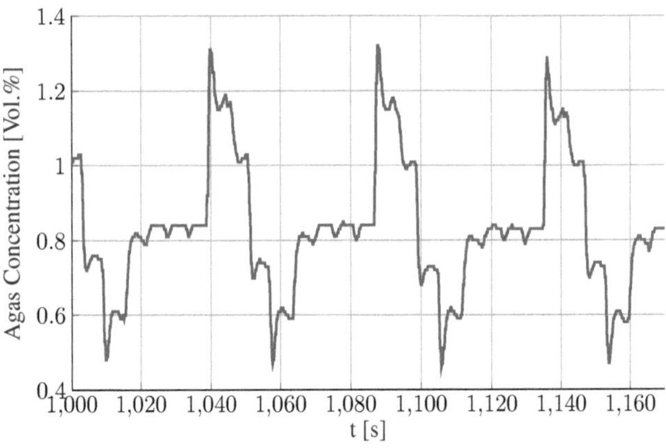

Figure 3: Raw data of anesthetic gas (Agas) concentration with fresh gas rise to $\dot{V} = 2 \frac{L}{min}$.

3.2 Bolus Maximum Depending on the Fresh Gas Rise

The raw data shown in Fig. 3 is analyzed regarding the bolus maximum over the time delay. In Fig. 4, the in (1) calculated dead time is used to compensate the time the fresh gas needs to reach the breathing system. This is the reason, the data does not start at a delay $\Delta t = 0$ s, but starts at around $\Delta t = 2$ s. As seen in (1), the dead time depends on the fresh gas flow. We can observe in Fig. 4 that the bolus has its largest maximum when it is triggered in the first quarter of the expiration. The higher the fresh gas flow, the higher the bolus. After the maximum in these curves, the bolus magnitude reaches its minimum in the second quarter of expiration and increases with increasing delay. In this phase, the bolus maximum is higher with smaller amounts of fresh gas than with larger amounts.

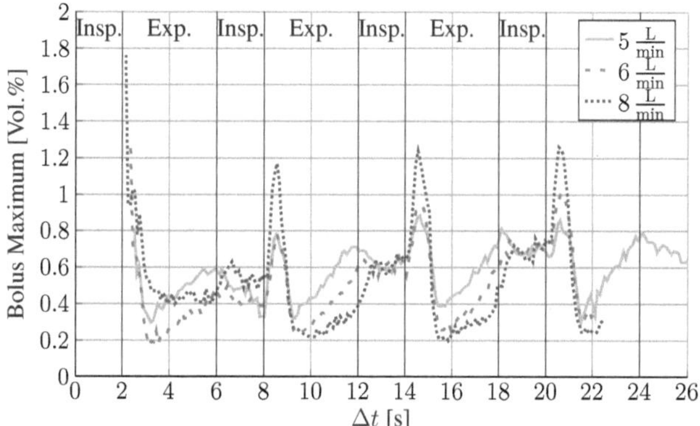

Figure 4: Bolus maximum at each time delay for different fresh gas levels.

3.3 Bolus Maximum Compared to Lower Fresh Gas Levels

While higher fresh gas values show a concrete point of maximal bolus magnitude in the breathing cycle, lower fresh gas increases as shown in Fig. 4 do not induce such a pronounced bolus strength. In all measurements the smallest bolus maximum is reached approximately in the second quarter of expiration during the time interval $t \in [9; 9.5]\ s$. Therefore, the dilute injection of anesthetic gases must depend on the level of fresh gas and an algorithm to reduce an unintended bolus should take this into account.

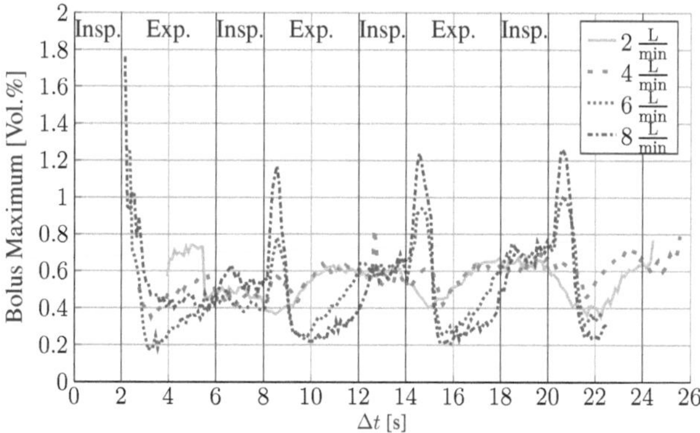

Figure 5: Bolus Maximum at each time delay for fresh gas levels from $\dot{V} = 2\ \frac{L}{min}$ to $\dot{V} = 8\ \frac{L}{min}$.

3.4 Bolus Maximum in Neonatal Ventilation

Since anesthesia devices are not only suitable for adults but also for ventilation of neonates, it is important to examine the bolus behavior with smaller tidal volumes and a higher respiratory rate. In this case, parameters for neonatal ventilation (RR = 30, $V_t = 30\,\text{mL}$ and inspiration to expiration ratio of 1:1) are used [5]. The results of this measurement are displayed in Fig. 6. The course of bolus maxima may lead to the assumption, that the bolus has its highest magnitude during the expiration, as in an adult ventilation. But in this case, the difference between the largest and lowest bolus maximum is only around $\Delta c = 0.2\,\%$. In addition to this, the bolus maxima are noisy so we cannot determine a specific point at which the highest or lowest magnitude of the bolus is reached. However, due to the small bolus size, it can be said that this bolus is negligible.

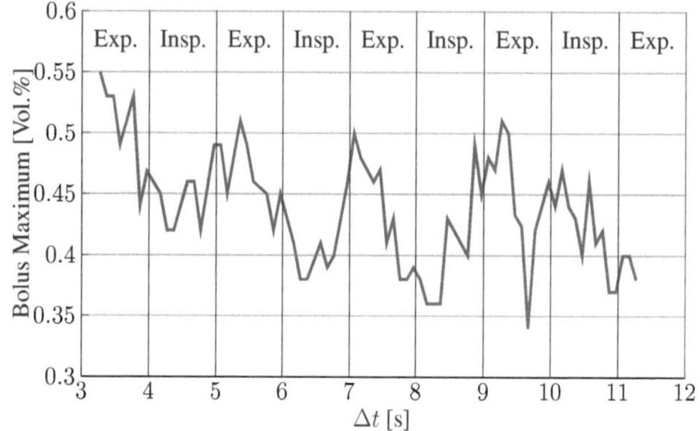

Figure 6: Bolus Maximum at delay time $\Delta t \in [3; 11.2]\ s$ for ventilation of a neonate at fresh gas level $\dot{V} = 3\ \frac{L}{min}$ with dead time compensation.

3.5 Time between fresh gas increase and bolus maximum

Because a dilute injection of anesthesic gas to avoid an unintended bolus should also consider the duration of the bolus and time until its maximum is reached, the analyzes in Fig. 7 and Fig. 8 are performed. Fig. 7 shows the time between fresh gas rise and the bolus maximum in dependence of the delay time for different fresh gas levels. The largest time between fresh gas rise and bolus maximum lies at the delay time that matches with the first quarter of the expiration. The shortest dead time is seen at the a delay time $\Delta t = 2\,\text{s}$. This means that the bolus maximum happens when the fresh gas is increased at the beginning of the expiration. Fig. 7 highlights the nearly linear decrease of dead time for a fresh gas level $\dot{V} = 2\ \frac{L}{min}$. The dead time then increases quickly and decreases again as the delay time increases. Higher fresh gas levels do not show such clear progress, but do show many outliers. These are attributed to the shape of the bolus. As we can see in Fig. 3, each bolus has one peak at its beginning and the data shows periodic fluctuations due to the breathing phases. As the time delay increases, the shape of the bolus changes to a bolus with two peaks that change in size. At the point, where both peaks are almost equal in size, the outliers appear when the first peak is higher than the second while the second should be higher.

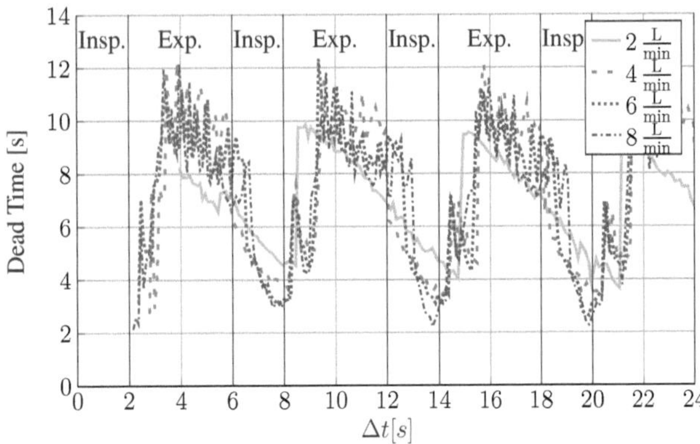

Figure 7: Time between fresh gas rise and bolus maximum at fresh gas levels from $\dot{V} = 2 \frac{\text{L}}{\text{min}}$ to $\dot{V} = 8 \frac{\text{L}}{\text{min}}$.

Fig. 8 shows a similar behavior of the bolus with reference to its duration. The graphs contain many outliers and are very noisy, so we cannot determine a specific bolus duration for each delay time. Furthermore the maximal bolus duration lies between 10 to 12 s, meaning it lasts for several breathing cycles. This has to be considered in an algorithm.

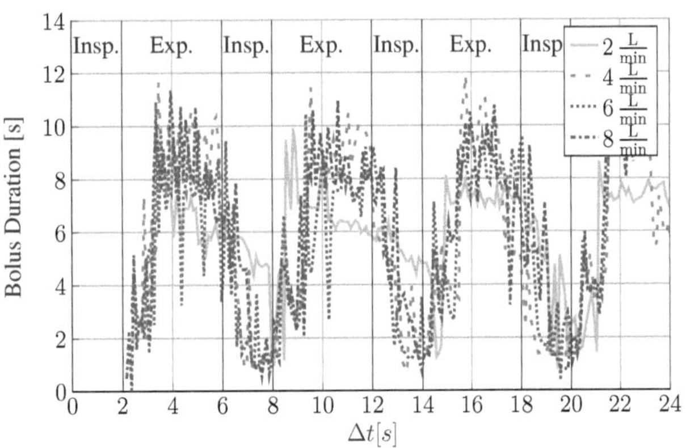

Figure 8: Bolus duration in dependence of delay time at fresh gas levels from $\dot{V} = 2 \frac{\text{L}}{\text{min}}$ to $\dot{V} = 8 \frac{\text{L}}{\text{min}}$.

4 Conclusion

The aim of this study is to determine a point in the breathing phase where an unintended bolus can be avoided or retarded. The data shows a periodic increase and decrease of the bolus height depending on the delay time over several breathing phases. For larger fresh gas flows, there is a maximum bolus when the calculated total delay time corresponds to the first half of expiration. This is the worst moment to increase the fresh gas flow. On the other hand the minimal bolus maximum is reached with a delay time that is only a little longer and at about $\Delta t_{total} \in [9; 9.5]\ s$. The results show a dependency between the delay time the bolus reaches its maximum concentration and the fresh gas flow. With a higher flow the highest bolus is produced in an earlier point of the breathing phase. While the bolus maximum can be easily determined, the time the bolus needs to reach its maximum cannot be determined precisely and neither can the duration of the bolus. In summary, the best time to increase fresh gas flow when ventilating an adult is in the second quarter of expiration. For neonatal ventilation, there would not be a concrete delay time, at which the bolus has its lowest strength. As a next research step we propose to implement an algorithm to reduce the unintended bolus and verify the results of this study.

Acknowledgement

The work has been carried out at Drägerwerk AG & Co. KGaA, Lübeck and supervised by the Institute of Software Engineering and Programming Languages, University of Lübeck.

Authors' Statement

Corinna Warnecke, Björn Bargstädt and Andreas Kater are working for Drägerwerk AG & Co. KGaA.

5 References

[1] German Federal Office of Statistics, "Fallpauschalen-bezogene Krankenhausstatistik (DRG-Statistik) Operationen und Prozeduren der vollstationären Patientinnen und Patienten in Krankenhäusern," 2022. [Online]. Available: https://www.destatis.de/DE/Themen/ Gesellschaft-Umwelt/Gesundheit/Krankenhaeuser/ Publikationen/Downloads-Krankenhaeuser/ operationen-prozeduren-5231401217014.pdf?__ blob=publicationFile

[2] M. Potdar, L. Kamat, and M. Save, "Cost efficiency of target-controlled inhalational anesthesia," *Journal of Anaesthesiology Clinical Pharmacology*, vol. 30, no. 2, p. 222, 2014.

[3] W. Oczenski, H. Andel, and A. Werba, *Atmen, Atemhilfen*, 9th ed. Stuttgart: Thieme, 2012.

[4] Drägerwerk & AG Co. KGaA, *Instruction for Use Atlan A300, A300 XL, A350, A350 XL*, 2nd ed., Lübeck, 2019. [Online]. Available: https://www.draeger.com/Content/Documents/Content/ atlan-a300-a300-xl-a350-a350-xl-ifu-9056001-en.pdf

[5] International Organization for Standardization, *ISO 80601-2-13 - Medical electrical equipment Part 2-13: Particular requirements for the basic safety and essential performance of an anaesthetic workstation*, 2nd ed., 2011, no. 1.

Implementation of a Declaration of Conformity (DoC) Process in the Process of Registration for Medical Devices in Compliance with European Regulations for a Medical Device Manufacturer

Ratul Hassan [1], Monika Schneider [3], Anna Natrup [3], and Stefan Müller [2]

[1] Biomedical Engineering, Luebeck University of Applied Sciences, ratul.hassan@stud.th-luebeck.de
[2] Medical Sensor and Device Technology Lab, Luebeck University of Applied Sciences, stefan.mueller@th-luebeck.de
[3] Eppendorf SE, Department of Global Regulatory Affairs Quality Management, schneider.m, natrup.a@eppendorf.de

Abstract

This research underscores the importance of implementing a robust Declaration of Conformity (DoC) process for a medical devices and In-vitro Diagnostic device manufacturer, aligned with European Medical Device Regulation (MDR): (Regulation (EU) 2017/745) and In-vitro Diagnostic Devices Regulation (IVDR) which is (Regulation (EU) 2017/746). Investigating the complex landscape of compliance in both product types, the study unveils the historical context and evolving role of the DoC as a pivotal instrument ensuring medical device safety information, efficacy, and global market access. Through a thorough examination of EU MDR and IVDR, it highlights the convergence and divergence of expectations, providing a comprehensive compliance framework. Drawing on literature, empirical data, and the research outlines essential components of an effective Declaration of Conformity (DoC) system. The study further offers practical recommendations for seamless implementation, emphasizing change management, resource allocation, and stakeholder empowerment with the help of this Declaration of Conformity (DoC).[4][5]

1 Introduction

Amidst this progress, safety, efficacy, and regulatory compliance are paramount, this internship navigates the intricacies of the industry, emphasizing the Declaration of Conformity (DoC) system implementation aligned with European Medical Device Regulation (MDR): (Regulation (EU) 2017/745) and In-vitro diagnostic Devices Regulation (IVDR). A declaration of conformity is a formal statement made by the manufacturer that their medical device conforms to the requirements set out in the regulation. This declaration is a crucial part of the conformity assessment process, development of it involves collaboration among individuals with diverse expertise, can be seen in Table 1. This includes oversight of regulatory aspects and key performer by the Regulatory Affairs (RA) Engineer from RD(Research and Development), management of technical documentation from a marketing perspective by the Product Life Cycle Manager (PLCM), planning and conformity management by the Product Manager (PM), change management and quality testing by the Quality Management (QM) Engineer, identification of commercial market requirements by the RA Manager, and consolidation of results and generation of ideas for finalizing the DoC template for the European Market by the RA Intern (Global RAQM). The internship aims to establish a robust DoC system by finding the sub process for it which is compliant to the EU MDR and IVDR. It seeks to provide valuable insights for regulatory compliance and global market access to industry professionals, regulators, and researchers. It addresses industry professionals, regulators, and researchers, providing insights for regulatory compliance and global market access in this dynamic field.[1][3] [CM: Change Management, CC:Change Control, QA: Quality Assurance, TD: Technical Documentation, Product Dev.: Product Development.]

Table 1: Responsibility diagram of different designations

Responsible Concern	Job Family	Responsibility
RA Engineer	Regulatory RD	RA aspects
PLC Manager	Marketing RD	TD
Product Manager	Product Dev.	Planning
QM Engineer	Quality RD	CM and CC
RA Manager	Commercial	Requirements
RA Intern	RAQM	Process creation

2 Material and Methods

The adopted method for this research is the comprehensive approach taken to investigate the implementation of a Declaration of Conformity (DoC) system for medical devices in line with European regulation. As a starting material to implement the (DoC), the EU MDR Article 19 and IVDR Annex IV and Article 17 has been thoroughly followed to be completely compliant to the European market. The software tool used, named BIC process design, commercially used in the industries to document different modeling processes. In Fig.1 the process (done via BIC) starts with the creation of "DoC Process" where all the sub processes will be stored. Input will be given initially by RA Engineer and PLCM. Final output will be the processed Declaration of Conformity.[4][5]

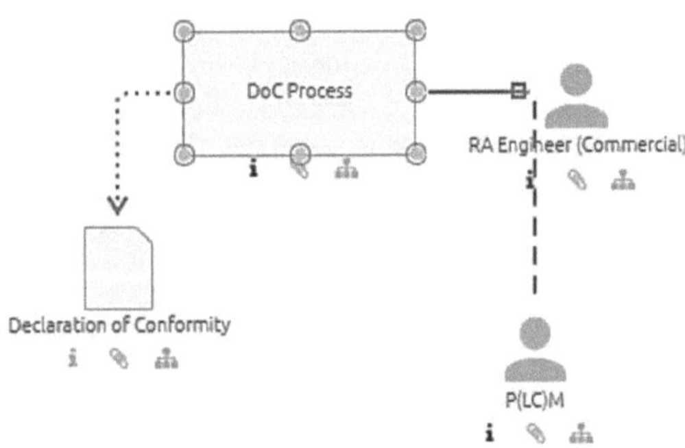

Figure 1: Storage of all sub process and the outcome.

A mixed-methods approach combining the qualitative and quantitative research methods with the overall research approach to create the DoC process where the main goal is to include the sub processes as follow: Country-wise Input collection, Fill out Template, Final Check, Get (new) DoC Number, Get signatures, Get Template, Save DoC, Upload or replace in Manufacturer Web Shop and Wait for finished Regulatory Conformity Review and establish the process.[2]

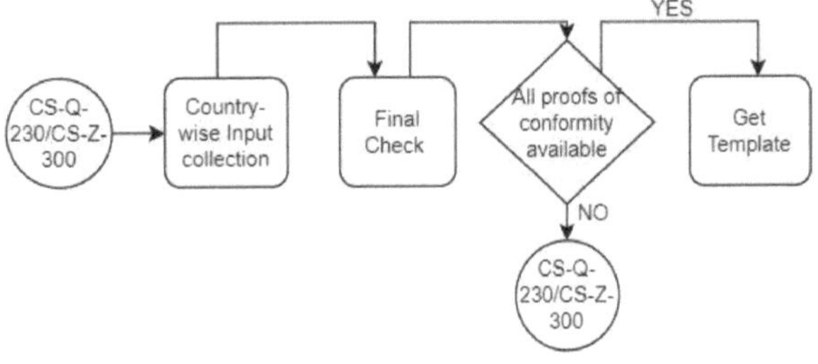

Figure 2: DoC sub processes flow chart (Phase -1).

Research Design: The investigation has been methodically formulated and structured in a predetermined sequence, with due consideration given to the insights provided by RA team while adhering to the Standard Operating Procedures (SOP) prescribed by the manufacturer in Fig.2 and Fig.3.

Qualitative Research: In-depth interviews with the key stakeholders which are the regulatory affairs personnel of the manufacturer's different branch offices in the target European countries, who provided insights like challenges, strategies, and experiences related to DoC process implementation.

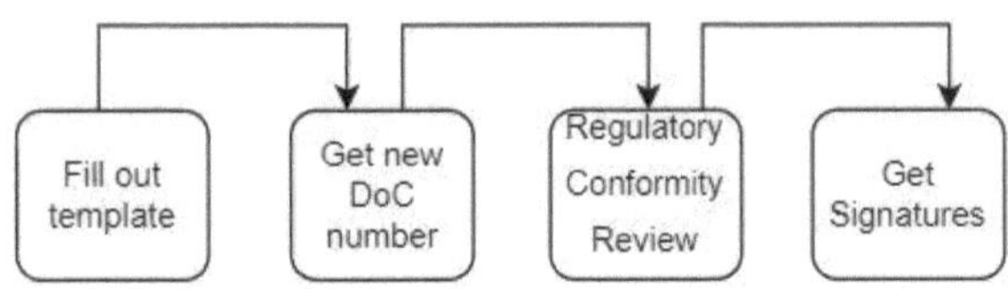

Figure 3: Flow chart of the sub processes flow chart (Phase -2).

Participant Selection:

Interviews: Key informants selected based on market, expertise, and involvement in the medical device industry, ensuring diversity in participants.

Surveys: Online surveys to Participants of the manufacturer's branch offices from ten countries of EU to collect quantitative data on various aspects of DoC system implementation.

Quantitative Research: Online surveys distributed to a larger sample of medical device manufacturers aim to gather quantitative data on experiences, practices, and perceptions regarding DoC system implementation.

Country-wise Input Collection: A Structured interview with 10 key informants from 10 different European countries (Germany, Austria, Switzerland, Netherlands, Italy, France, Denmark, Sweden, Belgium and Hungary) allow participants to provide detailed insights into their current DoC.

Data Analysis Techniques (Final Check): Thematic analysis of interview data using given template to be filled by RA personnel which identifies key themes, patterns, and insights related to DoC process implementation. Meanwhile the Corporate Standards (CS-Q-230 and CS-Z-300) from the manufacturer's Headquarters were analyzed and compared.

Quantitative Data Analysis (Get Template): Statistical analysis by employing descriptive and inferential statistics, and regression analysis to explore relationships and patterns.

Ethical Considerations with Informed Consent: Participants provided with informed consent forms detailing the research's purpose, data handling procedures, and confidentiality measures.

Research Approach:

Mixed-Methods Approach: Allows for triangulation, enhancing the validity and comprehensiveness of the research. This underscores the methodical approach to ensure a medical device manufacturer's adherence to regulations, standards, and quality management systems (QMS) in Element establishment. Key components involve regulatory understanding with the steps Country-wise Input collection, Fill out Template, Final Check, Get (new) DoC Number, Get signatures, Get Template, Save DoC, Upload or replace in manufacturer Web Shop and Wait for finished Regulatory Conformity Review.

3 Results and Discussion

3.1 Figures and Tables

Each step delineated the acquisition and provision of specific inputs, facilitated by the respective Regulatory Affairs personnel in their designated capacities as Table1.

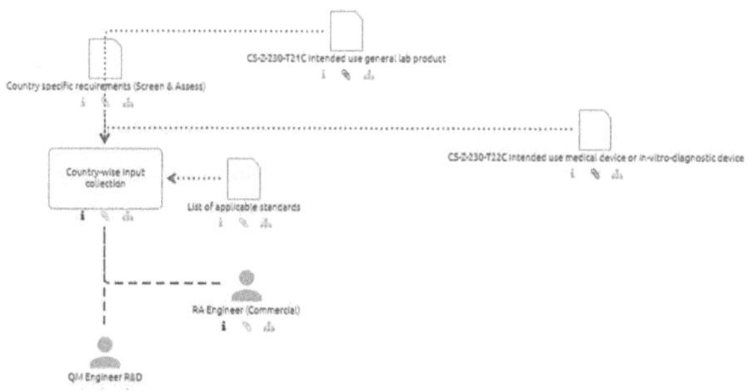

Figure 4: Country-wise Input collection.

In Fig.4, following the interview conducted with the ten affiliate countries, the initiation of the Declaration of Conformity (DoC) process commences if deemed necessary for a particular country and if all evidentiary documents demonstrating compliance with the Regulations are available as stipulated by corporate standard CS-Z-230. This prerequisite must be fulfilled prior to the commencement of the Regulatory Conformity Review. The specific procedure is delineated in CS-Q-026, specifically within the step denoted as "Create DoC". The event triggering the update of the DoC is outlined in CS-Z-300. The RA Engineer from Research and Development (R and D) undertakes the collection of requisite input, including a comprehensive list of all applicable standards, directives, regulations and checklists outlining essential requirements alongside accompanying evidence, pertinent to the countries under consideration. The resultant DoC is tailored to encompass all regulations necessitating such documentation within the designated country, with preference given to harmonized standards that align with the corresponding regulations.

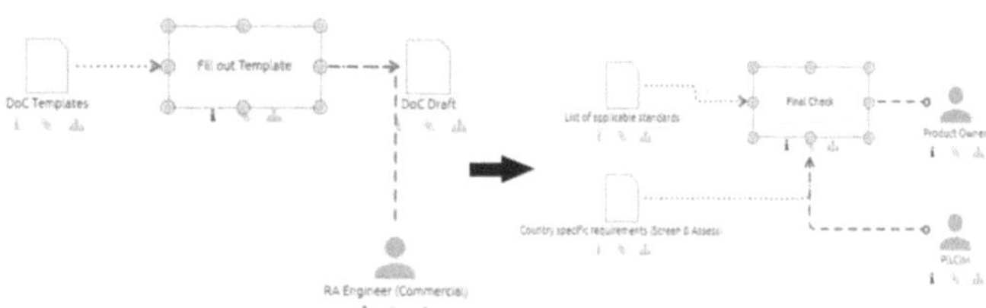

Figure 5: Fill Template and afterwards Final Checking.

In Fig.5. the templates are completed by the Responsible Authority (RA) personnel who participated in the interviews, resulting in the creation of a draft. The regulatory section of the Declaration of Conformity (DoC) must encompass the regulations necessitating such documentation. Preferably, harmonized standards corresponding to the regulations should be employed. During this final verification stage, the RA Engineer (R and D) is tasked with ensuring

the availability of all essential inputs pertinent to the specific country and the completion and finalization of the requisite checklists. This step entails a formal review and confirmation to ascertain the accuracy and compliance of the content, ensuring that all regulations and standards are referenced in their current iteration.

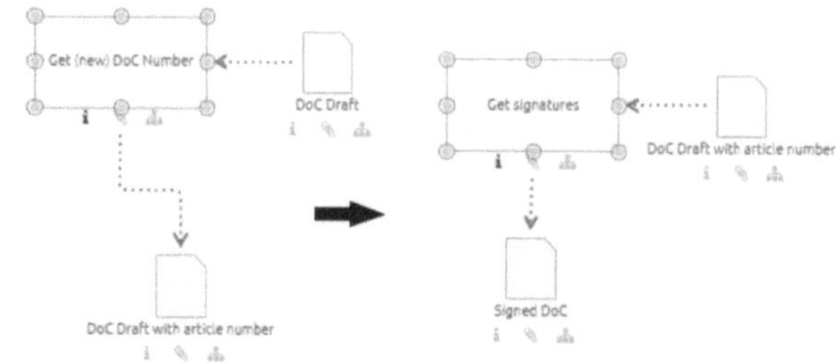

Figure 6: DoC Number (new) and corresponding signature collection.

In Fig.6. the DoC Number serves as a unique identifier assigned to the Declaration of Conformity (DoC), generated within the purview of the respective Technical Committees. Solicitation of the article number is initiated by the Responsible Authority (RA) Engineer within the Research and Development (R and D) department, directed to the designated division of the Technical Committee. It is imperative that the article number is integrated into the Bill of Material. The DoC necessitates physical signing and must also be made digitally accessible. Authentication of the DoC is executed by both the Head of Global Quality Management and Regulatory Affairs and the Head of Quality Management in accordance with the provisions set forth by the Technical Committee.

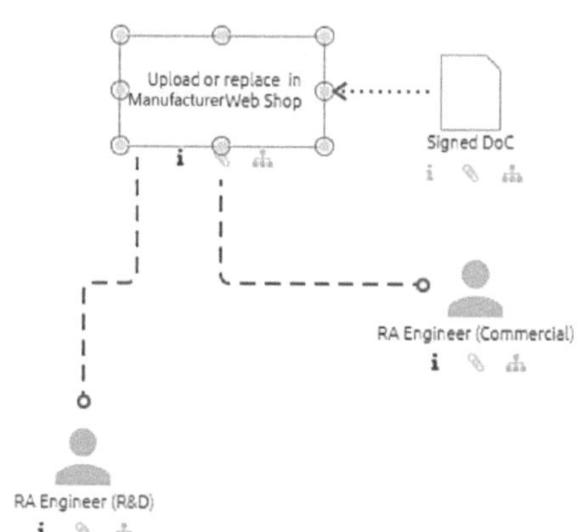

Figure 7: Save and uploaded in the web storage.

In Fig.7 the Declaration of Conformity (DoC) is archived within the Regulatory documentation package in compliance with CS-Q-026 guidelines. As per protocol, the file name must incorporate both the article number and version identification. Subsequently, the Product Life cycle Management (PLCM) team is responsible for the uploading of the updated or modified DoC onto Webshop, which is an internal cloud storage.

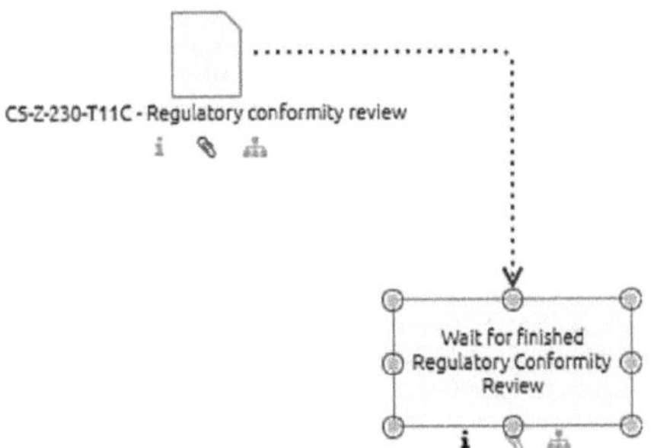

Figure 8: Awaiting of the process review and approval

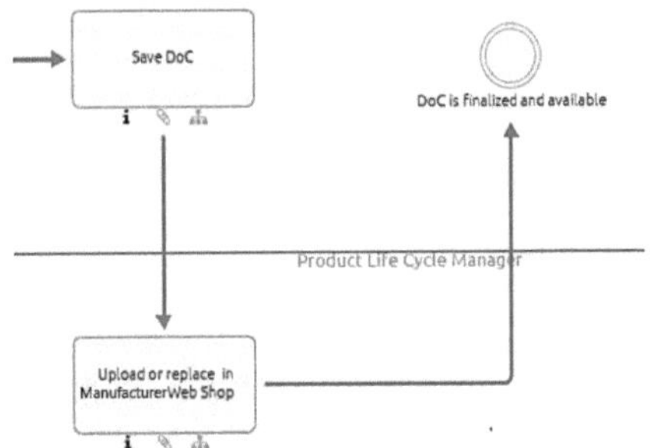

Figure 9: Finalized template of DoC.

In Fig.8 The draft versions of the documents are required to remain accessible until the conclusive Regulatory Conformity Review is undertaken. During this review process, the completeness and accuracy of the Declaration of Conformity (DoC) content, along with the availability of all requisite proofs of conformity, are meticulously evaluated. Upon successful confirmation of the Regulatory Conformity Review, the DoC is deemed eligible for release, facilitating the commencement of the signature acquisition process. In Fig.9 The template is finalized. It is kept and stored in the internal cloud storage of the manufacturer.

Finding of this analysis along with the process are, the regulatory requirements for (DoC) process for medical devices in compliance with European regulations, stressing the need for manufacturers to create and implement a robust DoC process. The DoC must be drafted by the manufacturer or their authorized representative, available in the official language(s) of the Member State and retained for at least ten years. It must affirm conformity to essential safety and performance requirements identify the device and its manufacturer, and explicitly declare compliance with EU MDR (2017/745) and IVDR (2017/746). Additionally, it must address post-market surveillance and vigilance reporting. The fundamental components of an effective DoC process, including risk assessment and management, quality management system alignment, clinical data and evaluation procedures, labeling, packaging, and traceability requirements, as well as harmonization of compliance efforts. It emphasizes the importance of periodic review and updates to maintain compliance and ensure the integrity of the DoC process, which is essential for regulatory compliance, patient safety, and market access in the European Union.

4   Conclusion

The future goal for the described Document of Conformity (DoC) process at Manufacturer is to continually enhance efficiency, agility, compliance and eradicate the limitations. To train the responsible concerns of the manufacturer and its affiliates. The Manufacturer aims to enhance the efficiency, agility, and compliance of its DoC process while eliminating limitations. This includes leveraging technology and automation for faster and more accurate DoC creation and fostering a culture of continuous learning. The detailed process description underscores the company's commitment to regulatory compliance, positioning it as a reliable and compliant industry leader. In conclusion, the detailed description of the established DoC process, responsibilities, tasks, and the involvement of the RA personnel provide a comprehensive view of Manufacturer's commitment to regulatory compliance. By aligning with technological advancements and maintaining a proactive approach Manufacturer aims to ensure the highest conformity.

Acknowledgement

The work has been carried out at Eppendorf SE and Laboratory for Medical Sensor and Device Technology, Luebeck University of Applied Sciences.

Authors' Statement

Authors state no conflict of interest.

5   References

[1] D. Shoukier, *Compendium The New European Medical Device Regulation MDR*. BellingsBooks Verlag, Dozwil, pp. 27-68, 2023.

[2] S. F. Amato, R. M. Ezzell Jr., *Regulatory Affairs for Biomaterials and Medical Devices*. Woodhead Publishing, Swaston, pp. 50-87, 2014.

[3] B. A. Fiedler, *Managing Medical Devices Within A Regulatory Framework*. Elsevier, Amsterdam, Chapter 8, section 3, pp. 215-242, 2016.

[4] Council of the European Union 2017 Brussels, *EU MDR*. Available: https://www.medical-device-regulation.eu/wpcontent/uploads/2019/05/CELEX-32017R0745-EN-TXT.pdf [last accessed on 2024-01-18].

[5] Council of the European Union 2017 Brussels, *EU IVDR*. Available: https://eur-lex.europa.eu/legal-content/EN/TXT/PDF/?uri=CELEX:32017R0746 [last accessed on 2024-01-18].

Construction of an endurance test rig for pressure regulators with associated pre-tests

Greta Göttling [1], Eduard Weber [2]

[1] Biomedical Engineering, Luebeck University of Applied Sciences, greta.goettling@stud.th-luebeck.de

[2] BU Pneumatics, Drägerwerk AG & Co. KGaA, Lübeck, eduard.weber@draeger.com

Abstract

In response to amendments in MDR 2017/745 pertaining to materials in contact with breathing gas, imperative adaptations are required for the pressure regulators employed by Dräger. These regulators, extensively employed in ventilation and anesthesia devices, mandate the substitution of two rubber materials, specifically acrylonitrile butadiene rubber (NBR) and chloroprene rubber (CR), utilized as seals. A dedicated endurance test rig is currently under development in anticipation of the upcoming endurance tests that will be administered to evaluate the performance of these pressure regulators under prolonged conditions. The test preparation encompassed preliminary assessments, confirming the suitability of proportional valves, and conducting a thorough comparison between the old and new pressure regulators. As the testing process is ongoing and the full complement of pressure regulators is yet to be evaluated, conclusive statements regarding the test outcomes are currently unavailable.

1 Introduction

This report focuses on pressure regulators purchased from the Knocks company, which Dräger rigorously tests to ensure their safety and compliance with the new guidelines in the MDR 2017/745. The endurance test rig is instrumental in evaluating the performance and long-term stability of the pressure regulators, especially after design and material changes. This work includes as preparation for the endurance test rig, tests like leakage, tightness, and response time. By identifying potential issues promptly, Dräger can ensure compliance with the strict requirements of the MDR and collaborate with the manufacturer for necessary improvements.

2 Material and Methods

The endurance test rig utilizes proportional valves, pressure regulators, and an Arduino to achieve precise regulation and control of flow and pressure in various applications. Proportional valves enable stepless flow control, making them essential for precise adjustments to achieve the desired pressure or flow. They can control the energy flow in terms of magnitude and direction, distinguishing them from switching valves that only switch between minimum and maximum values [1]. The performance of a proportional valve depends on the current flowing through the solenoid coil, which creates a linear relationship between current and the manipulated variable. However, hysteresis, caused by friction losses, affects the valve's ability to precisely reach the ideal setpoint [2]. Pressure regulators play a crucial role in monitoring and regulating the pressure in the system. They adjust the flow to maintain the desired pressure, working in conjunction with proportional valves to ensure precise control and stabilization [3]. Gas regulators maintain a constant outlet pressure by using a diaphragm, disc, and linkage system. They convert pressure energy into heat and lower regulated pressure, with an adjustable setpoint achieved through a wire spring and calibration screw [4]. The Arduino microcontroller is the controlling component in the experimental setup. It generates precise signals to control the proportional valves, allowing for the adjustment of pressure and flow according to specific requirements [5]. For this test an already existing Arduino program can be used, that is only minimally adapted for testing the pressure regulators.

A test setup, as shown in Fig. 1, was developed for pre-tests to individually test the components in continuous operation.

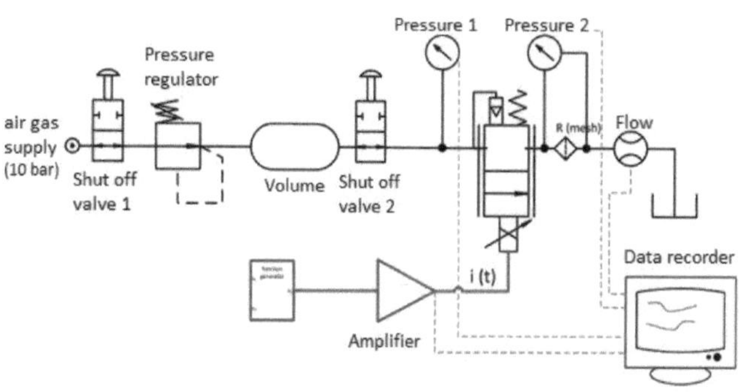

Figure 1: Test set up pre-tests

The main components of the test setup are as follows: 1. Gas supply: External compressed air is connected to the setup. 2. Pressure regulator and proportional valve: Multiple proportional valves (33 pcs, Norgren Fa. FAS) and new pressure regulator samples (4 pcs) are tested. 3. Buffer volume: Serves as a reservoir to compensate for pressure fluctuations and ensure a stable pressure flow, supporting continuous supply to the proportional valve. 4. Pressure gauge: Two gauges monitor primary and secondary pressure before and after the regulator. 5. Flow meter: Two different flow meters, for higher and lower flows, are used for precise measurement of gas flow. 6. Pressure gauge: Two gauges monitor primary and secondary pressure before and after the regulator.

2.1 Test Procedures

2.1.1 Characteristic curves of the proportional valves

The measurement of the proportional valves involves recording their characteristic curves, with 33 valves being tested. This serves as preparation for the endurance test of pressure regulators, while also checking the valves' leakage, which should not exceed 2 ml per measurement. The leakage is tested by pressurizing the system with closed valves, using a fine pressure gauge to detect any leaks. Subsequently, the desired values are set in Matlab software, including three consecutive pressure jumps from: 2 bar to 4 bar to 6 bar and again to and 4 bar. After connecting the valve to be tested and updating the file name through the software, specific current ranges (210 mA and 70 mA) and a stepsize of 1 mA are set, resulting in 140 steps. The current is used to precisely adjust the degree of opening and thus the flow. The measurement is then initiated.

2.1.2 Secondary pressure of the pressure regulator

Before starting the measurement, the pressure regulator is screwed into the measuring set-up. The secondary pressure in a pressure regulator is the constant output pressure maintained by the regulator, regardless of fluctuations in the input pressure or airflow. It is set to 2 bar using the pressure regulator. The inlet pressure is set manually to 3 bar, 5 bar and 7 bar in succession. The pressure is manually graded in steps, shown in the following Table 1. These steps were also reversed in the measurements.

Table 1: Flow steps for the secondary pressure of the pressure regulators

Flow steps in L/min											
0	0,1	1	2	3	4	5	6	7	8	9	10

2.1.3 Basic adjustment of the pressure regulators

The measurement focuses on determining the pulse length of the pressure regulators to ensure proper pressure switching. The pulse length is the duration it takes for the pressure regulator to self-adjust with each pulse. The objective is for the back pressure to settle within the first half of the pulse width. Inadequate pulse width can hinder correct pressure adjustment, leading to flow passing through the regulator. Both the pressure regulator and the pattern are measured and compared. A different control software is used for this pre-test, which sets a fixed current and pulse length for the proportional valves. The pulse length is minimized to 150 ms to save time, and various pulse lengths (200 ms, 300 ms, 400 ms) are tested to ensure back pressure settlement within the first half of the pulse. The test is conducted for three known pressure levels: 3 bar, 5 bar, and 7 bar.

3 Results and Discussion

3.1 Characteristic curves of the proportional valves

The diagrams below depict the recorded characteristic curves of the proportional valves, focusing on one measured valve. Fig. 2 displays selected hysteresis curves for four pressure jumps: 2 bar, 4 bar, 6 bar, and 4 bar again. The hysteresis curve in pressure valves describes the difference in pressure between the opening and closing of the valve, where the valve exhibits a varying pressure value for the same condition depending on whether it is in the process of opening or closing. Here, the flow is shown above the set current when different pressures are applied. In practice, not exactly 2, 4 and 6 bar are reached, but approximate values, as shown in Fig. 2. Not all valves are shown in the diagrams, due to the clarity. Similar curves are available for all 31 valves tested. The gradient above the flow

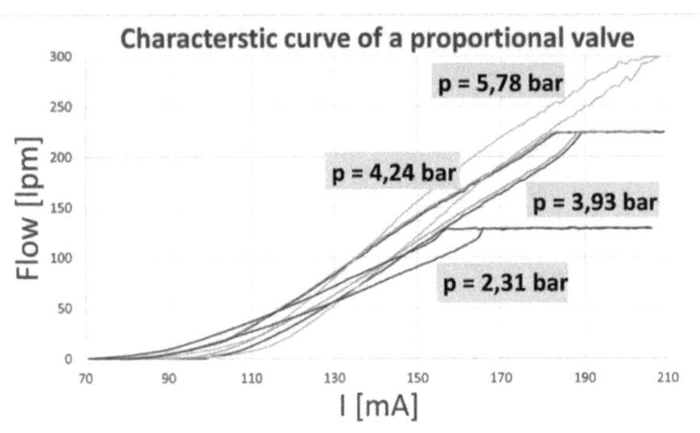

Figure 2: Characteristic curve of the proportional valve 10

is another important factor in assessing the valves. For the sake of clarity, only the gradient above the flow of the 2 bar measurement run is shown here in Fig. 3. The curves show a relatively flat course with minimal scatter, staying within the desired range indicated by dashed lines in the first two figures, Fig. 1 and Fig. 2. However, as the pressure increases, the scatter also increases, particularly in the upper flow range. This can be attributed to non-linear effects, friction and wear, and pressure-dependent leakage. Non-linear effects can cause deviations from ideal linear control, leading to increased dispersion of flow rate. High pressure can result in more friction and wear, affecting valve operation and increasing scatter. Additionally, leakage in seals and gaps can rise at higher pressures, diverting some fluid away from the desired channel and impacting the relationship between actual and target flow rates, thereby contributing to increased flow rate derivation.

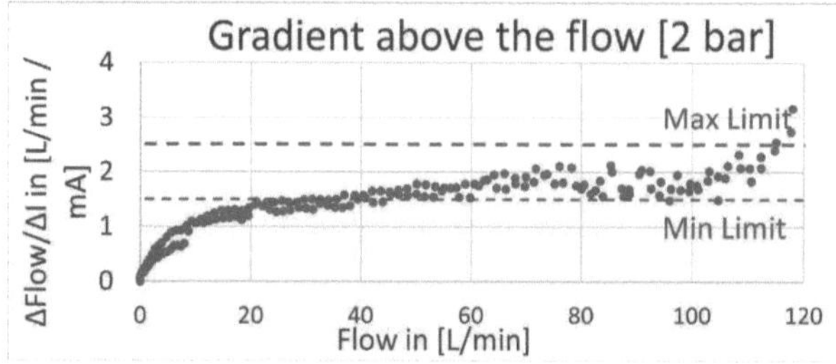

Figure 3: Characteristic curve of the proportional valve 10 at 2 bar

3.2 Secondary pressure of the pressure regulators

Fig. 4 shows the secondary flow behavior at 7 bar pressure, with one pattern displayed for clarity while the other two for 3 and 5 bar exhibit similar behavior. The example is shown for 7 bar, since the secondary pressure shows the greatest fluctuations here. The curves at 3 bar and 5 bar show fewer fluctuations and settle earlier. The new sample's pressure regulator consistently has higher values compared to the series, possibly due to user variations. The reworked pressure regulator pattern demonstrates a flatter curve and lower pressure difference, within the acceptable limit of 0,4 bar. The upper line represents the decreasing flow range, while the lower line represents the increasing flow range. This is attributed to the regulator's response to gas flow changes: the secondary pressure remains stable or slightly decreases during rising gas flow, as the regulator opens the flow path to maintain a constant outlet pressure. Conversely, during decreasing gas flow, the secondary pressure remains stable or slightly increases as the regulator narrows the flow path to maintain the desired outlet pressure.

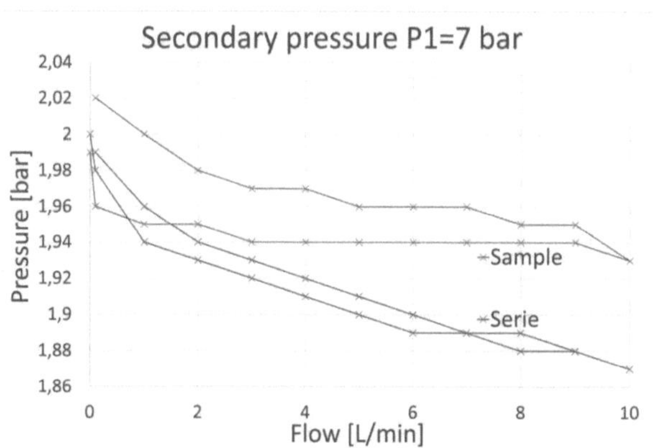

Figure 4: Secondary pressure P1 = 7 bar

This preliminary test served to check the behaviour of the secondary pressure of the pattern and compare it with that of the series. The result is a significantly more constant and flatter course of the new pattern.

3.3 Pulse length of pressure regulator

Results of the pulse length tests are shown in Fig. 5, displaying the finally selected frequency for clarity. Since the pulse length is most important for the highest frequency, because it shows the most irregularities and fluctuations, only the diagram for the highest frequency of 7 bar is shown. The selected frequency thereafter will also fit the lower frequencies. The known pressure jumps of 3 bar, 5 bar, and 7 bar are accompanied by their corresponding frequency and pulse width. A pulse width of 150 ms was chosen, resulting in a frequency of approximately 3,33 Hz. In Fig. 5, it can

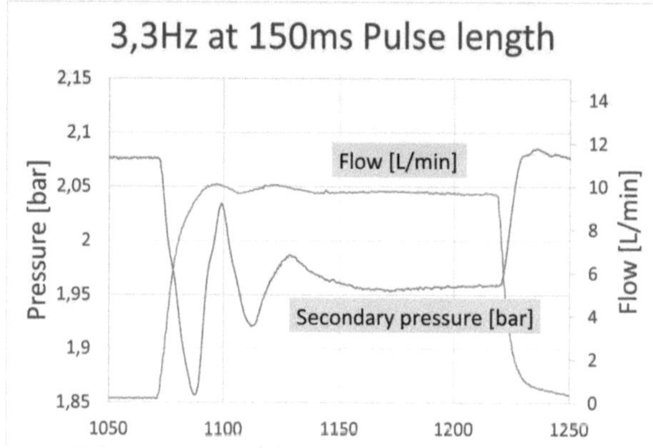

Figure 5: Secondary pressure at 3,3Hz and 150 ms Pulse length with 7 bar

be seen that rapid changes in gas flow can cause temporary fluctuations in the outlet pressure of the pressure regulators. These fluctuations occur due to the inertia of the regulator and the time required for it to adjust the control mechanism and stabilize the desired outlet pressure. When the gas flow increases rapidly, the pressure regulator opens the flow path to allow more gas flow and maintain a constant outlet pressure. However, there may be a slight delay before the regulator fully responds to the increased flow and stabilizes the pressure. The measurement demonstrates that at all three pressure levels, the output pressure stabilizes within half the pulse length as desired. The higher pressure range is particularly interesting because it can experience more fluctuations. However, with the chosen frequency of 3,33 Hz, the pressure stabilizes sufficiently before the next pulse begins. To maintain an even frequency, the next higher frequency of 4 Hz was selected for the continuous measurement setup.

3.4 Construction of Box and Holder

This chapter presents the construction of a specially designed holder for organizing 31 pressure regulators and proportional valves. The holder, made of aluminium, combines four components in each section. It fulfills various pre-defined requirements such as dimensions, stability, ergonomic hose connections, and air circulation. Aluminium was chosen for its optimal strength-to-weight ratio. The holder includes precise slots, secure mounting options, and strategically placed hose connections to minimize vibrations, movements, and potential leaks. Grouping four components in a holder saves space and simplifies the setup. Fig.6 shows the holder in an assembly with the associated other components, such as the hose connections and the proportional valves. The pressure regulators are attached to the top of the holder.

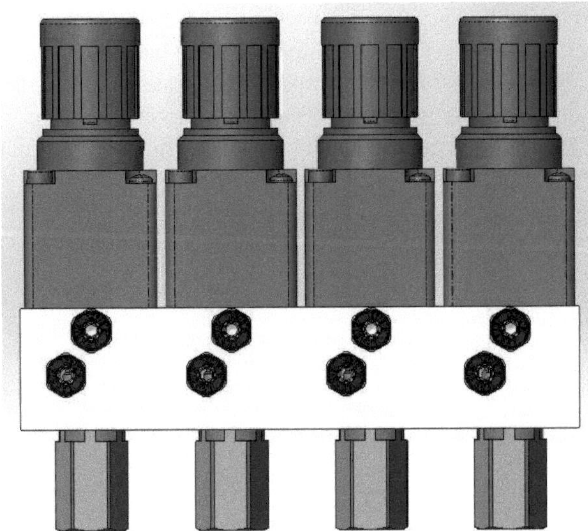

Figure 6: Graphic of the holder for pressure regulator, proportional valve and hose connections from above

The components of the endurance test have to be attached to a chipboard. For this purpose, a mounting system was designed for the holder. To maximize space utilization, the test stand is arranged with two rows of four blocks each, with the controlling printed circuit boards placed in between. The mounting design allows for the use of the holder from both sides, considering the stacked placement of the holders.

3.5 Continous test rig

The finished prepared endurance test stand, seen in Fig. 7, shows the chipboard on which the eight blocks, consisting of pressure regulators and proportional valves, are arranged. Between the blocks, the three printed circuit boards are strategically placed to allow connection and control of the proportional valves.

4 Conclusion

In response to MDR 2017/745 amendments, Dräger is adapting pressure regulators, replacing NBR and CR rubber seals. Rigorous testing, including a dedicated endurance test rig, ensures compliance and safety in ventilation and anesthesia devices. Preliminary assessments confirm proportional valve suitability and compare old and new regulators. Results highlight proportional valve characteris-

Figure 7: Finished measurement rig

tic curves and improved stability in the secondary pressure of the new regulator pattern. Pulse length tests indicate the ability of pressure regulators to self-adjust with precise pressure switching. The study emphasizes Dräger's commitment to advancing regulator performance and safety, addressing immediate regulatory needs. Findings offer insights for ongoing collaboration and potential improvements in meeting medical device regulations.

Acknowledgement

The work has been carried out at Drägerwerk AG & Co. KGaA, Intensive Care Area and supervised by Alexander Korff, Department of Computer Science and Economy, University of Applied Sciences Lübeck.

Author's Statement

Conflict of interest: Authors state no conflict of interest.

5 References

[1] D. G. N. S. H. Will, Hydraulik, Berlin, Heidelberg: Springer-Verlag, 2007.

[2] J. W. Norbert Gebhardt, Hydraulik – Fluid-Mechatronik: Grundlagen, Komponenten, Systeme, Messtechnik und virtuelles Engineering, Berlin, Heidelberg: Springer-Verlag, 2020.

[3] W. B. K.-H. Grote, Dubbel - Taschenbuch für den Maschinebau, Berlin, Heidelberg: Springer-Verlag, 1997.

[4] N. Z. G. Luecke, Stability of gas pressure regulators, Iowa State University, Ames: ScienceDirect, 2008.

[5] T. Brühlmann, Sensoren im Einsatz mit Arduino, Frechen: mitp, 2017.

Characterization of porphyrin-based sensor films to improve food safety of black tiger shrimps

Eleana Noti [1], Alexander Altmann[2], and Ramtin Rahmanzadeh [2]

[1] Medical Engineering, Universität zu Lübeck, eleana.noti@student.uni-luebeck.de

[2] Institut für Biomedizinische Optik, Universität zu Lübeck, {alexander.altmann, ramtin.rahmanzadeh}@uni-luebeck.de

Abstract

Millions of tons of food get thrown away every year. It is therefore of high interest for producers and consumers to be able to determine food spoilage continuously and non-invasively throughout the supply chain. In an attempt to reduce the amount of food waste and at the same time increase food safety, the reaction of porphyrin-based biosensor films to spoilage, occurring in black tiger shrimps, has been examined in this study. For that, shrimp spoilage in model packages was tested, with two different types of sensor films. These are able to detect amines, which are released during degradation. The pH values and bacterial growth within the samples were examined using two different methods. All sensor films successfully reacted with the amines, which could be seen in their fluorescence emission spectra, as well as color change with the eyes.

1 Introduction

As sustainability becomes increasingly relevant in today's society, addressing the problem of food waste and its detrimental consequences on the environment has become a priority. Yearly 1.3 billion tons of food get thrown away, with increasing numbers over the past years [1]. At the same time, ensuring food safety is important for both consumers and regulatory bodies. Although several optical sensors for determining food freshness have been developed over the last years [2], none of them have been applied yet. The biosensor film presented in this paper is able to detect food spoilage by the reaction with amines. Amines are released during food spoilage processes in protein-based food. That way, food freshness can be determined continuously and non-invasively, addressing both food safety and waste issues [3].

2 Material and Methods

2.1 Preparation of model packages with shrimp

Individually frozen, raw, peeled, deveined and glazed black tiger shrimps (Lat. Penaeus mondon) were purchased at a local supermarket and repacked in airtight plastic containers, with a volume of 280 ml, at 65-75 g portions. On the inside of the container lid, the circular porphyrin-based sensor film in shape, with an 8 mm diameter, was attached with adhesive tape. There were two different types of sensor films used for this experiment, *S1* with 500 ppm and *S2* with 1000 ppm phosphorus porphyrin molecules in polyethylene (PE)

film. After packaging, the containers were transported to the fridge for the shrimps to defrost. At each time point, one package was taken out of the fridge and analyzed by fluorescence spectroscopy and aerobic plate count. The fluorescence spectra were recorded at defined time intervals between 4 and 24 hours, while the samples were either stored at room temperature (experiment series I and II) or in the fridge between 2,5 °C and 7,5 °C (experiment series III and IV). After the fluorescence measurement was completed, the shrimps were taken out of the plastic container, homogenized with an immersion blender and filled in tubes at 5 g portions.

2.2 Measurement of fluorescence spectra

Directly after defrosting, a reference fluorescence spectrum of the sensor film was recorded using the FluoroMax Plus Horiba Scientific Spectrofluorometer, with an excitation wavelength of 422 nm. The end of the optical fiber was placed directly on the sensor film. The opening diameter during excitation was 2 mm and 9 mm during emission. The measurement was repeated three times and then averaged. The spectra were smoothed, with FFT filter (stepsize: 5), and normalised, with sum normalisation, using Origin (OriginLab Origin 2022).

2.3 Determination of pH value

To determine the pH value of the shrimps, 20 ml distilled water was mixed with 5 g homogenized mass and the sensor of the PCE Instruments pH Conductivity Meter PCE-BPH20 was inserted into the tube. The pH measurements were done directly after spectra acquisition and homogenization.

2.4 Determination of total viable count

Determination of bacterial growth was performed by the total viable count, in a down-scaled approach according to the FDA regulations [4]. For that, serial dilutions of up to 1:100000 of each sample were made by adding 45 ml buffer dilution solution to 5 g frozen homogenized shrimp mass, vortexing and then taking out 1 ml of the solution and mixing it with 9 ml buffer solution. One milliliter of the diluted sample solution was then pippeted onto a Petri dish and 10-12 ml of sterile Plate-Count Agar (APHA; ISO4833:2003) at 45 °C temperature were added. The Petri dish was gently tilted around, for the two liquids to combine. A duplicate was made for each sample and the Petri dishes were incubated for (48 ± 1) h at 35°C. After incubation, the colonies appearing on the Petri dish were counted. Each colony was marked and counted separately on Petri dishes with up to 250 colonies. For Petri dishes with more than 250 colonies, a 1 cm^2 square was marked on the dish, the colonies inside that square were counted and multiplied with the total area of the Petri dish, namely $56.75\,\mathrm{cm}^2$, as described by FDA regulations [4].

In addition to the above described plate count method, the drop down method was tested to determine the bacterial growth on the shrimps. For the drop down method, 20 ml of sterile Agar were pipetted into a square Petri dish, which was then put into the fridge for the Agar to solidify. After that, the dish was put up at a 60° angle and 10 µl of the dilutions were pipetted on the top of the plate, with a centimeter distance from each other. After the dilution had dropped down and reached the bottom of the plate, it was incubated for (48 ± 1) hours. When the incubation process was completed, the colonies were counted, according to ISO4833:2003.

3 Results and Discussion

The reaction of porphyrin-based biosensor films to detect spoilage of the shrimp samples was examined. The sensor showed a fluorescence and colour change, when food spoilage occured. This is a result of the reaction of the sensor molecules with the amines, that are set free during the degradation process [3].

The pH values and bacterial growth of the shrimps were monitored, to compare the sensor reaction with the actual spoilage of the shrimps.

3.1 Sensor film reaction

To quantify the sensor film reaction to the degradation processes taking place in the shrimps over time, the fluorescence emission spectra were recorded for sensor foils S1 and S2. Fig. 1 and Fig. 2. show the overlaid spectra for the room temperature experiments, in discrete time intervals. It

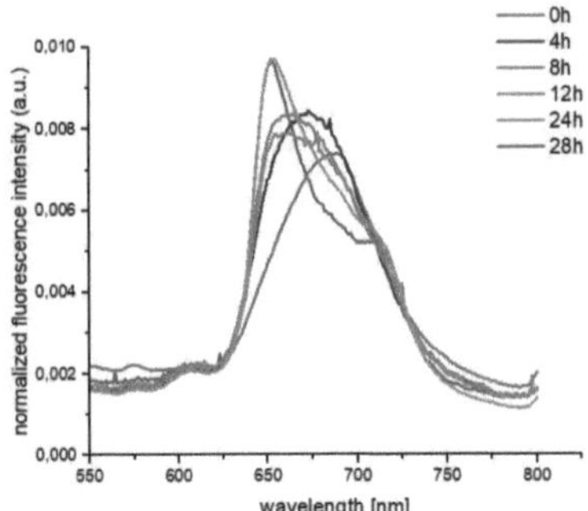

Figure 1: Fluorescence spectra for experiment series I (room temperature, S1 sensor film), 0 to 28 hours after defrosting.

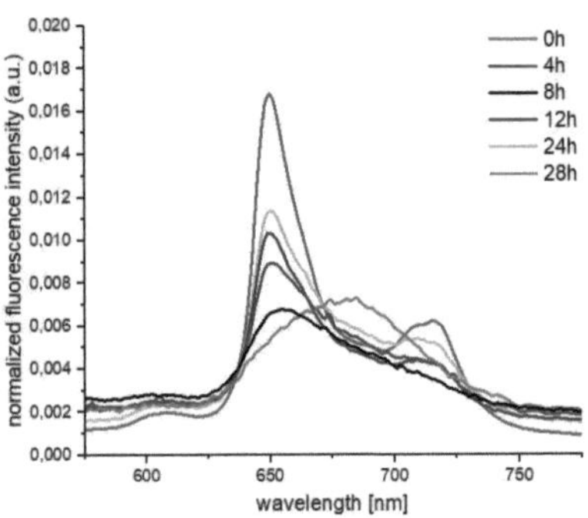

Figure 2: Fluorescence spectra for experiment series II (room temperature, S2 sensor film), 0 to 28 hours after defrosting.

can be seen that the maximum shifted towards lower wavelengths with time. Specifically, it shifts from 690 nm at 0 hours to 653 nm at 28 hours after defrosting, for the S1 sensor film. For the same parameters, the S2 sensor film emission peak shifts from 681 nm to 651 nm. Further, an increase in total fluorescence intensity can be observed over time, as well as a second, smaller peak forming on the right side of the spectrum, after 24 hours for the S1 film and 12 hours for the S2 film. Therefore, it can be concluded, that both sensor films are suitable for detecting the spoilage of shrimps through the gas phase.

Furthermore, the spoilage was investigated under more realistic conditions, i.e., under refrigeration. The fluorescence spectra for the refrigerated experiments in cooled conditions can be seen in Fig. 3 (S1 sensor film) and Fig. 4 (S2 sensor film). Compared to the room temperature spectra, it can be seen that the overall reaction of the sensor films took longer. A noticeable shift in wavelength and increase in intensity is visible for the first time after 96 hours, using S1 sensor film and 78 hours, using S2 sensor film. In this case, the shift is from 680 nm at 0 hours to 653 nm at 96 hours and from 680 nm at 0 hours to 651 nm at 144 hours after defrosting.

In all four experiments, a change in color of the sensor films, from bright green to a reddish brown, could be observed with naked eye (Fig.5). S1 and S2 sensor films, however show different spectroscopic reactions, due to their different porphyrin molecule concentrations. Spectral changes are more distinct, with the S2 sensor film, compared to S1. This becomes even more apparent in the refrigerated experiment setup, therefore the S2 sensor film should be chosen over S1 to evaluate shrimp freshness.

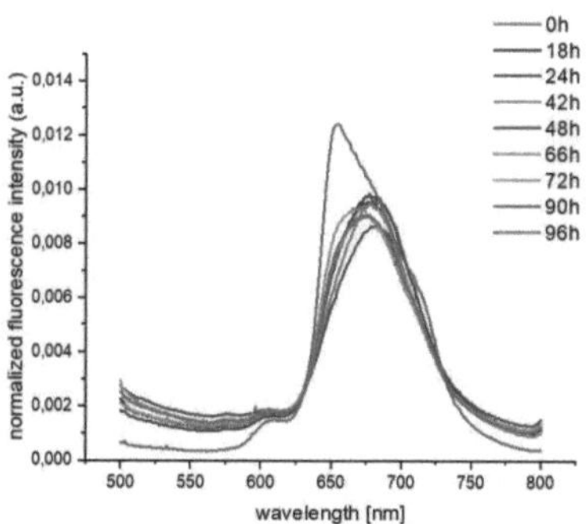

Figure 3: Fluorescence spectra for experiment series III (refrigerated, S1 sensor film), 0 to 96 hours after defrosting.

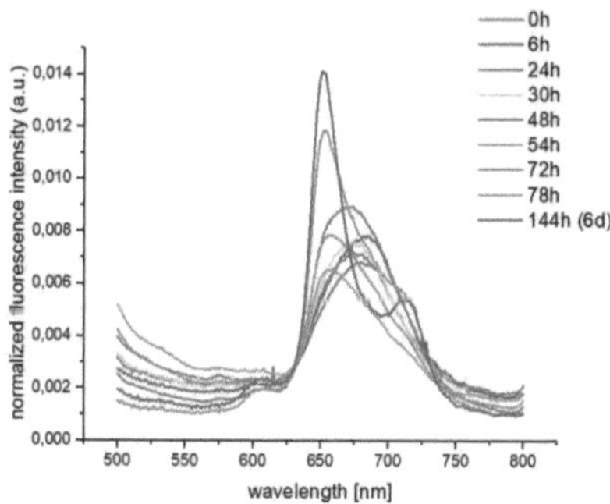

Figure 4: Fluorescence spectra for experiment series IV (refrigerated, S2 sensor film), 0 to 144 hours after defrosting.

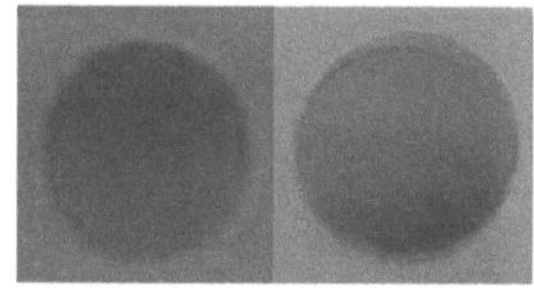

Figure 5: S2 Sensor film at 0 hours (left) and 28 hours (right).

3.2 Changes in pH value

As amines are bases and thus bind to acids, it is important to know, if the pH of the shrimps is high enough, so that the transfer of amines to the gas phase is not inhibited and they can be detected by the biosensor. Every experiment series showed a decrease in pH value with increasing storage time. For experiment series I, the pH value dropped from 8.34 to 7.86. In series II, pH value went from 7.82 to 7.65. The refrigerated experiment series, III and IV, also showed a decrease from 7.53 to 7.37 and from 8.52 to 7.41 accordingly.

As the pH values remain alkaline or neutral throughout the spoilage process, we can conclude that the amine should not be trapped on the shrimp surface during storage. Bacterial growth and pH value show an inverse proportional relation to each other.

3.3 Bacterial count

In order to quantify the degradation process in the packaged shrimp samples, the bacterial growth over time was observed. For that, the colonies on the shrimp were counted as total viable count. Generally, an exponential increase in the number of CFU/g can be observed, in all four experiments. Fig.6-9 show the bacterial growth in the samples for experiment series I-IV accordingly.

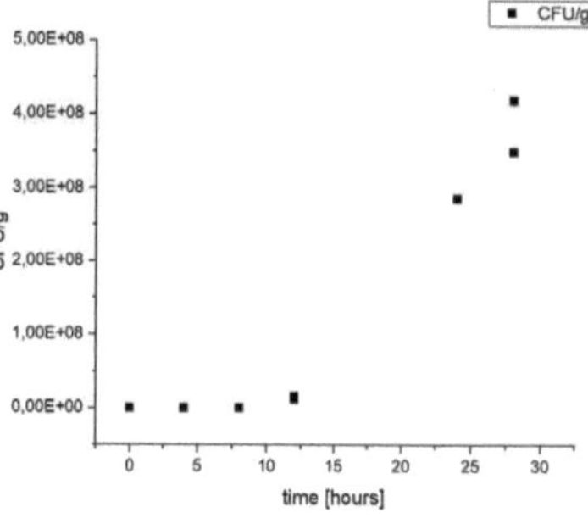

Figure 6: Bacterial growth over time for experiment series I (room temperature, S1 sensor film).

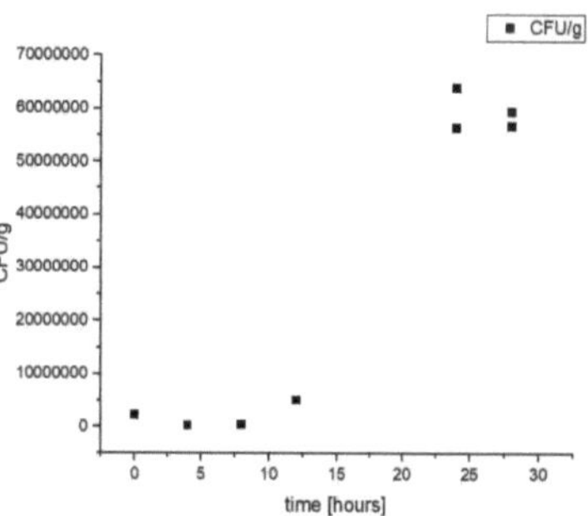

Figure 7: Bacterial growth over time for experiment series II (room temperature, S2 sensor film).

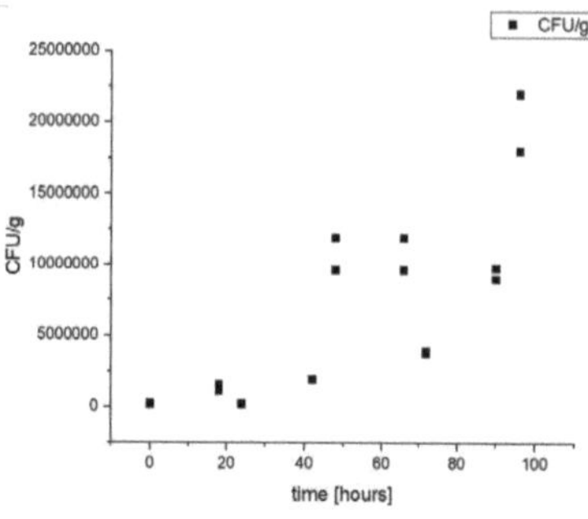

Figure 8: Bacterial growth over time for experiment series III (refrigerated, S1 sensor film).

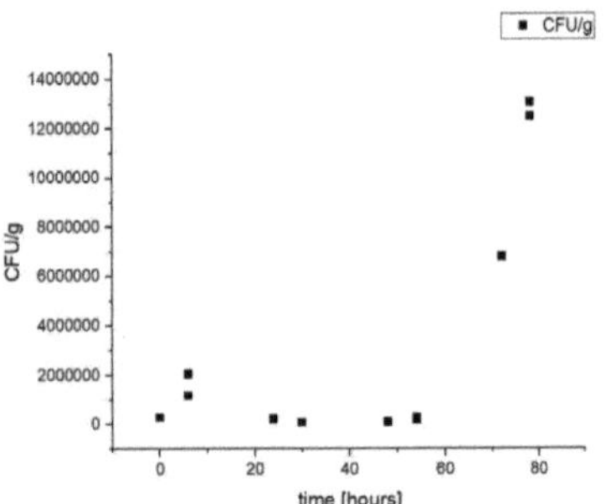

Figure 9: Bacterial growth over time for experiment series IV (refrigerated, S2 sensor film).

In the examined samples, the colony forming units counted, range from 84×10^3 to 419×10^6 CFU/g, thus exceeding the limit of 1.000.000 CFU/g for raw crustaceans, set by the International Commission on Microbiological Specifications for Foods (ICMSF) [5]. In total 59,32% of the samples exceeded this limit, during the controlled spoilage experiments. When comparing the bacterial growth with the S2 fluorescence spectra, it can be seen that the fluorescence intensity increases with the number of CFU/g. For both experiment series II and IV, the ICMSF limit is exceeded when the normalized fluorescence intensity is above 0,007 a.u..

3.4 Bacterial count with the drop down method

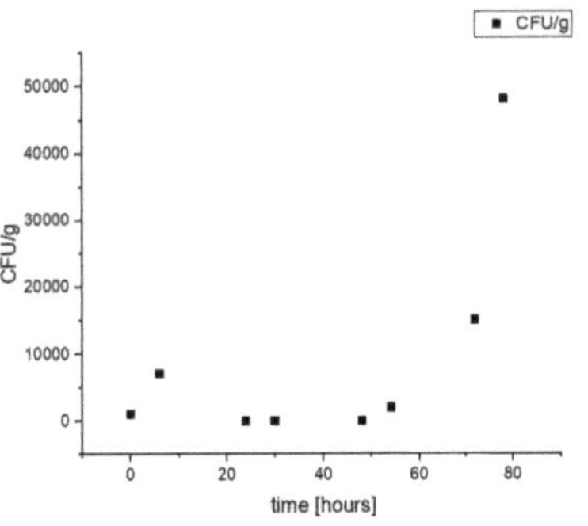

Figure 10: Bacterial growth for experiment series IV, determined with drop down.

For experiment series III and IV, the drop down method was used additionally to plate count in order to determine if it is a less time- and labour-intensive alternative. In both experiments the drop down method was shown to be comparable to plate count, as they both showed similar results in terms of bacterial growth over time for each sample (Fig. 10). We tested the recorded counts from both methods for correlation with each other. For that we applied rank correlation indices and found a Spearman coefficient of 0,57 for series III and 0,93 for series IV.

4 Conclusion

All sensor films successfully reacted, as degradation processes take place on the shrimps. This can be seen in an increase in intensity and peak shift in the fluorescence spectra.

For refrigerated experiment setups, the S2 sensor film is better suited, as spectral changes are more visible.

A significant decrease in the pH value of the homogenized shrimp mass can be observed over time.

The majority of the tested samples showed an up to 419 times higher bacterial concentration, than the ICMSF limit. This is visible in the S2 spectra, as an increase in fluorescence intensity.

The drop down method has been shown to be a reliable and equally accurate alternative method for determining the total viable count in the shrimp samples, while also being less time intensive compared to the plate count method.

Overall the biosensor films tested are able to contribute to an increased food safety and less food waste in black tiger shrimps.

Acknowledgement

The work was part of an internship, which has been carried out and supervised by the Institute of Biomedical Optics, Universität zu Lübeck.

Authors' Statement

Conflict of interest: Authors state no conflict of interest.

5 References

[1] N. Scialabba et al. , *Food Wastage Footprint: Impacts on Natural Resources. Summary Report.*, Food and Agriculture Organization of the United Nations, 2013.

[2] A. I. Danchuk, N. S. Komova, S. N. Mobarez, S. Y. Doronin, N. A. Burmistrova, A. V. Markin and A. Duerkop , *Optical sensors for determination of biogenic amines in food*, Analytical and Bioanalytical Chemistry 412, 4023–4036, 2020.

[3] A. Altmann, G. Hüttmann, S. Christian and R. Rahmanzadeh, *Porphyrin-based sensor films for monitoring food spoilage. Food Packaging and Shelf Life.*, 2023.

[4] L. Maturin and J. T. Peeler , *BAM Chapter 3: Aerobic Plate Count | FDA*, FDA Gov. 3 (January 2001), 1–11, 2020.

[5] ICMSF, *Microorganisms in Foods 2. Sampling for microbioloical analysis: Principles and specific applications.*, 2. Aufl., International Commitee on Microbiological Specifications for Food, 1986.

Exploring Cognitive Load in Surgical Skill Assessment using Event-Related Potentials

Lynn Yehia [1], Alexandra Eberenz [2], Georg Wolf [2], Georg Maennel [2], Marcus Heldmann [3], Freschta Malekzeda [4], Michael Thomaschewski [4], Jannis Hagenah [2]

[1] Biomedical Engineering, Lübeck University of Applied Sciences, lynn.yehia@stud.th-luebeck.de

[2] Fraunhofer Research Institution for Individualized and Cell-Based Medical Engineering IMTE, {alexandra.eberenz, georg.wolf, georg.maennel, jannis.hagenah} @imte.fraunhofer.de

[3] Department of Neurology, University of Lübeck, marcus.heldmann@uni-luebeck.de

[4] Department of Surgery, University Medical Center Schleswig-Holstein Campus Luebeck, {freschta.malekzada, michael.thomaschewski}@uksh.de

Abstract

This study explores cognitive aspects of surgical skills acquisition, emphasizing mental resource allocation and cognitive load in novice learners. Employing event-related potentials (ERPs) derived from electroencephalographic (EEG) data, the study introduces a dual-task scenario, combining a laparoscopic surgery exercise with an auditory detection task. Participants, medical students inexperienced in laparoscopy, underwent training of a 4-months laparoscopic training program. The ERPs reveal a significant increase in P300 component amplitude during the latter stages of laparoscopic training, indicative of heightened mental resources. This amplification aligns with the hypothesis that familiarity and structured training liberate mental resources, facilitating more efficient engagement in secondary tasks. The findings underscore the relevance of ERPs in tracking cognitive dynamics during surgical skills acquisition and its potential applications in refining training protocols for both novices and experts.

1 Introduction

Evaluating a learner's involvement and cognitive load during the process of learning is a crucial aspect of designing educational methods. This becomes particularly significant in the context of assessing surgical skills, where simulation-based training is employed for developing the motor skills essential for minimally invasive surgery (MIS). Novices often face limitations in their attentional resources due to the intricate nature of the task, particularly at the initial stages of the learning process. This limitation can impede their ability to handle disruptions or complications during exercises or actual surgical procedures [1].

To evaluate mental workload and fatigue, many studies rely on information from the frequency domain of the EEG [2]. Another approach involves the use of event-related potentials (ERPs) to estimate mental load. ERPs are voltage fluctuations in the human electroencephalogram (EEG) that can be recorded non-invasively from a volunteer's intact scalp. These fluctuations are induced by external events, such as sensory stimuli, providing valuable insights into the cognitive aspects of learning and task performance [3]. The ERPs are extracted from the EEG data using the method of signal averaging. The objective of signal averaging is to isolate the electrical potentials associated with specific events. This is achieved by assuming that events, such as stimuli produce similar waveforms across

trials, with random noise added to this waveform to produce the recorded EEG signal. By averaging EEG data across multiple events, the noise diminishes, and the signal will remain in the average [4].

Engaging in the simultaneous execution of two or more tasks, known as dual-tasking, is crucial for cognitive performance and demands attentional control to effectively allocate resources to each task [1]. Secondary tasks have been utilized to evaluate the processing capacity available to the secondary task under varying levels of difficulty in the primary task. Specifically, the amplitude of the P300 component has been employed as a marker for quantifying processing capacity [5]. The P300 component is a positive deflection observed in the EEG in response to an infrequent task-relevant stimulus presented among frequent non-target stimuli. It is consistently triggered in a dual-task paradigm [6]. Basically, in certain conditions, e.g. Alzheimer's disease, if two tasks share the same processing resources, then the allocation of resources are taken up by the primary task and are not sufficient to process the secondary task [7].

In this experimental study, a dual-task scenario was implemented, introducing an auditory detection task alongside a laparoscopic surgical task. The assessment of mental resources at the beginning and end of a laparoscopic training exercise was conducted through the analysis of ERPs.

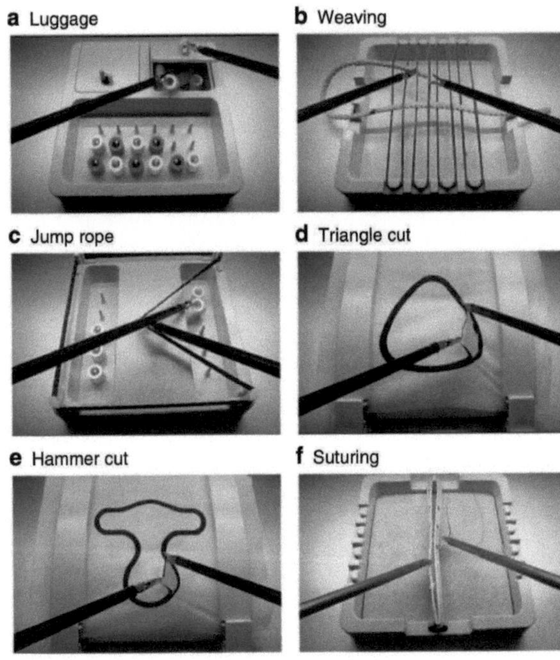

Figure 1: Six training exercises of the Lübecker Toolbox

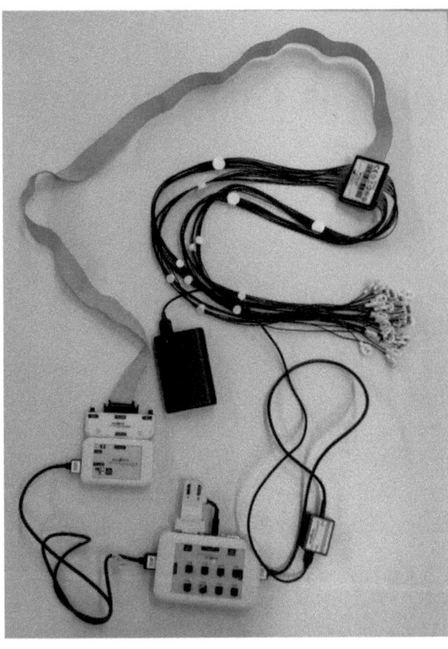

Figure 2: Brain Vision LIVEAMP USB amplifier system connected to the electrodes

The underlying hypothesis is that engagement in laparoscopic training along with the growing familiarity to laparoscopic tasks would lead to freeing up mental resources, which leads to more attention allocated to the secondary tasks.

2 Methods

Medical students without previous laparoscopic surgery experience participated in the study. All participants provided informed consent before participating in the study.

Participants underwent laparoscopic or robotic skills training using the Lübecker Toolbox (LTB) curriculum [8]. The curriculum includes six related exercises: pack your luggage, weaving, jump rope, triangle cut, hammer cut, and suturing as shown in Fig. 1. This study, however, focuses on the first exercise, "pack your luggage", where participants are required to alternately place pins of two different colors (white and blue) into their respective compartment by picking up each pin and inserting it into the corresponding slot, making sure it is placed correctly and the lid is closed, only then they can proceed to the next pin. Incorporating a dual-task paradigm, participants performed the exercise while simultaneously engaging in a secondary auditory detection task. The exercise "pack your luggage" was done over a certain number of trials, and every 5 trials, an auditory stimulus, represented by two alternating tones (800 and 880 Hz) of a beep sound, was introduced. This pattern will change to repeat a tone instead of presenting the alternating one. For every 20 alternating tones, a repeated tone is introduced, the position of which is selected at random. Participants were instructed to press a foot switch when they detect a sound repetition.

To capture the neural correlates of this dual-tasking, continuous EEG measurements were recorded using 32 electrodes, connected to a Brain Vision LIVEAMP USB amplifier system as shown in Fig. 2. The EEG electrodes were then affixed, based on the international 10-20 system, to a cap of a suitable size of the participant. To optimize conductivity and signal quality, a conductive gel was injected beneath the electrodes.

EEG data analysis was conducted using the EEGLAB and ERPLAB toolboxes [4]. Data were collected at a sampling rate of 250 Hz and subjected to bandpass filtering within the range of 0.1–30 Hz. The EEG data processing pipeline included a series of steps aimed at ensuring the integrity and reliability of the recorded data. Initially, the process involved correction and rejection of artifacts, such as blinks and eye movements through Independent Component Analysis (ICA), a widely employed statistical technique. The processing utilized EEGLAB's default ICA algorithm, Infomax, implemented through the runica routine.

Whenever an event such as a stimulus occurs, an event code is inserted into the EEG data file to denote the time of the event, i.e., the stimulus. A segment of EEG data surrounding each event, is called an epoch. Subsequently, epochs of the recorded EEG data correspond to the stimulus presentation window (-200 to 1200 ms) were extracted. Eventually, Individual ERPs for each trial were computed by averaging all the epochs together, resulting in an ERP waveform primarily reflecting the consistent ERP response. After all individual ERPs were computed, the grand average was found by averaging across trials and participants [4]. Due to excessive movement artifacts, three participants were excluded from the analysis, resulting in a final sample size of 16 subjects.

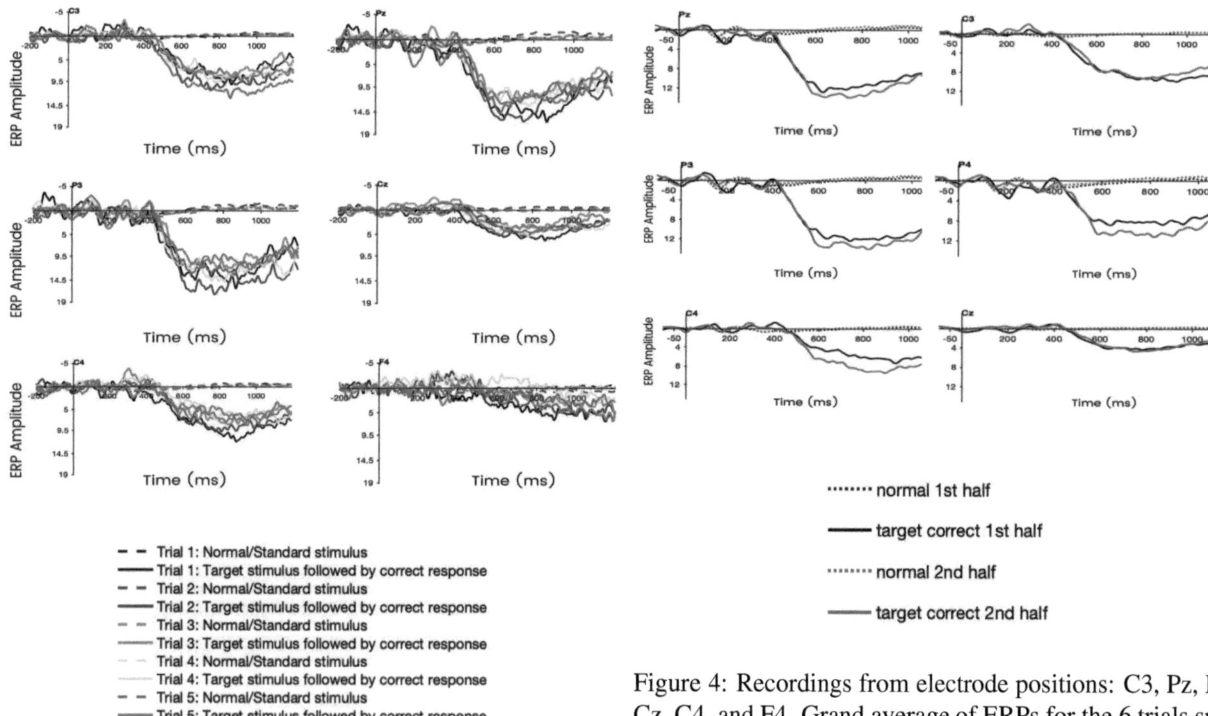

Figure 3: Recordings from electrode positions: C3, Pz, P3, Cz, C4, and F4. Grand average of ERPs for 6 trials shows the association of target tones with a positive deflection (P300 component).

Figure 4: Recordings from electrode positions: C3, Pz, P3, Cz, C4, and F4. Grand average of ERPs for the 6 trials split into 2 groups (first 3 trials and last 3 trials), shows target tones associated with a P300 component at the beginning of the training which was increased in the last sessions of training.

3 Results

A total of 31 trials were recorded for each participant throughout the laparoscopic and robotic training sessions, with dual-tasks occurring every 5 trials, leading to a cumulative count of 6 dual tasks for each subject.

3.1 Averaged ERPs for Standard and Target Tones

The first set of findings, shown in Fig. 3, offers an insight into the averaged ERPs associated with both the standard alternating tone and the target repeating tone across the six trials. A distinctive positivity prominent at central parietal electrode sites, namely C3, Pz, P3, Cz, C4, and F4, becomes apparent in response to the target tone (positive values downwards). This positivity represents the P300 component and manifests around 400 ms post-stimulus.

3.2 Early and Late Trials

Fig. 4 shows averaged ERPs split into 2 groups, one group is the first three trials and the other is the second half of the trials. This segmentation allows for a temporal analysis of the observed ERP responses. In the first half of trials, a P300 component manifests in response to the target tone, aligning with the pattern illustrated in Fig. 3. The second set of trials, representing the final phase of the training, reveals a notable amplification of the P300 component positivity.

4 Discussion

The subsequent analysis of ERPs yielded insightful patterns that provide a good understanding of cognitive processes during laparoscopic training.

The observed amplification of the P300 component in the later stages of laparoscopic training suggests a dynamic modulation in cognitive processes. The amplitude of the P300 component, which has been interpreted as an index for mental resources available for processing the task at hand, showed an increase.

This elevation in P300 amplitude implies that more mental resources are now being allocated to the secondary auditory detection task. These mental resources appear to have been liberated as a result of the laparoscopic training and the heightened familiarity acquired with the primary task.

The notion that the increase in P300 component amplitude signifies a rise in mental resources aligns with the expected outcome of the training program. As participants became more familiar with the laparoscopic tasks, the surplus mental resources could be efficiently directed towards the concurrent auditory detection task.

Previous studies have revealed mixed results. Thomaschewski et al [1], have observed a reduction in the

P300 amplitude in the secondary task over the course of training. This was argued to be the reason of a shift of mental resources towards the primary task. One key distinction which might be a reason to the discrepancy in results, lies in the methodology, where our study processed the EEG data of the dual-task recorded throughout one exercise, while Thomaschewski et al. compared the data taken from the first and last exercises of a 5 week training program. This difference in study design, with our data being based on a shorter, more focused course, may have contributed to the observed differences in the results. The longer course study design might have influenced the dynamics of mental resource allocation differently, possibly due to prolonged training effects. In addition, this analysis marks only the initial phase of an ongoing study. Subsequent phases will involve another exercise featuring a dual-task scenario at the end of the curriculum, which will hopefully present more insight. The contrasting results highlight the need for an understanding of the interplay between primary and secondary tasks in dual-task paradigms.

5 Conclusion

In conclusion, the study provides valuable insights into the dynamics of mental resource allocation during the acquisition of MIS skills. The increase in P300 component amplitude, indicates heightened mental resources, during the latter stages of laparoscopic training. This amplification aligned with the hypothesis that increasing familiarity and structured training in laparoscopic tasks facilitate the liberation of mental resources, enabling more efficient engagement in secondary tasks.

The observed amplification of the P300 component has direct relevance to surgical education, particularly in scenarios involving dual-task situations where cognitive performance and attentional control are of paramount importance. The findings imply that the enhanced mental resources may facilitate more effective multitasking during intricate surgical procedures, underscoring the practical implications of our research in the context of skill acquisition.

While our investigation focused on novice learners in the initial stages of skill acquisition, the broader implications of this research extend to potential applications in long-term studies involving expert surgeons. The ability of ERPs to differentiate between trained and untrained surgeons or serve as indicators of progress along a learning curve holds promise for the field. This suggests that neuroscientific measures could be instrumental in providing a comprehensive understanding of skill acquisition in surgical contexts.

Acknowledgement

The work has been carried out at the Fraunhofer Research Institution for Individualized and Cell-Based Medical Engineering IMTE and supervised by the Institute for Electrical Engineering in Medical, Universität zu Lübeck.

Research Funding

This work was partially funded by the European Union - European Regional Development Fund (ERDF), the Federal Government and Land Schleswig Holstein, Project No. 12420002 and No. 12422005.

Authors' Statement

Conflict of interest: Authors state no conflict of interest. Informed consent and ethical approval: Approved by ethics committee university of Luebeck, ref. No. 2023-626.

6 References

[1] Thomaschewski, M., Heldmann, M., Uter, J. C., Varbelow, D., Münte, T. F., & Keck, T. (2020). Changes in attentional resources during the acquisition of Laparoscopic Surgical Skills. BJS Open, 5(2). https://doi.org/10.1093/bjsopen/zraa012

[2] Käthner, I., Wriessnegger, S. C., Müller-Putz, G. R., Kübler, A., & Halder, S. (2014). Effects of mental workload and fatigue on the p300, alpha and theta band power during operation of an ERP (p300) brain–computer interface. Biological Psychology, 102, 118–129. https://doi.org/10.1016/j.biopsycho.2014.07.014

[3] Luck SJ. An Introduction to the Event-Related Potential Technique. Cambridge: MIT Press, (2014)

[4] Luck, S. J. (2022). Applied Event-Related Potential Data Analysis. LibreTexts. https://doi.org/10.18115/D5QG92

[5] Münte, T. F., Joppich, G., Däuper, J., Schrader, C., Dengler, R., & Heldmann, M. (2015). Random number generation and executive functions in parkinson's disease: An event-related brain potential study. Journal of Parkinson's Disease, 5(3), 613–620. https://doi.org/10.3233/jpd-150575

[6] Polich, J. (2007). Updating p300: An integrative theory of P3A and p3b. Clinical Neurophysiology, 118(10), 2128–2148. https://doi.org/10.1016/j.clinph.2007.04.019

[7] Camicioli R, Howieson D, Lehman S, Kaye J. Talking while walking: the effect of a dual task in aging and Alzheimer's disease. Neurology (1997);48:955–958

[8] Laubert T, Esnaashari H, Auerswald P, Höfer A, Thomaschewski M, Bruch HP, Keck T, Benecke C. Conception of the Lübeck Toolbox curriculum for basic minimally invasive surgery skills. Langenbecks Arch Surg. (2018) Mar;403(2):271-278. doi: 10.1007/s00423-017-1642-1. Epub 2017 Dec 1. PMID: 29196840.

Unveiling Cell-to-Cell Evolution by Raman Micro-Spectroscopy

Celina-Christin Schubbe [1], Javier Plou Izquierdo [2], Andreas Seifert [2,3] and Erhardt Barth [4]

[1] Medical Engineering Science, Universität zu Lübeck, celinachristin.schubbe@student.uni-luebeck.de
[2] CIC nanoGUNE BRTA, San Sebastian, Spain, {j.plou, a.seifert}@nanogune.eu
[3] IKERBASQUE, Basque Foundation for Science, Bilbao, Spain
[4] Institute for Neuro- and Bioinformatics, Universität zu Lübeck, erhardt.barth@uni-luebeck.de

Abstract

The investigation of the changes undergone by human cancer cell lines during culture has been traditionally overlooked. However, it is well-documented that such temporal changes in cellular states strongly influence the monitored behaviour which leads to reproducibility problems. These widely accepted limitations have in turn been reinforced by the lack of available techniques to routinely determine the fitness of cells over time. Herein, we aim to address this need by using Raman micro-spectroscopy for single-cell characterisation. In this way, we have developed a streamlined method to examine the Hek 293T cell population by acquiring multiple individual spectra and analysing their evolution throughout their in vitro growth. Our findings offer a rapid alternative to time-consuming sequencing techniques for monitoring cell lineage heterogeneity and evolution over time. Overall, our approach provides cellular information at a spectral level and thus a way to improve the reproducibility of in vitro assays.

1 Introduction

Human cancer cell lines are an indispensable tool in research, serving as a valuable resource for understanding cancer biology and drug testing. Despite their critical role, the evolution of these cell lines during culture and the resulting genetic and phenotypic heterogeneity are often insufficiently addressed [1]. This heterogeneity has significant implications, particularly in the context of drug response and the generalizability of research findings. The specified is considered to be the main cause of the frequent irreproducibility of in vitro assays [2]. However, current methods for monitoring cell changes over time, mostly sequencing-based, tend to be both time-consuming and costly. Said characteristics limit their routine application in favour of other quicker procedures. Those procedures are integral to routine cell culture tasks, such as the daily monitoring of cell density or cytotoxicity.

In this context, our study presents a novel application of Raman spectroscopy to overcome the challenges of cell screening [3], [4]. Renowned for its non-invasive, label-free, and precise molecular characterisation capabilities, Raman spectroscopy offers a rapid and cheaper alternative for monitoring the evolution of cell lines [5], [6]. Standard applications of Raman in cells often involve detailed hyperspectral images with extended acquisition times [7]. Whereas our method is tailored to expedite the collection of representative single-cell spectra from the culture population.

This innovative approach significantly reduces both the time and resources needed to monitor cell line evolution, making it viable for daily use. By swiftly acquiring the molecular fingerprints of individual cells, our method facilitated frequent and efficient tracking of cellular heterogeneity over time [4]. The proposed strategy not only monitors the changes and deviations from the original cell line but also promises to depict the progression of the cells over time.

2 Materials and Methods

2.1 Cell Culture

In this study, Hek 293T human embryonic kidney cells were cultured under controlled conditions. The incubator was set to maintain a constant temperature of $37\,^{\circ}\mathrm{C}$ and CO_2 concentration of $5\,\%$. Cell passaging was performed every 2 - 3 days. The passaging protocol involved rinsing the cells with 1 mL Phosphate-buffered saline (PBS), followed by the application of 1 mL Trypsin for cell detachment. The detached cells were then transferred into fresh culture medium. This medium consisted of Dulbecco's modified Eagle's medium (DMEM), enriched with 10% fetal bovine serum (FBS) and 1x penicillin-streptomycin, to form complete DMEM (cD-MEM).

2.2 Experimental Set-Up and Data Acquisition

Prior to any data acquisition, cells were detached using the aforementioned protocol and then diluted at a ratio of 1:4.

This dilution step was crucial to prevent cell clustering, which could otherwise impair single-cell data collection.

For data acquisition, Raman imaging was performed using a confocal Raman microscope (Renishaw® inVia™ confocal Raman microscope) equipped with a 785 nm laser, operating at an output of 100 mW. The laser was focused through a 50x/L NA microscope objective lens with a grating of 1200 l/mm, following protocols validated in previous studies [8].

For sample preparation, 1.5 - 2 μL of the prepared Hek 293T cells were placed between two Quartzglass Suprasil plates of a demountable cuvette from Hellma® Analytics, with a 0.01 mm light path. This cell-filled chamber was positioned on an aluminium coated microscope glass slide and placed under the Raman instrument, as illustrated in Fig. 1.

Raman mapping was conducted on multiple cells, each measured at a single point. These measurements were carried out with an integration time of 0.01 s and involved 5 - 20 acquisitions per point. Initially, the optical setup of the Raman microscope was critical for identifying individual cells. In the context of Raman mappings, a microscopic image was acquired across a region containing numerous cells, as shown in the left panel of Fig. 2. This setup enabled the selective collection of Raman spectra from specific points or areas on the cells. The resultant spectra are depicted spatially by overlaying square regions to the corresponding individual cells onto the microscopic image, as illustrated in the right panel of Fig. 2. This approach was crucial for obtaining representative spectra required for heterogeneity studies or for creating hyperspectral images.

Data acquisition was conducted over four days, approximately 30 min each day. The spectral range for the Raman spectra was set between 825 cm^{-1} and 1900 cm^{-1}, covering most biologically relevant spectral bands [4]. This range was selected based on insights gained from preliminary measurements.

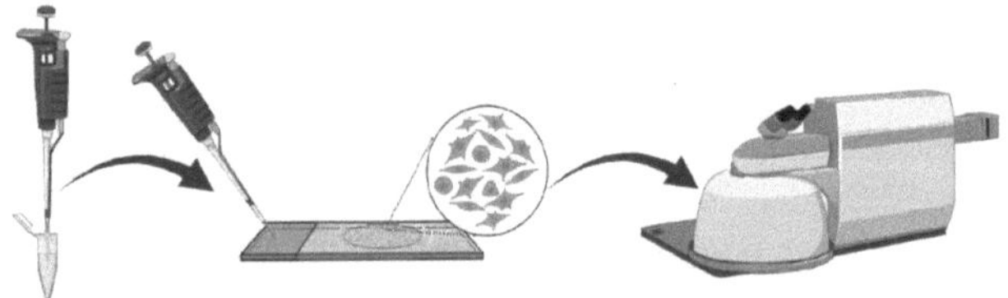

Figure 1: Workflow of the sample setup prior to data acquisition. The prepared Hek 293T cells are extracted from an Eppendorf tube using a pipette and positioned in a closed quartz chamber. Subsequently, the sample is mounted on an aluminium-coated microscope slide and inserted into the confocal Raman microscope, making the sample ready for measurements. (This figure was created with BioRender.com.)

2.3   Data-Preprocessing

The analysis of Raman data in biological samples poses significant challenges attributed to their heterogenous na-

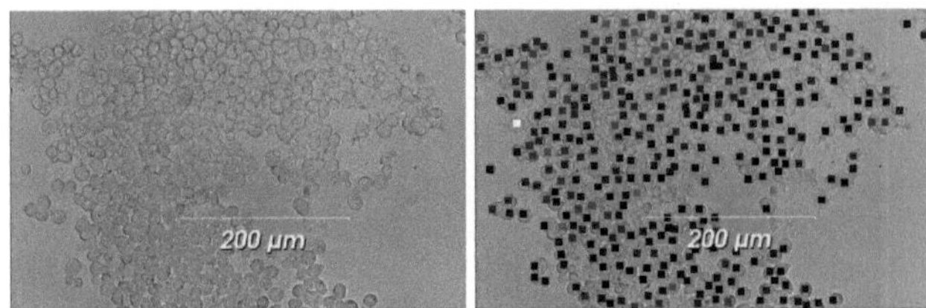

Figure 2: Light-microscopy image of a sample containing Hek 293T cells (left) compared to the light-microscopy image of the sample combined with the spectral Raman measurement of multiple Hek 293T cells (right). The Raman spectra are depicted as squares on the microscopic image.

ture. Undesired contributions arising from various background noise sources may hide weak Raman signals originating from the samples [9]. Therefore, a comprehensive data preprocessing was applied to the acquired raw Raman data using Python [10] to enable a reliable interpretation of the Raman spectra.

The first step involved cosmic ray removal. Subsequently, a 7th order polynomial was utilised for baseline correction, effectively differentiating the Raman signal from the cells. Following this, the baseline-corrected data underwent normalisation to facilitate comparison and visualisation. This normalisation process involved scaling the data to have a mean of 0 and a standard deviation of 1.

For feature extraction, Principal Component Analysis (PCA) was employed using the scikit-learn [11] PCA function. This process extracted main principal components from the preprocessed Raman data. Hereby the dimensionality is effectively reduced and the variance in the data is expressed while retaining the most significant features of the dataset.

3   Results and Discussion

To prove the concept and to reflect standard operating procedures in cell culture laboratories, we cultured Hek 293T cells over a period of one month. Over this time, we systematically collected Raman spectra from a representative sample of cells ($n = 100$) at four different time points. A key component of our experimental setup was a custom-made, closed quartz chamber with a capacity of 2.6 μL. This chamber allowed us to keep the cells viable inside during spectra acquisition. To identify the main Raman peaks associated with cellular components, we began our study by acquiring a hyperspectral image with a resolution of 5 μm of a representative single cell. PCA is applied to the preprocessed Raman data. In Fig. 3 the values of the first principal component (PC1) are represented. There are clear differences between the cell (light colours, low PC1 values) and its surroundings (dark colours, high PC1 values).

This observation led us to hypothesise that a single measurement with lower resolution ($> 20 \mu$m) could serve as a unique identifier for the phenotypic state of individual cells. The advantage of carrying out such measurements is that Raman spectra of several cells can be recorded in a shorter time. For the simple reason that only one spatial point per cell is measured. This finally allowed us to collect a larger number of individual measurements. This step was crucial for the robustness and depth of our subsequent statistical analysis of cell development. For better

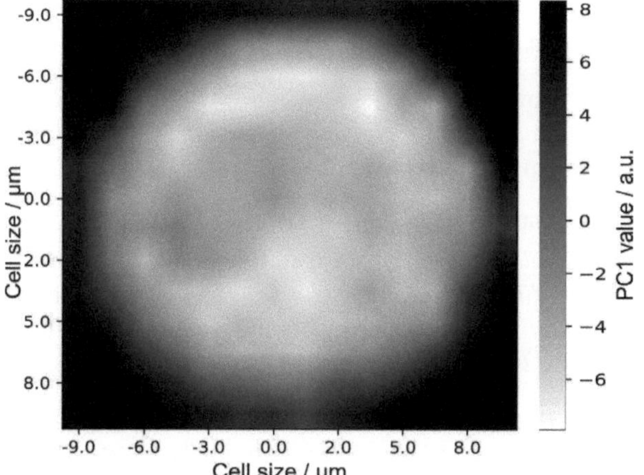

Figure 3: Values of the first principal component across the cell based on its hyperspectral Raman image. The relative x and y coordinates are proportional to the average Hek 293T cell size of $20\,\mu$m.

understanding, Fig. 4 compares the averaged spectra of the single-cell measurements of multiple cells. The spectra are categorised according to the days on which they were taken and shifted on the y-axis to provide a detailed view on each spectrum. The graphs in Fig. 4 show significant differences in the mean Raman spectra between the days. In particular, the spectra of the third and fourth day deviate more strongly from those of the first two days. For example, clearly visible deviations between the spectra from day 1 and day 3 can be seen in Fig. 4 in the peaks around 1300 and 1500 Raman shift / cm^{-1}. These observed deviations in the peaks indicate that changes of the cell states can be recognised over time. To examine and represent the spectral variations and heterogeneity at single-cell level, PCA was performed on the whole Raman dataset. The normalised density distribution of PC1 (80% of variability) across the four measurement days is illustrated in Fig. 5. The density values of the Raman signal highlight the heterogeneity both within and across different days. Particularly the signals of day 3 and day 4 show significant shifts in the distribution of the PC1 score. However, these noticeable shift in the spectral curves suggests a potential evolution of the cell line and significant variations in their Raman profiles. To comprehensively visualise the intercellular changes in cell states over the days, a violin plot in Fig. 6 was selected as another metric. This plot compares the overall distribution of PC1 across the days and accentuates the distinct characteristics of the four separate data groups. The pronounced differences in PC1 distribution suggest remarkable variations in spectral fingerprints or structures within each dataset. It should be emphasised that PC1 captures the most variance

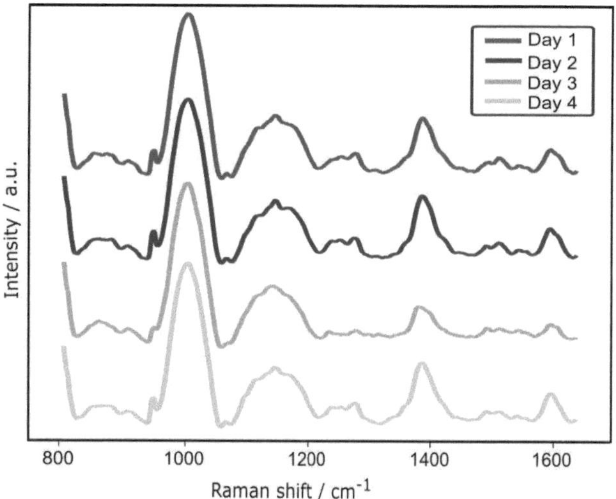

Figure 4: Averaged spectra of the single-cell measurements of multiple cells corresponding to each measurement day.

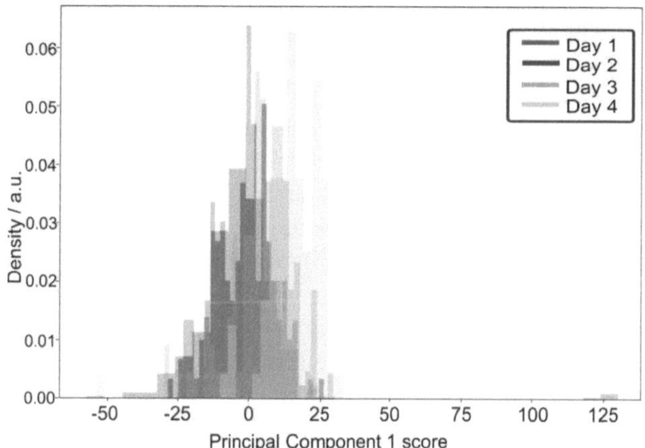

Figure 5: Normalised density distribution of the first principal component from single-cell measurements of multiple cells across the measurement days.

in the data. The distribution of PC1 provides compelling insights into the main sources of variability or patterns in the datasets.

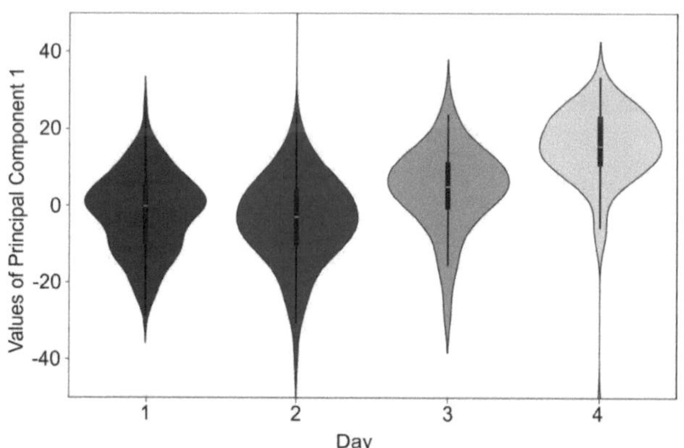

Figure 6: Comparison of the distribution of the first principal component along the single-cell measurements of multiple cells sorted by measurement days.

4 Conclusion

The observations based on our measurements suggest that the approach of obtaining representative single cell spectra from the cell culture population by Raman spectroscopy is a valid method to monitor the development of cell lines. And in like manner to obtain a molecular fingerprint of these cells. In addition, it has been shown that the acquisition of single cell measurements on several cells is a time-saving technique that provides valuable information about the condition of the cells. Moreover, our system, which includes data acquisition, processing and analysis, allows deviations from the original cell line to be detected. At the same time, it allows to monitor the dynamic processes that lead to these changes. Our method has proven effective in revealing the cell-to-cell heterogeneity of the Hek 293T cell line over time and thus the cell-to-cell evolution.

Possible subsequent steps could involve a more in-depth evaluation of our approach through the application of additional metrics to examine the Raman spectra on a more biologically meaningful level. Simultaneously, a comprehensive analysis of more principal components is necessary to uncover the underlying biochemical information they encrypt. Likewise, the application of other machine learning algorithms to the datasets is highly promising for maximising the utility of the available data. Moreover, acquiring more data at longer times is imperative to enhance the robustness and precision of our method. In this context, experiments involving cell media with diverse components could be designed to scrutinise the response of the cell culture under varying conditions. Following this target, cells can be exposed to different stressing conditions, as we partly already did, that force them to evolve and adapt. The uprising evolutions could then be monitored and a comparison between the different causes of stress and their impact on the cell states and thus cell-to-cell evolution could be drawn. Additionally, our approach can be easily extended to different cell lines to ensure its applicability across various cell types, providing further insights into their molecular fingerprints.

Acknowledgement

The work has been carried out at the Nanoengineering Group, CIC nanoGUNE BRTA, San Sebastian and supervised by the Institute for Neuro- and Bioinformatics, Universität zu Lübeck.
Furthermore, we thank Maitane Marquez Lopez for providing the initial Hek 293T cell line.

Authors' Statement

Conflict of interest: Authors state no conflict of interest.

5 References

[1] K. Iuchi, K. Oya, K. Hosoya et al., *Different morphologies of human embryonic kidney 293T cells in various types of culture dishes*. In: Cytotechnology, vol. 72, no. 1, pp. 131-140, 2020.

[2] U. Ben-David, B. Siranosian, G. Ha et al., *Genetic and transcriptional evolution alters cancer cell line drug response*. In: Nature, vol. 560, pp. 325–330, 2018.

[3] Z. Movasaghi, S. Rehman and I. U. Rehman, *Raman spectroscopy of biological tissues*. In: Applied Spectroscopy Reviews, vol. 42 no. 5, pp. 493-541, 2007.

[4] M. Sigle, AK. Rohlfing, M. Kenny et al. *Translating genomic tools to Raman spectroscopy analysis enables high-dimensional tissue characterization on molecular resolution*. In: Nature Communications vol. 14, pp. 5799, 2023.

[5] H. Butler, L. Ashton, B. Bird et al. *Using Raman spectroscopy to characterize biological materials*. In: Nature Protocols, vol. 11, no. 4, pp. 664–687, 2016.

[6] R. Pandey, S. Paidi, T. Valdez et al. *Noninvasive Monitoring of Blood Glucose with Raman Spectroscopy*. In: Accounts of chemical research. vol. 50 no. 2, pp. 264-272, 2017.

[7] C. C. Horgan, M. Jensen, A. Nagelkerke et al., *High-throughput molecular imaging via deep-learning-enabled Raman spectroscopy*. In: Analytical Chemistry, vol. 93, no. 48, pp. 15850–15860, 2021.

[8] C. Matthäus, T. Chernenko, JA. Newmark, CM. Warner and M. Diem, *Label-free detection of mitochondrial distribution in cells by nonresonant Raman microspectroscopy*. In: Biophysical Journal, vol. 93, no. 2, pp. 668-673, 2007.

[9] I. Azkarate-Askatsua *Photonic technology for diagnosis of perinatal asphyxia*, Ph.D. dissertation, University of the Basque Country, San Sebastián / Donostia, Spain, Apr. 2022.

[10] Python Software Foundation, *Python Language Reference*, version 3.10.12. Available: http://www.python.org [last accessed on 2024-02-07].

[11] F. Pedregosa, G. Varoquaux, A. Gramfort et al. *Scikit-learn: Machine Learning in Python*. In: JMLR vol. 12, pp. 2825-2830, 2011.

Development of a Cuvette Prototype for Infrared Temperature Measurement During Active Breathing Gas Conditioning

Joel-Maxim Gillner [1], Dominik Gaus [2], Ludger Tappehorn [2], and Norbert Linz [3]

[1] Medical Engineering Science, Universität zu Lübeck, joelmaxim.gillner@student.uni-luebeck.de
[2] Drägerwerk AG & Co. KGaA, {ludger.tappehorn, dominik.gaus}@draeger.com
[3] Institute of Biomedical Optics, Universität zu Lübeck, norbert.linz@uni-luebeck.de

Abstract

During active respiratory gas conditioning, it is important to humidify and heat the incoming air in order to maintain the self-cleaning function of the respiratory tract. For this purpose, sensors are used in the tubing system to regulate the heating capacity of the humidifier. To improve this and make this temperature measurement contactless, IR temperature measurement is introduced in this work. Therefore a blackbody radiator is integrated into the breathing gas, which enables to transfer in form of its radiation spectrum, information about the gas temperature to the IR sensor. In order to improve this process, four 3D models were constructed using SolidWorks and evaluated for flow dynamics and heat distribution through a simulation program. The simulations were carried out under the worst-case scenario of 4 L/min with dry air. The results show that the black body does not transfer the gas temperature optimally and the best design has a temperature difference of $\Delta T = 3.10\ °C$ to the set value.

1 Introduction

During normal respiration, the upper respiratory tract is responsible, among other functions, for warming and humidifying the inhaled air. If this process does not occur, mucociliary clearance cannot be maintained. Mucociliary clearance refers to the respiratory tract's ability to self-cleanse through the interaction of ciliated epithelium (cilia) and mucus-producing gland cells. The cilia move in a undulating motion, transporting mucus and foreign particles toward the throat, ensuring effective cleansing of the airways. Most foreign particles are thus removed within 24 hours [1].

In the case of intensive care patients, intubation disrupts the physiology of the upper respiratory tract. Therefore, artificial replacement of air warming and humidification is necessary since gases from the central supply have a humidity lower than 0.5 mg/L at ambient temperature. The gas passes through a heated water chamber, absorbing the required heat and moisture (active respiratory gas conditioning). At the patient side the gas shall have a temperature of 37 °C and a relative humidity of 100% (44 mg/L). After leaving the humidifier, the air passes through a heated inspiratory tube to maintain the temperature and prevent condensation. To ensure that the system provides the patient with the correct conditions, the temperature is measured in inspiratory tube using a sensor. If necessary, the heating power of the humidifier is adjusted. Currently available sensors on the market are either inserted or integrated into the tubing system. However,

insertable sensors have the disadvantage that, due to their sensitive electronics, they can only be disinfected instead of sterilized. This may results in cross contamination. Improper attachment can also result in system leaks, leading to inaccurate temperature measurements. Integrated temperature sensors are on a fixed position of the tubing system and are disposed after use, resulting in high waste generation and increased costs [2]. Therefore, this paper introduces a non-contact temperature measurement in the form of infrared temperature measurement to overcome the drawbacks of existing systems.

Every object with a temperature above absolute zero (-273.15 °C) emits electromagnetic radiation at its surface. Some of this radiation is infrared radiation, which can be used for temperature measurement. The infrared radiation range extends from 0.78 μm to 1 mm, with the temperature measurement utilizing the range from 0.7 μm to 12 μm, as longer wavelengths are insufficient for detection. A blackbody can be used to capture this infrared radiation. It is an ideal model that absorbs all radiation without reflection or transmission. According to Kirchhoff's radiation law, for black bodies, the absorption and emission coefficients are equal to one. This means they emit the maximum possible energy at every wavelength. It is important to note that the blackbody is a theoretical model. In practical applications, a grey body is assumed, which can have an emissivity in the range of zero to one (for maximum gray body emission, the goal should be to get as close to one as possible). This work employs a grey body to capture infrared temperature.

2 Material and Methods

2.1 Basic Structure of the System

The general test setup of the system is shown in Fig. 1. A pressure regulator and a TSI mass flow meter are connected to control pressure and flow, followed by the humidifier which heats and humidifies the air. After leaving the water chamber, the air is guided through the heated hose. Behind the hose is a cuvette where the IR temperature measurement takes place (see in Fig. 2). The measuring insert, which absorbs the heat from the gas flow and transfers it to the sensor, is located in the center of the cuvette. The IR sensor used is the MLX90614DCA (Melexis N.V., Belgium). The sensor has an accuracy of $\pm$ 0.5 °C and a resolution of 0.2 °C [3]. Reference sensors (NTC sensors) are located before and after the measuring insert. Since a linear temperature drop is assumed, the temperature measured by the IR sensor should be equal to the mean value of the two reference sensors. An Arduino is used to read the IR sensor and an Almemo is used to read the reference sensors.

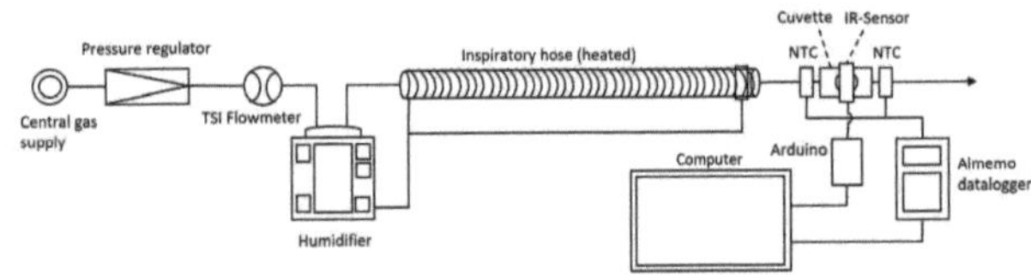

Figure 1: Schematic layout of the test system.

2.2 Cuvette - IR Temperature Measurement

This paper focuses on optimizing the cuvette and the measuring insert. A cross-section and a top view of a prototype is shown in Fig 2. The 3D-models were created using the Computer-Aided Design (CAD) program SolidWorks (Dassault Systèmes SolidWorks Corp., France). Subsequently, Computational Fluid Dynamics (CFD) simulations were conducted for analysis using Ansys (Ansys, Inc., Pennsylvania). The simulated CFD models are based on the 3D prototypes created with SolidWorks. The simulation model represents a complex three-dimensional structure, considering flow conditions under steady-state conditions. The underlying Reynolds Averaged Navier Stokes (RANS) equations served as the basis for a mean-based description of laminar flow. This approach allows for a precise analysis of velocity distributions, pressure profiles, flow patterns, and heat distribution within the model. In invasive mode the tubing system *VentStar Helix heated* is designed to be able to humidify flows ranging from 6 to 60 L/min [4]. In order to provide a buffer for the lowest flow rate, simulations in this paper were conducted with an even lower flow of 4 L/min using dry air. The objective is to enhance the worst-case scenario, as an improvement is anticipated with an increase in flow as well as humid air. To verify this thesis, a further simulation was carried out with a flow of 60 L/min.

2.2.1 Adjustment of Cuvette Geometry

According to current knowledge, restricting the cuvette in the horizontal direction proves to be less effective. Therefore, a cuvette was modelled with a constriction in the vertical plane. The opening was designed in the shape of a stadium geometry, with a side length of 9 mm and a radius of 5 mm. This specific geometry aims to maximize the distance between the measuring insert and the cool outer walls of the cuvette. Additionally, the cuvette's geometry is intended to create a central, large area with maximum heat development. The grey body was designed to be located at the center of this heat source. It consists of an elongated plate, which is intended to be maximally heated by the warm gas flow through a large surface area. This heat is transferred via a 1 mm diameter bridge to an upper plate, where the infrared radiation of the measuring insert is to be captured by an IR sensor. The upper plate has a diameter of 10 mm and is designed to have a small surface area to minimize energy losses.

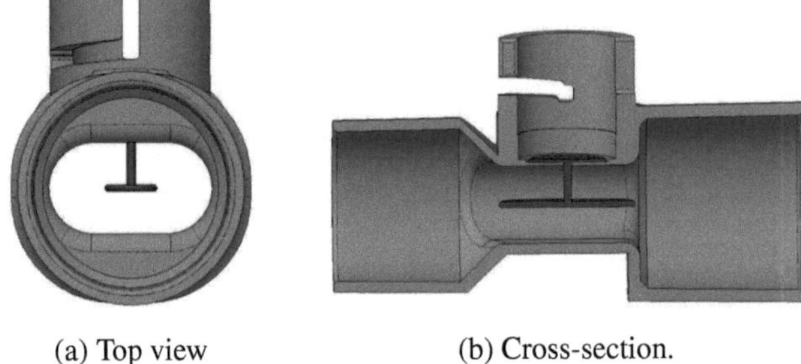

(a) Top view   (b) Cross-section.

Figure 2: Illustration of a prototype designed with Solid-Works: (a) top view of the cuvette, (b) cross-section through the cuvette.

Simulations have demonstrated that the bottom plate of the grey body, which is located in the gas flow is optimally heated and centrally positioned in the warm region of the flow. One issue is that the upper plate transmits a significant amount of heat, because of the colder cuvette material. This causes a decrease in the temperature of the entire component. This would potentially impact the IR temperature measurement. To ensure effective thermal insulation, four variants are discussed below. All concepts were designed to ensure that they could be manufactured using injection molding technology.

The first modification of the cuvette (K1) is illustrated in Fig. 3a. In this variant, the upper plate was integrated into the flow channel with the goal of allowing the plate to absorb additional heat from the airflow. Additionally, an attempt was made to create a minimal contact area of the grey body with the cool material of the cuvette. For this purpose, 0.15 mm of the plate was extended, so that this part is fixed in the cuvette wall. This is intended to secure the body in the vertical direction.

In the second variant (K2), an insulation layer is integrated between the cuvette material and the measuring insert. The cuvette material has a thermal conductivity of $0.23\frac{W}{mK}$, while the insulation layer, made of a thermoplastic elastomer (TPE) with an addition of 17% lime, has a thermal conductivity coefficient of $0.0209 \frac{W}{mK}$. To verify whether this effectively retains heat in the system, the initial analysis was performed using the thermal conductivity coefficient of air ($0.025 \frac{W}{mK}$). The expectation was that the low thermal conductivity coefficient would result in the heat remaining in the measuring insert and not being transferred to the cold cuvette wall by heat transfer. A potential drawback would be the need for a two-component injection molding process, leading to increased costs (more expensive tool, because it has to handle two different materials). In this process, the grey body would be initially encapsulated with the insulation material before the cuvette material encloses it.

A cost-effective variant (K3) was designed as a third option. The measuring insert is positioned on the cuvette material, which has a thickness of 0.4 mm. Similar to the first variant, the measuring insert is fixed in the vertical direction.

The fourth adaptation (K4) is based on the K3 cuvette. By reducing the vertical diameter, the flow velocity in the area of the grey body is to be increased. For this purpose, the diameter was reduced by a factor of 2 from 10 mm to 5 mm. A boundary layer forms in the immediate vicinity of a solid when a fluid flows over it. This boundary layer has a certain thickness that affects heat transfer. As the flow velocity of the fluid is increased, the boundary layer becomes thinner. A thinner boundary layer results in improved heat transfer between the fluid and the surface of the solid because the heat is transferred more efficiently. Therefore, the heat dissipation from the fluid to the body increases and the body is heated more effectively [5].

A short summary of the specific modifications of the four cuvettes is shown in Table 1.

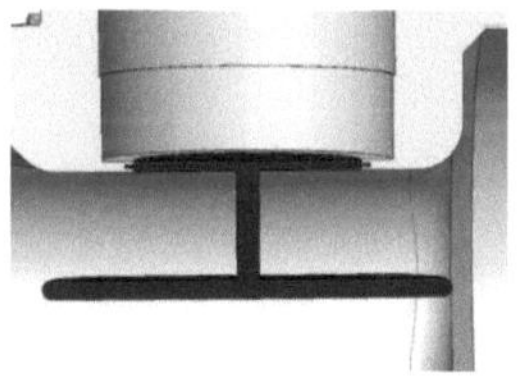

(a) Integration in the flow channel.

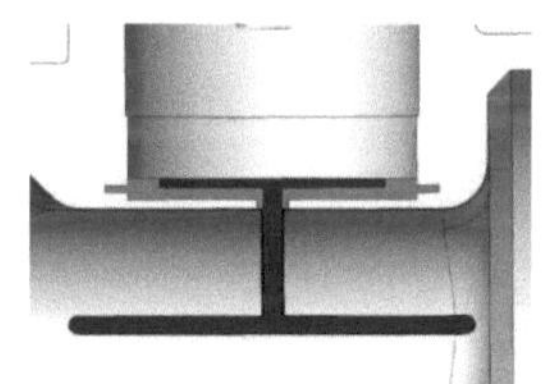

(b) Isolation of the material of the cuvette.

Figure 3: Different adjustments of cuvettes - (a): K1, (b): K2.

	Modification of the cuvettes
K1	Integration of the measuring insert into the flow channel. It is attached to the sides of the cuvette.
K2	Isolation of the grey body.
K3	Grey body placed on the cuvette material with fixation in vertical direction.
K4	Narrowed variant of K3.

Table 1: Different 3D-models of the cuvette.

2.3 Evaluation Method

In order to evaluate the heating of the measuring insert, a temperature curve was generated using Ansys, representing values in relation to the position in the system (solid line). The solid line depicts the temperature variation along the central axis of the system, as shown in Fig. 4. A linear interpolation was carried out for two points positioned 3 cm in front of and behind the midpoint (dotted line), where the reference sensors would be located in a real setup. The mean value was calculated from the interpolation, which is assumed to be the set value for the temperature of the grey body. The simulated temperature of the grey body was taken from the center of the solid line, and subsequently, the difference from the set value (ΔT) was determined and compared.

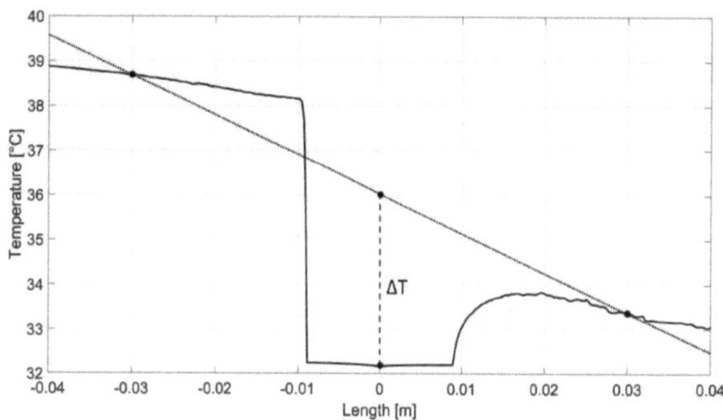

Figure 4: Determination of the temperature difference between the actual and set values.

3 Results and Discussion

3.1 Results

As described in Section 2, the average of the two temperature values, 3 cm before and after the measuring insert, was compared with the actual value of the simulated heating of the measuring insert along the central axis. In this section, the adjustments of K1-K3 are compared, as shown in Fig. 5. It can be observed that all three variants exhibit a similar trend. Once the measuring insert is reached, the temperature decreases. Behind the measuring insert, the temperature rises again before stabilizing at approximately 33 °C. In variant K3, the measuring insert has the lowest temperature among the three variants. While the grey body assumes a temperature of 32.18 °C, it warms up by an additional 0.76 °C in K2, reaching a temperature of 32.94 °C. The temperature difference between the actual and set values is $\Delta T = 3.84\,°C$ for K3 and $\Delta T = 3.18\,°C$ for K1. The grey body in variant K2 has a temperature of 32.44 °C and a temperature difference of $\Delta T = 3.62\,°C$.

Fig. 6 shows a comparison between K3 and the narrowed variant of K3 (K4). The temperature curve has a slightly different shape compared to the previous cuvettes. After the grey body, the curve rises minimally before falling again. The temperature of K4 is 0.87 °C lower than that of the K3

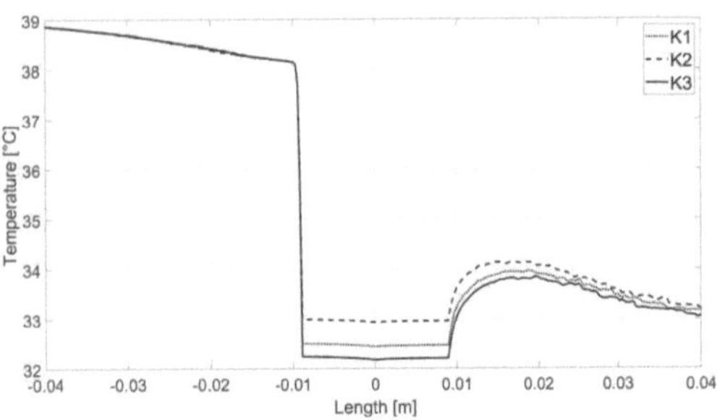

Figure 5: Temperature profiles of the different variants along the central axis (K1, K2, K3).

variant, as measured 3 cm behind the measuring insert. The measuring insert has a temperature of 32.84 °C, which is 0.67 °C higher than in K3. The temperature difference from the set value is $\Delta T = 3.13\,°C$, which is lower than the ΔT observed in cuvette K2. The upper temperature curve was simulated as a comparison with the K3 cuvette at a flow rate of 60 L/min. It can be seen that the curve has only a small temperature drop compared to the other variants. The grey body heats up to 37.8 °C and has a temperature difference to the set point of $\Delta T = 1.04\,°C$. In addition, the grey body heats up 5.61 °C more compared to a flow rate of 4 L/min. After leaving the grey body, the temperature rises as in the other examples, but does not lose as much heat and settles at about 38.6 °C.

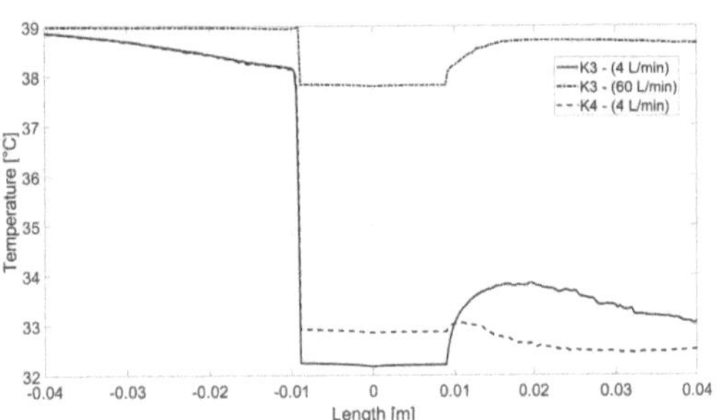

Figure 6: Temperature profiles of the different variants along the central axis (K3, K4). The upper temperature line was simulated at 60 L/min, the two lower ones at 4 L/min.

3.2 Discussion

The temperature difference between the best variant, where the measuring insert was insulated (K2), and the worst variant, where the grey body was placed on the cuvette (K3), was 19.87 %. In comparison, an even better result was obtained with K4 (narrowing of K3), which was 0.05 °C closer to the expected temperature. It should be discussed whether the result with the cheapest variant is sufficient for an accurate temperature measurement, or further simulations should be carried out with the insulated cuvette. A possible approach would be to reduce the diameter. Furthermore, a comparison of K3 and K4 shows that the temperature difference to the set value is reduced when the diameter of the cuvette is reduced. This result confirms our

initial assumption that a smaller diameter results in a higher flow velocity, which means that the grey body heats up better, shown in Fig. 6. By increasing the flow, the temperature difference to the set point is minimized to 1.04 °C. In addition, more heat remains in the system. The results show that at a distance of 3 cm from the center of the grey body, there is a temperature difference of 0.87 °C between K4 and K3. Due to the lower temperature, attention should be paid to the formation of condensation when performing the experiment in reality. Overall, it can be said that the modelled cuvettes deviate from the set temperature by at least 3.13 °C. The deviation could be explained by the large surface area of the cuvette, which causes the entire system to lose heat.

4 Conclusion

This study presented different prototypes of cuvettes for IR temperature measurement during active humidification. Therefore, four variants were modelled using SolidWorks and then simulated with a flow of 4 L/min and dry air. Overall, the results show that the vertically narrowed cuvette (K4) exhibited the smallest temperature difference from the set value, with $\Delta T = 3.13\,°C$. The next step is to compare the simulation results with the actual outcomes by testing the system in a real environment with the 3D-printed cuvettes.

Acknowledgement

The work has been carried out at Drägerwerk AG & Co. KGaA and supervised by Dr. rer. nat. N. Linz, Institute of Biomedical Optics, Universität zu Lübeck.

Authors' Statement

Conflict of interest: Authors state no conflict of interest.

5 References

[1] H. Lang, *Beatmung für Einsteiger*. Springer-Verlag, vol. 3, 2019.

[2] K. Kutschina, *Untersuchung einer berührungslosen Atemgastemperaturmessung mittels Infrarot als Eingangsgröße für die Regelung der Temperatur im Zusammenhang mit der Atemgaskonditionierung für die Langzeitbeatmung*. Diploma thesis, 2012.

[3] Melexis N.V., *MLX90614 family - Datasheet Single and Dual Zone - Infa Red Thermometer in TO-39*. Addison-Wesley, Harlow, 1999.

[4] Drägerwerk AG & Co. KGaA, *Instruction for use - VenStar Helix heated*, 2021.

[5] H. Schlichting, K. Gersten, *Grenzschicht-Theorie*. Springer Berlin, vol. 10, 2005.

3

Robotics

Latency Assessment of Console Input for a Virtualized Robotic Surgical Simulation

Gustav Krumnow [1], Tilo Jonathan Wöbber [2], Cem Adiyaman [2], Jannis Hagenah [2] and Georg Männel [2]

[1] Medical Engineering Science, Universität zu Lübeck, gustav.krumnow@student.uni-luebeck.de [2] Fraunhofer Research Institution for Individualized and Cell-Based Medical Engineering, Lübeck, {tilo.jonathan.woebber, cem.adiyaman, jannis.hagenah, georg.maennel}@imte.fraunhofer.de

Abstract

This study evaluates latency of an innovative surgical simulator which is designed to elevate surgeon training by replicating the entire operating theatre, including the surgical robot. Our focus centers on innovatively integrating the real robot's user console as an input device to enhance realism, thereby optimizing the efficacy of surgical training. The evaluation of system usability, conducted through latency measurements, reveals minimal computer-based simulation latency. However, the primary contributor to latency emerges from the interface-to-computer transfer. Despite current latency levels being inadequate for effective training, this paper identifies and discusses the challenges at hand, setting the stage for future research and improvements in latency. The outlined findings pave the way for a promising avenue in achieving realistic and efficient surgeon training for robotic-assisted procedures.

1 Introduction

The integration of robotic systems into modern surgical interventions has become a sector of rising importance necessitating specialized training paradigms for surgeons [1]. Simulators in robotic surgery serve as a crucial tool in this regard, providing a safe and controlled environment for surgeons to improve their skills without the risk of patient harm [2]. Departing from conventional approaches, this paper contributes towards a novel simulator approach that aims to virtualize the entire operating theatre. It incorporates not only the surgeon's view but also the complete surgical environment, processes, and personnel. This approach provides an immersive, realistic simulation setup without the need for real patients, phantoms, or the robot itself. However, The current simulator as illustrated in Fig. 1 operates similarly to existing robotic surgery simulators, using only the robot's instruments and performing tasks on a virtualized Lübecker Toolbox [3], which is a basic training concept consisting of a box with integrated camera and several tasks designed to practice and improve laparoscopic capabilities. The user's view is limited to the situs. A significant limitation of the current simulator is the lack of realism in the input tool which is a gamepad or the computer keyboard in present. These tools differ significantly in their handling from the actual console the surgeon has to operate with and therefore do not represent adequate input tools for training. Addressing this issue, the focus of this paper is to enhance the realism of the simulation by using the user console of the Dexter Surgical Robot, designed by Distalmotion [4], to control the virtual robot in three-dimensional space. A challenge in providing a cohesive experience is the delay between input and reaction of the virtual environment. Therefore, the latency within the implementation is evaluated. This work aims to contribute to the field of surgical training, especially towards more effective and efficient learning experiences for surgeons in the realm of robotic surgery. The content of the paper includes an introduction to the use of the user console as an input for controlling the simulation, a description of the technical realization, a demonstration of a proof of principle and a latency analysis.

2 Material and Methods

This study utilized the Blender Game Engine (BGE) [5] for the simulation of the robot. BGE is a powerful tool for creating interactive 3D applications, including simulations.

The choice of BGE for this simulation offers several advantages. Firstly, BGE is open-source and free to use, which makes it accessible for researchers and developers. Secondly, it provides a Python API, allowing for extensive customization and control over the simulation. This feature is particularly beneficial for complex simulations like the task at hand, where fine-tuning and adjustments, for example in positioning and behaviour of the virtual robot are required. The robot in the simulation was modeled using CAD data, ensuring a high degree of detail and accuracy.

The active component of the Dexter comprises four links, each of which is rotated by a distinct motor. These motors, referred to as the alpha-, beta-, gamma-, and theta-motors, respectively, function as joints.

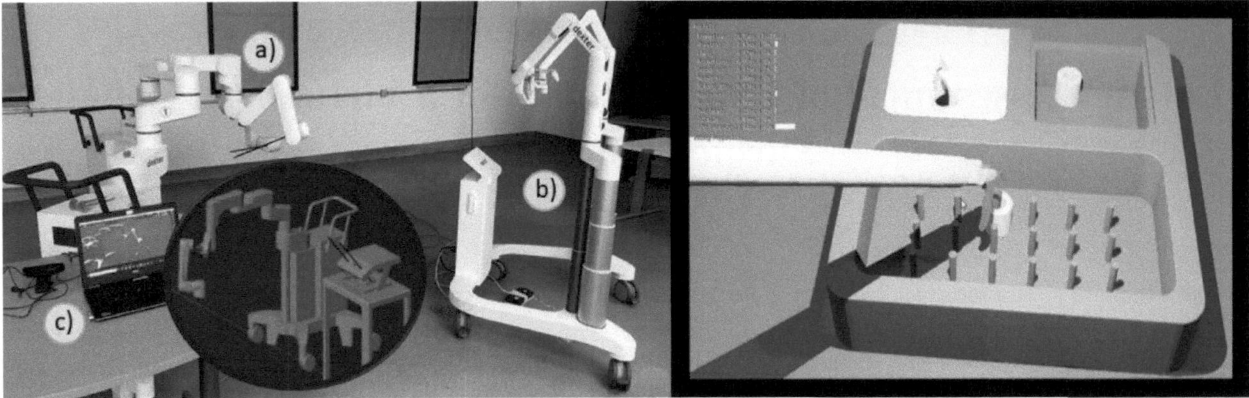

Figure 1: Left: The experimental setup of this work consisting of: a) The Dexter surgical robot utilized for this work. b) The user console, used as input device for the simulation. c) The laptop running the TCP/IP-client and the BGE/simulation. It is connected to the interface of the Dexter through ethernet cable. Right: a snapshot of the existing simulator, showcasing the surgical situs.

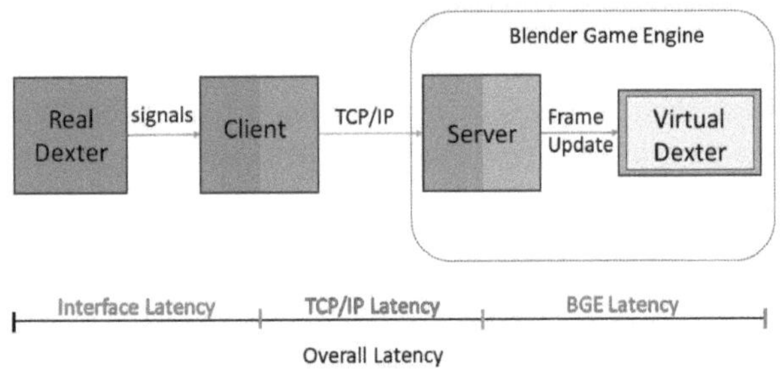

Figure 2: Schematic representation of the respective data transfer pipeline and the latency types to be measured. The overall latency consists of three parts, specifically the latency caused by the Dexters interface, the TCP/IP connection induced latency and BGE-related latency.

2.1 Experimental Setup

The experimental arrangement, as illustrated in Fig. 1, incorporates the Dexter Surgical Robot and its console. The Dexter robot is linked to a laptop via an ethernet cable, enabling real-time logging of various signals, including the robot's current angle values. These signals are transmitted to BGE via TCP/IP using a Python TCP/IP client implemented on the laptop. The Python API of BGE is used to receive these signals and apply them to the virtual model. Fig. 2 provides a schematic representation of the setup components along with their respective latencies.

2.2 TCP/IP Latency Measurement

To measure TCP/IP latency, signals from the robot, received at the client-side, are forwarded to the server. This server is implemented using BGE's Python API. Each signal is accompanied by a timestamp, denoted as the start time, which marks when it was sent. This process is not tied to a specific motion, as the goal is to replicate a package format similar to that used in the simulation, thereby creating a comparable situation for evaluation.

Upon receipt at the server-side, the timestamp is recorded as the end time. The latency for each transfer is then calculated by subtracting the start time from the end time for each of the 200 signals. This process is repeated ten times, with the mean latency calculated for each repetition. This method provides a comprehensive measure of the TCP/IP-connection's latency.

2.3 BGE Latency Measurement

The latency of BGE is measured using a log file that records an upward movement of the robotic arm. This log file encompasses 1329 signals in total, where each signal represents the angular value corresponding to the current position of the alpha-, beta-, gamma-, or theta-motor. Prior to applying the log file, it was downsampled to the aforementioned size. This downsampling was necessary because the signal sampling rate on the client side significantly exceeds the update rate of BGE, which stands at 60 Hz. For each call to update in the Python component, which manages changes in the simulation scene, a timestamp is recorded before the next signal from the log file is retrieved. This timestamp, referred to as the current timestamp, is used to calculate the latency of the last processing step. The current timestamp is also stored as the previous timestamp for the next call of the update function. The latency is then calculated by subtracting the current timestamp from the previous timestamp, yielding the latency for the last frame update.

This method was chosen to calculate the delay caused by an entire update cycle, not just by processing the signal. This is because a frame update in BGE consists of several phases, namely, the logic phase, physics phase, animation phase and rendering phase, each of which contributes differently to the delay caused by a frame update.

These calculations are performed for every signal in the log file, after which the mean latency is calculated. This entire process is repeated ten times to ensure accuracy and consistency in the measurements.

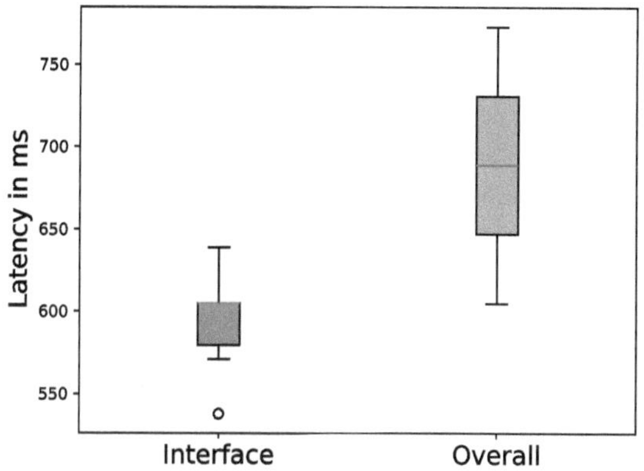

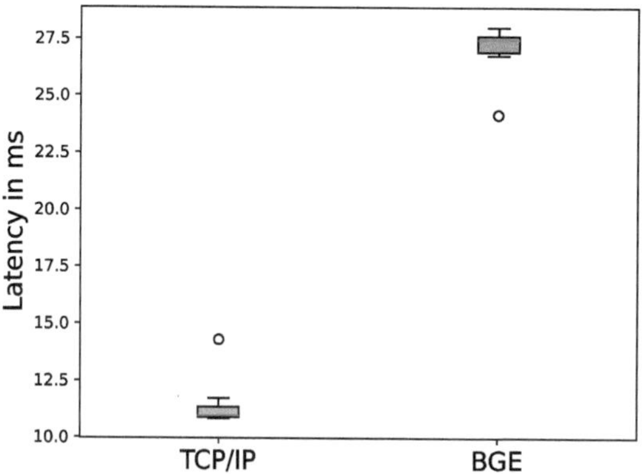

Figure 3: Interface latency and overall latency. Both determined by counting frames method explaining the broader interquartile range.

Figure 4: Latencies caused by the TCP/IP-connection and BGE. Both yield values within an acceptable range and a narrow interquartile range.

2.4 Overall and Interface Latency Measurement

The determination of the overall latency and the latency induced by the Dexter's interface involves capturing a video that simultaneously displays the upward movement of the real Dexter, the movement of the virtual Dexter, and a command prompt window. In the command prompt, a signal is printed as soon as the received signals on the client-side change by a certain value, indicating a one-millimeter arm movement of the real Dexter. This value was selected to surpass the noise threshold, which could result in a random print, and to link it to a visible motion step in the video. The video is recorded using a webcam, positioned in such a way that it simultaneously captures both the actual Dexter and the laptop screen, which displays the simulation and the command prompt.

The video is then divided into its constituent frames. The frames are counted from the start of the real Dexter's motion to the print in the terminal and to the start of the virtual motion, respectively. The latency of the interface and the overall latency are obtained by subtracting the number of frames at the real start from the number at the print and the number at the virtual motion, and then dividing this difference by the video's framerate.

This process is repeated ten times to ensure accuracy. The motion of the real Dexter is assumed to be the ground truth, as recognizing the start of the user motion in the console may not be as reliable, and the delay between console movement and the respective motion of the Dexter is minimal.

3 Results and Discussion

The performance of the surgical simulator, using the real user console of the robot, was evaluated by measuring the latencies of different components of the system, as shown in Fig. 3 and Fig. 4. The mean and standard deviation of

Table 1: Mean values and standard deviations for each measured latency type.

Type	Mean Latency in ms	Standard Deviation in ms
TCP/IP	11	1
BGE	27	1
Interface	600	29
Overall	686	55

the latency were calculated for each component based on ten repetitions, as can be seen in Table 1.

In this study, the TCP/IP component exhibited the lowest latency, with an average of 11 ms and a standard deviation of 1 ms. The BGE component had a slightly higher latency, averaging 27 ms with a standard deviation of 1 ms. Both components demonstrated acceptable and consistent performance for training purposes, as indicated by a narrow interquartile range (IQR). This suggests that the computational performance and the software used are capable of running the type of simulation under development.

However, the interface component had a significant impact on the overall system latency, with an average latency of 600 ms and a standard deviation of 29 ms. This led to an overall system latency of 686 ms, with a standard deviation of 55 ms. Despite the high delay observed between the actual motion and the reaction in terms of the change in signal values transferred by the interface, the client-side signal arrival rate remained fairly constant at 25 kHz. The broader IQR for the interface and overall latency measurements could be attributed to the discrete nature of the measurement method, which involved counting frames of a video with a framerate of 29.75 frames per second. This framerate could also vary for each recorded video. The overall latency could be influenced by the BGE and TCP/IP components. Moreover, as only 10 trials were conducted for the interface and overall latency measurements, outliers have a more pronounced impact on the observed signal variability. It is worth noting that in the TCP/IP and BGE measurements, the number of processed signals was significantly less than the number of signals transferred in the interface and overall latency measurements, which could also contribute to the small IQR observed in Fig. 4. In contrast, the BGE and TCP/IP latencies were measured by comparing timestamps, providing a more continuous and accurate method. The overall latency exceeded the threshold suitable for surgical training, suggesting that future efforts should be directed towards minimizing interface latency to enhance the user experience during training simulations.

One potential improvement could involve changing the programming language of the TCP/IP client, as Python is deemed relatively inefficient for real-time applications [6]. While the entire simulation is based on Python, this is particularly important on the client side because the signals arrive at a much higher rate of 25 kHz compared to the 60 Hz update rate of BGE, which does not necessarily require real-time capabilities. Reducing interface latency could make the simulator a more practical tool for surgical training, aligning with the goal of optimizing virtualized surgical environments for realism and effectiveness.

According to Xu's [7] assessment of latency effects on surgical performance, surgical performance deteriorates exponentially as latency increases. There are mild effects at 0–200 ms, increasing effects at 300–700 ms, and very large effects at 800–1000 ms. Latencies under 300 ms are suitable for telesurgery, 400–500 ms may be tolerable but tiring, and 600–700 ms are challenging and only acceptable for low-risk and simple procedures. Thus, the current system's overall latency, which averages 686 ms, may pose challenges for real training applications. This underscores the need for further optimization, particularly in reducing interface latency. While the methods for measuring TCP/IP and BGE latency were quite accurate, the methods used for measuring the overall latency and the interface latency were not very precise. This is because the time resolution of a video is not as accurate as timestamps, and only the start of real and virtual motion were compared, assuming that this delay is constant over time. Since there might be a nonlinear connection between both motions, the methods have to be adapted when reevaluating an improved version of the system. Also, factors like speed and complexity of the motion were not considered, which might also affect the BGE latency. However, for the observed delays and a preliminary evaluation of the proposed simulator, the used methods were accurate enough. They highlight the need for reducing the interface latency and serve as proof of the feasibility of using the real user console as an input device for a more immersive and realistic simulation.

4 Conclusion

This study aimed to assess the feasibility of a novel virtualized robotic surgical environment, with a special focus on integrating the real console as an input device. The latencies of different components of the system were measured and compared. The results showed that the TCP/IP communication and the BGE had acceptable latency, but the interface caused a significant delay that resulted in an overall latency unsuitable for effective training applications. The findings highlight the need to reduce interface latency to improve the user experience during training simulations. Furthermore, this research offers valuable insights for the development of realistic and responsive virtualized surgical training platforms. In line with the broader goals of advancing surgical robotics and improving training methods, the study lays the groundwork for future improvements, aiming to create more effective and immersive simulated surgical scenarios.

Acknowledgement

The work has been carried out at the Fraunhofer Research Institution for Individualized and Cell-Based Medical Engineering, Lübeck and supervised by P. Rostalski, Institute of Electrical Engineering in Medicine, Universität zu Lübeck.

Authors' Statement

Research funding: This work was partially funded by the European Union - European Regional Development Fund (ERDF), the Federal Government and Land Schleswig Holstein, Project No. 12420002 and No. 12422005. Conflict of interest: Authors state no conflict of interest.

References

[1] J. D. Bric, D. C Lumbard, M. J. Frelich, J. C Gould, *Current state of virtual reality simulation in robotic surgery training: a review. Surgical Endoscopy.* Surgical Endoscopy, vol. 30, no. 6, pp. 2169–78, 2016.

[2] M. W. Schmidt et al., *Virtual reality simulation in robot-assisted surgery: meta-analysis of skill transfer and predictability of skill*, BJS Open, vol. 5, no. 2, March 2021.

[3] T. Laubert et al., *Conception of the Lübeck Toolbox curriculum for basic minimally invasive surgery skills.* Langenbecks Arch Surg, vol. 403, pp. 271–278, 2018.

[4] Distalmotion. Dexter. Available: https://distalmotion.com/dexter/ [last accessed on 2024-01-19]

[5] UPBGE, Available: https://upbge.org [last accessed on 2024-01-22]

[6] Md. G. Rashed and R. Ahsan. "Python in computational science: applications and possibilities." International Journal of Computer Applications, vol. 46, no. 20, pp. 26-30 ,2012.

[7] S.Xu, M. Perez, K. Yang, *Determination of the latency effects on surgical performance and the acceptable latency levels in telesurgery using the dV-Trainer® simulator.* Surgical Endoscopy, vol. 28, pp. 2569–2576, 2014.

Mobile Robots for Machine Maintenance and Tending

Muhammad Junaid* [1], Jochen Lindermayr [2]

[1] Robotics and Autonomous Systems, Universität zu Lübeck, muhammad.junaid@student.uni-luebeck.de

[2] Department of Robot and Assistive Systems, Fraunhofer IPA, jochen.lindermayr@ipa.fraunhofer.de

Abstract

Mobile manipulators are used in industrial settings for mobile tending and maintenance tasks ensuring smooth processes and efficiency. Especially in industries with mass production where productivity is a core factor. With the increase in intelligence and flexibility of industrial robots, the implementation of industrial processes gets more complex and costly. This paper leverages behavior trees, which avoids this complexity in behavior design and allows us to design simple yet reusable nodes to design complex behaviors. Behavior trees are used to propose a complete pipeline for a mobile manipulator. The pipeline involves the detection and localization of an object in the workspace that needs processing. Then perform pick and place operation on the workpiece in simulation at the Fraunhofer IPA lab environment.

1 Introduction

In dynamic manufacturing industrial environments, the integration of robots is gaining popularity because of their prominent increase in efficiency, accuracy, and productivity. One of many applications of robotics in industries is mobile tending and machine maintenance. This study aims to shed some light on the design and practical applications of mobile manipulators in a simulation environment and to present the possible challenges encountered during the process. Robot mobile tending involves the deployment of a mobile manipulator equipped with modern and precise perception equipment to perform a variety of tasks including material handling, machine operation and even collaborating with workers in the industrial processes. Unlike the traditional static robotic systems deployed in industries, Mobile tending involves mobile robots along with robotic manipulators and end effectors with custom-designed tools for several applications. Due to the mobility of this robotic platform, the range of possibilities is huge [1].

In this work, a Mobile tending task pipeline is proposed for a pick and place task of a workpiece that requires further processing in the machine. We designed a Gazebo scenario in our lab environment. The mobile manipulator has to navigate autonomously to the table where the workpiece is placed to be processed in the machine. The robot has to detect the workpiece based on its shape and pick it up from the table in our lab. The robot has to bring this workpiece to the machine and place it inside the machine for further processing. The Robot operating system (ROS) framework was used for this project, which gave an additional advantage of using the wide variety of integrated tools provided by ROS, including the ROS navigation stack for navigation and Moveit for manipulating the mobile manipulator. To design the complex behavior of mobile manipulator behav-

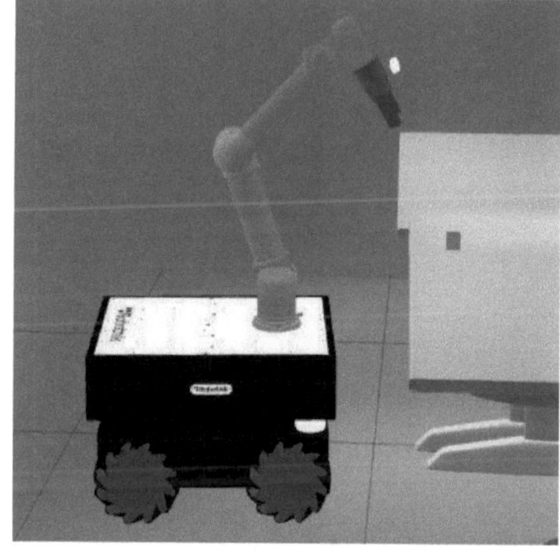

Figure 1: Mobile manipulator RB-KAIROS [2] in our simulation environment.

ior trees[3] is employed, which allows us to design complex behaviors using simple yet reusable nodes in the behavior trees.

2 Materials and Methods

The objective of this project was to design an algorithm to demonstrate the usability and effectiveness of using behavior trees for mobile manipulators to perform mobile tending tasks. To perform this task a scenario was designed in the Gazebo simulator similar to the real Fraunhofer IPA lab. The pipeline is designed to be implemented on the real robot after it is tested on the simulated robot. To perform this simulation the RB-KAIROS[2] robot was employed.

2.1 Mobile Manipulators

In robotics, mobile robots are used to allow mobility to the robots. In industrial settings where mobile tending is required, robot manipulation capabilities are a core need for its effective performance. Mobile manipulators are robotic systems that combine the best of both worlds, with a mobile robot base, a robot manipulator, and a gripper or an end effector. The kinematics of the mobile robot base may vary, however in industrial scenarios differential drive or omnidirectional kinematics is preferred given their wide opensource support and ease of mobility. Both have their pros and cons, hence kinematics are decided based on the respective scenario for which the robot is being designed. Robot arms or manipulators along with grippers provide robots the ability to interact with the environment. for example, picking and placing objects. The grippers can be designed to interact with the environment based on specific applications [4]. This combination of a mobile robot and a robotic arm makes a robot that allows diverse applications.

2.2 Robot Selection

To demonstrate the working of the mobile manipulator in the Gazebo simulator the official repository of Robotnik's RB-KAIROS robot [2] is employed. RB-KAIROS is a collaborative mobile manipulator designed specifically for industrial applications by Robotnik. The robot in simulation is equipped with two 2D lidar sensors - one on the front and one for the back - which ensures very high-quality occupancy grid maps when laser-based simultaneous localization and mapping (laser-based SLAM) is performed using the gmapping package [5] in the ROS Navigation stack. The robot is also equipped with two RGBD cameras. One of the cameras is attached to the front of the mobile robot base while the other is attached to the gripper to ensure visibility while performing the machine tending task.

The mobile robot base of the RB-KAIROS mobile manipulator is based on a mobile robot with omni-directional kinematics. To perform manipulation tasks, the robot is equipped with a Universal Robotics robotic arm. The robot is selected for simulation particularly because it almost mimics the design of the real robot manipulator in the lab. The simulated pipeline is designed basically for demonstration on the real mobile manipulator robot in lab [6]. The real robot manipulator is based on DeKonBot [4].

2.3 Behavior Trees

Graphical methods are usually preferred for designing behaviors in robotics. State machines are one of the traditional methods to simply design a particular behavior of a system. However, the state machines easily get very complex as the number of states increases in behavior design. Here come behavior trees, which have become recently popular as behavior designing tools replacing the traditional methods like state machines in robotics [3]. The library used to design behavior trees is *BehaviourTree.CPP*. The reason

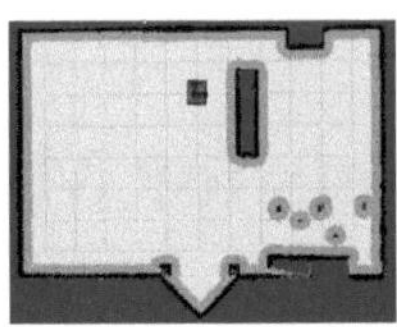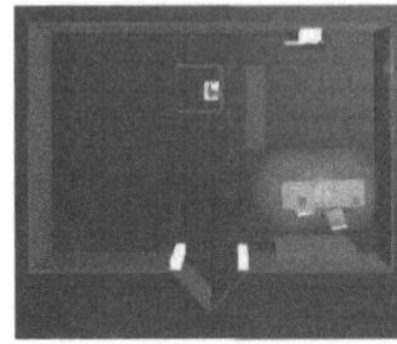

Figure 2: Left: Costmap of the scenario from Rviz. Right: Corresponding gazebo simulation environment

being their simplified hierarchical structure, high-level abstraction, reusability, and modularity. The core language to program behavior trees in the used *BehaviourTree.CPP* library is C++. So, we have all the powers of C++ along with ease in scripting the trees in XML language. The high reusability and modularity ensure that users to design very complex behaviors by reusing simpler components. Unlike state machines where we usually transition between different states, in behavior trees we focus on executing the predefined actions. Due to the high level of abstraction, a behavior designer does not need to know the fundamental workings of the algorithms.

2.4 Methodology

To simulate the experiment. we designed a simulation environment where the workpiece (in this case a box) was placed on the table and that required further processing in the machine. The task of the mobile manipulator is to navigate itself in front of the table, detect and localize the object based on shape, and plan a trajectory to pick the workpiece using Moveit. Then it has to navigate itself to the machine in which the workpiece has to be placed for further processing. Fig. (2) shows the cost map of the scenario generated by the ROS navigation stack along with the gazebo simulation environment of the scenario.

Fig. (3) shows an overview of the pipeline designed using behavior trees, where each step represents a subtree that consists of predefined functionality for that task. The following are the details of the components:

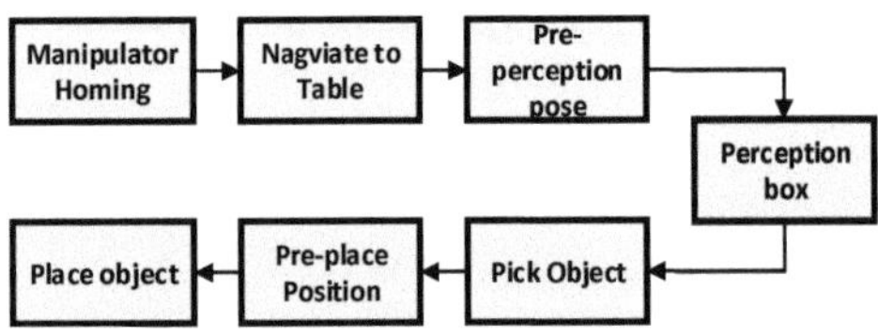

Figure 3: An overview of the pipeline in behavior trees.

- **Manipulator Homing:** This is the first step in the process. It consists of a series of behavior tree nodes using which the robot manipulator moves to a predefined home position using Moveit.

- **Navigate to Table:** After homing of the manipulator, the mobile robot navigates itself to the table, where the workpiece is placed using the ROS navigation stack according to the predefined goal provided to *move_base* node which is an implementation of a ROS action server. Once the robot reaches the goal we get a success message and we move to the next subtree.

- **Pre-perception pose:** Once the robot has reached the table, the robot manipulator has to move to a position so that the workpiece is easily visible using the arm-mounted RGBD camera's point cloud. This predefined manipulator position is achieved by planning trajectories using Moveit.

- **Perception box:** This is the subtree that contains the behavior tree nodes that detect and localize the object. It uses an object detection algorithm [7] that uses the point cloud information from the RGBD camera to detect, segment, and localize the object in the space relative to the robot base frame. This object detection algorithm uses the PCL library for manipulating point clouds. It is capable of detecting multiple objects of more than nine shapes and then separating the objects from the surfaces using a segmentation algorithm. Once the object and surfaces are detected and segmented, the algorithm publishes detailed information about the segmented point clouds concerning candidate object and surfaces, their location, and dimensions in a ROS message. The pose information of this detected object is set as an output of the behavior tree node, which is then used by the next sub-tree.

- **Pick Object:** Once the object is correctly localized, we have got the pose information about the object in space relative to the mobile robot base frame. We will use this pose information to create collision objects in the planning scene interface for the workpiece and table. When we want to pick the object we allow the collision of the end effector and the workpiece. Then we will use the pose information of the workpiece to plan a trajectory to pick the workpiece using Moveit. In the end, the object is picked.

- **Pre-place position:** In this subtree, the robot will navigate itself in front of the machine and adjust the pose of the robot manipulator to detect and localize the marker.

- **Place Object:** Finally, in this subtree the robot uses the Aruco marker detection algorithm [8] to detect and localize the Aruco marker. The spatial information of the Aruco marker helps the robot predict the position in the machine to place the object. Therefore, we finally use Moveit to place to object relative to the Aruco marker pose information and finally the object is placed inside the machine.

3 Results and Discussion

An experimental setup was created to simulate the robot in Gazebo and Rviz. The experiment was conducted 20

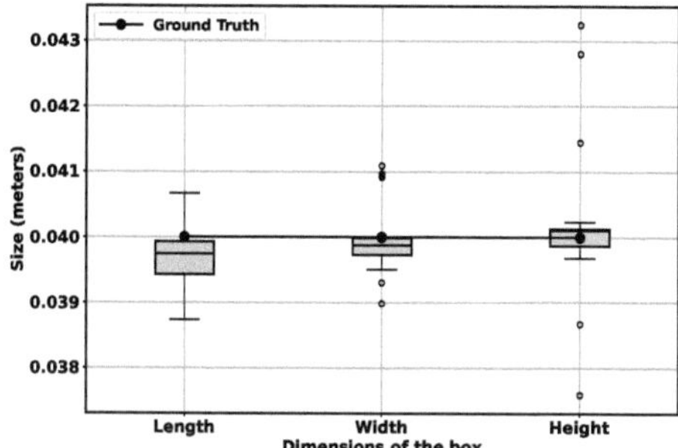

Figure 4: Estimated dimensions of the workpiece using the object detection algorithm [7].

times and important information was extracted. The statistical analysis of these experimental results revealed some interesting conclusions.

Firstly, the designed behavior tree used only six core behavior tree nodes which were repeated in the subtrees to design a complex behavior. This clearly shows how useful the behavior trees are for designing complex behaviors using small yet reusable nodes. The following table 1 shows the average time taken by each subtree during all of these 20 experiments.

Table 1: Mean and Median time taken by each subtree

Subtree Name	Mean (secs)	Median (secs)
Manipulator Homing	13.214	13.212
Navigate to Table	10.575	10.55
Pre-perception pose	19.022	18.839
Perception box	0.392	0.299
Pick Object	45.175	44.950
Pre-place position	13.137	13.133
Place Object	43.211	42.528

There are several anomalies in the time data of the *perception_box* subtree. In experiment numbers 16 and 20 the time taken by this subtree was 0.859 s and 1.033 s respectively. The reason for this anomaly is due to an issue with the perception algorithm itself as it does not detect an object right away once in a while, on the initial tick from the behavior tree. To mitigate this issue, the algorithm is designed in such a way that the behavior tree sends ticks to the node up to three times upon failure of the node which solves the issue. Hence, no failure is observed in the *perception_box* subtree in all 20 experiments.

Fig. 4 shows the box plot obtained using the dimensions of the detected workpiece box during the 20 experiments that used the detection algorithm [7]. From the results shown in Fig. 4, it can be seen that the object detection algorithm was able to detect the object correctly and the approximated size of these objects obtained is very near the ground truth. The median for the length, width, and height is 0.03974m, 0.03987m, and 0.0401m respectively. This certainly proves the high precision of the algorithm.

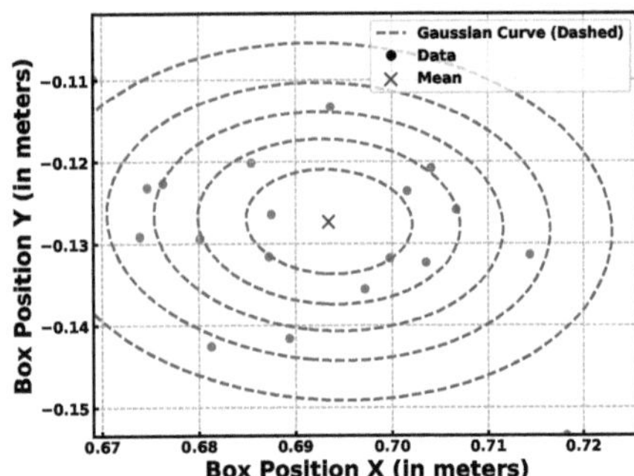

Figure 5: Scatter plot of the detected x and y axis position of the detected workpiece in robot base frame.

Lastly, Fig. 5 and Fig. 6 are the scatter plots obtained from the pose information obtained using the object detection and localization algorithm [7] and the Aruco marker detection and localization algorithm. The Fig. includes the Gaussian curve fitted in the obtained scatter plot. Here, it can be seen clearly that there is a variance in the pose information from the detection algorithm for the workpiece and Aruco marker. The pose information of both detections is with respect to the robot base frame which is constantly moving. Hence it may not be accurate due to thresholds in termination for the localization. Therefore, this variance can be because of inaccuracies in both localization and detection algorithms.

4   Conclusion

In conclusion, autonomous mobile manipulators have a huge potential in the field of mobile tending and maintenance of machines. Behavior trees can be used to design complex behaviors for such robots which allows a simple and reusable code.

In the proposed approach, we used a conventional algorithm for the detection of the objects. However, future work may address using a big 3D model dataset [9] to train specific deep learning models to detect and grasp the objects which

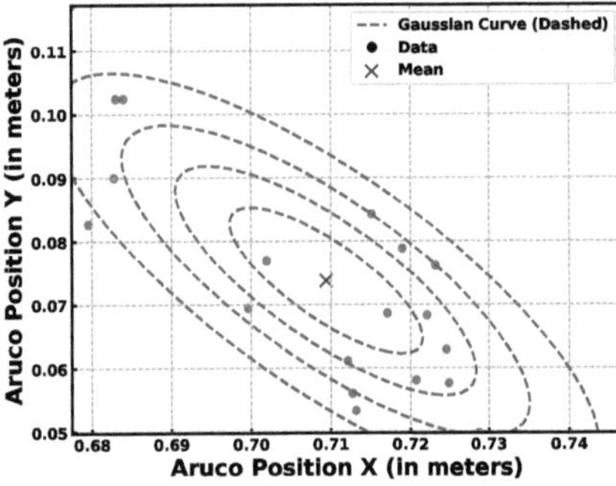

Figure 6: Scatter plot of the detected x and y axis position of the detected Aruco marker in robot base frame.

can increase the effectiveness of the system.

Acknowledgement

The work has been carried out at Fraunhofer IPA, Department of Robot and Assistive Systems, and supervised by Prof. Dr. Sebastian Otte from the Institute of Robotics and Cognitive Systems, Universität zu Lübeck.

Authors' Statement

Conflict of interest: Authors state no conflict of interest.

5   References

[1] L. Halt, F. Messmer, M. Hermann, T. Wochinger, M. Naumann, and A. Verl, "Amadeus - a robotic multi-purpose solution for intralogistics," in *ROBOTIK 2012; 7th German Conference on Robotics*, 2012, pp. 1–6.

[2] R. N. García, A. Arnal, M. Millet *et al.*, "rbkairos_sim," https://github.com/RobotnikAutomation/rbkairos_sim, 2018, online; accessed 23-Jan-2023.

[3] R. Ghzouli, T. Berger, E. B. Johnsen, A. Wasowski, and S. Dragule, "Behavior trees and state machines in robotics applications," 2023.

[4] S. Baumgarten, F. Jordan, M. Patel, M. Schmelzer, and B. Graf, "Development and evaluation of a mobile manipulation robot for surface disinfection," in *ISR Europe 2022; 54th International Symposium on Robotics*, 2022, pp. 1–8.

[5] G. Grisetti, C. Stachniss, and W. Burgard, "Improved techniques for grid mapping with rao-blackwellized particle filters," *IEEE Transactions on Robotics*, vol. 23, no. 1, pp. 34–46, 2007.

[6] J. Lindermayr and K. Predrag, "Ein mobiler manipulator hält maschinen selbstständig fit," https://www.maschinenmarkt.vogel.de/ein-mobiler-manipulator-haelt-maschinen-selbststaendig-fit-a-420df5e730c3bf5d6d75170c4aa50a86/, 2023.

[7] M. Ferguson, M. Arguedas, and C. Rauch, "simple_grasping," https://github.com/mikeferguson/simple_grasping, 2016, online; accessed 24-Apr-2023.

[8] H. Williams, J. McCulloch, B. Johnston *et al.*, "aruco_detector," https://github.com/maraatech/aruco_detector, 2020, online; accessed 25-Aug-2023.

[9] J. Lindermayr, C. Odabasi, F. Jordan, F. Graf, L. Knak, W. Kraus, R. Bormann, and M. F. Huber, "Ipa-3d1k: a large retail 3d model dataset for robot picking," in *2023 IEEE/RSJ International Conference on Intelligent Robots and Systems (IROS)*. IEEE, 2023, pp. 11 404–11 411.

Safe navigation with polytopic-based tools

Jakob Greten [1], and Ngoc Thinh Nguyen [2]

[1] Robotics and Autonomous Systems, Universität zu Lübeck, jakob.greten@student.uni-luebeck.de
[2] Institute for Robotics and Cognitive Systems, Universität zu Lübeck, ngocthinh.nguyen@uni-luebeck.de

Abstract

For mobile robot applications a robust navigation system is needed. This paper covers the development of a polytopic-based navigation tool. In order to simplify the path planning and control problem, an obstacle-free polytope is generated using the IRIS algorithm and a LiDAR sensor. A simple control pattern is sufficient for navigating towards the goal as long as both the goal and the robot are within the polytope. If the goal is not within the polytope, a temporary goal is chosen which is. New polytopes are dynamically generated based on the current pose and LiDAR data. The developed algorithm is validated using a high-fidelity simulation environment and intended for use in agriculture with the Jackal UGV.

1 Introduction

Any navigation systems performance depends heavily on the environment it is used in [1], [2]. For this project, the environment is well known and predictable. Developing a navigation system specifically geared towards performing well in the target environment is a promising area of research [3], [2].

In this case the environment the robot is supposed to operate in are fields with traversable corridors. This environment could easily be represented as a convex traversable polytope which is the goal of this project. Simplifying the environment in this way allows for the use of a simple controller with very limited path planning [2]. This is important because running more complex path planning and controlling systems in real-time can be difficult given hardware limitations.

The navigation tool developed for this project is comprised of the following two steps.

A LiDAR is used to detect obstacles in the robots vicinity. In order to fit larges possible polytopes in the traversable space, the IRIS algorithm [4] is used (Step 1).

This polytope is used to navigate to the intended destination (Step 2). New polytopes are generated dynamically based on the location of the robot, the location of the destination and the detection of new obstacles within the polytope. In order to navigate towards the intended destination a new temporary target is selected which is supposed to guide the robot towards its destination while also being located within the obstacle-free polytope. A simple controller is implemented to traverse within the obstacle-free polytope towards the temporary target and ultimately towards the selected destination.

The paper is organized as follows: Section 2 covers the materials and methods used, which include the robot platform, the IRIS algorithm, the localization system and the developed navigation process. Section 3 covers the results. Both the computing requirements as well as the simulation results are discussed. Section 4 covers the conclusion of the project.

2 Material and Methods

2.1 Jackal mobile Robot platform

The Jackal UGV (Fig. 2) by Clearpath Robotics is used in this project as a mobile robotics platform for all-terrain operation. The robot has a robust design for outdoor use and is weather- and dustproof. It utilizes skid steering meaning that both wheels on each side always rotate at the same speed but each side can be controlled individually. This allows the robot to turn on the spot. The Jackal is equipped with both a main computer and a NVIDIA Jetson. More complex projects with multiple individual processes can be split up between both computers.

2.1.1 ROS

The Robot Operating System(ROS) is used for this project. ROS is a framework which is based on a graph infrastructure where every running process takes place within a node. Nodes can publish messages to specific topics and subscribe to topics in order to receive the messages send by other nodes.

2.1.2 Ouster OS1 LiDAR

The Ouster OS1 LiDAR [5] is used to detect obstacles in the form of a 3D point cloud. The LiDAR has a vertical resolution of 16 channels and each channel can contain up to 2048 measurements. For this project a three dimensional representation of the environment is not required. Therefore the LiDAR data is converted into a simpler 2D Laserscan message using the *pointcloud_to_laserscan* [6] package.

Figure 1: The OS1 LiDAR generates a 3D point cloud which for the purpose of this project is converted to a 2D Laserscan message (white)

Figure 2: The Jackal mobile robot with the Ouster OS1 LiDAR attached

2.2 IRIS algorithm

The IRIS algorithm [4] is used to fit the largest possible polytopes into the traversable space based on the LiDAR data. The pseudocode for this process is shown in Algorithm 1. IRIS requires a list of obstacles O and a seed point which in this case is the position of the robot. The ellipsoid is represented as an image of the unit ball $E(C, d) = \{x = C\tilde{x} + d \mid \|\tilde{x}\| \leq 1\}$. For the initial ellipsoid (C_0, d_0), d_0 is set to the seed point s_0 and C_0 is set to the identity matrix in order to create a sphere around the seed point. For each obstacle a hyperplane (A, b) is generated which separates it from the ellipsoid. The hyperplanes are represented as linear constraints $P = \{x \mid Ax \leq b\}$. The intersection points between all of these hyperplanes form a polytope. A new ellipsoid is chosen by calculating the largest possible ellipsoid, which fits inside the polytope. This new ellipsoid can then be used to generate a new polytope. This process is repeated with the goal of increasing the volume of the ellipsoid. When the ellipsoids are not longer changing beyond a set tolerance, the process is finished and the polytope and ellipsoid are returned. Fig. 3 illustrates the iterative process of IRIS by showing the results after the first and the third iterations.

For more complex environments, multiple seed points and multiple polytopes and ellipsoid pairs have to be generated but for this project a single seed point is sufficient since only the immediate surroundings of the robot are of interest.

ALGORITHM 1
Pseudocode of the IRIS algorithm [4]

$$C_0 \leftarrow \epsilon I_{n \times n}$$
$$d_0 \leftarrow s_0$$
$$i \leftarrow 0$$
repeat
$$\quad (A_{i+1}, b_{i+1}) \leftarrow GenerateHyperplanes(C_i, d_i, O)$$
$$\quad (C_{i+1}, d_{i+1}) \leftarrow GenerateEllipsoid(A_{i+1}, b_{i+1})$$
$$\quad i \leftarrow i + 1$$
until $(detC_i - detC_{i-1}) < tolerance$
return (A_i, b_i, C_i, d_i)

The IRIS algorithm is designed to be used with obstacles in the form of polytopes. For this project the LiDAR data only shows individual points which are occupied. This is not especially problematic since a list of obstacle points can also be used for the IRIS algorithm, but computing this many obstacles is fairly slow. A possible solution would be to combine multiple obstacle points which belong to the same obstacle into a polytope before using this data to run IRIS but this is not implemented in this project as of yet.

In order to use the Laserscan data for IRIS, a conversion step is required. The LiDAR data comes in the form of range measurements, which correspond to specific angles at which the measurement was taken. This data needs to be transformed to the global coordinate system. Efficiency is very important in this step since these messages entail a lot of data and this conversion step is often running multiple times a second. Multiple conversion methods were tested and compared. The best method was to use the *laser_geometry* package to convert the Laserscan to a point cloud and then to iterated through those points to convert them into the for IRIS required format.

IRIS can generated three dimensional polytopes (polyhedrons) but for this project two dimensional polytopes (polygons) are sufficient. Therefore the generated polytopes will from now on be referred to as polygons and the ellipsoids will be referred to as ellipses.

IRIS maximizes the size of the generated polygon as much as possible given the obstacles. This means that the polygon borders and the borders of the obstacle often times are almost identical. In order to navigate safely on the polygon, a safety margin is needed. Generating a new polygon with a safety margin is a simple matter of downscaling the IRIS polygon.

2.3 Localization

In order to develop a robust navigation system, a reliable localization method is needed. The navigation system only requires an estimate of the position of the robot and not a map. However using a full Simultaneous Localization and Mapping (SLAM) system seems to yield the best results. In this case *gmapping* is used. When running the navigation system outside of the simulation environment, using a simpler localization system might be beneficial to improve the real-time performance.

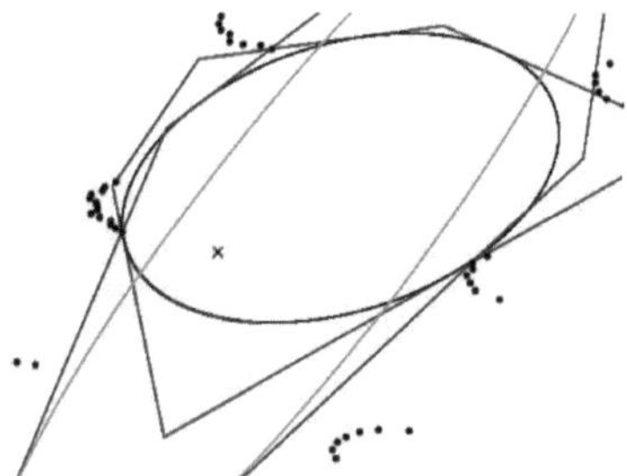

Figure 3: The polygon and the ellipse after one iteration as well as the polygon and ellipse after 3 iterations. The seed point is shown as a cross.

2.4 Safe control trajectories

In the following the program behaviour of the developed navigation system is explained and development decisions are discussed.

New IRIS obstacle-free polygons are generated whenever the robot is close to the border of the polygon or when an obstacle is detected within the old polygon. However a certain number of LiDAR measurements within the polygon are tolerated to compensate for sensor noise and prevent unnecessary regeneration of the polygon.

If the selected goal g is within the generated obstacle-free polygon then simply driving directly towards the goal is possible, since there cannot be any obstacles between the robot and the goal. However usually this is not the case at the beginning of the navigation process. Instead a second temporary goal g_{tmp} is chosen, which is leading the robot towards the actual goal. This is shown in Fig. 4. This temporary goal has to be within the obstacle-free polygon, which makes navigating towards it trivial. Choosing a g_{tmp} which is directly between g and the robot would not work because the robot would get stuck in dead ends. Instead g_{tmp} is chosen based on path planning from the *move_base* [7] package. Other less computationally expensive ways of selecting g_{tmp} could significantly improve overall performance but are not implemented as of yet.

A new g_{tmp} is continuously generated and as mentioned before navigating towards g_{tmp} is not very difficult. A simple control pattern is implemented, which turns on the spot if g_{tmp} is behind the robot and otherwise steers towards g_{tmp} while driving forward.

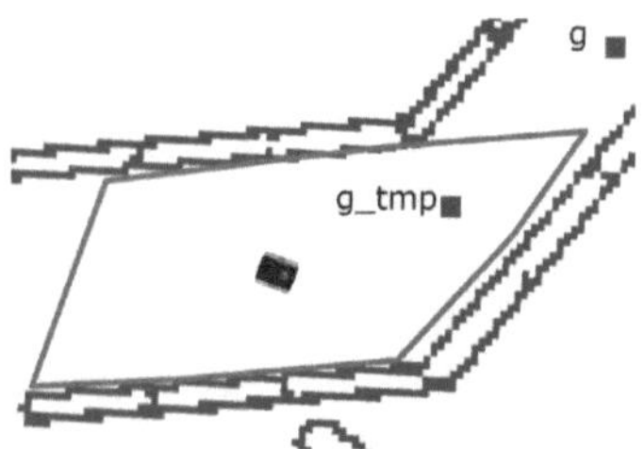

Figure 4: The selected goal g and the generated temporary goal g_{tmp} visualized in Rviz

3 Results and Discussion

3.1 Computing requirements

The computing requirements of the developed navigation system are very important for using the system outside of the simulation. The processing power of the Jackal robot is limited and processing delays should not exceed acceptable limits. As shown in Table 1 the longest processing step of the system is running the IRIS algorithm in order to generate an obstacle-free polygon. This takes 0.15691 seconds on average(N=100). This is why this step is only run when it is absolutely necessary. In order to check whether a new IRIS polygon has to be generated, the number of obstacle points in the polygon have to be counted which takes on average(N=4000) 0.011083 seconds.

The conversion of the LiDAR data to global coordinates takes 0.0035607 seconds on average(N=4000). Both the collision check as well as the conversion of the LiDAR data are run continuously. This should not be a problem since both of these steps require significantly less processing time compared to IRIS.

Table 1: Average computing time of individual steps of the navigation system. The average for the IRIS step was measured over 100 iterations and the average of the collision and data conversion steps were both measured over 4000 iterations.

IRIS	Checking for Collision	LiDAR data conversion
0.15691s	0.011083s	0.0035607s

3.2 Simulation results

The navigation system is tested in a Gazebo simulation. Here the robot can easily be placed in a maze-like environment which is useful for testing. The target destination can be selected using Rviz or using the */move_base_simple/goal* topic.

Fig. 5 shows both a generated IRIS polygon as well as the downscaled version of that polygon, which creates a safety margin. This downscaled version is used for the navigation system.

Fig. 6 and Fig. 7 show two paths taken by the robot in the simulation environment, along with generated IRIS polygons. There are too many polygons generated to show every single one but the shown polygons demonstrate that the overlapping polygons form a traversable corridor. Reference [2] shows that using the entire traversable corridor for navigation could have its benefits but for this project only the newest polygon is used. The navigation system almost always selects a path which leads to the intended destination but the path planning could be improved significantly. The robot takes turns which are usually sharper than they have to be which could be improved by implementing a better control system like Model Predictive Control(MPC).

Specifically a MPC system with tightening polytopic constraints [1] would benefit of the generated polygon. Reference [3] discusses a project similar to this one where a tube-based MPC system [8] is used along with IRIS.

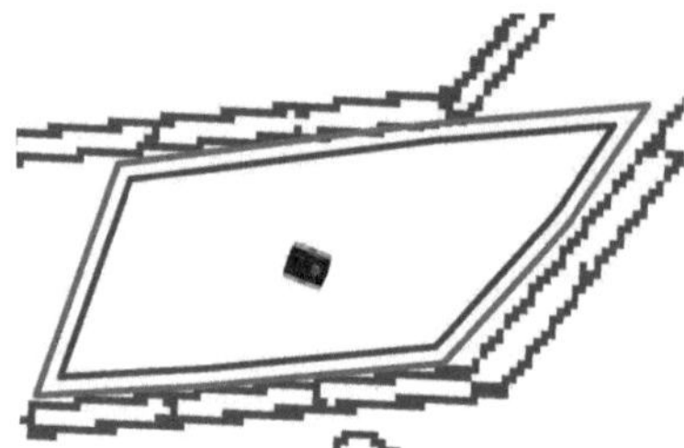

Figure 5: The generated IRIS polygon as well as the polygon with a safety margin(inner polygon).

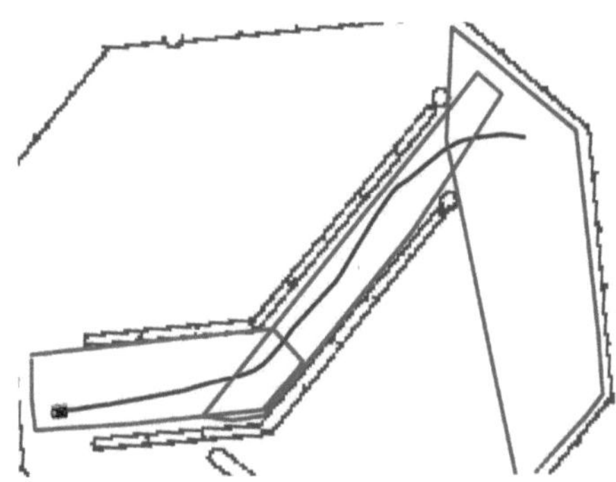

Figure 6: Path of the robot in simulation along with the generated polygon.

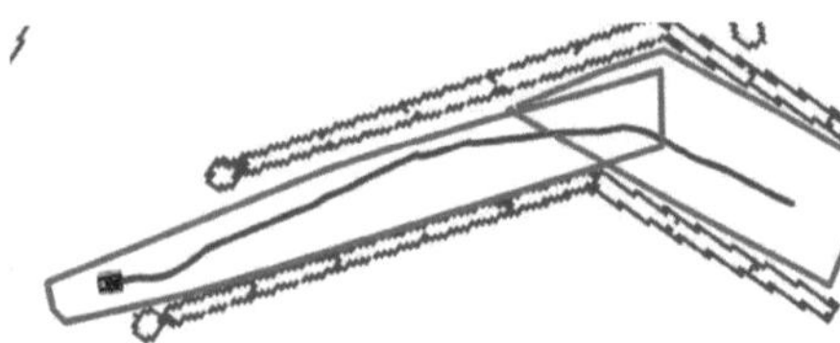

Figure 7: Path of the robot in a narrow corridor.

4   Conclusion

The navigation system presented in this paper successfully uses the existing IRIS system to generate obstacle-free polygons and and is able to navigate towards a goal using the dynamically generated polygons. By generating an obstacle-free polygon the control and path-planning problem can be simplified immensely. For the path-planning aspect the main difficulty remains to choose a useful temporary goal within the generated polygon. Implementing a controller which can navigate within the obstacle-free polygon towards the temporary goal is simple, although implementing a more efficient controller could be a promising extension of the project. Furthermore the implemented navigation system could not only be useful in the intended agricultural environment, but could be useful in any environment where there are obstacle-free corridors.

Acknowledgement

The work has been carried out at the Institute for Robotics and Cognitive Systems, University of Lübeck. We would like to thank many of our colleagues for fruitful discussions during the project.

Authors' Statement

Conflict of interest: Authors state no conflict of interest.

5   References

[1] N. T. Nguyen and G. Schildbach, "Tightening polytopic constraint in mpc designs for mobile robot navigation," in *2021 25th International Conference on System Theory, Control and Computing (ICSTCC)*, 2021, pp. 407–412.

[2] R. Geraerts and M. H. Overmars, "The corridor map method: Real-time high-quality path planning," in *Proceedings 2007 IEEE International Conference on Robotics and Automation*, 2007, pp. 1023–1028.

[3] S. Khaitan, Q. Lin, and J. M. Dolan, "Safe planning and control under uncertainty for self-driving," *IEEE Transactions on Vehicular Technology*, vol. 70, no. 10, pp. 9826–9837, 2021.

[4] R. Deits and R. Tedrake, *Computing Large Convex Regions of Obstacle-Free Space Through Semidefinite Programming*. Cham: Springer International Publishing, 2015, pp. 109–124. [Online]. Available: https://doi.org/10.1007/978-3-319-16595-0_7

[5] *OS1 Gen 1 (serial numbers starting with os1-) Mid-Range High-Resolution Imaging Lidar*, Ouster, Inc., 7 2020. [Online]. Available: https://ouster.com/downloads/

[6] "Github - ros-perception/pointcloud_to_laserscan: Converts a 3d point cloud into a 2d laser scan." https://github.com/ros-perception/pointcloud_to_laserscan, accessed: 2024-01-23.

[7] "Github - ros-planning/navigation: Ros navigation stack. code for finding where the robot is and how it can get somewhere else." https://github.com/ros-planning/navigation, accessed: 2024-01-23.

[8] D. Q. Mayne, E. C. Kerrigan, E. J. van Wyk, and P. Falugi, "Tube-based robust nonlinear model predictive control," *International Journal of Robust and Nonlinear Control*, vol. 21, no. 11, pp. 1341–1353, 2011. [Online]. Available: https://onlinelibrary.wiley.com/doi/abs/10.1002/rnc.1758

Refurbishment Scheme of a High Vacuum Tight Flange Seal with a Redundant Robot at ITER

Martin Rolfs [1], Emmanuel Brau [2], Floris Ernst [3], and David Hamilton [2]

[1] Robotics and Autonomous Systems, Universität zu Lübeck, martin.rolfs@student.uni-luebeck.de
[2] ITER Organization, {emmanuel.brau, david.hamilton}@iter.org
[3] Institute for Robotics and Cognitive Systems, Universität zu Lübeck, floris.ernst@uni-luebeck.de

Abstract

The vacuum of the ITER fusion reactor is contained by high performance vacuum tight seals. The project necessitates a refurbishment procedure for these seal surfaces around flanges. In this paper, a method to perform the refurbishment automatically with a redundant robotic arm is detailed. Also, since the operation has to be carried out in a constrained environment, an algorithm for obstacle avoidance is presented, combining a zero-order optimization algorithm with closed loop inverse kinematics. The refurbishment procedure - consisting of cleaning, inspecting and polishing - was tested on an experimental setup with a 7 degree of freedom (DOF) robotic arm. The obstacle avoidance was tested in a simulated environment. Both experiments produced satisfactory results and proved the concept of automatically refurbishing flange surfaces.

1 Introduction

1.1 The ITER Organization

The ITER Organization is an international project that aims to build the largest nuclear fusion reactor in the world. The organization is based in France and is organized by multiple different countries, including the EU, the USA, Japan, South Korea, the Russian Federation, India and China. The purpose of the fusion reactor is to strengthen the understanding of the science of nuclear fusion and plasma, as well as serve as an engineering example to aid in the construction of future power plants.

1.2 Problem Statement

Tokamak fusion reactors contain plasma by keeping it inside a high-pressure vacuum and a strong magnetic field. The pressure vessel at ITER aims to contain a high vacuum up to 10^{-6} Torr. This is possible due to the vacuum-tight seals on the port flanges that are fitted around the vacuum vessel. Whenever a flange has to be removed for maintenance purposes, the seal surface might be damaged in the process of the removal. Damage to the surface must be detected and corrected before the flange is refitted onto the vacuum vessel. The task is very labor-intensive since there is over 2 km of seal surface in total at ITER. Furthermore, once the reactor reaches first plasma, the vacuum vessel will be radiated, and consequently, manual refurbishments are challenging. For this reason, the Remote Handling Section at ITER aims to utilize robotic arms to perform refurbishments of these flange seals.

The vacuum vessel is fitted with a multitude of different flange types. One of the most important flanges is part of the so-called Test Blanket Module Port Plugs (TBM-PP). The TBM-PP can be seen in Fig. 1. These plugs form a casing around the Test Blanked Module (TBM), which is responsible for tritium breeding. Since the production of tritium is an essential step in the future of fusion energy, up to six different TBMs should be tested during the lifetime of ITER. This leads to the necessity, that the TBM-PP has to be detached multiple times in order to exchange the TBMs, which makes this type of flange a significant subject to be refurbished [1]. The process of detaching the TBM from the plug has been studied and tested in past years. The geometry of these flanges introduces motion constraints, as the seal to be refurbished is surrounded by the walls of the port plug. In order to avoid collisions, redundant robotic arms are utilized to perform this task. The redundancy of the robot enables it to find collision free configurations in this environment.

This paper discusses control schemes for the refurbishment maneuver of a redundant robot arm in a cluttered environment and presents the general setup of a prototype.

2 Related Work

This work is based on the work on flange seal refurbishment by the Technetics Group, an operating division of the company Enpro Industries. They developed a specialized polishing tool that can polish high vacuum seals. In their setup, the tool was mounted on a rail moving along the seal surface and applying constant pressure. In the presented work, however, the tool is mounted on a redundant robot - as seen in Fig 2a - that has to carry out displacements and force control.

The second part of this work is focusing on the development of an obstacle avoidance algorithm based on a Genetic Algorithm (GA). GAs are commonly used for the calculation of the inverse kinematics of redundant robots [2]. In the presented approach, the default algorithm is altered to fit the project's needs and decrease its computation time significantly.

3   Methods

3.1   Refurbishment Procedure

In this section, the refurbishment procedure is presented. The recoverable parts of the flanges will be detached and sent to a hot cell facility for refurbishment, and the fixed flange parts will be refurbished in situ.

The complete refurbishment scheme consists of three elementary steps. First, the surface has to be cleaned. This is because small metal shavings or similar dust particles might be on top of the surface. The second step is the inspection of the cleaned surface with the goal of finding damaged areas on the seal surface. This is done with a camera attached to the robot flange and an expert operator monitoring the video stream. The last step is polishing. Here, the previously detected points of interest are polished with a dedicated polishing tool. In this paper, only the polishing procedure is detailed.

3.1.1   Polishing

The seal surface is polished by a special polishing tool designed by Technetics and can be seen in Fig. 2a. This tool is attached to the robot and inherently comes with constraints. First, once the tool is activated, it should only be moved along one certain axis at a maximum velocity of 2 cm/min. This results in the robot having to perform orientational adjustments so that the tool only moves along the correct direction. Additionally, for the polishing to be effective, the tool should be applied at a set force. Technetics states that the tool tip has to be pushed onto the seal surface at a force of 10 N to yield acceptable results. Also, the downward force should not exceed 25 N, as the polishing might be too drastic, or the tool might even be damaged.

In order to control the downward force when using the polishing tool, two methods are utilized. Since the tool tip is fitted with a spring, the robot can apply force by moving the robot flange closer to the seal surface so that the spring contracts. The first method links the contraction of the spring to an applied downward force. This was possible by taking different measurements at different levels of spring contraction. This way, a relationship between the force exerted and the distance the robot move was established.

Another way to estimate the applied downward force is to infer it through the joint torques τ. With the geometric Jacobian $J \in R^{6 \times 7}$ the following relationship holds [3]:

$$\tau = J^T f_{wrench} \tag{1}$$

Here, f_{wrench} is the wrench force vector on the tool center point (TCP). Since the torques τ are known, (1) can be rewritten.

$$f_{wrench} = \left(J^T\right)^{-1} \tau \tag{2}$$

Since the robot comes with 7 degrees of freedom (DOF), the inverse of the transposed Jacobian cannot be calculated. As the polishing is always applied to a planar surface, the orientation of the robot to this plane of operation can be taken into account. In the currently defined setup of the robot and the flange, the screw axis of the first joint is supposed to be perpendicular to the plane of operation. This means that the first joint's torque value does not influence the force at the TCP in the direction of the normal of the plane of operation. This information can be used to solve (2).

$$f^*_{wrench} = \left(J^T_{6 \times 6}\right)^{-1} \tau_6 \tag{3}$$

Here, the $J_{6 \times 6} \in R^{6 \times 6}$ is the geometric Jacobian of the 7 DOF robot, where the first column is omitted, as it would establish the relationship between the joint torque of the first joint and the force at the TCP. But since only the force in the z direction of the TCP frame is of interest, this approximation can be used.

3.2   Obstacle Avoidance

As mentioned before, the refurbishment scheme has to be conducted, in some cases, in a constrained environment. Mostly the environment is cluttered with other pipes, which are to be avoided. The TBM-PP inherently constrains the movement of the robotic arm due to the geometry of the port plug. Here, the seal surface to be refurbished is below the outer flange and thus surrounded by four walls, as seen in Fig. 1.

Since there are no obstacles directly on the seal surface, no further path planning is necessary, and the TCP can follow a predefined trajectory. The robot body, on the other hand, must avoid collisions with the environment. Here the redundancy of the robotic arm comes into play, as there are, due to the extra DOF, infinite possible configurations to reach a desired pose [3]. The task of obstacle avoidance, therefore, is to find feasible configurations where the TCP is in the desired pose while the body of the robot does not collide with the environment.

In general, the issue of obstacle avoidance is an extension of the inverse kinematic problem (4). The configuration q is determined by the desired TCP pose x_d through the inverse kinematic function f^{-1}.

$$q = f^{-1}(x_d) \tag{4}$$

Obstacle avoidance can be regarded as an additional constraint. This can be expressed by a maximization of the distance from each joint position or link $T^O_{i,transl}$ to each obstacle o_j in the environment.

$$\max_q \sum_{i,j} \left\| o_j - T^O_{i,transl}(q) \right\| \tag{5}$$

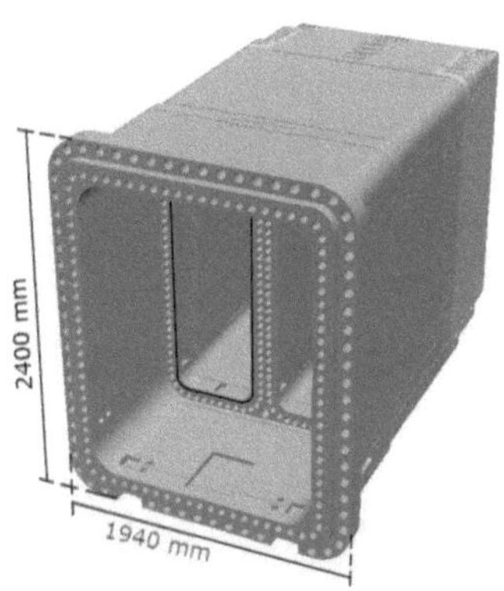

Figure 1: A render of the TBM-PP. Note that the seal surface to be polished is marked with a black line.

$T^O_{i,transl}$ is the translational part of the transformation matrix from the robot base O to the i-th joint or link. This can then be expressed as a single optimization problem.

$$q^* = \max_{q} \sum_{i,j} \left\| o_j - T^O_{i,transl}(q) \right\| \qquad (6)$$

$$\text{s.t. } x_d = f(q)$$

3.2.1 Modified GA

A common approach to solving the inverse kinematic problem of redundant robots is to use the inverse of the geometric Jacobian in order to retrieve the joint velocities from the target velocities. The Jacobian matrix of a redundant robot has a non-empty kernel. This means there are multiple different configurations to reach a desired end-effector pose [3]. In the presented approach, the idea is to use a GA to search this null space for *better* configurations by mutating one joint randomly so that the remaining 6 DOF can be used to find a unique solution.

Given the random mutation of joint i, the Jacobian that relates the joint velocities to the TCP velocities of the remaining joints can be constructed with the complete 7 DOF Jacobian matrix by removing the i-th column. The resulting matrix $J_i \in R^{6 \times 6}$ can now be inverted. With this, the change in configuration of the remaining 6 joints $\Delta q_{i,t}$ can be determined using the pose error e, a gain α and the time since the error estimation δ_t. In order to reduce the lag introduced by the integration of the joint velocities, the desired velocities of the trajectory $\begin{bmatrix} v & \omega \end{bmatrix}^T$ can be fed forward [4].

$$\Delta q_{i,t} = \delta_t \alpha J_i^{-1} \left(e + \begin{bmatrix} v \\ \omega \end{bmatrix} \right) \qquad (7)$$

From a given initial configuration q_0, a *population* of 100 configurations is generated by randomly mutating one joint from each *individual* configuration by 2 to 8 degrees. The mutation range from 2 to 8 degrees is chosen with the assumption that the initial configuration is close to the target configuration. The *individuals* are further altered using (7) in a closed loop. The pose error should at this point be minimal, since the robot should be capable of approaching the

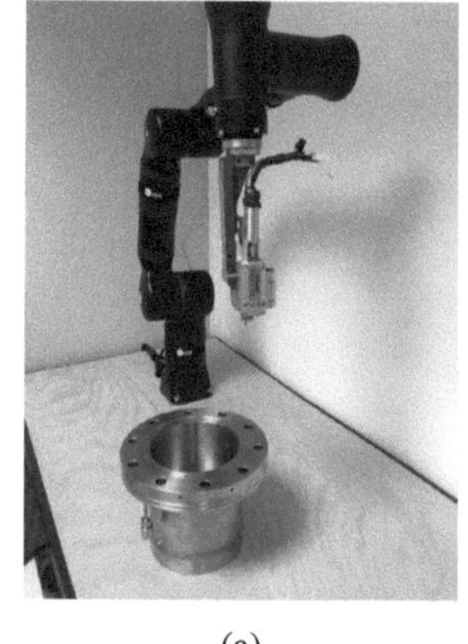
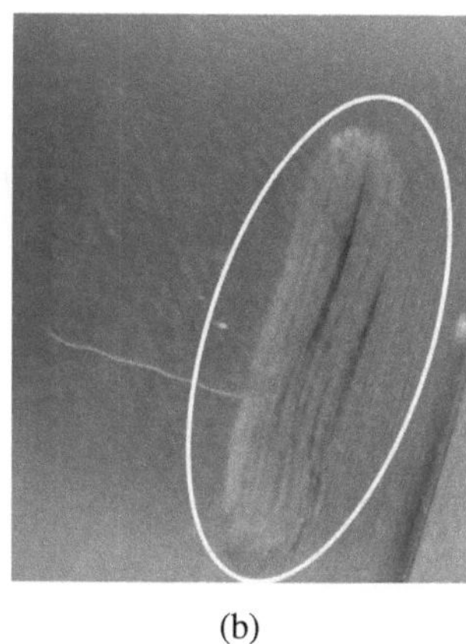

(a)   (b)

Figure 2: Fig. 2a shows the setup of the redundant robot arm, with the specialized tool designed by Technetics attached to the robot flange and the dummy flange bolted in place. Fig. 2b shows the scratch on the flange surface and the polishing result as seen on the inspection monitor. The latter is marked by the white ellipse.

target pose with the remaining 6 DOF. After that, for each *individual* in the population, its fitness is determined. The fitness is expressed as a scalar value that indicates the performance of each individual. In this case, lower fitness is preferred.

First, the minimum distance from the i-th joint given the configuration q to all the obstacles is determined. After doing that for all joints, the fitness r for each configuration q is computed. Here, the distances are squared in order to punish getting too close to obstacles.

$$r(q) = \sum_{i} \frac{1}{\left(\min_j \left\| o_j - T^O_{i,transl}(q) \right\| \right)^2} \qquad (8)$$

The individual q^* with the lowest fitness is selected and brought to the next *generation*. Now q^* is the new initial configuration from which the population is generated. The algorithm repeats this pattern until the fitness of the best individual from a generation exceeds a fitness threshold or the maximum number of generations is reached. Since the fitness depends heavily on the geometry of the obstacles in the environment and the target trajectory, the threshold to be chosen has to reflect an achievable fitness in that environment.

4 Evaluation

The polishing scheme is evaluated in an experimental setup. A 7 DOF KR810 Cobot developed by Kassow Robots is mounted in front of a dummy flange. The setup is shown in Fig. 2a. The obstacle avoidance algorithm is evaluated in a simulated environment created with the RoboDK software and MATLAB.

To measure the performance of the polishing scheme, small scratches were introduced to the seal surface of the dummy flange. The polishing tool was mounted on the robot, which

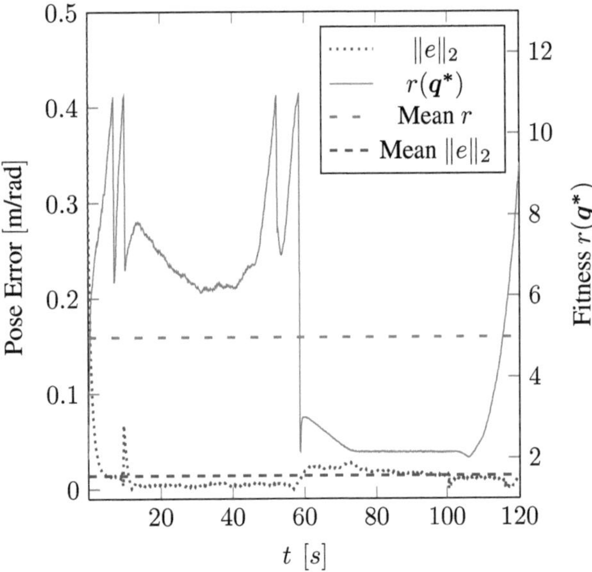

Figure 3: The fitness and pose error of the obstacle avoidance algorithm.

performed the polishing motion on the scratched surfaces. The tool moved at a tangential velocity of 0.33 mm/s and applied an estimated maximum downward force of 10 N to the surface. The force was controlled by the distance the TCP moved towards the surface and was verified by the mathematical estimation method described in Section 3.1.1. The polishing was deemed acceptable by an expert after multiple passes over the scratched area. The result of the polishing and the introduced scratch can be seen in Fig. 2b.

The obstacle avoidance algorithm was tested in simulation with the TBM-PP as the target flange for multiple different initial configurations. Since - especially in the first time step - the algorithm converged slowly onto the desired trajectory with an acceptable fitness value, the computation time was very high. Instead of computing the new configuration every time step with the evolutionary approach, a closed loop inverse kinematic (CLIK) algorithm, using the pseudo-inverse of the geometric Jacobian, was utilized when the fitness was below a threshold [5]. This decreased the computation time of the trajectory drastically. The downside of this is that the corrections made from the GA introduced fluctuations. Fig. 3 displays the fitness and the error over the course of the trajectory of a representative trial. The initial configuration in this particular trial was such, that the TCP was half a meter distant to the ingress point of the trajectory, which is evident by the large error at $t = 0$. The spikes in the fitness curve show the point where the threshold is encountered. The significant fluctuations of the fitness is a result from the geometry of the TBM-PP. The TCP has to move very close to the walls of the port plug, especially in the beginning of the trajectory. The fitness is very low between 60s and 110s as the TCP moves along the middle of the TBM-PP and maximizes the distance to the walls by doing that. Nonetheless, the robot body did not collide

with the environment in this test, and the fitness did not exceed the set threshold that ensures no collision.

Apart from a sudden increase in the magnitude of the pose error at $\sim$ 10 s, likely caused by the correction of the GA, the translational error was within a millimeter of accuracy.

5 Conclusion and Future Work

The results are promising, as the refurbishment scheme was successfully tested in a first experimental demonstration. Also, the presented obstacle avoidance algorithm that combines the concept of a GA with CLIK was tested successfully in a simulated environment.

The next step is to combine the obstacle avoidance algorithm with the complete refurbishment, ideally deploying it on a dummy TBM-PP. More experimentation on the effectiveness of the polishing scheme should also be conducted in order to find polishing schemes that yield the best results.

Acknowledgment

The work has been carried out at the ITER Organization in Saint-Paul-lez-Durance, France under supervision by the Institute of Robotics and Cognitive Systems, Universität zu Lübeck with great technical support by Soumik Sarkar, ITER Project Associate.

Authors' Statement

Authors state no conflict of interest.

6 References

[1] L. M. Giancarli *et al.*, "Overview of recent ITER TBM program activities," *Fusion Engineering and Design*, vol. 158, p. 111674, 2020. [Online]. Available: https://www.sciencedirect.com/science/article/pii/S0920379620302222

[2] A. C. Nearchou, "Solving the inverse kinematics problem of redundant robots operating in complex environments via a modified genetic algorithm," *Mechanism and Machine Theory*, vol. 33, no. 3, pp. 273–292, 1998.

[3] B. Siciliano and O. Khatib, *Springer Handbook of Robotics*, 2008th ed., ser. Springer Handbook of Robotics. Springer Berlin Heidelberg, 2008.

[4] M. Grotjahn and B. Heimann, "Model-based feedforward control in industrial robotics," *The International Journal of Robotics Research*, vol. 21, no. 1, pp. 45–60, 2002.

[5] A. Colomé and C. Torras, "Closed-loop inverse kinematics for redundant robots: Comparative assessment and two enhancements," *IEEE/ASME Transactions on Mechatronics*, vol. 20, no. 2, pp. 944–955, 2015.

Optical Pose Detection of a Cable-Driven Parallel Robot

Abeer Eisa Khalid Omer[1], Jonas Osburg [2]

[1] Robotics and Autonomous Systems, Universität zu Lübeck, abeer.eisakhalidomer@student.uni-luebeck.de
[2] Robotics and Cognitive Systems, Universität zu Lübeck, j.osburg@uni-luebeck.de

Abstract

This research addresses the challenge of determining the pose of the end effector in a cable-driven parallel robot by providing a simplified alternative to complex forward kinematics. Employing ArUco markers for pose detection, we conducted three experiments—robot repetition accuracy, ArUco marker repetition accuracy, and calibration. ArUco markers exhibited limited precision. Despite this drawback, they offer a lightweight and cost-effective option for pose detection. The findings underscore the potential need for post-processing in real-world applications. In conclusion, while ArUco markers present challenges, they remain a viable choice for scenarios prioritizing affordability and reduced weight in pose detection methods.

1   Introduction

Ultrasound technology has become an indispensable diagnostic tool due to its versatility and safety. Unlike other imaging modalities that involve ionizing radiation, ultrasound relies on sound waves, making it a radiation-free option suitable for repeated examinations, particularly during pregnancy. The real-time 3D imaging capability of ultrasound provides clinicians with immediate insights into anatomical structures and physiological processes, allowing for quick and accurate diagnoses. Furthermore, the accessibility and affordability of ultrasound equipment make it widely available, even in resource-constrained healthcare settings.

However, despite its numerous benefits, ultrasound does present challenges. Operator dependence, the difficulty in reproducing consistent views, and staff shortages in clinics are among the obstacles faced in routine clinical practice. These challenges emphasize the need for innovative solutions to enhance the efficiency and reliability of ultrasound procedures.

Recognizing the need for improvement, one promising solution lies in the integration of robotics into ultrasound systems [1] [2]. Robotic ultrasound systems offer several advantages, such as increased precision, repeatability, and the ability to access challenging anatomical locations. These robotic systems have found applications in diverse medical fields, including neurosurgery, cardiovascular interventions, and obstetrics.

Among the various robotic architectures are the cable-driven parallel robots (CDPRs) [3]. These robots employ a network of cables and pulleys to control the motion of the end-effector, allowing for a high degree of freedom and dexterity. The concept is that the patient would be positioned above the robot's frame and the scanning would be performed from under the patient.

The advantages of utilizing a cable-driven parallel robot for ultrasound diagnostics are manifold. Firstly, the cable-driven design offers a large working space, enabling the end-effector to reach anatomical structures with greater flexibility. Additionally, the relatively cost-effective manufacturing process of CDPRs makes them an attractive option for medical applications, potentially reducing the economic burden associated with advanced robotic systems. The fully automated diagnostic capabilities of a CDPR enable comprehensive checks of the entire abdominal cavity, leading to efficient and accurate examinations.

However, the inherent challenge in this setup arises from the over-constraining of the end-effector's forward kinematics [4], rendering analytical solutions unfeasible. Traditional approaches to solve such problems typically involve iterative methods or interval analysis [5]. As these methods are very complex, our research focuses on the approach of utilizing ArUco markers for pose detection.

The decision to employ ArUco markers stems from their versatility in computer vision applications. ArUco markers (Augmented Reality University of Cordoba, which is where it was developed) offer a lightweight and efficient means of detecting the pose of the end effector.

In this work, we delve into ArUco marker-based pose detection for the cable-driven parallel robot, with a particular emphasis on its application in robotic ultrasound. By addressing the challenges posed by the over-constrained forward kinematics, we aim to further develop the CDPR by allowing for the localization of the end-effector that will hold the ultrasound device. This results in three research questions for our paper: firstly, how accurate is the repeatability of the cable robot, secondly, how accurate is the repeatability of the ArUco marker detection and finally, how precisely can a hand-eye calibration be performed using the ArUco marker poses.

2   Material and Methods

2.1   The Cable-Driven Parallel Robot

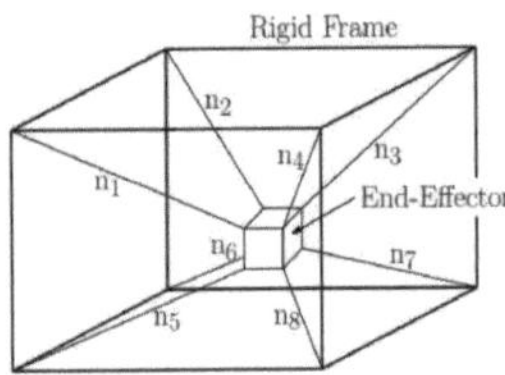

Figure 1: Cable-driven parallel robot with n = 8 cables [3].

As shown in Fig.1, the CDPR consists of n flexible cables that are firmly anchored at the end-effector and guided by pulleys to motorized winches strategically positioned within the robotic framework. The frame for the CDPR used measures 100 cm in length, 80 cm in width and 70 cm in height. The cables' endpoints are securely attached to the end-effector, forming a cohesive linkage between the rigid frame and the end-effector. The motorized winches serve as dynamic actuators, as these winches are actuated, the cables wind or unwind, resulting in an alteration of the cable lengths and subsequent movement of the end-effector. By precisely adjusting the lengths of the cables, the end-effector can traverse a three-dimensional Cartesian workspace, and the system manifests a a configuration space of n dimensions, offering remarkable versatility in positioning and orientation [3]. The design of the used system can be seen in Fig. 2.

Despite the capabilities of CDPRs, their application in practical scenarios remains relatively limited, owing to the specificity of their design and functionality. However, notable exceptions exist, with one of the most recognized derivatives of CDPRs being the Spidercam. Widely employed in stadiums, the Spidercam has garnered acclaim for its ability to capture cinematic video shots during sporting events or concerts. This showcases the unique niche that CDPRs occupy, offering specialized solutions in contexts where their distinctive attributes align with specific operational requirements.

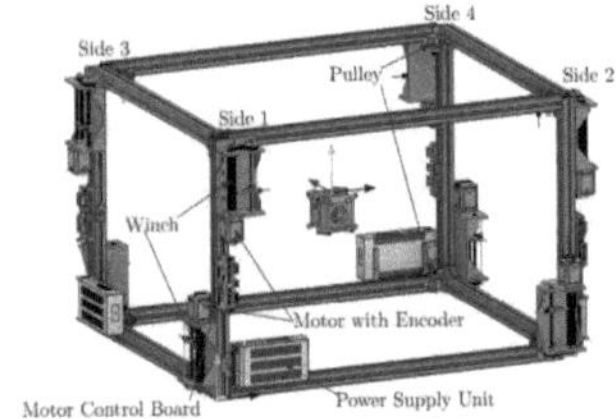

Figure 2: CAD design of the CDPR [3].

2.2   ArUco Markers

An ArUco marker is a square marker consisting of a wide black border and an inner binary matrix (Fig. 3) which determines its identifier (id). The marker size determines the size of the internal matrix. For example a marker size of 4x4 is composed by 16 bits [7].

In our work, we leverage ArUco markers to facilitate the detection of the pose of the end-effector in our cable-driven parallel robot. ArUco markers represent a subset of fiducial markers, these markers are designed to be visually distinctive and easily detectable, offering a lightweight yet powerful solution for vision-based tracking in various applications, including robotics.

The functionality of ArUco markers hinges on the principles wherein the camera system captures images of the environment, and specialized algorithms process these images to identify and interpret the markers. Each ArUco marker possesses a predefined code that, when detected, can be translated into a unique identifier. This identifier, along with the marker's position and orientation in the camera's field of view, are utilized to infer the pose of the end-effector.

A crucial advantage of ArUco markers lies in their versatility and adaptability to different scales and environments. By varying the size and complexity of the marker patterns, users can optimize the detection process for specific applications.

Figure 3: Examples of Aruco markers.

2.3   Setup and Experiments

The experimental apparatus consisted of the CDPR as the center, on one side, a webcam was positioned to capture images of the robot's workspace, focusing on the ArUco marker affixed to the end-effector (Fig. 4).

On the opposite side, the Polaris Spectra tracking system by NDI company served as the ground truth, providing high-precision tracking data for comparison and validation. 'Polaris,' a sophisticated tracking technology, employs a combination of optical and electromagnetic tracking to precisely monitor the position and orientation of objects in a three-dimensional space. It features a measurement accuracy of 0.35 mm, and was used alongside a tracker and the NDI tool tracker as an interface unit to read the measurements.

The cable-driven parallel robot was subjected to a series of 35 predefined poses (Fig. 5), 7 different translational poses that are repeated 5 times for 5 different rotations. The end-effector goes through positions that are 50 mm away from the workspace center in the three directions, while the rotation is chosen from 3 values (-10,0,10) degrees around the x and y coordinates. Rotation around the z axis (vertical axis) is not implemented because it's redundant, the system does not give the option to rotate around the z axis since it is not needed for the application. During these movements, the webcam captured images of the robot at regular intervals. Simultaneously, NDI's 'Polaris' system continuously tracked the robot's movements, providing a reference dataset representing the ground truth for comparison. This dual-setup allowed for the creation of a dataset comprising images from the webcam and corresponding tracking data from 'Polaris.'

The acquired images from the webcam were processed using detectMarkers function from OpenCV. The distinctive pattern of the ArUco marker, affixed to the end-effector, was identified and analyzed in each image frame. The ArUco marker size used was 120 mm. The resulting data, including the detected marker's position and orientation, constituted the basis for inferring the pose of the end-effector.

We conducted a series of experiments to assess the performance and reliability of the proposed robotic ultrasound system. Three key experiments were designed and executed: firstly, Robot Repetition Accuracy, which aimed to quantify the repeatability of the robotic ultrasound system by assessing the accuracy of robot movements during repetitive tasks. The focus was on evaluating how consistently the robot could replicate specific motions. Secondly, ArUco Marker Repetition Accuracy, which used the same poses as the first experiment to test the repeatability of ArUco markers. And finally, Calibration for both the ArUco marker poses and robot poses from Polaris.

Figure 4: Experiment setup with Webcam, CDPR with an ArUco marker and Polaris.

3 Results and Discussion

3.1 Robot Repetition Accuracy

The CDPR's accuracy is 17.31 mm/3.55 degrees [3]. To calculate its repetition accuracy, it went through the sequence of 35 poses 5 times recorded with Polaris to test the repetition accuracy of the robot. After the robot has moved for a while, the cables can become tighter or more relaxed due to a slight lateral deviations in cable placement around the cable winch when rewinding. This problem results in less accurate robot poses as time goes by with the robot functioning without turning it off and fixing the cables.

The 5 sequences were recorded one after the other without putting the cables back to their original state. The first recorded sequence was used as the reference and the following four sequences were compared to it. As can be seen from Table 1, the repetition error for the second to the fifth

sequence in comparison with the first sequence was calculated.

The translation error was calculated by getting the Euclidean distance between the position vectors of two respective poses from the first sequence and another sequence, the mean of the 35 errors corresponding to the 35 poses was calculated as the average translation error for that sequence. The rotation error was calculated using the following formula:

$$\theta = \arccos\left(\frac{tr(R_1 * R_2') - 1}{2}\right), \qquad (1)$$

where R_1 and R_2 are the rotation matrices of two respective poses from different sequences.

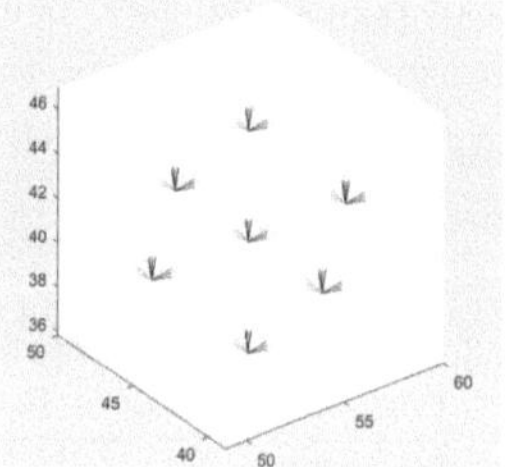

Figure 5: Predefined robot poses (cm).

Table 1: Average robot repetition accuracy (Polaris)

Repetition	Translation error	Rotation Error
2	1.22 mm	0.51°
3	2.03 mm	1.22°
4	2.39 mm	1.96°
5	2.69 mm	2.55°

3.2 ArUco Marker Repetition Accuracy

While Polaris was recording the robot poses from one side, the ArUco marker poses were being captured from the other side for the same sequences. The same procedure that was performed to calculate the robot repetition accuracy was used to calculate the ArUco marker repetition accuracy this time (Table 2). The translation errors are higher for the ArUco marker while the rotations errors are lower. This only measures the repeatability of the marker poses compared to the first pose, but it doesn't guarantee that the first pose is a good representation of the robot pose. Using the same logic as in the robot repeatability test, the error values increase as the commanded poses increase because of the cable's small entanglements.

Table 2: Average ArUco marker repetition accuracy

Repetition	Translation error	Rotation Error
2	6.00 mm	0.51°
3	8.2156 mm	0.93°
4	8.45 mm	0.60°
5	11.91 mm	1.50°

3.3 Calibration

The 35 poses were calibrated for both the robot poses from Polaris (Table 3) and the ArUco marker (Table 4), the calibration method used was QR24 [6]. Looking at the translation and rotation errors for the ArUco markers, it is clear that they are quite large to be useful in real life applications. Studying the errors for individual axes of the poses, we observed that rotation around the X axis led to large position errors along Y axis, and vice versa. The errors around the X and Y axis were reasonable otherwise. The translation calibration error for the Z axis was relatively good for all cases.

As can be seen from Fig. 6, the first 7 poses which are poses that rotate around the X axis, are shifted in the Y direction from the other set of poses, the same behavior can be seen for poses that rotate around the Y axis, for which the poses are shifted in the X axis, which explains the high calibration error.

Table 3: Mean calibration errors for Polaris

Repetition	Translation error	Rotation Error
1	2.13 mm	1.05°
2	1.90 mm	1.09°
3	1.97 mm	1.16°
4	2.56 mm	1.66°
5	2.47 mm	2.33°

Table 4: Mean calibration errors for ArUco markers

Repetition	Translation error	Rotation Error
1	38.60 mm	8.84°
2	38.41 mm	8.88°
3	36.95 mm	8.94°
4	39.35 mm	8.63°
5	44.74 mm	9.29°

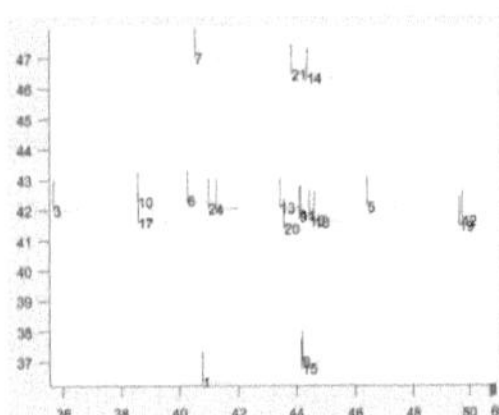

Figure 6: First 21 ArUco marker poses numbered and plotted along the Y and Z axes, the first 7 poses are shifted to the left (cm).

4 Conclusion

In conclusion, the accuracy of the cable-driven parallel robot declines during use due to variations in cable tension caused by entanglement around the winches. This issue is addressable by temporarily shutting down the robot and restoring the cable to its original state. On the other hand, the use of ArUco markers reveals a limitation in immediate accuracy, partly attributed to the modest dataset of 35 poses. The prospect of enhancing ArUco marker precision exists with a larger dataset and post-processing techniques. Acknowledging these challenges, our study emphasizes the need for periodic maintenance in the cable-driven robot's operation and anticipates improved accuracy in ArUco marker applications through dataset expansion and refinement.

Acknowledgement

The internship was completed at the University of Lübeck.

Authors' Statement

Conflict of interest: Authors state no conflict of interest.

5 References

[1] F. Najafi and n. Sepehri. *A novel hand-controller for remote ultrasound imaging.* In: Mechatronics, vol. 18, no. 10, pp. 578--590, 2008.

[2] P. Arbeille, A. Capri, J. Ayoub, V. Kieffer, M. Georgescu, G. and Poisson, *Use of a robotic arm to perform remote abdominal telesonography.* In: American journal of Roentgenology vol. 188, no. 4, 2007.

[3] R. Denz, F. Ernst and J. Osburg, *Development and Evaluation of a Cable-Driven Parallel Robot for Ultrasound Diagnosis.* University of Lübeck, Lübeck, 2022.

[4] A. Pott, and V. Schmidt, *On the forward kinematics of cable-driven parallel robots.* In: 2015 IEEE/RSJ International Conference on Intelligent Robots and Systems (IROS). IEEE. pp. 3182—3187, 2015.

[5] J. Merlet, *Solving the forward kinematics of a Gough-type parallel manipulator with interval analysis.* In: The International Journal of robotics research vol. 23, no. 3, 2004.

[6] F. Ernst *et al, Non-orthogonal tool/flange and robot/world calibration.* 2012

[7] *OpenCV: Detection of ArUco Markers.* https://t.ly/iUmJB.

Multi-Sensor Calibration on a Pan-Tilt-Unit System for Unmanned Aerial Vehicles

Hauke Budig [1] and Nikolaus Ammann [2]
[1] Robotics and Autonomous Systems, Universität zu Lübeck, hauke.budig@student.uni-luebeck.de
[2] German Aerospace Center, Institute of Flight Systems, Braunschweig, nikolaus.ammann@dlr.de

Abstract

Multi-sensor calibration is an important problem in the development of unmanned systems. All transformations between sensors must be known to interpret sensed data correctly. This applies especially when moving sensors are used. In this work, we build a calibration toolchain for future calibration tasks that includes camera-LiDAR, LiDAR-IMU, camera-IMU, and hand-eye calibration by evaluating preexisting calibration tools for each of these tasks. Afterwards, the toolchain is evaluated on a multi-sensor pan-tilt-unit (PTU) system. First, the resulting calibrations are separated into groups that form circular transformations. These transformations can be evaluated by calculating their deviation from the unit matrix. Second, we conduct a full system evaluation by testing all calibrations simultaneously on the PTU system. Full system evaluation shows a mean average error of $-3.14°$ in pan and $-4.88°$ in tilt direction. The outcome is a flexible calibration toolchain for multi-sensor calibration that will facilitate future calibration work.

1 Introduction

Autonomous systems rely heavily on sensors in terms of measuring their system state. They are designed to execute a certain behaviour without human intervention. Deviations from the measured state and the real state can impose potentially dangerous situations for the system, especially when it is airborne. Therefore, a complete and accurate intrinsic and extrinsic sensor calibration is needed. Due to the number of sensors and calibrations involved, and because of the amount of available calibration tools, we propose a toolchain of calibration methods for multi-sensor calibration, fulfilling the following requirements:

Open-Source License: The toolchain developed in this project will be used in future, potentially commercial-oriented projects. Therefore, we have to make sure that the licenses of the tools allow for such usage.

Good rotational and translational precision and accuracy: The calibration tool shall be precise and accurate. Comparing the given accuracies from the literature is difficult, because they use different metrics. Therefore, an individual evaluation for each tool that fulfills the criteria is necessary.

Good documentation: Good documentation saves time, especially when calibration is not the main objective. Understanding quickly how to use the software is desirable.

Easy to use: Besides being well documented, the tool should be intuitive. Highly complicated procedures are error-prone and could therefore be inefficient.

High degree of automation: Having to execute as few steps as possible is convenient for the user and makes the procedure less error-prone.

Fast computation: A quick execution time is a desirable property. Nevertheless, because sensor calibration consists of solving optimization problems, a reasonable amount of computation time should be expected.

System compatibility: The calibration software needs to have an interface to the Robot Operating System (ROS) and to run on Ubuntu systems.

The multi-sensor system used consists of two monocular cameras, a Light Detecting and Ranging (LiDAR) module, and an Inertial Measurement Unit (IMU). The cameras and the LiDAR are mounted on a pan-tilt-unit (PTU) which has two degrees of freedom (DOF). This makes hand-eye calibration necessary. The setup can be seen in Fig. 1.

In the next chapter, we will look into different calibration approaches and choose calibration tools accordingly. We describe our hardware setup and define evaluation methods for the resulting calibrations, The results of the calibration are presented in the third chapter. Here, calibration precision will be assessed, and the error distribution will be analyzed. The last chapter summarizes the work.

2 Material and Methods

In this work, we conduct four kinds of sensor calibrations: (1) Camera-LiDAR, (2) LiDAR-IMU, (3) Camera-IMU, and (4) Hand-eye calibration. The first three calibrations are "sensor-to-sensor"-calibrations, i.e. we obtain a transformation between two sensor frames. For the hand-eye calibration, we obtain a transformation between camera sensor and gripper, i.e., in our case, the PTU's rotational center.

Camera-LiDAR: Camera-LiDAR calibration computes the extrinsic parameters between a monocular camera and a

LiDAR scanner, using the camera images and the respective pointclouds. It requires intrinsic camera parameters to be known. These methods are mainly split into two approaches: target-based (see [1]) and targetless (see [2]). When using target-based approaches, a marker is necessary for the camera and LiDAR to detect and extract features. With targetless approaches, those features are extracted from the environment by special computer-vision algorithms such as feature point extraction [3].

LiDAR-IMU: LiDAR-IMU calibration computes the extrinsic parameters between a LiDAR scanner and an IMU, using changes in the pointcloud and the respective accelerations and angular rates. Methods for LiDAR-IMU calibration are described in [4] and [5]. Here, only one sensor gives visual information. In [4], LiDAR and IMU trajectories are represented using B-Splines and being matched. Reference [5] uses an Iterative Kalman Filter for solving the calibration problem.

Camera-IMU: The same caveat that applies to LiDAR-IMU calibration also applies to camera-IMU calibration. However, designing markers that a camera can detect is easier than designing markers that a LiDAR can detect. Therefore, most methods for camera-IMU calibration use checkerboards [6] [7]. Again, the problem is stated either using a Kalman filter [6] or using B-Splines [7].

Hand-eye: This type of calibration is different from the other three because the resulting transformation does not transform a frame from sensor to sensor. It computes the transformation from the rotational center of the PTU to the camera frame. Therefore, special techniques exist to solve the classic hand-eye calibration problem shown in (1), for example [8]. We assume, that the target is acquired at multiple poses. X represents the desired transformation between gripper and camera, and A and B denote the transformations from robot base to gripper and camera to world frame, respectively. Note that A and B consist of multiple transformations from several used poses [9].

$$AX = XB \tag{1}$$

There are also hand-eye calibration methods designed especially for PTUs, however, they assume pure rotation of the sensor [10]. This does not apply here, because the sensors move on circular paths when the PTU pans or tilts.

The hardware used in this work is built upon a PTU mounted on an aluminum frame. This PTU has 2-DOF and can rotate horizontally with an accuracy of $0.06°$ and vertically with an accuracy of $0.03°$. Two monocular cameras with fixed focal lengths are mounted left and right of the PTU. A LiDAR is mounted below. The LiDAR's field of view has been reduced internally to $90°$, i.e. from $45°$ to $-45°$ to match the cameras' field of views. Furthermore, an IMU is mounted on the aluminum frame behind the PTU. This IMU measures acceleration and angular velocity on the system with an update rate of $400\,\mathrm{Hz}$ and is fixed on the body frame of the unmanned aerial vehicle (UAV).

In the following, we list the calibration tools we chose ac-

Figure 1: The visual sensor array of the system. The cameras are attached on the sides of the PTU and the LiDAR is mounted at the bottom. The entire unit is attached below the UAV.

cording to the desired criteria:

Camera-LiDAR: For the camera-LiDAR calibration, we chose "direct_visual_lidar_calibration" [2]. This tool uses a targetless approach and is easy to use and well-documented. It promises deviations below $0.41°$ and $34\,\mathrm{mm}$. Furthermore, its accuracy proved itself in a flight test on a quadcopter. The only other tool taken into account was ATOM [11]. This tool fulfilled all named criteria, but lacked in speed and user-friendliness in the quadcopter test scenario.

LiDAR-IMU: The only preexisting tool found that complied with all criteria was "LiDAR_IMU_Init" [5]. Its main advantages are exceptional computation speed and a helpful visualization of the calibration process. It promises deviations of below $0.72°$ and $133\,\mathrm{mm}$. Other tools were found, however, they either failed during the test or lacked open-source licensing.

Camera-IMU: For camera-IMU calibration, we chose Kalibr [7], which is a calibration toolbox for camera and inertial calibrations and makes use of the B-Spline representation. Its expected accuracy is very good, with standard deviations around $1\,\mathrm{mm}$ and below $0.1°$. Three other tools were found, but they either failed during tests or needed an unreasonable amount of software adaptations.

Hand-eye: For hand-eye calibration, "easy_handeye" proved itself to be reliable and accurate. It implements several hand-eye calibration algorithms, but the one used was [8]. Documentation is good, but the calibration process is time-consuming. Furthermore, this tool was the only one to be well documented and to be tested on a current version of ROS.

For evaluating the calibration of the PTU-based hardware, we define two metrics:

- Deviation from the unit matrix after calculating loop transformations

- Deviation of calculated PTU pose from measured pose for rectangular paths in the configuration space

All matrices are given as 4x4 homogeneous transformation

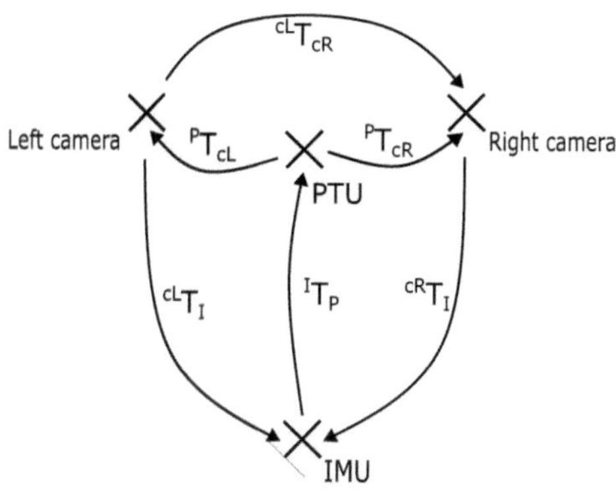

Figure 2: Overview of the transformation matrices in the system. Note that every frame is reachable from any other frame.

matrices. In the first step, we define transformations that form a loop from one frame onto itself using different pathways. An overview of the transformations can be found in Fig 2. All shown matrices are obtained from sensor calibration, except for $^{I}T_{P}$, which was known from the CAD model. Also note that the PTU includes the known transformation from its base to its rotational center from CAD. This transformation is considered, but not explicitly named. Now we define loops of transformations that map back to themselves to evaluate their accuracy. Here, I denotes the 4x4 unit matrix.

$$I = {}^{P}T_{cR}{}^{cR}T_{cL}{}^{cL}T_{P} \qquad (2)$$

$$I = {}^{I}T_{cR}{}^{cR}T_{P}{}^{P}T_{I} \qquad (3)$$

$$I = {}^{I}T_{cL}{}^{cL}T_{P}{}^{P}T_{I} \qquad (4)$$

$$I = {}^{I}T_{cL}{}^{cL}T_{cR}{}^{cR}T_{P}{}^{P}T_{I} \qquad (5)$$

Using these loops, we can measure the distance from the unit matrix, i.e. the error in the used calibration matrices. As a second step, we evaluate the entire calibrated system: We let the PTU move on a rectangular path around the $(0,0)$ position with a certain distance. Using a physical marker, we determine the position of the world frame. Using all calibrated frames and because of the PTU's high positional accuracy, we can calculate the calibrated pose of the PTU and compare it to the measured pose.

3 Results and Discussion

In this section, we will discuss the results achieved with the evaluation methods described in the last chapter. Unfortunately, LiDAR calibration could not be conducted due to a hardware failure of the LiDAR. Note that in the following, the asterisk in $^{cL}T_{P}^{*}$ indicates, that this matrix was acquired from a combination of the calibration with the lowest rotational and the calibration with the lowest translational error. It denotes a special, "hand-made" calibration.

Table 1: Relative deviations of evaluation matrices from the unit matrix obtained using the euclidean distance.

Transformation Loop	Deviation Rotation	Deviation Translation
$^{P}T_{cR}{}^{cR}T_{cL}{}^{cL}T_{P}$	2.47%	2.3 cm
$^{I}T_{cR}{}^{cR}T_{P}{}^{P}T_{I}$	6.49%	10.24 cm
$^{I}T_{cL}{}^{cL}T_{P}^{*}{}^{P}T_{I}$	6.61%	8.04 cm
$^{I}T_{cL}{}^{cL}T_{cR}{}^{cR}T_{P}{}^{P}T_{I}$	2.78%	5.99 cm

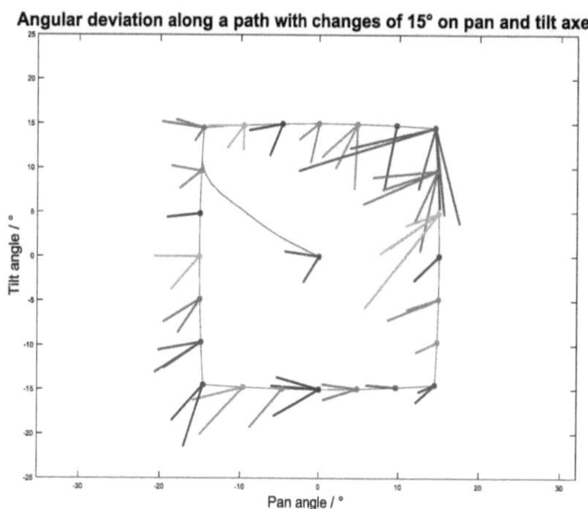

Figure 3: Rectangular path around the PTU zero position with step size of 5° and distance from the center of 15°. Different colors represent deviations from their respective true value

Table 1 shows the deviations from the unit matrix for the loop evaluations. The rotational deviations correspond to angles below 3°. However, we can not point on a single transformation to induce a higher error. When evaluating the full system with the PTU moving on a rectangular path, we can observe that most of the deviations tend towards negative pan and negative tilt direction, as can be seen in Fig. 3. The evaluation runs for 5° and 10° show the same behavior. This indicates a systematic deviation in one of the calibrations. This could be due to the fixed calibration between IMU and PTU that was taken from CAD and assumed zero rotational difference. When calculating this difference from the other calibrations, we obtain a rotation of $[0.01°, -2.83°, -2.11°]$ [XYZ] that could be due to manufacturing inaccuracies. The statistical analysis show a similar picture, with both pan and tilt deviations shifted to the negative with a mean average error of $-3.14°$ in pan and $-4.88°$ in tilt direction. When assuming a Gaussian distribution, we obtain the properties mentioned in Table 2. Note that the data is quite sparse, especially for the 5° and 10° runs.

4 Conclusion

In this work, we defined a toolchain for multi-sensor calibration on autonomous systems constisting of the fol-

Table 2: Statistical analysis of the error distribution. All distributions lean towards negative angles. For the five and ten degree run, the data is sparse. All values are given in [°]

P/T Angle		#	max	min	mean	median	std
5	pan	14	-0.938	-6.033	-3.697	-4.295	1.582
	tilt	14	-2.097	-7.581	-5.039	-5.215	1.880
	both	28	-0.938	-7.581	-4.368	-4.551	1.837
10	pan	17	3.702	-4.214	-1.692	-2.117	2.126
	tilt	17	-2.831	-11.993	-6.282	-6.001	2.217
	both	34	3.702	-11.993	-3.987	-3.773	3.163
15	pan	52	3.003	-17.128	-4.030	-3.762	3.209
	tilt	52	1.333	-10.763	-3.328	-2.879	3.075
	both	104	3.003	-17.128	-3.679	-3.425	3.147

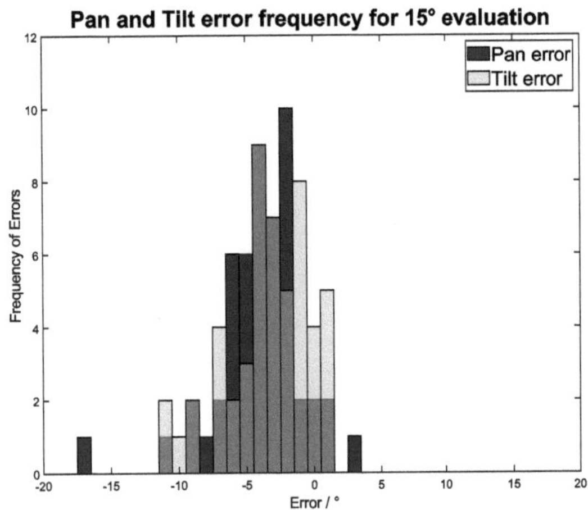

Figure 4: Error distribution from the $15°$ evaluation run. Both the pan and tilt error lean towards negative values.

lowing tools: (1) direct_visual_lidar_calibration (2) LiDAR_IMU_Init (3) Kalibr (4) easy_handeye. We applied this toolchain to a multi-sensor system mounted on a PTU and conducted a full calibration. Due to a hardware failure, the LiDAR-related calibrations could not be evaluated. The results show an average mean deviation of $-3.14°$ in pan and $-4.88°$ in tilt direction.. This could partially be explained due to the assumption of no rotation between IMU and PTU and needs further evaluation.

Acknowledgement

The work has been carried out at the German Aerospace Center, Institute of Flight Systems, and supervised by F. Ernst, Institute of Robotics and Cognitive Systems, Universität zu Lübeck.

Authors' Statement

Conflict of interest: Authors state no conflict of interest.

5 References

[1] M. Oliveira, A. Castro, T. Madeira, E. Pedrosa, P. Dias, and V. Santos, "A ros framework for the extrinsic calibration of intelligent vehicles: A multisensor, multi-modal approach," *Robotics and Autonomous Systems*, vol. 131, p. 103558, 2020.

[2] K. Koide, S. Oishi, M. Yokozuka, and A. Banno, "General, single-shot, target-less, and automatic lidar-camera extrinsic calibration toolbox," in *2023 IEEE International Conference on Robotics and Automation (ICRA)*, 2023, pp. 11 301–11 307.

[3] C. Yuan, X. Liu, X. Hong, and F. Zhang, "Pixel-level extrinsic self calibration of high resolution lidar and camera in targetless environments," *IEEE Robotics and Automation Letters*, vol. 6, no. 4, pp. 7517–7524, 2021.

[4] J. Lv, J. Xu, K. Hu, Y. Liu, and X. Zuo, "Targetless calibration of lidar-imu system based on continuous-time batch estimation," in *2020 IEEE/RSJ International Conference on Intelligent Robots and Systems (IROS)*. IEEE, 2020, pp. 9968–9975.

[5] F. Zhu, Y. Ren, and F. Zhang, "Robust real-time lidar-inertial initialization," in *2022 IEEE/RSJ International Conference on Intelligent Robots and Systems (IROS)*. IEEE, 2022, pp. 3948–3955.

[6] F. M. Mirzaei and S. I. Roumeliotis, "A kalman filter-based algorithm for imu-camera calibration: Observability analysis and performance evaluation," *IEEE transactions on robotics*, vol. 24, no. 5, pp. 1143–1156, 2008.

[7] P. Furgale, J. Rehder, and R. Siegwart, "Unified temporal and spatial calibration for multi-sensor systems," in *2013 IEEE/RSJ International Conference on Intelligent Robots and Systems*. IEEE, 2013, pp. 1280–1286.

[8] R. Y. Tsai and R. K. Lenz, "A new technique for fully autonomous and efficient 3 d robotics hand/eye calibration," *IEEE Transactions on robotics and automation*, vol. 5, no. 3, pp. 345–358, 1989.

[9] A. Schweikard and F. Ernst, *Medical robotics*, 1st ed. Springer, 2015.

[10] J. Li, G. Endo, and E. F. Fukushima, "Hand-eye calibration using stereo camera through pure rotations-fitting circular arc in 3d space with joint angle constraint," in *2015 IEEE International Conference on Advanced Intelligent Mechatronics (AIM)*. IEEE, 2015, pp. 1–6.

[11] "Atom: A general calibration framework for multi-modal, multi-sensor systems," *Expert Systems with Applications*, p. 118000, 2022.

Human-Robot Interaction in Industry: A Practical Approach on Integrating a Cobot into an Existing Workflow

Hannah Lück [1], Thomas Pisch [2], Javad Ghofrani [3]

[1] Robotics and Autonomous Systems, Universität zu Lübeck, hannah.lueck@student.uni-luebeck.de
[2] IBG Technology Hansestadt Lübeck GmbH, Lübeck, t.pisch@goeke-group.com
[3] Institute of Computer Engineering, Universität zu Lübeck, javad.ghofrani@uni-luebeck.de

Abstract

This paper explores the use of collaborative robots (Cobots) in industrial manufacturing, with a focus on efficient human-robot collaboration and safe hybrid mode operation. The study addresses a specific use case in an existing workflow, which includes the task of scanning two workpieces. For this, a program is created and safety measures are examined. To mitigate the impact on cycle time, a hybrid mode approach is adopted, allowing the Cobot to operate collaboratively selectively. A system of passive infrared sensors (PIR) and ultrasonic (US) sensors, controlled by an Arduino Uno Microcontroller, is integrated for human presence detection. This combines the movement detection of PIR sensors and the distance measuring of US sensors. Results highlight the potential benefits of Cobots in enhancing manufacturing efficiency, and the proposed multi-sensor system strategically utilizes false measurements induced by fabrics to further optimize its performance. Further exploration is recommended to refine the proposed sensor system.

1 Introduction

Industrial robots, from their beginning in the 1960s with the introduction of the Unimate [1], have progressed into sophisticated machines, incorporating advanced sensors and artificial intelligence. Unlike traditional industrial robots confined within safety enclosures, Collaborative Robots (Cobots) are designed to operate close to human workers. This characteristic enables flexible and dynamic production with advanced safety features, including external force-limiting technology, ensuring secure collaboration without safety fences[2].

While Cobots offer exciting prospects for human-robot collaboration, safety is still an important topic. Ensuring the safe coexistence of humans and Cobots requires careful planning, risk assessment, and adherence to safety regulations[3].

During this study, the goal was to investigate the integration of a Cobot in an existing industrial workflow, emphasizing efficient human-robot collaboration and safe hybrid mode operation, with the aim of enhancing productivity.

For this, a specific use case was chosen, for which a Cobot should be programmed. The selected application involved a task traditionally performed by humans: scanning two components of a workpiece, an exterior part with a matrix code and an interior part with a bar code. The scanned parts are then handed over to a worker for subsequent processing, potentially involving human-robot interaction. Due to this interaction, the robot needs to be in the collaborative mode. While a Cobot is in collaborative mode, it operates with limited speed and can be easily stopped by an external force. Although this is great for a collaborative application, it has a negative impact on cycle time.

Because of this, many collaborative robots like the YASKAWA HC10, which was used during this study, are so-called hybrid robots, that can operate in both, collaborative and normal modes. This leads to new opportunities since the robot only needs to be in collaborative mode when a human is nearby.

As a critical safety measure, the collaborative mode cannot be simply toggled through programming; instead, the signal to trigger this switch must come from an external device[7][8]. Because of this, we are going to add a system of passive infrared sensors (PIR) and ultrasonic (US) sensors. This Multi-Sensor fusion ensures the detection of human presence by detecting movements and ensuring that no human is standing in front of the system while it is not in collaborative mode.

In literature, PIR sensors are widely used. In [5] a dual PIR sensor system for human activity monitoring is used. In [4] a combination of PIR and US Sensors is used as a localization sensor network on a mobile social robot to track individuals within indoor spaces, estimating their movement direction and distance.

Like this work, our system combines the ability of the PIR to reliably detect human movements and the capacity of US sensors to measure the distance to an object. The sensors are controlled by an Arduino Uno Microcontroller, ensuring a cost-effective and user-friendly solution that can be easily adjusted.

2   Material and Methods

During this study, the Cobot YASKAWA MOTOMAN HC10DTP was used. To optimize the cycle time of the robot our aim was to implement a hybrid mode, wherein the Cobot operates collaboratively only when necessary.

The Cobot only needs to be in collaborative mode when a human is nearby. To achieve this, we explored employing PIR sensors and US sensors to reliably detect human presence.

2.1   Integration of the Cobot

Unlike traditional industrial robots, Cobots are designed with soft edges, minimizing the risk of a human getting caught between the robot's axes. This can be seen in Fig. 1, which shows the YASKAWA HC10DPT Cobot. The

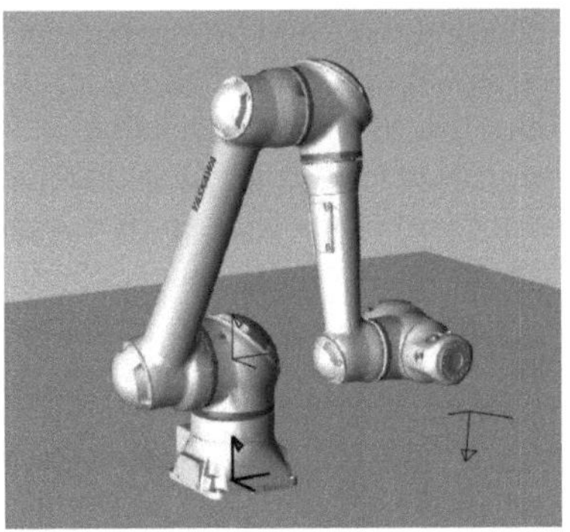

Figure 1: The Yaskawa HC10 Cobot in the MotoSim Simulation software.

Cobot's application involved picking up the interior part, scanning it, and placing it in front of the worker. Subsequently, the exterior part needed to be grabbed, scanned, and placed on top of the interior part with the matrix code facing the worker. Due to the robot's inability to perform a full rotation of the workpiece while gripped, there was a potential need for the robot to regrip the workpiece. Based on this understanding, the specifications for our test setup included: A feeding system for exterior and interior, a re-gripping location for both workpieces, and an area to place the scanned pieces as a symbolization of the worker's workplace.

With this information, we built a test setup that was suitable for our application while still being space-saving. The setup can be seen in Fig. 2.

To scan the codes on the exterior and interior parts, the Keyence Autofocus Code Reader was used. This scanner was mounted to match the height of the scan position. The scanner was connected via a PLC and Profinet.

2.2   Programming of the Cobot

For the scanning process, the robot rotates the workpieces around its z-axis, aiming to maximize the turning angle

Figure 2: Top down view of the test setup. 1 shows the feeding system for the exterior part, 2 the feeding system for the interior part, 3 the place for the stacked scanned parts, and 4 the regrip place.

while avoiding singularities that could lead to unpredictable behavior.

Predictability is crucial, to ensure comprehensive coverage of the entire workpiece area. A turn of 120 degrees was selected after extensive testing, as it proved to be unproblematic and accommodated the need to scan each workpiece a maximum of three times.

To prevent the omission of the matrix code or barcode, a total coverage of 126 degrees was chosen.

In terms of program planning, a master job is employed to call the sequence job, which does the further job calling. A variable indicates the selection between an exterior or interior for the next step. The program begins with the variable set to "Interior", indicating the initial scanning focus.

During the scan loop, the robot scans and regrips until the matrix code or barcode is successfully scanned, triggering an interrupt on the rising edge of the scan signal. Subsequently, the workpiece is transported to the worker.

It was challenging that the matrix code of the exterior needed to face the worker. An Interrupt Job was implemented to address this, triggered by the successful scanning interrupt. This Interrupt Job saved the current axes position, enabling the recreation of the scan position orientation.

Efficiency testing involved evaluating different positioning angles for the matrix code and barcode in eight different positions, 45 degrees apart. Results indicated that the barcode was scanned more efficiently and earlier than the matrix code. The barcode, although scanned earlier, lacked precise position control. In contrast, the matrix code was consistently scanned when directly in front of the sensor, leading to a more accurate reproduction of axis orientation.

2.3   Distance and Movement Measuring for a Safe Collaborative Usage

To optimize cycle time, the utilization of a robot in hybrid mode is proposed.

The hybrid mode dictates that the robot operates collaboratively only when necessary, particularly in the presence of a human. To achieve this, the integration of PIR sensors and

US sensors is explored for human presence confirmation.
PIR sensors are commonly employed for presence triggers.
The principle relies on the emission of heat by humans into
their surroundings with a wavelength of 9.4 μm. Given
that PIR detectors operate within a spectral range spanning
5-14 μm, they are suitable for detecting the presence of
humans[9].

They offer reliable motion detection, with minimal power
consumption and resilience to false triggers. However, their
limitations include the need for a direct line of sight. Ultrasonic sensors, providing distance measurement capabilities, complement PIR sensors by offering continuous coverage but face challenges with fabric sensitivity. The fusion
of these sensors aims to create a robust system for human
presence detection[6].

In the pursuit of optimizing our system for collaborative usage, we constructed a multi-sensor setup with US and PIR
sensors. The test configuration involved two PIR sensors
(HC-SR501) and three ultrasonic sensors (HC-SR04) controlled by an Arduino Uno Rev 3 microcontroller. This
selection was based on the components' affordability and
user-friendly characteristics, aligning well with the experimental nature of our testing. The test setup can be seen in
Fig. 3.

Figure 3: The test setup with the different sensors. The
three US sensors and the two PIR sensors are connected via
cables and bread boards with the Arduino Uno.

The PIR sensor was operated in repeatable trigger mode
to ensure continuous detection in the presence of human
movement. This mode allows the sensor to remain active
as long as human activity is detected, providing reliable
detection of ongoing movements. This choice is significant
for applications requiring continuous monitoring of the
environment for human presence, such as in our system
designed for collaborative functionality.

Moving to the code implementation, we leveraged the
distance data from the ultrasonic sensors to calculate
collaborative mode activation conditions. Collaborative
mode initiation depended on either PIR sensors detecting
human movement or ultrasonic sensors identifying an
object within a defined threshold. To enhance system
security, a timer was introduced, requiring a 5-second
absence of human activity before deactivating collaborative
mode. In this context, the US sensor was prone to generating inaccurate measurements when exposed to certain
fabric types. Consequently, we implemented a filtering
mechanism to exclude wrong readings from the dataset.

However, an inherent challenge arose when a person
remained stationary, leading to the PIR sensor failing
to detect them and the ultrasonic sensor struggling with
fabric detection, resulting in premature collaborative mode
deactivation.

To address this issue, we exploited the inherent high measurements produced by the ultrasonic sensor when faced
with certain fabrics. By establishing that these measurements only occurred due to human clothing, we introduced
a new detection threshold for high measurements as an
indicator of human presence. This adjustment increased the
system's robustness by mitigating the impact of stationary
individuals. While this occasionally resulted in detecting
humans farther away than the collaborative threshold,
testing indicated improved performance compared to
ignoring high measurements, facilitating earlier and faster
human presence detection.

3   Results and Discussion

The results from the collaborative robot (Cobot) programming reveal a meaningful addition to the manufacturing
process. By relieving humans of repetitive tasks, the
Cobot demonstrates its potential to enhance efficiency and
productivity. The integration of a Cobot into the work
environment optimizes the overall workflow, marking a
significant step towards a dynamic and flexible production
setting. To assess the sensor system's effectiveness, various
experiments were conducted.

The experiment involving a human walking from one side
to another showcased the advantages of a multi-sensor
system. The results can be seen in Fig.4. The PIR sensor
exhibits an earlier trigger response compared to the US
sensors. Additionally, the US sensors predominantly
register inaccurate readings during the walk, likely because
of the rapid movement surpassing the sensors' capacity
for relevant readings. Moreover, in the first graphic, it can
be seen, that the faulty measurements do indicate human
presence way earlier and better, than the US sensor would
normally detect them.

In situations where a person positioned themselves in the
proximity of the sensor system, the collaborative mode
would deactivate if the faulty sensor measurements were
ignored. But when we included the faulty measurements,
that occurred due to clothing, the collaborative mode
remained active despite the US sensors not measuring a
distance below the threshold.

Despite the challenges posed by clothing interference with
US sensor distance measurements, the system demonstrated
a robust ability to detect human presence, emphasizing the
effectiveness of a multi-sensor approach.

However, it is acknowledged that the current sensor system may not provide accurate distance measurements for
stationary individuals. Further exploration of alternative
sensor options or higher-quality ultrasonic sensors is recommended to address this limitation. The proposed Cobot

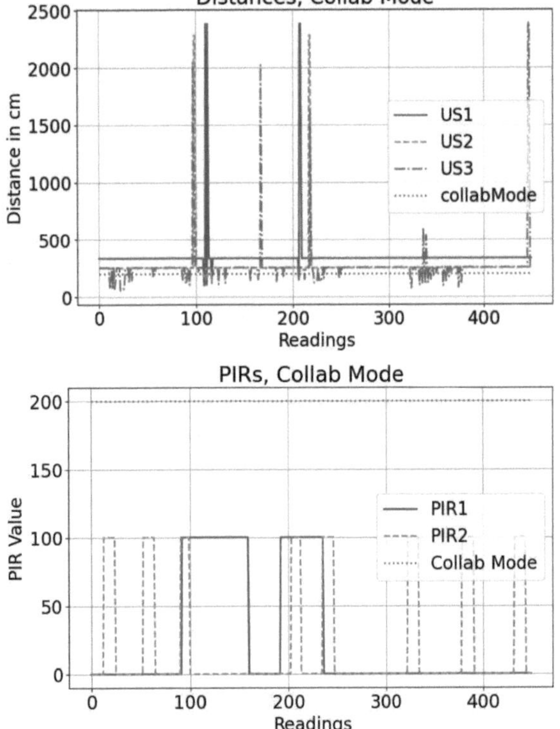

Figure 4: Measurements for the experiment where a human is walking side to side. The top graphic displays the three distance readings from the US sensors and the bottom graphic displays the readings from the PIR sensors.

programming approach presents a promising solution for safe and efficient human-robot collaboration. By leveraging human presence sensing, the system ensures a safe hybrid mode, prioritizing human safety while optimizing cycle time. While further testing and refinement are necessary, the concept holds great potential for application in critical industrial settings.

4 Conclusion

In conclusion, the integration of collaborative robots in industrial settings represents a significant advancement in manufacturing processes. Cobots, through their seamless collaboration with human workers, contribute to increased efficiency, reduced physical strain, and adaptability to evolving production demands.

While acknowledging the benefits of Cobots, their implementation requires careful consideration of safety measures, technical challenges, and the trade-off between cycle time and collaboration. The presented collaborative robot programming and sensor system for safe hybrid mode contributes to the ongoing exploration of effective human-robot collaboration in manufacturing environments.

The study demonstrates the promise of Cobots in addressing repetitive tasks. Additionally, the proposed sensor system, utilizing a combination of PIR and US sensors, shows potential for safe human-robot collaboration, although requiring further investigation and refinement.

As industries continue to evolve, the integration of Cobots, coupled with innovative programming and sensing solutions, holds the key to achieving enhanced productivity, safety, and efficiency in manufacturing processes.

Acknowledgement

The work has been carried out at IBG Technology Hansestadt Lübeck GmbH and supervised by the Institute of Computer Engineering, Universität zu Lübeck.

Authors' Statement

Conflict of interest: Authors state no conflict of interest.

5 References

[1] A. Gasparetto and L Scalera, *From the Unimate to the Delta Robot: the early Decades of Industrial Robotics.* In: Explorations in the History and Heritage of Machines and Mechanisms: Proceedings of the 2018 HMM IFToMM Symposium on History of Machines and Mechanisms, Springer, pp. 284–295, 2019

[2] SV. Sotnik, YS. Usenko and PV. Shakhov, "Safe cobots in development of industrial robotics", *Proceedings of VIII International Scientific and Practical Conference*, Barca Academy Publishing, 2023

[3] DGUV, *BG/BGIA risk assessment recommendations according to machinery directive: Design of workplaces with collaborative robots*, Deutsche Gesetzliche Unfallversicherung, 2011

[4] I. Ciuffreda, S. Casaccia and G. Revel, "A Multi-Sensor Fusion Approach Based on PIR and Ultrasonic Sensors Installed on a Robot to Localise People in Indoor Environments", *Sensors*, p.6963, 2023

[5] P. Hung, M. Tahir, R. Farrell, S. McLoone and T. McCarthy, "Wireless sensor networks for activity monitoring using multi-sensor multi-modal node architecture", *IET*, 2009

[6] L. Hodges, "Ultrasonic and Passive Infrared Sensor integration for dual technology user detection sensors", *Michigan State University*, p.5–6, 2009

[7] Yaskawa Robotics, *Anweisungen Kollaborierender Betrieb*,YASKAWA ROBOTICS, 2018

[8] Yaskawa Robotics, *Anweisungen für funktionale Sicherheitsfunktion*,YASKAWA ROBOTICS, 2018

[9] M. Shankar, J. Burchett, Q. Hao, B. Guenther and D. Brady, "Human-tracking systems using pyroelectric infrared detectors", *Optical engineering*, p.106401–106401, 2006

Towards the Development of a Robotic System for Automated Surgical Trocar Insertion

Sujay Mukherjee [1], Cem Adiyaman [2] Jannis Hagenah [3], Georg Männel [4], Dennis Kundrat [5] Dennis Wendt [6], and Floris Ernst [7]

[1] Medical Microtechnology, Technische Hochschule Lübeck & Universität zu Lübeck, sujay.mukherjee@stud.th-luebeck.de

[2] Fraunhofer Research Institution for Individualized and Cell-Based Medical Engineering, Lübeck, (cem.adiyaman; jannis.hagenah; georg.maennel; dennis.kundrat; dennis.wendt)@imte.fraunhofer.de

[7] Institute of Robotics and Cognitive Systems, Universität zu Lübeck, floris.ernst@uni-luebeck.de

Abstract

The global healthcare workforce shortage poses risks to patient well-being and increases the likelihood of medical errors. To address these challenges and mitigate the impact of demographic changes on labor shortages, automation in the medical industry, such as in surgeries, is essential. This research paper aims to demonstrate the feasibility of automation in surgical procedures, focusing on trocar insertion, achieved through the development of a gripper. Integrated into a robot designed for medical applications, the gripper securely holds the trocar. The system not only plans the robot's trajectory but also demonstrates the optimized surgical process in a phantom, simulating a laparoscopic surgery. Force analysis of the insertion has been performed, revealing similarities to actual surgeries. While recognizing the potential for refinement, this approach represents a pivotal step toward autonomous surgical processes even with a lack of trained personnel.

1 Introduction

The commencement of Laparoscopic surgery [3] typically involves the critical trocar insertion, marking the first exposure to the abdominal cavity. It is a type of minimally invasive surgery (MIS) that is performed through small incisions using small tubes, tiny cameras, and surgical tools. Skill-demanding primary procedures like trocar insertion require experienced surgeons, but healthcare understaffing and growing task complexity contribute to staff overload, jeopardizing patient well-being. Demographic shifts, particularly an aging population, further strain healthcare resources and increase the demand for health services.

In response, the use of robots in hospitals is expanding [4], with robot-assisted MIS gaining preference over traditional laparoscopy due to improved patient benefits and enhanced surgical capabilities in terms of dexterity, vision, and instrument handling. These advancements promotes the possibility for continued automation in specific surgical tasks to further enhance precision, consistency, and reduce surgeon workload [1]. Addressing these concerns holds the potential to optimize hospital throughput and reshape the future landscape of surgical procedures.

The vision of this research is to bring automation in surgeries, thus reducing staff workloads and increasing patient's safety in a long run. As a first step, this research aims to establish a demonstrator for a surgical process by optimizing its path and emulating the surgical procedures performed by the surgeons. It explores the controlled execution of the entire process within defined robot parameters, including velocities and forces, within an optimal time frame. The paper demonstrates automated trocar insertion in laboratory settings on abdomen phantoms using a fully functional, semi-automated gripper (for holding the trocar) integrated with a collaborative robot (cobot) – the LBR Med 14 R820 (KUKA AG, Augsburg, Germany). The conceptualization sets the groundwork for essential requirements in an autonomous surgical unit inside an operating room (OR) environment, representing a pivotal step towards potential solutions for fully autonomous surgeries.

2 Materials and Methods

This section delineates the hardware and software components employed and developed for the experiment, providing insight into their integration.

Trocar (a medical or veterinary device), essential in MIS, serves as a portal for the subsequent insertion of various endoscopic instruments, including graspers, scissors, staplers, electrocautery, and suction tips. It facilitates the passive evacuation of excess gas or fluid from internal organs. Modern trocars come in sizes ranging from 3mm to 12mm and fall into two main categories: *"Cutting Trocars"* and *"Dilating Trocars"*. For this project, an 11mm optical trocar, a type of dilating trocar has been used as the first trocar to provide primary entry point for endoscopic camera.

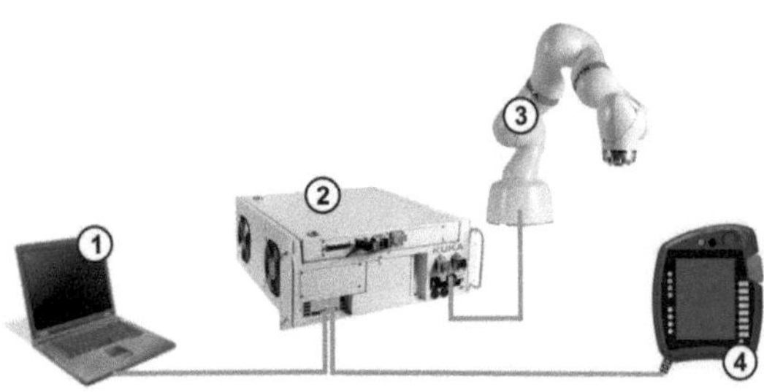

Figure 1: Overview of robot system 1)External computer for development, 2)KUKA Sunrise Cabinet, 3)Robot, and 4)KUKA Smartpad

Figure 2: Isometric view of the CAD model of gripper

Based on research, the most optimum method of insertion of a trocar is a rotating (torque acting around the trocar central axis) and pushing motion (force acting on the trocar central axis) motion [2]. This motion has been automated in this research project.

The robot used to perform this is LBR Med 14 R820 specially adapted to medical requirements. It is a seven-axis light-weight robot having seven degrees of freedom and comes with a versatile end effector interface making it easy for integration into medical products. It is designed as a safe, sensitive, precise, and flexible robot and is ideally suited for versatile assistance from diagnostics to treatment to surgical interventions. Fig. 1 shows the complete setup used to work with the robot.

This end effector interface provides connections for power supply, I/Os or Ethernet for various tools. The port X76 capable of providing a maximum of 60 V/ 5 A of external supply, present in the interface A1 which is located at the rear of the robot has been used to supply power and data (Pulse width modulation signals) to the media flange and thus to the gripper.

Grippers, a specific type of end-effectors are special devices fitted at the end of the robot, designed to help robots handle objects in the real world. In robotic surgery, grippers play a crucial role in performing delicate tasks such as suturing and tissue manipulation. In this project, a two-fingered electro-mechanical gripper has been developed to firmly hold trocars which when integrated into the robot allows it to smoothly penetrate the trocars in the predefined locations of the abdomen phantom.

The operating system used to operate the robot is KUKA Sunrise OS Med 2.6. It is a System Software package for robots in which programming and operator control tasks are strictly separated from one another. The robot applications are programmed with KUKA Sunrise Workbench. NEMA17-01, a bipolar stepper motor with 43.1 N cm rated torque and 1.8° step angle has been used to operate the gripper. To drive the motor by controlling the steps and direction, an A4988 stepper motor driver, an Arduino Mega 2560 microcontroller along with a 12 V-2 A external power supply were used. The stepper motor and stepper motor driver setup were connected with the end effector interface whereas the 12 V power supply and Arduino were connected to the other end i.e. with X76 port to control the

motor via an external microcontroller. Due to project time constraints, direct integration of the digital I/O pins present in the end effector interface, through the KUKA Sunrise Workbench was not successfully completed.

Several considerations were taken into account to ensure reliable experimental outcomes. The motor operated in a full-step mode with a limiting current of 1.2 A. Manual configuration of the X76 plug wiring facilitated integration into the A1 interface port of the robot for motor control. To avoid a collision, the experiment was conducted without obstacles around the robot's workspace. The insertion tests were performed on a static body (abdomen phantom) see Fig. 5, acknowledging that the dynamic changes observed in a human abdomen during laparoscopic procedures were not simulated.

2.1 Design of the Gripper

The primary objective was to achieve a lightweight design while optimizing grip performance and the intent was to control the gripper fingers by translating the motor power longitudinally. Various gripper design concepts were analyzed, considering the dimensions of the robot's media flange, and the motor's appropriate positioning and control. A detailed concept for the gripper's kinematics and mechanism was established, addressing housing space and strategic component placement. Initial sketches of components and their assembly, by incorporating major dimensions and addressing key considerations were made on paper to visualise the idea. A comprehensive Computer-Aided Design (CAD) model, created using Solidworks, further optimized through features such as collision detection and physical dynamics. The finalized CAD model served as the basis for 3D printing. To optimise the contact area, the components in direct contact with the trocar were printed with a soft material [6], Agilus (shore hardness 60, resolution 35 μm) as shown in Fig. 2 and Fig. 3. Remaining components were printed with PA2200 (resolution 100 μm), a strong, hard, and lightweight material commonly used for prototyping.

The design incorporated a slider between the gripper fingers and the motor, connected via a screw, flexible shaft coupling, and ball bearing. The motor's rotation translates the slider, enabling the gripper fingers to open and close. The final design was implemented for the 3D-printed proto-

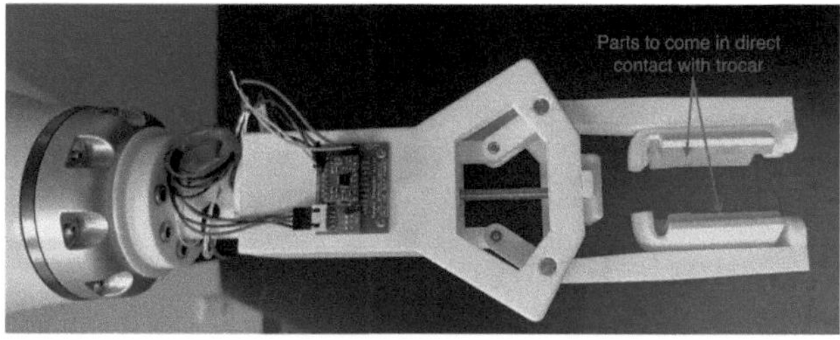

Figure 3: Front view of the 3D printed gripper (installed in the robot)

type. Figure 3 displays the initial prototype of a fully functional, two-fingered gripping unit with semi-automated capabilities, optimized through an angular gripping technique to maximize contact area with the cylindrical trocars.

The prototype measures 328 mm in height, 119 mm in maximum length, and 79.5 mm in maximum width, with a mass of 420g. The gripper can be easily attachable and detachable from the media flange, maintaining an open position to ensure a sterile interface and facilitate surgical draping.

2.2 Position Controlled Trajectory Planning and Force Analysis of the Robot

To operate the robot, a communication link was established between the KUKA LBR Med 14 R820 cabinet and an external computer. The Sunrise Workbench from KUKA was installed on the computer, and network settings were configured with the robot's cabinet. Once connected, the robot's trajectory was programmed in Java within a new project created under the Sunrise Workbench. Each Java file served as an application synchronized with the robot's cabinet by implementing the codes in it. The application could be executed using the KUKA smartpad to move the robot.

While planning, the objective was to assess the range of motion of the robot, to find a trajectory in point-to-point or linear motion, to mimic a surgeon's motions during trocar insertion in MIS, involving push and turn along the trocar's central axis, and to avoid potential singularities while performing the trajectory. The followed process has been explained via a flowchart in fig 4.

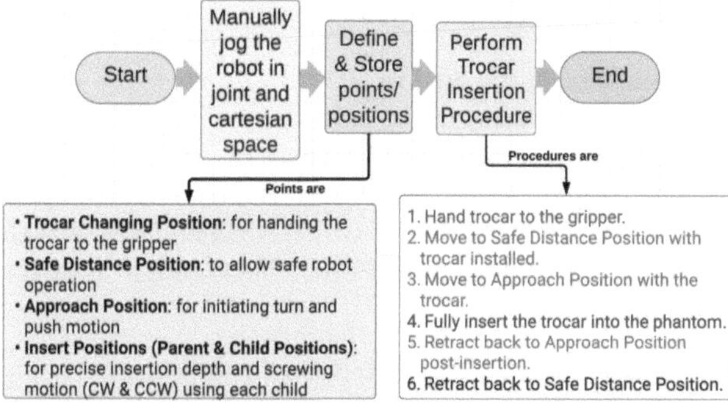

Figure 4: Flowchart of the entire trajectory planning

All points were defined in the WORLD coordinate frame, a fixed Cartesian coordinate frame with the robot's base as the

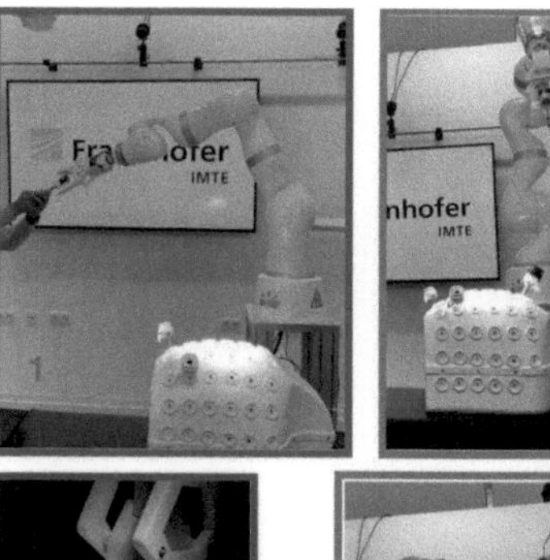
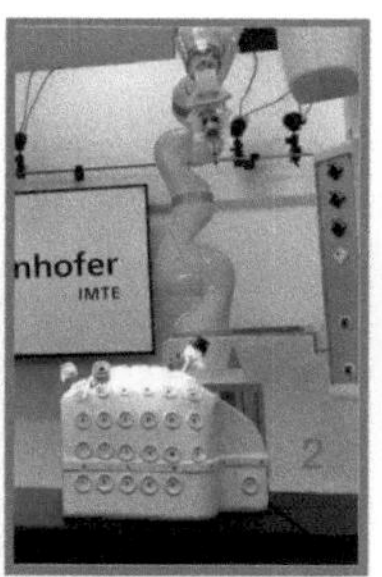
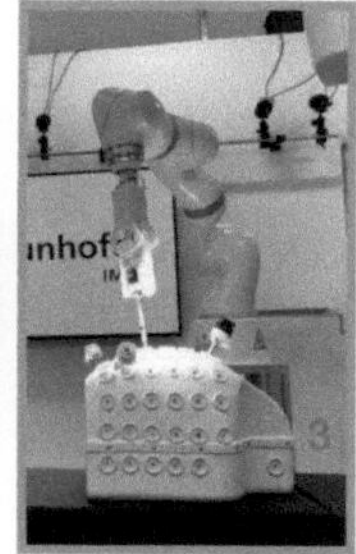
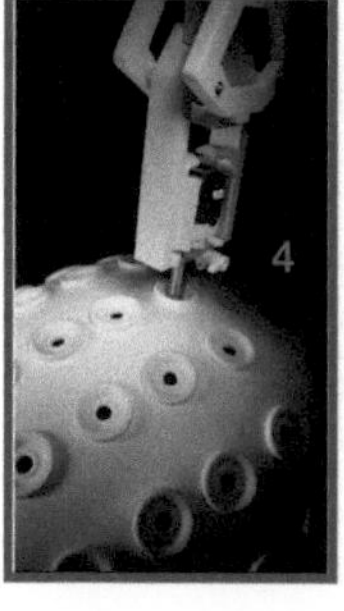
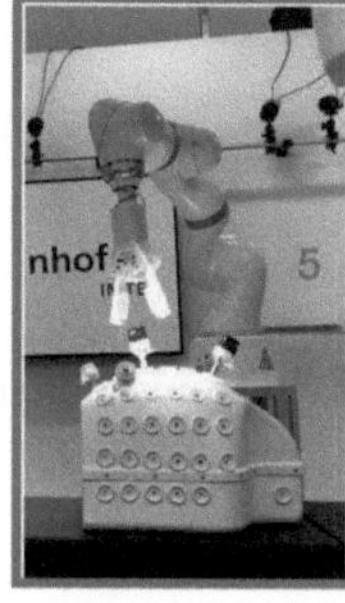

Figure 5: Robot performing the trocar insertion in the abdomen phantom (refer the numbers and color coding from flowchart in Fig. 4 to understand each frame)

center point. The WORLD frame uses (X, Y, Z, A, B, C) format, representing the robot's coordinates and Euler angles. The entire motion proceeded at a consistent velocity from point to point, with the lowest velocity during insertion set at 0.1×250 mm/s, which is the permissible jogging velocity of the robot for manual guidance.

Force plays a critical role in laparoscopic trocar insertion. Direct measurements of the force using tactile sensors in the gripper were not available in this project. Additionally, efforts to control the force exerted by the end effector pushing the trocar could also not be successfully completed within this preliminary investigation.

Instead, a comparative analysis was conducted by recording force data from the robot along the X, Y, and Z coordinates during trocar insertion inside the abdominal phantom. This data was then graphically compared to the process without actual trocar insertion. Despite the absence of direct force control, the analysis revealed a similarity in the maximum force during insertion, aligning with findings from two distinct research studies on trocar insertion.

3 Results and Discussion

Following the integration of the 3D printed gripper with the robot, along with meticulous wiring connections, the insertion procedure occurred in a laboratory with a realistic surgical environment. The robot, programmed with a defined trajectory of a surgical motion, was controlled using the smartpad and Arduino code simultaneously on an abdomen phantom positioned over a bed. Figure 5 visually captures the complete motion of the robot.

Trials were done through spatial points and linear velocity manipulations of the robot to optimise the process and achieve a comparable time to actual trocar insertion. For

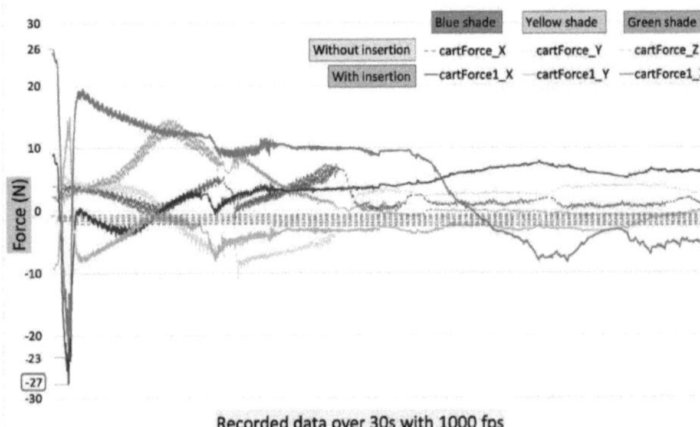

Figure 6: Line graph illustrating force comparison (dark solid lines represents the procedure with trocar insertion and light dashed lines represents without an actual insertion)

this project the final test recorded 36 secs for the complete insertion and further 10 seconds to retract the robotic arm back. This observed time closely aligns with findings from a related study utilizing the Visiport trocar system performed by surgeons taking a mean time of 37.7 ± 15.59 seconds for 50 patients [5].
A study indicates force fluctuations of 16.83 N to 61.86 N during trocar insertion in laparoscopic surgery in a human abdomen [2]. Another investigation reported a maximum force of around 50 N during insertion into streaky pork [3]. The force analysis results for this project, depicted in Fig. 6, exhibit spikes along the all three axes of the WORLD coordinate frame of the robot. The maximum force (calculated as the length of the resulting Vector) was 38.5 N. This force aligns with the range reported in the first study and is comparable to the latter.

4   Conclusion and Outlook

In conclusion, the development of the gripper prototype, its seamless integration with the robot, and the meticulous trajectory planning have effectively demonstrated the feasibility of automated trocar insertion. The research outlines an optimized MIS procedure, imitating the actions of skilled surgeons. The obtained results indicate significant similarities in terms of insertion force and trocar placement time.
This study illustrates the potential for creating an automated system that integrates seamlessly with surgical processes. It lays a foundational framework for addressing the shortage of healthcare personnel in the medical industry and marks significant progress toward the realization of fully autonomous surgical procedures in the future.

The project's scope encompasses the refinement and optimization of the gripper design to enhance its adaptability to various trocar types. The integration of image processing and sensor technologies could further extend the scope by enabling autonomous trajectory tracking, identification of anatomical landmarks, and real-time force feedback for dynamic force control, ultimately improving the precision of trocar insertion. Future exploration in this field, coupled with machine learning integration, may contribute to advanced capabilities such as collision detection and avoidance with both static and dynamic objects.

Acknowledgement

The work has been carried out at Fraunhofer Research Institution for Individualized and Cell-Based Medical Engineering, Lübeck.
Research funding: This work was partially funded by the European Union - European Regional Development Fund (ERDF), the Federal Government and Land Schleswig Holstein, Project No. 12420002 and No. 12422005.

Authors' Statement

Conflict of interest: Authors state no conflict of interest.

5   References

[1] Tyler J. Loftus, Amanda C. Filiberto, Jeremy Balch, Alexander L. Ayzengart, Patrick J. Tighe, et al., *Intelligent, Autonomous Machines in Surgery*, Journal of Surgical Research, Volume 253, 2020, Pages 92-99, ISSN 0022-4804

[2] Nillahoot, N., Pillai, B.M., Sharma, B., Wilasrusmee, C. and Suthakorn, J., *Interactive 3D Force/Torque Parameter Acquisition and Correlation Identification during Primary Trocar Insertion in Laparoscopic Abdominal Surgery: 5 Cases*.Sensors 2022, 22, 8970. https://doi.org/10.3390/s22228970

[3] Sun, Junpeng, Tadano and Kotaro, *Force Characteristics of Trocar Insertion Abdomen in Laparoscopic Surgery*. Booktitle:Proceedings of the 3rd World Congress on Electrical Engineering and Computer Systems and Science (EECSS'17), Rome, Italy. pages:4-6

[4] Paula Gomes, *Surgical robotics: Reviewing the past, analysing the present, imagining the future*. Volume 27, Issue 2, 2011, Pages 261-266, ISSN 0736-5845. https://doi.org/10.1016/j.rcim.2010.06.009.

[5] Mohammadi M, Shakiba B and Shirani M., *Comparison of two methods of laparoscopic trocar insertion (Hasson and Visiport) in terms of speed and complication in urologic surgery*. Biomedicine (Taipei). 2018 Dec;8(4):22. doi: 10.1051/bmdcn/2018080422. Epub 2018 Nov 26. PMID: 30474603; PMCID: PMC6254099.

[6] Giovanni Rateni, Matteo Cianchetti, Gastone Ciuti, Arianna Menciassi and Cecilia Laschi, *Design and development of a soft robotic gripper for manipulation in minimally invasive surgery: a proof of concept*. Meccanica 50, 2855–2863 (2015). https://doi.org/10.1007/s11012-015-0261-6

Simulation-based Fault Analysis in Object Detection Models

Alen James [1], Ahmed Mahmoudi [2], and Mladen Berekovic [2]

[1] Robotics and Autonomous Systems, Universität zu Lübeck, alen.james@student.uni-luebeck.de
[2] Institut für Technische Informatik, Universität zu Lübeck, {ahmed.mahmoudi, mladen.berekovic}@uni-luebeck.de

Abstract

The ubiquity of deep neural networks (DNNs) in critical applications demands a meticulous examination of their resilience to faults. This research investigates the sensitivity of DNNs, focusing on bit-flip faults induced by diverse environmental factors. Employing fault injection techniques on a pre-trained SSD MobileNet V2 FPNLite 640x640 model, we analyze the impact of weight perturbation on model accuracy. The study reveals nuanced fault tolerance patterns across layers and individual weights. Results highlight the importance of protecting hardware to mitigate unexpected errors. Object detection tests illustrate differences in outcomes on trained and faulty models, emphasizing the need for more reliable hardware.

1 Introduction

Artificial intelligence, especially machine learning and deep learning (DL), has emerged as a widely used solution in a variety of industries, to solve complex challenges and applications in areas such as automotive, aerospace, and healthcare. The success of deep neural networks (DNNs) over traditional machine learning algorithms is evident in tasks such as image and video recognition, text classification, and speech interpretation. The integration of deep learning systems in safety and security-critical domains, such as healthcare, financial systems, autonomous vehicles, etc., has raised concerns about potential malicious faults, leading to subtle data corruption.

Convolutional neural networks (CNNs) belong to a class of DNNs designed to automatically understand hierarchical representations of data, making them particularly effective for image-related tasks. In object detection, CNNs works by sequentially applying tensor multiplications to input images, utilizing a set of parameters referred to as weights. This complex process takes place in several layers, each with its own unique weight parameters. The resulting tensor undergoes successive multiplication by these weights, allowing the network to understand the complex patterns and hierarchical properties of the image. In such a framework, any fault occurring in the parameters of a specific layer within the CNNs has the potential to propagate through the network structure to subsequent layers, potentially impacting the network's ability to perform classification tasks, such as not identifying an incoming obstacle [1].

Neural networks, like any software, are susceptible to common random hardware faults, including bit-flips induced by factors like voltage fluctuations, increased heat, radiation, or electromagnetic induction. These environmental factors can corrupt the CPU or RAM, causing a change in the value of stored variables [2]. Consequently, it is important to understand how DNNs operate in the presence of random faults.

This research focuses on evaluating the sensitivity of DNNs to faults, with a specific emphasis on bit-flips, by employing fault injection techniques. We systematically analyze a pre-trained SSD MobileNet V2 FPNLite 640x640 model using the KITTI dataset. Through the strategic introduction of random weight modifications, we explore the nuanced impact on model accuracy. The toggling of weight bits for fault injection provides valuable insights into the model's fault tolerance and susceptibility, revealing potential weaknesses.

2 Related Work

Numerous researchers have explored the resilience of DNNs against faults, considering their diverse execution on various computing devices like GPUs, ASICs, and FPGAs [1]. Santos et al. [3] delved into the reliability of object detection algorithms, executing CNNs on various NVIDIA GPU architectures to assess the impact of transient faults on network reliability. Their findings indicated that a single fault in a GPU could propagate to multiple active threads, significantly diminishing CNNs reliability.

Further research focuses on analyzing faults and attacks within major memory-stored network parameters. Rakin et. al. [4] present a novel DNNs weight attack approach known as Bit-Flip Attack which can crush a neural network by maliciously flipping an extremely small amount of bits within its weight storage memory system.

Reagen et al. [5] propose a fault-injecting framework that conducts large-scale fault injection experiments. The authors investigate the fault tolerance of various neural networks and conclude that DNNs fault tolerance varies by orders of magnitude with respect to model, layer type, and structure. Li et al. [6] injected single and multi-bit faults into quantized network parameters, demonstrating their increased sensitivity to faults compared to floating point values.

Another line of related work is the study revolves around assessing the robustness of DNNs in the presence of bit errors, specifically when Error Correction Codes (ECCs) are deactivated [7]. This research critically examines the intricate trade-offs between storage, bandwidth considerations, and prediction accuracy, particularly in the context of neural networks residing in noisy media. Furthermore, they introduce a binary representation of parameters as an alternative that mitigates the distortions introduced by bit-flips.

Investigating fault tolerance in CNNs, an analysis [8] conducted through a series of fault-injecting experiments highlights the reliability increases from quantization. They analyze the most susceptible bits in the IEEE 754 format representation and also conduct an inspection of the network architecture to determine the most vulnerable layers at the layer level.

3   Methodology

In this section, we provide a comprehensive account of our approach for conducting fault analysis in the object detection model, dividing the methodology into three subsections. Commencing with the *Object Detection Model and Dataset*, an overview of the selected model and dataset upon which our fault analysis is conducted. Subsequently, we delve into the *Fault Injection Technique* employed to inject faults to simulate and analyze the model's response. This segment encompasses the type of fault considered and defines the function employed to toggle bits for fault injection. The final part of the section presents the *Implementation*, elucidating the execution of our experiment using the object detection model and how intentional faults are introduced within the network to study the potential impact on the model's performance.

3.1   Object Detection Model and Dataset

We have selected the pre-trained SSD MobileNet V2 FPN-Lite 640x640 model, sourced from the TensorFlow model zoo [9]. This variant of the Single Shot Multibox Detector (SSD) integrates the lightweight MobileNet V2 as its backbone feature extractor, augmented by a Feature Pyramid Network (FPN) for enhanced object detection capabilities. Its key components are the following:

- SSD is a popular object detection framework that aims to detect objects in images using a single forward pass of the neural network. It achieves this by predicting multiple bounding boxes and their corresponding class scores in a single pass.

- MobileNet V2 is a lightweight neural network architecture specifically designed for mobile and edge devices. It employs depthwise separable convolutions

to reduce the number of parameters and computations while maintaining good accuracy. This makes it suitable for real-time applications with resource constraints.

- FPN is a feature extraction technique that enhances the ability of a model to detect objects at different scales. It does so by creating a pyramid of feature maps with different resolutions, allowing the model to capture both high-level semantic features and fine-grained details.

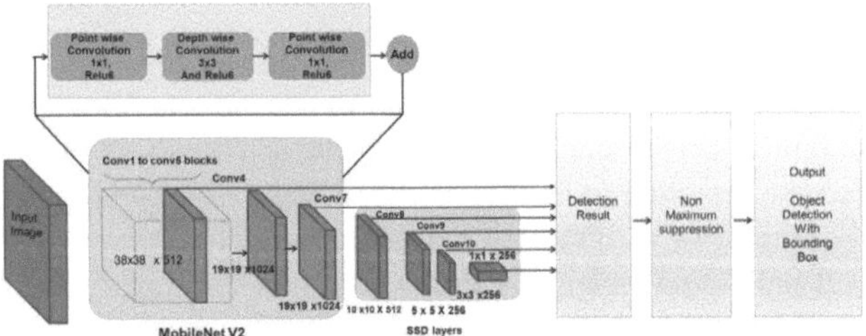

Figure 1:   Illustrative Representation of the SSD-MobileNetV2 Object Detection Framework.

The model is designed to process images with a resolution of 640x640 pixels. This resolution is a balance between capturing sufficient details for accurate object detection and maintaining computational efficiency. Fig. 1 provides a conceptual overview of the SSD MobileNetV2 object detection framework. This model is chosen for fault injection analysis due to its proficiency in efficient and accurate object detection, rendering it apt for deployment on devices with limited computational resources.

Moving to the dataset, our choice is the KITTI dataset sourced from the Object Detection Evaluation [10]. This dataset comprises 7481 training images and 7518 test images, encompassing a total of 80,256 labeled objects. The extensive annotations and variety of objects within the dataset contribute to a robust fault injection analysis, providing insights into the model's response across different scenarios.

The combined utilization of the SSD MobileNet V2 FPN-Lite 640x640 model and the KITTI dataset establishes a strong foundation for our fault analysis, enabling a comprehensive exploration of the model's resilience and susceptibility to faults in real-world conditions.

3.2   Fault Injection Technique

Bit-flip is a common type of fault that occurs in digital systems, including computer hardware and software. It refers to the unintentional change of a binary digit (bit) from 0 to 1 or vice versa. This phenomenon can occur due to various reasons, such as electrical noise, radiation, etc., that affect the stability of digital data storage. In the context of neural networks, when weights or parameters in the model are stored as binary data, a bit-flip can lead to an alteration of these values.

We use the concept of bit-flip fault injection as a controlled technique to intentionally introduce bit-flips into the binary representation of weights or biases in a neural network. This intentional injection allows us to study how the model responds to such faults and assess its resilience to errors that might occur in real-world hardware.

We define the *toggle_bit* function that efficiently flips the bit at a specified index in the binary representation. This operation involves converting the binary string into a list of individual characters, toggling the bit at the specified index, and subsequently rejoining the modified list into a new binary string. The function encapsulates the fundamental operation in fault injection processes, enabling the deliberate manipulation of individual bits within binary representations of neural network weights or biases.

3.3   Implementation

In this section, we present a comprehensive overview of the experimental execution involving the SSD MobileNet V2 FPNLite 640x640 model in conjunction with the KITTI dataset. The implementation process unfolds through several key phases, commencing with the creation of label maps, and TF records, and updating the configuration file to ensure seamless integration with the chosen object detection model. Subsequently, we trained the model on the KITTI dataset for 20,000 training steps to ensure robust learning. Post-training, the model underwent the evaluation phase to gauge its accuracy in object detection tasks. Further, we analyzed the *model.trainable_variables* to identify crucial layers for fault injection. Notably, layers 54, 146, and 228 were selected as the starting, middle, and ending layers for targeted fault injection, considering their critical roles in the overall architecture. Furthermore, we selected 100 weights from each layer and imposed a constraint on the bit positions within the range of 24 to 32. This selection ensures exploration of the impact of bit-flip faults within a controlled and meaningful range.

The fault injection process was characterized by the deliberate introduction of faults into each weight within the identified layers. This entailed the conversion of selected weights, typically represented as floating-point numbers, into their binary equivalents. The focus of our bit-flip injection methodology was the intentional toggling of specific bits in the binary representation. Subsequently, the *toggle_bit* function was applied to flip specific bits in the binary representation. The toggling function efficiently alters the state of the selected bit, simulating a bit-flip fault. The modified binary representation, now containing the introduced fault, is used to update the corresponding weight or bias in the neural network model. The model is then evaluated and tested with the injected fault to observe its impact on performance.

4   Results and Discussion

Here, we present the results obtained from our experiment involving the SSD MobileNet V2 FPNLite 640x640 model

and the intentional injection of faults to examine their impact on the model's performance.

To encapsulate our findings, we deliberately selected 100 random weights from three different layers, representing the beginning, middle, and ending layers of the neural network, for targeted fault injection. Each layer was studied independently. Initially, we injected faults into each weight using the *toggle_bit* function with randomly chosen bits in the binary representation. However, we observed that when the bit index ranged from 0 to 23, the fault injection did not yield significant changes in weight values, thereby impeding the feasibility of a meaningful accuracy analysis. To address this, we imposed a constraint on the bit positions, confining them to the range of 24 to 32 during the fault injection process. This targeted approach specifically focuses on the exponent and sign bits, introducing substantial changes to the float values.

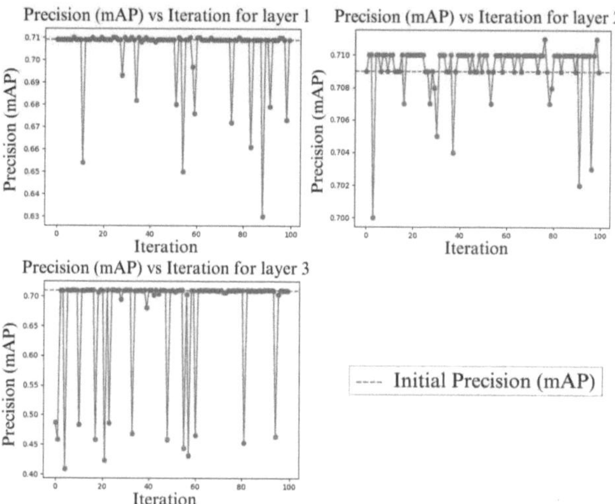

Figure 2: mAP vs Iteration across each layer.

Upon completion of fault injection, each layer was analyzed, and the Mean Average Precision (mAP) was compared against the iteration count, where each iteration represented a fault injection in the selected layers. Fig. 2 illustrates the fluctuations in mAP corresponding to each injected fault in the starting, middle, and ending layers, respectively. The observations revealed that layers, in the beginning, were less affected by injected faults in accuracy changes compared to later layers. The middle layer exhibited an increase in accuracy for certain faults in the weights and demonstrated moderate sensitivity to injected faults. Conversely, the ending layer displayed greater fluctuations in accuracy, with a notable decrease in mAP values for specific faults.

Additionally, the experiment highlighted the varying sensitivity of individual weights within each layer, providing insights into the significance of specific weights. This understanding sheds light on the pivotal role played by specific weights, unraveling their varying degrees of importance within the network's architecture. Such insights contribute to a deeper comprehension of the intricate dynamics of weight sensitivity across different layers.

Figure 3: Comparison of Detection Tests: Top image: Trained Model, Bottom image: Faulty Model.

Furthermore, we conducted object detection assessments on the same set of test images using the trained model and the model injected with faults. For the sake of comparison, we chose the fault model that exhibited diminished accuracy in our experimentation. In Fig. 3, the top image illustrates the output of the detection test on the trained model, while the bottom image displays the results for the model subjected to fault injection. The results, illustrated in the figures, showcase the differences in detection outcomes attributed to the injected faults.

Our experiment underscores the impact of bit-flip faults in the binary representation of weights, revealing changes in trained parameters that affect the network's task performance. This emphasizes the critical need to secure hardware data from faults, as observed in the experiment.

5 Conclusion

In conclusion, our study underscores the significance of hardware reliability for deep neural networks, emphasizing the potential impact of bit-flip faults on model performance. The experiment reveals layer-specific and weight-dependent fault sensitivity, contributing to a comprehensive understanding of neural network robustness. Future work involves extending this analysis to transformer-based models, aiming to broaden insights into diverse architectures and enhance overall model resilience. Protecting neural network hardware against faults is imperative for ensuring reliable AI applications in critical domains.

Acknowledgement

The work has been carried out at the Institute of Computer Engineering (ITI), under the supervision of Prof. Dr. Mladen Berekovic and M.Sc. Ahmed Mahmoudi, Universität zu Lübeck.

Authors' Statement

Conflict of interest: Authors state no conflict of interest.

6 References

[1] S. Ribes, A. Malek, P. Trancoso, and I. Sourdis, "Reliability analysis of compressed cnns," *Secur. Hardw.-Softw. Arch. Robust Comput. Syst*, 2021.

[2] M. Beyer, A. Morozov, E. Valiev, C. Schorn, L. Gauerhof, K. Ding, and K. Janschek, "Fault injectors for tensorflow: evaluation of the impact of random hardware faults on deep cnns," *arXiv preprint arXiv:2012.07037*, 2020.

[3] F. F. dos Santos, P. F. Pimenta, C. Lunardi, L. Draghetti, L. Carro, D. Kaeli, and P. Rech, "Analyzing and increasing the reliability of convolutional neural networks on gpus," *IEEE Transactions on Reliability*, vol. 68, no. 2, pp. 663–677, 2018.

[4] A. S. Rakin, Z. He, and D. Fan, "Bit-flip attack: Crushing neural network with progressive bit search," in *Proceedings of the IEEE/CVF International Conference on Computer Vision*, 2019, pp. 1211–1220.

[5] B. Reagen, U. Gupta, L. Pentecost, P. Whatmough, S. K. Lee, N. Mulholland, D. Brooks, and G.-Y. Wei, "Ares: A framework for quantifying the resilience of deep neural networks," in *Proceedings of the 55th Annual Design Automation Conference*, 2018, pp. 1–6.

[6] G. Li, S. K. S. Hari, M. Sullivan, T. Tsai, K. Pattabiraman, J. Emer, and S. W. Keckler, "Understanding error propagation in deep learning neural network (dnn) accelerators and applications," in *Proceedings of the International Conference for High Performance Computing, Networking, Storage and Analysis*, 2017, pp. 1–12.

[7] M. Qin, C. Sun, and D. Vucinic, "Robustness of neural networks against storage media errors," *arXiv preprint arXiv:1709.06173*, 2017.

[8] M. A. Neggaz, I. Alouani, S. Niar, and F. Kurdahi, "Are cnns reliable enough for critical applications? an exploratory study," *IEEE Design & Test*, vol. 37, no. 2, pp. 76–83, 2019.

[9] Tensorflow model zoo. Accessed: August 18, 2023. [Online]. Available: https://github.com/tensorflow/models/blob/master/research/object_detection/g3doc/tf2_detection_zoo.md

[10] The kitti vision benchmark suite. Accessed: August 22, 2023. [Online]. Available: https://www.cvlibs.net/datasets/kitti/eval_object.php?obj_benchmark=2d

Practical Palletizing
with Real World Considerations

Max-Ole Bastian von Waldow[1], Floris Ernst[2], and Eduard Marti Bigorra[3]

[1] Robotics and Autonomous Systems, Universität zu Lübeck, maxole.vonwaldow@student.uni-luebeck.de
[2] Institute for Robotics and Cognitive Systems, Universität zu Lübeck, floris.ernst@uni-luebeck.de
[3] Discharge and Dismantling, Northvolt Revolt AB, Västerås, eduard.marti.b@northvolt.com

Abstract

We explore automated palletizing in a factory setting where a robot picks items and stacks them onto pallets to reduce storage and shipment costs and to simplify item handling. Offline 3D bin packing as well as batched online 3D bin packing are examined for the palletizing calculation. Knowledge of future items is added to the offline heuristic, improving the packing density, and transforming the algorithm into the batched online version of bin packing. A one-pass heuristic and a metaheuristic are employed to solve the bin packing problems. Practical issues are considered by adding various constraints to the algorithms that tackle stability of the items and collision avoidance of the robotic gripper with items on the pallet as well as with external obstacles around the pallet. Our efforts result in a safe and robust stacking of items.

1 Introduction

As part of any material processing flow, items need to be transported, stored, and handled. Items are unpacked and then packed onto standard pallets – a step called palletizing in the following – which makes them easily accessible. The pallets can be transported and stored in a standardized, easy and economic fashion. A good packing technique reduces wasted space on each pallet. The goal is to minimize the pallets needed for a given set of items or to maximize the items packed on each pallet. These requirements boil down to optimizing the packing density by intelligently choosing the position and orientation of each item on the pallet and by assigning suitable items to the right pallets. The technique must also respect different constraints to ensure a safe and robust packing regarding collisions and stability. In literature, palletizing falls under the extensive field of *bin packing* or *nesting* problems.

2 Related Work

The goal in bin packing settings is to maximize the (worth of) items fitting in a container or to minimize the number of containers needed or the size of the smallest container that fits all items. Bin packing can be done in varying dimensions. The 1D case coincides with the classical knapsack problem and optimizes which items are packed [1]. 2D bin packing additionally positions each item on a plane to optimize the area usage. This can reduce material waste e.g. in laser cutting [2] or memory space when storing digital textures in a bit map [3]. We consider the 3D case where items are also stacked on top of each other.

Bin packing variations exist regarding the available knowledge and accessibility of items. We distinguish the offline and the online case. In *offline bin packing*, the algorithm has full knowledge of all items to be packed. The input (the items to be packed) as well as the output (the containers to allocate the items into) are freely accessible. On the other hand, *online bin packing* is used when only one item is known and accessible at a time. It may still be placed freely in any output container [4].

The constraints of the *online bin packing* problem can be relaxed to form further variations. *Closed online bin packing* is a relaxation where the accessibility is unaltered, but a queue of n incoming items is known instead of just the next item. This additional knowledge can be used to improve the packing density by keeping room for not yet available items. A further relaxation is *batched online bin packing* [5]. In this case, again n items are known and a batch of $m \leq n$ items are accessible freely.

2.1 Approximations

A common assumption in bin packing is that all items or all bins are of the same size. For many methods it is also common not to consider different orientations of an item. Some only allow rotations by 90 degrees. Allowing more options quickly increases the search space.

Approximating items as rectangles or cuboids and keeping them aligned through limited rotation reduces the complexity without too much impact on the packing quality. While this is a promising simplification, there are cases where optimal packing requires non-orthogonal placements.

2.2   Different Approaches to Palletizing

Optimization solvers find the optimal solution to bin packing problems but require large computation times and efforts need to be taken to keep constraints linear [6]. This approach is useful when the task can be written as a mixed integer programming (MIP) problem and only deals with few items to be packed due to the computational effort.

Heuristic methods place items using empirically proven rules. They are approximate algorithms but terminate very quickly and are thus useful for large problem instances. *One-pass heuristics* do not revisit placed items to optimize the packing. The quality of the result depends on the order in which items are placed as they are not changed if a following one cannot fit properly. In online bin packing, the order of incoming items is fixed, but in offline bin packing, the items can be sorted by different criteria, commonly by size, that empirically improve the packing.

In exchange for longer computations, variations of heuristics can bring quality improvements. *Iterative heuristics* such as simulated annealing begin using an initial solution and iteratively improve upon it [7].

Unlike the previous constructive heuristics, a *metaheuristic* does not generate the solution itself. Rather a metaheuristic fine-tunes a constructive heuristic, calling it repeatedly. An example is the *Greedy Adaptive Search Procedure (GASP)*, which attempts to optimize the order of items for a one-pass heuristic [6]. Metaheuristics can also be based on evolutionary algorithms [8] or swarm intelligence and organic behavior [9].

3   Problem Formulation and Algorithm Selection

3.1   Defining and Categorizing the Problem

The problem can be defined as an offline 3D bin packing problem where all items are approximately cuboid shaped. Full knowledge of the active bins and of every item in the active batch is available and they are freely accessible. This packing is performed many times with new active items and bins. There is an input stream of batches that arrive sequentially and there is an output stream of empty pallets. The problem can thus be expanded to a batched online bin packing. Additional knowledge of incoming packs is available to further optimize the individual offline packings and make them better fit together. We have access to an active batch of m items as well as knowledge of n items consisting of the active batch and a single additional batch of potentially different size.

3.2   Heuristics and Sorting

We choose an offline technique and later modify it into a batched online bin packing technique using future knowledge. The offline method we choose is sardine-can [6], a well written package of heuristics and optimization solvers.

We choose *Extreme Point Insertion* as a constructive one-pass heuristic which will be referred to as the *algorithm*.

3.3   GASP - Random Ordering

The order in which the items are sorted by the offline heuristic is a best guess and typically yields better results than using unordered input. Still, it is smarter to occasionally deviate from the order of items and to shuffle the orientations in which items are attempted to be added first. GASP is a meta-heuristic from the sardine-can package which we will use together with Extreme Point Insertion. While GASP supports a score-based order, even random shuffling can improve the results.

4   Stability

4.1   Constraints and Considerations

The packing solution must be such that it can be performed by a robot with a gripper. In the considered case, the gripper grips each item from two opposing sides. The large gripper also extends over the edges of the top of the item. To prevent the gripper from colliding with the floor, the item can only be placed on its bottom surface but be freely rotated around the upwards pointing z-axis. For easier computation, we restrict the orientations to multiples of 90 degrees. Because only the bounding box of each item is important for nesting, it does not matter whether the orientation is 0 degrees or 180 degrees or similarly 90 degrees or 270 degrees. This leaves only two distinct options.

Besides the rotation, a margin around each piece needs to be considered. The goal of this margin is to (1) have a safe tolerance when placing items for robustness against inaccuracies. Another margin is added on just two sides to (2) leave space for the gripper opening motion.

4.2   Mechanical Principles of Stability

To ensure stability, each item in the packing must be properly supported. It is assumed that every item has the same uniform density and is cuboid-shaped with a level surface for an item on top. Some error tolerance is necessary due to these approximations and to handle external accelerations when handling the final pallet.

Error margins in the stability considerations might reduce packing density but help to avoid accidents. We consider three different stability cases with and without margins of error.

(1) **Direct support:** The center of gravity (CoG) of an item is supported directly by another item. As a margin, a region around the center of gravity is required to be supported. This still allows for an item to safely overhang over another.

(2) **Indirect support:** The CoG might be free hanging when an item bridges the gap between two or more supporting items. The item is stable if a level line through the CoG exists from one supporting region to an opposing one [10]. This can also be extended to a region around the CoG.

(3) **Item groups:** By adding weight on the overhanging side of a previously stable item, the combined CoG shifts which might lead to the group of items tipping [10]. This is a rare scenario and will not be explored in this work. The previous error margins in addition to a limited stacking height help to mitigate instabilities related to item groups.

5 Multi-Batch Knowledge

5.1 Chaining Bin Packing Operations

In a factory scenario, multiple bin packing operations are chained to handle a continuous material inflow and a continuous outflow of full pallets. A pallet is replaced once full and a batch is replaced once empty. As these are not synchronized, bin packing must be able to be performed on a non-empty pallet. The remaining free space is used as a non-cuboidal container for the next packing operation by adding virtual obstacles to the standard cuboidal container.

5.2 Stackability of Next-Batch Items

While only a single batch is accessible to the robot at a given time, knowledge about queued batches can improve the packing quality for a pallet that will contain items from multiple batches. With the added information, the current batch can be packed in a way that might be less dense on its own, but that leaves space to accommodate future items.

A future batch is not available to be picked and placed until the former one is complete. Therefore, a future batch item may not be buried by items from the current batch. We take inspiration from [6], where fragile items can be marked as *non-stackable*, prohibiting any other item from being packed on top to avoid crushing the item. In our case, marking future batch items as *non-stackable* ensures that these items are placed last. We ignore the stackability flag if both the top and the supporting item are part of the same future batch. The rule is enforced by modifying the aforementioned stability consideration. A placement that violates the rule is deemed unstable and rejected.

These modifications transform the offline algorithm into a batched online bin packing algorithm. Each item is used by the palletizing algorithm multiple times and is part of a rolling window of varying size over the input and output streams. The flow of an item is visualized in Fig. 1. It is first used as a future item, then as an item to be packed - which might cover multiple pallets - and finally as an obstacle. The quick runtime of the heuristic makes this approach possible. Instead of a single complex algorithm that might take a long time to run, the task is split and performed through multiple quick runs of the modified heuristic.

5.3 Knowledge Horizon Length

We use a horizon length of $h = 1$ and thus have a single batch $b_i = b_1$ as future knowledge, where $0 \geq i \geq h;\, h, i \in \mathbb{N}$. This is extendable to multiple batches, but the usefulness of a larger horizon length depends on the

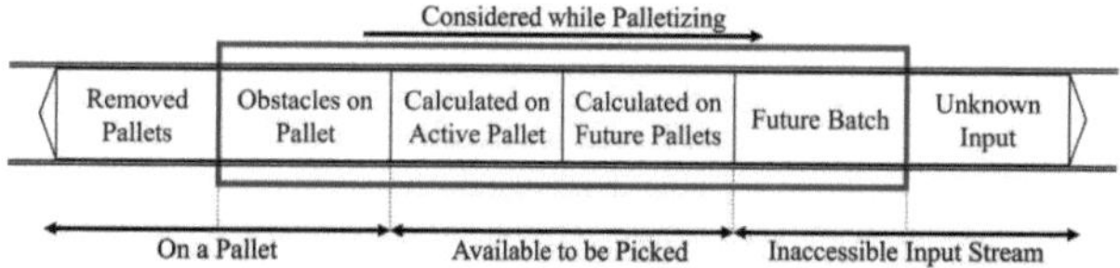

Figure 1: A window moving forward in time of items included in the palletizing. Items start on the right as unknown input and are part of the steps in between before becoming part of pallets that are then removed and stored.

relation between batch size and pallet size. An item from a batch b_i must not be stacked on one from a batch b_{i+1}. If the area of the container is small in relation to the space needed by an average batch, then the items mostly get stacked vertically and a batch b_{i+1} might be separated from the current batch b_0 by a batch b_i. This reduces the influence the batch b_{i+1} has on current items. At that point the gain from the added knowledge becomes minimal and is not worth the additional computation effort of palletizing with more items.

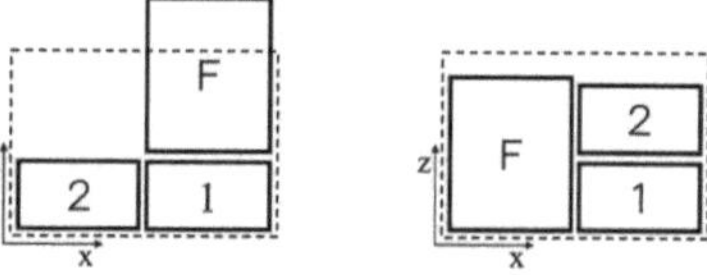

Figure 2: A side view of a packing without (left) and with (right) future knowledge enabled. Two active items $(1, 2)$ are rearranged to accommodate a tall future item (F) that would otherwise not fit the height limit (dotted boundary).

6 Gripper Collision Avoidance

6.1 Internal Items

When placing an item, the large gripper must not collide with any of the other previously placed pieces. These might be taller than the current item, in which case the gripper needs to move past them to place the item without a high drop. If a collision is anticipated, the current item is moved over to clear the gripper from the obstacle. A new palletizing solution is generated that includes a virtual obstacle in the shape of the required spacing. This spacing is lost space on the final pallet and is calculated to be just as big as necessary. In case the new solution still collides, another spacing obstacle is added. The process continues, iteratively getting closer to a collision free packing.

6.2 External Obstacles

Items must be palletized in a way that the overhanging gripper stays clear of any external obstacles outside of the pallet, such as fencing around the robot cell. If an item is to be placed too close to the obstacle, its location is rejected and the algorithm must find a new location.

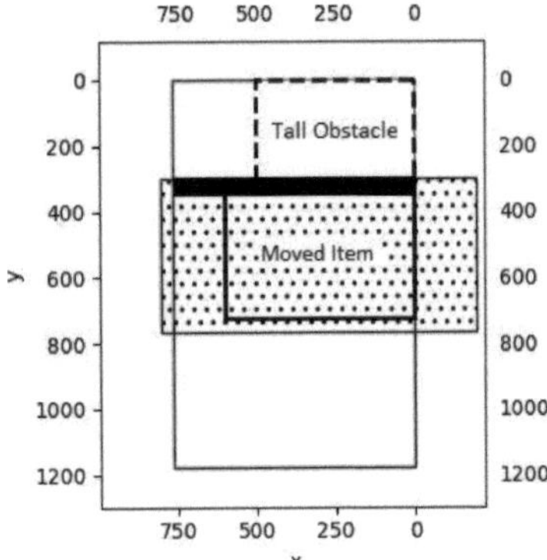

Figure 3: Virtual spacing (black rectangle) moves the new item to avoid a collision with the gripper (dotted).

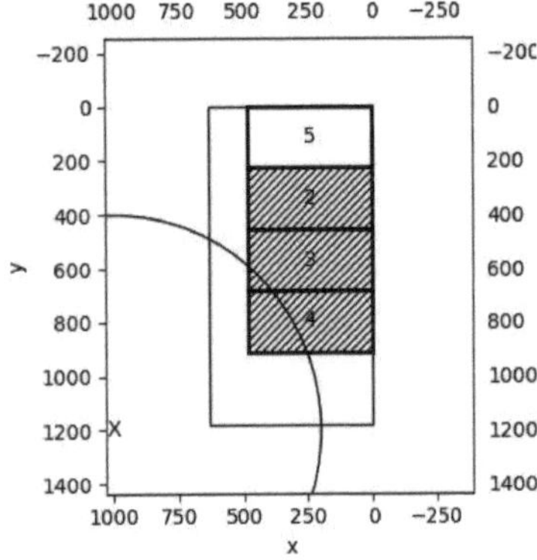

Figure 4: Item 5 would fit at the bottom but was moved out of the avoidance radius of an external obstacle marked X (checking the item center) and on top of four hatched items.

7 Results

Fig. 3 and Fig. 4 show a top view of the pallet and demonstrate successful internal and external obstacle avoidance respectively. In the example in Fig. 2, future knowledge rearranges two active items to accommodate a tall future item that would not fit otherwise. Improvements in packing densities depend on combinations of item sizes and are not further quantified here.

8 Conclusion

In this work, offline 3D bin packing and batched online 3D bin packing are used for robotic palletizing in a production setting. Practical issues are considered such as stability and collisions of the gripper with internal and external obstacles. The use of future knowledge is demonstrated using an example where an additional item can be packed.

Acknowledgement

The work has been carried out at Northvolt Revolt AB and supervised by the Institute for Robotics and Cognitive Systems, Universität zu Lübeck. The project was sponsored in part by Vinnova, Sweden's innovation agency.

Authors' Statement

Conflict of interest: Authors state no conflict of interest.

9 References

[1] J. Puchinger, G. R. Raidl, and U. Pferschy, "The Core Concept for the Multidimensional Knapsack Problem," in *Evolutionary Computation in Combinatorial Optimization: 6th European Conference, EvoCOP 2006, Budapest, Hungary, April 10-12, 2006. Proceedings 6*. Springer, 2006, pp. 195–208.

[2] W. Ao, G. Zhang, Y. Li, and D. Jin, "Learning to Solve Grouped 2D Bin Packing Problems in the Manufacturing Industry," in *Proceedings of the 29th ACM SIGKDD Conference on Knowledge Discovery and Data Mining*, 2023, pp. 3713–3723.

[3] Z. Yang, Z. Pan, M. Li, K. Wu, and X. Gao, "Learning based 2D Irregular Shape Packing," *ACM Transactions on Graphics (TOG)*, vol. 42, no. 6, pp. 1–16, 2023.

[4] S. Liu and X. Li, "Online Bin Packing with Known T," *arXiv preprint arXiv:2112.03200*, 2021.

[5] G. Gutin, T. Jensen, and A. Yeo, "Batched Bin Packing," *Discrete Optimization*, vol. 2, no. 1, pp. 71–82, 2005.

[6] M. Merschformann, "Models and Algorithms for Extended Three-Dimensional Bin Packing Problems in Air Cargo Applications," Master thesis, University of Paderborn, 2014.

[7] R. Rao and S. Iyengar, "Bin-Packing by Simulated Annealing," *Computers & Mathematics with Applications*, vol. 27, no. 5, pp. 71–82, 1994.

[8] X. Wang, L. Gong, H. Zhao, B. Li, and M. Tian, "A 3D Offline Packing Algorithm considering Cargo Orientation and Stability," *International Journal of Distributed Sensor Networks*, vol. 2023, 2023, Article ID 5299891.

[9] S. Rajesh Kanna and K. Udaiyakumar, "A Complete Framework for Multi-Constrained 3D Bin Packing Optimization using Firefly Algorithm," *International Journal of Pure and Applied Mathematics*, vol. 114, no. 6, pp. 267–282, 2017.

[10] J. de Castro Silva, N. Soma, and N. Maculan, "A Greedy Search for the Three-Dimensional Bin Packing Problem: The Packing Static Stability Case," *International Transactions in Operational Research*, vol. 10, no. 2, pp. 141–153, 2003.

4

Sensor Data Analysis

Realtime assessment of food spoilage using multiple sensors

Samira Assarzadeh [1], Alexander Altmann [2], Martin Leucker [3], Mohammad Khodaygani [3], and Ramtin Rahmanzadeh [2]

[1] Biomedical Engineering, Luebeck University of Applied Sciences, samira.assarzadeh@stud.th-luebeck.de

[2] Institute of Biomedical Optics, Universität zu Lübeck, {alexander.altmann, ramtin.rahmanzadeh} @uni-luebeck.de

[3] Institute for Software Engineering and Programming Languages, Universität zu Lübeck, {Leucker, Moham-mad.Khodaygani} @isp.uni-luebeck.de

Abstract

This study aims to develop a method for monitoring fish spoilage at various temperatures using a Raspberry Pi and connected sensors. The study measured different parameters affected by fish spoilage, such as oxygen, humidity, pressure, temperature, Total Volatile Organic Compounds (TVOC), and equivalent Carbon Dioxide (eCO2). The data was compared with the fluorescencespectrum of Prophyrin-based sensor films using two spectrometers. The films react upon amine exposure and change the fluorescence and color based on the amine amount present. A high-quality Raspberry Pi camera captured RGB images of the sensor films. The sensor film's RGB values and spectra were then transformed into the u ´v ´ chromaticity values for correlation analysis. The study conducted bacterial growth experiments and analyzed the sensor data to establish a method for predicting food freshness using only images from standard mobile cameras, which will be a valuable tool for waste reduction.

1 Introduction

In the food industry, ensuring the quality and safety of food is a top priority. It is crucial for consumer welfare, as well as for the sustainability of the supply chain. The industry is dedicated to maintaining high standards beyond regulatory requirements to ensure public trust. One of the biggest challenges in this process is detecting and managing food spoilage, especially in perishable items like fruits, vegetables, and meats that can degrade quickly.[1][2]

To address this issue, sensor technologies have been developed that use electrical and chemical sensors to detect gases like O2, NH3, H2S, CO2, and TVOC, which are spoilage indicators [1]. Colorimetric pH sensor foils have also been introduced, which change color in response to specific gases or amines, providing a quantifiable way to measure food spoilage [2]. These tools are essential for a proactive food safety system, allowing for early detection and intervention.

Our research takes this one step further with a new approach to detecting fish spoilage at ambient temperature. We have designed an integrated system that incorporates cutting-edge technology and analytical methods. The cornerstone of our design is a Raspberry Pi equipped with sensors that monitor vital environmental factors, recording data on the conditions that lead to spoilage.

2 Material and Methods

2.1 Setup for Fish Experiments

A setup comprising a glass box with a polypropylene (PP) lid stored in a cooling box (VEVOR refrigerator) with a defined temperature was used to investigate fish spoilage. For the experiment, the glass box was equipped with an absorbing sheet (Fresh pads) to decrease the humidity in the box. The box contained approximately 200 g of Alaska Pollock fillets purchased from a local retailer. Before the experiments, the fillets were thawed in a standard refrigerator at 5 degrees Celsius for eight hours. The lid of the box was equipped with different sensors to observe different parameters over time. The sensors were controlled by a Raspberry Pi 4. A DFRobot Gravity I2C Oxygen Sensor (SEN0322, detection range: 0-25%) was manually calibrated and used to measure oxygen electrochemically. The Pimoroni PIM480 SGP30 Sensor was used to measure eCO2 (0-60,000 ppm) and TVOC (400-60,000 ppb), and the results were normalized to a 0-100% scale for ease of understanding. A BME/BMP 280 sensor was used to measure temperature, humidity, and barometric pressure. A porphyrin-based sensor film (750 ppm porphyrin in polyethylene, including 100 ppm titanium dioxide) is used for a color read-out [3]. For color measurement indication, a Raspberry Pi High-Quality Camera, coupled with a 35mm Teleobjective Lens, captures high-resolution images of the sensor film, aided by an SK6812 LED string to provide illumination during image capture. the second film per package was also used to measure the fluorescence change. The sensor film was attached to the lid by adhesive tape or parafilm. Horiba FlouromaxPlus and ATTO 3 were used to measure fluorescence spectra. Fig. 1 displays the experimental setup, including the VEVOR refrigerator, Sensors, Raspberry Pi 4, Sensor Foils, and ATTO 3 Spectrometer.

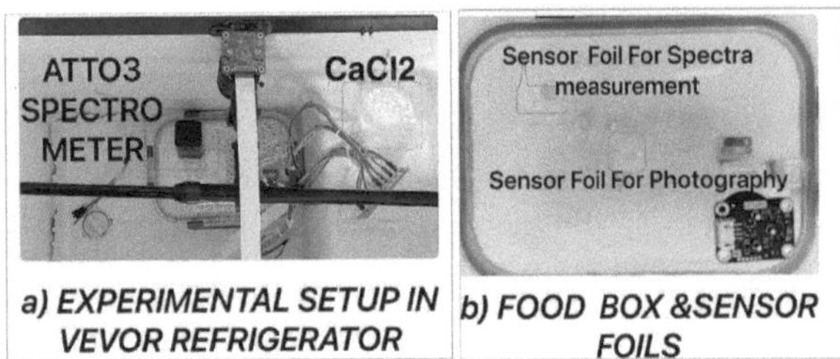

Figure 1: Experimental Setup including VEVOR refrigerator, Sensors, Raspberry pi 4, Sensor Foils, ATTO 3 Spectrometer

2.2 Acquisition and Processing of Color and Spectral Data

During each experimental set, four images were analyzed at 0, 12, 24, and 40 hours at room temperature. To assess the sensor foil images, ImageJ software, was utilized to determine the mode RGB values, which were first normalized by dividing by 255, then transformed to u´v´Chromaticity values using equations (1), (2) and (3). The spectrometer readings were normalized by dividing the intensity values by the maximum intensity recorded during the experiments. The Fast Fourier Transform (FFT) smoothing function was applied using Origin Pro 2024 to minimize unwanted fluctuations and noise. The Origin Chromaticity Diagram app was used to transform wavelengths ranging from 600 nm to 780 nm into the CIE 1976 u'v' color space.

2.3 Transformation Equations of Normalized RGB to u´v´ Chromaticity Values

First, the Gamma correction will be used to align color and light representation with human perception and convert non-linear to linear scales [4](1).

$$R', G', B' = \begin{cases} \frac{R_n, G_n, B_n}{12.92}, \text{if } R_n, G_n, B_n \leq 0.04045 \\ \left(\frac{R_n, G_n, B_n + 0.055}{1.055}\right)^{2.4}, \text{otherwise} \end{cases}$$
(1)

Second, the gamma-corrected RGB values will be converted to XYZ [4](2).

$$\begin{bmatrix} X \\ Y \\ Z \end{bmatrix} = \begin{bmatrix} 0.4124 & 0.3576 & 0.1805 \\ 0.2126 & 0.7152 & 0.0722 \\ 0.0193 & 0.1192 & 0.9505 \end{bmatrix} \begin{bmatrix} R' \\ G' \\ B' \end{bmatrix}$$
(2)

Finally, the XYZ value will be converted to the u´v´ chromaticity coordinates [4](3).

$$u' = \frac{4X}{X + 15Y + 3Z}, v' = \frac{9Y}{X + 15Y + 3Z}$$
(3)

2.4 Detection of total viable count (TVC) of bacteria

To determine the TVC of microorganisms present in the fish samples, plate-count agar (PCA, NutriSelect grade), monopotassium phosphate (reagent grade), and sodium hydroxide (reagent grade) were purchased from Merck (Darmstadt, Germany). We followed U.S. Food and Drug Administration (FDA) guidelines to analyze microbial growth on fish fillets [5] [6]. After homogenizing the fillets, dilutions were made, and 1 ml of the dilution was pipetted into a Petri dish. 10-12 mL Plate Count Agar was added to the dilution and mixed by hand. The total viable count was counted, and the bacterial growth was calculated as colony-forming units (CFU) per gram, according to FDA guideline [6](4). Only Petri dishes with 250 or fewer bacterial colonies were used to calculate each sample set's median CFU/ gram to determine the number of aerobic bacteria present[6].

$$\frac{CFU}{ml \text{ or gram}} = \frac{TVC}{\text{Volume of sample per (ml or gram)} \times \text{Dilution}}$$
(4)

3 Results and Discussion

3.1 Sensor values

An experiment was conducted at a stable temperature of 24°C to understand fish spoilage at stressed storage conditions. The aim was to measure TVOC, eCO2, and O2 levels to detect spoilage onset, as depicted in fig. 2. The experiment revealed that spoilage began after 24 hours when oxygen levels fell below sharply from 20 %Vol and later got stable at 13 % Vol at 40 hours, showing intense microbial respiration. This was marked by an exponential increase in TVOC levels, indicating heightened bacterial degradation of fish Tissue. The peak of TVOC was at 100 % ppb at 31 hours before plateauing. At the same time, eCO2 levels surged, peaking at 100 % ppm at 30 hours before reducing to around 45 % ppm at 37 hours. This shift might indicate a change from aerobic to anaerobic microbial activity. TVOC levels remained high at 100 % ppb, and eCO2 eventually stabilized at around 45 % ppm after declining. The peak levels of TVOC and eCO2 aligned with detecting a spoiled odor, indicating a highly contaminated environment within the experiment's confines.

3.2 Images of the sensor foil over 40 hours

As can be seen from table 1 and 2, smartphone and Raspberry Pi cameras had different reactions to colors. The u' value derived from iPhone images shifted from 0.189 to 0.211, which means an absolute delta value of 0.022 and indicates a move towards the red axis. On the other hand, the Raspberry Pi camera shows less significant u' value change from 0.1809 to 0.190, with a delta of 0.0091, showing a slighter redshift. The iPhone's measurements closely match

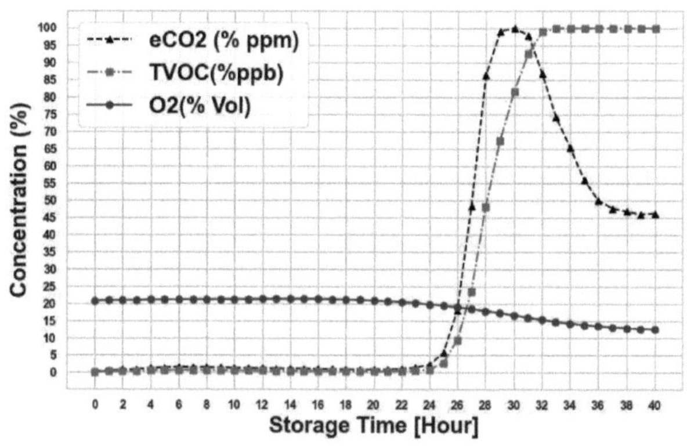

Figure 2: TVOC, eCo2 and O2 measurement over 40 hours at room temperature

hours, indicating an advanced reaction. A consistent rise in u´ values from 0.212 to 0.242 confirm a change in the foil's colorimetric reaction.

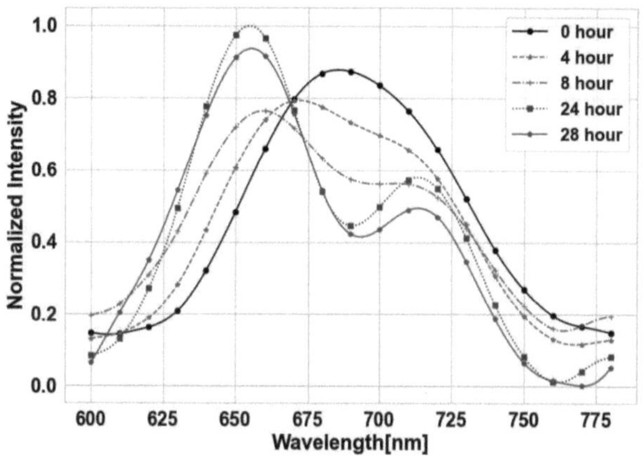

Figure 3: Normalized intensity against wavelength for sensor foil at room temperature measured by ATTO3

those of human vision, while the differences between the smartphone and Raspberry Pi camera results could be due to the lighting in the experimental setup. The LED light that was used gave a greenish tint, which altered the color detection and perception. Furthermore, the light's low intensity failed to light the inside of the refrigerator sufficiently.

Table 1: RGB and u´v´ values of Sensor Foil taken with Raspberry Pi camera at room temperature

Time	RGB	u'	v'
0 h	(96,117,102)	0.181	0.483
12 h	(106,120,105)	0.182	0.477
24 h	(102,122,112)	0.187	0.484
40 h	(93,102,87)	0.190	0.487

Table 2: RGB and u´ v ´values of Sensor Foil taken with iPhone camera at room temperature

Time	RGB	u'	v'
0 h	(125,141,58)	0.189	0.537
40 h	(147,135,97)	0.211	0.507

3.3 Spectra readout taken from sensor foil over 28h

As it can be seen from fig.3, and table 3, the spectral response of a sensor foil measured by ATTO3 changes over 3me, with initial peaks at 680 nm (intensity 0.85) shifting to 670 nm (intensity 0.8) in four hours and forming dual peaks at 660 and 720 nm (intensities 1 and 0.6) after a day. A change in u' values from 0.115 to 0.211 suggests a color transition towards red/magenta. The u' value fluctuates from 4 to 24 hours may be due to the P-P-lid used between the spectrometer and the foil. This setup could have caused unwanted reflections and noise, affecting the readings. On the other hand, as it can be seen from fig.4, and table 4, shows that initial peaks at 670 nm (intensity 0.6), moving to 660 nm (intensity 0.5) in four hours and forming peaks at 655 and 710 nm (intensities 1 and 0.4) after 24

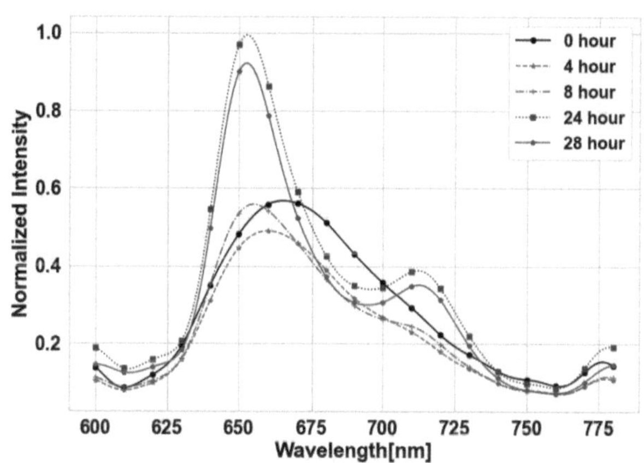

Figure 4: Normalized intensity against wavelength for sensor foil at room temperature measured by Flouromax

Table 3: u' and v' values from spectra by ATTO3

ATTO3	u'	v'
0 h	0.115	0.572
4 h	0.113	0.571
8 h	0.108	0.572
24 h	0.102	0.572
28 h	0.211	0.558

Table 4: u' and v' values from spectra by Flouromax

Flouromax	u'	v'
0 h	0.212	0.560
4 h	0.209	0.561
8 h	0.227	0.559
24 h	0.235	0.558
28 h	0.242	0.557

3.4 Detection of total viable count

Fig.5 shows the TVC levels over 28 hours, which indicates a significant increase in microbial count over time. The low microbial count initially suggests freshness or effective preservatives, but TVC levels increased rapidly, indicating potential spoilage. The steep increase in TVC levels after 12 hours shows an alarming microbial growth rate and sample deterioration. After 28 hours, the point of full spoilage $(5 * 10^5)$ was not observed [7]. However, bacteria growth increased exponentially after 12 hours, confirming the trend of O2, eCo2, and TVOC values.

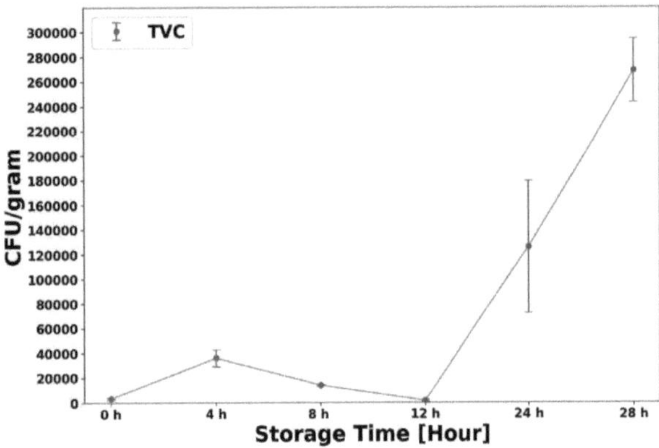

Figure 5: TVC at room temperature over 28 hours

4 Conclusion

In our experiment conducted at room temperature, we observed a consistent pattern in sensor readings. We found that a reduction of oxygen levels below 20 % volume caused an increase in both Total Volatile Organic Compounds (TVOC) and equivalent Carbon Dioxide (eCO2) levels. On the other hand, when oxygen levels stabilized or plateaued, the levels of TVOC and eCO2 either remained stable or decreased. Images captured with an iPhone revealed a correlation between sensor color change and chromaticity coordinate u', which trended towards the red axis. Spectrometer results showed a reaction of the sensor foil, resulting in wavelength peak transformations. After 24 hours, bacterial growth increased exponentially, approaching the reference threshold for food spoilage at 28 hours. Additionally, we aim to extend our investigation to include experiments at lower temperatures, such as 5 and 10 degrees Celsius to thoroughly examine the stability of our results under different thermal conditions. We intend to simplify our equipment by utilizing a commonly available mobile phone paired with a lightbox, which offers a more user-friendly and "Real-life experience method" compared to the previous setup with a Raspberry Pi camera and LED string. This adjustment in our experimental tools is anticipated to not only streamline the process but also enhance the accessibility of our method. Collecting a wider range of data, we aim to establish and validate our experimental outcomes.

Acknowledgement

The work has been carried out at the Institute for Biomedical Optics(BMO) ,University of Lübeck.

Authors' Statement

Authors state no conflict of interest.

5 References

[1] K. Lee, S. Baek, D. Kim J. Seo, *A freshness indicator for monitoring chicken-breast spoilage using a Tyvek® sheet and RGB color analysis*. Food Packaging and Shelf Life, 19, pp. 40–46, 2019.

[2] E. Osmólska, M. Stoma, A. Starek-Wójcicka, *Application of Biosensors, Sensors, and Tags in Intelligent Packaging Used for Food Products—A Review*. Sensors, 22, 9956, 2022.

[3] A. Altmann, M. Eden, G. Hürmann, C. Schell, R. Rahmanzadeh, *Porphyrin-based sensor films for monitoring food spoilage*. Food Packaging and Shelf Life, 38, 101105, 2023.

[4] D. Malacara, *Color Vision and Colorimetry: Theory and Applications*. 2nd ed., Press Monograph, vol. 204, ISBN 978-0-8194-8397-3, 2011.

[5] U.S. Food and Drug Administration, *BAM Chapter 1: Food Sampling/Preparation of Sample Homogenate*.Available online: https://www.fda.gov/food/laboratory-methods-food/bam-chapter-1-food-samplingpreparation-sample-homogenate.

[6] U.S. Food and Drug Administration, *BAM Chapter 3: "BAM Chapter 3:Aerobic Plate Count*.Available online: https://www.fda.gov/food/laboratory-methods-food/bam-chapter-3-aerobic-plate-count.

[7] J. A. Sciortino, R. Ravikumar, *Fishery Harbour Manual on the Prevention of Pollution - Bay of Bengal Programme*. Fish Quality Assurance, Chapter 5, 1999.

Advancing Robotic Perception:
Extracting Gyroscopic Data from Camera Streams

Steffen Häusler [1],

[1] Robotics and Autonomous Systems, Universität zu Lübeck, steffen.haeusler@student.uni-luebeck.de

Abstract

This paper presents a novel approach to enhancing robotic perception by extracting gyroscopic data (pitch, roll, and yaw) from camera streams, particularly focusing on its application in autonomous robots with hardware limitations for additional sensors. The project, executed during an internship at STIHL, involved various methods, starting with feature extraction using OpenCV's Scale-Invariant Feature Transform (SIFT) detector and culminating in a hardware-based approach using motion vector analysis of H.264 encoded camera streams. This method proved superior due to its real-time processing capabilities and minimal jitter, making it ideal for dynamic and challenging outdoor environments. The research also explored the adaptability of these methods to different cameras, particularly with a Raspberry Pi camera. The findings indicate significant potential for enhancing the spatial awareness and operational efficiency of autonomous robots, providing insights into future developments in robotic navigation and perception.

1 Introduction

This paper aims to be a report of a project undertaken during an internship at STIHL, a company renowned for its expertise in outdoor craftsmanship tools, as well as robotics applications, particularly in battery-powered devices such as lawnmower robots. These robots have a specific selection of on-board sensors, that help them navigate their environment. The goal of this project is to expand the sensors available to the robot without changing its hardware, but rather by processing already available data. Specifically, this report will cover the extraction of gyroscopic rotational data in terms of the pitch, roll, and yaw rotation from a video stream to improve the navigation of robots, that are equipped with a camera, but not a gyroscope.

To achieve this goal, this report will cover multiple approaches, ordered in phases. They are all executed in Python, aiming to improve future Roboter Operating System (ROS) compatibility. Initially, current state-of-the-art approaches have to be understood for this project. The following approaches were considered for this:

1. Analyzing the optical flow of single image frames [1]

2. Training an artificial neural network on sensor data [2]

3. Creating a feature detector to compare successive frames [3] [4]

4. Exploiting the hardware encoding of the camera for extracting the motion vector [5]

Before an approach can be pursued further, it is important to understand the limitations of this project. The script has to run on low-powered hardware and should not require significant storage space. Also, the gyroscopic data should be created in real-time, in the case of a lawnmower robot, that means at least 10Hz. Finally, the reading should be reliable and robust against vibrations.

For these reasons, in the section 2, Material and Methods, this report will cover the creation of a feature extractor and extraction of the motion vector of an H.264-encoded camera. Due to hardware limitations, the usage of an artificial neural network was not further pursued. Also, due to reliability issues, especially regarding vibrations, optical flow analysis of frames was not an option.

The following chapter 3, Results and Discussion, will cover the results gained from both approaches implemented for this specific use case. Both proved successful, however, the motion vector approach, running primarily on hardware, showcased real-time capabilities, making it a more viable solution for practical applications. Additionally, future work is discussed to improve each approach, especially for a different hardware setting.

In the final chapter 4, Conclusion, this project is summarized. The results are used to determine the performance of each approach and their ability to satisfy the initial goal.

2   Material and Methods

The project was implemented in phases using a prototype-based approach. The following sections show the development of the project for each phase.

Preliminary Testing and Basic Approaches

Initially, the project commenced with fundamental tools, using a standard webcam and a laptop. The primary objective during this phase was to explore and test various basic approaches for gyroscopic data extraction. These approaches included simple image processing techniques, experimenting with different types of sensors, and analyzing the feasibility of using machine learning algorithms for pattern recognition. However, after extensive testing, the decision was made to focus on feature extraction as the most promising method for this application.

Feature Extraction and Implementation

Feature extraction was then implemented using OpenCV, a popular computer vision library, and the Scale-Invariant Feature Transform (SIFT) detector. The SIFT detector was chosen for its robustness in detecting and describing local features in images, which is crucial for accurately determining the orientation of the robot. To integrate this approach with the Robot Operating System (ROS), a target framerate of at least 10 Hz was set. Through meticulous adjustments of parameters such as the number of features, the use of masks, and other optimization techniques, the system achieved an impressive framerate of up to 50 Hz, processing nearly every captured image effectively. The following images show examples of the SIFT detector, highlighting detected features in two frames.

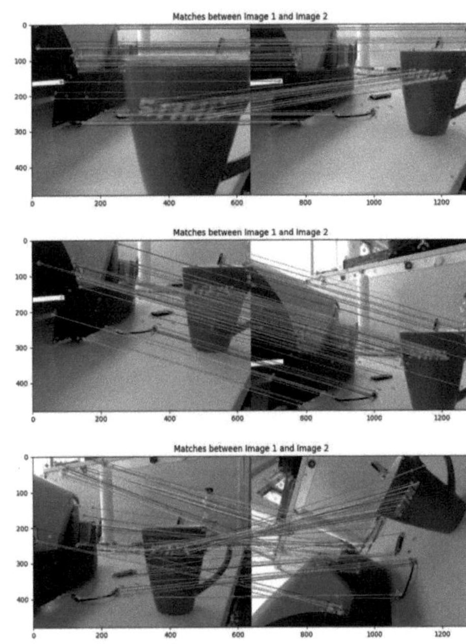

Figure 1: Testing of the SIFT detector.

Portability with Raspberry Pi 3B

The next challenge was portability, crucial for real-world applications. The system was transitioned to a Raspberry Pi 3B with a Pi camera, a more realistic setup for a robotic application. However, this transition was not straightforward. Several OpenCV operations that were seamless on a more powerful laptop proved too demanding for the Raspberry Pi, necessitating significant alterations to the script. Ultimately, the adapted script achieved a maximum processing speed of 2 Hz on the Raspberry Pi.

Additionally, another issue emerged: The new camera in combination with the already tested SIFT detector showed unreliable at detecting rotations around a single axis. Plotting and comparing the images revealed, that the 170-degree field of view of this camera caused a fish-eye distortion, so that features at the corner of the image were displayed closer to the center than they are in reality.

The initial webcam was internally corrected for this, but this step has to be done manually for the new camera. To achieve this, the camera needed to be calibrated using a chess pattern. Figure 2 shows the distortion of the original image and the compensation calculated by the calibration tool, provided by OpenCV.

Figure 2: Pattern used to calibrate the PiCamera.
Before calibration (left) and after calibration (right).

Field Testing with a Remote-Controlled Car

To simulate real-world conditions, the Raspberry Pi setup was mounted on a remote-controlled car which can be seen in Figure 3. This phase was critical for assessing the performance of the detector under dynamic conditions. The SIFT detector, however, struggled with maintaining consistent tracking at speeds above $1\frac{m}{s}$. The reduced framerate further exacerbated this issue.

Figure 3: RC-Car used for practical testing.

Motion Vector Extraction

The final phase of the project addressed the framerate limitations by leveraging the motion vector of an H.264-encoded camera. This approach has the advantage of being hardware-accelerated, enabling real-time processing. By transforming the 2D flow into a 3D rotation using the camera's rotation matrix, calibrated with a chess pattern, the system could accurately translate this information into roll, pitch, and yaw data. This method not only matched the camera's framerate but also allowed the video stream to be concurrently processed by other programs, demonstrating exceptional compatibility and flexibility.

A script was created, that can extract the gyroscope data in real time while the camera is operating. Also, another script can later process the data with the corresponding frames and display them at the same time. The following Figure 4 shows the same test run, recorded at 30 Hz, while the detected rotation of the car can be seen below each image. Only the yaw rotation is regarded here since it can be easily affected by the steering angle of the car.

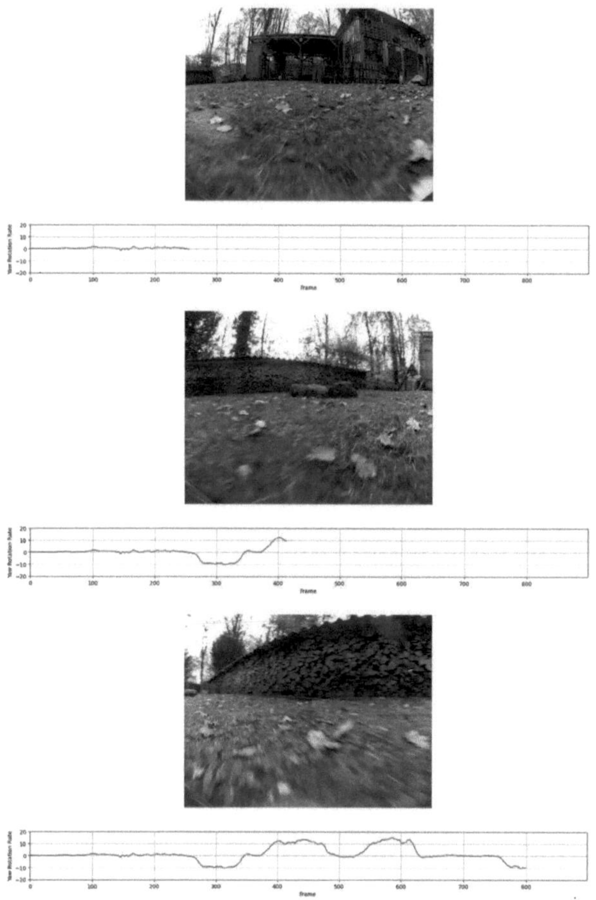

Figure 4: Frames and corresponding rotation below, ranging from +20 to -20 deg. Steering is straight (top), right (center), and left (bottom).

3   Results and Discussion

Project Evolution and Diverse Approaches

This project, aimed at extracting gyroscopic data from camera streams for robotic applications, has evolved significantly from its inception. The journey involved a series of explorations and trials, each contributing to the project's ultimate success. Various approaches, ranging from basic image processing techniques to advanced feature detection methods, were thoroughly investigated. The evaluation of the project will mainly compare the mentioned approaches as a proof of concept, due to time constraints an evaluation against a regular gyroscope was not possible.

Superiority of Motion Flow Extraction

The conclusive phase of the project, focusing on extracting the motion flow from H.264 encoded camera streams, emerged as the superior technique. Its primary advantage lies in its hardware-based operation, which significantly minimizes jitter and artifacts, ensuring a smooth and reliable flow of data. This method's real-time processing capability stands out, particularly when compared to other techniques that were more computationally demanding and less efficient in dynamic environments.

It is important to note, that this approach is not suitable for all applications. Only cameras with integrated encoders for video compression work, like the H.264 encoder used for the MPEG format. Additionally, the motion vector needs to be extractable from this hardware. The PiCamera supports this and even provides a library, but for other systems, this might be more difficult or impossible. Another aspect to keep in mind is, that the motion vector can be understood as the sum of all pixel movements between frames. That means, that a motion is detected, even if only a small part of the frame moves. For example, a person walking in front of the camera, while the robot remains stationary.

Comparative Analysis with Feature Extraction

While motion flow extraction demonstrated clear advantages, feature extraction with the SIFT detector in OpenCV also showed significant potential. Its primary benefit is its universality – it can be applied to virtually any camera, regardless of the encoding format. This flexibility makes the feature extraction method a valuable alternative, especially in scenarios where hardware-accelerated motion flow extraction is not feasible.

The practical implementation of the SIFT detector showed, that most features were detected on the ground, directly in front of the robot. This presents an issue since the robot is expected to move forward, so most features can not be found in the next frame. Also, similar-looking objects, like leaves on the ground caused additional false detections. To resolve this, masks can be applied to blacklist areas of the image for feature detection, but this might not be ideal for dynamic use cases.

Working with the Pi Camera

The Raspberry Pi camera's integration into the project was a critical step, providing a practical and realistic test environment. Its ease of use and compatibility with the Raspberry Pi made it an ideal choice for this application. However, the challenges encountered with processing speed and robustness under varying operational conditions highlighted areas for improvement. Future work could focus on enhancing this approach to be compatible with a broader range of cameras, potentially increasing the system's applicability in various robotic contexts.

Future Prospects and Considerations

Looking forward, this project lays the groundwork for further exploration and development in the field of robotics, particularly in autonomous systems like lawnmower robots. The potential to adapt and optimize these techniques for a wide range of camera systems opens up numerous possibilities for enhancing robotic perception and navigation capabilities. Testing against other sensors and regular gyroscopes can help to further evaluate the accuracy. Additionally, refining the current methodologies to improve processing speed and accuracy, especially in diverse environmental conditions, will be a crucial area of focus for future research.

4 Conclusion

The primary goal of this research was to explore and develop a method for extracting gyroscopic data from camera streams, a key component in enhancing the spatial awareness and operational efficiency of autonomous lawn mowing robots. This endeavor, undertaken during an internship at STIHL, was driven by the need to compensate for the lack of gyroscopic sensors in some of these robots, using the available sensory data from a standard camera.

The research was anchored on the hypothesis that gyroscopic data, crucial for robotic orientation and navigation, could be effectively derived from camera streams. This was explored through various methods, beginning with feature extraction using OpenCV and SIFT, and culminating in a novel approach using the motion vector from H.264 encoded camera streams. The latter proved particularly promising due to its hardware-based processing, ensuring real-time data extraction with minimal jitter.

The results demonstrated the feasibility and effectiveness of both approaches. The feature extraction method showed versatility across different camera types, though it was limited by processing speed and environmental conditions. In contrast, the motion vector approach, while more limited in camera compatibility, offered superior performance in terms of speed and reliability, making it a more viable solution for real-time applications in autonomous systems.

The project successfully achieved its initial goal of gyroscopic data extraction through multiple methodologies. The motion vector approach, in particular, represents a significant advancement in the field, offering a practical solution for enhancing the capabilities of autonomous lawn mowing robots. However, there remains room for improvement, further evaluation, and exploration to improve the approaches presented. The findings from this research not only contribute to the field of robotics and autonomous systems but also lay the groundwork for future innovations and advancements in robotic perception and navigation.

Acknowledgement

The work has been carried out in Waiblingen, at the company STIHL, and supervised by Ralf Bruder, Institute of Robotics, Universität zu Lübeck.

Authors' Statement

Conflict of interest: The authors state no conflict of interest regarding the publication of this paper.

Informed consent: Not applicable. This research did not involve human participants, human data, or human tissue.

5 References

[1] L. Haipeng, K. Luo, and S. Liu, *Gyroflow: gyroscope-guided unsupervised optical flow learning.* Proceedings of the IEEE/CVF International Conference on Computer Vision, 2021.

[2] S. Bonnabel, and A. Barrau, *Denoising imu gyroscopes with deep learning for open-loop attitude estimation.* IEEE Robotics and Automation Letters 5.3, 2020

[3] S. Battiato, G. Gallo, G. Puglisi, and S. Scellato, *SIFT features tracking for video stabilization.* 14th international conference on image analysis and processing, IEEE, 2007.

[4] Y. Durdu and A. Durdu, *Performance and Trade-off Evaluation of SIFT, SURF, FAST, STAR and ORB feature detection algorithms in Visual Odometry.* Avrupa Bilim ve Teknoloji Dergisi, 2020.

[5] B. Zhang, L. Wang, z. Wang, Y. Qiao, *Real-time action recognition with enhanced motion vector CNNs.* Proceedings of the IEEE conference on computer vision and pattern recognition, 2016.

Use of RGB and Depth Information for AI-assisted Estimation of an optimal Patient Position

Isabel Haasler [1], Christian Seitzer [2], and Sebastian Fudickar [2]

[1] Medical Informatics, Universität zu Lübeck, isabel.haasler@student.uni-luebeck.de

[2] Institute of Medical Informatics, Universität zu Lübeck,{c.seitzer, sebastian.fudickar}@uni-luebeck.de

Abstract

Voice therapy is a widely accepted therapy method for voice disorders. The LAOLA (proper name) project aims to develop a digital assistance app to support voice therapy. Video data analysis is an essential part of this project, using RGB and depth data and AI to assess the performance of the exercises performed. This paper explores the influence of patient-to-camera distance on depth data acquisition using MediaPipe face mesh and the Microsoft Azure Kinect (depth) camera. From the face mesh, defined face landmarks were used to read out the corresponding depth values. In the study, subjects performed voice exercises at three different distance ranges, with an additional orthogonal camera for manual distance measurement. The results indicate that the medium distance range (45-65 cm) provides both acceptable error values and encourages committed exercise performance, suggesting its optimal use. This finding can be used for the real-time feedback in the LAOLA app.

1 Introduction

Voice disorders have a serious impact on quality of life and cause work absences and financial losses. According to research by [1], the prevalence of voice disorders is 6.6% and the lifetime prevalence is 29.9% in Germany. The cost of voice therapy amounted to 636 million euros in 2015, and the need for therapy is continuously increasing. A voice disorder (dyphonia) is characterized by limited vocal performance (phonation) [2]. Dysphonia can be divided into organic, functional and psychogenic disorders, whereby combined disorders are possible [3]. An organic voice disorder is often due to a pathological change in the larynx. If there is no organic cause for hoarseness, it is referred to as a functional voice disorder. This is often based on overuse of the voice [4]. If organic and functional causes have been ruled out, there is usually a psychogenic voice disorder. This can result from psychological stress in the form of voice disorders or even voice loss (aphonia) [5]. Studies such as [6] have proven the effectiveness of voice therapies for non-organic dyphonia.

The LAOLA (proper name derived from the OLA (Oldenburger Logopädie App) app) project aims to develop a digital assistance system in the form of an app to support functional voice therapy. The predecessor of this project is the Oldenburger Voice Therapy App OLA, which provides digital voice therapy videos for the rehabilitation of patients after thyroid surgery. The LAOLA app is designed to make it easy to carry out training sessions and to document and evaluate the training simultaneously. Biofeedback is provided in real-time [7]. The smartphone should therefore be placed in front of the patient on a tripod so that the patient has both hands free. A part of the project is investigating the extent to which machine learning methods can measure the performance of vocal exercises using visual data. A face mesh algorithm based on RGB and depth data are used for this purpose.

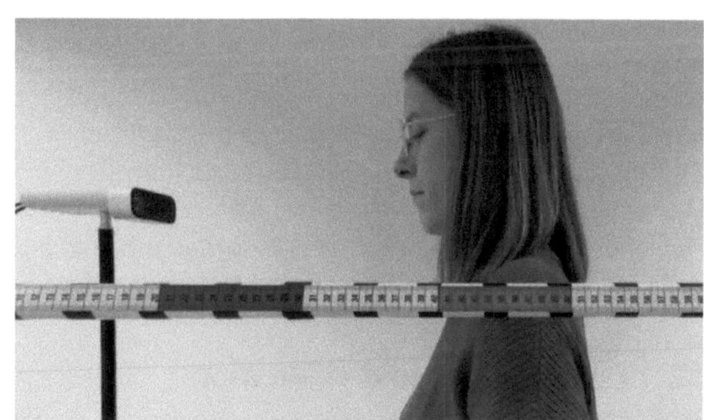

Figure 1: Section from the view point of the control camera in the experiment set-up, which is aligned orthogonally to the subject's face with the measuring tape in front of the side profile.

In this study, a Mircosoft Azure Kinect was used. Studies such as [8] have examined the accuracy of the Azure Kinect and confirmed the officially stated values, but without an underlying neural network which determines certain landmarks from which the depth data is read out. There, a planar surface fixed on a tripod was used as the measurement object. A laser scanner and defined distances to the camera were deployed to validate the distance measurement of the

Azure Kinect. In [9], the influence of the camera viewing angle of the Azure Kinect on the recognition of kinematic gait patterns, which is based on the integrated body tracking algorithm, was investigated. It was found that the camera viewing angle has a significant effect on the kinematic gait analysis. The angle of the camera to the face could therefore theoretically also be relevant when performing vocal exercises. However, the angle of the face to the camera is not investigated in this study.

This paper examines whether there is an optimal distance range from patient to camera and whether the distance has a relevant impact on the performance of the depth data based on the face mesh. Therefore, RGB data (for the face mesh) forms the basis for measuring the distance using the depth data. The face mesh algorithm detects facial landmarks, from which two are selected to measure the distance.

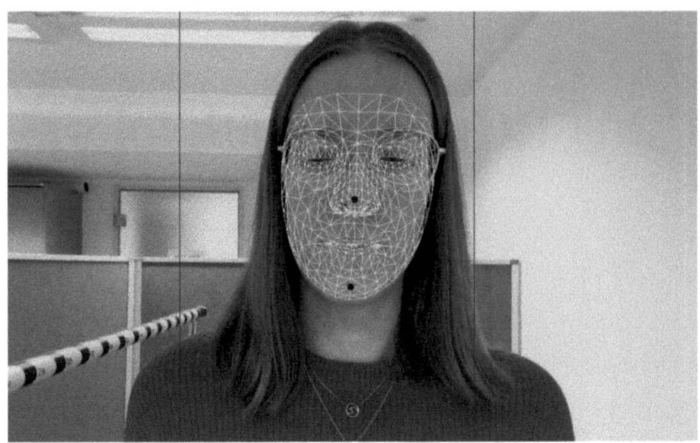

Figure 2: The field of view of the Azure Kinect, here with the face mesh from MediaPipe. The upper black dot is the selected landmark for the nose tip, the lower one for the chin.

2 Material and Methods

2.1 Data aquisition

Seven test subjects took part in the data collection. There were no exclusion criteria. The test subjects were standing in front of the Microsoft Azure Kinect. They were instructed to repeat the exercises from a voice training video at each of three different distance ranges (25-45cm (short); 45-65cm (medium); 65-85 cm (large)) to the camera. These distance ranges were chosen as they are not considered too close or too far away for patients when performing the exercises on their own. Meanwhile, RGB and depth data were recorded with the Azure Kinect at a frame rate of 10fps. The chosen depth mode was narrow field-of-view (NFOV) without binning. The data was recorded continuously for 2:30 minutes. In addition, a second (control) camera simultaneously recorded the subject orthogonally from the side. A bar with an attached measuring tape was positioned in front of the second camera so that the distance between the patient's nose tip and chin could be manually read from the video data as seen in Fig. 1. The values were annotated every 4 seconds (every 40 frames). This data was used for later comparison of the depth data recorded at the same time.

2.2 Data analysis and evaluation

Two Python applications were developed: one for data capturing and the other for data processing. In the capturing application, the video data is captured using the Microsoft Azure Kinect and stored as RGB video, colorized depth video and numpy array for the depth data. A control camera was also started simultaneously from the same application (sufficient synchronisation can therefore be assumed) for verification purposes. In the processing application, face landmarks are first detected on the basis of the RGB data. The face landmark detector used was Face Mesh from MediaPipe, which detects 478 3D face landmarks in real time [10]. Other face landmark detectors such as [11] and [12] only calculate 27 and 68 face landmarks respectively. Since a high density of face landmarks is useful for

a later analysis of the performance of voice exercises, the face mesh from MediaPipe was chosen in this study. Of these face landmarks, the landmark of the nose tip (no. 1) and the chin (no. 199) were selected. The face mesh and the selected face landmarks are visualized in Fig. 2.

In order to find a potential optimal distance range from camera to patient for the later analysis of the voice exercises and to test the depth sensor in combination with the face mesh, subject recordings were taken. The landmarks for the nose tip and chin and the difference between them were used to check whether the same depth values are given at different distance ranges. To evaluate the measured values, the Mean Absolute Errors (MAE) are calculated. A measuring deviation of up to 30 mm is considered acceptable.

3 Results and Discussion

No face landmarks were detected in the recordings with the greatest distance range (large), so no depth data could be read out. Possible causes for this are a too low resolution or that the face mesh model is not trained for such a distance. For this reason, there are no results for this distance range. This leads to the conclusion that a camera-to-patient distance of more than 65 cm results in unreliable face mesh performance and that the analysis of the execution of the voice exercises is not possible. If MediaPipe is also used in the LAOLA app, it is recommended that the patient should be a maximum of 65 cm away from the camera. If a different face mesh algorithm is used, a reliable distance should be examined.

For the other two distance ranges, the Mean Absolute Error (MAE) was calculated for the chin-nose tip depth distance. The distance between the two landmarks was used to reduce the basic error of the measurement. The MAE is 8 mm for the short distance range and 9 mm for the medium distance range. Both are within the accepted tolerance range of 30 mm. In addition, the values of the nose tip and the chin were normalized to the range 0 to 1 for the calculation of the normalized MAE in order to reduce the basic error in this way. This resulted in values of 0.11 and 0.10 for the nose tip and 0.11 and 0.20 for the chin for the short and medium distance ranges respectively. The MAEs for the chin-nose tip depth distance and for the nose tip only show a small difference for both distance ranges as can be seen in Table 1. But there is a difference in the MAE for the chin, which is almost twice as large at the medium distance range.

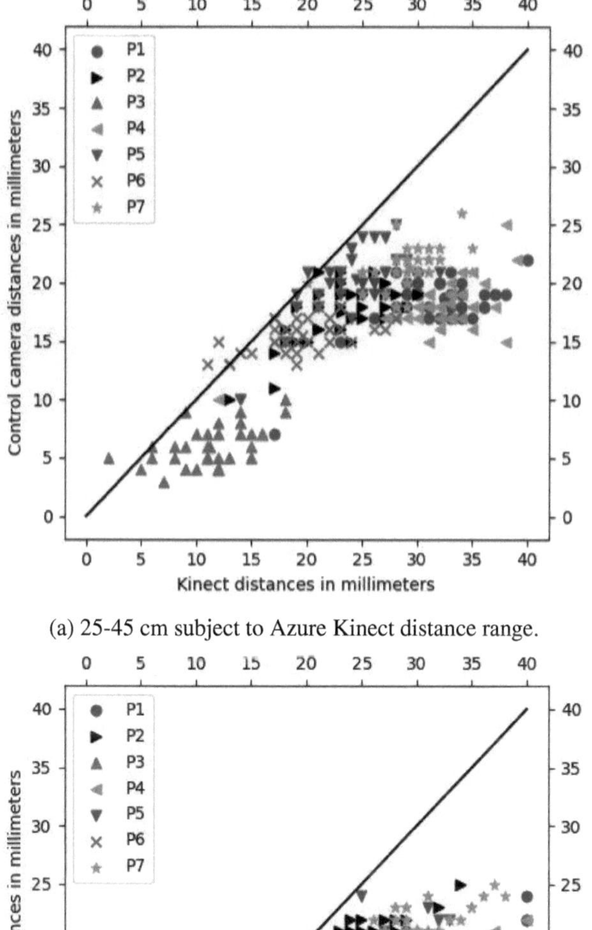

(a) 25-45 cm subject to Azure Kinect distance range.

(b) 45-65 cm subject to Azure Kinect distance range.

Figure 3: The chin-nose tip depth distances measured from the Azure Kinect and the control camera for each subject-to-Azure Kinect distance ranges. P1 to P7 represent the seven subjects.

Furthermore, the average movement spans of the nose tip and chin were calculated. In Table 1 can be seen that the Kinect values are approximately twice as large as the

control values. Fig. 3 also shows that the distances measured with the Kinect tend to be greater than the control values for both the short (see Fig. 3a) and the medium distance range (see Fig. 3b). The generally greater values of the Azure Kinect are consistent with the fact that the difference between two points also tend to be greater than the control values. The deviations between the measured values of the Azure Kinect and the control camera lie also in the accepted range of tolerance. In Table 1, it is interesting to note that the movement spans are always greater at the medium distance than at the shorter distance. This indicates that the subjects performed the voice exercises with more commitment at a greater distance to the camera and that a shorter distance to the camera tended to inhibit the subjects. But since the aim is to achieve a dedicated exercise performance, as this has a greater training effect, the medium distance is assumed to be better.

Table 1: Comparison between the short (25-45 cm) and the medium (45-65 cm) distance range. MAE = mean absolute error; SD = standard deviation.

	short distance	medium distance
MAE chin-nose tip depth distance	8 mm	9 mm
Normalized MAE nose tip	0.11	0.10
Normalized MAE chin	0.11	0.20
Mean span from chin (Kinect); SD	48 mm; 13 mm	61 mm; 17 mm
Mean span from chin (control camera); SD	26 mm; 8 mm	31 mm; 10 mm
Mean span from nose tip (Kinect); SD	50 mm; 14 mm	63 mm; 22 mm
Mean span from nose tip (control camera); SD	25 mm; 7 mm	34 mm; 9 mm

The presented study has some limitations. First, the depth data was tested with an orthogonally positioned control camera and a bar with attached measuring tape (see Fig. 1). The control camera was placed in a way that minimized distortion during annotation, but a certain degree of perspective distortion cannot be ruled out. Future studies could, for example, use a laser for distance measurement instead, like [8]. Additionally, the points from which the distances were read were based on the calculated face landmarks. This means that a landmark may slip slightly for a moment and the position from which the depth data is read may differ slightly between Azure Kinect and the control camera, resulting in deviations. As described in other studies, the exact accuracy can be measured using a planar, fixed surface [8]. Fixing people to carry out a study like this is not possible and would not be appropriate for the application. Moreover, the larger ranges of movement at the medium distance range mentioned earlier are likely to have a negative effect on the face mesh and the accuracy of the location of the facial landmarks, as the MAE for the chin, which moved the most, was considerably larger than for less movement. This can also lead to differences between the two distance ranges. It would also be interesting to investigate the effect of camera viewing angles on the Face Landmark Detector and the resulting depth measurement, as [9] did for kinematic gait patterns. Besides, the here presented study was conducted with seven test subjects. A larger number of subjects could lead to more precise and meaningful results.

4 Conclusion

The paper provides a recommendation for a distance range the patient should maintain from the camera when performing the functional voice exercises in the LAOLA app. The study was conducted using a Microsoft Azure Kinect as a depth camera and the MediaPipe Face Mesh with seven subjects. Two landmarks (chin and nose tip) were selected from which the depth data was obtained. Three different distances to the camera were tested. At the largest distance (65-85 cm), no face mesh was created due to the possible limitations of MediaPipe, so no depth data could be captured. As the performance of the voice exercises cannot be analyzed in this way, a distance of more than 65 cm is not recommended. At the short (25-45 cm) and medium distance (45-65 cm), face meshes were always created and the MAE of the chin-nose tip depth distance (8 mm vs. 9 mm) and the normalized MAE for the nose tip (0.11 vs. 0.10) were relatively similar. At the medium distance range, however, the test subjects moved more than at the short distance range. This is intentional, because a committed execution of the exercises, which results in more movement, is important for the effect of the exercise. For this reason, a camera-patient distance range of 45-65 cm is recommended. It was also found that the distance has no relevant impact on the performance of the depth data based on the face mesh, given the defined deviation tolerance of 30 mm. This knowledge can be used to give the patient real-time feedback whether they are too far or too close to the camera. In the future, the influence of the angle of the face to the camera should also be examined.

Acknowledgement

The work has been carried out and supervised by the Institute of Medical Informatics, Universität zu Lübeck in the research group MoveGroup. The LAOLA project is funded by the Bundesministerium für Bildung und Forschung (BMBF) (16SV8958).

Authors' Statement

Conflict of interest: Authors state no conflict of interest.
Informed consent: Informed consent has been obtained from all individuals included in this study.
Ethical approval: The research related to human use complies with all the relevant national regulations, institutional policies and was performed in accordance with the tenets of the Helsinki Declaration, and has been approved by the authors' institutional review board or equivalent committee.

5 References

[1] S. Jung, *Systematisches Review und Metaanalyse zur konservativen Behandlung von funktionellen Stimmstörungen und einseitigen Stimmlippenparesen*. Homburg/Saar, 2018. doi: 10.22028/D291-27839.

[2] U. Beushausen, *Intensität in der Stimmtherapie*. In: H. Grötzbach (Ed.), Therapieintensität in der Sprachtherapie/Logopädie, Schulz-Kirchner Verlag, Idstein, p. 107, 2017.

[3] S. Fleischer and M. Hess, *Diagnostik und Therapie von Stimmstörungen*. HNO Nachrichten, vol. 46, pp. 26–33, 2016, doi: 10.1007/s00060-016-5169-9.

[4] F. Hofer, *Stimmstörungen*. Wehrmedizinische Monatsschrift, vol. 66, no. 8, pp. 295–300, 2022, doi: 10.48701/opus4-36.

[5] C. Grüber and O. Arndt, *Was hilft bei psychogenen Stimmstörungen?*. HNO Nachrichten, vol. 47, p. 50, 2017, doi: 10.1007/s00060-017-5392-z.

[6] P. N. Carding, I. A. Horsley and G. J. Docherty, *The effectiveness of voice therapy for patients with non-organic dyphonia*. Clinical otolaryngology and allied sciences, vol. 23, no. 4, pp. 310–318, 2001, doi: 10.1046/j.1365-2273.1998.00147.x.

[7] LAOLA 2023 *Die Vision: Stimmtraining mit dem Handy in der Hand*. Available: https://laola-projekt.de/#projekt [last accessed on 2023-10-21].

[8] G. Kurillo, E. Hemingway, M. L. Cheng and L. Cheng, *Evaluating the Accuracy of the Azure Kinect and Kinect v2*. Sensors (Basel), vol. 22, no. 7, 2022, doi: 10.3390/s22072469.

[9] L. Yeung, Z. Yang, K. Cheng, D. Du and R. Kai-Yu, *Effects of camera viewing angles on tracking kinematic gait patterns using Azure Kinect, Kinect v2 and Orbbec Astra Pro v2*. Gait & Posture, vol. 87, pp. 19–26, 2021, doi: 10.1016/j.gaitpost.2021.04.005.

[10] MediaPipe 2023 *Face landmark detection guide*. Available: https://developers.google.com/mediapipe/solutions/vision/face_landmarker [last accessed on 2023-11-15].

[11] Microsoft 2023, *Face detection, attributes, and input data*. Available: https://learn.microsoft.com/en-us/azure/ai-services/computer-vision/concept-face-detection [last accessed on 2023-12-30].

[12] A. Rosenbrock 2017 *Facial landmarks with dlib, OpenCV, and Python*. Available: https://pyimagesearch.com/2017/04/03/facial-landmarks-dlib-opencv-python/ [last accessed on 2023-12-30].

Development of a Modular Zigbee/LoRaWAN Wireless Sensor Network for Environmental Monitoring with Biohybrid Sensors

Tim-Lucas Rabbel [1], Marko Križmančić [2]

[1] Robotics and Autonomous Systems, Universität zu Lübeck, timlucas.rabbel@student.uni-luebeck.de

[2] University of Zagreb, Faculty of Electrical Engineering and Computing, Croatia, marko.krizmancic@fer.hr

Abstract

The Watchplant project develops a biohybrid wireless sensor network (WSN) for environmental monitoring. Sensor nodes are exposed to nature and wilderness, making WSNs commonly suffer from node failure and data loss. We propose a novel approach to increase the network's robustness: Using two different communicaion technologies. Zigbee is used for fast, short-range communication of measurement data, and LoRaWAN for slow, long-range communication of configuration messages and error detection. Experiments were conducted with 15 sensor nodes in a controlled lab environment with simulated node failures. The WSN displayed its ability to effectively self-regulate node failures and shows that the heterogeneous communication approach is a viable alternative to conventional WSNs.

1 Introduction

Watchplant [1] is an EU Horizon 2020 project concerned with environmental monitoring for climate change and urban air pollution research. Watchplant develops a novel approach based on a wireless sensor network (WSN) of biohybrid devices, so-called *phytonodes*. Each phytonode consists of a plant with attached electronics. Electrodes are inserted into the plant's stem to read the electrical potential, which changes depending on external stimuli like temperature, light, and wind. These signals are processed with statistical methods, e.g. feature extraction and discriminant analysis, and machine learning methods to classify the triggering stimulus [2]. The proposed setup aims to be a long-lasting, energy-efficient, cheaper alternative to conventional sensor stations.

This paper's contribution is a prototype WSN that was deployed in lab conditions. One single sensor can not give an accurate representation of a larger area like a city. A WSN allows the deployment of multiple sensor nodes evenly distributed over a target area. This is especially important for phytosensors, as a single phytosensor has a higher variance compared to a conventional sensor. As sensor nodes are prone to failure, WSNs have to be developed with robustness in mind. Previous approaches sacrifice bandwith for data messages for network management messages. We develop a heterogeneous WSN using Zigbee and LoRaWAN communication technologies. The hybrid Zigbee/LoRaWAN network provides a novel alternative to WSNs which rely on one communication technology. Based on our literature overview, very little research is done in the area of WSNs with multiple communication technologies. Examples of Zigbee and LoRa combinations are sensor nodes that can switch the technologies, but only

after user request [3], and Zigbee and LoRa clusters that communicate over a Zigbee/LoRa bridge [4]. Our goal is to design a network that takes advantage of the strengths of both communication technologies: Zigbee for communication of measurement data, and LoRaWAN for network management. Initial experiments show that our approach to increase the robustness is effective without reducing communication bandwith for network management.

2 Technologies

Wireless Sensor Networks use a variety of communication technologies to transmit data wirelessly. The most commonly used communication technologies in WSNs include Zigbee, Bluetooth Low Energy (BLE), Wi-Fi, and LoRa (Long-Range).

Zigbee [5] is a wireless communication protocol designed for high data rate, low power consumption, and long battery life applications. Zigbee operates in the 2.4 GHz band and uses a mesh networking topology to enable communication between nodes. It is based on the IEEE 802.15.4 standard, which specifies the physical layer and medium access control layer for low-rate wireless personal area networks (LR-WPANs). Zigbee supports a maximum range of up to 100 meters and a maximum network size of 65,536 devices. It is suitable for applications such as smart homes, industrial automation, and building automation. Computer chips that implement Zigbee communication can easily be integrated into smart devices or attached externally using, e.g., an XBee module.

LoRaWAN (Long Range Wide Area Network, see
`https://lora-alliance.org/resource_hub/`
`lorawan-specification-v1-1/`) is a low-power,

long-range wireless communication protocol designed for IoT applications. The network topology consists of low-power end devices connected to gateways with internet access. The gateways are connected via a network server back-end like The Things Network (TTN, see `https://www.thethingsnetwork.org/`) forming a star-of-stars topology and thus increasing coverage. Using the sub-GHz spectrum, LoRaWAN can provide long-range communication capabilities of up to 10 km in rural areas and 2-3 km in urban environments. However, it has a low data rate (typically in the range of a few kilobits per second).

LoRa (Long Range) [6] is the physical layer technology used in LoRaWAN that provides long-range communication capabilities with low power consumption. LoRa uses chirp spread spectrum modulation, which enables robust communication with low sensitivity to noise and interference.

3 Zigbee/LoRaWAN Hybrid Network

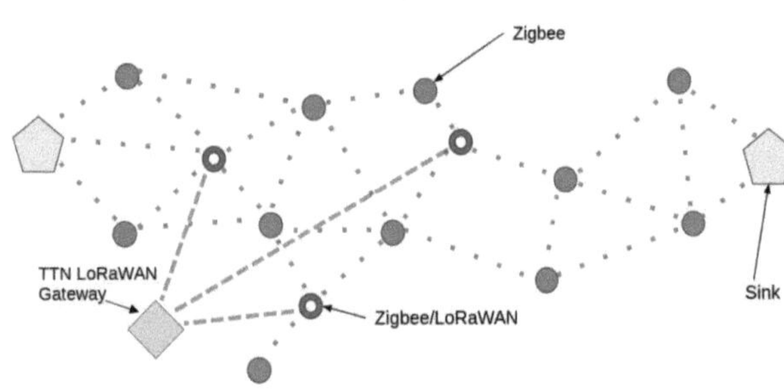

Figure 1: The schematic of the proposed hybrid Zigbee/LoRaWAN network.

For the prototype of our heterogeneous sensor network, we decided to use Zigbee and LoRaWAN. The main differences between these technologies are communication range and bandwidth. Zigbee can transfer data fast over a short range, while LoRa can transmit data slow, but over a far longer range. The network uses the advantages of both technologies and uses Zigbee to transmit measurement data and LoRaWAN to transmit control flow messages. During long-term network deployment, we expect more regular and long data messages, but fewer, shorter control messages. In a bigger network, data messages that are sent via Zigbee do not have the range to be directly transmitted to a data sink. Instead, Zigbee will utilize intermediary nodes for multi-hop routing. This can be problematic if intermediary nodes fail. LoRa, on the other hand, has a far higher range of multiple kilometers. Using the LoRaWAN network, a LoRaWAN node can always directly communicate with a LoRaWAN gateway without dependence on intermediary nodes.

The network design is displayed in Fig. 1. Zigbee nodes collect data and send it via Zigbee to the sink. Zig-bee/LoRaWAN nodes do the same, but they also have additional tasks. In contrast to pure Zigbee nodes, they request an acknowledgment from the sink. They count over a fixed period how many messages failed to get through and send the count to the nearest public TTN LoRaWAN gateway. A server-side Python program evaluates the data and, if necessary, changes the network configuration. The behavior of Zigbee and Zigbee/LoRaWAN nodes is summarized in Fig. 2 and 3.

For example, if only one of the LoRaWAN nodes reports failed messages, the Python controller can assume that the network is damaged, but also that the damage is contained in a relatively small area, i.e., the Zigbee/LoRaWAN node itself or a few intermediary Zigbee nodes. The majority of the network (majority of Zigbee nodes and also the sink) is still intact. Otherwise, more Zigbee/LoRaWAN nodes would have reported failed transmissions. As long as the majority of the network is still operating, there is no need to react and change the configuration. The controller can also be used to send out a notification to the operators running the network (via email) to fix the broken nodes.

A bigger problem is a failing data sink because the network loses all data from all nodes. In this case, a majority of Zigbee/LoRaWAN nodes report failed transmissions. Now the controller reacts and sends out a new network address to change the data flow to the backup sink. The Zigbee/LoRaWAN nodes receive the new address as a LoRaWAN message and reconfigure their sink addresses. Then they broadcast the change to all Zigbee nodes in the network. In this way, only a small portion of the information is lost.

Additionally, the operators can use the controller to change network settings remotely. We implemented changing the measurement frequency of all nodes, but there are more possibilities depending on the sensor loadout and the requirements.

3.1 Zigbee Node

The Zigbee node is the base of a sensor node. It has only Zigbee communication capabilities through the use of the attached XBee module. We use STM32WB55 NU-CLEO boards and USB dongles for processing the data. For the power supply, we use a solar power manager board DFR0559 from DFRobot connected to a 1500 mAh battery and a solar panel.

3.2 Zigbee/LoRaWAN Node

Zigbee/LoRaWAN nodes are extensions of Zigbee nodes. They do the same task - measuring data and sending it to the sink. In addition to their XBee modules, they also have LoRa modules for LoRaWAN communication. They periodically send LoRaWAN messages over the nearest LoRaWAN gateway to the server. Using Zigbee/LoRaWAN nodes with TTN allows the use of any connected community gateway. In this way, the whole network is easier to deploy because it can use public infrastructure for commu-

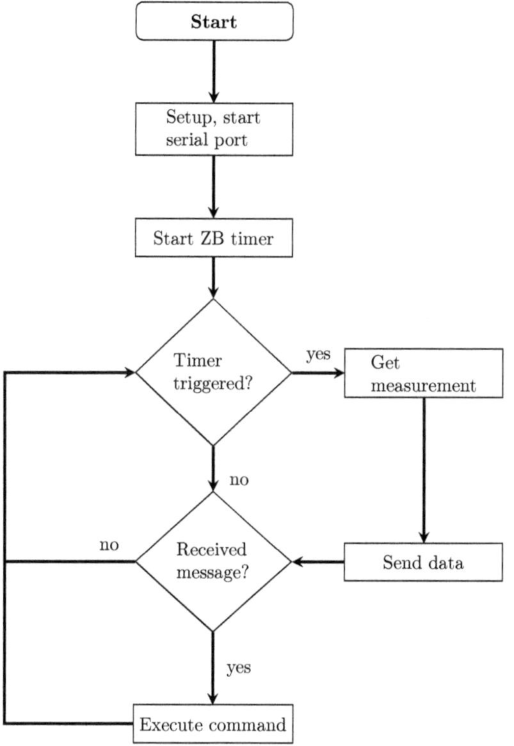

Figure 2: Flowchart explaining the operation of Zigbee nodes.

nication. The used hardware consists of a Heltec Wireless Stick v2.1, which includes a LoRa module, and a connected XBee. These devices are also powered by a battery and a solar panel.

3.3 Data sink

The primary sink used in the experiment was the custom-made, waterproof, autonomous data acquisition unit called the Orange Box. The main processing unit inside the box is a Linux-based mini-computer Rock Pi S, similar to the Raspberry Pi. The second sink was built out of a Raspberry Pi 4 and an XBee module. Every device that can interface with an XBee can be used as a sink. However, the program of the sink uses Python and relies on a conventional file system, so it is recommended to use mini-computers instead of microcontrollers. The program on the sink listens to the data and creates new folders and files for each transmitting sensor node. The names of the files are based on the 64-bit MAC addresses of the XBee modules.

3.4 LoRaWAN Controller

Using LoRaWAN and TTN, the Zigbee/LoRaWAN nodes send their status messages to the nearest TTN gateway. The controller downloads the messages from the TTN server. This requires that the controller code is running on a device that has internet access. The controller is used for periodically checking the network status and sending user commands.

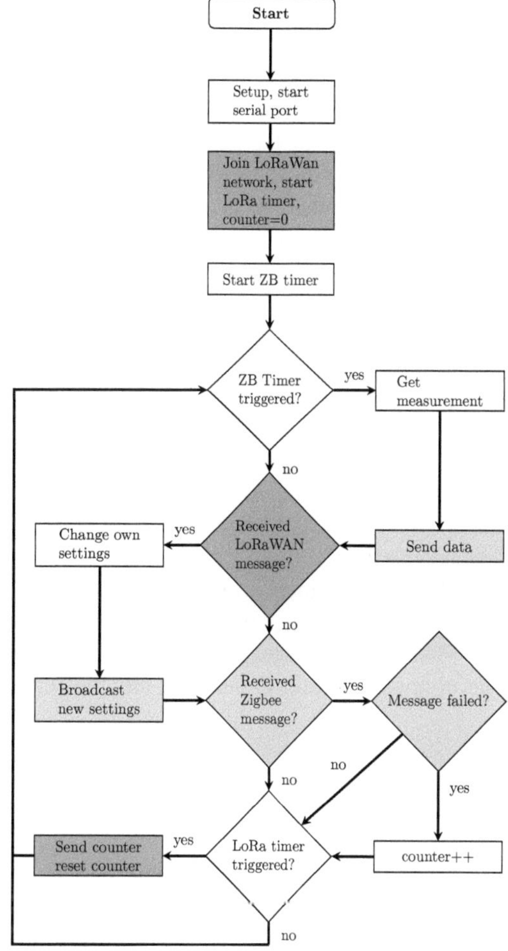

Figure 3: Flowchart explaining the operation of Zigbee/LoRaWAN nodes. Blue nodes highlight LoRaWAN communication, while orange nodes highlight Zigbee communication.

To check the network status, it relies on received messages that notify the controller about failed transmissions between Zigbee/LoRaWAN nodes and the sink. In our practical test, we used three Zigbee/LoRaWAN nodes. In this scenario, the controller implements the following rule: "If the majority of Zigbee/LoRaWAN nodes report one or more failed transmissions, the sink has failed." The controller then sends a LoRaWAN message back to the network containing an alternative sink address. The controller cycles through all known link addresses if a change is necessary. Selecting a good threshold of failed transmissions and reporting devices is a design decision and depends on the size and the structure of the network. In general, some messages can be expected to fail. The second task of the controller is to react to human input. An operator working with the network might want to change network settings, like the measurement frequency of the sensor nodes. This is possible over the internet using the controller, so that the network can be controlled and monitored worldwide, as long as the operator has internet access. Depending on the task the network should perform, a multitude of options can easily be added to the system.

4 Results and Discussion

In the laboratory test setup shown in Fig. 4, we used 15 devices: two sinks, three Zigbee/LoRaWAN nodes, and ten Zigbee nodes. Failures of the sink and other nodes were simulated by disconnecting their XBee modules, effectively disabling their communication. In our experiments, the network fulfilled the set expectations and proved that our heterogeneous networking concept has great potential for further development and future use in the project. The network can transmit messages as fast as a native Zigbee network but has more reliability and robustness. In contrast to other approaches for increasing the robustness and reliability of WSNs, our network is keeping its performance because it does not need to sacrifice communication bandwidth for network management; this task is performed by LoRaWAN communication. Moreover, the network can be easily extended with additional types of devices. We have developed the code for interfacing with the XBee module for the Arduino framework, Arduino through PlatformIO, MBed OS, and Python. As long as a device can run any of these four options, it is easy to integrate into the network.

Figure 4: Test network with 15 nodes: Two sinks (RPi on the left side, open Orange Box on the right side), three Zigbee/LoRaWAN nodes, and 10 Zigbee nodes.

While the first test was promising, there are still drawbacks to consider. One disadvantage is the increasing cost of sensor nodes: Each additional communication technology adds at minimum a new transceiver. This was especially notable in the prototype stage. A second disadvantage could be the power consumption. The used Heltec Wireless Sticks did not perform as promised by the manufacturer, so we could not do an in-depth investigation of the additional power consumption due to the use of LoRaWAN.

5 Conclusion and Future Work

We developed a heterogeneous Wireless Sensor Network with a hybrid communication system and successfully deployed the prototype in our lab. Due to the use of two different communication technologies, we were able to decouple the infrastructure for sending data messages and the infrastructure for sending status and reconfiguration messages. In a conventional WSN, the sink (base station) has two responsibilities: receiving sensor data and sending out reconfiguration messages when it detects errors. Using two infrastructures, we were able to split these responsibilities so that the LoRaWAN controller is responsible for error detection and reconfiguration, and the sink is used only for data collection. This reduces the bottlenecks in the network and increases overall robustness since the critical roles are distributed to multiple devices. Furthermore, the LoRaWAN range spans multiple kilometers, so as long as there is a gateway close enough, the network can still be monitored and configured even without direct access to the internet. As a next step, we deploy the prototype in outdoor conditions for a longer time and compare the performance to networks with a single communication technology.

Acknowledgement

The work has been carried out at the University of Zagreb and supervised by Prof. Stjepan Bogdan, Ph.D., LARICS, University of Zagreb, Prof. Dr.-Ing. Heiko Hamann, ITI, Universität zu Lübeck, Universität Konstanz, and Dr.-Ing. Kristian Ehlers, ITI, Universität zu Lübeck.

Authors' Statement

Conflict of interest: Authors state no conflict of interest.

6 References

[1] H. Hamann, S. Bogdan, A. Diaz-Espejo, L. García-Carmona, V. Hernandez-Santana, S. Kernbach, A. Kernbach, A. Quijano-López, B. Salamat, and M. Wahby, "Watchplant: Networked bio-hybrid systems for pollution monitoring of urban areas," in *ALIFE 2021: The 2021 Conference on Artificial Life*. MIT Press, 2021.

[2] E. Buss, T. Aust, M. Wahby, T.-L. Rabbel, S. Kernbach, and H. Hamann, "Stimulus classification with electrical potential and impedance of living plants: comparing discriminant analysis and deep-learning methods," *Bioinspiration & Biomimetics*, vol. 18, no. 2, p. 025003, 2023.

[3] A. I. Ali, S. Z. Partal, S. Kepke, and H. P. Partal, "Zigbee and lora based wireless sensors for smart environment and iot applications," in *2019 1st Global Power, Energy and Communication Conference (GPECOM)*. IEEE, 2019, pp. 19–23.

[4] V.-T. Truong, A. Nayyar, and S. A. Lone, "System performance of wireless sensor network using lora–zigbee hybrid communication," *Computers, Materials & Continua*, vol. 68, no. 2, pp. 1615–1635, 2021.

[5] S. C. Ergen, "Zigbee/ieee 802.15. 4 summary," *UC Berkeley, September*, vol. 10, no. 17, p. 11, 2004.

[6] D. Croce, M. Gucciardo, S. Mangione, G. Santaromita, and I. Tinnirello, "Lora technology demystified: From link behavior to cell-level performance," *IEEE Transactions on Wireless Communications*, vol. 19, no. 2, pp. 822–834, 2019.

Impact of Dietary Choices on Human Gut Microbiome Composition

Anna Osipjan [1], Lena Best [2], Christoph Kaleta [2], and Marcin Grzegorzek [3]

[1] Medical Informatics, Universität zu Lübeck, anna.osipjan@student.uni-luebeck.de
[2] Institute of Experimental Medicine, Kiel University, {l.best, c.kaleta}@iem.uni-kiel.de
[3] Institute of Medical Informatics, Universität zu Lübeck, marcin.grzegorzek@uni-luebeck.de

Abstract

A person's diet and the associated impact on the gut microbiome composition has a major influence on health. However, the exact effect of specific diets and dietary patterns on microbial composition remains incompletely understood. In this study, microbial, nutritional, and phenotypical data from the German population cohort FoCus with a total sample of 1561 subjects were examined. The focus was on analyzing the meat/vegetable ratio using the programming language R. The Wilcoxon test, linear model and PERMANOVA were used as statistical methods. In addition, the data was analyzed with regard to age, BMI and gender. Different diets lead to different abundance of bacteria in the gut microbiome: *Eubacterium eligens* was associated with a vegetable-rich and *Eubacterium rectale* with a meat-rich diet. Gender, age and BMI were significant parameters, although they only explained a very small proportion of the variance in the data.

1 Introduction

The gut microbiome plays a crucial role in a person's health. A disturbance in its composition, a so-called dysbiosis, is associated with various metabolic disorders such as obesity, type 2 diabetes and dyslipidemia. These disorders increase the risk of cardiovascular diseases and thus the number of deaths in industrialized countries [1]. Various factors influence the composition of the microbiome and its activity. Firstly, the composition of the gut microbiome changes with age. Considering the aging society, the research field of healthy aging is increasingly focusing on the role of the gut microbiome. Furthermore, it not only varies within an individual, but also shows changes between individuals. These interindividual variations are mainly due to enterotypes, BMI (body mass index) and external factors such as lifestyle, exercise frequency and ethnicity [2].

However, diet has one of the greatest influences on changes in the gut microbiome. The gut microbiome is involved in the absorption and synthesis of various nutrients and the release of their metabolic products. These can provide a variety of growth-promoting and growth-inhibiting factors that influence human health either directly or indirectly [3]. Numerous recent studies have focused on the influence of diet, i.e. the intake of macronutrients such as proteins, carbohydrates and fats as well as micronutrients such as vitamins, minerals and trace elements, on the composition of the gut microbiome. Despite growing knowledge of these interactions, the exact impact of specific diets and dietary patterns on microbial composition and functionality remains incompletely understood. This study aims to contribute to the understanding of the relationship between dietary habits and the gut microbiome and provides new insights into the design of effective dietary strategies for better health.

2 Material and Methods

The data analyzed were taken from the German population cohort FoCus [4]. Individuals with a known diagnosis of inflammatory bowel disease or inflammatory bowel syndrome, chronic diarrhoea, kidney disease or any type of diabetes were excluded in order to minimize potential biases due to these diseases. This resulted in a total sample of 1561 subjects. For each participant, the daily intake of nutrient groups was inferred from a food frequency questionnaire. The nutritional data was normalized. This ensures comparability of the diet between the subjects based on the relative proportions of foods and food groups in their total diet. Furthermore, the normalized data per subject was aggregated into the meat/vegetable ratio. As the data was obtained by 16S sequencing, the read counts were normalized in a sample-wise manner to obtain relative abundances of operational taxonomic units (OTU) per subject. This normalization is important to ensure correct interpretation and comparability in order to compensate for sequencing depths between the samples. The R programming language (version 4.3.0) in the RStudio development environment with the libraries dplyr for data manipulation and transformation and vegan for calculating similarity and distance measures was used as the analysis tool. Statistical analyses were performed by using the Wilcoxon rank-sum test and the linear model. The two samples for the Wilcoxon test were extracted from the dataset by filtering the subjects with the lowest and highest 25% of the respective food consumption. For the analysis using a linear model, the entire sample was used to examine the linear relationships between the variables. Test results with a p-value smaller than or equal to 0.05 were considered statistically significant. Furthermore, a Permutational Multivariate Analysis of Variance (PERMANOVA) was performed using the R package vegan. A distance matrix containing the differences between the objects was used as the database. It was calculated using the vegdist method on the normalized abundance data of the gut microbiome composition with the Euclidean distance as metric. In the PERMANOVA analysis using the adonis2 method, the variables gender, BMI and age were taken into account to investigate how these variables can explain the variation in the distance matrix. For the statistical tests, 999 permutations were performed. By setting the by argument to margin, the marginal effects of the independent variables were analyzed.

3   Results

First, the gut microbiome compositions of the two groups low meat/vegetable ratio, i.e. a diet low in meat and rich in vegetables, and high meat/vegetable ratio, i.e. a diet rich in meat and low in vegetables, were analyzed using the Wilcoxon test. The bacterium *Eubacterium rectale ATCC 33656*, hereinafter referred to as *Eubacterium rectale*, occurs significantly more frequently in the gut microbiome composition within the high meat/vegetable ratio group. The boxplot in Fig. 1 shows the relative abundance of *Eubacterium rectale* in both groups.

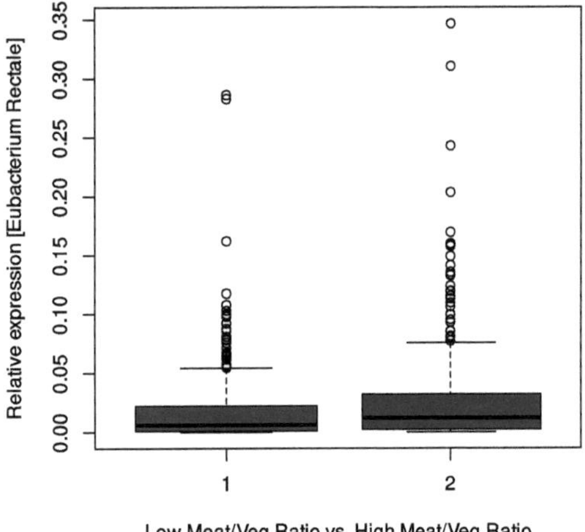

Figure 1: Boxplot showing the comparision between the groups low (left) and high (right) meat/vegetable ratio. The increased abundance of *Eubacterium rectale* in the high meat/vegetable group was significantly different with a Wilcoxon rank-sum test adj. p-value of 8.87×10^{-4}.

For Group 1 (low meat/vegetable ratio), the median is 0.0059, with the interquartile range from 0.0008 (Q1) to 0.0222 (Q3). The whiskers range from 0 to 0.0541 and there are 33 outliers. For Group 2 (high meat/vegetable ratio), the median is 0.0119, with the interquartile range from 0.0022 (Q1) to 0.0318 (Q3). The whiskers range from 0 to 0.0755 and there are 31 outliers.

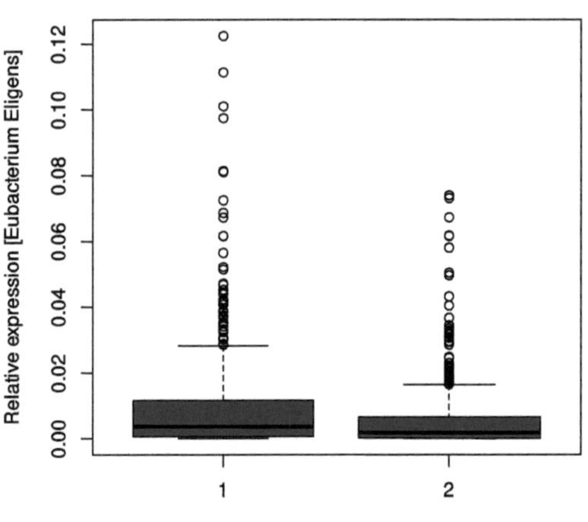

Figure 2: Boxplot showing the comparision between the groups low and high meat/vegetable ratio. The increased abundance of *Eubacterium eligens* in the low meat/vegetable group was significantly different with a Wilcoxon rank-sum test adj. p-value of 1.43×10^{-3}.

In contrast to this finding, the bacterium *Eubacterium eligens ATCC 27750*, hereinafter referred to as *Eubacterium eligens*, occurs significantly more frequently in the gut microbiome composition within the low meat/vegetable ratio group as can be seen in Fig. 2. For Group 1, the median is 0.0036, with the interquartile range from 0.0006 (Q1) to 0.0117 (Q3). The whiskers range from 0 to 0.0282 and there are 42 outliers. For Group 2, the median is 0.0018, with the interquartile range from 0.0002 (Q1) to 0.0067 (Q3). The whiskers range from 0 to 0.0164 and there are 51 outliers.

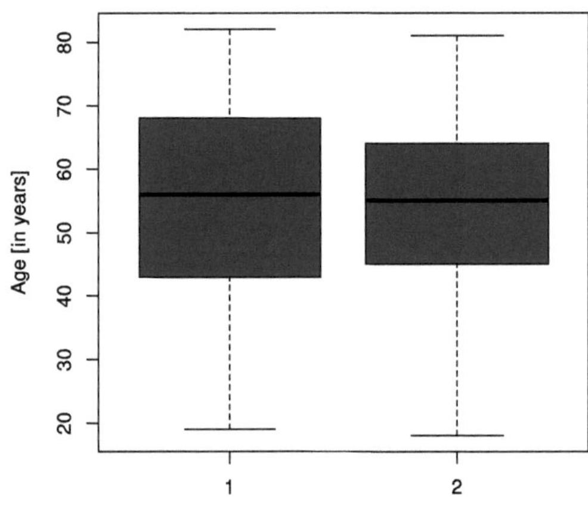

Figure 3: The boxplot compares the age distribution between the low and high meat/vegetable ratio groups.

Additionally, this data was analyzed with regard to age, BMI and gender. In Fig. 3, the phenotypic parameter age is considered in more detail. For low meat/vegetable ratio, the median is 56 years, with the interquartile range from 43 (Q1) to 68 (Q3) years. The whiskers range from 19 to 82 years and there are no outliers. For high meat/vegetable ratio, the median is 55 years, with the interquartile range from 45 (Q1) to 64 (Q3) years. The whiskers range from 18 to 81 years and there are also no outliers. The boxplot shows that both groups are very similar both on average and in the scatter range. Fig. 4 shows the phenotypic parameter BMI

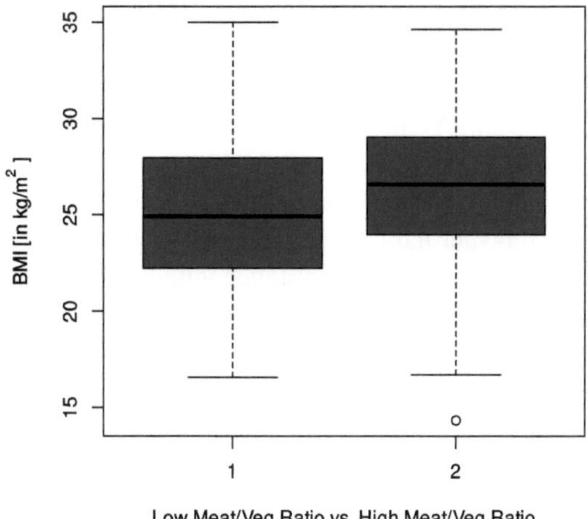

Figure 4: The boxplot compares the BMI distribution between the low and high meat/vegetable ratio groups.

in more detail. For Group 1, the median is 24.91 $\frac{kg}{m^2}$, with the interquartile range from 22.24 (Q1) to 27.96 (Q3) $\frac{kg}{m^2}$. The whiskers range from 16.56 to 34.99 $\frac{kg}{m^2}$ and there are no outliers. For Group 2, the median is 26.57 $\frac{kg}{m^2}$, with the interquartile range from 23.97 (Q1) to 29.03 (Q3) $\frac{kg}{m^2}$. The

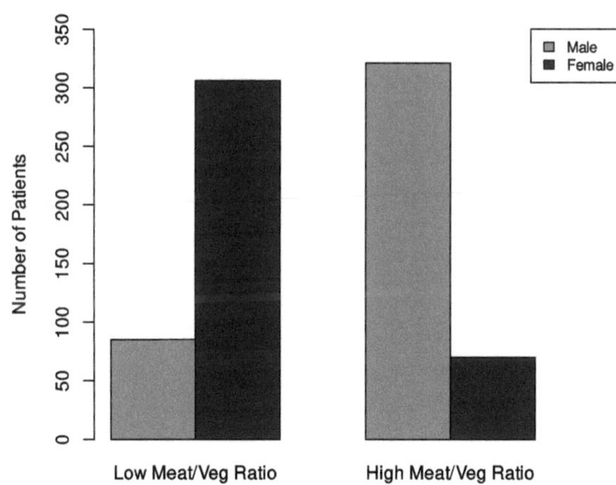

Figure 5: The barplot shows the gender distribution within the low and high meat/vegetable ratio groups.

whiskers range from 16.69 to 34.62 $\frac{kg}{m^2}$ and there is an outlier at 14.33 $\frac{kg}{m^2}$. Fig. 4 shows that the high meat/vegetable

ratio group tends to have a slightly higher median and wider BMI distribution. Fig. 5 depicts the gender composition of the two dietary groups. The low meat/vegetable ratio group is made up of 85 men and 306 women, whereas the high meat/vegetable ratio group contains 321 men and 70 women. This shows very clearly that the group composition is unbalanced in terms of gender. There are predominantly women in the low-meat, vegetable-rich diet group and vice versa. Furthermore, a Wilcoxon analysis was performed for the two groups in relation to age, BMI and gender. Table 1 shows the results, which include the log2FC and the adjusted p-value for these parameters.

Table 1: Wilcoxon analysis results for low vs. high meat/vegetable ratio groups concerning Age, BMI, and Gender

	log2FC	p.adj
Age	6.771×10^{-4}	6.354×10^{-1}
BMI	6.579×10^{-2}	3.337×10^{-6}
Gender	-2.128	0

The high p-value of 6.354×10^{-1} for age shows that there is no siginificant difference between the two groups. In contrast, the small p-value of BMI (3.337×10^{-6}) shows a difference between the groups, although the effect size is only weakly pronounced. However, the gender parameter shows the most striking difference, which is statistically highly significant with a p-value of 0. The log2FC value of -2.128 shows that the effect size indicates an approximately fourfold difference between the groups in terms of their gender composition.

Table 2: PERMANOVA analysis on Age, BMI, Gender, and MeatVegRatio

	R2	Pr(>F)
Age	3.148×10^{-3}	0.006
Gender	3.264×10^{-3}	0.007
BMI	1.407×10^{-3}	0.345
MeatVegRatio	1.230×10^{-3}	0.436

A PERMANOVA was carried out to further investigate the influence of these parameters. In addition, the respective group association (low or high meat/vegetable ratio) is taken into account in the independent variable MeatVegRatio. In Table 2 can be seen that the parameters BMI and MeatVegRatio with high p-values show no significance. Only the parameters age and gender with p-values of 0.006 and 0.007 significantly influence the distances between the subjects. However, the coefficient of determination R^2 also shows that they only explain the variations in the data to a very small extent. Around 0.315 % of the total variation is explained by the age parameter and 0.326 % by the gender parameter.

4   Discussion

The aim of the study was to investigate the influence of diet on the gut microbiome composition. The data shows that a vegetable- or meat-rich diet leads to different bacterial abundance. It was shown that *Eubacterium rectale* has a higher relative abundance in the high meat/vegetable ratio group. This observation is additionally supported by the study by [5], in which *Eubacterium rectale* is linked to a meat-rich diet. Moreover, [6] show an association between *Eubacterium rectale* and a higher BMI. This relationship can also be seen in the analyzed data based on the higher median BMI, see Fig. 4.

Furthermore, the analyses showed a higher relative abundance of *Eubacterium eligens* in the low-meat, vegetable-rich diet. A thai population study highlights a correlation between *Eubacterium eligens* and a vegetarian diet [7]. Similar results were found in a spanish population for a diet rich in lettuce, green beans and spinach [8]. In addition, it was observed that the phenotypic characteristics of gender and, to a lesser extent, BMI were significant parameters for the differences between the low and high meat/vegetable ratio groups. The PERMANOVA analysis, on the other hand, showed that gender and age were significant parameters. In particular, it can be seen that gender is one of the significant factors in both analyses and that there is also a link between gender and eating habits, see Fig. 5.

However, there are various limitations that make it difficult to interpret the results. As these are clinical and biological data from humans, they have a high variance. Especially compared to data from animal studies, which can be conducted in controlled environments over the entire lifespan, human data has a greater unpredictability, which makes the evaluation and recognition of causal relationships more challenging. In addition, the subjects' nutritional information was collected via a questionnaire. Since this data collection instrument is dependent on self-reporting, various biases of the data can occur due to, e.g., misunderstandings, lack of memory and limited response options. Furthermore, this is a cross-sectional study. This study design represents a snapshot of data collected at a specific point in time. However, this snapshot offers insufficient opportunity to investigate causal as well as cause-effect relationships.

5   Conclusion

The study aimed to deepen the understanding of the effects between dietary choices and the gut microbiome. It was possible to extract that a vegetable- or meat-rich diet leads to different bacterial abundance. *Eubacterium eligens* was associated with a more vegetable-rich diet and *Eubacterium rectale* with a more meat-rich diet. *Eubacterium rectale* was associated with a higher BMI, which was also seen in the analysis of this group of subjects. In addition, it was found that gender and, to a lesser extent, BMI were significant parameters for the differences between the low and high meat/vegetable ratio groups as phenotypic characteristics. The PERMANOVA analysis, on the other hand, showed

that gender and age were significant parameters, although they only explained a very small proportion of the variance in the data. In particular, it was found that there is a correlation between gender and dietary choices and habits.

Acknowledgement

The work has been carried out at the Institute of Experimental Medicine, Kiel University and the Institute of Medical Informatics, Universität zu Lübeck.

Authors' Statement

Conflict of interest: Authors state no conflict of interest. Informed consent: Written informed consent has been obtained from all individuals included in this study. Ethical approval: The research related to human use complies with all the relevant national regulations, institutional policies and was performed in accordance with the tenets of the Helsinki Declaration, and has been approved by the ethical review board of the Medical Faculty of Kiel University.

6   References

[1] M. Moszak, M. Szulinska, and P. Bogdanski, "You are what you eat—the relationship between diet, microbiota, and metabolic disorders—a review," *Nutrients*, vol. 12, p. 1096, 04 2020.

[2] E. Rinninella *et al.*, "What is the healthy gut microbiota composition? a changing ecosystem across age, environment, diet, and diseases," *Microorganisms*, vol. 7, no. 1, p. 14, 2019.

[3] S. Mansour, M. Moustafa, B. Saad, R. Hamed, and A.-R. Moustafa, "Impact of diet on human gut microbiome and disease risk," *New Microbes and New Infections*, vol. 41, p. 100845, 2021.

[4] C. Geisler *et al.*, "Cohort profile: the food chain plus (focus) cohort," *European Journal of Epidemiology*, vol. 37, no. 10, pp. 1087–1105, 2022.

[5] J. Tap *et al.*, "Diet and gut microbiome interactions of relevance for symptoms in irritable bowel syndrome," *Microbiome*, vol. 9, no. 1, p. 74, Mar. 2021.

[6] P. I. Costea *et al.*, "Subspecies in the global human gut microbiome," *Molecular systems biology*, vol. 13, no. 12, p. 960, 2017.

[7] S. Ruengsomwong *et al.*, "Microbial community of healthy thai vegetarians and non-vegetarians, their core gut microbiota, and pathogen risk," *Journal of microbiology and biotechnology*, vol. 26, 07 2016.

[8] A. Latorre-Pérez *et al.*, "The spanish gut microbiome reveals links between microorganisms and mediterranean diet," *Scientific Reports*, vol. 11, 11 2021.

5

Control technology

Creating a Graphical User Interface as Clickdummy with Simple Directmedia Layer Library in C++

Liesa Kunert [1]

[1] Medical Engineering Science, Universität zu Lübeck, liesa.kunert@student.uni-luebeck.de

Abstract

Graphical user interfaces (GUI) are omnipresent in our everyday lives today and make it much easier. However, the development of a GUI is a very extensive process that needs to be well-planned. Clickdummies often help to guide structural and conceptual thinking, mediating between the product owner's expectations and the implementer's understanding. Although these are often created with the help of existing toolkits, they were not used in this project and the GUI was implemented directly with the Simple DirectMedia Layer (SDL) in C++. However, this method was chosen without knowing how time-consuming and realistic this implementation would be for the client's desired project with 150 different pages. While this approach presents time challenges and design constraints, it offers a major advantage in terms of compatibility with other platforms. A clickdummy is created that can later be integrated directly into the planned device as a GUI.

1 Introduction

Graphical user interfaces (GUIs) are an important part of today's highly modern, technological society. We encounter them several times a day on our cell phones and computers, but also at home, on the road, or in the laboratory. Wherever human-machine interaction takes place, they enable simple and straightforward operation of the machine without the need for extensive background knowledge of the system. The first theories about a GUI were published in 1945 by Vannevar Bush in his article "As We May Think" [1] and adapted by Douglas Englebart and first presented publicly in 1968 [2]. A recent review is given by Martinez [3]. They should support the user and not replace them. As GUIs are event-based, the user can navigate through the application and operate the machine simply by clicking the mouse, gesturing in front of the camera, or moving the touchpad.

In the practical project on which this article is based, the original intention was to design a graphical user interface for a laboratory measuring device that was also to be developed first. However, in consultation with all those involved and to protect the confidentiality of the developers, this work is based on a general example familiar to all readers, such as a fully automatic coffee machine. This should have a large number of functions and settings as well as status displays for WLAN or Bluetooth and thus comprise a GUI of about 150 pages.

It starts with the basic steps of GUI development, the special project requirements, the choice of the Simple DirectMedia Layer Library (SDL), and the subsequent implementation, before finally evaluating the results.

2 Material and Methods

The following sections discuss general aspects of GUI development, specific requirements for the GUI discussed in this paper, details of the hardware and software used, and finally the implementation of the GUI.

2.1 General development of a GUI

The first step is to create an overview with answers to the following questions:

- What functions should the GUI have?

- Who should be able to use it?

- What problems should it solve?

- How can it best help the user?

- How should the individual functions be operated and accessed?

The answers to these questions should then be refined to such an extent that a structure map can be derived. This should provide an overview of individual pages and their interrelationships. With the aim of understanding which button to press to navigate from one page to another. In addition to the buttons, icons, the definition of the toolbar, other dialog windows, lists, drop-down menus, checkboxes, and text boxes are also part of a GUI [3]. The structure of the pages and their interrelationships can then be displayed in a wireframe or, somewhat more dynamically, in a clickdummy. This is a good way to check the structure again at an early stage, keep possible incorrect development costs low, and avoid duplicate development steps because such a simplified representation makes it possible to quickly incorporate new ideas into the concept and to obtain further changes and feedback in tests with users or product owners. Possible tools to create a mockup would be:

- *Balsamiq* (https://balsamiq.com/wireframes/),

- *Axure rp* (https://www.axure.com/),

- *figma* (https://www.figma.com/de/prototyping/),

- *adobe indesign* (https://www.adobe.com/de/products/indesign/),

- *adobe illustrator* (https://www.adobe.com/de/products/illustrator/c).

To make full use of these tools, you often need a subscription, without knowing where the copyright lies in such cases and how well your own designs and developments of a device that is not yet on the market are protected.

Once you have a rough concept, you can start developing the actual GUI directly or in parallel. There are usually various toolkits or programming interfaces (API) to facilitate the creation, but some of these are only compatible with one platform (macOS, Windows, ...), like *Cocoa* (macOS, http://www.cocoa-coding.de) and *Universal Windows Platform* (UWP, Windows) or the cross-platform *Toolkit* (TK, https://wiki.tcl-lang.org). Overall, a GUI should always take the following criteria into account:

- requires little user knowledge

- should be easy and convenient to use and understand

- should react quickly and sensibly to an event

A comprehensive literature review of menu requirements is given by Murano and Sanders [4].

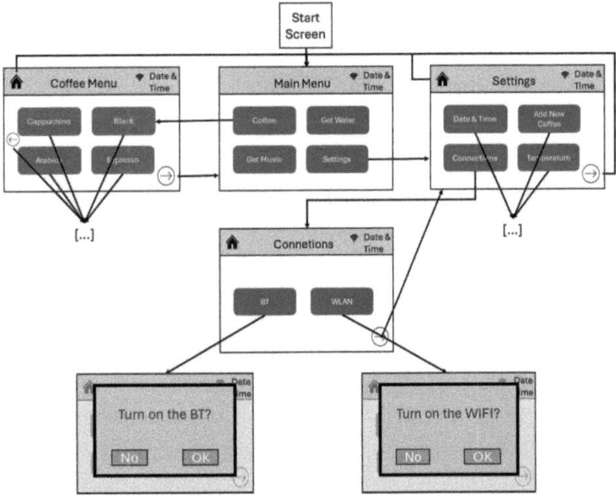

Figure 1: Structure Overview: Small excerpt from the rough structure of GUI of the coffee machine. Actually, over 150 pages, simplified here.

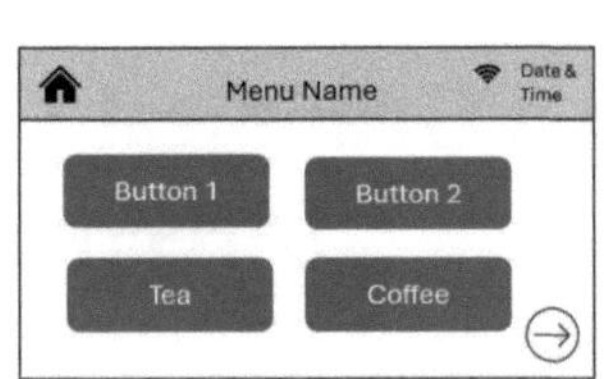

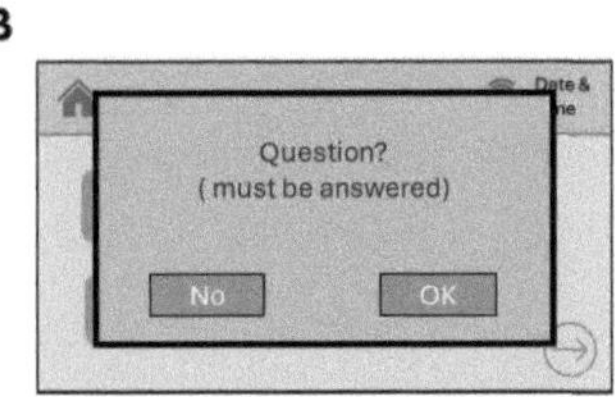

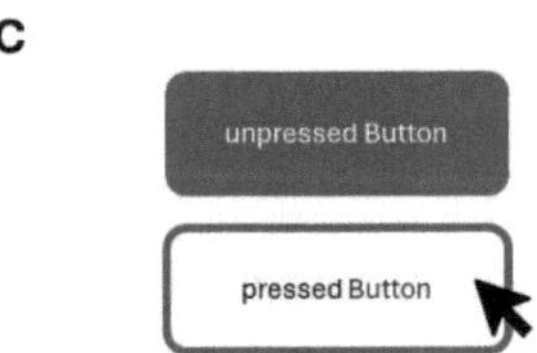

Figure 2: Structure of various pages. A) The main window with a toolbar and back button. B) Modal dialog window with locked background window. C) Pressed and unpressed button.

2.2 Requirements for this GUI

In this case, the project owner's desired user interface for the coffee machine should comprise around 150 pages. The rough structure had already been determined. Nevertheless, details were repeatedly clarified and adjusted. In addition, the question of cost-effective and timely implementation remained. Fig. 1 shows a small sample excerpt of the structure map of a GUI. There are two groups of pages, as shown in Fig. 2. One is the group of main pages. They have a toolbar, various buttons, a back button, and a home button. The second group is the dialog boxes or so-called modal windows. They require interaction from the user and block the rest of the screen. With our coffee machine, for example, this could be the warning "No more coffee. Please refill beans." The buttons should also change their appearance when clicked. The following were also defined as further important criteria for the clickdummy:

- The clickdummy should be executed directly on the end module, but also as easily as possible on a computer.

- It should not create any license or data protection problems.

- The clickdummy should be created with as little effort as possible.

- The creation should generate as little to no additional costs as possible.

- It should be easy to handle so that it can be presented to the customer quickly and without great effort if necessary.

- It should enable quick and targeted feedback.

2.3 Technical Conditions

For our coffee machine, it was specified that we use a Raspberry Pi with the Compute Module 4 [5]. This is a single-board computer that runs on Linux and initially has all the features to allow smooth communication between the individual modules within the USB architecture. A resistive 5 inch touchpad was chosen as the platform for the GUI.

2.4 Simple Directmedia Layer

The Simple DirectMedia Layer (SDL) is a portable open-source multimedia library [6]. Since the SDL has a zlib license, it is free to use and can be combined with any software [6]. The SDL makes it possible, for example, to process hardware input such as mouse movements, clicking buttons, or keyboard input and to enable audio output. It is written in C and therefore also works very well in C++. It runs cross-platform on Windows, macOS, Linux, iOS, Android and others. It is also sufficiently well documented, as it has already been used successfully in games such as *OpenTTD* (https://www.openttd.org) [6]. A list of other games based on the SDL can be found in the LibreGameWiki [7]. To implement the GUI, standard programming tools such as an Integrated Development Environment (IDE) e.g. *Visual Studion Code*, a version management program with git repository, a builder *cmake* (Windows and Linux) and a compiler *minGw-w64* are used. This enabled an executable file to be created as an artifact.

2.5 The implementation

The SDL provides tools such as drawing graphic shapes, but also important functions such as a timer or recognizing events, e.g. operating the touch. Based on the class *Widgets* with the functions *set* and *fill*, several widgets such as buttons are created. Each *widget* contains information on how an event is processed or what triggers it and is then redrawn. Here, the current mouse position and the *clicked or not* state are compared with the position of the button and, if necessary, the button is set to *activated* mode. Entire pages or modal dialogs can be assembled from the widgets. Each resulting page is assigned an enum with which it can be called via the Lambda function. This makes it possible to implement links between the pages. If you start the device, it starts the function *run* in the *GUI* class and creates a message loop that waits for input, i.e. events. This prevents multiple runs. In an event, all widgets on a page are painted new, whereby the last trumps all others and is displayed on top of the others.

The modal dialogs are also based on the *GUI* and create a *runModal*, which is stacked and then destacked to return to the original page. The structure and sequence of the dialog boxes are similar to the normal pages but differ in the way they are called. The main image is painted and the modal is placed over it blocking everything else on the screen e.g. the home button.

3 Results and Discussion

By using SDL, it was possible to implement a comprehensive GUI. The originally planned clickdummy thus became an essential part of the GUI, which minimized the time factor as hoped. Other clickdummy tools such as Balsamiq, Figma, or the Adobe products were initially considered but were not taken into account for the time being due to the additional costs involved in ensuring that data protection was maintained and the possibilities that could not be implemented in such detail. As the GUI concept already appeared to be very mature, this simplified direct programming with SDL. In addition, access to feedback and rapid deployment on the various platforms was a deciding factor.

Thanks to the multi-platform capabilities of SDL, the GUI could be programmed in various developer environments. It is also very practical to test the GUI on the target device (5 inch display) during development and then modify it directly on the PC. An additional scaling function allows a rough approximation of the final image on the desktop. This multi-platform application therefore proved to be particularly helpful when adjusting details such as font size or button size and positioning.

With its extensive geometric functions, SDL provided a good basis for all kinds of representations. The color scheme was also very flexible. The only thing that took a little more time was creating more complex designs such as rounded corners on the buttons. These had to be assembled using drawings of a circle and several rectangles, as shown in Fig. 3. At the same time, however, this also shows the diverse possibilities offered by the SDL.

Figure 3: Concept of buttons with round corners: A) construction of unpressed button B) pressed button with round corners

The drop-down menus, which are important for GUIs, were not to be implemented due to the display size and would have taken a little more time to develop in SDL programming compared to a GUI toolkit. Nevertheless, the combination of individual smaller widgets made it possible to implement more complex widgets such as a keyboard or display brightness settings using a slider. Both could be realized by combining various elements already used in the button concept (see Fig. 3), if not the buttons themselves.

The SDL also offered many different possible event trackings and therefore many options for activating the GUI. My colleague was even able to implement a kinetic scrolling function for tables with SDL.

All in all, the criteria listed under 2.2 were all met. This clickdummy did not generate any additional costs for Software, time expenditure, or license problems. Data protection was also not jeopardised during development, as the work was carried out locally, mostly offline, and, if necessary, with already validated tools such as Git repositories. Furthermore, it can be easily presented on various media through an executable file. The only thing that has not been considered yet is the simple feedback function. This is achieved in the dummy by assigning a page number to each page in the schematic before conversion, and this is also displayed on each page in the GUI. It helps with simple communication about individual pages and changes but must be removed before the GUI is published.

Unfortunately, I was unable to continue working on this project after my internship ended, but I know that later connections to databases such as *SQLite* (https://www.sqlite.org) have been implemented by my colleague in line with the model view controller principle. In addition, other aspects already considered in the GUI were made possible by the quick integration of further libraries, e.g. the connection of a camera. Other useful functions, such as the translation of the GUI into another language, have already been prepared in the GUI. Furthermore, the drop-down menus are to be integrated into the GUI at some point. In addition, the page numbering, which previously only served to simplify feedback, has been given a new function. It is now also used to call up help texts.

4   Conclusion

In my project and with the experience of my colleagues, it was a unique advantage to program the GUI directly in this project. With this structure, all 150 pages and modal dialogs could be created functionally. Furthermore, it was possible to implement special functions such as kinetic scrolls, keyboards, sliders, and even drop-down menus. Using a library such as SDL, which offers a wide range of components, significantly reduced the development time for this project and ensured a consistent look and feel throughout the application. As well as saving time, the library comes with extensive documentation, making it easier for programmers

to learn. The open-source nature of the software encourages a wider developer community and can facilitate continuous improvement and future enhancements. In addition, SDL's cross-platform compatibility allows the GUI to be used on multiple operating systems. For less complex design projects, however, ready-made toolboxes remain the faster and simpler choice. This is exactly what limits them. However, anyone who enjoys structured planning, programming, and unlimited creativity will understand why there are already so many games with the SDL. Not to mention it has many versatile functions and possibilities.

Acknowledgement

The work has been carried out at who Ingenieurgesellschaft mbH, Lübeck and supervised by the Institute of Physics, Universität zu Lübeck.

I would like to express my gratitude to all my colleagues at who Ingenieurgesellschaft mbH, Lübeck for their willingness to help me and share their knowledge and experience during my internship. Special thanks to Hartmut Quast, Head of Software Development and Stefan Kanarski, Software Developer, both from the who Ingenieurgesellschaft mbH.

Authors' Statement

Conflict of interest: Authors state no conflict of interest.

5   References

[1] V. Bush, *As we may think*. Atlantic Monthly, pp. 101–108, 1945.

[2] DC. Engelbart, WK. English, *A research center for augmenting human intellect*. AFIPS Conference Proceedings Fall Joint Computer Conference; pp. 395–410, 1968.

[3] WL. Martinez, *Graphical user interfaces*. Wiley Interdisciplinary Reviews: Computational Statistics; 2011.

[4] P. Murano, M. Sander, *User Interface Menu Design Performance and User Preferences: A Review and Ways Forward*. International Journal of Advanced Computer Science and Applications(IJACSA); 7(4), 2016.

[5] *Raspberry Pi-Compute Module 4*. Available: https://www.raspberrypi.com/products/compute-module-4/ [last accessed on 2024-02-05].

[6] *SDL Website*. Available: https://www.libsdl.org/ [last accessed on 2024-02-05].

[7] *Games Website*. Available: https://libregamewiki.org/SDL_games [last accessed on 2024-02-05].

Implementation of an interface between a foot switch and an augmented reality device based on the SDC standard

Manuel Herbst [1] , Martin Leucker[2] and Lennart Landt [3]

[1] Student of Medical Informatics, Universität zu Lübeck, manuel.herbst@student.uni-luebeck.de
[2] Director of Institute for Software Engineering and Programming Languages, Universität zu Lübeck, leucker@isp.uni-luebeck.de
[3] Projektmanagement KI-Med. Ökosystem, UniTransferKlinik GmbH, llandt@unitransferklinik.de

Abstract

This research deals with the implementation of an interface between a foot switch and an augmented reality device based on the Service-oriented Device Connectivity (SDC) standard, making the HoloLens 2 SDC-ready. While the progress in augmented reality is creating new application possibilities in medicine, the recently developed communication standard SDC offers the possibility to integrate the HoloLens 2 interoperably into a medical scenario. The research was implemented with the SDC JAVA library (SDCLib/J) and contributes to exploring the possibilities of SDC.

1 Introduction

Interoperability between devices ensures high levels of security and consistency for the transferred data. It is especially vital in the medical context which relies on the protection and correctness of personal information. Because of the ever-increasing amount of data and digitalization, it is also important that the data transfer is reliable and fast. Service-oriented Device Connectivity (SDC) is an IEEE standard for machine-to-machine communication, which deals precisely with this interoperability between medical devices. It enables bidirectional communication and remote control between devices [1], [2]. As it is fairly new there are not many scientific papers available to study, although some medical tech companies are already using it.

This research is based on the extension of interoperability to an Augmented Reality (AR) Headset which allows for new possibilities to access data during medical procedures [3]. Such an approach has the advantage of a simplified and personally adjustable interface that makes complex data more accessible, providing reliable support to medical professionals. The Microsoft HoloLens 2 is an AR Headset that can be used in various work scenarios that prove to be very beneficial, especially in medicine [3]. It can reduce the number of devices needed while providing more variable information for the user.

The main idea behind this research is to use a surgical foot switch that is SDC-ready and to connect it to the HoloLens 2 to show the integration of the HoloLens 2 into surgical devices is possible by using SDC. With that, one should be able to control the headset with the foot switch, for example taking a photograph mid-surgery for documentation purposes. Furthermore, this can be used to possibly classify objects such as organs in real-time and provide the information in the field of view of the user. The implementation of this use case presents one of many possibilities for the integration of AR in medicine.

Considering how recently SDC was developed and AR starting to be adapted into research, this project also aims to address the lack of theoretical foundations in this area.

This paper begins with the chapter on materials and methods which describes the hardware used in this research, as well as the approach toward upcoming challenges. Results and discussion deals with unexpected changes, the experimental setup and functional outputs. Moreover, the limitations and challenges and future work are discussed. The last chapter contains the conclusion of the research.

2 Materials and Methods

2.1 Microsoft HoloLens 2

The Microsoft HoloLens 2, as seen in Fig. 1, is an AR Headset. Specifically, it is an Optical See-Through Head Mounted Display (OST-HMD) with the ability to display images without restricting the view of the surrounding area. Different interaction mechanisms can be used, such as hand interaction or voice commands. In a medical context, it could be used as a supporting tool for medical professionals, helping by displaying data. The field of view can be modified and adapted to personal preferences. There is no lag between what the user sees as it is a see-through device. This is relevant because a fast reaction time can be critical. The HoloLens 2 is used because it is currently

Figure 1: Image of the HoloLens 2 [6].

the leading product on the market. With the ability to scan the surrounding area constantly it is possible to place and attach objects in space and on different objects.

2.2 Foot Switch by .steute

The foot switch is provided by .steute and consists of three different parts. The foot switch itself, a receiver and a Raspberry Pi. .steute provides different kinds of foot switches with a different number of pedals. For this research, the foot switch with three pedals is used, as seen in Fig. 2. It acts as the User Interface, for example by pressing the left pedal the HoloLens 2 takes a photograph. To prepare the foot switch for use, it is necessary to connect it to a receiver, as seen in Fig. 3, which is connected to a RapberryPi by a USB cable. The Raspberry Pi transforms the signal into an SDC message and transmits the signal into the SDC network. SDC is described in Section 2.3. In the following, the three parts (foot switch, receiver and Raspberry Pi) are referred to as the foot switch.

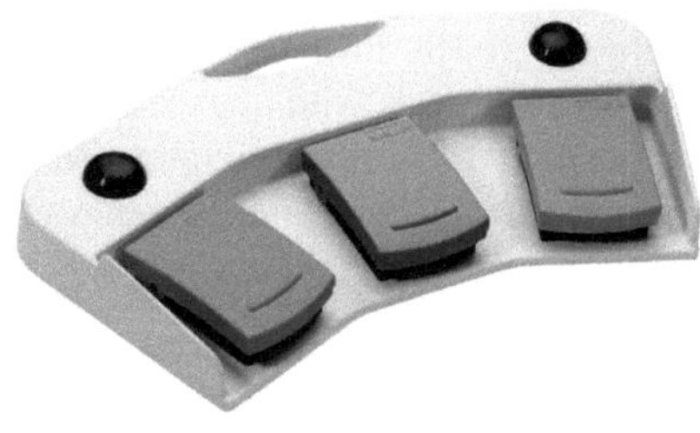

Figure 2: The foot switch used in this research [7].

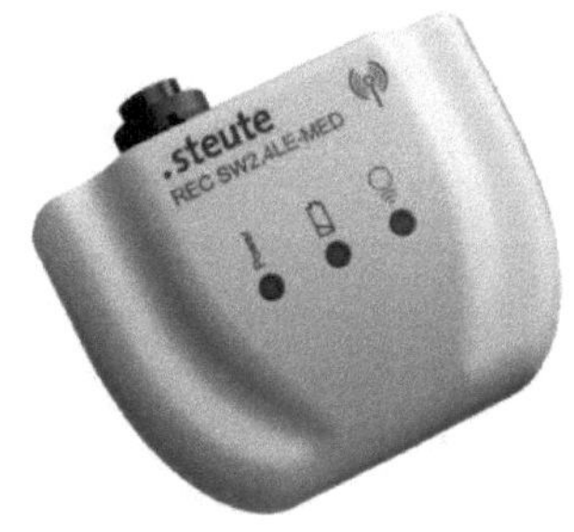

Figure 3: The receiver used in this research [8].

2.3 Service-Oriented Device Connectivity (SDC)

The research is focused on the implementation of SDC, the Service-oriented Device Connectivity framework. SDC is supposed to address the lack of interoperability between devices in the medical sector. Standardized communication is already implemented for medical imaging and patient data using DICOM or HL7 FHIR [1]. But for the exchange of data between medical devices, there are just obsolete standards of the old ISO/IEEE 11073 standard family [1]. SDC enables bidirectional data transfer between medical devices from different manufacturers and even enables remote control of these devices while maintaining high quality and consistency. The framework is based on the service-oriented architecture (SOA) and is available in different OpenSource libraries.

There are software libraries available in C++ (SDCLib/C) and Java (SDCLib/J), e.g. at [4]. There is also a reference implementation at [5]. SDC uses provider and consumer objects. Provider objects can send metrics and alerts into the network, while the consumer objects listen if any signals are received and react accordingly.

2.4 Implementation of the Connection

To be able to transmit signals from the foot switch to the HoloLens 2, a connection needed to be established. For this, the SDCLib/J is used, which connects the foot switch to a computer. The computer sends the received signal to the HoloLens 2. The signal is processed and triggers a reaction on the headset e.g. taking a picture. The implemented configuration is further explained in Section 3.2. As there is no code documentation available for SDCLib/J, the implementation was realized step by step. At first, assumptions were made, for example, the foot switch sends flags whenever a pedal is pressed. The corresponding functions in the code were then searched. If the appropriate function was found it was implemented in the work. Using this method the consumer was applied properly and able to receive signals from the foot switch.

3 Results and Discussion

3.1 Unexpected changes and Adjustments

In the context of this research, the HoloLens 2 should be the consumer that gets controlled by the provider, in this case, the foot switch. Since the original goal was to use the HoloLens 2 as an SDC consumer device, the application needed to run natively on the HoloLens 2. For a better understanding Fig. 4 is provided. As SDLib/C failed to install due to the failure of a dependency during project setup, SDCLib/J is currently used as an alternative in this research. The consumer cannot run on the HoloLens 2 natively, because the headset does not support JAVA applications. Therefore an external computer had to be used instead to act as the consumer. Moreover, another connection had to be established, connecting the computer with the HoloLens 2 using sockets.

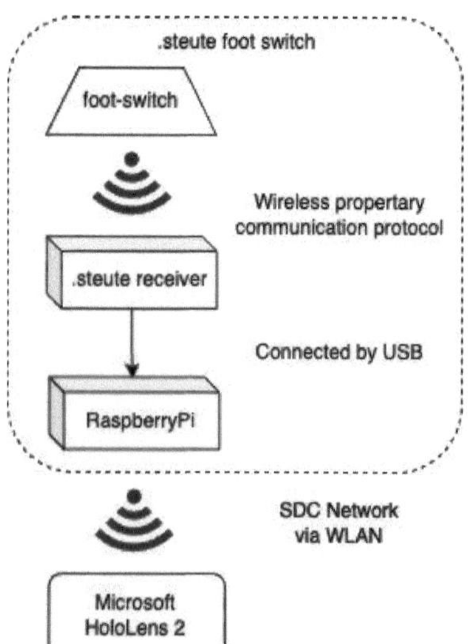

Figure 4: This graphic shows the initial idea for the test setup. This includes the setup of the foot switch from .steute, as well as the HoloLens 2. The various types of transmission media are also illustrated. The signal originates from the foot switch and is sent downwards to the HoloLens 2.

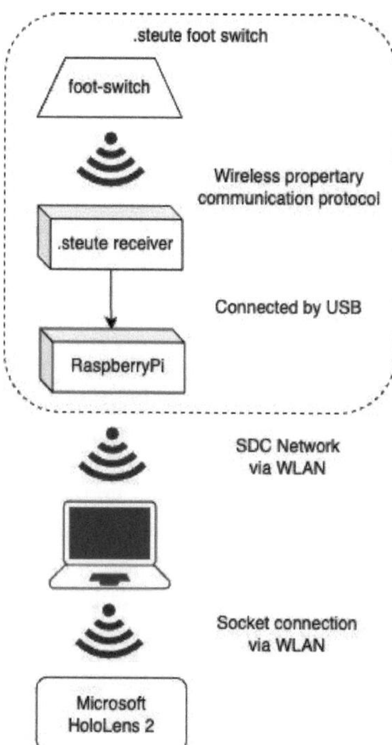

Figure 5: This graphic shows the test setup. This includes the setup of the foot switch from .steute, as well as the computer and the HoloLens 2. The various types of transmission media are also illustrated. The signal originates from the foot switch and is sent downwards to the HoloLens 2.

3.2   Implemented Configuration

In this section, the connection between the foot switch and the HoloLens 2 is described in detail. For a better understanding an illustration is provided, see Fig. 5. The activation of the foot switch is the trigger for a run through the connection. If the user activates the foot switch, a signal is sent into the SDC network, which runs on the Wireless Local Area Network (WLAN). The configuration of the foot switch is already explained in Section 2.3. As a consumer object, the computer can listen, if a provider is sending messages. To achieve this, the computer has to be connected to the SDC network and a listener function needs to be executed. If a message is received, the computer starts a socket and sends a message to the HoloLens 2. This message contains information on the state of the foot switch, e.g. the pedal ID ("1", "2" or "3") and the state ("pressed" or "released").

The HoloLens 2 runs an application that starts a socket. Whenever a signal is received from this socket, the message is processed and analyzed which pedal the signal is coming from and reads the state of the pedal. Depending on the state and the ID, the HoloLens 2 reacts accordingly. For now, the headset captures a photograph, if the pedal with id="1" is pressed.

3.3   Functional outputs

During implementation, console outputs were used to check whether a connection was successful. The connection exists between the foot switch and the HoloLens 2 and consists of the connection between the foot switch and the computer, as well as between the computer and the HoloLens 2. For example, all the different messages sent by the foot switch to the consumer were printed. This output was used to help implement the listener function and to be able to capture the correct message. Another console output was made to validate the successful connection between the foot switch and the HoloLens 2. Here the output just printed the incoming message, containing the id of the pedal and the state of the pedal. This console output was later exchanged by the function to trigger the capture of a photograph. The capture of a photograph after pressing a pedal indicates that the connection between the foot switch and the HoloLens 2 is successful.

3.4   Implications of code implementations

This research is concerned with testing the possibilities of SDC and showing its potential, but also its limitations (Section 3.6). In general, the idea of bringing AR glasses into the medical context is promising [3]. Maybe SDC will be further developed in the future so that it can be used in similar projects and unfold that potential.

As the documentation of SDC is not very well developed, this research also helps to expand the documentation, as the implementation was documented in parallel to the research.

3.5 Limitations and Challenges

The major limitation was that SDCLib/C was not available for this project (Section 3.1). This led to the use of SDCLib/J instead, which added additional complexity, due to the use of an external computer. The additional complexity can lead to instability and inconsistencies. This step also adds a delay between the physical input of the foot switch and the corresponding response, in this case, the capture of a photograph. Ideally, the delay should be brought to a minimum, which can only be reached through the use of SDCLib/C running natively on the HoloLens 2. Furthermore, it might be helpful to build the library in C#. This would make it even more applicable for the HoloLens 2 because the development for the HoloLens 2 is done in C#. As already stated in Section 2.4, a big challenge in this research was the insufficient documentation of SDCLib/J. This slowed down the progress significantly and was also caused by SDCLib/C not working properly.

3.6 Future work

A future use case could also be to connect an endoscope to the SDC network. This would open more possibilities for HoloLens 2 implementations, such as using the HoloLens 2 as the display for the video. This could make external monitors obsolete, as the video would already be displayed in the headset. Furthermore, other information, such as medical images or vital parameters could be displayed in the headset. This would lead to a reduction of the number of displays necessary and provide a more variable surrounding for the user.

4 Conclusion

The adoption of socket-based communication was not the original objective, as it does not correspond to the idea of SDC. The idea behind SDC, as mentioned before, is the interoperability of different manufacturers and the standardization of the communication between medical devices. However, if SDCLib/C is set up, it might be possible to use SDC on the HoloLens 2 natively, the way it is intended.

In the end, it can be said that it might still be possible to run SDC natively on the HoloLens 2. With the workaround created in this project, it is still possible to transmit signals from the foot switch to the HoloLens 2 and to use them as user input.

Ideally, the foot switch should contain only the set of pedals without the receiver and the Raspberry Pi to minimize the amount of hardware. This leads to fewer connections and therefore, less vulnerability regarding the transmission processes.

Many possible applications are coming to mind, whether displaying X-ray or CT images, vital data or analyzing what tissue is malign or healthy. With that being said, there is a lot to discover in the field of AR in medicine, but also a lot of issues that have to be addressed. In this case, implementing the communication using SDC and the resulting interoperability. This research demonstrates what SDC may be capable of in the future.

Acknowledgement

The work has been carried out at the Institute of Software Development and Programming Languages at the University of Lübeck and supervised by the Institute of Software Development and Programming Languages at the University of Lübeck.

Authors' Statement

Conflict of interest: Authors state no conflict of interest.

5 References

[1] OR.NET e.V., *11.01.2024*. Available: https://ornet.org/services-2-2/ [last accessed on 2024-01-15].

[2] N. Kraus, M. Viertel and O. Burgert, *Control of KNX devices over IEEE 11073 service-oriented device connectivity.* 2020 IEEE Conference on Industrial Cyberphysical Systems (ICPS), Tampere, Finland, 2020, pp. 421-424, doi: 10.1109/ICPS48405.2020.9274729.

[3] Palumbo A. *Microsoft HoloLens 2 in Medical and Healthcare Context: State of the Art and Future Prospects.* Sensors (Basel). 2022 Oct 11;22(20):7709. doi: 10.3390/s22207709. PMID: 36298059; PMCID: PMC9611914.

[4] OR.NET e.V., *Software Stacks (Libraries).* Available: https://ornet.org/en/arbeitsgruppe_standardisierung-und-internationalisierung-2-2-2/#1554298685114-41719706-5b29 [last accessed on 2024-02-02].

[5] sdc-suite, *SDC Reference Implementation.* Available: https://gitlab.com/sdc-suite/sdc-ri [last accessed on 2024-02-02].

[6] Microsoft HoloLens 2, *02.02.2024.* Available: https://www.microsoft.com/de-de/d/hololens-2-industrial-edition/8mqn5pzp01x5?activetab=pivot:%C3%BCbersichttab [last accessed on 2024-02-02].

[7] .steute foot-switch, *11.01.2024.* Available: https://www.steute-meditec.com/en/products/mkf-3-sw24le-med-gp34.html [last accessed on 2024-01-12].

[8] .steute receiver, *11.01.2024.* Available: https://www.steute-meditec.com/en/products/rec-sw24le-med-ag43.html [last accessed on 2024-01-12].

Pressure Control of a Blower-Based Valve Test Bench

Malin-Berit Kluge [1], Robin Brütt [2], and Georg Männel [2]

[1] Medical Engineering Science, Universität zu Lübeck, malinberit.kluge@student.uni-luebeck.de
[2] Fraunhofer Research Institution for Individualized and Cell-Based Medical Engineering, Lübeck,
{robin.bruett,georg.maennel}@imte.fraunhofer.de

Abstract

For the development of a lung simulator, which should be capable of simulating fast lung mechanics, valves and blowers are required that offer the necessary dynamics. A test bench for simultaneous development of the simulator's blower and valve combination is used. This paper focuses on the control of the generated upstream pressure by the blower. The nonlinearity of the blower was largely compensated by recording its static characteristic field and integrating it into the model as a lookup table (LUT). A system for the blower and LUT was identified and used for tuning a PI-controller with a method based on the frequency response with requirements for rise time and overshoot. To analyse the performance of the controller, its behaviour was simulated and measured. The results show that the requirements are met for most operation points and the upstream pressure can be controlled.

1 Introduction

A physical lung simulator can be used to test ventilators in critical scenarios [1]. For the development of such a simulator blowers and valves are used to simulate the lung mechanics (e.g. the pressure) at the device interface to a ventilator as realistically as possible [2]. The valve should be proportional controllable and should have high dynamics [2], therefore a variety of valve designs should be tested to determine a valve with the desired behaviour. This is done on a separate test bench, whereby the blower used is identical to the blower of the lung simulator.

The blower shall be controlled to be able to generate a specific pressure in front of the valve. There are various approaches for controlling this blower based on variable-gain control and adaptive control, among others, but in these the blower is used within mechanical ventilators [3], [4].

In this paper, a PI-controller is designed to set a specific upstream pressure. The nonlinear dynamic behaviour of the blower is compensated with the knowledge of the static characteristic and a model is identified for the almost linear system. Based on this model, a controller is tuned with special requirements regarding the dynamics that arise from the fact that not only the valve, but also the blower should be used within the lung simulator.

2 Material and Methods

The turbine blower is specific for mechanical ventilators. A model of it is identified, for which a black-box model approach is chosen. As the blower is nonlinear, the static characteristic is recorded and a compensator is implemented to linearise the system, that is used for tuning a PI-controller.

2.1 Test Bench

The test bench, as shown in Fig. 1, is an updated version of the test bench used for the modelling of a positive end-expiratory pressure (PEEP) valve by Brütt et al. [5]. A blower (Macawi turbine blower, Demcon) with faster dynamics than the original blower, new pressure sensors (MPXV7025, NXP) with a wider measuring range, and a new flow meter (SFM3000, Sensirion) are used. A real time target machine, that is integrated in Simulink (2023a), controls the test bench and processes the data. For the characterization of the blower different flow resistances are set by using a PEEP valve and a ball valve, with the second having less influence on the dynamics. The opening of the PEEP valve can be adjusted with a screw or a voice-coil actuator, which is controlled with a pulse width modulation signal.

The test bench is limited by the blower and the flow meter, because the maximum measurable flow is $200\,\mathrm{L\,min^{-1}}$ and the maximum speed of the blower decreases with high load.

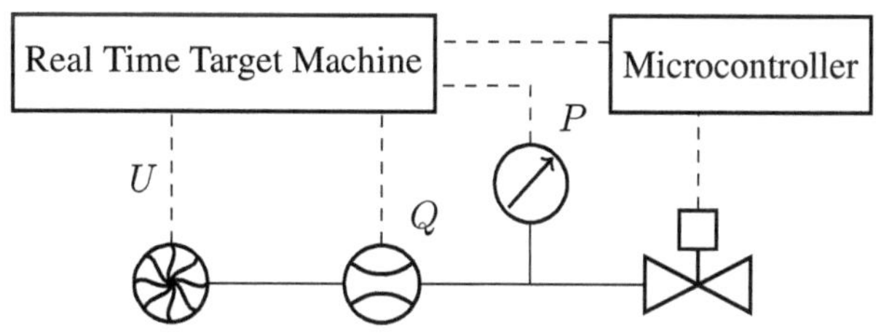

Figure 1: Test bench, where U is the blower control voltage, Q the flow, and P the upstream pressure. Next to the blower are the flow meter, the pressure sensor, and the valve. The solid lines indicate the tubes of the test bench and the dashed lines the data transfer.

2.2 System Identification

In order to apply standard linear control theory, the nonlinearity must be compensated. A lookup table (LUT), which selects a blower control voltage depending on the pressure and flow, is recorded with the PEEP valve and the adjusting screw. A ramp-shaped control voltage is used where the voltage is increased up to 5 V in 100 s, and both flow and pressure are measured. This measurement is repeated for different positions of the screw to take different flow resistances into account. The signal is filtered and smoothed to reduce the noise.

A black-box model approach is chosen to determine a model for the blower, so only the inputs and outputs of the system are considered but not the exact processes of the system [6]. To obtain such a blower model, the pressure and flow in the test bench are measured for different openings of a ball valve. It replaces the PEEP valve to avoid dynamic disturbance through fluttering of the valve membrane. The LUT is included in the system to achieve a linear system. Its inputs are the measured flow and the reference pressure $P_{\text{ref,LUT}}$, which is changed following an ascending and descending step function signal with different step heights. A step function can be used as an input for the system identification when the system order and the noise are low [6]. To reduce the noise, the data are averaged over five measurements. With the MATLAB (2023a) System Identification Toolbox a second order transfer function

$$G(s) = \frac{P(s)}{P_{\text{ref,LUT}}(s)} \tag{1}$$

is calculated for each step, where P is the upstream pressure of the valve.

Within all transfer functions, one is determined to describe the behaviour of the blower. Therefore, the mean value of the gains is used as the gain of this transfer function. As the poles vary, it is difficult to determine one transfer function. Transfer functions with poles with a small real part correspond to a slow system dynamic compared to those with a larger real part [7]. Choosing a complex conjugate pole pair which is further away from the imaginary axis causes instability in the pressure regions where the dynamics are slower because the poles are moved to the right-half plane. Following that, a compromise has to be found to avoid instability, but also describe the behaviour in the other pressure regions as good as possible and have a signal with low noise.

2.3 Controller

The transfer function of the system identification is used to tune a controller. For the controller a PI-controller is chosen, where a proportional and integral term is used to reduce the control error [7]. Fig. 2 shows the block diagram of the blower feedback control. Here the reference pressure P_{ref} is the pressure which should be achieved and the controller adjusts the pressure for the input of the LUT $P_{\text{ref,LUT}}$.

To obtain a sufficiently good controller, requirements for the performance are set for the tuning process. The rise time is defined as the time it takes for the pressure to rise

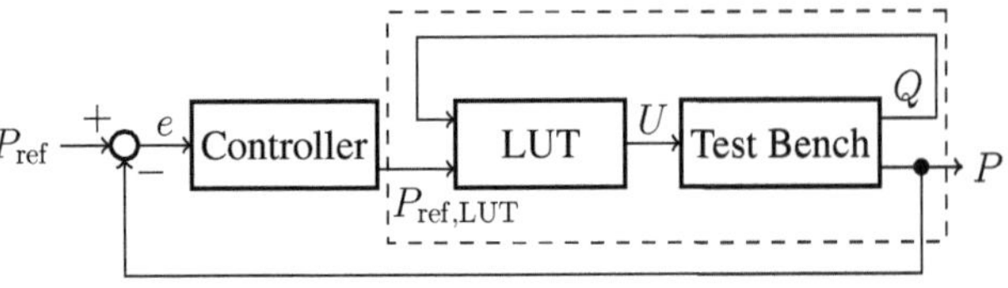

Figure 2: Block diagram of the feedback control system. The dashed rectangle is the new system where the LUT is included in the system. (P_{ref}: Reference pressure, e: Control error, $P_{\text{ref,LUT}}$: Reference pressure for LUT, U: Blower control voltage, P: Pressure, Q: Flow)

from 10 % to 90 % of the difference between the steps [7]. It should be less than 15 ms with regard to the lung simulator dynamics. The maximum overshoot should be below 20 %. Following that, the phase margin is 50° which is in the acceptable range over 30° to achieve stability [7], [8]. With respect to these requirements a controller should be designed. Therefore, a method based on the frequency response of the system is chosen, as described in [9]. The controller is a PI-controller

$$K(s) = K_{\text{P}} + \frac{K_{\text{I}}}{s} = \frac{K_{\text{I}}(sT_{\text{I}} + 1)}{s} \tag{2}$$

with a proportional and an integral controller parameter K_{P} and K_{I}. The crossover frequency

$$\omega_{\text{C}} \approx 1.5 \cdot t_{\text{r}} \tag{3}$$

depends on the system rise time t_{r} and the phase margin

$$\phi_{\text{m}} \approx 70 - o \tag{4}$$

on the maximum overshoot o. With the transfer function of the open-loop

$$L(s) = K(s) \cdot G(s) \tag{5}$$

the proportional and integral controller parameters are calculated. At first only the known parameters are considered:

$$L_1(s) = \frac{1}{s} \cdot G(s). \tag{6}$$

By calculating the phase of this system L_1, it is possible to determine how much the phase needs to be increased to achieve the required phase margin:

$$\Phi = (-180° + \phi_{\text{m}}) - \arg(L_1(i\,\omega_{\text{C}})). \tag{7}$$

With the value for the phase increase Φ and the crossover frequency the integration time

$$T_{\text{I}} = \frac{\tan(\Phi)}{\omega_{\text{C}}}. \tag{8}$$

can be calculated. Because the magnitude of the transfer function $L(s)$ must be 1 at the crossover frequency, the value for the integral parameter results from

$$K_{\text{I}} = \frac{1}{|(T_{\text{I}}i\omega_{\text{C}} + 1) \cdot G(i\omega_{\text{C}})|}. \tag{9}$$

3 Results and Discussion

3.1 Static Characteristic of the Blower

The behaviour of the blower is nonlinear. A LUT, as seen in Fig. 3, is used to compensate this.

For testing the accuracy of the LUT a ramped-shaped signal for the reference pressure is used. The measurement was carried out for different flow resistances. The relative deviation is constant for different pressures while the flow resistance is constant, but increases with decreasing maximum pressure if the flow resistance varies. The mean deviation over all measured openings of the valve is 0.76 mbar and the maximum deviation is 5.66 mbar.

It is possible to select values for the blower control voltage with the LUT so that the pressure is within the range of the reference pressure, but the deviation is too high, hence a controller is necessary to reduce this error.

The gain of the identified transfer functions is almost constant, see Fig. 4, thus it seems possible to use only one controller for the whole system. At regions with low pressure and high flow the gain is significantly smaller compared to the gain of other regions because the limits of the blower are reached and it is not capable to generate the reference pressure. Unlike the gain, the poles, shown in Fig. 5, still vary considerably. The complex conjugate pole pairs and real poles are all in the left-half plane. The pole locations selected to describe the system lead to relatively slow dynamics and low attenuation compared to all determined poles.

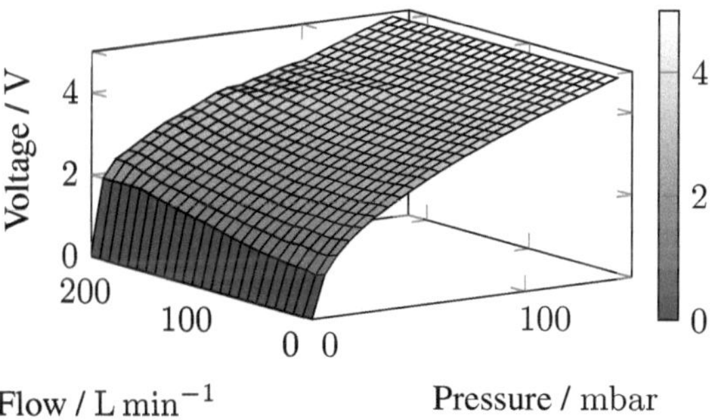

Figure 3: Lookup table for the control voltage of the blower depending on the pressure and the flow in front of the valve. The scale indicates the control voltage.

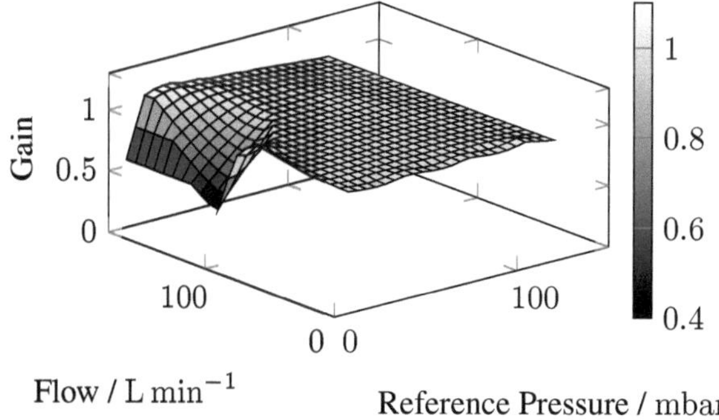

Figure 4: Gains of the transfer function of the identified system. The scale indicates the gain.

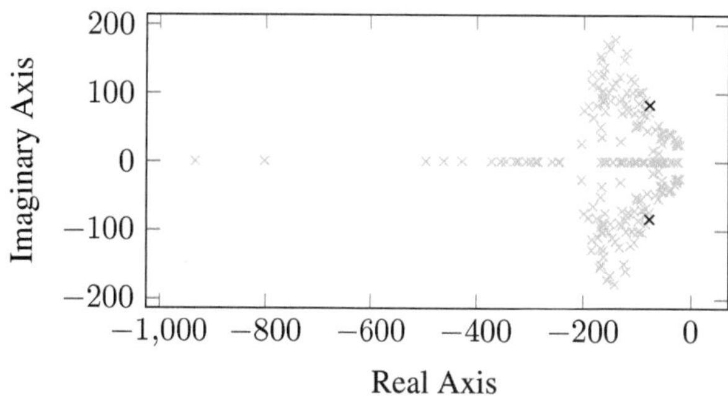

Figure 5: Poles of the transfer functions. The black pole pair with the value $-77.96 \pm 83.81i$ is the one chosen for the transfer function.

The chosen transfer function to describe the blower

$$G(s) = \frac{1.02}{7.6324 \cdot 10^{-5}s^2 + 0.0119s + 1} \quad (10)$$

is used to tune a PI-controller according to (2) - (9) with the parameters $K_\mathrm{P} = 0.7445$ and $K_\mathrm{I} = 92.7731$.

3.2 Closed-loop Evaluation

Step responses are simulated and measured to evaluate the controller performance. The rise time, the settling time with a tolerance of 5%, and the overshoot is analysed [8]. For the simulation the step responses of closed-loop systems with the identified transfer functions and the tuned controller are used. A total of 96 step responses were simulated. For the second test 120 step responses were measured.

In the simulation the system becomes unstable at two operation points with low pressure. Moreover, for the steps from 0 mbar to a higher pressure the oscillation is high and takes a long time to be damped. For all measured step responses there is no detectable steady-state offset and no instability could be determined.

The results for the requirements, as described in 2.3, are shown in Table 1. The rise time in the simulation is slightly higher than the required time whereas the rise time of the measurement is significantly higher. The settling time is sufficiently accurate for both. However, in the simulation the step responses of the steps from 0 mbar to a higher pressure are removed because the settling time is much higher, up to 43.50 s with a mean value of 583.50 ms. It is particularly high for the transfer functions of larger steps. In the measurements there are steps where noise peaks distort the settling time and in some case it cannot be determined at all because the noise is larger than 5 % of the signal. The mean overshoot meets the requirements for both the simulation and the measurement, but there are overshoots over 20 % which are too high.

The instability and oscillation at some steps in the simulation are caused by the actual system dynamics being slower than the transfer function used for the controller design. Following that, the poles are moved to the right-half plane. This can be avoided when using the slowest poles, however, this leads to a measured signal with strong noise, which is the reason why these were not chosen.

Table 1: Simulated (S) and Measured (M) Step Responses

Criteria	Mean	Max	Min
Rise Time (S)	17.47 ms	27.86 ms	11.47 ms
Rise Time (M)	33.76 ms	150.46 ms	14.48 ms
Settling Time (S)	65.22 ms	420.25 ms	18.22 ms
Settling Time (M)	75.07 ms	576.32 ms	18.93 ms
Overshoot (S)	13.67 %	79.24 %	0.00 %
Overshoot (M)	16.21 %	93.90 %	0.05 %

The rise time is still acceptable because the requirement was chosen strictly with regard to the lung simulator, which requires fast dynamics, while the test bench is also functional with slower dynamics. The longer rise time in the measurements is partly caused by the fact that at the step to the maximum pressure the overshoot but not the actual reference pressure is limited by the performance of the blower. Therefore, the rise is much slower compared to the other steps. Excluding the highest steps, the mean rise time is 29.50 ms and the longest rise time is 53.77 ms, which is still higher than in the simulation and has to be improved.

In the simulation the excluded values for the settling time are due to the slower system dynamics during these steps, where the selected controller adjusts the pressure value too strongly. The higher values for the measured settling time above 100 ms are caused by the step to the highest pressure where the increase is slower, as described for the rise time.

The high overshoot is caused by the system dynamics, that vary too much to be controlled with a single PI-controller. In the measurements the large overshoots appear at the steps from 0 mbar to a higher pressure. Here the chosen transfer function has a faster dynamic than the actual system which worsens the performance. Without these steps, the mean overshoot is 11.62 % and the maximum overshoot is 43.98 %. The maximum overshoot is still too high, but 72.50 % of all operating points meet the requirement.

When comparing the simulated and measured step responses, there is a significant difference in the rise time but most of the values are in a similar range, which is acceptable for the use of the controller.

4   Conclusion

The dynamic behaviour of a blower was identified and linearised with a LUT. Hence it was possible to identify a system where the gain is nearly constant over the whole pressure and flow range, but the poles still varied greatly and therefore the transient dynamics. A mean transfer function for this system was selected as a compromise for the pole location which was used for a PI-controller design. Overall, the pressure in front of a valve can be controlled with the tuned PI-controller.

While not all requirements were yet met with the current controller tuning, the approach shows great potential. Employing loop shaping to better handle the uncertainty in the system poles, should further improve the given results.

Acknowledgement

The work has been carried out at the Fraunhofer Research Institution for Individualized and Cell-Based Medical Engineering, Lübeck and supervised by P. Rostalski, Institute for Electrical Engineering in Medicine, Universität zu Lübeck.

Authors' Statement

Research funding: This work was partially funded by the European Union - European Regional Development Fund (ERDF), the Federal Government and Land Schleswig Holstein, Project No. 125 23 009.
Conflict of interest: Authors state no conflict of interest.

5   References

[1] S. Henn et al., *Concept for the testing of automated functions in therapeutic medical devices.* at - Automatisierungstechnik, vol. 70, no. 11, pp. 946-956, 2022.

[2] F. Bautsch, G. Männel, and P. Rostalski, *Development of a Novel Low-cost Lung Function Simulator.* Current Directions in Biomedical Engineering, vol. 5, no. 1, pp. 557-560, 2019.

[3] B. Hunnekens, S. Kamps, and N. van de Wouw, *Variable-Gain Control for Respiratory Systems.* IEEE Transactions on Control Systems Technology, vol. 28, no. 1, pp. 163-171, 2020.

[4] J. Reinders, B. Hunnekens, F. Heck, T. Oomen, and N. van de Wouw, *Adaptive Control for Mechanical Ventilation for Improved Pressure Support.* IEEE Transactions on Control Systems Technology, vol. 29, no. 1, pp. 180-193, 2021.

[5] R. Brütt, G. Männel, C. Brendle, and P. Rostalski, *Modelling of a Positive End-Expiratory Pressure Valve with a Voice-coil Actuator.* Current Directions in Biomedical Engineering, vol. 8, no. 2, pp. 516-519, 2022.

[6] C. Bohn, *Systemidentifikation.* In: M. Hennecke and B. Skrotzki (eds) HÜTTE - Das Ingenieurwesen, Springer, Berlin, Heidelberg, 2021.

[7] G. F. Franklin, J. D. Powell, and A. Emami-Naeini, *Feedback Control of Dynamic Systems.* Pearson, Upper Saddle River, 2010.

[8] S. Skogestad and I. Postlethwaite, *Multivariable Feedback Control: Analysis and design.* John Wiley & Sons, Chichester, 2001.

[9] A. Kugi, *5. Das Frequenzkennlinienverfahren, Vorlesung und Übung Automatisierung (Wintersemester 2013/2014).* Institut für Automatisierungs- und Regelungstechnik, TU Wien. Available: https://www.acin.tuwien.ac.at/fileadmin/cds/lehre/aut /Archiv/WS1314/Kapitel5.pdf [last accessed on 2024-01-16].

Balancing and Trajectory Tracking of a Ballbot based on Linear Parameter-Varying Modeling and Control

Kishore Kumar Mariappan [1], Hossam S. Abbas [2], and Dimitrios S. Karachalios [2]

[1] Robotics and Autonomous Systems, Universität zu Lübeck, kishore.mariappan@student.uni-luebeck.de

[2] Institute for Electrical Engineering in Medicine, Universität zu Lübeck, {h.abbas, dimitrios.karachalios} @uni-luebeck.de

Abstract

In this paper, we study and implement a novel method for controlling a ballbot nonlinear system in simulation. A ballbot is a mobile robot with a spherical wheel, i.e., a ball driven by actuator wheels that make contact with the top of the ball. The single point of contact of the spherical wheel with the ground makes it omnidirectional and highly maneuverable. The linear parameter-varying (LPV) model of the ballbot equivalently represents the nonlinear ballbot model. The parameters of the nonlinear ballbot model were identified through known parameters of a linearized model from a past study. We use the ballbot LPV model combined with model predictive control (MPC) for balancing and trajectory tracking. The ballbot LPV-model predictive control (LPVMPC) leads to an optimization problem that can be solved efficiently with quadratic programming (QP). The ballbot simultaneously balances and follows a predefined trajectory, such as a Lissajous curve.

1 Introduction

A ballbot [1]-[4], is an omnidirectional, single-wheel agile robot system where the ball acts as the wheel, making a single point of contact with the ground, while the actuators driving the ball are wheels on top of the ball, making contact with the ball as shown in Fig. 1. They can be considered as an extension of an inverted pendulum. The use cases of a ballbot include healthcare, virtual assistants, etc., as service robots, where maneuverability in closed environments while maintaining stability is important.

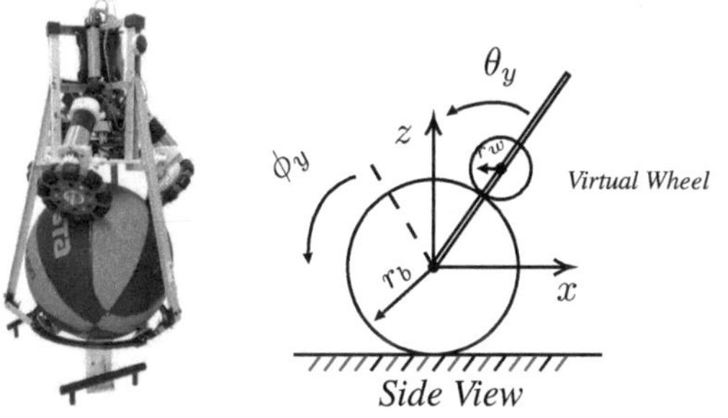

Figure 1: The Ballbot constructed at IME (left), side view of ballbot on xz-plane (right), r_b, and r_w are radii of the ball and actuator wheel respectively.

Ballbot system's stability implies balancing and simultaneous trajectory tracking through a predetermined path. As the use cases, the environments mentioned above need good control stability in crowded or busy environments, which is the motivation to improve ballbot control. Linear control of a ballbot [1] is proven insufficient, and the LPV model,

being much closer to the nonlinear model, has shown great potential in applications such as autonomous vehicles [5]. The novelty of this control method is that it uses the nonlinear ballbot system instead of relying only on a linear one. Therefore, the accuracy is comparable with nonlinear MPC frameworks, and the computational cost of LPVMPC can be compared with linear control performance as detailed in [5]. The nonlinear system parameters can be discovered after solving a set of nonlinear equations that connect the linearized model parameters from [1] with the nonlinear model. Such a solution can be achieved after utilizing iterative methods i.e., Netwon.

2 Preliminaries and Definitions

2.1 The Ballbot Nonlinear Model

The ballbot model is thoroughly discussed regarding the Euler-Lagrange method of formulating the dynamics in [1]-[4]. The ballbot model used here [1] is the one in which system identification was performed to identify a linearized model used to balance the ballbot.

The ballbot is a step ahead of an inverted pendulum, which is a 1-degree of freedom (DoF) system, while a Ballbot is a 5-DoF system, where three are the rotational motion of the body and the two are the translation in the xy-plane. We will see that θ is mainly used for balancing angle, and ϕ is mainly used for tracking angle.

The symmetry of the ballbot control problem allows us to focus on just the xz-plane [1] to simulate and understand its stability and controllability. The ballbot dynamics as given in (1) gives us the two observable and controllable states, $q = [\phi_y, \theta_y]^T$, and τ_y, being the control input torque of the virtual wheel actuator of a ballbot in the xz-plane as seen in Fig. 1.

M, C, D, G and $\tilde{B}$ denote the mass matrix, and the vectors of Coriolis force, centrifugal forces, frictional torque, gravitational forces, and input vector, respectively, ε is the residual error term.

$$M(q)\ddot{q} + C(q,\dot{q}) + D(\dot{q}) + G(q) + \varepsilon(q,\dot{q}) = \tilde{B}\tau_y. \quad (1)$$

The matrix and vectors in (1) are defined as:

$$M = \begin{bmatrix} b_1 & -b_2 + \ell r_b \cos(\theta_y) \\ -b_2 + \ell r_b \cos(\theta_y) & b_3 \end{bmatrix},$$

$$C = \begin{bmatrix} -\ell r_b \sin(\theta_y)\dot{\theta}_y{}^2 \\ 0 \end{bmatrix}, \quad D = \begin{bmatrix} b_4\dot{\phi}_y \\ 0 \end{bmatrix}, \quad (2)$$

$$G = \begin{bmatrix} 0 \\ -\ell g \sin(\theta_y) \end{bmatrix}, \quad \tilde{B} = \begin{bmatrix} \frac{r_b}{r_w} \\ -\frac{r_b}{r_w} \end{bmatrix}, \quad \varepsilon = \begin{bmatrix} -b_e\dot{\phi}_y \\ 0 \end{bmatrix}.$$

ℓ, r_b, and r_w are the location of the center of gravity, the radius of the ball and the radius of the actuator wheels, respectively. The expanded form of b-parameters can be found in [1].

The state-space of a planar ballbot equation is given in (3).

$$\underbrace{\begin{bmatrix} \dot{q} \\ \ddot{q} \end{bmatrix}}_{\dot{x}} = \underbrace{\begin{bmatrix} \dot{q} \\ M^{-1}(-C - D - G - \varepsilon) \end{bmatrix}}_{f(x)} + \underbrace{\begin{bmatrix} 0 \\ M^{-1}\tilde{B} \end{bmatrix}}_{g(x)} \tau_y. \quad (3)$$

where $x = [\phi_y, \theta_y, \dot{\phi}_y, \dot{\theta}_y]^T$ are the states of the system.

2.2 The Linear Parameter-Varying (LPV) Embedding

Linear parameter-varying embeddings as defined in [5] refer to an adaptive linear state-space model with scheduling of time-varying parameters such as θ_y and $\dot{\theta}_y$. These scheduling parameters are used to control the ballbot nonlinear model.

The LPV formulation of (3) can be represented in the continuous time domain as denoted in the matrix A_c and vector B_c as in (4)

$$\begin{cases} \dot{x} = A_c(\rho)x + B_c(\rho)u \equiv f(x) + g(x)\tau_y, \\ \rho = \sigma(x), \ x_0 = x(0), \ t \geq 0, \end{cases} \quad (4)$$

where the parameter ρ is the scheduling parameter, which depends on the mapping function $\sigma: \mathbb{R}^{n_x} \longrightarrow \mathbb{R}^{n_\rho}$, between the four states of the nonlinear ballbot systems, $n_x = 4$, and two scheduling parameters of the LPV model, $n_\rho = 2$. The input dimension is $n_u = 1$ with $u = \tau_y$.

The final LPV form of continuous time state-space matrices is given as (5), the nonzero elements of the matrices A_c and B_c are given in Table 1.

$$A_c(\rho) = \begin{bmatrix} 0 & 0 & 1 & 0 \\ 0 & 0 & 0 & 1 \\ 0 & A_{32} & A_{33} & A_{34} \\ 0 & A_{42} & A_{43} & A_{44} \end{bmatrix}, \quad B_c(\rho) = \begin{bmatrix} 0 \\ 0 \\ B_3 \\ B_4 \end{bmatrix}. \quad (5)$$

Table 1: Elements of A_c and B_c, with their respective scheduling parameters given as the scalar entries $A_{ij}(\rho) \in \mathbb{R}$ and $B_i(\rho) \in \mathbb{R}$. Their scheduling signals are given as a function of $\rho = \sigma(\theta_y, \dot{\theta}_y)$. The common denominator $d(\theta_y) = b_1 b_3 - (-b_2 + \ell r_b \cos(\theta_y))^2$.

$\rho = \sigma(\theta_y, \dot{\theta}_y)$	Scheduling signals
$A_{32}(\rho)$	$\dfrac{\ell g \operatorname{sinc}(\theta_y)(b_2 - \ell r_b \cos(\theta_y))}{d(\theta_y)}$
$A_{33}(\rho)$	$-\dfrac{b_3(b_4 - b_e)}{d(\theta_y)}$
$A_{34}(\rho)$	$\dfrac{b_3 \ell r_b \sin(\theta_y)\dot{\theta}_y}{d(\theta_y)}$
$A_{42}(\rho)$	$\dfrac{b_1 \ell g \operatorname{sinc}(\theta_y)}{d(\theta_y)}$
$A_{43}(\rho)$	$-\dfrac{(b_4 - b_e)(b_2 - \ell r_b \cos(\theta_y))}{d(\theta_y)}$
$A_{44}(\rho)$	$\dfrac{\ell r_b \sin(\theta_y)\dot{\theta}_y(b_2 - \ell r_b \cos(\theta_y))}{d(\theta_y)}$
$B_3(\rho)$	$\dfrac{r_b(b_3 - (b_2 - \ell r_b \cos(\theta_y))}{r_w d(\theta_y)}$
$B_4(\rho)$	$\dfrac{r_b((b_2 - \ell r_b \cos(\theta_y)) - b_1)}{r_w d(\theta_y)}$

2.3 Stabilization Control

The ballbot is stabilized in [1] through a proportional integral derivative (PID) controller. During stabilization in [1], a multi-harmonic excitation provided measurements to identify a linear model with respect to the p-parameters as in Table 2. The linearized model can stabilize the ballbot with optimal feedback gain after using a linear quadratic regulator (LQR).

In this study, after identifying the nonlinear b-parameters, the LPVMPC control method will simultaneously take care of the stabilization and reference tracking.

2.4 Model Predictive Control with Linear Parameter-Varying Embedding

Model predictive control with linear parameter-varying embeddings implies that the MPC optimally controls the underlying time-varying parameters based on the scheduling of the parameters as shown in the LPV formulation (4).

The standard MPC form begins with the energy function defined as $J_k(u)$ as seen in (6), where $x_{i|k}$ denotes the discretized state vector at time k and prediction i. The i varies in length equal to the prediction horizon. The reference state vector is given as $x_{i|k}^{\text{ref}} = \begin{bmatrix} \phi_{i|k}^{ref} & 0 & 0 & 0 \end{bmatrix}^\top$.

$$J_k(u) = \sum_{i=0}^{N-1} (\|x_{i|k} - x_{i|k}^{\text{ref}}\|_Q^2 + \|u_{i|k}\|_R^2), \quad (6)$$

where Q and R are quadratic costs, they can be written as weighted norms i.e., $\|x_{i|k}\|_Q^2 = x_{i|k}^\top Q x_{i|k}$.

The energy function is used to calculate all the future states for N steps. We use a modified backward Euler method to discretize (4), with discretization step $x(t) = x(kt_s) = x_k$, where t_s is the sampling time. The discrete LPV system is:

$$\begin{cases} x_{k+1} = A(\rho_k)x_k + B(\rho_k)u_k, \\ \rho_k = \sigma(x_k), k = 0, \ldots, \infty. \end{cases} \quad (7)$$

The aim is to minimize the quadratic cost function J in (6) under the constraints given in (8), with $G^x \in \mathbb{R}^{n_{cx} \times n_x}$, $h^x \in \mathbb{R}^{n_{cx} \times 1}$, $G^u \in \mathbb{R}^{n_{cu} \times n_u}$, and $h^u \in \mathbb{R}^{n_{cu} \times 1}$, $n_{cx} = 8$ and $n_{cu} = 2$ are the total number of upper and lower constraint bounds of states and input respectively.

$$\begin{aligned} \mathcal{X} &= \{x_k \in \mathbb{R}^{n_x} | G^x x_k \le h^x\} \\ \mathcal{U} &= \{u_k \in \mathbb{R}^{n_u} | G^u u_k \le h^u\}, \end{aligned} \quad (8)$$

With the parameterization $\hat{p}_{i|k}$ of the states due to LPV we get the quadratic program optimization given as $\mathrm{QP}(\hat{p}_{i|k}, \hat{x}_{i|k}, x_{i|k}^{\mathrm{ref}})$ in (9) and for $i = 0, \ldots, N - 1$:

$$\begin{aligned} \mathrm{QP} : \min_{u_{i|k}} &\sum_{i=0}^{N-1} \left(\|\hat{x}_{i|k} - x_{i|k}^{\mathrm{ref}}\|_Q^2 + \|\hat{u}_{i|k}\|_R^2 \right), \\ \text{s.t. } &\hat{x}_{i+1|k} = A(\hat{p}_{i|k})\hat{x}_{i|k} + Bu_{i|k}, \\ &\hat{x}_{0|k} = x_{0|k} = x_k, \\ &\hat{x}_{i|k} \in \mathcal{X}, \forall i = 0, 1, \ldots, N, \\ &\hat{u}_{i|k} \in \mathcal{U}, \forall i = 0, 1, \ldots, N - 1. \end{aligned} \quad (9)$$

Starting with $u_{0|k}$, $x_{0|k}$, and initializing with $\hat{p}_{i|k} = \sigma(x_{0|k})$, $i = 1, \ldots, N - 1$, we can compute $\hat{x}_{i+1|k}$ with respect to the decision variables $\hat{u}_{i|k}$. After solving the QP in (9), for every future prediction horizon, we infer the scheduling signal as $\hat{p}_{i|k+1} = \hat{p}_{i+1|k}$, $i = 0, \ldots, N - 1$.

3 Results and Discussion

3.1 Identification of Nonlinear Parameters

According to [1], the linearization of the continuous state-space model (3) is to be performed on its equilibrium state point of $x_e = [0, 0, 0, 0]^T$ as shown in (10).

$$\dot{x} = f(x) + g(x)\tau_y \overset{\text{linearization}}{\underset{x_e=0}{\approx}} \underbrace{\nabla_x f(0)}_{A_{lin}} x + \underbrace{g(0)}_{B_{lin}} \tau_y. \quad (10)$$

The linearized continuous state-space model remains with the following parameterized matrix A_{lin} and vector B_{lin},

$$A_{lin} = \begin{bmatrix} 0 & 0 & 1 & 0 \\ 0 & 0 & 0 & 1 \\ 0 & p_1 & p_2 & 0 \\ 0 & p_4 & p_5 & 0 \end{bmatrix}, \quad B_{lin} = \begin{bmatrix} 0 \\ 0 \\ p_3 \\ p_6 \end{bmatrix}. \quad (11)$$

Table 2 shows the values of the nonlinear b-parameters identified through the Newton method, which is just one solution amongst many solutions that can explain the transient behavior of the linearized model close to the equilibrium state point.

Table 2: Physical parameters of the ballbot model (2), with p being the linearized model parameters from [1], and b being the discovered nonlinear model parameters.

p_1	p_2	p_3	p_4	p_5
-342.6038	-52.8301	-1425.9	-36.0734	-9.1477
p_6	$\ell[m]$	$r_b[m]$	$r_w[m]$	$g[m/s^2]$
251.8476	0.2978	0.12	0.05	9.81
b_1	b_2	b_3	b_4	b_e
0.002483	0.059325	0.143093	-0.933813	-0.859453

Multisine excitation is given as input τ_y, to simulate and understand the behavior of nonlinear (3), LPV (4) and linear (11) models based on the newly identified parameters in Table 2, and validated in Fig. 2, showing us both the angular displacements in $\phi(t)$ and $\theta(t)$ in rad, and corresponding angular velocities in $\dot{\phi}(t)$ and $\dot{\theta}(t)$ in rad/s.

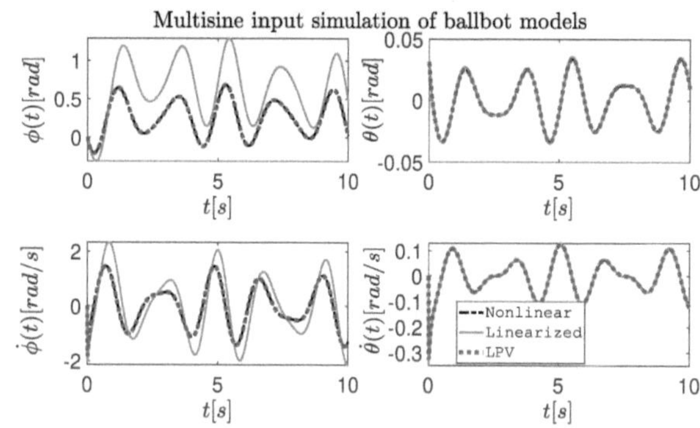

Figure 2: Multisine input τ_y is given to all the ballbot models, nonlinear in black dashed lines, linearized in gray line, and LPV in dotted gray line. All simulations were performed in ode45 in MATLAB. The angular displacement (1st row) and angular velocity (2nd row) of the ballbot's rotation on the xz-plane is given by $\phi_y(t), \theta_y(t), \dot{\phi}_y(t), \dot{\theta}_y(t)$ respectively.

The simulations show that the LPV model formulation is the same as the nonlinear model. Further, the ϕ value of the linearized model starts to differ only away from the linearization operational point compared to the LPV and nonlinear models that justify the expected performance.

3.2 LPVMPC for Reference Tracking

Reference tracking for a ballbot refers to the stabilized motion of the ballbot in both 1D in x or y axis or 2D in xy plane, by the LPVMPC controller that can maneuver with any given reference states $x_{i|k}^{\mathrm{ref}}$ (9) with $x_{0|0} = [0, 0, 0, 0]^T$. The tuning matrices are; $Q = diag([10, 1, 0.1, 0.1])$ and $R = 1000$ at any given instance of discrete time k; the sampling time is $t_s = 0.1 \ (s)$; the state and input constraints can be found in Table 3.

For the 1D reference tracking, as shown in Fig. 3, traveling in y direction, the ballbot system also balances itself simultaneously, as shown with θ being between ranges of $-0.65 \ (rad)$ to $0.62 \ (rad)$, as it is referenced to $0 \ (rad)$,

Table 3: MPC Constraints

Variable	Lower bound	Upper bound	Units
ϕ_k	$-\infty$	$+\infty$	[rad]
θ_k	$-\pi/5$	$+\pi/5$	[rad]
$\dot{\phi}_k$	-15π	$+15\pi$	$\left[\frac{\text{rad}}{\text{s}}\right]$
$\dot{\theta}_k$	-10π	$+10\pi$	$\left[\frac{\text{rad}}{\text{s}}\right]$
u	-1	$+1$	[Nm]

while traveling a distance according to ϕ, which is referenced to move from $2\pi\,(rad)$ to $-4\pi\,(rad)$, and angular velocities of both being referenced to $0\,(rad/s)$, are shown in $\dot{\theta}$, and $\dot{\phi}$ respectively. They show the agile and robust movements of the ballbot while following the reference. In addition, the virtual torque τ_y and energy function J cost are shown along with the optimal control input $u = \tau_y$.

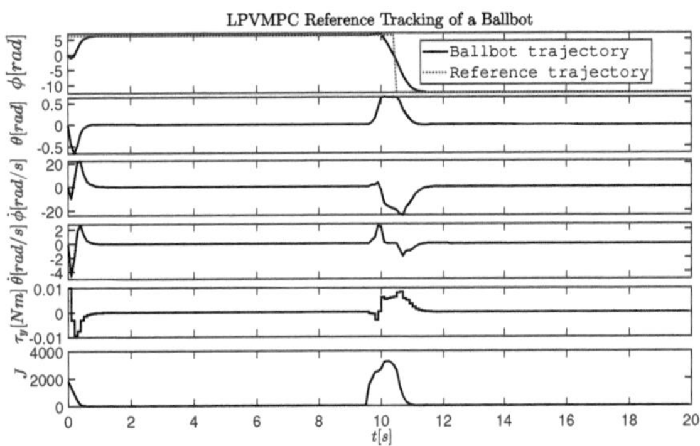

Figure 3: LPVMPC reference tracking rolling to reference of ϕ traveling from $2\pi\,(rad)$ to $-4\pi\,(rad)$ in y direction, θ measuring balancing of ballbot, $\dot{\phi}$, $\dot{\theta}$ measuring the angular velocities, τ_y is the virtual torque, J is the energy function.

Solving the LPVMPC reference tracking problem in both yz-plane, solving for x-direction and xz-plane, solving for y-direction, we can drive the ballbot in xy plane with Lissajous curves as a reference as shown in Fig. 4 starting at $(x, y) = (\phi_x r_b, \phi_y r_b) = (0, 0)$, and following a path of $\phi_x = 2\pi \sin 0.3t$ and $\phi_y = 2\pi \sin 0.4t$, and under the constraints in Table 3. With $Q = diag([1000, 1, 0.1, 0.1])$ to improve position accuracy.

4 Conclusion

The simulations illustrate that applying LPVMPC-based control is feasible for a ballbot nonlinear system and has an advantage over the control of a linearized ballbot model, as it can robustly balance while simultaneously tracking the trajectory close to the references, as shown in the Lissajous curve tracking. The ballbot system is not only stable but also highly maneuverable.

The LPV model being closer to the nonlinear model gives an added advantage of computation time where MPC solves the prediction problem much faster as a quadratic program is solved, unlike the nonlinear MPC (NMPC), where computation time to predict is larger, as shown in [5]. Finally,

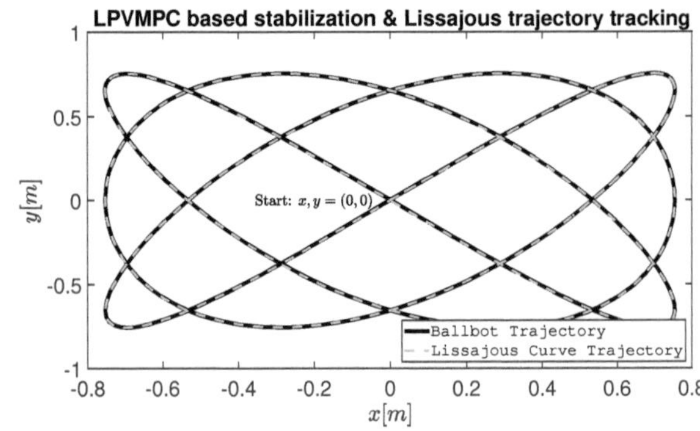

Figure 4: Stabilization and Lissajous curve trajectory tracking of the ballbot in 2D with LPVMPC method solved in both yz and xz planes. Covering an area $\approx 2\ m^2$ in $70\ s$.

implementing this method on the physical ballbot system in a real environment is left for future endeavors.

Acknowledgement

The work has been carried out and supervised by the Institute for Electrical Engineering in Medicine, Universität zu Lübeck. We thank Mr. Ing.-Ievgen Zhavzharov who constructed the actual ballbot in Fig. 1.

Authors' Statement

Conflict of interest: Authors state no conflict of interest.

5 References

[1] M. Studt, I. Zhavzharov and H. Abbas, *Parameter Identification and LQR/MPC Balancing Control of a Ballbot* . European Control Conference, pp. 1315-1321, 2022.

[2] M. A. Alyousify, H. S. Abbas, M. M. M. Hassan and M. H. Amin, *Parameter Identification and Control of a Ball Balancing Robot*. 2022 8th International Conference on Mechatronics and Robotics Engineering (ICMRE), Munich, Germany, pp. 91-97, 2022.

[3] U. Nagarajan, G. Kantor, and R. Hollis, *The Ballbot: An Omnidirectional Balancing Mobile Robot*. The International Journal of Robotics Research, vol. 33, no. 6, pp. 917–930, 2014.

[4] P. Frankhauser and C. Gwerder, *Modeling and control of a ballbot*. Master's thesis, Eidgenössische Technische Hochschule Zürich, ETH - Swiss Federal Institute of Technology Zürich, 2010.

[5] M. Nezami, D. Karachalios, G. Schildbach, and H. Abbas, *On the Design of Nonlinear MPC and LPVMPC for Obstacle Avoidance in Autonomous Driving*. International Conference on Control, Decision and Information Technologies (CoDIT), Italy, 2023.

Looking Ahead: Comparison of Stanley and Model Predictive Control for Miniature Sized Cars Using Lookahead Point

Miklos Libak [1], Robin Kensbock [2], Fabian Domberg [2] and Georg Schildbach [2]

[1] Robotics and Autonomous Systems, Universität zu Lübeck, miklos.libak@student.uni-luebeck.de

[2] Institute for Electrical Engineering in Medicine, Universität zu Lübeck, {r.kensbock, f.domberg, georg.schildbach}@uni-luebeck.de

Abstract

In this study, the performance of the Stanley Controller and Model Predictive Control (MPC) for miniature-sized race car, focusing on the utilization of a lookahead point is evaluated. The controllers are assessed in terms of lap time and track error on an unknown track, considering varying velocities and lookahead distances. Experimental results indicate that both controllers demonstrate competitive performance at higher speeds with a moderate lookahead distance. However, challenges arise at lower velocities and shorter lookahead distances, highlighting the importance of these parameters in controller design.

1 Introduction

Rapid progress has been observed in the field of autonomous driving in recent years. Path tracking is a crucial task for autonomous vehicles, especially in racing scenarios requiring high speed and accuracy. One of the most challenging tasks is developing a stable vehicle controller to realize the function of tracking the desired paths automatically using different speeds. This work compares two of them: Stanley Control and Model Predictive Control (MPC). In this paper, both approaches are customized to utilize a lookahead point and used for path tracking of a miniature race car, which is a one-seventh-scaled electric car. The goal is to assess the performance regarding lap time and lateral tracking error of the previously mentioned controllers, incorporating a lookahead point, applied to an unfamiliar track.

1.1 Related Work

Recent studies from K et al. [1] and Liu et al. [2] have compared MPC and Stanley control for path tracking scenarios but have not evaluated their performance in racing scenarios. Both found that MPC demonstrated the most favorable performance in path tracking. The superior performance of MPC was attributed to its optimization properties, which resulted in better trajectory tracking. In both cases, the performance assessments are conducted without the utilization of a lookahead point.

2 Background

2.1 Stanley Controller

The Stanley Controller is a widely adopted control strategy utilized in autonomous vehicle path tracking. This controller was initially proposed by Thrun et al. [3]. Its primary objective is to enable accurate tracking of the desired path, even in challenging conditions such as rapidly changing or rough terrains.

A crucial component of the Stanley Controller is the cross-track error, denoted as $e(t)$. Defined by Thrun et al. [3], this metric quantifies the lateral distance between the center of the vehicle's front wheels and the nearest point on the reference path. In this paper, a so-called lookahead point p was considered for cross-track error calculation. This point lies ahead of the vehicle's reference point by a distance D, as depicted in Fig. 1. The lookahead distance determines how far ahead the controller looks. It allows the controller to anticipate upcoming changes in the path and proactively adjust the steering angle. Mathematically, the lookahead point's position can be expressed as:

$$\begin{bmatrix} p_x \\ p_y \end{bmatrix} = \begin{bmatrix} x \\ y \end{bmatrix} + D \begin{bmatrix} \cos(\theta) \\ \sin(\theta) \end{bmatrix}, \qquad (1)$$

where $[x, y]^T$ is the vehicle reference point and θ is the yaw angle of the vehicle.

The angle ψ represents the orientation on the closest path segment relative to the vehicle's orientation. When there are no lateral errors, the control law aligns the front wheels parallel to the reference path.

The fundamental control law for the steering angle is given by:

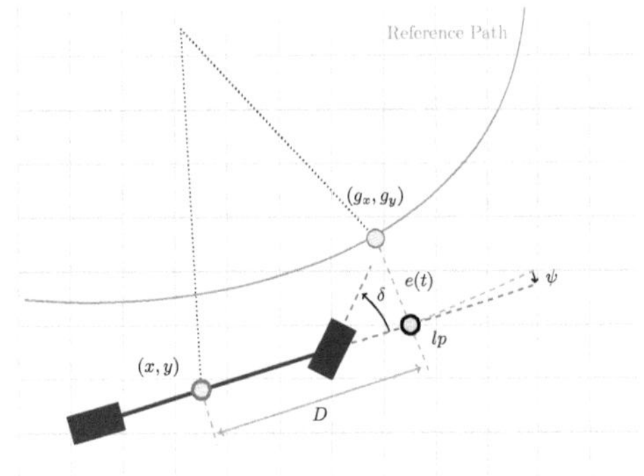

Figure 1: Illustration of the Stanley Controller

$$u_{\text{steer}}(t) = \psi(t) + \arctan\frac{ke(t)}{v(t)}. \qquad (2)$$

Here, k denotes a gain parameter and $v(t)$ is the velocity of the vehicle at time t. u_{steer} is the input to the steering system of the car and thus can be regarded as the steering angle. The second term of the equation adjusts the steering angle in (nonlinear) proportion to the cross-track error $e(t)$. As the error increases, the steering response toward the path becomes stronger [3].

While, the Stanley Controller controls the steering of the car, a Proportional-Integral-Derivative (PID) controller was implemented for keeping a constant velocity. The basic update amounts to the following equation, where u_{vel} is the input that controls the throttle of the car and therefore influences the acceleration of the vehicle:

$$u_{\text{vel}}(t) = K_p e_v(t) + K_i \int_0^t e_v(\tau)\mathrm{d}\tau + K_d \frac{\mathrm{d}e_v(t)}{\mathrm{d}t}, \qquad (3)$$

where $e(t) = v_{\text{ref}} - v(t)$ is a measured velocity error and K_p, K_i and K_d are proportional, integral and differential gains, respectively.

2.2 Model Predictive Controller

MPC operates on the principle of solving an optimization problem over a finite time horizon, considering given constraints on inputs and states. The goal is to determine optimal control signals, guided by a cost function, to steer the system along a desired trajectory. This process involves forecasting the system's behavior based on calculated inputs, employing a prediction model. At each time step, the MPC solves an Optimal Control Problem (OCP) across the shifted horizon, with the first computed control input subsequently applied to the vehicle. The OCP is structured as a constrained least squares minimization, as documented in the ACADO manual [4]:

$$\min_{\mathbb{X},\mathbb{U}} \sum_{k=0}^{N-1} \|h(\mathbf{x}_k,\mathbf{u}_k) - \mathbf{x}_{\text{ref},k}\|_{\mathbf{Q}}^2 + \|h_N(\mathbf{x}_N) - \mathbf{x}_{\text{ref},N}\|_{\mathbf{P}}^2$$

$$(4a)$$

$$\text{s.\,t.} \quad \mathbf{x}_0 = \mathbf{x}(t_0) \qquad (4b)$$

$$\mathbf{x}_{k+1} = F(\mathbf{x}_k,\,\mathbf{u}_k),\; k = 0,\ldots,N-1 \qquad (4c)$$

$$\mathbf{x}_k^{\min} \leq \mathbf{x}_k \leq \mathbf{x}_k^{\max},\; k = 0,\ldots,N \qquad (4d)$$

$$\mathbf{u}_k^{\min} \leq \mathbf{u}_k \leq \mathbf{u}_k^{\max},\; k = 0,\ldots,N-1 \qquad (4e)$$

$$\mathbf{x}_k \in \mathbb{X} \qquad (4f)$$

$$\mathbf{u}_k \in \mathbb{U}. \qquad (4g)$$

In (4a), $\mathbf{Q} \in \mathbb{R}_+^{6\times6}$ and $\mathbf{R} \in \mathbb{R}_+^{2\times2}$ denote positive semi-definite weighting matrices. N represents the length of the prediction horizon. The state vector $\mathbf{x}_k$ is expressed as $[x,y,v,\psi]^T$, while $\mathbf{u}_k$ contains the controls $[\delta_{\text{f}},a]^T$. Both state and control vectors are subject to upper and lower bounds. Here x_0 refers to the current state and $x_{\text{ref},k}$ denotes the reference vector $[x_{\text{ref}}, y_{\text{ref}}, v_{\text{ref}}, \psi_{\text{ref}}, \delta_{\text{f,ref}}, a_{\text{ref}}]^T$. The least squares problem is formulated by providing the states to the reference function h.

The effectiveness of MPC relies on having a mathematical model that accurately reflects the dynamic behavior of the system. In this paper, a kinematic bicycle model (Fig. 2) serves as a fundamental representation of the miniature race car's kinematics. It simplifies dynamics by assuming a point mass and focuses on essential aspects such as position, velocity and steering angle [5]. The equations that describe the dynamics of the vehicle are

$$\dot{x} = v\cos(\psi + \beta), \qquad (5a)$$

$$\dot{y} = v\sin(\psi + \beta), \qquad (5b)$$

$$\dot{\psi} = \frac{v}{l_{\text{r}}}\sin(\beta), \qquad (5c)$$

$$\dot{v} = a, \qquad (5d)$$

$$\beta = \tan^{-1}\left(\frac{l_{\text{r}}}{l_{\text{f}} + l_{\text{r}}}\tan(\delta_{\text{f}})\right), \qquad (5e)$$

where x and y are the coordinates of the center of mass in an inertial frame (X,Y), while ψ describes the orientation of the vehicle. The parameters l_{f} and l_{r} represent the distance from the center of mass of the vehicle to the front and rear axles, respectively. v is the speed of the vehicle and β is the angle of the current velocity of the center of mass of the vehicle with respect to the longitudinal axis of the car. a is the acceleration of the center of mass in the same direction as the velocity. The control inputs are the front steering angles δ_{f} and acceleration a [6]. The MPC Controller, as described by Kensbock et al. [7], was implemented using ACADO [8], an open-source toolkit for optimal control, employing the qpOASES solver. ACADO generates C code incorporating all these properties. This C code is then integrated into a ROS program that communicates with the simulation outlined in Section 3.1. ACADO is provided with the dif-

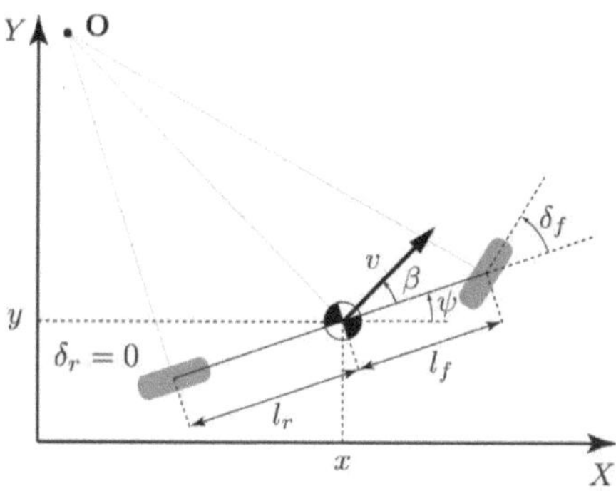

Figure 2: Kinematic Bicycle Model [6]

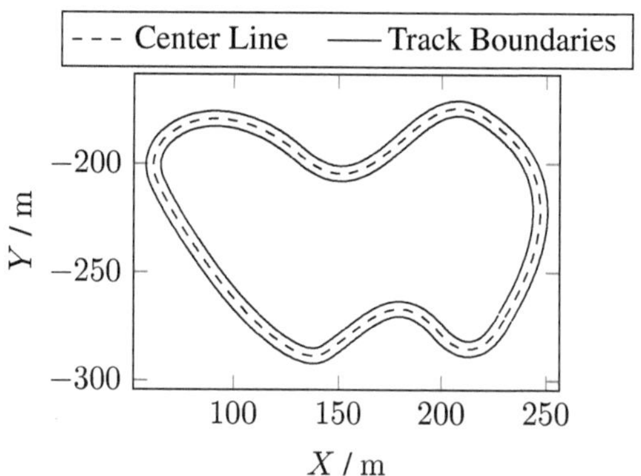

Figure 3: A visualization of the racing track including the center line and lane boundaries.

ferential equations, constraints and MPC algorithm properties. The constraints include hardware limitations, such as the maximum steering angle of the front wheels. The MPC operates at a frequency of $50\,\mathrm{Hz}$, with a prediction horizon length of 3.0 seconds, resulting in $N = 30$ for a discretization time of 0.1 seconds. Constraints on the steering angle δ_f and the acceleration a are set as:

$$-28° \leq \delta_\mathrm{f} \leq 28° \tag{6}$$

$$-10\,\mathrm{m/s^2} \leq a \leq 10\,\mathrm{m/s^2}. \tag{7}$$

The MPC incorporates a lookahead point by setting $[x_\mathrm{ref}, y_\mathrm{ref}]$ to a single point across the entire horizon, specifically the closest point on the center line of the track to the lookahead point. The lookahead point is calculated as in (1). The reference velocity v_ref is set to a constant value as well.

3 Results and Discussion

3.1 Simulation Environment

The Unity-based [9] simulation environment serves as the testing ground for the miniature race car controllers. Representation of the vehicle behavior is achieved through the utilization of the Vehicle Physics Pro [10] plugin for Unity. This plugin enhances the simulation's realism, ensuring accurate modeling of the miniature race car's dynamics. The racing track (Fig. 3) is constructed using Unity's Splines feature, allowing for the creation of a customizable track layout. Subsequently, the center line of the track (along with the track boundaries), characterized by discrete points, is extracted.

Communication with the simulation is done using (ROS), which connects the simulation to the control loop. An input the vehicle in the simulation accepts for throttle ($\in [0, 1]$), brake ($\in [0, 1]$) and steering ($\in [-1, 1]$).

3.2 Experiments & Discussion

To analyze the performance of the controllers, the evaluation metrics have to be determined. In this paper, the average tracking error, the maximum tracking error, and the lap time were used.

With a given reference path $p_\mathrm{ref}(s)$, the tracking error can be regarded as the distance between the vehicle reference point $[x, y]^T$ and the closest point on $p_\mathrm{ref}(s)$. Mathematically, this can be expressed as $s_\mathrm{close} = \arg\min_s ||[x_\mathrm{ref}, y_\mathrm{ref}] - [x, y]|| t$. The closest point is then $p_\mathrm{close} = p_\mathrm{ref}(s_\mathrm{close})$ and the tracking error is $e_\mathrm{tracking} = ||[p_\mathrm{close}] - [x, y]||$.

Experiments are performed using three reference velocities v_ref ($30\,\mathrm{km/h}$, $50\,\mathrm{km/h}$ and $70\,\mathrm{km/h}$) and three lookahead distances D ($5\,\mathrm{m}$, $10\,\mathrm{m}$ and $15\,\mathrm{m}$). Path tracking has been successfully accomplished using both controllers and the results are provided in the tables below.

Stanley Controller			
v_ref	$30\,\mathrm{km/h}$	$50\,\mathrm{km/h}$	$70\,\mathrm{km/h}$
$D = 5\,\mathrm{m}$	$67.27\,\mathrm{s}$	$40.42\,\mathrm{s}$	$32.39\,\mathrm{s}$
$D = 10\,\mathrm{m}$	$67.82\,\mathrm{s}$	$39.98\,\mathrm{s}$	$31.30\,\mathrm{s}$
$D = 15\,\mathrm{m}$	-	-	$30.11\,\mathrm{s}$

MPC			
v_ref	$30\,\mathrm{km/h}$	$50\,\mathrm{km/h}$	$70\,\mathrm{km/h}$
$D = 5\,\mathrm{m}$	$60.32\,\mathrm{s}$	$40.04\,\mathrm{s}$	$38.45\,\mathrm{s}$
$D = 10\,\mathrm{m}$	$50.74\,\mathrm{s}$	$38.40\,\mathrm{s}$	$31.40\,\mathrm{s}$
$D = 15\,\mathrm{m}$	-	$33.16\,\mathrm{s}$	$29.40\,\mathrm{s}$

Table 1: Lap times of the Stanley Controller and MPC using different lookahead distances D and reference velocities v_ref.

As indicated in Table 1, superior lap times are observed with higher velocities when employing a longer lookahead distance. However, neither the Stanley Controller nor the MPC successfully completed a lap while adhering to the minimum reference velocity of $30\,\mathrm{km/h}$ and employing a lookahead distance of $15\,\mathrm{m}$. Both controllers significantly undershot the curve during entry due to the placement of the lookahead point at the curve's exit (or even further away), leading to this outcome.

Analyzing the average and maximum tracking errors presented in Table 2 and Table 3 respectively, it is evident that both controllers demonstrate minimal deviation from the center line when operating at the maximum velocity of

Stanley Controller			
v_{ref}	30 km/h	50 km/h	70 km/h
$D = 5\,\mathrm{m}$	1.35 m	1.34 m	1.25 m
$D = 10\,\mathrm{m}$	1.27 m	1.27 m	1.21 m
$D = 15\,\mathrm{m}$	-	-	1.27 m

MPC			
v_{ref}	30 km/h	50 km/h	70 km/h
$D = 5\,\mathrm{m}$	1.36 m	1.78 m	1.30 m
$D = 10\,\mathrm{m}$	9.18 m	1.96 m	0.33 m
$D = 15\,\mathrm{m}$	-	1.25 m	1.28 m

Table 2: Average tracking errors of the Stanley Controller and MPC using different lookahead distances D and reference velocities v_{ref}.

Stanley Controller			
v_{ref}	30 km/h	50 km/h	70 km/h
$D = 5\,\mathrm{m}$	3.37 m	2.91 m	3.27 m
$D = 10\,\mathrm{m}$	3.18 m	3.01 m	3.59 m
$D = 15\,\mathrm{m}$	-	-	3.18 m

MPC			
v_{ref}	30 km/h	50 km/h	70 km/h
$D = 5\,\mathrm{m}$	2.95 m	4.03 m	3.18 m
$D = 10\,\mathrm{m}$	30.64 m	8.97 m	0.74 m
$D = 15\,\mathrm{m}$	-	2.80 m	2.87 m

Table 3: Maximum tracking errors of the Stanley Controller and MPC using different lookahead distances D and reference velocities v_{ref}.

70 km/h and utilizing a lookahead distance of 10 m. Considering both lap time and tracking error, both controllers exhibit similar challenges and no clear superiority is evident between them based on the results obtained. However, better tuning of the controllers could potentially improve their performance under varying conditions.

4 Conclusion

This paper compares Stanley and Model Predictive Control, both with a lookahead point, for path tracking on an unfamiliar track. The behaviors are examined utilizing different reference velocities and lookahead distances. The findings indicate that both controllers exhibit satisfactory performance at higher speeds with an intermediate lookahead distance.

Acknowledgment

The work has been carried out and supervised by the Institute for Electrical Engineering in Medicine (IME), Universität zu Lübeck. We thank Ievgen Zhavzharov for the development of the 3D model of the car for the simulation.

Authors' Statement

Conflict of interest: Authors state no conflict of interest.

5 References

[1] V. K, M. Ambalal Sheta, and V. Gumtapure, "A Comparative Study of Stanley, LQR and MPC Controllers for Path Tracking Application (ADAS/AD)," in *2019 IEEE International Conference on Intelligent Systems and Green Technology (ICISGT)*. Visakhapatnam, India: IEEE, Jun. 2019, pp. 67–674.

[2] J. Liu, Z. Yang, Z. Huang, W. Li, S. Dang, and H. Li, "Simulation Performance Evaluation of Pure Pursuit, Stanley, LQR, MPC Controller for Autonomous Vehicles," in *2021 IEEE International Conference on Real-time Computing and Robotics (RCAR)*. Xining, China: IEEE, Jul. 2021, pp. 1444–1449.

[3] S. Thrun, M. Montemerlo, H. Dahlkamp, D. Stavens, A. Aron, J. Diebel, P. Fong, J. Gale, M. Halpenny, G. Hoffmann *et al.*, "Stanley: The robot that won the DARPA Grand Challenge," *Journal of Field Robotics*, vol. 23, no. 9, pp. 661–692, Sep. 2006.

[4] B. Houska, H. Ferreau, M. Vukov, and R. Quirynen, "ACADO Toolkit User's Manual," http://www.acadotoolkit.org, 2009–2013.

[5] R. Rajamani, *Vehicle Dynamics and Control*, ser. Mechanical Engineering Series. Boston, MA: Springer US, 2012.

[6] J. Kong, M. Pfeiffer, G. Schildbach, and F. Borrelli, "Kinematic and dynamic vehicle models for autonomous driving control design," in *2015 IEEE Intelligent Vehicles Symposium (IV)*. Seoul, South Korea: IEEE, Jun. 2015, pp. 1094–1099.

[7] R. Kensbock, P. Huß, and G. Schildbach, "Trajectory Control with Lane Boundary Constraints for Autonomous Vehicles Using Model Predictive Control," in *Student Conference Proceedings 2020*. Lübeck, Germany: Infinite Science Publishing, Mar. 2020, pp. 285–288.

[8] B. Houska, H. Ferreau, and M. Diehl, "ACADO Toolkit – An Open Source Framework for Automatic Control and Dynamic Optimization," *Optimal Control Applications and Methods*, vol. 32, no. 3, pp. 298–312, 2011.

[9] "Unity Real-Time Development Platform | 3D, 2D, VR & AR Engine." [Online]. Available: https://unity.com

[10] "Vehicle Physics Pro." [Online]. Available: https://vehiclephysics.com/

6

Safety and Quality

Advancing Automotive ECU Software Validation: Automated Testing in a Software-in-the-Loop Environment for Enhanced Efficiency and Robustness

Michael Kurenkov [1,2], Tobias Klass [2] and Georg Schildbach [3]

[1] Robotics and Autonomous Systems, Universität zu Lübeck, michael.kurenkov@student.uni-luebeck.de

[2] Powertrain Development of Plug-In Hybrid Electrical Vehicles, Mercedes-Benz AG

[3] Institute for Electrical Engineering in Medicine, Universität zu Lübeck, georg.schildbach@uni-luebeck.de

Abstract

This study introduces an advanced automated testing framework tailored for the comprehensive assessment and validation of automotive software. Employing a Software-in-the-Loop approach, the framework allows for the early testing of software in the development cycle. An automated pipeline organizes the creation of Software-in-the-Loop environments and the execution of test cases, enhancing efficiency. The dashboard provided by test.guide aids in monitoring test case outcomes across software updates, supplemented by email notifications for prompt intervention in case of failures. In essence, the automated testing framework offers a comprehensive solution to guarantee the reliability of automotive software. It is well-positioned to tackle the evolving challenges inherent in the ongoing development and testing processes within the automotive industry.

1 Introduction

In the evolving landscape of vehicle development, the rise of software has become a major player alongside traditional mechanical components. As software, especially in electronic control units (ECUs), takes on an increasingly crucial role, ensuring its reliability has become a top priority. Modern vehicles heavily rely on software to manage various functions, from basic engine control to advanced features like driver-assistance systems and safety features. The heart of this software-driven intelligence lies in electronic control units. As vehicles become more like computers on wheels, the performance and dependability of these ECUs become critical and therefore a growing challenge. Traditional validation methods, like manual testing or code review, struggle to keep pace with the complexity of automotive software. Often Hardware-in-the-Loop (HIL) enviroments are used for automated testing, which can result in a slower develompment cycle [1],[2]. As shown in Fig. 1, testing in HIL is one step before implementation on real systems. Assisted by a Software-in-the-Loop (SIL) enviroment, the code can be tested directly after the implementation, resulting in several advantages [3]. To leverage these adavantages, this paper focuses on a solution: using automated testing in a SIL environment. This approach, which simulates real-world scenarios in a controlled virtual space, aims to make the validation process more efficient and robust. The following sections show the challenges of ECU software validation today, discuss the benefits of automated

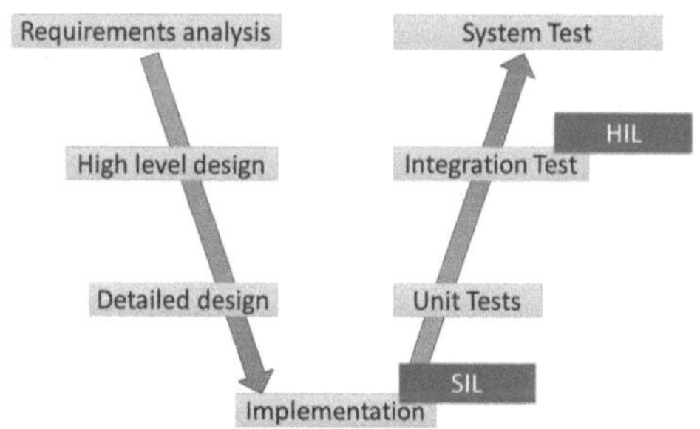

Figure 1: Development V-Cycle with added timing for SIL and HIL testing

testing in a SIL environment, and present test examples to illustrate the advantages of this approach.

2 Material and Methods

2.1 Simulation Platform

In the virtual powertrain development area, a reliable tool is Synopsys Inc.'s Silver — a Software-in-the-Loop platform [4]. Specifically tailored for powertrain development, Silver implements the interaction between virtual Electronic Control Units (vECUs), engines, and vital vehicle components through advanced simulations [5] as shown in Fig. 2. Some further key features of Silver are:

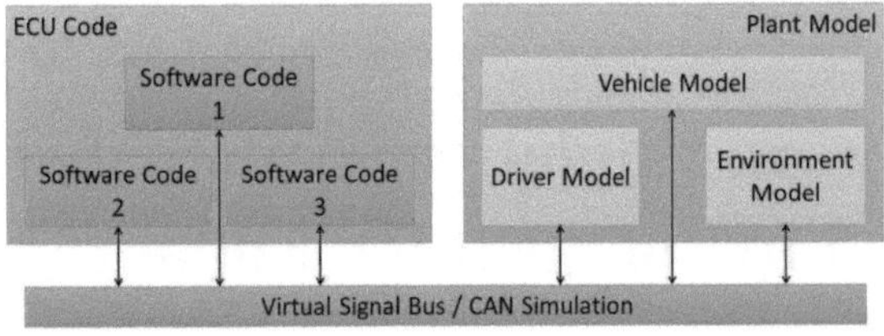

Figure 2: Silver SIL showing its underlying infrastructure with the plant model on the right side and the ECU software to be tested on the left side.

- **Comprehensive Signal Manipulation:** Whether through an intuitive GUI or a Python scripting API, Silver allows developers to wield control over one or multiple signals. This flexibility is paramount in tailoring simulations to specific scenarios, ensuring a validation of powertrain functionalities.

- **Behavior Analysis:** The platform facilitates in-depth analysis of powertrain behavior under diverse conditions. Through simulation, developers can inspect the responses of ECUs, engines, and associated components, providing valuable insights into system communications.

- **Modularity Advantages:** Silver introduces a modular approach through Silver Functional Units (SFU), enabling seamless component replacement. This modularity facilitates the effortless substitution of components, such as engines, fostering adaptability for diverse testing scenarios and system configurations.

- **Pipeline Integration Excellence:** Silver seamlessly integrates into software development pipelines, automating the creation of new simulations for each software version. This streamlined integration enhances efficiency by ensuring that every software update undergoes thorough simulation testing, contributing to a robust validation process.

2.2 Simulated Vehicle and Powertrain Components

The simulated vehicle in this methodology is a Plugin Hybrid Electric Vehicle (PHEV). A simplified schematic for a PHEV is given in Fig. 3. The SIL environment can be categorized into two different types — the plant model and the ECU software. For the simulation the PHEV is configured with multiple SFUs. Each SFU represents one critical element like Internal Combustion Engine (ICE), Transmission Control Module (TCM), Battery, Electric Motor (EM) and more. To ensure that the simulation closely approximates real-world vehicles, cycles such as the Worldwide harmonized Light Duty Test Cycle (WLTC) are executed to achieve an as comparable system as possible.

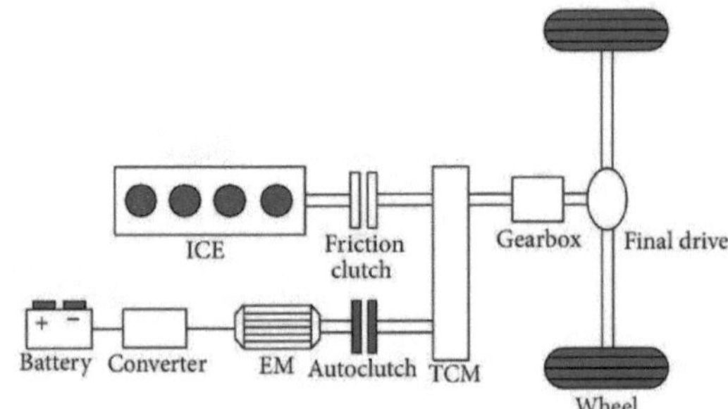

Figure 3: Blockdiagram of a PHEV with a parallel hybrid powertrain [6]

2.3 Application and Diagnose Tool

Central to the testing approach is INCA by ETAS [7], a versatile software crucial for application and diagnosis in testing vehicles. Although INCA's focus is the real world application, it's flexibility spans the entire development process, from desk analysis to on-road evaluations and seamlessly adapts to the simulation platform. The software allows for precise adjustments to functional parameters, characteristic maps, and tables, enabling offline and online modifications while the control unit actively regulates the vehicle. INCA's data acquisition capabilities from control units and vehicle buses, including CAN, LIN, FlexRay, or Ethernet, contribute to a detailed analysis of electronic control system behavior.

2.4 Testing Platform

At the core of the testing framework lies ecu.test by tracetronic GmbH [8], a robust platform meticulously designed to automate the testing of ECU software, complemented by the automated generation of comprehensive test reports. One of the standout features of ecu.test is its ability to facilitate automated testing and significantly enhancing the efficiency of ECU software evaluation. This capability is further boosted by its adaptability, supporting interfaces like ARXML, DBC, A2L and ODX, ensuring a fluid connection between the simulation and testing platform. The platform also introduces the invaluable feature of measurement recording during test stimulations, empowering developers with a improved understanding of software behavior under diverse conditions. Beyond automation, ecu.test offers versatility with two distinct testing modes — synchronous and asynchronous. In synchronous testing, signals are stimulated in real-time, allowing for a simultaneous examination of the plant model's new states. On the other hand, asynchronous testing relies on measurements at the conclusion of a testing loop, initiating a trace analysis. When using the asynchronous mehtod, it is also possible to apply trace analysis to real vehicle data and test different functions. This flexibility accommodates a spectrum of testing scenarios tailored to specific validation requirements. The asynchronous testing method further incorporates trace analysis, requiring the definition of start and end triggers, and the identification of various states in between. This approach offers a detailed examination of the software's response to changing conditions.

Adding a layer of accessibility and user-friendly interaction, tracetronic GmbH introduces the test.guide — a web-based dashboard that centralizes and simplifies the analysis of test outcomes. This dashboard becomes a pivotal tool for efficient monitoring and interpretation of testing results. Additionally, it facillitates the creation of pipelines for automated testing.

2.5 Testing process

The testing process follows a well-defined sequence, as shown in Fig. 4, aimed at ensuring a thorough evaluation and validation of the automotive software. The journey begins with the "Creation of Test Specification" phase, which involves laying out the specifics and objectives of the tests to be conducted. The subsequent step involves the "Modeling of Test Cases", where intricate test scenarios are conceptualized and shaped. These models serve as the blueprint for the actual testing process. The "Creation of Test Cases" phase then sees the transformation of these models into executable test cases, ready for execution. In the next phase test.guide executes the created software checks using ecu.test. This step involves running all tests for any changes in the software, including Functional Testing, First Run Tests, and Generic Test Stimulation, which employs the WLTC. The results obtained from these tests undergo comprehensive analysis during the "Evaluation of Test Cases" phase. This critical step involves analyzing the performance of the software under various conditions, ensuring that it meets the required standards and safety regulations. Upon successful completion of the evaluation, the "Approval" phase is reached, marking the official approval and clearance of the tested software. The SIL test-

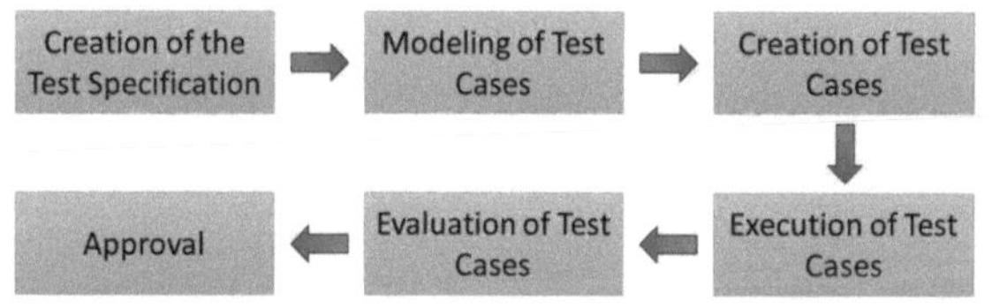

Figure 4: ECU software testing process

ing is just the inital part of the entire testing chain and is insufficient alone to make definitive statements about the software's functionality. It lacks certain aspects that differ in the real vehicle, apart from the inaccurately represented hardware. These include for example noise on the signals, latency times, and transmision dropouts. However, it allows for the early and very time efficient detection and resolution of inital errors in the development process.

2.6 Test example

The testchain with a SIL environment is illustrated using a simple example. For this purpose, a typical greatly simplified function from a operating strategy of a PHEV is presented. It involves the engine starting when the Battery SOC is low. The simple model receives the current Battery State-of-Charge (SOC) via a CAN signal and decides whether the EM and/or the ICE should be used. To further address some boundary conditions, we are also examining effects of a battery SOC outside the range of 0% - 100% or when no signal is received. These signals would also be sent to the respective control unit via the CAN. The following tests would be specified for this function:

Description					
Battery SOC greater equal 3.5% and lower equal 100%	T	-	-	-	-
Battery SOC between 3% and 3.5%	-	T	-	-	-
Battery SOC lower equal 3% and greater equal 0%	-	-	T	-	-
Battery SOC lower 0% or higher 100%	F	F	-	T	-
Battery SOC Signal not availible	F	F	-	-	T
Actions	A1	A2	A3	A4	A5

Action	Description
A1	only EM active
A2	EM active & ICE active
A3	only ICE active
A4	SOC invalid range & EM inactive & ICE active
A5	SOC error & EM inactive & ICE active

Table 1: Decision Table for the test example

It results in five different cases, that need to be tested. These are implemented in ecu.test and integrated into the automated testing system.

3 Results and Discussion

The results are summarized in two parts, the test cases themselves and the underlying pipline for the automatic creation of the simulation enviroment and the execution of the tests.

3.1 Automated testing pipeline

The established automated pipeline successfully generates a new SIL with each software update. Subsequently, test.guide automatically triggers the previously implemented test cases and presents the results through a user-friendly web interface. The dashboard's clear layout allows for tracking whether test cases that failed in the past have passed with a new software update. It allows to subscribe to different events and test cases to ensure a tailored information flow, where only relevant email notifications are sent. Following this, it is possible to use the interface to view the test sequences through the recorded signals and analyze any potential errors.

3.2 Test cases

After implementing the test cases in ecu.test, it is easily possible to examine the software for any problems. While creating the test cases, both minor errors in the software and issues with the Silver Simulation environment were identified. These problems were subsequently resolved through corrections. A subsequent run of the same test cases revealed no further issues. The overview of the tests in ecu.test allows for a straightforward analysis of potential problems, as shown in Fig. 5. It is possible to inspect the

5	⊟ 🗐 Action	
6		⏱ Wait
7	⊞ 🗐	1) em active, when battery SOC between 3% and 100%
14	⊞ 🗐	2) em active & ice active, when battery SOC between 3% and 3.5%
21	⊞ 🗐	3) em not active & ice active, when battery SOC between 3% and 0%
28	⊞ 🗐	4) em not active & ice active & soc_inv_range, when battery SOC lower 0% or higher 100%
40	⊞ 🗐	5) em not active & ice active & soc_error, when no signal

Figure 5: Testreport in ecu.test with successfull cases

measurement data related to the tests. Especially in the case of errors, conducting a quick data analysis is helpful. Additionally, it is possible to test edge cases that are not easily reproducible in a real vehicle. In the case of the exemplary test described above, one can test SOCs that are not typically reachable in a car (e.g., SOC above 100%). Also, special edge cases such as a non-arriving SOC signal are not easily reproducible in a vehicle. Here, testing in a SIL environment demonstrates its strengths. Additionally, the asynchronous trace analysis provides a comprehensive overview of individual test cases. Relevant signals are plotted and displayed alongside the set triggers. In Fig. 6, it was verified whether the ICE is active when the Battery SOC is between 0% and 3%. These conducted trace analyses can later be applied to real experimental drives, making it easy and automated to identify issues in a real vehicle. Given that ecu.test is not solely dedicated to SIL testing, the test cases formulated within its framework are versatile and can be seamlessly transitioned for later HIL implementations.

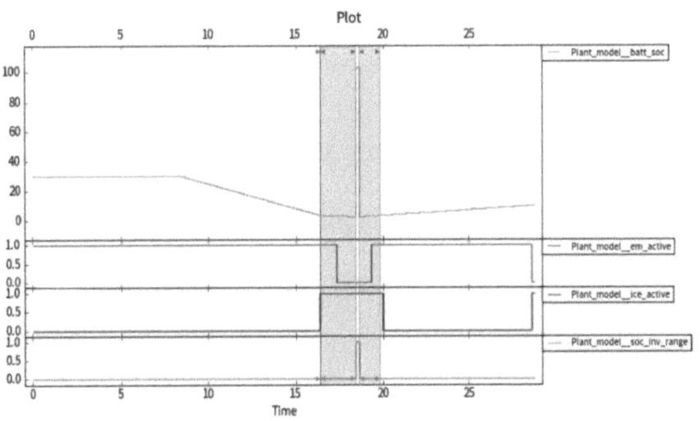

Figure 6: Trace analysis in ecu.test: Trigger is set to Battery SOC between 0% and 3%, checking if ICE is active

4 Conclusion

In conclusion, the automated testing pipeline, seamlessly integrates with test.guide, consistently generates SIL environments for thorough evaluations. The user-friendly interface and email notifications enhance result visibility and prompt intervention. Test cases implemented in ecu.test proved effective in identifying and resolving software and simulation environment issues. The framework's versatility in testing edge cases and its potential for real-world application demonstrate its robustness in ensuring automotive software reliability. It provides a high-quality approach to addressing the challenges in ECU software development by contributing a fast and comprehensive testing method.

Acknowledgement

The work has been carried out at Mercedes-Benz AG at the department of powertrain development of Plug-in Hybrid Electric Vehicles and supervised by the Institute for Electrical Engineering in Medicine, Universität zu Lübeck.

Authors' Statement

Conflict of interest: Authors state no conflict of interest.

5 References

[1] A. Mouzakitis, D. Copp, R. Parke and K. Burnham. (2009). *Hardware-in-the-Loop System for Testing Automotive Ecu Diagnostic Software.* Measurement and Control.

[2] N. Brayanov and A. Stoynova. (2019). *Review of hardware-in-the-loop -a hundred years progress in the pseudo-real testing.*

[3] V. Jaikamal and T. Zurawka. (2010). *Advanced Techniques for Simulating ECU C-code on the PC.*

[4] Synopsys Inc., [Online]. Available: https://www.synopsys.com/verification/virtual-prototyping/silver.html [last accessed on 2023-12-04].

[5] N. Amringer and P. Asemann. (2022). *Simulation of Virtual ECUs in the context of ECU Consolidation.*

[6] A. Gao, X. Deng, Zhang, M. Zhang and Z. Fu. (2017). *Design and Validation of Real-Time Optimal Control with ECMS to Minimize Energy Consumption for Parallel Hybrid Electric Vehicles.* Mathematical Problems in Engineering.

[7] ETAS GmbH., [Online]. Available: https://www.etas.com/de/portfolio/inca.php [last accessed on 2023-12-05].

[8] tracetronic GmbH., [Online]. Available: https://www.tracetronic.de/produkte/ecu-test/ [last accessed on 2023-12-05].

Literature Research on Medical Technology Products for Monitoring Perioperative Processes in Anesthesia

Julia Röller [1], Karsten Hiltawsky [2]

[1] Medical Engineering Science, Universität zu Lübeck, julia.roeller@student.uni-luebeck.de
[2] Drägerwerk AG & Co. KGaA - Corporate Technology & Innovation, Karsten.Hiltawsky@draeger.com

Abstract

Accurate anesthetic administration is essential, as both underdose and overdose can lead to complications for the patient. The use of quantitative monitoring systems is expected to support the anesthesiologist in carrying out individualized treatment.

This work was intended to research monitoring systems and determine their clinical evidence. An approach for standardized evidence assessment was developed using literature research and Model-based System Engineering (MBSE), which focuses on data extraction and study evaluation.

MBSE enabled the management of large amounts of data: 15 hypnosis depth systems were identified, for which 102 studies were evaluated. 83 literature references were used for 15 quantitative systems focusing on analgesia and 10 relaxation systems were identified for which 16 studies were processed. Currently there is no monitoring system available which shows superior clinical evidence. This might be a reason, why there is no single monitoring system yet, which is commonly used in clinical practice.

1 Introduction

The three main components of anesthesia are analgesia, hypnosis and muscle relaxation, which characterize the perioperative condition of a patient. Different medications are required to eliminate consciousness (hypnosis) and the perception of pain (analgesia) as well as to prevent the transmission of stimuli to the motor end plate (neuromuscular relaxation). By temporarily eliminating consciousness and the perception of pain, muscle relaxation is initiated. On the other hand, muscle relaxation caused by neuromuscular blocking agents (NMBA) has no influence on the analgesia and hypnosis state. The use of NMBA ensures that muscle responses arising from stimuli are suppressed, thereby impairing the assessment of the hypnosis and analgesia components [1]. Nevertheless, NMBAs are used because of the significant surgical advantages [2]. Even for an experienced anesthesiologist, optimal dosing of anesthetics can be challenging because:

1. The combination of anesthetics triggers synergistic effects. On the one hand, additive effects can be desirable, but on the other hand, the incidence of drug interaction increases exponentially with the number of drugs administered [3].

2. When selecting and dosing anesthetics, individual characteristics such as the patient's age, weight, gender and previous illnesses must be taken into account.

Underdoses can lead to awareness [4], pain [5] and difficult surgical conditions due to body movements. Overdoses can lead to delirium, vomiting and the risk of opiate addiction as well as residual blockages [2]. In the transition zones between underdose, optimal dose and overdose, the risk of adverse events is particularly high, since, for example, the anesthesia levels *still awake* and *already awake* are difficult to differentiate. Non-specific physiological parameters such as blood pressure and heart rate as well as clinical symptoms (sweating, tearing and body movements) are used to monitor the patient during general anesthesia [4]. A variety of quantitative monitoring devices and parameters, most of which focus on one of the three main components, have already been developed. The methods aim to individualize the administered drug doses during perioperative anesthesia in order to avoid adverse events. It has already been proven that with the help of specialized monitoring methods, significant improvements in the anesthesia process can be achieved, such as reduced anesthetic dose and shorter recovery times [6], [7]. Despite this, it has not been established in clinical routine.

In medical technology, it is necessary to confirm safety, validity and social relevance with comprehensive clinical data, taking into account a cost-benefit analysis. This work deals with the evaluation of studies and relevant articles using Model-based Systems Engineering (MBSE) to determine the level of evidence of quantitative monitoring systems in a standardized manner.

2 Material and Methods

The methods consist of literature research and literature management with MBSE. The components hypnosis, analgesia and muscle relaxation were processed independently.

2.1 Literature Research

Fig. 1 illustrates the literature research procedure as an example for hypnosis monitoring systems.

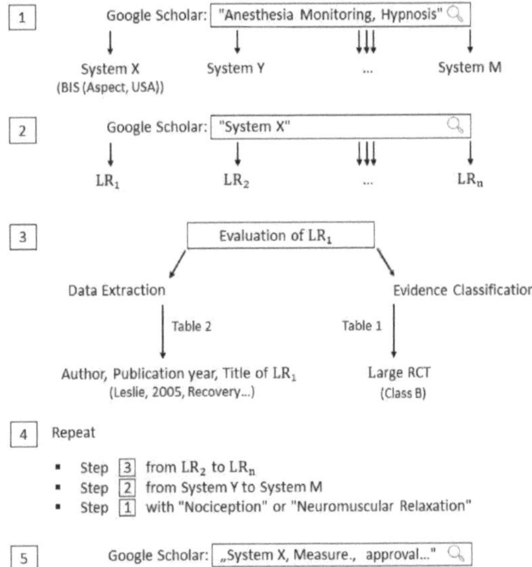

Figure 1: Process of literature research exemplary for hypnosis monitoring systems.

The digital database Google Scholar was used to identify available monitoring systems in perioperative anesthesia. Keywords were *anesthesia monitoring* combined with *hypnosis* (Step 1). For a System X (e.g. *Bispectral Index, BIS* (Aspect Medical Systems, USA)), various literature references (LR_1, LR_2,..., LR_n) are filtered out in the Google Scholar search (Step 2). The evaluation begins with LR_1 and consists of data extraction and evidence classification (Step 3). For the latter, Table 1, which has already been determined in advance, is used. Class A to Class E defines the decreasing evaluation potential of scientific studies in alphabetical order (e.g. LR_1 is a large randomized controlled trial (RCT), so it is assigned to Class B). Class F represents a special case as a computer-simulated study.

The author, the year of publication, the title of LR_1 and other data are recorded (Leslie, 2005, Recovery...) in the MagicDraw database (details in Table 2).

Step 3 is repeated for all subsequent literature references (LR_2,..., LR_n). The process is then repeated from Step 2 for subsequent monitoring systems (System Y,..., System M). From Step 1 onwards, the procedure is repeated analogously with the keywords *Nociception* or *Neuromuscular Relaxation* instead of *Hypnosis* (Step 4).

For each monitoring system, the measuring principle, medical technology approvals and technical data are also researched (Step 5).

Table 1: Evidence classification

Level	Definition
Class A	Multiple large studies & low risk of error
Class B	One large study with low risk of error
Class C	Randomized study with moderate to low risk of error
Class D	Not randomizes studies
Class E	Other human studies than Class A to D or expert opinions
Class F: In Silico	Computer-simulated studies, using special programs or algorithms

2.2 Literature Management

To efficiently collect the data volumes, Model-based Systems Engineering is applied using the MagicDraw software (No Magic, Inc., USA) in the modeling language SysML. Table 2 provides an overview of the data extraction from literature references. As an example, the concept is demonstrated for [8].

From each literature reference, the elements of basis information, the main topic and detailed information are entered into the MagiDraw database using generic tables and allocation matrices. Generic tables are implemented in order to present elements and their properties in a manageable manner. Allocation matrices are required for detailed analysis and show dependencies between elements. Important results of the study are summarized in the Diagram Documentation.

Table 2: Data Extraction from Literature References

	Function	Elements	Example [8]
Basis	Table	1. Author Public.Year Publisher Title Patients Study Type	Leslie, K. 2005 Anaesthesia and intensive care Recovery From Bispectral Index-guided Anaesthesia in a Large Randomized Controlled Trial of Patients at High Risk of Awareness 2463 Class B
Topic	Matrix	Hypnosis Analgesia Relaxation	Hypnosis
Details	Matrix	System Drugs Models	BIS Not specified Not specified
Result	Diagram	Docum.	BIS: No affect on time to tracheal extubation

3   Results and Discussion

In order to determine the level of evidence of a monitoring system, we decided to introduce *Evidence Assessment (EA)* as follows:

$$EA_{MS_X} = \sum_{Class A}^{Class F} \sum_{i=1}^{n} LR_{Class,i}.$$ (1)

Literature References $(LR_1...LR_n)$ depending on their level of evidence $(Class\ A...Class\ F)$ are used to assess the level of evidence of a Monitoring System$_X$ (MS$_X$).

3.1   Literature Research: Hypnosis

The literature research resulted in 15 monitoring systems which focus on assessing the depth of hypnosis during general anesthesia. A total of 102 studies and relevant articles were evaluated and integrated in the MagicDraw database. Quantitative systems are primarily based on electroencephalography, using complex algorithms to generate one, two or more parameters that numerically represent a patient's level of consciousness. Examples of researched developments and their evidence are:

- *Bispectral Index* (Aspect Medical Systems, USA)
  $EA_{BIS} = 49$ LR (Class A: 1; Class B: 5; Class C: 16; Class D: 23; Class E: 1; Class F: 3)

- *Entropy™ module* (GE Healthcare, Finland)
  $EA_{Entr.} = 13$ LR (Class C: 4; Class D: 8; Class E: 1)

- *Conox® QM-7000M* (Fresenius Kabi AG, Germany)
  $EA_{Conox} = 6$ LR (Class D: 6).

With the exception of the *Bispectral Index*, the level of evidence varies from Class C to Class E. Due to the extensive studies, the *BIS* is often used as a reference method and has become the most established monitoring system in everyday clinical practice, although it has not been proven that it can measure the depth of hypnosis better than other systems.

3.2   Literature Research: Analgesia

15 monitoring systems that focus on the analgesia component were identified and a total of 83 studies and relevant articles were evaluated.

Nociception monitors are based on the detection of passive changes in effect organs that are triggered by painful stimuli [9]. Measurement techniques for assessing nociception and anti-nociception balance are based on skin conductivity, plethysmography, pupil dilation reflex, heart rate variability as well as electroenzephalography and electromyography. Examples of researched nociception systems and their evidence are:

- *Analgesia Nociception Index* (Mdoloris, France)
  $EA_{ANI} = 13$ LR (Class C: 3; Class D: 8; Class E: 2)

- *Surgical Pleth Index* (GE Healthcare, Finland)
  $EA_{SPI} = 12$ LR (Class C: 8; Class D: 4)

- *Nociception Level Index* (Medasense, Israel)
  $EA_{NOL} = 6$ LR (Class C: 6)

Despite various developments, monitoring analgesia in the perioperative context is a relatively new area of research. The level of evidence for quantitative analgesia monitoring systems varies only from Class C to Class E. Although many systems reflect nociception better than non-specific parameters, no system has been able to demonstrate a convincing and clinically relevant benefit for routine use.

Quantitative monitoring of hypnosis and analgesia requires innovative systems, because complex processes in the autonomic nervous system as well as the central nervous system must be taken into account [1].

The developed evidence procedure provides a basis for assessing the current state of technology for both components.

3.3   Literature Research: NM. Relaxation

Through the literature research, 10 monitoring systems that focus on measuring neuromuscular relaxation during general anesthesia were identified. A total of 16 studies on these were evaluated and recorded in the MagicDraw database.

The most important methods of quantitative relaxation measurement include electromyography, accerlography and mechanomyography. Examples of relaxometry devices and their evidence are:

- *TwitchView™* (Blink Device Company, USA)
  $EA_{Twitch} = 2$ LR (Class D: 2)

- *NMT ElectroSensor* (GE Healthcare, USA)
  $EA_{NMT} = 1$ LR (Class C: 1)

- *Stimpod* (Xavant Technology, South Africa)
  $EA_{Stimpod} = 2$ LR (Class D: 2)

Researched studies only reach the level of evidence from Class C to Class D. In contrast to hypnosis and analgesia, the principle of neuromuscular monitoring using peripheral nerve stimulation is well known. However, this evidence level does not reflect the extensive state of knowledge. Instead of evaluating quantitative relaxation devices, studies focus primarily on comparing methods. Therefore, literature references that focus on relaxometry devices only reach a maximum evidence level of Class C.

It should also be noted that the evidence classification from Table 1 takes into account the study type and the number of patients, but is based on subjective assumptions, for example through the determination of a "large study".

3.4   Literature Management

By selecting component-specific data and exporting it as an Excel file, tabular overviews of perioperative anesthesia monitoring systems were generated. Fig. 2 visualizes the table content developed through Model-based Systems Engineering as an example for the hypnosis depth system

Bispectral Index. The manufacturer, measurement principles and values as well as medical technology approvals are represented. The column *Evidence [Literature Reference]* lists the studies evaluated in MagicDraw in descending order of their evidence classification as a result of (1). For the literature reference name, the last name of the first author is combined with the year of publication of the reference. General information and technical data are entered in the last column.

The table overview does not serve as a rigid snapshot, but can be updated as the database is expanded and modified to reflect the current level of evidence from monitoring systems.

System	Manu-facturer	Meausurement [Score]	Approval	Evidence [Literature Reference]	Information
Bispectral Index™ (BIS™) Monitoring System	Aspect Medical System, USA	Electroencephalography Bispectral Index: Range 0...100 • 0 represents the absence of brain activity • Target Value: 40...60 during General Anesthesia • BIS < 40 represents a deep hypnotic state	CE & FDA	**Class A** [Oliveira2017] **Class B** [Chen2011, Leslie2005, Leslie2010, Liu2004, Wang2021] **Class C** [Gan1997, Gruber2022, Kertai2011, Liu2006, Petrun2013, Rüsch2018, Wu2017, Aimé2006, Kreuer2003, Kreuer2005, Delfino2009, Shalnovic2014, Bruhn2005, Liao2011] **Class D** [Sadhasivam2006, Haenggi2008, Laitio2008, Kreuer2004, Panousis2007, Christ2014, Hoymork2007, Cortinez2007, Revuelta2008, Fuentes2008, Hrelec2010, Bergese2017, Springman2010, Chen2002, Schneider2003, Soehle2008, Nishiyama2013, Lu2008, Jensen2006, Kreuer2006, Nishiyama2004, Vereecke2003, Vereecke2006] **Class E** [Kissin2000] **Class F** [Struys2004, Pilge2006, Zanner2009]	*Intended Use:* BIS processes EEG information to provide a measurement of the patient's level of consciousness. BIS uses EEG signals between 0.5 up to 47 Hz; 2-channel and 4-channel monitors.

Figure 2: Data and clinical evidence of the BIS.
The results of the literature research, evidence categorization and data extraction from various studies using *Bispectral Index* as an example are presented as a tabular overview. The table content was structured and managed in the MagicDraw software using MBSE.

Model-based Systems Engineering represents an efficient foundation for managing large amounts of data in an interdisciplinary manner and is able to capture the resulting complexity, which typically increases over time. Although numerous literature references have been integrated into the MagicDraw database, a limitation of this work is the incompleteness: The database requires continuous expansion in order to adequately represent the current state of technology and clinical evidence.

4 Conclusion

This work revealed that many different quantitative monitoring systems are available for each anesthesia component. Overall, the study situation is mediocre. It is very likely that specialized systems are not used as clinical standard yet, because very few have a high level of evidence.
Model-based Systems Engineering enabled the development of a standardized evidence process and the management of large amounts of data. The evidence assessment (EA) offers a solid basis for presenting the current state of a certain monitoring system and its evidence level.
In future approaches, it could be weighted according to the level of evidence of literature references: Meta-analyses of randomized controlled trials (Class A) are particularly suitable for confirming the validity of medical treatment methods and should therefore be given a higher weight than small non-randomized studies (Class E).

Acknowledgement

The work has been carried out at Drägerwerk AG & Co. KGaA, Corporate Technology & Innovation and was supervised by Prof. Dr.-Ing. Dr. med. Karsten Hiltawsky.

Authors' Statement

The authors declare that they have no conflicts of interest.

5 References

[1] P. Martinez-Vazquez and E. W. Jensen, *Different perspectives for monitoring nociception during general anesthesia.* Korean Journal of Anesthesiology 75.2, pp. 112-123, 2022.

[2] S. J. Brull and A. F. Kopman, *Current status of neuromuscular reversal and monitoring: challenges and opportunities.* Anesthesiology 126.1, pp. 173-190, 2017.

[3] A. S. Milde and J. Motsch, *Medikamenteninteraktionen für den Anästhesisten.* Weiterbildung für Anästhesisten 2003, Springer Berlin Heidelberg, Berlin, pp. 117-135, 2004.

[4] P. Bischoff and I. Rundshagen, *Awareness under general anesthesia.* Deutsches Ärzteblatt International 108.1-2, pp. 1-7, 2011.

[5] M. Gruenewald, C. Ilies, J. Herz, T. Schoenherr, A. Fudickar, J. Höcker, B. Bein.*Influence of nociceptive stimulation on analgesia nociception index (ANI) during propofol–remifentanil anaesthesia.* British journal of anaesthesia, 110(6), pp. 1024-1030, 2013

[6] A. Vakkuri, et al., *Spectral entropy monitoring is associated with reduced propofol use and faster emergence in propofol–nitrous oxide–alfentanil anesthesia.* The Journal of the American Society of Anesthesiologists 103.2, pp. 274-279, 2005.

[7] S. Liu, *Effects of Bispectral Index monitoring on ambulatory anesthesia: a meta-analysis of randomized controlled trials and a cost analysis.* The Journal of the American Society of Anesthesiologists 101.2 pp. 311-315, 2004.

[8] K. Leslie, et al., *Recovery from bispectral index-guided anaesthesia in a large randomized controlled trial of patients at high risk of awareness.* Anaesthesia and intensive care 33.4, pp. 443-451, 2005.

[9] F. von Dincklage, *Monitoring von Schmerz, Nozizeption und Analgesie unter Allgemeinanästhesie.* Der Anaesthesist, 10(64), pp. 758-764, 2015

Humidification system of the incubator: An investigation of possible methods and the development of potential concepts

Stjepan Radic [1], Peter Baltin [2], Ulf Pilz [3]
[1] Biomedical Engineering, Luebeck University of Applied Sciences, stjepan.radic@stud.th-luebeck.de
[2] Drägerwerk AG & Co. KGaA, Peter.Baltin@draeger.com
[3] Department of Applied Natural Sciences; Lübeck University of Applied Sciences, ulf.pilz@th-luebeck.de

Abstract

Premature babies require special care for their body temperature. The evaporation of water from the baby's body influences the heat balance, which can be minimised by high humidity in the incubator. This work investigates methods for generating humidity in the incubator in order to develop promising humidification system concepts. The VDI 2221 process model was used up to the concept phase. A list of requirements, system analysis, derivation of functions and a morphological box led to five humidification concepts based on evaporation and vaporisation. The vaporisation method is used in the humidification system of the Dräger Babyleo. The question of whether evaporation is a practicable and cost-effective option for humidity control remains. A test carried out with evaporation-based humidity generation in the incubator showed a relative humidity of 85% under worst-case conditions. Further optimisation could enable more efficient humidity control. Five concepts are therefore available for a new development of the humidification system.

1 Introduction

Every year around 15 million children worldwide are born prematurely, more than one in ten newborns. Four million die before the first month. A normal pregnancy lasts 40 weeks [1]. Premature babies, born before 37 weeks of pregnancy, require special care. A stable microclimate, non-invasive therapies and external factors such as a reduction in noise, light and contact are crucial for long-term healthy development.[2]. Body temperature is the most important parameter for premature babies as they are unable to regulate their own body temperature. They lack the brown fatty tissue that produces heat [3]. Premature babies have no mechanisms such as sweating and shivering, and their metabolism is restricted. Even small temperature differences of 1°C mean stress and loss of energy, energy that they urgently need for healthy development [4]. It is important to keep the body temperature of premature babies at an optimum level to avoid cold stress. Heat is lost in premature babies through evaporation, convection, radiation and conduction [5]. Most of all, heat loss through evaporation is particularly high as their skin is not yet fully developed. When liquid evaporates, heat is lost through evaporative heat [2]. The thin skin dries out quickly and cracks appear. Together with the limited ability of the skin vessels to constrict in the immature state, the skin becomes more permeable [3]. The younger the baby, the greater the effect of water loss and the greater the heat loss. Fig. 1 shows the water loss in premature babies as a function of the month of gestation and the first days after birth. It shows an exponential relationship between week of gestation and transepidermal water loss in the first 4 weeks. The effect of water loss can be minimized by high ambient humidity [4]. Humidity is created by the mixing of dry air and water. The capacity of the air to absorb humidity depends on the temperature and increases with higher temperatures.

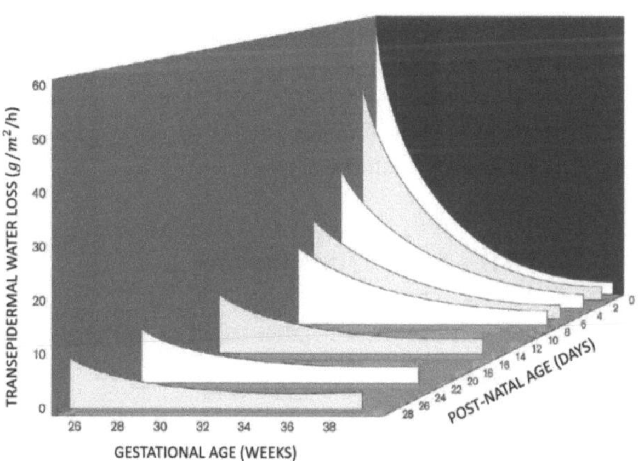

Figure 1: Premature babies with a low gestational age exhibit excessive water loss and therefore high heat loss in the first few days after birth [4]

Premature babies are treated in incubators to adjust their temperature and humidity according to their physiological needs. For babies under 500 g, the temperature is maintained between 36 and 38°C, with around 85% relative humidity. A drop from 80% to 60% humidity leads to a 1°C drop in body core temperature within 5 minutes for very premature babies [3]. Therefore the humidification system can be crucial for the survival of a premature baby. The humidification systems used in incubators operate under special conditions such as high temperatures, water vapour and limited installation space. The high level of safety, reliability and thermodynamic conditions make the humidification system a highly complex component in the incubator. Humidification systems face a variety of challenges. Besides the reliable regulation and control of humidity and the requirement to achieve a relative humidity of up to 99% in the closed incubator at temperatures of up to 39°C, the increasing demands on service life, simplified installation for worldwide use, germ free maintenance through reprocessing and the reduction of noise levels in the patient room must also be taken into account. These are challenges that the humidification system must meet. This paper therefore examines concepts for humidification in the incubator that could be considered in the future.

2   Material and Methods

Firstly, the concept of a new design is analysed and a concept phase according to VDI (Verein Deutscher Ingenieure) 2221 is carried out in order to obtain an overview of the possibilities and scope for change that can be considered for moisture generation in the incubator. Based on the analysis, a decision is then made as to whether optimisation or a new design should be sought in the future. VDI Guideline 2221 is a generally recognised procedural guideline for development engineers in product development. The systematic application of methods, techniques and guidelines is intended to increase innovative strength, save time, avoid wrong decisions and objectively evaluate the results. The aim is to efficiently find a creative solution that is not accidentally achieved. In this work, the focus was placed on phases 1 and 2 of VDI 2221. In phase 1, objectives are defined and requirements are analysed. In phase 2, functions and solutions are evaluated, variants are developed and the optimum solution is identified [6]-[7].

3   Results and Discussion

3.1   Methodical development of a humidification system in the incubator

3.1.1   List of requirements

In order to develop the humidification system, various departments had to work together to identify requirements. Based on experience with existing solutions and the identification of upcoming requirements, a comprehensive catalog of requirements was drawn up. The analysis shows that most of the requirements relate to the overall system and can therefore only be tested in the system. In the concept phase, technical subsystem requirements (TSSR) relating only to the humidification system are primarily relevant. Fifteen such subsystem requirements were created in customer, service and function categories. Customer and safety requirements are used for the subsequent assessment

of partial solutions. Functional requirements are crucial at this stage and are listed in Table 1. Humidification of the air is required, which can be achieved by various principles based on nebulisation, evaporation or vaporization. Furthermore, an external water supply is required as well as the sterility of the humidification system during use in order to prevent infections in an infant.

TSSR		
Functional requirments		
No:	Type:	Attribute
1	Shall	The humidification system shall produce humidified air
2	Shall	An external water supply shall be available for the humidification system.
3	Shall	The humidification system shall be germ-free when used for air humidification.

Table 1: The functional requirements for the humidification system

3.1.2   Function analysis

The black box analysis of the humidification system defines clear system boundaries, input and output variables and the main function, see Fig. 2. Water and air as external input variables should be used to generate 99% relative humidity in a closed room, thereby also creating a certain amount of noise. The required thermal energy dissolves the molecular bond of the water, whereby a portion leaves the system as heated, humid air. External influences such as air pressure and temperature, as well as possible leaks, affect the conversion of 99% relative humidity in the closed room. By analysing the input and output variables and the functional requirements in detail, the main function is defined as "to humidify the air in the incubator compartment germ-free".

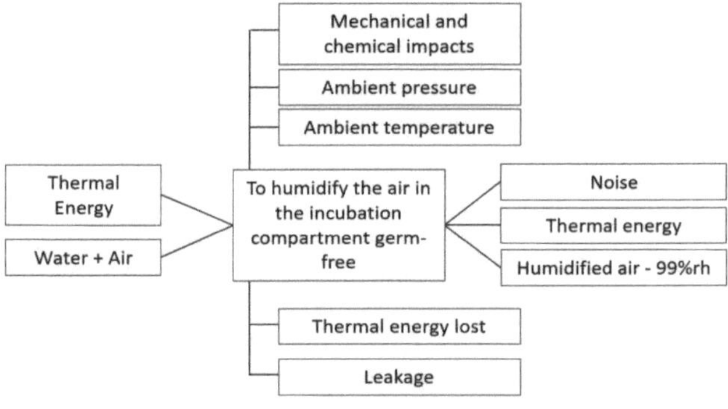

Figure 2: Black box representation of the humidification system in an incubator

Sub-functions were derived from the overall function in order to fulfill it. These include "to supply water to the humidification system", "to provide moisture to the air" and "to reprocess the humidification system". The first two sub-functions can be further subdivided to refine the functional structure. Fig. 3 illustrates the resulting functional structure. The sub-functions on the second level and the function "to reprocess the humidification system" form the direct basis for the systematic search for partial solutions.

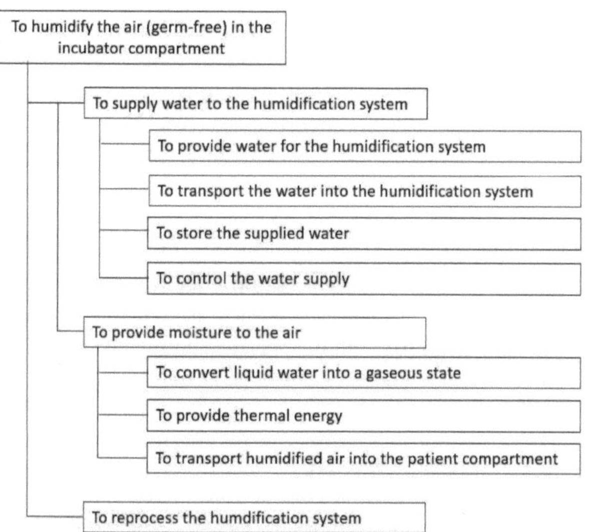

Figure 3: Functional structure of a humidification system

3.1.3 Morphological box

By applying the conventional, intuitive and discursive methods, possible partial solutions for partial functions were identified and summarised in a morphological box. A total of 54 partial solutions for the sub-functions were found, resulting in 3.5 million possible concept combinations. However, many of the partial solutions or combinations were not suitable for a humidification system in an incubator because they were classified as unsuitable for technical, price, safety, installation space or service-related reasons. For this reason, the morphological box was reduced to meaningful partial solutions in a three-stage process. Firstly, independent subtasks were set to "hold", then utopian and risky partial solutions were excluded. In the third step, the partial solutions were classified if they followed similar solution principles. The reduction was carried out carefully in several iterations, taking into account the list of requirements. The analysis of the morphological box and the created compatibility matrix finally resulted in 5 suitable concepts with 5 partial solutions. Fig. 4 shows the complete morphological matrix with all sub-solutions and the reduced morphological matrix. The current humidification system used in the Babyleo T500 from Dräger is based on water evaporation in combination with hydrostatics, which is one of the five concepts. Four further concepts can be derived from the morphological box, three of which are based on evaporation and another on vaporization. An experimental set-up is intended to clarify whether the evaporation principle is suitable for humidifying incubators.

3.2 Experimental test of evaporation

The experiment on incubator humidification included an evaporation chamber already available on the market. Originally designed for humidifying the respiratory tract, the

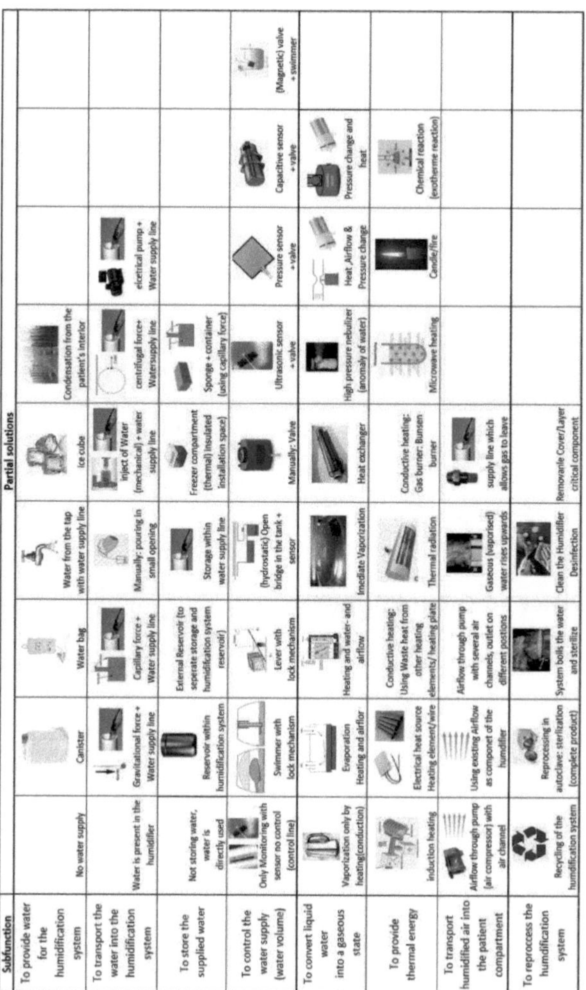

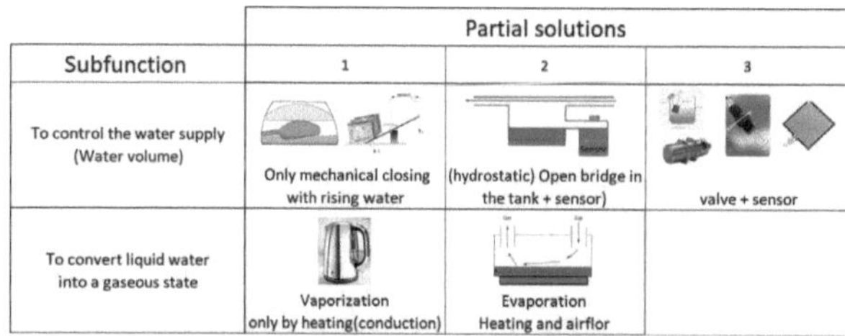

Figure 4: The morphological box of a humidification system (top), the reduced matrix with the meaningful partial solutions (bottom)

chamber enabled an initial analysis of its general suitability in the incubator. The test setup included the chamber, a water tank for the water supply, a blower for generating airflow, a heating base for heating the water and hoses for guiding air into the incubator, see Fig. 5. Firstly, dry air was drawn in by the fan and directed into the evaporation chamber via a hose. There, the air was humidified through contact with the heated water. The chamber was designed in such a way that the supplied air comes into contact with as much of the water surface as possible. The humidified air was directed into the incubator through a second hose and placed there in such a way that it was optimally mixed with the internal circulating airflow of the incubator. The incubator humidity was monitored using a capacitive humidity sensor. The experiment was carried out under worst-case conditions at an air temperature of 39 °C and an initial humidification of 20% in the incubator.

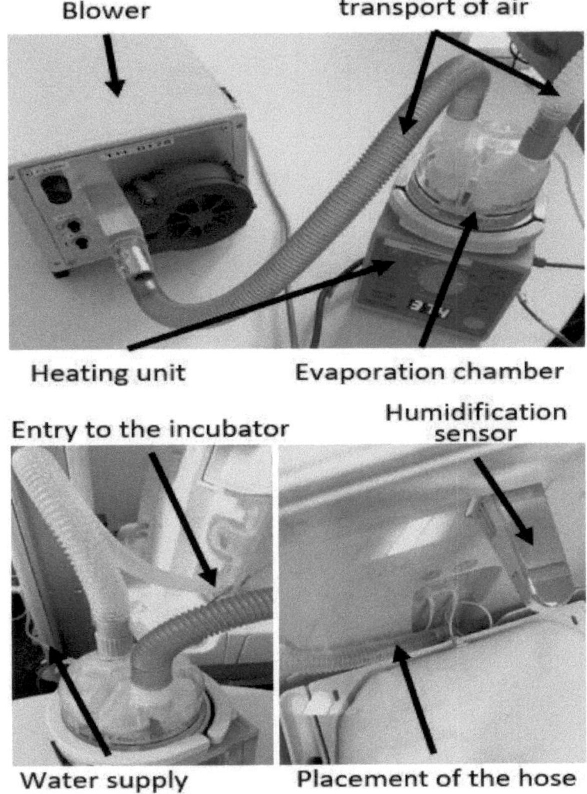

Figure 5: Experimental setup

The first experiment was carried out with a constant airflow and a heating plate of approx. 77°C. The humidity in the incubator stabilized at approx. 68% after 20 minutes and reached around 85% after an hour and remained at this level. The experiment was then repeated, but with a gradually increasing airflow. Initially, the humidity increased faster than in the previous experiment. However, a too high airflow led to turbulence in the chamber, which reduced evaporation. A relative humidity of over 85% could not be achieved despite adjusting the airflow. It was not possible to generate a relative humidity of 99% within an hour. Challenges such as the slow formation of humidity, the noise of the fan, condensation in the hoses and leaks in the system were identified. From the test setup, it can be concluded that a humidity of up to 85% is possible. An optimized evaporation chamber and experimental setup could allow a higher humidity in the incubator, but requires further research on the ideal design, airflow rates and temperatures.

4   Conclusion

Using the methodical development according to VDI 2221, for a neutral view of the product through a list of requirements and functional analysis, the concept phase led to a morphological matrix. Many solutions were subsequently classified as unsuitable. The matrix was iteratively reduced until only 5 concepts based on evaporation or vaporization remained. For further insights, an experimental setup was then developed to analyze the principle of evaporation in the incubator. The evaporation principle was tested with an existing chamber in the incubator at 39°C; 85% humidity was achieved, but not the required 99%. This could be due to the worst case conditions in the non-optimized mechanical setup of the experiment. An optimization of the setup and process considering temperature, surface area and airflow could lead to a suitable humidification system based on evaporation. Further experiments are necessary. Therefore, a new setup for both evaporation and vaporization can be considered for the future.

Acknowledgement

The work was carried out at Drägerwerk AG & Co. KGaA and supervised by Professor Pilz from the Technical University of Applied Sciences, Lübeck. I would like to take this opportunity to thank my supervisor and all other employees of the company for their cooperation and support during the project.

Authors' Statement

The authors declare that there are no conflicts of interest in connection with the research work.

5   References

[1] Drägerwerk AG & Co. KGaA, *Alles, was ein Frühgeborenes für den optimalen Start braucht.* Drägerwerk AG & Co. KGaA, Lübeck, 2017.

[2] P. P. Andrew Lyon, *ThermoMonitoring A step forward in neonatal intensive care.* Dräger, Lübeck, 2015.

[3] S. Avenarius, *Wärmepflege bei Frühgeborenen.* Georg Thieme, Stuttgart, 2017.

[4] G. Sedin, K. Hammarlund, B. Strömberg, *Transepidermal water loss in newborn infants. VII. Relation to postnatal age in very pre-term and full-term appropriate for gestational age infants.* Acta Paediatr Scand.,1983.

[5] Drägerwerk AG & Co. KGaA, *A simulation training program for heat gain and loss of premature infants.* Dräger, Lübeck, 2015.

[6] B. B. K. Gericke *Pahl/Beitz Konstruktionslehre - Methoden und Anwendung erfolgreicher Produktentwicklung.* Springer Vieweg, Bochum, 2020.

[7] P. Naefe, *Methodisches Konstruieren - Auf den Punkt gebracht.* Springer Vieweg, Aachen, 2018.

Optimizing the Endoscope Bending Rubber Adhesive Process

Negar Sedaghatimonavar[1], Kevin Vaughan[2] and Stefan Müller[3]

[1] Biomedical Engineering, Lübeck University of Applied Science, negar.sedaghatimonavar@stud.th-luebeck.de
[2] Medical Repair Service, Olympus Winter & Ibe GmbH, kevin.vaughan@olympus.com
[3] Biomedical Engineering, Lübeck University of Applied Sciences, stefan.mueller@th-luebeck.de

Abstract

Internal company processes evolve and advance as technology does. Finding the inefficiencies in the process and continuous improvement are necessary for a successful business. This scientific paper investigates the optimization of the A-rubber gluing process on a flexible endoscope for the Olympus Medical Repair Service (MRS). After having identified A-rubber gluing as a major bottleneck in the repair process, the study analyses the impact of temperature and drying time on adhesive bonding. The service manual specification indicates room for improvement, particularly in the second coating adhesive drying step. Experimental findings suggest a reduction of second coat drying time by one-third, thus speeding up processing. Additionally, the study addresses non-uniform temperature distribution in oven chambers and proposes precise temperature control for consistent adhesive quality. Overall, the paper provides practical insights for improving efficiency and enhancing the quality of flexible endoscope repairs in the Europe, Middle East, and Africa (EMEA) region.

1 Introduction

The flexible endoscope has a bending part (Fig. 1) that is covered by a so-called A-rubber. This rubber must be sealed to the endoscope at two ends using adhesive. In Olympus MRS, A-rubber gluing is one of the three most problematic steps causing a high amount of rework and also is a time-consuming step in the whole process. Through this, the gluing step becomes a bottleneck in repair for the whole EMEA.

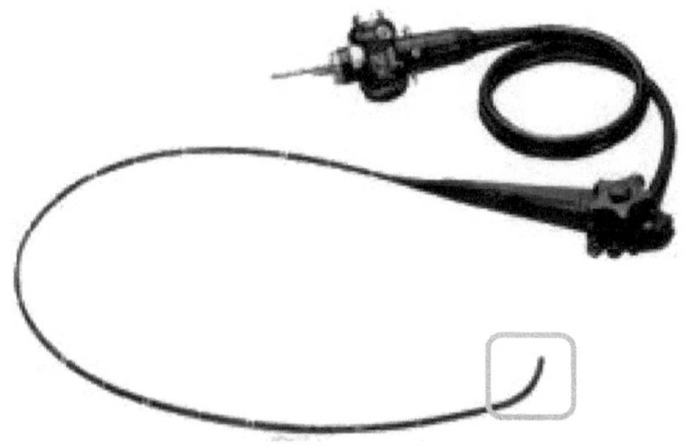

Figure 1: Flexible Endoscope with the bending part highlighted (courtesy of Olympus)

The process of gluing the A-rubber is done by a technician following the service manual which is a company-wide standard. Gluing in the current process involves four steps: preparing and applying the adhesive, curing the adhesive in the oven, and quality control.

The A-rubber is inserted around the bending part of the endoscope and the adhesive is applied in two steps. After applying the first coat of adhesive, the endoscope is placed in an oven for 30 minutes for curing. Then the second coat is applied. It ensures a secure bond and leads to the final appearance of the adhesive and its accurate diameter. Once the second coat of adhesive is applied, the endoscope is again placed in the oven. The service manual requires a minimum of 30 minutes for this step. At many sites, endoscopes are dried for 60 minutes in the oven.

This project aims to reduce the A-rubber gluing repair time in the Olympus MRS.

As a preliminary work of this study environmental variables with a significant impact were identified for each of the A-rubber gluing steps as so-called *impacting factors*. For example, humidity and temperature influence the adhesive bonding process and thus count as impacting factors [1]–[2]. After having excluded factors and steps of the whole repair process which were out of the scope of this study, *temperature* and *drying time* were left as impacting factors. These two factors affect each other as increasing temperature reduces drying time [3].

Comparing the service manual to the actual process shows that the second coating adhesive drying time differs from the specifications by 30 minutes. At the same time oven drying time is the most time-consuming step in the process. Therefore, there lies significant optimization potential in reducing drying time by increasing drying temperature.

Temperature control and uniform heating are two key performance indicators of the oven. Ovens used for adhesive drying should have precise temperature control capabilities and ensure uniform drying of the adhesive. Temperature variations can affect the drying process and potentially compromise the quality of the adhesive bond. [4]

In the Olympus service manual, the drying temperature range for this specific adhesive is 60–70 °C. However, some of the ovens in the repair section have an opening window which causes the temperature to drop in the chamber. In addition, the temperature is currently recorded only at one point in the oven (at the back wall of the oven chamber).

This study addresses these issues by measuring temperature uniformity and oven performance to find the minimum drying time and achieve consistent quality across the products.

2 Material and Methods

This study analysed by how much the drying time can be reduced. It started with measuring the oven characteristics first and verified that the adhesive is being dried at a temperature within range of the service manual specification. The temperature was then raised so that the adhesive was exposed to a temperature close to the upper boundary of the specification. This was to determine the so-called *optimal drying temperature*. Finally, the ovens of repair lines at different sites in EMEA were set to the optimal drying temperature to examine whether the drying time can be reduced.

The selected ovens were 3 models that can be considered essentially similar in terms of their primary features. This includes having an open window, a fan at the top, drying one endoscope at a time, and having rotational arms for rotating the endoscope while drying. These oven models were chosen because they are the most commonly used models in the repair lines. The temperature uniformity has never been checked on them. Calibrated thermocouples and ALMEMO® 710 data logger were used to record temperature.

2.1 Temperature Characteristics of the Ovens (Test 1)

Temperature sensors were installed at four points in the oven chamber. Fig. 2 illustrates the positions of the sensors installed for this test.

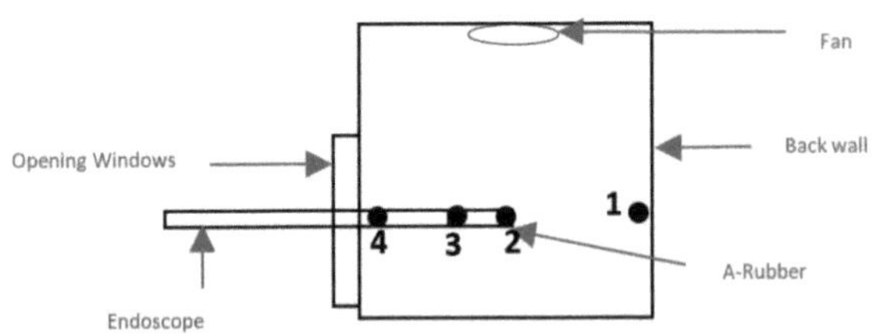

Figure 2: Side-view of an oven chamber and location of temperature sensors

There is a pre-installed sensor at position 1 used to monitor the oven temperature in the current MRS repair lines. An additional sensor was installed at the same spot as a control. To monitor the temperature at the exact location of the adhesives, a thermometer was installed at each positions, 2 and 3 in Fig. 2. This allowed insertion of the sensors into the oven along with the endoscope through the opening window.

The ovens were set by adjusting the temperature dial to 65 °C. This is the current operating procedure in repair lines. All sensor values were logged with a temperature data logger over 6 hours every 10 seconds.

2.2 Optimal Drying Temperature (Test 2)

The temperature of the oven was increased in steps until the maximum temperature at the potential position of an adhesive (positions 2 and 3 Fig. 2) neared but did not exceed the maximum allowed temperature of 70 °C. After having reached this temperature a new recording was started to determine the median temperature that is just as high as to not exceed 70 °C while the temperature in the oven chamber is oscillating. The resulting median temperature was then used as the *optimal drying temperature*.

2.3 Minimum Drying Time (Test 3)

The newfound optimal drying temperature was then used to determine if the drying time can be reduced. On a set of dummies, the new drying time was found. A dummy consisted of a cylindrical tube that closely matched the size and shape of the bending end of the highest diameter of the gastrointestinal (GI) endoscope with an A-rubber attached to it and the application of an adhesive on both sides.

By increasing the drying time in five-minute steps each dummy was exposed for a different duration. First starting at the minimum allowed drying time of 30 minutes and each dummy was quality checked afterwards. On the first sample at which the adhesive bond passed the quality check the time was recorded as the *minimum second coat adhesive drying time*.

After cleaning the adhesive with alcohol, the quality was examined to ensure the bonding. This quality check consisted of a visual inspection of the color and the texture of the adhesive bond. A dull surface after cleaning with alcohol is an indicator of a not thoroughly cured adhesive. Those checks resemble the quality inspection in the MRS repair lines.

After testing on dummies, the minimum second coat adhesive drying time with the optimal temperature was verified under realistic repair environment conditions. Tests were carried out on real endoscopes in the real repair lines across the different EMEA sites to account for variations in environment parameters, equipment, and other factors. The tests were done on the thickest diameter model of the gastrointestinal endoscope.

The first step was to adjust the oven temperature in a way that the condition of having 67 °C at the two adhesives is true but at the same time not letting the maximum step over the 70 °C maximum. After the adhesive was applied to the endoscope, the endoscope was placed into the oven and the drying time was set to 40 minutes. During this time the temperature was recorded at the location of the two adhesives and at the back wall where the temperature is being recorded to show the difference between them.

At some sites, the endoscopes were taken out of the oven after 5 minutes to do the final correction. This 5 minutes was also included in the 40 minutes. After the 40-minute drying cycle was completed, the endoscope had to cool down at room temperature for 5 minutes and then had to be cleaned with alcohol of 80–90 % concentration. After that, the quality of the adhesive bond, its uniformity, and visual defect were carefully assessed. The quality parameters of the adhesive such as color and texture were examined after each drying test.

The sample size was determined using the Bayes Success-Run Theorem [5].

$$n = \frac{\ln\left(1 - C\right)}{\ln R} \tag{1}$$

Equation 1 yields that $n = 59$ defect free samples are required with a probability of success $R = 0.95$ and confidence level $C = 0.95$. For the test a total sample size of 65 was evenly assigned to each of the 5 repair sites in EMEA. A probability of success of

$$R = \sqrt[n]{1 - C} = 0.955 \tag{2}$$

was expected with a confidence $C = 0.95$ if all of $n = 65$ are successful. Every site was instructed to choose its 15 endoscopes randomly.

3 Results and Discussion

Results from monitoring temperature, finding optimal temperature, and minimum drying time are presented and discussed below.

3.1 Temperature Characteristics of the Ovens (Test 1)

Data collected from five thermometers are shown in Fig. 3 as a box and whisker plot. All values except the minimum temperature at sensor location 4 $t_{4,min} = 59,80$ – close to the opening window – stay within the service manual specification when the temperature is set to the usual 65 °C. The temperature spread at each location in Fig. 3 is caused by the heat cycling behavior of the oven models.

The results show that the temperature is not uniformly distributed in the oven chamber. Sensor 4 near the opening window consistently experiences the lowest temperature values whereas sensor 1 (at the back) experiences the highest. Furthermore the temperature at the adhesive joints, locations 2 and 3 in Fig. 2, experience a median temperature lower than the temperature the oven is adjusted to. Also, the maximum temperature readings at all of the sensor locations range below the allowed maximum of 70 °C specified by the service manual. This indicates that by increasing the set temperature the drying time can be reduced.

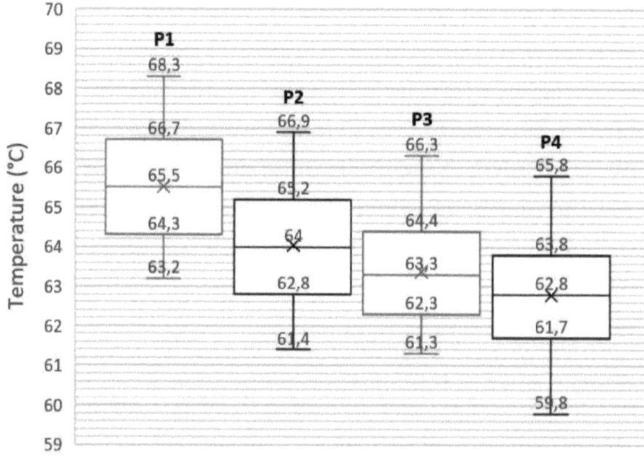

Figure 3: Oven temperature at the four locations with temperature set to 65 °C

3.2 Optimal Drying Temperature (Test 2)

The temperature at each 5 points had an overall upward shift compared to Fig. 3. The result indicated that the optimal drying temperature for adhesive is around 67 °C. Temperatures above 67 °C should be avoided, because then the temperature at the adhesive joints surpasses the recommended maximum limit.

3.3 Minimum Drying Time (Test 3)

Tests on the dummies indicate that 40 minutes is sufficient for drying the adhesive. According to the calculations based on the Bayes Success-Run Theorem in (Equation 1).

Table 1: Samples and Acceptance Ratio of Dried A-rubber on Endoscopes

Sites	No.Samples	Dried & Quality accepted	Acceptance Ratio
Site 1	5	4	80 %
Site 2	15	15	100 %
Site 3	15	0	0 %
Site 4	15	15	100 %
Site 5	15	15	100 %

Table 1 summarizes the result of the quality check of the endoscopes after 40 minutes of drying. 40 minutes is the minimum drying time without compromising the quality of the adhesive. The results indicate that in three out of the five sites, a drying time of 40 minutes is sufficient to achieve the desired outcome. Here all samples pass the quality check because the surface of the adhesive did not become dull after cleaning with alcohol.

At site 3 the 40 minute drying results in failure, but the site assures that by letting the endoscopes rest in the oven for another 5 minutes, the quality is acceptable. Site 1 could not complete the test on 15 samples and only 5 samples were tested but 40 minutes of drying time is sufficient for 4 samples out of 5.

The expected probability of success of $R = 0.955$ in Equation 2 lies in stark contrast to the overall success rate of 75 % as shown in Table 2. On the other hand the low acceptance ratio mostly concentrates around Site 3 where all of the tests failed. Notably the temperature readings at this site did not show the same characteristics as at other sites or in the preliminary tests on the dummies, suggesting there might be a site or process specific issue. Without taking Site 3 into account the size of $n = 49$ samples without error still yields a probability of success of $R = 0.94$ with a confidence of $C = 0.95$ analog to Equation 2.

Table 2: Median and Maximum Temperature of the Tests Conducted by the Author at Site 4

Sensor	1	2	3
Median	66.1	65.0	65.8
Max.	70.0	68.9	69.98

As an example Table 2 shows the temperature readings from the tests conducted at site 4. Even though the maximum temperature at the glue position reached closely to 70 °C, the median value did not reach the 67 °C determined to be the optimal drying temperature from the previous tests, even though the same oven model was used. Still the drying time of 40 minutes is acceptable at site 4.

4 Conclusion

Longer drying time in repairing endoscopes results in higher work-in-progress and higher costs. Increasing the temperature usually decreases the drying time, but the compatibility of the adhesive must be considered. The Olympus MRS service manual already defines a lower boundary of 30 minutes, however in many sites, the adhesive is dried for 60 minutes.

Without compromising the quality of the adhesive bond the drying time can be safely reduced to 45 minutes. If site 3 can be regarded as an outlier, the drying time can be even reduced further. Thus the duration of this step can be reduced by one-third. Even if there were no current efforts to also address shortcomings in the repair section, a reduction of this most time-consuming step leads to significantly faster feedback in case a redo of the adhesive bond is necessary. This reduction is common across EMEA based on the assumption that the repair process is standardized across all sites in that region.

However, it cannot be ruled out that there lies further optimization potential at specific sites: The performance analysis of the ovens compared with the results from Table 2 already shows different characteristics even across ovens of the same model. Pointing out the non-uniform temperature distribution in the oven chamber yields a new approach to measuring current optimization efforts. Apart from the reduction in time, quality improvement regarding the curing of the adhesive bond can be achieved by measuring the temperature at the exact location of the adhesive and ensuring it lies in the upper range of the allowed temperature spectrum.

Acknowledgement

The work has been carried out at Olympus Winter & Ibe GmbH and supervised by Dr. Müller at Technische Hochschule Lübeck.

Authors' Statement

Conflict of interest: Authors state no conflict of interest.

5 References

[1] A. Gursel, *Fundamentals in Adhesive Bonding Design for Complex Structures and Conditions*. In: Journal of Advanced Technology Sciences, vol. 8, Düzce Üniversitesi, Düzce, pp. 1–10, 2019.

[2] B. Duncan, S. Abbott, R. Court, R. Roberts and D. Leatherdale, *A Review of Adhesive Bonding Assembly Processes and Measurement Methods*, 2003, NPL, Teddington.

[3] S. Ebnesajjad and A. H. Landrock, *Adhesive Applications and Bonding Processes*. Adhesives Technology Handbook (Third Edition), Boston, 2015.

[4] M. Banea and L. F. M. Silva, *The Effect of Temperature on the Mechanical Properties of Adhesives for the Automotive Industry*. In: Proceedings of The Institution of Mechanical Engineers Part L-journal of Materials-design and Applications, vol. 224, no. 4, 2010.

[5] M. Durivage, *Sample Sizes: How Many Do I Need?*. Available: https://qscompliance.com/wp-content/uploads/2017/10/Sample-Sizes-How-Many-Do-I-Need.pdf [last accessed on 2024-01-05].

Development of a Microcontroller-Based Test Device for Air Temperature Sensor Probe of a Neonatal Incubator

Sherbeen Vadakken [1], Igor Knor [2], Luca Belloi [2], Ulf Pilz [3]

[1] Biomedical Engineering, Lübeck University of Applied Sciences , sherbeen.vadakken@stud.th-luebeck.de

[2] LMT Medical Systems GmbH, {knor, belloi}@lmt-medicalsystems.com

[3] Department of Applied Natural Sciences, Lübeck University of Applied Sciences, ulf.pilz@th-luebeck.de

Abstract

Neonatal incubators are vital for infants and premature babies, providing a precisely controlled thermal environment to effectively protect and regulate their body temperature [1]. This paper details the development of a test device for the air temperature sensor probes in the neonatal incubator nomag®IC Advanced from the requirements stage to testing a functional prototype. Additionally, this paper discusses the importance of post-manufacturing sensor probes verification. The device allows visualization of current temperatures in degrees Celsius, the tolerance threshold based on the negative temperature coefficient (NTC) manufacturer's datasheet and the temperature difference between two NTC sensors. Furthermore, it indicates the success or failure of the probe verification according to the internally defined tolerance threshold. The developed device was tested on 29 sensor probes, demonstrating high reliability in determining their production quality, offering optimism for the implementation of the device in routine quality control processes.

1 Introduction

The nomag®IC Advanced is a magnetic resonance (MR) compatible transport incubator produced by LMT Medical Systems GmbH for seamless examination of premature and neonatal infants in intensive care with the non-invasive imaging technique of magnetic resonance imaging. Neonatal incubators are developed to offer a thermoneutral environment for neonates and premature infants who are unable to regulate body temperature on their own [1]. They are vulnerable to both hypothermia (dangerously below normal body temperature) and hyperthermia (overheating) due to their unstable body temperatures [2]. Therefore, a neonatal incubator should ensure that the infant's body core temperature remains within a safe range, typically defined as between 35°C and 37°C , and that the patient does not overheat, typically considered to be above 40°C [3].

The air temperature of the patient cabin in the nomag®IC Advanced is measured using a sensor probe consisting of two identical GA2.2K3A1 NTCs manufactured by TE Connectivity (Fig.1). An NTC thermistor is a type of temperature sensor that shows a decrease in electrical resistance as temperature increases [4]. The user interface of the nomag®IC Advanced displays the temperature measured by the NTC1. To control the measurement of NTC1 and verify that its value corresponds to the current air temperature, a second NTC of the same type is positioned next to NTC1. As both NTCs are placed next to each other on the same airflow line within the incubator, this allows them to monitor the measurements of each other. The incubator is designed to trigger an alarm state if the temperature measurement be-

tween the two NTCs differs by more than 1K.

Sensor probes manufacturing, including soldering of NTC thermistors, is done manually by a supplier. The NTC thermistors are highly sensitive components that can be easily damaged during production due to improper soldering [4]. In this case, if one of the NTCs was not soldered correctly, e.g. causing a short circuit, the incubator will go into alarm state. However, connecting the sensor probe to the incubator system during the production can take place weeks after receiving the batch of sensor probes. Hence, the main aim of this work was to create an in-house test device to control the air temperature sensor after its manufacturing. The test device must display the current air temperature in Celsius, the temperature difference between the two NTCs

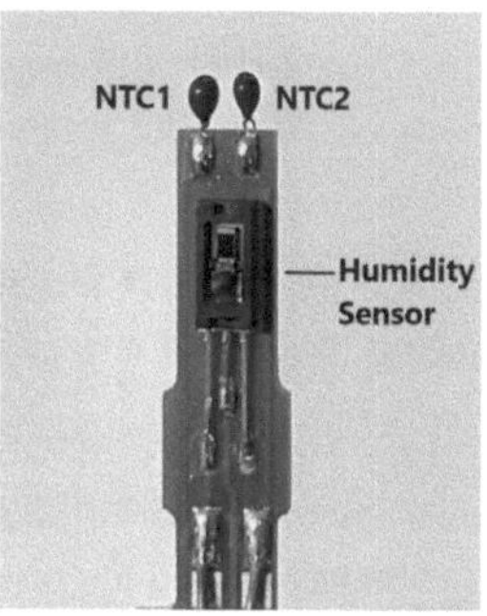

Figure 1: Air temperature sensor probe of the neonatal incubator nomag®IC Advanced without the protective housing. Using an additional NTC allows to control the measurement of the primary NTC, thus increasing patient safety.

and the acceptable deviation threshold for this difference. Additionally, the maximum temperature difference between the two NTCs must be displayed during testing to help document the results. Furthermore, an environment appropriate to the incubator working conditions must be provided. The second aim of this work was to investigate whether manual soldering, as opposed to automated methods, could have an effect on the final temperature measurement.

2 Material and Methods

This following section begins with the theoretical background explaining the concepts of development. It is followed by the setup of the two developed prototypes, including the initial breadboard design (Prototype 1) and its subsesquent refined version made using the printed circuit board (PCB). Finally, the testing of the second prototype according to the specified requirements is described.

2.1 Theoretical Background

The simplified principle of measuring the air temperature in the incubator cabin is explained as follows. As mentioned above, the resistance value of the NTCs changes according to the temperature value. The signal from each NTC goes to the analog-to-digital converter on the incubator's mainboard, where the microcontroller converts the resistance value to a temperature value in Celsius/Fahrenheit, including necessary steps such as filtering and calibrating the values. The final temperature value is then sent to the incubator user interface. A similar measurement scheme was chosen for development of the test device. The air temperature sensor probe should be connected to the test device and the following data should be displayed to the user to identify the probe quality:

- air temperature reading of the first NTC (NTC1) in degree Celsius;

- air temperature reading of the second NTC (NTC2) in degree Celsius;

- difference in temperature values between the two NTCs in kelvin;

- the maximum acceptable difference between two NTCs in kelvin based on the manufacturer's specifications;

- allowed tolerance for the NTCs defined by the company, considering the manufacturer's specifications in kelvin.

In addition, the user should be informed in an automated way whether the probe has passed the test or not.
The temperature readings measured by the NTC thermistors was calculated by using the beta parameter Equation

$$ T = \frac{1}{\frac{1}{T_0} + \frac{1}{\beta} \ln\left(\frac{R}{R_0}\right)}, \qquad (1) $$

where T is the temperature in kelvin for the temperature at that instant, T_0 is the reference temperature in kelvin, R is the resistance at temperature T, R_0 is the resistance at reference temperature T_0 and β is the beta value of a thermistor available in the manufacturer's datasheet.
The calculations of the resistance tolerance were made using a simplified equation derived from the Beta exponential model by calculating its partial derivatives and is given by

$$ \frac{\Delta R}{R} = |0.25\%| + \left| 0.5\% \cdot 3976 \cdot \left(\frac{1}{T} - \frac{1}{T_0}\right) \right|, \qquad (2) $$

where $\frac{\Delta R}{R}$ represents the maximum resistance tolerance in % at temperature T, with 0.25% as the rated tolerance of resistance value at 25°C , 0.5% as the tolerance on beta value 25/85 and 3976 as the β value from datasheet [5].

2.2 Prototype 1

For processing operations and taking signals from air temperature sensor probes, an STM32 Nucleo-64 development board with STM32F103RB MCU was chosen. This development board offers sufficient functionality to perform the tasks described above for this work.
The electrical circuit of the test device was designed, as shown in Fig.2. According to this design, two NTC thermistors with a nominal resistance of $R = 2.252k\,\Omega$ at 25°C were connected to the ADC pins of the microcontroller via series resistors of $R = 2.2k\,\Omega$. The value for these resistors was selected as close as possible to the nominal resistance of the thermistors for better sensitivity and linearity [6]. The analog signal which is received at the ADC pin of the microcontroller is converted to a readable output voltage. The real input voltage was measured with a multimeter and showed 3.17 V, which differs from the 3.3 V indicated on the development board due to a voltage drop. The obtained output voltage was then converted to resistance using the voltage divider equation

$$ R_{\text{NTC}} = \frac{R_{\text{ref}} \cdot V_{\text{out}}}{V_{\text{in}} - V_{\text{out}}}, \qquad (3) $$

where R_{NTC} represents the NTC resistance, R_{ref} is the reference resistance with a value of $R = 2.2\,k\Omega$, V_{in} is the input

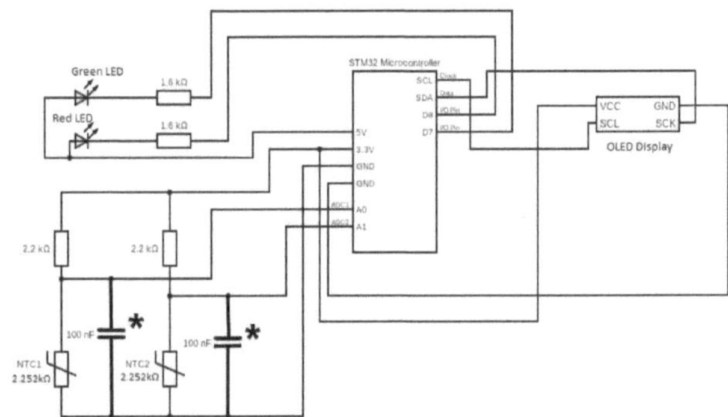

Figure 2: Circuit diagram of prototype 1 and 2. The asterisk indicates the added shunt capacitors in prototype 2 for noise filtering.

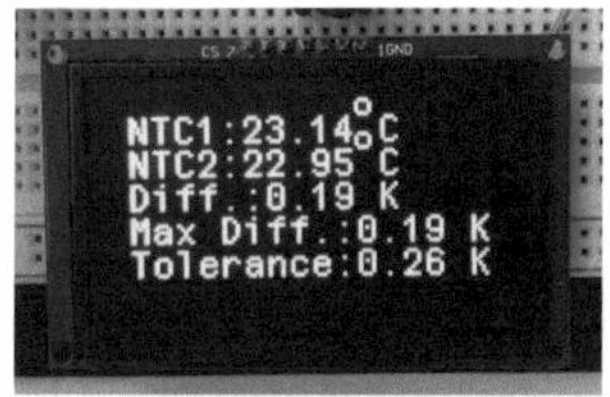

Figure 3: Prototype 1, showing the specified values during the functional test.

voltage of $3.17V$ and V_{out} is the output voltage. Solving this equation gives the value of the unknown resistance.

After that, the NTC resistance value was converted to a temperature value in Celsius using (1), modified in order to display the output values in Celsius, and is given by

$$T = \frac{T_0 \cdot \beta}{T_0 \cdot \left(\ln \left(\frac{R_{\text{NTC}}}{R_{\text{ref}}} \right) + \beta - 273.15 \right)}, \qquad (4)$$

where T_0 is the reference temperature in kelvin at 25 °C, which is 298.15 K, 273.15 is the conversion factor from kelvin to degree Celsius and β is 3976 for the GA2.2K3A1 Series I BetaTherm thermistors.

The resistance tolerance for NTC1 was calculated using (2) based on the current air temperature. The resulting value was doubled, as in a worst-case scenario, NTC1 could have the minimum tolerance value and NTC2 the maximum, which was used as the test pass criterion to determine the quality of the sensor probes. The maximum temperature difference between the two NTCs during the measurement procedure was also calculated for future completion of the test report.

To display the values required for sensor probe verification, a 2.4-inch 128x64 OLED display from DollaTek was used. This display enables the use of open access software libraries and support the Inter-Integrated Circuit (I2C) data exchange protocol. The I2C protocol uses the Serial Data (SDA) and Serial Clock Line (SCL) signals to send signals to the display module. The final output on the display, based on the requirements described in Chapter 2.1, is shown in Fig.3. For displaying values the microcontroller reads and sends data every 2 seconds.

As shown in Fig.2, in order to inform the user and indicate if a probe passed the test or not, it was decided to use two light emitting diodes (LEDs) of green and red light, for successful and unsuccessful result respectively. The LEDs were connected via current limiting resistors of $R = 1.6k\,\Omega$ each to the microcontroller. Activation of one of the two LEDs was based on the acceptable tolerance criterion described above.

To connect the circuit elements of the test device, initially a TRU COMPONENTS EID-102B breadbord was chosen, which has a sufficient size to arrange the electronic components, including OLED display. To verify the required functionality, the NTC probe was positioned in an environment with stable air temperature using a closed cardboard box to ensure that the measurement was not influenced by parasitic parameters (e.g., air flow, sunlight, direct light, etc.).

A reference probe, a calibrated AMR dual NTC sensor fitted with a ZA 9040-FS2 ALMEMO® connector was used for the verification of the test. ALMEMO 2390-3 is a calibrated measuring instrument and was used to visualise the reference probe values. During testing, it was observed that loose wires in the breadboard caused circuit instability and additional noise at the ADC channel's receiving end, resulting in erratic values. However, this problem should be eliminated when creating the PCB prototype.

2.3 Prototype 2

Compared to prototype 1, the following changes were made for the new prototype. Bypass capacitors of 100 nF were placed between the output voltage signal from the NTCs and ground to filter out interference and improve the desired device behaviour (see Fig.2). This prototype was also programmed to take the median of every 10 ADC values that were measured to improve accuracy.

To transfer the circuit diagram to PCB, Altium Designer software was used. The required electronic components were assembled and soldered manually to create the printed circuit board assembly (PCBA). The STM32 microcontroller was positioned beneath the PCBA, forming an integrated assembly as shown in Fig.4. For user convenience, an ON/OFF switch connected to the microcontroller's power supply jumper was integrated into the device. The PCBA was cut accordingly to expose the microcontroller's reset button, facilitating easy access in case of system freezing.

The calibration procedure for the output temperature values was conducted similar to the calibration procedure in the incubator. The prototype 2 was calibrated using a $5\,\text{k}\Omega$ trimmer potentiometer by setting it to different values and observing the displayed temperature readings. This calibrated setup recorded temperature values displayed by the device across nine measurements within the functional range of $20\,°C$ to $40\,°C$. When plotted, the results exhibited a deviation ($y = 0.9927x + 1.5133$) from the theoretically expected values ($y = x$). To align with theoretical expectations, the correction factor was programmed into the system, ensuring results closer to the anticipated values. To verify the required functionality, the prototype 2 was tested identically to the prototype 1. The functionality test showed no deviations. Therefore, it was decided to test the available sensor probes before proceeding to the stage of enclosing the test device in a housing.

Figure 4: Prototype 2, showing the PCBA with integrated circuit mounted on the STM32 development board.

2.4 Testing the setup

As test samples, 29 air temperature sensor probes were used. Prior to enclosing the PCBA prototype in its designated housing and authorising it for production, a testing trial was conducted to identify any potential anomalies, bugs, or deviations from the specifications. The calibrated PCBA prototype was subjected to testing according to the functional test for prototype 1 described in Chapter 2.2. The measurements were taken in the temperature range of 20°C to 21 °C. Each sensor was monitored for a minimum set duration of 20 seconds before recording the values. All values were written in a prepared Microsoft Excel table for further evaluation of the test results.

3 Results and Discussion

Testing of the prototype 2 showed the following results (see Table 1). All 29 sensor probes met the evaluation criteria, demonstrating a maximum temperature deviation of 0.17 K, well within the defined average resistance tolerance of 0.30 K. The average temperature difference between NTC1 and the reference sensor was 0.12 K, confirming alignment with the calibrated ALMEMO sensor. Similarly, NTC2 displayed minimal deviation with an average temperature difference of 0.09 K. The average temperature differance of 0.05 K showed the close correlation between the NTCs. This showed that even after the sensor probe was manufactured, the NTCs still gave correct data according to their datasheet. However, before releasing the device to production, a proper housing should be developed to avoid electrostatic discharge (ESD) problems during use and damage to individual PCBA elements, which could result in erratic values. Also, a usability test should be done with the people in the production to see if they will be able to use the device or not.

Manual soldering introduces the possibility of variations in soldering quality, such as differences in solder joint consistency, potential solder bridges, or the application of excessive heat during the soldering process. However, the observed temperature deviations between the NTCs were minor, suggesting that while manual soldering does have some impact on temperature readings, it is not substantial. This

issue can be investigated in more detail when more tests are performed as part of the sensor control process.

4 Conclusion

The sensor test device, currently in its PCBA model without housing, has been successfully developed to assess the functionality of the air temperature sensor probes employed in the nomag®IC Advanced incubator. This test device demonstrated the capability to identify NTCs deviating from the tolerance range and to detect functional problems or damages during production. It must be noted that the tolerance of reference resistances was not taken into account in determining the boundary of allowed tolerance, resulting in a more stringent test. This decision on the strictness is planned before the device is released.

The current PCBA prototype without housing, is slated for future modification. The modifications will include testing of humidity sensor modules in incubator probes and the development of an ESD protective housing by the company's production department.

Acknowledgement

The work has been carried out at the research and development department at LMT Medical Systems GmbH and supervised by the Department of Applied Natural Sciences, Lübeck University of Applied Sciences.

Authors' Statement

Conflict of interest: Authors state no conflict of interest.

5 References

[1] J. de Júnior, J. M. P. de Menezes Júnior, A. A. de Albuquerque, O. Almeida, F. M. de Araújo, *Assessment and Certification of Neonatal Incubator Sensors through an Inferential Neural Network*. Sensors, 2013.

[2] S. A. Ringer, *Core Concepts: Thermoregulation in the Newborn, Part II:Prevention of Aberrant Body Temperature*. NeoReviews, vol.14, no.5, 2012.

[3] 60601-2-20, *Medical Electrical Equipment-Part 2-20: Particular requirements for the basic safety and essential performance of infant transport incubators*. International Electrotechnical Commission, 2020.

[4] R. K. Kamat and G. M. Naik, *Thermistors-in search of new applications, manufacturers cultivate advanced NTC techniques*. Sensor Review, vol. 22, no.4, 2002.

[5] TDK, *NTC Thermistors*. EPCOS AG, 2018.

[6] B. C. Baker, *Thermistors in Single Supply Temperature Sensing Circuits*. Microchip Technology Inc., pp.5, 1999.

Table 1: Test results of 29 air temperature sensor probes. Here Δ NTC1-2 is the average temperature difference between the NTCs, Δmax NTC1-2 is the maximum temperature difference between the NTCs across all measured sensors, Δavg NTC1-Ref and Δavg NTC2-Ref is the average temperature difference between the corresponding NTCs and the reference sensor and Δavg R tolerance is the average of maximum accepted tolerance between NTCs (see Chapter 2.2).

Δavg NTC1-2	Δmax NTC1-2	Δavg R tolerance	Δavg NTC1-Ref	Δavg NTC2-Ref
0.05 K	0.17 K	0.30 K	0.12 K	0.09 K

7

Machine Learning

A Comparison of Imputation Techniques to Improve Lung Cancer Survival Estimation

Nina Cassandra Wiegers[1], Sebastian Germer[2], Christiane Rudolph[3], Katharina Rausch[4], Natalie Rath[4], Heinz Handels[2, 5]

[1]Medical Informatics, University of Lübeck, nina.wiegers@student.uni-luebeck.de

[2]German Research Center for Artifical Intelligence (DFKI), Lübeck, {sebastian.germer, heinz.handels}@dfki.de

[3]Institute for Cancer Epidemiology, University of Lübeck, christiane.rudolph@uksh.de

[4]Saarland Cancer Registry, {k.rausch, n.rath}@soziales.saarland.de

[5]Institute of Medical Informatics, University of Lübeck

Abstract

Survival analysis in oncology is important, e.g. for tracking and comparing the success of therapeutic regimens. Cancer registry data include sociodemographic information, tumor histology and progression information, and overall patient survival time, but tend to have a high percentage of missing values in some variables, which complicates data analysis and model fitting. This study investigates the impact of data imputation on survival prediction for lung cancer patients using data from the Schleswig-Holstein Cancer Registry. We use three methods to impute the data: Simple Imputer, Multivariate Imputation by Chained Equations (MICE) and Generation of Realistic Tabular Data (GReaT). We then estimate patients survival in a classification task with above and below one year. GReaT outperforms MICE, showing an increase in f1-score of 3%. Despite not imputing every missing value, GReaT's higher performance suggests, that imputing all missing values, as MICE does, may lead to misleading results.

1 Introduction

Cancer registries record the course of disease for cancer patients. Through the analysis of these data, statements about the mortality and survival period can be made. Additionally, different treatments in relation to age, health condition and other relevant attributes can be analyzed with the data from the cancer registry. Unfortunately, these registries contain missing values, which are a result of incomplete reporting by clinical staff to the cancer registry. This can lead to biased results in statistical analyses [1]. Here the imputation methods come into play. Through imputation, missing values are eventually predicted based on a given data distribution. In the following three different imputation methods, Simple Imputer, Multivariate Imputation by Chained Equations (MICE) and the large language model (LLM) approach Generation of Realistic Tabular Data (GReaT), will be evaluated. Imputation is a challenging task, because no ground truth is available. The aim of this paper is to analyze whether the prognosis of the survival time can be more accurate on an imputed dataset. The evaluation of imputation methods will be conducted on a cancer registry dataset from Schleswig-Holstein containing patients diagnosed with lung cancer. This is especially important, because lung cancer is often diagnosed in a late stage of the disease. In improving the estimation of the survival time of these patients, the aim is to provide suitable therapies based on the measured survival time at the time of the first diagnosis.

2 Material and Methods

2.1 Dataset

The dataset consists of tabular data related to the diagnosis of patients with lung cancer. There are 11,641 cancer patients of which, 6,820 are male and 4,821 are female. Patients range in age from 21 to 97 years, with an average age of approximately 69 years. Most of the data is categorical data, but there are some attributes with numerical and date time type. The dataset includes the TNM-classification, information to the histology of the tumor and the health condition of the patient. To evaluate the performance of the imputation methods, attributes with no missing values are needed. Due to their completeness and their representation of numerical and categorical data, the patient's *age* and *histology groups* are selected as primary attributes. Randomly selected values of both attributes will be deleted and subsequently imputed. The deleted values serve as a ground truth to compare the estimated values of the imputation methods against. As training data will the other incomplete attributes be of assistance, so that the imputation is based on the data distribution and correlations between the attributes of the whole dataset. The amount of deleted values will also be chosen as 10%, 20%, 30% and 40% to test the influence of the amount of missing values and the robustness of the chosen imputation method. In Fig. 1 the distribution of the missing values per patient is shown. Most of the patients have 18 or 19 missing values, which correspond to a third of all attributes. Every patient has at least four and at most

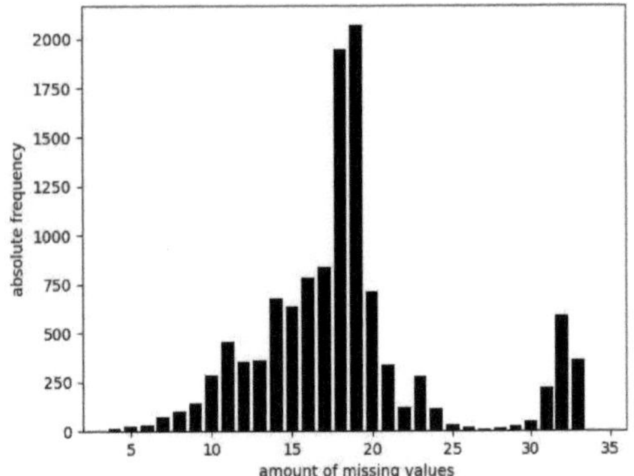

Figure 1: Distribution of missing values per patient

34 missing values, while overall 81% have between 19% and 39% missing values. This amount of missing data per patient is high enough to make an imputation worthwhile, whilst not being close to the limit of 40%. If this limit is exceeded, the error of the unknown imputed values can lead to high bias in the data, which may not be valid [2].

2.2 Imputation Methods

As a baseline method for comparing the other imputation methods in performance, a simple method called Simple Imputer is chosen. It replaces every missing value with the median of the original data distribution for the attribute *age*. For the categorical attribute, all missing values will be replaced with the *most frequent* category. The MICE algorithm is an iterative multiple imputation algorithm. It models each missing value of each attribute as a function of the other attributes. Depending on the type of the missing values, the MICE algorithm selects either a classifier for categorical values or a regressor for numerical values. It is based on the scheduling method *Round Robin*, which is a scheduling method for competing processes. For each column y of the dataset, the respective classifier or regressor is fitted based on the other columns X as input and y as output. After the fitting on X, the estimator predicts every missing value in y. The estimated results of the already imputed columns serve as new training data for the next iteration and the missing values of the selected attribute will be predicted again. The algorithm repeats this process for every column y, until the imputed values converge. The results of the last predicted output serve as the imputation result [3]. GReaT uses a large language model (LLM), which is based on a pre-trained transformer model. After fine-tuning on a dataset, the model can be used to synthesize new data similar to the training dataset. The imputation is based on the acquired ability of the LLM to model the data distribution of the data set. Afterward, it can predict the missing values based on the learned correlation between the attributes. As LLMs require text as input, each row of the tabular data is transformed into a string with the pattern "attribute-is-value" for each attribute-value pair inside the row. Now each patient is modeled by a sentence, which contains these attribute-value pairs. Because there is no dependency on

the attribute order in tabular data, the attribute-value pairs in each row are also shuffled [4].

2.3 Evaluation Metrics

To evaluate the imputation on the numerical attribute *age* and the categorical attribute *histology groups*, we choose different error measurements. The *error quotient* (EQ) will be evaluated for both attributes and is defined as the amount of wrong imputations divided by the amount of missing values (1).

$$EQ = \frac{Amount\ of\ wrong\ imputations}{Amount\ of\ missing\ values} \qquad (1)$$

For the attribute *age* the *mean absolute error* (MAE) (ref. 2), the *mean squared error* (MSE) (3) and the *normalized root mean squared error* (NRMSE) will be calculated (4). The *error quotient* for this attribute is calculated by rounding the estimated value and comparing it to the ground truth, which are the artificial-deleted values.

$$MAE = \frac{1}{n}\sum_{i=1}^{n}|\hat{y}_i - y_i| \qquad (2)$$

$$MSE = \frac{1}{n}\sum_{i=1}^{n}(\hat{y}_i - y_i)^2 \qquad (3)$$

$$NRMSE = \frac{1}{\sigma_y}\sqrt{\frac{1}{n}\sum_{i=1}^{n}(\hat{y}_i - y_i)^2} \qquad (4)$$

In all three of the equations above, n describes the amount of samples, $\hat{y}$ the predicted value and y the ground truth of the missing value. i symbolizes an index for all missing values and σ is the standard deviation of the ground truth y. As classification metrics for the categorical attribute *histology groups* the *balanced accuracy* is chosen.

$$acc_{bal} = \frac{1}{n}\sum_{i=0}^{n}recall_i \qquad (5)$$

$$recall = \frac{TP}{TP + FN} \qquad (6)$$

The *balanced accuracy*, as shown in (5), is the average of the recall (6) for each category. For the survival time analysis the *f1-score* (7) of both classes *under one* and *over one* year survival time are evaluated. *TP* stands for True Positives, *FP* represents the False Positives, *TN* includes only the True Negatives and *FN* denotes the False Negatives. In addition, the *accuracy* (9) of the survival time classification will be calculated.

$$f1\text{-}score = 2 \cdot \frac{precision \cdot recall}{precision + recall} \qquad (7)$$

$$precision = \frac{TP}{TP + FP} \qquad (8)$$

$$accuracy = \frac{TP + TN}{TP + TN + FP + FN} \qquad (9)$$

2.4   Preprocessing

All results are cross validated on five different splits and averaged over these. Also, the data set is pre-processed for MICE. All attributes with the datatype *categorical* and *string* are transformed by an *ordinal encoder* to integer values. Additionally, all date time attributes are scaled with the help of a *MinMaxScaler* in the value range between 0 and 1. As GReaT has its own transformation by converting every patient row into a string, the whole tabular data can be transformed at once. The survival time of the patients with the diagnosis of lung cancer is estimated through one binary classification. Therefore, a binary classifier was implemented, which tries to separate patients who survived less than one year from those who survived more than one year. This separation was selected, because the median survival time in days corresponds to 330 days for this dataset. Because the imputation methods MICE and GReaT performed best on the complete attributes *age* and *histology groups*, they are chosen to impute the whole dataset as a preprocessing step for fitting the classificator *Light Gradient Boosting Machine* (LGBM) ([5]) on the dataset. LGBM was selected, because it can handle data with missing values. As GReaT does not impute every missing value, it was additionally combined with MICE, so that MICE imputes the remaining missing values of GReaT.

3   Results and Discussion

3.1   Evaluation of Imputation Methods

In the following, the results of the attribute *age* and *histology groups* will be shown to get an overview of the imputation method's performance. The aim of this experiment is to select the best imputation methods for the preprocessing of the dataset for the survival time classification based on the results of the two complete attributes. Table 1 shows the error metrics results for the Simple Imputer. The *error quotient* is around approx. 97%. The other metrics increase proportional to the amount of deleted values. For the imputation of the MICE algorithm, the best performance was achieved using the *Random Forest* classifier and the *HistGradientBoost* regressor. The error for every regression metric increases for the attribute *age* as the amount of deleted values increases. The *error quotient* drops by max. 42% in comparison to the Simple Imputer (see Table 2). GReaT also reduces the regression metrics for the numerical attribute (ref. Table 3). The error quotient can be cut down by approx. 60% in comparison to the Simple Imputer and by 18% compared to MICE. For the categorical attribute, the results for all three imputation methods are summarized in Figure 2. The *balanced accuracy* can be improved by approx. 28% with MICE and by approx. 76% with GReaT compared to the Simple Imputer. Compared to the results of the numerical attribute, the error metrics do not increase with the amount of deleted values. This is because, the categorical attribute is highly imbalanced.

Table 1: Results of Simple Imputation for attribute *age*

amount of deleted values	EQ	MAE	MSE	NRMSE
10 %	96.56 %	0.8023	9.8296	0.0455
20 %	96.67 %	1.5834	18.9846	0.0632
30 %	96.44 %	2.3750	28.3858	0.0773
40 %	96.47 %	3.1719	38.0282	0.0894

Table 2: Results for MICE for attribute *age*

amount of deleted values	EQ	MAE	MSE	NRMSE
10 %	54.28 %	0.0967	0.4319	0.0095
20 %	55.91 %	0.2061	0.9468	0.0422
30 %	59.09 %	0.3086	1.5026	0.0177
40 %	61.96 %	0.4917	2.2707	0.0218

Table 3: Results for GReaT Imputation for attribute *age*

amount of deleted values	EQ	MAE	MSE	NRMSE
10 %	36.49 %	0.0403	0.0835	0.0041
20 %	35.72 %	0.0762	0.1158	0.0048
30 %	36.12 %	0.1172	0.2110	0.0066
40 %	35.59 %	0.1542	0.2570	0.0072

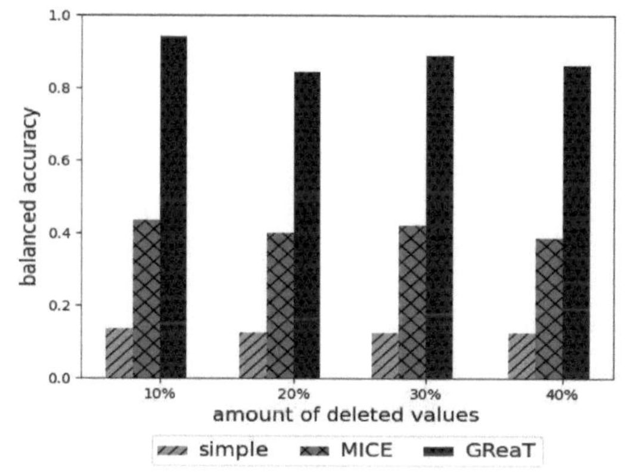

Figure 2: Summarized Results for attribute *histology groups*

3.2   Binary Classification of Survival Time

The f1-scores improve for the class *over one year* by 3% with the prior imputation using only GReaT. The other class remains constant and only improves by max. 1% with GReaT. The classification based on MICE alone performs worse than the baseline for the class *over one year* and decreases by 1%. The performance of the combination of MICE and GReaT is between the baseline and GReaT. The accuracy also increases with the imputation of GReaT by 2% and with the combination of both methods by 1%.

Table 4: Results for Binary Classification of Survival Time below and above one year

method	f1-score under 1	f1-score over 1	accuracy
no method	76 %	57 %	69 %
MICE	76 %	56 %	69 %
GReaT + MICE	76 %	58 %	70 %
GReaT	**77 %**	**60 %**	**71 %**

3.3 Discussion

The evaluation of the three imputation methods on both attributes shows, that the error rates can be significantly decreased for the numerical attribute. Also, the f1-scores and accuracy can be immensely improved. Based on these results, it can be concluded, that the imputation of both attributes works fine. As for the binary classification of the survival time, there is also an improvement by using the evaluated imputation methods. A large difference between the MICE and the LLM GReaT is, that GReaT does not impute every missing value and leaves some at the original value NONE. Moreover, the calculation time of both methods differ a lot. As MICE only needs 10 seconds to impute, GReaT must fit the values to the data set for three hours. Furthermore, the LLM requires an additional time of three hours to impute. This circumstance is based on the problem, that GReaT tries to impute every missing value for a limited amount of times defined by the user. It also does not perform the imputation parallel, so there is room for improvement. Besides, the LLM GReaT learns the data distribution and does not recognize the value NONE as an invalid value. As every value is transformed as a string for the input to the LLM, the NONE value is not distinguishable from the others. However, as the classification results with GReaT imputation are better than with MICE, this does not seem to be a major problem. This indicates, that an incomplete imputation may be better than an imputation over all missing values. The imputation of missing values always bears the risk of imputing misleading values. As the amount of the missing values and the importance of the attribute for the classification increases, the risk of imputing misleading values rises proportionally. As the data set is incomplete, there is no ground truth to check, whether the imputed values are realistic or not. On top of that, a high amount of missing values (over 40 %) complicates the learning process for GReaT to recognize underlying causal connections between the different attributes of the data set [2]. Additionally, MICE could be improved by analyzing, which attributes can be derived by others. Those will undergo another imputation following the imputation of their source attributes. This leads to derived variables, that are consistent with the derivation of their source attributes [6]. The standard method would be to impute each attribute simultaneously, without considering the correlations between them.

4 Conclusion

Based on the discussion, the task of estimating the survival time as a binary classification can be improved with the help of imputation as a pre-processing step. Despite the high computing time and the still existing proportion of missing values, GReaT still improves the classification performance. With paralleling the imputation of GReaT, the calculation time could be decreased. Besides, the data imputation is typically done only once before data analysis and does therefore not necessarily need to be optimized in terms of runtime. Considering that imputation and survival time analysis are difficult problems, the improvement of the binary classification of survival time by 3 % for the f1-score of the class *over one year* and the accuracy by 2 % is nevertheless a success.

Acknowledgement

The work has been carried out at the DFKI and supervised by the Institute of Medical Informatics, University of Lübeck. This work was done as part of the AI-CARE project[1], funded by the German Ministry of Health (BMG), grant number (Förderkennzeichen) ZMI5-2522DAT13J. The dataset analyzed for this study was retrieved from the Cancer Registry Schleswig-Holstein and can be obtained there upon reasonable request. We thank them for their support.

Authors' Statement

Authors state no conflict of interest.

5 References

[1] Stegmaier, C., Hentschel, S., Hofstädter, F., Katalinic, A., Tillack, A. und Klinkhammer- Schalke, M., Hrsg. (2019). *Das Manual der Krebsregistrierung*. München: W. Zuckschwerdt Verlag. 219 S.

[2] Jakobsen, J.C., Gluud, C., Wetterslev, J. und Winkel, P. (Dez. 2017). When and how should multiple imputation be used for handling missing data in randomised clinical trials – a practical guide with flowcharts. BMC Medical Research Methodology 17. Number: 1 Publisher: BioMed Central, 1–10. (last accessed on 2023-12-19).

[3] Buuren, S.V. und Groothuis-Oudshoorn, K. (2011). *mice : Multivariate Imputation by Chained Equations in R*. Journal of Statistical Software 45. (last accessed on 2023-10-25).

[4] Borisov, V., Sessler, K., Leemann, T., Pawelczyk, M. und Kasneci, G. (29. Sep. 2022). *Language Models are Realistic Tabular Data Generators*. In. The Eleventh International Conference on Learning Representations. (last accessed on 2023-10-23)

[5] Ke, G., Meng, Q., Finley, T., Wang, T., Chen, W., Ma, W., Ye, Q. und Liu, T.-Y. (o. D.). *LightGBM: A Highly Efficient Gradient Boosting Decision Tree*.

[6] Austin, P.C., White, I.R., Lee, D.S. und Buuren, S. van (1. Sep. 2021). Missing Data in Clinical Research: A Tutorial on Multiple Imputation. Canadian Journal of Cardiology 37, 1322–1331. (last accessed on 2023-09-27)

---

[1]https://ai-care-cancer.de

Numerical- vs NN-based Solutions for Solving Differential Equations

Amira Moualhi [1] , Saleh Mulhem [2] , Mladen Berekovic [2]

[1] Robotics and Autonomous Systems, Universität zu Lübeck, amira.moualhi@student.uni-luebeck.de

[2] Institute for Computer Engineering, Universität zu Lübeck, {saleh.mulhem, mladen.berekovic}@uni-luebeck.de

Abstract

Machine Learning has become increasingly popular due to its ability to offer efficient solutions that replace conventional methods with more accurate ones. This paper presents modern solutions for an engineering problem, which replaces the well-known conventional solution in approximating first and second order differential equations with the Euler method using a Neural Network (NN) and compares its performance with the exact and numerical solution provided by the Euler method. The results show that a simple shallow NN is able to outperform the Euler method approximation when the equation is converging.

1 Introduction

In recent years, Artificial Intelligence (AI) has revolutionised many fields. With the release of ChatGPT by OpenAI, the world is relying on AI more than ever. This has led to exploring AI as a substitute for conventional mathematical methods, as AI may offer even greater performance.

Applying AI to conventional problems further allows us to explore benefits regarding computation time and accuracy of the solution. Autonomous system is one field that tries to investigate the impact of integrating AI. Here, Ordinary Differential Equations (ODEs) present a profound problem for many physical systems. They are used to predict the continuous change of a system [1]. For example, ODEs are used in autonomous driving to describe the vehicle's motion [2]. For this reason, we need to find an exact solution to these ODEs, but not all have an analytical solution [3]. For this reason, we resort to another solution, Approximation, which we will refer to as the numerical solution. Despite the high accuracy of numerical solutions, they still present some drawbacks, as the error grows when the step size is increased. However, a larger step size is more computationally beneficial, resulting in a tradeoff. In the case of non-linear ODE, the error accumulates when moving away from the initial condition [4]. In some autonomous systems, having the analytical/exact solution is critical, and an approximate solution is insufficient. For this reason, new approaches that are able to outperform numerical solutions are highly interesting for researchers.

In this paper, we solve a ODE using a simple Multi-Layered Perceptron (MLP) and compare it to the exact and numerical solution. We solve first- and second-order ODEs and investigate the limitations of Neural Networks (NNs) in solving this problem.

2 Background: Analytical and Numerical Solution using the Euler Method

There are two methods to solve ODEs mathematically. If the ODE has an exact solution, it can be solved analytically. This presents the exact solution. Alternatively, if no exact solution exists or finding it would be too computationally expensive, numerical solutions can be used to find an approximation.

2.1 Exact Solution: Analytical

To solve an ODE analytically, there are different methods depending on the order of the ODEs. The solution of the exact method presented in this paper is computed using Matlab [5].

2.2 Approximate Solution: Numerical

Many ODEs can be solved analytically, but not all equations are solvable. In such cases, we need to approximate the solution.

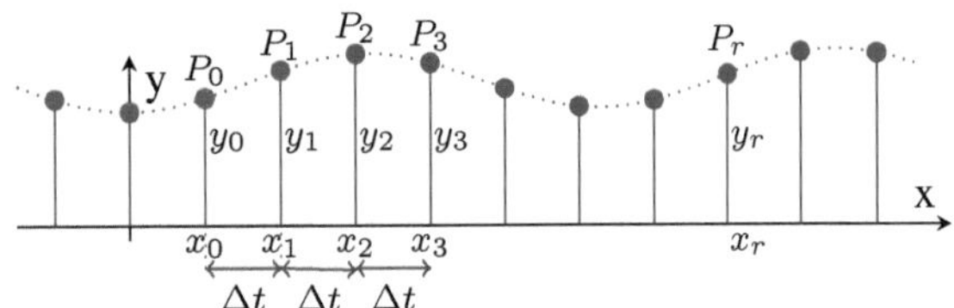

Figure 1: Illustration of the numerical solution using the Euler method (adapted from [6]).

The numerical solution is a step-by-step method as shown in Fig. 1. It takes the initial condition and tries to estimate the next point based on the function's rate of change, the previous result and the step size. There are different methods for finding the numerical solution. In this paper, we focus on the Euler method. With $\frac{dy}{dx} = f(x, y)$ and P_r a point on the curve with coordinate (x_r, y_r) and $f(x_r, y_r)$ is the gradient at P_r. The Euler approximation is as follows:

$$\frac{y_{r+1} - y_r}{\Delta t} = f(x_r, y_r) \rightarrow y_{r+1} = y_r + f(x_r, y_r) \cdot \Delta t,$$

with Δt as the step size [6].

3 Solving ODEs using NNs

The core idea is to solve ODEs without using any other conventional solution. This is done using a custom training loop [5] because there is no dataset. We start by generating an interval of x for the training data, where x is defined as our input. The training process is defined as a custom loop and illustrated in Fig. 2. Starting with a forward pass through the NN to get the NN's prediction with initialized weights. After that, we will calculate the loss and start back-propagation to calculate the NN gradients in order to update the weights [7]. The loss function plays the most important role. It is a custom loss function depending on the equation and the initial condition. In this case, a traditional loss function is not the right choice because we do not have ground truth labels we can use to define the loss as needed for mean squared error or absolute error. To solve an Ordinary ODE without knowing the exact solution of the equation, we need to update the weights based on a custom loss function that penalizes the deviation of the predicted function without solving it. The loss function used in this experiment is adapted from [8] and defined as follows: We consider the ODE:

$$\frac{dy(x)}{dx} = f(x, y),$$

where:

$$\mathcal{F} = \sum_{i=0}^{n} \left(\frac{dy_t(x_i)}{dx} - f(x, y_t(x_i)) \right),$$

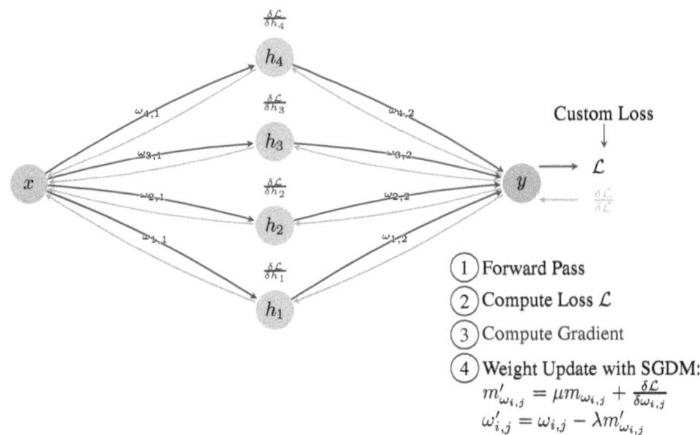

Figure 2: NN Training Process.

with y_t the forward pass of the NN, $\frac{dy_t(x)}{dx}$ its gradient and n the batch size of the training.

$$\mathcal{L} = \left| \frac{(\mathcal{F})^2 + \alpha \cdot (y_t(0) - y(0)^2)}{2} \right|,$$

with $\mathcal{L}$ the loss, α as a hyperparameter, $\mathcal{F}$ refers to the equation that we want to solve, and $y_t(0) - y(0)$ is the MLP prediction at 0 against the initial conditions at 0. The choice of the loss function, all NN hyperparameters, the learning rate λ, the learning drop rate and the momentum μ are tuned manually in order to reach better performance. Here, the performance is defined by how close out NN prediction is to the exact solution.

3.1 Solving First Order ODE using NNs

The equation used in this experiment is:

$$y' + y = 2 \cdot e^x,$$

with $y' = \frac{dy}{dx} = f(x, y)$ and initial condition $y(0) = 1$. The exact solution to this first-order ODE is solved analytically:

$$y = e^x.$$

3.1.1 Multi-Layer Perceptron (MLP)

Solving first-order ODEs using NNs shows that a simple MLP with just one layer is sufficient if the activation function and the hyperparameters are tuned correctly. The architecture of the MLP is shown in Table 1.

Table 1: The architecture of MLP for the first experiment.

Layer	Hyperparameter	Activation Function
Input	1	
Dense	2	Softplus
Output	1	Softplus
Parameters	7	

3.1.2 Results and Comparaison

The results in Fig. 3 show that a simple MLP is able to outperform the numerical solution computed using the Euler method with step size $\Delta t = 0.1$. The MLP is trained on the interval of [0 .. 2]. Therefore, these results show the performance on the training data. It also shows that the error

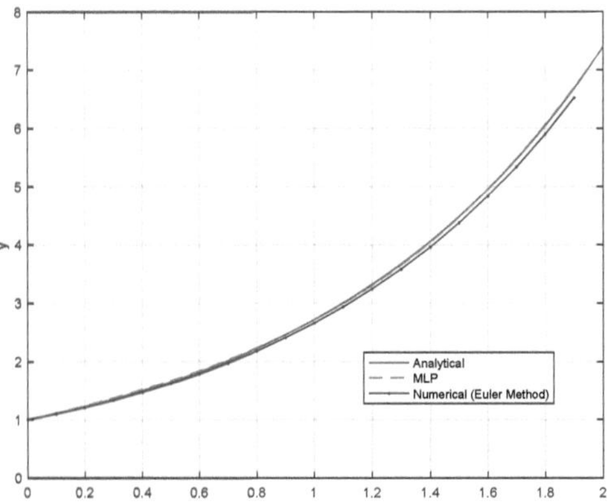

Figure 3: Results of solving first order ODE.

of the Euler method increases when moving too far from the initial condition, while the error of the MLP does not diverge.

3.2 Solving Second Order ODE

The equation used in this experiment is:

$$y'' + 6 \cdot y + 5 \cdot y = 10,$$

with $y' = \frac{dy}{dx} = f(x, y)$ and the initial conditions $y(0) = 0$ and $y'(0) = 5$. The Exact solution to this second-order ODE is solved analytically:

$$y = -3/4e^{-5x} - 5/4e^{-5x} + 2.$$

3.2.1 Multi-Layer Perceptron (MLP)

A simple shallow NN is insufficient for solving second-order ODEs. For this experiment, we optimized an MLP architecture with four dense layers and optimized hyperparameters, as shown in Table 2. The experiment of finding optimized activation functions shows that the traditional activation functions are not suitable for this equation; for this reason, custom activation functions are chosen based on empirical methods to find the optimal ones.

3.2.2 Results and Comparaison

The MLP used to solve the second-order ODE is more complex and uses the custom activation functions. However, the results do not match the analytical solution. The Euler method clearly outperforms the MLP on the interval of [0 ..

Table 2: The architecture of MLP for experiment 2.

Layer	Hyperparameter	Activation Function
Input	1	
Dense	10	x^2
Dense	5	$\frac{3}{1+e^{-3x}}$
Dense	2	$\frac{3}{e^{-3x}}$
Dense	10	$2 \cdot \tanh 3x$
Output	1	
Parameters		**128**

4], while in the interval of [5 .. 7], the results of the three proposed methods are similar when the function converges to a real number as illustrated in Fig. 4.

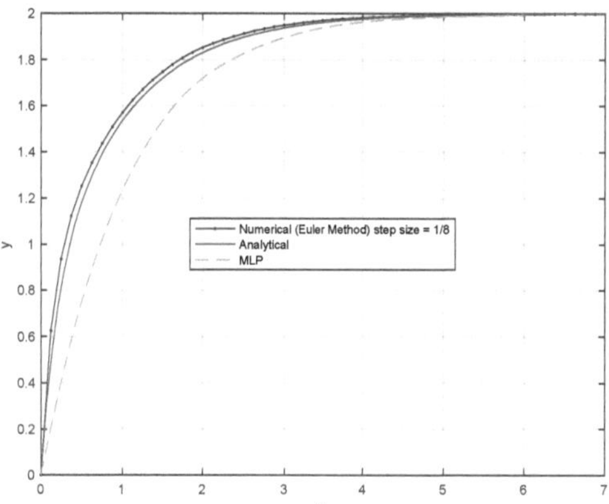

Figure 4: Results for solving second order ODE.

3.3 Limitation of NN in Solving ODEs

The previous two sections show using NNs to solve ODEs is not always an advantage. In this section, we highlight some of the limitations of NNs when applied to solving ODEs. In this part, we used the same NN with the same activation functions from Table 1 with fine-tuning the hyperparameter and the loss. Fig. 5 presents a NN for solving the following equation:

$$y' + 4 \cdot y = e^{-t} \,, \; y(0) = 1.$$

The exact solution of the ODE:

$$y(t) = \frac{e^{-t}}{3} + \frac{2 \cdot e^{-4 \cdot t}}{3},$$

where the NN is trained on the close interval of [0 .. 2], but it was tested on the interval of [0 .. 5]. Fig. 5 shows that the NN is a good replacement for the numerical solution even on the testing data.

Fig. 6 presents results for a NN for solving the following equation:

$$y' = y \cdot t \,, \; y(0) = 2.$$

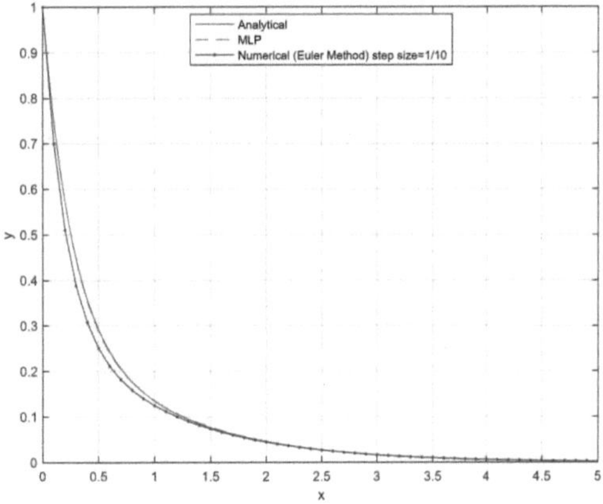

Figure 5: Results of solving first order ODEs on train- and test data.

The exact solution of the ODE:

$$y(t) = 2 \cdot e^{\frac{t^2}{2}}.$$

The NN is trained on the close interval from [0 .. 1], but it was tested on the interval of [0 .. 3]. It shows that the NN is similar to a numerical solution with step size $\Delta t = 0.1$ on the training data, but on the test data, the MLP prediction clearly deviates from the exact solution. Here, the numerical solution is superior.

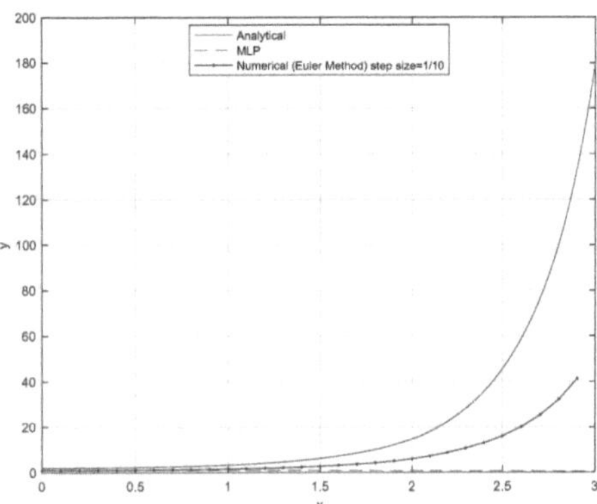

Figure 6: Results of solving first order ODEs on train- and test data.

In conclusion, the current limitation of solving ODEs is that the numerical solution may outperform the NN on certain equations. However, the biggest limitation of this modern method is that the equation might just be solved in the defined training interval, and the NN might not be able to generalise outside the training data.

4   Conclusion

Conventional methods have been a relevant solution for solving ODEs for years. However, this paper demonstrates that a simple MLP with only seven trainable parameters can outperform the Euler method's approximation on the trained interval without solving the equation analytically. This conclusion is particularly relevant for first-order ODEs. Additionally, our findings indicate that the MLP method falls short of the Euler method in the second-order ODE experiment. Furthermore, it is challenging to generalize beyond the training interval. Although the current methodology has some limitations, this work holds promising implications for future research in this domain.

Acknowledgement

The Institute of Computer Engineering (ITI), Universität zu Lübeck, has carried out the work and was supervised by Prof. Dr.-Ing. Mladen Berekovic and Dr.-Ing. Saleh Mulhem.

Authors' Statement

Conflict of interest: Authors state no conflict of interest.

5   References

[1] S. J. Farlow, *An introduction to differential equations and their applications.* Courier Corporation, 2006.

[2] J. Kong, M. Pfeiffer, G. Schildbach, and F. Borrelli, "Kinematic and dynamic vehicle models for autonomous driving control design," in *2015 IEEE intelligent vehicles symposium (IV).* IEEE, 2015.

[3] M. L. Piscopo, M. Spannowsky, and P. Waite, "Solving differential equations with neural networks: Applications to the calculation of cosmological phase transitions," *Physical Review D*, 2019.

[4] B. Biswas, S. Chatterjee, S. Mukherjee, and S. Pal, "A discussion on euler method: A review," *Electronic Journal of Mathematical Analysis and Applications*, 2013.

[5] T. M. Inc., "Matlab version: 9.13.0 (r2022b)," Natick, Massachusetts, United States, 2022. [Online]. Available: https://www.mathworks.com

[6] B. Denis, "An overview of numerical and analytical methods for solving ordinary differential equations," *arXiv preprint arXiv:2012.07558*, 2020.

[7] I. Goodfellow, Y. Bengio, and A. Courville, *Deep Learning.* MIT Press, 2016, http://www.deeplearningbook.org.

[8] I. Lagaris, A. Likas, and D. Fotiadis, "Artificial neural networks for solving ordinary and partial differential equations," *IEEE Transactions on Neural Networks*, 1998.

Using Contrastive Language-Image Pretraining Models for Medical Image Retrieval

Caterina Grundt [1]
[1] Medical Engineering Science, Universität zu Lübeck, caterina.grundt@student.uni-luebeck.de

Abstract

The interpretation of medical images presents a challenging task and requires considerable experience. To facilitate the learning process for future radiologists, being able to view similar radiographs for specific diagnosis represents considerable benefits. This requires a model retrieving suitable radiographs for particular reports containing the diagnosis. Recently, neural network models have proven to be successful in *Vision Language Processing* (VLP) through contrastive pre-training. However successful in the general domain, recent models show limitations in medical context. To improve performance in biomedical VLP, we pre-processed a private dataset of medical image-text pairs and fine-tuned existing neural networks on it. While first results only reach a 0.58% Top 1 retrieval rate, they show, that similar inputs lead to similar embeddings. Differentiation of body parts on radiographs showed promising results with correct body part retrieval of 84.22%. The current results leave room for improvement while providing a good starting point for further experiments.

1 Introduction

Interpreting a medical radiograph and determining an appropriate diagnosis is a challenging task in the hospital environment and requires a significant level of experience and training. To facilitate the learning process, especially when educating students or inexperienced radiologists, it can be beneficial to retrieve and examine similar radiographs for a particular radiology report. However, preparing such data manually requires extensive effort and cannot be scaled. To enable this learning approach with minimum effort required from the lecturer, a model for radiograph retrieval based on radiology reports can provide significant benefits.

In the field of self-supervised *Vision Language Processing* (VLP), neural network models have achieved great success in the past years by using contrastive pretraining on image-text datasets, as demonstrated by the *Contrastive Language-Image Pre-training* (CLIP) [1] model.

CLIP is a neural network model trained to learn visual concepts from natural language supervision. The approach of contrastive pretraining is to simultaneously train an image encoder and text encoder aiming to minimize the distances between matching image-text pairs and maximize distances between non-matching pairs in the embedding space during training. With such pretraining on an image-text dataset, CLIP was able to achieve competitive and partially even higher accuracies compared to state of the art results without further data-specific training [1].

Previous research, including CLIP, has often used publicly available data from the web for training, with the result that respectively trained models are successful in the general domain but often not suitable for use in a medical context due to the considerable difference of medical images [2].

To improve performance in biomedical VLP on common medical imaging tasks such as retrieval and classification, BiomedCLIP [2] conducted large-scale domain-specific pretraining on image-text data and achieved results matching and even outperforming state of the art models on radiology-specific tasks. With these results, Biomed-CLIP has demonstrated the promising potential of domain-specific pre-training in biomedical VLP.

In contrast to the general domain, domain-specific training in medical VLP is still limited by the availability of large datasets of medical image-text pairs that are publicly accessible [2]. Although some medical datasets are publicly available, they are unsuitable for our application.

For instance, there is MIMIC-CXR [3], a publicly available dataset of chest radiographs and corresponding radiology reports consisting of 377,110 images. However, it is limited to chest radiographs and contains no further anatomies. Another publicly available dataset is PMC-15M [2], which was used to train BiomedCLIP and comprises 15 million biomedical image-text pairs. However, since it was created by mining image-caption pairs from PubMed Central [4] articles, it contains not only radiography and microscopy images but also generic biomedical illustrations such as tables, charts, and statistical figures. As texts, it contains figure captions instead of radiology reports.

2 Approach

To obtain a dataset suitable for our objective of training a model for medical image retrieval based on radiology reports, we used a private biomedical image-text dataset containing pairs of anonymized radiographs and the corresponding radiology reports and conducted pre-processing to prepare the dataset for fine-tuning purposes. On this dataset, we then fine-tuned pretrained neural network models previously proven successful.

2.1 Preprocessing of the Dataset

The data is provided anonymized in DICOM format. We extracted the radiographs and associated radiology reports and saved them as individual PNG and TXT files.

2.1.1 Extracting Radiographs and Radiology Reports

During extraction, the radiographs were inspected to ensure that the DICOM files were readable. If necessary, the radiographs were resized to match the dimension which the model expects as input while keeping the aspect ratio. No further image processing was performed on the images.

For each image, additional meta-data was provided in the DICOM image. The following fields were extracted from the DICOM file and saved in individual JSON files: StudyDescription, SeriesDescription, BodyPartExamined, ImageLaterality and Laterality. The extracted metadata was analyzed to determine the body part depicted in the radiograph as well as the laterality and view position.

The report texts were processed as follows: First, the signature field was removed as it contained information about the radiologist conducting the assessment. Additionally, using a set of regular expressions, all dates, times and names of patients or doctors were identified and removed or replaced with placeholders to maintain the sentence's structure.

2.1.2 Matching Radiographs and Radiology Reports

The extracted PNG, TXT and JSON files are uniquely identifiable based on the anonymized patient ID, study ID and series ID. For each patient, only those studies containing one report and one or more associated images were included for training. Studies containing no report or several reports were excluded due to lack of reliable assignability. The resulting dataset contains 866,750 image-text pairs with 866,750 images and 549,459 reports, some of which match several images.

2.2 Architectures and Models

Our implementation is based on the CLIP model [1] which aims to simultaneously train an image and text encoder to maximize the cosine similarity of matching image and text embeddings and minimize the similarity of non-matching pairs. This is achieved by optimizing the symmetric cross entropy loss over the similarity scores during training.

On the image side we use BiomedCLIP [2]. BiomedCLIP is a vision language model using contrastive learning which is pretrained on PMC-15M and therefore specifically suitable for the biomedical domain.

Because the radiology reports are written in German, we replace the original text encoder and tokenizer used in CLIP with a pre-trained German language model, German medBERT.de [5]. Based on the *Bidirectional Encoder Representations from Transformers* (BERT) [6] architecture, medBERT.de [5] is a German *Natural Language Processing* (NLP) model for the medical domain. It has been trained on 4.7 million German medical documents to perform various NLP tasks in the medical domain. The model achieved new state of the art performance on medical benchmarks.

2.3 Model Training

To fine-tune, we freeze the weights of BiomedCLIP as image encoder and medBERT.de as text encoder and add two projection layers which we fine-tune on the pre-processed dataset. For optimization we use the Adam [7] optimizer. For the tokenizer, we increase the context length of 77 tokens implemented by CLIP to 512 tokens to be able to cover the token lengths of the more extensive radiology reports. With regards to the training hyper parameters, we choose a learning rate of 1e-04, batch size of 64 and dropout of 0.5. We trained the projection layers for 10 epochs on one GPU (RTX 2080 Ti, 11GB). The training took three days.

3 Results and Discussion

To evaluate suitability of the fine-tuned model for medical image retrieval, Top 1 and Top 5 accuracies as metrics are presented. Additionally, two retrieval examples are evaluated qualitatively and generated embeddings are visualized.

3.1 Accuracies

The metrics used to assess the trained model are Top 1 and Top 5 accuracies on training and test dataset. Top 1 accuracy measures how often the model retrieved the correct radiograph for a given radiology report. Top 5 accuracy measures how often the correct radiograph is among the five retrieved images that the network embedded most closely. The achieved accuracies are summarized in Table 1.

Although the fine-tuned model achieves a Top 1 accuracy on the training and test set of 60.94% and 29.34% within a batch respectively, the Top 1 accuracy on the total test set only stands at 0.58%. The considerable gap between the accuracies can be explained by the different measurements used. While the batchwise accuracies are only evaluated within a training or test batch containing 64 pairs, the total accuracy is assessed across the entire test set containing around 87,000 pairs. The network therefore has a greater probability of retrieving the correct radiograph simply by guessing at a smaller batch size. Therefore, a higher batch size would reduce the random chance to guess the correct radiograph compared to a smaller batch size. Looking at the total test accuracy, the probability of retrieving the correct image out of the test set containing around 87,000 image-text-pairs by chance is at 0.0011%. Therefore, the fine-tuned model performs 525 times better than random chance. The Top 5 accuracy amounts to 2.27%. With respect to the objective of retrieving radiographs based on radiology reports, these results are still insufficient.

Top 1 and Top 5 accuracies per body part and laterality were assessed on the test set (see Table 2). In contrast to previously discussed accuracies, Top 5 accuracies per body part and laterality indicate that all five retrieved images match the correct label. In 84.22% of cases the model retrieved a radiograph depicting the correct body part. In 74.34% of cases, all five retrieved radiographs contained the correct body part. The Top 1 and Top 5 accuracies for laterality show lower results at 79.41% and 59.92%.

Table 1: Top 1 and Top 5 retrieval accuracies on training and test dataset compared to Top 1 random chance, given batchwise (bw) and on the total testset (total). Top 5 accuracy indicates that **at least one** of the five retrieved images matches the correct report.

	Train (bw)	Test (bw)	Test (total)
Top 1 random	1.56%	1.56%	0.0011%
Top 1 accuracy	60.94%	29.34%	0.5774%
Top 5 accuracy	-	-	2.2733%

Table 2: Top 1 and Top 5 accuracies for retrieving the correct body parts and lateralities, given on the total testset. Top 5 accuracies indicate that **all** five retrieved images match the correct label.

	Test set (total)
Top 1 accuracy body part	84.22%
Top 5 accuracy body part	74.34%
Top 1 accuracy laterality	79.41%
Top 5 accuracy laterality	59.92%

3.2 Retrieval

In Fig. 1 we show what the two projection layers have learned during fine-tuning taking the example of a retrieval with Top 1 match in the first row and the example of a retrieval without Top 1 or Top 5 match in the second row. This is done by computing the pairwise euclidean distances between the embeddings of all radiographs and radiology reports and retrieving the five images with the smallest distances. In the following, the contents of individual radiology reports are referred to as diagnostic findings.

In the Top 1 match retrieval in the first row, the original report (see the corresponding radiograph in Fig. 1.1a) matches the retrieved image with the smallest euclidean distance (1.1b). Fig. 1.1c, e and f show images corresponding to different reports, however, containing similar diagnostic findings such as "No fracture/dislocation". In contrast, 1.1d shows a radiograph corresponding to a report containing a fracture and therefore is not a suitable retrieval result.

In the retrieval without Top 1 or Top 5 match seen in the second row, the original report (see in Fig 1.2a the associated radiograph) does not match any of the retrieved images (1.2b-f). However, the four closest images (1.2b-e) correspond to reports containing similar diagnostic findings such

as "No fracture/dislocation" or "Normal representation". In contrast, the fifth closest image (1.2f) shows a fracture and therefore is an inappropriate match for the original report. Although Top 5 picks for both retrieval examples contain one unsuitable radiograph, it is noticeable, that all Top 5 radiographs show the same body part. This demonstrates that retrieving of radiographs containing correct body parts is quite successful. However, retrieving radiographs containing the same diagnostic findings does not perform adequately, even looking at Top 1 match retrievals.

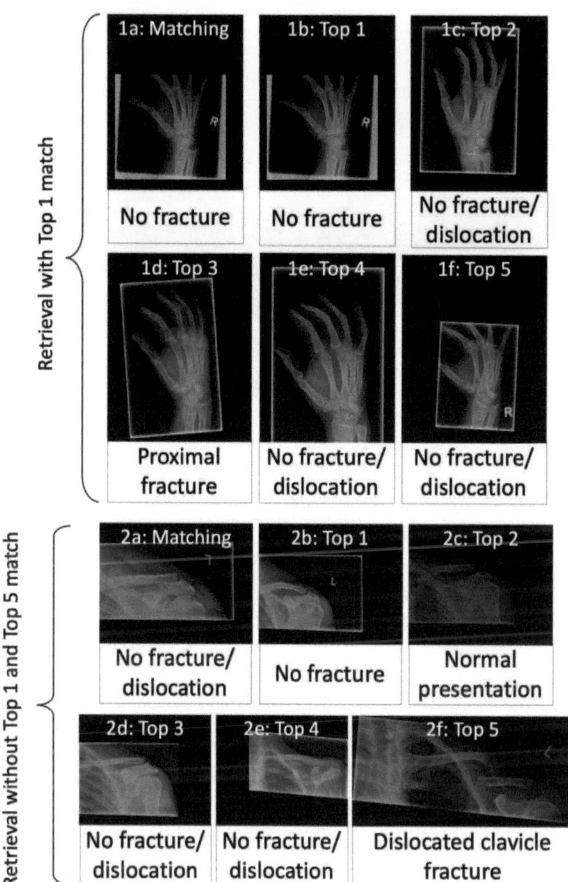

Figure 1: 1a and 2a each show the matching radiograph to the report examined while 1b-f and 2b-f show the radiographs with smallest pairwise distances to the report, in ascending order. While reports of four retrieved radiographs contain similar diagnostic findings, one radiograph each (1d and 2f) shows different diagnostic findings.

3.3 Embedding Visualization

Fig. 2 shows a qualitative visualization of the embeddings generated by the two fine-tuned projection layers. To visualize we use the *T-distributed Stochastic Neighbor Embedding* (t-SNE) [8] algorithm. By applying t-SNE we reduce the 512-dimensional embedding space to a 2-dimensional space. In order to maintain differentiation between all categories, only the 5 most common body parts are included: Thorax, hand, spine, knee and pelvis.

The map produced by t-SNE reveals the natural classes of body parts in the data. However, some radiographs are located outside of the clusters containing points with the same body parts. Those could either be outliers in the data, images that were clustered inaccurately by t-SNE or radiographs which received unsuitable embeddings by the network. Although the objects are not all cleanly separated from each other, in general a differentiation of body parts is still noticeable in the visualization. This observation is supported by the achieved body part accuracies.

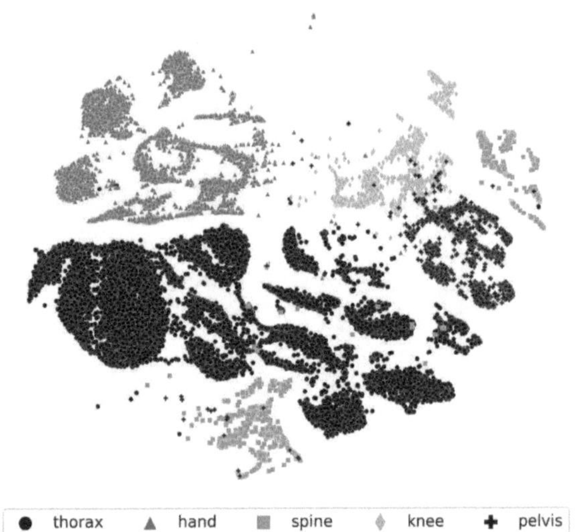

Figure 2: t-SNE plot visualizing embeddings. Classes of body parts are clearly recognizable, underlining the network's ability to distinguish body parts in radiographs.

4   Conclusion

In this study, we fine-tuned existing pre-trained neural network models on a medical dataset containing roughly 870,000 pairs of radiographs and corresponding radiology reports. Top 1 and Top 5 accuracies of 0.58% and 2.27% on the test set show insufficient results for retrieval of the identical radiographs associated with radiology reports.

However, given the objective to facilitate education of young doctors, it may rather be sufficient to retrieve similar radiographs for reports instead of identical matches. Therefore, the accuracies only represent lower bounds while in reality the performance might be even higher.

Retrieval of similar radiographs in terms of retrieving correct body parts and laterality already shows promising results with a retrieval rate of 84.22% and 79.40%. Unfortunately, retrieval of similar radiographs in terms of similar diagnostic results cannot be evaluated due to the lack of standardized information on radiology report content.

To improve the accuracy, further experiments need to be conducted. Different experimental approaches varying the fine-tuning parameters, such as the batch size, learning rate, dropout, and epochs, could potentially increase the accuracies. Reducing the dropout to 0.00 and increasing the learning rate in particular could lead to better performance.

These experiments have not yet been carried out due to the long training duration and lack of time.

Another approach to improve the network´s performance could be to train the whole model from scratch instead of freezing the weights and only fine-tuning projection layers. To conduct such an experiment, more computational resources are required, since fine-tuning of the two projection layers already took three days to train.

The initial results from training on the extensive medical dataset already show considerable potential and are a promising starting point for further experiments.

Acknowledgement

The work has been carried out at the Institute of Neuro- and Bioinformatics, Universität zu Lübeck and supervised by Dominik Mairhoefer and Manuel Laufer.

Authors' Statement

Conflict of interest: Author states no conflict of interest.

5   References

[1] A. Radford et al., *Learning Transferable Visual Models From Natural Language Supervision*. In: Proceedings of the 38th International Conference on Machine Learning, PMLR 139, pp. 8748–8763, 2021.

[2] S. Zhang et al., *BiomedCLIP: a multimodal biomedical foundation model pretrained from fifteen million scientific image-text pairs*. 2023.

[3] A. E. Johnson et al., *MIMIC-CXR, a de-identified publicly available database of chest radiographs with free-text reports*. In: Scientific Data, vol. 6, 2019.

[4] PubMed Central (PMC). [Internet] (2024). URL https://www.ncbi.nlm.nih.gov/pmc/.

[5] K. K. Bressem et al., *MEDBERT.de: A Comprehensive German BERT Model for the Medical Domain*. In: Expert Systems with Applications, vol. 237, p. 121598, 2024.

[6] J. Devlin, M.-W. Chang, K. Lee, and K. Toutanova, *BERT: Pre-training of Deep Bidirectional Transformers for Language Understanding*. In: Proceedings of the 2019 Conference of the North American Chapter of the Association for Computational Linguistics: Human Language Technologies, vol. 1, pp. 4171–4186, 2018.

[7] D. P. Kingma and J. Ba, *Adam: A Method for Stochastic Optimization*. In: Proceedings of the 3rd International Conference on Learning Representations (ICLR 2015), 2015.

[8] L. V. D. Maaten and G. Hinton, *Visualizing Data using t-SNE*. In: Journal of Machine Learning Research, vol. 9, pp. 2579–2605, 2008.

Fine-Tuning Pre-Trained Image and Language Models on a Large German Radiological Dataset for Image-to-Text Retrieval

Konrad von Kügelgen [1]

[1] Medical Informatics, Universität zu Lübeck, konrad.vonkuegelgen@student.uni-luebeck.de

Abstract

Time is a valuable resource in the medical field. Creating diagnostic reports is a time-consuming task. Radiologists could benefit from pre-selected, radiograph-specific reports, only adapting them as needed for efficiency. Through image-to-text retrieval, radiologists could optimize their workflow by receiving automated report templates, minimizing the time allocated to the reporting process. This paper explores the adaptation of Contrastive Language-Image Pre-training (CLIP) to the German medical domain, utilizing a large private radiological dataset and pre-trained models. The proposed model combines the German language model MedBERT.de with BiomedCLIP's vision transformer. A projection is fine-tuned for image-to-text retrieval. The study outlines the methodology, dataset structure, pre-processing steps and preliminary results of ongoing research. However, the initial evaluation reveals a limited recall performance with a recall@1 of 0.18% and recall@5 at 0.79%, emphasizing the need for further refinement.

1 Introduction

The past years have been highly influenced by the progress of artificial intelligence. Specifically Natural Language Processing (NLP) like Bard and ChatGPT took NLP to new standards and spawned more commercial interest, political debates and general awareness of the AI capabilities.

With the abundance of general domain images and descriptive texts, vision-language processing (VLP) has also made great progress, not least due to improvements in NLP. In 2021 Contrastive Language-Image Pre-training (CLIP) [1] was introduced which presented a scalable method for learning from natural language supervision, surpassing previous VLP systems in performance. This self-supervised neural network encodes each an image and a text into a shared embedding space. A loss function ensures the shortest distance between corresponding image and text embeddings and larger distances for non-related pairs. It can be used for downstream tasks such as cross-modal retrieval.

In this paper we utilized a large private radiological dataset for *fine-tuning* a CLIP model, specifically tailored for the medical domain, as *pre-training* of CLIP or NLP models is very computationally intensive. For this we combined medBERT.de [2], a German language model, with Biomed-CLIP's [3] vision transformer thus creating a new CLIP-like model. We froze their weights and fine-tuned a Multi-Layer Perceptron (projection) and used it for image-to-text retrieval. Both of these very topical model releases of 2023 were pre-trained on data of the medical domain. The objective is to expedite the reporting process for radiologists by providing report templates through image-to-text retrieval. This paper details our methodology, including the radiolog-ical image-text dataset, pre-processing steps, and the architecture of the CLIP-like model. The evaluation section outlines the model's performance metrics, focusing on image-to-text retrieval accuracy and training dynamics. Section 3 and 4 discuss the quality of the results, the limitations and describe further steps of improvements in medical VLP.

2 Material and Methods

2.1 Radiological Image-Text Dataset

Labeled (training) data is often scarce in medical research. This is reflected in the availability and scope of public medical datasets, especially in the German medical domain. The CLIP architecture in particular does not require class labels but descriptive text, making the search for applicable datasets even narrower. We were able to receive large amounts of anonymized radiological data (private at present) from a German hospital. Our dataset consists of X-ray images and their respective diagnostic reports, spanning over ten years. The diagram in Fig. 1 shows the overall composure of our dataset. The raw data which is stored as DICOM files has a storage size of almost 9 TB. It is apparent that pre-processing is required in order to use the data for training.

2.2 Pre-Processing

The DICOM file format stores metadata that contains valuable information about the patient, imaging device parameters, and acquisition settings, providing a comprehensive context for each image. We discarded faulty pixel arrays,

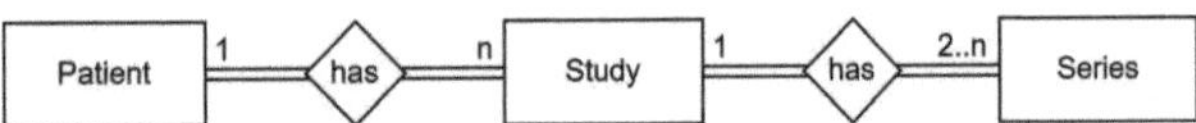

Figure 1: Entity-relationship model (Chen-Notation) of the dataset structure. Series can either be a report or an X-ray image. A study has exactly one report and has one or more radiographs.

handled different photometric interpretations to get an overall conformity, resized the images to 512 pixels at the shortest side and then stored the images as 16 bit grayscale PNGs. When used for training, the images are normalized with the dataset's mean and standard deviation and then receive the same augmentation as in BiomedCLIP: random resized crop for training images, resize plus a center crop for validation and test images. For the Structured Reports (SR) we extracted the field findings (German: "Befund"), removed dates and names through regular expression filters and then saved them as TXT files. Upon loading and using a TXT report in the tokenizer, it is condensed and cleaned by more regular expressions and functions on-the-fly, enhancing its overall clarity and readability for tokenization. In general a study has to have exactly one report and at least one image to be relevant for our use case. Only studies meeting these criteria were used for our training. This results in 866.750 image-text pairs, featuring 549,459 distinct reports. In cases with multiple images per study, each image is paired with the one report from the study. Thus, shuffling the dataset during training, validation and testing reduces the consecutive occurrences of the same report in one batch, improving pair diversity per batch. Our strategy reduces the dataset from 9TB to approximately 500GB, resulting in more feasible data handling. The split into training, validation and test sets is approximately 80/10/10 - no patient is part of multiple sets.

2.3 Model architecture (CLIP)

CLIP-like models comprise two primary components: the image encoder and the text encoder. The image encoder can take the form of a CNN, such as ResNet, or, as in our case, a vision transformer. The text encoder again consists of two parts: a tokenizer and the encoder. Tokenization, an integral step in NLP, involves breaking down a text into smaller units, like words or characters. This process provides NLP systems with a structured representation, enabling the conversion of tokens into numerical representations through tasks like vectorization. These algorithms are thus the first step before the text encoder can leverage the tokens to encode the textual information. The original CLIP uses GPT-2 [4] as their text encoder, BiomedCLIP and ulimately we use a variant of Bidirectional Encoder Representations from Transformers (BERT) [5]. BERT captures contextual information from both left and right context, making it well-suited for tasks requiring an understanding of word relationships and context. Figure 2 shows an overview of the CLIP pre-training procedure. The loss function in CLIP is based on a contrastive learning approach, where the model

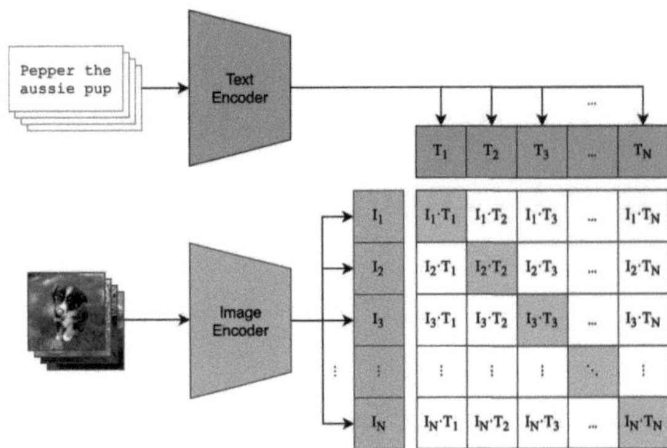

Figure 2: Overview of contrastive pre-training from [1] where the diagonal corresponds to matching image-text pairs.

is trained to pull together positive pairs (correct image-text pairs, in Fig. 2 the diagonal) and push apart negative pairs (incorrect image-text pairs). The loss is calculated using cosine similarity that measures the agreement between the representations of the image and text. In Figure 2, N refers to the amount of image-text pairs in the batch. The diagonal corresponds to matching image-text pairs. The network learns to associate semantic meanings between visual and textual information. Reference [1] showed that their CLIP model can perform many different zero-shot tasks on datasets, and is even able to outperform supervised ImageNet models. This all concerns the general, not the medical domain. Domain-specific pre-training is essential for a biomedical application, as images as well as the language used in the medical field greatly differ from general domain data. We are bound to find fitting models for our radiological dataset, as training from scratch exceeds available computational capacities. We extracted the pre-trained image encoder from BiomedCLIP, discarded the corresponding English text encoder and used a German one instead. The weights of these pre-trained models were frozen. We only trained a multi-layer perceptron as a projection of the model outputs to a shared embedding space. Figure 3 shows the architecture of our CLIP-like model.

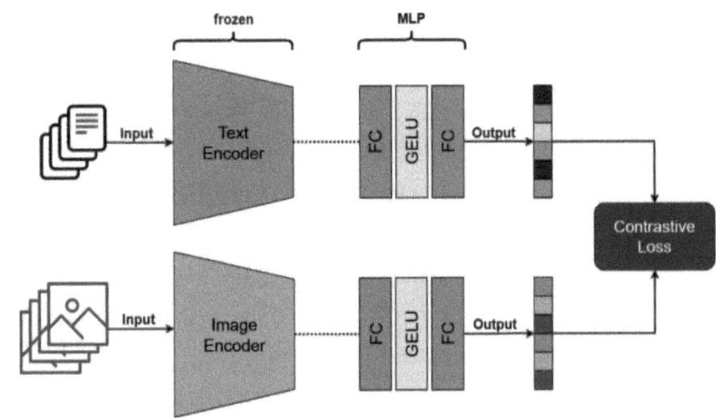

Figure 3: The model architecture of our approach. Only the MLP is trained. The text encoder is MedBERT.de [2] and the image encoder is extracted from BiomedCLIP [3].

BiomedCLIP [3] is a domain-specific model for biomedical VLP. It was pre-trained on a massive dataset, PMC-

15M [3], which contains 15 million biomedical image-text pairs, making it significantly larger than existing biomedical image-text datasets. This extensive pre-training on diverse biomedical image types has enabled BiomedCLIP to outperform prior VLP approaches including radiology-specific models like BioVil [6].

MedBERT.de [2], a specialized German BERT model for the medical domain, has been trained on an extensive corpus comprising 4.7 million German medical documents. The model's effectiveness is attributed to its comprehensive training data, including scientific texts, medical books, and hospital records from various medical domains. The results emphasize the model's utility for longer texts and underscore the importance of a domain-specific approach in medical language modeling. Whilst CLIP's tokenizer uses a context length of 77 and BiomedCLIP's uses one of 256, medBERT.de introduced a length of 512 maximum tokens.

2.4 Model Training and Evaluation

We acquired our models from the transformers library of the Hugging Face Model Hub [7]. We used the hub as a streamlining integration into our pipeline but it revealed its complexity as we were challenged with versioning and compatibility issues over time. The models were saved locally and most of the hugging face wrappers were stripped, implementing and correcting parts of their functionality ourselves. BiomedCLIP was trained on RGB images which defines the input channels of the image encoder. We repeated our grayscale image in each channel so as not to change the model due to the frozen weights and possibly expected activations. We employed PyTorch Lightning as the foundational framework for our experiments. While most hyperparameters were tuned to align with the recommended values of our two models, our batch size was not able to match the 64k from BiomedCLIP. Our feasible maximum was 512, the first training runs were conducted with a batch size of 64. Training our model with a batch size of 64 for 10 epochs on one GPU (RTX 2080 Ti, 11GB) is a time investment of more than three days. Creating the embedding for every image and text with the trained model takes 15 hours.

Evaluation (Image-to-Text Retrieval)

To evaluate the performance of our CLIP-like model, we focused on image-to-text retrieval, a crucial aspect of its utility. Image-to-text retrieval involves presenting the model with an image query and assessing its ability to retrieve relevant textual descriptions, here the matching report. We employed embeddings generated by our model to measure the recall at one ($R@1$) and recall at five ($R@5$). The recall measures the model's ability to correctly retrieve the most relevant report for a given image query. $R@1$ quantifies the percentage of queries for which the correct match is found in the top-ranked result, for $R@5$ within top-five-ranked result respectively. While the primary focus is on the model's performance in image-to-text retrieval, understanding its behavior during training is essential. Training

accuracy metrics shed light on how well the model is learning and adapting to the dataset.

3 Results and Discussion

Figure 4-5 show the image-to-text accuracy during training and validation of three exemplary models (Table 1). Training time and resources are limiting factors in result acquisition. We have trained over ten models with different parameters determining promising parameters for longer training sessions. In Table 1 we show three exemplary models. The model with the largest batch size of 512 shows potential for a continuous accuracy increase and needs more available GPU hours which could not be pursued as of now. The models with the batch size of 64 both reach an approximate training accuracy of 60% after the first few epochs. There is a correlation between batch size and accuracy: the smaller the batch size the larger the accuracy. This is expected as choosing the correct counterpart from 64 instead of 512 is more likely (batch-wise computation of accuracy). As described in 2.2, shuffling the data loaders is important (training is always shuffled). Figure 5 illustrates this accuracy improvement in the model with smaller batch size. Although validation shuffling improves the accuracy, it does not remove the error of the current metric when the correct report (same study) is returned but is counted a mismatch as that report is paired with another image of the study. Furthermore, even if the report of another study is returned, semantically it could describe the same diagnosis. This highlights problems in finding a suitable concise metric without validating half a million unstructured texts by hand.

Table 1: Information of three exemplary models also occurring in Fig. 4 and 5. Average Recall@1 and Recall@5 for 88k (validation set) images tested on image-to-text retrieval against the whole dataset ($1/549459 = 1.8e{-}6$ random chance).

Epochs	Runtime	Batch Size	Shuffled Val.	R@1	R@5
17	4d 11h	64	No	0.13%	0.54%
10	3d 6h	64	Yes	0.14%	0.65%
20	4d 11h	512	Yes	0.18%	0.79%

The current recall performance (Table 1) of the models falls below anticipated standards. Retrieving the correct text for an image has an average probability of at least $1.3e{-}3$, random guessing $1.8e{-}6$. Reasons for this poor performance are the current metric but also the the lack of hyperparameter tuning due to time constraints. Time constraints also apply to parameters in changing dataset characteristics. For example, increasing the resolution entails the necessity of sufficient storage capacities, time for the export (approximately 5 days) and GPU memory when loading them for training (in comparable batch sizes of the reference models). There are multiple structural bottlenecks that hinder the fluent hyperparameter tuning - it is yet to be optimized. The nature of the clinical reports poses a further challenge. Doctors do not always describe a single X-ray image, but combine findings from several radiographs in one report.

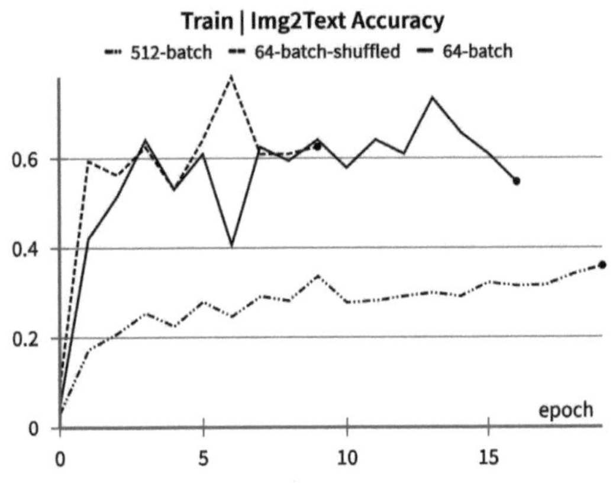

Figure 4: Batch-wise image-to-text top-1-accuracy during training, averaged after each epoch.

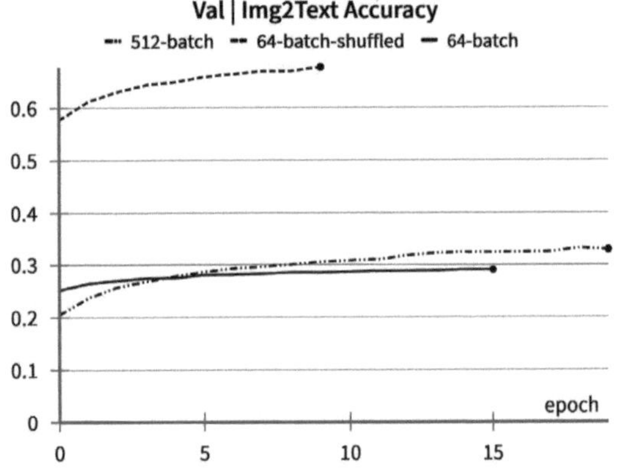

Figure 5: Batch-wise image-to-text top-1-accuracy during validation, averaged after each epoch. Note the increase in accuracy of the 64-batch model after the validation shuffle.

This leads to a blurred description of the report paired to the image. These fuzzy or even non-semantically usable descriptions (e.g. German: "Ohne Befund gültig.") are fed into the network in the expectation that it will learn a perfect representation from imperfect data of the real world.

4 Conclusion and Outlook

In this paper, we presented an approach to enhance image-to-text retrieval in the German medical domain through the adaptation of CLIP. Leveraging a large private radiological dataset, we fine-tuned a CLIP-like model, merging the vision transformer from BiomedCLIP and the German language model medBERT.de. Despite the promising potential and novelty of the individual encoders, our initial evaluation results have shown a limited recall performance. Several factors contribute to this outcome, including constraints on hyperparameter tuning, suitable metrics, and the inherent complexity of the dataset characteristics. The challenges encountered point towards many approaches of future research and improvement. While our current results may not

meet the desired performance standards, we believe that we can significantly improve the outcome with continuous tuning, process optimization and an improved metric.

Generative reports could be of interest, eliminating the finite return of a report and thus rendering a synthetic report template more closely correlating to the input image, showing the models semantic interpretation of a radiograph.

Acknowledgement

The work has been carried out at University of Lübeck in the Neuro- and Bioinformatics and supervised by Dominik Mairhöfer and Manuel Laufer. I also thank Caterina Grundt for co-working and contributions on this topic.

Authors' Statement

The author states no conflict of interest.

5 References

[1] A. Radford et al., "Learning transferable visual models from natural language supervision," in *Proceedings of the 38th International Conference on Machine Learning*, ser. Proceedings of Machine Learning Research, vol. 139. PMLR, 18–24 Jul 2021, pp. 8748–8763.

[2] K. K. Bressem et al., "medbert.de: A comprehensive german bert model for the medical domain," *Expert Systems with Applications*, vol. 237, p. 121598, Mar. 2024.

[3] S. Zhang et al., "Biomedclip: a multimodal biomedical foundation model pretrained from fifteen million scientific image-text pairs," 2024.

[4] A. Radford, J. Wu, R. Child, D. Luan, D. Amodei, and I. Sutskever, "Language models are unsupervised multitask learners," 2019.

[5] J. Devlin, M.-W. Chang, K. Lee, and K. Toutanova, "BERT: Pre-training of deep bidirectional transformers for language understanding," in *Proceedings of the 2019 Conference of the North American Chapter of the Association for Computational Linguistics: Human Language Technologies, Volume 1 (Long and Short Papers)*. Minneapolis, Minnesota: Association for Computational Linguistics, Jun. 2019, pp. 4171–4186.

[6] B. Boecking et al., "Making the most of text semantics to improve biomedical vision–language processing," in *Computer Vision – ECCV 2022*. Cham: Springer Nature Switzerland, 2022, pp. 1–21.

[7] T. Wolf et al., "Transformers: State-of-the-art natural language processing," in *Proceedings of the 2020 Conference on Empirical Methods in Natural Language Processing: System Demonstrations*. Association for Computational Linguistics, Oct. 2020, pp. 38–45.

Optimizing Planetary Rover Simulation Dynamics with Real Trajectories: A Comparative Study

Tim-Louis Lewejohann [1]

[1] Robotics and Autonomous Systems, Universität zu Lübeck, tim.lewejohann@student.uni-luebeck.de

Abstract

The evolving landscape of planetary exploration requires the development of advanced learning-based path planning and control algorithms. To overcome the challenges and expenses associated with physical prototyping, robust and accurate simulation environments are required. This paper examines two simulation approaches: (1) Augmenting the Gazebo simulator with a wheel slip and wheel plowing plugin requiring parameter optimization. (2) A comparative runtime analysis with the particle-based simulator Project Chrono. The findings indicate that although Project Chrono is more realistic, it is unsuitable due to a runtime 25 times slower than real-time. The extended Gazebo simulation, with optimized parameters, demonstrates efficient performance, achieving a real-time factor of 1.8. This research provides valuable insights into practical considerations and trade-offs when selecting a simulation environment for developing and testing advanced rover algorithms.

1 Introduction

Extensive testing and validation are necessary when developing planning and control software for planetary rovers. Especially when adding learning-based features, like Regression and Neural Networks to control tasks [1]. However, implementing these features on actual hardware can be costly in terms of both time and money and may not be the optimal choice in the early design stages. Realistic simulation environments can help bridge this gap and decrease both development and testing times. Consequently, this research investigates two different simulation approaches and presents a method to improve the simulation using real rover data collected with the Lightweight Rover Unit (LRU) of the German Aerospace Center (DLR) shown in Fig. 1. The article [2] presents a Gazebo plugin that utilizes a mathematical approach to incorporate deformable soil interaction for a rover. The appropriate parameters are determined from a wheel test stand, and the approach is validated using a rover prototype. Similar research has been proposed by [3]. They implemented the Gazebo Wheel Slip Plugin successfully and determined parameters using real rover data. However, the data only includes driving orthogonal to a slope, and there has been no testing to determine the slip ratio during parallel sideways driving.

2 Material and Methods

This section introduces two distinct simulation environments for comparison: Project Chrono, emphasizing realistic particle-based wheel-soil interaction, and Gazebo, prioritizing nearly real-time capabilities with a simple Coulomb-

Figure 1: Lightweight Rover Unit from the German Aerospace Center (DLR), a real-world counterpart to the simulated planetary rover. The depicted image was generated within the scope of the research project DeLeMIS.

like friction model. Notably, Gazebo's realism is augmented through the incorporation of additional plugins, necessitating the identification of suitable parameters. Subsequently, the methodology for generating these parameters using real rover data is elaborated upon.

Project Chrono

Project Chrono is an open-source physics-based simulation framework designed for modeling and simulating multibody dynamics [4]. It is particularly useful for simulating complex mechanical systems, including vehicles, robots, and machinery [5]. The framework provides a platform for simulating the interaction of rigid bodies, deformable bod-

Figure 2: Simulated planetary rover navigating uneven terrain using Project Chrono.

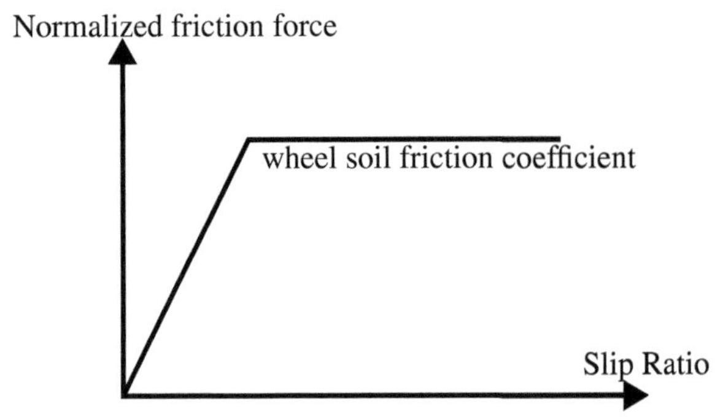

Figure 3: Wheel soil interaction model employed through the Gazebo wheel slip plugin. (Adapted from [6])

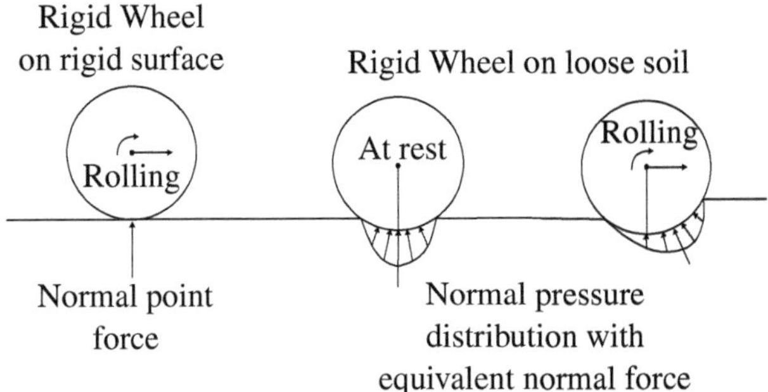

Figure 4: Illustration of the wheel plowing effect that counteracts the movement with increased wheel speed. (Adapted from [7])

ies, and various constraints, enabling the simulation of realistic physical behaviors.

In the context of modeling the wheel-soil interaction of a planetary rover, Project Chrono can be employed for the modeling of deformable terrain, such as sand and gravel, making it suitable for simulating the interaction between rover wheels and the planetary surface. Furthermore, Chrono provides tools for defining contact and friction models, which are crucial for simulating the wheel-soil interaction accurately.

Gazebo Simulator with Plugins

The Gazebo simulation environment is currently used in the DeLeMIS project and is closely linked to the Robot Operating System (ROS). URDF files can be used to adapt the behavior of a robot. In addition, the functionality of Gazebo can be extended via plugins. In the following, the two plugins for improved ground interaction modeling between tires and terrain are detailed. These plugins improve upon the simple default friction coefficient model.

Wheel Slip Plugin

Wheel slip is one of the major tire soil interactions present in planetary rovers. However, in contrast to cars where asphalt provides a nearly uniform terrain, sand and rocklike soil is prone to variations in the amount of slip. Furthermore, most rover tires have shovel-like attachments on the circumference. Thus making it easier to grip in loose sand. However, they do not have rubber-like deformation to adapt to the soil. Even with the help of plugins, Gazebo cannot simulate loose sand. A simple force-based friction model is employed through the use of the Gazebo plugin. The wheel slip plugin allows the definition of a slope angle which based on the slip ratio determines the maximum wheel force. Additionally, a maximum slip angle is defined through the friction coefficient, which states the maximum wheel force possible. An illustration of the slip ratio and wheel force dependency is shown in Fig. 3.

Wheel Plowing Plugin

An additional effect that occurs when driving through loose soil is wheel plowing. At a standstill, a tire will sink into the soil relative to the wheel's normal force, contact surface area, and the soil's properties. When the vehicle starts to move, the soil starts to build up in front of the tire thus providing a force that counteracts the movement. The plowing angle perpendicular to the tire plane on stand still rotates with the velocity and acts opposed to the direction of travel. This is illustrated in Fig. 4. A plugin implementing these mechanics has already been developed by NASA for the Gazebo simulation [2].

Parameter Optimisation

While the aforementioned plugins can greatly enhance the degree of realism regarding the tire-soil interaction, one problem that remains is the estimation of appropriate parameters. It is desirable to have the simulated rover behave as closely related to the LRU as possible. Therefore a method is proposed that through the help of Bayesian optimization minimizes the deviation of the simulated rover from the real rover using Bayesian optimization. As a starting point, the positional data of the real rover was collected

Figure 5: The top image shows the Planetary Exploration Lab at the Institute of Robotics and Mechatronics, German Aerospace Center (DLR). The lower image shows the same test site modeled in the Gazebo environment with integrated wheel slip and plowing plug-ins.

through multiple runs of trajectory following scenarios in the Planetary Exploration Lab. The same trajectories are then imported into the Gazebo simulation and followed, employing the same controller used on the real rover. A loss function is then used to numerically describe between deviation between the real rover trajectory and the simulated one. It has been found empirically, that the summed squared error can better model this deviation compared to the mean squared error. The plugin parameters, consisting of 7 distinct values across the two plugins, are then used in the Bayesian optimization while minimizing the loss function.

In Fig. 5, the rover is depicted both in the Planetary Exploration Lab at the DLR's test facility and in a simulated environment in Gazebo, as shown in the unified representation of the two scenarios.

Testing Environment

The Planetary Exploration Lab at the German Institute of Aerospace (DLR) serves as a testing ground for simulating diverse terrains. A notable feature of this facility is its adjustable slope, allowing for the simulation of various inclines and challenging obstacles. The Lightweight Rover Unit, developed by the DLR, plays a pivotal role in the simulation process [6]. Weighing approximately 30 kg, this rover boasts a payload capacity of 5 kg, with the payload utilized for a computer vision head. The inclusion of visual odometry enables the rover to localize itself within its environment autonomously.

Data Collection

To bridge the gap between simulation and reality, positional data from the rover is collected. This includes both the rover position and orientation. Multiple runs of trajectory following within the Planetary Exploration Lab serve as the basis for this data collection. The same trajectories are then replicated in the Gazebo simulation, employing an identical control structure as that used on the real rover.

Control Structure

To ensure repeatability and comparability of the results, the optimization process cannot consider any adaptive control laws. Therefore, a non-linear controller from [1] is used as the controller structure to steer the rover along the trajectory. This controller is a full-state feedback controller that operates around the linearized trajectory. The controller's gain parameters are set identically between both the real rover and the simulation.

Gaussian Process Regression

The parameter estimation process employs Bayesian optimization, specifically Gaussian process regression [8]. This technique minimizes the deviation between the behavior of the simulated rover and the Lightweight Rover Unit. As previously mentioned, the summed squared error is used as the loss function. At the beginning, the parameters are initialized randomly. Then the predefined trajectory is run with the applied parameters and the deviation is determined. A new iteration of the optimization is then started and the resulting parameters are evaluated again. This process is repeated until a predefined threshold is reached.

3   Results

Performance Results

Performance assessments were conducted on Project Chrono using the built-in Curiosity rover scenario across diverse hardware platforms. By measuring the step time, the real-time factor can be calculated as:

$$\text{Real-time Factor} = \frac{1}{\text{Step Time (ms)} \times \frac{\text{steps per second}}{1000}}.$$

On the i7 system, a step time of approximately 260 ms was measured, while the Ryzen 9 system exhibited improved performance with a step time reduced to 170 ms. Detailed performance metrics are presented in the accompanying Table 1. Gazebo simulations, tested exclusively on the i7 system, achieved real-time factors of 1.8 with and without the implemented plugins.

Optimisation Results

The results of the parameter optimization are shown in Fig. 6. Here, the largest deviations occurred on the ramp due to

Platform	CPUs	Step Time	RT Factor
Intel Xeon 6134	12	430ms	0.023
Core i7 11850H	8	260ms	0.038
Ryzen 7 5900X	12	170ms	0.058
NVIDIA DGX	96	303ms	0.033

Table 1: Results are obtained from simulations using Project Chrono with the Curiosity rover scenario. Each row corresponds to a specific hardware platform, providing details on the number of CPU cores, step time (in milliseconds), and the calculated real-time factor for the respective step time.

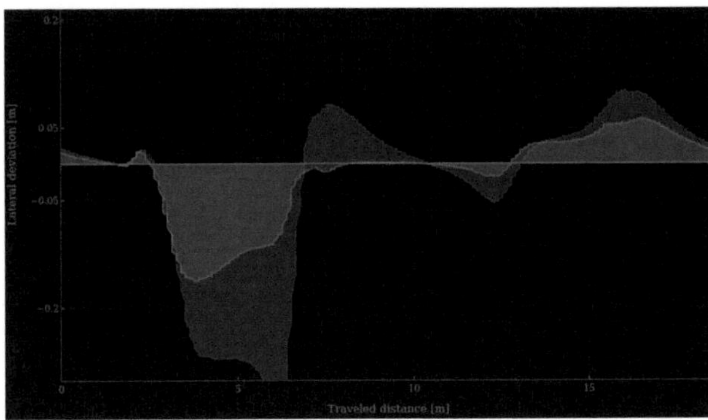

Figure 6: The Lateral Deviation between the simulated rover and the real rover reference data. Results before the optimisation process are shown in red and with optimised parameters in green.

strong slippage. By optimizing the parameters, the deviation in this section was reduced by several centimeters.

4 Discussion

Although Gazebo continued to run at a real-time factor of 1.8 with the plugin enhancements, Project Chrono's runtime results were not very useful for the intended purpose of testing different control strategies before deploying them on real hardware. However, a limitation of the particle calculation in the immediate vicinity of the tire contact area is currently under development. Additionally, the use of real rover data improved the realism of the Gazebo simulation through the use of plugins, meeting the project requirements. In the future, additional plugins may be used, which could include a contact model based on particles.

5 Conclusion

This research addresses two critical aspects: conducting a runtime comparison between two rover terrain interaction simulations and deriving simulation parameters from real rover data. The runtime comparison revealed that Project Chrono, despite featuring a more realistic soil interaction model, is unsuitable for the development of different learning-based control strategies when compared to Gazebo. The optimization process showed that using actual rover data improves the slip and plowing behavior of the simulated rover. However, there are still opportunities for additional research to create plugins that increase realism without significantly slowing down the simulation.

Acknowledgement

The work has been carried out at e:fs TechHub GmbH and supervised by Prof. Dr. Georg Schildbach, Institute of Electrical Engineering in Medicine, Universität zu Lübeck.

Authors' Statement

Conflict of interest: Authors state no conflict of interest.

6 References

[1] N. Baldauf, A. Turnwald, *Iterative learning-based model predictive control for mobile robots in space applications*, 27th International Conference on Methods and Models in Automation and Robotics (MMAR), 2023

[2] M. Allan et al., *Planetary Rover Simulation for Lunar Exploration Missions*, IEEE Aerospace Conference, 2019

[3] R. Zhou, W. Feng, L. Ding, H. Yang, H. Gao, G. Liu, Z. Deng, *MarsSim: A High-Fidelity Physical and Visual Simulation for Mars Rovers*, IEEE Transactions on Aerospace and Electronic Systems, 2023

[4] A. Tasora, R. Serban, H. Mazhar, and M. Pazouki, *Chrono: An Open Source Multi-physics Dynamics Engine*, High Performance Computing in Science and Engineering, Springer International Publishing, 2016

[5] R. Serban, M. Taylor, A. Tasora, and D. Negrut, *Chrono::Vehicle: template-based ground vehicle modelling and simulation*, International Journal of Vehicle Performance, 2019

[6] M. J. Schuster et al., *Towards Autonomous Planetary Exploration*, Journal of Intelligent & Robotic Systems, 2019

[7] S. C. Peters, *Gazebo Wheel Plowing Physics Plugin*, Functionality described in Pull Request 3164., https://github.com/gazebosim/gazebo-classic/pull/3164, 2019

[8] E. Brochu and V. M. Cora, N. de Freitas, *A Tutorial on Bayesian Optimization of Expensive Cost Functions, with Application to Active User Modeling and Hierarchical Reinforcement Learning*, ArXiv, 2010

Pattern Recognition and Machine Learning-driven Classification of Solid-State Polymer Signals in Continuous Wave NMR

Agastya Heryudhanto [1], Jörg Schroeter [2], Max Urban [2].
[1] Biomedical Engineering, Luebeck University of Applied Sciences, agastya.heryudhanto@stud.th-luebeck.de
[2] Medical Sensors and Devices Laboratory, Luebeck University of Applied Sciences, {joerg.schroeter, max.urban}@th-luebeck.de

Abstract

This study explores material identification using Continuous Wave (CW) Nuclear Magnetic Resonance (NMR) spectroscopy signals acquired from a CW-NMR device. Focusing on polystyrene, glycerol, and teflon, the study examines their raw NMR data using pattern recognition techniques. Raw signals obtained from the Leybold-NMR device were quantified, and the dissimilarity between datasets was assessed using euclidean distance and cosine similarity. The most similar material showed a euclidean distance of 3.3 and a cosine similarity of 0.6. Classification models: logistic regression, naive Bayes, and support vector machine, achieved 100% accuracy with five principal components. This consistency in results is attributed to the uniform modulation magnetic strength (5 Vpp) and frequency (21 Hz) of the raw data, which yielded highly repeatable signals. Future works should further assess the model by varying the CW modulation field.

1   Introduction

This paper builds upon several prior milestones in the overarching project conducted within our laboratory for constructing a benchtop Continuous Wave (CW) Nuclear Magnetic Resonance (NMR) machine for mobile spectroscopy. The initial study began with the work by Grundman, which explored the feasibility of constructing a benchtop Continuous Wave (CW) Nuclear Magnetic Resonance (NMR) machine for mobile spectroscopy. Grundman's exploratory research laid the groundwork for the current pursuit of pattern recognition within the field of NMR technology to address the low signal-to-noise ratio in CW-NMR[1].

Subsequently, Mehta's contributions propelled the project forward by enabling the initial digitization of analog CW signals with an STM32 microcontroller and subsequently improving the speed of digitizing these raw analog signals. This advancement allowed the ability to capture and process NMR data efficiently for signal processing [2].

Following Mehta's work, Subedi's research focused on automating signal acquisition and analysis by the CW-NMR device with the STM32 microcontroller. By developing a digital control system, Subedi enabled the automated data acquisition of analog CW-NMR signals, allowing efficient data collection and analysis [3].

Building upon these advancements, the study aims to establish a reliable classification model for material identification. Through the application of classical machine learning techniques such as logistic regression, naïve bayes, and support vector machine (SVM), the effectiveness of machine learning models on raw CW-NMR signals is examined.

2   Material and Methods

2.1   Hardware Setup

2.1.1   CASSY Module for Signal Acquisition

Analog signals of glycerol, polystyrene and polytetrafluoroethylene (PTFE) from the Benchtop NMR device were acquired using a Sensor-CASSY. CASSY Lab 2 software facilitated the recording and analysis of NMR signals, providing a graphical representation with time on the X-axis and voltage on the Y-axis (refer to Figure 1).

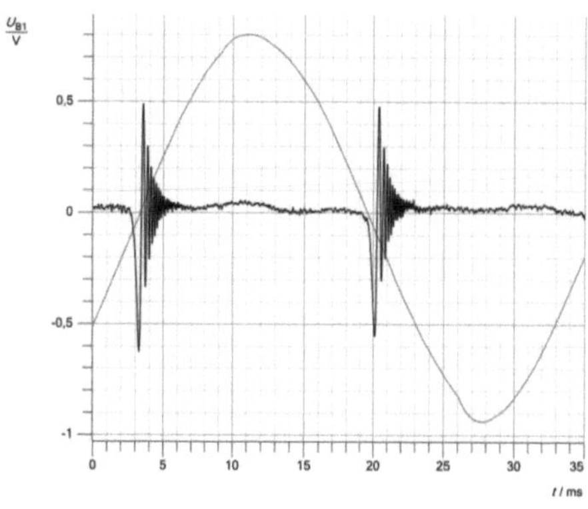

Figure 1: Clean NMR Signals.

2.1.2   NMR Apparatus Setup

The NMR apparatus utilized the nuclear magnetic resonance technique, with the U-core housing 10 A coils for generating a consistent magnetic field. Modulation coils provided the necessary magnetic modulation field during experiments.

2.1.3 Function Generator

A function generator modulated the magnetic field during experiments, capable of generating various waveforms across a broad frequency range. In this project, a sinusoidal signal with a frequency of 21 Hz and an amplitude of 5 Vpp was employed.

2.2 Software Setup

2.2.1 Virtual Machine and Machine Learning Frameworks

For development and training, a virtual machine on Google Colab was utilized with Python and libraries such as NumPy, Pandas, Matplotlib, and Scikit-learn (Sklearn), which significantly reduced training time.

2.2.2 Data Splitting for Training and Testing

Data for glycerol, polystyrene, and polytetrafluoroethylene (PTFE) was divided into training and testing sets. This splitting is vital for training and testing classification. It prevent overfitting and ensures accurate evaluation of the model's generalization ability.

2.2.3 Model Training and Validation

Validation of machine learning models during training is a necessary step to prevent overfitting. It uses testing dataset to ensure models generalize well to new and unknown data.K-cross-validation and evaluation metrics: confusion matrix and F1 score, assess model performance.

2.2.4 Principal Component Analysis (PCA)

Principal Component Analysis (PCA) was applied to analyze and interpret the raw data. This statistical method simplifies the complexity in high-dimensional data while retaining trends and patterns through decomposition. PCA helps in extracting the most significant features and patterns within the NMR signals, contributing to the overall analysis and material identification process [4].

2.2.5 K-fold Cross-Validation

K-fold Cross-Validation quantified model performance and generalization capability by dividing the dataset into K subsets, training on K-1 folds, and validating on the remaining fold, providing a reliable estimation of effectiveness[5].

2.2.6 Confusion Matrix and F1 Score

The Confusion Matrix provides a detailed breakdown of the model's performance in true positives and negatives. The F1 Score, the average of precision and recall, gives a balanced metric on the model's accuracy.[5].

3 Results and Discussion

3.1 Raw Signal Analysis

The raw data obtained from the CASSY module displayed distinct signals for the materials glycerol, polystyrene, and PTFE (see Figure 2 - 4).

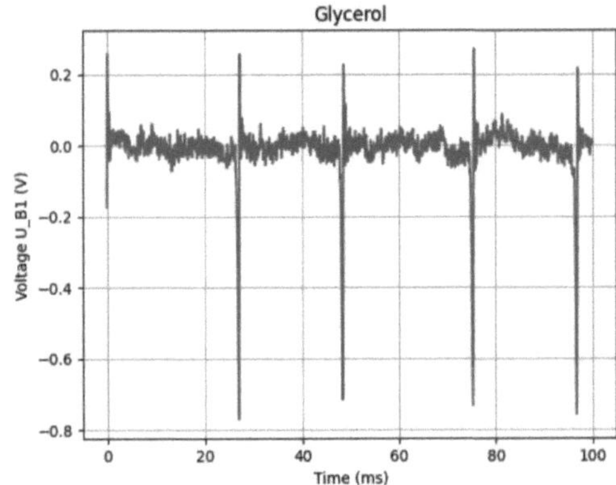

Figure 2: Raw signals for glycerol.

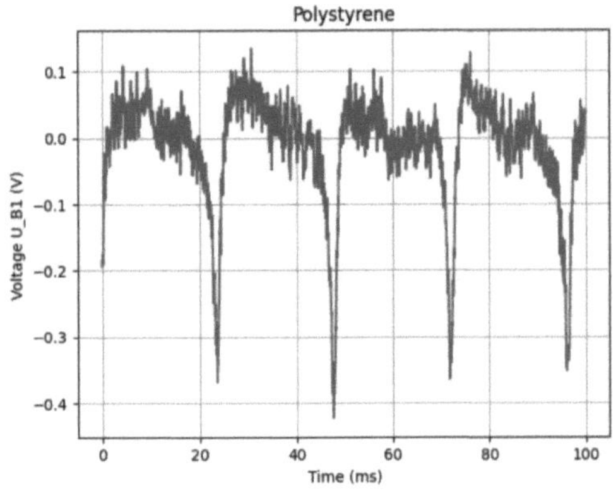

Figure 3: Raw signals for polystyrene.

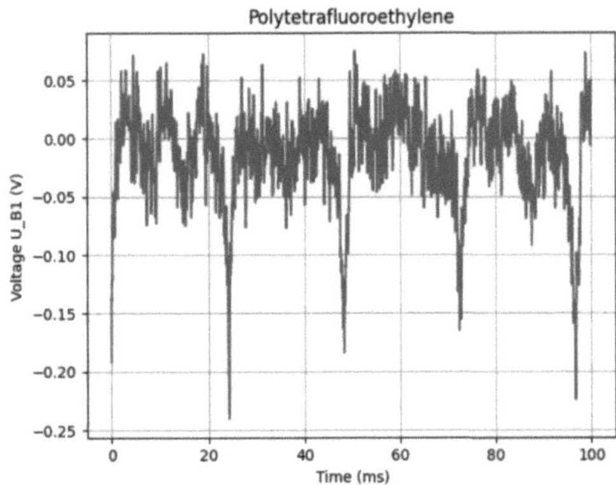

Figure 4: Raw signals for Tetrafluoroethylene.

3.1.1 Interpretation of Similarity Measurements

The euclidean distances and cosine similarities quantified correlation in terms of straight-line distance (euclidean) and cosine angle (cosine similarity), extracting shape similarity due to identical frequency and phase. For reference, the pairs (1, 2), (1, 3), and (2, 3) correspond to glycerol, polystyrene, and PTFE (teflon), respectively.

- **Euclidean Distance**:
    - Pair (1, 2): 5.78
    - Pair (1, 3): 4.33
    - Pair (2, 3): 3.27

- **Cosine Similarity**:
    - Pair (1, 2): -0.0017
    - Pair (1, 3): 0.1023
    - Pair (2, 3): 0.6076

- Signals of polystyrene and glycerol exhibit the greatest difference.

- Signals of polystyrene and teflon show moderate, yet still significant, difference.

- Signals of glycerol and teflon are relatively similar compared to glycerol and polystyrene.

The observed signals clearly vary for each material. Cosine similarity shows relatively high positive correlation between signals of glycerol and teflon, and nearly orthogonal directions for signals of polystyrene and glycerol. Both metrics paired together show the similarity between the solid state materials teflon and polystyrene and their distinction from the liquid material glycerol.

3.2 Dimensionality Reduction Strategy

Dimensionality reduction is a critical step in preparing data for training machine learning models and neural networks. Principal Component Analysis (PCA) is the most commonly used technique for reducing dimensions. PCA extracts the most important features in the data and represents them as a set of new variables, as principal components.

The cumulative explained variance graph, depicted in Figure 5, shows the effectiveness of PCA in capturing the variance in the raw data. With as few as one component, PCA captures 95% of the variance in the entire dataset(glycerol,polystyrene and teflon), while with ten principle components(PCs), 99% of the variance is captured.

However, despite the results shown in the cumulative explained variance graph, there are considerations to using PCA for our dataset. The presence of significant noise near the time base of relevant signals and a signal-to-noise ratio that could impede PCA. Undesired signals are abundantly present before PCA, the resulting principle components will represent them in proportion to the desired signals.

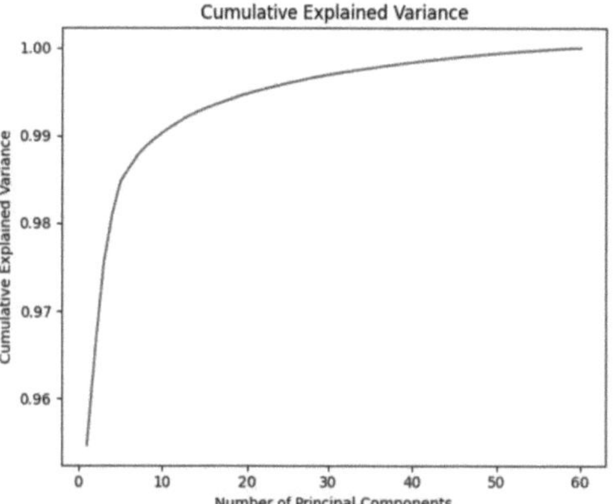

Figure 5: Explained Variance.

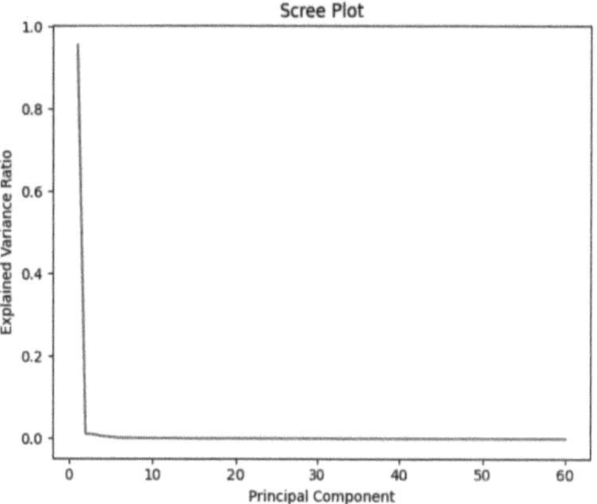

Figure 6: Scree plot.

Figure 6 depicts a scree plot, which displays the explained variance ratio when compared to the principle components chosen. Typically, an 'elbow' will form, and indicates the compromise of data loss through simplification and effective capturing of signals. The results are components here are that the elbow occurs between 2 and 10 PCs could be sufficient components in order to retain the relevant information to classify correctly.

3.3 Model Results and Identical Outputs

600 thousand datapoints represent 20 samples of each material in the dataset used. Without PCA, 100% was achieved at great computational cost. The results with PCA for the three classification methods SVM, Naive Bayes(NB), and Logistic Regression(LR) yielded identical outputs for 2 PCs for 87.5% (see figure 7 -9). The graphs highlight the feasibility of visualisation when smaller PCs are chosen, yet leads to information loss and difficulty for models to perform. The models produces identical 100 percent accuracy when 5 principles components(PCs) were chosen. However logistic regression(LR) and naive bayes(NB) performs well as principle components are increased, SVM perform unreliably as the number of principle components increases, with accuracy reducing at 3 PCs when compared when 2 PCs were chosen, as well as performing poorer at 3 PCs

compared to LR and NB (see Table 1).

Method	None	2 PC	3 PC	4 PC	5 PC
LR	100%	87.5%	87.5%	93.75%	100.0%
NB	100%	87.5%	87.5%	93.75%	100.0%
SVM	100%	87.5%	85.4%	91.66%	100.0%

Table 1: Accuracy of models with different numbers of principal components.

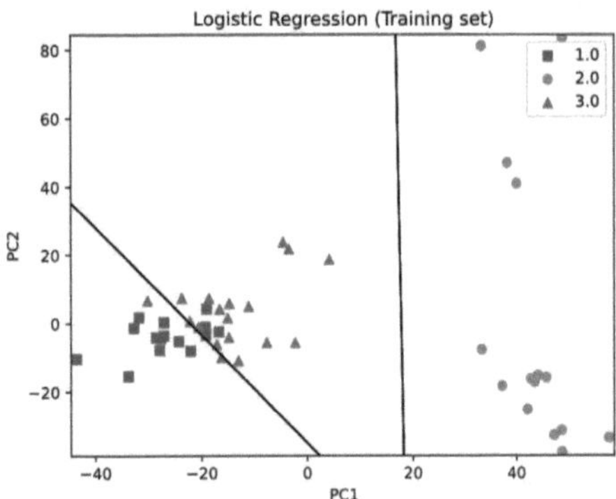

Figure 7: Logistic Regression with 2 PCs.

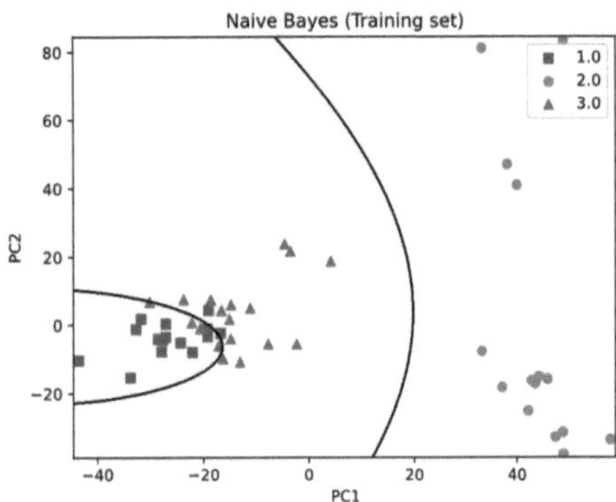

Figure 8: Naive Bayes with 2 PCs.

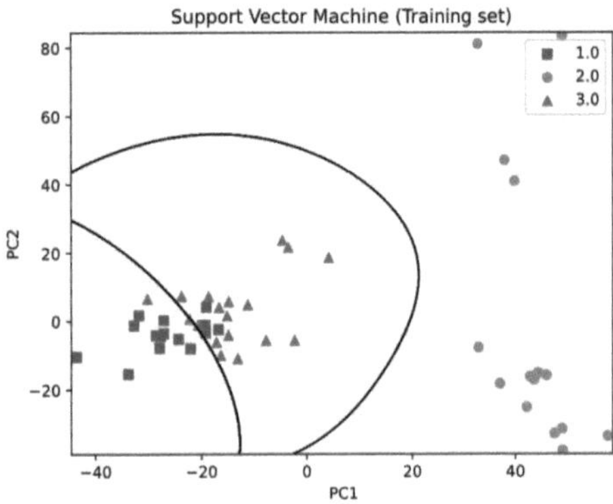

Figure 9: Support Vector Machine with 2 PCs.

4 Conclusion

The exploration into material identification using CW NMR signals revealed insights into the trade-off between dimensionality reduction and efficacy of models. Clear distinctions in raw signals for glycerol, polystyrene, and PTFE shows the potential of benchtop NMR devices for applications involving materials with distinct resonance signals, quantified with high euclidean distance of 3.3 amongst the most similar datasets, a cosine similarity of 0.6 which indicates significant differences. Achieving 100% accuracy in classification methods was feasible with five principal components, due to the near-ideal nature of the raw data. The consistency of the results were the result of a single modulation magnetic strength of 5 Vpp and a modulation frequency of 21 Hz, yielding highly repeatable signals. Pairwise comparisons revealed varying degrees of similarity, with cosine similarity showing correlations between similar solid-state datasets, indicating the need for robust classification methods when similar CW-NMR samples are to be distinguished from each other.

Moving forward, future works should explore the robustness of classification methods under more realistic conditions, considering variations in modulation magnetic strength, frequency and introduce non-ideal conditions.

Acknowledgement

This work was carried out in the Labor für Sensor-und Gerätetechnik in the institute of Biomedical engineering, Universität zu Lübeck.

Authors' Statement

Conflict of interest: Authors state no conflict of interest

5 References

[1] C.Grundman, *Digitization of Continuous Wave NMR.* In: BioMedTec Studierendentagung 2023, Lübeck, 2023.

[2] S.Mehta, *Development of fast digitalization for analog NMR signals of a benchtop NMR system.* Bachelor Thesis, Technische Hochschule, Lübeck, 2023.

[3] M. Subedi, *Development of a Control System for Digital Benchtop Nuclear Magnetic Resonance (NMR) Spectrometer.* Bachelor Thesis, Technische Hochschule, Lübeck, 2023.

[4] C.M. Bishop, *Pattern Recognition and Machine Learning.* New York, NY: Springer, 2006.

[5] scikit-learn Machine Learning in Python *Machine Learning in Phyton.* Available: https://scikit-learn.org/stable/ [last accessed on 2024-01-23].

8

Biochemical Physics

Structures and dynamics of DNA three-way junctions: The role of interrupts in hairpins formed by CAG repeats

Svea J. Wilken [1], Gillian Cadden [2], and Steven W. Magennis [2]

[1] Biophysics, Universität zu Lübeck, svea.wilken@student.uni-luebeck.de
[2] School of Chemistry, University of Glasgow, {gillian.cadden, steven.magennis}@glasgow.ac.uk

Abstract

It is known that triplet repeat (TR) expansion causes various neurodegenerative diseases, but the underlying mechanism is not yet fully understood. A more detailed understanding could support the development of a treatment that inhibits expansion. The formation of DNA secondary structures and their dynamics are a key factor in the expansion mechanism. We used a combination of multiparameter fluorescence detection (MFD) and total internal reflection fluorescence microscopy (TIRF) to analyse the structure and dynamics of mobile DNA three-way junctions (3WJs) with interrupted CAG repeats in the hairpin. We compared our results with previously measured uninterrupted hairpin structures. Our findings suggest interrupts in 3WJs suppress branch migration. The reduced branch migration in interrupted repeat structures might be one reason, why genes with interrupts expand less compared to uninterrupted ones.

1   Introduction

With its basic sequence, the coding regions of a deoxyribonucleic acid (DNA) define the amino acid arrangement of all synthesised proteins. Erroneous base sequences thus can cause erroneous proteins. This can be very hazardous to organisms and is the reason for several diseases. One such is Huntington's disease (HD), a hereditary disease caused by expansion of the CAG tract on the huntingtin (HTT) gene [1]. In healthy individuals, the HTT gene contains 26 CAG repeats or less [2]. Genes with repeats greater than 26 become unstable and elongation of the CAG tract from one generation to the next may occur (germline expansion) [2]. This increases the risk for HD in the next generation, as a DNA with more than 39 CAG repeats on the HTT gene reliably causes HD [3]. In diseased organisms, elongation of the CAG tract during a lifetime (somatic expansion) contributes to progression of HD [4]. A higher number of CAG repeats increases the propensity for the somatic expansion [4] and thus accounts for an earlier onset of disease and for a higher severity [2].

The secondary structure of DNA and its rearrangements influences the mismatch repair (MMR) process and with this the probability of triplet repeat (TR) expansion [4], [5], [6]. This opens up one opportunity of tackling HD and reinforces the relevance of researching the structure and dynamics of DNA segments with CAG repeats. It has been previously shown that interruptions in CAG repeats impact DNA secondary structures [6]. The influence of interrupts is of particular interest, because only the number of uninterrupted CAGs determines the age of onset in Huntington's disease [1], [6].

In vitro experiments showed, that non-canonical base pairing within TR slipped-strand structures facilitates the formation of stable intrastrand hairpin-loop structures [6]. Mobile three-way junctions (mobile 3WJ) with repeats protruding from the hairpin show two different kinds of dynamics, "local conformational motions" at the branchpoint and "longer-range branch migration" [5] (Fig. 1).

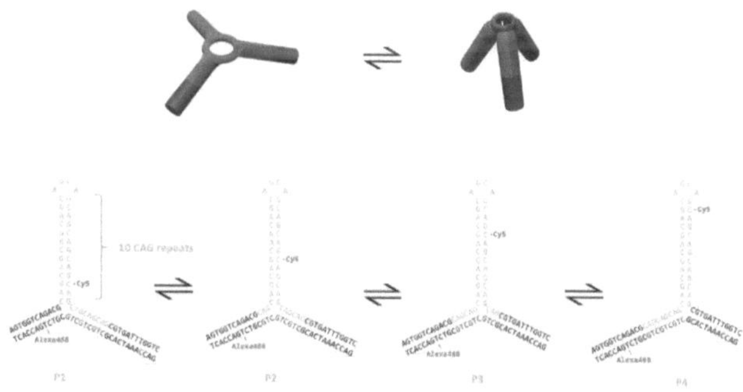

Figure 1: Dynamics of mobile three-way junctions: Local conformational motions (top) and longer-range branch migration with four positional isomers (bottom). The figure is taken from [5].

The migration dynamics of a 3WJ are influenced by the hairpin length, as small loops migrate faster than long hairpins [5]. Experiments with CAG repeats of different parity that were attached to a biotin anchor on one side and free on the other indicated, that slipping dynamics of hairpins are influenced by repeat parity. Weninger *et al.* demonstrated that 5'-AGCA-3' tetraloops, which were only formed by even-numbered repeats without having hanging trinucleotides on the free side of the hairpin strand, are more stable than 5'-CAG-3' triloops, causing structures with even numbered repeats to be less dynamic [6]. It was also shown that interruptions near the hairpin loop could reduce strand slipping and stabilise the triloops [6].

It has been previously proposed that loop migration may influence the probability for TR expansion [5]. The migration might allow two complementary hairpins on opposite strands to travel apart, diminishing the opportunity to collapse back into a DNA double strand [6]. MMR machinery may then recognise mismatched bases in the hairpins and introduce nicks, causing hairpin opening and a subsequent refilling with complementary nucleotides resulting in TR expansion [6]. MMR enzymes have been shown to bind to small loop structures [7]. Long hairpins may facilitate TR expansion by blocking smaller loops, delaying their merging and allowing more time for MMR machinery to incorrectly process small loops [5].

We used a combination of multiparameter fluorescence detection (MFD) [8] and total internal reflection fluorescence microscopy (TIRF) [9] to build on the experiments by Hu *et al.* [10] and analysed the structure and dynamics of further DNA 3WJs containing TRs. We introduced interrupts in hairpins that were formed by even or odd numbers of CAG repeats to investigate their impact on slipping dynamics. Our findings are relevant for further understanding of TR expansion in HD and the development of treatments, as there are no FDA-approved disease-modifying treatments until today [2].

2   Material and Methods

The structure and dynamics of two different mobile DNA 3WJs were probed. Interrupts were introduced in 3WJs with 10 (iCAG$_{10}$) or 11 (iCAG$_{11}$) CAG repeats forming the hairpin. The results were compared with previously measured static and mobile CAG$_{10}$ and CAG$_{11}$.

The annealing of the acceptor strand which formed the hairpin (end concentration of 15 µM, ATDBio) with the complementary donor strand (5 µM, ATDBio) was executed in a buffer solution (20 mM Tris (Sigma-Aldrich), 50 mM NaCl (Fluka), pH 7.5). For MFD, a buffer (40 mM Tris, 15 mM NaCl, 1 mM ascorbic acid (Fluka), pH 7.5) was used to dilute the annealed DNA to a single molecule concentration. For TIRF, a slightly different buffer solution (20 mM Tris, 15 mM NaCl, 6 % glucose (Sigma-Aldrich), pH 7.5) was used to dilute the annealed DNA to a concentration of 10-30 pM. A solution of 97 µL TIRF buffer, 2 µL Trolox (2 mM) and 1 µL Gloxy (10 mg oxidase (Sigma-Aldrich) and 20 µL glucose catalase (Sigma-Aldrich) in 100 µL 20 mM Tris at pH 8.0) was flushed through the slide after the DNA was injected to reduce photobleaching and blinking.

For the MFD measurements, a home-built system was used [10]. Pulsed, linearly polarized laser light with a wavelength of 483 nm was focused into the sample to excite the donor (Alexa 488) [10]. The fluorescence from the donor and from the acceptor (Cy5) emitted light was separated into parallel and perpendicular and this light into donor and acceptor components and detected via four avalanche photodiodes [10]. For the data analysis, software provided by Prof. Claus Seidel, Heinrich-Heine-Universität Düsseldorf, was used.

The TIRF experiments were carried out on a custom-built objective-type TIRF microscope [10]. The DNA was immobilized in a microchannel via a neutravidin-biotin interaction and then excited using an evanescent wave of a 488 nm laser beam via total internal reflection [10]. The fluorescence from the donor and from the acceptor emitted light was separated and then detected via an EMCCD camera [10]. Emission intensities were saved in movies of 25 seconds via the ImagePro-Plus v7.0 software and then extracted using TwoTone (v3.1) [10]. The time traces were further analysed using HaMMy v4.0 and MASH-FRET 1.3.2.

3   Results and Discussion

We did MFD experiments with the mobile iCAG$_{10}$ and compared our results with the MFD plots of mobile and static CAG$_{10}$ from Hu *et al.* [10]. The MFD plots are shown in Fig. 2. Besides the donor-only peak at zero FRET, we only observed high FRET states (the FRET efficiency (E_{FRET}) was distributed around 0.82) for the mobile iCAG$_{10}$ (Fig. 2c). The graph is similar to the one measured for static CAG$_{10}$, which shows two high FRET states ($E_{FRET, 1} \approx 0.75$ and $E_{FRET, 2} \approx 0.88$) because of the local conformational motions (Fig. 2b) [10]. This similarity of the FRET states suggests that the majority of iCAG$_{10}$ molecules perform similar motions as the static CAG$_{10}$ (no slipping) and allows conclusions about the most favoured positional isomer.

To analyse the dynamics, we carried out TIRF experiments with mobile iCAG$_{10}$ and measured the FRET signal of individual immobilized molecules over a time of 25 seconds (Fig. 4a). We observed good correlation with our MFD results (compare E_{FRET} histograms of Figs. 2c and 4b), with the majority of transitions occurring between two high FRET states ($E_{FRET} \approx 0.69$ and 0.84) (Fig. 4c), indicating the local conformational motions. We observed a very small proportion of molecules undergoing transitions between the high and one medium FRET state ($E_{FRET} \approx 0.35$) (Fig. 4c), which may result from minimal slipping. Our results suggest the addition of the interrupt greatly suppresses branch migration compared to mobile CAG$_{10}$ by stabilising one positional isomer. Slipping into the next positional isomer is still possible but less probable. This behaviour can be explained by assessment of the base-pairing of different positional isomers. The most favoured structure is the one with the lowest number of mismatched base pairs.

We also conducted experiments with mobile iCAG$_{11}$ to assess the influence of parity on interrupted structure dynamics. The MFD plots of the static and mobile CAG$_{11}$

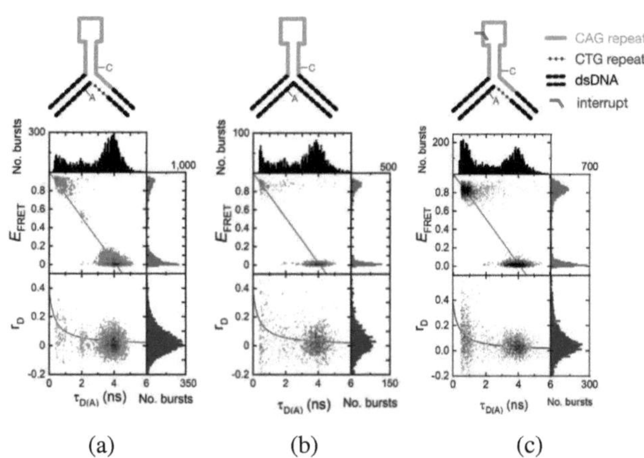

(a) (b) (c)

Figure 2: MFD plots of (a) mobile CAG_{10}, (b) static CAG_{10} and (c) mobile $iCAG_{10}$. Sketched are the FRET efficiency E_{FRET} or the donor anisotropy r_D against the donor lifetime $\tau_{D(A)}$, the grey level indicates the number of single-molecule bursts (increasing from white to black). The line in the upper graphs is the theoretical FRET relationship, $E = 1 + \tau_D / \tau_{D(A)}$, the line in the lower graphs is the Perrin equation, $r_D = r_0 (1 + \tau_{D(A)} / \rho_D)$. The plot of the mobile $iCAG_{10}$ is similar to the plot of the static CAG_{10}. This indicates that an interrupt reduces slipping of mobile 3WJs by stabilising one positional isomer. The MFD plots in Figs. 2a and b are taken from [10].

and mobile $iCAG_{11}$ are shown in Fig. 3. Besides the donor-only peak at zero FRET, we observed multiple FRET states with overlapping populations for the mobile $iCAG_{11}$, the most prominent being the mid-FRET population with $E_{mid\text{-}FRET} \approx 0.45$ (Fig. 3c). Unlike the static CAG_{11} structure in the non-slipped configuration ($E_{mid\text{-}FRET} \approx 0.48$, Fig. 3b), the mid-FRET state is more populated than all other states. One possibility is that the proportion of the different local conformational states changes with the addition of the interrupt. However, since the static CAG_{11} in the slipped-strand configuration (dye being shifted three bases) also contains a mid-FRET population ($E_{mid\text{-}FRET} \approx 0.38$), and we observe small populations with low FRET efficiency ($E_{low\ FRET} \approx 0.07$), this could indicate that slipping is more tolerated for odd numbered hairpins.

With TIRF we measured a FRET efficiency distribution that was similar to that one measured with the MFD, with a mid-FRET population being most prominent ($E_{mid\text{-}FRET} \approx 0.29$) (Fig. 4e). The TIRF analysis via MASH-FRET [11] differentiates between five different FRET states ($E_{FRET} \approx 0.11$, 0.29, 0.44, 0.69 and 0.88) (Fig. 4f), which is one state more than observed for the static CAG_{11} and one state less than observed for the mobile CAG_{11}. There were transitions between all states observable using MASH-FRET (Fig. 4f). Besides the expected local conformational motions, the least populated fifth FRET state with the lowest FRET efficiency indicates that there are a few additional slipping dynamics.
A comparison between data measured for even (Figs. 2

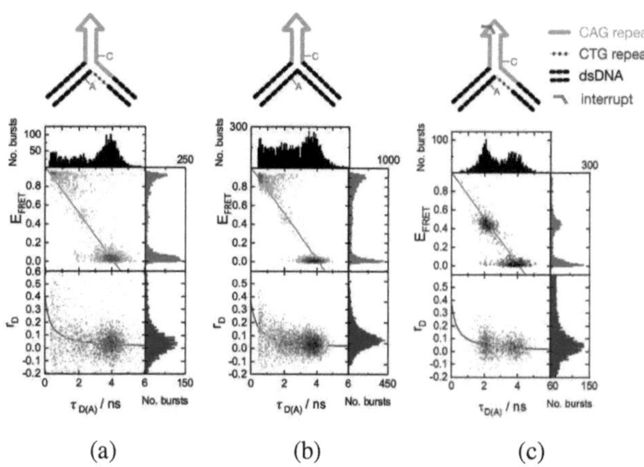

(a) (b) (c)

Figure 3: MFD plots of (a) mobile CAG_{11}, (b) static CAG_{11} and (c) mobile $iCAG_{11}$. Compared to the non-interrupted mobile 3WJ, slipping is reduced with the addition of the interrupt.

and 4a - c) and odd (Figs. 3 and 4d - f) numbered hairpin structures provides evidence that dynamics are affected by parity. Odd numbered samples seem to permit more slipping. Probably, this is because of a frustration of the triloops, which are less stable than the tetraloops that are being formed in even numbered samples [6].

4 Conclusion

We have carried out MFD and TIRF experiments on DNA 3WJs that are believed to be formed in TRs containing genes [6]. The hairpins were formed by CAG repeats of different parity with introduced interrupts to work out their effects on structure and dynamics of 3WJs. A comparison of the behaviour of uninterrupted and interrupted structures could be relevant for a more detailed understanding of TR expansion, because those genes with interrupts expand less compared to uninterrupted ones [6].

We have shown that interrupts in 3WJs reduce the branch migration of even and odd numbered hairpins and increase the number of molecules in the positional isomer with the lowest number of mismatched base pairs. Nevertheless, there are few slipping dynamics occurring. We have also shown that dynamics of interrupted structures are influenced by parity. Odd numbered structures are more dynamic than even numbered ones because of the less stable triloop structure. The reduced branch migration in interrupted repeat structures might be one reason, why genes with interrupts expand less compared to uninterrupted ones.

In future projects, a dwell time analysis of the TIRF data could be carried out to get more detailed information about the slipping kinetics which then could be used to work out more differences between even and odd numbered hairpins.

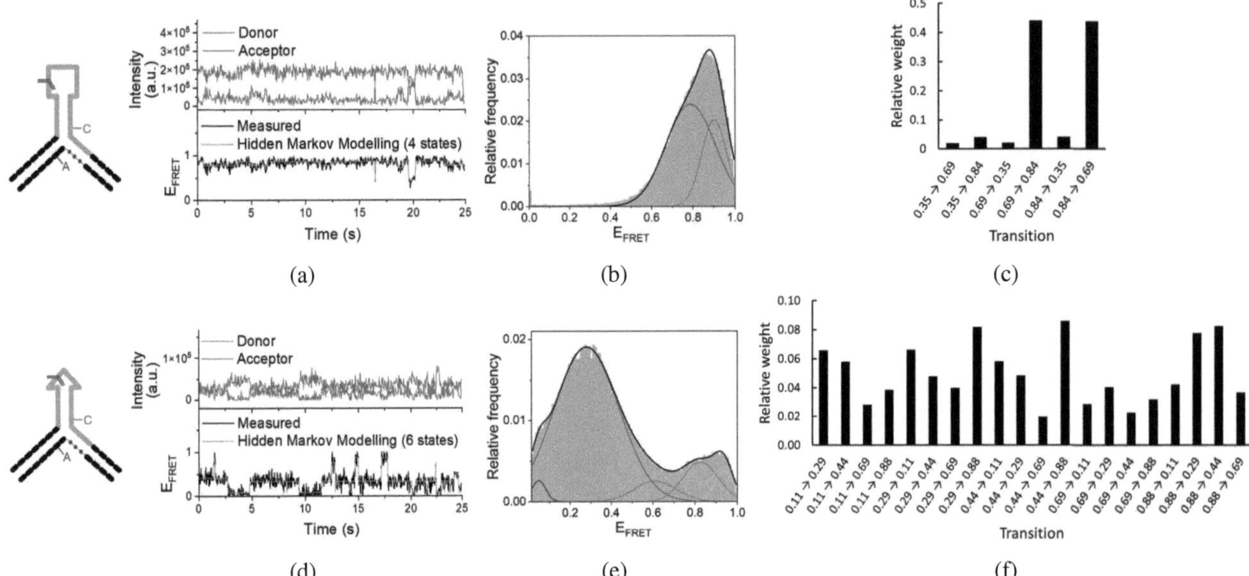

Figure 4: Results of TIRF experiments with the even (top row) and odd (bottom row) numbered 3WJ. (a, d) Exemplary TIRF signal (top) with corresponding FRET efficiency E_{FRET} (bottom). (b, e) E_{FRET} histogram combining all dynamic traces. (c, f) Relative weight of transitions between the observed FRET states.

Acknowledgement

The work has been carried out in the laboratory of Steven Magennis at the University of Glasgow, School of Chemistry and supervised by Y.-H. Song, Institute of Physics, Universität zu Lübeck.

Authors' Statement

Conflict of interest: Authors state no conflict of interest.

5 References

[1] G. M. of Huntington's Disease (GeM-HD) Consortium, "CAG repeat not polyglutamine length determines timing of Huntington's disease onset," *Cell*, vol. 178, no. 4, pp. 887–900, August 2019.

[2] S. Ahamad and S. A. Bhat, "The emerging landscape of small-molecule therapeutics for the treatment of Huntington's disease," *Journal of Medical Chemistry*, vol. 65, no. 24, pp. 15 993–16 032, December 2022.

[3] P. McColgan and S. J. Tabrizi, "Huntington's disease: a clinical review," *European Journal of Neurology*, vol. 25, no. 1, pp. 24–34, January 2018.

[4] L. Jones, H. Houlden, and S. J. Tabrizi, "DNA repair in the trinucleotide repeat disorders," *Lancet Neurol*, vol. 16, pp. 88–96, January 2017.

[5] S. Bianko, T. Hu, O. Henrich, and S. W. Magennis, "Heterogeneous migration routes of DNA triplet repeat slip-outs," *Biophysical Reports*, vol. 2, no. 3, p. 100070, September 2022.

[6] P. Xu, F. Pan, C. Roland, C. Sagui, and K. Weninger, "Dynamics of strand slippage in DNA hairpins formed by CAG repeats: roles of sequence parity and trinucleotide interrupts," *Nucleic Acids Research*, vol. 48, no. 5, pp. 2232–2245, January 2020.

[7] R. R. Iyer and A. Pluciennik, "DNA mismatch repair and its role in Huntington's disease," *Journal of Huntington's Disease*, vol. 10, no. 1, pp. 75–94, February 2021.

[8] C. Eggeling, S. Berger, L. Brand, J. Fries, J. Schaffer, A. Volkmer, and C. Seidel, "Data registration and selective single-molecule analysis using multiparameter fluorescence detection," *Journal of Biotechnology*, vol. 86, no. 3, pp. 163–180, April 2001.

[9] D. Axelrod, "Total internal reflection fluorescence microscopy," *Methods in Cell Biology*, vol. 89, pp. 169–221, 2008.

[10] T. Hu, M. J. Morten, and S. W. Magennis, "Conformational and migrational dynamics of slipped-strand DNA three-way junctions containing trinucleotide repeats," *Nature Communications*, vol. 12, no. 204, January 2021.

[11] M. C. A. S. Hadzic, R. Börner, S. L. B. König, D. Kowerko, and R. K. O. Sigel, "Reliable state identification and state transition detection in fluorescence intensity-based single-molecule Förster resonance energy-transfer data," *J. Phys. Chem. B*, vol. 122, no. 23, pp. 6134–6147, May 2018.

Investigation of different reaction coordinates in MD simulations of the Ala$_9$ peptide at different pressures and temperatures

Yannik Kasprzak[1], Jessica Rückert[2], Niclas Ludolph[3], and Hauke Paulsen[4]

[1] Biophysics, Universität zu Lübeck, yannik.kasprzak@student.uni-luebeck.de
[2] Medical Laser Center Lübeck, Lübeck, j.rueckert@uni-luebeck.de
[3] Institute of Physics, Universität zu Lübeck, n.ludolph@student.uni-luebeck.de
[4] Institute of Physics, Universität zu Lübeck, hauke.paulsen@uni-luebeck.de

Abstract

Protein unfolding commonly happens at high temperatures or pressures or in the presence of a denaturants. *Molecular dynamics* (MD) simulations offer valuable insights into the unfolding and stabilisation processes and can help to identify suitable *reaction coordinates*, such as the *root mean square deviation, hydrogen bond distances*, and the *end-to-end distance*. To investigate the suitability of these reaction coordinates, the peptide Ala$_9$—the smallest peptide with a stable α-helix—was simulated at different pressures (0.001, 1, 2 and 4 kbar) and temperatures (300, 320, 350 and 370 K). From the resulting trajectories *energy landscapes*, were constructed that can be used to characterise the folding and unfolding processes. All three reaction coordinates investigated here proved to be suitable, but in view of larger proteins the root mean square deviation and the hydrogen bond distances turned out to be particularly promising.

1 Introduction

Proteins are fundamental biomolecules that exist universally in all cells and perform essential tasks such as transporting substances and catalysing chemical reactions in cellular processes [1]. The study of proteins encompasses a vast area of research, with particular interest in the tertiary structure and function of these molecules and the mechanisms involved in protein folding and unfolding, the latter resulting in a decrement of their efficacy [1]. To investigate this phenomenon, variations of temperature, pressure and incorporation of denaturants such as urea can be used in laboratory experiments and in *molecular dynamics* (MD) simulations.

MD simulations allow in-depth analyses of key fold-defining properties of the unfolding process, like unfolding times, rates, and molecular interactions [2]. Given that biomolecules function in an aqueous environment, simulations usually incorporate water-based environments and, therefore, demand challenging large amounts of computing time. One way to circumvent this problem are simulations at high temperatures and a subsequent extrapolation of the data to ambient temperature. Vital to this procedure is raising the pressure in order to maintain the liquid state and prevent the simulated water from evaporating.

The process of a protein unfolding from its folded *native state* can be described using so called *reaction coordinates* [3]. The basic idea behind these reaction coordinates is the following: Most details of atomic motion in the protein are not relevant for the folding process [4]. Therefore, it should be possible without losing much of accuracy to reduce the complexity of an all atom description of this process to one or two degrees of freedom that are called reaction coordinates. Candidates for such coordinates are (i) the *root mean square deviation* (RMSD) between equivalent atoms in different conformations of the protein, (ii) the average value of the *hydrogen bond distances* that contribute to the stabilisation of the native state of proteins, and (iii) the distance between the first and the last atom of the polypeptide chain. The root mean square deviation, defined by

$$\mathrm{RMSD} = \sqrt{\frac{1}{N} \sum_{i=1}^{N} (\mathbf{r}_i - \mathbf{r}_i')^2}\,, \qquad (1)$$

compares the simulated structures (given by vectors $\mathbf{r}_i$ for the N atoms of the protein) with the structure of the native state (given by the vectors $\mathbf{r}_i'$) after aligning both structures. The RMSD can be used to measure the similarity of two different conformations of the same protein [5].

In a previous study [6] the characteristic of the reaction coordinates RMSD and hydrogen bond distance were explored for the protein Ubiquitin. In this study, the suitability of the above mentioned reaction coordinates was investigated at the example of the peptide Ala$_9$, which is one of the smallest proteins that forms a stable α-helix [7].

2 Material and Methods

The peptide Ala$_9$, the molecule under study, consists of nine consecutive alanine residues. The initial structure was created in a previous work [8] using the software *Avogadro* and is shown in Fig. 1. At ambient temperature and pressure the native and the unfolded conformation of Ala_9 can coexist and at elevated pressure a stabilisation of the α-helix is observed as has been shown earlier by MD simulations and experiments [9].

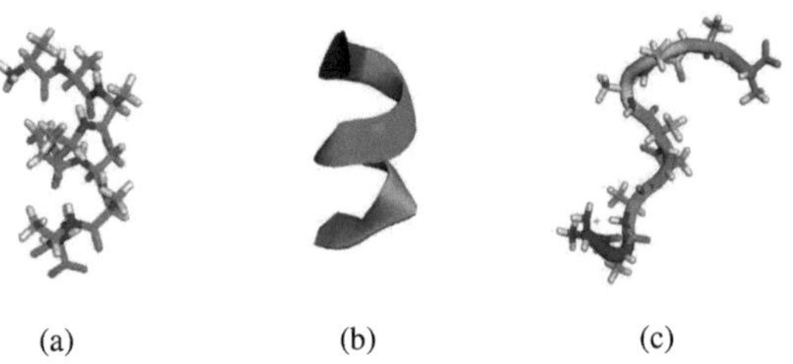

(a)   (b)   (c)

Figure 1: The native folded state of Ala_9 in (a) stick presentation and (b) in cartoon presentation to emphasise the α-helix and (c) a possible partially unfolded state.

Several *trajectories* of about 5 μs length for Ala_9 in water were simulated for different sets of pressure and temperature (1, 1000, 2000 and 4000 bar at 300 K as well as 320, 350 and 370 K at 1 bar) with the help of GROMACS version 2022.4 (Groningen Machine for Chemical Simulations), a software, which enables the simulation of the dynamics of large biomolecules, together with the *force field* CHARMM27 (Chemistry at HARvard Macromolecular Mechanics 27) and the TIP3P (transferable intermolecular potential with 3 points) *water model*. For the visualisation of the protein, the graphic software PyMOL (Version 1.7.2.1) was utilised.

While the time dependent values for the reaction coordinates RMSD and end-to-end distance could be automatically extracted from the simulated trajectories using tools provided by the GROMACS program package, the calculation of the reaction coordinate average hydrogen bond distance required the deliberate selection of three hydrogen bonds in the native Ala_9 peptide that have been determined in a previous study [8]. The donor and acceptor atoms (numbered as in the structure file constructed with the *Avogadro* software) involved in the three hydrogen bonds are given in Fig. 2.

The axes of the three reaction coordinates investigated here, RMSD, hydrogen bond length and end-to-end distance, were divided into intervals of constant width labelled with i. All peptide conformations extracted from the simulated trajectories were assigned to these intervals, and from the frequencies of conformations in each interval i the corresponding probabilities w_i were calculated. From these probabilities the Gibbs energies G_i of the intervals could be calculated according to the relation

$$G_i = -k_B T \ln(w_i) \,. \tag{2}$$

In this way, energy landscapes were constructed that depict the Gibbs energy as functions of the respective reaction coordinates. The obtained energy landscapes depend crucially on the interval width. A probability interval of 0.01 nm was chosen because it provides an acceptable statistical noise

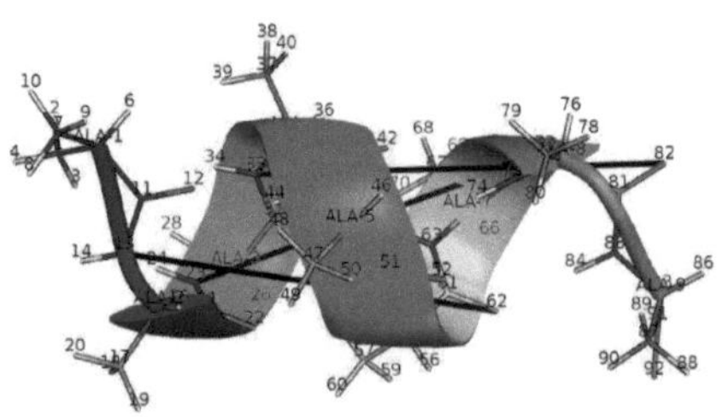

Figure 2: Initial conformation of Ala_9 with the three native hydrogen bonds (black lines) for a left-handed direction of rotation. The pairs of atoms are (13,62), (23,72) and (33,82). The figure is taken from [8] and modified.

without losing structural information. An interval width of 0.1 nm leads to a much lower statistical noise and the energy landscapes are smoothed, but features of folded and unfolded states are lost.

3   Results and Discussion

In the following figures, the energy landscapes of the different reaction coordinates are displayed for different pressure values at room temperature (Fig. 3-5) and for different temperatures at ambient pressure (Fig. 6-8). In each case, the energy landscapes were shifted in a way that the Gibbs energy of the native conformation of the peptide equals zero.

3.1   Influence of pressure

At 300 K an RMSD value of about 0.107 nm corresponds to the native state (Fig. 1). For all pressure values, a maximum at around 0.15 nm and 6 kJ mol^{-1} is attributed to a partially unfolded first transition state that separates the native state from unfolded states.

The Gibbs energy difference between the native and an unfolded state at about 0.4 nm increases from about 0.6 kJ · mol^{-1} at 1 bar to about 4 kJ mol^{-1} at 4000 bar. This observation is consistent with the observation that α-helices can be stabilised under pressure [9]. The energy landscape for the commonly used reaction coordinate of hydrogen bond distances (Fig. 4) appears to be much smoother in comparison to the energy landscape for the RMSD. However the stabilisation of the secondary structure is also visible here. The first transition state is at around 0.35 nm and 6 kJ mol^{-1} and the folded native state is at 0.3 nm.

The third reaction coordinate to be evaluated is the end-to-end distance of the Ala_9-peptide. As shown in Fig. 5, the stabilisation of the α-helix can also be observed here. The folded native state is at around 1.35 nm and the transition state is at 1.25 nm with an activation energy ΔG of about 4 kJ mol^{-1}.

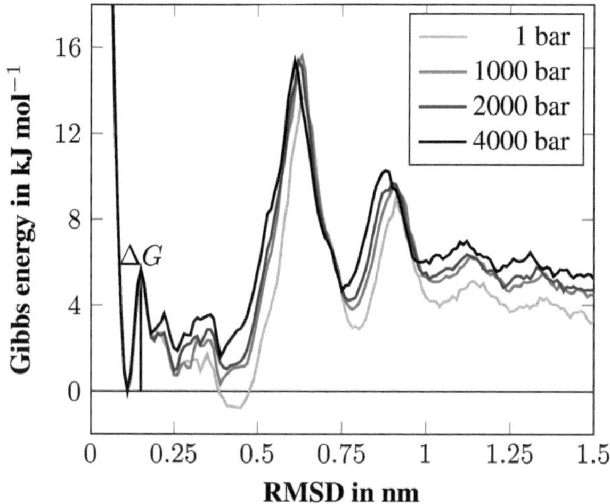

Figure 3: Gibbs energy as a function of the RMSD at 300 K and at pressure values of 1 [8], 1000, 2000 and 4000 bar. An activation energy of ΔG for the first transition state around 6 kJ mol^{-1} is represented by a grey vertical bar. The probability interval is 0.01 nm.

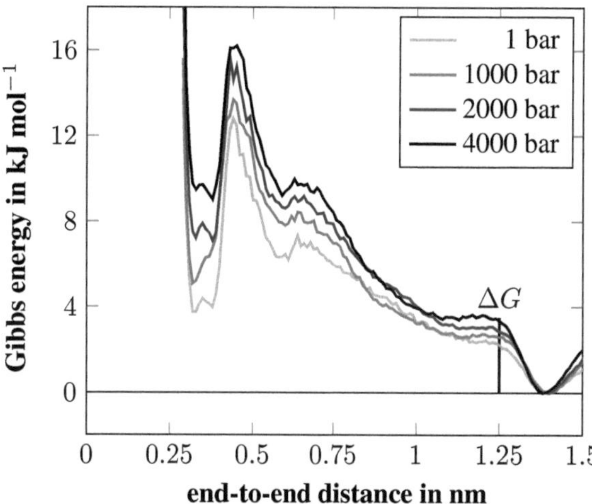

Figure 5: Energy landscape at pressures of 1 [8], 1000, 2000 and 4000 bar, all measured at a temperature of 300 K for the reaction coordinates of the end-to-end distance. ΔG for the first transition state is at 4 kJ mol^{-1} and at 1.25 nm. The probability interval is 0.01 nm.

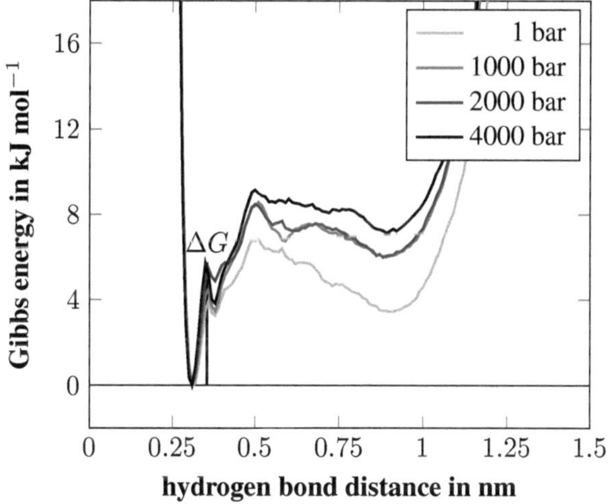

Figure 4: Energy landscape at pressure values of 1 [8], 1000, 2000 and 4000 bar, all simulated at a temperature of 300 K for the reaction coordinates of the hydrogen bond distance. An activation energy of ΔG for the first transition state around 6 kJ mol^{-1} is represented by a grey vertical bar. The probability interval is 0.01 nm.

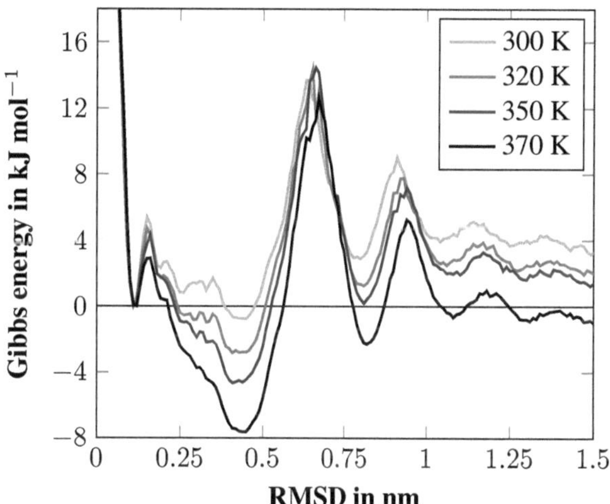

Figure 6: Energy landscape at atmospheric pressure and temperatures of 300 [8], 320, 350 and 370 K for the reaction coordinate RMSD. The probability interval is 0.01 nm.

3.2 Influence of temperature

After the evaluation of the suitability of different reaction coordinates for different pressures, we take a look into simulations with increasing temperatures. For these simulations, the pressure was fixed at atmospheric pressure. As before, the Gibbs free energy was calculated and plotted against the different reaction coordinates with the native state at an energy of 0 kJ mol^{-1}. For the reaction coordinate RMSD, shown in Fig. 6, the native state is at a value of 0.11 nm and the first transition state is at 0.16 nm for all simulated temperatures. The graph of the often used reaction coordinate of the hydrogen bond distance is somewhat

smoother than the graph of the RSMD as shown in Fig. 7. The native folded state can be considered at 0.31 nm and the first transition is at around 0.36 nm. The reaction coordinate end-to-end distance appears to be a little bit noisier than the other reaction coordinates. However the influence of increasing temperature can be seen here as well. The native folded state is at 1.41 nm and the first transition state is at 1.25 nm.

4 Conclusion

We find a stabilisation of the α-helix with increasing pressure in agreement with similar results for other peptides reported in literature [9]. This stabilisation is observed in all three reaction coordinates used in this study. While hydrogen bond distances are widely used as a reaction coordi-

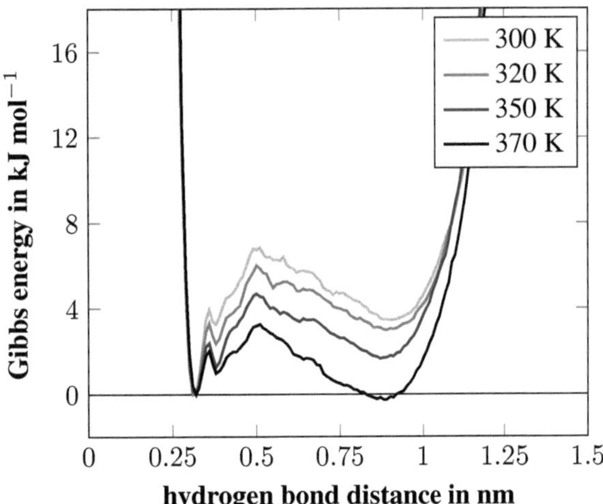

Figure 7: Energy landscape at atmospheric pressure and measured at a temperature of 300 [8], 320, 350 and 370 K for the reaction coordinates of the hydrogen bond distance. The probability interval is 0.01 nm.

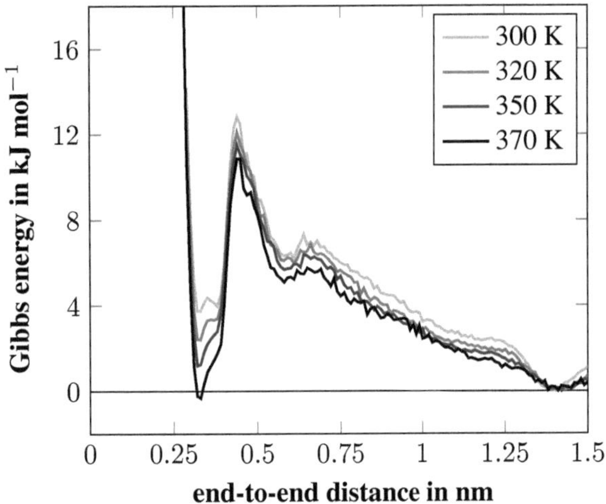

Figure 8: Energy landscape at atmospheric pressure and measured at a temperature of 300 [8], 320, 350 and 370 K for the reaction coordinates of the end-to-end distance. The probability interval is 0.01 nm.

nates in MD simulations, the RMSD turns out to be a viable alternative. Also the reaction coordinate end-to-end distance reflects the stabilisation of the α-helix under pressure. In terms of the energy landscapes for the Ala$_9$, an increase of temperature also has an effect on the folding and unfolding of the peptide. The results indicate that the reaction coordinate RMSD is a suitable choice for MD simulations at temperatures between 300 K and 370 K. To somewhat less extent this is also true for the end-to-end distance.

The root mean square deviation and the hydrogen bond distance as reaction coordinates turned out to be comparably powerful methods for MD simulations of larger molecules like Ubiquitin.

Acknowledgement

The work has been carried out and supervised at the Institute of Physics, Universität zu Lübeck.

Authors' Statement

Conflict of interest: Authors state no conflict of interest.

5 References

[1] J.M. Berg, J.L. Tymoczko, L. Stryer. *Biochemie.* Springer Spektrum, 8th ed., pp. 32-73, 2018.

[2] B.J. Alder, T.E. Wainwright. *Studies in molecular dynamics. I. General method.* The Journal of Chemical Physics 31(2), pp. 459-866, 1959.

[3] J. Rogal. *Reaction coordinates in complex systems- a perspective.* European Physical Journal B 94(11), p. 223, 2021.

[4] H. Grubmüller, P. Tavan. *Multiple time step algorithms for molecular dynamics simulations of proteins: How good are they?* Journal of Computational Chemistry 19(13), pp. 1534-52, 1998.

[5] U. Fechner, S. Renner, M. Schmuker. *RMS, RMSD,* Available: https://roempp.thieme.de/lexicon/RD-18-02334 [last accessed on 2023-12-11]

[6] J. Rückert, N. Ludolph, H. Paulsen. *Reaction coordinates of protein folding: Comparison between the number of native hydrogen bonds and the RMSD of the crystal structure.* In: Student Conference Proceedings 2023, Infinite Science Publishing, pp. 393-396, 2023.

[7] C. Ayaz et. al., *Non-Markovian modeling of protein folding.* Proceedings of the National Academy of Sciences, vol. 118, no. 31, p. e2023856118, 2021.

[8] J. Rückert, *Reaktionskoordinaten der Proteinentfaltung: MD-Simulationen für das 9-Alanin-Peptid,* Master thesis, Universität zu Lübeck, 2023.

[9] H. Hata, M. Nishiyama, A. Kitao. *Molecular dynamics simulation of proteins under high pressure: Structure, function and thermodynamics.* Biochimica et Biophysica Acta (BBA) - General Subjects, vol. 1864, pp. 129395, 2020.

Construction of a short dead-time small volume stopped-flow system for kinetics reactions

Thore Viemann [1], Janosch Kappel [2], Christian Hübner [2]

[1] Medical Engineering Science, Universität zu Lübeck, thore.viemann@student.uni-luebeck.de
[2] Institute of Physics, Universität zu Lübeck, {kappel, huebner}@physik.uni-luebeck.de

Abstract

The study of the dynamic interactions of biomolecules provides fundamental insights into molecular processes such as protein unfolding or enzymatic reaction kinetics. This knowledge contributes significantly to advances in medicine and biophysics. The stopped-flow method is an approach that enables precise study of such dynamic interactions. This article presents the design and the development of a stand-alone stopped-flow system using fluorescence microscopy for the investigation of reaction kinetics. This system allows for more efficient handling and minimizes potential interference when adapting to different microscopic configurations. A microfluidic mixing device enables very short mixing times and small sample volume, but requires efficient microscopic light detection. The system is thus based on an epifluorescence microscope with a fast sensitive CCD camera. We present details of the construction and a first image from the microfluidic mixer.

1 Introduction

Investigating the dynamic interactions of biomolecules can aid in elucidating reaction mechanisms, leading to a better understanding of diseases caused by misfolding and aggregation of proteins, such as Parkinson's disease. One way to investigate the reaction kinetics of molecular processes is the stopped-flow method. The reaction rate is a key parameter in reaction kinetics studies, measuring the change in reactant concentration over time [1]. It provides insights into reaction mechanisms and transition states. An example of a first-order reaction is protein unfolding, where the conversion from the native state (N) to the unfolded state (U) is defined by the rate constant k_u:

$$N \xrightarrow{k_u} U. \tag{1}$$

Mixing two reactants in a stopped-flow system initiates the conversion process. The speed rate v of a molecular process can be expressed mathematically as:

$$v = \left[-\frac{d[C]}{dt} \right], \tag{2}$$

where [C] represents the concentration of a protein in its native state. In an unfolding reaction, the concentration of the native protein decreases exponentially with time, while the concentration of the unfolded protein increases. In order to drive this reaction, the protein is mixed with a denaturant.

The reaction rate can be described mathematically by solving the differential equation, which represents the dependence of the concentration $[C_0]$ on the rate constant of the reactants involved.

$$C(t) = C_0 e^{-k_u t}. \tag{3}$$

The reaction kinetics can be recorded optically, e.g. by detecting emitted fluorescence, or electrochemically, by measuring conductivity [2]. The most suitable method for a specific reaction varies, as each technique has its individual advantages and disadvantages. The stopped-flow method has significant advantages, such as short dead-times and low sample consumption [3].

2 Material and Methods

This section explains the elementary principles of the stopped-flow method and describes its integration on a fluorescence microscope. Additionally, it provides a detailed description of the microfluidic chip geometry.

2.1 Stopped-flow method

A proven technique for analyzing rapid biochemical reactions is the stopped-flow method. It involves rapidly mixing reaction components and then abruptly stopping the flow for precise kinetic analysis. The stopped-flow method was first introduced in 1937 by B. Chance [4] and has been continuously improved and modified over the last few decades. In a mixing chamber, two or more reactants are mixed at defined speeds. The flow stops rapidly within an extremely short period of time as soon as the observation chamber is completely filled. The reaction can be monitored as a change in signal over time by a data acquisition system with high temporal resolution, allowing for accurate analysis. The method enables for precise analysis and characterization of rapid biochemical reactions by recording their temporal dynamics. The dead-time, which marks the beginning of data acquisition, is defined as the time period between the first contact of the reactants and the arrival of the completely mixed solution in the observation chamber. The length of the dead-time is primarily influenced by several factors such as mixing time, flow rate, and mixer geometry. Its precise determination can be achieved by analyzing signals using fluorescence measurements as first-order responses [5]. Another significant advantage of the stopped-flow method is its low sample consumption. After the flow has been stopped, the mixed solution can be observed without the need for additional sample consumption. The stopped-flow method is an efficient and exact technique for investigating fast biochemical reactions.

A pump is necessary in a stopped-flow system as it allows an adjustable speed and flow rate. This controlled and precise introduction of reactants into the microfluidic chip shortens the mixing process. The customized syringe drive, which is integrated into the stand-alone system, is described in [9]. It uses two high-precision glass syringes (Hamilton, United States) with a volume of 250 µl. Two stepper motors (NEMA 23) enable the exact dosing of two reagents into a specifically taylored microfluidic chip.

2.2 Microfluidic mixers

Modern stopped-flow systems utilize the concept of microfluidics to minimize reaction volumes and optimize mixing times. This miniaturization not only enables for accurate mixing conditions but also contributes to the reduction of reaction losses [6]. A herringbone mixer (Fludic 187, Microfluidic Chip Shop, Germany) was used for the tests as it is inexpensive and widely used. However, the final setup is designed for a helical structure.

The herringbone mixer is a significant innovation in microfluidics that enables efficient mixing of liquids on a microscopic scale. It was designed by Stroock et al. [7] and is characterized by a structured arrangement of obstacles within the micro-channels. These obstacles are typically arranged in a wavy pattern, similar to the bones of a fish. The unique structure of the herringbone mixer leads to effective mixing of the liquids. The wave-shaped obstacles induce vortices in the flow, which improve the diffusion and mixing of the liquids. The herringbone mixer is known for its ability to create fast and homogeneous mixtures on a microscopic level. Compared to traditional methods, this mixer requires smaller volumes and produces highly reproducible results. The dimensions of the herringbone mixer are crucial parameters that significantly impact the system's behavior and performance. The channel depth, which determines the channel's volume capacity, is 200 µm. The channel width at the inlet is 300 µm. In the mixer area of the herringbone mixer, the channel width

is increased to 600 µm. The width of the mixing area contributes to homogeneous mixing. The channel width at the outlet remains constant at 600 µm.

The helical mixer mentioned earlier is a custom design [8] that allows mixing of small quantities of liquid in the sub millisecond range. This design has the advantage of a short distance between the mixing chamber and the observation chamber, resulting in shorter transport distances and a shorter dead-time. The mixer is made of fused silica produced by selective laser induced etching (ISLE). The dimensions of the helical mixer are critical to its functionality and performance. Both the mixing chamber and the observation chamber have a capacity of 0.0136 µl. The inlet and outlet channels of the helical mixer have a diameter of 0.5 mm, thus drastically reducing the sample volume.

3 Results and Discussion

The following sections describe the stand-alone stopped-flow system that has been developed in this work. In addition, the advantages of a stand-alone system with stopped-flow are explained.

3.1 Epifluorescence microscopy in combination with a stopped-flow apparatus

Combining the stopped-flow method with epifluorescence microscopy broadens the range of investigation possibilities. Epifluorescence microscopy allows real-time observation of fluorescent proteins and other fluorescently labelled biomolecules. By integrating this technique with the stopped-flow method, accurate spatial and temporal resolution is achieved, allowing detailed analysis of biochemical reactions. Fig. 1 shows the schematic setup of the stopped-flow device in combination with an epifluorescence microscope. The two stepper motors enable the exact dosing of two reagents into the microfluidic chip. The reactants are mixed in the mixing chamber of the microfluidic chip before being transferred to the observation chamber. The positioning of the observation chamber is adjusted by a custom 3D printed XY translation stage. Both components are placed on the table of the microscope stand (Eclipse TS100-F, Nikon, Japan). Excitation of the sample in the observation chamber is performed using a 488 nm laser (Sapphire 488 SF NX, Coherent, Germany), which acts as the light source for the system. The laser beam is expanded by a plano-concave lens with a focal length of 50 mm (LC1715, Thorlabs, Germany) and focused at the back focal plane of the objective (Plan Flour 10x, Nikon, Japan) by a plano-convex lens with a focal length of 125 mm (LA1986-A, Thorlabs, Germany). A dualband dichroic mirror (Z488/568rpc, Chroma Technology Corp, United States) separates the excitation light from the emission light. The emitted light is then recorded by a CCD camera with a resolution of 1376x1040 pixel (PCO Sensicam qe, Excelitas, United States). The recorded signal can be analyzed via PCI frame grabber board (PCO PCI-Interface-Board, Excelitas, United States) on a dedicated computer using appropriate software (PCO CamWare, Excelitas, United States).

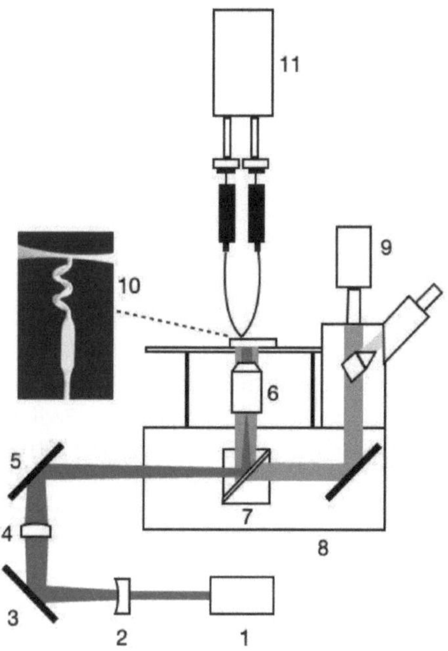

Figure 1: Schematic representation of the planned stop-flow system in combination with the fluorescence microscopy setup. The Laser (1) emits blue light with a wavelength of 488 nm. The laser beam is expanded by a plano-concave lens (2). A plano-convex lens (4) and two mirrors (3, 5) are used to focus the expanded beam into the rear aperture of the microscope objective (6). This allows the excitation beam to exit the objective at a larger angle. Before entering the objective lens, the focused excitation beam is reflected by a dichroic beam splitter (7), which separates the excitation light from the the emitted light of the sample. The objective and the dichroic beam splitter are located in a microscope stand (8). The emission light is directed through the optics onto the CCD camera (9) through the optical structure of the microscope stand. The sample can then be analyzed using software on an external computer. The microfluidic mixer (10) is located on the stage of the microscope stand. The observation chamber of the microfluidic mixer is illuminated with a widened beam path through the objective. The syringe drive (11) is connected to the mixer.

3.2   Optical characterization

The developed stand-alone stopped-flow system captured a representative image (Fig. 2) displaying the structures of the helical chip using its detection system. The image was taken using an external white light source and shows both the mixing chamber and the observation chamber. The acquisition of this image allows the conclusion that reaction kinetics measurements are possible, paving the way for further studies.

Figure 2: The helical structure of the mixer can be seen in the upper part of the picture. The observation chamber is located below the mixing channels in the image. The structures shown were captured with the detection system of the developed stand-alone setup. It is important to note that only the observation chamber is relevant for data recording, not the mixing process. The microscope's magnification is limited to a specific area. Therefore, we visualized the geometric structure of the helical mixer by combining two images.

3.3   Benefits of a stand-alone stopped-flow system

To obtain reliable data, precise and efficient systems are necessary for conducting stopped-flow measurements and reaction kinetic analyses. Previous systems had deficiencies in setup, dismantling, and microscopic settings. The integration of the existing stopped-flow pump and microfluidic chips into existing setups was not possible without significant modifications. This integration of the existing components made the entire experimental setup time-consuming before each measurement. Fig. 3 shows the developed stand-alone stopped-flow system that is specifically tailored to the needs of reaction kinetic measurements.

The stand-alone stopped-flow system represents a significant improvement as it eliminates the need for complicated and time consuming assembly before each measurement. This not only simplifies handling for users but also saves valuable time. The system requires fewer adjustments, resulting in more efficient assembly and disassembly compared to previous systems.

A significant improvement is the reduction in the number of adjustments required in the overall system. Previous systems required frequent adjustments to ensure optimum measurement conditions. The stand-alone stopped-flow system minimizes this need, which not only reduces the workload but will also improve the reproducibility of the experiments. The designed system allows precise customization and adjustment to perform stopped-flow measurements and reaction kinetics studies. Compared to previous systems used with different confocal microscope setups, the improved

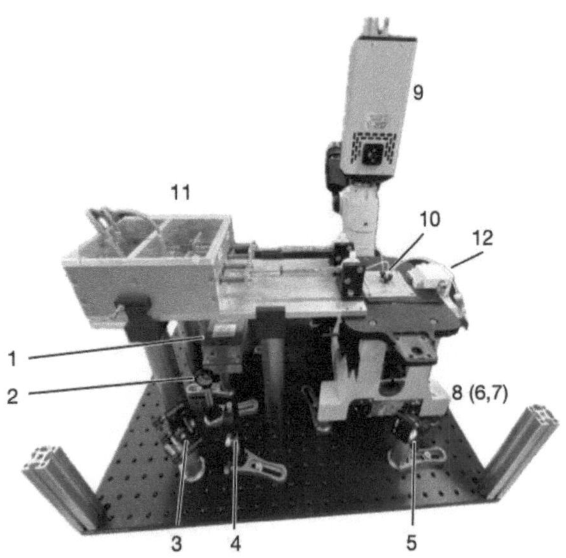

Figure 3: Final assembly of the designed stop-flow system in combination with a fluorescence microscope. The assembly includes the excitation laser (1) which is located in front of the optical components such as lenses (2,4) and mirrors (3,5). Not visible in the picture are the objective (6) and the dichroic beam splitter (7), which are inserted into the microscope stand (8). The CCD camera (9) is also mounted on the microscope stand. The stand holds the mixer chip (10) and the self-made 3D-printed XY stage (12). The tip drive (11) is mounted horizontally on specially manufactured 3D-printed brackets, which in turn are attached to the mounting posts.

system eliminates compromises in matching the two systems. The beam path and data acquisition are now easily adaptable, facilitating for the use of different objectives and optical components.

Another key advantage of the stand-alone stopped-flow system is the use of an epifluorescence excitation beam path as opposed to confocal fluorescence microscopy. This configuration allows illumination and detection of the entire channel cross-section or observation chamber. In contrast, the confocal configuration is limited to excitation and detection in a small partial volume of the channel cross-section or observation chamber.

4 Conclusion

The developed stand-alone stopped-flow system offers significant advantages in terms of assembly, adjustment and customization for reaction kinetics measurements. The integration of the epifluorescence microscope not only enables more accurate measurements, but also avoids time-consuming compromises when adapting different systems. This makes the optimized stopped-flow system a powerful tool for studying reaction kinetics in biochemical analysis. Further steps include testing the whole system with a real reaction kinetics experiment using the epifluorescence con-

figuration.

Acknowledgement

This work was carried out at the Institute of Physics at the Universität zu Lübeck under the supervision of Janosch Kappel and Prof. Dr. Christian Hübner.

Authors' Statement

Conflict of interest: Authors state no conflict of interest.

5 References

[1] Lechner, M. Physikalische Reaktionskinetik. *Einführung In Die Kinetik: Chemische Reaktionskinetik Und Transporteigenschaften*. pp. 29-52 (2018)

[2] Kathuria, S., Chan, A., Graceffa, R., Paul Nobrega, R., Robert Matthews, C., Irving, T., Perot, B. & Bilsel, O. Advances in turbulent mixing techniques to study microsecond protein folding reactions. *Biopolymers*. **99**, 888-896 (2013)

[3] Khorassani, S., Ebrahimi, A., Maghsoodlou, M., Shahraki, M., Price, D. & Paknahad, A. A novel high performance stopped-flow apparatus equipped with a special constructed mixing chamber containing a plunger under inert condition with a very short dead-time to investigate very rapid reactions. *Arabian Journal Of Chemistry*. **8**, 873-884 (2015)

[4] Chance, B. The accelerated flow method for rapid reactions. *Journal Of The Franklin Institute*. **229**, 737-766 (1940)

[5] Tonomura, B., Nakatani, H., Ohnishi, M., Yamaguchi-Ito, J. & Hiromi, K. Test reactions for a stopped-flow apparatus: Reduction of 2, 6-dichlorophenolindophenol and potassium ferricyanide by L-ascorbic acid. *Analytical Biochemistry*. **84**, 370-383 (1978)

[6] Song, H., Tice, J. & Ismagilov, R. A microfluidic system for controlling reaction networks in time. *Angewandte Chemie*. **115**, 792-796 (2003)

[7] Stroock, A., Dertinger, S., Ajdari, A., Mezic, I., Stone, H. & Whitesides, G. Chaotic mixer for microchannels. *Science*. **295**, 647-651 (2002)

[8] H. Mueller, *Design and simulation of a passive micromixer for rapid mixing of small fluid volumes in reaction kinetic studies*. Master Thesis, Universität zu Lübeck, 2016.

[9] J. Kappel, *Development of a stopped-flow apparatus for rapid mixing of small fluid volumes in reaction kinetic studies*. Master Thesis, Universität zu Lübeck, 2018

An optimised small-scale ninhydrin-based assay for the quantification of amines on aminosilanised coverslips

Elena Romiana Schütte[1], Christian Rönnau[2], Christian Hübner[2]

[1] Biophysics, Universität zu Lübeck, elena.schuette@student.uni-luebeck.de

[2] Institute of Physics, Universität zu Lübeck, c.roennau@uni-luebeck.de, christian.huebner@physik.uni-luebeck.de

Abstract

For many biological applications, surface modifications, e.g. aminosilanisation of coverslips, are essential for tailoring their properties. A small-scale ninhydrin-based assay was developed and tested for amine quantification on aminosilanised coverglasses. A high reproducibility as well as a more than 10-times higher sensitivity compared to the literature could be demonstrated.

1 Introduction

Surface modification of coverslips (CS) through aminosilanisation is a crucial process for tailoring their properties for various biological applications. Aminosilanisation techniques provide a versatile method for introducing functional amino groups onto glass coverslips. These groups can then be used to attach biomolecules like (i) proteins using glutaraldehyde as a linker or (ii) polymers for targeted applications [1], [2].

In order to use surface functionalisation reliably, it is important to be able to apply a thin, homogeneous film of aminosilane. The standard procedure for aminosilanisation of coverslips is liquid-phase silanisation, in which the coverslips are incubated in aminosilane dilutions prepared with organic solvents, such as methanol, ethanol, acetone, or toluol [2], [3], [4]. One disadvantage of liquid-phase silanisation is the large amount of silanising agent that is required as well as the plethora of liquid, organic waste that is produced. Vapor-phase silanisation, on the other hand, produces significantly less waste and requires less silanising agent [5]. Another difference between the methods is the type of silane layer produced. In contrast to vapor-phase silanisation, liquid-phase silanisation can produce multilayers instead of monolayers, resulting in a significantly thicker silane layer [6]. For applications like label-free biosensing or some optical measurements such as total internal reflection fluorescence microscopy, a thin silane layer is crucial for a homogeneous amine layer. As this layer is the foundation for later modifications, it has significant influence on the background signals and therefore the sensitivity of low-intensity measurements (e.g. single-molecule measurements). [6].

The silanisation agent used also has an impact on the type of silane layer. 3-aminopropyltriethoxysilane (APTES) is the commonly used aminosilane for coverslip silanisation [7]. The corresponding trimethoxysilane is 3-aminopropyltrimethoxysilane (APTMS). The latter is less complex to handle due to its lower hydrolysis rate [7].

Various methods exist for amine determination, including high pressure liquid chromatography (HPLC), hydrophilic interaction liquid chromatography (HILIC), and potentiometric titration. However, these are only suitable for amine quantification in solution and therefore cannot be used to quantify amines on coverslips. In addition, HILIC and HPLC are time-consuming, while relatively large sample volumes are required for potentiometric titration. In contrast, the ninhydrin-based amine quantification according to Starcher [8] is simple to implement, time-saving, and cost-effective. It is a spectroscopic approach for the quantification of amines in liquid samples in the ml range. This method is a bit modified also suitable for the detection of immobilised amine groups on surfaces.

In this work, the method described by Starcher [8] was adapted for small volumes to minimise sample consumption. The developed method was just characterised for the detection of amine groups on coverslips.

2 Material and Methods

2.1 Aminosilanisation of coverslips

Coverslip pretreatment: The coverslips were incubated at 480 °C for 5 h to remove any remaining impurities. To etch and activate the surface, the CS were sonicated in 1 M KOH for 30 minutes. To test the influence of the KOH etching, CS were also prepared without incubation in KOH. Hereafter, CS were washed at least five times with

MP-water while sonicating for 2 minutes. After washing, the pH of the solution was evaluated using pH strips to ensure neutral or slightly acidic conditions.

Vapor-phase silanisation: Washed CS were dried in a desiccator for 60-90 minutes. As shown in Fig. 1, the dried CS were placed in a rack on top of two 5 ml glass beakers containing either 200 µl APTES or 200 µl APTMS each. Next, the assembly was placed in a vacuum chamber that was subsequently evacuated to around $2 * 10^{-2}$ mbar for 3 h. Hereafter, the CS were first washed with acetone followed by three times rinsing with MP-water, each step involving sonication for 5 minutes. Finally, CS were dried for 60-90 minutes in a desiccator.

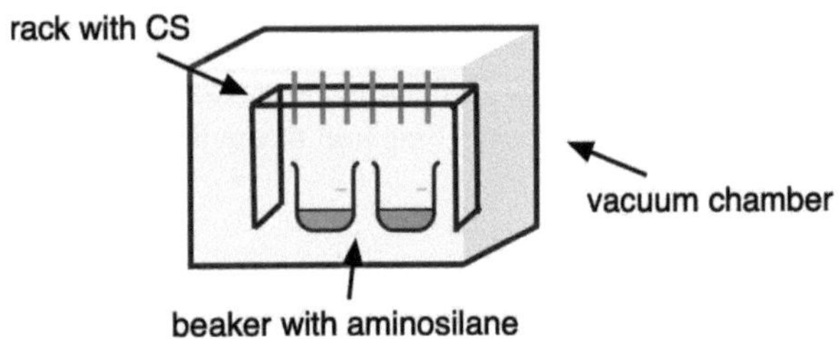

Figure 1: Setup for vacuum-phase silanisation. The coverslips (CS) were placed in a vacuum chamber in a rack above two 5 ml glass beakers each containing 200 µl aminosilane agent (APTES or APTMS).

2.2 Quantification of amines

Ninhydrin-based assay for CS: The CS were crushed, treated with the ninhydrin reagent, and the resulting extinction (E) was measured using a UV-VIS-spectrometer (NanoDrop, Peqlab). The ninhydrin solution was prepared by dissolving 20 mg of ninhydrin in 750 µl ethylene glycol, 250 µl 4 M sodium acetate buffer pH 5.4, and 25 µl of stannous chloride solution (20 mg $Sn(II)Cl_2$ in 10 µl ethylene glycol). The coverslips were placed in a 50 ml centrifuge tube and crushed using a vortex mixer. Four samples containing CS fragments between 20 mg and 140 mg were weighed into 500 µl PCR tubes. Next, 50 µl of ninhydrin reagent was added to each tube. The samples were incubated in a thermal cycler at 90 °C for 10 minutes and then cooled for 2 minutes at 15 °C. The glass debris was filtered out and the liquid was collected. The extinction was measured at 570 nm using 5 µl of the collected liquid.

Calibration measurement: The amino acid L-alanine was used for calibration. For this purpose, a 11.8 mM stock solution was first prepared by dissolving 10.55 mg L-alanine in 10 ml 4M sodium acetate buffer. This was used to prepare a 472 µM and then a 236 µM dilution. Subsequently, five samples of 50 µl were prepared according to the Table 1. Next, 50 µl of ninhydrin solution was added to each, vortexed and incubated at 90 °C for 10 minutes. After cooling at 15 °C for 2 minutes, vortexing was repeated and

Table 1: L-alanine concentrations and volumes for calibration measurements. The starting solution was a 236 µM L-alanine solution (L-alanine-dilution) containing L-alanine and 4M sodium acetate buffer (NaOAc).

[L-alanine] / µM	$V_{\text{L-alanine-dilution}}$ / µl	V_{NaOAc} / µl
14.77	3.13	46.9
29.5	6.25	43.8
59	12.5	37.5
118	25.0	25.0
177	37.5	12.5

the extinction was measured at 570 nm.

Data analysis: The extinction at 570 nm was extracted from the spectra and a line fit was created for each measurement. Next, the slope of the fit function was extracted and corrected for the slope of the control - burnt CS for non-KOH-treated silanised CS and KOH-treated CS for KOH-treated silanised CS. Subsequently, the density of amines on a CS (ρ_{NH_2}) was determined using the following equation:

$$\rho_{\text{NH}_2} = \frac{N_\text{A} * V_{\text{Ninh}} * w_{\text{CS}} * m_{\text{CS}}}{m_{\text{cal}} * A}, \qquad (1)$$

where N_A is the Avogadro constant, V_{Ninh} is the volume of the ninhydrin solution added (i.e. 50 µl), w_{CS} is the weight of a CS ($\approx$209 mg), A is the corresponding surface area of the CS ($\approx$1166 mm^2), m_{CS} represents the corrected slope of the fitted data, and m_{cal} is the mean slope of the calibration measurements.

3 Results and Discussion

3.1 Sensitivity and reproducibility

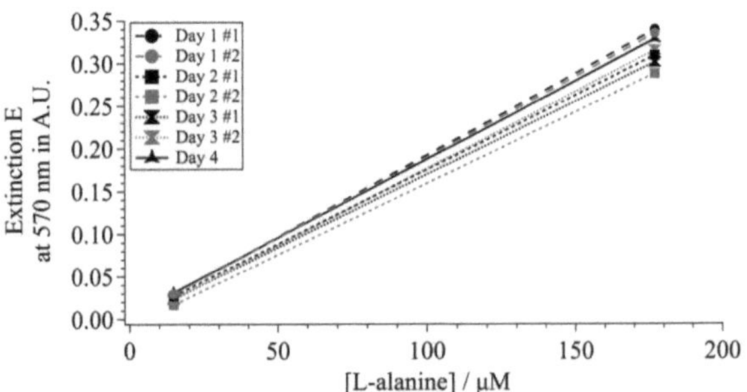

Figure 2: Linear fits of calibration measurements for the ninhydrin assay, using the same L-alanine concentrations (five data points per experiment) between 14 µM and 180 µM. Duplicates (indicated by black and grey) recorded on the same day using the same ninhydrin solution have the same marks and line styles.

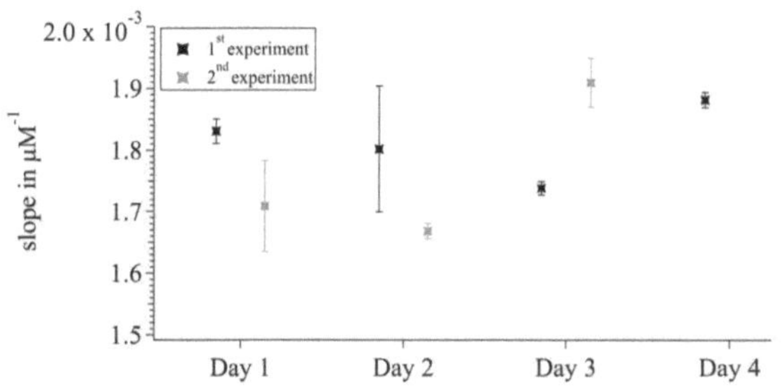

Figure 3: Slope of the linear fit for each of the seven calibration series (Fig. 2). Duplicates (black and grey) are grouped by day. Error bars indicate fitting errors.

A ninhydrin-based assay for quantification of amines in solutions was adapted for small volumes and determination of amines on coverslips.

The sensitivity of the assay is given by the slope of line-fitted independent calibration curves recorded for the same L-alanine concentrations using different ninhydrin solutions at different days (see Fig. 2 and 3). Increasing the incubation time to 15 minutes at 90 °C or increasing the temperature to 95 °C for 10 minutes had no effect on colour development and thus sensitivity (data not shown). The average sensitivity obtained was $1.7913\,\mathrm{E\,mM^{-1}}$ (SD: $\pm\,0.09\,\mathrm{E\,mM^{-1}}$). The deviation between the measurements ($N = 7$) was 5.02 % and the SEM was 1.90%, which is in good agreement with the original publication by Starcher (3%, 16 samples using the same ninhydrin reagent and 16 samples using different ninhydrin reagents) [8].

3.2   Quantification of amines on CS

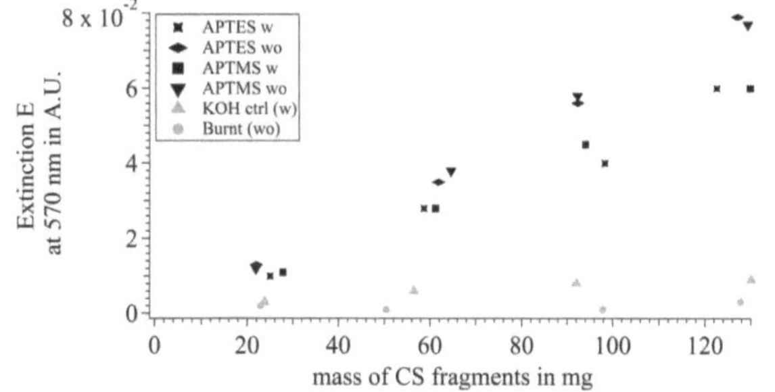

Figure 4: Extinction at 570 nm plotted against the mass of the CS fragments. KOH treatment is labelled 'w', untreated CS are labelled 'wo' while silanised CS are depicted in black marks and controls are light grey.

As the method by Starcher was adapted for the quantification of amines on coverslips, the amine density on vapour-phase silanised coverslips was also determined. Therefore, four samples with different masses were measured, the slope was determined from the line-fitted data and the amine density $\rho_{\mathrm{NH_2}}$ was calculated using (1).

The extinction at 570 nm depending on the mass of the

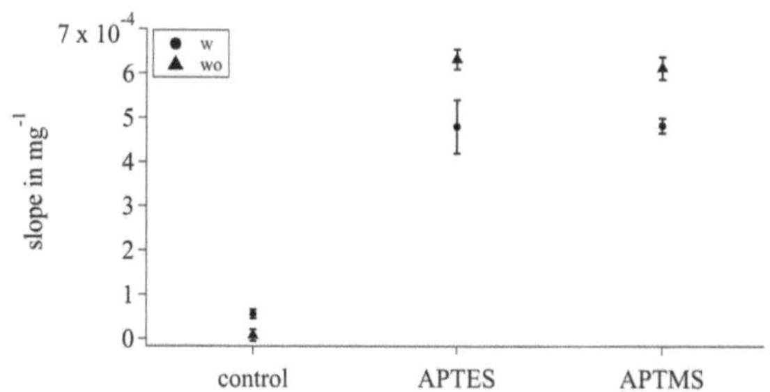

Figure 5: Slopes extracted from line fitting the data shown in Fig. 4, samples treated with KOH (w) are indicated by circle marks, triangular marks indicate no treatment with KOH (wo). Fitting errors are represented by error bars.

Table 2: Extracted slopes from line fitting the data shown in Fig. 4, slopes corrected for the control (m_{CS}), KOH-treated (w) or burnt (wo), and estimated amine density ($\rho_{\mathrm{NH_2}}$) on coverslips per $\mathrm{nm^2}$.

condition	slope in $10^{-4}\,\mathrm{mg^{-1}}$	m_{CS} in $10^{-4}\,\mathrm{mg^{-1}}$	$\rho_{\mathrm{NH_2}}$ in $\mathrm{nm^{-2}}$
APTES w	4.799	4.239	1.278
APTES wo	6.316	6.234	1.876
APTMS w	4.834	4.274	1.286
APTMS wo	6.131	6.049	1.820

CS fragments is shown in Fig. 4. Only a slight slope is recognisable for burnt and KOH-treated CS, as they both contain very few amine groups, as expected. There is a significant difference in the slope of the silanised coverslips. However, there is no discernible difference when using APTES or APTMS as silanising agent. It should be noted that neither the thickness nor the homogeneity of the silane film is assessed in this study. Fig. 5 displays the slope with error bars indicating fitting errors for the different coverslip conditions. The graph illustrates that the slope is lower for KOH-treated CS than for untreated ones. This suggests that treated CS have fewer amine groups than untreated ones, despite having a larger surface area. The extracted slopes, as well as the slopes corrected (m_{CS}) and the estimated density of amines ($\rho_{\mathrm{NH_2}}$) on the aminosilanised coverslips are shown in Table 2.

4   Conclusion

In this paper, an adapted, small-scale ninhydrin-based assay was established, providing high reproducibility (5 % between independent calibrations) and 11.7-times higher sensitivity, achieved by utilising a NanoDrop UV-VIS-photospectrometer, compared to the literature [9], [10]. Furthermore, it was demonstrated that the assay is capable of detecting amines on CS that were aminosilanised using a vapor-deposition method.   These preliminary results indicate that KOH-pretreatment of CS reduce the amount of deposited amines whereas the two tested aminosilanes (APTMS and APTES) did not have an influence on the amount of deposited amines. For a recommendation regarding the silanisation agent used, both the homogeneity and the thickness of the silane layer applied would have to be investigated. To ensure reliable determination of the influence of KOH treatment prior to silanisation, additional measurements of silanised cover glasses are necessary.

The presented assay enables quantification of small amounts of amines on coverslips, allowing to compare different silanisation methods and conditions with respect to amine-density. This facilitates both the optimisation and the quality control of the aminosilanisation procedure.

Acknowledgement

The work has been carried out at the Institute of Physics, Universität zu Lübeck.

Authors' Statement

Conflict of interest: Authors state no conflict of interest.

5   References

[1] N. S. K. Gunda, M. Singh, L. Norman, K. Kaur, and S. K. Mitra, "Optimization and characterization of biomolecule immobilization on silicon substrates using (3-aminopropyl)triethoxysilane (APTES) and glutaraldehyde linker," *Applied Surface Science*, vol. 305, pp. 522–530, 6 2014.

[2] S. D. Chandradoss, A. C. Haagsma, Y. Kwang Lee, J.-H. Hwang, J.-M. Nam, C. Joo, and C. Surface, "Surface passivation for single-molecule protein studies," *J. Vis. Exp*, no. 86, p. 50549, 4 2014.

[3] N. Graf, E. Yeğen, A. Lippitz, D. Treu, T. Wirth, and W. E. Unger, "Optimization of cleaning and aminosilanization protocols for Si wafers to be used as platforms for biochip microarrays by surface analysis (XPS, ToF-SIMS and NEXAFS spectroscopy)," *Surface and Interface Analysis*, vol. 40, no. 3-4, pp. 180–183, 3 2008.

[4] G. Jakša, B. Štefane, and J. Kovač, "Influence of different solvents on the morphology of APTMS-modified silicon surfaces," *Applied Surface Science*, vol. 315, no. 1, pp. 516–522, 10 2014.

[5] F. Zhang, K. Sautter, A. M. Larsen, D. A. Findley, R. C. Davis, H. Samha, and M. R. Linford, "Chemical vapor deposition of three aminosilanes on silicon dioxide: Surface characterization, stability, effects of silane concentration, and cyanine dye adsorption," *Langmuir*, vol. 26, no. 18, pp. 14648–14654, 9 2010.

[6] A. R. Yadav, R. Sriram, J. A. Carter, and B. L. Miller, "Comparative study of solution-phase and vapor-phase deposition of aminosilanes on silicon dioxide surfaces," *Materials science & engineering. C, Materials for biological applications*, vol. 35, no. 1, pp. 283–290, 2 2014.

[7] M. Zhu, M. Z. Lerum, and W. Chen, "How to prepare reproducible, homogeneous, and hydrolytically stable aminosilane-derived layers on silica," *Langmuir*, vol. 28, no. 1, p. 416, 1 2012.

[8] B. Starcher, "A ninhydrin-based assay to quantitate the total protein content of tissue samples," *Analytical Biochemistry*, vol. 292, no. 1, pp. 125–129, 5 2001.

[9] G. Haeger, J. Bongaerts, and P. Siegert, "A convenient ninhydrin assay in 96-well format for amino acid-releasing enzymes using an air-stable reagent," *Analytical Biochemistry*, vol. 654, p. 114819, 2022.

[10] Y. Zhang, Y. Fu, S. Zhou, L. Kang, and C. Li, "A straightforward ninhydrin-based method for collagenase activity and inhibitor screening of collagenase using spectrophotometry," *Analytical Biochemistry*, vol. 437, no. 1, pp. 46–48, 6 2013.

Microfluidic production of water-in-oil droplets

Marie C. Wetzel [1], Janina Nandy [2], and Christian Nehls [2]

[1] Medical Engineering Science, Universität zu Lübeck, marie.wetzel@student.uni-luebeck.de
[2] Research Center Borstel - Leibniz Lung Center, Division of Biophysics, Borstel, Germany, {jnandy,cnehls}@fz-borstel.de

Abstract

There is still much to learn about entrapped intracellular pathogens, including their production and impact on the compartments in which they are enclosed. A model system for pathogen-enclosing compartments can be water-in-oil droplets surrounded by a lipid monolayer. In this study, we investigated the conditions under which stable droplets can be produced and the influence of the bonding of the chip. Four different oils were compared and the producibility of a lipid monolayer around the oil droplets was analysed. We have shown that stable droplets can be formed four days after chip production with a combination of mineral oil and the surfactant Span 80. Furthermore, we encapsulated bacteria in the droplets, these should serve as model systems for intracellular compartments in which pathogens are enclosed.

1 Introduction

Pathogens are phagocytized by phagocytes in the body and end up in intracellular compartments. Because many factors are involved, investigating the interactions between pathogen and host factors is very complex. The aim of this project is to develop a model system in a microfluidic setup that allows to study individual processes in such isolated compartments.

The production of W/O (water-in-oil) droplets is used for more than 20 years and can be used to show how the interaction between two immiscible liquids can be utilized to create a model system for pathogen-enclosing compartments. When water and oil collide in a microfluidic chip, droplet formation occurs in the junction of the chip. The water flow is partially obstructed by the oil and the droplet formation is achieved by high shear forces generated at the leading edge of the water perpendicular to the oil flow, creating droplets in the picoliter range [1]. The size of the droplets depends on the inlet pressures of water and oil. To create stable droplets, a surfactant is often used. The term 'surfactant' derives from 'surface active agent'. A surfactant is an amphiphilic compound that reduces the surface tension between oil and water phases [4]. It is composed of a hydrophilic and a hydrophobic part, which allows it to adsorb at the interfaces. Furthermore the surrounding layer of an intracellular compartment can be simulated, which can be added to the W/O droplets by mixing lipids into the oil phase. The microfluidic system allows for easy manipulation, immobilization and examination. This work uses microfluidically generated droplets as model systems for intracellular compartments to lay the foundations for enclosing substances and bacteria in droplets.

2 Material and Methods

2.1 Material

The microfluidic flow controller (Flow EZ) was obtained from Fluigent, (Jena, Germany). Fluorinert FC-40, mineral oil, octanol, dodecan and Span 80 are from Sigma-Aldrich, (Merck, Darmstadt, Germany). For the chip production a Sylgard 184 Silicone Elastomer Kit was used, (Dow Corning, MI, USA). DOPC and Liss Rhod PE are from Avanti Polar Lipids (Merck, Darmstadt, Germany).

2.2 Methods

The microfluidic chips were produced from polydimethylsiloxane (PDMS) by mixing an elastomer component and curing agent in a 10:1 weight ratio. Subsequently, the PDMS was degased and poured into the mold of the silicon wafer master, which contained the structure for the chip design. At the same time, cover glasses were cleaned and coated with PDMS. Afterwards the PDMS and the cover glasses hardened for 140 minutes at 65 °C and was cut out of the mold. Then the outlet and the inlets of the aqueous phase and the oil were punched. The microfluidic channels of the structure are between 50 μm and 150 μm wide and the design for the structure was taken from [2].

After punching, the chip was cleaned with ethanol. Subsequently, the chip and cover glass with a PDMS surface were bonded in a plasma cleaner.

The aqueous solution was phosphate-buffered saline (PBS) with 15% Glycerol and 400 mM Sucrose. We compared the oils octanol, dodecan, FC-40 and mineral oil and used the surfactant Span 80 in combination with mineral oil based on [5]. To form droplets, air pressure controllers pushed the oil and buffer from the sample reservoirs through tubes into

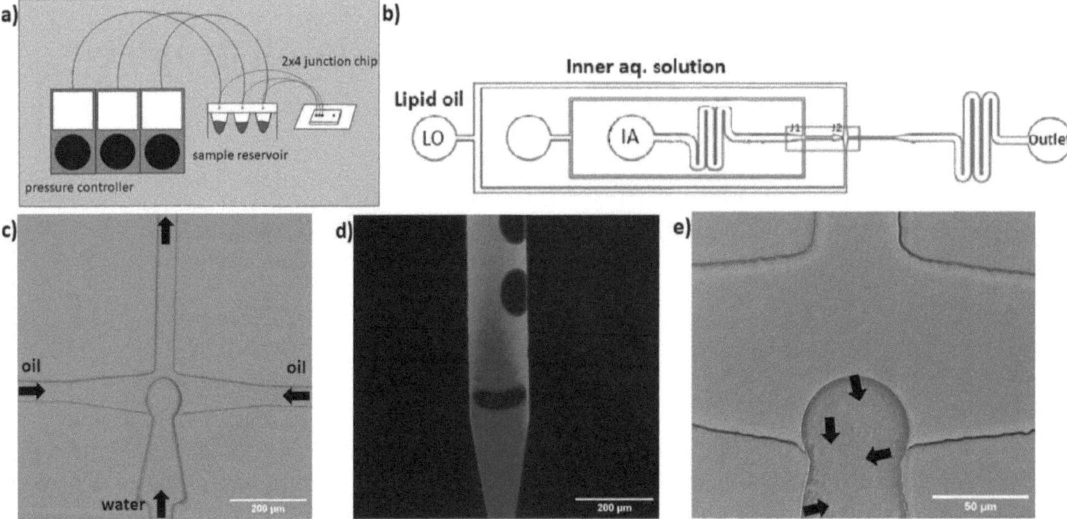

Figure 1: The microfluidic setup is shown in a), while b) depicts the structure of the chip which was taken from [2]. The droplet formation at junction 2 is shown in c), the aqueous solution flows from the lower channel and the oil from the left and right channel into the outlet. The beginning of the outlet and droplets are shown in d) with 0.1 µg/ml rhodamine in mineral oil. Bacteria were enclosed in W/O droplets through the water phase in e).

the chip, as can be seen in Fig. 1a. No more than 100µl was used per experiment. The entire chip was flushed with oil. Injection of buffer then lead to the formation of W/O droplets upon arriving at junction 2, which is shown in Fig. 1c. The W/O droplets can then be observed in the beginning of the outlet channel as can be seen in Fig. 1d. Because the oil cuts off the aqueous phase when it flows into the outlet, the pressures of oil and aqueous phase determine the size of the droplets and their formation rate. The maximal pressure was 1000 mbar, but usually not more than 200 mbar are needed to produce droplets. When the pressure rises and the droplets become smaller, the distance between them increases. The droplets were observed with an objective lens with 10x magnification. Except for Fig. 1e which was observed with an 32x magnification. It shows bacteria in the aqueous phase.

3 Results and Discussion

3.1 Bonding of the chip

When producing the chips, the plasma cleaner generates radicals that react with the methyl groups, resulting in substitution with hydroxyl groups. This leaves silanol (SiOH) groups on the surface, making the surface more hydrophilic. The PDMS returns to the initial state after a few days and is hydrophobic under normal condition. To be able to use the chip it must be hydrophobic so that the oil adheres to the walls and droplets can form. This is necessary due to the hydrophilicity of the aqueous buffer solution. By doing so, unstable droplets or oil-in-water droplets can be avoided. To investigate the influence of plasma treatment and the as-

sociated hydrophilisation of the PDMS chips on the formation of W/O droplets, 4 different time points were investigated. Fig. 2 shows the formation of droplets in octanol over the course of four days. On the first day, oil-in-water droplets formed at the junction, because the chips were too hydrophilic. However, the droplets could not maintain their form in the outlet, and the buffer adhered to the walls of channels as shown in Fig 2a. On the second day, as shown in Fig. 2b, W/O droplets formed, but the chip remained too hydrophilic and the buffer continued to stick to the walls of the channel. In Fig 2c, a production of W/O droplets is shown on the third day, as the buffer no longer adhered so strongly to the walls. However, the droplets fused immediatly when they came into contact. As shown in [6], it is usually sufficient to form a localised bridge between two adjacent drops to fuse them. The formation of this liquid bridge leads to a region of the drop with concave curvature, which corresponds to a low pressure region within the droplets. This low pressure draws liquid into the bridge and thus increases its size [6]. For this reason, surfactants are often added to the solution to stabilize the droplets against merging.

After the fourth day, as shown in Fig. 2d, the buffer no longer adhered to the walls of the channel. The same test was additionally carried out with dodecan, FC-40 and mineral oil. The chip was also too hydrophilic during the first three days. Therefore, the experiments for all oils were carried out at least 4 days after plasma bonding.

3.2 Droplet formation with different oils

Octanol, dodecane, FC-40 and mineral oil were compared for their ability to create stable droplets. Stable droplet for-

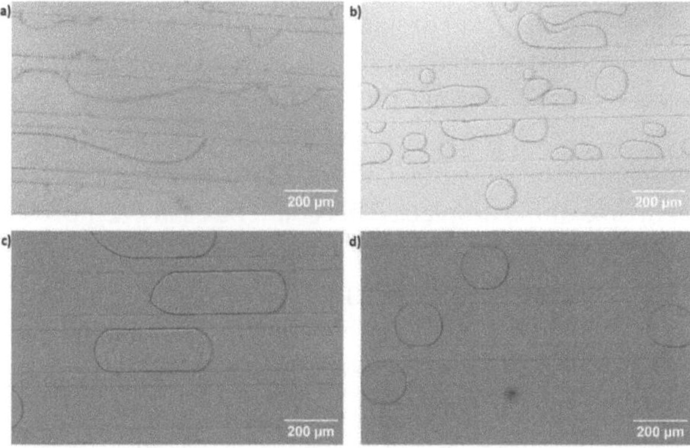

Figure 2: Droplet formation with octanol at different times after plasma bonding. Each picture illustrates the outlet of the chip and the relationship between PDMS hydrophobicity and droplet formation a) one day, b) two days, c) three days and d) four days after plasma bonding.

mation was not possible with dodecan, although Fig. 3b shows a droplet formation in the junction, the pressed-out buffer did not retain its shape in the outlet. It required a pressure of over 800 mbar to push the dodecane to the junction. It was also not possible to produce droplets with FC-40, as shown in Fig. 3c . The oil was pressed jerkily through the channels and the buffer persisted in adhering to the channel walls, even four days after plasma bonding. A surfactant might improve the droplet formation with the oil, as suggested in [4].

On the contrary, Fig. 3a and Fig. 3d show that both octanol and mineral oil can be used to produce droplets four days after plasma bonding. However, when observing the W/O droplets in the channel, it can be seen that the droplets are too unstable and fuse when they come into contact, this happened more frequently with octanol droplets. To further stabilize the droplets the surfactant Span 80 was added.

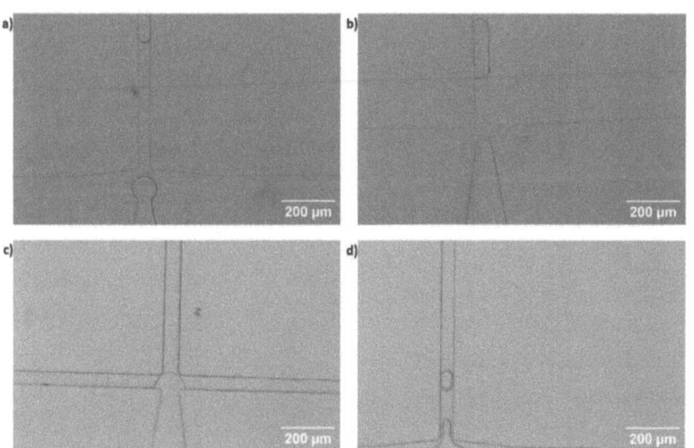

Figure 3: Four different types of oil were used: a) octanol, b) dodecane, c) FC 40 and d) mineral oil.

3.3 Stabilizing the droplets with surfactant and lipid monolayer

Due to the high stability of the droplets, we decided to use mineral oil for our further experiments. Based on [3], the surfactant Span 80 was added to the oil to further improve the stability and was mixed with the oil and put in an ultrasonic bath for 15 minutes. Fig. 4a shows that a concentration of 1% Span 80 is not enough for a robust droplet formation, although there is a clear droplet formation, most droplets still fuse on contact. The unstable behavior may be due to the fact that the surfactant concentration does not cover the entire surface of the droplet. Therefore Fig. 4b shows the concentration increased to 5%. This lead to a strong improvement with droplets that do not fuse. To create a lipid monolayer surrounding the stable droplets 5% Span 80 was combined with 10 mg/ml DOPC (dipalmitoylphosphatidylcholine). The resulting stabilized droplets can be seen in Fig. 4c. However, as it is not yet clear how the lipids are distributed in the droplets, a lower Span 80 concentration would be desirable. While the monolayer has a positiv influence on the stability of the droplets, Fig. 4d indicates that even with a combination of mineral oil and 10 mg/ml DOPC, the droplets still merge on contact. This did not change even when the lipid concentration was increased to 20 mg/ml DOPC, so a surfactant is essential. If the lipids form a monolayer, the question is whether the surfactant displaces them from the surface of the W/O droplets. It is also not clear, if the mixture of lipids and Span 80 is less stable than the surfactant alone. The distribution of Span 80 and lipids does not seem to be uniform, because although most droplets do not fuse with each other, it does happen occasionally.

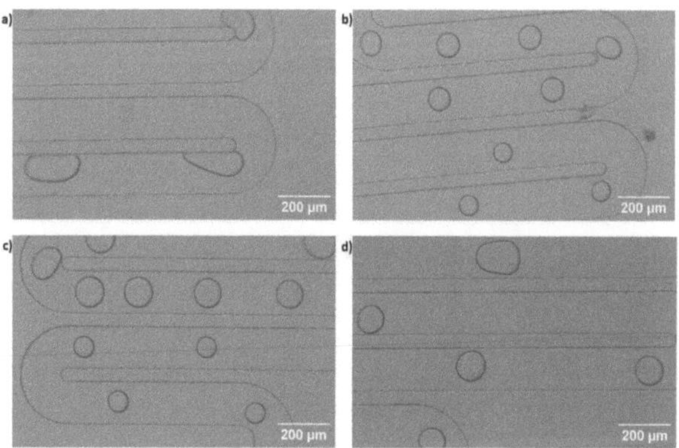

Figure 4: This figure illustrates W/O droplets in microfluidic channels, produced with different oil phases. While a) shows the usage of mineral oil with 1% Span 80, b) uses 5% Span 80. In c) a combination of mineral oil, 10 mg/ml DOPC and 5% Span 80 is shown and d) shows a combination of mineral oil and 10 mg/ml DOPC without a surfactant.

For this reason, an experiment with fluorescence labelled lipids would be advantageous to see where the lipids are located. A first attempt is shown in Fig. 1d, where the mineral oil was labeled with 0.1 µg/ml rhodamine. However, it was not possible to distinguish between the drop surface and the oil phase. In addition, a fluorescently labelled Span 80 would be helpful to distinguish between the surfactant and the monolayer.

3.4 Production of different droplet sizes

Mineral oil, 5% Span 80 and 10 mg/ml DOPC were used to form stable droplets and to test for different droplet sizes. Fig. 5a shows the biggest droplets were derived at 30 mbar oil pressure and 40.9 mbar water pressure with 140 μm in diameter. With the same water pressure, an oil pressure of 60 mbar let to droplet diameters of 100 μm, which can be seen in Fig. 5b. An increase of the oil pressure to 82,9 mbar led to droplet diameters of 77 μm, as shown in Fig. 5c. The minimum with about 56 μm was reached at an oil pressure of 134.5 mbar and a water pressure of 57.8 mbar. After increasing the presure even further it was no longer possible to tell whether there was a difference in size. There are deviations in the relationship between pressure and size due to differences in the chips, which lead to changes in resistance and thus change the required pressure. In addition, the ratio between organic and aqueous solution may be relevant and not the specific values of the pressure. The exact size values also deviate due to the calibration of the microscope and the measurement of the individual droplets.

To produce even smaller droplets, the size of the junction in which the droplets are formed, would have to be reduced.

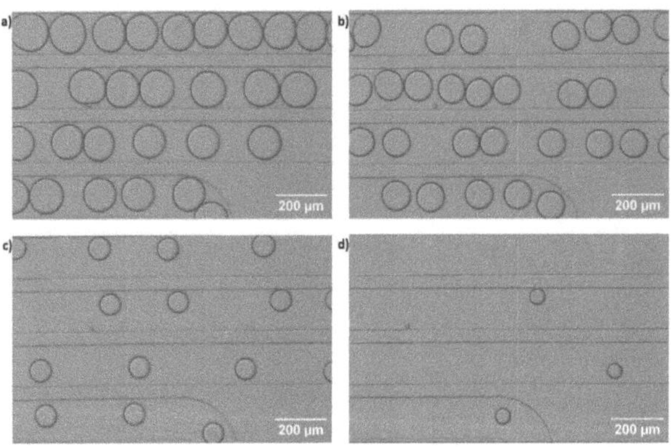

Figure 5: The size of the droplets was adjusted through changing the pressure of the water and oil channels. In a) the size of the droplet is around 140 μm, in b) 100 μm, in c) 77 μm and in d) 56 μm.

4   Conclusion

In this work the microfluidic model of W/O droplets was modified through changes in the oil medium. It was shown that the production of the chip has an influence on the hydrophobicity and consequently on the immediate usability of the chip. It was shown that the use of oils alone does not allow stable droplet formation and that the four oils differ significantly regarding droplet formation. Neither dodecan nor FC-40 could form droplets at the junction. Furthermore the application of lipids in the oil increased the stability of the droplets, but did not prevent them from fusing. Therfore a surfactant was added and improved the stability. The combination of oil, surfactant and lipids facilitated the control of size of the droplets. This could be easily achieved by changing the pressure in the channels.

In the continuation of the project, we plan to use the droplet-enclosed bacteria to study intracellular interaction processes between bacteria and effector molecules.

Acknowledgement

This work has been carried out at the Research Center Borstel with the help of Franziska Kiesow, Kerstin Stephan, Sabrina Groth and Thomas Gutsmann, who I would like to thank for their support during this project.

Authors' Statement

All authors state no conflict of interest.

5   References

[1] T. Thorsen, R. W. Roberts, F. H. Arnold, S. R. Quake. *Dynamic pattern formation in a vesicle-generating microfluidic device*, 2001. Physical review letters, 86(18), pp. 4163.

[2] N. Yandrapalli, J. Petit, O. Bäumchen, T. Robinson, *Surfactant-free production of biomimetic giant unilamellar vesicles using PDMS-based microfluidics*, 2021. Commun Chem 4, 100. https://doi.org/10.1038/s42004-021-00530-1

[3] S. L. Anna, N. Bontoux, and H. A. Stone, *Formation of dispersions using "flow focusing" in microchannels*, 2003. American Institute of Physics. https://doi.org/10.1063/1.1537519

[4] J.-C. Baret, *Surfactants in droplet-based microfluidics*, 2012. DOI: 10.1039/C1LC20582J, Critical Review, Lab Chip, 12, 422-433

[5] M. Chabert, J.-L. Viovy, *Microfluidic high-throughput encapsulation and hydrodynamic self-sorting of single cells*, 2008. https://doi.org/10.1073/pnas.07083211

[6] C. N. Baroud, F. Gallaire and R. Dangla, *Dynamics of microfluidic droplets*, 2010. DOI: 10.1039/C001191F, Critical Review, Lab Chip, 10, 2032-2045

Wounding induces sequential pinna movement in Mimosa Pudica

Jasmin Handel [1], Shouguang Huang [2], Rainer Hedrich [3] and Young-Hwa Song [4]

[1] Biophysics, Universität zu Lübeck, jasmin.handel@student.uni-luebeck.de
[2] Molecular Plant Physiology and Biophysics, Julius-Maximilians-Universität Würzburg, shouguang.huang@uni-wuerzburg.de
[3] Molecular Plant Physiology and Biophysics, Julius-Maximilians-Universität Würzburg, hedrich@botanik.uni-wuerzburg.de
[4] Institut für Physik, Universität zu Lübeck, song@physik.uni-luebeck.de

Abstract

Leaf movements of *Mimosa Pudica* are triggered by electrical and calcium signals, however, the mechanism by which these signals are transmitted within the pinnae of a leaf remains exclusive. In this study, the action potential and the calcium signal were recorded in all four pinnae of a leaf. When touch or wounding were imposed in the pinna closest to the petiole, the actions potential as well as the Ca^{2+}-signal is not transmitted directly to the opposite pinna, despite the physical closeness between them. Instead, they first move counterclockwise or clockwise to the other two pinna, despite the further physical distance between them. We thus characterized a novel signal transmission mode that leads to a sequential pinna movement in *Mimosa Pudica* for a better understanding of the secondary pulvinus.

1 Introduction

Mimosa Pudica, the sensitive plant, is known for its intriguing response to touch or wounding. A recent study has highlighted that rapid leaf folding or drooping is a self-defence manner against herbivorous insects [1]. With a gentle touch, a rapid electrical signal, called action potential (AP) is provoked, which can basipetally propagate to the base of the pinna. If the leaf gets a stronger stimulation, like wounding, in addition to the AP, a variation potential (VP) is evoked, which is a long-lasting depolarization [1]. It was reported that touch only induces leaflet closure within the stimulated pinna, whereas all pinnae of the entire leaf close upon wounding stimulation [1], [5]. The AP is also accompanied by a Ca^{2+} signal, which can be recorded using a genetically encoded Ca^{2+} indicator called GCaMP6f, which belongs to the single fluorophore sensors [6]. However, it remains unknown how the electrical and Ca^{2+}-signals travel from one pinna to the others.

Leaflet closure is due to the loss of turgor pressure in the motor cell of the pulvinus which is a result of K^+ efflux. Loss of the K^+ leads to pulvinus shrinking while uptake of the K^+ results in swelling of the cell. Up to 60 % of the K^+ moves from the flexor side to the extensor side during the folding process [4], [1], [5]. This type of movement is only possible due to the pulvinus, which is located at the end of each leaflet, rachilla and petiole, with the primary and tertiary pulvinus being the more sensitive two [3].
Touch and wounding effects on the *Mimosa Pudica* are described as so-called rapid movements but there is also the slow movement. This so-called nyctinastic movement acts like a biological clock of the plant. The leaves open during

the day and close at night or darkness, whereby the primary pulvinus does not lose their turgor pressure [2].

2 Material and Methods

Plants
The *Mimosa Pudica*, both the wild type and the GCaMP6f-manipulated plants, are cultivated with a 14-h/10-h light/dark time and a temperature of 26°C/24°C in the greenhouse, with a humidity of 60 %. The GCaMP6f-manipulated plant seeds were obtained from [1] and cultivated using the same protocol.

Action potential
The AP is measured with electrodes that are made out of silver and were galvanized using a potassium-chloride solution, which creates a silverchloride layer on their surface. They were connected to a headstage of a custom-made amplifier (input impedance $> 10^{11}\,\Omega$).
The electrodes are connected to the plant via a 15 % gelatin from bovine skin (G9391, *Sigma-Aldrich*) solution, containing 0.1 mol/l of KCl for conductivity. The ground-electrode was attached to the stem of the plant.

Calcium-Signal
To determine whether the Ca^{2+}-signal and the AP correlate with each other, the Ca^{2+}-signal was monitored in GCaMP6f expressing *Mimosa Pudica*. This fluorescence signals were detected in the center of the leaf of a GCamp6f plant. The fluorescence signals were monitored with a fluorescence stereomicroscope (*Leica MZFLIII*), with a 1.0x

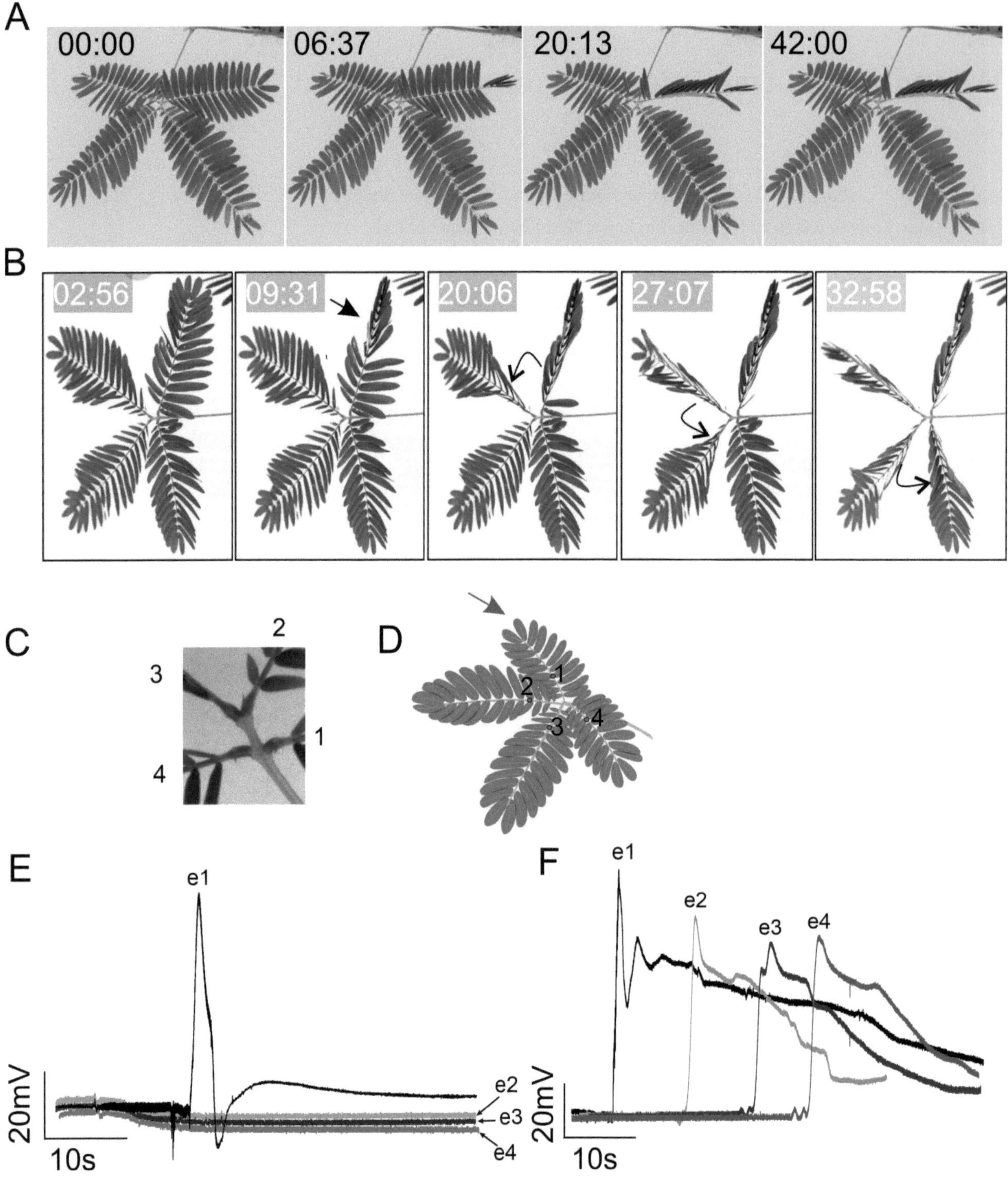

Figure 1: **Wounding induces a sequential leaflet movement in *Mimosa Pudica*. A-B** Touch (**A**) and wounding (**B**) induced leaflet movement. **A** Touch only induced leaflet closure within the stimulated pinna. **B** Wounding due to fire (arrow) induced leaflet closure in a sequential order. **C** Secondary pulvina with the numbered pinnae. **D** Leaf of *Mimosa Pudica* and its numbered attached electrodes for recording the AP by touch or wounding (arrow). **E** AP by touch at leaflet of pinna 1. **F** AP through a wounding effect, due to fire on the tip of the leaflet of pinna 1.

plan objective lens, equipped with a sCMOS microscope camera (Prime BSI Scientific CMOS *Teledyne Photometrics*). It was illuminated via an integrated light source with an excitation wavelength of $\lambda_{Ex} = 470\,nm/40\,nm$ and a emission wavelength of $\lambda_{Ex} = 525\,nm/50\,nm$. All experiments were made with an exposure time of 150 ms and with a binning of 2x2. The interval between the frames was 500 ms. The images were analyzed with the *IMAGE-J/FIJI* software.

For the nyctinastic movement, the secondary pulvinus was

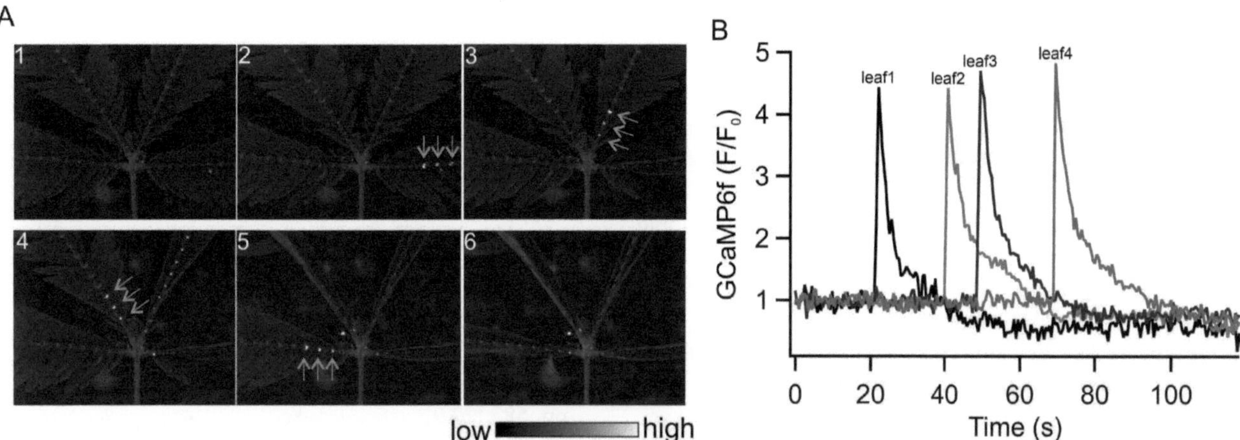

Figure 2: Ca^{2+}-signal was provoked sequentially in different leaves by a wounding stimulation. **A** Wounding through scissors at the tip of the leaflet of pinna 1 which triggers a Ca^{2+}-signal that travels through all leaflets (arrows). **B** Fluorescence signal of the Ca^{2+}-signal in all four pinnae through wounding, due to cut at the tip of the leaflet of pinna 1.

positioned under UV-light, as in the measurement of the Ca^{2+}-signal of *Mimosa Pudica*, and the rest of the room was darkened so that individual leaflets were in the dark. Thus, the plant was partially illuminated for one hour.

3 Results and Discussion

3.1 Action Potential

Mimosa Pudica has a different response to touch and wounding. When a tertiary pulvinus is touched at the tip of the leaflet, as can be seen in Fig. 1A, touch only induced leaflet closure within the stimulated pinna, while the rests had no response. Given that leaflet closure is induced by the electrical signal, the AP was simultaneously recorded in four pinnae of a leaf. Electrodes attached to the tertiary pulvini were displayed in Fig. 1D. In line with the Fig. 1A and Fig. 1E the AP was constrained to the leaflet that was touched but could not travel to the rest of the pinnae. Here it can be shown that the rapid movement by touch has a very steep rise in the AP as well as a steep fall again, which shows that no VP is present.

When wounding occures, by cutting the tip of the leaflet or by putting a flame at the tip of the leaflet for a duration of t=1 s, all the pinnae connected to the secondary pulvinus close, as illustrated in Fig. 1B. The AP travels through the individual pinnae as shown in Fig. 1F. The closing of the leaflets by wounding occures in a specific order (n=5). When the pinna 1 gets wounded, then all the other leaflets close counterclockwise. If the wounding occurs on the leaflet of pinna 2 or pinna 3, then the two pinnae next to them close after the wounded one. When the pinna 4 gets wounded, all leaflets close clockwise. From this it can be concluded that the secondary pulvinus has a barrier within it that separates pinna 1 and pinna 4, as these two are the pinnae that are physically closest to each other as Fig. 1C shows.

The determination of the speed for pinna 1, as well as the speed for the transmission of the AP from electrode 1 to the other three electrodes shows, that the speed slows down. Pinna 1 exhibits a speed of $v_{AP} = 1.00\pm0.36mm/s$ (n=5). However, the speed of e1 to e2 is $v_{e1,e2} = 0.28\pm0.08mm/s$, the speed of e1 to e3 is $v_{e1,e3} = 0.18\pm0.03mm/s$ and the speed of e1 to e4 is $v_{e1,e4} = 0.13\pm0.06mm/s$. The speed of pinna 1 is comparable to the speed from [1]. The reduction in speed can be explained by the fact that the AP has to overcome fewer barriers from pinna 1 to pinna 2 than from pinna 1 to pinna 4.

3.2 Calcium-Signal

In line with the AP, the Ca^{2+}-signal was first detected in the wounded pinna, with 10-20 s delayed, the signal went to the other pinnae.

In Fig. 2A the fluorescence signal of the Ca^{2+}-signal for *Mimosa Pudica* is illustrated, where the tip of the leaflet of pinna 1 was cut with a scissor. The Ca^{2+}-signal travels, as well as the AP, counterclockwise through the plant. The Ca^{2+}-signal travels from the tip of the leaflet to the secondary pulvinus and from there to the other pinnae. The signal is mainly visible in the pulvini. If the wounding is very severe, however, the signal is so strong that it is also visible in the individual leaves. It is then also visible in the petiole. This is because more Ca^{2+} is then released, which is distributed further into the leaves

Fig. 2B shows the fluorescence signal recorded at the tertiary pulvinus, at the same level as the electrodes used to record the AP, as a function of time. There is illustrated that there is a larger time interval between pinna 1 and 2 and between pinna 3 and pinna 4 than between pinna 2 and pinna 3. The time between pinna 1 and pinna 2 is $t_{leaf1,leaf2} = 15.2\,s$, while the time between pinna 2 and pinna 3 is $t_{leaf2,leaf3} = 6.8\,s$. In contrast, the time between pinna 3 and pinna 4 is greater again with a value of $t_{leaf3,leaf4} = 16.0\,s$. This is due to the fact that there is a greater distance between pinna 1 and pinna 2 and between pinna 3 and 4 than between pinna 2 and pinna 3 and thus the Ca^{2+}-signal can be transmitted faster. Using the length of the pinna and the time it takes for the Ca^{2+}-signal to reach the secondary pulvina, the speed of pinna 1 can be calculated, resulting in $v_{Ca^{2+}} = 0.30 \pm 0.18$ mm/s. This velocity compared to that of the AP infers that the AP is generated due to the Ca^{2+}-signal.

When illuminating the secondary pulvina and individual leaflets around it with UV-light, it can be discovered, that the individual leaves that lie in the UV-light remain open, but the others all close over time due to the darkness. This is due to the fact that blue light activates the H^+ pump, which is important for opening respectively keeping open the pulvini.

4   Conclusion

Mimosa Pudica tries to protect itself from predators by closing its leaves, inducing an AP and a Ca^{2+}-signal when touched or wounded. The AP at touch is not strong enough to overcome the secondary pulvinus, but in the case of a wounding effect, the signal is transmitted to the other pinnae, which then also close. The AP moves at a speed of $v_{AP}=1.00\pm0.36$ mm/s whereas the Ca^{2+}-signal moves at a speed of $v_{Ca^{2+}}=0.30\pm0.18$ mm/s in the wounded leaf.

The leaf closure sequence, which starts when the leaves are wounded, was also determined. From this it can be concluded that the secondary pulvinus has a internal barrier separating pinna 1 and pinna 4. This assumption is based on fact that when pinna 1 is wounded, pinna 4 is the last to receive an AP and Ca^{2+}-signal. This will be further investigated in the near future using sectioning techniques.

As seen in the previous experiments, there is a lot to find out about *Mimosa Pudica*. So far only the fast movement has been considered, so one of the next steps would be to look at the slow movement, the so called nyctinastic movement. It has already been observed that the individual leaves, which are irradiated by blue light, remain open and thus the tertiary pulvini have no loss of turgor pressure. This is due to the fact that the H^+-pump is activated by blue light.

The other next step is to find out in which way the glutamate receptor-like channels play a role in the transmission of the AP and the Ca^{2+}-signal.

Acknowledgement

The work has been carried out at Julius-Maximilians-Universität Würzburg at the Institute Molecular Plant Physiology & Biophysics and supervised by Dr. Shouguang Huang and supervised by Dr. Young-Hwa Song from the physics institute, Universität zu Lübeck.

Authors' Statement

Conflict of interest: Authors state no conflict of interest

5   References

[1] T. Hagihara, H. Mano, T. Miura, M. Hasebe and M. Toyota, *Calcium-mediated rapid movements defend against herbivorous insects in Mimosa pudica.* Nature communications, vol. 13, no. 1, pp. 6412, 2022.

[2] M. Ueda and S. Yamamura, *The chemistry of leaf-movement in Mimosa pudica L.* Tetrahedron, vol. 55, no. 36, pp. 10937–10948, 1999.

[3] T. Hagihara and M. Toyota, *Mechanical Signaling in the Sensitive Plant Mimosa pudica L..* Plants, vol. 9, pp. 587, 2020.

[4] G.G. Cote, *Signal transduction in leaf movement.* Plant Physiology, vol. 109, no. 3, pp. 729, 1995.

[5] T. Sibaoka, *Rapid plant movements triggered by action potentials.* The botanical magazine= Shokubutsu-gaku-zasshi, vol. 104, pp. 73–95, 1991.

[6] T. Chen, et al. *Ultrasensitive fluorescent proteins for imaging neuronal activity.* Nature, vol. 499, no. 7458, pp. 295–300, 2013.

9

E-Health

A web-based listening test suite for an immersive listening lab

Robin Holland [1], Gabriel Gomez [2], and Hendrik Husstedt [3]

[1] Auditory Technology, Universität zu Lübeck, robin.holland@student.uni-luebeck.de
[2] WSAudiology, gabriel.gomez@wsa.com
[3] Deutsches Hörgeräte Insititut, h.husstedt@dhi-online.de

Abstract

Speech testing plays a crucial role in evaluating and improving the performance of hearing aids. Especially in complex acoustic situations where background noise and reverberation can affect speech intelligibility, accurate and reliable speech tests are of significant use. At the hearing aid manufacturer WSAudiology, an immersive listening lab has been created that aims to be as easy to use as possible by investigators and test subjects. This paper presents a web-based approach to a test suite that can be controlled in the laboratory via a tablet. The web app implements a multichannel audio stream on the server side and offers a standard speech intelligibility test and a calibration routine for loudspeakers. The realized framework can be extended by various audiological tests and experiments and shows great added value, especially with regard to the work with test subjects, due to its high accessibility.

1 Introduction

In modern audiology, the evaluation of the effectiveness of hearing aids and their algorithms relies heavily on robust speech test procedures. In everyday life, hearing aid wearers are exposed to difficult acoustic situations such as reverberation and background noise, which can significantly impair the intelligibility of speech. It is therefore important for hearing aid manufacturers to take such situations into account when evaluating hearing aids. At WSAudiology, the pursuit for improved evaluation techniques has led to the creation of an immersive hearing laboratory that can virtually represent acoustic scenes.

Virtual acoustic environments are digital replications of real listening scenarios in which reverberation is simulated and the (dynamic) sound sources are reproduced or simulated [1].

The "Wonderful Sound Lab" takes the approach of using hidden equipment and loudspeakers to create the most everyday environment possible, as can be seen in Fig. 1. This approach, combined with a home-like choice of colors and furnishings, can be assumed to make test subjects behave more authentically in experiments.

Speech tests and other listening experiments usually require additional hardware such as a sound card and computer in the laboratory. The use of additional hardware creates a lab-like environment and can reduce the high ecological validity. Therefore, a web-based test suite was developed for listening labs such as the Wonderful Sound Lab, which can be operated with various end devices like tablets and is easily expandable.

Figure 1: Picture of the Wonderful Sound Lab. Most of the speakers and microphones are placed behind the panels. In the center of the room is a table with eight seats.

2 Material and Methods

2.1 Laboratory

The Wonderful Sound Lab is primarily operated via a tablet that controls the digital signal processor (DSP) server in a seperate room via Hypertext Transfer Protocol (http). The DSP server convolves the sounds to be played back and the sounds occurring in the room with the desired impulse responses. This allows the lab to be used for various acoustic scenes that can be flexibly adapted in terms of their reverberation. 24 equalized, directional microphones ensure that sounds in the room can be convolved in real time. The reverberation and sound material is reproduced via 39 loudspeakers. As shown in Fig. 2, there are 9 loudspeaker on the ceiling, 10 on the upper wall, 18 on the lower of the wall and 2 subwoofers in two corners. In addition, 6 loudspeakers on the lower wall that can be moved in one dimension (x-axis) are used for additional playback material such as speech. All inputs and outputs of the lab are passed through the DSP server.

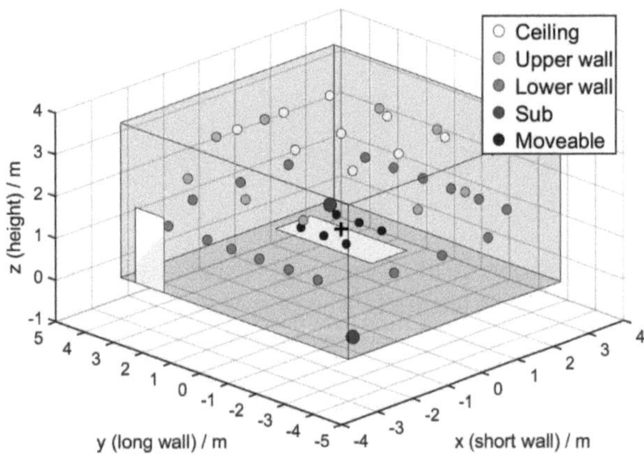

Figure 2: Schematic representation of the laboratory with its loudspeakers. The loudspeakers of the lab are plotted as dots. A plus sign is shown in the center of the room. The white plane in the room represents the lab's table.

2.2 Implementation of the web app

The concept of the laboratory envisions that all tests and settings can be controlled via a tablet and therefore no additional hardware is required. A reasonable independence from operating systems, convenient deployment and simple revisions can be achieved with web applications. The audio stream should take place on the lab computer, which is connected to the audio interface of the DSP server. Django was used as the web framework, which allows for managing databases (models) in addition to the web-templating system [4]. Examples of the models are the test settings or audio devices and their channels. As Django, the entire backend is implemented in Python.

The audio stream was realized with the Sounddevice library [5]. The audio stream must be executed in parallel in a second process so that it runs continuously across different templates and functions. Communication between the server and the audio process was ensured by message queuing. The audio stream signal is transmitted to the DSP. Because DSP server and test suite server are separate from each other, the DSP can be used during this process to reverberate the signals in real time, enabling complex audiological tests.

A brief representation of the web app is shown in Fig. 3. The client, i.e. the end device, renders the web-page (templates) and executes Javascript functions (statics) on the client side, for example to send commands to the server or plot graphs on the browser window. The server is responsible for providing the templates and statics to the presen-

tation layer. The view functions are executed as soon as a client visits an address or fetches data. Furthermore, the views send commands to the audio process. The callback functions of the audio process communicate with the hardware of the sound device.

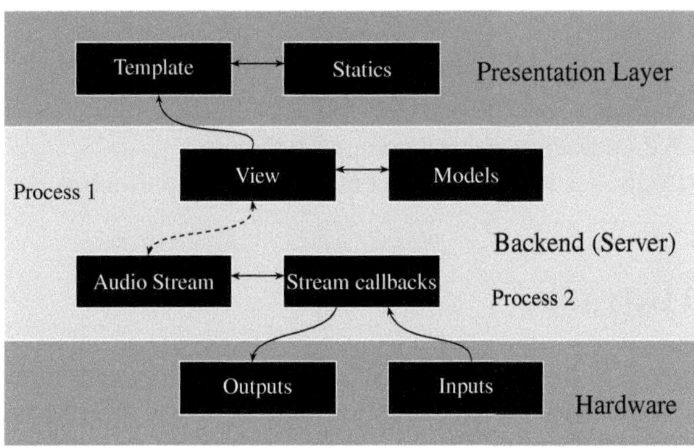

Figure 3: Diagram of the layers of the web application.

2.3 Calibration of the speakers

Correct levels are essential for hearing experiments and especially for determining the speech reception threshold (SRT). There are eight different seats for test persons in the lab. The amplitude and phase of the same playback vary at the different seats. The speaker settings never change, neither within the DSP software nor on the hardware. It can therefore be assumed that amplitude and phase depend only on the input signal and seat.

The loudspeakers in the laboratory are of two different types, stationary and movable loudspeakers. A reference point in dB FS was established by playing pink noise on each stationary speaker and recorded at each seat using a calibrated microphone. The difference between the output level and the input level was used to determine the reference points for the speaker-seat pairs. These reference values now make it possible to determine the sound pressure level for each speaker-seat pair before the signal is reproduced. The level L_{total} of a multichannel signal with n incoherent sound sources with levels L_i can be determined by level addition (1) [3]. However, the reference values for the moving speakers must be determined after each movement, for which the microphones could be used.

$$L_{\text{total}} = 10 \log_{10} \sum_{i=0}^{n} 10^{L_i/10}. \qquad (1)$$

Neither the stationary nor the movable loudspeakers were equalized when the laboratory was set up. It can be assumed that the equalization depends too much on the listening position and, in the case of the movable loudspeakers, on the position of the loudspeaker.

To simplify matters, the seat-dependent equalization of the loudspeakers was not taken into account in this work. In addition, it is technically not possible to access the microphone signals outside the DSP in the current state of the lab, which is why the level of the moving speakers currently has to be determined manually.

2.4 OLSA

The Oldenburg sentence test (OLSA) is a matrix-based speech intelligibility test that primarily measures the SRT in noise. The sentences consisting of name-verb-number-word-adjective-object are meaningless, resulting in only a small learning effect [2].

The version of OLSA implemented here adaptively adjusts the level of the speech material based on the procedure presented in [6]. If more than 50 % of the words are understood, the speech level is reduced and thus also the rendered signal-to-noise ratio (SNR). If less than 50 % of the words are understood, the speech level is increased, which also increases the rendered SNR. The level of the background noise remains the same. The SRT is estimated at the end of a test run.

The used speech material is in german language. The graphical control system was realized as a web interface with HTML and Javascript. The graphs for the course plot and the scatter plot were created with the Javascript library plotly.js [7].

A mapping function makes it possible to route each channel of a sound file (in a 1:1 relationship) to the loudspeaker channels. Using the calibrated sensitivities of the speakers, the average dB FS of each channel and level addition, the test suite can determine the average level in the lab for each routed signal.

3 Results and Discussion

In this work, a test suite for the Wonderful Sound Lab was developed. The test suite is web-based and can be controlled solely with a terminal device. The server backend was implemented on the lab computer, which is connected to the lab's DSP. The OLSA speech intelligibility test has also been implemented in the test suite. An example of a running OLSA test is shown in Fig. 4. The test suite can be controlled on various devices such as tablets or computers. Speaker calibrations for the stationary speakers are retrieved automatically. A multichannel audio stream has been implemented. Examples of input signals in the app are white noise, speech signals, but also everyday situations recorded and played back. In case of the OLSA, sound-channels can be mapped as required.

An important factor in this project was accessibility for users. Thanks to the implementation as a web interface, no separate application is required on the tablet. It is sufficient to call up a URL in the browser. This approach not only ensures easy deployment, but also enables use on different devices in the lab network. For some test subjects use of a laptop with a mouse and keyboard may be more accessible. Older people in particular may have problems finding their way around a tablet. For open-set speech intelligibility testing using the OLSA, which is controlled by the test instructor, this factor does not yet play a role. Though with other

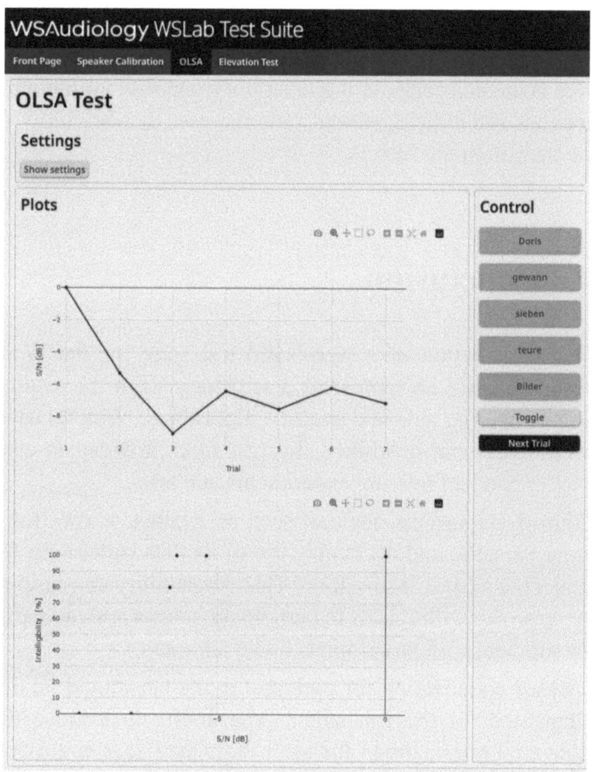

Figure 4: Screenshot of a running open-set OLSA test within the test suite. On the right is the control panel with the words of the played sentence. The upper course plot displays the trial number on the x-axis and the SNR in dB on the y-axis. The lower scatter plot contains the SNR in dB on the x-axis and the discrimination values in % on the y-axis.

experiments and tests in mind, this could be a decisive argument. In addition, a certain robustness is to be expected, as the application is installed on the labs computer that is not directly used by test instructors and test subjects.

Up to now, Matlab implementations have mainly been used for listening experiments and speech intelligibility tests. The correct functioning of these Matlab implementations often depends on the version used, the installed libraries and the connected sound card. This requires a certain level of expertise from the user, which is not required in the web environment. In addition, the implemented test suite including audio stream is expected to allow other tests and listening experiments for the listening lab to be implemented more quickly.

A key advantage of the web-based test suite is its higher ecological validity compared to the status quo. The need for additional hardware is eliminated and subjects experience tests in an environment closer to real-life situations. These aspects could help to enable a more realistic evaluation of the performance of hearing aids or algorithms in different acoustic environments.

It should be noted that reference values for speech intelligibility tests cannot simply be transferred to this room. For this reason, reference values must be determined for the laboratory in order to obtain valid results. The loudspeaker layout and the playback in the laboratory do not follow the approach of a physically correct sound field. Therefore, complex acoustic scenes in the laboratory generate an inhomogeneous sound field, which could increase the variability of the measurement results.

4 Conclusion

The introduction of a web-based test suite for the Wonderful Sound Lab represents a step forward in the evaluation of hearing aids and acoustic algorithms. With its user-friendliness and flexibility, this test suite provides an easy way to conduct hearing experiments and tests.

Control via various devices, such as a tablet, allows for a more versatile and accessible use of the test suite, both for subjects and test instructors. This adaptability can help to increase the willingness to participate in tests and increases the efficiency of experiments in the laboratory.

Speaker equalization for each seat in the lab should be implemented for the test suite. The ability to retrieve the acoustic optimizations for each individual seat could further improve the comparability, accuracy and reliability of the test results. In addition, the calibration routine for the movable speakers is not yet usable due to the lack of access to the microphone signals.

There is also the potential to integrate further tests and experiments into this environment. This expansion could offer the efficiency of hearing aid evaluation and audiology in the laboratory. This would allow a wider range of audiological tests to be integrated into this test suite, optimizing workflows and providing more convenience for researchers.

Another important step would be to verify the OLSA and investigate the potential effects of increased ecological validity on the test results. Such an analysis could help to understand the effects of the more natural laboratory environment on the test subjects. With the OLSA of the test suite, the artificial reverberation of the laboratory can also be conveniently used to test effects on speech intelligibility.

Overall, the web-based test suite is one possible approach to conducting device-independent audiological experiments in laboratories. This approach is particularly promising if the laboratory equipment is concealed and a homely atmosphere is to be created in order to obtain the most authentic test results possible. The remaining improvements and aspects to be investigated could further strengthen the applicability and informative value of the platform in audiology research and practice.

Acknowledgement

The work has been carried out at WSAudiology, Erlangen and supervised by the German Institute of Hearing Aids, Luebeck.

Authors' Statement

Conflict of interest: Authors state no conflict of interest.

5 References

[1] L. Savioja, J. Huopaniemi, T. Lokki and R. Väänänen, *Creating Interactive Virtual Acoustic Environments.* Journal of the Audio Engineering Society, vol. 47, pp. 675–705, 1999.

[2] K. Wagener, V. Kühnel and B. Kollmeier, *Entwicklung und Evaluation eines Satztests für die deutsche Sprache I: Design des Oldenburger Satztests.* Zeitschrift für Audiologie, vol. 38, pp. 4–15, 1999.

[3] M. Möser, *Technische Akustik.* Springer, vol. 8, pp. 475–477, 2009.

[4] Django Software Foundation, *Django.* Available: https://www.djangoproject.com/ [last accessed on 2024-01-11].

[5] M. Geier, *Python-Sounddevice.* Available: https://github.com/spatialaudio/python-sounddevice [last accessed on 2024-02-6].

[6] T. Brand and B. Kollmeier, *Efficient adaptive procedures for threshold and concurrent slope estimates for psychophysics and speech intelligibility tests.* Acoustical Society of America, vol. 111, pp. 2801—2810, 2002.

[7] Plotly Inc, *Plotly Javascript.* Available: https://plotly.com/javascript/ [last accessed on 2024-01-23].

Pathwalker - a tool for supporting interactive analytics on FHIR data

Bettina Uliczka [1], Josef Ingenerf [2], Joshua Wiedekopf [3], and Lorenz Rosenau [3]

[1] Medical Informatics, Universität zu Lübeck, bettina.uliczka@student.uni-luebeck.de

[2] Institute of Medical Informatics, Universität zu Lübeck, josef.ingenerf@uni-luebeck.de

[3] IT Center for Clinical Research, Universität zu Lübeck, {j.wiedekopf, lorenz.rosenau}@uni-luebeck.de

Abstract

Pathling is a powerful tool to analyse large amounts of Fast Healthcare Interoperability Resource (FHIR) data, extract certain information and prepare FHIR data for further use. As input, it uses FHIRPath expressions. FHIRPath is a specialized query language tailored explicitly for FHIR resources and profiles. To enhance the accessibility of Pathling, Pathwalker, a tool that provides support for creating FHIRPath expressions, was developed. It is a web-based syntax-supported visual editor that visualizes the selected resources as a tree, shows function that are allowed to be added and possible parameters that can be added. Users can explore the available functions, including those establishing connections to a terminology server. This functionality, for instance, enables users to directly visualize value sets, if they are allowed, enhancing the efficiency and user-friendliness of the data analysis process.

1 Introduction

Pathling is a tool, developed by the Australian e-Health Research Center, CSIRO, to analyse Fast Healthcare Interoperability Resources (FHIR) data [1]. FHIR is a health data standard that supports data exchange and interoperability in the health sector [2]. Pathling is able to analyse a large amounts of FHIR data, for example big patient cohort studies. It supports among others three operations: `Aggregate`, `Search` and `Extract` that take FHIRPath expressions as parameters. Pathling, in comparison to other tools that analyse large amounts of electronic health data like the Forschungsdatenportal für Gesundheit (FDPG), gives data scientists a lot of freedom to express what they want through FHIRPath expressions. The FDPG is build for medical researches and thus limits the parameters, e.g., the user can not analyse encounters. Something that Pathling is able to do [3].

To use Pathling one must first learn the language FHIRPath which requires a deep understanding of FHIR and its profiles and also an understanding of hierarchical structures, query languages, data types and functions. This is challenging for all users, independent of their domain knowledge. However, it is important to emphasise that mastery of the quite unknown but complex FHIRPath language is a basic requirement for using Pathling [1]. To deal with that challenge Pathwalker was created.

2 Material and Methods

For understanding the resulting tool, FHIR, FHIRPath, Pathling, and Synthea are explained.

2.1 FHIR

FHIR is a healthcare standard crafted by the Standard Developing Organization Health Level Seven International (HL7) [2]. It builds upon prior HL7 standards and combines the advantages and modernizes it. Its goal is to support interoperability and data exchange in the health sector. FHIR is structured in various resources like patient, condition, observation and many more. They are hierarchical structured data models. The detailed structure of each resource including elements, types and constraints is represented uniformly based on the meta the resource StructureDefinition.

Resources or profiles (as modified resources) can be represented as trees, where each element of the resource functions as a node in the tree and optionally has children. In the context of a FHIR resource, the root is the resource's name, such as Patient or Observation. All nodes can be searched through a process that can be based on various parameters. Either searching based on the data of the node, results in non-unique matches with several possible leafs and FHIRPaths. Alternatively, searching by the nodes unique identifier ensures an exact solution. Profiles are extensions or constraints of a FHIR resource. They can be used if their requirements are not covered by the FHIR base standard.

2.2 FHIRPath

FHIRPath is a language for navigating through FHIR data models and extracting useful data for use and interpretation. It is similar to XPath, but it is adapted to work with FHIR data [2]. The simplest case of FHIRPath is chaining. Chaining involves recursively combining all the parent elements of a child. There are different functions, that can be added in different cases. The functions that are implemented here are `where`, `exists`, `count`, `memberOf`, `resolve` and `reverseResolve`. There are functions that are unique to the Pathling implementation and not part of the FHIRPath specification, for example `reverseResolve` [5].

where can be added to any collection regardless of the type. A criteria is added as a parameter, which evaluates with or without an expression to a Boolean value. After the function any other parts of the FHIRPath of the resource can be added. Example: `Patient.name.where(given='Thomas').birthDate`

exists can be added to any node. This can work without an argument, but also with a criteria with an optional expression, similar to the where-function. No further elements are allowed to be added. Example: `Patient.where(gender='male').photo.exists()`

count can be added to any node, no further elements are allowed to be added. Example: `Patient.count()`

memberOf can be added to nodes that are of type `Coding` or `CodeableConcept`. It takes the URL of a value set as a parameter. By default it is the value set that is deposited in the StructureDefinition of the reference. Other value sets can also be accessed if there is a connection to a terminology server. Example: `Patient.gender.memberOf('https://www.netzwerk-universitaets medizin.de/fhir/ValueSet/birth-sex')`

resolve can be added to a node, which has the type reference and loads in the elements from the reference, offering several different possible references. Further elements can be added to the FHIRPath. Example: `Encounter.subject.resolve().name.family`

reverseResolve takes the reference as a parameter and continues with the original resource. The reference is the root of the FHIRPath. Example: `Encounter.reverseResolve(Observation.encounter).has-Member`

2.3 Pathling

Pathling supports six operations in total, here the focus is on the three that take FHIRPath expressions as input each having different parameters. The search operation has only one parameter, filter, which takes a FHIRPath as input that evaluates to a Boolean value. The aggregate-function has three input parameters: aggregation, grouping and filter. Filter is a Boolean expression, while grouping and aggregation must return a so called materializable type. Materializable types include Boolean, String, Integer, Decimal, Date, DateTime, Time and Coding. The types can also be returned by a function, `count()`, which returns an Integer or `exists()`, which returns a Boolean value. The extract-function requires three different parameters: column, filter and limit. Similar to aggregation and grouping the column input must evaluate to a materializable type. Filter must produce a Boolean value and limit must evaluate to an integer.

2.4 Synthea

Synthea is an open-source tool designed to address the challenges for developers to access electronic health records (EHR), because of legal, privacy and security reasons [6]. Synthea provides the possibilities to create realistic but not real patient data as HL7 FHIR data. It allows the developers to tailor the created data to specific parameters, for example the age of the patients, the gender, as well as the clinical module can be selected. The modules include medical conditions like asthma, injuries and life situations like homelessness and sexual activities. As output ndjson files are created which include a specific resource type. Like real data the resource types are connected to each other by references and references to value sets and code systems.

3 Results and Discussion

In this internship a web-based syntax-supported visual editor (called Pathwalker) was developed, empowering users to craft FHIRPaths. It visualizes a selected FHIR resource as a tree. The tree structure is primarily implemented in Java, for the representation HTML is used. Specifically lists are used recursively to build the trees. Each node is a button, in most of the cases when its clicked a FHIRPath for that specific node is build, based on the id in the background. The FHIRPath is created by recursively walking the tree up to the root.

The structure of a FHIR resource is defined in the StructureDefinition. To get further information, such as the type of an element, FHIRPath can be used. Every time a FHIRPath is created the type of the selected node is searched for. With the FHIRPath expression `StructureDefinition.descendants().where(path=FHIRPath ).type.code` the type can be found, if the FHIRPath is part of the StructureDefinition. Based on different types, there are different possibilities to continue; for example, the function `resolve` can only be used if the type of the node is `Reference`. In some cases the FHIRPath is not part of the StructureDefinition, , this can occur when more trees are added because of references or for other elements with non-primitive data types (like `CodeableConcept`). In that case, the tool searches for the part of the FHIRPath that is included in the StructureDefinition, then uses the type of that node as the root for the next part and checks again if is part of the next StructureDefinition. Even though the root of the elements that are added are not part of the FHIRPath at all layers the type can be detected, for example the expression `Condition.code.coding.display`.

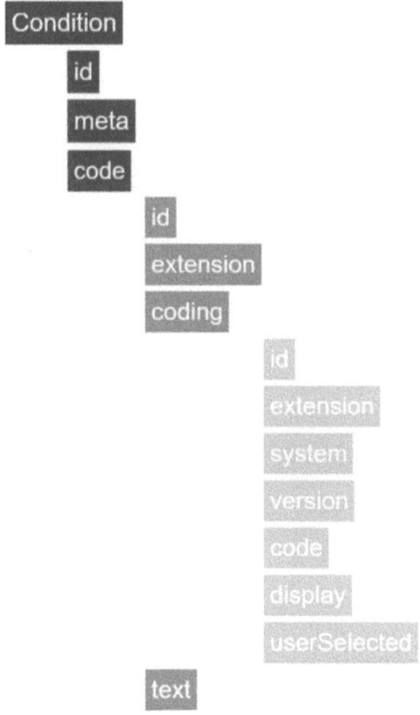

Figure 1: Dark gray is part of the `Condition` resource. Gray is part of the `CodeableConcept`. Light gray is part of `Coding`.

As seen in Fig. 1 the first part of the FHIRPath is part of the Condition resource. `Condition.code` is of type `CodeableConcept`, so the elements of CodeableConcept, which are `id`, `extension`, `coding` and `text` are added, in the figure in gray. When `coding` is selected the type of `CodeableConcept.coding` must be searched for, which is `Coding`, then `Coding` is loaded in, in the figure in light gray.

The elements of `CodeableConcept` and `Coding` are currently loaded in automatically. When a node is of type reference it is possible to select the function resolve. When it is selected the possible references are searched for, that is again being done by a FHIRPath expression `StructureDefinition.descendants().where (path=FHIRPath).type.where(code= 'Refe-rence').targetProfile`. The possible references are represented as buttons, as seen in Fig. 2. When a node is selected the FHIRPath is for example `Encounter.subject.resolve().gender`, where gender is part of the Patient resource. In the background, because of the tree, the FHIRPath `Encounter.subject.Patient.gender` is created, which is wrong, but helps to find the type of the node that is selected.

Another possibility of working with a reference is the function reverseResolve. Again the possible references are presented, but in this case the tree is not being loaded in

Profile

The fhir path you are looking for is **Encounter.subject**.

The type of the selected node: Reference

Add function here:

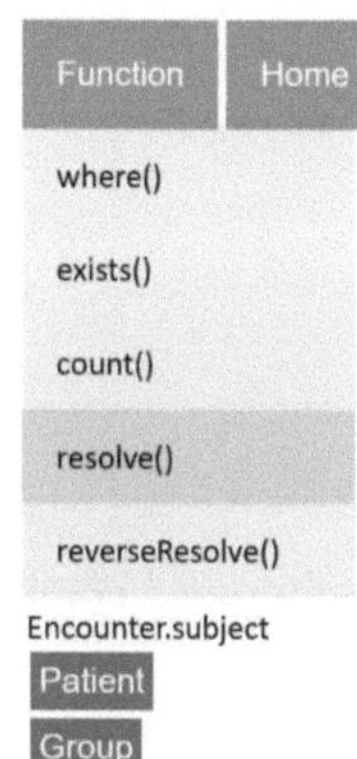

Figure 2: `Encounter.subject` is of type Reference, the possible resources are Patient and Group. If one of them is selected the resource is loaded in and added to the FHIRPath.

but the referenced resources name is the new root of the FHIRPath. Then further elements can be added. Fig. 3 shows the final FHIRPath, where first an element of type reference was selected (here: `Observation.subject`), then the referenced resource was selected (`Patient`) and then further elements of the observation resource were added (`component.code`). When the where-function is

Profile

The fhir path you are looking for is **Patient.reverseResolve(Observation.subject).component.code**.

The type of the selected node: CodeableConcept

Add function here:

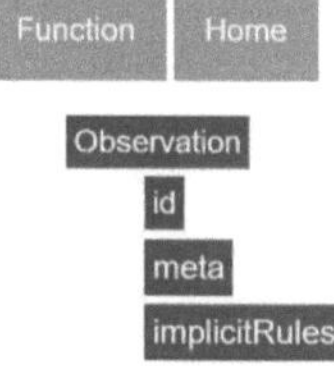

Figure 3: The reverseResolve-function used for the observation resource.

selected the buttons that are not allowed to be selected are disabled, e.g. when `Patient` is selected only the direct children are active and can be selected as criteria, not for example the children of `contact`. When a node is selected as criteria there are three options to continue. The comparison operator and the expression can be added as free text, seen in Fig. 4 the upper part. The comparison operators $[\leq, \geq, <, >, =]$ are allowed.

The other two options are only available if the selected criteria can have bindings, which include elements of types code, `CodeableConcept` or `Coding`. When the selected criteria node is of those types it is possible to se-

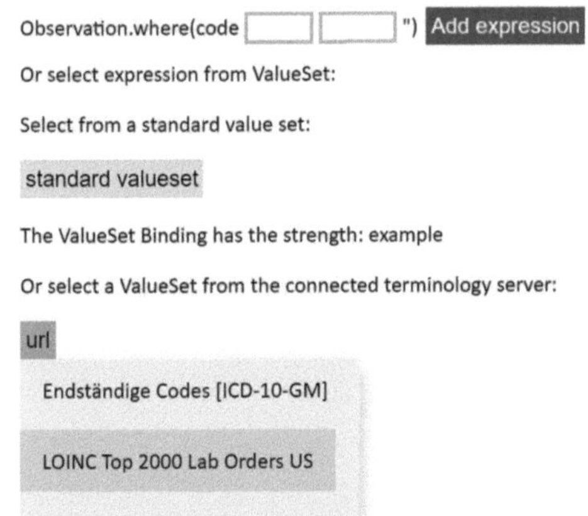

Figure 4: A value set can be selected from a connected terminology server.

lect a value set either from a terminology server or from the default or the required value set that references a code system from HL7. The value sets for either option can be selected with a dropdown menu. For the terminology server, all of the URLs of the value sets from the connected terminology server are loaded in, as seen in Fig. 4. When a value set is selected, it is expanded and loaded based on the URL. The value sets also have StructureDefinitions in the JSON format and the display of the values can be accessed by `ValueSet.descendants().display` and are loaded in another dropdown menu. Besides working with resources and profiles the tool is also able to load in instances of those resources. For testing purposes the instances were created using Synthea, including 12 different resources and in total 100 instances. Instances result in different issues, e.g. in a lot of cases multiple elements or bundles of elements can be part of a node. Example: a patient can have multiple names (e.g. official and maiden name), which result in multiple HumanNames, which result in multiple elements of HumanName. To visualize that nodes numbers were added to differentiate those bundles, e.g. `Patient.name.0.given.1."Roosevelt"`, which can be translate to `Patient.name.where(given= "Roosevelt")`.

The Pathling implementation includes 24 different FHIRPath functions and not all of them are implemented in Pathwalker. Only six of them are implemented, so potential future works could be focused on that. Additionally a certain computer science background is still helpful for example when datatypes are compared to each other, especially because sometimes there is the option of freetext, as seen in Fig. 4.

4 Conclusion

To improve the users experience with Pathling a tool to easily create syntactically correct FHIRPath expression is needed. Pathwalker fulfills this need by helping the user to craft simple FHIRPath, add functions, add parameters to functions and further more expand the FHIRPath based on FHIR resources, profiles and instances. Because FHIRPath is a complex language not all of the function and possibilities are implemented, which leaves room for improvement.

Acknowledgement

This work has been carried out at the IT Center for Clinical Research and supervised by the Institute of Medical Informatics, Universität zu Lübeck.

Authors' Statement

Conflict of interest: Authors state no conflict of interest.

Availability

The code of Pathwalker is available at: https://git.imi.uni-luebeck.de/ehStudents/pathwalker-on-fhir

5 References

[1] Grimes, J., Szul, P., Metke-Jimenez, A., Lawley, M. Loi, K. *Pathling: analytics on FHIR*. J Biomed Semant 13, 23 (2022). https://doi.org/10.1186/s13326-022-00277-1

[2] HL7. (n.d.). FHIR - Fast Healthcare Interoperability Resources. Retrieved January 8, 2024, from https://www.hl7.org/fhir/

[3] Gruendner J, Deppenwiese N, Folz M, Köhler T, Kroll B, Prokosch H-U, Rosenau L, et al. *The Architecture of a Feasibility Query Portal for Distributed COVID-19 Fast Healthcare Interoperability Resources (FHIR) Patient Data Repositories: Design and Implementation Study*. JMIR Med Inform. 2022 May 25;10(5):e36709. doi: 10.2196/36709. PMID: 35486893; PMCID: PMC9135115.

[4] Vorisek C, Lehne M, Klopfenstein S, Mayer P, Bartschke A, Haese T, Thun S, *Fast Healthcare Interoperability Resources (FHIR) for Interoperability in Health Research: Systematic Review*, JMIR Med Inform 2022;10(7):e35724, https://doi.org/10.2196/35724

[5] CSIRO. (n.d.). Pathling Documentation: FHIRPath. Retrieved January 3, 2024, from https://pathling.csiro.au/docs/fhirpath

[6] Walonoski, J., Kramer, M., Nichols, J., Quina, A., Moesel, C., Hall, C., A. et al. *Synthea: An approach, method, and software mechanism for generating synthetic patients and the synthetic electronic health care record*, Journal of the American Medical Informatics Association, Volume 25, Issue 3, March 2018, Pages 230–238, https://doi.org/10.1093/jamia/ocx079

Integration of the Oncologic Base Data Set into the UKSH Medical Data Integration Center

Herwig Nicolaus [1], Michael Anywar [2,3], Santiago Pazmino Pinto [2,3], Hannes Ulrich [2], Jan Schladetzky [3,4], Ann-Kristin Kock-Schoppenhauer [3,4], Josef Ingenerf [4,5]

[1] Medical Informatics, Universität zu Lübeck, herwig.nicolaus@student.uni-luebeck.de

[2] Institute for Medical Informatics and Statistics (IMIS), Kiel University and University Hospital Schleswig-Holstein, {michael.anywar, santiagodavid.pinto, hannes.ulrich}@uksh.de

[3] Medical Data Integration Center, University Hospital Schleswig-Holstein

[4] IT Center for Clinical Research, Lübeck, University of Luebeck, 23562 Lübeck, Germany, {jan.schladetzky, ann-kristin.kock-schoppenhauer}@uksh.de

[5] Institute of Medical Informatics (IMI), Universität zu Lübeck, josef.ingenerf@uni-luebeck.de

Abstract

Interoperability between present data standards is becoming increasingly crucial due to the growing number of distributed systems and data standards. As the Medical Data Integration Center (*MeDIC*) uses *openEHR* as the standard for storing clinical data and the *oncologic base data set* is the national standard for tumor documentation, a bridge between these two standards is necessary. This paper discusses and illustrates the creation of openEHR templates, data processing pipelines and mapping between these two standards. It also presents additional information about the tools used, the standards, and the Information Technology (*IT*) architecture of the University Hospital Schleswig-Holstein (*UKSH*).

1 Introduction

Modern healthcare informatics emphasizes standardized approaches to ensure interoperability and sustainable data management. OpenEHR is an open source health information standard enabling the modelling of interoperable Electronic Health Records (*EHRs*) [1]. Additionally, the HIGHmed Consortium has chosen to use openEHR as their primary approach for storing clinical data, enabling cross-institutional data analysis at participating locations [2]. On the other side, the oBDS is a national standard for the documentation of oncologic data within the clinical context [3]. As a member of the HiGHmed consortium [4], the UKSH relies on mapping the oBDS to openEHR templates in order to be able to use oncological data for secondary purposes. Various facilities in Germany are in the process of integrating more and more data into the Data Integration Center. Together, the data can be used to create new knowledge and added value. This paper describes the abovementioned standards, the tools used to create a pipeline for data processing and the mapping of the oBDS to openEHR templates.

The approach, initially referred to as two-level modelling [5], separates information, represented by the reference model (*RM*), from knowledge, represented by archetypes and templates [6]. They enable domain-specific customization, ensuring adaptability to diverse clinical contexts [1]. The RM exists for logical data structures and attributes required to express and persist data. The model is stable and serves as a basic framework for the construction of concrete medical information models. The combination of the reference model and the archetype model has the powerful ability to evolve information systems and adapt to the dynamic evolution of clinical knowledge [1]. While meeting the complex requirements of the clinical environment, the clinical information models represented by archetypes are likely to change in the future, but the described data will always remain interoperable [6].

The Clinical Knowledge Manager (*CKM*) is a well-known tool for sharing archetypes and templates nationally and internationally. The HIGHmed Consortium utilises openEHR as a standard for exchanging clinical data [2].

2 Material and Methods

2.1 OpenEHR Standard

OpenEHR provides a robust and flexible multi-level modelling framework for representing and exchanging EHRs.

2.2 Oncologic Base Data Set

The Working Group of German Tumor Centers (*ADT*) and the Society of Epidemiological Cancer Registries in Germany (*GEKID*) are responsible for defining and maintaining the nationally standardized oBDS, including specific

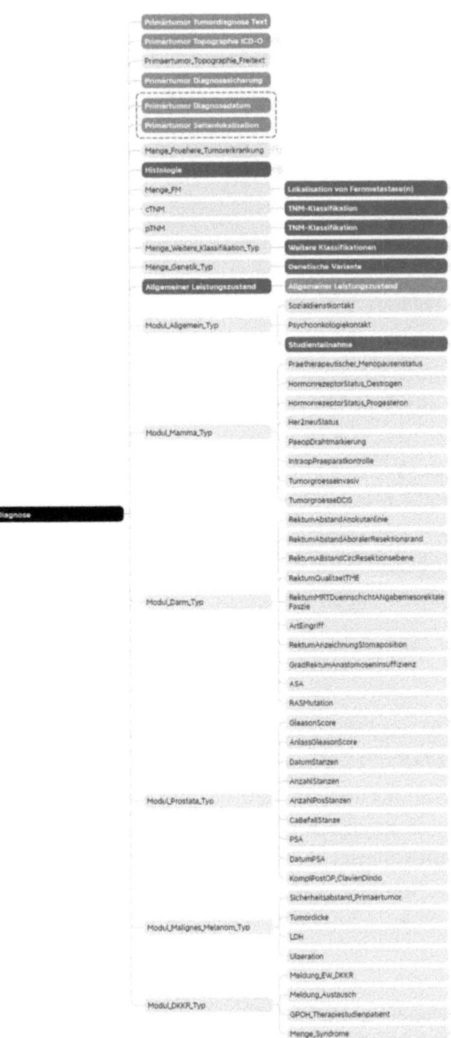

Figure 1: A representation of the oBDS XML Schema file for the *Diagnosis* use case. Subnodes have been collapsed for clarity.

```
□ oBDS_Diagnose  NAME (from: 'Bericht')
  □ → context
    □ → other_context
        T Bericht ID
        T Status
      □ Erweiterung
        ⊞ Modul_Allgemein  Δ [0..*] to [0..1]
        ⊞ Modul_Mamma  Δ [0..*] to [0..1]
        ⊞ Modul_Darm  Δ [0..*] to [0..1]
        ⊞ Modul_Prostata  Δ [0..*] to [0..1]
        ⊞ Modul_Malignes_Melanom  Δ [0..*] to [0..1]
        ⊞ Modul_DKKR  Δ [0..*] to [0..1]
  □ → content
    □ Problem/Diagnose
      □ → data
          T Primaertumor_Diagnosetext  NAME (from: 'Name des Problems/ der Diagnose')
          T Klinische Beschreibung
          T Körperstelle
          T Ursache
          Datum/ Zeitpunkt des Auftretens/ der Erstdiagnose
          Datum/Zeitpunkt der klinischen Feststellung
          Schweregrad
        □ Spezifische Angaben
          □ Tumorklassifikation ICD-O  Δ [0..*] to [0..1]
              T Morphologie Code ICD-O
              T Morphologie Codebeschreibung
              Primaertumor_Topographie_ICD_O  NAME (from: 'Topographie Code ICD-O')  Δ Values changed
              T Primaertumor_Topographie_Freitext  NAME (from: 'Topographie Codebeschreibung')
          T Beschreibung des Verlaufs
          Datum/Zeitpunkt der Genesung
          Diagnosesicherung  NAME (from: 'Diagnostische Sicherheit')
          T Kommentar
      □ → protocol
          Zuletzt aktualisiert
        □ Erweiterung
          ⊞ Klinische TNM-Klassifikation  Δ [0..*] to [0..1]
          ⊞ Pathologische TNM-Klassifikation  Δ [0..*] to [0..1]
          □ Evidenzgraduierung  Δ [0..*] to [0..1]
            ⊞ Therapieansatz
            ⊞ Genetische Variation
                T Zellulärer Mechanismus
                T Kommentar zur Evidenzgraduierung
    ⊞ Fruehere_Tumorerkrankung  NAME (from: 'Ad hoc Überschrift')
    ⊞ Histologie  NAME (from: 'Ad hoc Überschrift')
    ⊞ Fernmetastase  NAME (from: 'Ad hoc Überschrift')
    ⊞ Weitere_Klassifikation  NAME (from: 'Ad hoc Überschrift')
```

Figure 2: Overview of the template "obdsDiagnose" for the report case "Diagnosis".

modules.

The national standard framework for all types of tumors is defined by the oBDS, which lays the foundation for comparable data collection and analysis, according to § 65c SGB V [3]. The oBDS is applicable to all types of cancer and is continuously expanded with organ-specific modules [7]. It is the basis for all further data transformation, processing and mapping. Its purpose is to reduce and standardise cancer treatment data improving quality assurance, population-based analysis and scientific evaluation of treatment strategies [8].

Based on the structures defined in the schema file, a completed XML file is generated within the UKSH domain. The dataset schema includes pre-defined categories and organ-specific supplementary modules. The completed files contain sensible patient data such as location and health insurance information, as well as specific details related to a medical procedure or case in the context of oncology. Each file contains multiple patient records, each of which may contain one or more reports regarding the same patient. The eight different report cases include *Diagnosis*, *Pathology*, *OP*, *Radio Therapy*, *Systemic Therapy*, *History*, *Death*, and *Tumorconference*. The main nodes of one of the report cases, namely the *Diagnosis* case, are shown in Fig. 1. Each report case requires an individual openEHR template file and unique manual mapping, making them crucial for later mapping.

2.3 Data Source and Aquisition

The architecture of the UKSH Medical Data Integration Center is highly complex, incorporating multiple healthcare data sources, as well as diverse knowledge management and research capabilities, using a wide range of systems and tools for integration. The data acquisition process begins with the design and implementation of a custom Extract, Transform, Load (*ETL*) job for the UKSH. Raw data from various sources such as clinical information systems etc. are collected in a block-storage-based data lake. Data lakes are designed to store large volumes of data in any available format and structure [9]. The ETL process is then used to transfer the medical data to the openEHR repository, representing the data warehouse architecture [10].

Due to the complexity, a light variant of the MeDIC stack was used in this internship. This light variant did not include all the components of MeDIC and used different but comparable operating systems for some of the structures. In

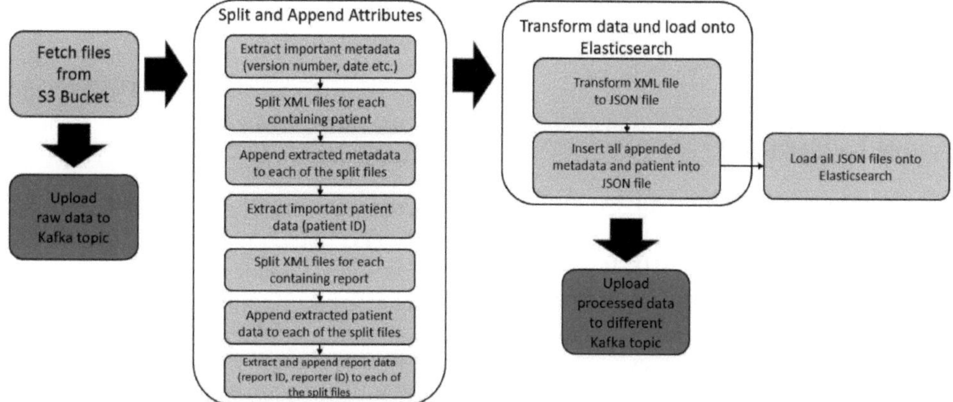

Figure 3: Simplified Pipeline for data transformation and storage that was constructed in Apache NiFi.

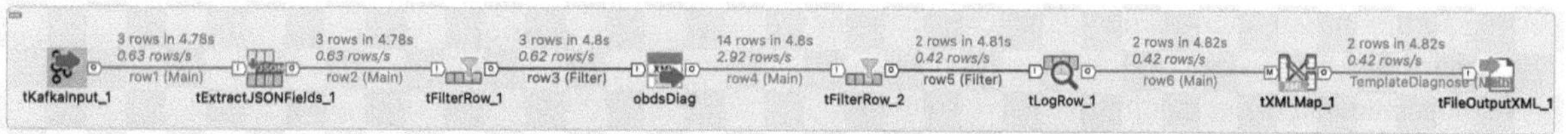

Figure 4: Developed Pipeline for data transformation and extraction in Talend for the use case *Diagnosis*.

this case, a MinIO Simple Storage Service (*S3*) bucket was used to make the raw data available internally for further processing. The S3 bucket ensures a scalable cloud storage for data without the loss of performance [11]. The local use of this light version made the working process very flexible, allowing the settings to be quickly adjusted. On the other hand, it is not a very accurate representation of the structure and workflow of MeDIC at the UKSH.

3 Results and Discussion

3.1 Designing openEHR Templates

Using the web-based open source visual design tool *Archetype Designer* created by Better [12], eight different openEHR templates were created. The oBDS report cases were manually mapped to an openEHR template using existing and verified archetypes from sources such as the Use Case Oncology from the HIGHmed CKM, where possible. However, some data could not be modelled with present archetypes, so new ones had to be created, such as for the multiple oncologic modules.

The template for the 'Diagnosis' example is shown in Fig. 2 using the Archetype Designer. The *Gleason Score* archetype was not originally available in German and has to be manually translated and officially accepted by the openEHR community. As of today, this task is still ongoing.

The limitations of each data standard make designing openEHR templates for specific XML data structures a difficult task. Due to the requirement to consistently use existing archetypes and the fact that even minor modifications result in a new archetype ID, severing any connection to the pre-existing archetype, the data cannot always be transposed in a one-to-one manner.

3.2 Data Processing and Transformation

The ETL workflow was implemented through a robust pipeline built using Apache Niagara Files (*NiFi*), see Fig. 3. This pipeline was designed to process raw XML data following the structure of the oBDS and pass it on to later components.

Given the diverse nature of the information within each file, different mappings are required. Using Apache NiFi, the data undergoes an annotation process where essential metadata such as IDs and timestamps from the metadata and patient data are appended. The data is then segmented and loaded into Elasticsearch for efficient storage and retrieval.

The processed and unprocessed raw data are directed to separate Kafka topics, each acting as a designated channel for data organization. This step in the workflow ensures the seamless transfer and availability of the annotated and segmented data to subsequent steps in the data processing pipeline, including other steps that require the base structure of the raw data.

3.3 Mapping to openEHR

To start the mapping task, the data is first retrieved from the associated Kafka topic. Then, the data needs to be processed and the appropriate fields extracted for the required report case. Fig. 4 shows the pipeline for the *Diagnosis* report case. The other cases proceed in a similar manner.

After the extraction process in the *obdsDiag* component, empty reports are filtered out. The data is then mapped to the openEHR templates and the resulting completed template is saved as an XML file for further work in the MeDIC repository.

Despite its adaptability, making even minor adjustment in the openEHR template results in a new export from the

Archetype Designer, which leads to a new import into Talend, which means the mapping task has to be done all over again.

3.4 Results and Evaluation

The evaluation of the templates generated by the constructed pipeline was carried out at the MeDIC in Kiel. Although the pipeline produced valid openEHR templates, a number of issues arose that offer potential for future work. Due to the different structures of the oBDS XML files and the openEHR templates, individual attributes of the XML file containing multiple elements cannot currently be translated on a one-to-one basis, resulting in either data redundancy or data loss. In addition, references to terminology systems and value sets in the template could not be resolved, suggesting further minor adjustments to the template structure. After further discussions and testing, it appears this task is manageable and offers a comprehensive solution.

4 Conclusion

Interoperability between different data standards is critical in clinical environments [10], but mapping between these standards can be a time-consuming and complex process.
Designing a functional template for a specific use case is challenging within the constraints of the openEHR standard. Editing templates within your local environment is possible, but it is not always guaranteed to be valid and conform to the openEHR guidelines. Additionally, the official Archetype Designer is not connected to the CKM, which can make it laborious to work with.
However, processing the oBDS data with tools like Apache NiFi or Talend works seamlessly and can be easily adapted to different needs or future changes in the data set. In the end, a functional pipeline for data transformation and subsequent mapping from one standard to another was successfully created, allowing further customisation to better fit existing structures.

Acknowledgement

This work has been carried out at the IT Center for Clinical Research in close collaboration with the MeDIC of the UKSH and supervised by the Institute of Medical Informatics, University of Lübeck.

Authors' Statement

Conflict of interest: Authors state no conflict of interest.

5 References

[1] openEHR Foundation, *Architecture Overview*, Available: https://specifications.openehr.org/releases /BASE/latest/architecture_overview.html [last accessed on 2024-02-07].

[2] HiGHmed e.V., *HIGHmed Medical Informatics*, Available: https://www.highmed.org/en/home [last accessed on 2024-02-07].

[3] Arbeitsgemeinschaft Deutscher Tumorzentren e.V. (ADT), *Bundeseinheitlicher Onkologischer Basisdatensatz*, Available: https://basisdatensatz.de [last accessed on 2024-02-07].

[4] Vernetzen. Forschen. Heilen., *Medizin Informatik Initiative*, Available: https://www.medizininformatik-initiative.de/de/start [last accessed on 2024-02-07].

[5] Beale, Thomas, (2002), *Archetypes: Constraint-based Domain Models for Future-proof Information Systems*, Eleventh OOPSLA Workshop on Behavioral Semantics, Serving the Customer.

[6] International Standardization Organization (ISO), *The ISO 13606 standard*, Available: http://www.en13606.org/information.html [last accessed on 2024-02-07].

[7] Klinisches Krebsregister Niedersachsen, *Onkologischer Basisdatensatz – oBDS*, Available: https://www.kk-n.de/melder-aerzte/onkologischer-basisdatensatz-obds/ [last accessed on 2024-02-07].

[8] Bundesministerium für Gesundheit, *Querschnittsthema: Datensparsame einheitliche Tumordokumentation*, Available: https://www.bundesgesundheitsministerium.de/themen/ praevention/nationaler-krebsplan/was-haben-wir-bisher-erreicht/querschnittsthema-datensparsame-einheitliche-tumordokumentation [last accessed on 2024-02-07].

[9] Madera, Cedrine and Laurent, Anne, *The next Information Architecture Evolution: The Data Lake Wave*, 2016, Association for Computing Machinery, https://doi.org/10.1145/3012071.3012077, 174–180, 7, Biarritz, France, MEDES.

[10] Kock-Schoppenhauer, A. K., Schreiweis, B., Ulrich, H., Reimer, N., Wiedekopf, J., Kinast, B., Busch, H., Bergh, B., Ingenerf, J., *Medical Data Engineering – Theory and Practice. In Communications in Computer and Information Science*, 2021, vol 1481., Springer, Cham., (pp. 269-284)., https://doi.org/10.1007/978-3-030-87657-9_21

[11] MinIO, *MinIO in Healthcare*, Available: https://min.io/resources/docs/MinIO-in-Healthcare.pdf [last accessed on 2024-02-07].

[12] openEHR, *Modelling Tools*, Available: https://openehr.org/products$_t ools/modelling_tools/$ [last accessed on 2024-02-07].

Structured data for the legal electronic health record using a FHIR-based clinical data repository

Laura Meyenborg [1], Josef Ingenerf [2], Jonas Schön [3]

[1] Medical Informatics, Universität zu Lübeck, laura.meyenborg@student.uni-luebeck.de
[2] Institute of Medical Informatics, Universität zu Lübeck, josef.ingenerf@uni-luebeck.de
[3] Tiplu GmbH, Hamburg, j.schoen@tiplu.de

Abstract

With the introduction of the electronic health record (ePA) in Germany it is inevitable for healthcare and research to have interoperable patient data across system boundaries. The different use cases of the ePA are structured in Medical Information Objects (MIOs) based on the international standard Health Level 7 Fast Healthcare Interoperability Resources (HL7 FHIR). The newly created MIOs for laboratory findings and hospital discharge letters are analysed in terms of their feasibility based on a pre-structured clinical data repository. As a result, a detailed table of chances and challenges was created and a converter service for generating the two MIOs was implemented. The main findings are that FHIR servers that are based on accepted standards offer great opportunities for providing structured data for the electronic health record. Nevertheless, the structure of the MIOs and the ePA itself severely restricts the interoperability because the ePA works as a bundle of non-related MIO files that are not evaluable without high effort.

1   Introduction

Since January 1, 2021, all Statutory Health Insurances in Germany have to provide their patients with an electronic patient record, often referred to as the Elektronische Patientenakte (ePA) in German. The ePA is a key feature of the Digital Act that will promote and support the exchange and utilisation of health data in Germany [1]. It represents a repository of information extracted from medical documents such as medical letters or findings so that health-related information is available when it is needed for treatment [2]. Since the ePA should be able to interact with many different systems in the healthcare sector, its data has to be interoperable. In other words, it must be interpreted correctly regardless of the software system [3]. To achieve this, the National Association of Statutory Health Insurance Physicians (Kassenärztliche Bundesvereinigung, KBV) developed the concept of Medical Information Objects (MIOs) for documenting medical data in a standardised form based on international standards. A MIO can be seen as a digital information component which can be used and combined universally [4]. While the MIO concept assumes the data to be in a structured, standardised form, reality shows that the required data is not necessarily available in hospitals in a structured and interoperable form [5].

The research and implementation was conducted at the company Tiplu GmbH which develops a clinical data repository represented by a proprietary patient record. The project includes a feasibility study for filling the newly created MIOs *Hospital Discharge Letter* (Krankenhausent-lassbrief) and *Laboratory Report* (Laborbefund), and the development of a converter that generates valid MIOs from Tiplu patient records.

2   Materials and Methods

The project subjective is based on the data exchange architecture of Tiplu GmbH (Hamburg, Germany), which is aligned with international standards and German profile definitions.

2.1   Data representation

As interoperability can only be achieved through binding standards, the National Agency for Digital Medicine (gematik) has the legal mandate to develop national specifications for the exchange of medical data, e.g. ISiK (Informationstechnische Systeme in Krankenhäusern) for fast and secure exchange of treatment data. On June 30, 2021, the gematik published their first requirements specification on the base of HL7 FHIR (Fast Healthcare Interoperability Resources) which are to be further developed in several expansion stages [6]. HL7 FHIR is an open-source standard that is oriented towards modern web technology that uses the REST paradigm and is therefore easy to implement, scalable and customisable. The data objects are exchanged as resources that can be manipulated by the HTTP methods GET, PUT, POST and DELETE and be serialized in JSON or XML format. FHIR defined a set of about 150 resource types for relevant medical objects and their relations such as *Patient* for master data of the patient,

Condition for diagnoses or *Observation* for measurements and findings [7]. With the use of profiles, these resources can be extended to cover a certain use case. The 80/20 rule applies here, which states that the resources should cover around 80% of the data elements while the remaining 20% must be covered by custom extensions [10]. In addition to the data format, the use of harmonised terminology is an important aspect of achieving interoperability since clinical data is only comparable if the terms used have been defined, e.g. Logical Observation Identifiers Names and Codes (LOINC) [6]. A commonly used online platform to publish these profile definitions is called Simplifier.net [8]. Besides, it offers a variety of tools including validation of resources against their specification [9].

In FHIR one MIO is represented as a Bundle resource that holds all related resources that are needed to cover the necessary data. All resources are specified by the KBV for a particular use case (e.g. documentation of laboratory findings). For the MIO *Laboratory Report*, the main component of the bundle is the DiagnosticReport resource that references all other resources. The MIO *Hospital Discharge Letter* contains a Composition resource that represents a document with several sections [4].

2.2 Implementation and Validation

The project data, coming from a hospital information system was already stored on a FHIR server developed by Tiplu referred to as TipluDB. The profiles used are profiles of, or at least based on, ISiK or Medizininformatik Initiative (MMI). The data is read from the database of a hospital information system and mapped to the FHIR model. This mapping is an extensive process that on one hand analyses the documents to code their sections dependent on their content and on the other hand, standardises observations, diagnoses, procedures and medication. For documents, Tiplu developed a code system to identify document types (e.g. discharge letters) and their individual document sections (e.g. admission diagnosis) since the existing ones from LOINC and SNOMED-CT are not as fine-grained as needed. In this project, the data was analysed about their value for the considered MIOs as well as the MIOs were checked for fulfilment and limits. As a proof of concept, a converter service was implemented using the .NET SDK for HL7 FHIR by Firely which provides a set of tools and libraries to work with FHIR resources. The converter reads a JSON File containing a test FHIR Bundle resource with all patient record-related resources, scans it for either information about laboratory findings or hospital discharge letters and creates compliant MIOs from it, also in JSON format. This JSON is uploaded to the simplifier validation tool and in an iterative process, the errors were eliminated. Fig. 1 demonstrates the setup described. The correctness and completeness of the generated resources are checked by the validation tool on Simplifier.net. For this, pseudonymised real patient records from the Universitätsklinikum Schleswig-Holstein (UKSH) are randomly selected from their TipluDB

server to generate the requested MIOs from it. The resulting instances are validated. If the implementation is technically correct it should not throw any errors. The practicability and feasibility of the actual content of the MIOs are validated by presenting and discussing the results in a workshop in the Department of Cybersecurity and Interoperability at the Ministry of Health to employees of the department and the gematik.

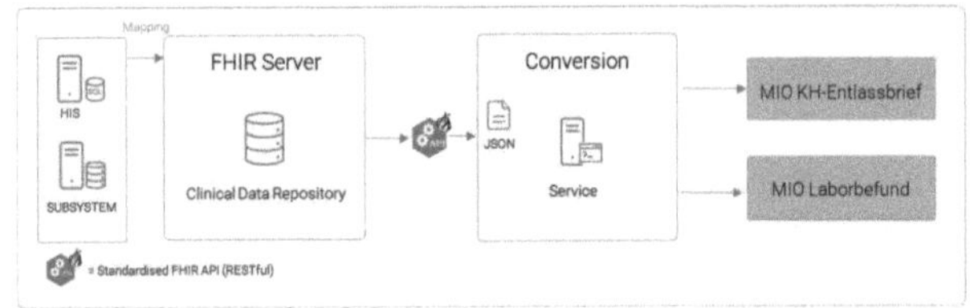

Figure 1: The data is collected from the hospital system and its subsystems, structured and made available in FHIR by TipluDB. To create a MIO a patient record is requested and the resulting JSON is read by the converter service that builds corresponding MIOs based on the medical information.

3 Results and Discussion

For both use cases (Documentation of laboratory findings and preparation of a hospital discharge letter) it was analysed which data must be provided and to what extent they must be structured (free text or highly structured with coded values and standardised code systems).

3.1 MIO Laboratory Result

Table 1 demonstrates the feasibility of creating the MIO *Laboratory Result*. The most relevant section of the laboratory findings is the laboratory study. There is a free text option if the result cannot be provided in a standardised form but due to the extensive mapping process in advance, the result is assigned with a LOINC code and the measurement result has a standardised physical unit as well as a reference range. Only by using standardised code systems such as LOINC can a value be interpreted in the same way by different systems, which is highly relevant both for patient care and for the procurement of research data. However, there are a few limitations as not all data is currently available on the source FHIR server. Many of them could be added by technical extensions, while others are more content-related problems. The MIO specification expects the identifier of the diagnostic report to be a Universal Unique Identifier (UUID). This UUID has to be assigned by the laboratory and as this technical interface may not yet be available in practice, it is questionable whether another form of identification is insufficient until the technical requirement is implemented in all the laboratories. Furthermore, the time of laboratory sampling is not necessarily valid in the hospital context, so this information should not be treated as mandatory in the specimen resource. In addition, the MIO should currently contain the original findings in PDF format. Since the use of TipluDB intends to move away from unstructured data, this can not be provided. In times of the changeover to electronic patient files it might help to have the original findings document additionally, but it should not be a mandatory field as it works in the opposite direction to the approach of having structured data in the ePA.

Table 1: Laboratory report with regards to its conformity and feasibility

Element	Conformity	realisable
Patient data	mandatory	yes
Contracting laboratory	required	partly
Order data	required pt	partly
Unique identification	mandatory	no
Laboratory study group	mandatory	no
Laboratory study	mandatory	partly
Status of laboratory report	mandatory	yes
Timestamp laboratory report	mandatory	yes
Approval of findings	required	no
Overall rating	optional	no
Additional information	optional	no
Attachment overall findings	mandatory	no

3.2 MIO Hospital Discharge Letter

The structure of the MIO *Hospital Discharge Letter* is oriented towards the discharge letter on paper. The information from the header and footer concerning administrative data of discharging institution, receiver, patient data and case data are mandatory information in the MIO (see Table 2). The rest of the letter contains the medical information whereby the minimum discharge letter must include the admission, epicrisis, discharge findings, therapy recommendations and discharge medication. As the clinical documents were analysed and mapped onto a document section code system before being stored on the FHIR server, this mandatory information can be filled with texts. In addition, optional sections such as diagnoses, procedures, medication and care level can be created in a structured way with ICD-10 codes for diagnoses and OPS codes for procedures and care level as well as ATC codes for medication. In summary, this leads to a higher value of this letter in terms of interoperability. Currently, the discharge letter is a mixture of structured and free text data which results in a few inconsistencies. If the data is not in the expected standard a few sections offer an alternative profile for free texts, for example, the resource for documenting allergies and intolerances. Instead of a code, it is also allowed to enter a free text, however, the clinical status and verification status of the allergy has to be recorded as a coding, which means that this resource can still not be used. A general question concerning the discharge letter is the point in time when it is made available to the ePA. For a valid MIO it is currently mandatory to add a discharge reason and date which would mean in most cases the report can only be made available once the case has been finalised and approved by the head

physician. This process, depending on the hospital, can take days or weeks to finalise. Conversely, this means that no temporary report can be generated which may be helpful for the patient immediately after discharge. This is one of the cracking points of the ePA. Since the data is encrypted on the client side it is not possible to reference another document (e.g. a preliminary doctor's letter) or replace it with a newer version.

Table 2: Components of the discharge letter with regards to their conformaty and feasibility

Element	Conformity	realisable
Discharging institution	mandatory	partly
Receiving person	mandatory	partly
Patient data	mandatory	yes
Case data	mandatory	partly
Referral	optional	no
Admission	mandatory	yes
Anamnesis	required	yes
Diagnoses	optional	yes
Procedures	optional	yes
Implants	optional	no
Epicrisis	mandatory	yes
Discharge findings	mandatory	yes
Therapy recommendations	mandatory	yes
Care level	optional	yes
Discharge medication	mandatory	yes
Incapacity for work	optional	no
Documentation	optional	no

3.3 General issues concerning MIO profiles

A problem that affects all already developed and newly created MIOs is the severe restriction of the used FHIR resources. For example, as seen in Fig. 2, the cardinality of the supported profiles is limited to one, which means that resources that support other profiles can never be used in a MIO, even if their structure is otherwise compliant. The MIO profiles are therefore not compatible with the ISiK standard. This behaviour is contradictory and hinders the interoperability concept of FHIR. Moreover, the profiles have a certain redundancy that also violates the 80/20 rule. For example, there are two profiles based on the Observation resource, one for documenting the discharge report and one for the inpatient stay. Both profiles have exactly the same attributes and constraints and only differ in their profile URL. Following the 80/20 rule all data could be covered by a single observation profile per MIO. It would increase the interoperability and implementability to loosen the restrictions (by allowing more attributes and having fewer mandatory fields) and instead focus on a detailed implementation guide.

3.4 Validation of results

For the technical validation, 10 different real cases were selected to generate the MIOs *Hospital Discharge Letter*

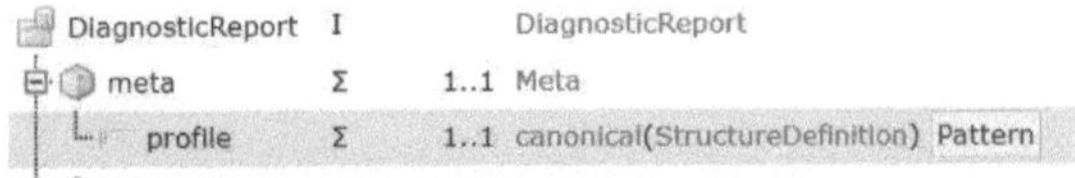

Figure 2: Cardinality of supported profiles is restricted to 1

and *Laboratory Report* in separate processes. All missing information on the TipluDB such as personal information were complemented with default values. As the validation showed no errors it could be shown that the implementation can create valuable MIOs. The results were discussed in a workshop format and the criticism should now be directed to the gematic with their form for submitting interoperability problems. In particular, the comment that too many fields of the MIOs are mandatory was a new point of criticism for the gematik participants. The described problems resulting from this restriction were recognised.

4   Conclusion

In summary, it can be said that the requirements of the MIOs for the electronic health record mean high development effort for information systems that do not operate in FHIR yet as the transmitted data should be already largely structured. Using the example of two different use cases for medical data in the ePA, this project showed the benefit of having a standardised data repository in the form of a FHIR server as the mapping of the data takes place only once and in one place and thereby reduces the number of interfaces and enhances the interoperability. Besides, it was elaborated that the structure of the MIOs is rather restrictive and complicates the implementation of them. The use of more general resources would lead to higher flexibility and simplify the process of satisfying other MIOs. A general problem that has to be further discussed is the structure of the ePA itself. As mentioned in its current form it is a bundle of single encrypted MIO files. To generate added value for patient care and research all these files need to be decrypted, merged and cleaned from redundancies (e.g. each bundle holds the administrative patient data). This means a high effort, especially regarding the fact, that the data has been structured when they are transferred to the ePA. This topic will have to be a focus in the upcoming discussions about the ePA and its further development. The implemented service is focused on the development of the MIOs *Laboratory Report* and *Hospital Discharge Letter* but with the data basis, it could be extended to support other MIOs as well. In addition, the service could be integrated into the infrastructure by adding an API service that requests the needed information instead of interpreting a JSON file with all case-related information. This measure would also increase the performance of the service as irrelevant data for the MIO in question would not be processed.

Acknowledgement

The work has been carried out at Tiplu GmbH, Hamburg and supervised by J. Ingenerf, Institute of Medical Informatics, Universität zu Lübeck.

Authors' Statement

Authors state no conflict of interest.

5   References

[1] Bundesministerium für Gesundheit, *Die elektronische Patientenakte (ePA)*. Available: https://www.bundesgesundheitsministerium.de /elektronische-patientenakte.html [last accessed on 2023-01-08]

[2] gematik, *ePA*. Available: https://www.gematik.de/anwendungen/e-patientenakte [last accessed on 2023-01-08].

[3] gematik, *Medizinische Informationsobjekte*. Available: https://www.ina.gematik.de/themenbereiche /medizinische-informationsobjekte [last accessed on 2023-01-08].

[4] Kassenärztliche Bundesvereinigung, *Medizinische Informationsobjekte (MIO)*. Available: https://www.kbv.de/html/mio.php [last accessed on 2023-01-08].

[5] Hoffmann, L. (2023) *Wenig Datenfluss in Kliniken*. Available: https://www.handelsblatt.com/inside/digital-health/isik-wenig-datenfluss-in-kliniken/29395926.html [last accessed on 2023-01-22].

[6] Heckmann, S. (2022), *HL7 FHIR als Grundlage für moderne Digitalstrategien*. In: Henke, V., Hülsken, G., Meier, PM., Beß, A. (eds) Digitalstrategie im Krankenhaus. Springer Gabler, Wiesbaden. pp. 307–331, 2022

[7] Ploner, N. (2022), *Akzeptanz und Architektur einer vertrauenswürdigen Patientenanbindung auf Basis des FHIR-Standards*, Doctoral dissertation, Friedrich-Alexander-Universität Erlangen-Nürnberg(FAU)

[8] SIMPLIFIER.NET, *About Simplifier.net*. Available: https://simplifier.net/about [last accessed on 2023-01-11].

[9] Toonstra, A. (2020), *How to Validate FHIR Resources Like a Boss*. Available: https://fire.ly/blog/validate-fhir-resources-like-a-boss/ [last accessed on 2023-01-16].

[10] Yesakov, E. (2023), *FHIR Components & Resources: Everything You Need to Know*. Available: https://kodjin.com/blog/understanding-fhir-components-fhir-resources [last accessed on 2023-01-16].

SWOT Analysis of Utilizing Medical Device Log Data into Post-Market Activities According to the EU Medical Device Regulation

Rghia Fakrun [1]
[1] Biomedical Engineering, Luebeck University of Applied Sciences, Rghia.fakrun@student.th-luebeck.de

Abstract

Emerging trends such as digital transformation and the Internet of Things are impacting many industries, including the medical device industry. Digital, real-world device data is therefore expected to provide benefits which will enable the medical device industry to better meet the global post-market surveillance requirements for medical devices ensuring the safety and performance of the devices and protecting the health of all stakeholders. This paper addresses the benefits and risks of using real-world medical device log data for post-market surveillance activities required by the EU MDR by conducting a SWOT analysis on the data collected from the relevant internal and external stakeholders. The results show that medical device data can provide a better quality of clinical data, especially nowadays in times of interoperability and connectivity solutions. The analysis further shows the gaps and challenges that the manufacturer should consider to optimally meet the requirements of the upcoming digital era.

1   Introduction and Background

The European data strategy aims towards making the European Union (EU) a leader in the data-driven society, which will enable a dynamic flow of data within the EU and bring more benefits to businesses, researchers and public administrations [1]. The National Institute of Standards and Technology (NIST) has stated that "the lack of interoperability between medical devices can lead to preventable medical errors and potentially serious inefficiencies that would otherwise be avoided" [2]. Therefore, as the medical devices industry is growing, it needs to improve its use of data to optimize the overall quality of the interaction with the surrounding ecosystem while complying to the regulations on medical device. This paper analyzes the utilization of real-world medical device data into post-market surveillance.

1.1   Regulatory framework

The new market strategy of the European Economic Community (EEC) influenced the medical device regulations [3]. As a consequence, the previously existing directives on Medical Device (MDD) and Active Implantable Medical Devices (AIMDD) were replaced by the Regulation (EU) 2017/745 on Medical Devices (MDR). The MDR is legally binding and valid for each member state within the EU since May 2021. The purpose of the new approach for the medical devices sector is to "establish a robust, transparent, predictable and sustainable regulatory framework for medical devices which ensures a high level of safety and health whilst supporting innovation" [4]. The MDR covers all phases of a product's life cycle, from develop-ment and placing on the market to post-market surveillance. Moreover, it concentrates on transparency regarding safety and performance of the devices after placing on the market. Since the MDR demands the manufacturer to strictly monitor the collection and analysis of clinically relevant data from Post Market Surveillance (PMS); clinical evidence, clinical data and clinical evaluation are described more comprehensively.

1.2   Post market surveillance

1.2.1   Post market surveillance according to the MDR

According to Article 83 of the MDR, "the post-market surveillance system shall be suited to actively and systematically gathering, recording and analysing relevant data on the quality, performance and safety of a device throughout its entire lifetime, and to drawing the necessary conclusions and to determining, implementing and monitoring any preventive and corrective actions" [4]. PMS is a combination of reactive and proactive activities that the manufacturer must carry out. Vigilance is the reporting of serious incidents and field safety corrective actions, which implies a reactive process. While, the proactive type of PMS is the Post-market clinical follow-up (PMCF), a continuous process that extends throughout the entire life cycle of the product. It aims to provide clinical evidence through the clinical evaluation of post-marketing experience [4]. Clinical evaluation is performed based on the collected clinical data which in the MDR means: "information concerning safety or performance that is generated from the use of a device" [4] and can be sourced from many resources such as clinical investigations, studies reported in scientific literature, or clinically relevant information coming from post-market surveillance, in particular the PMCF [4].

1.2.2 PMCF activities in real-world settings

In addition to the regulations of the MDR, the guidance information mentioned from the Medical Device Coordination Group (MDCG) documents and in the guidance MEDDEV 2.7/1 rev. 4 on medical devices regarding clinical evaluation, provide supportive and informative resources for the manufacturer to comply with the regulatory requirements. One relevant aspect from these guidelines is that the follow-up period should be long enough for outcomes to occur, and it should be frequent enough to detect temporary side effects and complications. Another relevant point is that the methods used for quantifying the outcomes (including validation of the methods) should be reliable [5].

The choice of PMCF activities type is the responsibility of the manufacturer and depends on various criteria. General PMCF activities can be in the form of a literature research or feedback gathered from the users, while specific PMCF activities mean that PMCF studies are performed or registers are evaluated [4]. An example of a specific PMCF activity is to perform a PMCF study on specific functions of the product to ensure the safety and performance of a medical device in the everyday clinical practice. This PMCF study starts with a defined scope, plan, procedure, tools and methodology. To ensure the study design's validity and reliability, it is initially implemented at the manufacturer's test laboratories. Once the results are deemed satisfactory, the planned study can be carried out on the basis of the customer's device in the real-world environment (e.g., in a hospital) which represents the intended normal conditions of use of the device. The findings of this study are documented on the clinical evaluation report and expected to reflect the safety and performance state of the device from the post-market phase. As mentioned in the MEDDEV 2.7/1 rev. 4; "the follow-up should be frequent enough", which means that this procedure is repeated over the whole lifetime of the product at regular intervals. Such studies pose major challenges in means of planning, implementation effort, time and costs. In addition to these challenges, the assessment of the data during the entire PMCF study can be influenced by many factors, such as bias, statistical errors or data misinterpretation.

1.3 Complexity of medical device definition

According to the Article 2 of the MDR, "a 'medical device' means any instrument, apparatus, appliance, software, implant, reagent, material or other article intended by the manufacturer to be used, alone or in combination, for human beings for one or more of the following specific medical purposes: diagnosis, prevention, monitoring, prediction, prognosis, treatment or alleviation of disease [4]". This broad definition covers a wide range of devices, ranging from the very simple examples such as blasters to the more sophisticated devices found in modern intensive care units and operating rooms, including ventilators, monitors, anesthesia workstations and infusion pumps. These medical devices contain microprocessors, sensors and actuators running continuously, performing operations as data acquisition, measurements, signal processing and control to fulfill its intended purpose [6].

2 Materials and Method

The aim of this research paper is the analysis of the benefits and risks of utilizing medical device real-world log data into post-market surveillance activities. The used method to fulfill this goal is a SWOT analysis, a strategic planning and management technique used to help identify strengths, weaknesses, opportunities, and threats related to certain project plans. Strengths are the internal advantages which characterize this project, while weaknesses represent the internal factors that place the project in a disadvantageous situation and need to be improved. Opportunities are factors in the external environment which can be utilized for the project's benefit. Threats are factors also in the external environment that could pose risks for the project [7]. The analysis began with the identification of the involved stakeholders from the internal and external environment. The identification criteria are based on the factors that a manufacturer can or cannot influence. The internal environment includes the Software Research and Development (SW R&D) division that is responsible for the development of software projects, and the Clinical Affairs (CA) division which is responsible for the clinical evaluation process and its compliance to the regulations. On the other hand, the external environment includes the users of the device (Hospitals), the international and national regulations and laws, and the trends. The second phase focused on collecting and analyzing the data from relevant stakeholders through the following steps:

Internal factors:

1. Medical device data characteristics from the point of view of SW R&D.

2. Areas of improvement from the point of view of CA.

External factors:

1. Opportunities from the external environment such as trends, strategies, and new laws.

2. Threats regarding the implementation efforts at the hospitals

Finally, the data is then evaluated and interpreted in the SWOT matrix.

3 Results and Discussion

As shown in Figure 1 the matrix presents the four main aspects, in which the first row shows the internal strengths and weaknesses while the second row shows the external opportunities and threats.

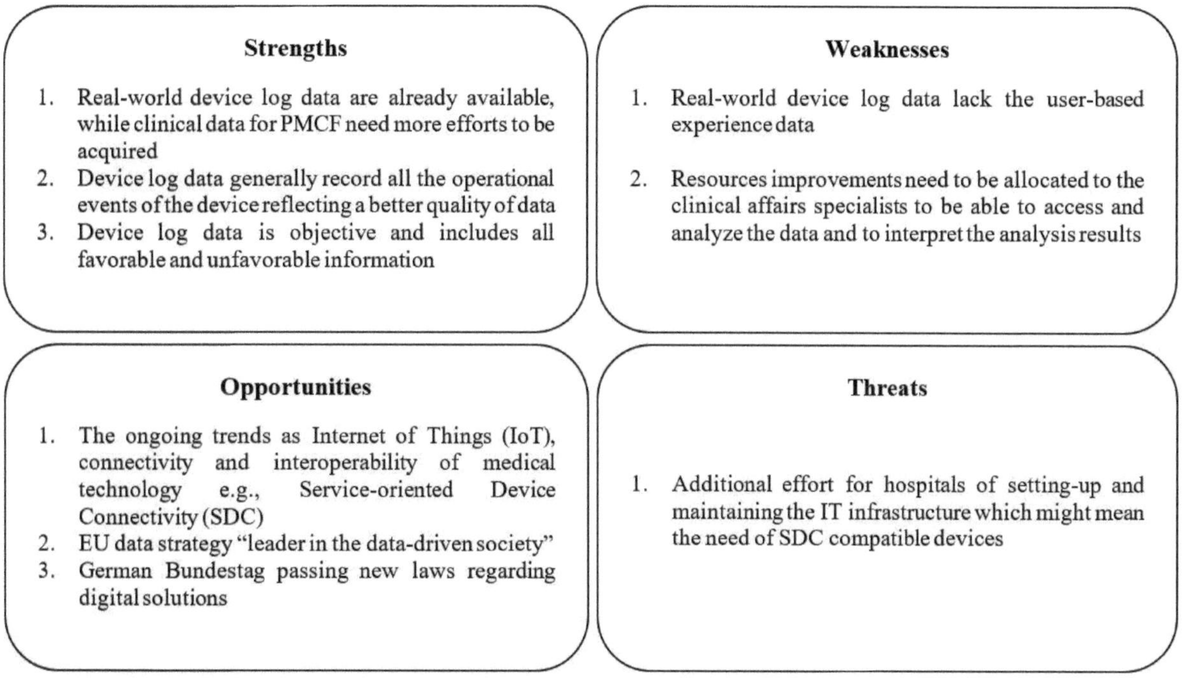

Strengths

1. Real-world device log data are already available, while clinical data for PMCF need more efforts to be acquired
2. Device log data generally record all the operational events of the device reflecting a better quality of data
3. Device log data is objective and includes all favorable and unfavorable information

Weaknesses

1. Real-world device log data lack the user-based experience data
2. Resources improvements need to be allocated to the clinical affairs specialists to be able to access and analyze the data and to interpret the analysis results

Opportunities

1. The ongoing trends as Internet of Things (IoT), connectivity and interoperability of medical technology e.g., Service-oriented Device Connectivity (SDC)
2. EU data strategy "leader in the data-driven society"
3. German Bundestag passing new laws regarding digital solutions

Threats

1. Additional effort for hospitals of setting-up and maintaining the IT infrastructure which might mean the need of SDC compatible devices

Figure 1: SWOT matrix illustrating the internal and external factors (Strengths, Weaknesses, Opportunities, and Threats) of using logged device data for post-market surveillance activities.

3.1 Internal factors analysis

From the perspective of the SW R&D division, and based on the information provided in (section 1.3) medical devices store their operational data in many forms, one example is the log book. A log book is a chronological historical record of the events of the operating time of the device e.g. the measured value of the temperature sensor. Each file contains several entries, with each entry representing an event that occurred at a specific point in time. The data in the log files is systematically recorded according to a standard structure, and comprises all events as well as measurement data. This in fact is used for regular maintenance and repair of faulty sensors or actuators, because it provides a historical overview of the functional mode of these microprocessors. It is assumed that such data will also provide information on the safety and performance of the device. This is because, according to the MDR, clinical performance is defined by the ability of a device to fulfill its intended purpose as stated by the manufacturer in relation to its technical or functional characteristics [4].

Comparing the availability of current clinical data against device data, the existing PMCF studies demand significant efforts to acquire necessary clinical data. In contrast, device data provide a more sustainable source of clinical data due to them being logged continuously and stored in the medical device. All of these aspects characterize the device's log data and considered as strengths. The only aspect which will be missing is the user-based constructive experience. This is because the user's perception of the functionality of the device is no longer taken into account.

Another area for improvement is the fact that the log files contain raw data that includes both clinically relevant and non-relevant events. Extracting the relevant data requires considerable technical knowledge. In addition, in the context of PMCF studies, the tools used are designed for specific purpose of data collection and are only suitable for the respective study scope. However, the tools which will be required for extracting the relevant data from such a huge amount of raw data should be set up accordingly. For this reason, the current situation calls for more resources to be made available to the CA specialists e.g., an appropriate tool (Software) to acquire the relevant data and to analyze , and means to understand and interpret the analysis results.

3.2 External factors analysis

Given the new EU data strategy [1], decisions are being made to improve the everyday healthcare and research opportunities in Germany. December 2023, the German Bundestag has passed new laws for digital solutions, namely the "Act to Accelerate the Digitization of the Healthcare System" (Digitization Act - DigiG) and the "Act on the Improved Use of Health Data" (Health Data Use Act - GDNG) in the second and third readings. The law contains a new procedure for the creation and definition of data standards, ensuring that the requirements for interoperability have high quality and are binding [8].

In addition, current trends such as the "Internet of Things" (IoT), connectivity and interoperability of medical technology as well as the pace of innovation within the medical devices industry are another opportunities to this research's use case. One example is Service-oriented Device Connectivity (SDC), which is standardized by the Institute of

Electrical and Electronics Engineers (IEEE) for the interoperability of medical devices. It supports clinicians by enabling secure exchange between the devices, clinical and hospital information systems. The integration of all relevant devices, whether in the Operating Room (OR), Intensive Care Unit (ICU) or Neonatal Intensive Care Unit (NICU), requires a secure hospital's IT infrastructure. SDC ensures reliable multi-directional data transmission, remote control, high level of data-security and provides medical-grade data quality, which in turn reduces the operating and administrative effort and time [9].

A device connected to the SDC is a secure device due to the authentication concepts defined in the standard, where only the connected devices can exchange data. The transmitted data is encrypted and protected. Ensuring that hospital data and patient records are protected from unauthorized use or manipulation [9].

The system requirements for SDC are minimal, it allows for implementation into most existing hospital IT infrastructures. However, there are a few challenges to consider. The effort required for the initial setup to implement the network in the hospital, and the cost of this implementation, and the compatibility of the hospital's devices to SDC [9]. If the device is connected to a third-party device, such as a patient monitor, the hospital may need to replace its devices to make them SDC-compatible, which in turn incurs additional costs.

4   Conclusion

In this paper, a SWOT analysis is carried out for the utilization of of medical device real-world log data for post-market activities under the EU MDR.

The SWOT analysis results show that medical device real-world log data have a potential advantage in improving the quality and availability of clinical data, because they are continuously recording all favorable and unfavorable device events, unfortunately, only lacking user-based experience data. Overall, logged device data facilitates the systematic, proactive and continuous data gathering process of the PMCF activities.

The EU strategy and the latest national laws and current trends in digital solutions require better use of digital data to respond to the future market needs. Interoperability and connectivity solutions such as SDC offer a secure, multi-directional data transmission, and high quality medical data. It requires, however, an initial implementation effort for the entire system, thereby incurring potential costs for the manufacturer. Additionally, the manufacturers internal IT resources for CA specialists need to be improved so that they can efficiently draw conclusions from the acquired data.

Weighing up all the factors leads to the conclusion that the use of log data from medical devices is generally advantageous. The full integration of medical device data into post-market surveillance activities requires more effort, and detailed assessment to ensure the compliance to all the relevant regulations.

Acknowledgement

The work has been carried out at Drägerwerk AG & Co. KGaA, Lübeck and supervised by the Institute of Applied Natural Sciences, Technische Hochschule Lübeck. Furthermore, I would like to thank all my supervisors for their continuous guidance during the internship: Ulrich Behrje, Isabel Wagner, and Angela Schober, as well as to my university supervisor for his valuable knowledge and academic support, Prof. Dr. sc. hum. Folker Spitzenberger.

Authors' Statement

The author states that there is no conflict of interest.

5   References

[1] European Commission. *European data strategy*. Available: https://commission.europa.eu/strategy-and-pol icy/priorities-20192024/europefit-digital-age/europea n-data-strategy_en [Last accessed on 2024-01-19]

[2] National Institute of Standards and Technology. *Medical Device Interoperability, Fact Sheet*. Available: http s://www.nist.gov/programs-projects/emerging-technol ogies-healthcare/medical-device-interoperability [Last accessed on 2024-01-22]

[3] Wikipedia. *The EEC single market strategy*. Available: https://en.wikipedia.org/wiki/European_single_market [Last accessed on 2024-01-22]

[4] European Parliament and council, 2017. *Regulation (EU) 2017/745 on medical devices*.

[5] European Commission. *MEDDEV 2.7/1 revision 4, Guidance on Clinical Evaluation*. Available: https: //ec.europa.eu/docsroom/documents/17522/attach ments/1/translations/ [Last accessed on 2024-02-02]

[6] Reinaldo J. Perez, *Design of Medical Electronic Devices*. San Diego, California: Academic Press,vol. 279, pp. 75-78, 2002.

[7] Wikipedia. *SWOT analysis*. Available: https://en.wikip edia.org/wiki/SWOT-analysis [Last accessed on 2024-01-19]

[8] Federal Ministry of Health, Germany. *Parliament passes digital laws for better care and research in the healthcare sector. (Bundestag verabschiedet Digitalgesetze für bessere Versorgung und Forschung im Gesundheitswesen)*. Available: https://www.bundesge sundheitsministerium.de/presse/pressemitteilungen/ bundestag-verabschiedet-digitalgesetze-pm-14-12-23 [Last accessed on 2024-02-01]

[9] Drägerwerk AG & Co. KGaA. *Medical device interoperability*. Available: https://www.draeger.com/en_s ea/Hospital/Medical-Device-Interoperability [Last accessed on 2024-01-23]

10

Artificial Intelligence

Self-Supervised DEnoising UltraSound Network (DEUS-Net)

Jan Meyer [1] and Christian Janorschke [2]

[1] Medical Informatics, Universität zu Lübeck, ja.meyer@student.uni-luebeck.de
[2] Institute for Robotics and Cognitive Systems, Universität zu Lübeck, c.janorschke@uni-luebeck.de

Abstract

Ultrasound is one of the most widespread technologies for medical imaging. This is due to its many benefits, like being non-invasive, radiation free, portable and relatively low cost, compared to e.g. computed tomography and magnetic resonance imaging. However, one of the biggest challenges for ultrasound imaging is image quality. This is mainly caused by speckle noise, which interferes with most image processing algorithms. Many mathematical methods are traditionally used to improve the image quality, but modern neural networks offer a new perspective to this issue. In this paper we propose a self-supervised neural network approach that seeks to enhance image quality for ultrasound images from different anatomical domains and stations, thus achieving domain translation and real-time denoising. We trained a denoising network on four domains that is on par with the best traditional denoising algorithms in terms of performance.

1 Introduction

Ultrasound (US) imaging is very common in medical diagnosis because it is non-invasive, radiation free and relatively low cost while being mobile, real-time capable and quick to use for examinations. Yet, there are still some major problems for automatic processing as low contrast and the presence of speckle noise degrades the image quality. Speckle noise is multiplicative and follows a gamma distribution [1]. This corruption interferes with commonly used algorithms that assume low variance of brightness for specific types of tissue and strong edges between structures. Many mathematical methods are in use, but most cannot fully reduce speckle noise while retaining edges. This has lead to increased interest in deep learning methods for US image denoising, that promise to reduce noise while preserving image details. The block-matching and 3D filtering algorithm (BM3D) is a state-of-the-art denoising algorithm, however it is slightly outperformed by deep leaning models in general denoising tasks [2]. Many deep learning approaches for US noise removal have been discussed including generative adversarial networks [3] and autoencoders [4]. While some have tried to extract the noise pattern directly from US images [4], others simply try to replicate mathematical methods, that are computationally expensive, to improve runtime [5]. However, most papers do not look at the domain translation problem present with US. As there is no standardized way to configure US stations, every physician uses his own individual settings, affecting the image. Anatomical domains are also a factor as different tissues and organs look different, leading to a great variety of contrast changes in combination with the stations settings. This needs to be considered when using deep learning approaches to avoid problems regarding generalization.

2 Material and Methods

Starting with some commonly used mathematical filters, the following sections describe the used dataset, evaluation metrics and the network structure.

2.1 Mathematical Methods

There are many mathematical filters to use for image denoising. The following were chosen to compare to the network based on their prevalence in the literature:

1. Scalar (low pass) filters are a simple way to remove high frequency noise, however they also filter out edges from the image. Common examples are the mean and gaussian filter [1].

2. Adaptive filters smooth the image without removing edge information, thus preserving the structural information. The wiener filter is one common example that aims to minimize the mean square error while the lee filter accounts for neighboring pixels [1], [5].

3. The wavelet filter uses a discrete wavelet transform to decompose the image and then reconstruct it with the inverse wavelet transform, omitting all coefficients below a certain threshold. Thus, the image is compressed to the most important information, removing unimportant noise. If the threshold is too high, this can lead to reconstruction artifacts and information loss [1], [5].

4. Low-rank filters work on a similar assumption, but use for example a singular value decomposition. Small singular values are assumed to be noise and set to zero before reconstructing the image, thus removing potential noise from the image [2], [5].

5. Total variance (TV) filters assume that the noise has high variance and that reducing the variance lowers the amount of noise present in the image. This is usually achieved by solving an optimization problem [2].

6. The BM3D algorithm is a two-stage non-locally collaborative filtering method that computes and stacks similar patches of the image into 3D groups by block matching [2]. Each stack is processed with a wavelet filter. The denoised patches are aggregated by a collaborative filter [5]. This reduces the computing time, and effectively suppresses noise by grouping 3D data arrays of similar 2D image fragments. However, it often struggles to maintain a balance between noise suppression and image detail preservation [3].

2.2 Dataset

The used dataset from the Institute for Robotics and Cognitive Systems consists of four domains (Heart, Liver, Kidney and Prostate), each with data from multiple subjects recorded on two different US systems. The original data consists of 3D volumes from which 2D slices were extracted around the center to generate the image data. The number of volumes per domain also varied drastically, as seen in Table 1, leading to only a percentage of volumes from the liver and prostate data being used. The data from the subject with the least amount of volumes per domain was used as the test set and the rest as training data.

Table 1: Number of subjects, volumes, training and test images for each domain of the dataset

Domain	Subjects	Volumes	Train	Test
Heart	6	1128	20000	2400
Liver	3	15719	20000	2400
Kidney	8	1668	20000	2400
Prostate	3	18829	20000	2400

2.3 Evaluation Metrics

The evaluation of denoising methods requires metrics for image quality. Based on the current literature, four commonly used metrics were chosen for this:

1. Mean Square Error (MSE) is calculated as follows:

$$\text{MSE} = \frac{1}{M \cdot N} \sum_{i=0}^{N-1} \sum_{j=0}^{M-1} \left(\mathbf{X}(i,j) - \mathbf{Y}(i,j) \right)^2, \quad (1)$$

with $\mathbf{X}$ and $\mathbf{Y}$ being two different images $M \times N$ in size, meaning a low MSE indicates high similarity [1].

2. Signal-to-Noise-Ratio (SNR) describes the difference in power of the signal compared to the noise in image $\mathbf{I}$, meaning a higher SNR indicates less noise [1]:

$$\text{SNR} = 10 \log_{10} \left(\frac{1}{M \cdot N \cdot \text{MSE}} \sum_{i=0}^{N-1} \sum_{j=0}^{M-1} \mathbf{I}(i,j)^2 \right). \quad (2)$$

3. Peak SNR (PSNR) indicates the size of the error proportionate to the peak value of the signal. A high value specifies a better reduction in noise. It can be calculated as follows:

$$\text{PSNR} = 10 \log_{10} \left(\frac{(2^B - 1)^2}{\text{MSE}} \right), \quad (3)$$

with B being the bits ($B = 8$ for the used data) used for encoding the grey values in the image [1]-[3].

4. The Structural Similarity Index Measure (SSIM) does not directly compare pixel values, but tries to compare structural components of the images:

$$\text{SSIM} = \frac{(2\mu_{\mathbf{X}}\mu_{\mathbf{Y}} + c_1) \cdot (2\sigma_{\mathbf{XY}} + c_2)}{(\mu_{\mathbf{X}}^2 + \mu_{\mathbf{Y}}^2 + c_1) \cdot (\sigma_{\mathbf{X}}^2 + \sigma_{\mathbf{Y}}^2 + c_2)}, \quad (4)$$

where $\mu_{\mathbf{X}}, \mu_{\mathbf{Y}}$ are the mean values, $\sigma_{\mathbf{X}}^2, \sigma_{\mathbf{Y}}^2$ the variances and $\sigma_{\mathbf{XY}}$ the covariance of images $\mathbf{X}$ and $\mathbf{Y}$, with c_1, c_2 being constants derived from the dynamic range of the images. This makes it more robust against contrast changes, compared to the MSE [2], [3], [5].

2.4 Neural Network

In this section, the network implementation, architecture, loss function and training target are described.

2.4.1 Network Architecture

The denoising US network (DEUS-Net) is a U-net [6] like autoencoder with skip connections between two blocks each for encoding and decoding as well as one block in the middle for the latent space, as seen in Fig. 1. The encoding module on the left consists of two smaller blocks. It extracts important features from the image using convolution layers and max pooling, which are then stored in the latent space. From there it is reconstructed using upsampling and convolutions in the decoding module on the right, which also consists of two smaller blocks. The idea is to achieve an encoding, from which the image is reconstructed without the noise similar to wavelet and low-rank approaches, but at the same time preserve edge information in the process. Additionally, skip connections are used, as they were shown to improve the spatial resolution of the original U-net.

2.4.2 Training Target

Since there is no ground truth in terms of noise free US images, a training target has to be chosen in order to calculate the loss function. One could average multiple consecutive US images, but this is not possible for e.g. heart data due to the organs movement. Other approaches try to train a network with just noisy images [4], but DEUS-Net uses a make-shift ground truth generated from the filters introduced in section 2.1, making it a self-supervised approach. We chose multiple different filters (mean, wiener, lee, TV and bilateral), from which one is randomly selected for each batch during training. The idea is that DEUS-Net learns the best aspects from multiple filters instead of just replicating the characteristic of a single filter.

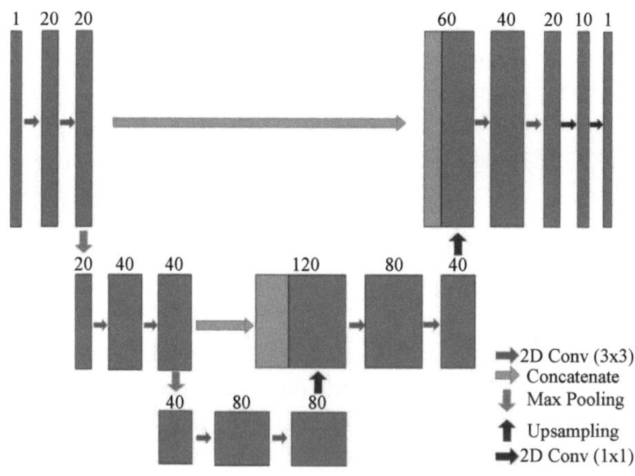

Figure 1: A visual overview of the DEUS-Net structure with the encoding module on the left, latent space in the middle and decoding module on the right.

2.4.3 Loss Function and Optimizer

As a loss function the MSE between the predicted image and the target image is calculated. This loss function performed best during preparation test runs on a smaller dataset. The Adam optimizer was used with a learning rate of 0.1 that is reduced by a scheduler once a plateau is reached during training. To save runtime, early stopping was implemented, terminating the training process once the network has not improved significantly for three epochs.

3 Results

In this section, DEUS-Nets training and runtime, denoising performance and domain translation are evaluated.

3.1 Network Training and Runtime

DEUS-Net was trained for 12 epochs using the complete dataset with 80,000 images, a batch size of 256 and no data augmentation. One training epoch took about 7 hours on a Nvidia GeForce RTX 3060 GPU. Furthermore, the average speed of filtering over 100 images was taken for all evaluated methods (see Table 2).

Table 2: Computation times for all mathematical filters and DEUS-Net averaged over 100 images

Method	Time	Method	Time
Mean	9.06 ms	Gaussian	0.47 ms
Wiener	2.19 ms	Lee	1110 ms
Wavelet	1.72 ms	Low-rank	2.34 ms
TV	39.1 ms	Bilateral	17.5 ms
BM3D	1040 ms	DEUS-Net	1.56 ms

3.2 Denoising Performance

Due to lack of a ground truth, the denoising performance of DEUS-Net is evaluated by comparison with mathematical filters. Therefore, artificial speckle noise was added with three different noise levels (multiplicative gaussian noise with $\sigma = 25, \sigma = 50$ and $\sigma = 75$) to the 9600 test images. Afterwards, the four metrics discussed in section 2.3 were calculated between the original and the denoised images. As a reference, the metrics were also calculated for the original and noisy images to allow for better interpretation of the results. We also used some more advanced filters, like the bilateral filter and BM3D algorithm as state-of-the-art methods. The results can be seen in Table 3 with the best result per metric highlighted bold.

Surprisingly, the noise image performs best in terms of MSE for $\sigma = 25$. The wiener and wavelet filter as well as DEUS-Net surpass it on higher noise levels. The low-rank filter and BM3D perform best in terms of SNR across all noise levels. DEUS-Net performs best in PSNR, expect for $\sigma = 25$ were the noise image is better, followed by wiener and lee filter. The gaussian and wiener filter performed best in SSIM for $\sigma = 25$ while DEUS-Net performs better on higher noise levels. Overall, DEUS-Net performs best in three out of four metrics for $\sigma = 50$ and $\sigma = 75$.

3.3 Domain Generalization

One important aspect of DEUS-Net is its ability to adapt to different domains within US imaging. As the dataset contains images from multiple organs, patients and US stations DEUS-Net is trained to adapt to these. Different characteristics like contrast shifts, noise patterns, etc., can hinder other approaches and some classical filters that rely on user-defined and problem-specific parameters.

Table 3: Evaluation of different methods for three different noise levels

Methods	$\sigma = 25$				$\sigma = 50$				$\sigma = 75$			
	MSE	SNR	PSNR	SSIM	MSE	SNR	PSNR	SSIM	MSE	SNR	PSNR	SSIM
Noise	**31.38**	17.31	**29.78**	0.823	44.04	18.95	23.83	0.691	49.23	19.49	20.49	0.616
Mean	37.41	18.16	28.30	0.823	43.62	18.91	23.87	0.696	48.64	19.43	20.62	0.621
Gaussian	52.05	19.38	24.97	**0.891**	51.24	19.40	25.09	0.829	50.20	19.37	24.72	0.772
Wiener	39.86	18.22	29.47	0.890	37.67	18.12	27.93	0.811	43.04	18.61	24.77	0.724
Lee	49.13	19.25	26.27	0.865	38.51	18.28	28.26	0.838	40.16	18.55	26.02	0.748
Wavelet	37.01	18.02	27.97	0.815	37.83	18.15	26.24	0.746	41.71	18.50	24.69	0.704
Low-rank	108.4	21.37	19.46	0.418	104.5	21.24	19.31	0.346	106.9	21.32	18.34	0.283
TV	62.11	20.26	22.37	0.849	60.43	20.14	23.15	0.822	58.48	20.01	23.24	0.769
Bilateral	55.86	19.79	21.91	0.728	51.37	19.45	23.27	0.707	47.63	19.16	23.60	0.669
BM3D	99.98	**21.50**	20.38	0.480	104.4	**21.61**	19.91	0.474	107.3	**21.66**	19.45	0.464
DEUS-Net	39.72	18.24	29.17	0.880	**34.29**	17.80	**29.55**	**0.842**	**35.83**	18.01	**28.68**	**0.807**

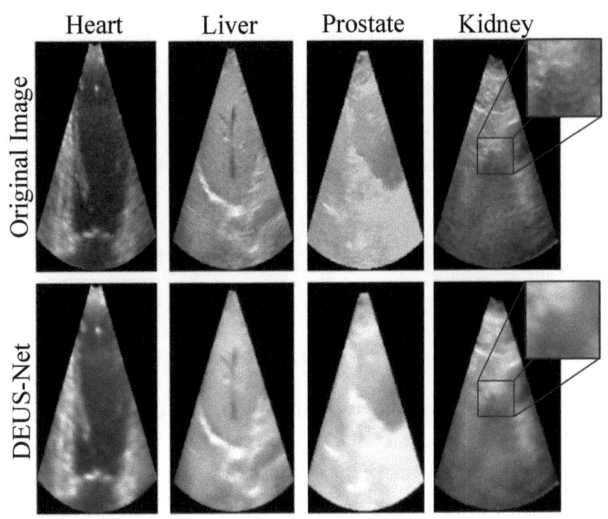

Figure 2: Clean images from all four domains (top row) and the corresponding network outputs (bottom row).

4   Discussion

One of the most important aspects is the runtime of DEUS-Net. As seen in Table 2, DEUS-Net is among the fastest denoising methods with only the gaussian filter being significantly faster. This allows for real-time applications, filtering and processing data while streaming.

The results seen in Table 3 provide some insight into the challenge of evaluation since, from a numeric point of view, added noise results in a better MSE and PSNR than any denoising method for $\sigma = 25$. This changes for higher noise levels, underlining why multiple evaluation metrics should always be considered to assess the results. The simple gaussian filter shows great resilience towards added noise in the SSIM metric for lower noise levels, probably due the averaging effect while other filters, like BM3D, blur the edges too much resulting in a loss of structural similarity. While DEUS-Net also performs well for $\sigma = 25$, it performs best on higher noise levels for most metrics. The results of DEUS-Net also do not deteriorate as much for higher noise levels as for many of the other filters.

As stated above the MSE loss was used in DEUS-Nets training. However, the loss function could be extended to further improve the performance or specific denoising behavior considering specific use-cases and subjective visual evaluation. For instance, the images of DEUS-Net seen in Fig. 2 show a slight increase in brightness as well as strong smoothing effect. Specific characteristics like these could be altered by adding e.g. the SSIM to the loss function, by pre-processing the brightness level of the input or by implementing data augmentation varying brightness.

To avoid misleading numeric or subjective evaluations, the performance of denoising could also be assessed depending on the use-case. For example, the output of a segmentation algorithm could be used to evaluate denoising performance comparing both filtered and unfiltered inputs against a ground truth.

Though further testing is warranted for domains not encompassed in the dataset, DEUS-Net can handle four vastly different input images as shown in Fig. 2. None of the four domains utilized correlated with a deterioration in terms of the (MSE) loss. Therefore, a robust generalization can be inferred.

5   Conclusion

In a comparative test using added noise, DEUS-Net performs best for most metrics on higher noise levels. It is among the fastest denoising methods, making it viable for real-time applications. Domain translation for four different domains was achieved and will be further evaluated in future work using new data. Additionally, ongoing efforts are dedicated to enhance both performance and evaluation methods.

Acknowledgement

The work was conducted at the Institute for Robotics and Cognitive Systems, Universität zu Lübeck, under supervision of Prof. Dr.-Ing. Achim Schweikard.

Authors' Statement

Conflict of interest: Authors state no conflict of interest.

6   References

[1] K. T. Dilna and D. J. Hemanth, "Novel image enhancement approaches for despeckling in ultrasound images for fibroid detection in human uterus," *Open Computer Science*, vol. 11, no. 1, pp. 399–410, 2021.

[2] L. Fan, F. Zhang, H. Fan, and C. Zhang, "Brief review of image denoising techniques," *Visual Computing for Industry, Biomedicine, and Art*, vol. 2, no. 7, 2019.

[3] L. Zhang and J. Zhang, "Ultrasound image denoising using generative adversarial networks with residual dense connectivity and weighted joint loss," *PeerJ Computer Science*, 8:e873, 2022.

[4] S. Goudarzi and H. Rivaz, "Deep ultrasound denoising without clean data," *Medical Imaging 2023: Ultrasonic Imaging and Tomography*, vol. 12470, 124700Q, 2023.

[5] S. Cammarasana, P. Nicolardi, and G. Patanè, "Real-time denoising of ultrasound images based on deep learning," *Medical & Biological Engineering & Computing*, vol. 60, pp. 2229–2244, 2022.

[6] O. Ronneberger, P. Fischer, and T. Brox, "U-net: Convolutional networks for biomedical image segmentation," *Medical Image Computing and Computer-Assisted Intervention*, vol. 9351, pp. 234–241, 2015.

Stress Classification with a Deep Neural Network using Electrodermal Activity and Blood Volume Pulse

Michelle Diez [1], Philip Gouverneur [2], Marco Bazzani [3], and Marcin Grzegorzek [2]

[1] Medical Informatics, Universität zu Lübeck, michelle.diez@student.uni-luebeck.de

[2] Institute of Medical Informatics, Universität zu Lübeck, {philip.gouverneur, marcin.grzegorzek}@uni-luebeck.de

[3] Teoresi Group, Department of Innovation, Turin, Italy, marco.bazzani@teoresigroup.com

Abstract

The feeling of stress is an ordinary part of life, however, excessive and chronic stress can lead to severe health risks. Detecting stress effectively is crucial for preventing related illnesses. Machine learning, and deep neural networks in particular, show promise in this area. The current lack of standardized architectures and data engineering hinders accurate comparisons of stress classification approaches. This research addresses this gap by evaluating data processing methods in deep learning to identify optimal practices. Using the WESAD dataset, the focus lies on evaluating electrodermal activity (EDA) and blood volume pulse (BVP) signals, which are known for strong stress correlations. A multi-layer perceptron architecture is modified regarding used signals, pre-processing methods, and feature extraction. The results emphasize the importance of EDA components over BVP, recommend a 5 Hz low-pass filtering for EDA, 0.5–8 Hz band-pass filtering for BVP, and z-score normalization.

1   Introduction

Stress is characterized by the release of cortisol in 'fight or flight' situations and can be beneficial to some extent by providing energy for acute stressors [1]. However, continuous stressful situations signalize the body to be in a constant state of threat awareness, leading to consistently high levels of cortisol and chronic stress. This can have a serious impact on the overall health, leading to physical consequences such as cardiovascular diseases and mental illnesses like depression [2]. Therefore, stress detection has become a widely researched area to reduce chronic stress and its effects. In the field of machine learning, techniques for detecting and classifying stress are increasingly being developed. Common methods analyze various physiological signals through public databases or self-acquired data while using machine learning algorithms like random forest classification, k-nearest neighbours clustering, support vector machines, and neural networks. Studies have shown, that deep neural networks can reliably classify stress and often show even better results than traditional machine learning methods [3, 4]. However, the current challenge lies in comparing different deep learning approaches due to the variability of which physiological data and sensors are used and which architectures and pre-processing operations are applied to the raw data. Consequently, a general architecture and the best data processing approach for stress classification are yet to be determined.

This research paper aims to provide a first impetus for an optimal deep learning approach for stress classification, using a simple neural network architecture to focus on the assessment of the best possible data engineering practices. Since heart rate variability (HRV) and electrodermal activity (EDA) evidently correlate with cortisol levels and are thus considered reliable indicators for stress classification [5], this work focuses on the processing of the corresponding EDA and blood volume pulse (BVP) signals.

2   Material and Methods

2.1   Public Dataset

In the field of stress recognition, several public datasets have been released throughout the years and used for machine learning classification. To achieve the best comparability with other research articles, the WESAD dataset [6] was selected for this work, which has been extensively used in the context of stress detection in recent years.

The dataset consists of recorded stress and amusement data from both the Empatica E4 wristband and the RespiBAN chestband. 15 participants were undergoing the Trier Social Stress Test as a stimulus for the label 'stressed', which consists of a public speaking exercise and an arithmetic exercise. The relaxed state is measured without any stimulation and labeled as 'baseline'. Furthermore, participants were shown several funny video clips for the detection of the label 'amusement', which is a commonly chosen third label in emotion recognition datasets. Electrocardiography, EDA, electromyography, and respiratory data were acquired by the chest-worn device, including electrodes attached to the chest to derive ECG data and to the upper *trapezius* muscle for EDA data acquisition. The Empatica E4 device includes built-in sensors for measuring BVP, EDA, skin temperature, and accelerometer signals. Regarding this project, only EDA and BVP signals from the Empatica E4 were considered.

2.2 Baseline Implementation

Since this research focuses on the optimization of data engineering methods, the neural network was chosen to be a simple multi-layer perceptron architecture as the baseline architecture. The architecture was adapted according to Eren and Navruz [7], which has shown high accuracy for a binary stress classification task and, therefore, being competitive with current state-of-the-art approaches. Regarding the initial implementation of the neural network by Eren and Navruz [7], the raw data underwent similar pre-processing and feature extraction operations to establish a baseline for the optimization task. After data loading, the signals and labels first had to undergo resampling to 50 Hz, since the original sampling rates for EDA, BVP, and labels were 4, 64, and 700 Hz respectively. The network was used for a binary classification differentiating between the labels 'baseline' and 'stressed', as well as a three-class classification with the inclusion of the label 'amusement'.

A 5 Hz Butterworth low-pass filter was applied to the EDA signal to remove high-frequency noise and focus on relevant frequency ranges suitable for the physiological components of the signal. The raw EDA signal consists of a tonic and phasic component, which were parsed by using the convex optimization tool for EDA (cvxEDA). Regarding the BVP signal, a heartbeat count algorithm was implemented, which extracts the signal peaks to obtain information about the HRV. Feature extraction was conducted by applying a sliding window of 60 seconds to the signals with a shift of 1 second [7]. For every window, the signals' mean, minimum, maximum, and standard deviation (STD) were calculated and normalized using z-score normalization from the sklearn package. Thus, 4 features for each signal were derived from the EDA raw signal, the EDA phasic and tonic components, the BVP signal, and 2 features of the extracted heartbeats, namely mean and standard deviation. In total, 18 features were extracted from the signals and given into the first neural network layer.

2.3 Data Processing Variations

The baseline implementation of data processing operations was altered regarding three different aspects, containing the used sensors, pre-processing methods, and extracted features. Various research papers were analyzed to compare the most commonly used data processing methods, including Bobade & Vani [3], Ninh et al. [4], Schmidt et al. [6], Eren & Navruz [7], and Khan & Sarkar [8].

Firstly, the performances of separate EDA and BVP signals were compared, including all signals that were derived from the raw data. Therefore, regarding EDA, the tonic and phasic components as well as the sparse sudomotor nerve activity (SMNA) driver of the phasic component were considered. Since the EDA tonic and phasic components can be acquired using different methods, both performances of the packages cvxEDA and the NeuroKit2 were analyzed. The extracted heartbeats of the BVP raw signal were also included in the analysis.

The comparison of pre-processing methods contained different filtering methods for EDA and BVP raw signals as well as normalization methods. The filtering methods for the EDA signal comprised of a 5 Hz Butterworth low-pass filter of the fourth order according to the baseline implementation, a 3 Hz Butterworth low-pass filter of the fourth order using NeuroKit2, as well as a low-pass filter of the fourth order with a cutoff frequency of 0.5 Hz from Ninh et al. [4]. The BVP raw signal performance was compared to the usage of a zero-phase second-order Butterworth band-pass filter, with cutoff frequencies of 0.5 and 8 Hz using Neurokit2.Regarding normalization methods, z-score normalization with and without normalization using the baseline label [8] was analyzed, as well as min-max normalization [4] using the sklearn package.

Finally, to find the most relevant features for an optimal classification, different features derived from the mentioned studies were included and their weights were analyzed. The feature weights were obtained by extracting the network's gradients after calculating a mean saliency map over all participants and labels. The classification performance was evaluated with the mean accuracy and F1-score from the LOSO cross-validation.

2.4 Neural Network Architecture

The network architecture consists of five hidden layers between the input and output layers. The ReLU activation function was used for all hidden layers, whereas for the binary classification of the output layer, the sigmoid activation function was used. For a 3-label classification, the output layer was altered to three channels and a softmax activation function.

The architecture was implemented with the stochastic gradient descent optimizer (SGD), due to the network having the best results with SGD compared to other optimizers [7]. In order to optimize training time and prevent fluctuation around the minimum loss function, the starting learning rate was set to 0.1 and decreased exponentially with more epochs. Binary cross-entropy was selected as the loss function since it also provided the best classification results compared to other loss functions. Regarding the 3-class problem, the categorical cross-entropy loss function was used. Furthermore, a leave-one-subject-out (LOSO) cross-validation was implemented to increase the network's ability to generalize data from different subjects as well as to ensure an evaluation that is not subject-dependent. The network is trained 15 times, respective to the number of subjects. For every training cycle, one subject is left out of training and used for validation. After a grid search analysis of the batch size and the number of epochs, the model was implemented with a batch size of 32 and 60 epochs per iteration.

3 Results and Discussion

3.1 Performance of Signals

Figure 1 shows the neural network's F1-scores for each analyzed signal regarding a binary and 3-class classification. Especially the phasic component of the EDA signal seems to greatly impact high 2-class classification performances. However, while conducting a 3-class prediction, the tonic component of the EDA signal shows increased F1-scores. The performance also depends on which method is used to extract the components, since the cvxEDA method has a more successful phasic component for 2-class predictions, whereas the neurokit method performs better regarding the tonic component. All signals acquired from EDA generally surpass signals, which are acquired from BVP. Signals derived from BVP appear to have stronger performances in a 3-class setting than in 2-class predictions.

In general, it is debatable which results are to be considered more relevant based on the number of included classes. In three-class classifications, the network has the opportunity to learn more patterns relevant to distinguishing between different classes, resulting in better class separation and improved generalization. Therefore, the results may be more expressive for an overall network evaluation. However, since this research specifically focuses on stress classification, the results of the binary classification task are considered more significant for this topic.

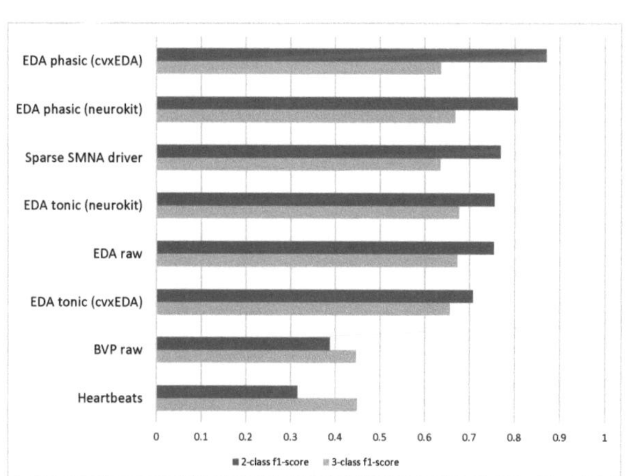

Figure 1: F1-scores for 2-class and 3-class predictions of every signal derived from EDA and BVP raw signals.

3.2 Comparison of Data Pre-processing Methods

The analysis of different pre-processing methods can be categorized into filtering methods and normalization methods.

The F1-scores achieved with various filtering methods are shown in Figure 2. The EDA signal was compared to three low-pass filtering methods with different cutoff frequencies. Overall, the diagram shows that for both classification tasks, the EDA performance increases with a growing cutoff frequency of the filter. It appears that the filter has to have a cutoff frequency of at least 3 Hz to have a positive impact on the classification performance. The performance of the BVP signal could be significantly improved by using a 0.5–8 Hz band-pass filter, both for a 2-class and 3-class classification. Thus, the following comparisons were performed with a filtered BVP signal and a 5 Hz low-pass filter for the EDA signal according to the original baseline implementation.

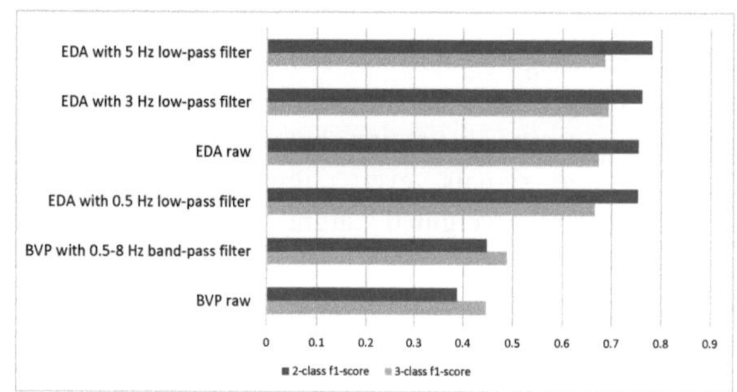

Figure 2: F1-scores for 2-class and 3-class predictions of various EDA and BVP filtering methods.

Performances from the variation of normalization methods are depicted in Figure 3 and show, that the best performance for a 2-class prediction could be reached by using z-score normalization. When considering a 3-class prediction, the F1-score can be improved using normalization by subtracting the mean value of the baseline label from all signals in addition to z-score. The following feature analysis was conducted using only z-score normalization.

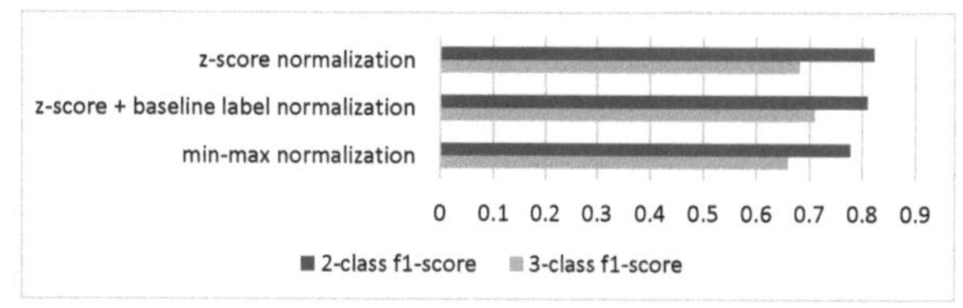

Figure 3: F1-scores for 2-class and 3-class predictions of different normalization methods.

3.3 Feature Variations

In order to find features, that have the most impact on the classification performance, various features for EDA, BVP, and HRV signals were derived according to Ninh et al. [4], Schmidt et al. [6], Bobade & Vani [3], and the NeuroKit2 package. A total of 56 features were included and are depicted in Table 1. After training the network for a binary classification, the individual weights of the features were calculated and analyzed.

Table 1: Used features per signal. HRV features are abbreviated following Neurokit2. Positively weighted features are marked in bold.

Signal	Features
EDA	**mean, min, max, STD** of raw EDA; **mean, min, max, STD** of tonic component; **mean, min, max**, STD of phasic component; **mean, min, max**, STD of SMNA; mean, **STD, counts** of SCR onsets; mean, STD, counts of SCR peaks; **mean, STD, counts** of SCR recovery
BVP	**mean, min**, max, STD of raw BVP; **mean**, min, max, **STD** of heartrate; mean, STD, counts of heartbeats
HRV	mean, STD, RMSSD, SDSD, **CVNN**, CVSD, MedianNN, **MadNN, MCVNN, IQRNN, pNN50, pNN20**, TINN, HTI, $\sum(ULF, VLF, LF, HF, VHF)$, TP, **LFHF, LFn**, HFn, **LnHF**

The feature weights analysis resulted in 30 positively weighted features and 26 negatively weighted features. There is no specific distinction that can be identified between the used signals, every signal is equally represented in positively weighted features as well as negatively weighted features. Features that are rated the most important for classification include the HRV median absolute deviation of the intervals between peaks divided by the median of the intervals (MCVNN), as well as the mean values of the EDA components SMNA, tonic, phasic and heart rate. Negatively rated features overall comprise STD values of the BVP raw signal, the EDA SMNA component, and heartbeats, among others. Other negatively weighted features include single features like the minimum heart rate, the heartbeat counts, and the HRV square root of the mean squared successive differences between adjacent peak intervals (RMSSD), thus, no specific patterns can be identified. It could be argued whether signal features such as mean should generally be favoured more, whereas STD values may not need to be considered at all.

4 Conclusion

In summary, this study attempts to find the best practices for stress classification in the context of deep learning. The research utilizes the WESAD dataset, emphasizing EDA and BVP signals known for stress correlation. A multi-layer perceptron architecture adapted from Eren & Navruz [7] undergoes modifications for binary and three-class predictions in terms of used signals, pre-processing methods and extracted features. The results highlight the large impact of EDA components on both binary and three-class classifications and recommend the choice of a 5 Hz low-pass filter for EDA and a 0.5-8 Hz band-pass filter for BVP signals. Moreover, z-score normalization using the sklearn package is recommended. The weights analysis of various features illustrates the balanced significance of all signals, where especially mean values show high importance in addition to the HRV MCVNN feature. Future works should focus on analyzing further methods from various studies as well as standardizing data processing procedures to enable more cross-study comparisons.

Acknowledgement

The work has been carried out at the Teoresi Group, Department of Innovation in Turin, Italy and supervised by the Institute of Medical Informatics, Universität zu Lübeck.

Authors' Statement

Conflict of interest: Authors state no conflict of interest.

5 References

[1] D. Y. Lee, E. Kim, and M. H. Choi, "Technical and clinical aspects of cortisol as a biochemical marker of chronic stress," *BMB reports*, vol. 48, no. 4, p. 209, 2015.

[2] S. Cohen, D. Janicki-Deverts, and G. E. Miller, "Psychological stress and disease," *Jama*, vol. 298, no. 14, pp. 1685–1687, 2007.

[3] P. Bobade and M. Vani, "Stress detection with machine learning and deep learning using multimodal physiological data," in *2020 Second International Conference on Inventive Research in Computing Applications (ICIRCA)*. IEEE, 2020, pp. 51–57.

[4] V.-T. Ninh, M.-D. Nguyen, S. Smyth, M.-T. Tran, G. Healy, B. T. Nguyen, and C. Gurrin, "An improved subject-independent stress detection model applied to consumer-grade wearable devices," in *International Conference on Industrial, Engineering and Other Applications of Applied Intelligent Systems*. Springer, 2022, pp. 907–919.

[5] B. Mahesh, T. Hassan, E. Prassler, and J.-U. Garbas, "Requirements for a reference dataset for multimodal human stress detection," in *2019 IEEE International Conference on Pervasive Computing and Communications Workshops (PerCom Workshops)*. IEEE, 2019, pp. 492–498.

[6] P. Schmidt, A. Reiss, R. Duerichen, C. Marberger, and K. Van Laerhoven, "Introducing wesad, a multimodal dataset for wearable stress and affect detection," in *Proceedings of the 20th ACM international conference on multimodal interaction*, 2018, pp. 400–408.

[7] E. Eren and T. S. Navruz, "Stress detection with deep learning using bvp and eda signals," in *2022 International Congress on Human-Computer Interaction, Optimization and Robotic Applications (HORA)*. IEEE, 2022, pp. 1–7.

[8] N. Khan and N. Sarkar, "Semi-supervised generative adversarial network for stress detection using partially labeled physiological data," *arXiv preprint arXiv:2206.14976*, 2022.

A Template-Based Region Proposal Approach for Two-Stage Object Detection in Point Clouds

Lara Presser [1], Jasper Diesel [2], and Mattias Heinrich [3]

[1] Medical Informatics, Universität zu Lübeck, lara.presser@student.uni-luebeck.de
[2] Corporate Technology & Innovation, Drägerwerk AG & Co. KGaA, jasper.diesel@draeger.com
[3] Institute of Medical Informatics, Universität zu Lübeck, mattias.heinrich@uni-luebeck.de

Abstract

3D object detection is a crucial first step for various visual processing tasks in point clouds. It can be approached in two stages: a region proposal and region classification stage. Explainable approaches that are adaptable to novel objects without large amounts of training data are of high relevance, especially in the medical domain. The present paper presents a template-based region proposal algorithm that identifies candidate regions of target objects in point clouds. The evaluation on the task of generating bed region proposals showed promising results with a stable prediction quality and an average best overlap (ABO3d) of 0.74 ± 0.09. However, future work is needed to enhance the recall rate of 0.72, obtained using an ABO3d threshold of 0.7 and to explore the approach's applicability to novel domains and target objects.

1 Introduction

Point clouds provide a detailed representation of spatial information and are widely used in domains such as robotics, autonomous driving, or 3D medical imaging.

To gain a comprehensive scene understanding, 3D object detection is a crucial first step for various visual processing tasks involving point clouds [1]. It comprises the 3D localization, classification and often pose estimation of object instances and results in the prediction of oriented bounding boxes (OBBs) enclosing the target objects.

One major approach towards object detection in general divides the task into two stages: a region proposal stage and an object classification stage. The first stage determines a set of promising candidate regions that potentially contain a target object and the second stage identifies the presence of an object and its class membership. Thereby the region proposal stage replaces a computationally expensive exhaustive search across the whole image by means of a sliding window approach [2]. Essential properties of a region proposal algorithm are a high recall, as missed object regions cannot be recovered in the second stage, computational efficiency and a good trade-off between proposal quality and quantity [3], [4].

Traditionally, two-stage object detection was applied to image-data, where candidate regions were formed based on segment-merging or window scoring methods based on hand-crafted features [3], [4]. More recently, the use of deep learning-based approaches has become common practice for 2D and 3D object detection tasks [5].

However, challenges of deep learning approaches such as the need for large amounts of training data, insufficient ex-

plainability of predictions and vulnerabilities against adversarial attacks are a major drawback especially in medical applications [6]. Moreover, these models often lack the flexibility to detect novel objects, as this would require retraining.

The present paper proposes a template-based region proposal algorithm for detecting candidate regions of specific target objects in point clouds. It is intended to be used in conjunction with an object classifier as part of a two-stage 3D object detection framework. Template matching is employed as an interpretable and computationally efficient method to determine the candidate regions based on their similarity to a set of 2D templates from birds' eye view (BEV). Unlike deep learning-based approaches, the proposed algorithm can be easily adjusted to novel target objects without requiring training data. Instead, only limited prior scene knowledge and simplified BEV representations of the target object are needed.

2 Methods

2.1 Region Proposal Algorithm

The proposed template-based region proposal algorithm aims to identify a set of candidate regions in form of OBBs of a given target object. Thereby target objects are described as templates in the 2D BEV and their corresponding instances are found using a template matching scheme on the BEV representation of the point cloud together with basic prior information about the scene.

The key components of the current approach are the transformation of point clouds into a compact 2D grid repre-

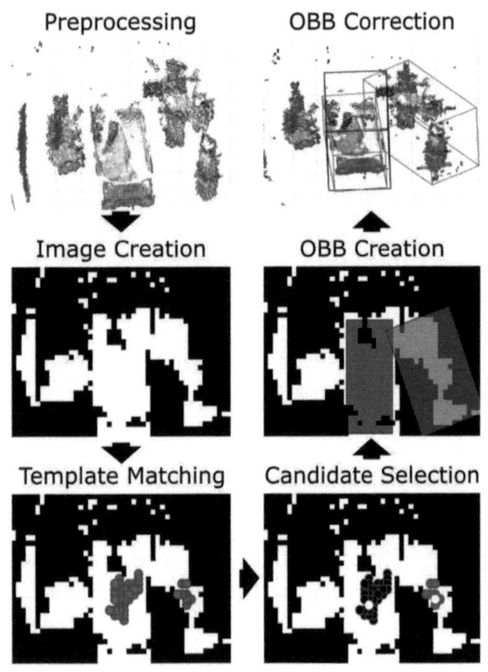

Figure 1: Illustration of the processing stages of the template-based region proposal algorithm.

sentation (Section 2.1.2), the efficient identification of candidate locations using a template matching and -selection scheme (Section 2.1.3 and 2.1.4) and the subsequent candidate location refinement in the original 3D space (Section 2.1.5). An overview of the processing pipeline is provided in Fig. 1.

2.1.1 Initialization and Point Cloud Pre-processing

The region proposal algorithm is initialized with a set of templates T and prior knowledge about the scene including the floor plane location, approximate axes of the target object and the expected height bounds of the object search space. As a point cloud pre-processing step, the points outside the given height bounds are excluded.

2.1.2 Image Creation

Next, the point cloud is transformed into a BEV representation, obtained from an orthogonal projection of points onto the gridded floor plane. Each pixel in the grid denotes the occupancy of points in its corresponding region. The pixel size, affecting the grid granularity, is controlled by a parameter. The process results in a binary pseudo image that efficiently represents the point cloud and enables the use of established image processing methods.

2.1.3 Template Matching

Afterwards, the best matching image location for each template is determined. The templates T are binary images that encode possible shapes of the target object in BEV in various rotations and sizes. To prevent border effects of

outer pixels during the matching process, they are padded with NaN values. During template matching, similarity scores $S_{T_i}(x, y)$ are computed between each image location $I(x, y)$ and every template T_i, utilizing the method of summed squared differences. Subsequently, the image location $I(x', y')$ with the best matching score $S_{T_i}(x', y')$ is selected for each template. It is referred to as match in the following.

2.1.4 Candidate Selection

Following the template matching, the next step aims to select a subset of matches as candidate regions. Thereby candidate regions should be the best matching largest templates within an image proximity and have minimal spatial overlap with neighbouring matches.

To achieve this, every match is first scored based on the previous template matching score $S_{T_i}(x', y')$ and its corresponding template size $size(T_i)$, resulting in the candidate region score $S_{C_i}(x', y')$ defined in (1). It introduces a positive bias towards the selection of larger templates, weighted by w.

$$S_{C_i}(x', y') = S_{T_i}(x', y') - size(T_i) \cdot w \qquad (1)$$

Next, the matches are clustered in a bottom-up hierarchical manner with the complete linkage Euclidean cluster distance as the merge criterion. Finally, for each cluster, the match with the best candidate score $S_{C_i}(x', y')$ is selected. The bias towards the selection of larger templates accounts for the problem of partial object occlusions and aims to prevent choosing matches of small templates that only partially cover the target objects. Conceptually, each cluster can be thought of as representing a candidate region hypothesis, and selecting the match with the best candidate score from a cluster yields a similar effect like non-maximum suppression [7].

2.1.5 OBB Creation and Refinement

After the best candidate regions were found in 2D BEV, they are subsequently transformed to 3D OBBs. For each candidate region, floor plane aligned OBBs are created, that encompass all points, whose BEV projection lie within the respective candidate region. While an OBB's ground-plane corresponds to the candidate region shape, its height is defined by the largest floor plane distance of its points. The height of an OBB is therefore highly susceptible to outliers and requires a subsequent height correction. Points are identified as outliers, if the mean height difference towards their neighbours exceeds a certain threshold. They are removed by linear interpolation before the final floor plane aligned OBB is calculated from the smoothed points.

2.2 Evaluation

The region proposal algorithm was evaluated on the task of generating region proposals for beds as target objects. The templates represented the beds in the BEV in a simplified way as rectangles of different orientations and sizes.

2.2.1 Evaluation Metrics

The quality of the proposals for a scene was evaluated using the average best overlap in 3D (ABO3d). The metric is an adaptation of the average best overlap (ABO) for region proposals in images [3]. ABO3d describes the best 3D intersection over union (IoU3d) between region proposals $RP_j^s \in RP^s$ and the ground truth GT^s for a scene s as in (2).

$$ABO3d = \frac{1}{S} \sum_{s=1}^{S} \max_{RP_j^s \in RP^s} IoU3d(RP_j^s, GT^s) \qquad (2)$$

It is common evaluation practice to consider only the top n proposals in order to punish algorithms that output a high number of proposals with low quality [3]. For the current scenario the top 10 proposals, ranked by their candidate score S_C, were chosen.

To evaluate the recall of the region proposal algorithm, true positives are defined as having an ABO3d greater than a threshold value. The threshold is set to 0.7, which is consistent with ABO3d values for detecting specific object instances in a common benchmark [8].

2.2.2 Hyperparameter Optimization

Tuning the hyperparameters of the region proposal algorithm was crucial to achieve a good performance. To align with the goals of a region proposal algorithm (Section 1), a multi-objective Tree-structured Parzen Estimator was employed that simultaneously aimed to minimize computation time of the algorithm, measured as CPU time, and maximize ABO3d of the proposals. Minimizing the computation time also regulated the quantity of proposals, as both were positively related. The best hyperparameter configuration was selected by visual assessment of the Pareto front.

2.2.3 Dataset

Hyperparameter tuning and inference was performed on a private indoor dataset with a $40\%/60\%$ data split for tuning and testing. The complete dataset contained a total of 4952 scenes emulating hospital environments with manually annotated OBBs for people, beds and various medical devices. Thereby a bed, as the target object, was visible in every scene. It was displayed in various backrest height configurations and was either empty or with a person in it. The dataset was acquired by a PrimeSense Camine 1.08 depth camera that was positioned at 2.2m height in varying angles and distances to the bed. As a pre-processing step, background and floor were removed from all final scenes.

3 Results and Discussion

3.1 Hyperparameter Tuning

Hyperparameter tuning revealed the four most important parameters: the number of different templates provided (Section 2.1.1), the pixel size of the BEV grid (Section 2.1.2), the template size weight (Section 2.1.3) and the linkage-distance threshold as cluster merge criterion (Section 2.1.4).

The best hyperparameter configuration had an ABO3d of 0.75 ± 0.08, an average computation time of $0.3 \pm 0.08s$ and an average number of proposed regions of 3 ± 1.46 on the training set. It was subsequently evaluated on the test dataset.

3.2 Evaluation

The region proposal algorithm achieved an ABO3d of 0.74 ± 0.09 on the test set with an average number of region proposals of 2.98 ± 1.46.

Fig. 2 illustrates the relation between recall and ABO3d threshold. The steepness of the curve implies a low dispersion of ABO3d values, which is underlined by the ABO3d interquantile range of 0.11 around the median of 0.75. The maximum ABO3d value was 0.94. With the chosen threshold, a recall of 0.72 was achieved.

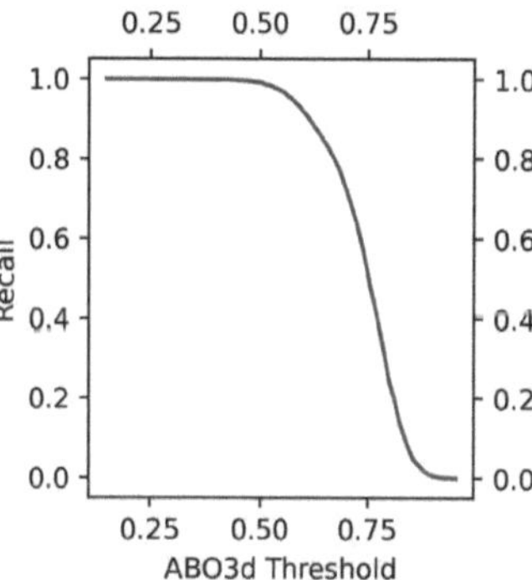

Figure 2: The recall versus different IoU3d thresholds.

Overall, the evaluation results indicate a good trade-off between quality and quantity of proposals, and the low dispersion of ABO3d values suggests a stable quality of region proposals across different scenes. Further, the results of hyperparameter tuning and testing were consistent, implying a good generalization of the selected parameter configuration.

Placing the current approach in context of existing methods is challenging. Most two-stage 3D object detection paper focus solely on the final object classification evaluation by reporting metrics such as Average Precision (AP). However, calculating the AP of region proposals algorithms is not meaningful, as they intentionally compromise precision in favor of a high recall. Moreover, the current approach is evaluated on a private dataset. Potential disparities in point cloud densities and task complexities have likely influenced computation times and performance, rending a comparison with existing methods difficult.

State of the art 3D object detection approaches achieve an inference speed of up to 16ms [1], indicating superior computational performance than the current approach, despite comparison difficulties.

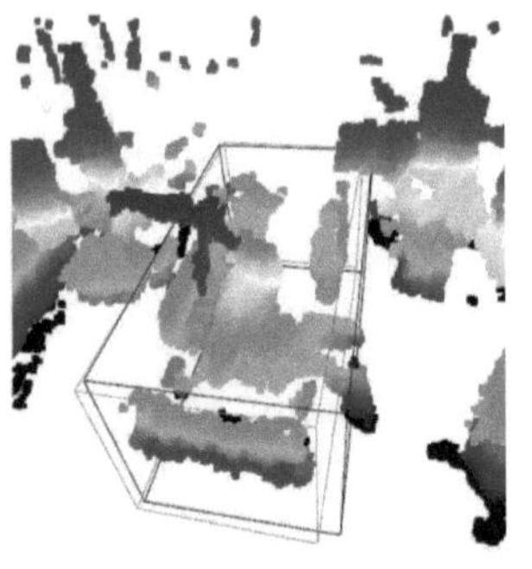 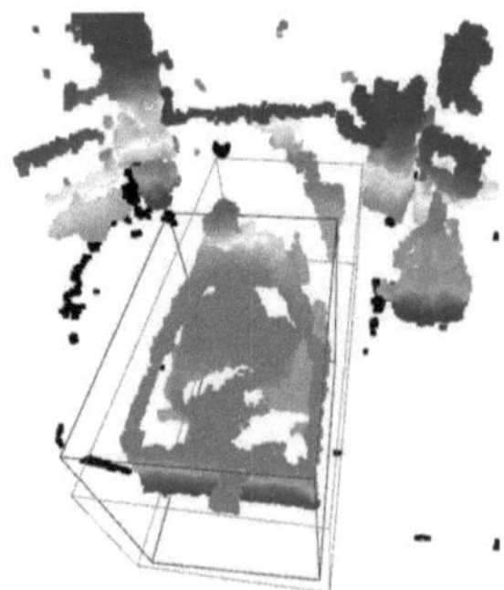

(a) Successful Region Proposal   (b) Failed Region Proposal

Figure 3: Examples of region proposals. The best OBB proposal of a scene is depicted in black and the ground truth OBB in gray. In Fig. 3a the bed was estimated correctly ($IoU3d = 0.85$), while in Fig. 3b, it was overestimated in height and not captured with its entire length ($IoU3d = 0.43$).

Compared to traditional 2D region proposal algorithms with recall rates of up to 0.87 [4], the current recall rate of 0.72 is comparatively poor. However, these algorithms generate hundreds of object proposals per image, while the current approach generated only 2.98 proposals in average.

Reasons for the lower ABO3d, resulting in a low recall, could be twofold. First, IoU3d deceases rapidly in the presence of slight inaccuracies, making it more difficult to achieve high IoU scores in 3D compared to IoU in 2D [9]. Second, the dataset contained annotation inaccuracies, from small to medium extents, potentially exacerbating the effect of decreasing IoUs in 3D. An annotation inaccuracy affecting the length of the OBB is visible in Fig. 3b.

One limitation of the approach is the neglect of height information during candidate selection. It could render template matching in highly cluttered scenes with overlapping object shapes unfeasible. Future work could explore the usage of height maps that preserves this information, while holding all advantages of an 2D point cloud representation. Another point of improvement is the susceptibility of template matching to occlusion, often present in point cloud data (see Fig. 3b). While the problem is approached using the template size weight, the exact value of the parameter could be dataset and context dependent.

4 Conclusion

This paper presented a novel approach to identify object-specific region proposals in an explainable and adaptable way. The evaluation on a specific target object class shows promising results regarding the robustness and trade-off between quality and quantity of proposals. However, future work is needed to improve the ABO3d score and the overall recall by addressing the discussed limitations. Further, conducting evaluations on public datasets is essential to enhance comparability with existing approaches and explore its applicability to novel domains and target objects with more precise annotations.

Acknowledgement

The work has been carried out at the department Corporate Technology & Innovation, Drägerwerk AG & Co. KGaA and supervised by the Institute of Medical Informatics, Universität zu Lübeck.

Authors' Statement

Conflict of interest: Authors state no conflict of interest.

5 References

[1] J. Mao, S. Shi, X. Wang, and H. Li, "3d object detection for autonomous driving: A comprehensive survey," *International Journal of Computer Vision*, pp. 1–55, 2023.

[2] N. Chavali, H. Agrawal, A. Mahendru, and D. Batra, "Object-Proposal Evaluation Protocol is 'Gameable'," in *2016 IEEE Conference on Computer Vision and Pattern Recognition (CVPR)*. Las Vegas, NV, USA: IEEE, Jun. 2016, pp. 835–844.

[3] J. R. R. Uijlings, K. E. A. van de Sande, T. Gevers, and A. W. M. Smeulders, "Selective Search for Object Recognition," *International Journal of Computer Vision*, vol. 104, no. 2, pp. 154–171, Sep. 2013.

[4] C. L. Zitnick and P. Dollár, "Edge boxes: Locating object proposals from edges," in *Computer Vision–ECCV 2014: 13th European Conference, Zurich, Switzerland, September 6-12, 2014, Proceedings, Part V 13*. Springer, 2014, pp. 391–405.

[5] W. Liu, J. Sun, W. Li, T. Hu, and P. Wang, "Deep Learning on Point Clouds and Its Application: A Survey," *Sensors*, vol. 19, no. 19, p. 4188, Jan. 2019.

[6] T. Dhar, N. Dey, S. Borra, and R. S. Sherratt, "Challenges of Deep Learning in Medical Image Analysis—Improving Explainability and Trust," *IEEE Transactions on Technology and Society*, vol. 4, no. 1, pp. 68–75, Mar. 2023.

[7] R. He, J. Rojas, and Y. Guan, "A 3d object detection and pose estimation pipeline using rgb-d images," in *2017 IEEE International Conference on Robotics and Biomimetics (ROBIO)*. IEEE, 2017, pp. 1527–1532.

[8] "The KITTI Vision Benchmark Suite." [Online]. Available: https://www.cvlibs.net/datasets/kitti/eval_object.php?obj_benchmark=3d

[9] G. Brazil, A. Kumar, J. Straub, N. Ravi, J. Johnson, and G. Gkioxari, "Omni3D: A Large Benchmark and Model for 3D Object Detection in the Wild," in *2023 IEEE/CVF Conference on Computer Vision and Pattern Recognition (CVPR)*. Vancouver, BC, Canada: IEEE, Jun. 2023, pp. 13 154–13 164.

Reward Function Design for End-to-end Off-road Autonomous Simulator

Sazid Rahman Simanto [1], Fabian Domberg [2], and Georg Schildbach [2],

[1] Robotics and Autonomous Systems, Universität zu Lübeck, sazid.simanto@student.uni-luebeck.de

[2] Institute for Electrical Engineering in Medicine, Universität zu Lübeck, {f.domberg, georg.schildbach}@uni-luebeck.de

Abstract

This paper presents the reward function designing method of a new off-road simulation environment that drives a 4-wheel drive vehicle to reach its destination autonomously starting from an initial location. Both the initial and goal locations are chosen randomly inside the terrain. The vehicle acts here as a naive blind agent and it observes the environment using some measurements related to the vehicle dynamics. A simple multilayer perceptron is used as a baseline model which is trained using the Proximal Policy Optimization (PPO) algorithm to determine the best-performing reward function for this environment.

1 Introduction

In recent decades, autonomous navigation for structured environments has been a key focus of research. However, the off-road environment has not been extensively explored. In earlier work [1], [2], [3], [4] each of them has handled the part of perception, planning, and control separately. Initially, in these papers, based on the data from the perception module (LiDAR, RGB, Camera), an elevation 2.5d map was created. Then using that map, some of them used classical planning algorithms like A*, D* or probabilistic models (e.g., Gaussian regression model) or even Reinforcement Learning (RL) to generate the optimal trajectory. Afterward, Model Predictive Control (MPC) or classical controller (e.g., Proportional Integral Derivative controller) is used to follow the trajectory for navigation. However, one of the papers [5] has used a single neural network in an end-to-end fashion using Deep Reinforcement Learning (DRL) without computing the perception, planning, and control blocks separately. This method takes the front camera RGB image and vehicle state as input and then directly predicts the throttle and steering angle of the vehicle. It significantly reduces computational overhead but also poses challenges in its use in real-world environments. They have published their environment but there is little information about the design of the environment's reward function in the paper. The main inspiration for this project has been drawn from this paper and a new simulation environment has been created which is easy to set up. After creating a simulation environment the very essential task is to design an appropriate reward function for the environment. For this task, it is very convenient to start with a very basic approach because then it becomes easy to track down the occurred problems. For this reason, only the vehicle state data from the observation is considered as input for this experiment by excluding the front camera RGB image for simplicity. This procedure can be compared with a blind human being who can not perceive the surroundings but has an idea of his current state and can gradually achieve his own map representation of a frequently visiting place or a place where he had spent most of the time. The content of this paper is to make an available off-road simulation environment to facilitate other's contributions in this domain.

2 Material and Methods

In this project, a unity simulation environment [7] was created to train a 4-wheel drive vehicle [6] to reach its goal position autonomously starting from an initial location. The environment is a 256m × 256m uneven surface area which consists of large trees and stones as obstacles and the goal point is indicated by a sign. Following the DRL framework to solve this problem, the vehicle acts here as an agent and it observes the environment using an RGB camera and some measurements related to the vehicle dynamics. As an action if the vehicle applies throttle $(0, 1.0)$ and the steering $(-40°, 40°)$ then the environment gives a reward based on the evaluation of the action. However, designing the reward function is the most difficult task for an RL environment because if it is not designed appropriately, the learning will not be effective. For this setup, as there are no publicly published reward functions available for uneven terrain, inspiration has been taken from [8] where an evaluation has been performed on 10 different reward functions, which are for structured road environments. During the time of testing these reward functions for this environment, it was visible that most of the conditions for structured environments do not equally impact the uneven terrain environment, e.g., lane keeping, reaching maximum speed by satisfying physical constraints, following traffic rules, overtaking other vehicles, etc. But reaching the goal state, collision with an obstacle, and considering the stable vehicle dynamics are common conditions between these two environments. A well-designed reward function is a prerequisite to learning a better policy. For this purpose, a simple multilayer perceptron is used as a baseline model that is trained using the Proximal Policy Optimization (PPO) algorithm [9] to measure the performance of that particular reward function. The vehicle state data which is the only input to the policy learning algorithm is a latent vector of size 19 and table 1 briefly describes the attributes that are represented by the vehicle state data. Here the designed reward functions for this unstructured environment are now listed below.

1.
   - +5000 if the goal is reached.
   - -2000 if the vehicle goes outside the terrain, gets flipped, and collides with a stone or tree.
   - Reward at each nonterminal timestep:

$$R_t = (1.0 - \frac{|v - 20.0|}{20.0}) + (1 - d_{goal}) \quad (1)$$

2.
   - +5000 if the goal is reached.
   - -2000 if the vehicle goes outside the terrain, gets flipped, or collides with a stone or tree.
   - Reward at each nonterminal timestep:

$$R_t = (1.0 - \frac{|v - 20.0|}{20.0}) * (1 - d_{goal}) \quad (2)$$

3.
   - +5000 if the goal is reached.
   - -2000 if the vehicle goes outside the terrain, gets flipped, or collides with a stone or tree.
   - Reward at each nonterminal timestep:

$$R_t = -2.5 + (1.0 - \frac{|v - 20.0|}{20.0}) * (1 - d_{goal}) \quad (3)$$

4.
   - +5000 if the goal is reached.
   - -2000 if the vehicle goes outside the terrain, gets flipped, and collides with a stone or tree.
   - 0 Reward at each nonterminal timestep.

$$R_t = 0 \quad (4)$$

5.
   - +10 if the goal is reached.
   - -5 if the vehicle goes outside the terrain, gets flipped, and collides with a stone or tree.
   - Reward at each nonterminal timestep:

$$R_t = -0.01 * \frac{|v - 20.0|}{20.0} \quad (5)$$

Table 1: Vehcile state

Index	Attributes
state[0-2]	Goal location relative to vehicle(X, Y, Z)
state[3-5]	Vehicles velocity(X, Y, Z)
state[6-8]	Vehicles acceleration(X, Y, Z)
state[9-11]	Vehicles angular velocity(X, Y, Z)
state[12-14]	Vehicles angular acceleration(X, Y, Z)
state[15]	Vehicles heading
state[16]	Heading difference to target location
state[17]	Vehicles RPM
state[18]	Vehicles gear

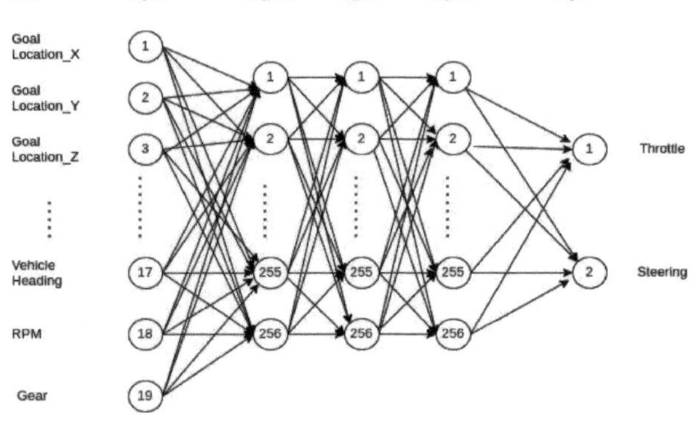

Figure 1: Network Architecture of PPO Algorithm

3   Results and Discussion

The widely used open-source implementation repository stable-baselines3[1] of RL algorithms have been used to train the policy model. As the configuration of the baseline PPO model, "MlpPolicy" is set as the policy model, which only consists of 3 hidden layers (256, 256, 256), and Fig. 1 represents the network architecture for better understanding. The discount factor is $\gamma = 1.0$, which indicates that the expected cumulative reward is not discounted, which means that all actions are equally important across the time steps. The total time step for training the model has been set to 10^6 and to facilitate the training speed, the environment has been parally trained on 4 instances and the simulation timescale has been set to 5.0 which makes the simulation run 5 times faster. This simulation environment is complex regarding driving the vehicle through uneven surfaces, and initially, the vehicle has to face many difficulties to learn how to drive, control, and adjust the speed in that toughest hilly area. Even to realize its purpose, it should get the easier tasks at the beginning, and by successfully reaching those goals, it encourages learning. On the other hand, at the beginning of learning if it does not reach the goal then it can not be well aware of its objective. This whole idea is based on this paper [10] and to incorporate this method during learning time, some changes have been made in the environment. The curriculum step (50% of total timesteps) has been set, which will gradually make the environment surface hilly. Also in the initial steps, the location of the goal will be very close to the vehicle, and after exceeding the curriculum step it will also gradually move away from the vehicle's initial position. After setting up the environment like this, the next stage of the experiment is to evaluate the reward functions. In the case of the first two reward functions (1), (2), during the time of nonterminal timesteps the vehicle is rewarded if it tries to maintain a constant speed near $20ms^{-1}$ and reduces the distance from the goal as much as possible. But in both cases, the vehicle tries to stay in the same place (velocity=0; Fig. 2) and collects the reward by making the episode longer.

It is interesting to note that for the third reward function (3), things started to improve. As a base value, -2.5 is added to the reward function, which always punishes at least a negative value of -0.5 for each nonterminal timestep. From Fig. 3, it is clear that even due to curriculum learning, though the environment is becoming harder, the mean episode length is decreasing and the mean episode reward is increasing at the same time. In the case of the fourth reward function (4), the results improve significantly when the intermediate rewards are set to zero. From Fig. 4, it can be noticed that the mean episode length converges to a particular range (~ 500) and also the mean episode reward has converged to a positive value (~ 2000), though the negative rewards for termination are -2000, this indicates that there is a performance boost here but it takes longer time to reach the goal.

The best average cumulative reward has been achieved by the fifth reward function (5), although it is very simple, but provides a very elegant solution to this problem. Here, the velocity constraint is very important for driving the vehicle on an uneven track and examining the direct control of the vehicle, it has been observed that at the maximum speed of $20ms^{-1}$, the vehicle can handle all bumpy or steep scenarios pretty smoothly. Multiplying the value by a negative constant factor of 0.01 adds a very little negative reward for each nonterminal state by keeping the velocity close to that constant value, and this small negative reward also forces the vehicle to reach the goal as early as possible. For better convergence and multiplication with that small constant factor, the maximum and minimum rewards have been set to +10 and -5 consecutively. From Fig. 5, the significance of the improvement is visible, and also in the case of evaluation (Fig. 6) steps, it successfully reached the goal 8/10 times.

Figure 2: The vehicle is not moving(left indicator in the image showing speed = 0).

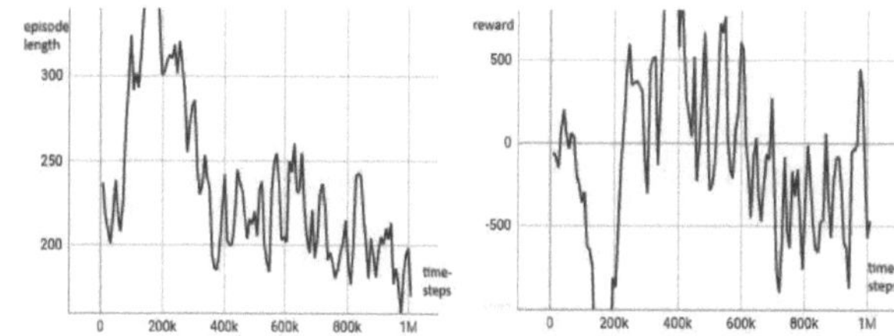

Figure 3: Left: Vehicle's episode length is decreasing, Right: The mean episode reward is also increasing at the same time for the third reward function (3).

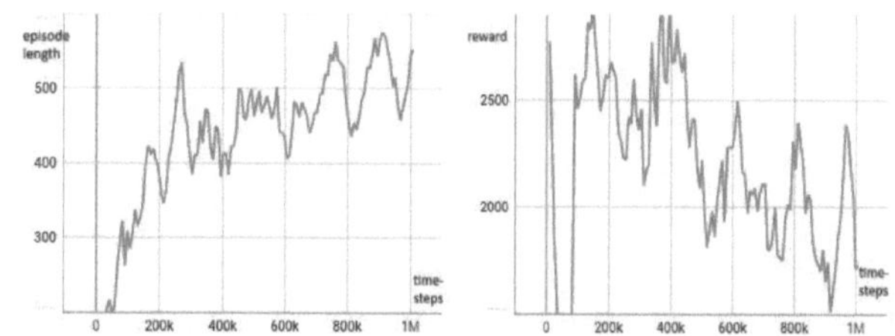

Figure 4: Left: Vehicle's episode length has converged near 500, Right: The mean episode reward has reached near 2000 for the fourth reward function (4).

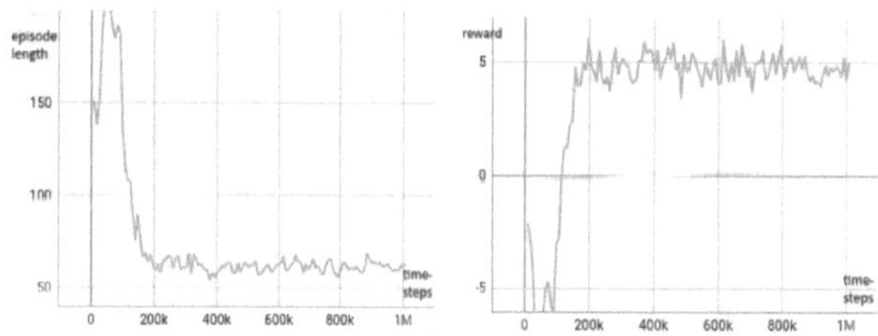

Figure 5: Left: Vehicle's average episode length has converged near 60, Right: The mean episode reward has also converged to 5.0 for the fifth reward function (5).

Figure 6: Vehicle successfully reaching the destinations.

4 Conclusion

To obtain the optimal reward function that is most suitable for the simulation environment, one has to start with a basic baseline model for learning the policy. Now after finding a well-performed reward function the future direction of this project will be to focus on visual input processing and then combine it with the vehicle state data, which could lead to a better policy. This model can then be deployed to a real mobile robot and have to check how it performs in a real-world environment.

Acknowledgement

The work has been carried out at the Institute for Electrical Engineering in Medicine, Universität zu Lübeck.

Authors' Statement

Conflict of interest: Authors state no conflict of interest.

5 References

[1] Hojin Lee, Junsung Kwon, Cheolhyeon Kwon, *Learning-based Uncertainty-aware Navigation in 3D Off-Road Terrains*, CoRR, vol. abs/2209.09177, 2022, https://doi.org/10.48550/arXiv.2209.09177

[2] David M. Bradley, David Silver, Boris Sofman, *Learning Rough-Terrain Autonomous Navigation*, 2009, https://api.semanticscholar.org/CorpusID:14305087,

[3] Kasun Weerakoon, Adarsh Jagan Sathyamoorthy, Utsav Patel, Dinesh Manocha, *TERP: Reliable Planning in Uneven Outdoor Environments using Deep Reinforcement Learning*, CoRR, vol. abs/2109.05120, 2021, https://arxiv.org/abs/2109.05120.

[4] Stepan Dergachev, Kirill Muravyev, Konstantin Yakovlev, *2.5D Mapping, Pathfinding and Path Following For Navigation Of A Differential Drive Robot In Uneven Terrain*, IFAC-PapersOnLine, vol. 38, pp. 80–85, 2022, https://www.sciencedirect.com/science/article/pii/S2405896323001441

[5] Simone Benatti, Aaron Young, Asher Elmquist, Jay Taves, Alessandro Tasora, Radu Serban, Dan Negrut, *End-to-end learning for off-road terrain navigation using the Chrono open-source simulation platform*, Multibody System Dynamics, vol. 54, pp. 399–414, 2022, https://doi.org/10.1007/s11044-022-09816-1.

[6] *Vehicle Physics*, https://vehiclephysics.com/

[7] Arthur Juliani, Vincent-Pierre Berges, Ervin Teng, Andrew Cohen, Jonathan Harper, Chris Elion, Chris Goy, Yuan Gao, Hunter Henry, Marwan Mattar, Danny Lange, *Unity: A general platform for intelligent agents*, CoRR, 2020, http://arxiv.org/abs/1809.02627.

[8] W. Bradley Knox, Alessandro Allievi, Holger Banzhaf, Felix Schmitt, and Peter Stone, *Reward (Mis)design for Autonomous Driving*, CoRR, vol. abs/2104.13906, 2023, https://arxiv.org/abs/2104.13906.

[9] J. Schulman, F. Wolski, P. Dhariwal, A. Radford, and O. Klimov, *Proximal Policy Optimization Algorithms*, CoRR, vol. abs/1707.06347, 2017, http://arxiv.org/abs/1707.06347 [last accessed on August 13, 2018].

[10] Y. Bengio, J. Louradour, R. Collobert, and J. Weston, *Curriculum Learning*, Proceedings of the 26th Annual International Conference on Machine Learning (ICML '09), Association for Computing Machinery, New York, NY, USA, pp. 41–48, 2009, https://doi.org/10.1145/1553374.1553380.

Obstacle Avoidance of MONSUN AUV Using SONAR

Suneesh Omanakuttan Sumesh Nivas [1]

[1] Robotics and Autonomous Systems, Universität zu Lübeck, suneesh.omanakuttansumeshnivas@student.uni-luebeck.de

Abstract

Underwater robotics has gained significant attention in various fields, including marine exploration, underwater inspections, and underwater infrastructure maintenance. One critical challenge in these applications is obstacle avoidance, as underwater environments often contain complex and dynamic obstacles. SONAR (Sound Navigation And Ranging) based sensing techniques offer a promising solution for obstacle detection and avoidance due to their ability to operate effectively in the underwater domain. Obstacle avoidance was achieved by employing a sonar positioned at the front of the underwater robot. The implementation of this principle utilized MONSUN underwater robots, Robot Operating System (ROS), and the programming language Python.

1   Introduction

Obstacle avoidance in Autonomous Underwater Vehicles (AUVs) using sonar technology is a crucial area of research. The primary objective of obstacle avoidance in AUVs is to enable safe navigation and prevent collisions with underwater structures, marine organisms, and other potential obstacles. To get inspiration for this project, we must first understand the current state-of-the-art approaches. The following approaches for obstacle avoidance were considered:

- An intelligent single-beam echo sounder designed for collision avoidance[1].

- Segmenting sonar images, monitoring underwater objects, and estimating motion[2].

- A control framework, drawing inspiration from neurobiology[3].

- An expert system that actively avoids obstacles in real time[4].

- An obstacle detection system for collision avoidance of AUV tested in a real environmet[5].

The MOVE project at the University of Lübeck, also known as the Monitoring of Vegetation and Water Quality in Lakes with Underwater Robot Swarms has the primary objective of observing the ecosystem of aquatic plants and animals to address issues caused by climate change, pollution, and overexploitation and restore the water bodies to their optimal ecological state. The automation of data collection and analysis in water bodies is achieved through the utilization of an autonomous underwater robot called MONSUN, as shown in Fig. 1. MONSUN's objective is to gather data on the characteristics of macrophytes and other relevant information, which will subsequently be processed according to

the assessment guidelines offline. The objectives of the paper is to incorporate obstacle avoidance techniques utilizing a pinger sonar altimeter and scanning sonar, both positioned at the front of the MONSUN micro AUV. By utilizing these sensors, MONSUN can detect and navigate around potential obstacles on its way.

Figure 1: The AUV MONSUN is fitted with a Blueview M900-2250 imaging sonar for inspection (bottom left), a GoPro HERO7 camera with an LED lamp (top), and a Tritech Micron DST scanning sonar for navigation (bottom right)[6].

Figure 2: Pinger Sonar from Blue Robotics (left)[7] Scanning sonar from Tritech (right)[8].

2 Material and Methods

2.1 MONSUN AUV

The inception of the MONSUN Autonomous Underwater Vehicle project commenced in 2009 at the Institute of Computer Engineering, University of Lübeck. The robot's architecture is scalable, facilitating its utilization within a coordinated swarm of independent AUVs. The fundamental design of the AUV is characterized by its robust and modular structure, enabling significant adaptability for a wide range of applications related to environmental monitoring and inspections. Its comparatively tiny size makes it easily deployable by one person. Six brushless motors compose the actuation system. Four of the six thrusters are positioned vertically and are utilized for attitude control and diving, while the other two are used for forward thrust, providing the AUV with differential kinematics. The Robot Operating System (ROS) is running on a Raspberry Pi [6, 9].

2.2 Pinger And Scanning Sonar

A Pinger Sonar Altimeter is a device used to measure the distance between the device and a solid object, typically the seabed or the bottom of a body of water. It operates based on the principle of echolocation, similar to how bats use sound waves to navigate their surroundings. By knowing the speed of sound in water and the time it takes for the echo to return, the pinger sonar altimeter can calculate the distance between the device and the object.

The scanning sonar system operates on the principle of acoustic waves and employs advanced scanning techniques to provide detailed images of the underwater environment. It uses a single rotating transducer that emits acoustic pulses in a fan-shaped beam pattern. Here, we use a Blue Robotics pinger sonar and a Tritech Micron scanning sonar to gauge the distance to submerged entities. Pinger sonar can operate with a 50-meter range, a 30-degree beam width, and the ability to function at depths of up to 300 meters. One of the key features of the Micron scanning sonar is its ability to provide a real time image of the underwater environment. The scanning sonar used for navigation emits an acoustic pulse extended by 3° horizontally and 35° vertically with a maximum range of 75 m and measures the intensity of the received echo. The sonar beam can be rotated mechanically, allowing for a 360-degree field of view. Pinger sonar from

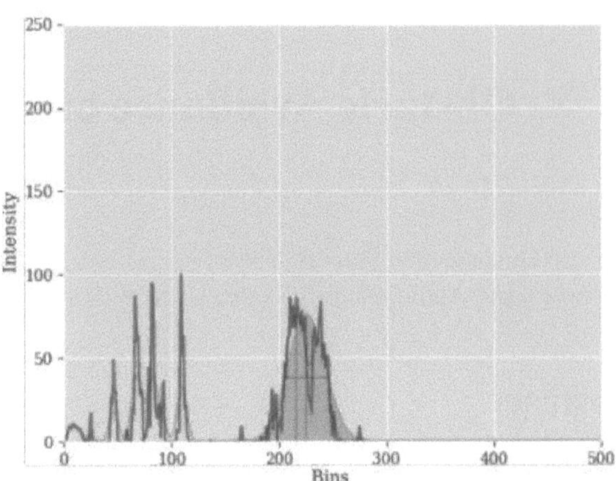

Figure 3: The scanning sonar's raw data, represented in blue, its filtered data, represented in red, and its peak detection with a bell curve, represented in green, are used to identify the most powerful echo indicating an obstacle[6].

Blue Robotics and scanning sonar from Tritech are depicted in Fig. 2. Fig. 3 illustrates how the distance to an obstacle is computed for a beam from the raw data.

2.3 Obstacle Avoidance Node

Creating an obstacle avoidance algorithm for an autonomous underwater robot using ROS and sonar sensors involves several key steps and components. This feature is programmed using Python and ROS scripting. Fig. 4 illustrates the software architecture with the obstacle avoidance node using pinger sonar. The obstacle avoidance node subscribes to the sonar node to acquire distance measurements and then communicates with the engine node (engine), which is already integrated within the MONSUN software. The engine node governs the actions of the robot's six thrusters: four vertical and two horizontal. It subscribes to linear and angular velocity inputs encompassing drive(x), dive (z), roll (x), pitch (y), and yaw (z). The heading_ctrl node is used to drive along a given compass direction and the depth_ctrl node to dive at a specified depth. The gps_navigation node allows to drive along a given path described by a waypoint list. For this, it subscribes to the GPS (Global Positioning System) data from the xsens node and publishes the course to the heading_ctrl node. This orchestration empowers the AUV to navigate around the detected obstacle effectively.

2.4 Avoid Behavior

This behavior involves mechanisms that allow the vehicle to swiftly respond to detected obstacles and navigate around them to ensure safe and effective operation. The following two simple algorithms describe the avoid behaviors for the MONSUN AUV using a pinger and a scanning sonar respectively.

In Algorithm 1, which is used by pinger sonar, firstly, MONSUN AUV will rotate in the right direction when the

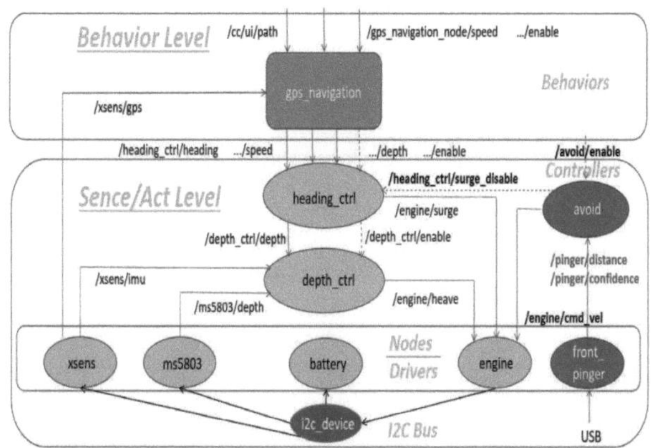

Figure 4: Software architecture of the obstacle avoidance node using pinger sonar.

detected distance is lower than the threshold distance and confidence greater than the threshold confidence. Then it will complete a 90-degree rotation. This can be ensured with the help of yaw angle from the IMU (Inertial Measurement Unit). After the 90-degree rotation towards the right, the AUV will again check the distance and confidence value. If it does not meet the condition, the AUV will rotate 180-degrees in the left direction. The mission will abort if the obstacle is present after the 180-degree rotation.

In Algorithm 2, the vector of segmented output from the scanning sonar is divided into three sectors. The algorithm initializes variables for counting the number of obstacles in each sector. It loops through each sector and increments the corresponding variable if the vector element is greater than 0 indicating an obstacle. The AUV can move without any problem when there are no obstacles in the second sector. If there are obstacles in the second sector, the AUV must navigate to another sector that is free of obstacles. Firstly, the algorithm checks if there are any obstacles in the second sector. If there are, it disables depth_control and enables heading_control. If there are free spaces in the first sector, it sets the final angle to -35-degrees. It publishes these values to the engine node and checks if the IMU angle matches the final angle. If there are no free spaces in the second sector and the first sector, the algorithm checks if there are any free spaces in the third sector. If there are, it follows a similar process as before, but with a final angle of 35-degrees. If there are no free spaces in either the second, third, or first sector, the algorithm assumes that there is a wall present in front of the vehicle and aborts the mission.

3   Results and Discussion

This project, which aims to obstacle avoidance of MON-SUN AUV using sonar, has progressed greatly since its inception. Various methods, ranging from basic to advanced obstacle avoidance techniques, were investigated, including improved algorithms. It turned out that simple reactive avoid algorithms as Algorithm 1 or 2 described in the

---

Algorithm 1 Algorithm for pinger sonar

Require: $distance\_threshold, x\_dir\_speed,$
$\quad z\_dir\_speed$
$\quad pinger\,sonar\,subs \leftarrow subscribe\,to\,pinger\,sonar\,node$
$\quad$ If distance from sonar $<=$ distance_ threshold AND
$\quad$ confidence $>=$ confidence_ threshold
$\qquad heading/ctrl \leftarrow disable$
$\qquad depth/ctrl \leftarrow enable$
$\qquad angular\_z \leftarrow z\_dir\_speed$
$\qquad linear\_x \leftarrow x\_dir\_speed$
$\qquad pub \leftarrow publish\,to\,engine\,node$
$\qquad check\,imu\_angle\,is\,update\,to\,90\,degree$
$\quad$ If distance from sonar $<=$ distance_ threshold AND
$\quad$ confidence $>=$ confidence_ threshold
$\qquad heading/ctrl \leftarrow disable$
$\qquad depth/ctrl \leftarrow enable$
$\qquad angular\_z \leftarrow -z\_dir\_speed$
$\qquad linear\_x \leftarrow x\_dir\_speed$
$\qquad pub \leftarrow publish\,to\,engine\,node$
$\qquad check\,imu\_angle\,is\,update\,to\,180\,degree$
$\quad$ If distance from sonar $<=$ distance_ threshold AND
$\quad$ confidence $>=$ confidence_
$\qquad mission \leftarrow abort$

Algorithm 2 Algorithm for scanning sonar

Require: x_dir_speed, z_dir_speed
$\quad scanning\,sonar\,subs \leftarrow$
$\quad subscribe\,to\,scanning\,sonar\,node$
Ensure: $Data\,from\,sonar\,filtered\,with\,median\,filter$
$\quad segmented\,with\,threshold\,segmentation$
$\quad$ For $k\,from\,1\,to\,length(Sec\_1)$
$\qquad$ If Sec_1[k]> 0
$\qquad\quad$ Sec_1_free $=$ Sec_1_free $+1$
$\quad$ For $l\,from\,1\,to\,length(Sec\_2)$
$\qquad$ If Sec_2[l]> 0
$\qquad\quad$ Sec_2_free $=$ Sec_2_free $+1$
$\quad$ For $m\,from\,1\,to\,length(Sec\_3)$
$\qquad$ If Sec_3[m]> 0
$\qquad\quad$ Sec_3_free $=$ Sec_3_free $+1$
$\quad$ If Sec_2_free> 0 $\qquad\qquad$ ▷ checking obstacle presence
$\qquad heading/ctrl \leftarrow disable$
$\qquad depth/ctrl \leftarrow enable$
$\qquad$ If Sec_1_free$== 0$
$\qquad\quad$ Final_Angle$= -35$
$\qquad\quad angular\_z \leftarrow z\_dir\_speed$
$\qquad\quad linear\_x \leftarrow x\_dir\_speed$
$\qquad\quad pub \leftarrow publish\,to\,engine\,node$
$\qquad\quad check\,imu\_angle\,is\,equal\,to\,final\_angle$
$\qquad$ Elif Sec_3_free$== 0$
$\qquad\quad$ Final_Angle$= 35$
$\qquad\quad angular\_z \leftarrow z\_dir\_speed$
$\qquad\quad linear\_x \leftarrow x\_dir\_speed$
$\qquad\quad pub \leftarrow publish\,to\,engine\,node$
$\qquad\quad check\,imu\_angle\,is\,equal\,to\,final\_angle$
$\qquad$ Else $\qquad\qquad$ ▷ wall present in front of vehicle
$\qquad\quad$ mission $\leftarrow abort$

---

previous chapter are sufficient in the context of the MOVE project, since here MONSUN has to drive along a transect in a straight line where usually no obstacles are present. In case of an unexpected obstacle, the robot should try to drive around it and then continue its path. If this is not possible, it shall simply stop and emerge. More sophisticated strategies like using maps and path planning are not required for MOVE project.

During MONSUN AUV's operation for the MOVE project, the primary purpose is to drive along a given waypoint list, often in a straight line, utilizing node gps_navigation, which uses heading_ctrl and depth_ctrl. A GPS buoy on the surface provides access to underwater GPS. In the absence of obstacles which is the most usual case, the waypoints on the waypoint list are visited one by one while driving in a straight line between them. If the sonar distance readings fall below the predetermined threshold, signifying the presence of an obstacle ahead, the obstacle avoidance node becomes active, which stops heading_ctrl and takes direct control of the engines to drive the robot around the obstacle. Finally, gps_navigation is taking over control again to resume driving to the next waypoint.

To prove the capability of the presented realization, we conducted simulations using dummy messages in different scenarios for pinger and scanning sonar. Different locations of the obstacles relative to the robot were investigated and checked if the algorithms reacts correctly to avoid them. Also, various parameter settings were studied. In these tests, the collision avoidance proved to work successfully.

When comparing the avoidance behavior of scanning and pinger sonar, scanning sonar provides a comprehensive coverage of the surrounding environment. This allows for a more complete and detailed perception of obstacles in all directions. Scanning sonar typically offers higher resolution compared to pinger sonar. This higher resolution allows for better discrimination of objects and a more accurate representation of the surroundings. Scanning sonar is more adaptable to different underwater environments, including complex terrains. It can effectively handle scenarios with irregular obstacles and varying depths. It can minimize blind spots, providing a more reliable perception of the surroundings. A drawback is, however, that it is much more expensive than the pinger sonar and consumes more energy. Further research including real application tests in water will show if the pinger sonar is sufficient for the MOVE project.

4 Conclusion and Future Work

Through the integration of sonar-based obstacle detection and avoidance mechanisms, AUVs can effectively navigate complex and challenging underwater environments while minimizing the risk of collisions and potential damage. This paper has presented two simple reactive avoid algorithms that are embedded in the ROS software of the MONSUN AUV using a pinger sonar or a scanning sonar respectively. The conducted tests utilizing ROS-based simulations have provided valuable insights into the functionality and performance of the implemented obstacle avoidance technique. Real in-water tests with MONSUN are planned for the future. Prospective research should focus on refining the algorithm's accuracy and robustness.

Acknowledgement

The work has been carried out at the Institute of Computer Engineering, Universität zu Lübeck, and supervised by Prof .Dr.-Ing. Erik Maehle.

Authors' Statement

Conflict of interest: Authors state no conflict of interest.

5 References

[1] P. Calado, R. Gomes, M. Nogueira, J. Cardoso, P. Teixeira, P. Sujit, and J. Sousa, "Obstacle avoidance using echo sounder sonar," in *OCEANS 2011 IEEE-Spain*. IEEE, 2011, pp. 1–6.

[2] Y. Petillot, I. T. Ruiz, and D. M. Lane, "Underwater vehicle obstacle avoidance and path planning using a multi-beam forward looking sonar," *IEEE journal of oceanic engineering*, vol. 26, no. 2, pp. 240–251, 2001.

[3] A. Guerrero-González, F. García-Córdova, and J. Gilabert, "A biologically inspired neural network for navigation with obstacle avoidance in autonomous underwater and surface vehicles," in *OCEANS 2011 IEEE-Spain*. IEEE, 2011, pp. 1–8.

[4] H. Zou, X. Bian, and Z. Chang, "A real-time obstacle avoidance method for auv using a multibeam forward looking sonar," *Robot*, vol. 29, no. 1, pp. 82–87, 2007.

[5] R. Kot, "Review of obstacle detection systems for collision avoidance of autonomous underwater vehicles tested in a real environment," *Electronics*, vol. 11, no. 21, p. 3615, 2022.

[6] E. Maehle, B. Meyer, C. Isokeit, and U. Behrje, "Monsun: a swarm auv for environmental monitoring and inspection," in *Autonomous Underwater Vehicles: Design and Practice*. IET, 2020, pp. 329–357.

[7] Ping echosounder sonar user manual. Accessed: January 23, 2024. [Online]. Available: https://bluerobotics.com/learn/ping-sonar-technical-guide/

[8] M. Zhou, R. Bachmayer, and B. de Young, "Initial performance analysis on underside iceberg profiling with autonomous underwater vehicle," in *2014 Oceans-St. John's*. IEEE, 2014, pp. 1–6.

[9] U. Behrje, C. Isokeit, B. Meyer, K. Ehlers, and E. Maehle, "Auv-based quay wall inspection using a scanning sonar-based wall following algorithm," in *OCEANS 2022-Chennai*. IEEE, 2022, pp. 1–9.

Leveraging complementary labels in breast cancer segmentation on mammography data: An in-depth analysis

Solveig Thrun [1], Stine Hansen [2], Robert Jenssen [2], Michael Kampffmeyer [2] and Maik Stille [3,4]

[1] Medical Engineering Science, Universität zu Lübeck, solveig.thrun@student.uni-luebeck.de

[2] Department of Physics and Technology, UiT The Arctic University of Norway, Tromsø, Norway, {s.hansen, robert.jenssen, michael.c.kampffmeyer}@uit.no

[3] Institute of Medical Enginnering, Universität zu Lübeck, stille@imt.uni-luebeck.de

[4] Fraunhofer IMTE, Research Institution for Individualized and Cell-Based Medical Engineering, Lübeck, Germany, maik.stille@imte.fraunhofer.de

Abstract

The inclusion of complementary, seemingly uninformative but easy to obtain labels has recently been linked to increased performance in deep-learning based segmentation. In this work, we provide an in-depth analysis of this phenomenon in the setting of breast cancer segmentation on mammography images to further investigate how to select these labels. The DeepLabV3 model was trained on different subsets of complementary labels to analyse which labels improve the cancer segmentation. Results demonstrate that leveraging complementary labels in breast cancer segmentation leads to a statistically significant improvement ($p < 0.05$) compared to a training without complementary labels. Classes with a high number of pixels improve the segmentation, whereas noisy labels or classes with few pixels peform worse.

1 Introduction

Breast cancer is the most diagnosed cancer in women and the leading cause of cancer deaths among women. In 2020, among all women worldwide, 24.5% of all new cancer diagnoses were breast cancer, and 15.5% of cancer deaths were caused by breast cancer [10]. Early diagnosis of breast cancer is crucial for effective treatment and a positive prognosis. Mammography is frequently used for breast cancer screening and diagnosis due to its high spatial resolution, cost efficiency, and low dose of radiation [1].

Automatic segmentation of mammograms helps physicians detect breast cancer and quantify the volume of the tumour in the breast, which is an important step in treatment planning since the dose for breast cancer treatment depends strongly on the size of the tumour [8]. Most of deep learning-based methods in semantic medical image segmentation rely on numerous images with pixel-wise annotations of the class of interest made by medical experts. However, in a recent study C. Matsoukas et al. [7] indicate that including complementary label classes (e.g. calcification, lymph nodes, and skin) can improve the segmentation of breast cancer. Nevertheless, the study does not address which complementary labels are helpful. In this work, we therefore analyse the influence of different complementary labels on the cancer segmentation.

Our main contribution is an in-depth analysis of complementary labels in breast cancer segmentation on mammography images. We trained a Deep Convolutional Neural Network, and performed various training scenarios where we subsequently added complementary labels to analyse which labels improve the cancer segmentation.

Our evaluation indicates that not all complementary labels are helpful. The results of our analysis showed that labels with a high pixel count have a positive influence on the segmentation result, and labels with few pixels or with noisy annotations have a negative influence.

2 Material and Methods

2.1 Dataset

The CSAW-S dataset [7] is a publicly available subset of the Swedish CSAW dataset [3], which is a large dataset with mammography images from breast cancer screenings at the Karolinska University Hospital, Stockholm, Sweden.

In total, the CSAW-S dataset contains 338 mammograms with annotations for semantic segmentation of both mediolateral oblique (MLO) and craniocaudal (CC) views from 172 women. Included in the dataset are annotations of 12 classes, which we will also refer to as labels in this paper: the cancer class, 10 complementary classes of breast anatomy (mammary gland, pectoral muscle, skin, thick vessels, non-mammary tissue, nipple, axillary lymph nodes, text, foreign objects, calcification) and the background class (see Fig. 1). The dataset is split into training, validation and test sets. The training set contains 263 images from 130 patients, the validation set contains 49 images from 20 patients, and the test set contains 26 images from 23 patients. For the training and validation sets, the cancer annotations are provided by a radiology expert (annotator1), and the complementary labels are provided by non-experts. Three

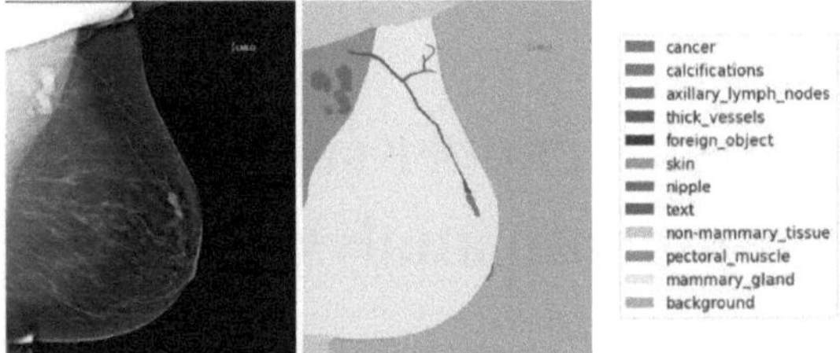

Figure 1: Example mammogram with corresponding ground truth segmentation map from the CSAW-S dataset.

	Class	Number of pixels	% of pixels
1	mammary gland	2 033 129 037	0.45757
-	background	1 571 497 258	0.35367
2	pectoral muscle	425 958 074	0.09586
-	cancer	104 745 737	0.02357
3	skin	77 695 137	0.01749
4	thick vessels	76 165 444	0.01714
5	non-mammary tissue	61 619 308	0.01387
6	nipple	43 618 537	0.00982
7	axillary lymph nodes	25 438 224	0.00573
8	text	13 781 930	0.00310
9	foreign object	8 923 528	0.00201
10	calcification	768 586	0.00017

Table 1: The classes in the training dataset are imbalanced. The numbers correspond to the x-axis in Figs 2 and 3. Cancer and background have no number because they are included in every training scenario.

radiology experts (annotator1, annotator2, annotator3) provided the cancer annotations for the test set, and two of them also provided complementary labels in the test set.

The complementary classes are of different size. For each class in the training dataset, we calculated the sum of all pixels across the training dataset and the percentage of all pixels (see Table 1).

2.2   Pre-Processing and Patch Extraction

The mammography images are provided as pre-processed 8-bit PNG files. For each image, each class label is provided as binary mask.

Due to memory limitations, we cropped the mammograms into patches of size 512x512. To extract patches from the mammograms and the corresponding ground truth segmentation, the procedure proposed by C. Matsoukas et al. [7] was followed. For every mammogram in the training set, 10 patches were extracted from each class (cancer, pectoral muscle, skin, thick vessels, non-mammary tissue, nipple, axillary lymph nodes, text, foreign objects, calcification) if the class was present in the mammogram. The 10 patches belong to different, randomly selected locations of each label. In total, there are 16950 image patches and corresponding segmentation patches to train the models.

2.3   Implementation Details

We used the PyTorch 2.0.1 deep learning framework [9] for all implementations. A DeepLabV3 model [2] with a ResNet50 [5] as backbone was trained to perform semantic segmentation of the patches because the segmentation prediction of the DeepLabV3 model is better when compared with other techniques [2]. The backbone was initialized with ImageNet [4] pretrained weights. For the DeeplabV3 model, the atrous rates are set to 6, 12 and 18. Following C. Matsoukas et al. [7], all BatchNorm layers were replaced with GroupNorm layers (num groups = 32). The Cross-Entropy loss without weights was selected as the loss function. We optimize the loss using the ADAM optimizer [6] with a learning rate of 1e-4 and a weight decay of 1e-6. If the validation Intersection over Union (IoU) for cancer does not improve within 10 epochs, the learning rate is reduced by a factor of 10. The number of epochs was set to 80, the batch size to 12, and the number of workers to 6. To prevent overfitting, data augmentation including RandomRotate90, ColorJitter(brightness = 0.2, contrast 0.2), Flip and Normalization was applied. Training takes 35 hours on a Nvidia RTX A6000 and Nvidia GeForce RTX 3090 GPU.

2.4   Experimental Setup

C. Matsoukas et al. [7] indicate that the segmentation of breast cancer in mammography images is improved when leveraging complementary labels for training. We provide more insight into the influence of different complementary labels on breast cancer segmentation in order to find out which complementary labels improve the segmentation.

Therefore, we defined one training setting without complementary labels (only the cancer label is used to train the network) and multiple training settings based on subsets of complementary labels. Due to the high computational cost of training the network, it is not feasible to test all possible combinations of complementary labels. Therefore, we chose the approach of adding the complementary classes in descending order of the number of pixels in the training dataset (see Table 1). In each training scenario, the number of complementary labels is increased by one, resulting in a total number of 11 training settings. For each setting, the model was trained four times and the results were averaged over the runs.

2.5   Evaluation Metrics

To quantitatively evaluate the performance of the trained models, we calculated the IoU, the sensitivity (S), and the precision (P) between the model predictions for the test set and the ground truth. The equations for calculating these evaluation measures are as follows:

$$IoU = \frac{TP}{TP + FP + FN}, \tag{1}$$

$$S = \frac{TP}{TP + FN}, \tag{2}$$

$$P = \frac{TP}{TP + FP}, \tag{3}$$

where TP denotes the number of true positive pixels, TN denotes the number of true negative pixels, FP denotes the number of false positive pixels, and FN denotes the number of false negative pixels.

3 Results and Discussion

We performed an in-depth analysis of complementary labels in breast cancer segmentation by training a DeepLabV3 model on different subsets of complementary labels.

Quantitative results: Figs. 2 and 3 show how the IoU, sensitivity, and precision for the cancer segmentation are influenced by the number of complementary labels. The mean value is plotted, and the standard deviation is shaded.

By looking at these graphs, it is visible that not all labels have a positive effect on the segmentation result. The performance drops significantly at 2 points: When the number of labels was increased from 3 to 4 (adding the label thick vessels), and when increasing the number of labels from 8 to 9 (adding the label foreign objects). The thick vessels have a relatively high number of pixels and one would assume that they improve the segmentation result and do not worsen it. But by looking at Fig. 1 it is apparent that these labels are quite noisy because not all vessels that are visible in the mammogram have been annotated by the expert. Moreover, foreign objects have few pixels and therefore probably reduce the performance.

We defined another training scenario where we included those labels that in the used order of labels had a positive influence on the IoU of cancer. This means, the classes mammary gland, pectoral muscle, skin, non-mammary tissue, nipple and calcification were used for the training. By comparing the results from the training based on these 6 labels with the other results, it is visible that the training is about as good as the training with 3 complementary labels (mammary gland, pectoral muscle and skin).

To indicate statistically significant improvements by using complementary labels in breast cancer segmentation, one-sided T-tests are performed to compare the mean IoU, mean sensitivity and mean precision between the training without complementary labels and the training based on the 6 good complementary labels. Results indicate that there is a statistically significant improvement by leveraging these 6 complementary labels in breast cancer segmentation ($p < 0.05$). We categorise the labels into three different categories: category A are classes that improve the segmentation, category B are classes which worsen the segmentation and category C are classes that only improve the segmentation if we also use classes from category B. Looking at the labels of the various training settings, it can be concluded that the labels mammary gland, pectoral muscle and skin improve the performance (they belong to category A), the labels thick vessels, axillary lymph nodes, text and foreign objects reduce the performance (they belong to category B) and the labels non-mammary tissue, nipple and calcification belong to category C. A closer look at Table 1 shows that labels with a high number of pixels have a positive influence (mammary gland, pectoral muscle and skin). Labels with a low pixel

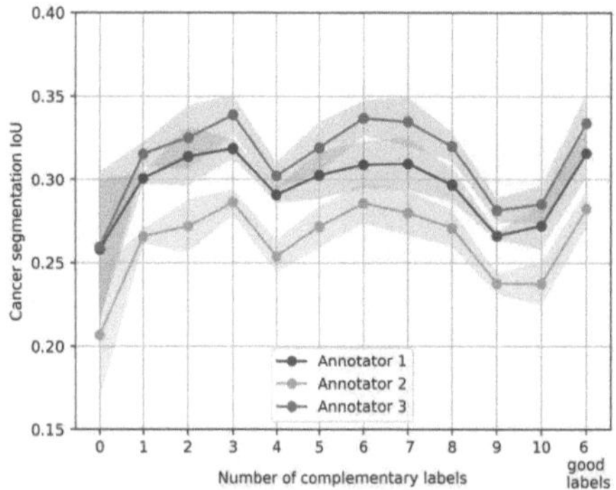

Figure 2: In each training setting we increased the number of complementary labels by one and analysed how the IoU for cancer changed. "6 good labels" corresponds to the training setting based on the complementary classes that improve the cancer IoU (mammary gland, pectoral muscle, skin, non-mammary tissue, nipple and calcification).

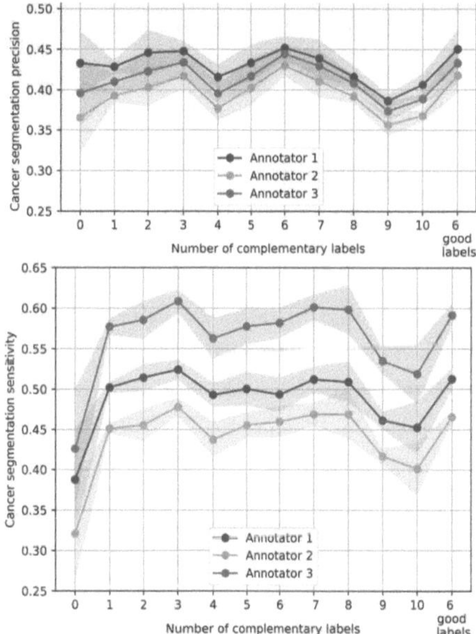

Figure 3: The effect of the number of complementary labels on the precision and the sensitivity is shown.

count (text, foreign object and axillary lymph nodes) reduce the segmentation performance. Among these, only the label calcification has no negative influence, even though it has the lowest number of pixels. Probably because they are easy to distinguish from the other anatomy tissue, as they are very bright on the mammogram and have a clear edge. It can also be assumed that some complementary labels are more helpful than others. When we only use one complementary label (mammary gland), the metrics increase quite a lot. The tumour is mainly located in the mammary gland, which suggests that this class is very helpful in localizing the tumour in the right location of the patches.

Qualitative results: Fig. 4 shows that the cancer segmentation looks very similar for the training based on 3 labels and the training based on the 6 good complementary labels. However, the rate of FP cancer segmentation could be slightly reduced in the training based on 6 good

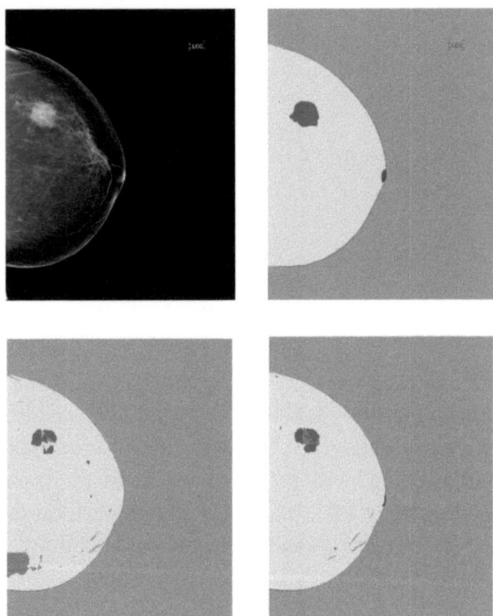

Figure 4: Top left image: one mammogram from the test set. Top right image: corresponding ground truth segmentation from annotator 1. The segmentation were reconstructed from the segmented patches (bottom left image: model trained on 3 complementary labels, bottom right image: model trained on the 6 good complementary labels).

labels. The visual analysis also showed that the segmentation of complementary classes is better in the training setting with 6 good labels. By looking at further segmentation results, it was visible that for some mammograms, the tumour was labelled as axillary lymph node. Since the axillary lymph nodes are also brighter on the mammogram, the network most likely has problems to distinguish between these classes.

4   Conclusion

In this work, we proposed an in-depth analysis of complementary labels in breast cancer segmentation on mammography images. We analysed several training settings with a different number of complementary labels. Results demonstrated that leveraging complementary labels increases segmentation performance compared to training without complementary labels. But not all labels have a positive influence on breast cancer segmentation. Labels such as foreign objects and thick vessels worsen the segmentation, whereas the labels mammary gland and pectoral muscle improve the segmentation. Thus, it can be concluded that the choice of complementary labels is important. Labels with a small number of pixels or noisy labels reduce the cancer segmentation performance, whereas labels with a high number of pixels have a positive influence on the segmentation. This analysis is also helpful when creating new datasets. The thick vessels for example reduce the performance and are time-consuming to annotate, which is why they do not need to be annotated by experts for future datasets.

One limitation of our chosen approach is that there could be an interdependence of label contributions. In each training

scenario, we add a label, and it may be that the positive or negative contribution of the newly added label depend on the other labels included before, and the newly added label alone might have a different influence on the segmentation performance. In future work, it would be interesting to explore feature selection methods, e.g. forward selection as a new strategy to combine the different classes.

Acknowledgement

The work has been carried out at the UiT Machine Learning Group, Department of Physics and Technology, UiT The Arctic University of Norway, Tromsø, Norway and supervised by Fraunhofer IMTE, Research Institution for Individualized and Cell-Based Medical Engineering, Lübeck.

Authors' Statement

Conflict of interest: Authors state no conflict of interest.

5   References

[1] A. Bhushan, A. Gonsalves and J. U. Menon, *Current State of Breast Cancer Diagnosis, Treatment, and Theranostics*, Pharmaceutics, vol. 13, no. 5, 2021.

[2] L.-C. Chen, G. Papandreou, F. Schroff and H. Adam, *Rethinking Atrous Convolution for Semantic Image Segmentation*, arXiv, arXiv: 1706.05587, 2017.

[3] K. Dembrower, P. Lindholm and F. Strand, *A Multimillion Mammography Image Dataset and Population-Based Screening Cohort for the Training and Evaluation of Deep Neural Networks —the Cohort of Screen-Aged Women (CSAW)*. In: Journal of Digital Imaging, vol. 33, no. 2, 2019.

[4] J. Deng, W. Dong, R. Socher, L.-J. Li, K. Li and L. Fei-Fei, *ImageNet: A large-scale hierarchical image database*, In: CVPR, 2009.

[5] K. He, X. Zhang, S. Ren and J. Sun, *Deep Residual Learning for Image Recognition*, In: CVPR, 2016.

[6] D. P. Kingma and J. L. Ba *Adam: A method for stochastic optimization*. In: ICLR, 2015

[7] C. Matsoukas et al., *Adding seemingly uninformative labels helps in low data regimes*. In: ICML, 2020

[8] E. Michael, H. Ma, H. Li, F. Kulwa and J. Li, *Breast Cancer Segmentation Methods: Current Status and Future Potentials*, BioMed Research International, vol. 2021, 2021.

[9] A. Paszke et al., *PyTorch: An Imperative Style, High-Performance Deep Learning Library*, In: NeurIPS, 2019.

[10] H. Sung et al., *Global Cancer Statistics 2020: GLOBOCAN Estimates of Incidence and Mortality Worldwide for 36 Cancers in 185 Countries*, CA: A Cancer Journal for Clinicians, vol. 71, no. 3, 2021.

Efficient Processing of Multi-Channel sEMG With Neural Networks on Mobile Hardware

Benedikt Stepanek [1], and Lukas Boudnik [2]
[1] Medical Microtechnology, Universität zu Lübeck, b.stepanek@student.uni-luebeck.de
[2] Fraunhofer Research Institution for Individualized and Cell-Based Medical Engineering IMTE,
lukas.boudnik@imte.fraunhofer.de

Abstract

Measuring a patient's natural breathing effort is crucial for mechanical ventilation. Recent research has demonstrated that surface electromyography is an appropriate non-invasive method for this purpose. To remove the overlying electrocardiogram, artificial intelligence (AI) methods can be used. In this study, a dry electrode thorax belt, connected to a low-power mobile device, was used to investigate the feasibility of implementing a temporal convolutional network to continuously measure and separate the signals. Signal processing was implemented fast enough for continuous processing through vectorization of arithmetic operations. By optimizing the model with TensorRT and quantizing its weights to INT8, inference was accelerated by a factor of 4. The findings of this study demonstrate that AI methods can be used for continuous signal processing with high precision while consuming little power.

1 Introduction

Mechanical ventilation maintains a vital function of the human body in seriously injured and sedated patients. However, it poses a major technical challenge as the activity of the ventilator must closely match the patient's natural spontaneous breathing if it is still functioning. Mechanical ventilation can permanently damage the lungs due to excessive air pressure or volume, as well as the diaphragm, which can suffer muscle atrophy and structural damage to the muscle fibers [1].
To avoid such damage, the patient's natural breathing activity can be measured. Traditional methods such as measuring air flow or esophageal pressure are invasive and unreliable [2], [3]. Current research therefore focuses on directly measuring the activity of the respiratory muscles. Surface electromyography (sEMG) is a non-invasive technique that can be used to reliably measure respiratory activity [3], [4], while putting the patient at low risk. After recording the muscle activity signals, various signal processing methods can be applied. In addition to traditional filters that reduce noise and common modes from the mains, it is necessary to remove the electrocardiogram (ECG) from the signal. The activity of the heart muscle is significantly stronger and overlays the signals of the respiratory muscles [5].
Since medical devices require reliable data processing, [2] and [5] developed a flexible sEMG thorax belt with 32 leads to investigate whether combining multiple channels can reduce noise and improve signal quality. In [6], a *temporal convolutional network* (TCN) was used to separate the ECG from the sEMG of the diaphragm.
The present research aims to combine these two studies and investigate whether it is possible to efficiently implement neural networks like TCNs on low power mobile hardware

for online signal processing and appropriate result visualization. This research aims to

- establish a connection between the sEMG belt and an Nvidia® Jetson™ system-on-chip,
- efficiently implement an online signal processing chain,
- quantize the model's weight to the INT8 datatype, and
- evaluate the performance of different AI inference frameworks.

2 Materials and Methods

In this section, the structure of the measurement system, the recording and pre-processing of the signals, and the implementation of the inference frameworks is explained. The measurement system includes a dry electrode thorax belt with 64 round gold-plated electrodes, four A/D converters, a reference driver circuit for common mode rejection, a power supply circuit, a microcontroller, and an Ethernet connector kit. A Jetson Orin Nano 8 GB system-on-chip (Nvidia Corporation, Santa Clara, CA, USA) serves as the primary system for AI-based signal processing and visualization. This device combines a low-power mobile processor with an Nvidia graphics card and 8 GB of shared memory. It consumes only 15 W of power, making it ideal for running AI models directly on end devices, also known as *Edge AI*. The two systems are connected via an Ethernet connection. To ensure electrical safety, the devices are galvanically separated by an Ethernet isolator (MI 1005 from Baaske Medical, Lübbecke, Germany).

Signal Transmission and Reconstruction The microcontroller continuously sends 10 samples for 36 channels each, including the 32 EMG channels and four status channels as

User Datagram Protocol (UDP) packets. The system samples the electrodes at a rate of 4000 Hz, resulting in 400 packets being sent per second or one packet every 2.5 ms. To prevent the graphical user interface from being blocked by accepting such a large number of packets, the application was divided into three threads. The main thread initializes the application and processes input to the graphical user interface (GUI). The second thread opens a UDP socket, continuously receives packets, and places them in a thread-safe queue. As one packet contains too few samples for further processing, the third thread waits until the data queue contains an adjustable number of $N = 100$ packets. These packets are then processed and forwarded to the GUI for visualization. Given that 400 packets arrive per second, this results in a 250 ms window to process these 100 packets before new data arrives.

To begin signal processing, the 24-bit samples must first be reconstructed from the UDP packets. These are 1085 bytes in size and include a 5 byte identifier followed by the samples. The signal processing software developed by [2] and [5] was implemented in MATLAB (The Mathworks, Natick, MA, USA) for offline post-processing. This means that the entire signal is recorded and then analyzed after the recording has ended. In this study, however, the system shall continuously measure, process and visualize signals. Python was chosen as the programming language for its ease of integration with various machine learning frameworks. To minimize signal processing latency, vectorization was used extensively in the NumPy and SciPy computations to reconstruct and filter the signals.

Preprocessing Using Conventional Methods   The signal processing chain consists of six stages:

1. Aggregate Packets   4. Filter Signal
2. Remove Status Channels   5. Check Plausibility
3. Downsample Signal   6. Detrend Signal

Butterworth filters can attenuate frequencies more efficiently with increasing order. However, with increasing filter order, more input data is required to obtain a valid result. Thus, as previously described, several packets are bundled. Then, the four status channels are removed as they do not provide any relevant information. The TCN was trained using measurements with a sample rate of 1000 Hz. Since the measurement system records with 4000 Hz, the signal has to be downsampled by a factor of 4. To avoid resulting artifacts, the signal is zero-phase filtered with a Chebyshev type 1 IIR filter ($N = 8$, $\varepsilon = 0.05$, $\omega_0 = 0.2$) beforehand [7].

Since the expected frequency components of the EMG and the ECG are unlikely below 0.5 Hz, the filter pipeline begins with a high-pass filter ($N = 5$, $f_c = 0.5$ Hz) to reduce baseline wandering, as suggested by [3] and [8]. The remaining filter steps are based on [2]. The 50 Hz noise from the mains and its harmonics is filtered using three IIR notch filters for 50 Hz, 100 Hz, and 150 Hz. Unwanted high-frequency components, that also do not occur in this application, are then removed with a low-pass filter ($N = 5$, $f_c = 350$ Hz). Once the signal has been filtered, a plausibility check is performed. The electrodes of the belt do not always lie perfectly on the skin surface and might even float. To detect this, an approach based on root-mean-square (RMS) error was implemented by [2] to set the channel to zero if the RMS of one electrode is greater than twice the mean RMS over all electrodes. Ultimately, the signal has to be detrended as it drifts significantly. This is removed by subtracting a linear least-squares fit of the data from the data itself.

Implementation of the Neural Network   In [6], a TCN was trained to remove the sEMG of the diaphragm from the Einthoven II lead of the ECG. Here, a modified version of the network was trained with a sample rate of 1000 Hz and chest wall leads were included in the ECG training data. This closely resembles an actual application with the dry electrode belt. The network was implemented by [6] using PyTorch (The Linux Foundation, San Francisco, CA, USA), which therefore was also used as a reference.

PyTorch can perform mathematical operations for neural networks on multiple hardware platforms. It also supports different floating point data types for network weights. However, some functions have limited use and may not be implemented for all data types or computing units. For example, the network used in this work cannot be run with FP16 weights when the CPU backend is used. Additionally, PyTorch offers experimental support for INT8 quantization. Quantizing the model weights to the INT8 data type reduces the model size and can speed up calculations if supported by the computing hardware. But at the time of writing, only the CPU can be used for that. PyTorch does not support INT8 quantization on GPUs.

Alongside PyTorch, the *Open Neural Network Exchange* (ONNX) format and its promising runtime are used. Many deep learning frameworks can export their models to this format. The ONNX runtime (Microsoft, Redmond, WA, USA) offers execution providers for various hardware platforms and operating systems. This allows for the efficient execution of AI models on mobile hardware as well as embedded systems. Among these execution providers, the Nvidia TensorRT execution provider is of particular interest for the present work. TensorRT optimizes models in a thorough process for the underlying hardware to optimize inference speed, throughput, and power consumption. This is especially beneficial for low-power mobile devices.

Running ONNX Models with TensorRT   To export a PyTorch model to the ONNX format, the model must be prepared for export. The PyTorch documentation lists several unsupported operations, such as negative tensor indices, that must be modified to make them exportable. The export can then be performed using a test tensor. If a model allows input tensors with variable sizes, the ONNX export with PyTorch 1.11 and older fixes the size because not all execution providers of the ONNX runtime support dynamic axes. Starting from PyTorch 1.12, this behavior changed, requiring the model to be sanitized with the polygraphy tool of TensorRT before it can be used. To avoid this additional preprocessing step, PyTorch 1.11 was used in this work.

The optimization process of TensorRT can take anywhere from a few minutes to several hours, depending on the hardware and the model size. Unfortunately, running this process on a high-performance graphics card and then transferring the optimized model to a Jetson device is not supported. Furthermore, the process must be repeated whenever the configuration changes. The TensorRT execution provider of the ONNX runtime automatically triggers this process. Note that the execution provider does not save or cache the optimized model by default. This must be requested explicitly.

Quantizing Model Weights to INT8   Several methods exist to quantize model weights to the INT8 datatype. The major challenge is that INT8 has a greatly reduced value range compared to conventional floating point data types, leading to inaccuracies during inference. While *quantization-aware training* typically yields the best results in terms of accuracy, it requires a modified training process. In this work, a static *post-training quantization* was used instead. This means that all weights are converted using sample data prior to inference. While the theoretical principles of quantization may seem simple, practical implementation can be challenging due to the need to consider quality metrics to prevent result degradation. However, the ONNX runtime takes care of this.

The process involves creating a calibration table for TensorRT. This cannot be done with TensorRT itself. Instead, the Nvidia CUDA execution provider is used to generate this table using a one-minute EMG measurement as a reference data set. The TensorRT execution provider can then be set to INT8 mode and use the calibration table for future inferences. It is important to note that no new ONNX model with INT8 weights will be generated. Instead, the original model will be loaded along with the calibration table. The execution provider will then apply the table to the model weights just in time. While this has the advantage that one model file can be used with all weight precisions, it has the disadvantage that the quantized INT8 weights cannot be extracted and viewed easily.

3   Results and Discussion

In this section, the results of experiments are presented, in which the speed of the AI inference frameworks and the accuracy of the results were determined. Automated performance tests were conducted to compare PyTorch with the ONNX runtime (henceforth referred as ORT). The average inference time on the CPU and with CUDA was averaged over 10 runs each with PyTorch and the ORT, as well as with TensorRT only with the ORT. The tests ran using various artificially enlarged datasets with up to 96 channels with 1000 samples each, with both FP16 and FP32 weight precision. Larger datasets could not be processed due to the limited amount of memory.

As previously mentioned, TensorRT does not support dynamic axes. Therefore, an ONNX model had to be exported for each configuration and then converted by TensorRT. As this process is already exceptionally time-intensive, it was decided not to quantize the weights into the INT8 datatype for each model. Instead, only two models were quantized: One with support for 32 channels with 250 samples, resembling the use case with the dry electrode belt, and another one

with 64 channels with 1000 samples. Larger input caused TensorRT to skip certain optimization strategies due to insufficient memory.

Results of the Performance Tests   Figure 1 shows the measurements with FP32 weights and a batch size of 1. As expected, PyTorch's CPU runtime is slow, taking 538 ms to process one channel with 20 samples. This is unsuitable for the present use case as only 250 ms are available to process 32 channels with more samples.

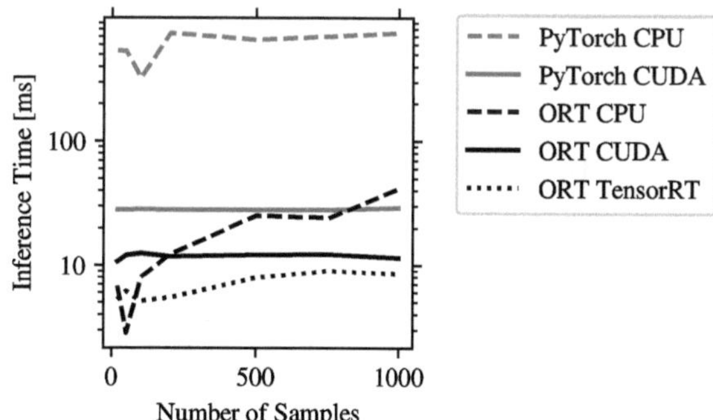

Figure 1: Inference time with FP32 weigths and one data batch, measured with different execution providers.

In contrast, the ORT's CPU execution provider is much more efficient, requiring only about 7 ms for the same amount of data. The inference time with this execution provider is acceptable for data volumes of up to four channels with 1000 samples each. However, it increases rapidly beyond that point, making it again unsuitable. Despite this, even with the largest data set measured, execution is still approximately ten times faster than with PyTorch.

Regardless of the configuration, PyTorch with CUDA, however, is up to 70 ms faster than the ORT with both the CUDA and the TensorRT execution providers. The efficiency of PyTorch with CUDA is especially notable when measuring warm-up time. PyTorch takes between 3 and 4 seconds, regardless of the model size. TensorRT even takes a little less with only 1 to 2 seconds. However, the ORT's CUDA execution provider requires up to 20 seconds. The ORT only seems to be faster when processing very small amounts of data. The recordings show a clear trend that PyTorch consistently takes at least 28 ms, while there is no apparent lower limit for the ORT.

Quantizing to INT8 proved to be successful. However, on the Nvidia Jetson, the difference is only marginally measurable. Both TensorRT with INT8 weights and PyTorch with FP16 weights run with an average execution time of 30 ms equally fast on the small model. On the larger model, TensorRT with INT8 weights only takes around 89 ms, while PyTorch requires 93 ms with FP16 weights and 116 ms with FP32 weights. Although the difference is not as significant as anticipated, it seems to increase with larger data volumes. Additional measurements that were conducted to verify the accuracy of the results, confirm the efficiency of using INT8 weights.

Calculation of the Accuracy of Results   To evaluate accuracy, [6] developed an evaluation script that calculates the improvement in signal-to-noise ratio (SNRI) and root mean square error (RMSE) for both, the EMG and the ECG. The program artificially mixed EMG and ECG measurements and ran inference on over 6200 measurements requiring up to 18 GB of memory. Thus, the tests were performed on a Nvidia RTX A6000. The tests, the conversion process to TensorRT, and quantization to INT8 weights were carried out under the same conditions as on the Jetson device.

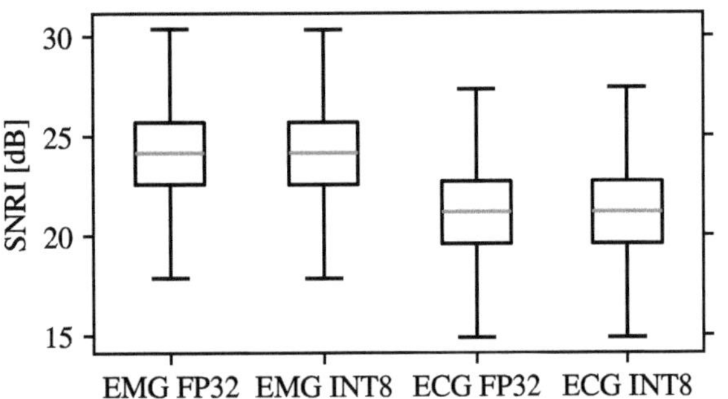

Figure 2: Average improvement in signal-to-noise ratio with different model weights.

Figure 2 shows the average SNRI with different model weights. The results indicate that the lower resolution data types do not affect the accuracy of the result. With the FP32 and FP16 weights as well as with INT8 weights, an average SNRI of 24 dB for the EMG and 21 dB for the ECG could be achieved. An average RMSE of 0.011 mV was calculated over all samples. These results are even slightly better than those reported by [6], which can be attributed to the addition of the chest wall leads for the ECG and the higher sampling rate.

Due to the graphic card's higher performance, coupled with a power consumption of up to 300 W, a significant decrease in inference speed was expected. Processing 10,000 samples with PyTorch and FP32 weights took 9 ms. Surprisingly, using FP16 weights increased the time to almost 12 ms. The inference with INT8 weights with the ORT's TensorRT execution provider, however, took only 2 ms.

4 Summary and Outlook

This research demonstrated the viability of implementing AI models such as temporal convolutional networks on low-power mobile hardware to process measurements continuously and with minimal delay. As the signal processing with the TCN takes only around 30 ms, the 250 ms window available until new data arrives is sufficient for reconstructing, filtering, and processing incoming data using AI on an Nvidia Jetson. The experiments show that more complex measurement systems with more leads can be used as well. The ONNX format and its associated runtime facilitate the porting of neural networks to various hardware platforms and enable efficient execution on mobile devices. As the measurements show, an efficient implementation of mathematical operations for neural networks enables fast inference even on a CPU. Inference can also be significantly accelerated by quantizing model weights to lower resolution integer datatypes, as done here with TensorRT. Future research

should focus on implementing computational operations for neural networks on a low level close to the hardware. Additionally, further research could explore the quantization of model weights to INT8 on non-Nvidia hardware. It may also be worthwhile to investigate the use of even lower-resolution data types.

Acknowledgements

This work was carried out at the Fraunhofer Research Institution for Individualized and Cell-Based Medical Engineering IMTE, Lübeck and supervised by Prof. Dr. Philipp Rostalski, Institute for Electrical Engineering in Medicine, Universität zu Lübeck. The authors would like to thank Felix Vollmer for his support and technical advice.

Authors' Statement

Conflict of interest: Authors state no conflict of interest.

5 References

[1] U. Hintzenstern and T. Bein, *Praxisbuch Beatmung.* Urban & Fischer, Munich, 2019.

[2] R. Kusche, J. Graßhoff, A. Oltmann, and P. Rostalski, *A multichannel EMG system for spatial measurement of diaphragm activities.* In: IEEE Sensors Journal, vol. 22, no. 23, pp. 23393–23402, Dec. 2022.

[3] G. Bellani et al., *Measurement of diaphragmatic electrical activity by surface electromyography in intubated subjects and its relationship with inspiratory effort.* In: Respiratory Care, vol. 63, no. 11, pp. 1341–1349, Nov. 2018.

[4] J. Graßhoff et al., *Surface EMG-based quantification of inspiratory effort: a quantitative comparison with P_{es}.* In: Critical Care, vol. 25, no. 1, Dec. 2021.

[5] R. Kusche, J. Graßhoff, A. Oltmann, L. Boudnik, and P. Rostalski, *A robust Multi-Channel EMG system for lower back and abdominal muscles training.* In: Current Directions in Biomedical Engineering, vol. 7, no. 2, pp. 159–162, Oct. 2021.

[6] L. Boudnik, J. Graßhoff, F. Vollmer, and P. Rostalski, *Muscle Artifact Removal in Single-Channel Electrocardiograms using Temporal Convolutional Networks.* In: 2022 44th Annual International Conference of the IEEE Engineering in Medicine & Biology Society (EMBC), Jul. 2022.

[7] P. Virtanen et al., *SciPy 1.0: fundamental algorithms for scientific computing in Python.* In: Nature Methods, vol. 17, no. 3, pp. 261–272, Feb. 2020.

[8] P. Kligfield et al., *Recommendations for the standardization and interpretation of the electrocardiogram.* In: Circulation, vol. 115, no. 10, pp. 1306–1324, Mar. 2007.

Using Computer Vision for Inventory Management

Eike Schultz[1]

[1] Institute of Computer Engineering, Universität zu Lübeck, eikelukas.schultz@student.uni-luebeck.de

Abstract

Inventory management in retail stores is traditionally done by synchronizing the inventory, which consists of counting the number of items and comparing them to a list of expected stock. The project described in this paper was proposed to partially automate this task with the help of computer vision software. For this, an image recognition pipeline was developed that can interpret a photo of a display case and automatically register visible items in a central databank. This pipeline can be accessed by users via a web application built for it, which proved to reduce the time needed for an inventory synchronization by up to 50% compared to existing processes in benchmark and live-environment tests.

1 Introduction

The Amor Group GmbH is a jewelry wholesaler and retail chain that operates sales floors across the globe. In order to allow to accurately plan operations and to prevent losses, the inventory of these sales floors needs to be managed uniformly in all stores, which is a non-trivial challenge in multinational networks [1]. The goal of the Computer Vision Synchronization project is to replace manual counting and bookkeeping steps with a tool that allows creation of digital records of physical stocks with minimal organizational overhead. This use of computer vision does not stem from practical necessity, but from exploring novel approaches. While comparable concepts have been proposed [4], [5], no fully-developed solutions were found that could be feasibly adopted.

The project development was undertaken over the course of two years in three phases. The first phase saw the development of a working prototype of a service recognizing merchandise displays from images, which constitutes the first two modules of the image recognition pipeline described below. The second and third phases iterated on this prototype to add more features and refine the underlying process, adding the other two modules as well as the overall user-facing side of the web application.

This paper will first highlight the use case for this new process and detail some of the deciding considerations. It will then describe the image recognition pipeline and its components, followed by a short explanation of the web application used by the sales floor operators. Finally, the results and performance of the final iteration of the process will be outlined together with research opportunities that became apparent during development.

2 Use Case and Considerations

The Computer Vision Synchronization project has the goal of optimizing conventional inventory management processes with image recognition-based automation. In jewelry retail, inventory is predominantly both shallow and front-facing, meaning that most items are stored visible to the public and only very limited stock of each item is available on location. Following this, the deciding factor on how items are stored is the presentation to potential customers, and logistical considerations are secondary. This means that automation-friendly practices like standardized boxes and prominently placed, machine-readable barcodes are not possible, which in turn leads to most inventory management still being conducted manually.

The two reference processes for the project, which are both still in use at Amor in various locations, were an electronic and a paper-based inventory system. The electronic process called "MAI Synchronization" is only viable for a limited number of locations, as it needs specialized tablets and handheld scanners. These cannot be issued to all locations due to logistical constraints especially common in international locations. The paper process can be conducted at all locations, but is known to be error-prone and very labor-intensive, as it relies on multiple manual counting and data processing steps.

The use case for the web application mostly consists of locations that rely on the paper process and cannot feasibly make use of the MAI Synchronization system. Because of this, the application cannot require any dedicated hardware and needs to be relatively platform-agnostic. This suggests a web-based application, as the vast majority of modern handheld devices can be expected to come with a web browser to run the application. As image recognition is too computationally intensive to be conducted on a hardware-agnostic web application, it requires a near-constant high speed internet connection in order to send the necessary image data to a central server. To account for this, a fall-back mode of operation is needed, in which products can be registered as present or missing without the need to transmit images. While an internet connection is still required to access the inventory data, this secondary mode greatly reduces the minimum requirements to make use of the application.

3   Image Recognition Pipeline

The image recognition of trays and products is the core component of the process. It is organized as a pipeline with multiple modules that separately edit and analyze the image sent from the web client to the server. This allows for the fine-tuning and re-training of each module on its own, adopting to new display cases, tray layouts, or product types.

It can be divided into four parts: Image Warping, Tray Recognition, Product Recognition and Synchronization. The first three parts are built around deep neural networks trained for their individual tasks. Each module also includes automated image manipulation algorithms that preprocesses the image for the neural network. The fourth part consists of data transfers to log the recognized information in the central system and display the information to the operator.

Between steps, information from the central system is used in tandem with the previous classification to restrict the search space of possible results. By reducing the choices available to the next module, for example only allowing classification from trays physically present in a location, the classification rate can be significantly increased.

After passing the pipeline, the augmented image is saved to the archive data bank. The entire process takes between 3 and 10 seconds, with a large portion of the delay being attributed to data transmission between the mobile device and the server.

3.1   Image Warper

In the first module, the image is normalized to only show the relevant tray in a standardized format. If there are multiple trays in the photo, the centermost tray in the image gets recognized as the one to focus on. By finding the four corner points of the tray, it can then be extracted as a quadrilateral. This is aided by the assumption that all trays are square in form and visually distinct from the surrounding display cabinet.

The extracted quadrilateral is then stretched to a standardized square format using the recognized corner points as anchors. This is done by using a conventional image manipulation algorithm. In this new form factor, the image can get segmented and an overlay of the stored tray layouts and position numbers can be superimposed on it for the user.

3.2   Tray Recognition

Since there is only a relatively small number of different layouts for a large number of individual trays, the tray

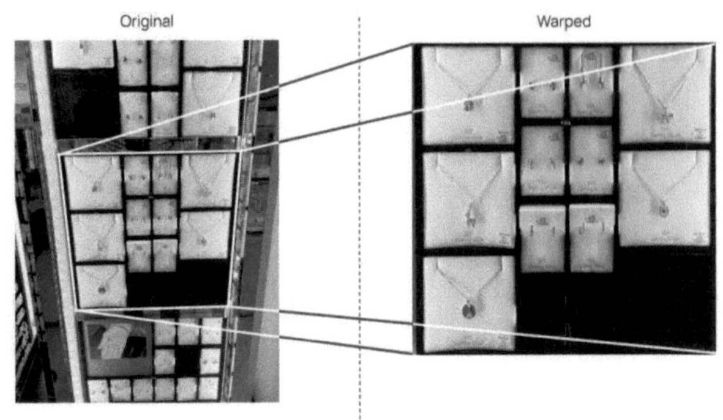

Figure 1: Example result of the image warping algorithm, extracting the tray and stretching it to the normalized format.

recognition relies on reading the tray ID number present on product labels and the tray itself. The ID labels are not in any standardized position on the picture, so the network identifies as many labels (both on the tray and on product cushions) as possible to poll a consensus decision.

This is done with a traditional OCR algorithm after the text position is identified. The read labels are compared with the list of known tray IDs to reduce the search space. The layout design of each tray is known and is stored in the data bank as a vector image. With this, the tray can be segmented into separate product position images. In most positions, products sit on a cushion that has a significantly lighter color than the tray background color. This allows for the identification of empty positions that do not have this color difference and marks the item expected to be there as missing. In trays holding bracelets and similar products, these cushions are not present, so the absence of items is not as easily detectable. The network needed to be fine-tuned to also recognize these items against the similarly colored background to improve the success rate to an acceptable level.

Positions that do not hold any products and instead show artwork or advertising images are not passed along to the product recognition module. They are instead always considered to be present, as they regularly change and do not hold visible position labels, making them impractical to be trained for with each change.

3.3   Product Recognition

For this step to function properly, products need to have a visible label specifying the tray and product position on their cushion. This is required for the module to utilize text detection and recognition in order as the previous does.

The product recognition module works in three successive steps. First, a text detection algorithm finds the label on the tray position segments from the previous module. These are not always at the same position on the cushion, and may also be rotated from a normal orientation.

With the detected label, a text recognition module using OCR can translate the label text on the image into a machine readable format. All labels currently in use feature the tray and position numbers in the last row of the label text. This means that they can easily be extracted from the recognized label irrespective of the rest of the text. The tray and position numbers are then compared to the expected values within the database.

Since the tray and tray layout were identified in the second step of the pipeline, an exact match is marked as "correctly placed". If the label is not readable or the read text does not match the expected value, the module will mark the position as "misplaced". This ensures that the pipeline will be more likely to produce false negatives than false positives.

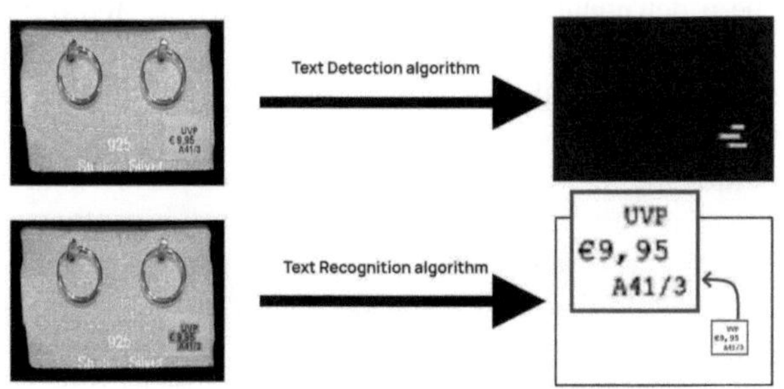

Figure 2: Visualization of the Product Recognition module. On each tray position segment, it detects where on the product position a label text can be read, and an OCR algorithm will read the label for further use.

3.4 Synchronization

After running all tray position segments through the third module, a list of all products in a tray with their respective status can be compiled. This list is then processed and sent back to the user. There, it is displayed as a model representation of the tray.

This representation uses color coding for the individual positions, with correct positions being highlighted in green, misplaced or unrecognized ones in yellow, and empty positions in grey. In each position that is not correct, the position number is being displayed to allow operators to quickly check positions themselves if needed.

The operator is able to manually update the presented information if a misclassification has taken place. When they are satisfied with the result, they can confirm the tray synchronization, which prompts the application to save the registered data. Once all products in a location have been recorded, this preview is sent to the central inventory system as a fulfilled synchronization to be processed from there.

The central inventory system receives the information as a list of missing items, encoded by their position and tray numbers. Everything not included in the list is treated as being present. Notably, the ID numbers of the items are not being sent, because the central system might work with different product lists than the store. This is relevant in countries that don't use the EAN format, which is the standard in Germany.

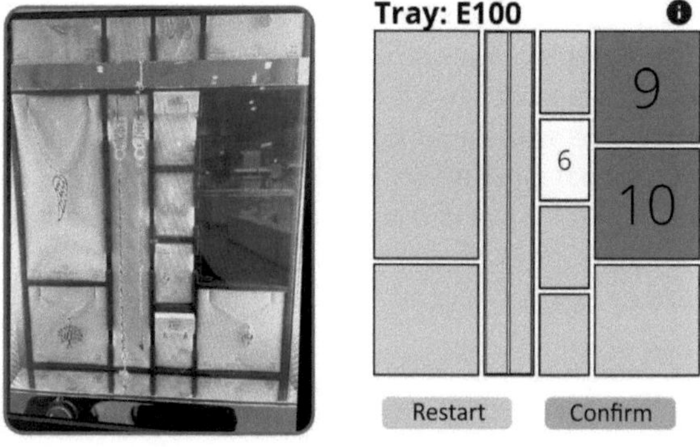

Figure 3: Comparison between an image sent to the image recognition pipeline and the resulting visualization in the web application.

4 Web Application

The web application "wraps around" the image recognition pipeline and allows operators to make use of it. After logging into an operator's account, the web application also provides acceess to a history of past sessions as well as statistics synthesized from collected information. When an inventory synchronization session is started, an operator can choose between different operating modes, either structured arrangement, ring presenters, or free arrangements.

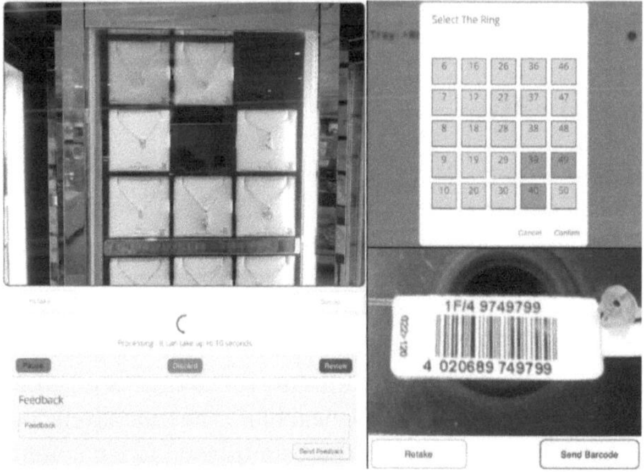

Figure 4: Overview of the different operation modes. Structured arrangements on the left, ring presenters on the upper right, free arrangement on the lower right.

Structured arrangements are processed with the image recognition pipeline by default, but can switch to a manual mode analogue to the ring presenters. Once a photo is processed, the camera view is replaced with a schematic view of the tray. From here, the status of each position can be changed by the operator, allowing to correct errors.

As rings are not displayed with visible labels on their presentation displays, they only allow for manual processing. The manual mode only shows the schematic view of the tray, skipping the image recognition step. The tray can be selected from a drop down menu, which also allows for searching for trays by text name.

Free arrangements are items that are not presented in a standardized display. The corresponding functionality reads the product ID encoded in a bar code visible on a product label and registers it in the inventory list. The interface also offers a manual input mode, allowing to enter the ID number directly. The system will compare the product ID to a list of all available products to prevent recognition mistakes, but otherwise doesn't offer any automation.

5   Testing

The performance of the image recognition pipeline was tested in mock inventory sessions, both in the controlled environment of a show room as well as on live sales floors. During each test, disruptive errors would be marked down, and the time needed to conduct the session would be noted. Since these tests aimed to provide comparisons with the paper and MAI processes while assessing, the processes needed to be conducted in parallel. This provided benchmarks for time spent and error rates to improve upon.

To train the neural networks, an initial training set of 2322 reference images of display cases was used. These were augmented with randomized artifacts like glare to harden the network against this kind of fault. In addition, all photos taken during prototyping and testing were included in the training set for fine-tuning the model. As these did not feature extreme angles or light conditions, they represented the expected prediction conditions.

6   Results

While the paper process synchronization takes between 90 and 120 minutes in the show room, the MAI process needs 45 minutes on average. At the end of the second development phase, a full synchronization with the computer vision process needed about 60 minutes with first-time users and 30 minutes with experienced operators. This difference can be attributed to familiarity with the requirements regarding acceptable levels of glare, optical blur. The time saved compared to the paper process can be attributed to the automation of the product identification, which vastly outweighs the computation time of the image recognition.

While being hardened against these artifacts, these proved to be the most disruptive influences. While this didn't meaningfully affects the image warper module, the tray and product recognition modules suffered from these, as labels could become unreadable. Overall lighting conditions played little part in the recognition rates due to being loosely standardized in the sales locations. They only noticeably affected performance in cases where the operator created a shadow (e.g. from an overhead spot light). All three issues were included in the training material, as they were easy to control by the operator.

At the end of development, the project was deemed a success and the web application is scheduled to be gradually rolled out to sales floors over the course of a year. During this time, the reliability and performance will be monitored in a testing phase to ensure the new process continues to meet the existing standards in a live environment.

6.1   Additional Research Opportunities

During development, it was found that using a great number of different devices was important for improving classification success rates. As most devices possess unique camera hardware and image encoding configurations, they create images with noticeably differing recognition rates. It was also found that the use of multiple cameras lenses, which is becoming more common in consumer smartphones, showed to be a detriment for recognition rates, as the algorithms were trained on single lens cameras. These are both phenomena that are already documented [2], but not yet exhaustively explored.

Additionally, it was found that optical blur played a direct role in recognition rates. An initial assumption was that any blur, if it was below an accounted-for threshold, would not be a deciding factor in the overall performance [3]. The severity of the optical blur seemed to have a proportional relationship to the recognition rate of images across the explored range even after corresponding fine-tuning.

Acknowledgement

The work has been carried out at Amor Group GmbH in Hanau, Germany. It has been supervised by Prof. Dr.-Ing. Erhardt Barth at the Institute of Neuro- and Bioinformatics, Universität zu Lübeck.

Authors' Statement

Conflict of interest: Authors state no conflict of interest.

7   References

[1] F. Chen, A. Federgruen and Y.-S. Zheng, *Near-Optimal Pricing and Replenishment Strategies for a Retail/Distribution System.* Operations Research, vol. 49, no. 61, pp. 839–53, 2001.

[2] S. Dodge, L. Karam, *Understanding How Image Quality Affects Deep Neural Networks.* arXiv:1604.04004, 2016.

[3] I. Vasiljevic, A. Chakrabarti, G. Shakhnarovich, *Examining the Impact of Blur on Recognition by Convolutional Networks.* arXiv:1611.05760, 2017.

[4] R. Goonatilake, S. Maldonado, *Essentials of Novel Inventory Management Systems.* Engineering Management Research, vol. 7, No. 1, pp. 31-45, 2018.

[5] M. Fiorucci et al, *A Computer Vision System for the Automatic Inventory of a Cooler.* CIAP 2017, Part I, LNCS 10484, pp. 575–585, 2017, 2017.

Developing a Camera-based Hands-on Detection System for the Steering Wheel Using OpenPose Library

Sadra Nasiri [1], Foti Coleca [2], and Erhardt Barth [3]

[1] Robotics and Autonomous Systems, Universität zu Lübeck, m.nasiri@student.uni-luebeck.de
[2] Gestigon GmbH - a Valeo brand, foti.coleca@valeo.com
[3] Institut für Neuro- und Bioinformatik, Universität zu Lübeck, erhardt.barth@uni-luebeck.de

Abstract

In this research, the primary motivation is to address the crucial need for a camera-based hands-on detection system in autonomous vehicles, specifically focusing on the safety and functionality of advanced driver assistance systems (ADAS). The gap this paper aims to fill lies in ensuring precise detection of the driver's hand position on the steering wheel. To achieve this, OpenPose library is employed for accurate hand localization, leveraging its spatial relationships understanding. Additionally, a dedicated hand detection model is developed using the YOLO (You Only Look Once) object detector as a prerequisite for OpenPose. Overcoming the challenge of limited datasets, the HoD dataset is created by sampling images from in-car camera streams and annotating them. This paper assesses effectiveness and reliability of the system through evaluating the metrics like accuracy, precision, and recall. This research contributes practical solutions to enhance safety and facilitate the implementation of advanced driver assistance systems in autonomous vehicles.

1 Introduction

1.1 Overview

This research addresses the vital safety concern in autonomous vehicles by developing a hands-on detection system. Focused on advanced driver assistance systems (ADAS) and achieving Level 4 and 5 autonomy, the proposed camera-only-based system offers cost-effective advantages over existing solutions like capacitive sensors. As vehicles increasingly integrate cameras for driver monitoring, the system provides detailed information on hand position and orientation, crucial for implementing advanced ADAS like lane-keeping assistance (LKA) and adapting to diverse driving scenarios.

1.2 Contributions

The main contribution of this paper is to:

- **Propose a Prototype of a Hands-on Detection System:** The main goal of this paper project is to develop a prototype Hands-on Detection (HoD) system, comprising two phases. In the initial phase, crucial information such as the position and orientation of hand key points and the steering wheel's location is extracted from images. The subsequent phase involves inputting this information into the hands-on detection block to determine if the driver's hands are positioned on the steering wheel.

- **Develop a Hand Detection model:** To enable OpenPose to detect hand key points accurately, a crucial initial step involves annotating hands through bounding boxes. Before utilizing OpenPose, a dedicated hand detection block is developed. This block is employed specifically to facilitate the initial stage of hand detection.

- **Study and Comprehend the OpenPose Approaches:** OpenPose is vital for precisely locating the driver's hands on the steering wheel in our project, addressing challenges like occlusions and varied hand poses. It plays a central role in estimating hand key points in the complex HoD system. This research actively studies and experiments to optimize OpenPose's application, refining its role in hand detection for correct functioning within the HoD system.

- **Prepare HoD dataset:** Obtaining suitable data for training the initial hand detection model was a challenge, as existing datasets did not align with our camera's specifications. Consequently, we opted to sample a sufficient number of images from various in-car recorded streams and annotated them, creating the HoD dataset.

- **Evaluate Performance and Analyze Metrics:** Ultimately, performance of the developed block is assessed. This evaluation will utilize key metrics like accuracy, precision, and recall, offering valuable insights into the system's effectiveness and reliability.

2 Material and Methods

2.1 HoD Architecture

The proposed Hands-on Detection (HoD) system prototype (fig. 1) is structured into two sequential phases. The first phase involves extracting relevant information from images, with a focus on hand key points estimation. Two components, the "Hands detection" block (utilizing Yolov8 model [3]) and the "Pose estimation" block (employing OpenPose [1]), collaborate to detect hands and estimate their spatial coordinates and orientations. Simultaneously, this phase includes a branch to detect the steering wheel position.

The second phase integrates information from the first part, combining hand key points, their estimated orientation, and steering wheel location. This unified information is directed into the hands-on detection block to determine if the driver's hands are correctly positioned on the steering wheel.

The detailed exploration of the steering wheel position detection and further elaboration on the second phase are beyond the scope of this research.

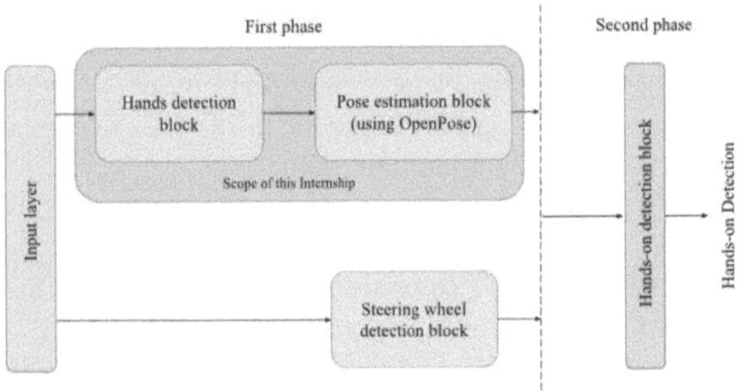

Figure 1: The Hands-on Detection (HoD) system, in two phases, extracts hand key points and steering wheel position from images. It then determines if the driver's hands are on the steering wheel. The paper focuses on the highlighted phases.

2.2 Pose Estimation

OpenPose [1], is the first real-time system specializing in detecting the human body, foot, hand, and facial key points (in total 135 key points) and spatial relationships between key points like fingers and palms in one image. OpenPose plays a crucial role in precisely locating the driver's hands on the steering wheel in the project.

As it's shown in fig. 2, OpenPose is able to accurately identify body and hand key points in RGB images under optimal conditions. However, it faces challenges when it is integrated with our images. Firstly, OpenPose relies on the arm's key points to estimate hand locations. Consequently, if the body arm key points are not detectable, it becomes unable to identify hand key points. Unfortunately, our camera's perspective, hindered arm key points detection, as the torso remained unobservable. To address this, hands were manually annotated with bounding boxes in experiments. However, a dedicated hand detection model is later devel-

oped to overcome this issue. Additionally, the use of an infrared camera in our images led to pixel variance, with some areas excessively illuminated and others dimmed. This high variance, coupled with gray scale images, increased complexity for OpenPose in detecting hand key points.

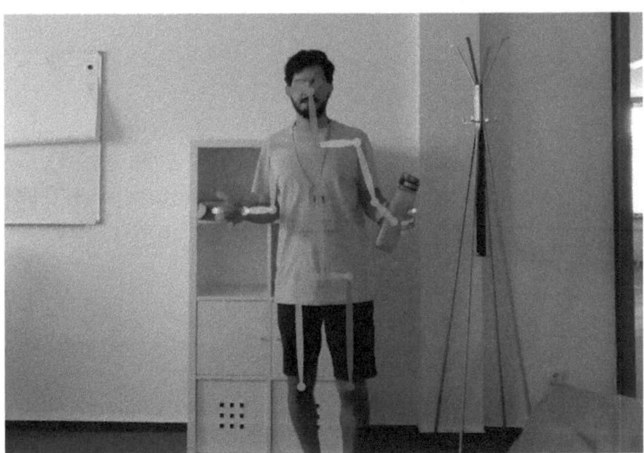

Figure 2: OpenPose has accurately detected body, legs, hands, and face key points in real-time.

Lacking labeled data for the target dataset, we experimented with image manipulation techniques on sampled images to align them better with the expected inputs for OpenPose. In Experiment 1, as shown in fig. 3(a), an original sample image (232×220 pixels) led to unsatisfactory results with OpenPose, possibly due to the image size or significant pixel variance. In Experiment 2, to address potential obstacles, images were scaled up by a factor of three (928×880 pixels) using the FSRCNN model by Dong et al [4]. This scaling produced more favorable outcomes, as depicted in fig. 3(b). However, challenges persisted, particularly in detecting keypoints for the hand farther from the camera, which remained in darkness in many images.

Experiment 3 involved applying "Clipped Min-Max normalization" and upscaling (fig. 3(c)) to address potential outliers in the image. This normalization technique ensures that the scaled values within the dataset represent the majority of the data while minimizing the impact of outliers.

Experiment 4 applied Contrast-Limited Adaptive Histogram Equalization (CLAHE) from [5] to a scaled-up image. While CLAHE improved visibility and key point detection for the right hand, its high contrast may have caused the left hand (closer to the camera) to go unrecognized

In experiment 5, we applied "Gamma Correction" and an "Unsharp Mask" to address the high contrast generated by CLAHE after upscaling. Gamma correction adjusts perceived brightness or contrast, while the unsharp mask enhances edges and details. Despite these adjustments, the results in fig. 3(e) did not show a significant improvement in OpenPose estimation, possibly due to the persistent high contrast in the image.

Experiment 6 employs a non-linear method, using a gamma-controlled mask and Gaussian blur, to selectively reduce image sharpness. This enhances OpenPose's accuracy in detecting key points, achieving balanced sharpness between hands. The approach strategically reduces sharpness in overexposed areas while maintaining clarity where needed. Results in Figure 6 demonstrate improved OpenPose performance, particularly in frames with significant exposure variations.

(a) Experiment 1: OpenPose output on the original image (232×220 pixels)

(b) Experiment 2: The image is scaled up by a factor of 3. The new image size is 928×880

(c) Experiment 3: Clipped Min-Max normalization is applied on the scaled-up image

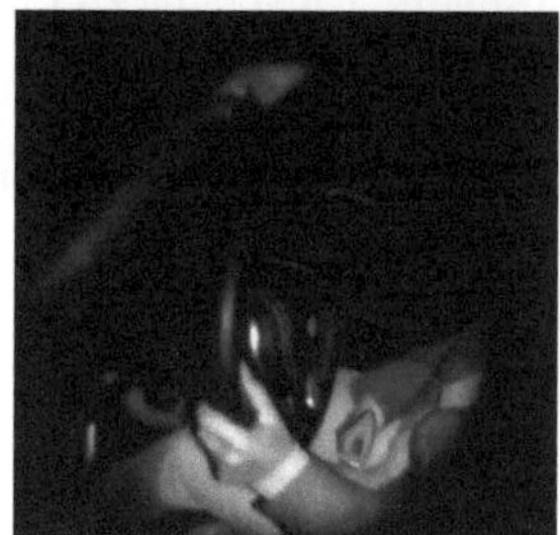

(d) Experiment 3: CLAHE method is applied on the scaled-up image

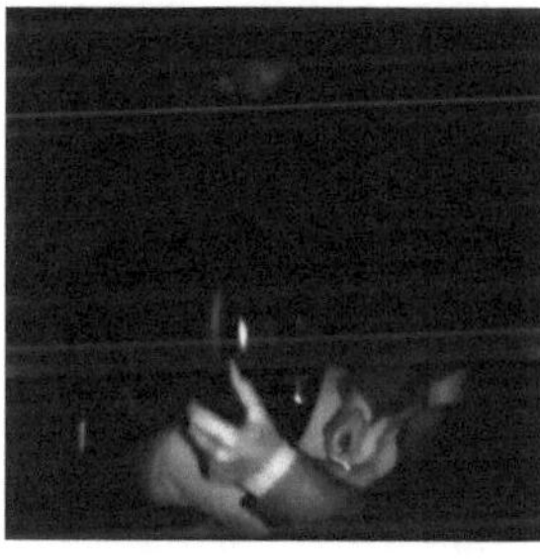

(e) Experiment 5: Up-scaling and CLAHE, followed by Gamma correction and unsharp mask sequentially are performed on the image

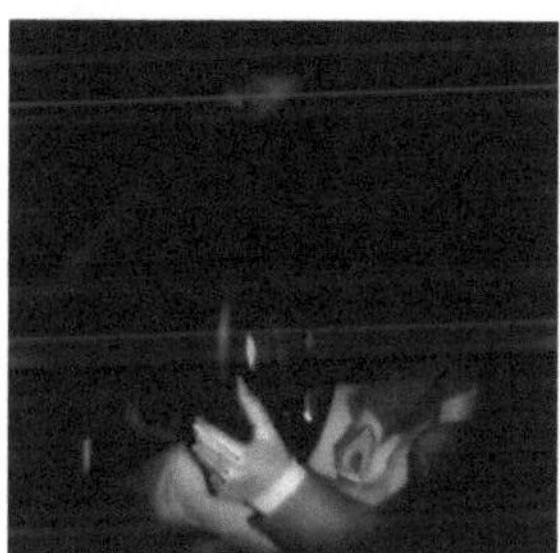

(f) Experiment 6: Upscaling, sharpness reduction, adaptive histogram equalization, gamma correction, and unsharp mask are applied sequentially

Figure 3: Outcomes from the executed experiments on a sample image

2.3 Hand Detection

OpenPose relies heavily on results from the body key points detector to estimate hand positions. However, when body key points are unclear, particularly in our over-the-shoulder camera perspective capturing arms and laps, OpenPose struggles to identify hand key points. To overcome this, a dedicated hand detection block was developed to enhance OpenPose's capability in detecting hand key points, especially when body key points are indiscernible. This development aims to improve the overall performance and robustness of the system.

In this paper, YOLO [3] is employed for hand detection due to its speed, treating detection as a regression problem without requiring a complex pipeline. YOLOv8n (nano) pre-trained model [2] served as the starting point, and is retrained using the HoD dataset for this project.

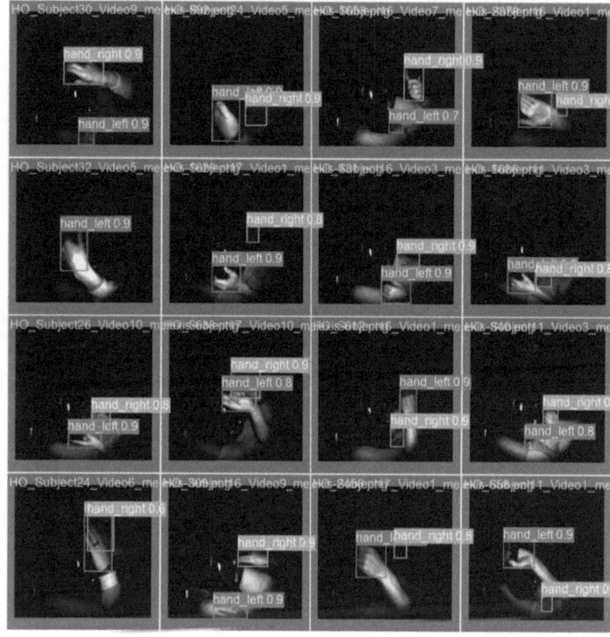

Figure 4: Some example predictions of the hand detection model.

3 Results and Discussion

3.1 Pose Estimation

The assessment of the experiment setups involves evaluating the mean confidence of key points detected by OpenPose, focusing on 20 key points for each hand. Through experiments with 25 sample images, key insights emerged. Experiment 2 addressed the initial struggles in effective hand detection by upscaling the image, notably improving left hand confidence. Despite slight improvements with clipped Min-Max normalization in experiment 3, a substantial gap persisted between left and right hand detection confidences. Introducing the CLAHE method in experiment 4 significantly boosted confidence for detecting the right hand but impacted the left hand negatively. Attempting to mitigate high image contrast with gamma correction and unsharp mask methods yielded improved confidence for both hands in the experiment 5 but maintained the gap. In the final experiment, acceptable confidence levels for both hands were achieved using a sharpness reduction method, showcasing progressive improvement.

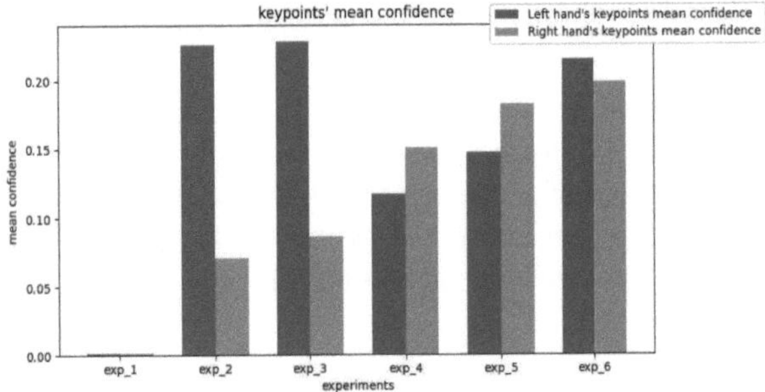

Figure 5: Left and right hands detection confidence over different experiment setups illustrates the progressive improvement in confidence for both left and right hand detection across our six experiments

3.2 Hand Detection

To assess the performance of our hand detection model, three different loss functions including box-loss, cls-loss, and dfl-loss are evaluated across validation iterations, observing convergence in all three losses. The hand detection model classifies instances into Left Hand (LH), Right Hand (RH), and Background (BG). Confusion matrix analysis reveals true positives, false positives, and false negatives for LH and RH. Figure 6 demonstrates recall and precision for LH and RH, indicating effective identification and accurate detection with minimal false positives low false-positive rate for both hands. These recall and precision values highlight the model's robust performance in distinguishing between LH, RH, and backgrounds, achieving a balance between accurate detection and minimizing false positives.

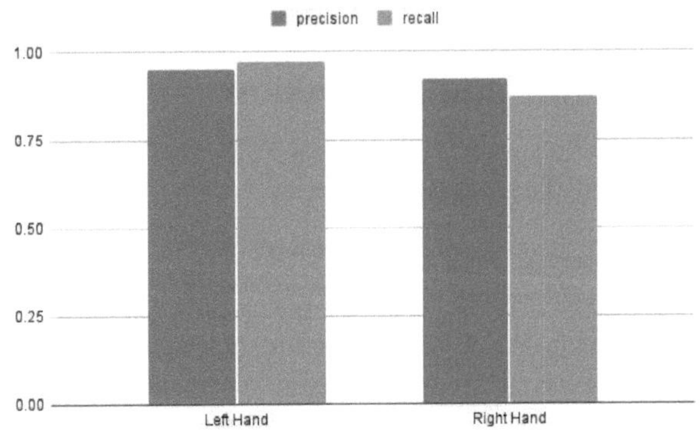

Figure 6: Approximately 0.97 recall and around 0.95 precision for LH and a recall of approximately 0.87 and precision of around 0.92 for RH

4 Conclusion

During the research, the focus is on developing a Hands-on Detection (HoD) system with a dedicated architecture comprising pose estimation and hand detection elements. The pose estimation block is optimized through extensive exploration of the OpenPose library and experiments. The HoD dataset is prepared to fulfill the requirements of the hand detection block. Subsequently, a hand detection model is trained and evaluated using the HoD dataset, enabling independent identification of hands in images.

Future tasks include implementing a steering wheel detection block, potentially using object detection or instance segmentation techniques, and integrating this information with the hands' key points and orientations. The final phase involves constructing a dedicated block to determine the driver's hand positions in relation to the steering wheel, bringing the HoD system to full functionality.

Acknowledgement

The work has been carried out at Gestigon GmbH - a Valeo brand, and supervised by the Institute of Neuro- und Bioinformatik, Universität zu Lübeck.

Authors' Statement

Conflict of interest: Authors state no conflict of interest.

5 References

[1] Zhe Cao, Gines Hidalgo, Tomas Simon, Shih-En Wei, and Yaser Sheikh, *OpenPose: Realtime Multi-Person 2D Pose Estimation using Part Affinity Fields*, arXiv preprint arXiv:1812.08008 (2019), `https://arxiv.org/abs/1812.08008`, primaryClass: cs.CV.

[2] Glenn Jocher, Ayush Chaurasia, Jing Qiu, *YOLO by Ultralytics*, Version 8.0.0, Released under AGPL-3.0 License, January 2023, `https://github.com/ultralytics/ultralytics`.

[3] Joseph Redmon, Santosh Divvala, Ross Girshick, Ali Farhadi, *You Only Look Once: Unified, Real-Time Object Detection*, In *Proceedings of the IEEE Conference on Computer Vision and Pattern Recognition (CVPR)*, June 2016.

[4] Chao Dong, Chen Change Loy, Xiaoou Tang, *Accelerating the Super-Resolution Convolutional Neural Network*, arXiv preprint arXiv:1608.00367 (2016), `https://arxiv.org/abs/1608.00367`, primaryClass: cs.CV.

[5] Garima Yadav, Saurabh Maheshwari, Anjali Agarwal, *Contrast limited adaptive histogram equalization based enhancement for real-time video system*, In *2014 International Conference on Advances in Computing, Communications and Informatics (ICACCI)*, 2014, Pages 2392-2397, DOI: `https://doi.org/10.1109/ICACCI.2014.6968381,`

Exploring neural network acceleration using Gemmini

Leon Dietrich [1], Ahmed Mahmoudi [2] and Mladen Berekovic [2]

[1] Robotics and Autonomous Systems, Universität zu Lübeck, leon.dietrich@student.uni-luebeck.de

[2] Institute of Computer Engineering, Universität zu Lübeck, [ahmed.mahmoudi, mladen.berekovic]@uni-luebeck.de

Abstract

Neural Networks have acquired an important role within modern automated decision-making systems such as computer vision or intent recognition. While these algorithms are quite powerful, their processing demand is huge, requiring acceleration. Furthermore, energy efficiency and continuous advances in throughput are desired, resulting in the development of specialized hardware accelerators. As the primary work load within these networks can be reduced to matrix multiplication, systolic arrays, augmented with post-processing options, are one of the technologies with a natural fitting. As this technology, combined with this use case is quite novel, the software stacks required for operation are lacking behind. This is expressed by the fact that performance losses, energy inefficiencies and unsupported operations are still common. This work focuses on the adaptation of a specific framework in order to better support modern neural network architectures. Furthermore, various opportunities for performance improvements have been identified of which some have been implemented.

1   Introduction

While convolutional neural networks are increasingly important for computer vision tasks in autonomous systems, their inference, meaning the evaluation of a given network with the corresponding inputs, remains computationally challenging. General purpose devices, like GPUs or CPUs, impose significant power demand increase and die space overhead at a given performance point. Therefore, these devices are usually not suited for inference in embedded application contexts.

The primary task that needs to be handled is matrix multiplication with post-processing. While there are some accelerators targeting this work load, only few of them are capable of targeting lower power scenarios [7]. Systolic arrays, like the one found within Gemmini [2], are particularly well suited for this task. In addition, Gemmini features some scratch pad memory, a transposing engine, data stream processing filters in order to perform further operations, such as activation or aggregation functions, on the outputs of the systolic and finally control logic for its operation. The accelerator is attached to a RISC-V processor using the RoCC interface.

In order to operate Gemmini from a software perspective, there is libgemmini, which offers some low level control over the accelerator as well as a version of onnxruntime modified for RISC-V based processors. Without modification, onnxruntime is capable of performing inference on basically any model given to it. In order to do so, there is a CPU targeting software implementation for every operation (called kernel) that might occur within such model. These however are notoriously slow. As a consequence one can define special hardware accelerated handlers for selected kernels in order to speed them up. This has already been done for selected functions required during training, but inference requirements have generally not been met.

The contribution of this paper is the enablement of onnxruntime for RISC-V to perform inference on arbitrary opset 10 generation networks. While doing so we also identified multiple optimization opportunities, both in software as in hardware, of which we implemented some. This research is aligned with further research at the institute targeting computer vision for autonomous vehicles.

2   Overview on Gemmini Hardware

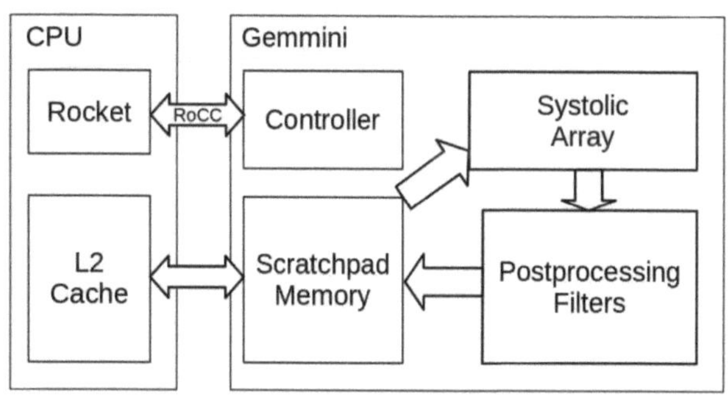

Figure 1: Dataflow within Gemmini. The RoCC interface carries only instructions.

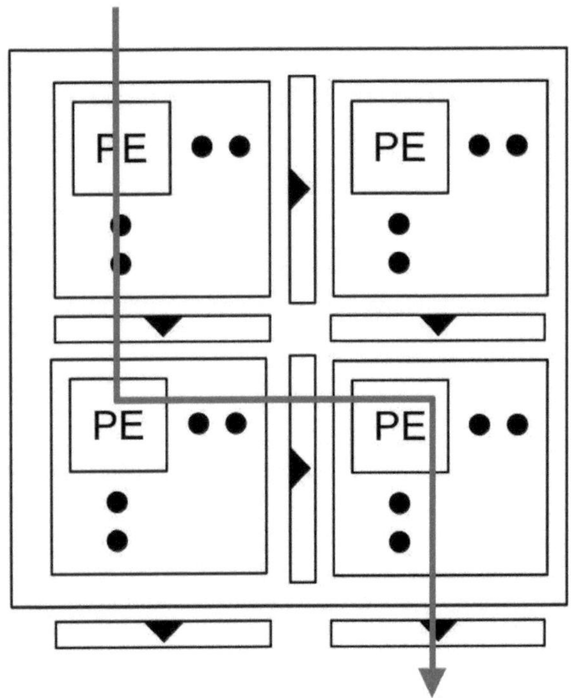

Figure 2: Explanation of layout and critical path within Gemmini systolic array. The red arrow indicates the critical path.

Gemmini is part of the chipyard [1] ecosystem, which is a toolkit to compose processing cores to be utilized within systems on a chip, primarily designed at Berkely University. In order to use this module one has to instantiate at least one Rocket Core [3]. The usage of other included processor options were not tested.

As displayed in fig. 1, the accelerator is integrated using their RoCC interface, resulting in an instruction set extension for RISC-V. Furthermore, it contains a control unit handling scheduling, arbiting, data movement and sequence control. Gemminis core is made out of a systolic array which data is fed by a scratchpad memory (basically a large register file), and post-processing filters, consisting out of accumulators, multipliers, ReLU processing units and various other optional elements. The provided instructions range from bare metal instructions targeting data movement and cache management, compute pipeline configuration and execution triggers up to quite complex instructions for issuing of complete compute sequences usable for tasks such as matrix multiplication.

Gemini issues only one interrupt meant for handling page faults during background data transfers. All other operations are carries out in a synchronous manner after the main processors commit stage. While this is an obvious drawback, no further speedup can be achieved here without redesigning the entire RoCC interface as all instructions issued using this interface stall the entire pipeline.

Unlike most compute accelerators for neural network inference, Gemmini expects the data presented to it to be NHWC (here referring to associated data being right next to each other)format for memory transactions and output filtering such as activation functions. There is no accelerator available for data format conversion, rendering this operation expensive. We therefore reduced the number of required conversions by chaining subsequent Gemmini calls inside onnxruntime.

The configuration of the systolic array used for testing is based on an eight by eight grid of compute tiles where each tile only features a single processing element. This layout is visualized in fig. 2. Our critical path therefore exists only inside a processing element. We clocked the test setup at 100MHz. We opted to use this configuration in order to maintain comparability with results from Berkely [6].

First, we used the default 32 bit floating point unit as our processing element in order to verify the existing software. Later, we switched to eight bit signed and unsigned integer computation within the processing element. Using Gemmini with multiple compatible data types still requires multiple array instances. Fixing this was left to future work.

3 How Onnx handles neural networks

The open neural network exchange (Onnx) specification is the only one available for vendor-independent interchange of neural network models. Accompanying, there is also onnxruntime, which is a framework to perform inference on models stored in Onnx representation. A fork of this project, called onnxruntime for RISC-V, exists and enables execution on RISC-V based devices including available accelerators.

Onnx organizes neural networks as nodes on a directed graph whereas each node represents an operation that is performed on the data. A node does not necessarily represent a layer of the network. Instead, a layer can be formed out of different nodes. Onnxruntime is a framework that can perform inference using models saved in the Onnx representation. As time goes by, newer versions of the file format specification (operation sets, later called OpSet), have emerged. Older OpSet kept convolutions or matrix multiplications separate from their surrounding activation functions, whereas newer versions (such as OpSet 10 – our target operation set version) usually combines these steps into single nodes in order to enable better optimization capabilities. As only a selection of these older operators were previously implemented in onnxruntime for RISC-V, there is still a lot of potential to implement further accelerator integration without introducing new use cases for the systolic array and its data filters.

Last but not least, a node might use multiple input edges and might provide multiple output edges and possibilities to activate or deactivate some of these edges based on data being processed exist. While this enables options such as data feedback or branching this also introduces further complexity in regard to data reuse on a given accelerator. As a result, such optimizations were not previously enabled within onnxruntime for RISC-V.

4 Overview on Onnxruntime Software Stack

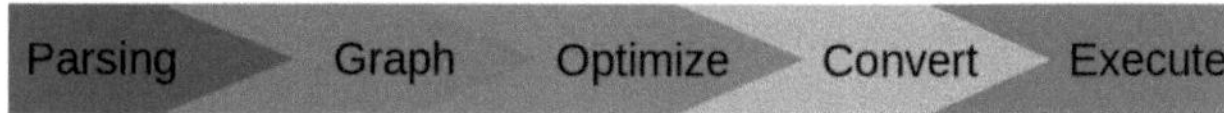

Figure 3: Data flow within onnxruntime

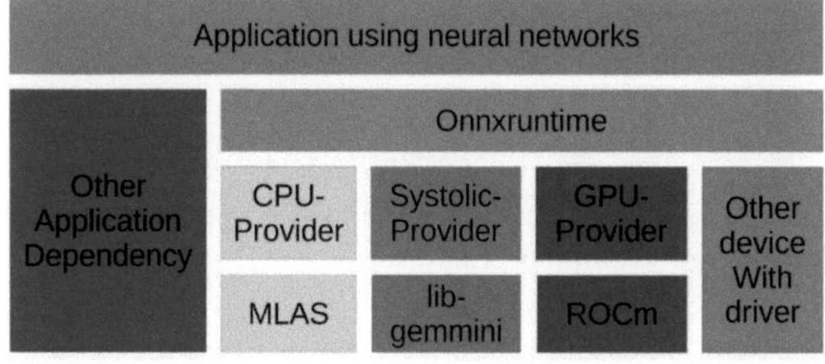

Figure 4: Software stack involved in inference using onnxruntime.

Primarily, onnxruntime [4] consists out of modules in charge of parsing the model, generation of execution graphs, optimizing these graphs, converting the used data structures to a format the local machine understands and execution or code generation. These modules basically hand their results out to each other as visualized in fig. 3. While the first three stages only need to be executed once per session, the last two will be executed on every inference.

While the parsing module is not of particular interest as all it really does is reading in flatbuffer objects concatenated within the target .onnx files, all other modules had to be modified in order to obtain working acceleration support.

In the next stage, onnxruntime performs optimizations on the computational graph. The scope of these optimizations is defined by graph optimizations, optimizations regarding the mathematical structure as well as ones improving the execution characteristics of the particular accelerators available to the system. Regarding Gemmini there have been only few modifications involving the data layout transformation compatibility as onnxruntime internally works with NCHW instead of NHWC. The user might choose to export the optimized graph at this point in order to save the optimized graph to a new .onnx file, resulting in faster start up times in the following sessions.

Afterwards, accelerators claim the kernels they may accelerate and there is a dedicated stage where accelerators can perform transformations on the data of their acclaimed kernels such as converting their data to NHWC if required. As the modified library already performed these conversions, there is limited compatibility with other accelerators. Fixing this issue would be part of future work though.

The final module of onnxruntime defines the execution of the loaded model. Fig. 4 visualizes the distribution across the corresponding model. Each accelerator uses this stage to initialize his corresponding devices, using his own libraries. In the case of Gemmini, libgemmini will be used

to reset the accelerator. Afterwards onnxruntime simply iterates over all kernels, calling their registered implementation. As the RISC-V version of onnxruntime does not perform its data conversion using the predetermined layer module and onnxruntime allows multiple inference run per session, the kernels for Gemmini check the conversion requirement again prior to their work dispatch.

Last but not least, onnxruntime will gather all result tensors, converts the data back to the frameworks representation and provides the results to the next kernel or back to the calling application. Data pre- and post-processing are not part of the framework and need to be handled by the implementing user.

5 Testing Methodology and Results

In order to benchmark the achieved performance with the accelerator, we used different techniques.

Perhaps, the most accurate performance measurements are possible, running the integrated System on a Chipon an FPGA. These measurements should be performed without a lot of other processes running and using direct tools. Unfortunately, this process is quite error-prone and time-consuming which is why it is not used for rapid prototyping and development. Other methods are discussed later.

Most of the testing was done using simulators. Simulation results have been verified using a Xilinx ZCU104 FPGA using a relatively small systolic array of eight by eight processing elements without a TLB and only a single Rocket CPU core. Other settings have been copied from the Arty-100T definition. Our FPGA is smaller than the default (Xilinx VCU118) configuration provided. Nevertheless, we are confident that our results are comparable as onnxruntime does not benefit from multiple CPU cores.

There are two kinds of simulators. First, there are cycle-accurate hardware simulators. These are the most precise way to measure performance (even more than running on real hardware). Nevertheless, they are unsuited for development as simulating the execution of a medium-sized network takes weeks. The reason for this is that these pieces of software simulate every hardware state change within the system as the software executes.

The RISC-V ecosystem provides a functional simulator called spike which has been modified by the chipyard project to support their hardware. Together with a syscall proxy (called pk) we're able to run risc-v software on arbitrary hardware. In order to be able to run instructions targeting Gemmini, the simulator needs to be patched for the specific behavior of rocket and needs to be augmented with parts of libgemmini that perform the simulation.

Using this simulator for profiling would be a bad idea most of the time though. This is due to the fact, that userland profiling tools query the system time, which is proxied to the host system. In other words: timing the execution of software running in spike is a benchmark of the simulator, not of the actual software being emulated. The instruction count registers on the other hand yield exact numbers as there is no scheduler or interrupt handler running inside the simulator. While the instruction count is not useful as a performance metric as various instructions take different amount of time and energy to execute in modern processors [5], it proved to be useful as an indication whether Gemmini was utilized. Performing a matrix multiplication or convolution on Gemmini requires a handful of instructions (10^1) whereas performing these operations in software usually requires 10^5 to 10^6 instructions on the networks we used. Some networks use many small scale GEMM operations. The data layout conversion overhead might counteract the efficiency and speed gains from Gemmini. Therefore, this might not be worth it for small matrices. Checking this would be a task for future work.

A different option on gaining performance data is using the internal profiling system of Onnxruntime. This can be enabled by specifying a directory to save the report to. Once enabled, they log every kernel being executed together with its total system execution time. Unfortunately, this cannot be used for performance measurements. However, it provides some useful information about which kernels are executed in general and at which frequency. Using this information in order to decide how much time to spend on which optimization turned out to be crucial.

Last but not least, in order to check which kernel versions were acclaimed, we simply inserted corresponding print statements inside the kernel registry and filtered the output. The information acquired this way was then used to further investigate bugs causing Onnxruntime to not accelerate operations that Gemmini would otherwise support.

We used various types of pretrained neural networks for testing. We sourced our networks from the Onnx model zoo github repository. We started with vgg-ImageNet based networks as they were already working. These networks had to be converted back to Opset 10 in order to run with the base version of Onnxruntime for RISC-V. Later we introduced support for RCNN and resnet50based networks. Finally we got MobileNet anf Googlenet based networksworking. For Opset 10, there should be no unsupported Onnx operations anymore. However, as breaking changes to the Onnx Opset specification are common and not all newer versions can be backported, using a new network type will cause engineering work at this stage.

6 Conclusion

Execution of neural networks requires specialized hardware acceleration in conjunction with corresponding drivers and frameworks. The latter ones are still lacking behind. In order to improve on this issue, we analyzed the runtime behavior of Onnxruntime for RISC-V based CPUs using the Gemmini accelerator. Various possibilities for improvements have been found, both in the way, Gemmini is used, the utilization within Onnxruntime and Gemmini itself as well as fixing issues causing most networks to fail on attempted inference. Non of our changes degraded the inference accuracy compared to a pure CPU execution using the same data type. Furthermore, we established methods to systematically analyze the performance improvements var-

ious optimizations provided. Last but not least, scripts have been implemented in order to automatically set up the build and (software based) testing environment.

While some improvements have been made, there is still a lot of potential for further ones. Future work would also include efforts towards automatic benchmarking and testing on real hardware. As the RISC-V version of Onnxruntime has never been integrated with the upstream project and its features are currently distributed across many (largely incompatible) repositories and versions, the task to maintain and consolidate these efforts also still persists.

Acknowledgement

The work has been carried out and was supervised at the Institute of Computer Engineering (ITI), Universität zu Lübeck.

Authors' Statement

Conflict of interest: All authors state no conflict of interest.

7 References

[1] A. Amid and D. Biancolin, A. Gonzalez, D. Grubb, S. Karandikar, H. Liew, A. Magyar, H. Mao, A. Ou, N. Pemberton and others, *Chipyard: Integrated design, simulation, and implementation framework for custom socs.* In: IEEE Micro, vol 40, no. 4, pp. 10–21, 2020.

[2] H. Genc and A. Haj-Ali, V. Iyer, A. Amid, H. Mao, J. Wright, C. Schmidt, J. Zhao, A. Ou, M. Banister and others, *Gemmini: An agile systolic array generator enabling systematic evaluations of deep-learning architectures.* In: arXiv preprint, vol 3, pp. 25, 2019.

[3] K. Asanovic, and R. Avizienis, J. Bachrach, S. Beamer, D. Biancolin, C. Celio, H. Cook, D. Dabbelt, J. Hauser, A. Izraelevitz and others, *The rocket chip generator.* In: EECS Department, University of California, Berkeley, Tech. Rep. UCB/EECS-2016-17, vol 4, pp. 6–2, 2016.

[4] Microsoft Coorporation, *Onnx Runtime Framework*, Available: https://onnxruntime.ai/ and https://github.com/microsoft/onnxruntime [last accessed on 2024-01-12].

[5] D. H. Albonesi, *Dynamic IPC/clock rate optimization.* In: Proceedings on 25th Annual International Symposium on Computer Architecture, pp. 282–292, 1998.

[6] P. Prakash, *End-to-end Model Inference and Training on Gemmini.* MA thesis. EECS Department, University of California, Berkeley, 2021.

[7] Y. Chen and Y. Xie, L. Song, F. Chen and T. Tang, *A Survey of Accelerator Architectures for Deep Neural Networks.* In: Engineering, vol 6, no. 3, pp. 264–274, 2020.

Efficient Inference Hardware Acceleration of the Yolov5s Object Detection Model at the Edge

Lukas Groth [1], Saleh Mulhem [2], and Mladen Berekovic [2]

[1] Robotics and Autonomous Systems, Universität zu Lübeck, lukas.groth@student.uni-luebeck.de

[2] Institute of Computer Engineering, Universität zu Lübeck, {saleh.mulhem, mladen.berekovic}@uni-luebeck.de

Abstract

Many energy-constrained edge devices only feature CPUs, making them unsuitable for artificial intelligence workloads. This limitation can be addressed by adding dedicated hardware accelerators such as systolic arrays for core computations of DNNs (convolutions, matrix multiplications). Modern DNN architectures such as Yolo also rely on additional compute-intensive operations. These would introduce severe bottlenecks when having to run on the CPU. The type of required operation varies between DNN architectures. In this paper, we implement the Sigmoid activation in hardware with High-Level Synthesis. The resulting accelerator is attached to a RISC-V core coupled with a systolic array. The complete system is implemented on FPGA. Our approach boosts Yolov5s inference speed by $2.27\times$ compared to systolic array alone. Moreover, compared to CPU-only execution of the Sigmoid activation, we achieve a $58.78\times$ speedup while requiring <3% of additional hardware resources.

1   Introduction

Over recent years, the demand for artificial intelligence (AI) solutions has pushed deep neural network (DNN) inference needs further to the edge. Especially in power-constrained or privacy-focused applications, the constant streaming of possibly sensitive sensor readings to the cloud for processing is highly undesirable or even impossible [1]. The complexity of modern deep learning algorithms necessitates increasingly powerful inference hardware. In this regard, dedicated hardware architectures have been proposed for efficient acceleration of compute-intensive DNN operations at the edge [2] [3]. Convolutions and matrix multiplications often account for more than 90% of the total required computation time during inference [4]. When deploying DNN models on low-power and resource-constrained platforms such as RISC-V, convolution and matrix multiplication operations might not be the only bottleneck. One challenging problem in computer vision is object detection. Object detection refers to the classification and localization of objects in an image. The You Only Look Once (Yolo) [5] family of DNN models addresses this challenge. Fig. 1 shows the number of cycles for different operations in the Yolov5s model [6] when running on a single RISC-V CPU core. Depicted is the sum of cycles for each operation during the inference for a single sample with the Yolov5s DNN. As expected, convolution operations account for the majority of overall inference cycles, contributing $\sim$95.9% to the total run time. With $\sim$3.5%, the Sigmoid activation functions contribute significantly to the total number of cycles ($> 19\times$ the cycles of all other single operations).

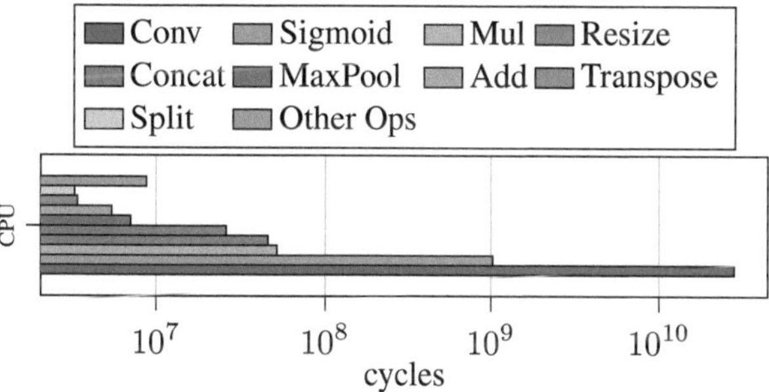

Figure 1: Profiling of DNN operations in the Yolov5s model regarding their execution time in cycles when running on a single core RISC-V CPU.

In brief, this paper makes the following contributions: to address the bottleneck of running Sigmoid on CPU, this paper proposes a custom Sigmoid hardware Unit that can be attached to a RISC-V CPU via Direct Memory Access (DMA). We generate a 64-bit RISC-V core with systolic array accelerator and use it as the baseline for performance analysis when adding our Sigmoid hardware accelerator. We implement the complete hardware system and evaluate it on a Xilinx ZCU104 FPGA.

2   Background

2.1   Rocket Chip RISC-V

RISC-V is a license free Instruction Set Architecture (ISA) for CPUs. Due to the open nature of the ISA, it can be customized to fit a large variety of application domains, such as AI [7]. Rocket [8] is a System-on-Chip (SoC) generator developed by UC Berkeley that uses the hardware construction language Chisel [9] for generating customizable general-purpose processor cores based on RISC-V. The openness of the RISC-V ISA combined with the flexibility provided by Rocket and Chisel make them well suited for computer architecture research in domains ranging from traditional computing to emerging technologies such as AI.

2.2 Gemmini Accelerator

Gemmini [2] is the systolic array-based matrix multiplication and convolution accelerator developed by UC Berkeley. Gemmini makes use of the Rocket Custom Coprocessor interface (RoCC) to integrate with the Rocket core. RoCC commands extend the RISC-V ISA with custom commands, in this case for executing matrix or tensor operations in parallel on the systolic array. Furthermore, Gemmini supports the commonly used Rectified Linear Unit (ReLU) activation function. It would be possible to add other element-wise operations to Gemmini. However, Gemmini applies ReLU directly to the result of a preceding computation and then writes the result after ReLU back to the main memory. Using Gemmini to apply the ReLU activation function without fusing it to a convolution or matrix multiplication operation would be less efficient as a large part of total latency stems from data movement and not the computation itself. In this paper, Gemmini is configured with a 4×4 systolic array and float as the datatype for all computations.

2.3 The Yolov5s Object Detection Model

The Yolo object detection family of models [5] implements a single shot detection mechanism. This makes it more lightweight than multi-stage detectors such as MaskRCNN [10]. The Yolo model takes an image as input and directly computes bounding boxes and classes of all objects the model was trained on contained in the image. MaskRCNN on the other hand would first calculate regions of interest (RoIs), then crop these regions from the image and run a bounding box regressor and classifier on each region. The light-weight single-stage approach implemented in Yolo makes it more suitable for resource constrained platforms. Yolov5 offers multiple versions varying in size. In this paper, we focus on Yolov5s, the second smallest variant with 7.2M trainable parameters and mean average precision (mAP) of 37.4 on the COCO val2017 dataset [11]. To run inference, the Yolov5s model is implemented in C. Fig. 2 shows a block consisting of convolution, Sigmoid activation, and element-wise multiplication used throughout the Yolov5s architecture. The output tensor produced by the convolution is used as input to the Sigmoid activation as well as to the multiplication. Even if Gemmini supported the Sigmoid activation, it could not be applied in the same way as the ReLU activation. The result of the convolution is needed before modification so it can not be produced during write-back to main memory.

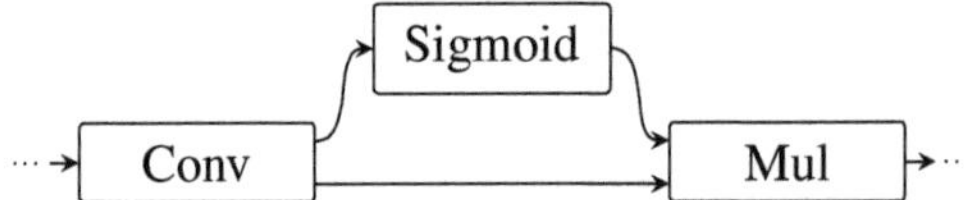

Figure 2: Common block in the Yolov5s architecture consisting of convolution, Sigmoid, and element-wise multiplication.

3 Material and Methods

3.1 Sigmoid Unit

The Sigmoid activation function for an input tensor $\mathcal{X}$ of arbitrary dimension is computed element-wise as per

$$\mathcal{Y} = \mathrm{Sigmoid}(\mathcal{X}) \text{ with } \mathcal{Y}_i = \frac{1}{1 + e^{-\mathcal{X}_i}}, i = 0, ..., n$$

with $\mathcal{X}, \mathcal{Y} \in \mathbb{R}^n$ the input- and output tensor and n the total number of elements in the tensor. For hardware implementation, we choose float as the input/output data type to maintain compatibility with the remaining operations during inference. We then apply the Sigmoid function over 256 elements in a pipelined way. Initiation of the pipeline takes 36 cycles, afterwards a new result is produced every cycle, leading to 292 cycles total for computing the 256 elements on the Sigmoid Unit, not considering any overhead introduced by communication between RISC-V core and the Sigmoid Unit. The algorithm is implemented with C++ in Xilinx Vitis HLS 2022.2 using the M_AXI AXI memory mapped interface for I/O. Therefore, both the Sigmoid Unit and the RISC-V CPU core can exchange data directly via DMA. All transactions with the Sigmoid Unit are instigated by the Rocket core via an AXI Lite control interface. This interface allows setting address pointers to the input/output data the Sigmoid Unit will then work on directly as well as starting/observing the computation state of the Sigmoid Unit.

Vitis HLS compiles the C++ algorithm to hardware description language (HDL) files. After compilation, the HDL source files are bundled as a single instance that can be used in Xilinx Vivado and connected to the Rocket RISC-V core. As the Sigmoid Unit is separate from the RISC-V CPU, it can operate at a different, higher, clock speed. In the implementation presented here, it is set to 100 MHz.

The Sigmoid Unit is capable of processing 256 elements at a time. All tensors in the Yolov5s object detection model are larger than 256, so the data is split into chunks of 256 elements. During inference, a pointer to each chunk of the input tensor X and output tensor Y is passed to the Sigmoid Unit by the RISC-V CPU. Then the execution is started. This computes the element-wise Sigmoid activation for 256 elements in X and writes 256 results to Y. After waiting for the computation to be done, the next chunk of data is processed. This is illustrated in Fig. 3, showing the C-Code running on the RISC-V CPU for computing Sigmoid over a tensor with dimension $1 \times 32 \times 320 \times 320$. Here, the functions SIGM_SetIn and SIGM_SetOut write the address of the input/output data to the AXI Lite Control interface.

```c
static inline void model_0_Sigmoid(
    const float X[1][32][320][320],
    float Y[1][32][320][320])
{
 const float *X_ptr = X;
 float *Y_ptr = Y;
 for (unsigned i=0; i<12800; i++) {
    SIGM_SetIn(&sigm, &X_ptr[i*256]);
    SIGM_SetOut(&sigm, &Y_ptr[i*256]);
    SIGM_Start(&sigm);
    while(!SIGM_IsDone(&sigm));
 }
}
```

Figure 3: C-Code for Sigmoid Activation running on the RISC-V CPU for communicating with the Sigmoid Unit.

`SIGM_Start` starts the Sigmoid Unit computation. The Rocket core then waits for the `SIGM_IsDone` flag to be set in the AXI Lite interface before writing the next address chunk. Inference performance could be improved further by increasing the dimension of the Sigmoid Unit to more than 256 or by increasing the clock speed of the Sigmoid Unit.

3.2 New Hardware Accelerator Architecture

To evaluate the performance in a real-world scenario, the complete system consisting of Rocket core RISC-V CPU, the Gemmini accelerator, and our proposed Sigmoid Unit is implemented in Xilinx Vivado and synthesized for the Xilinx ZCU104 FPGA. The Chipyard [12] Framework is used to generate the SoC with the Rocket core and Gemmini according to the configuration from Sections 2.1 and 2.2. The resulting HDL modules are placed in Vivado. A DDR4 SDRAM Controller is added to serve as the DRAM for the complete system. The physical DDR4 memory bank on the ZCU104 FPGA connected to the programmable logic is populated with a 16GB module. The Sigmoid Unit generated by Vitis HLS is connected in Vivado to the Rocket core via the AXI Lite interface and to DRAM via AXI. Additionally, an AXI UART for serial communication between the SoC and a host computer is added and connected to the Rocket core via AXI Lite. Because of the tight integration between the Rocket core and Gemmini, they use the same clock source which is set to 60 MHz. The Sigmoid Unit clock is independent and can thus be set higher at 100 MHz. For Rocket+Gemmini, 60 MHz is the maximum attainable clock speed on our target hardware. It might be possible to further increase the Sigmoid Unit clock speed at the cost of additional routing resources. Fig. 4 shows the simplified SoC hardware design with our Sigmoid Unit in the bottom left. It is synthesized for the ZCU104 FPGA using Vivado 2022.2. Any Software running on the prototype is written in C and compiled for RISC-V using the toolchain provided by Chipyard. After compilation, the binary is loaded into DRAM via JTag using Xilinx XSDB. Performance results

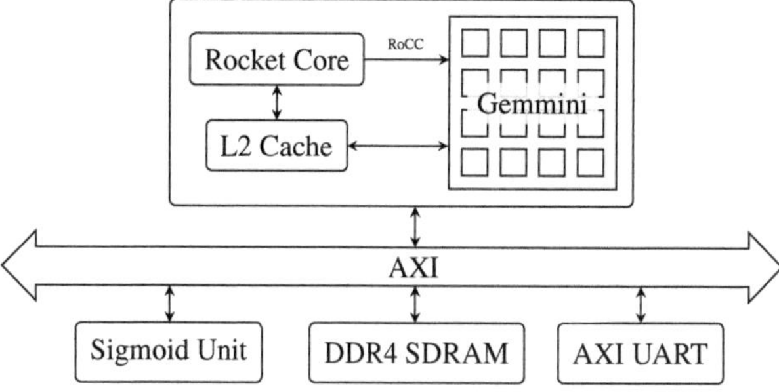

Figure 4: Gemmini accelerator [2] with Sigmoid Unit: new edge hardware SoC accelerator architecture.

are gathered via the UART serial interface between RISC-V CPU and a host computer.

4 Results and Discussion

Performance metrics are captured by counting the number of cycles each operation takes as seen by the RISC-V CPU running on hardware. This is independent of whether the operation is running on the CPU, on Gemmini, or the Sigmoid Unit. Fig. 5 shows the performance profiling of different operations in the Yolov5s model with the number of cycles an operation takes on the x-axis. The y-axis indicates

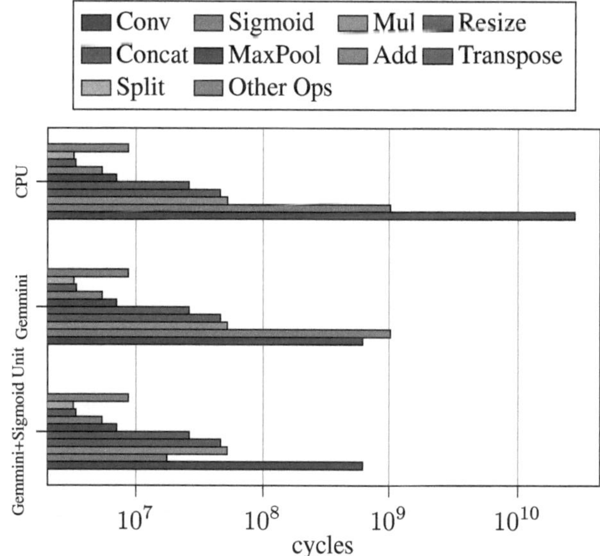

Figure 5: Profiling of DNN operations in the Yolov5s model regarding their execution time in cycles depending on the inference hardware.

the hardware component the execution was run on. CPU means all convolutions and Sigmoid operations are running on the RISC-V CPU, while Gemmini+Sigmoid Unit means convolutions are running on Gemmini and Sigmoid on our proposed Sigmoid hardware Unit. All remaining operations are running on the CPU. These results show a significant increase in inference performance when using the dedicated hardware accelerators. When running on the CPU, the Sigmoid activations alone take $\sim$1025.95M cycles in total on

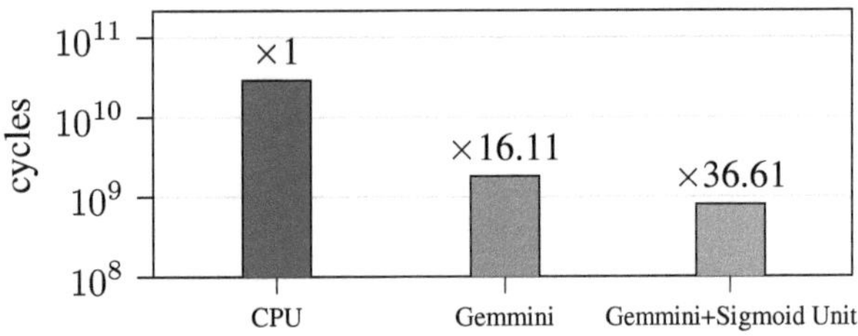

Figure 6: Execution time in cycles for one sample w.r.t inference hardware and the total speedup over CPU.

the CPU, versus $\sim$17.45M cycles on the Sigmoid Unit. This results in a 58.78× speedup in Sigmoid performance. Fig. 6 plots the total number of inference cycles for the aforementioned configurations of CPU/Gemmini/Sigmoid Unit on the y-axis and the respective speedup compared to CPU on top. Adding the Gemmini convolution accelerator boosts performance by 16.11× over the CPU implementation. We achieve an additional 2.27× boost when adding our proposed Sigmoid Unit. Thus, with both accelerators working together, a total speedup of 36.61× is achieved over the baseline CPU implementation.

	LUTs	FFs	BRAM	DSPs
Rocket RISC-V	42.223 (18.3%)	20.120 (4.4%)	142 (45.5%)	15 (0.9%)
+ Gemmini	130.767 (56.8%)	30.039 (6.5%)	80 (25.6%)	352 (20.4%)
+ Sigmoid (256)	6.449 (2.8%)	4.714 (1.0%)	3 (1.0%)	18 (1.0%)

Table 1: FPGA resource usage of Rocket RISC-V CPU, the additional Gemmini accelerator, and the Sigmoid Unit with percentage of total ZCU104 resources in brackets.

Table 1 shows the FPGA resource utilization of the Rocket RISC-V CPU, the additional Gemmini accelerator, and the Sigmoid Unit on the FPGA. The Gemmini accelerator significantly increases resource utilization with an overhead of 6.5% to 56.8% over just the CPU. This is justified by the significant performance improvement of 16.11×. The Sigmoid Unit is more lightweight, incurring a <3% resource overhead on top of the Rocket RISC-V core while boosting performance by an additional 2.27× as previously mentioned.

5   Conclusion

This paper proposes extending RISC-V-based edge devices used for AI inference with a Sigmoid Unit. We demonstrate the viability of our approach by implementing a complete SoC prototype on the Xilinx ZCU104 FPGA. This prototype boosts Yolov5s inference performance by 58.78× for Sigmoid layers over the baseline RISC-V CPU implementation and an additional 2.27× overall compared to convolution acceleration alone leading to a 36.31× improvement in inference cycles for the complete model.

These results support the validity of integrating specialized units like the Sigmoid Unit into RISC-V-based edge devices. The performance gains hold promising implications for real-world applications across diverse domains.

Acknowledgement

The work has been carried out at the Institute of Computer Engineering, Universität zu Lübeck and supervised by Prof. Dr.-Ing. Mladen Berekovic.

Authors' Statement

Conflict of interest: Authors state no conflict of interest.

6   References

[1] S. Deng, H. Zhao, J. Yin, S. Dustdar, and A. Y. Zomaya, "Edge intelligence: the confluence of edge computing and artificial intelligence," *CoRR*, vol. abs/1909.00560, 2019.

[2] H. Genc *et al.*, "Gemmini: Enabling systematic deep-learning architecture evaluation via full-stack integration," in *Proceedings of the 58th Annual Design Automation Conference (DAC)*, 2021.

[3] Nvidia, "Nvdla: Nvidia deep learning accelerator," http://nvdla.org/primer.html, 2018.

[4] A. G. Howard *et al.*, "Mobilenets: Efficient convolutional neural networks for mobile vision applications," 2017.

[5] J. Redmon, S. Divvala, R. Girshick, and A. Farhadi, "You only look once: Unified, real-time object detection," in *Proceedings of the IEEE conference on computer vision and pattern recognition*, 2016.

[6] Ultralytics, "YOLOv5: A state-of-the-art real-time object detection system," https://docs.ultralytics.com, 2021, accessed: 03.01.2024.

[7] S. Kalapothas, M. Galetakis, G. Flamis, F. Plessas, and P. Kitsos, "A survey on risc-v-based machine learning ecosystem," *Information*, 2023.

[8] K. Asanović *et al.*, "The Rocket Chip Generator," EECS Department, University of California, Berkeley, Tech. Rep. UCB/EECS-2016-17, April 2016.

[9] J. Bachrach *et al.*, "Chisel: Constructing hardware in a scala embedded language," in *Proceedings of the 49th Annual Design Automation Conference*, ser. DAC '12. New York, NY, USA: Association for Computing Machinery, 2012.

[10] K. He, G. Gkioxari, P. Dollár, and R. Girshick, "Mask r-cnn," in *2017 IEEE International Conference on Computer Vision (ICCV)*, 2017.

[11] T.-Y. Lin *et al.*, "Microsoft coco: Common objects in context," 2015.

[12] A. Amid *et al.*, "Chipyard: Integrated design, simulation, and implementation framework for custom socs," *IEEE Micro*, 2020.

11

Psychology

Researching the Effect of Collective Efficacy on Team Flow in High Responsibility Teams: A Focus on Firefighting Operations

Svea Kaduk [1], Fabienne Aust [2], Lena Heinemann [3], Maik Holtz [4], Vera Hagemann [3], and Corinna Peifer [2]

[1] Psychology - Cognitive Systems, Universität zu Lübeck, svea.kaduk@student.uni-luebeck.de

[2] Research Group Work and Health, Department of Psychology, Universität zu Lübeck, {fabienne.aust, corinna.peifer}@uni-luebeck.de

[3] Business Psychology and Human Resource Management, Faculty of Business Studies and Economics, Universität Bremen, {lena.heinemann, vhagemann}@uni-bremen.de

[4] Cologne Professional Fire Department, Institute for Security Science and Rescue Technology (ISR), maik.holtz@stadt-koeln.de

Abstract

Investigating the interaction of collective efficacy and team flow in firefighting teams, the present research aims to fill a gap in understanding how collective psychological conditions correlate with team performance under high-stress conditions. Utilizing quantitative methods, the study involved surveying firefighter trainees to assess their collective efficacy beliefs before and experiences of team flow after a firefighting related exercise. The results revealed a significant positive correlation, showing that teams with higher collective efficacy beliefs tend to report greater experiences of team flow. These findings suggest that supporting and strengthening collective efficacy through targeted training could lead to improved team dynamics and performance in the field. In addition, the promotion of high collective efficacy and the resulting team flow could reduce stress levels within the team and thus contribute to the well-being of team members.

1 Introduction

To work in a fire brigade means to work in a so-called High Responsibility Team (HRT) [1]. The working conditions are demanding: high pressure, high responsibility for other people's lives and the environment, managing challenging tasks continuously, without breaks, stopping or adjusting their actions [1]. In the context of fire brigades, the defining characteristics of HRTs include not only the necessity for rapid decision-making, but also the ability to operate effectively under extreme uncertainty and risk.

The concept of flow, initially explored at the individual level by [2], describes a state of being fully immersed by a task that is challenging but still under control. This occurs when individuals face a balance between the challenge of a task and their own skill level [2]. It is often encountered as a state in situations that are relevant to stress. In addition, however, flow can lead to stress reactions being alleviated and thus increase job satisfaction in the work context [3]. Recent studies have expanded this concept to encompass team flow, a collective state where team members experience a deep sense of engagement and focus while performing tasks that rely on each other, all for the benefit of the team [4]. In high-stress professions like firefighting, achieving flow could enhance focus and performance, reducing the perception of stress.

Self-efficacy, a construct reflecting an individual's beliefs in their own competence to successfully plan, organize and carry out necessary actions to achieve specific goals or attainments [5], is often integrated with the concept of flow, serving as an antecedent that facilitates the emergence of flow states [6]. According to [5] it does not only empower individuals to face challenging tasks but also enhances their ability to cope with stress and pressure, which is a crucial aspect in high responsibility environments like firefighting. Perceived collective efficacy refers to the common beliefs within a group regarding its ability to effectively organize and work together to achieve specific goals or desired results [5].

Reference [7] showed in a longitudinal, controlled laboratory setting that collective efficacy beliefs could influence the way a group and its team members feel. Additionally, high collective efficacy beliefs were a predictor for team flow. These findings are based on a study with university students who participated in three laboratory tasks (three measurement points at intervals of 1 week). Collective efficacy beliefs were measured by using the scale as in [8] developed. Collective flow, which is another term for team flow, was measured using a group task absorption scale, a group task enjoyment scale as well as the group challenge and group skills with two self-constructed items. Limitations of the study included the fact that it was only tested under laboratory conditions and the fact that the cohort consisted mainly of women [7]. Therefore, this study aims to close this gap in understanding these dynamics in real-world HRT contexts, particularly within firefighting teams that often operate in unpredictable and high-stress environments. In addition, this study examines a cohort consisting primarily of male subjects.

Based on previous research, the present study aims to investigate the predictive relationship between collective efficacy beliefs and team flow within the context of firefighting operations. It is hypothesized that higher collective efficacy beliefs in firefighting teams will correlate with an increased experience of team flow during operations. This research not only contributes to the theoretical foundations of team dynamics in HRTs, but also has practical implications for training and development within such teams, ultimately enhancing their effectiveness and well-being in high-stakes environments.

Hypothesis 1: The more collective efficacy beliefs at T1, the higher the level of experienced team flow at T2.

2 Material and Methods

2.1 Participants and Design

The sample consisted of trainees from the Cologne professional fire department and the full-time Frechen fire station. The experiment took place between September 28, 2020, and July 20, 2022. Questionnaires were filled in at two different time points. The first questionnaire at time 1 (T1) could be filled out at home or at work. The second questionnaire at time 2 (T2) was filled out immediately after participating in an exercise which involved fire extinguishing as well as rescuing a human. 168 firefighters answered the questionnaires and participated in the exercise. 29 participants were excluded due to missing values on the study variables. The final sample consisted of 135 participants, 128 men, 7 women and 0 diverse. Ages ranged from 20 to 40 years ($M = 27.0$; $SD = 4.67$).

2.2 Measures

The experience of team flow was measured using 12 items based on the Flow-Short-Scale [9] and the Team-Flow-Monitor [10], translated into German. The resulting scale has four subcategories: being absorbed, fluency, sense of unity and trust and control. Participants rated items on a 7-point Likert scale (1 = *strongly disagree* to 7 = *strongly agree*). An example of an item is 'We felt optimally challenged as a team'. For the analysis, a total score using the mean of the items was calculated because according to [11] the team flow experience embodies a combination of the various elements and is only presumed to be a team flow experience if they happen concurrently.

The collective efficacy beliefs were measured using a scale developed by [8], translated into German. The 4 items were rated on a 5-point Likert scale (1 = *disagree at all* to 5 = *fully agree*). An example of an item is 'I feel confident about the capability of my group to perform this task very well'. Cronbach's α for the scale measuring team flow was .90.

The Cronbach's α of the subcategories ranged between .64 and .80 (see Table 1). Cronbach's α for the scale measuring collective efficacy beliefs was .87.

2.3 Data Analysis

The statistical software jamovi was used to analyze the data. First, internal consistencies of the two scales were computed. Then the descriptive analyses were performed. A linear regression was calculated to investigate whether the collective efficacy beliefs predicted team flow.

3 Results and Discussion

3.1 Descriptive Results

Table 1 displays the means, standard deviations and the internal consistencies of the study variables. Noticeably, high mean values for collective efficacy beliefs and the team flow experience can be seen.

Table 1: Means (M), standard deviations (SD) and internal consistencies (Cronbach's α).

Variable	n	M	SD	α
Age	165	27.01	4.67	–
Team flow (Tf)	139	5.70	0.76	.90
Tf being absorbed	156	5.20	0.75	.64
Tf fluency	157	4.66	0.86	.77
Tf sense of unity	155	5.22	0.83	.80
Tf trust and control	156	5.42	0.87	.74
Collective efficacy beliefs	155	4.36	0.55	.87

3.2 Linear Regression

To test if there is a relationship between the collective efficacy beliefs and team flow, a linear regression was computed. The criteria necessary for the analysis, including outlier identification, linearity, absence of multicollinearity, normal distribution, and homoscedasticity, were successfully fulfilled. The linear model, which examines the relationship between team flow and collective efficacy beliefs, proved to be significant, $F(1, 133) = 12.66, p < .001$. The R^2 for the model was 0.09 (adjusted $R^2 = 0.08$). This indicates a significant, albeit small, relationship. A scatter plot with a linear regression line was constructed to quantify their relationship, stating a correlation coefficient of $\beta = .295, p < .001$. This moderate positive correlation suggests that as collective efficacy beliefs in teams increase, there is a corresponding rise in team flow experiences, though the relationship is not strongly marked. Similarly, the four subcategories within the team flow scale demonstrated a consistent trend, each exhibiting a moderate positive correlation with self-efficacy, indicating that an increase in the collective efficacy beliefs is also associated with a corresponding rise in the specific aspects of team flow, albeit the relationship remains moderate. The visual presentation as in Fig. 1 of the relationship between collective efficacy beliefs and team flow provides an overview of the data trends in addition to the numerical analysis and offers a better understanding of the interaction between the two variables.

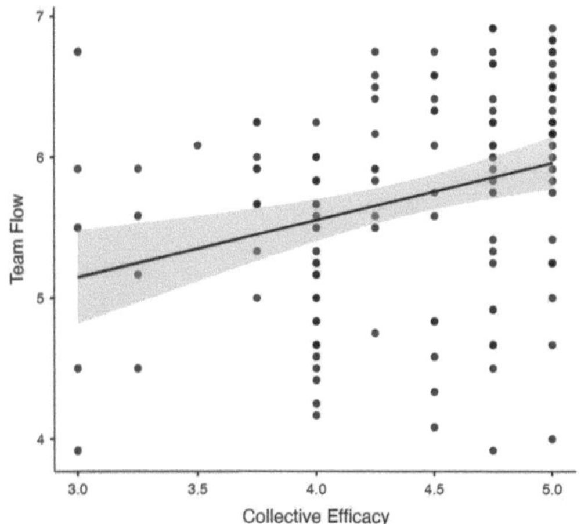

Figure 1: Relationship between Collective Efficacy and Team Flow with Regression Line.

3.3 Discussion

The aim of the present study was to investigate the relationship between the collective efficacy beliefs recorded before a firefighting related exercise and the team flow recorded after the exercise. The firefighters' assessments of the perceived collective efficacy beliefs and the experienced team flow were collected using questionnaires at two different points in time. Through the results of the linear regression, Hypothesis 1, "The more collective efficacy beliefs at T1, the higher the level of experienced team flow at T2." was supported by the data, meaning that the collective efficacy beliefs in the team were a significant predictor for experienced team flow. These findings align with those of [7], who discovered positive correlations between the collective efficacy beliefs in teams and the experienced team flow, thereby enriching the understanding of group dynamics in high-stress environments like firefighting.

The present study observed only a weak yet significant effect of collective efficacy on team flow. This highlights the complexity of real-world operational exercises and the multitude of uncontrolled confounding factors inherent in such settings. Despite these challenges, the significant predictive relationship found is noteworthy and underscores the interaction between collective efficacy and team flow in operational environments.

In firefighting operations, where rapid decision-making and effective operation under extreme uncertainty and risk are essential [1], the presence of a strong collective efficacy belief can be beneficial. This study extends the application of these concepts from a controlled laboratory environment, predominantly involving female participants, to a real-world setting with male-dominated firefighter trainees. Such a shift in demographic and environment offers a new perspective, contributing to the literature on team flow and collective efficacy in diverse contexts.

In accordance with the findings of [11], this study suggests that targeted training programs focusing on building collective efficacy could be instrumental in enhancing team flow experiences. Training that simulates the high-stress conditions of firefighting can prepare teams to develop a stronger sense of collective capability, which is essential for achieving optimal flow states in real-world operations.

In conclusion, this study not only underlines the findings of [7] regarding the influence of collective efficacy beliefs on team flow but also expands this understanding to the context of firefighter trainees in real-world scenarios. The significance of these findings lies in their applicability to developing training programs and interventions aimed at enhancing team performance in high-stress, high-responsibility environments.

3.4 Strengths, Limitations and Future Research

This study extends the application of the concepts of collective efficacy and team flow from controlled laboratory settings to real-world contexts, specifically within firefighter trainees, offering new insights into team dynamics in high-stress environments. The significant correlation found between collective efficacy beliefs and team flow, despite the complex and unpredictable nature of firefighting, underlines the robustness of these psychological constructs in operational settings. This study also contributes to diversifying the demographic scope of existing studies, which predominantly involved female participants, by utilizing a sample primarily consisting of male firefighter trainees. The methodological approach, involving the assessment of collective efficacy beliefs prior to a firefighting exercise and the measurement of team flow afterwards, provides a structured and sequential understanding of these interrelated concepts. A limitation of the study is that it did not control for other potential confounders that occur in real-life situations. However, due to the relatively small effect size observed in the predictive relationship between collective efficacy and team flow, this is suggested. The study's reliance on self-reported questionnaires may lead to common method biases, especially in the assessment of complex constructs as team flow and collective efficacy are. Given the specific context of firefighting, the generalizability of the findings to other high-responsibility teams operating in different environments remains to be established. Finally, due to the cross-sectional design of the study, no causal relationship between the collective efficacy beliefs and team flow can be assumed.

This study opens fields for further research, for example in the realm of collective emotions and their contagion in high-responsibility settings. The theory of emotional contagion, suggesting that individuals naturally imitate and synchronize their emotions with others [12], could be a potential mechanism underlying the spread of flow experiences within teams. In firefighting, where mimic expressions are often constrained by equipment, understanding how non-verbal cues and collective emotional states contribute to team flow becomes an interesting area for future inquiry.

4 Conclusion

The results presented in this paper has brought new insights into the phenomena of collective efficacy and team flow within real workgroups. It was shown, that the collective efficacy beliefs within a team have a significant correlation with the experienced team flow during a realistic operational exercise. These findings play a role in deepening the understanding of team dynamics and their influence on workgroup performance. This study suggests that when firefighter teams have strong beliefs in their collective capabilities, they are more likely to experience a state of deep engagement and synchronization in their tasks, known as team flow. This state is not just beneficial for their performance during emergency responses, but may also be relevant for maintaining high morale and team cohesion. Given the challenging and unpredictable nature of firefighting, fostering this cycle of collective efficacy and flow may be vital for both the safety and the success of the teams. By nurturing an environment where firefighter teams can cultivate strong collective efficacy beliefs and experience team flow, organizations could enhance the efficiency, safety, and well-being of their teams, ultimately leading to more effective emergency response and community safety. However, this study reveals just a moderate correlation between collective efficacy and team flow, indicating that flow experiences in workgroups, especially among firefighters, are multifaceted and can be affected by numerous antecedents. Given the unpredictable and challenging nature of firefighting, this highlights the complexity of fostering flow under conditions of high responsibility and uncertainty.

Acknowledgement

The work has been carried out and was supervised by the Research Group Work and Health of the Department of Psychology, Universität zu Lübeck. The data was collected as part of the 'Gemeinsam stark — Teamtraining Brandbekämpfung [Strong Together - Team Training Firefighting]' project.

Authors' Statement

Conflict of interest: Authors state no conflict of interest. Informed consent: Informed consent has been obtained from all individuals included in this study. Ethical approval: The research related to human use complies with all the relevant national regulations, institutional policies and was performed in accordance with the tenets of the Helsinki Declaration, and has been approved by the authors' institutional review board or equivalent committee.

5 References

[1] V. Hagemann, A. Kluge and S. Ritzmann, *High responsibility teams - a systematic analysis of teamwork contexts for effective skills acquisition [High Responsibility Teams - Eine systematische Analyse von Teamarbeitskontexten für einen effektiven Kompetenzerwerb]*. In: Journal Psychologie des Alltagshandelns, vol. 4, no. 1, pp. 22–42, 2011.

[2] M. Csikszentmihalyi, *Beyond boredom and anxiety.* Jossey-bass, 1975.

[3] C. Peifer and G. Wolters, *Flow in the context of work.* In: C. Peifer and S. Engeser (Eds.), *Advances in flow research,* Springer International Publishing, pp. 287–321, 2021.

[4] J. J. J. van den Hout and O. C. Davis, *Team flow: The psychology of optimal collaboration.* Springer International Publishing, 2019.

[5] A. Bandura, *Self-efficacy: The exercise of control.* NY: Freeman, New York, 1997.

[6] A. Rodríguez-Sánchez, M. Salanova, E. Cifre and W. B. Schaufeli, *When good is good: A virtuous circle of self-efficacy and flow at work among teachers.* In: International Journal of Social Psychology, vol. 26, no. 3, pp. 427–441, 2011.

[7] M. Salanova, A. M. Rodríguez-Sánchez, W. B. Schaufeli and E. Cifre, *Flowing together: A longitudinal study of collective efficacy and collective flow among workgroups.* In: The Journal of psychology, vol. 148, no. 4, pp. 435–455, 2014.

[8] M. Salanova, S. Llorens, E. Cifre, I. M. Martínez and W. B. Schaufeli, *Perceived collective efficacy, subjective well-being and task performance among electronic work groups: An experimental study.* In: Small Group Research, vol. 34, no. 1, pp. 43–73, 2003.

[9] F. Rheinberg, R. Vollmeyer and S. Engeser, *Measuring flow-experience [Die Erfassung des Flow-Erlebens] .* In: J. Stiensmeier-Pelster and F. Rheinberg (Eds.), Diagnostik von Motivation und Selbstkonzept, Hogrefe, Göttingen, pp. 261–279, 2003.

[10] J. J. J. van den Hout, J. M. P. Gevers, O. C. Davis and M. C. D. P. Weggeman, *Developing and Testing the Team Flow Monitor (TFM).* In: Cogent Psychology, vol. 6, no. 1, 1643962, 2019.

[11] F. Aust, L. Heinemann, M. Holtz, V. Hagemann and C. Peifer, *Team Flow Among Firefighters: Associations with Collective Orientation, Teamwork-Related Stressors, and Resources.* In: International Journal of Applied Positive Psychology, vol. 8, pp. 339–363, 2023.

[12] E. Hatfield, J. T. Cacioppo and R. L. Rapson, *Emotional contagion.* Cambridge University Press, Cambridge, 1993.

Influence of a multimodal stimulus presentation on emotion perception

Ricus Gattermann [1] and Sarah Jessen [2],

[1] Psychology Cognitive Systems, Universität zu Lübeck, ricus.gattermann@student.uni-luebeck.de

[2] Institute of Medical Psychology, Universität zu Lübeck, sarah.jessen@neuro.uni-luebeck.de

Abstract

The ability to perceive emotions correctly is important especially for social behavior. There are several different methods to measure emotion perception. In this paper, we want to investigate to what extent the multimodal presentation of stimuli affects the performance of emotion recognition, the perceived strength and naturalness of the emotion, and the perceived arousal compared to a unimodal presentation. For this purpose, we divided participants into three groups and showed them either a video with sound, a muted video, or an audio file of a semantically neutral sentence. This resulted in the conditions audiovisual (AV), auditory (A), and visual (V). It was shown that a multimodal presentation positively influences performance and perceived strength. A general influence of the multimodal presentation on perceived naturalness or a positive influence on arousal could not be demonstrated. However, further studies, for example with a more diverse sample composition, could provide more insight in this regard.

1 Introduction

Emotion perception is an important skill, cross-cultural and of great importance in social behavior [1]. It plays a major role in early childhood and infant development, but also in adolescence and adulthood and is also associated with the development of a range of psychopathologies [2].

In recent years, various studies have been published on the perception of emotions, particularly in facial expressions, using imaging techniques, EEG or behavioral observation. It has been shown that emotion perception is a complex process in which several brain regions (e.g. amygdala, orbitofrontal cortex) are involved [1]. It can be assumed that the representation of emotions is at least partially amodal [3].

In addition, perception of amodal information is enhanced by multimodal stimulation, as information presented simultaneously and synchronously in two sensory modalities attracts attention and facilitates perceptual learning of amodal information to a greater extent than when the same information is presented in only one sensory modality [4]. EEG studies also show that multimodal processing plays an important role in the perception of emotions [5].

Therefore, the question is whether subjects' emotion perception performance is positively influenced when emotions are presented with multimodal stimuli compared to unimodal stimuli.

It can be assumed that performance is improved when the stimuli are audiovisual compared to auditory or visual stimuli. It is also expected that the emotions will be perceived as stronger and more intense and that the perceived naturalness of the presented emotions will be influenced.

Our hypotheses are therefore as follows:

Hypothesis 1a (H1a): The accuracy of emotion perception is significantly higher with multimodal presentation of stimuli than with unimodal presentation.

Hypothesis 1b (H1b): The intensity of perceived emotion is significantly stronger with multimodal presentation of stimuli than with unimodal presentation.

Hypothesis 2 (H2): The perceived naturalness of emotion differs significantly between multimodal and unimodal presentation of stimuli.

Hypothesis 3 (H3): Perceived arousal (intensity) is significantly higher with multimodal presentation of stimuli than with unimodal presentation.

2 Material and Methods

2.1 Participants

The participants were recruited from the student environment via the mailing list. A total of $n = 67$ participants took part and completed the survey. $N = 59$ were female, $n = 8$ were male. The average age was $M = 22.51$ years, ranging from 18 to 41 years ($SD = 4.61$). Most of the participants were psychology students.

2.2 Materials

The participants watched short videos or listened to audios that depicted various emotions: fear, joy, and anger. There were five different semantically neutral sentences that were presented by six different actresses in a neutral setting.

In total, the participants were presented with 90 different videos or audios.

2.3 Design

The study used a between-subject design, where the participants were divided into three groups: one group that only heard the auditory emotions (A with n = 21), one group that saw/heard the emotions audiovisually (AV with n = 27), and one group that only saw the visual emotions (V with n = 19). The assignment was made at random.

2.4 Procedure

The survey was conducted online via the Tivian platform. The participants were asked to answer a series of questions after watching or listening to each video or audio. They were asked which emotion they perceived, how strongly they perceived the emotion (on a Likert scale from 1 to 7, from very weak to very strong), how naturally they perceived the emotion (on a Likert scale from 1 to 7, from very unnatural to very natural), and how intensely they perceived the emotion (according to the arousal panel of the SAM scale (SELF-ASSESSMENT MANIKIN)) [6].

2.5 Statistical Analysis

The statistical analysis was carried out with R. A Chi-square test was conducted to verify the first hypothesis (H1a). For the verification of hypotheses H1b, H2, and H3, an ANOVA was conducted for each, as well as post hoc tests. (Tukey multiple comparisons of means; 95% family-wise confidence level).

3   Results

3.1   Hypothesis 1a

To investigate the extent to which the mode of presentation of stimuli influences the hit rate in recognized emotions (H1a), the individual success rates for each group were first calculated. In the case of the AV condition, 2210 out of a total of 2430 played items (90 videos with 27 test persons) were correctly recognized. This results in a hit rate of 90.1%. In the A condition, 1381 out of 1890 were correctly recognized. This corresponds to a hit rate of 73.1%. In the V condition, a hit rate of 1447 out of 1710 was achieved, 84.6%. To check for significant differences for these results, a Pearson's Chi-squared test was calculated χ^2 (2, n = [6,030]) = 249.21, p <.001. It can therefore be assumed that the hit rate is significantly higher under the AV condition than under the unimodal conditions. The individual quotas for the different emotions were also evaluated. It is shown that in the V condition joy (93.7%) and anger (93.7%) were recognized very well, while in the A condition only joy (86.8%) stands out with a higher hit rate and anger (65.9%) was recognized as poorly as fear (66.5%). As detailed in Table 1. Pearson's Chi-squared tests also show significant differences between the conditions for the individual emotions: Fear: χ^2 (2, n = [2,010]) = 78.01, p <.001, Joy: χ^2 (2, n = [2,010]) = 38.99, p <.001, Anger: χ^2 (2, n = [2,010]) = 254.11, p <.001.

Table 1: Hit rate for the perception of emotions

Group	Fear	Joy	Anger	Total
A	419/630	547/630	415/630	1381/1890
	66,5%	86,8%	65,9%	73,1%
AV	682/810	773/810	755/810	2210/2430
	84,2%	95,4%	93,2%	90,1%
V	379/570	534/570	534/570	1447/1710
	66,5%	93,7%	93,7%	84,6%

3.2   Hypothesis 1b

To investigate whether emotions are perceived more strongly when presented with a multimodal stimulus (H1b), an ANOVA was calculated (3x1 one-way ANOVA with the between-subject factor group). The mean for the queried intensity in the AV condition is M = 4.2, for the A condition M = 4.02, and for the V condition M = 4.06. Thus, the perceived intensity of the emotion was reported as slightly higher under the multimodal presentation of stimuli. The ANOVA revealed a significant effect of: F(2, 6027) = 7.687, p <.001. A group differences can therefore be assumed to be significant. To specify more precisely whether this also applies to the difference between the multimodal and the two unimodal conditions, a post hoc test was conducted (Tukey multiple comparisons of means; 95% family-wise confidence level). The difference between the AV group and the A group was statistically significant, with a mean difference of 0.17 (p <.001). This suggests that the Participants in the AV group judged the emotions as significant more intense compared to the A group. The comparison between the V group and the AV group showed a significant negative mean difference of -0.13 (p <.013). This suggests that the V group has a significant lower mean than the AV group. As depicted in Fig. 1. The A group showed no significant difference to the V group in the evaluation of the intensity of emotions. It can therefore be assumed that emotions are perceived as significantly stronger under multimodal conditions.

3.3   Hypothesis 2

For the question of whether a multimodal presentation influences the perceived naturalness (H2), an ANOVA was conducted too (3x1 one-way ANOVA with the between-subject factor group). The mean for perceived naturalness in the AV condition is M = 3.5, for the A condition M = 3.57, and for the V condition M = 3.92. The ANOVA revealed a significant effect of: F(2, 6027) = 42.85, p <.001. The post hoc test (Tukey multiple comparisons of means; 95% family-wise confidence level) shows that there was no

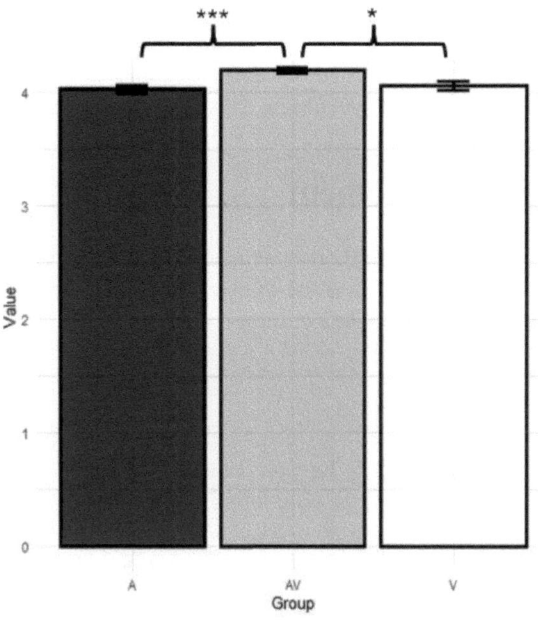

Figure 1: Perceived strength of emotion The mean differences between the AV and A groups show a very high level of significance, the mean differences between the AV and V groups show a moderate level of significance. Visualised with SEM. * p<.05, *** p<.001

significant difference between the AV and A condition, but there was between the V group and the A group (mean difference 0.35, p <.001) and between the V group and the AV group (mean difference 0.42, p <.001). As Fig. 2 illustrates. Given these data, the hypothesis that emotions are perceived as significantly different natural under a multimodal condition is not confirmed. Rather, the perceived naturalness seems to be significantly higher under a purely visual condition than under a purely auditory or an audiovisual condition.

3.4   Hypothesis 3

To investigate the question of whether a multimodal presentation of stimuli leads to higher arousal (H3), an ANOVA was also calculated (3x1 one-way ANOVA with the between-subject factor group). The mean for the queried arousal in the AV condition is M = 3.02, for the A condition M = 2.99, and for the V condition M = 2.94. It should be noted that the higher the indicated value, the lower the arousal was rated (according to the arousal panel of the SAM scale (SELF-ASSESSMENT MANIKIN)). Therefore, the participants indicate the least arousal under the AV condition, although the differences are quite small. The ANOVA revealed a significant effect of $F(2, 6027) = 3.163$, p <.042. The post hoc test (Tukey multiple comparisons of means; 95% family-wise confidence level) only shows a slight significance of p <.032 in the comparison between the V group and the AV group with a mean difference of 0.084. As can be seen in Fig. 3. The assumption formulated in the hypothesis could therefore not be confirmed.

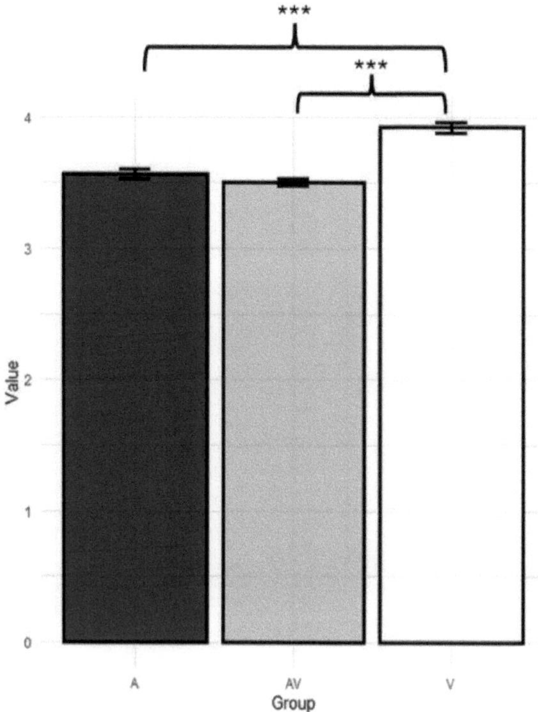

Figure 2: Perceived naturalness of emotion. The mean differences between the V and A group and between the V and AV group show a very high level of significance. No significant difference was found between the AV and A groups. Visualised with SEM. *** p<.001

4   Discussion

In this study, we investigated to what extent the multimodal presentation of stimuli affects the performance of emotion perception, the perceived strength of emotion and naturalness of emotion, and the perceived arousal compared to a unimodal presentation. Our hypothesis was that a multimodal presentation of stimuli would enhance the participants' ability to discern the emotion being conveyed to them. We also hypothesised that the multimodal presentation would lead to a higher perceived strength of the emotion, a higher perceived naturalness and a higher arousal. The results showed that a multimodal presentation positively influences the performance of emotion perception and the perceived strength of emotion. This observation is consistent with previous findings on the effect of multimodal stimuli presentation on emotion perception, particularly in the effects observed in EEG [5]. A general influence of the multimodal presentation on the perceived naturalness or a positive influence on arousal could not be demonstrated. It is important to consider that the participants in our sample exhibit little diversity in terms of both gender and age. Approximately 88% of the participants were female and the average age was only M = 22.51 years. Moreover, most of the participants were psychology students. Young adults do not benefit as much from multimodal stimuli in emotion perception as do infants or older adults [7]. Moreover women have a higher ability to perceive emotions in faces than men

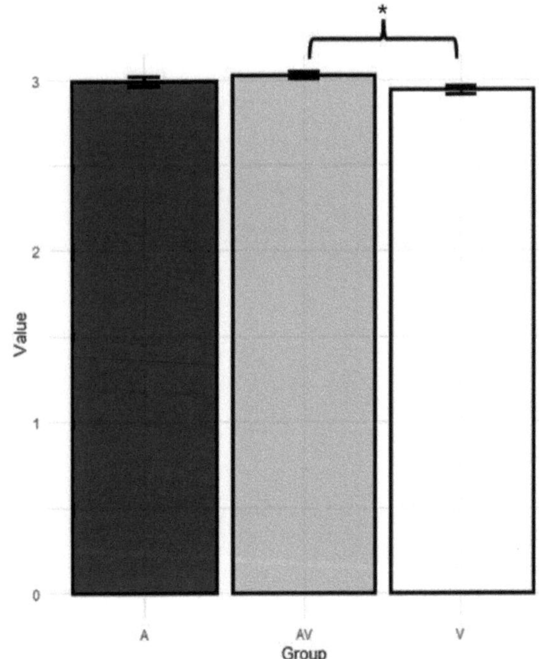

Figure 3: Perceived arousal of emotion. The mean differences between the AV and V groups show a low level of significance. On the other hand, no significant difference was found between the AV and A groups. It should be noted that a higher value means a low level of arousal. Visualised with SEM. * p<.05

[1]. However, for the first hypothesis (H1a, H1b), this study provides evidence that seems to confirm that participants recognize emotions better and perceive them as stronger when they are presented with multimodal stimuli. For perceived naturalness, a purely visual presentation seems to be more advantageous. Why this is the case is questionable. Perhaps the participants rated the acting performance of the actresses as more natural in facial expressions, while the emphasis, tonality of the sentences contradicted this impression. No clear difference between the conditions could be identified for the indicated arousal (slight significance in the V-AV group).

5 Conclusion

At least for the hit rate and the perceived intensity in emotion perception, a multimodal presentation seems to have a significant, positive influence. The research questions offer opportunities for further and more in-depth research, also with adapted procedures or methods. The assumption that perceived naturalness and arousal are also positively influenced could certainly be further investigated in further, adapted studies. For example, a survey with a more diverse sample (especially in terms of age and gender) would be conceivable. In summary, it can be stated that emotion perception can indeed be influenced by a multimodal presentation style. However, further research in this field is needed to investigate in more detail which aspects it influences.

6 Acknowledgement

The work has been carried out and was supervised by the Institute of Medical Psychology, Universität zu Lübeck.

Authors' Statement

Conflict of interest: Authors state no conflict of interest. Informed consent: Informed consent has been obtained from all individuals included in this study.

7 References

[1] B. Hoheisel and I. Kryspin-Exner, *"Emotionserkennung in Gesichtern und emotionales Gesichtergedächtnis,"* Zeitschrift Fur Neuropsychologie, vol. 16, no. 2, pp. 77–87, Jan. 2005 *"Emotion recognition in faces and emotional face memory."* Journal of Neuropsychopsychology, vol. 16, no. 2, pp. 77–87, Jan. 2005, doi: https://doi.org/10.1024/1016-264x.16.2.77.

[2] N. C. Lee et al., *"Do you see what I see? Sex differences in the discrimination of facial emotions during adolescence.,"* Emotion, vol. 13, no. 6, pp. 1030–1040, 2013, doi: https://doi.org/10.1037/a0033560.

[3] B. Kaup et al., *"Modal and amodal cognition: an overarching principle in various domains of psychology,"* Psychological Research, Oct. 2023, doi: https://doi.org/10.1007/s00426-023-01878-w.

[4] R. Flom and L. E. Bahrick, *"The development of infant discrimination of affect in multimodal and unimodal stimulation: The role of intersensory redundancy.,"* Developmental Psychology, vol. 43, no. 1, pp. 238–252, 2007, doi: https://doi.org/10.1037/0012-1649.43.1.238.

[5] S. Jessen and S. A. Kotz, *"The temporal dynamics of processing emotions from vocal, facial, and bodily expressions,"* NeuroImage, vol. 58, no. 2, pp. 665–674, Sep. 2011, doi: https://doi.org/10.1016/j.neuroimage.2011.06.035.

[6] M. M. Bradley and P. J. Lang, *"Measuring emotion: The self-assessment manikin and the semantic differential,"* Journal of Behavior Therapy and Experimental Psychiatry, vol. 25, no. 1, pp. 49–59, Mar. 1994, doi: https://doi.org/10.1016/0005-7916(94)90063-9.

[7] M. Murugappan, M. R. B. M. Juhari, R. Nagarajan, and S. Yaacob, *"An investigation on visual and audiovisual stimulus based emotion recognition using EEG,"* International Journal of Medical Engineering and Informatics, vol. 1, no. 3, p. 342, 2009, doi: https://doi.org/10.1504/ijmei.2009.022645.

The Relationship between Work-Related Stressors, Flow and the Role of Strengths Use

Karl Benjamin Bieber [1], Kirsten Handschuch [2], Fabienne Aust [2], and Corinna Peifer [2]

[1] Psychology - Cognitive Systems, Universität zu Lübeck, karl.bieber@student.uni-luebeck.de
[2] Research Group Work and Health, Department of Psychology, Universität zu Lübeck, {kirsten.handschuch, fabienne.aust, corinna.peifer}@uni-luebeck.de

Abstract

The relationship between different types of work-related stressors and their potential impact on the experience of flow has received limited scientific attention. This study aims to investigate the connection between the challenge-hindrance stressors framework and work-related flow, with a focus on the moderating role of personal strengths. A total of 132 working adults from diverse occupational backgrounds completed a questionnaire assessing challenge/hindrance stressors, flow frequency, and the utilization of personal strengths. Linear regression and moderation analysis revealed a significant negative relationship between hindrance stressors and flow, moderated by strengths use. No significant relationship was found between challenge stressors and flow, and moderation analysis yielded non-significant results. This study contributes evidence indicating that hindrance stressors negatively impact employees' frequency of flow experience, emphasizing the need for their minimization. Additionally, encouraging workers to cultivate and regularly apply their strengths at work is recommended.

1 Theoretical Background

1.1 Introduction

Since Mihaly Csikszentmihalyi described it in his book "Beyond Boredom and Anxiety" [1] , the concept of flow has been extensively researched as a promising state in leisure and the work context. It can be described as "a state in which an individual is completely absorbed in activity without reflective self-consciousness but with a deep sense of control." [1]. Although Csikszentmihalyi was not the only researcher investigating the phenomenon of being fully immersed in a given task, he coined the term and initiated a promising field of research. Reaching flow necessitates certain prerequisites [2]. Firstly, flow requires a skill-based task or activity, for example, sporting activities like rock climbing or musical endeavors like playing the piano. Secondly, the flow state is characterized by a feeling of control, enjoyment, and deep involvement in the task, a state of high concentration, a feeling of time standing still, a loss of self-awareness and the perception of the task as being rewarding. As outlined in flow theory [3], there are three important antecedents for the flow state, namely a clear goal or task instruction, direct and unequivocal feedback and a balance between the skills and demands regarding a certain task or activity. Barthelmäs and Keller [2] argue that the perception of a balance between a set of skills and the task demands seem to be the most prominent and important prerequisite of reaching flow. This is because it can be argued that receiving clear feedback and having a clear goal in a given task are relevant preconditions for a skill-demand-balance necessary for flow. Assessing if one has the relevant skills and if those skills are sufficient for completing the task in regard to the demands, in and of itself requires sufficient feedback and having a clear goal at hand, therefore making it possible to reduce the antecedents of flow to a perceived skill-demand balance.

1.2 Flow at work

Looking at the work environment, flow has several positive implications. Peifer and Wolters [2] divide antecedents and consequences of experiencing flow at work in three spheres, namely an individual sphere, a social/organizational and a job/task sphere. Regarding the individual sphere, people who experience flow more often show better performance at work. This is likely due to heightened focus during the flow state and the intrinsic motivation of repeating a certain task because of the joy experienced after completing it. Furthermore, flow shows associations with well-being and job satisfaction [2]. When looking at the consequences of flow on the social or organizational sphere, there are two important mechanisms worthy to mention. Firstly, there are studies showing that flow can spread from one person to another, indicating that flow can be 'contagious'. Furthermore, research suggests that flow fosters team engagement which in turn influences team performance. Finally, looking at the task sphere, experiencing flow can lead individuals to build resources regarding the task and may even be related to job crafting, i.e. actively changing one's own work environment [2].

In conclusion, work-related flow has numerous positive implications for the workplace. Since work demands are changing and people experience more stress regarding their work, flow can be seen "as a Healthy Path to Productivity." [2].

1.3   Stress and Flow

While research on flow is constantly growing, there are still unanswered questions that need to be investigated. According to Peifer and Wolters [2] for example, we need a better understanding of different types of work-related stressors and their influence on the experience of flow at the workplace. As for now, there is some laboratory evidence regarding the intensity of stress in the form of physiological arousal and flow.

In their study, Peifer et al. [4] experimentally manipulated flow via a computer program and induced stress via the Trier social stress test (TSST). They measured flow experience, cortisol levels and sympathetic and parasympathetic activation as indicators for stress and found evidence for an inverted U-shaped relationship between sympathetic arousal and flow experience. This adds support to the broader observation of Csiksentmihalyi [1], who discovered that flow often occurred during stressful situations. Flow can therefore not be characterized as a relaxed state but by a certain level of arousal that should not be exceeded or fallen short of [1].

But the intensity of a stressor has no informative value about the types of stressors that influence the flow state. A widely used categorization of work-related stressors is the challenge-hindrance stressor framework [5] which distinguishes between stressors that are more or less controllable (e.g. time pressure or workload) and stressors that are not under one's control (i.e. administrative hassles or conflicts with coworkers). Unsurprisingly, those stressors go along with different appraisal processes, challenge stressors being viewed as something to grow from and hindrance stressors as a factor limiting one's own scope of action. This falls in line with the transactional model of stress and flow [2], in which a stressor appraised as a challenge will lead to a flow state, while stress occurs when a stressor is appraised as a threat or a loss.

One of the first studies to investigate the relationship between work-related flow and job stressors was carried out by Feng [6]. Feng conducted a longitudinal survey with 344 employees in the tech industry in China. They included cognitive processes like problem solving pondering and affective rumination in the study as mediators. Results showed that challenge and hindrance stressors are related to different cognitive processes which influence flow. Their first finding was that affective rumination shows to be flow-hindering, while problem-solving pondering is flow-fostering. The process of problem- solving pondering furthermore fully mediated the relationship between hindrance stressors and flow, while partially mediating the effect of challenge stressors on flow. Affective rumination played a

masking role between challenge stressors and flow, while fully mediating the relationship between hindrance stressors and flow. The study overall shows the importance of how people evaluate different types of stressors, therefore underlining the relevance of appraisal processes in the relationship of stress and flow.

Since one important component of reaching flow is a sense of control and because hindrance stressors (like office politics or conflicts with coworkers) constitute stressors that are likely appraised as out of one's own control, we hypothesize that there is a negative association between hindrance stressors and flow at work. Challenge Stressors (like time pressure or skill variety) however fit in well with the antecedents of flow at work. Because time pressure for instance is probably more likely to be appraised as a challenge, we hypothesize that challenge stressors are positively related to flow experience.

Hypothesis 1: There is a positive association between challenge stressors and flow experience.

Hypothesis 2: There is a negative association between hindrance stressors and flow experience.

1.4   Strengths Use

As a counterbalance to the pathology oriented psychology, the strengths-based approach emerged within the positive psychology movement over 20 years ago. Widely studied in the mental health sector, the research of using strengths in organizations is still limited. There are many approaches to defining the construct of personal strengths, see Bakker and van Woerkom for an overview [7]. Here, we use Govindji and Linleys definition of strengths, which states that "strengths are understood to be natural capacities that we yearn to use, that enable authentic expression, and that energise us." [8]. Using one's own strengths in daily work can have several positive consequences [7]. Research found positive associations with work engagement, well-being and self-efficacy [7]. Also, Wood et al. [9] did a study investigating the influence of strengths use on different outcomes. They measured strengths use, positive and negative affect, perceived stress, self-esteem and vitality. Results showed that strengths use was related to a significantly higher well-being as well as lower perceived stress levels.

The influence of strengths on flow is also one research gap that has not been extensively investigated according to Peifer and Wolters [2]. Liu, van der Linden & Bakker [10] did a weekly diary study examining the episodic fluctuations of work related flow and the use of personal strengths. In addition, risk taking behavior and attentional performance was measured. 164 Chinese workers filled out the questionnaires on five consecutive working days. Findings showed that strengths use is positively related to flow at work, which indicates that using one's own strengths can lead to more flow. Taking together the negative relationship of strengths use with stress and the positive association of strengths use with flow, it is conceivable that using strengths regularly at work may moderate the relationship of stressors and flow, resulting in the following hypotheses:

Hypothesis 3: The positive association between challenge stressors and flow experience is moderated by regular strengths use.

Hypothesis 4: The negative association between hindrance stressors and flow experience is moderated by regular strengths use.

2 Material and Methods

2.1 Procedure

The survey was distributed via social media and Survey-Circle and there was no monetary incentive or course credit given for participation. After information about data protection and the voluntary nature of participation, informed consent was obtained. Afterwards the participants filled out demographic information and subsequently the questionnaire.

2.2 Measures

Flow frequency was measured using the 10- Item- scale developed by Bartzik and Peifer [11]. Items included for example: "How often have you experienced in the last two weeks that...you were completely absorbed in an activity?". Participants were instructed to indicate their answer on a 5-point-scale ranging from "never" to "frequently". Cronbach's α was 0.89. The correlations are shown in Table 1. Challenge and hindrance Stressors were measured using the scale by Lepine et al [12]. Challenge stressors were measured with 10 Items and included for example "How often do you have time pressure?" or "How often do you have to perform complex tasks?". Hindrance Stressors were measured with 9 Items, example items include "How often are you confronted with unclear work tasks?" or "How often are you confronted with office politics?". Participants were instructed to indicate on a 5- point Likert- scale ranging from "never" to "very often". Cronbach's α for the scales were 0.82 and 0.77, respectively. The use of personal strengths was measured adapting the 10- item- scale from Govindji and Linley [8], with example items including "In my daily work...I can use my strengths in many situations." or "In my daily work...I find it easy to use my strengths". The scale ranged from 1 ("do not agree at all") to 7 ("fully agree") and showed high reliability with a Cronbach's α of 0.94. The data analysis was performed using the "Stats" package in R using the lm() function.

Table 1: Correlations (Pearsons r)

Variables	1	2	3	4
1 Flow Frequency				
2 Challenge Stressors	.12			
3 Hindrance Stressors	-.43	.13		
4 Strength Use	.60	.31	-.30	

2.3 Sample

The initial sample included 147 participants. After excluding participants who did not answer the questionnaire (13) and who indicated working less than 20 hours per week (2) the final sample consisted of 132 participants with 62% being female (n=82). The mean age was 32.25 years (SD=9.54) with minimum age being 21 and the maximum age being 63. The average working hours per week were 35.08 hours (SD=10.91) with the lowest score being 20 hours and the highest score being 70 hours per week. On average, the working experience was 11.63 years (SD = 9.88) and the majority held an university degree (45%) or A- Levels (30%).

3 Results

Descriptive statistics of the scales are shown in table 2. To test hypotheses 1 and 2, linear regression was used. The first linear regression showed that challenge stressors did not have a significant positive effect on flow experience, $F(1,130)=1.884$, p>.05 (β=.13, adj. R^2=0.006), therefore hypothesis 1 is not supported. However, hindrance stressors did have a significant negative effect on flow experience, $F(1,130)=30.47$, p<.05 (β=-.49, adj. R^2=), supporting hypothesis 2.

Table 2: Descriptive statistics

Variable	n	M	SD	Md
Flow Frequency	132	3.22	0.71	3.15
Challenge Stressors	132	3.75	0.62	3.78
Hindrance Stressors	132	2.61	0.62	2.6
Strength Use	132	5.41	1.06	5.6

To investigate the influence of strengths use, a moderation analysis was used to test hypothesis 3 and 4. Regarding hypothesis 3, though the overall model was significant, there was no significant interaction between challenge stressors and strengths use when predicting flow frequency, $F(3,128)=25.45$, p<.05 (R^2=0.37), therefore rejecting hypothesis 3. Strength use was, however, a significant moderator in the relationship between hindrance stressors and flow frequency, therefore rendering support for hypothesis 4, $F(3,128)=37.90$, p<.05 (R^2=0.47).

4 Discussion

This study explored the relationship of hindrance and challenge stressors with the experience of flow at work and the moderating role of regular strengths use at work. Challenge stressors and flow showed no significant relationship. In addition, strengths use was no significant moderator in this relationship. On the other hand, hindrance stressors were significantly negatively related to flow and regular strengths use significantly moderated this relationship. Considering the results obtained by Feng [6] this is somewhat surprising. Regarding the direct relationship between challenge/hindrance stressors and flow, Feng found exactly the opposite pattern of significance. This may be explainable by the differences in the samples used in the studies. Since Feng [6] reports recruiting only tech workers, the appraisal process of challenge stressors may be somewhat homogeneous. Because of the diversity regarding occupations in the present sample with the majority working in health (21%), sales/ marketing (17%), and administration (16%) the appraisal processes may differentiate between the samples. It is conceivable that people working in the health or sales sector appraise challenge stressors as less positive. Having to multitask or experiencing time pressure might thereby trigger stress responses which in turn hinders flow. In the present study, not including the appraisal processes connected to different stressors should therefore be considered a significant flaw. Nevertheless, having a diverse sample should be mentioned as a strong point as well. In addition, the fact that the participants were all working individuals should also be mentioned positively, since many studies usually draw on student samples. The results point to the practical implication that minimizing hindrance stressors at work have positive consequences on employees in the sense that they are able to experience flow more often. Employers should also encourage employees to identify and use their strengths, since using them shows to have a buffering effect of the negative influence of hindrance stressors on flow. Due to the cross-sectional design of this study, there are no causal inferences possible. Future research should focus more on experimentally manipulating different stressors to be able to draw causal inferences. Also, as mentioned above, the appraisal processes should not be neglected in investigating stress and flow.

Acknoledgement

The work has been carried out and was supervised by the Research Group Work and Health of the Department of Psychology, Universität zu Lübeck. The data was collected as part of the Research Internship. Special thanks to Emily Fitzgibbon for helping with the translation of the questionnaires.

Authors' Statement

Conflict of interest: Authors state no conflict of interest. Informed consent: Informed consent has been obtained from all individuals included in this study.

5 References

[1] M. Csikszentmihalyi and I. Csikszentmihalyi, Beyond boredom and anxiety, 1. ed., [Nachdr.]. in The Jossey-Bass behavioral science series. San Francisco, Calif.: Jossey-Bass Publ, 1975.

[2] C. Peifer and S. Engeser, Eds., Advances in Flow Research. Cham: Springer International Publishing, 2021.

[3] J. Nakamura and M. Csikszentmihalyi, "Flow Theory and Research," in The Oxford Handbook of Positive Psychology, S. J. Lopez and C. R. Snyder, Eds., Oxford University Press, 2009, pp. 194–206. doi: 10.1093/oxfordhb/9780195187243.013.0018.

[4] C. Peifer, A. Schulz, H. Schächinger, N. Baumann, and C. H. Antoni, "The relation of flow-experience and physiological arousal under stress — Can u shape it?," Journal of Experimental Social Psychology, vol. 53, pp. 62–69, Jul. 2014, doi: 10.1016/j.jesp.2014.01.009.

[5] J. A. Lepine, N. P. Podsakoff, and M. A. Lepine, "A Meta-Analytic Test of the Challenge Stressor–Hindrance Stressor Framework: An Explanation for Inconsistent Relationships Among Stressors and Performance," AMJ, vol. 48, no. 5, pp. 764–775, Oct. 2005, doi: 10.5465/amj.2005.18803921.

[6] X. Feng, "How Job Stress Affect Flow Experience at Work: The Masking and Mediating Effect of Work-Related Rumination," Psychol Rep, p. 003329412211228, Aug. 2022, doi: 10.1177/00332941221122881.

[7] A. B. Bakker and M. Van Woerkom, "Strengths use in organizations: A positive approach of occupational health.," Canadian Psychology / Psychologie canadienne, vol. 59, no. 1, pp. 38–46, Feb. 2018, doi: 10.1037/cap0000120.

[8] Govindji, R., and Linley, P. A., "Strengths use, self-concordance and well-being: Implications for strengths coaching and coaching psychologists", International Coaching Psychology Review, 2(2), 143-153, Jul. 2007

[9] A. M. Wood, P. A. Linley, J. Maltby, T. B. Kashdan, and R. Hurling, "Using personal and psychological strengths leads to increases in well-being over time: A longitudinal study and the development of the strengths use questionnaire," Personality and Individual Differences, vol. 50, no. 1, pp. 15–19, Jan. 2011, doi: 10.1016/j.paid.2010.08.004.

[10] W. Liu, D. Van Der Linden, and A. B. Bakker, "Strengths use and work-related flow: an experience sampling study on implications for risk taking and attentional behaviors", JMP, vol. 37, no. 1, pp. 47–60, Jan. 2022, doi: 10.1108/JMP-07-2020-0403.

[11] M. Bartzik, C. Peifer, "Flow Frequency Scale", in preperation.

[12] M. A. LePine, Y. Zhang, E. R. Crawford, and B. L. Rich, "Turning their Pain to Gain: Charismatic Leader Influence on Follower Stress Appraisal and Job Performance," AMJ, vol. 59, no. 3, pp. 1036–1059, Jun. 2016, doi: 10.5465/amj.2013.0778.

Factorial structure in Executive Functions: A replication study

Maja Fritzi Saidl [1], Marcus Heldmann[2, 4], Baldrun Köhncke [3]

[1] Psychologie - Cognitive Systems, Universität zu Lübeck, maja.saidl@student.uni-luebeck.de
[2] Klinik für Neurologie, Universität zu Lübeck, marcus.heldmann@uni-luebeck.de
[3] Klinik für Neurologie, Universität zu Lübeck, b.koehncke@uni-luebeck.de
[4] Center of Brain, Behavior and Metabolism (CBBM), Campus Lübeck

Abstract

Executive functions is an umbrella term that summarises cognitive processes that are necessary for the control of actions and the adaptation of action goals. Miyake postulated a 3-factorial structure of executive functions based on factor analysis [1]. In the present paper, we tried to replicate this three-factorial structure by investigating 15 students (undergraduate and Ph.D.) using 10 different executive tasks from the Miyake study [1]. Due to violated assumptions, an exploratory factor analysis assuming two different executive factors had to be calculated instead of a confirmatory one. The analyses will be re-examined once the intended sample size of 25 is reached.

1 Introduction

Executive functions (EF) enable us to adapt our actions and behavior to a constantly changing environment. Depending on the circumstances, we can actively influence our actions or let automatic processes take control. We can rely on existing habits or act in a goal-oriented manner (i.e. with behavioral self-control) [2], [3]. Attentional factors have been closely linked to the performance in executive tasks, so they explained 30% of the variance in self-regulation as measured by behavioral ratings and was a general factor measured by working memory capacity, response inhibition, and general intelligence [4]. EF thus include "broad-based attentional control or self-regulation [...] across a variety of tasks and domains" [5].

Previous and recent research assumes that EF comprise three central factors: inhibition (which includes inhibitory and inference control, as well as behavioral inhibition) working memory (its "functions to maintain and manipulate information over brief periods of time" [4], basically its capacity and permanent updating), and cognitive flexibility, also known as (mental) set-shifting. EF are thus part of our overall cognitive abilities as lower-level cognitive functions add up to more complex performances [2] in everyday life. Mental and physical health, school readiness and success, job success, marital harmony, and public safety are common examples [6]. Psychological constructs are often invented ad hoc and ad lib, which results in a lack of sharp boundaries and detailed definitions [7]. It is thus important to verify our conceptual understanding by replicating existing research. As part of a study with Parkinson's patients and healthy controls, the factor solutions of this behavioral study will be used to relate them to EEG parameters that are intended to map executive functions.

The goal of this study is to confirm the mentioned three-factor structure in EF.

2 Material and Methods

2.1 Inclusion/exclusion

Inclusion criteria were subjects aged 18 and older who could independently agree to informed consent. Individuals with relevant neurological or psychiatric diagnoses, regular alcohol or drug consumption or abuse, and those taking medication could not take part in the study. Furthermore, left-handedness was an exclusion criterion to avoid or reduce EEG irregularities. Participants were required to have non-impaired or correctable hearing and vision as the tasks included visual and auditory stimuli. Two participants had to be excluded due to a BDI depression score ≥ 18.

2.2 Participant characteristics

Participants were undergraduate (n = 13) and PhD students (n = 2) at the University of Luebeck. The age of the final sample ranged from 19 to 41 years (M = 24.00, SD = 5.67). 12 of them were female, and 3 male, resulting in a total sample of 15 test subjects. The intended sample size is 25.

2.3 Sampling Procedures

Data were collected between November 2023 and January 2024 in a behavioral laboratory of the CBBM on the campus of the University of Lübeck while participants sat at a computer with an 24" screen and completed the tasks either by pressing keys on the keyboard or by responding verbally. Participants received either course credits or monetary compensation of €10 per hour. The superordinate multisectional study was approved by the ethics board of the University of Lübeck.

2.4 Measures and covariates

Beck-Depression Inventory (BDI, revision II) scores were only used to exclude participants with hints about or tendencies toward a mild to moderate depressive episode. Depression is associated with executive dysfunction and could therefore lead to bias [8]. For the respective EF factors inhibition (antisaccade, stop signal, Stroop), updating of working memory (keep track, spatial 2-back, letter memory), and set-shifting (category switch, color-shape, number letter), we used the behavioral tasks, stimuli, and dependent measures from the study by Feng et al. [9]. For detailed information about the dependent variables see Table 1.

In addition to the original study, subjects completed the digit ordering task (dot) as part of the working memory factor. First, participants were presented with five numerals, which they had to arrange in ascending order so they could later decide whether the numeral corresponded to a place marked in red in a line of five dots. The dependent measure was the proportion of incorrect assignments.

Table 1: Executive tasks and their dependent variables

Task	Dependent Measure
antis and dot	proportion of errors (%)
cat-sw, c-hsape, num-let	RT differences (switch - repeat)
stroop	RT difference (congruent - incongruent)
keep	recalled words (%)
let-mem	recalled 3-letter lists (%)
sp2back	detected repeated blinks (%)
ssrt	Stop-Signal Reaction Time

Antisaccade (antis):
In each trial, a gaze point "+" was first presented in the center of the screen. Then a visual cue (a black square) was presented on one side of the screen, followed by the target (an arrow within a square), which appeared on the screen's opposite side and was shaded by a gray square. The participants had to control their attention to the target, not the cue, and pressed the left or top or right key to judge the arrow's direction.

Stop-Signal (ssrt):
In each trial a green arrow occurred in the middle of the screen, right after a gaze point "+" was presented (go trials). Participants should react based on the arrow's pointing direction. Randomly selected, after the green arrow, a red cross could occur, requiring the participants to inhibit an answer (pressing left or right button; stop signal).

Stroop:
We adapted this task from a classic one. Participants had to press a button according to the color of a written word, which was either green, red, yellow, or blue. Thus, two conditions were present: incongruent stimuli (word meaning and printed color did not match) and congruent stimuli (word meaning corresponded to printed color or had a completely different meaning, e. g. lake).

Category switch (cat-sw):
In each trial, participants had to judge the animacy or size of a stimulus (written word). Depending on whether a scale symbol or a heart appeared in the middle of the screen, they had to press a button. All trials were either assigned to the repeat or switch condition, depending on the task of the previous trial.

Color-shape (c-shape):
The color-shape switching task is similar to the category switch task, but presents different stimuli. In each trial, a cue ("F" or "S") was presented on top of the screen. If "F" for "Farbe" (color) occurred, participants had to decide, whether the stimulus was red or blue. If "S" for "Symbol" (shape) occurred, they should decide between triangle and circle.

Number letter (num-let):
In each trial of this task, a number-letter pair appeared on the screen. If they appeared at the top of the screen, the participants were asked to judge if the number was odd or even and press the button as quickly and correctly as possible. If they appeared at the bottom of the screen, the participants were asked to judge if the letters were vowels or consonants. The appearance of stimuli (top or bottom) was pseudo-randomized.

Keep track (keep):
In each trial, 2–4 target categories of animals, relatives, furniture, countries, metals, or colors appeared on the bottom of a screen. A total list of 15 words (either part of a category or a random other word) occurred one by one. At the end of a trial, participants should remember the most recent word according to the target categories.

Letter memory (let-mem):
In this task, letters were presented in the middle of a screen one by one, each followed by a gaze point. Participants were asked to update a list of three letters as a new letter was presented. At the end of a trial, they were asked to recall the most recent three letters.

Spatial 2-back (sp2back):
In this task, 10 squares were presented on a screen. The task was to recall, whether the recently flashing square had also been blinking two items before.

2.5 Quality of measurements

At the beginning of each task, participants performed a training trial. For the stop signal, dot, and color-shape task the subjects performed three experimental blocks consisting of 62, 31, and 191 trials, respectively. Subjects were not assigned to groups, as all completed the ten tasks. Due to a mistake in task order permutation, the first six participants did not perform the dot and two participants' data from stop-signal were not gathered correctly. To transform outputs from stimulus presentation software ("Presentation"), we used MATLAB_R2023b to generate the read-out, which was then transmitted into CSV files to do the data cleaning and actual data analysis in RStudio (Version 9.3.191269). We cleaned our data by excluding response times <100 ms, averaging them per subject and condition. Cat-sw, c-shape, num-let, Stroop, and ssrt tasks used the RTs to calculate the dependent variables, for the rest, we used correctly recalled (keep, let-mem), correctly detected stimuli (sp2back) and errors in stimulus detection (antis) and recalled word order (dot), respectively (also, see Table 1).

3 Results and Discussion

The missing data in the ssrt and dot tasks were replaced by the mean values of the total sample. Before the main analyses were calculated, the data were inspected. It became clear that the response time (RT) differences from the category switch, color-shape, and number letter tasks were strongly skewed to the left. On average, the 10 variables correlated at .18 (range: -.52 - .89), with correlations often being non-linear. Overall, participants performed well above average, esp. in the antisaccade, dot, letter memory, and spatial 2-back tasks (mean error proportions being .06 and .04, and mean accuracy being .76 and .99, respectively). When calculating the confirmatory factor analysis (CFA), it was noticed that the variances of the factors were negative, which violates the requirements for a CFA. It is possible that the correlations of variables outside their assumed factors are too strong or that a so-called "Heywood case" exists. In this case, only two items load on one or more factors.

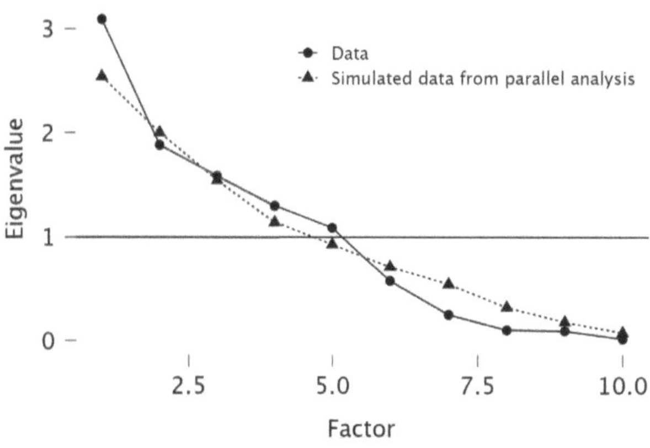

Figure 1: Scree Plot based on factorial analysis.

Requirements for a CFA are violated. In order to reveal other structures in the data and despite the result of Bartlett's test for sphericity (p = .0011), an exploratory factor analysis (EFA) estimated using maximum likelihood (ML) was conducted. The scree plot of a parallel analysis based on FA indicated five executive factors present (see Fig. 1). However, when evaluating the 5-factor EFA, the Root Mean Square Error of Approximation (RMSEA) value was large (df = 5 χ^2 = 24.22, RMSEA =.11, TLI = -12.69, BIC = -9.73, p = .58)., indicating a poor model fit. The best model fit was achieved by a 2-factor geominQ-rotated

solution (df = 26, χ^2 = 3.81, RMSEA =.11, TLI = 1.24, BIC = -46.19, p = .56). Neither of the models was significant, nor had a good model fit. Fig. 2 shows the result of the EFA with a 2-factor structure. As there are no specific confirmatory factors, they were named ML 1 and ML 2 according to the estimation method. Interestingly, the inhibition tasks (antisaccade, stop-signal, Stroop) are divided from the shifting tasks (category switch, color shape, and number letter). The working memory tasks, on the other hand, do not belong to a clear factor: while letter memory and keep track are part of the second factor (ML 2), the dot task is contained in the first factor (ML 1). The loading of the individual tasks on their respective factors ranges from .20 to .80 or .90, with the loading of the antisaccade task on the second factor being negative (-.40) as it contains the proportion of incorrect answers.

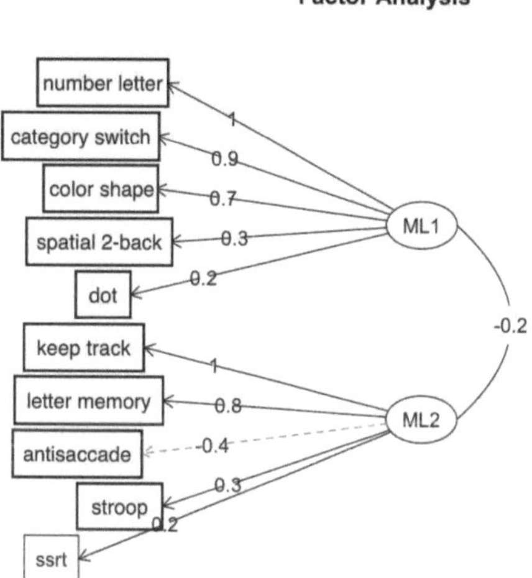

Figure 2: Executive tasks and their loadings on two factors, retrieved from EFA.

This study aimed to replicate the 3-factor structure of executive functions from previous research [9]. Due to negative factor variances running a CFA was not possible. In order to approach the factor structure of our data two EFAs were calculated exploratively. Notice that, although the 10 tasks did not load on three factors, set-shifting and inhibition tasks were completely separated on both calculated factors in the bifactor model.

The factor analyses may not have been significant for various reasons, e.g. relating to the structure of the data. According to this, Bartlett's test result was hardly significant, which indicates both of the model fits (for two and five factors) were poor and not significant. The sample size might be the main issue resulting in a high variability of factorial structure. General limitations could have led to bias and thus corrupted statistical analyses. The sample only consisted of students and was thus homogeneous. It does not reflect the overall population. This could result in lower external validity. Because students were recruited via monetary or course credit compensation, a selection bias concerning executive performances could have occurred. Since the number and allocation of different executive tasks varies across studies, it remains unclear whether there is possibly a higher-level general and other distinct factors. In a previous study, the (performance in) executive tasks correlated more strongly in older adults than in younger ones. However, the bifactor EFA's results would be more in line with dedifferentiation across the lifespan. Furthermore, the authors concluded that "executive functions are better described as a partly overlapping rather than a factorial structure" [10]. Regarding how well the sample performed overall, but especially in four out of ten tasks, ceiling effects could have affected calculations. As proposed in the literature, the variance reduction problem could be tackled by multiplying a weight matrix by the sample covariance matrix [11]. As soon as the data from the other 10 planned test subjects are available, the factor structure should be re-examined, using the proposed weighting method.

4   Conclusion

This study could not confirm the 3-factor structure in EF. This is likely due to the small sample size and other methodological limitations. Nevertheless, research into EF remains an important field of science. If we understand the interindividual differences and factor structure of EF, this understanding may help us to explain changes in EF. Our cognitive abilities change throughout our lives or during clinical diagnoses (e.g. Parkinson's Disease, frontotemporal dementia, psychiatric diseases), for example. Thus, their preservation and protection rely on the findings of basic research.

Acknowledgement

The work has been carried out at Center of Brain, Behavior and Metabolism, Campus Lübeck. It is part of the EFProfile study by Marcus Heldmann and his research group.

Authors' Statement

Conflict of interest: Authors state no conflict of interest. Informed consent: Informed consent has been given by all participants.

5   References

[1] A. Miyake, N. P. Friedman, M. J. Emerson, A. H. Witzki, A. Howerter, and T. D. Wager, "The Unity and Diversity of Executive Functions and Their Contributions to Complex 'Frontal Lobe' Tasks: A Latent Variable Analysis," *Cognitive Psychology*, vol. 41, no. 1, pp. 49–100, 2000, doi: https://doi.org/10.1006/cogp.1999.0734.

[2] N. P. Friedman, A. Miyake, S. E. Young, J. C. DeFries, R. P. Corley, and J. K. Hewitt, "Individual differences in executive functions are almost entirely genetic in origin.," *Journal of Experimental Psychology: General*, vol. 137, no. 2, pp. 201–225, 2008, doi: https://doi.org/10.1037/0096-3445.137.2.201.

[3] N. P. Friedman and T. W. Robbins, "The role of prefrontal cortex in cognitive control and executive function," *Neuropsychopharmacology*, vol. 47, pp. 1–18, 2021, doi: https://doi.org/10.1038/s41386-021-01132-0.

[4] J. Tiego, M. A. Bellgrove, S. Whittle, C. Pantelis, and R. Testa, "Common mechanisms of executive attention underlie executive function and effortful control in children," *Developmental Science*, vol. 23, no. 3, 2020, doi: https://doi.org/10.1111/desc.12918.

[5] E. L. Glisky *et al.*, "Differences between young and older adults in unity and diversity of executive functions," *Aging, Neuropsychology, and Cognition*, pp. 1–26, 2020, doi: https://doi.org/10.1080/13825585.2020.1830936.

[6] A. Diamond, "Executive Functions," *Annual Review of Psychology*, vol. 64, no. 1, pp. 135–168, 2013, doi: https://doi.org/10.1146/annurev-psych-113011-143750.

[7] R. A. Poldrack and T. Yarkoni, "From Brain Maps to Cognitive Ontologies: Informatics and the Search for Mental Structure," *Annual Review of Psychology*, vol. 67, no. 1, pp. 587–612, 2016, doi: https://doi.org/10.1146/annurev-psych-122414-033729.

[8] S. L. Warren, W. Heller, and G. A. Miller, "The Structure of Executive Dysfunction in Depression and Anxiety," *Journal of Affective Disorders*, vol. 279, pp. 208–216, 2021, doi: https://doi.org/10.1016/j.jad.2020.09.132.

[9] J. Feng *et al.*, "A cognitive neurogenetic approach to uncovering the structure of executive functions," *Nature Communications, vol. 13*, no. 1, 2022, doi: https://doi.org/10.1038/s41467-022-32383-0.

[10] O. Bock, M. Haeger, and C. Voelcker-Rehage, "Structure of executive functions in young and in older persons," *PLOS ONE*, vol. 14, no. 5, 2019, doi: https://doi.org/10.1371/journal.pone.0216149.

[11] Q. Liu and L. Wang, "T-Test and ANOVA for data with ceiling and/or floor effects," *Behavior Research Methods*, vol. 53, 2020, doi: https://doi.org/10.3758/s13428-020-01407-2.

Analysing the neural foundations of ego- and altercentricity bias and their association with judgement of emotional states

Gerwin Akpadji [1], Ulrike Krämer [2], Martin Göttlich [3], and Emily Fitzgibbon [1]

[1] Psychology - Cognitive Systems, Universität zu Lübeck, {gerwin.akpadji, emily.fitzgibbon}@student.uni-luebeck.de
[2] Medical Psychology, Universität zu Lübeck, {ulrike.kraemer, martin,goettlich}@uni-luebeck.de

Abstract

In social interactions of two or more individuals, incongruent emotional states can lead to biases in judgement. One's own emotional state can influence the judgement of the other persons emotional state. Likewise, the perceived emotional state of another person can influence one's own emotional state. The former has previously been defined as the egocentricity bias, the latter as the altercentricity bias. In this study we applied a newly developed paradigm using direct facial feedback to assess the influence of these biases on behavior and analysed their neural foundations using fMRI. We found that the paradigm successfully induced emotional biases reflected in neural activity and that specific brain regions were directly correlated to behavioral bias scores.

1 Introduction

In any social interaction, humans will attempt to understand their interaction partners' thinking, decision-making and their emotional state of mind [1]. Distinguishing between one's own emotions and those of another person is vital to accurately judge situations and perspectives. This becomes especially important if the perspectives of two individuals are incongruent. In such situations, judgement of another's emotional state can be biased towards the own state (*emotional egocentricity bias*; EEB) [2]. Conversely, a person's emotional state may be influenced by the perceived emotional state of another (*emotional altercentricity bias*; EAB) [3]. To induce real emotional states reliably and create an incongruent emotional state between the subject and the target ("other"), we used the Food-EEB paradigm. Previous work has mainly used visuotactile stimulation to create incongruent emotional states (e.g. the touch of rose petals vs. the touch of slimy worms) [2]. Mohr et al. [4] introduced a new audiovisual paradigm to better control the exact valence of the stimuli presented (e.g. using aversive or pleasant sounds that remain consistant and visual cues to indicate the targets experience). This study used the *Food emotional egocentricity bias* paradigm (Food-EEB or FEEB), which uses direct social-cues in the form of face-stimuli displaying the targets emotion, ensuring correct assessment of the target's emotional state[5]. Functional brain imaging (fMRI) was used to identify the brain regions associated with the task. We expected significant correlation of both the neural activity in the right temporoparietal junction (TPJ), which is involved in self-other distinction and theory-of-mind processes [6], and the right supramarginal gyrus (rSMG), which has previously been associated with the underlying neural processes of the EEB [2]), with the extent of the induced emotional biases.

2 Material and Methods

2.1 Participants

A total of 40 participants was recruited. Due to outliers and exclusion criteria, 26 remaining participants (19 female, 7 male) were included in this study. All participants were self-reportedly healthy and had no history of neurological or psychiatric diseases. Participants were recruited via university mailing lists and compensated for their participation (monetary or hourly credit as experimental participants). All participants were neurotypical and within an age range of 18 to 35 years (mean age 23 ± 2.8 years).

2.2 Material

The food stimuli were compiled from various databases. A pilot study with N= 41 participants verified a mild dichotomous classification of the food pictures into appetitive and aversive for the expected western-culturized sample. The videos showing emotional displays were taken from the Jerusalem Facial Emotion Expression Set [7]. Four actors (three male and one female) displaying once a happy face and once a disgusted face were picked for the current task based on subjective criteria fitting into the restaurant setting implied in the FEEB task.

2.3   FEEB-Paradigm

The experimental design includes three factors with two levels each: Congruency (congruent vs incongruent), Target (self vs other), and Valence (appetitive vs aversive). Participants are instructed to imagine sitting in a restaurant, with the food served being the dish pictured in the food stimulus. They look up and see the emotional expression of the person at their table (i.e., a video of a person smiling or displaying disgust). Afterwards they are asked to rate their own feeling towards the food or the feeling of the other person towards the food. The FEEB setup has two parts: (i) a prerating of each food stimulus alone (n=20), and (ii) the main task, where the setting in a restaurant is implied and the food stimulus and video is shown (n=80). For the prerating, all trials are self-paced without a time limit. The food pictures are shown in a randomized fashion on the screen and the participant is asked to rate each on a 9-point scale from -4 (very aversive) to +4 (very appetitive) with a neutral point 0. For the main task (see trial example in Fig. 1), first a fixation cross was shown, followed by a food picture for two seconds, then the presentation of a video in the middle of the screen and the food picture in the lower part of the screen for 3 to 4 seconds, and a colored circle indicating the target cue for two seconds. Afterwards, the rating scale was presented until the subject gave either their own judgement of the food, their estimate of the other person's judgement or until a time-out of five seconds. The next trial was started by clicking on a 'Next trial' button and did not include further breaks. This design creates multiple conditions in which emotional biases are induced. When the subject is asked to give their own rating of the presented food stimuli and the emotional expression of the face stimuli does not match the presented food, emotional altercentricity bias is induced. This happens when the food stimulus is appetitive and the emotional expression shows disgust, or when the food stimulus is aversive and the emotional expression is happy. In these conditions, the subjects rating may be influenced by the perceived rating of the other person. Conversely, when asked to rate the other person's judgement of a food stimulus, the subject's own rating of the food stimulus may alter their jugdement of the other person's emotional state (emotional egocentricity bias). This happens in trials in which the presented stimulus is appetitive, but the other person shows disgust, or in trials in which the food stimulus is aversive and the other person shows happiness.When comparing these ratings to those in conditions in which self and other rating are congruent, the extent of the biases can be measured.

2.4   Imaging

Structural and functional MR imaging was performed at the CBBM Core Facility Magnetic Resonance Imaging using a 3-T Siemens Magnetom Skyra scanner equipped with a 64-channel head-coil. Functional images were acquired applying a single-shot gradient-recalled echo-planar imaging (GRE-EPI) sequence sensitive to blood oxygen level dependent (BOLD) contrast (TR=980 ms; TE=30 ms; flip

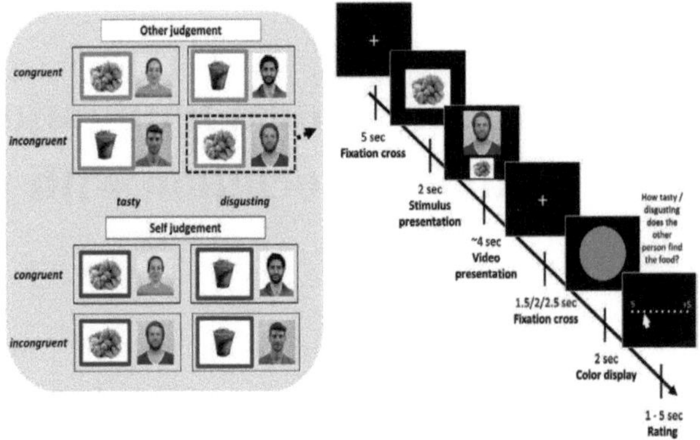

Figure 1: Example of the main task in the FEEB-paradigm. Left side shows the 3 factors: target (self vs. other), valence (tasty vs. disgusting), congruence (congruent vs. incongruent). Right side shows timeline of a single trial.

angle=65°; in-plane resolution 3×3 mm^2; 204×204 mm^2 field of view; 56 transversal slices; 3 mm slice thickness; simultaneous multi-slice factor 4). Additionally, structural images of the whole brain using a 3D T1-weighted MP-RAGE sequence were acquired (TR=1900 ms; TE=2.44 ms; TI=900 ms; flip angle 9°; 1×1×1 mm^3 resolution; 192×256×256 mm^3 field of view; acquisition time 4.5 minutes). The average total scanning time was 37 minutes.

2.5   Analysis

Analysis was conducted using MATLAB and SPM12. Two one-sample t-tests were performed to establish whether there were brain regions showing changed activity in conditions inducing EEB or EAB. To measure a correlation between the subject's bias and their neural activity, a behavioral EEB and EAB score was calculated from the rating data. The emotional egocentricity bias was calculated as [(-1* d1 + d3) /2] and the emotional altercentricity bias as [(-1* d2 + d4) /2], where d1 = other-rating pleasant incongruent – congruent, d2 = self-rating pleasant incongruent – congruent, d3 = other-rating unpleasant incongruent – congruent, and d4 = self-rating unpleasant incongruent– congruent. These scores were then used to perform a multiple regression with contrast images for EEB and EAB. For both t-tests and regression, whole brain analysis was used.

3   Results

3.1   T-Test

Both in conditions inducing egocentricity and altercentricity bias, the t-test results showed significant differences in neural activity (see Fig. 2 and 3). All p-values are cluster-level family-wise error (FWE) corrected. k describes the extent threshold for voxels forming a cluster. During induced EEB, neural activity was increased in the right inferior frontal gyrus (rIFG)(p=.009, k=134), the right supplementary motor cortex (SMC)(p=.003, k=134), the thalamus (p=.005, k=134) and the left posterior orbital gyrus (left POrG)(p=.002, k=134). During EAB, neural activity

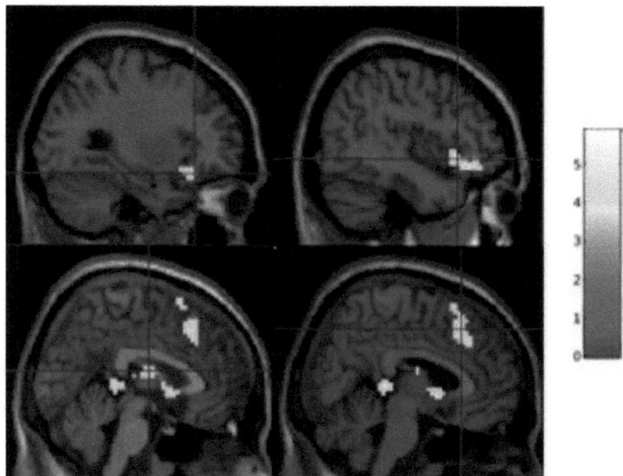

Figure 2: Images displaying brain regions with significant t-test result (EEB). Coordinates set at (from left to right, top to bottom): left POrG, rIFG, Thalamus, rSMC. Color legend show t-level.

was increased in the right inferior frontal gyrus (p=.002, k=153).

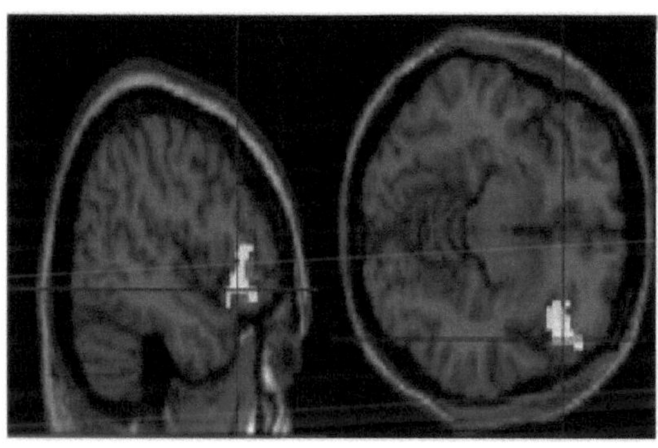

Figure 3: Significant brain region in EAB t-test: rIFG

3.2 Regression

Multiple regression of the EEB-contrast images with the individual EEB-scores showed a positive correlation of the EEB-score with the left posterior orbital gyrus (p=.001, k=93), the rIFG (p=.006,k=93), the right superior frontal gyrus (rSFG)(p=.025, k=93), the rSMC (p<.001, k=93) and the thalamus (p=.008, k=93). See Fig. 4. For the EEB, individual scores correlated positively with activity in the right inferior frontal gyrus (p=.002, k=140). See Fig. 5.

4 Discussion

This study aimed to evaluate whether data collected using the FEEB-paradigm could be used not only to highlight differences in brain activity based on t-tests, but also to calculate correlations of specific brain regions, specifically the

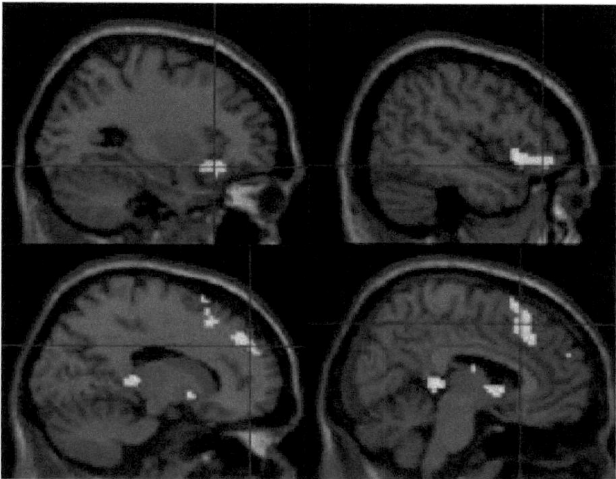

Figure 4: Regions correlating with EEB-score. Top left to bottom right: Left POrG, rIFG, rSFG, rSMC

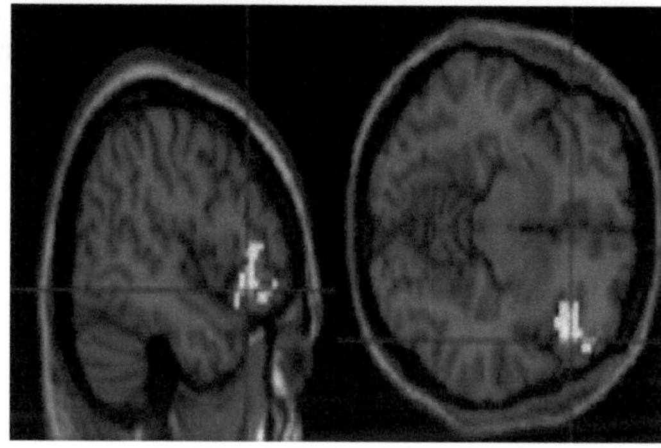

Figure 5: Region correlated with EAB-score: rIFG

rTPJ and rSMG, with behavioral bias scores. The results of the t-tests show, that the paradigm successfully creates an emotional egocentricity and altercentricity bias that is reflected in the neural activity. The activity in the right inferior frontal gyrus is in line with its involvement in a network processing surprising or unexpected information and inhibiting responses [8]. Activity in the thalamus and rSMC may be due to the processing and encoding of sensory information and motor-response selection. Involvement of the left posterior orbital gyrus may indicate its involvement due to a role in social cognition [9] but will have to be further explored in the future. The results of the regression analyses show that a direct correlation between behavior-scores and neural activity is possible in this FEEB-paradigm. The areas corresponding to the bias-scores where the same as in the t-tests, indicating that there is a linear connection between neural activity and the extent of bias. The suspected brain regions rTPJ and rSMG did not show significant activity or correlation in either model. This may be an indication of insuffient sample size. A post-hoc power analysis revealed a power of 68, indicating a relatively high chance of type-2 errors and suggesting a sample size of at least 36 for further studies. As a comparably large standard error can lead to false negatives in smaller sample sizes [10], an exploratory regression was performed using k=10 as the extent threshold for cluster size. In this uncorrected regres-

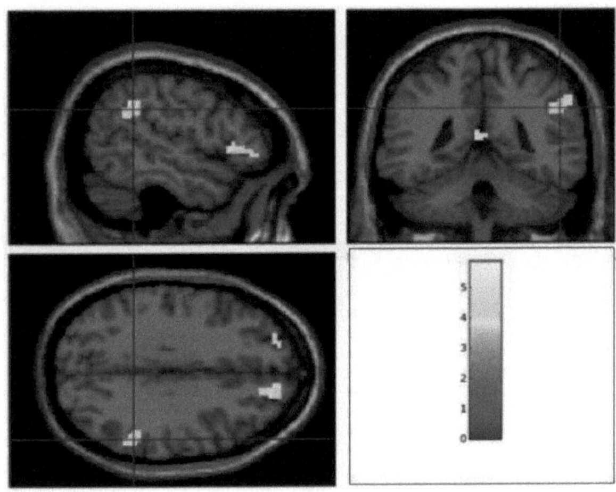

Figure 6: Correlated activity of rSMG with EEB-score as discovered in exploratory regression analysis

sion, the right supramarginal gyrus did show a correlation to the EEB-score (as seen in Fig. 6), which supports the hypotheses, that a larger sample size may reveal more accurate results. Future research could focus on exploring the correlation between bias extent and brain activity and aim for specific ROI-analyses with larger sample size, that include the regions identified in this study.

Acknowledgement

The work has been carried out at University of Lübeck and supervised by the Cognitive Neuroscience Group at the Institute of Medical Psychology. Due to communication issues regarding the participation of the Cognitive Systems Master in the student conference module, recruitment and testing of participants was conducted outside of the internship time frame.

Authors' Statement

Conflict of interest: Authors state no conflict of interest. Informed consent: Informed consent has been obtained from all individuals included in this study. Ethical approval: The research related to human use complies with all the relevant national regulations, institutional policies and was performed in accordance with the tenets of the Helsinki Declaration, and has been approved by the authors' institutional review board or equivalent committee.

5 References

[1] Decety, J. & Jackson, P. L. (2004). The functional architecture of human empathy. *Behavioral and cognitive neuroscience reviews*, 3(2), 71–100. https://doi.org/10.1177/1534582304267187

[2] Silani, G, Lamm, C, Ruff, C. C & Singer, T. (2013). Right supramarginal gyrus is crucial to overcome emotional egocentricity bias in social judgments. *The Journal of neuroscience : the official journal of the Society for Neuroscience*, 33(39), 15466–15476. https://doi.org/10.1523/JNEUROSCI.1488-13.2013

[3] Bukowski, H., Tik, M., Silani, G, Ruff, C., Windischberger, C. & Lamm, C (2020). When differences matter: rTMS/fMRI reveals how differences in dispositional empathy translate to distinct neural underpinnings of self-other distinction in empathy. *Cortex; a journal devoted to the study of the nervous system and behavior*, 128, 143–161. https://doi.org/10.1016/j.cortex.2020.03.009

[4] Mohr, M. von, Finotti, G., Ambroziak, K. B. & Tsakiris, M. (2020). Do you hear what I see? An audio-visual paradigm to assess emotional egocentricity bias. *Cognition and emotion*, 34(4), 756–770. https://doi.org/10.1080/02699931.2019.1683516

[5] Goregliad Fjaellingsdal, T., Makowka, N. & Krämer, U. M. (2023). Studying trait-characteristics and neural correlates of the emotional ego- and altercentric bias using an audiovisual paradigm. *Cognition & emotion*, 37(4), 818–834. https://doi.org/10.1080/02699931.2023.2211253

[6] Corbetta, M., Patel, G. & Shulman, G. L. (2008). The reorienting system of the human brain: from environment to theory of mind. *Neuron*, 58(3), 306–324. https://doi.org/10.1016/j.neuron.2008.04.017

[7] Yitzhak, N., Giladi, N., Gurevich, T., Messinger, D. S., Prince, E. B., Martin, K., & Aviezer, H. (2017). Gently does it: Humans outperform a software classifier in recognizing subtle, nonstereotypical facial expressions. *Emotion*, 17(8), 1187–1198. https://doi.org/10.1037/emo0000287

[8] Lopez-Sosa, F. et al. (2021). Nucleus Accumbens Stimulation Modulates Inhibitory Control by Right Prefrontal Cortex Activation in Obsessive-Compulsive Disorder. *Cerebral cortex* (New York, N.Y. : 1991), 31(5), 2742–2758. https://doi.org/10.1093/cercor/bhaa397

[9] Nestor, P. G., et al. (2013). In search of the functional neuroanatomy of sociality: MRI subdivisions of orbital frontal cortex and social cognition. *Social cognitive and affective neuroscience*, 8(4), 460–467. https://doi.org/10.1093/scan/nss018

[10] Gnambs, T. (2023). A Brief Note on the Standard Error of the Pearson Correlation. *Collabra: Psychology*, 9(1), Artikel 87615. https://doi.org/10.1525/collabra.87615

Local variations in ultrafine particles
– A pilot study –

Vivian Mohr [1], Lisa Marshall [2, 3]

[1] Psychology - Cognitive Systems, Universität zu Lübeck, Vivian.Mohr@student.uni-luebeck.de
[2] Institute of Experimental and Clinical Pharmacology and Toxicology, Universität zu Lübeck, Lisa.Marshall@uni-luebeck.de
[3] Center of Brain, Behavior and Metabolism (CBBM), Universität zu Lübeck

Abstract

Air pollution is a topic, that affects us all. Recently, ultrafine particles (UFPs) have come into focus, since their association with negative health outcomes has been underscored. Although the World Health Organization issued regulations for larger particulate matter none are established for UFPs. This is in part due to the lack of standardized measurement procedures and the use of various instruments for UFP detection. This study presents a first step toward establishing a standardized measurement procedure for UFPs with a mobile device. Measurements were taken in three different locations (rural, urban, control condition). We analyzed differences in UFP concentrations between the three locations and compared current weather conditions. The results showed, that UFP concentrations differed significantly between the rural and urban areas and correlations with weather conditions exist. It is concluded, that future studies and analysis should include environmental factors when establishing a protocol for UFP measurement.

1 Introduction

Ultrafine particles (UFPs) are defined as particulate matter (PM) that have a diameter of ≤ 0.1 µm. In contrast, fine particles either have a diameter of ≤ 10 µm (PM_{10}) or ≤ 2.5 µm ($PM_{2.5}$). The different sizes of particulate matter navigate the small bronchioles and lung defense system differently and lead to diverse health effects [1]. In comparison, UFPs show more delayed effects and a greater influence on mortality whereas $PM_{2.5}$ show more immediate effects [1]. Environmental airborne pollutants and in particular ultrafine particles are taken in through the olfactory pathway as well as the lungs [2]. Studies have reported major associations between UFPs and diseases regarding the respiratory and cardiovascular system, metal fume fever and polymer fume fever, diabetes, cancer, effects on embryos during pregnancies (i.e. increased risk of low birthweight, possible changes in mRNA associated with growth factors, cell migration and more) and the central nervous system [1]. When UFPs penetrate the nasal mucosa, they can overcome the nose-to-brain barrier [2]. Studies have shown that there is a great brain uptake of ultrafine particles into the olfactory bulb [3]-[2]. The uptake of UFPs into the brain by transport across the blood-brain barrier was mainly shown in rats [3]. However, analysis of case studies for example in Mexico City strongly suggest that long exposure and long-term accumulation of UFPs together with brain toxicants can result in neurodegenerative diseases (i.e. Alzheimer Disease, Parkinson Disease) [4]. Neuropathological examinations showed that not only adults (29.2 ± 6.8 years age) but also young children (12.89 ± 4.9 years age) displayed neurodegenerative pathology of Alzheimer and Parkinson diseases when living in areas with a high UFP concentration. Cognitive tests indicated an earlier beginning of dementia and a general decline in cognitive abilities [4]. Air pollution is also associated with symptoms and behavior of anxiety or depression and changes in brain structure and function [5]. A recent review by Zundel and colleagues [5] showed that only a small percentage of studies on air pollution focused on UFPs. Their review also revealed that early life, childhood and adolescence are time-periods not closely looked at. That is unfortunate since children's brains are in development and may be more susceptible to air pollution [5]. They state a general need for more human studies (i.e. neuroimaging assessments), to establish a connection to the already existing literature of animal studies [5].

UFPs have natural and anthropogenic sources. For example, volcanic eruptions and wildfires produce UFPs naturally [6]. A major anthropogenic source of UFPs is combustion (machines burning fossil fuels, i.e. the whole transportation sector with vehicles, ships and planes) [6]-[7]. Synthetic sources frequently come from engineered products used for microtechnology or toner pigment [1]. Major occupation exposures occur during welding and blast furnace operation [1]. The World Health Organization [8] issued Air quality guidelines (AQG) which regulate the annual and 24-hour emission of PM_{10} µm/m^3 and $PM_{2.5}$ µm/m^3. However, there are no regulations for UFPs [7].

An average concentration of UFPs measured in rural areas was 2610 particles/cm^3 (pt/cc), whereas for roadsides it amounts to 48.180 pt/cc [1]. However, there are also associations between high UFP concentrations and different ambient conditions, for example high humidity, low air movement, increased number of diesel vehicles and traffic accelerations. Additionally, there are also changes in UFP numbers, because of season, location and weekday. For example, the UFP concentrations in winter are twice as high and in summer half as much as the annual average. There are also daily peak levels between 7 am and 10 am [1]. Yet, many studies do not control for meteorological conditions [5].

Moreover, through a lack of standardized measuring instruments for UFPs, methodologies and protocols vary and are not coherent [1]. The lack of standardization prohibits cohesive conclusions, which impairs the establishment of guidelines or regulations for measuring UFP emissions. However, UFP levels and various health outcomes are associated, as mentioned above, thus it is utmost relevant to facilitate UFP investigations. To help close the gap regarding UFP measurements, results from a mobile measurement instrument were compared to literature reports. This study also contributes to establishing a standardization procedure and encourages further research on the topic of ultrafine particles. Our working hypothesis is that there are differences in UFP concentration between the different locations. The second hypothesis is that there are correlations between UFP concentrations and meteorological conditions.

2   Material and Methods

UFP measurements were taken in three different locations: Marie-Curie-Straße (in front of the Center of Brain, Behavior and Metabolism, CBBM), Carlebach-Park (Max-Linde-Weg 8) and at Ratzeburger Allee (in front of the 'Well you fitness studio'). For measurements the P-Trak Ultrafine Particle Counter Model 8525 (TSI Incorporated, USA) was used, which counted particles per cubic centimeter (pt/cc). Measurements took place between 7 and 10 am in November 2023. One measurement took 30 minutes with 15 measurement points, of one minute each every other minute. Since the Particle Counter is not water resistant measurements could not always be conducted as planned but were dependent on the weather conditions. Detailed weather parameters (temperature, air pressure, relative humidity, wind velocity, wind direction and horizontal solar radiation) were taken from the website of the Technische Hochschule Lübeck (https://wetter.th-luebeck.de/ [9]) and recorded in a protocol. Since data were not normally distributed values at the three different locations were compared using Kruskal-Wallis and pairwise Wilcoxon-Tests. Furthermore, correlations with the weather conditions were conducted. All tests were conducted with the software RStudio.

3   Results and Discussion

3.1   Results

First, data were explored: UFP at Marie-Curie-Straße, Carlebach-Park and Ratzeburger Allee were compared in a graph that showed the concentration on a timeline for each location per day separately (Figures 1-3). In the Figures for the location Ratzeburger Allee as well as the Marie-Curie-Straße noticeable peaks were visible. After looking into the protocol to see if these peaks could be explained through environmental changes (i.e. someone smoking in near proximity), environmental explanations could be ruled out. Descriptive data analyses showed that these two datapoints were over 3 standard deviations from the respective mean and were therefore labeled as outliers. As a result, the datapoints at 8:48 am on the 09.11. for the Marie-Curie-Straße and at 7:29 am on the 07.11. for Ratzeburger Allee were excluded ($N = 14$) in Figures 1-3. As mentioned before, the data does not follow a normal distribution. For the Marie-Curie-Straße the general mean $\pm$ standard deviation (SD) was 4232.72 pt/cc $\pm$ 3677.13 pt/cc SD. For Carlebach-Park the mean and SD are 3637 pt/cc $\pm$ 2471.85 pt/cc, and for Ratzeburger Allee, 5115.27 pt/cc $\pm$ 3270.48 pt/cc.

The Kruskal-Wallis-Test revealed a significant difference between the three groups with $H(2) = 26.74$, $p < .001$. Further analysis with a pairwise Wilcoxon-Test revealed significant differences between the location Carlebach-Park and Ratzeburger Allee ($p < .001$) as well as Marie-Curie-Straße and Ratzeburger Allee ($p < .001$), with small effect sizes ($r = 0.28$, $r = 0.23$). The descriptive statistics as well as the Figures 1 to 3 indicate that Ratzeburger Allee displayed higher UFP concentrations than Carlebach-Park and Marie-Curie-Straße, whereas Carlebach-Park and Marie-Curie-Straße do not differ significantly from each other. To investigate for susceptibility regarding weather conditions, correlations were conducted with the UFP concentrations for each day at every location. The correlated weather conditions were: temperature, air pressure, relative humidity, wind velocity, wind direction and horizontal solar radiation. Except temperature and wind velocity none of the other correlations were significant. We found strong correlations between the temperature and UFP concentrations at the Carlebach-Park location $r(9) = -.68$, $p = .022$ as well as at the Marie-Curie-Straße location $r(8) = -.69$, $p = .028$. However, there was no significant correlation for Ratzeburger Allee, $r(8) = .02$, $p = .95$. Figure 4 shows the different correlations. The other significant correlations were for wind velocity. UFP concentrations in Carlebach-Park had a strong correlation with wind velocity $r(9) = -.61$, $p = .046$ together with the Marie-Curie-Straße location $r(8) = -.64$, $p = .047$. Ratzeburger Allee again had no significant correlation $r(8) = .31$, $p = .38$. Correlations with wind velocity can be found in Figure 5. These pilot results were not corrected for multiple comparisons.

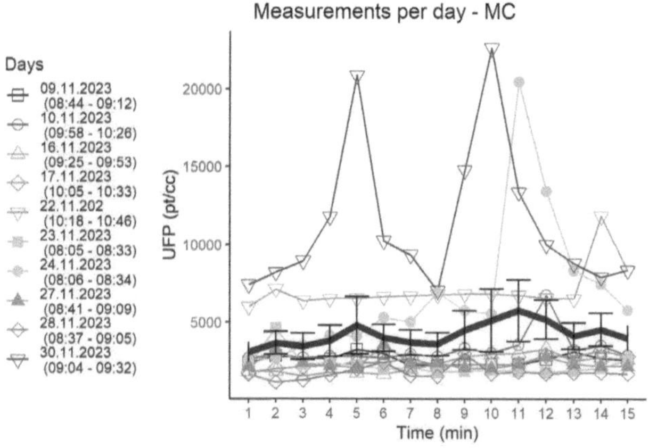

Figure 1: Measurements per day for the Marie-Curie-Straße (MC). Each measurement point represents the average UFP value over the one minute interval. Outliers are not included.

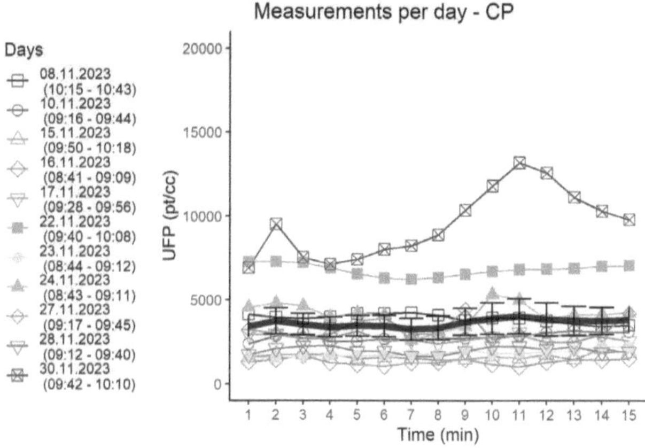

Figure 2: Measurements per day for the Carlebach-Park (CP). Each measurement point represents the average UFP value over the one minute interval.

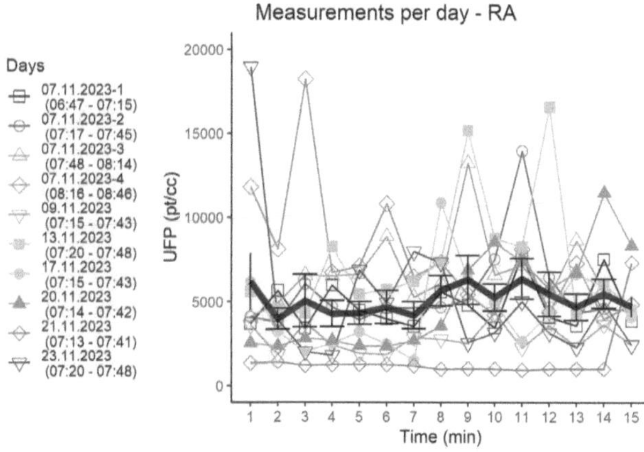

Figure 3: Measurements per day for the Ratzeburger Allee (RA). Each measurement point represents the average UFP value over the one minute interval. Outliers are not included.

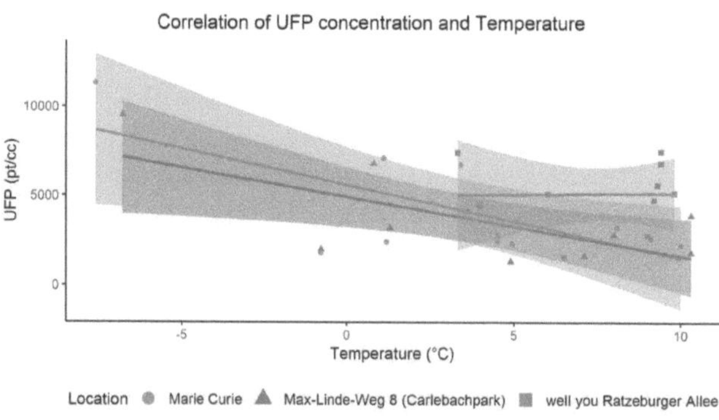

Figure 4: The correlation between UFP concentration and temperature for each location.

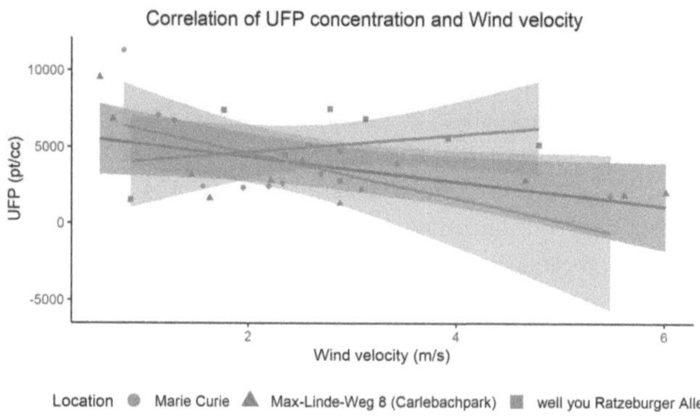

Figure 5: The correlation between UFP concentration and wind velocity for each location.

3.2 Discussion

The significant differences between the locations Marie-Curie-Straße (control condition), Carlebach-Park (rural area) and Ratzeburger Allee (urban area, roadside) validate that the instrument can differentiate UFP concentrations successfully and confirm that the UFP concentration is higher in urban areas, especially at the roadside. Particularly between 7 and 10 am there are many vehicles on the road due to the rush hour. Since that is also the time when children are on their way to school, it represents a sensitive time window. Children whose school way is along the roadside could be affected by UFPs. This is an important result considering children, with developing body and brain are generally more susceptible towards environmental factors such as air pollution [5]. More longitudinal human studies should be conducted in this research area, to ascertain possible delayed effects resulting from child exposure [5]. Equally important are other time windows that have been neglected (i.e. adolescence) [5]. In addition to school routes future studies should measure UFPs concentration at schools and see how the values are in classrooms and school environments (i.e. in school yards).

Correlations of temperature and wind velocity with UFP concentrations illustrate, that measurements are weather dependent and should be taken into account during assessments. Interestingly, the ambient temperature during the last measurements for Marie-Curie-Straße and Carlebach-Park was really low, -7.6 degree Celsius and -6.8 degree Celsius, respectively. On that day it was also snowing; very high UFP concentrations were measured on these cold days. This relationship is in accordance with the literature [1]. With increased wind velocity UFP may be blown away faster and may not stay in place long enough to be measured (enter the sensor tube).

In summary, our first hypothesis regarding differences in UFP concentration in the different locations was confirmed. In regard to the second hypothesis, significant correlations were found only for temperature and wind velocity.

3.2.1 Comparison with the literature

In rural areas a mean value of 2610 pt/cc was reported in the literature. We obtained a mean value of 3637 pt/cc. Both values are in the same order of magnitude [1].
At Ratzeburger Allee we measured 5115.37 pt/cc whereas a roadside value of 48.180 pt/cc was reported in the literature. Our value was thus way below this average [1].
Details of the locations per se as well as the above mentioned weather conditions may account for these differences. In addition, these were our first measurements with the instrument. Future studies should control for weather conditions and should consider them in their analysis. Many previous studies didn't control for meteorological conditions [5]. Furthermore, as previously mentioned we were limited in measurements due to the fact that the measurements could not be conducted on rainy days. Subsequent studies within this project will continue to develop standards for the assessment of UFP concentrations including protocols for taking meteorological conditions into account. UFP measurement standards are a key step toward establishing UFP thresholds. We believe this study has made a small contribution to this undertaking.

4 Conclusion

Taken together, these first measurements demonstrate the importance of establishing a standardized process for measuring ultrafine particles. Once a preliminary guideline for measurements exists, studies are more likely to provide coherent or at least comparable results. This would facilitate scientific investigations into adverse health effects of UFPs. Finally, this could support the World Health Organization in issuing thresholds for not only PM_{10} and $PM_{2.5}$ but also for ultrafine particles.

Acknowledgements

The work has been carried out at and supervised by the Institute of Experimental and Clinical Pharmacology and Toxicology, Universität zu Lübeck. Special thanks go to Lina Johanna Steinlein, student of Molecular Life Science and Pia Witt, student of Nutritional Medicine, for assisting in measurements of the ultrafine particles. Special thanks go to Lübeck Hoch 3 (LH3) for supporting the project and financing the P-Trak Ultrafine Particle Counter. We furthermore acknowledge Claas Heymann, Technische Hochschule Lübeck as collaborator in the LH3 project. The Technische Hochschule Lübeck provided the website for weather conditions.

Authors' Statement

Conflict of interest: Authors state no conflict of interest.

5 References

[1] D. E. Schraufnagel, *The health effects of ultrafine particles.* In: Experimental & Molecular Medicine, vol. 52, no. 3, pp. 311–317, 2020.

[2] Y. Shang, R. Chen, R. Bai, J. Tu and L. Tian, *Quantification of long-term accumulation of inhaled ultrafine particles via human olfactory-brain pathway due to environmental emissions – a pilot study.* In: NanoImpact, vol. 22, 100322, 2021.

[3] G. Oberdörster, Z. D. Sharp, V. Atudorei, A. Elder, R. Gelein, W. G. Kreyling and C. Cox, *Translocation of inhaled ultrafine particles to the brain.* In: Inhalation Toxicology, vol.16, no. (6–7), pp. 437–445, 2004.

[4] L. Calderón-Garcidueñas and A. Ayala, *Air pollution, ultrafine particles, and your brain: Are combustion nanoparticle emissions and engineered nanoparticles causing preventable fatal neurodegenerative diseases and common neuropsychiatric outcomes?* In: Environmental Science & Technology, vol. 56, no. 11, pp. 6847–6856, 2022.

[5] C. G. Zundel, P. Ryan, C. Brokamp, A. Heeter, Y. Huang, J.R. Strawn and H.A. Marusak, *Air Pollution, depressive and Anxiety Disorders, and Brain Effects: A Systematic review.* In: NeuroToxicology, vol. 93, pp. 272–300, 2022.

[6] A. L. Moreno-Ríos, L. P. T. Benítez and C. Bustillo-Lecompte, *Sources, Characteristics, toxicity, and Control of Ultrafine Particles: An Overview.* In: Geoscience Frontiers, vol. 13, no. 1, 101147, 2022.

[7] World Health Organization, *WHO Global Air Quality Guidelines: particulate matter (PM2.5 and PM10), ozone, nitrogen dioxide, sulfur dioxide and carbon monoxide.* World Health Organization, Geneva, 2021.

[8] World Health Organization, *What are the WHO air quality guidelines?* Available: https://www.who.int/news-room/feature-stories/detail/what-are-the-who-air-quality-guidelines [last accessed on 2024-02-02].

[9] Technische Hochschule Lübeck, *THL Campus Wetterseite* Available: https://wetter.th-luebeck.de/ [last accessed on 2024-02-02].

Assessing Driving Situation Complexity: Development and Pilot Evaluation of a Short Questionnaire with Traffic Psychology Experts

Elise Sophie Banach [1] and Markus Gödker [2]

[1] Psychology - Cognitive Systems, Universität zu Lübeck, elise.banach@student.uni-luebeck.de

[2] Institute of Multimedia and Interactive Systems, Universität zu Lübeck, markus.goedker@uni-luebeck.de

Abstract

For field or simulator driving studies in the field of traffic and engineering psychology, it is necessary but difficult to accurately classify driving situations in terms of their situational complexity. One approach is to regard driving tasks as similar to a work task making a requirements analysis possible. Based on the "Situational analysis of behavioural requirements of driving tasks" (SAFE) [1], a short questionnaire for assessing situational complexity was created in this study. The questionnaire was used in an expert study with professors and research assistants ($N = 8$) from the field of traffic and engineering psychology from various German universities in order to obtain a rating of driving sectors from the DFG-funded AMORi project. The video-based online study resulted in an assessment of complexity that was consistent with the planned design of the driving sectors. The newly designed short questionnaire resulted in sufficient to high internal consistency and allowed a differentiation of the driving scenes with regard to their situational complexity in comparison to each other.

1   Introduction

The situational complexity of driving scenes is difficult to measure, describe and manipulate as many different factors determine it and the subjectively experienced load of a driving situation does not directly result from objective features of a situation. Influencing factors are, for example, the driver themself with their characteristics, their perception and behaviour but also the environment including factors as the traffic situation [2]. A higher complexity for the driver could "maybe [mean] more disturbing or distracting objects or events. [...] But the complexity of a driver situation cannot be quantified generally, as the driver himself in his current state is a variable which cannot be disregarded." [3, p. 48]. One factor that is used in many studies to manipulate complexity is the area in which a driving task is performed. A distinction is usually made between urban and rural scenarios (e.g. [4]), whereby the urban area should lead to greater complexity than the rural one. Other factors are the visibility and weather conditions of a scene, whereby worse conditions lead to a higher situational complexity [5]. In order to be able to classify driving situations, one can understand the driving task as similar to a work task [5]. This is because driving behaviour and working behaviour can both be viewed as performance in specific situations. On this basis the SAFE [1] was developed. SAFE is a detailed expert situation assessment tool comprised of rating and coding schemes to assess situation requirements of driving scenes. In the original approach, the driving task is

divided into spatial and temporal segments, to which individual subtasks are assigned. Based on a model of information processing [6], an analysis of behavioural requirements is then carried out in twelve categories. As the situational complexity of driving scenes consists mostly of information processing requirements and those relating to vehicle operation [5], the first six categories of the SAFE cover precisely these factors. The categories one to five, named "Perception", "Expectations", "Judgement", "Memory" and "Decision & Planning" include the factors of information processing. The sixth category consists of the vehicle operation requirements and is named "Car Handling & Driver Action". Category eight deals with complexity assessment and the information from the previous categories is analysed. In the present study, we focus on the categories just mentioned, since they are most relevant for the aspect of situational complexity. Other factors used in the analysis to enable classification and have an impact on complexity are the time available to successfully complete a task and the required accuracy performing it. If the driver has to act under time pressure or particularly accurate, the task gets more difficult [1]. The more demands are placed on the driver when driving, the more resources he has to use and the more complex the situation is perceived. Therefore, the number of demands is also an important factor to regard in the assessment of situational complexity [1]. The driver's advanced cognitive functions on a conscious level need to be utilised more when a situation is more complex [5]. That means "to what extent single requirements are processed either consciously or unconsciously is a significant indicator for assessing task complexity." [1, p.969]. This study is part of the DFG-funded AMORi project at the Institute of Interactive and Multimedia Systems, which deals with the "action-integrated modelling and optimisation of energy-related driver-vehicle interaction." [7]. The objective of the project is to examine novel ecodriving displays in the Institute's driving simulator. In order to understand how the situational complexity moderates the effect of the displays, driving scenes of varying complexity are needed. To this end, an assessment of the complexity of the designed driving scenes is required beforehand. In the present research, we first want to develop a suitable new method to economically capture the situational complexity of driving scenes as the smallest examination unit, since the SAFE is highly detailed. Second, we have the objective to obtain ratings of the scenes from experts on the field of traffic and engineering psychology to classify the scenes. Third, we want to compare the predefined complexity of the scenes from the design phase in the AMORi project with the ratings of the experts.

2 Materials and Methods

2.1 Questionnaire Development

The SAFE is not a completed and standardized procedure, but it "could give a theoretical and methodological framework for analysing safe driver behaviour in different contexts." [1, p.976]. To conduct a complete SAFE for all seven driving scenes is not a suitable approach for our purpose, as it takes much time for multiple raters. We rather focused on an overall assessment to rank situations and sort them into coarse categories, hence, the extremely detailed SAFE appears not economical. Therefore, our goal was to develop a short subjective rating questionnaire based on and adapted from the theoretical and methodological basis of SAFE. To this end, the previously mentioned factors influencing situational complexity were used as items. The instruction in the beginning of the questionnaire was: "Please answer the following questions and refer them to the video you have just watched." The following items were included with a 11-point rating scale (0 - very low or few, 10 - very high or many): time pressure when completing the driving task, degree to which the driving task had to be carried out particularly accurately, the number of requirements to successfully complete the driving task and the situational complexity. Also, the percentage of conscious and automatic actions to fulfil the driving task had to be divided to 100% as this was also the scale in SAFE. The *functions of information processing and vehicle operation* were also included in the questionnaire. Here, the subscale *functions of information processing* included the items *perception*, different *expectation processes, temporal spatial constellations, risk and situation assessments, retrieval of information from long-term memory, reading and retaining information in working memory, integrating information from different sources* and *decision processes*, which were the first five categories

of SAFE. The items included in the subscale *vehicle operations*, coming from category six of SAFE, were *keeping track and speed, indicators and signalling, operating activities, compensating for incorrect behaviour of others* and *information intake from inside of the vehicle*. All these items had to be rated on a scale from zero to ten.

2.2 Description of the Expert Study

An explorative expert video-based online study was conducted, that was supervised via webconference. The experts ($N = 8$) were professors and doctoral students from various German universities. Six worked as research assistants in the field of engineering or traffic psychology, one as a professor of engineering psychology and cognitive ergonomics and one as a professor of traffic psychology. On average, they had practised their current profession for 3.38 years ($SD = 2.31$) and were 32.4 years old ($SD = 5.36$). Of all experts, 25% were male, 75% female. The seven driving sectors were designed in [8] by students in a media informatics project and driving through at the recommended speed took an average of 1.2 minutes. An overview of the sectors can be found in Table 1. Each sector was driven through in the simulator and a screen video was recorded which also included the speed display. These videos were used in the questionnaire with the instruction to put oneself in the situation of a person driving through the scene in the driving simulator. A list of definitions of the used constructs was gone through in detail so that all experts had the same idea of the developed items and every video was watched once.

Table 1: Description of the sectors

Sector	Description
2	urban area, intersection and stop sign, no other traffic
3	rural area, departure from four-lane motorway, no other traffic
6	rural area, downhill and roadworks on opposite lane, no other traffic
7	urban area, uphill and downhill, no other traffic
8	urban area, three bus stops, bus and other cars on own lane
9	rural area, intersection and stop sign, no other traffic
10	rural area, sharp turn and stop sign, no other traffic

In the course of the driving scene design phase in the AMORi project, complexity blocks were formed which gave the map designers information about the required complexity of the driving scene. Sectors 3, 6, 8 and 10 were supposed to become the more complex block in comparison to 2,7 and 9. Our research questions are, therefore, whether the newly designed questionnaire based on the SAFE is suitable for determining situational complexity, how the sectors are evaluated by experts in general, and whether the ratings of the eight experts agree with the previously planned complexity. All statistical analyses in this study were carried out with [9].

3 Results and Discussion

3.1 Descriptives and Internal Consistency

In order to find out how the experts assessed the individual sectors and whether these assessments correspond to the planning in the AMORi project, the descriptives were first analysed. Looking at the mean values of the *information processing functions*, the *perception function* had the highest value with a mean of 5.98 (SD = 2.32). The other mean values of the *information processing functions* range from 2.29 (SD = 2.31; *decision making under uncertainty*) to 4.45 (SD = 2.66; *Temporal spatial constellations*). For the *vehicle operating functions*, *maintaining speed limit* had the highest mean value with 5.88 (SD = 2.62) and *compensating for incorrect behaviour of others* the lowest (M = 1.16; SD = 2.03). Cronbach's α was then calculated to determine the internal consistency of the subscales *information processing functions* and *vehicle operation functions*. According to the interpretation of [10], the internal consistency of the *information processing functions* subscale was high (α = .87) and for the *vehicle operation functions* it was acceptable (α = .74). For this reason, in all further analyses the mean of the respective values of these subscales was used. The calculation of the single measures Intraclass Correlation Coefficient (ICC) results in a value of .83.

The mean values of the items per sector can be found in Table 2. The highest mean value was 6.00 (SD = 2.39) on the item *accuracy* in sector 10. When rating conscious and automated actions, the highest average value was 43.13 % (SD = 24.04 %), also in sector 10. The last two items sector 10 was ranked highest compared to all other sectors, were *number of requirements* (M = 5.00; SD = 2.00) and *situational complexity* with a mean value of 5.29 (SD = 2.75). Sector 9 has the lowest values on four items, which are required *accuracy* (M = 3.25; SD = 1.58), *number of requirements* (M = 2.63; SD = 1.30), *consciousness of actions* (M = 25.00 %; SD = 16.69 %) and *vehicle operation functions* (M = 2.05; SD = 1.06). On the item *situational complexity* sector 9 (M = 2.50, SD = 1.77) is rated as minimally more complex than sector 7, which has the lowest mean value with 2.25 (SD = 1.39). For comparison with the complexity blocks defined in the AMORi project, an average value was calculated for all items per sector. For this purpose, the item *consciousness of actions* was also transformed to the zero to ten scale. The overall mean values of sector 3 (M = 4.14), sector 6 (M = 3.87), sector 8 (M = 3.93) and sector 10 (M = 4.41) were the highest. Sectors 2 (M = 2.81), 7 (M = 2.86) and 9 (M = 2.50) were therefore rated as less complex by the experts, compared to the other sectors.

3.2 Correlations

Next, the Spearman correlations of the items that should measure complexity (*time pressure, accuracy, number of requirements, consciousness of actions, information processing functions* and *vehicle operation functions*) and the item *situational complexity* were calculated. All included influencing factors on situational complexity correlate significantly with the item *situational complexity*. According to [11], the strength of the correlation is moderate for *vehicle operation functions* (r = .27, p = .049), high for *information processing functions* (r = .54, p < .001), *time pressure* (r = .52, p < .001), *accuracy* (r = .57, p < .001) and *consciousness of actions* (r = .69, p < .001 and highest for *number of requirements* (r = .83, p < .001).

3.3 Interpretation

First, we had the objective to develop a new method to economically capture the situational complexity of driving scenes. The values of Cronbach's alpha are high and acceptable, which indicates that the subscales of the two most important influencing factors *functions of information processing* and *functions of vehicle operation* on *situational complexity* can be captured well using the designed items on the basis of the SAFE. The high correlations are a first indicator for an existing convergent validity, which would imply that the influencing factors can be used to get a ranking of the complexity of different scenes. The ICC shows that there is a high interrater reliability and that the experts are consistent in their assessments. From the results it can be concluded that it is possible to differentiate driving scenes from one another in terms of their situational complexity and that the questionnaire developed here can be used for this purpose.

Our second research question, how complex the experts rated the sectors, can be answered by means of the descriptives. One conclusion that can be drawn, is that no sector was categorised as particularly complex. Thus, for example the highest overall mean value was only 4.41 on a scale of zero to ten. This is also shown by the predominant *functions of perception* of the item *functions of information processing*, which play a greater role in less complex situations than, for example, judgement or decision processes [5]. The highest mean value on the *consciousness of actions* item is also rather low with 43.13%. Lower figures indicate a smaller proportion of conscious actions and therefore a larger proportion of unconscious and automated actions, which require fewer resources and ensure less perceived complexity. The overall mean value would make sector 9 the least complex sector and sector 10 the most complex one. Sector 10 includes a sharp unexpected turn and curves which require a high accuracy compared to the other sectors. Probably because of that, the number of requirements is also highest in this sector. In sector 9, this item is lowest since it is a rural area with no traffic and no turns. This might be the reason that also the *consciousness of actions* is lowest compared to the other sectors, since many actions take place automatically. The last research question, regarding the comparison of the complexity of the sectors with the previously defined blocks in the AMORi project shows that the experts' ratings correspond to the requirements placed

Table 2: Mean Values

Sector	Time Pressure	Accuracy	Number Requirements	Consciousness	Information Processing	Vehicle Operation	Situational Complexity
2	1.50	3.50	3.25	29.38	2.79	2.18	3.50
3	3.38	5.38	4.63	40.00	3.44	3.75	4.38
6	3.13	4.38	4.00	40.00	3.80	2.88	4.88
7	1.75	3.50	3.25	38.75	2.61	2.77	2.23
8	3.75	4.63	4.00	41.25	4.13	2.75	4.13
9	1.75	3.25	2.63	25.00	2.82	2.05	2.50
10	3.00	6.00	5.00	43.13	3.94	3.30	5.29

on the map designers beforehand. Sectors 3, 6, 8 and 10 were the higher ranked sectors in the overall average and can therefore be used in the project as the more complex block compared to sectors 2, 7 and 9.

4   Conclusion

In the present research, the newly developed short questionnaire on the basis of the SAFE enabled a differentiation and an assessment of the complexity of scenes in the driving simulator that corresponds to the predefined specifications. In agreement with [5], according to whom uncomplex situations often occur on long routes without oncoming traffic, this study resulted in rather low complexity ratings. With the exception of sector 8, no other traffic was seen in any of the driving scenes. In addition, visibility and weather conditions are good in every scene, which is also reflected in the low complexity ratings. In order to create more complexity in driving scenes, more traffic could be added to increase the number of requirements or the time pressure to take a decision. When adding other weather conditions, that lead to a worse visibility the required accuracy of the driving task could increase, which could also result in a higher complexity, since the visibility and weather conditions are a crucial factor regarding the complexity [5]. The limited sample size of eight experts requires caution in interpreting the results, as it could affect the external generalizability of the results. The short questionnaire can be tested on subjects driving through the sectors to check how similar the ratings of the subjects would be to those of experts. In general, this study created the basis for an economic method of an assessment of the complexity of driving scenes.

Acknowledgement

The work has been carried out at Institute of Multimedia and Interactive Systems (IMIS), Universität zu Lübeck and supervised by T. Franke, IMIS, Universität zu Lübeck.

Authors' Statement

Conflict of interest: Authors state no conflict of interest. Informed consent: Informed consent has been obtained via the online survey tool LimeSurvey from every participant.

References

[1] W. Fastenmeier and H. Gstalter, "Driving task analysis as a tool in traffic safety research and practice," *Safety Science*, vol. 45, no. 9, pp. 952–979, 2007.

[2] M. Plavsic, "Analysis and modeling of driver behavior for assistance systems at road intersections," Ph.D. dissertation, Technical University of Munich, 2010.

[3] C. P. Rommerskirchen, M. Helmbrecht, and K. J. Bengler, "The impact of an anticipatory eco-driver assistant system in different complex driving situations on the driver behavior," *IEEE Intelligent Transportation Systems Magazine*, vol. 6, no. 2, pp. 45–56, 2014.

[4] D. Kaber, Y. Zhang, S. Jin, P. Mosaly, and M. Garner, "Effects of hazard exposure and roadway complexity on young and older driver situation awareness and performance," *Transportation research part F: traffic psychology and behaviour*, vol. 15, no. 5, pp. 600–611, 2012.

[5] W. Fastenmeier *et al.*, *Autofahrer und Verkehrssituation. Neue Wege zur Bewertung von Sicherheit und Zuverlässigkeit moderner Strassenverkehrssysteme.* 1995.

[6] J Rasmussen, *Abduction hierarchy, information processing and human-machine interaction*, 1986.

[7] Institute of Multimedia and Interactive Systems, Universität zu Lübeck. "AMORi," imis.uni-luebeck.de. last accessed on 20 January, 2024. (2022), [Online]. Available: https://www.imis.uni-luebeck.de/de/forschung/projekte/amori.

[8] BeamNG GmbH, *beamNG.drive*, https://www.beamng.com, last accessed on 20 January 2024.

[9] Jamovi Project, *Jamovi (version 2.3) computer software*, 2022.

[10] M. Blanz, *Forschungsmethoden und Statistik für die Soziale Arbeit: Grundlagen und Anwendungen.* Kohlhammer Verlag, 2021.

[11] J. Cohen, *Statistical power analysis for the behavioral sciences.* Lawrence Erlbaum Associates, 1988.

Development and Validation of the Preference for Automation Adaption Scale (PAAS)

Anne Tichy [1] and Mourad Zoubir [2]

[1] Psychology - Cognitive Systems, Universität zu Lübeck, anne.tichy@student.uni-luebeck.de
[2] Institute for Multimedia and Interactive Systems, Universität zu Lübeck, m.zoubir@uni-luebeck.de

Abstract

Without optimizing for users' behaviour and preferences, automated systems cannot achieve their full potential. This is especially crucial for systems where changes in levels of automation are necessary for optimal performance. However, it can be unclear whether users prefer these changes to be manual or automated. The aim of this work was the development and validation of a scale to capture the adaptation preferences in the context of system automation. The scale was developed in an iterative process, including cognitive interviews with experts and potential users. An online validation study with students ($N = 86$) found a poor model fit, but good internal consistency and correlation analysis indicated at least partial construct validity. This study highlights PAAS' potential to assess preferences for a new adaption-based automated system in advance, preventing resource waste on a system with uncertain user adoption.

1 Introduction

Automated systems are present in almost every aspect of modern lives. The benefits of human-machine systems depend on both technical potential and user behaviour. It is therefore important to not only consider technical challenges, but also gain an understanding of individual preferences regarding automated systems.

Parasuraman [1] suggests that a fixed application of automation to reduce workload may lead to diminished awareness, trust issues, complacency, and skill decline. Conversely, adopting an approach involving flexible automation, where the level or type can adapt during system operations, has been shown to mitigate out-of-the-loop problems and enhance overall human-automation system performance. This adaption-based automation concept extends beyond low and high levels of automation (LOAs), allowing the assignment of different LOAs to individual tasks within system operation.

Various systems to differentiate between multiple levels of automation have been developed. Parasuraman et al. [2] created a model for types and levels of human interaction with automation that is divided into four classes of function: 1) information acquisition (detection and recording of input data); 2) information analysis (further processing of the data); 3) decision selection (deciding multiple action alternatives); and 4) action implementation (executing the previously made choice). Within each of these types, automation can be applied in different LOAs. This raises the question of who or what is in charge of assigning the LOAs. Two options are adaptive and adaptable automation.

1.1 Current state of research

Calhoun [3] reviewed existing research pertaining to adaptive and adaptable automation and found that so far researchers focused more on adaptive than on adaptable automation, and that in particular, little research has been done on the direct comparison between adaptive and adaptable automation. The term "adaptive automation" is used for technology enabling flexible recommendations and/or initiation of LOA changes based on specific triggers. Although adaptive automation has demonstrated to improve task performance and reduce attention management demands compared to fixed automation, human acceptance may decrease when the automation initiates changes in LOAs [4].

In contrast, with adaptable automation, the authority for invoking or changing LOAs remains with the human operator. This approach keeps the human operator in the loop and simultaneously utilizes the human's ability to anticipate and prepare for changes, offering needed flexibility in new situations. Studies have shown that adaptable automation fosters higher human operator acceptance, appropriate levels of trust in automation and improved situation awareness [5]. However, entrusting humans with the management of automation levels raises concerns about additional management workload and potential preoccupation with changing LOAs, potentially leading to people failing to make necessary LOA changes or choosing inappropriate levels of automation [6]. To establish the optimal adaption type for a given system, a direct comparative analysis of the two types should be carried out by systematically evaluating user preferences. A subjective measure is particularly useful to identify requirements for a system before its (potential) development begins to conserve resources, and can be applied to diverse fields to examine preferences particular to that context.

1.2   Research questions

The objective of this study was to validate the developed Preference for Automation Adaption Scale (PAAS). This involved extracting factors from the collected data and exploring potential associations by examining correlations between PAAS scores and other relevant variables. Therefore, the following research questions were formulated: RQ 1: To what extent does the PAAS fulfil convergent and discriminant construct validity? RQ 2: To what extent does the PAAS fulfil internal consistency? RQ 3: To what extent does an exploratory factor analysis reveal an underlying factor structure of the PAAS?

2   Material and Methods

2.1   Scale development

Based on the types of functions by Parasuraman et al. [2], the Preference for Automation Types Scale (PATS) was created by a research team at the Institute for Multimedia and Interactive Systems, University of Lübeck. The PATS assesses preferences for tasks that can be performed by both humans and automated systems. The scale encompasses a spectrum from completely human to fully automatic control, with four items for each of the four types of functions [7]. Building on the PATS, the PAAS was developed in an iterative way, including cognitive interviews with experts and potential users. The PAAS captures the user's preferences on adaptive versus adaptable automation on a 6-point Likert scale ranging from "completely prefer adaptable automation" (1) to "completely prefer adaptive automation" (6) on the same 16 items as the PATS. The scale is designed to be applied to specific functions (information acquisition, information analysis, decision selection and action implementation) of a chosen technical system utilizing automation. The ability of this system to take over a specific task should be given in the instruction statement. Therefore, the PATS evaluates the individual preference for the fixed specific task.

2.2   Additional scales

Included in the survey were four questionnaires to measure convergent and one subscale to measure divergent construct validity.
Affinity for Technology Interaction (ATI) was measured with the unidimensional scale from Franke et al., [8] (9 items on a scale from 1-6, 1 being "completely disagree" and 6 being "completely agree"). Propensity to Trust (PTT) was measured with the unidimensional scale translated into German [9] (6 items on a scale from 1-5, 1 being "does not apply at all" and 5 being "fully applies"). Basic Psychological Need Satisfaction for Technology Use (BPN-TU) was measured with the scale from Moradbakhti et al.

[10] (12 items on a scale from 1-5, 1 being "does not apply at all" and 5 being "fully applies"). Agreeableness and Openness to Experience were measured with the Big-Five-Inventory-10 [11] (4 items on a scale from 1-5, 1 being "disagree strongly" and 5 being "agree strongly"). Neuroticism was measured with the same Big-Five-Inventory-10 [11] as agreeableness and openness. This subscale was used for discriminant validity.

Research has shown that most of these constructs correlate with each other and, apart from the Big Five, altogether form a theoretical foundation in the context of automation via definition [7]. Consequently, it was reasonable to anticipate correlations with the PAAS assuming construct validity.

2.3   Participants

Participants were recruited through a student forum of the University of Lübeck. There were no inclusion criteria. Students of psychology and media informatics from the university of Lübeck were given course credit for participation. 118 participants started the survey, but we excluded due to incomplete submissions ($n = 23$) or those incorrectly answering the attention check ($n = 9$). The average age was 22,5 years old ($SD = 3,85$, $Mdn = 21$), ranging from $18 - 37$ years old. 65 participants identified as women, 19 as men and 1 as diverse.

2.4   Procedure

The survey was performed with an online survey tool and for all questionnaires German versions were used. After filling out sociodemographic questions, the participants completed the PAAS. For this study, the context given in the PAAS consisted of an automated driving system that was able to take over the collection of environmental data, determination of the optimum speed, selection of the most energy-efficient driving style and adoption of acceleration and speed maintenance. Lastly, the participants filled out the four questionnaires for construct validation described above.

2.5   Statistical analysis

For construct validity, averaged sum scores of the PAAS and its subscales (information acquisition, information analysis, decision selection and action implementation) were correlated with the averaged sum score of additional scales. Spearman correlation was selected due to Shapiro-Wilk Tests for normality indicating deviating distributions for several measured constructs. Cronbach's alpha was assessed for both the entire PAAS and each type of function to evaluate internal consistency, which is acceptable when $.60 < \alpha < .90$. To assess if the data gathered with the PAAS was suitable for factor analyses, Bartlett's Test of Sphericity, which should turn out significant for suitable data, and the Kaiser-Meyer-Olkin (KMO) test, with KMO

Table 1: Correlation matrix

PAAS		BPN-TU						Big Five Inventory		
	PTT	Total	Autonomy	Competence	Rel to Others	Rel to Tech	ATI	Agreeableness	Openness	Neuroticism
Total	0.23*	0.21	0.27*	0.19	0.07	-0.01	0.21	0.04	-0.03	-0.11
Acq	0.32**	0.23*	0.23*	0.19	0.06	0.04	0.12	0.00	-0.07	-0.04
Ana	0.15	0.08	0.24*	0.05	0.05	-0.13	-0.02	0.03	-0.08	-0.21*
Dec	0.22*	0.27*	0.27*	0.33**	0.07	-0.06	0.14	-0.07	0.08	0.02
Act	0.08	0.14	0.17	0.12	-0.03	0.09	0.17	0.06	0.02	-0.09

Note. * $p < .05$, ** $p < .01$
Acq = Information Acquisition, Ana = Information Analysis, Dec = Decision Selection, Act = Action Implementation, PTT = Propensity to Trust, Rel to Others = Relatedness to Others, Rel to Tech = Relatedness to Technology, ATI = Affinity for Technology Interaction

above .5 being acceptable, were performed. Next, an exploratory factor analysis (EFA) was carried out. As correlations between factors were assumed, an EFA with oblimin rotation was carried out. To test for internal consistency of the indicated factors and examine the interrelatedness of their items, Cronbach's alpha was calculated. To evaluate model fit $\chi2$ and the Root-Mean Square Error of Approximation (RMSEA) were reported. For a good model fit, $\chi2$ should be non-significant with $p > .05$ and RMSEA should score low (< .05). The computer software jamovi was used for statistical analysis [12].

3 Results and Discussion

3.1 Construct validity

For convergent construct validity, significantly positive Spearman correlations were found between global PAAS and BPN-TU autonomy and PTT; PAAS information acquisition and global BPN-TU, BPN-TU autonomy and PTT; PAAS information analysis and BPN-TU autonomy; PAAS decision selection and global BPN-TU, BPN-TU autonomy, BPN-TU competence and PTT. In other words, PAAS and most of its subscales correlated with Autonomy need fulfilment, and Propensity to Trust. For discriminant construct validity, a significantly negative Spearman correlation was found between PAAS information analysis and neuroticism. See Table 1 for all correlations.

3.2 Internal consistency

The whole PAAS (Cronbach's α= .80) and its dimension action implementation (Cronbach's α= .81) had good internal consistency, while information acquisition (Cronbach's α= .69), information analysis (Cronbach's α= .66) and decision selection (Cronbach's α= .65) were acceptable.

3.3 Exploratory factor analysis

Bartlett's Test of Sphericity gave a significant result with $p < .001$. The KMO ranged from .62 till .83 across all items of the PAAS. Scree plot indicated three factors. All items were included in the model because all primary factor loadings were > .30 (see Table 2). Cronbach's alpha was acceptable for all factors. This model explained a total of 43,4% of the variance. Since RMSEA = .07 and $\chi2$ =

103, $p < .05$ showed a significant difference between most likely data matrix and the postulated one, the model should be rejected.

Table 2: EFA of the Items of the PAAS

Item	Factor loading		
	1	2	3
Acq1	.69		
Acq2	.56		
Acq3	.59		
Acq4	.45		
Ana1	.66		
Ana2	.67		
Ana3	.36		
Ana4	.46		
Dec1	.52		
Dec2			.52
Dec3			.86
Dec4			.79
Act1		.62	
Act2		.75	
Act3		.91	
Act4		.49	
Cronbach's alpha	.71	.76	.82
% of variance	17.8	13.5	12.1

Note. EFA with oblimin rotation. Values above cutoff .3 displayed.

3.4 Implications

An EFA model was tested with three factors. Factor loadings indicated a different combination of items than proposed with the types of function. Information acquisition and information analysis loaded together on one factor. This could mean that those two types of function are too closely related to each other, at least when it comes to driving a vehicle. One could argue that in other contexts, like medical diagnostics, the difference between data acquisition and data analysis is more profound and transitioning from one to the other is less intuitive. Further studies with other contexts could therefore show different results. The first item of decision selection also loaded on the first factor. Looking at the wording of this item ("Present options of what to do based on available information.") shows that it depicts the display of options rather than selecting a decision and therefore matches the contents of the second type of function more. Consequently, this item should be rewritten or removed. With regard to the other two factors, they represent decision selection and action implementation adequately. While Cronbach's alpha was acceptable for all factors in the model, it was rejected as $\chi2$ indicated the data matrices did not support the model. Concerning convergent validity, it is possible that the limited number of significant correlations with other measured constructs may stem from substantial differences between these constructs and PAAS. For instance, PAAS consistently pertains to a specific system, potentially yielding varied results in diverse contexts. In contrast, the Big Five are recognized as traits, exhibiting relative stability across time and various situations. Despite these differences, the identified correlations provide some support for construct validity, suggesting that PAAS likely measures a construct associated with automation.

3.5 Limitations

The determination of sample size in factor analysis is a controversial subject in literature with varying guidelines. To ensure robust statistical power, it is recommended to involve a minimum of 100 participants for an EFA. Falling short of the required sample size for adequate power increases the risk of Type II errors, potentially leading to the oversight of actual effects and compromising parameter estimation accuracy. Consequently, results should be interpreted cautiously, serving more as guidance for future studies with appropriate sample sizes.

4 Conclusion

This study underlined the potential for the PAAS to be used in the future to assess preferences for a new automated system before development, thereby avoiding a waste of resources when creating an already decided upon adaption-based automated system no one would use. PAAS could also be compared before and after the usage of an automated system to measure its potential effect. Future work should not only include performing more studies, but also specifically revising and improving the PAAS. Items might need to be rewritten to cover each type of function more specifically.

Acknowledgement

This work was supervised by Prof. Dr. rer. nat. Thomas Franke, Institute of Multimedia and Interactive Systems, Universität zu Lübeck.

Authors' Statement

Conflict of interest: Authors state no conflict of interest. Informed consent: Informed consent has been obtained from all individuals included in this study.

References

[1] R. Parasuraman, M. Mouloua, and R. Molloy, "Effects of adaptive task allocation on monitoring of automated systems.," *Occupational Health and Industrial Medicine*, vol. 3, no. 36, p. 101, 1997.

[2] R. Parasuraman, T. B. Sheridan, and C. D. Wickens, "A model for types and levels of human interaction with automation," *IEEE Transactions on systems, man, and cybernetics-Part A: Systems and Humans*, vol. 30, no. 3, pp. 286–297, 2000.

[3] G. Calhoun, "Adaptable (not adaptive) automation: Forefront of human–automation teaming," *Human factors*, vol. 64, no. 2, pp. 269–277, 2022.

[4] G. Calhoun, V. B. R. Ward, and H. A. Ruff, "Performance-based adaptive automation for supervisory control," in *Proceedings of the Human Factors and Ergonomics Society Annual Meeting*, vol. 55, SAGE Publications Sage CA: Los Angeles, CA, 2011, pp. 2059–2063.

[5] R. Parasuraman and C. D. Wickens, "Humans: Still vital after all these years of automation," in *Decision Making in Aviation*, Routledge, 2017, pp. 251–260.

[6] D. B. Kaber and J. M. Riley, "Adaptive automation of a dynamic control task based on secondary task workload measurement," *International journal of cognitive ergonomics*, vol. 3, no. 3, pp. 169–187, 1999.

[7] A.-M. Warzecha and M. Zoubir, "Towards validation of the preference for automation types scale in the context of chatbots writing essays," 2023.

[8] T. Franke, C. Attig, and D. Wessel, "A personal resource for technology interaction: Development and validation of the affinity for technology interaction (ati) scale," *International Journal of Human–Computer Interaction*, vol. 35, no. 6, pp. 456–467, 2019.

[9] B. E. Holthausen, P. Wintersberger, B. N. Walker, and A. Riener, "Situational trust scale for automated driving (sts-ad): Development and initial validation," in *12th International Conference on Automotive User Interfaces and Interactive Vehicular Applications*, 2020, pp. 40–47.

[10] L. Moradbakhti, B. Leichtmann, and M. Mara, "Development and validation of a basic psychological needs scale for technology use," *Psychological test adaptation and development*, 2023.

[11] B. Rammstedt and O. P. John, "Measuring personality in one minute or less: A 10-item short version of the big five inventory in english and german," *Journal of research in Personality*, vol. 41, no. 1, pp. 203–212, 2007.

[12] JamoviProject, *Jamovi (version 2.3)[computer software]*, 2022.

Assessing the Impact of AI Capability Feedback and Reliability in LLM-Assisted Error Detection

Luisa Winzer [1], and Tim Schrills [2]

[1] Student of Psychology - Cognitive Systems, Universität zu Lübeck, luisa.winzer@student.uni-luebeck.de
[2] Institute of Multimedia and interactive Systems, Universität zu Lübeck, tim.schrills@uni-luebeck.de

Abstract

Artificial Intelligence (AI) technologies, especially Large Language Models (LLM), are increasingly used in various tasks. This study investigates the impact of ability feedback and AI reliability in LLMs on user experience and performance in tasks focusing on spelling error detection. We conducted a 2x2 mixed design experiment with $N = 43$ participants interacting with either a LLM of high (90%) or low (60%) reliability, providing ability feedback in one task and not providing feedback in another task. Ability feedback impacted how users perceived the system in high vs. low AI reliability conditions, being beneficial when the AI is reliable, but detrimental when the AI is unreliable. This finding is critical, as the varying effects on user experience depending on system reliability highlight the potential risks associated with eXplainable AI (XAI) in LLM contexts.

1 Introduction

Systems based on AI profoundly impact our world by analyzing vast amounts of data, offering real-time decision support, and conducting predictive analyses. The widespread adoption of LLMs, like ChatGPT, has markedly enhanced the accessibility of AI systems to a broad audience [1]. AI's proficiency in handling complex cognitive tasks has significantly improved efficiency in task management, establishing it as a valuable collaborator in everyday tasks [2]. While Algorithms can perform remarkably, complete reliance on Machine Learning (ML) models is often not desirable, since they can act probabilistic, which does not guarantee the correctness of a specific decision. Additionally, ML models can be influenced by errors in the input data and hidden biases, or be brittle when encountering new situations [3]. ML models can assist users in decision-making to produce a joint decision outcome, that is desirably better than what could be produced by either the model or the user alone. Efficient collaboration requires users of AI systems to understand the AI's error boundaries and discern when to trust its conclusions [4]. XAI refers to AI systems that provide insights into their functionality and decision-making processes in a way that can be understood by humans. The goal of XAI is to create a transparent and trustable relationship between AI systems and their users by explaining how the AI models make their decisions, especially in complex systems where the decision-making process is not inherently obvious [5]. This can be used to understand when to trust or distrust AI, enabling users to apply their knowledge appropriately and improve decision outcomes, especially in scenarios where the model is likely to perform poorly [4].

Like other LLMs, ChatGPT is designed to perform a wide range of tasks with a high degree of competence, but it is important to remember that it is not infallible. In the example of spelling error detection, ChatGPT is generally effective but may struggle with uncommon terms, proper nouns, context-dependent words, and homophones. By integrating ability feedback in the task through XAI, users can enhance their ability to evaluate ChatGPT's output by enabling mutual predictability, detectability, and maintaining a common ground [6]. This encourages vigilance in areas prone to errors, such as uncommon terms or homophones, and prompting users to cross-check these instances more thoroughly [7]. This collaboration may then lead to more accurate outcomes by leveraging ChatGPT's strengths in identifying common errors while compensating for its limitations with the user's contextual knowledge. Furthermore, transparency about the model's limitations may not only improve task outcomes but also establish trust by recognizing the model's shortcomings thus nurturing a durable trust between users and technology.

The concept of AI systems providing self-disclosure about their capabilities as a basis for cooperation remains largely unexplored. The objective of the present research was to examine the influence of ability feedback in high and low-reliability AI systems on performance and user experience. To this end, we programmed a LLM, similar to ChatGPT, which should assist in a joint error detection task. Based on the issues presented, several hypotheses were derived for this study. Firstly, The presence or absence of ability feedback serves as the independent variable, while traceability of the system (H1a), joint performance (H1b), subjective measures of trustworthiness (H2), human-likeness

(H3), and cooperativeness (H4) are the dependent variables. We hypothesize an increase in all four dependent variables in the feedback condition compared to the no feedback condition.

Furthermore, we assume that AI-reliability as a second independent variable, operationalized as 60% or 90% accuracy in finding spelling errors, increases the same dependent variables as well (H5 - H8).

Lastly, we hypothesize a significant interaction between ability feedback and AI reliability in all dependent variables, suggesting that ability feedback may compensate for reduced AI reliability in the task (H9).

2 Material and Methods

Participants 65 participants completed the experiment. We recruited participants via mailing lists and Prolific. We excluded participants who failed to answer the dependent variable. The final sample consisted of $N = 43$ participants, ranging from 19 to 72 years old ($M = 34.7$, $SD = 14.4$). 23 participants identified themselves as female, 18 as male, one person as neither, and one person preferred not to say.

Experimental Environment To create an experimental environment, we used Gradio, a Python library that allows for the easy creation and sharing of ML models through user-friendly interfaces. To develop the chatbots, we integrated an NLP model using GPT, designed an interface for users to interact with, and manipulated which responses the user received from the chatbot in terms of the research question (see Fig. 1). The Gradio interface was then embedded into the LimeSurvey environment in which the study was hosted.

Measures We asked participants to rate the system's traceability after the interaction since feedback offers additional information about the AI's hidden processes (Subjective Information Processing Scale; [8]). Further, we captured a trust measure (Trust in Automation; [9]), as feedback offers transparency and could calibrate appropriate trust in a given task. Since chatbots are designed to mimic humanlike conversation and the addition of explainability in the form of ability feedback could enhance this perception, we also conducted a human-likeness scale (Godspeed Scale I; [10]). Finally, as the participants needed to cooperate with an AI, we also assessed an unpublished scale of the perceived cooperativeness of the agent.

Procedure The study was conducted in English. We used a 2x2 mixed design experiment, with ability feedback (feedback vs. no feedback) as a within-subject variable and AI reliability (60% vs. 90%) as a between-subjects variable. Initially, all participants were randomly assigned to either the low AI reliability (60%) or the high AI reliability (90%) condition. Participants were then tasked with detecting spelling errors in a text of approximately 350 words while seeking assistance from the chatbot. The spelling errors included five homophones — words that sound identical to another word but differ in meaning and often in spelling — and five other typical types of spelling mistakes. Participants were then instructed to list the identified errors in

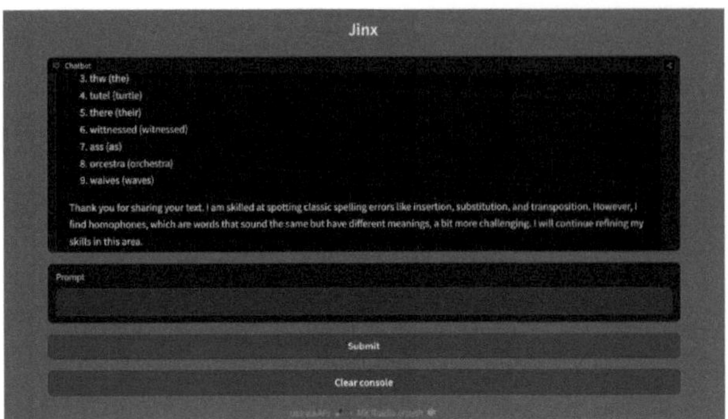

Figure 1: Chatbot "Jinx": Feedback - High AI Reliability Condition.

a textbox located below the chatbot interface. They had 4 minutes to complete this task. Depending on the experimental condition, the chatbot failed to identify 4 or 1 out of 10 spelling errors, which were all homophones. Regardless of the condition, participants interacted with two versions of the chatbot: one that provided ability feedback (Version A: "Jinx") and one that did not (Version B: "Loom"). The ability feedback was phrased as follows: "Thank you for sharing your text. I am skilled at spotting classic spelling errors like insertion, substitution, and transposition errors. However, I find homophones, which are words that sound the same but have different meanings, somewhat more challenging. I am continually refining my skills in this area." After completing the task with one chatbot version, the perceived traceability, trustworthiness, human-likeness, and cooperativeness of the agent were assessed.

3 Results and Discussion

3.1 Results

Comparison of Ability Feedback Conditions Hypotheses H1 – H4 related to the effect of ability feedback on the number of identified spelling errors, as well as subjective measures of system traceability, trustworthiness, human-likeness, and cooperativeness. Given that feedback was evaluated within the same participants, paired t-tests (or Wilcoxon tests for non-parametric data) were employed, anticipating a more pronounced effect of feedback versus no feedback across all dependent variables. Table 1 presents the results of the within-group analysis employing Wilcoxon tests (one-sided).

Table 1: Results of Wilcoxon Tests (one-sided)

	Median (SD)				
	F	No F	V	p	d
Performance	7.4 (2.2)	7.7 (2.5)	181	.791	.08
Traceability	4.1 (1.2)	4.0 (1.3)	264	.619	.06
Trust	3.5 (0.7)	3.6 (0.6)	298.5	.957	.21
Humanness	2.2 (0.8)	2.2 (0.8)	176	.628	.05
Cooperation	3.8 (1.0)	3.8 (1.0)	234.5	.940	.15

Note. F = Feedback, NF = No Feedback, d = Cohen's d.

Participants did not identify a significantly higher number of spelling errors when they received ability feedback from the chatbot, compared to when they did not receive such feedback. They also did not rate the traceability of the system higher, trust the chatbot more, find the agent to be more humanlike, nor find the cooperativeness of the system better when receiving feedback compared to receiving no feedback. In summary, these observations collectively suggest that feedback had minimal influence, if any, on users' performance or perception of subjective measures.

Comparison of AI Reliability Conditions We next report the effects of the independent variable AI reliability on our dependent variables (H5 - H8). Again, we expected an increase in all dependent variables in the high AI reliability condition compared to a low AI reliability condition. We ran independent t-tests for each of our five measures since AI reliability is our grouping factor which we conducted between participants. In case the assumptions for calculating independent t-tests were not met, we instead ran Mann-Whitney U Tests. Firstly, we found that high AI reliability ($M = 8.2$; $SD = 2.5$) leads to a significant increase in performance compared to low AI reliability ($M = 7.0$; $SD = 2.0$), $W = 1284$, $p = .001$; $d = .53$. Likewise, participants in the high AI reliability conditions ($M = 3.7$; $SD = 0.6$) significantly trusted the system more compared to participants in the low AI reliability conditions ($M = 3.4$; $SD = 0.6$), $W = 1186$, $p = .021$; $d = .50$. No such effects could be found for system tracability, $W = 1012.5$, $p = .422$, $d = .11$, humanlikeness, $t(84) = 0.99$, $p = .161$, $d = .21$, or cooperativeness, $W = 1062$, $p = .218$, $d = .31$.

Interactions of Ability Feedback and AI Reliability Lastly, we ran two-way mixed ANOVAs for our five measures (H9). We found no interaction effect for ability feedback and AI reliability for joint performance. Yet, we found interaction effects for ability feedback and AI reliability for all four user experience measures. The interaction plots are depicted in Fig. 2. There was no significant difference in how humanlike, trustworthy, or cooperative participants subjectively perceived the chatbot when no ability feedback was presented between the AI reliability conditions. For information processing awareness, participants in the low AI reliability condition benefited significantly more from not receiving feedback than those in the high AI reliability condition. In contrast, when feedback was presented, the impact of AI reliability became present for all subjective measures of user experience. Participants had increased subjective information processing awareness, trusted the system more, and perceived the interaction with the system as more cooperative and humanlike in the high AI reliability condition compared to the low AI reliability condition.

3.2   Discussion

In our study, we examined the impact of capability feedback and AI reliability on user experience and performance in tasks assisted by LLMs. One key observation is the significant improvement in task performance and trust when the AI is more reliable, regardless of feedback. It high-

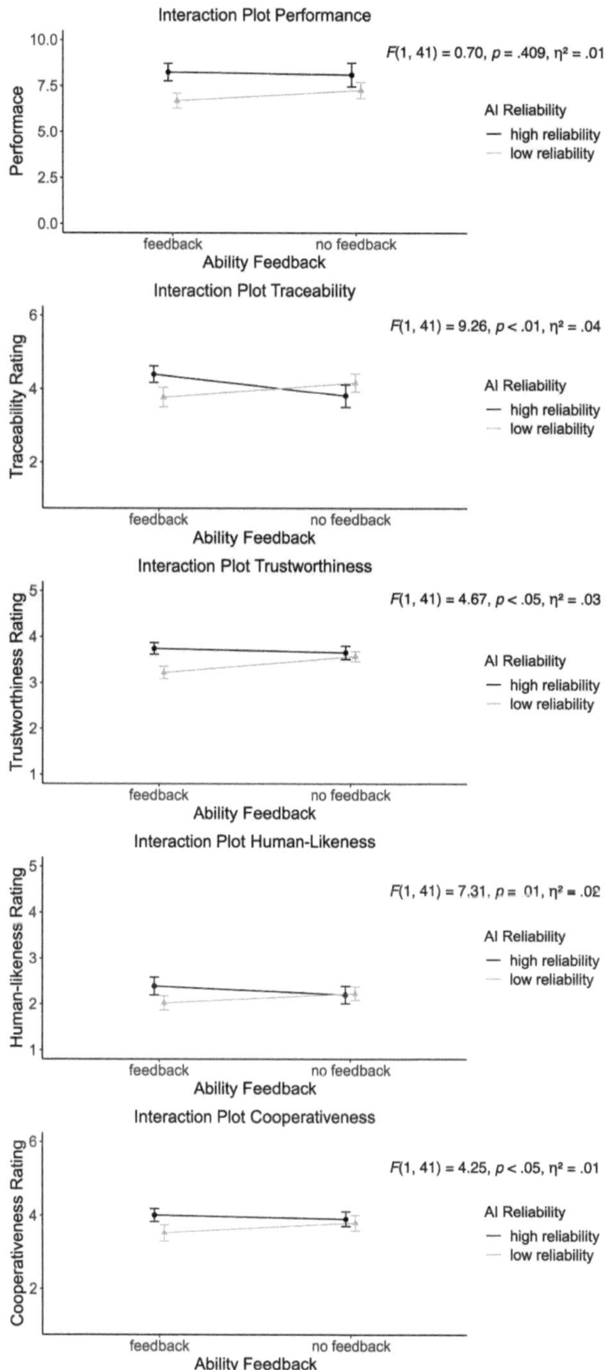

Figure 2: Interaction Plots for Ability Feedback and AI Reliability on Joint Performance and User Experience Measures

lights the importance of AI's functional accuracy in shaping user trust and reliance. Another highlight is the significant interactions which allow a more nuanced insight into understanding the complex relationship between XAI in scenarios of varying AI reliability and the impact on user experience in the context of LLMs. They suggest that in the absence of feedback, there are no significant differences between the AI reliability conditions for human-likeness, trustworthiness, and cooperativeness. Only, when feedback is provided, we can see that these measures of user experience benefit in scenarios where the AI is reliable, while it diminishes user experience in scenarios where the AI is unreliable. It shows that feedback enables a subjective perceivable difference in varying AI reliability contexts. This might indicate a potential mismatch between the explanations provided and the users' expectations or comprehension levels and inadvertently highlight the AI's limitations. In the case of the system's traceability, there also was a significant difference in conditions where no feedback was provided. Subjective information processing increases when AI reliability is high and decreases when it is low. These findings are pivotal for AI research, indicating that while transparency is crucial, it must be carefully balanced with the AI's actual performance capabilities to avoid diminishing user trust and cooperation. It underscores a critical aspect of AI-user interaction: the potential risks inherent in XAI when dealing with LLMs, which often have accuracy rates lower than 90%. This aligns with Eiband & Chromik's observations about the potential backfiring of explanations in less reliable systems [11]. This suggests a need for adaptive explainability strategies that consider the AI's reliability and the user's context and experience level.

4 Conclusion

In summary, this research not only contributes to our understanding of how users interact with LLMs but also raises crucial questions about the design and implementation of explainable AI features. It emphasizes the need for a careful and context-sensitive approach to developing explainable AI systems, particularly in applications where AI reliability may vary. The findings underscore the importance of ongoing research into user-centric and adaptive explainability in AI, which respects the balance between transparency, user trust, and effective collaboration.

Acknowledgement

The work has been carried out at the Institute for Multimedia and interactive Systems (IMIS) and additionally supervised by Prof. Thomas Franke, IMIS, Universität zu Lübeck.

Authors' Statement

The authors state no conflict of interest. Informed consent has been obtained from all participants. The research has been approved by the authors' institutional review board or equivalent committee.

5 References

[1] Y. Xu et al., "Artificial intelligence: A powerful paradigm for scientific research," *The Innovation*, vol. 2, no. 4, p. 100179, Nov. 2021. DOI:10.1016/j.xinn.2021.100179.

[2] P. P. Ray, "ChatGPT: A comprehensive review on background, applications, key challenges, bias, ethics, limitations and future scope," *Internet of Things and Cyber-Physical Systems*, vol. 3, pp. 121–154, 2023. DOI:10.1016/j.iotcps.2023.04.003.

[3] M. R. Endsley, "Supporting Human-AI Teams: Transparency, explainability, and situation awareness," *Computers in Human Behavior*, vol. 140, p. 107574, Mar. 2023. DOI:10.1016/j.chb.2022.107574.

[4] Y. Zhang, Q. V. Liao, and R. K. Bellamy, "Effect of confidence and explanation on accuracy and trust calibration in AI-assisted decision making," in *Proc. 2020 Conference on Fairness, Accountability, and Transparency*, Jan. 2020, pp. 295–305. DOI:10.1145/3351095.3372852.

[5] T. Miller, "Explainable AI is Dead, Long Live Explainable AI! Hypothesis-driven Decision Support using Evaluative AI," in *Proc. 2023 ACM Conference on Fairness, Accountability, and Transparency*, Jun. 2023, pp. 333–342. DOI:10.1145/3593013.3594001.

[6] G. Klein, D. D. Woods, J. M. Bradshaw, R. R. Hoffman, and P. J. Feltovich, "Ten challenges for making automation a 'team player' in joint human-agent activity," *IEEE Intelligent Systems*, vol. 19, no. 6, pp. 91–95, Feb. 2004. DOI:10.1109/MIS.2004.74.

[7] E. K. Chiou and J. D. Lee, "Trusting automation: Designing for responsivity and resilience," *Human Factors*, vol. 65, no. 1, pp. 137–165, Feb. 2023. DOI:10.1177/0018720821100999.

[8] T. Schrills and T. Franke, "How Do Users Experience Traceability of AI Systems? Examining Subjective Information Processing Awareness in Automated Insulin Delivery (AID) Systems," A*CM Trans. Interact. Intell. Syst.*, Mar. 2023. DOI:10.1145/3588594.

[9] M. Körber, "Theoretical considerations and development of a questionnaire to measure trust in automation," in *Proc. 20th Congress of the International Ergonomics Association (IEA 2018): Volume VI: Transport Ergonomics and Human Factors (TEHF), Aerospace Human Factors and Ergonomics*, S. Bagnara, R. Tartaglia, S. Albolino, T. Alexander, and Y. Fujita, Eds., 1st ed., Springer, 2019, pp. 13–30.

[10] C. Bartneck, E. Croft, D. Kulic, and S. Zoghbi, "Measurement instruments for the anthropomorphism, animacy, likeability, perceived intelligence, and perceived safety of robots," *International Journal of Social Robotics*, vol. 1, no. 1, pp. 71–81, Jan. 2009. DOI:10.1007/s12369-008-0001-3.

[11] M. Chromik, M. Eiband, F. Buchner, A. Krüger, and A. Butz, "I think I get your point, AI! The illusion of explanatory depth in explainable AI," in *Proc. 26th International Conference on Intelligent User Interfaces*, Apr. 2021, pp. 307–317. DOI:10.1145/3397481.3450644.

Investigating right supramarginal gyrus activation and behavioral scores in the FFEEB-paradigm

Emily Leah Fitzgibbon [1], Ulrike Krämer [2], Martin Göttlich [2], and Gerwin Akpadji [1]

[1] Psychology - Cognitive Systems, Universität zu Lübeck, emily.fitzgibbon, gerwin.akpadji@student.uni-luebeck.de
[2] Cognitive Neurosciences, Universität zu Lübeck, ulrike.kraemer, martin.goettlich@.uni-luebeck.de

Abstract

In our study, we researched emotional egocentricity bias and emotional alterocentricity bias. We employed the Food Emotional Egocentricity Bias (FEEB) paradigm, utilizing food stimuli during functional magnetic resonance imaging (fMRI) to elicit these biases and investigate their neural and behavioral basis. The correlation analysis between the activation of the right supramarginal gyrus under conditions inducing EEB and EAB and their respective behavioral scores did not yield significance. The ROI activations for EEB and EAB conditions were not significantly different. The ROI activation in the EEB condition and the behavioral biases were correlated with the subscales of the IRI, finding that only one subscale correlated with the EEB behavioral score. This underscores the relation between behavioral biases, self-perceived empathic abilities, and activation in the brain being more complex than expected. In the future, functional connectivity analyses should be employed to investigate the underlying neural networks of these biases.

1 Introduction

Self-other distinction is the ability to discern between the experience of oneself and others. This concept plays a crucial role in social cognition, as individuals constantly navigate social situations, seeking to interpret the emotions and thoughts of others [1]. These estimations can be biased in two important ways: emotional egocentricity bias (EEB) and emotional alterocentricity bias (EAB). The emotional egocentricity bias describes the concept of one's state influencing how one perceives the state of the other [2]. In contrast, the alterocentricity bias describes the concept of the state of the other influencing the way one perceives one's own emotions [3]. Goregliad Flaellingsdal et al [4] developed a food-stimulus-based paradigm to investigate activations in the brain linked to EEB and EAB using fMRI (FEEB – food emotional egocentricity bias). The experimental manipulation was successful as EEB and EAB were induced on a behavioral level and significant activations for the biases were found in the brain. A region of interest (ROI) analysis of the right supramarginal gyrus (rSMG) informed by previous literature pointing towards it being activated during EEB conditions with other paradigms [2] was conducted. Significant activations in EEB and incongruent conditions were found. The Interpersonal Reactivity Index (IRI) questionnaire [5] was also given to the participants to measure self-perceived empathy. This paper aims to deepen the understanding of the relation between behavioral and brain-level activation in emotional egocentricity and alterocentricity bias. To do this, we investigated the differences in ROI activation in EEB and EAB conditions and checked for behavioral and ROI activation correlations. We also ex-plored if the self-perceived empathic sensitivity (measured through the subscales of the IRI) correlated with the amount of bias the participants showed.

2 Material and Methods

2.1 Participants

40 participants were recruited using the university email distribution list and flyers were hung up around the university. There was monetary compensation or compensation in the form of student credit. Sixteen participants had to be excluded due to head movements, behavioral criteria, and other exclusion criteria leaving data of 24 participants to be used of which 18 were female and 6 were male. The participants were all neurotypical, right-handed, and aged 18 to 36 with a mean age of 23 (SD=2.9).

2.2 FEEB Paradigm

The FEEB paradigm is a 2x2x2 design. The factors are as follows: One factor is the valence of the food stimulus, with the conditions tasty or disgusting. The next is the target, with the conditions self and other, indicating whose emotions the participant evaluates in a given trial. The last is congruence with the conditions congruent and incongruent, which indicates whether the valence of the food matches the emotion depicted in the subsequent video (see Figure 1). Participants were instructed to envision themselves sitting in a restaurant, with the presented food stimulus representing the dish placed before them. The person sitting across from them is represented by the video of the actor and they were told to imagine how they or their opposite would feel receiving this food. In any given trial, the participant would first see the food stimulus for two seconds, then the video of the actor presenting a happy or disgusted expression, followed by a fixation cross. Afterward, they would see a color, which indicated to them whose emotions they would rate, followed by another fixation cross, and finally, the scale ranging from minus five to plus five to which the participant would then give their rating with a time-out of five seconds (see Figure 1). The experiment started with a prerating, in which the participants rated the 20 stimuli in a randomized order on the same minus five to plus five scales without any other given context. There were two runs of the experiment, with 80 runs each.

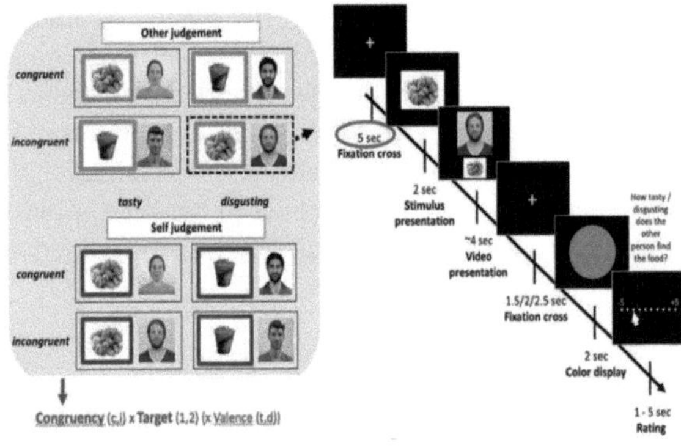

Figure 1: Conditions and exemplary trial of FEEB experiment.

2.3 Material

The tasty and disgusting stimuli were sourced from various food databases and validated in a pilot study where participants classified them [4]. The results indicated a mild dichotomous classification. Videos depicting facial expressions (happy and disgusted) were taken from the Jerusalem Facial Emotion Expression Set [6], featuring four actors (three male, one female).

2.4 Imaging

Structural and functional magnetic resonance imaging to investigate EEB and EAB was performed using a 3-T Siemens Magnetom Skyra scanner. The voxel size was set to 3x3x3 mm3, with a repetition time (TR) of 1000 ms, and there were 800 total scans measured per run.

2.5 Data Analysis

Data preprocessing was performed using SPM12 [7] running on MATLAB [8]. The region of interest was extracted using MarsBaR [9] (see Figure 2) and contrast maps for EEB and EAB conditions were set up in SPM. The coding

of the contrasts was as follows: For the Emotional Egocentricity Bias (EEB) contrast, negative weights were assigned to both congruent conditions, where participants decided for others, and positive weights were assigned to both incongruent conditions, where participants decided for others. Similarly, for the Emotional Alterocentricity Bias (EAB) contrast, the same coding was applied, but this time, the conditions where participants decided for themselves were used. The results were one value of ROI activation in procent difference of BOLD activity compared to the resting state per participant and condition. One behavioral score per participant of the size of the EEB and EAB was calculated in RStudio [10]. All other tests were conducted via MATLAB [8]. A Mann-Whitny U test was performed to determine the difference in activation in EEB conditions compared to EAB. Multiple Spearman correlations were performed between EEB and EAB ROI activation and behavioral scores. They were also used to correlate IRI scores with the behavioral bias scores as well as with EEB and EAB ROI activation.

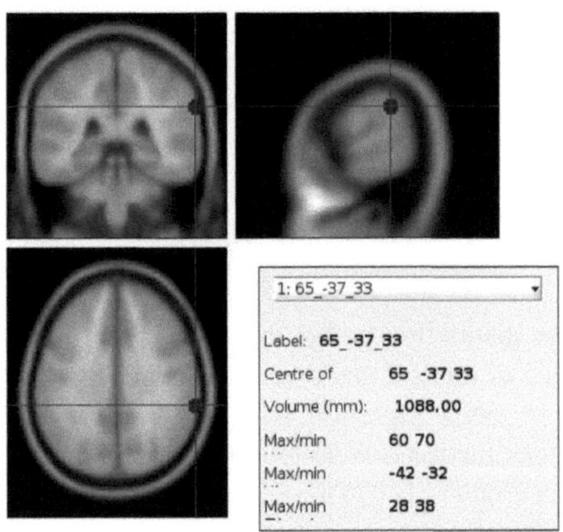

Figure 2: Region of interest rSMG extracted with MarsBaR.

3   Results and Discussion

3.1 Kolmogorov-Smirnov test

All but one Kolmogorov-Smirnov test (EAB rSMG activation) showed that the data was not normally distributed.

3.2 Difference between EEB and EAB ROI activation

The Mann-Whitney U test between rSMG activation during EEB conditions and EAB conditions was not significant (p=0.31, z= 1.02), see Figure 3.

3.3 Correlations between ROI activation and behavioral biases

The Spearman correlations between rSMG activation and EEB (r(24) = .05, p = .82) as seen in Figure 4, as well as

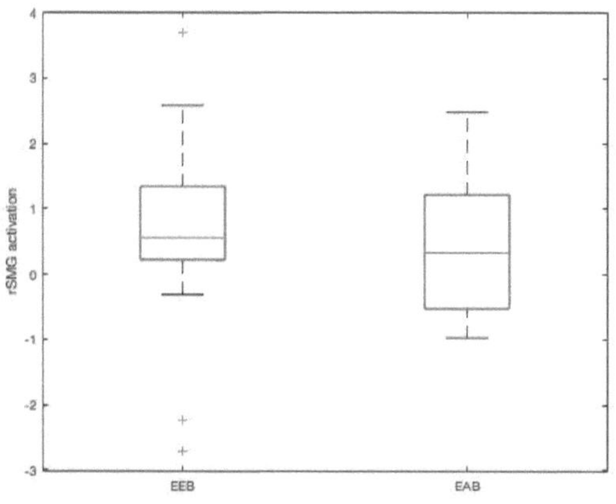

Figure 3: EEB and EAB conditions rSMG activation with no significant difference, red crosses are outliers.

EAB (r(24) = .17, p = .41) as seen in Figure 5, were not significant.

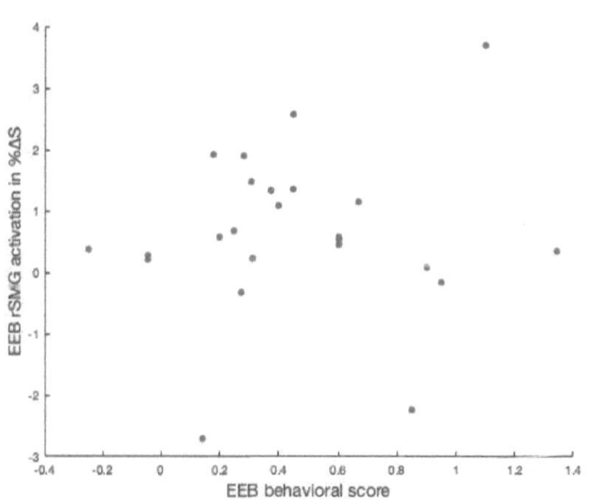

Figure 4: No significant correlation of ROI activation and EEB.

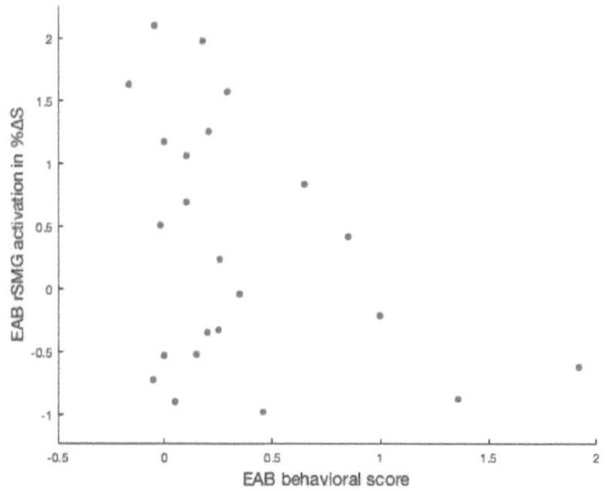

Figure 5: No significant correlation of ROI activation and EAB.

3.4 Correlations between IRI scores and ROI activation

The Spearman correlations between the rSMG activation in the EEB condition and the IRI subscales empathic concern r(24) = .13, p = .56), personal distress (r(24) = .05, p = .82), perspective taking (r(24) = .04, p = .86) and fantasy (r(24) =- .15, p = .50) were not significant.

3.5 Correlations between IRI scores and behavioral EEB

The Spearman correlations between the behavioral emotional egocentricity bias and the IRI subscales empathic concern (r(24) = .13, p = .56), perspective taking (r(24) =- .04, p = .84) and fantasy (r(24) = .30, p = .16) were not significant. The correlation between behavioral EEB and personal distress was significant (p = .03) with a moderate correlation (r(24) = .43), see Figure 6.

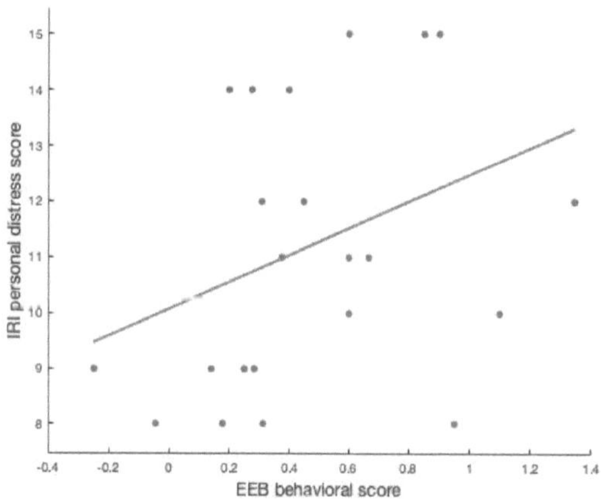

Figure 6: Significant correlation between EEB behavioral score and IRI personal distress with regression line.

3.6 Correlations between IRI scores and behavioral EAB

All correlations between behavioral EAB and the IRI subscales (empathic concern: r(24) = .22, p = .31; personal distress: r(24) = .38, p = .07; perspective taking: r(24) = .03, p = .90 as well as fantasy: r(24) = .28, p = .19) were not significant.

3.7 Discussion

The non-significant results of the Mann-Whitney-U test between the rSMG activations of EEB and EAB might point to the rSMG not only being involved in EEB processing but might instead more broadly process incongruence. The fact that rSMG activations and the size of the biases do not correlate points in the direction of the rSMG not being the only area of the brain involved in the processing of these biases. The IRI scales neither interact with the ROI activation nor the EEB and EAB scores (except for one) which implies that self-perceived empathy might not inform measured biases. The correlation between personal distress and EEB scores is rather unexpected, as personal distress has been linked to EAB, not EEB in other studies [11]. Yet the correlation between EAB and personal distress, while not quite significant (r(24) = .38, p = .07), hints at a potential statistical connection between incongruence and personal distress rather than a unique association with EEB.

4 Conclusion

The results of this paper show that the rSMG most likely is not the only area involved in the processing of the EEB and EAB. Therefore, more whole-brain analyses and in particular functional connectivity analyses such as gPPI [12] or beta-series correlation should be implemented. Further research should also look into the correlation between incongruency and personal distress scores to further the understanding of self-perceived empathy and self-other distinction effects.

Acknowledgment

The work has been carried out at the University of Lübeck and supervised by the Institute of Institute of Medical Psychology (Cognitive Neuroscience Lab), University of Lübeck.

Authors' Statement

Conflict of interest: The authors state no conflict of interest. Informed consent: Informed consent has been obtained from all individuals included in this study. Ethical approval: The research related to human use complies with all the relevant national regulations, institutional policies and was performed in accordance with the tenets of the Helsinki Declaration, and has been approved by the author's institutional review board or equivalent committee.

5 References

[1] A. Schütz and K. Rentzsch, *Dorsch – Lexikon der Psychologie*. 2020. doi: 10.1024/85914-000.

[2] G. Silani, C. Lamm, C. C. Ruff, and T. Singer, "Right supramarginal gyrus is crucial to overcome emotional egocentricity bias in social judgments," *The Journal of Neuroscience*, vol. 33, no. 39, pp. 15466–15476, Sep. 2013, doi: 10.1523/jneurosci.1488-13.2013.

[3] C. Lamm, H. Bukowski, and G. Silani, "From shared to distinct self–other representations in empathy: evidence from neurotypical function and socio-cognitive disorders," *Philosophical Transactions of the Royal Society B*, vol. 371, no. 1686, p. 20150083, Jan. 2016, doi: 10.1098/rstb.2015.0083.

[4] T. G. Fjaellingsdal, N. Makowka, and U. M. Krämer, "Studying trait-characteristics and neural correlates of the emotional ego- and altercentric bias using an audiovisual paradigm," *Cognition & Emotion*, vol. 37, no. 4, pp. 818–834, May 2023, doi: 10.1080/02699931.2023.2211253.

[5] Davis, M. H. (1980). "A multidimensional approach to individual differences in empathy" *JSAS Catalog of Selected Documents in Psychology*, 10, 85–104.

[6] N. Yitzhak *et al.*, "Gently does it: Humans outperform a software classifier in recognizing subtle, nonstereotypical facial expressions.," *Emotion*, vol. 17, no. 8, pp. 1187–1198, Dec. 2017, doi: 10.1037/emo0000287.

[7] K. J. Friston, *Statistical Parametric mapping*. 2007. doi: 10.1016/b978-0-12-372560-8.x5000-1.

[8] The MathWorks, Inc. (2022). MATLAB version: 9.13.0 (R2022b). Accessed: January 01, 2023. Available: https://www.mathworks.com.

[9] M. Brett, "Region of interest analysis using an SPM toolbox," *NeuroImage*, vol. 16, p. 497, Jan. 2002, [Online]. Available: https://ci.nii.ac.jp/naid/10029427927

[10] RStudio Team (2020). RStudio: Integrated Development for R. RStudio, PBC, Boston, MA URL http://www.rstudio.com/.

[11] F. Hoffmann, C. Banzhaf, P. Kanske, M. Gärtner, F. Bermpohl, and T. Singer, "Empathy in depression: Egocentric and altercentric biases and the role of alexithymia," *Journal of Affective Disorders*, vol. 199, pp. 23–29, Jul. 2016, doi: 10.1016/j.jad.2016.03.007.

[12] D. G. McLaren, M. L. Ries, G. Xu, and S. C. Johnson, "A generalized form of context-dependent psychophysiological interactions (gPPI): A comparison to standard approaches," *NeuroImage*, vol. 61, no. 4, pp. 1277–1286, Jul. 2012, doi: 10.1016/j.neuroimage.2012.03.068.

12

Medical imaging

Design, assembly, and validation of a magnetic particle imaging low-noise amplifier

Frauke H. Niebel [1], Jorge Chacon-Caldera [2,4], Alex C. Barksdale [2,3], Monika Śliwiak [2], Lawrence L. Wald [2,4], and Matthias Graeser [5,6]

[1] Medical Engineering Science, Universität zu Lübeck, Lübeck, Germany, frauke.niebel@student.uni-luebeck.de

[2] MGH/HST Athinoula A. Martinos Center for Biomedical Imaging, Massachusetts General Hospital, Charlestown, MA, USA, {jchaconcaldera, msliwiak, lwald}@mgh.harvard.edu

[3] Massachusetts Institute of Technology, Cambridge, MA, USA, alexbark@mit.edu

[4] Harvard Medical School, Boston, MA, USA

[5] Fraunhofer Research Institution for Individualized and Cell-Based Medical Engineering IMTE, Lübeck, Germany, matthias.graeser@imte.fraunhofer.de

[6] Institute of Medical Engineering, Universität zu Lübeck, Lübeck, Germany

Abstract

Magnetic particle imaging (MPI) is an emerging imaging technique that maps the local concentration of superparamagnetic iron-oxide nanoparticles (SPION) by exploiting their non-linear magnetization response to externally applied magnetic fields. A major challenge in MPI signal recording is that the detected particle signals have a very low amplitude, requiring a low-noise amplifier (LNA) to fit them to the analog-to-digital converter's (ADC) finite input range with minimum degradation of the noise floor. This work presents an LNA based on parallelization to reduce the noise contribution of the amplification process itself. The prototyped LNA features an input rated voltage noise of $373\,\mathrm{pV}/\sqrt{\mathrm{Hz}}$, which is 3.2 times less compared to the LNA employed in the present scanner setup. With only a factor of 2.1 away from coil noise dominance, the presented LNA holds the potential to substantially enhance the scanner's sensitivity.

1 Introduction

The following sections outline the importance of the low-noise amplifier (LNA) in magnetic particle imaging (MPI) and introduce the low-noise design technique used for the LNA proposed in this work.

1.1 MPI Low-noise Amplifier

Magnetic particle imaging (MPI) measures the non-linear magnetization response of nanoparticles based on their physical interactions with oscillating magnetic fields [1]. In order to detect minute particle concentrations, signal detection needs to be highly sensitive. However, achieving high sensitivity is challenging since the time-varying particle magnetization induces a very low signal of only a few microvolts amplitude. Without amplification, the signal becomes submerged in the quantization noise of the analog-to-digital converter (ADC). Therefore, the MPI receive chain requires a low-noise amplifier (LNA) to raise the signal amplitude above the ADC's noise floor [2]. The purpose of low-noise amplification is to amplify the signal amplitude with minimum noise floor degradation, preventing the LNA from becoming the dominant source of noise. Whereas the ultimate goal is patient noise dominance, the intermediate goal is to be dominated by the thermal noise of the receive

coil. The LNA introduced in this work involves taking a further step towards coil noise dominance by reducing the noise arising from the amplification process itself.

1.2 Operational Amplifier Noise Theory

An amplifier's noise performance is usually characterized as noise density (root-mean-square (RMS) noise amplitude in a $1\,\mathrm{Hz}$ band of frequency), specifically, voltage noise density e_n in units of $\mathrm{V}/\sqrt{\mathrm{Hz}}$ and current noise density i_n in units of $\mathrm{A}/\sqrt{\mathrm{Hz}}$. Since the output noise of an amplifier depends on the amplifier's gain, it is customary to refer all noise to the input [3]. Another standard metric for characterizing noise performance is the noise figure (NF). The noise figure of an amplifier is defined as the logarithmic ratio between the noise output of a real amplifier and the noise output of an ideal amplifier with the same gain and a resistor R_s connected across the amplifiers' input terminals:

$$\mathrm{NF} = 10\log_{10}\left(1 + \frac{e_n^2 + i_n^2 R_s^2}{4k_B T R_s}\right) \quad (1)$$

The denominator denotes the resistor's mean squared noise density, where k_B is the Boltzmann constant and T is the absolute temperature of the load (usually $290\,\mathrm{K}$) [3], [4]. The goal of low-noise design is to reduce the noise arising from the amplification process itself so that the signal is

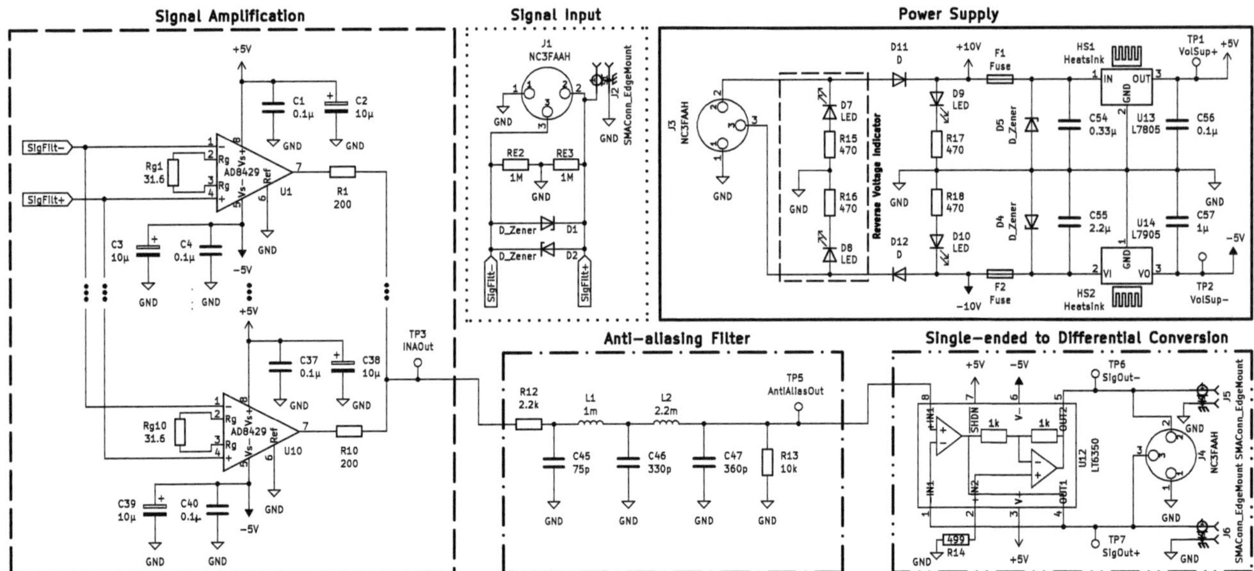

Figure 1: Simplified circuit schematic of the prototyped LNA. The signal amplification stage (dashed outline) is followed by an anti-aliasing filter (dash-dot outline) and a single-ended to differential converter (dash-dot-dot outline). The power supply (solid outline) and signal input (dotted outline) are shown for completion.

amplified with minimum noise floor degradation. One effective low-noise design technique involves the parallelization of multiple amplification devices [3]-[5]. While in a configuration of N parallelized amplifiers, the input signal (assuming correlation) sums linearly ($V_{total} = NV_{in}$), the current noise (assuming uncorrelation) adds with the square root of the number of devices in parallel ($i_{n,total} = \sqrt{N}i_n$). The voltage noise decreases by $1/\sqrt{N}$. However, this holds only true for zero source resistance since a noise current at an amplifier's input flowing through the signal's source resistance R_s generates its own voltage noise $e_n = R_s i_n$ [3]. The LNA proposed in this work employs the idea of parallelization to reduce the input rated voltage noise. The LNA's noise performance is characterized and compared to the previous LNA design employed in the present scanner setup.

2 Material and Methods

The following sections define the design goal of the prototyped LNA, present its circuitry and printed circuit board, and describe the tests performed for its validation.

2.1 LNA Design Goal

The present scanner setup implements an LNA based on an ultralow noise instrumentation amplifier (INA) (AD8429, Analog Devices, Wilmington, MA, USA) with a voltage noise density e_n of $1\,\mathrm{nV}/\sqrt{\mathrm{Hz}}$ and current noise density i_n of $1.5\,\mathrm{pV}/\sqrt{\mathrm{Hz}}$. The INA is operated with a gain resistor R_g of $31.6\,\Omega$ resulting in a gain G of 191. Taking the gain resistor's thermal noise into account, the total input rated voltage noise e_n is $1.25\,\mathrm{V}/\sqrt{\mathrm{Hz}}$. The proposed LNA design is based on this previous design but aims to reduce the input rated voltage noise by placing ten AD8429 INAs (equal gain of 191) in parallel. Since this work is intended

to demonstrate proof of concept, the number of parallelized amplifiers is independent of a specific source resistance. In total, the input rated voltage noise is supposed to decrease by $1/\sqrt{10}$ giving $e_n = 395\,\mathrm{pV}/\sqrt{\mathrm{Hz}}$ which is an improvement factor of about 3 compared to the previous design.

2.2 LNA Circuit Design

Fig. 1 shows a simplified circuit schematic of the proposed LNA subdivided into three stages (signal amplification, anti-aliasing filtering, and single-ended to differential conversion). The signal amplification stage (dashed outline) contains the parallelized ten INAs. The input signal from the receive filter is fed to each of the INAs' inputs, amplified by 191, and recombined in TP3. A summing amplifier was omitted as the INAs' output voltages are assumed equal. The amplification stage is followed by an anti-aliasing filter (dash-dot outline) that limits the bandwidth of the noise at the ADC input and thereby prevents out-of-band noise from wrapping around. The proposed LNA implements a fifth-order Butterworth low-pass filter with a cutoff frequency of $320\,\mathrm{kHz}$. The final stage of the LNA involves a differential amplifier (LT6350, Analog Devices, Wilmington, MA, USA) that converts the single-ended signal into a differential signal (dash-dot-dot outline). The output signals are of equal amplitude but $180°$ out-of-phase, providing a low impedance differential drive for the ADC. The circuit is supplied by $\pm 12\,\mathrm{V}$ (solid outline). Two linear voltage regulators (L7805/L7905, STMicroelectronics, Plan-les-Outes, Switzerland) are used to provide a stable split-voltage supply of $\pm 5\,\mathrm{V}$. Assuming a current of $100\,\mathrm{mA}$ and a voltage drop of $7\,\mathrm{V}$, the temperature is expected to rise by $35\,°\mathrm{C}$ above ambient temperature (given a thermal resistance of $50\,°\mathrm{C/W}$). Therefore, the voltage regulators are equipped with heat sinks (thermal resistance of $24.4\,°\mathrm{C/W}$) reducing the temperature rise by half. The power inputs are protected from reverse polarity and overvoltage by fuses and diodes. Furthermore, the LNA input (dotted outline) is protected by diodes limiting the input voltage to $\pm 0.5\,\text{V}$.

2.3 LNA Printed Circuit Board Layout

For the LNA, a custom printed circuit board (PCB) was designed in KiCad 7.0.8. The PCB has a four-layer board stack-up with top and bottom layers intended for signal traces and two internal layers serving as ground and power planes. The ten INAs are placed on a circle with tracks routed in a star-like pattern from the circle's center to the INAs' inputs to ensure equal signal phases at the INA input ports. Decoupling capacitors with capacitance values $0.1\,\mu\text{F}$ and $10\,\mu\text{F}$ are placed closely to each IC's power pins that shorten current return paths to mitigate signal coupling on the board. Moreover, they keep high-frequency noise away from the chip and ensure a stable supply voltage during any transient load spikes. The power plane is split up into two contiguous $+5\,\text{V}$ and $-5\,\text{V}$ areas with a minimum clearance of $0.5\,\text{mm}$ for electrical isolation. In addition to the internal ground plane, the external PCB layers feature dedicated ground copper pours that contribute to the mitigation of electromagnetic interference (EMI) and maintenance of signal integrity. To minimize the impedance between the separate ground planes, they are tied together by multiple vias (plated-through holes that interconnect layers) placed around the PCB perimeter. These so-called stitching vias are arranged in a picket fence pattern, acting as a Faraday cage that effectively shields the internal components from EMI when operating the board outside its aluminum enclosure. The board was manufactured by JLCPCB (JL-CPCB, Shenzen, Guangdong, Republic of China). Fig. 2 shows a photograph of the populated PCB.

2.4 LNA Validation Tests

The prototyped LNA and the previous LNA were characterized and compared through simulations and measurements. Firstly, the input voltage noise of both LNA designs was simulated in LTSpice. To verify a stable operation of the ten parallelized INAs on the PCB, a sine wave with a frequency of $100\,\text{kHz}$ and an amplitude of $10\,\text{mV}$ was applied to the LNA's input and detected at the inverting (TP6) and non-inverting (TP7) output of the ADC driver with an oscilloscope. In order to validate the simulation results, the output noise voltages of both LNAs with inputs shorted were determined using a RIGOL DSA815 spectrum analyzer ($100\,\text{Hz}$ resolution bandwidth). To relate the output rated noise levels to the input, they must be divided by the LNAs' gains. Therefore, the LNAs' transfer functions (TF) were determined by measuring their S21 parameters with a vector network analyzer (VNA). The S21 parameter describes the ratio between a transmitted wave passing into the LNA input to the received wave generated by the LNA at the output [3]. Lastly, noise figures were calculated according to (1) assuming a $2\,\Omega$ resistor (resistance of the receive coil) connected across the input terminals.

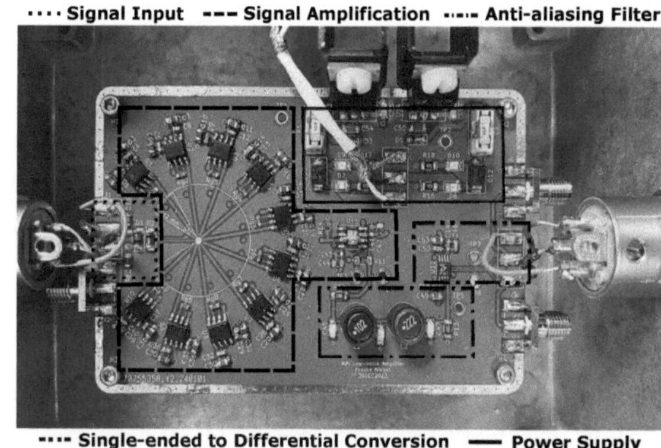

Figure 2: Photograph of the populated PCB within its aluminum shielding enclosure. The individual circuit parts are labeled in accordance with the schematic in Fig. 1.

3 Results and Discussion

Fig. 3 shows the results of the LTSpice noise analysis. At a frequency of $100\,\text{kHz}$, the simulated input voltage noise density is $e_n = 1.26\,\text{nV}/\sqrt{\text{Hz}}$ for the previous LNA design (left) and $e_n = 398\,\text{pV}/\sqrt{\text{Hz}}$ for the new LNA design (right), which matches well with the theoretical predictions. Fig. 4 depicts the waveforms detected at the inverting (TP6) and non-inverting (TP7) output of the ADC driver. The output signals are stable and $180°$ out-of-phase. With a peak amplitude of $1.6\,\text{V}$, the amplifier's gain is 160 and thereby below the aimed gain of 191. However, the overall noise characteristic of the LNA remains unchanged since the output noise is proportionally less. Moreover, the loss in gain could be remedied by increasing the differential gain of the ADC driver in the last stage. Fig. 5 shows the measurement based input rated noise levels for both LNAs obtained by the division of the measured output rated noise levels by each LNA's TF (representatively shown for the prototyped LNA in Fig. 6). In a frequency range from $100\,\text{kHz}$ to $350\,\text{kHz}$, the mean input voltage noise level of the proposed LNA is $373\,\text{pV}/\sqrt{\text{Hz}}$ and $1.2\,\text{nV}/\sqrt{\text{Hz}}$ for the previous LNA. Above $350\,\text{kHz}$, the input noise of both amplifiers increases due to the low-pass characteristic of their TFs. In comparison, the prototyped LNA has 3.2 times less input noise than the previous LNA which meets the set design goal. Compared to the resistive noise of the human-head sized receive coil ($324\,\mu\text{H}$, $2\,\Omega$) of $179\,\text{pV}/\sqrt{\text{Hz}}$, the prototyped LNA's input noise is only 2.1 times higher (6.7 times for the previous LNA). As noted before, this improvement holds only true for zero source impedance. Since the used INA has a relatively high current noise of $1.5\,\text{pA}/\sqrt{\text{Hz}}$, its created voltage noise will exceed the measured input noise of $373\,\text{pV}/\sqrt{\text{Hz}}$ for any source impedance greater than $79\,\Omega$. Therefore, the predicted improvement factor of about 3 might not be reached within the system as the receive coil has an impedance of $203\,\Omega$. Furthermore, the parallelization of amplification devices results in an addition of their input capacitances, which affects the resonance with the receive coil [4]. Table 1 summarizes the input rated noise levels and calculated noise figures.

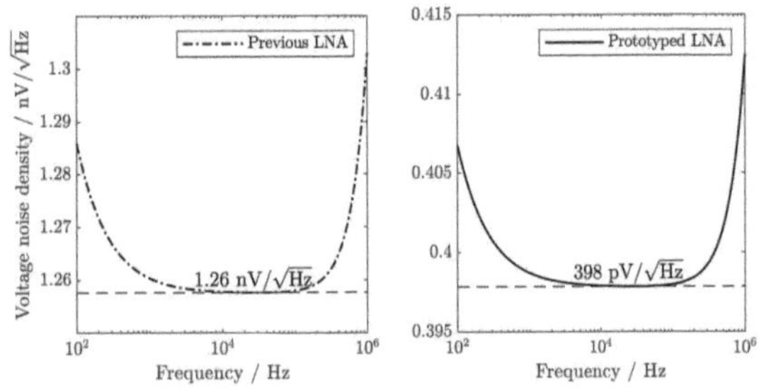

Figure 3: Simulated input rated noise levels for the previous LNA (left) and the prototyped LNA (right).

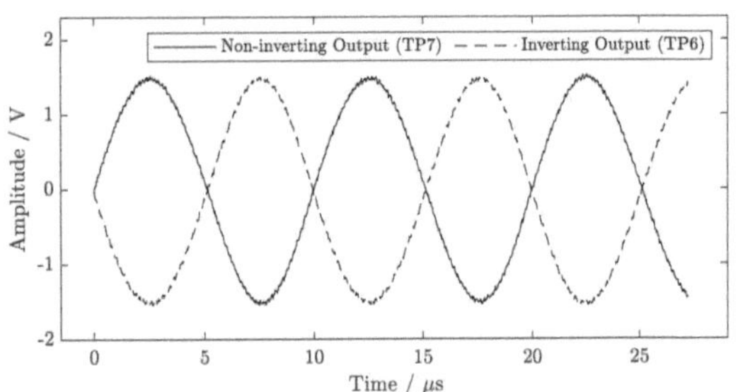

Figure 4: Differential output signal of the prototyped LNA measured at the inverting (TP6) and non-inverting (TP7) output of the ADC driver.

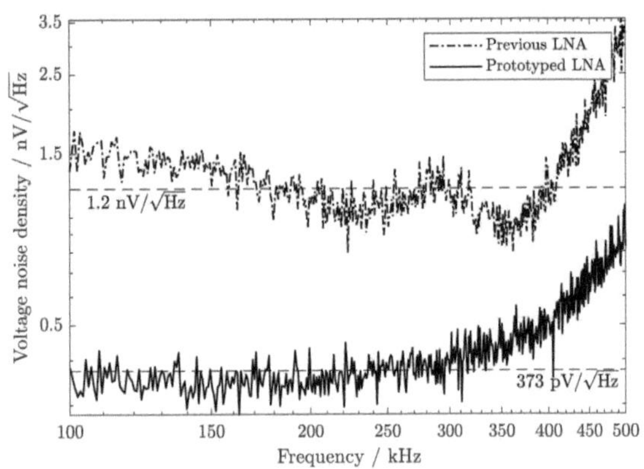

Figure 5: Measurement based input rated noise levels for the previous LNA (dash-dot line) and the prototyped LNA (solid line). The dashed line represents the mean noise level from 100 kHz to 350 kHz.

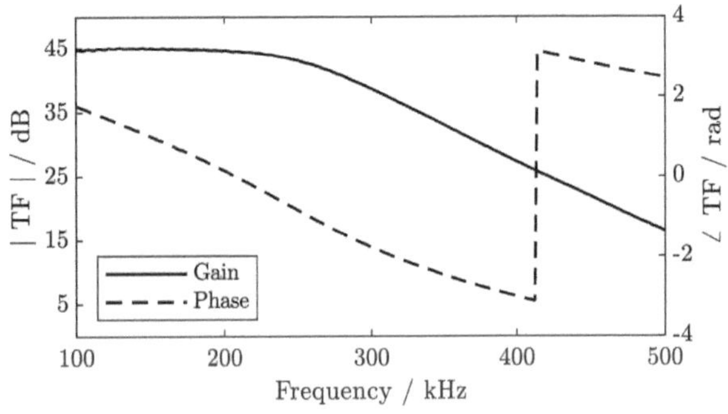

Figure 6: TF of the prototyped LNA represented by the magnitude and phase of the measured S21 parameter.

Table 1: Input noise and noise figures for both LNAs.

	Input noise	Noise figure
Previous LNA	$1.2\,\mathrm{nV}/\sqrt{\mathrm{Hz}}$	$16.6\,\mathrm{dB}$
Prototyped LNA	$373\,\mathrm{pV}/\sqrt{\mathrm{Hz}}$	$6.4\,\mathrm{dB}$

4 Conclusion

In this work, an LNA based on the parallelization of multiple amplifiers has been presented and compared to the LNA of the present scanner setup. The prototyped LNA has an input rated voltage noise density of $373\,\mathrm{pV}/\sqrt{\mathrm{Hz}}$, which is a 3.2-fold improvement compared to the previous design. With a noise figure of $6.4\,\mathrm{dB}$, the LNA's input noise is only 2.1 times above the resistive coil noise. Future work will focus on the implementation of the prototyped LNA in the scanner setup. Furthermore, the number of paralleled INAs can be increased in order to match the coil noise. In this way, the presented LNA might significantly enhance the imager's SNR by bringing coil noise dominance within reach.

Acknowledgement

The work has been carried out at the MGH/HST A. A. Martinos Center for Biomedical Imaging, and supervised by the Institute of Medical Engineering, University zu Lübeck.

Authors' Statement

Conflict of interest: Authors state no conflict of interest.

5 References

[1] B. Gleich and J. Weizenecker, *Tomographic imaging using the nonlinear response of magnetic particles.* Nature, vol. 435, no. 7046, pp. 1214–1217, 2005.

[2] T. Knopp and T. M. Buzug, *Magnetic particle imaging: an introduction to imaging principles and scanner instrumentation.* Springer, 2012.

[3] P. Horowitz and W. Hill, *The art of electronics.* Cambridge University Press, 2015.

[4] B. Zheng, P. W. Goodwill, N. Dixit, D. Xiao, W. Zhang, B. Gunel et al., *Optimal broadband noise matching to inductive sensors: Application to magnetic particle imaging.* IEEE transactions on biomedical circuits and systems, vol. 11, no. 5, pp. 1041–1052, 2017.

[5] A. Malhotra and T. M. Buzug, *A summing configuration based low noise amplifier for MPI and MPS.* Current Directions in Biomedical Engineering, vol. 4, no. 1, pp. 83–86, 2018.

Pilot Study for Coating of Superparamagnetic Iron Oxide Nanoparticles using Single Emulsion

Nitish Panchpor [1], Ankit Malhotra [2], Agnes Weimer [2], and Kerstin Lüdtke-Buzug [3]

[1] Medical Microtechnology, Lübeck University of Applied Sciences, nitish.panchpor@student.uni-luebeck.de

[2] Fraunhofer IMTE, Lübeck, {ankit.malhotra, agnes.weimer}@imte.fraunhofer.de

[3] Institut für Medizintechnik, Universität zu Lübeck, kerstin.luedtkebuzug@uni-luebeck.de

Abstract

This research explores the coating of superparamagnetic iron oxide nanoparticles (SPIONs) using an advanced microfluidic system. Unlike conventional methods, this approach offers precision, scalability, and enhanced control over key parameters. The microfluidic system, particularly with a single emulsion device, which provides better fluid management and reproducibility. The experiment successfully demonstrated oil-in-water emulsion, showcasing the system's ability to produce high-quality, monodisperse droplets at a viable production rate. Subsequent experimentation involved coating SPIONs with Dodecyltrimethylammonium bromide (D-TAB), revealing challenges in the nozzle and material compatibility. These findings pave the way for improved SPION synthesis and coating methodologies.

1 Introduction

In recent years tomographic imaging techniques have emerged strongly in medical imaging, diagnostic, and therapeutic applications. Magnetic Particle Imaging (MPI) is a technique developed by Gleich and Weizenecker in the year 2005 [1]. MPI quantitatively measures the spatial distribution of nanoparticles, with high sensitivity and spatial as well as temporal resolution. Magnetic nanoparticles, especially superparamagnetic iron oxide nanoparticles (SPIONs) have non-linear magnetization characteristics for which they are subjected to the magnetic field, a static selection field, and an oscillatory magnetic field. As the particles react in a non-linear manner, the resultant signal is picked up by a pick-up coil also known as a receiver coil. Due to the non-linearity, the resultant signal consists of the fundamental excitation frequency as well as harmonics. These higher harmonics are generated by nanoparticles and can be easily separated through appropriate filtering. The spatial distribution of the nanoparticles is realized with the static selection field that produces a field-free point, which is moved through the field of view by the drive fields [2]. The frequency response of SPIONs can be seen in Fig.1 in which an oscillating magnetic field (H, modulation field) is applied to the magnetic material at a single frequency f. As the magnetization curve (M) is nonlinear, resulting in time-dependent magnetization (yellow curve) as shown in Fig.1 (a).

To characterize the SPIONs, Magnetic Particle Spectroscopy (MPS) is a potent tool regarding applicability for MPI. MPS devices are also referred to as "0D-MPI" as it is based on the same principles as MPI. The use of this device

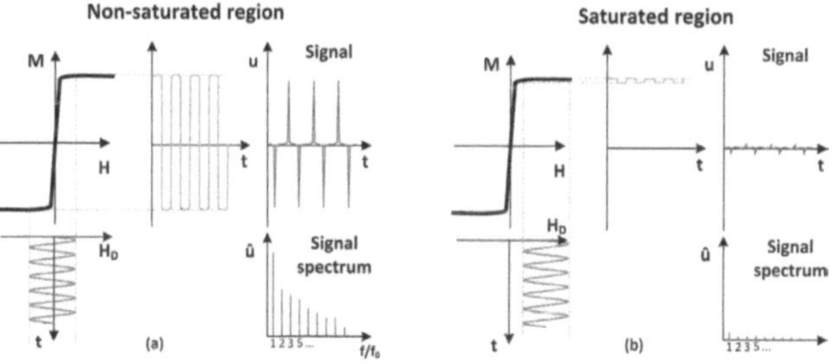

Figure 1: (a) response of nanoparticles within the FFP/FFL. The response consists of the excitation frequency f and higher harmonics of it.   (b) nanoparticles outside the FFP/FFL are magnetically saturated and do not respond to the excitation field [1].

enables the characterization of SPIONs in the units of magnetic moments (Am^2). By calibrating the measurements, quantitative results can be made based on the linear relation between the magnetic moment of the nanoparticles and the concentration of SPIONs in the sample.

In this paper, the coating of superparamagnetic nanoparticles is discussed using an advanced microfluidic system rather than conventional methods such as chemical coprecipitation [3], thermal oxidation [9], and reverse microemulsion [3]. Microfluidic devices can also be used for the production of nanoparticles which allows precision, scalability as well as control over parameters for example, stoichiometry, temperature dependence, and reaction time in comparison to conventional methods.

Magnetic nanoparticles

A variety of SPIONs (usually magnetite, Fe_3O_4, or

maghemite, $\gamma-Fe_2O_3$) has been developed to deliver drugs to specific target sites in vivo with different functional biocompatible layers. Biocompatible SPIONs have been widely used for in vivo biomedical applications as mentioned below.

Coatings, typically composed of biocompatible substances such as polymers or surfactants, play a crucial role in enhancing the performance and versatility of nanoparticles. These coatings prevent particle clumping, ensuring colloidal stability, and can be tailored to enhance targeted drug delivery and biocompatibility for in vivo applications. Surface characteristics and size of SPIONs, carefully adjusted through coatings, provide stability, biocompatibility, and the ability to traverse biological barriers [9]. The fundamental structure of a SPION with a coating layer is depicted in Fig. 2.

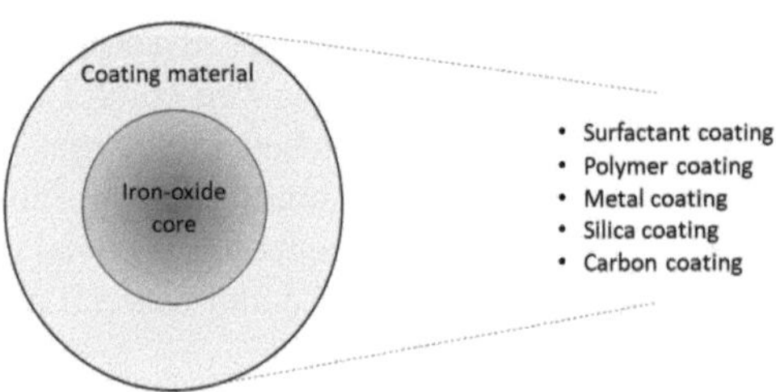

Figure 2: Fundamental structure of SPIONS

SPIONs are used for various interesting applications such as:

1. *Magnetic Resonance Imaging (MRI)*: MNPs, particularly superparamagnetic iron oxide nanoparticles (SPIONs), significantly enhance the sensitivity and resolution of MRI scans, providing detailed anatomical information for biomedical imaging [5].

2. *Targeted Drug Delivery*: SPIONs play a pivotal role in drug delivery, enabling precision medicine. The surface modification allows for the attachment of therapeutic agents, creating nanocarriers that can be guided to specific tissues or cells through an external magnetic field. This minimizes systemic side effects and enhances drug efficacy [6].

3. *Magnetic Hyperthermia*: In cancer therapy, magnetic hyperthermia emerges as a promising application. Effects of nanoparticle size on cellular uptake and liver MRI, detailing how SPIONs, when exposed to an alternating magnetic field, generate controlled hyperthermia within tumor tissues. This leads to localized thermal damage to cancer cells, which showcases the potential of SPIONs in cancer treatment.[7]

2 Material and Methods

The coating of the particles can be achieved through different means however in this paper, emulsion using a microfluidic system is discussed below.

The particles can be emulsified either with a single layer or a double layer. To achieve this, Fluigent's Raydrop (FLUIGENT Deutschland GmbH, Jena, Germany) is used, which is a microfluidic single-emulsion device. The single emulsion here refers to the mixture of one liquid containing a dispersion of the other liquid. This essentially means to coat a layer of a material around an object. The RayDrop is composed of three main fully removable parts: two inserts on each side, a center section containing a nozzle, and an outlet capillary. The Co-flow mechanism which is used in RayDrop usually consists of two phases, the first is the continuous phase and the other is the dispersed phase. The continuous phase is referred to as the material that is emulsifying the other material that is coming out from the nozzle of the dispersed phase. This design presents both the characteristics of a co-flow (axisymmetric geometry) and a flow focusing (local accelerations of the continuous phase), thereby called non-embedded co-flow focusing. The nozzle and outlet capillary are aligned in the continuous phase chamber, the dispersed phase comes through the nozzle to create the microparticles in the continuous phase and exits by the outlet capillary [8]. Refer to the Fig. 3.

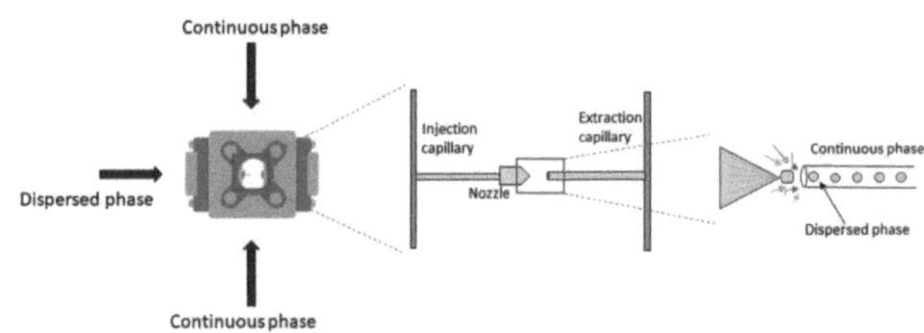

Figure 3: Co-flow mechanism of RayDrop [8]

Moreover, the system consists of various components that provide remarkable control over the flow rates. In this setup, there are low noise pressure generator (FLPG+), microfluidic pumps (Flow EZ), microfluidic valve controller for flow redirection (Switch EZ), and Microfluidic Sampling Valve (2-Switch), flow sensors (Flow unit), RayDrop for the emulsion, and tubing with the inner diameter of 1/16 and 1/32 respectively. These instruments were provided by the Fluigent Deutschland GmbH. The illustrative representation of the microfluidic experimental setup can be seen in Fig. 4.

2.1 Experimental protocol

The initial goal of the experiment was to establish a microfluidic system that can produce the emulsions as well as give better control over the experimental parameters such as flow rates, resistances in the system, pressure input, and also valve control for fluid management.

The experimental setup involved in preparing the RayDrop chamber for emulsion trials is shown in Fig. 4. Deionized water mixed with Rose Bengal B dye (CI 45440) (1.5 g, Carl Roth GmbH Karlsruhe, Germany) formed the continuous phase, while Fluosurf-C 2% w/w in HFE 7500 (10

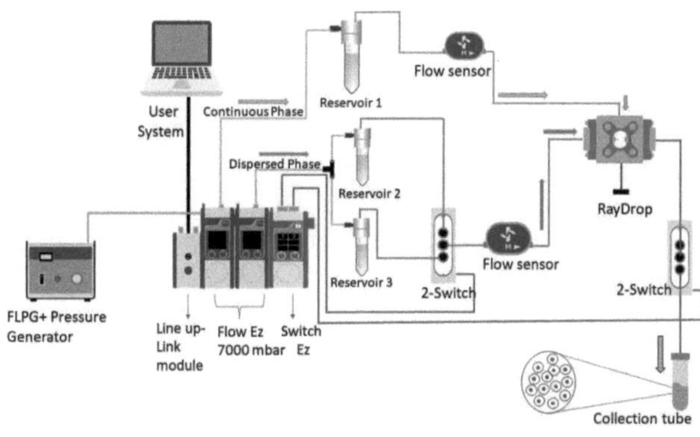

Figure 4: Experimental setup using microfluidic system

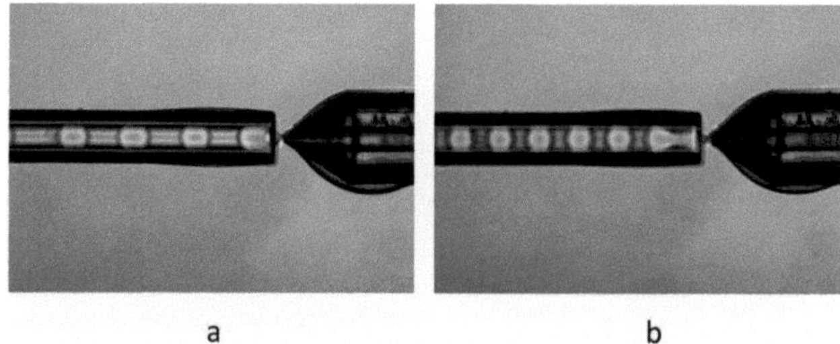

Figure 5: (a) Continuous phase: 30 μL/min, dispersed phase: 12 μL/min, (b) Continuous phase: 35 μL/min, dispersed phase: 13μL/min

ml, Emulso, Pessac, France) served as the dispersed phase. Isopropanol (20 ml, 99.9%, Höfer Chemie GmbH, Kleinblittersdorf, Germany) was used for cleaning. The cleaning process is necessary after each experiment to clear out the residuals in the tubing and the instruments.

The RayDrop chamber was primed with dyed water, and the air was purged. The continuous phase pressure was set (500-600 mbar), and pressure ($\approx$100 mbar) was applied to the dispersed phase. Pressure adjustments were done to establish the droplet regime.

For cleaning purposes, isopropanol was run for a minute through the setup, its pressure was reduced to 0 mbar, and the water inlet was flushed for at least a minute.

3 Results and Discussion

3.1 Oil in water emulsion

The Fig. 5 shows oil in water emulsion. In the outlet capillary, the oil is encapsulated by the dyed water. Adjusting and changing the flow rates of both the phases gives ability to produce high-quality, monodisperse droplets and the ability for reproducibly and at a viable production rate. The size and the production rate were visually assessed by using an inverted optical microscope (ZEISS Axio, Jena, Germany).

The results were obtained with the flow manipulation of continuous phase and dispersed phases. Two of the droplet regimes can be seen in Fig. 5. These pictures were obtained using an inverted optical microscope. In Fig. 5 the triangular nozzle is the output of the dispersed phase which contains the oil and emulsions of the same size at two different flow rates can be seen in the output capillary.

3.2 Further experimentation

After the successful completion of oil in water emulsification, the next goal was to coat the SPIONs with D-TAB (Dodecyltrimethylammonium bromide, $C_{15}H_{34}BrN$), which is a versatile quaternary ammonium surfactant with a wide range of applications in cell biology and biochemical research. Its unique properties make it a valuable tool for solubilizing proteins and peptides, extracting DNA, and lysing cells.

Preparation of chemicals

The SPIONs were synthesized at Hamburg University, Hamburg, Germany by Feld et al. [9] using the thermal oxidation method. Afterward, these SPIONs were mixed with Trichloromethane/Chloroform (99.9%, Carl Roth GmbH Karlsruhe, Germany) which contributes to the stability of the SPIONs in the solution. Firstly, an ultrasonication bath (Fisherbrand FB15065, Germany) was performed on the SPIONs for around 30 seconds. Afterward, 300 ml of ultrasonicated SPION suspension was added to the 600 ml of Chloroform to produce a 10 g/L concentration which was used in the dispersed phase of the system. Moreover, for the continuous phase, 1.5 ml of deionized water was added to the 1.5 ml DTAB (concentration-20 mg/ml) solution to make a 30 g/L concentrated solution.

Challenges in the later experiment

1. The RayDrop nozzle is 3D printed in a Nanoscribe printer (Nanoscribe, Eggenstein-Leopoldshafen, Germany) using IP-L resin. According to the RayDrop datasheet the use of Trichloromethane ($CHCl_3$) is not recommended. However, the datasheet states that Dichloromethane (CH_2Cl_2) can be used for the short run (approximately for one hour without any damage to the nozzle). To check the compatibility of the Trichloromethane for emulsification, several test cylinders (refer Fig.6) were printed using an in-house nanoscribe printer with the above-mentioned resin which is used in the fabrication of the nozzle in RayDrop. The test was divided into two different experiments. In the first experiment, the test cylinders were immersed in the solution of Trichloromethane for a short duration of one hour. In the second experiment, the duration of the immersion was changed to 24 hours. After the experiments, the test cylinders were assessed for any deformities with the microscope. The same tests were performed on Dichloromethane however, the emphasis was on the long experiment.

2. The same situation occurred for cleaning purposes Hydrochloric acid (32%, Carl Roth GmbH Karlsruhe, Germany) that was supposed to be used. However, the RayDrop datasheet didn't mention the usage of HCl

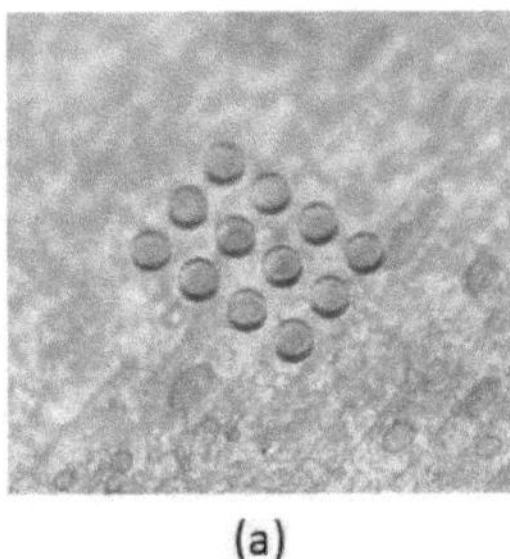 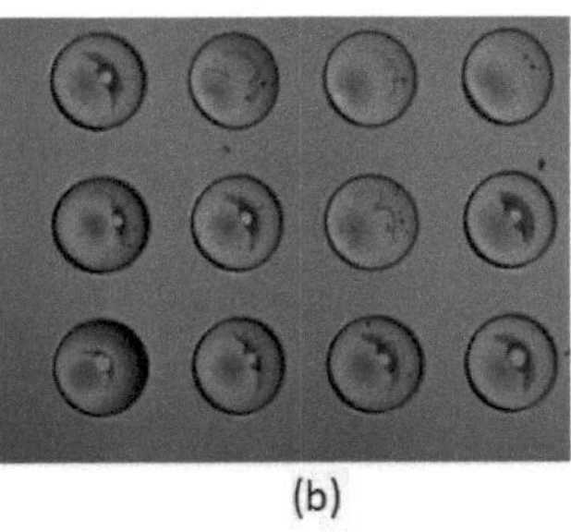

Figure 6: Cylinders printed with IP-L resin in Nanoscribe with 250 μm height and 500 μm diameter. (a) Final image captured via Nanoscribe after the completion of 3D printing process (b) Image obtained by an external microscope.

for both short and long periods. To verify whether the cleaning solution consisting of different HCl concentrations can damage the nozzle, the cylinders were immersed in HCl solution (5%, and 10%) for short (< 1 hr) and long (24 hrs) duration.

The results obtained from the previous experiments are shown in Table 1. The "Normal" condition represents the resin material that did not react with the chemical and when the resin material reacted with the chemical "Deformation" condition is stated. The normal condition represents that the geometry of the cylinder is not changed (not deformed) however if there is structural damage/change seen then that is referred to as a deformed condition. The inspection for the deformity was done using a microscope.

Table 1: Results for long and short experiments

Chemicals	Short experiment	Long experiment
$CHCl_3$	Normal	Deformation
CH_2Cl_2	Normal	Deformation
HCl (5%)	Normal	Normal
HCl (10%)	Normal	Deformation

4   Conclusion

In conclusion, the utilization of a microfluidic system, specifically Fluigent's RayDrop, can be effective in achieving precise and reproducible emulsions. The successful generation of oil-in-water emulsion highlights the system's capabilities. Challenges in material compatibility, especially regarding nozzle resilience, were identified, emphasizing the need for further optimization. The results provide valuable insight for improving the synthesis as well as the coating process of SPIONs through emulsification. In the future experiments will be conducted using different continuous and dispering phases.

Acknowledgement

The author expresses their sincere gratitude for the funding provided by the project, Diagnose und Therapieverfahren für die Individualisierte Medizintechnik (IMTE), Antrags-Nr.: 124 20 002.

Author's Statement

The authors declare no conflict of interest in connection with this research paper.

5   References

[1] Gleich B., Weizenecker J., *Tomographic imaging using the nonlinear response of magnetic particles*, Nature, 435(7046), 1214-1217, 2005.

[2] Thorsten B. et al., *Magnetic particle imaging: Introduction to imaging and hardware realization*, Zeitschrift für Medizinische Physik, 22-4, 323-334, 2012.

[3] Sun S., Zeng, H., *Chemical synthesis of magnetic nanoparticles*, Chemical reviews, 102(4), 1026-1071, 2002.

[4] Morteza M., Shilpa S., Ben W., Sophie L., Tapas S., *Superparamagnetic iron oxide nanoparticles (SPIONs): Development, surface modification and applications in chemotherapy*, Advanced Drug Delivery Reviews, 63-1, 2011.

[5] Laurent S., Forge D., Port M., Roch A., Robic C., Vander L., and Muller R., *Magnetic iron oxide nanoparticles: synthesis, stabilization, vectorization, physicochemical characterizations, and biological applications*, Chemical Reviews, 108(6), 2064-2110, 2008.

[6] Huang, J. et al., *Effects of nanoparticle size on cellular uptake and liver MRI with polyvinylpyrrolidone-coated iron oxide nanoparticles*, ACS Nano, 4(12), 7151-7160, 2010.

[7] Pankhurst A., Connolly J., Jones K., and Dobson. *Applications of magnetic nanoparticles in biomedicine*, Journal of Physics: Applied Physics, 36(13), R167, 2003.

[8] Dewandre A., Rivero-Rodriguez, J., and Vitry Y., Sobac B., Scheid, B., *Microfluidic droplet generation based on non-embedded co-flow-focusing using 3D printed nozzle*, Scientific Reports, 10, 21616, 2020.

[9] Artur F. et al., *Chemistry of Shape-Controlled Iron Oxide Nanocrystal Formation*, ACS Nano 13 (1), 152-162, DOI: 10.1021/acsnano.8b05032

Detector Efficiency Correction for the MERMAID Scanner Prototype

Rebecca Kantorek[1], Hong Phuc Vo[2], Steven Seeger[2], Jorge Roser [2], and Magdalena Rafecas[2]

[1] Medical Engineering Science, Universität zu Lübeck, rebecca.kantorek@student.uni-luebeck.de

[2] Institute of Medical Engineering, Universität zu Lübeck, {h.vo, s.seeger, jorge.rosermartinez, magdalena.rafecas} @uni-luebeck.de

Abstract

To improve uniformity and spatial resolution of a positron emission tomography scanner prototype dedicated to small aquatic animals, detector efficiency corrections were implemented into image reconstruction based on experimental measurements. The detector efficiency was measured using a planar V-shaped phantom. To evaluate the image quality, the homogeneous region and active rods of a downscaled NEMA NU4-2008 image quality phantom were measured and reconstructed using the *Maximum Likelihood Expectation Maximization* algorithm with and without detector efficiency correction factors. The correction improved the system sensitivity, which results in more uniform reconstructed images. To obtain conclusive results, further measurements should be conducted, both to obtain robust efficiency factors as well as to evaluate the correction.

1 Introduction

The Multi-Emission Radioisotopes - Marine Animal Imaging Device (MERMAID) scanner prototype has been designed to enable positron emission tomography (PET) of small aquatic animals, e.g. zebrafish. The latter are used in biomedical research to develop and study models of human diseases [1]. The motivation for this work is to improve the quality of the reconstructed images by compensating for individual detector responses.

PET is based on the β^+ decay of specific radionuclides. During the decay a positron is emitted, which after annihilation typically produces two antiparallel photons, each with an energy of $511\,keV$. When two detector elements of the PET scanner register two $511\,keV$ photons almost simultaneously, the pair of detected photons is referred to as a coincidence. The line of response (LOR) is defined as a straight line connecting the two involved detector elements. The annihilation origin is assumed to be along this LOR [2][3].

There are different physical phenomena which affect the quality of the reconstructed images. These include scattering, attenuation, or positron range. The image quality is influenced by the specific scanner configuration and the detector size. Additionally, variations in the detector efficiency of single detector elements further affect the image. They are caused by variations in the components of the readout electronics as well as differences or defects in the detector elements [2]. These variations can be corrected by factors, which can be determined for each specific scanner [2]. In this study, the detector efficiency for each crystal is experimentally determined and implemented as detector efficiency factors into the reconstruction of the image. Finally, the effect of the efficiency correction to the reconstructed image is evaluated.

2 Material and Methods

2.1 PET Scanner

Within the scope of the MERMAID project, a PET scanner prototype was developed by the Nuclear Imaging Group of the Universität zu Lübeck [4]. The scanner consists of four modules, each one comprising 16×8 LYSO scintillator crystals coupled to two 8×8 silicon photomultipliers. Each crystal has a volume of $1.12 \times 1.12 \times 15\,mm^3$.

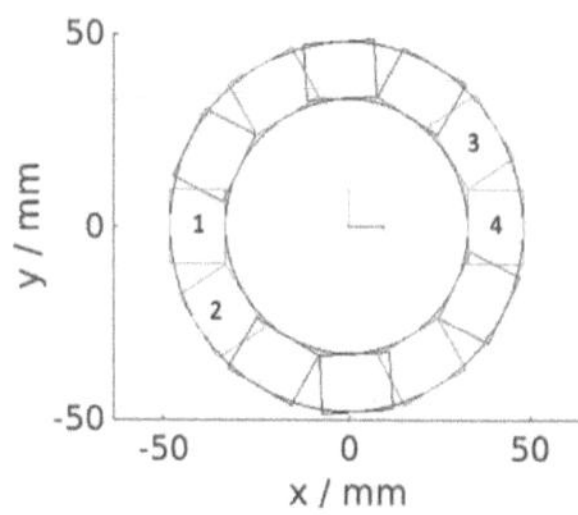

Figure 1: Geometry of the MERMAID detector modules and their rotated positions of 60° steps

The modules are constructed in pairs of two, with an angle of 33° between the two neighbouring modules. The two pairs are mounted on a rotating disc on opposite sides of a circle with a radius of $33\,mm$. For this work, a scan consisted of three acquisition steps with a rotation of the module setting by 60° to cover the full ring (see Fig. 1). The single measured photons (single events) are processed using Matlab after detection. Two single events within a time window of $2\,ns$ and an energy window of $450 - 550\,keV$ are considered a coincidence. These coincidences are sorted according to their acquisition time and labelled with the detector positions depending on the step and rotational angle.

2.2 Detector Efficiency

To measure the detector efficiency, a planar V-shaped phantom was used as shown in Fig. 2. The phantom was specifically constructed and 3D-printed with resin. The phantom has a depth of $3\,mm$ and a length of $29.5\,mm$ to cover two neighboring detector modules each with one wing. The

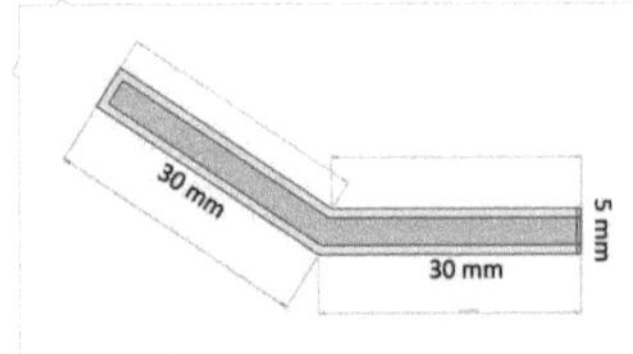

Figure 2: Measurements of the V-shaped phantom

phantom has an active volume of $2.7\,ml$. For the experimental data, ^{18}F-FDG (fluorodeoxyglucose) with an initial activity of $4.74\,MBq$ was used. Modules one and two were measured first with an acquisition time of thirty minutes. After that, data for module three and four were acquired for forty minutes, extending the acquisition time for decay correction. To theoretically match the number of emitted events from the first measurement, a correction factor for the number of single events was used.

The efficiency of the n-th detector channel ϵ_n was calculated as follows:

$$\epsilon_n = \frac{N_n}{N_{max}} \tag{1}$$

where N_n is the number of measured single events at the detector channel n and N_{max} is the maximum number of measured single events at any detector channel. N_{max} was used instead of N_{mean} to calculate the efficiency, because previous measurements show several low efficiency channels influencing the mean value. Noise of N_{max} can be excluded due to the type of the phantom and the long measurement period, although statistical fluctuations cannot be ruled out. For the reconstruction the efficiency for each LOR, so each possible detector pairing, is required. These factors were calculated by multiplying the efficiency factor of the two involved detectors.

2.3 Image Reconstruction

The reconstruction relies on the Customizable and Advanced software for Tomographic Reconstruction (CASToR [5]. CASToR is an open-source reconstruction software based on C++ which offers various reconstruction techniques. Here we used the *Maximum-Likelihood Expectation Maximization* (MLEM) algorithm [6][7], previously

adapted and implemented for the MERMAID scanner. The method models the radiotracer distribution as an algebraic function of a system matrix element a_{ij} and the 3D image, which is calculated as follows,

$$\tilde{f}_j^{(k+1)} = \frac{\tilde{f}_j^{(k)}}{S_{normj}} \sum_{i=1}^{M} \frac{1}{\sum_{j'=1}^{J} a_{ij'}\tilde{f}_{j'}^{(k)}} a_{ij} \tag{2}$$

with event number i, voxel number j, image of iteration k $f(k)$, system matrix element a_{ij} and number of coincidences M. The efficiency correction factors n_i are included into the calculation of the sensitivity S_{normj} is calculated in the following way:

$$S_{normj} = \sum_{i=1}^{I} n_i a_{ij} \tag{3}$$

2.4 Metrics

To evaluate the image quality and compare the proposed reconstruction with and without the calculated efficiency correction factor, a downscaled NEMA NU4-2008 image quality (IQ) phantom [8] was used. Two regions of the phantom were considered, the cylindrical homogeneous region ($d = 15\,mm, l = 10\,mm$) as well as the active rods. Both are shown with their specifics in Fig. 3. For the measurement, the phantom contained an activity concentration of $4.24\,MBq/ml$. Based on a preliminary study of the metrics, the images were reconstructed with 32 iterations for the homogeneous region and 128 iterations for the rod region. The different numbers of iterations are necessary to ensure that the reconstruction converges for the data of the specific phantom. The voxel size of the reconstructed image is $0.25 \times 0.25 \times 0.25\,mm^3$.

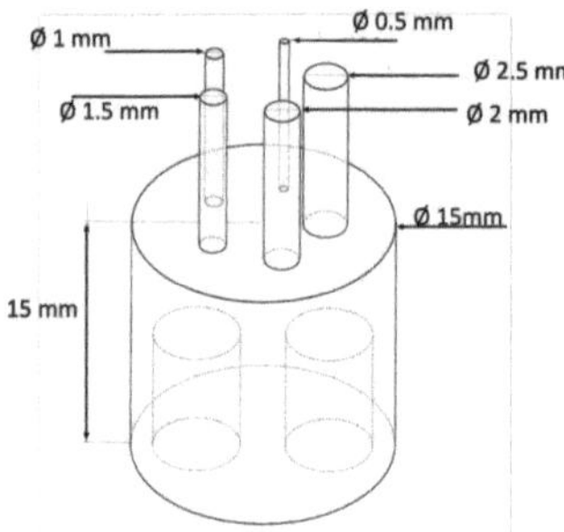

Figure 3: Image of the downscaled NEMA NU4-2008 IQ phantom

Sensitivity: The sensitivity was calculated with, as in (3), and without efficiency correction factors. To examine the differences between the two sensitivities, they were normalized and a quotient image was determined by calculating the ratio between these sensitivities.

Uniformity: The uniformity is a measurement to evaluate possible fluctuations in sensitivity. A cylindrical volume of interest with a diameter of $13.5\,mm$ and a depth of $6\,mm$ was centrally allocated within the homogeneous phantom and the percentage standard deviation (%STD) in each iteration was calculated.

Recovery Coefficient: The recovery coefficient (RC) replicates the system's capacity to reconstruct small features and therefore quantifies the spatial resolution. The RC is determined by dividing the mean of the central region (75 % diameter) of each rod by the mean of the uniform region, characterized as the inner central region of the largest rod (50 % diameter).

3 Results and Discussion

3.1 Detector Efficiency

Fig. 4 shows the measured efficiencies for module two, and can be considered representative for the differences encountered in the other modules. These measurements were acquired by placing the V-shaped phantom directly on top of module two covering the surface of the crystals. However, since this phantom is constructed with the second wing for the neighboring module on the right (Fig. 2), more counts are expected towards the right edge of this module. In future studies, this effect can be compensated with additional simulation and experimental data.

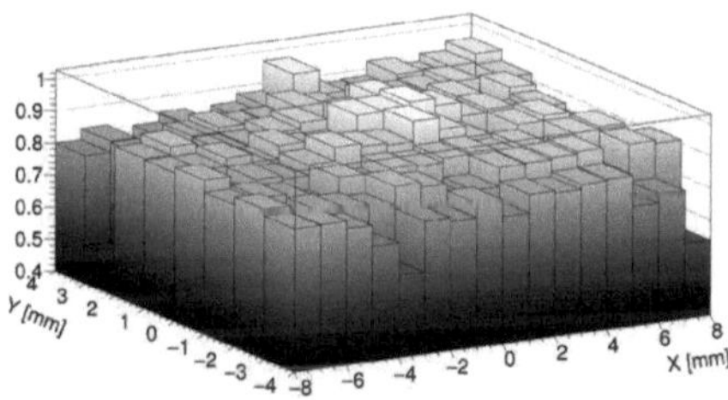

Figure 4: Efficiency of the detector crystals for module two

Table 1: Statistics for the measured detector efficiency

Module	Max	Min	Mean	Median	STD
1	1	0.7342	0.89	0.89	0.04
2	0.99	0.06	0.86	0.88	0.10
3	0.81	0.04	0.74	0.75	0.07
4	0.81	0.58	0.74	0.75	0.04
LORs	1	0.0017	0.65	0.66	0.11

To evaluate the variations in the efficiencies, the maximum (max), minimum (min), mean, median and standard deviation (STD) of the calculated efficiencies of each module as well as all possible LORs are reported in Table 1. The minimum values for modules two and three originate from the low detection channels mentioned in 2.2. The lower values for modules three and four could originate from either actual detector differences or an inaccuracy in the decay correction, which could be determined with further testing. Since there are only small deviations in detector efficiency in each module, we expect only small changes in the reconstructed images compared to the non-corrected ones.

3.2 Evaluation of Image Quality

3.2.1 Sensitivity

Fig. 5a shows a central transaxial slice of the sensitivity image with an implemented efficiency correction factor. Fig. 5b points out the quotient of two sensitivity images with and without efficiency correction. The quotient shows

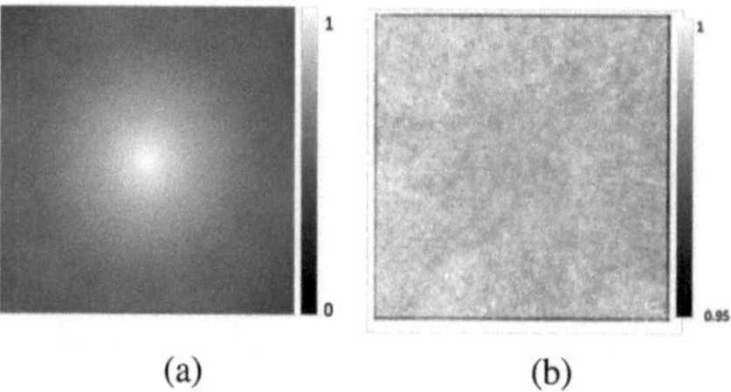

(a)   (b)

Figure 5: (a) Central transaxial slice of the sensitivity image with efficiency correction (b) Quotient of the central transaxial sensitivity

that there are minimal differences in the sensitivity. The colorbar is set from 0.95 to 1 to highlight the slight changes. Due to the relatively low STD observed across all LORs (Table 1), the efficiency correction factors have only slight fluctuations.

3.2.2 Uniformity

To evaluate the uniformity of the image, the homogeneous part of the phantom was scanned. Fig. 7a) and b) show the reconstructed transaxial and coronal slices with and without the efficiency correction. The phantom was positioned with a slight offset towards the edge of the field-of-view (FOV), which causes the cylinder to fade out towards the right side of the image. The two images are visually similar. These observations are confirmed with the comparison of the %STD over all 32 iterations, see Fig. 6. The minimum is located for both reconstructions at iteration 14 with a value of 14.81 % without efficiency correction and 14.64 % with efficiency correction. As discussed above, since there are slight differences in detector efficiencies, there is only a slight improvement in terms of uniformity.

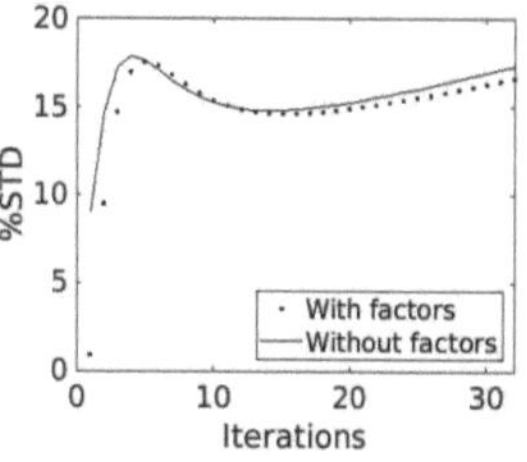

Figure 6: %STD of the homogeneous phantom over 32 iterations, with and without efficiency correction

3.2.3 Recovery Coefficient

To quantify the change in spatial resolution two measurements of the active rod region of the IQ phantom were reconstructed and evaluated. The phantom is placed with the rods on the edge of the FOV, Fig. 7c) and at the center of the FOV, Fig. 7d). The four biggest rods are visually resolved.

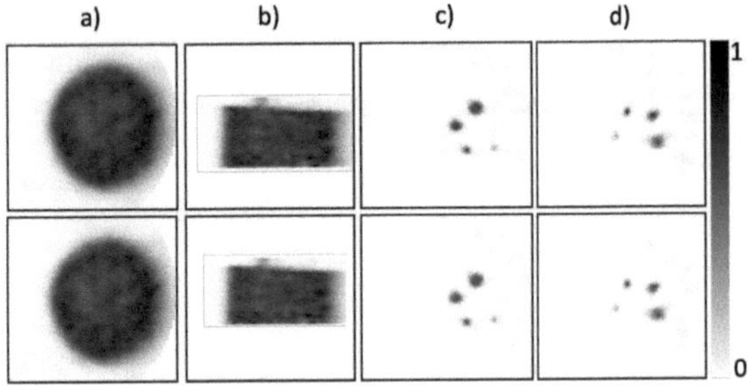

Figure 7: Reconstructed images without (top) and with (bottom) efficiency correction after iteration 32, a) transaxial and b) coronal slice of the homogeneous region of the NEMA phantom, c) and d) transaxial slice of the NEMA phantom rods

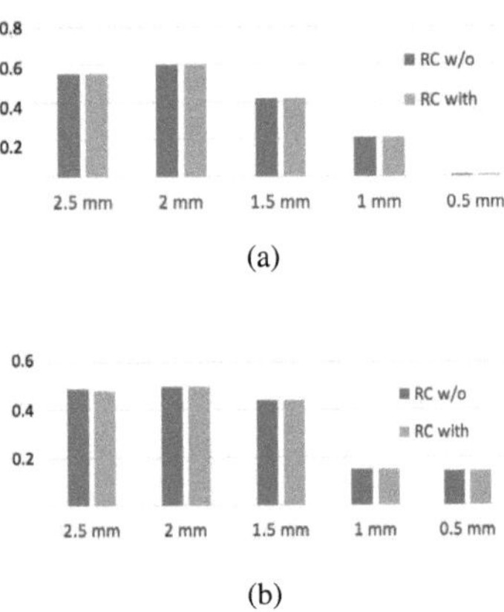

Figure 8: (a) RC of the first measurement, (b) RC of the second measurement

The RC does not show any improvement of the resolution as seen in Fig. 8, corresponding that the influence of the variations in sensitivity are minimal. The scanner setup for this study also affects image quality. The intrinsic resolution is limited by the size of the crystals and the reconstructed spatial resolution is affected by the partial volume effect due to the voxel size. Additionally, the reconstruction does not include any correction for scattering or attenuation in the object. Because of these limitations, the smallest rod is not resolved. The RC is very low, making the evaluation minor. Furthermore, the evaluated regions of interest were selected by visual observation and might be imprecise. To evaluate possible improvement of the uniformity, this measurement needs to be repeated.

4   Conclusion

The detector efficiency was experimentally obtained and successfully integrated into the CASToR's reconstruction algorithm as additional efficiency correction coefficients. With this result, this work lays out a framework for future automatic optimization of the reconstruction process. From the given data, there is a small visible improvement in the obtained uniformity metric.

To further investigate the influence of the detector efficiency correction, the efficiency could be modelled through a Monte Carlo simulated system. Since only one measurement of the detector efficiency was conducted there are several limitations, including temperature sensitive characteristic of the detectors, choice of phantom, decay compensation and also limited data to study the effectiveness of implementing the efficiency correction factor into the reconstruction. Therefore, more measurements of the detector efficiency would help to reduce the statistical noise and allow robust efficiency factors to be calculated. Additional scans of the IQ and other extended phantoms should be carried out to further evaluate of the image quality improvement.

Acknowledgement

The work has been carried out at the Institute for Medical Engineering, Universität zu Lübeck.

Authors' Statement

Conflict of interest: Authors state no conflict of interest.

5   References

[1] TY. Choi et al., *Zebrafish as an animal model for biomedical research*, Exp Mol Med, vol. 53, no. 3, pp. 310-317, 2021.

[2] D. Bailey et al., *Positron Emission Tomography - Basic Science*. Springer-Verlag, London, 2005.

[3] A. Karellas et al., *Physics of PET and SPECT Imaging*, CRC Press, Boca Raton, USA, 2017.

[4] S.Seeger et al., *First Images from MERMAID, a Small Aquatic Animal PET Scanner Prototype*, 2023 IEEE NSS MIC, Vancouver, Canada, 2023.

[5] T. Merlin et al., *CASToR: a generic data organization and processing code framework for multi-modal and multi-dimensional tomographic reconstruction*, Phys. Med. Biol., vol. 63, no. 18, 2018.

[6] LA. Shepp et al., *Maximum likelihood reconstruction for emission tomography*, IEEE Trans Med Imaging, 1982.

[7] A. J. Reader et al., *Accelerated list-mode EM algorithm*, IEEE Trans Nucl Sci, vol. 49, no. 1, pp. 42-49, 2002.

[8] *NEMA standard publication NU4-2008: performance measurements of small animal positron emission tomographs*, National Electrical Manufacturers Association, Rosslyn, VA, 2001.

X-Ray Spectrometry with Photon Counting Detectors

Simeon Placzek [1,2], Andreas Deresch [2]
[1] Medical Engineering Science, Universität zu Lübeck, simeon.placzek@student.uni-luebeck.de
[2] Comet Yxlon GmbH, Hamburg, andreas.deresch@comet.tech

Abstract

The energy sensitive measurement of X-rays is not yet very common in non-destructive testing (NDT). This is mainly due to the fact that a spectroscopic setup in an X-ray system with limited space entails several difficulties. Since spectrometers have a dead time and industrial X-ray tubes emit a large number of photons, this must be reduced to receive reliable results. Photon counting flat panel detectors convert X-ray photons directly into an electrical signal by ionizing atoms in a semiconductor. Due to the much smaller measuring volume of the pixels, semiconductor detectors enable higher counting rates than conventional spectrometers, but have a significantly lower probability of full photon energy transfer as a drawback. In this study one silicon (Si) and one cadmium telluride (CdTe) semiconductor detector were tested regarding their specific properties. The experiments show that semiconductor detectors are capable of recording X-ray spectra at high X-ray intensities.

1 Introduction

In X-ray spectrometry using indirect converting spectrometers, X-rays are first converted into visible light. Afterwards, a photomultiplier converts the light into an electric signal and amplifies it. The resulting signal is proportional to the transferred energy of the incident x-ray photon, and by careful design the probability of a full energy transfer can be maximized. The time the spectrometer needs to detect a photon is the so-called dead time. No other X-ray photon may enter the measurement volume within this time. Otherwise, the energies of the photons would be combined into one event and the dead time is extended additively [1].

X-ray tubes that are used in NDT usually emit high intensities. To reduce dead time losses, the intensity arriving at the detector must be limited. The simplest method is to increase the focus detector distance (FDD). According to the inverse square law, the intensity of radiation on a certain area decreases quadratically with increasing distance form the source [1]. Besides that, the use of heavy shielding with a small entry window also reduces the incident intensity. Precise alignment of the entry window to the source is then required. In order to determine the energy of the photons accurately, they must be completely absorbed in the measurement volume. Therefore, the measuring volume of a spectrometer is usually in the order of cubic centimetres [2]. These requirements are difficult to achieve in practice, especially when space in an X-ray system is limited. Photon-counting flat panel detectors convert X-rays directly into an electrical signal. The ionization of a semiconductor produces free charge carriers. These are separated by the applied bias voltage and cause a measurable current. Each individual pixel has its own evaluation electronics and represents a measurement volume, which enables high counting rates. The signal is proportional to the energy transferred during the interaction(s). This makes it possible to measure an energy for each photon in addition to the spatial information [3].

The advantage of this technology is that each photon is assigned to an energy channel. In this way, only photons above a certain energy threshold are taken into account. The offset signal known from scintillation detectors is therefore completely eliminated [3]. This paper investigated the feasibility to which photon counting flat panel detectors are a viable alternative to conventional spectrometers. First, spectra were recorded to increase the energy resolution and efficiency by optimizing the settings. Further measurements were carried out to specifically measure characteristic radiation, which is caused by atomic excitation either by electrons or by photons in order to evaluate the achievable energy resolution. In addition, comparative measurements were made using the detector with a CdTe layer.

2 Material and Methods

2.1 Hardware

Two different detectors from the company Advacam were used to acquire X-ray spectra. The MiniPIX, with a 300 µm thick Si layer and the ModuPIX with 1000 µm of CdTe. Both detectors have 256 x 256 pixels with a pixel size of 55 µm. Two different X-ray systems were used to irradiate the detectors. A Cheetah made by Comet Yxlon, this system uses a microfocus X-ray tube with a maximum voltage of 160 kV. In this system, work was carried out at an FDD of 0.7 m. The other system used is a walk-in cabinet using a MXR-225HP11 tube made by Comet, with an FDD of

2.1 m. This tube is a minifocus X-ray tube with a maximum voltage of 225 kV.

2.2 Measurements

The setup in the Cheetah was used to find the best possible settings for the semiconductor detector to measure energy-resolved spectra efficiently in time. Measurements were made at 120 kV and 5 µA target current. A 100 µm copper (Cu) filter was positioned on the tube. X-ray spectra were then recorded at different bias voltages and integration times. The setup with the MXR-225HP11 tube was used to maximize the detected amount of characteristic radiation from the tube. A voltage of 160 kV at a target current of 10 µA was used with a 500 µm Cu filter in front of the tube. With this setup among other things a measurement was taken over 14 days, in which the data were analyzed pixel by pixel. Finally, a comparative measurement was taken with the CdTe detector and the same setup.

2.3 Clustering

The spectra in this paper were recorded using the *Clustering* plugin of the software PixetPro (v1.8.1) from Advacam. The *Clustering* plugin writes every event that takes place within the semiconductor frame by frame to a file. This contains the x- and y-coordinates of the event as well as the corresponding energy in keV. Events that extend over several adjacent pixels are recognized as so-called *clusters* and written in a line. This effect is known as charge sharing. Each line of raw data represents an event or cluster. To convert the raw data into an energy spectrum, the width of the energy channels must first be chosen. Each cluster is then assigned to a channel. The individual energies within clusters of several hit pixels are added together. Energies below 5 keV count as offset noise and are ignored. This can lead to a slight shift of the spectrum into the low-energy range. Another effect that leads to a down-shift of the spectrum is the incomplete energy transfer during photon scattering in the sensitive volume. In addition to that, pileup must be taken into account. Pileup describes the effect of two or more photons hitting the same or adjacent pixels within one readout cycle. As a result, the energies of the pixel or cluster are added together. This leads to an overestimation of the photon energy. [4]

2.4 Evaluation of the Spectra

The spectra were evaluated using a Bash script wich was written by *A. Deresch* at the *Comet Yxlon GmbH* specifically for evaluating the raw data of spectra recorded with PixetPro. An additional version of this script evaluates a spectrum for each pixel. This script defines the width of the energy channels and accumulates events in the corresponding energy channels. In addition to that, the spectra are also filtered according several criteria. First, the events are divided into single, double, triple and quad events according to the number of pixels hit in one event. In addition, the events are divided into valid and invalid. Double

events with diagonally connected pixels are considered invalid. For triples and quads anything outside a 2 x 2 grid is considered invalid. Any events with more than four pixels hit are always considered invalid. The ratio of valid to invalid events provides information about possible overexposure of the detector. In the spectrum of single pixel events the counting rate is reduced, but the influence of charge sharing and pileup can also be reduced. For this reason, single spectra are used in this paper for illustrations and evaluations.

XRayGUI was used to evaluate the spectra. This is a program that is primarily used to simulate characteristic X-ray spectra and continuous X-ray spectra for a wide variety of anode materials. In addition, XRayGUI offers the possibility to simulate different filters and detector reactions. XRayGUI is based on the model described in *A general spectrum model for X-ray generators* [5].

3   Results and Discussion

3.1   Optimization of the detector parameters

Fig. 1 shows X-ray spectra with the same X-ray parameters and different bias voltages. It can be clearly seen that the intensity of the spectrum increases with increasing bias voltage. This effect increases in the higher energy range. Consequently, more high-energy photons are measured when a high bias voltage is set.

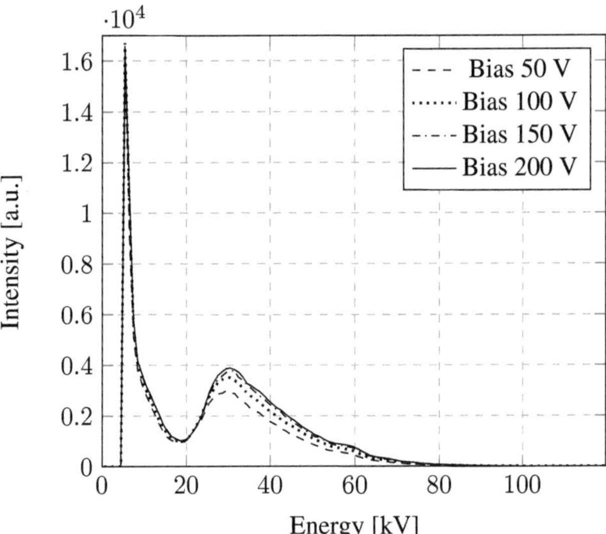

Figure 1: Comparison of X-ray spectra at different bias voltages.

As already mentioned, the bias voltage is responsible for separating pairs of charge carriers generated by ionization. The voltage generates an electric field in which the charge carriers are located. Since pairs of charge carriers are oppositely charged, the Coulomb force of the electric field pulls them in different directions. The higher the bias voltage, the stronger the separating force and the less likely they recombine due to the attraction of opposite charges. At some point, the bias voltage is high enough to separate all charge carriers. From this point onwards, increasing the bias voltage no longer results in an increase of the yield. The measurements indicate such a plateau region above 150 V. As the yield nevertheless increases slightly, 200 V is selected as the bias voltage for all further measurements.

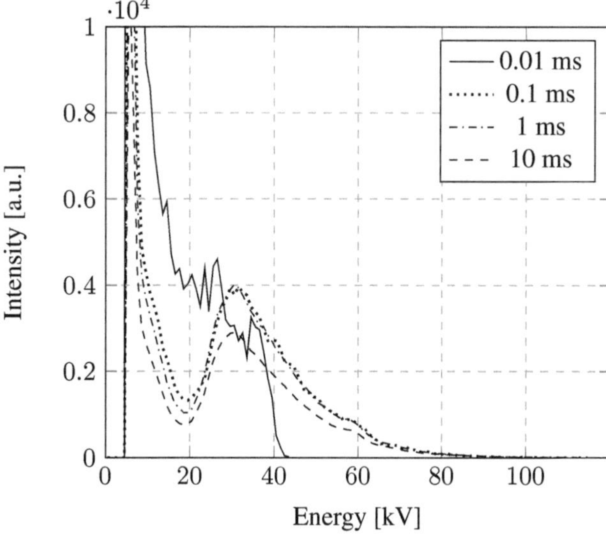

Figure 2: Comparison of X-ray spectra with frame count of 5000 frames and different integration times. The intensity axis was cut off at 1.0e+04 in order to better display the spectra in the high-energy range.

The spectra shown in Fig. 2 were recorded at different integration times. A short integration time of 0.01 ms results in noisy spectra with low statistics. This is because, as Table 1 shows, a total acquisition time of only 0.05 s results for 5000 images. This period is not long enough to measure enough X-ray photons. Nevertheless, in Fig. 2 the intensity of the spectrum at 0.01 ms integration time seems similar to the others because the flux is normalized to one second. At an integration time of 0.1 ms, the spectrum is still slightly noisy, but the continuous *Bremsstrahlung* spectrum, caused by continuous deceleration of electrons in the target, can be clearly identified. The reason for this is that the acquisition time here is longer by a factor of 10. If this is increased again by a factor of 10, the spectrum is similarly shaped but smoother. Apparently, enough events were measured here to form a low-noise spectrum. From an integration time of 10 ms, the intensity of the spectrum decreases visibly. This is because longer integration times mean that many more pixels are hit within one readout cycle. This increases the probability of adjacent pixels being hit. Since only single spectra are used in the evaluation, many events are filtered out. This is also shown by the ratio of invalid to valid events in Table 1 of 11.5 %. This rate does not include the case that a pixel is hit by several photons within a readout cycle, which is why the rate in Table 1 underestimates the true rate. To keep this error as low as possible the integration time must be shortened. If the integration time is too short, the required measurement time increases, as the frame rate does not exceed 27.5 fps. This limitation can be caused by

readout, data transfer or clustering. The most efficient integration time is therefore a compromise between time efficiency and error frequency.

Table 1: Effect of different integration times

Integration t.	Acq. t.	Frame rate	Invalid/Valid
0.01 ms	0.05 s	27.0 fps	0.01 %
0.1 ms	0.5 s	27.3 fps	0.26 %
1 ms	5 s	27.5 fps	1.12 %
10 ms	50 s	21.4 fps	11.5 %

3.2 Spectra per pixel

This measurement was made to check whether the energy resolution for single pixels is better or different than averaged over the whole detector. The pixel spectra shown in Fig. 3 have many positive and negative peaks. These peaks are not part of the characteristic spectrum. A possible cause for these peaks is that the calibration file supplied by the manufacturer is not correct for each individual pixel. Only averaging over all pixels smoothes out this error. This obscures other information such as characteristic lines in the spectra per pixel. Consequently the comparison between single-pixel and averaged spectra could not be carried out.

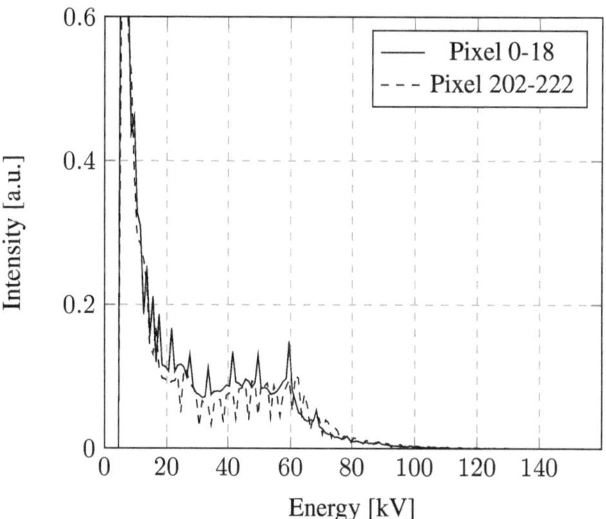

Figure 3: Spectra of two randomly selected pixels. The intensity axis was cut off at 0.6 in order to better display the spectra in the high-energy range.

3.3 Comparison of Si and CdTe

The spectra shown in Fig. 4 start with high intensities. Real X-ray spectra do not have this peak, as it is the result of the interaction probability of radiation decreasing with increasing energy and incomplete energy transfer. Much of the low-energy part of the X-ray radiation is usually already absorbed in the target and filters. Since the 300 µm thick silicon layer of the detector has a low interaction probability,

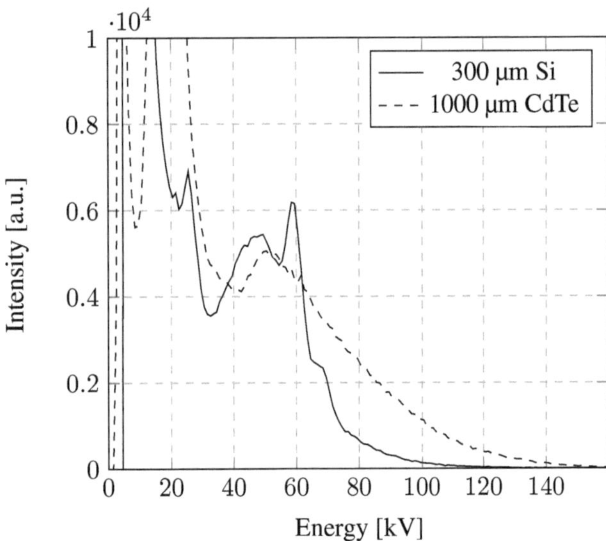

Figure 4: Comparison of spectra recorded with Si detector and CdTe detector (CdTe spectrum was scaled by a factor of 13). The intensity axis was cut off at 1.0e+04 in order to better display the spectra in the high-energy range.

especially at higher energies, many photons are not completely absorbed in the detector. The continuous part of the spectrum is the *Bremsstrahlung*. In addition, the characteristic radiation can be seen in the 300 µm Si spectrum in Fig. 4. For tungsten (W), the Kα_1 line is located at 59.3 kV and the Kα_2 line at 58.0 kV. Due to the low energy resolution, these lines blur into one peak. The following peak in the spectrum is the Kβ_1 line of W which is located at 67.2 kV [6]. The characteristic lines are more visible here than in previous measurements because they are more strongly excited at 160 kV and much of the lower-energy radiation has been intentionally attenuated. Since CdTe detectors have a lower energy resolution than Si detectors, no K lines can be seen here. The CdTe detector has a higher density than Si and a higher layer thickness in this experiment, which increases the probability of interactions. Therefore, significantly more events were recorded with the CdTe detector, so that many charge sharing events occurred that were filtered out. Consequently, the intensity of the CdTe spectrum is lower and was scaled in Fig. 4. In the range of higher-energy X-rays, the spectrum recorded with the Si detector drops significantly, so that barely any events are measured above 100 kV. This is due to the low absorption of 300 µm Si, which is already below 2.5 % at 60 keV (simulated in XRayGUI). Additionally the absorption of a high-energy photon in just one pixel is unlikely and therefore rarely occurs in a spectrum of single-pixel events. In contrast, the CdTe detector here measures events up to 160 kV due to its thickness and density.

4   Conclusion

The present work shows that semiconductor flat panel detectors are quite capable of recording X-ray spectra with comparatively small effort in terms of setup and time. The Si semiconductor detector was able to measure X-ray spectra without shielding and with low pre-filtering in an X-ray system. Due to the high intensity and the high-counting rate, spectra could be recorded within 3 minutes. This was more difficult with the CdTe detector, as it has a higher absorption rate. In terms of energy resolution, the Si performed better and was able to show various K lines of the target material. Nevertheless, these could not be seen separately from each other. So the Si detector has advantages in energy resolution but does not perform as well at energies above 60 keV, when compared to the CdTe detector. Which detector is a better fit depends on the individual measurement task. Due to the parallel measurement of spatial and energy information, semiconductor detectors offer advantages such as the elimination of offset noise and the possible use of an energy threshold. Semiconductor flat panel detectors show a higher probability of photon escape and incomplete energy transfer than spectrometers with large measurement volumes, making the direct evaluation of spectra more complex. The measurement of spectra per pixel could not provide any further insights due to calibration artifacts.

Acknowledgement

The work has been carried out at Comet YXLON GmbH, Hamburg, and was supervised by M. Stille, Medical Engineering IMTE, Lübeck.

Authors' Statement

Conflict of interest: Authors state no conflict of interest.

5   References

[1] H. Krieger and P. W. Daly, *Strahlungmessung und Dosimetrie*. Springer Spektrum, Wiesbaden, 2012.

[2] W. Schlegel, C.P. Karger, O. Jaekel, *Medizinische Physik*. Springer Spektrum, Heidelberg, 2018.

[3] D. Walter, U. Zscherpel, U. Ewert, *Photon Counting and Energy Discriminating X-Ray Detector - Benefits and Applications*. e-Journal of Non-Destructive Testing, Vol21(7),pp.1-10, 2016.

[4] C. Ullberg, M. Urech, et al., *Measurement of a Dual-Energy Fast Photon Counting CdTe Detector with integrated Charge Sharing Correction*. Medical Imaging, 2013.

[5] A. Deresch, C. Bellon, and G.-R. Jaenisch, *A general spectrum model for X-ray generators*. NDT and E International, Vol 79, pp. 92-97, 2016.

[6] Lawrence Berkeley National Laboratory, 2009 Berkeley, X-Ray Data Booklet. Available: `https://xdb.lbl.gov/Section1/Table_1-2.pdf` [last accessed on 2024-01-19].

Cinematic Rendering Methods of Dental-CBCT-Data using Visualization Toolkit

David Melenberg [1,2], Nele Blum [2], Thorsten M. Buzug [2], Maik Stille [2]

[1] Medical Engineering Science, Universität zu Lübeck, david.melenberg@student.uni-luebeck.de

[2] Fraunhofer-Einrichtung für Individualisierte und Zellbasierte Medizintechnik IMTE {david.melenberg, nele.blum, thorsten.buzug, maik.stille}@imte.fraunhofer.de

Abstract

Cinematic Volume Rendering (CVR) is gaining increasing importance in the medical field, particularly within dentistry. It enables photorealistic 3D representations of teeth and jaws, which are invaluable for treatment planning and educational purposes. In this study a real-time rendering algorithm for cone-beam CT in dentistry using the Visualization Toolkit (VTK) was developed. The proposed pipeline features global ambient occlusion, 2D transfer function generation, and post-rendering tone mapping. Various methods for creating 2D transfer functions where evaluated and high-quality renderings generated. Future developments may explore additional effects and optimizations, including the integration of stochastic ray casting and the automated generation of 2D transfer functions.

1 Introduction

In recent decades, medical imaging has made significant strides and is pivotal for diagnosing and treating diseases. Cone-beam CT is among the promising techniques used in this field that provide high spatial resolution with low radiation dose. These high-quality medical images allow renderings with exceptional detail and realism. CVR has been particularly successful in producing such renderings. Inspired by the animation film industry, CVR enables photorealistic rendering of medical volume data. Compared to traditional direct volume rendering (DVR), CVR provides the advantage of improved recognition of shapes and depth structures [1]. Following Siemens Healthineers' implementation of this method, there is currently a high demand for open-source solutions in the field [2]. Ray casting is a common method of DVR in which rays are cast from each pixel into the volume, sampling equidistant points. Based on Max's absorption/emission model [3], which is the most widely used in DVR, each sampling point must be assigned appropriate properties. This is done by a so-called transfer function [4]. The mapping to an RGBA (Red, Green, Blue, Alpha) tuple can be achieved through a one-dimensional transfer function (TF) based on a scalar value or a multi-dimensional TF based on a scalar value and a gradient amount. Compositing combines the color and alpha values of all sampling points along a ray. To shade the volume, DVR conventionally employs the gradient-based Phong shading method. In recent years, a range of volumetric shading techniques have been developed in addition to the gradient-based method. Jönsson et al. [5] introduced some of these techniques. As lighting complexity increases, multiple techniques are available, such as gradient-based shading, local ambient occlusion (LAO), shadow mapping,

and Monte Carlo Ray Tracing (MCRT). Considered state-of-the-art for generating photorealistic volumetric data renderings, MCRT creates global illumination by simulating random light paths across numerous possibilities and paths. Kroes et. al. [6] demonstrate that the utilization of a camera model, hybrid shading techniques, and incorporation of direct and indirect lighting can lead to highly realistic outcomes [1, 6]. Xu et. al [7] proposed a hybrid shading model based on linear interpolation between gradient-based shading and volumetric shading. This approach differs from the stochastic mixing approach used in [6], where a user-defined mixing coefficient is used to determine the weight of each model. The volumetric shading model, which is based on the local ambient occlusion method, uses a Heney-Greenstein phase function and a secondary ray at each sample point to determine occlusion [7].

This work focused on creating transfer functions, which play a crucial role in rendering and its quality.

2 Material and Methods

2.1 Data foundation

Our data set consists of 16 CT head images with a size of $547 \times 421 \times 547$ and a pixel spacing of 0.2mm from a test group of predominantly females aged between 40 and 50 years. The dental data were captured in the cranial region extending from the chin to the zygomatic arch area but some of the studied volumes include only a section of the mandible. The data were provided in anonymized DICOM format by our cooperation partner Düerr Dental. The Vista-Soft software was used to gather the data with a tube power of 83.7kW (0.9A·93kV) and an exposure time of 16.4s or

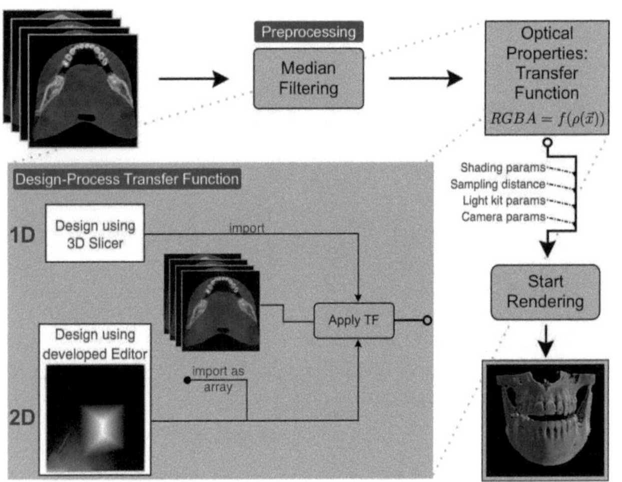

Figure 1: Pipeline of the proposed Algorithm.

117.6kW (1.2A·98kV) for 11s, respectively. The SNR of the dataset ranges from -0.82dB - 1.54dB.

2.2 Implementation

Fig. 1 illustrates our developed volume rendering pipeline based on the Visualization Toolkit (VTK) [8]. To obtain an improved image quality, an edge-preserving smoothing was first introduced by applying a median filter with a kernel size of $5 \times 5 \times 5$.

After preprocessing of the volume the TF was defined. In a first approach we designed 1D TFs using 3D Slicer, an open-source program based on VTK [9]. 3D Slicer delivers an interactive GUI to design a 1D TF with live preview. With the export feature the TF can be imported in our pipeline, where it can be read as two piecewise defined functions for opacity- and color-classification.

The second approach employs the complete design loop of a 2D TF. Our developed editor allows for the design of a 2D TF based on polygonal and rectangular Regions of Interest (ROI) with the ability to assign color and predefined intensity gradients. The designed TF can both be exported and re-imported as a numpy array (see. Fig. 1). To define a ROI in the scalar-gradient-magnitude domain, we used Matplotlib. To ensure a smooth mapping from the data domain to an RGBA tuple, we calculated several intensity modulations for the alpha channel that can be applied to any polygonal shape, producing highly visually appealing renderings. Fig. 2 presents a visual demonstration of the various techniques. The red channel of the TF is superimposed with the alpha channel of the respective TF to visualize its definition range. Fig. 2a displays a TF constructed using the distance transform technique, enabling the creation of adaptable TFs. By utilizing the fast-marching method extension of the Scikit Python module, this method calculates the scalar value of the distance from every point in the mask to its contour. Additionally, the TF can be considered as a coordinate system. Suitable 1D functions, such as the square root function, can be mapped onto this coordinate system. This can be observed in Fig. 2b, where the

square root function $f(\vec{x}) = \sqrt{\vec{x}}$ is mapped onto the distance map $d(\vec{x})$. The third method erodes the binary mask first then smooths the result with a gaussian filter (see Fig. 2c). The kernels for both operations are being calculated automatically. Fig. 2d shows the last method, in which a shape-dependent gaussian intensity modeling was applied. To achieve this, a convolution with a gaussian kernel σ was executed within the mask. However, it is possible to obtain non-zero intensities outside the mask. To avoid black color assignments, the method fills the entire bounding box of the ROI with valid values if any non-zero intensities are detected outside of it.

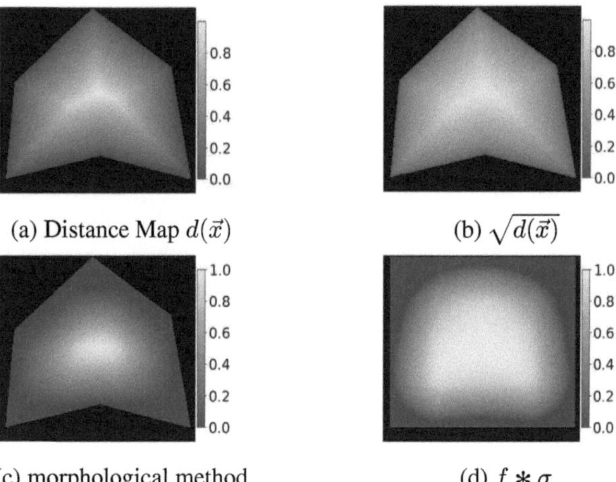

(a) Distance Map $d(\vec{x})$

(b) $\sqrt{d(\vec{x})}$

(c) morphological method

(d) $f * \sigma$

Figure 2: Four different transfer functions. **a)** shows the distance map. **b)** shows the mapping of the square-root onto the distance map. **c)** shows the morphological method. **d)** shows method of convolution within the mask.

After the application of the TF, the system configures several rendering parameters. The VTK shading model is based on the hybrid scattering model presented by Xu et. al. [7]. The model allows for user-specified mixing of local and volumetric shading. Our standard configuration was full volumetric shading unless otherwise specified. Next the components of the gradient based Phong shading model can be adjusted. To achieve a glossy appearance of bone and teeth of the data, setting the specular component high and the ambient component

Further the sampling parameter defines the distance of the sampling points during the ray casting process. A small parameter leads to fewer artifacts, such as the wood-grain effect, caused by under-sampling. However, this also reduces the frame rate. This issue can be addressed using an internal feature of the *vtkRenderer*, which jitters the sampling rays to replace the wood-grain effect with a noise texture. Running the standard configuration the sample distance is set to 1 and ray jittering is enabled during rendering. But when saving a rendering the sample distance is set to 0.1 to reduce artifacts.

Another crucial factor for a high-quality rendering result is the design of virtual light sources. Our lighting configurations were pre-set and utilize the VTK LightKit, which consists of four components: key-, fill-, back-, and headlight. This provides simple but high quality lighting. Furthermore, we developed a sub-application to adjust the parameters of the LightKit interactively. Our second preset is inspired by the work of Zhang and Ma [10]. Adjusting the backlight color enhances the perceived depth and emphasizes the significance of lighting design within cinematic rendering approaches [10].

In the last step we performed the rendering using the *vtkG-PUVolumeRayCastMapper* as an interface between the data and the graphics pipeline, as it supports 2D TFs unlike the *vtkSmartVolumeMapper*.

To test the capabilities of the VTK scattering model in the context of realism, we used a translucency test, by placing a light source with 100% intensity on the dorsal side of the volume and the camera on the lateral side of the volume. The volume was rendered using a TF with simple material assignment without smoothed alpha channel and with an anisotropy of zero and truncated the volume laterally by half. We doubled the transparency of the medium.

All renderings were performed on a Windows machine with Nvidia GeForce RTX 3080 Ti GPU and AMD Ryzen 9 5900X 12-core CPU. All renderings depicted are saved with high sampling rate, which does not correspond to the rendering seen during interaction.

3 Results

3.1 Translucency test

Fig. 3 shows the result of the translucency test where Fig. 3a and 3b where rendered with an opacity of 100% and 50% respectively. It can be seen that when the material is fully opaque (Fig. 3a), no light penetrates into the bone interior. Furthermore, the wood-grain effect can be observed (Fig. 3a bottom right). In Fig. 3b and 3c the *Ray Jittering* (see. 2.2) was applied. At an opacity of 50%, light can penetrate the volume and is scattered forward. Halving the opacity

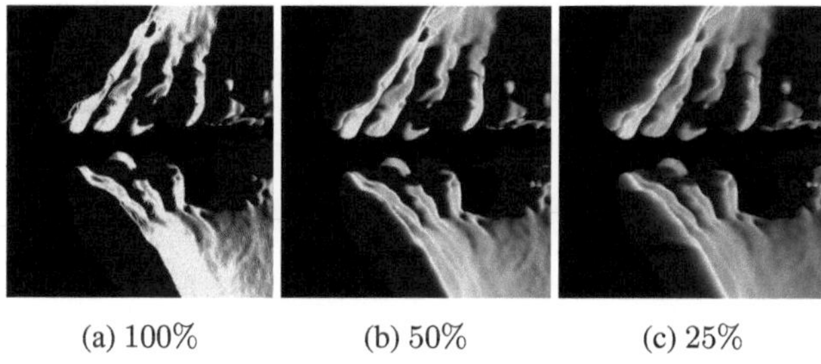

(a) 100% (b) 50% (c) 25%

Figure 3: Renderings of the translucency test of upper right and lower right incisor. **a)** Rendering with fully opaque volume. **b)** Rendering with 50% volumetric opacity. **c)** Rendering with 25% volumetric opacity.

to 25% (Fig. 3c) further intensifies the forward scattering (transmission) and softens its appearance. Moreover, we observe that regardless of the ray jittering activated, the image quality is affected by both the wood-grain artifacts and the noise texture. In our observations, we were able to reduce these artifacts by decreasing the sampling distance.

3.2 Renderings

Fig. 4 shows several renderings obtained by applying our previously described pipeline (see Fig. 1) on a sample image from the data set. The subject is a 21-year-old woman. Fig. 4a illustrates the result of a 1D TF that offers a fast color assignment for the design process of bones and teeth. In contrast, Fig. 4b shows a 2D TF that utilizes the morphological method to achieve accurate classification of bones and teeth. Similar to this, Fig. 4c demonstrates a 2D TF with an additional tone mapping operator during post-rendering, resulting in a clearer representation of colors. Fig. 4d shows the same TF with a higher weighting of local shading, thus producing secular highlights on the image. Fig. 4e illustrates the distance map TF, allowing better highlights of tooth features while displaying the teeth very clearly.

4 Discussion and Conclusion

Overall, our pipeline was designed to facilitate high-quality renderings within a modular framework. The presented pipeline cannot be fully classified as a cinematic renderer, as it lacks fully global illumination and multiple scattering features. Achieving real-world effects, such as color bleeding, subsurface scattering, depth-of-field, and environmental lighting, requires a stochastic ray casting framework that simulates millions of possible light paths. This type of framework is presented by [6]. These effects create slight color gradients that enhance the authenticity of the rendered image. Although VTK supports environmental lighting and depth-of-field features, they are not yet supported by the volume mapper and are only available for surface objects. Further developments could involve integrating render passes and pre-existing stochastic ray casting engines into our pipeline. Additionally, it is possible to automate the tedious process of designing a TF using a convolutional neural network.

In the context of cinematic rendering, we achieved promising results working with the standard VTK library and adapt it to our requirements. The developed TFs appear promising for utilizing the appearance of the renderings.

Acknowledgement

The work has been carried out at Fraunhofer IMTE in cooperation with Düerr Dental.

Author's Statement

Conflict of interest: Authors state no conflict of interest. Informed consent: Informed consent has been obtained.

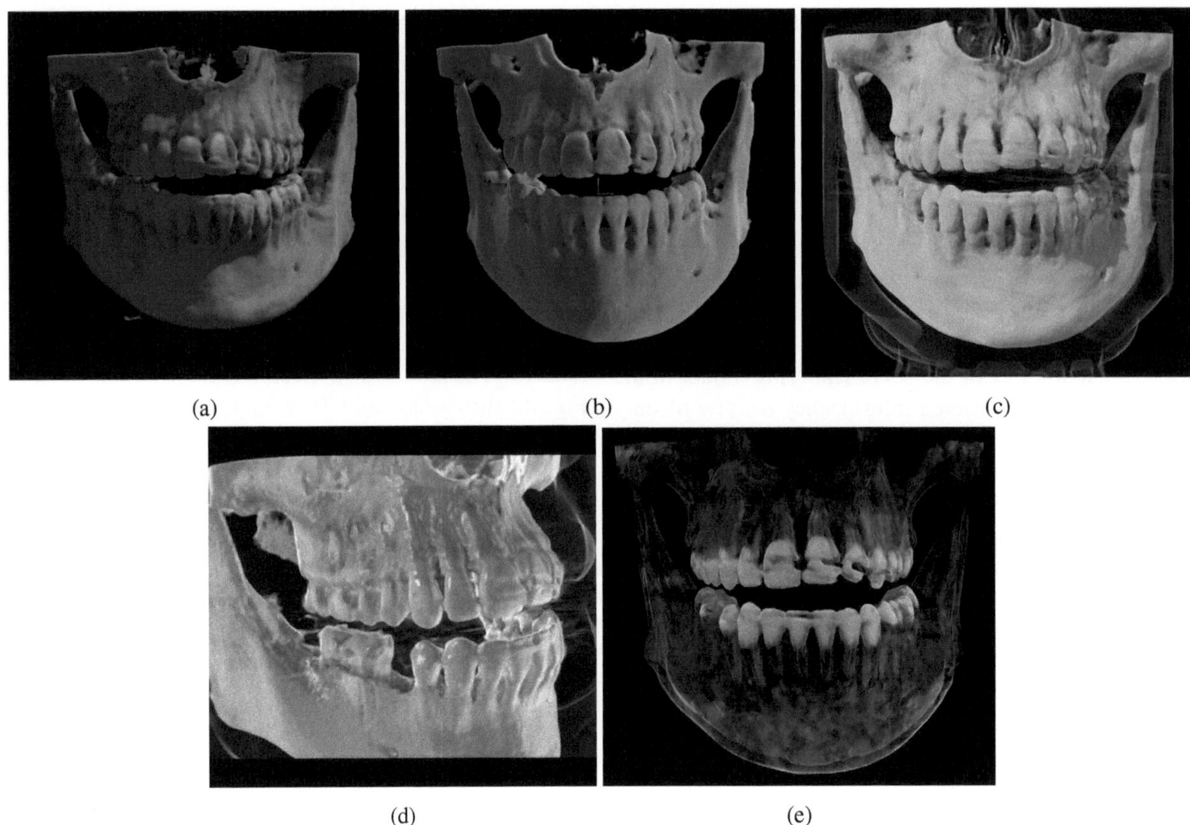

(a)   (b)   (c)

(d)   (e)

Figure 4: The figure displays various renderings of the proposed Pipeline, each with different configurations of TF, shading, lighting parameters, and post-processing. In **a**), a 1D TF is used. In contrast, **b**) uses a 2D TF in combination with the pre-set LightKit configuration. In **c**), a rendering with a 2D TF and tone mapping applied is shown. A similar configuration is used in **d**), where the shader is weighted with local shading. Finally, **e**) displays a different 2D TF.

5   References

[1] T. Steffen, S. Winklhofer, F. Starz, D. Wiedemeier, U. Ahmadli, and B. Stadlinger, "Three-dimensional perception of cinematic rendering versus conventional volume rendering using ct and cbct data of the facial skeleton," *Annals of Anatomy*, vol. 241, 2022.

[2] D. Comaniciu, K. Engel, B. Georgescu, and T. Mansi, "Shaping the future through innovations: From medical imaging to precision medicine," *Medical Image Analysis*, vol. 33, pp. 19–26, 2016.

[3] N. Max, "Optical models for direct volume rendering," *IEEE Transactions on Visualization and Computer Graphics*, vol. 1, no. 2, pp. 99–108, 1995.

[4] M. Levoy, "Display of surfaces from volume data," *IEEE Computer Graphics and Applications*, vol. 8, no. 3, pp. 29–37, 1988.

[5] D. Jönsson, E. Sundén, A. Ynnerman, and T. Ropinski, "A survey of volumetric illumination techniques for interactive volume rendering," *Computer Graphics Forum*, vol. 33, no. 1, pp. 27–51, 2014.

[6] T. Kroes, F. H. Post, and C. P. Botha, "Exposure render: An interactive photo-realistic volume rendering framework," *PLOS ONE*, vol. 7, no. 7, p. e38586, 2012.

[7] J. Xu, G. Thevenon, T. Chabat, M. McCormick, F. Li, T. Birdsong, K. Martin, Y. Lee, and S. Aylward, "Interactive, in-browser cinematic volume rendering of medical images," *Computer Methods in Biomechanics and Biomedical Engineering: Imaging & Visualization*, vol. 11, no. 4, pp. 1019–1026, 2023.

[8] W. Schroeder, K. Martin, and B. Lorensen, *The Visualization Toolkit (4th ed.).*   Kitware, 2006.

[9] A. Fedorov, R. Beichel, J. Kalpathy-Cramer, J. Finet, J.-C. Fillion-Robin, S. Pujol, C. Bauer, D. Jennings, F. Fennessy, M. Sonka, J. Buatti, S. Aylward, J. V. Miller, S. Pieper, and R. Kikinis, "3d slicer as an image computing platform for the quantitative imaging network," *Magnetic Resonance Imaging*, vol. 30, no. 9, pp. 1323–1341, 2012.

[10] Y. Zhang and K. L. Ma, "Lighting design for globally illuminated volume rendering," *IEEE Transactions on Visualization and Computer Graphics*, vol. 19, no. 12, pp. 2946–2955, 2013.

VTK-Enabled Depth Rendering for Enhanced Bronchoscopy Guidance Using CT Images

Sai Pavan Kumar Reddy Nusuma[1], Marian Himstedt [2]
[1] Robotics and Autonomous Systems, Universität zu Lübeck, sai.nusuma@student.uni-luebeck.de
[2] Institute for Medical Informatics, University of Lübeck, marian.himstedt@uni-luebeck.de

Abstract

Bronchoscophy is a medical procedure used to examine and visualize the airways of the lungs for lesion detection. Depth image rendering plays a vital role in enhancing the accuracy and effectiveness of bronchoscophy. However, traditional methods like stereo camera, structure from motion (SFM) and simultaneous localization and mapping (SLAM) proved to impractical due to the limitations with size and accuracy. To address these limitations, we use Visualization Toolkit (VTK) a powerful visualization library which provides a robust environment for Visualizing complex 3D scenes making it suitable for depth image rendering. By integrating CT scans with VTK it provides detailed anatomical information, allowing for more accurate depth image rendering.

1 Introduction

Lung cancer is a significant global health concern, being the leading cause of cancer-related deaths worldwide. According to the World Health Organization (WHO), almost 1.8 million people died from lung cancer in 2020, accounting for the highest mortality rates[1]. The International Agency for Research on Cancer (IARC) also highlights that lung cancer has been and remains the most common cause of cancer death globally.However, most of the lesions found are benign and it is important to distinguish between benign/cancer lesions. Nodule biopsy is needed for the diagnosis confirmation. Bronchoscopy, also known as bronchoscopic navigation is a minimally invasive procedure that is commonly performed to diagnose and treat various pulmonary disorders. The procedure involves navigating a flexible bronchoscope through intricate airway structures to visualize and examine different regions of the lung as shown in Fig 1. Accurate depth information is essential for precise interventions, enhanced spatial perception, and improved diagnostic accuracy, making depth image rendering a critical aspect of bronchoscopy.

Rendering depth images involves generating images where pixel values represent the distance from the camera to the corresponding points in the scene. This is often done using depth maps, where each pixel's intensity is proportional to its distance from the camera. The lack of ground truth depth measurement in procedures like bronchoscopy, poses a significant challenge, inherently lacks the provision of depth information. Traditional imaging modalities used in bronchoscopy, such as RGB images, do not capture the third dimension, hindering the ability to precisely determine distances and spatial relationships within the airways. The absence of ground truth depth measurements complicates tasks like navigation through complex anatomical structures, accurate intervention targeting, and comprehensive spatial understanding. To address this limitation, alternative methods, such as depth image rendering through computational techniques or the incorporation of prior knowledge from imaging modalities like CT scans, become essential for overcoming the inherent challenges associated with the lack of ground truth depth in procedures like bronchoscopy.

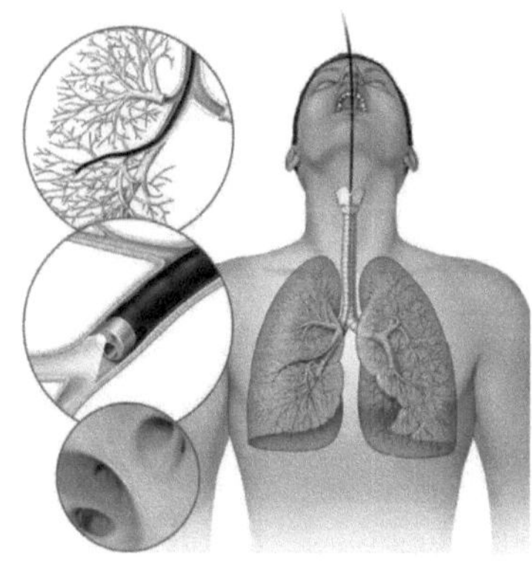

Figure 1: Bronoscophy[2]
Visualizing the airways by using Broncoscophy

2 Material and Methods

2.1 Traditional Methods to Capture Depth

2.1.1 Normal Camera

The projection of a 3D object onto the image plane involves capturing the spatial information of the object and representing it in two dimensions. This process simulates how light rays from the 3D object interact with a camera, where

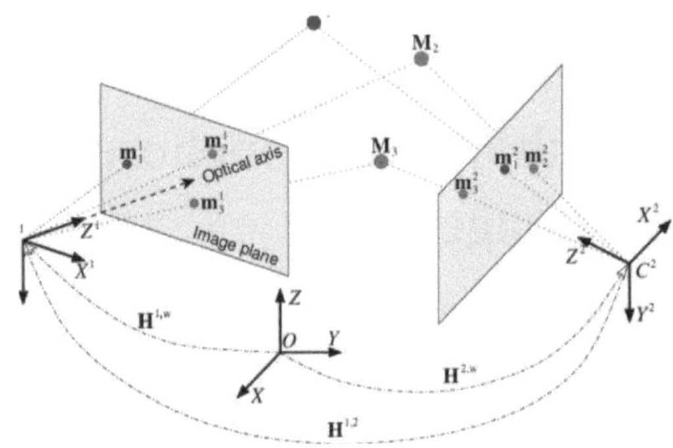

Figure 2: Camera Projection Model [3]
showing the projections of a point from different view points
similar to stereo camera.

each point on the object corresponds to a point on the image plane. Mathematical transformations, such as *perspective projection*, are used to perform this projection, wherein points closer to the camera appear larger than those farther away. The image sensor records the light intensity at each pixel, creating a mirrored representation of the 3D scene. Depths are inferred from the size and position of objects in the image, with objects closer to the camera appearing larger and those farther away appearing smaller. This information helps determine the relative depths of different objects within the scene, and it is a crucial parameter for creating accurate and realistic 3D models by understanding how objects are positioned in relation to the camera[3]. The distance from the object to the camera is a key factor in 3D reconstruction, influencing the size and position of objects in the 2D image. By capturing images from multiple viewpoints, the differences in perspectives can be used to triangulate the 3D position of points in the scene. Triangulation, a fundamental technique in 3D reconstruction, involves determining the 3D position of a point by finding the intersection of rays from multiple camera perspectives as shown in Fig 2. which created a basis for using *Stereo camera* and *Structure From Motion* setups, This technique enables accurate localization of points and forms the basis for reconstructing the geometry of a scene. Cameras with planar motion can adversely impact 3D reconstruction from multiple views by causing a lack of scene parallax, leading to the loss of depth information and the introduction of ambiguities in the reconstruction process. To mitigate these challenges, techniques such as using non-planar camera motions, incorporating additional depth cues, or leveraging prior knowledge about the scene geometry can be employed to improve the accuracy of 3D reconstruction from multiple views[3].

2.1.2 Stereo Camera

Stereo vision involves the use a pair of cameras positioned at a certain baseline distance from each other. This configuration allows the cameras to capture slightly different perspectives of the same scene. By analyzing the disparities (the differences in the images), depth information can be computed through triangulation.However, its application in medical settings, such as bronchoscopy, is constrained by the limited space within bronchoscopes. The small diameter of these instruments restricts the distance between the two cameras, reducing the baseline and consequently impacting the accuracy and effectiveness of stereo vision.

2.1.3 Structure from Motion(SFM)

Structure from Motion (SfM) is a related technique that deals with the reconstruction of 3D structures from 2D images. and its working involves reconstructing 3D structures from 2D image sequences. The limitation of SFM is that it is inaccurate at similar kind of environment(dynamic environment) like airway anatomy in bronchoscophy[4].

2.2 VTK(Visualization Toolkit)

2.2.1 VTK Environment

The Visualization Toolkit (VTK) is an open-source software system for 3D computer graphics, modeling, image processing, volume rendering, scientific visualization, and information visualization. It provides a large C++ class library and supports interpreted interfaces in languages such as Tcl/Tk, Python, and Java, making it user-friendly for programming and interactive controls. VTK includes ancillary support for 3D interaction widgets, two- and three-dimensional annotation, and parallel computing. The core functionality of VTK is written in C++ and Python, and it is wrapped into other language bindings to maximize efficiency and expose it to a wider audience. VTK is used by academicians for teaching and research, as well as by government research institutions and commercial firms. It offers a comprehensive API reference section for C++, Python, and CMake API, along with a User's Guide and a textbook for further in-depth understanding. VTK's impact is highlighted by its role as a fundamental tool for visualization, with applications ranging from teaching to specialized visualization tasks, such as Depth Image rendering in bronchoscopy[5].

VTK (Visualization Toolkit) offers significant advantages over stereo cameras, structure-from-motion (SfM) methods. VTK provides specialized visualization capabilities tailored for 3D scene rendering and processing, offering advanced techniques which are not found in stereo cameras or SfM. Its flexibility allows precise control over camera parameters and rendering techniques, enabling tailored solutions for diverse applications. Additionally, VTK's scripting capabilities streamline complex tasks and integration into larger systems, supported by extensive documentation and community support. With cross-platform compatibility and seamless integration with other tools, VTK offers advanced rendering techniques for visualizing complex 3D data sets and interpreting depth maps effectively.

2.2.2 VTK Camera Model

The camera model used in VTK is represented by the vtkCamera class, which is a virtual camera for 3D rendering. It

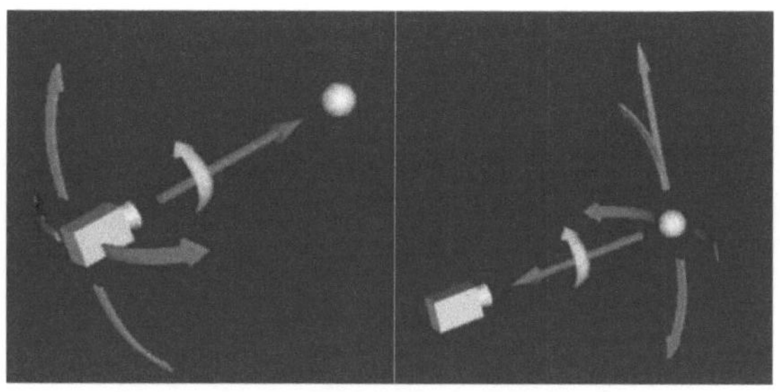

Figure 3: VTK camera model [5]
Left: Camera movements around focal point.
Right: Camera movements centered at camera position

provides methods to position and orient the view point and focal point. The camera model includes support for more complex methods such as the manipulation of the computer graphics model, view up vector, clipping planes, and camera perspective. Additionally, it offers convenience methods for moving about the focal point and allows the application of transforms to the camera as in Fig 3. The vtkCamera class is essential for controlling the viewpoint and perspective in 3D scenes, making it a fundamental component of the VTK rendering pipeline[5].

2.3 Pipeline Description

2.3.1 Data Set

To delve further into Depth Image Rendering in bronchoscopy, it is essential to acknowledge challenges. The scarcity of data, particularly the absence of ground truth depth on video data and limited availability of paired data with corresponding CT scans, presents a significant hurdle. We Start with the acquisition of high-resolution CT images of the anatomical region of interest, which serves as a prior to Generate Depth images using VTK.

2.3.2 Airway Segmentation

Implementing an airway segmentation algorithm to isolate the airway structures from the CT scans. This involves thresholding, region growing and other segmentation techniques tailored to airway identification as shown in Fig 3.

2.3.3 Centerline Extraction

The VTK library is used to extract center line for the segmented airways and plotting a graph from the extracted centerline, The AirwayCenterlineExtraction algorithm[6] is used to extract the centerline from the segmented airway as shown in Fig 5. It utilizes various libraries such as VTK, scikit-image, scipy, nibabel, numpy, igraph, and networkx for image processing, graph generation, and analysis. The algorithm includes methods for extracting the centerline, updating the extraction, generating paths, performing random walks, and other related functionalities for airway centerline extraction.

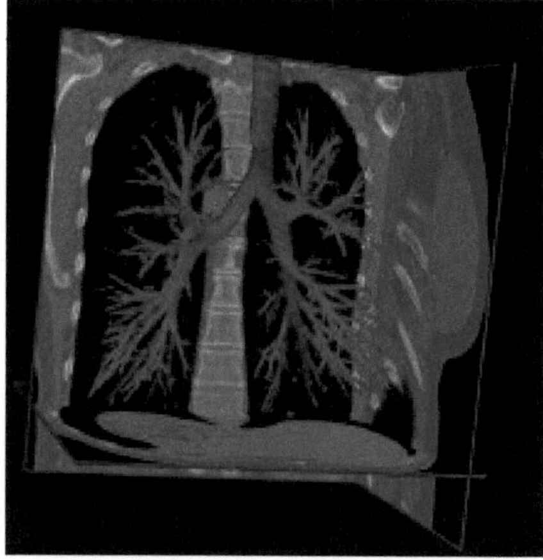

Figure 4: Segmented Airways
Airways Segmented from 3D CT

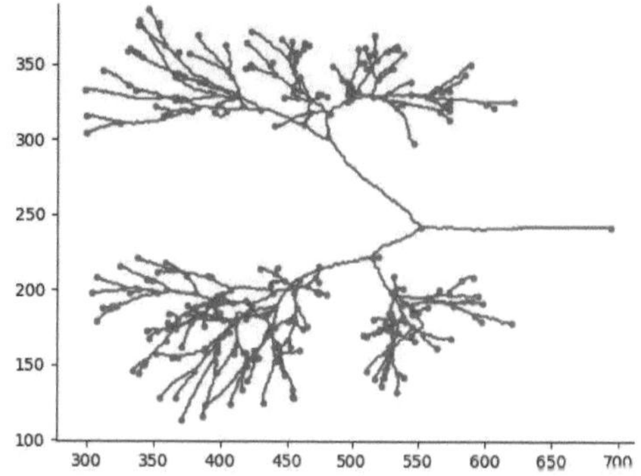

Figure 5: Center line along the Airways[6]
The Graph indicates the Center line in the airways and points indicate the bifurcations and egdes along the path.

2.3.4 Path Generation for Camera

Generating the path or vtkcamera path coordinates starts from manipulating the obtained Centerline path. The path generation algorithm[6] is used to create a path along the extracted centerline, with customizable parameters. The algorithm also includes functionality for visualizing the generated path and airway mesh using existing libraries and tools such as Matplotlib and 3D scatter plots, Additionally, the generated path is aligned with the coordinate frame of the given airway object(3D) as shown in Fig 6, involving translation, rotation, and manual offset of the path points. It is part of a broader framework for airway segmentation and centerline extraction. This path represents the trajectory the camera will follow during the simulation to generate Depth maps along the path of a given coordinate to a particular node in the airways. Considerations may include avoiding collisions with airway walls and ensuring coverage of the entire airway system.

3 Results and Discussion

Simulating the movement of the VTK camera along the generated path, depth maps are generated at each coordinate

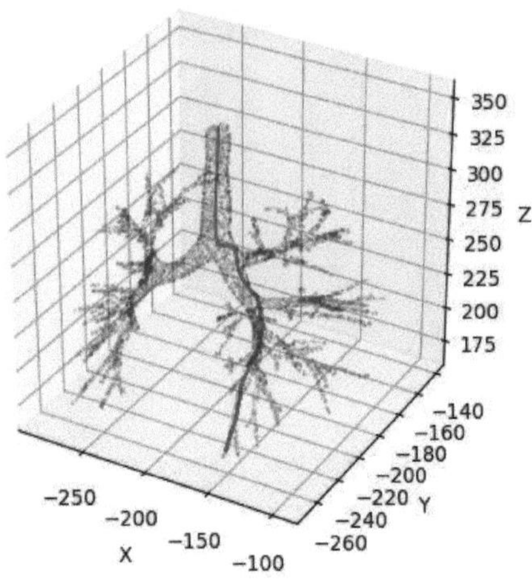

Figure 6: Camera path inside the airways[6]
Graph represents the camera path along the the airways to a
particular node.

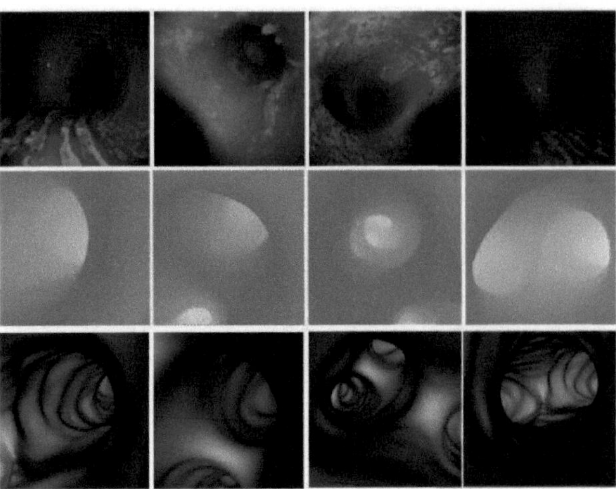

Figure 7: Depth Maps
Depth maps generated by vtk along the path inside the airways
generated from CT images and compared with RGB images.
Note:Depth maps(2nd row) and virtual bronchoscophy
images(3rd row) are paired but not RGB images(1st row).

along the airway mesh's trajectory, as depicted in Fig. 7. Depth image rendering using VTK on CT images represents a significant advancement in enhancing the accuracy and effectiveness of bronchoscopy compared to existing methods. The integration of VTK with CT scans enables the utilization of detailed anatomical information, resulting in more precise depth image rendering. This innovative aspect of VTK's performance distinguishes it from conventional methods. Unlike traditional techniques that may lack the depth and accuracy required for effective bronchoscopy, VTK leverages advanced visualization techniques and customizable rendering pipelines to provide real-time interactive visualization of 3D data sets. Moreover, VTK's scripting capabilities and integration with machine learning and AI frameworks enhance its capabilities for data analysis and visualization. Overall, the integration of VTK with CT scans offers a novel approach to depth image rendering in bronchoscopy, providing clinicians with detailed anatomical insights and improving diagnostic accuracy. Through its innovative features and advanced visualization techniques, VTK sets a new standard for depth image rendering in medical imaging applications, paving the way for more precise and effective diagnostic procedures.

4   Conclusion

The use of VTK for depth image rendering has proven instrumental in generating precise depth maps within the context of bronchoscopy.This approach overcomes the limitations of traditional methods such as Stereo Camera and Structure from Motion (SFM), which were found to be impractical due to size and accuracy constraints. Leveraging VTK's capabilities, we successfully translated the intricate airway anatomy into detailed depth information, providing a foundation for Deep learning-based depth estimation is simplified through providing paired image data (RGB images and their corresponding depth maps).

Acknowledgement

The work has been carried out at the Institute of Medical Informatics, and supervised by Dr. Marian Himstedt, Institute of Medical Informatics, Universität zu Lübeck

Author's Statement

Conflict of interest: Authors state that they have no conflict of interest.

5   References

[1] World Health Organization, "Lung cancer," 2023. Accessed: January 10, 2023.

[2] G. A. Lillington, "Bronchoscopy," *Mayo Clinic Proceedings*, vol. 70, no. 2, p. 201, 1995.

[3] R. Hartley and A. Zisserman, *Multiple View Geometry in Computer Vision*. Cambridge University Press, 2 ed., 2004.

[4] M. Visentini-Scarzanella, T. Sugiura, T. Kaneko, and S. Koto, "Deep monocular 3d reconstruction for assisted navigation in bronchoscopy," *International journal of computer assisted radiology and surgery*, vol. 12, 05 2017.

[5] W. Schroeder, K. Martin, and B. Lorensen, *The Visualization Toolkit (4th ed.)*. Kitware, 2006.

[6] E. Smistad, A. C. Elster, and F. Lindseth, "Gpu accelerated segmentation and centerline extraction of tubular structures from medical images," *International Journal of Computer Assisted Radiology and Surgery*, vol. 9, pp. 561–575, July 2014. Epub 2013 Nov 1.

Examining the synthesization of physically plausible MRI images

Hassan Shayea [1], Christian Weihsbach [2], Ziad Al-Haj Hemidi [2], and Mattias P. Heinrich [2]

[1] Robotics and Autonomous Systems, University of Lübeck, h.shayea@student.uni-luebeck.de

[2] Institute of Medical Informatics, University of Lübeck, {christian.weihsbach, z.alhajhemidi, mattias.heinrich}@uni-luebeck.de

Abstract

Magnetic Resonance Imaging (MRI) offers excellent soft tissue contrast, adjustable through different magnetic gradient sequences, enhancing visualization of specific structures. Deep learning methods, now standard in MRI analysis, require extensive datasets for tasks like image registration, segmentation, and reconstruction. This study introduces a synthesis method for MR images addressing data scarcity, leveraging Bloch equations for creating neuro and cardiac images from semantic labels and T_1/T_2 maps. We simulated four distinct MR sequences and achieved high Structural Similarity Index Measure (SSIM) values when synthesizing brain images from label maps compared to reference images. However, cardiac imaging synthesis could have been more successful, reaching an SSIM of 60%, due to dataset constraints and insufficient structural content in label maps. Future work is needed to improve inputs with more structural content, like utilizing edge maps and obtaining better ground truth by examining existing data or acquiring data with known sequence properties.

1 Introduction

Magnetic Resonance Imaging (MRI) is a non-invasive diagnostic tool enabling high-resolution imaging that avoids ionizing radiation [1]. It is ideal for studying brain structures, detecting tumors and chronic diseases [1]. MRI is also essential in cardiology for capturing detailed heart images [2]. Deep learning (DL) is used in the context of MR imaging to solve downstream tasks, such as image registration, segmentation, and enhancing reconstruction from undersampled k-space data [3]. Yet, DL faces challenges like overfitting due to limited data and the need for accurate, labor-intensive data labeling [3]. Models trained on repetitive datasets may not properly represent the wide range of possible MR sequences with different weightings. Considering these challenges, synthetic datasets have become a promising solution. Synthetic datasets can expand training sets by augmenting data, as demonstrated by BrainWeb's simulated images [4] and MRXCAT tools [5]. These advancements are considered in this study, where we combine physical principles with DL techniques to gain more control over synthetic image generation. This study proposes an approach to MR image synthesis that combines physics-based modeling and deep learning. The paper presents a method for reconstructing brain MR images from semantic labels using the Bloch equation, adjusted to the specific parameters of different brain tissues. Additionally, the study explores integrating DL techniques to improve the realism of synthesized cardiac images. Unlike the existing synthesization pipelines such as MRXCAT 2.0, our approach targets the prediction of separable tissue parameter maps of T_1 (longitudinal relaxation time) and T_2 (transverse relaxation time) that enable the dynamic synthesization of different MR contrast settings afterwards.

2 Material and Methods

In this work, we investigate the generation of intensity images using the principles of MRI physics. This process involves two primary approaches. In the first method, we create images using broad tissue labels based on tissue properties. In the second scenario, we aim to refine the initial results by addressing inaccuracies in the first phase through a voxel-label T_1/T_2 mapping technique. This mapping technique is crucial in enhancing image accuracy by adjusting the initial labels based on the intrinsic properties of tissues. Further details regarding these methods are provided in the following sections.

2.1 MRI synthesis based on semantic labels

We follow a pipeline similar to BrainWeb [4] for the physics-only approach, as depicted in Fig. 1. We begin with the coarse semantic labels from the *twenty anatomical models of the twenty normal brain* dataset [4], a dataset constructed from MRI scans of 20 healthy adults to capture variations in anatomy [4]. These labels serve as templates where tissue properties are mapped to each corresponding label. Using the Bloch equation Eq. 1, we then calculate the intensity values for these labels. Given the challenge of precise T_1, T_2, and Proton Density (PD) parameter acquisition, we apply histogram matching in postprocessing to align the synthesized images more closely with ground truth data. We used histogram matching in postprocessing due to the lack of ground-truth tissue parameters.

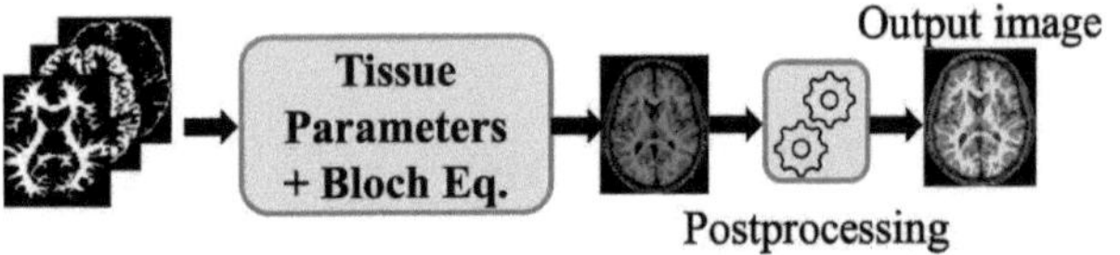

Figure 1: The process of generating images involves using the Bloch equation with weighting specific parameters, followed by post-processing techniques.

The image synthesis process leverages the Bloch equation [6], which models the behavior of magnetization vectors as fundamental to generating MRI signals. By adjusting parameters like Repetition Time (TR), Echo Time (TE), and flip angle, we can synthesize a variety of images, each highlighting different tissue contrasts. In our synthesis, we employ the following imaging techniques: Spoiled Gradient Echo (Spoiled GRE), Balanced Steady-State Free Precession (bSSFP), Inversion Recovery (IR), and Spin Echo (SE) to generate T_1 and T_2 weighted images. To generalize across these techniques, we use the following representation of the Bloch equation without modeling diffusion. [6]:

$$\frac{d\mathbf{M}}{dt} = \gamma(\mathbf{M} \times \mathbf{B}) - \frac{M - M_0}{T_2} - \frac{M - M_0}{T_1} \quad (1)$$

where M is the magnetization vector, M_0 is the equilibrium magnetization vector, γ is the gyromagnetic ratio, B is the magnetic field.

2.2 MRI synthesis based on fine T1/T2 maps

In contrast to the only-physics approach, where the tissue values are set, we explore the prediction of such parameters using maps. To generate T_1 and T_2 maps, a set of images with varying T_1 or T_2 weightings is captured. For T_1 mapping, this involves capturing images at different inversion times, which is essential for observing the recovery process of longitudinal magnetization. On the other hand, in the case of T_2 mapping, images are taken at different echo times to observe the decay of transverse magnetization. After capturing these images, a curve fitting process [9] is carried out to calculate the T_1 and T_2 relaxation times. Once the relaxation times are calculated, T_1 and T_2 maps are created by assigning the corresponding T_1 and T_2 values to each voxel. The modified Look-Locker inversion recovery (MOLLI) sequence is used for T_1 mapping, while the Flash sequence is used for T_2 mapping, as suggested by [7]. Further details on the mapping techniques can be found in the referenced literature [7] [8]. The method for reconstructing T_1- and T_2-weighted cardiac images is based on a dual-model approach. This strategy is summarised in Fig. 2. Here, the two parts of the method are separated by the Bloch equation acting as a connector, which translates the parameter output from the first model into an initial image, which the second model then refines. Following the implementation of MOLLI and FLASH sequences, we estimated T_1 and T_2 target maps to train our first model. The chosen model for

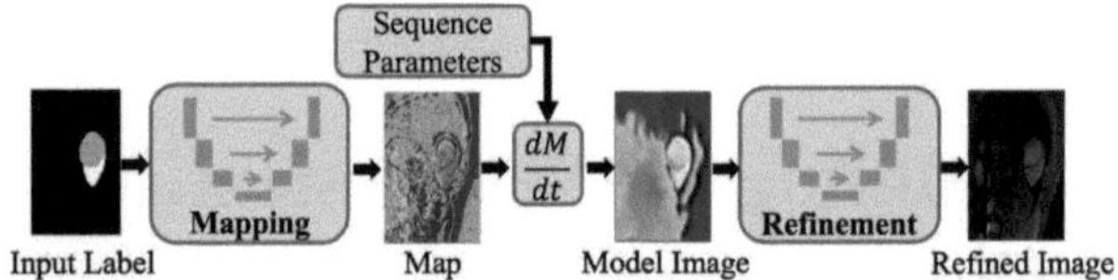

Figure 2: The mapping model takes coarse label masks as input, generating T_1 or T_2 maps depending on the sequence. Along with sequence parameters, these maps are integrated into the Bloch equation $\frac{dM}{dt}$, producing initial images which are then further processed by the refinement model.

MRI synthesization was a U-Net, starting with 32 channels doubled in each layer. The model consists of four pooling layers for depth, incorporating convolution, batch normalization, leaky ReLU activation, and dropout for regularization. Up-sampling halves the channels to return to the initial size, with final dimension adjustments. The model is trained using a weighted combination of Mean Squared Error (MSE) and Structural Similarity Index Measure (SSIM) losses. These losses were calculated between the predicted and ground truth intensity images and between the predicted and generated maps with curve fitting, as seen in Eq. 2.

$$L = \alpha \cdot \frac{1}{2}\|P - I\|_2^2 + (1 - \alpha) \cdot (1 - \text{SSIM}(P, I)) \quad (2)$$

The equation defines a weighted loss combining MSE between predictions P and ground truth I with the addition of SSIM, balanced by a parameter α. The first model translates coarse labels into detailed T_1 and T_2 maps, using MSE for accuracy and SSIM for structural integrity. This process improves the quality of the input for further processing. The Bloch equation then generates initial images using these maps and sequence parameters. A second phase refines the initial images for higher quality, using MSE and SSIM again. This end-to-end methodology ensures an integrated process from initial mapping to refined output.

3 Experiments and Results

The analysis is divided into two parts. Firstly, we will explore the synthesis of brain images generated from semantic labels. We have generated T_1 and T_2-weighted images using different techniques. Spoiled GRE and bSSFP techniques synthesize T_1-weighted images, while both bSSFP and SE sequences synthesize T_2-weighted images. The optimal parameters for synthesizing weighted MRI images are determined through a two-step process involving parameter space exploration and selection based on SSIM value compared to the original images. The Spoiled GRE method used parameters of TR: 780 ms, TE: 15 ms, and flip angle: $80°$, while bSSFP was optimized using TR: 6 ms, TE: 1 ms, and flip angle: $20°$. For T2-weighted imaging, we adjusted the bSSFP settings to TR: 180 ms, TE: 100 ms, flip angle: $3°$ and used a TR of 5000 ms, TE of 120 ms, and flip angle: $6°$ for the SE sequence of T2-weighted imaging. The quality synthesis of all techniques was evaluated through SSIM and PSNR metrics.

Secondly, we will discuss synthesizing cardiac images using fine T_1 and T_2 maps. To train the models, we combined different loss functions. We integrated MSE to target pixel-level precision and SSIM to evaluate structural similarity with parameter α equals 0.5, as discussed in Sec. 2.2. The models are trained for 700 epochs to ensure it is learned thoroughly from the dataset. We used Prodigy optimizer with an initial learning rate of 1 [10]. The optimizer dynamically adjusts learning rates to improve performance. To prevent overfitting and enhance the models generalization capability, we applied a dropout rate of 0.2. Following the preparation of T_1/T_2 maps, as discussed in Sec. 2.2, the *CMRxRecon* dataset of 120 scans was partitioned into three subsets: 70% for training, 20% for validation, and 10% for testing. Afterward, the end-to-end model is trained to achieve optimal results. This evaluation involved comparing the Mapping and Refinement models. We utilized SSIM and PSNR to evaluate the accuracy of model predictions compared to ground truth images.

3.1 MRI synthesis based on semantic labels

The SSIM, indicating structural similarity, consistently exceeded 88% for both T_1 and T_2-weighted images, with the highest reaching 96%. The PSNR values underlined the effectiveness of the synthesis process, especially with peaks at 73.45 dB for T_1 and 71.75 dB for T_2-weighted images, as illustrated in Fig. 3. These metrics collectively demonstrate

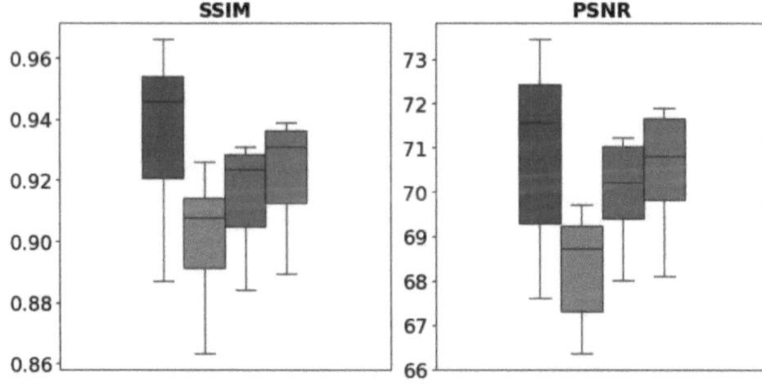

Figure 3: SSIM and PSNR evaluation metrics across ten subjects form the BrainWeb dataset. Different image types are represented by colors, emphasizing the variability of the image synthesis process across ■ Spoiled GRE T1, ■ bSSFP T1, ■ Spin Echo T2, and ■ bSSFP T2 techniques.

the capability of the methods employed to synthesize MRI images, as further illustrated in the following Fig. 4, which compares synthesized and target images for Subjects 2 and 9 for T_1 and T_2-weighted MRI images, respectively.

3.2 MRI synthesis based on fine T1/T2 maps

For T_1-weighted images, Fig. 5 illustrates that the Mapping model's outcomes improve with the Refinement model. This enhancement is evidenced by a higher average SSIM in

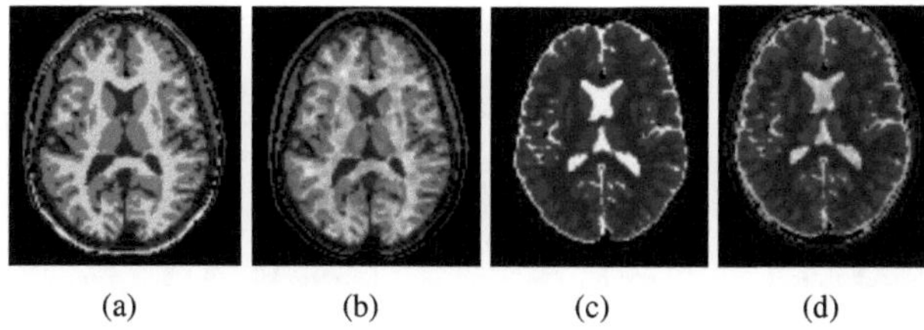

Figure 4: (a) Subject 2's Spoiled GRE synthesized vs. (b) target T_1-weighted, and (c) Subject 9's bSSFP synthesized vs. (d) target T_2-weighted images, illustrating the effectiveness of the synthesis process.

the Refinement model, indicating improved structural similarity. However, outliers, denoted by small circles in Fig. 5, show variability in SSIM scores across some samples. Additionally, the Refinement model achieved a PSNR increase to 29.0 dB, up by +17.1 dB, representing image quality.

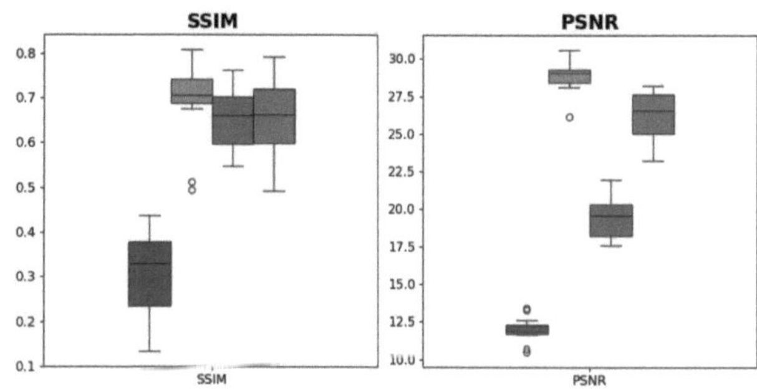

Figure 5: Evaluation metrics for the T_1 and T_2-weighted reconstructions across samples using both models. ■ represent the T_1 Mapping Model, ■ represents the T_1 Refinement Model, ■ represent the T_2 Mapping Model, and ■ represents the T_2 Refinement Model.

On T_2-weighted images, we observed improved image quality through higher SSIM and PSNR values, as seen in Fig. 5. The Mapping model's SSIM values showed variability, ranging from 55% to 76%. However, the Refinement model enhanced these scores, particularly in some samples. The Refinement model also showed its ability to enhance image quality by reducing noise, as it increased the PSNR values from 17.6 dB to as high as 28.2 dB.

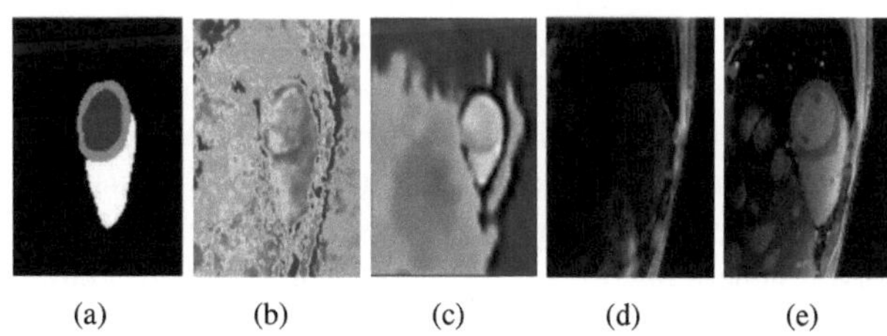

Figure 6: (a) Mapping Model input. (b) Mapping Model Output. (c) Initial image input for the Refinement Model was reconstructed using the Bloch equation. (d) Output of the Refinement Model. (e) Target T_1-weighted image.

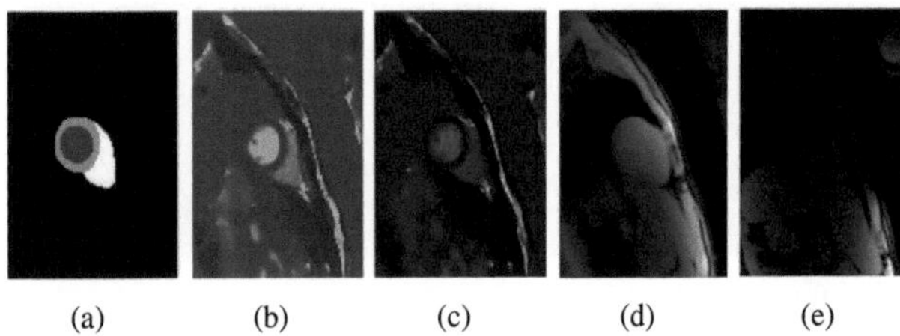

(a)   (b)   (c)   (d)   (e)

Figure 7: (a) Mapping Model input. (b) Mapping Model Output. (c) Initial image input for the Refinement Model was reconstructed using the Bloch equation. (d) Output of the Refinement Model. (e) Target T_2-weighted image, which shows the transformation process from label to image in the test dataset, displaying the results achieved with the Refinement model.

4 Discussion

The performance difference stems from the Mapping model's less efficient process of transforming semantic labels into T1 maps. In contrast, by leveraging actual images for training, the Refinement model yields more accurate results than the Mapping model's reliance on approximations. Fig. 6 shows this difference by illustrating the transformation process from label to image in the test dataset, displaying the results achieved with the Refinement model. In evaluating T2-weighted images, the Refinement model improved the outcome of the mapping model regarding SSIM values. However, the performance varied across samples, as depicted in Fig. 5. This figure illustrates the comparison of evaluation metrics for both models. Some samples showed the Mapping model performing better than the Refinement model, indicating an interaction between their functionalities, as further evidenced in Fig. 7. The effectiveness of the Mapping model in generating T2 maps influences the outcomes of the Refinement model. Accurately predicted T2 maps by the Mapping model tend to benefit the Refinement model's performance, highlighting that deficiencies in the Mapping model's outputs are not correctable by the Refinement model as illustrated in Fig. 7.

5 Conclusion

The results demonstrated that synthesizing brain images from coarse semantic labels was effective. We generated optimal T_1 and T_2-weighted images using four MR sequences, which we verified against the BrainWeb dataset. However, the challenge of determining accurate T_1, T_2, and PD values led us to the project's next phase. This phase involved generating T_1 and T_2 maps, which helped address the absence of specific tissue parameters and further improved the synthesis with pixel-level T_1 and T_2 tissue property maps. These maps enhanced semantic masks, particularly in cardiac areas. The iterative learning between models revealed that early-stage discrepancies negatively affect final image quality and highlighted the importance of accurate mapping. For future work, we aim to synthesize images from a real dataset and assess our model's accuracy against known parameters, validating our synthesis techniques. We will also use Generative Adversarial Networks (GANs) to improve image synthesis.

Acknowledgement

This work has been carried out in the scope of an internship at the Institute of Medical Informatics, University of Lübeck.

Authors' Statement

Conflict of interest: Authors state no conflict of interest.

6 References

[1] C. Wang et al., Recommendation for cardiac magnetic resonance imaging-based phenotypic study: imaging part, Phenomics, vol. 1, pp. 151170, 2021

[2] J. P. Ridgway, Cardiovascular magnetic resonance physics for clinicians: part I, Journal of cardiovascular magnetic resonance, vol. 12, no. 1, pp. 128, 2010.

[3] H. Zhang and Y. Qie, Applying Deep Learning to Medical Imaging: A Review, Applied Sciences, vol. 13, no. 18, p. 10521, 2023.

[4] C. A. Cocosco, online interface to a 3D MRI simulated brain database, NeuroImage, vol. 5, p. S425, 1997.

[5] S. Buoso, T. Joyce, N. Schulthess, and S. Kozerke, MRXCAT2. 0: Synthesis of realistic numerical phantoms by combining left-ventricular shape learning, biophysical simulations, and tissue texture generation, Journal of Cardiovascular Magnetic Resonance, vol. 25, no. 1, p. 25, 2023.

[6] M. A. Brown and R. C. Semelka, MRI: basic principles and applications. John Wiley & Sons, 2011.

[7] C. Wang et al., CMRxRecon: An open cardiac MRI dataset for the competition of accelerated image reconstruction, arXiv preprint arXiv:2309.10836, 2023.

[8] A. J. Taylor, M. Salerno, R. Dharmakumar, and M. Jerosch-Herold, T1 mapping: basic techniques and clinical applications, JACC: Cardiovascular Imaging, vol. 9, no. 1, pp. 6781, 2016.

[9] I. A. Dekkers et al., Consensus-based technical recommendations for clinical translation of renal T1 and T2 mapping MRI, Magnetic Resonance Materials in Physics, Biology, and Medicine, vol. 33, pp. 163176, 2020.

[10] K. Mishchenko and A. Defazio, Prodigy: An Expeditiously Adaptive Parameter-Free Learner, arXiv preprint arXiv:2306.06101, 2023.

Enhancing 3D Ultrasound Reconstruction: A Comparative Analysis Using IMU Data

Ahmad Mohammad [1], Lasse Hansen [2], and Mattias P. Heinrich [3],

[1] Medical Informatics, Universität zu Lübeck, ahmad.mohammad@student.uni-luebeck.de

[2] EchoScout GmbH, Lübeck, Germany, lasse@echoscout.ai

[3] Institute of Medical Informatics, University of Lübeck, Lübeck, Germany, mattias.heinrich@uni-luebeck.de

Abstract

Ultrasound's efficiency and safety make it a valuable imaging modality, yet 3D ultrasound adoption is hindered by cost-intensive equipment. This research addresses this limitation by exploring the feasibility of reconstructing 3D ultrasound images from economical 2D scans. Leveraging a publicly available optically tracked 3D ultrasound dataset, our study focuses on evaluating existing reconstruction algorithms and determining the impact of incorporating Inertial Measurement Unit (IMU) data. We simulate IMU data on the dataset, re-implement algorithms, and assess their performance with and without IMU input. Our work establish a foundation for comparing reconstruction algorithms, shedding light on the potential enhancements achievable through IMU integration.

1 Introduction

In the past years, the task of transformation estimation in the realm of ultrasound image analysis, specifically within the context of freehand ultrasound scans has been a formidable challenge for a lot of researchers. Historically, the primary focus of research endeavors has centered on understanding the transformation between individual image pairs. In contrast to this traditional approach, our work introduces a shift in perspective. We not only concentrate on the transformation between individual image pairs but also extend our focus to incorporate Inertial Measurement Unit (IMU) data. This additional dimension allows us to simulate IMU data on the dataset and explore the potential enhancements it brings to reconstruction results. By broadening the scope of our investigation, we aim to uncover new insights that go beyond the conventional understanding of transformation estimation in ultrasound image analysis.

2 Related Work

Freehand 3d Ultrasound scanning, along with mechanical scanning [1] and scanning 2D-array Ultrasound probe [2] are the three main approaches categorized in the task of 3D Ultrasound reconstruction. The reliance on external spatial trackers, despite their positive contribution, has prompted an interest in developing trackerless methods.

Prominently, trackerless reconstruction methods based on deep-learning have shown huge potential of surpassing previous techniques.

3 Materials

The dataset utilized in our study was derived from the data presented by Q. Li et al. [3]. The Ultrasound data were gathered from 19 volunteers, encompassing both their right and left arms. Three distinct trajectories, namely, *straight, c-shape*, and *s-shape*, were captured in the distal-to-proximal direction for each forearm. Within each trajectory, two scans were acquired by orienting the transducer (probe) either *perpendicular* or *parallel* to the scanning direction.

Our analysis focused exclusively on scans obtained with the ultrasound probe positioned **perpendicularly** to the scanning direction. In total, the dataset utilized for this study comprises 114 scans (19 participants x 6 scans per participant), each constituting a 3D image with dimensions [Depth, Height, Width]. Each frame within a scan is associated with a transformation matrix representing the relationship between the ultrasound probe and the tracker. Furthermore, the dataset encompasses a calibration matrix, signifying the transformation from an image to tool coordinates. This calibration matrix is multiplied by each frame transformation matrix from the probe to the tracker, resulting a transformation matrix for each frame. These matrices collectively capture the transformation from the image coordinate system to the world coordinate system.

For the trajectory of the IMU, the simulation utilizes a Splined Trajectory to represent IMU's position and orientation over time.

Gyroscope and acceleration data is simulated using the timestamps and the transformations presented in the US dataset. Random noise is introduced to the gyroscope data to mimic real-world sensor noise. The gyroscope measurements are then interpolated to match the timestamps of the acquired ultrasound images, ensuring temporal coherence between the simulated gyroscope data and the corresponding ultrasound frames.

Rotation matrices are simulated by employing an Attitude and Heading Reference System (AHRS) algorithm [4]. The AHRS algorithm processes gyroscope and accelerometer measurements to estimate the orientation quaternion, linear acceleration, and earth acceleration. The resulting quaternion is converted into a rotation matrix, providing the simulated rotations for the IMU over time.

All algorithms and libraries used in simulating the IMU data were used from this paper [5].

4 Method

We define a set of 2D ultrasound image frames as $\mathcal{S}$, comprising an image sequence with a length of M. The sequence is represented as $S = \{I_m\}, m = 1, 2, ..., M$, where $S \subseteq \mathcal{S}$ and m signifies consecutively increasing time-steps at which frames are acquired.

For a given sequence, a spatial transformation $T_{j \leftarrow i}$, where $1 \leq i < j \leq M$, characterizes the relative translation and rotation between the ith and jth frames.

The goal is to predict the spatial transformation $T_{j^* \leftarrow i^*}$ between a specific pair of frames (j^*, i^*)

4.1 Network development and implemenation

This work focuses on establishing comparability among existing reconstruction algorithms and explores the potential improvement in results through the integration of IMU data. The primary objective is not centered around further architecture optimization.

A feed-forward EfficientNet (b1) f_{fwd} [6] parametrized by θ takes the image frames in sequence to predict $T_{j^* \leftarrow i^*}$ at the end of each sequence. The architecture undergoes a slight modification to accommodate future IMU data utilization. It produces a 32-dimensional feature vector at the last fully connected layer, which is subsequently fed into a new feed-forward neural network. This secondary network takes the produced feature vector, concatenates it with IMU data, and applies a ReLU activation function. The final output is generated through a second linear layer, resulting in a 9-dimensional vector.

The models in this analysis are tested using two adjacent frames as input and output transformations between them. A minibatch size of 32 and the Adam optimizer with learning rate 1^{-4} were used to train each model for 50,000 epochs. Three distinct scenarios are examined:

- Network with no additional input (i.e. only images)

- Network with additional rotations (predicted) as input

- Network with gyroscope data as input

4.2 Loss function

To address the challenge of empirical tuning between rotational and translational contributions, this work employs the Mean Squared Error (MSE) loss function. The loss is computed based on the distance between the predicted transformed points $\hat{p}_n^{(j)}, n = 1, ..., N$ and the ground truth transformed points $p_n^{(j)}, n = 1, ..., N$ by the predicted $\hat{T}_{j \leftarrow i}$ and the ground truth $T_{j \leftarrow i}^{(gt)}$ respectively. In this work N = 4 corner points in the jth image are utilized, represented in their homogeneous *image space*.

$$Loss = \frac{1}{N} \sum_{i=1}^{N} D(p_n^{(j)}, \hat{p}_n^{(j)}) \qquad (1)$$

Here $D(.)$ denotes the Mean Squared Error between x, y and z coordinates of the two points sets. The predicted transformed points $\hat{p}_n^{(j)}$ are computed as $\hat{T}_{j \leftarrow i} \cdot p_n^{(i)}$ and the ground truth transformed points $p_n^{(j)}$ are given by $T_{j \leftarrow i}^{(gt)} \cdot p_n^{(i)}$. the common sets of points $p_n^{(i)}$ represents the same points in the ith *image space*.

The ground truth transformation is composed by two image-to-world transformation, $T_{j \leftarrow i}^{(gt)} = (T_{world \leftarrow i}^{(gt)})^{-1} \cdot T_{world \leftarrow j}^{(gt)}$, at time-steps i and j.

This formulation allows the loss function to capture the disparity between the predicted and ground truth transformations, facilitating the training of the model to accurately predict the spatial transformation between consecutive image frames.

4.3 Evaluation Metrics

In this research two methods have been used to quantitatively evaluate the accuracy of the spatial transformations predicted during a reconstruction scan to assess model generalisation.

For each reconstructed scan, two reconstruction errors are reported:

4.3.1 Euclidean Distance

Euclidean Distance/Euclidean norm (L2 norm) is utilized to measure the translation error between the predicted and target spatial transformations in mm. The error is calculated along three axis, the x, y and z axis as follows: $||t_x||$, $||t_y||$ and $||t_z||$, where $|| \cdot ||$ represents the L2 norm between the predicted and ground truth points.

4.3.2 Rotation Error

Rotation Error is employed to evaluate the accuracy of the rotational component in the predicted spatial transformations. The error is expressed in terms of Euler angles (in degrees) along the three axes as well (x,y and z) [7].

The rotation matrices R_{target} and R_{pred} are converted to Euler angles $(\theta_x, \theta_y, \theta_z)$ and the rotational error is computed as the absolute difference between corresponding Euler angles.

These metrics, which offer a comprehensive evaluation of the model's performance by capturing both translational and rotational aspects of spatial transformations, are reported on the hold-out test set—unseen during model training and development. The Euclidean distance error provides insights into the accuracy of translation, while the rotation error assesses the fidelity of the predicted rotations during the reconstruction scan.

Evaluation metric	images	gyroscope	rotations
MSE	0.48	0.42	0.42
$t_x(mm)$	0.28	0.27	0.27
$t_y(mm)$	0.38	0.35	0.35
$t_z(mm)$	0.56	0.52	0.51
$\theta_x(degree)$	0.11	0.10	0.11
$\theta_y(degree)$	0.09	0.08	0.08
$\theta_z(degree)$	0.14	0.12	0.14

Table 1: Comparative Analysis of Reconstruction Performance Across Three Scenarios on Unseen Test Data

5 Results and Discussion

As it has been mentioned before, the objective of this work is to establish comparability among different scenarios using a released dataset, with a focus on quantifying the impact of varying input configurations on reconstruction performance through ablation studies.

The hold-out test set is employed for evaluation, specifically investigating the influence of 1) having images as the sole input to the model and 2) incorporating Inertial Measurement Unit (IMU) data into the input.

Table 1 presents the outcomes of the ablation studies, revealing crucial insights into the performance metrics:

MSE Loss results demonstrate a noteworthy enhancement in overall reconstruction accuracy when incorporating IMU data. As we can see from Table 1 the overall MSE loss of 0.48 for the images input only. This improvement is observed consistently across scenarios involving additional (predicted) rotation and gyroscope data in the input, as it drops down to 0.42. The findings strongly suggest the beneficial impact of IMU data on spatial transformation prediction.

Translation error (t_x, t_y, t_z) consistently exhibit reductions with the inclusion of IMU data, 0.38mm being the case along all axis (x,y and z). This consistent improvement implies that the integration of IMU input significantly aids in refining translation predictions.

Rotational error $(\theta_x, \theta_y, \theta_z)$ reveal nuanced improvements. While gyroscope data contributes positively to rotational accuracy (0.1°along x,y and z axes), the impact on specific rotation components may vary. This underscores the importance of further investigation into the nuanced effects of IMU input on rotational predictions.

The results from fig. 1 emphasize the significance of IMU data in enhancing reconstruction algorithms, particularly in refining translation predictions. The nuanced impact on rotational components warrants further investigation. These findings contribute valuable insights for the continued development and optimization of reconstruction algorithms in the presence of IMU input.

6 Further Research Directions

Exploring more sophisticated IMU simulation techniques could be further studied to get more insight into the analysis. Due also to the importance of understanding the dynamics of ultrasound image transformation over time, this work could be extend to include a new paradigm that accepts more than two adjacent frames to estimate the transformation.

7 Conclusion

In conclusion, this research advances our understanding of the interplay between IMU data and reconstruction algorithms. The positive impact observed on translation predictions, along with nuanced improvements in rotations, highlights the potential for further advancements in refining spatial transformation predictions. These insights contribute to the broader field of medical computer vision and offer valuable directions for the continued development and optimization of reconstruction algorithms.

Acknowledgement

The work has been carried out at EchoScout, and supervised by the Institute of Medical Informatics, Universität zu Lübeck.

Authors' Statement

Conflict of interest: Authors state no conflict of interest.

8 References

[1] L. Mercier et al., "A review of calibration techniques for freehand 3-d ultrasound systems," Ultrasound in medicine biology, vol. 31, no. 2, pp. 143–165, 2005.

[2] E. D. Light et al., "Progress in two-dimensional arrays for real-time volumetric imaging," Ultrasonic imaging, vol. 20, no. 1, pp. 1–15, 1998.

[3] Q. Li et al., "Trackerless Freehand Ultrasound with Sequence Modelling and Auxiliary Transformation Over Past and Future Frames," 2023 IEEE 20th International Symposium on Biomedical Imaging (ISBI), Cartagena, Colombia, 2023, pp. 1-5

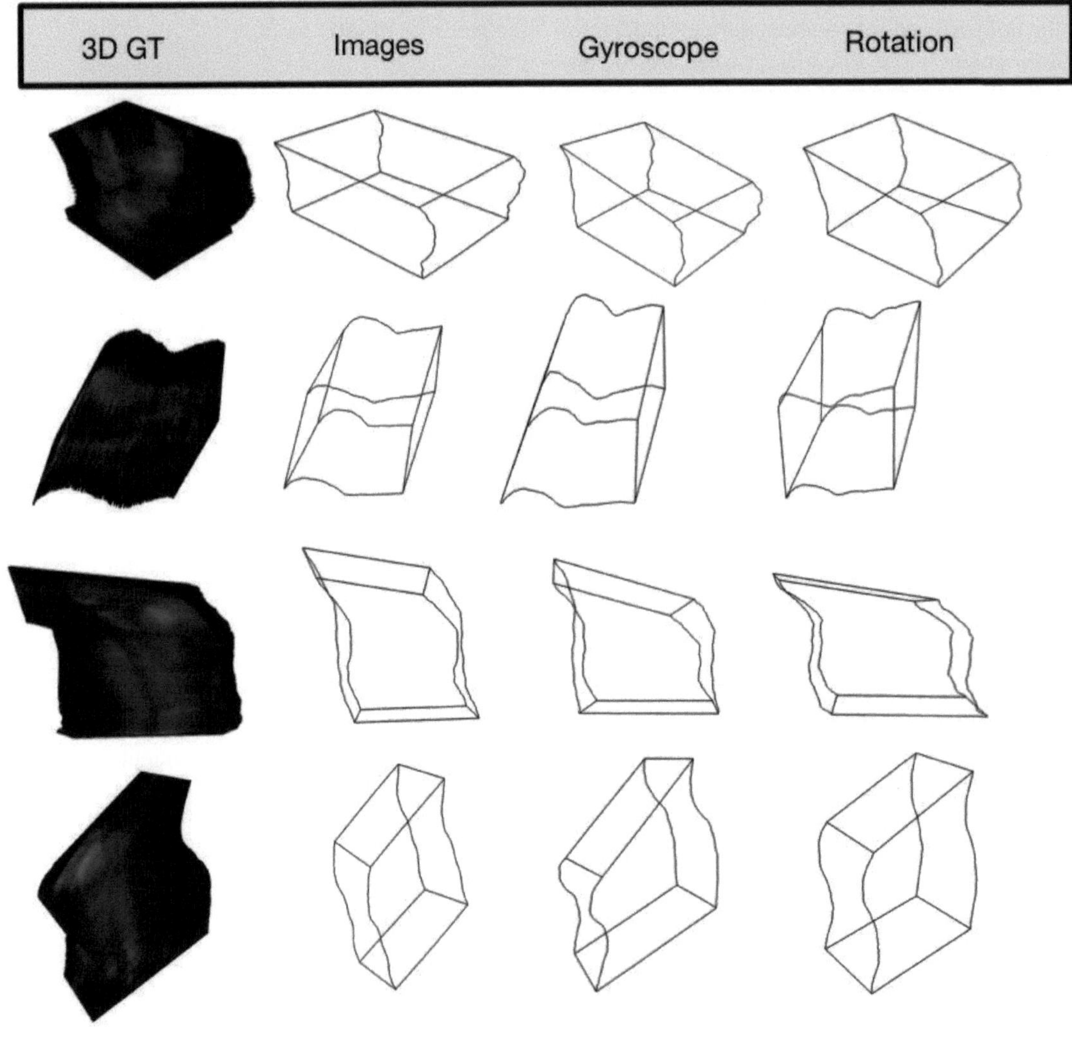

Figure 1: Reconstruction results of the three scenarios

[4] Tomaszewski, Dariusz, Jacek Rapiński, and Renata Pelc-Mieczkowska. "Concept of AHRS algorithm designed for platform independent IMU attitude alignment." Reports on Geodesy and Geoinformatics 104.1 (2017): 33-47.

[5] IMUSim: A Simulation Environment for Inertial Sensing Algorithm Design and Evaluation" by A. D. Young, M. J. Ling and D. K. Arvind, in Proceedings of the 10th ACM/IEEE International Conference on Information Processing in Sensor Networks (IPSN 2011), pp. 199-210. ACM, 2011.

[6] Tan, M., Le, Q.: Efficientnet: Rethinking model scaling for convolutional neural networks. In: International conference on machine learning. pp. 6105–6114. PMLR (2019)

[7] Pio, Richard. "Euler angle transformations." IEEE Transactions on automatic control 11.4 (1966): 707-715.

Detection of Pathological Mediastinal Lymph Nodes with Multiple-Instance Learning: A Proof-of-concept using Synthetic Data

Jasmin Katz [1], Sofija Engelson [2], Nele Brügge [3] and Heinz Handels [2]

[1] Medical Informatics, Universität zu Lübeck, jasmin.katz@student.uni-luebeck.de
[2] Institute of Medical Informatics (IMI), Universität zu Lübeck, {sofija.engelson, heinz.handels}@uni-luebeck.de
[3] German Research Center for Artificial Intelligence, nele.bruegge@dfki.de

Abstract

Detecting pathological lymph nodes has important prognostic and therapeutic significance for cancer staging. The main challenges in lymph node evaluation are low contrast to surrounding tissue, high morphological variability, and only image-level labels being available for a diverse patient cohort. This study presents a proof-of-concept using multiple-instance learning on synthetic data with the aim to assess to what extent the challenges faced in lymph node detection with image-level annotations can be adressed. The classification tasks and training data are designed to imitate the concepts that distinguish pathological lymph nodes, that is, form and shape changes. The proposed network architecture is based on the clustering-constrained-attention multiple-instance learning (CLAM) model [1]. Hyperparameter settings such as patch size, top k values and the number of patches are examined. Achieving accuracy ranges of 0.86 to 1.00 for detecting various shapes demonstrates the model's ability to detect size and shape changes in a simplified experimental setup.

1 Introduction

Lymph node detection is a crucial aspect of cancer diagnosis and the decision for treatment options. Tumour diseases are classified according to the TNM staging system, which assesses the anatomical extent of malignant tumour spread to recommend a treatment. N-staging involves categorizing tumours based on lymph-node involvement [2]. In the clinical workflow for lymph node assessment, the invasive method is called lymph node biopsy, which involves identifying, removing, and examining lymph nodes tissue for cancer cells [3]. The gold standard of non-invasive methods is the combination of CT scans with positron emission tomography (PET) scans [3]. Invasive methods are time-consuming and increase the risk of complications for the patient. PET scans are harmful to the patient's health due to exposure to radioactive radiation and cause high costs. Using CT images only for lymph node detection and classification is less expensive and resource-intensive in retrieval. However, this is a difficult task even for experienced physicians because lymph nodes strongly vary in shape, size as well as other morphological factors. Additionally, their borders are ill-defined and the intensity contrasts to surrounding soft tissue are minimal. Consequently, there is a pressing need for automatic lymph node staging including a visualization of which lymph nodes are pathological. Thus, the challenges of minimal size and shape changes are examined in this study with synthetic data to serve as a proof-of-concept. Obtaining images with segmentation masks of pathological lymph nodes can be challenging because it demands a high level of expertise and time commitment and may be subject to inter-observer variability. Thus, a model has been developed, consisting of a feature extractor and an attention-based multiple-instance learning (MIL) model. Training with MIL gives the opportunity to train on a large and diverse range of CT images collected from multiple centres, where pixel-level annotations are not available. MIL is a variation of supervised training and provides a heatmap of important image regions given only image-level labels.

1.1 Related Work

For the detection of mediastinal lymph nodes in CT images, various methods, such as fully convolutional neural networks [4] and a modified U-Net [5], have already been employed. However, these methods utilize images with pixel-level annotations of a comparably small dataset. Notably, MIL has only been employed for various other CT-related challenges, such as in diagnosing gastric cancer [6] and identifying intracranial haemorrhage [7] in CT images. This study aims to apply multiple-instance learning to detect mediastinal lymph nodes in CT images. Focussing on Whole Slide Images (WSI) rather than CT images, the detection of lymph node metastasis, was also addressed by a study [1]. Building upon this research, this study customizes their CLAM model for a proof-of-concept using synthetic data.

2   Material and Methods

2.1   Data

Synthetic data was used for training and evaluating the model to mimic challenges in CT scans, focusing on size and shape differences. Although synthetic data may not fully capture CT image complexity, it enables the evaluation of the model under ideal conditions, with unlimited data availability and arbitrary task complexity. Table 1 illustrates diverse classification tasks on the synthetic data, comprising a Gaussian noise background and encompassing five shapes selected from circles, ellipses, rectangles, and triangles. The spatial arrangement of these shapes is

Table 1: Different segmentation challenges

Task	Target	Other forms	Variation
1	Triangle	Rectangle, Circle	–
1.1	Triangle	Rectangle, Circle	Darker colour
2	Circle	Rectangle, Triangle	–
2.1	Circle	Rectangle, Triangle	Noise
3	Ellipse	Triangle, Circle	–
4	Ellipse	Circle	–
5	Large Ellipse	Ellipses of smaller size	–
6	Ellipse	Circle	2D ChestMNIST

intended to mirror the distribution of lymph nodes in the mediastinum. The complexity of the task is gradually increased by increasing the noise and changing colour. The initial task was to discriminate triangles and circles from other shapes, progressing in complexity up to distinguishing between small and large ellipses. To further increase the contextual complexity, the medical 2D ChestMNIST dataset [8] was used as background.

2.2   Multiple-Instance Learning

Multiple-instance learning organizes training data into bags, each containing multiple instances. In image processing, these instances are usually referred to as patches. The key characteristic in MIL is that labels are assigned at the bag level rather than at the instance level. Images are categorized into positive bags, which include at least one patch containing the target shape, and negative bags, which lack the target shape. In this study, a variant of the MIL method called top-k attention-based multiple-instance learning (top-k AMIL) method is used. Top-k AMIL is a method that assigns an attention score to each patch and focuses on identifying the top k instances within positive bags that contribute the most to the positive label. This way, the model needs to learn to concentrate on the most informative parts of the data without requiring specific instance-level labels. This is particularly useful when there is variability in the importance of instances within a bag. To aggregate image-level labels from patch-level representations for each class, an attention-based pooling function is used [1].

2.3   Model

This study adapts the CLAM model, originally designed for WSIs of $100,000 \times 100,000$ pixels per slice. As CT images have smaller dimensions (512×512 pixels per slice) yielding to fewer patches, the batch size could be increased to speed up computations. The training of CLAM involves extracting patches from the WSI, obtaining features using a pretrained ResNet50, passing the feature vector to the CLAM model, and using an attention network to generate scores for each patch, visualized in heatmaps [1]. Adapting this approach for lymph node detection in synthetic data, the feature extractor was adjusted to a Convolutional Neural Network (CNN) with fewer layers due to the significantly smaller size of the patches and the lower complexity of the synthetic data. The model architecture, including the feature extractor and the MIL is illustrated in Fig. 1. The MIL

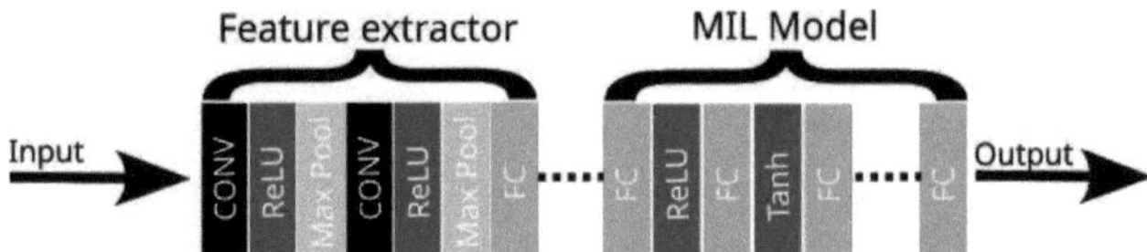

Figure 1: The model's architecture, consisting of feature extractor and MIL model.

model consists of an attention network and a classifier. The attention network, consisting of three fully connected layers with ReLU and Tanh activations, discerns the contributions of each patch to decision-making. The classifier, a single fully connected layer, determines the image label.

2.4   Model Training

During training, images were randomly divided into patches and normalized. For inference, a sliding window approach was used to extract patches. These patches were then processed by a feature extractor and the attention network in the MIL model assigned attention scores to each patch's feature vector. The final image label is computed by averaging all patches in an image, weighted by their respective attention score. Training supervision involves calculating a total loss, consisting of an image-level and instance-level loss. For the image-level loss, the predicted image labels are compared with the ground-truth image-level labels using the cross-entropy loss. To compute the instance loss, a binary clustering method focuses on the k most- and least-attended patches in each iteration, with pseudo labels generated from the attention network's output due to the absence of patch-level labels. Pseudo labels for images containing the target shape are generated by assigning a negative cluster label to the patches with the k lowest attention scores and patches with the highest attention scores receive a positive cluster label. An instance classifier, consisting of a fully connected layer, assigns instance-level clustering prediction scores to patches, classifying them as either containing the target form (class 1) or not (class 0). The instance loss is computed by comparing these scores to corresponding pseudo-cluster labels using cross-entropy loss. The goal of this training strategy is that, for images containing the target shape, the patches with the highest k attention scores are indicators for class 1.

3   Results and Discussion

The model training parameters are depicted in Table 2 and remained consistent, but for each task, a model was trained using different training data. Table 3 outlines outcomes from the tasks presented in Table 1. In medical applications, high sensitivity is often prioritized because false negative predictions, such as truly pathological lymph nodes being classified as normal, carry severe consequences.

Table 2: Hyperparameters

Parameter	Value
Patch size	16
Learning rate	0.0005
Number of top k	8
Number of patches	450
Number of training samples	500
Number of validation samples	100
Number of testing samples	100
Batch size	16

The model achieves perfect accuracy in detecting basic shapes, including triangles and circles. The inclusion of darker coloured shapes as an additional challenge does not adversely affect the model's accuracy or other performance metrics. Despite the overlay of Gaussian noise, the model maintains robust performance, achieving high accuracy (0.98) and F1-score (0.99). The model is good at detecting ellipses, with accuracy and F1-score above 0.90.

Table 3: Results of segmentation challenges of Table 1

Task	Accuracy	F1-Score	Sensitivity	Specificity
1	1.00	1.00	1.00	1.00
1.1	1.00	1.00	1.00	1.00
2	0.99	0.99	1.00	0.97
2.1	0.98	0.99	0.97	1.00
3	0.93	0.94	0.90	0.97
4	0.91	0.89	0.81	0.98
5	0.88	0.91	0.94	0.75
6	0.86	0.81	0.74	0.93

Detecting larger ellipses presents a slightly lower accuracy (0.88) and sensitivity (0.94) but the model still maintains adequate performance. Fig. 2 shows an example with reasonable feature visualization. The generated heatmap on the right shows that the model has learned to prioritize the elongated sides of an ellipse to distinguish it from a circle. In Fig. 3 it is evident that there are images where the model concentrates on patches that have no informative value. The patches with high attention fail to capture the distinctive features distinguishing circles from ellipses. Consequently,

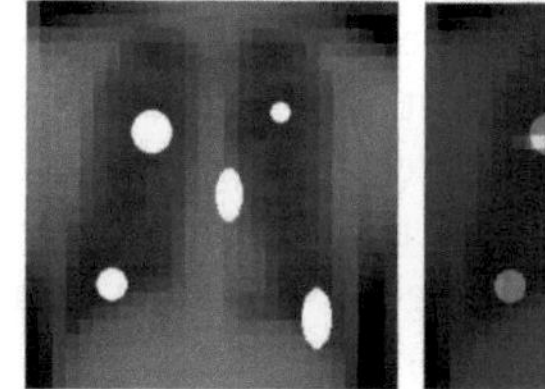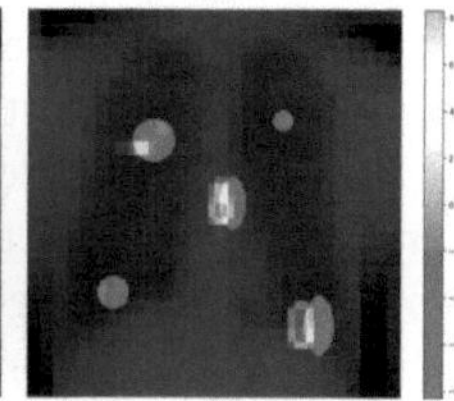

Figure 2: The left image shows the original synthetic training image and the right displays the heatmap after training to distinguish between ellipses and circles. Patches with high attention scores are represented in white or light grey, while low scores are in dark grey.

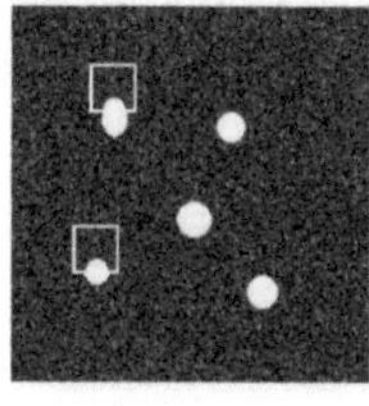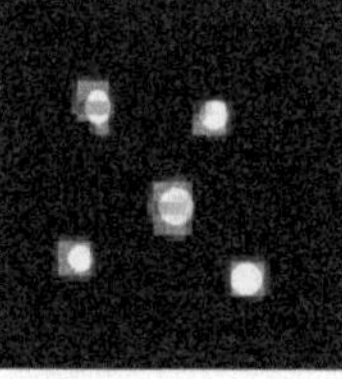

Figure 3: The original synthetic training image is on the left, the heatmap after training is in the middle. The task was to distinguish between ellipses and circles. High attention scores are in light grey. On the right, a patch is displayed and two potential extraction locations in the original image are indicated by white rectangles.

the highlighted areas erroneously mark not only ellipses but also circles. In Fig. 3 one can see, that small patches do not always capture the differences between circles and ellipses.

3.1   Distinction of Ellipses and Circles

Distinguishing ellipses from circles shows the ability of the model to recognize small changes in size and shape and thus the ability to differentiate between pathological and healthy lymph nodes. To assess the impact of patch size and variations in ellipse shape, patch sizes of 16, 24, 32 and 64 were tested. Different ellipse shapes, ranging from easily distin-

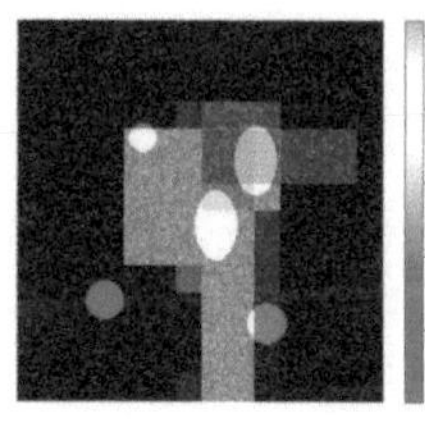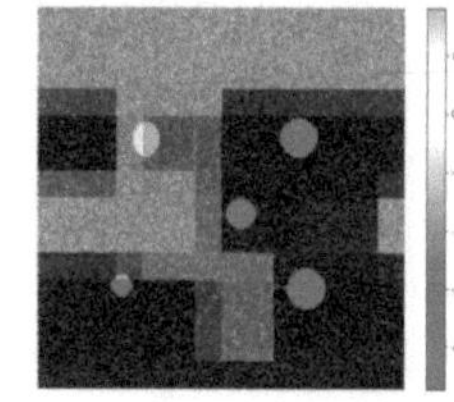

Figure 4: The left heatmap reflects low similarity between circles and ellipses in training, with a patch size of 64. On the right, an image with high similarity, using the same patch size, is displayed.

guishable to high similarities with circles, were considered, as illustrated in Fig. 4. This shows that the quality of the heatmaps decreases with higher task complexity. Starting with a patch size of 64 and decreasing to 16, the values of

k were set to 4, 8, 12 or 16. The correlation between patch size and k becomes apparent, with smaller patches requiring a larger k as they may contain less discriminative information. By increasing k, the model considers a larger number of patches, potentially capturing more relevant details. The results in Table 4 show that the model consistently performs well across all metrics for medium and low similarity, while the performance slightly decreases for high similarity. Generally, high similarity tasks exhibit accuracies ranging from 0.87 to 0.97, while low similarity tasks, excluding patches of size 32, achieved an accuracy of 1.00. Medium similarity tasks have an accuracy of above 0.96. Patch size 64 consistently performs well across all similarity levels, likely due to its ability to cover entire shapes, facilitating easier differentiation. For a patch size of 16, effective performance

Table 4: Results of distinguishing ellipses from circles

Patch size	Similarity to circle	Accuracy	F1-Score	Sensitivity	Specificity
16	high	0.87	0.86	0.80	0.94
16	medium	1.00	1.00	1.00	1.00
16	low	1.00	1.00	1.00	1.00
24	high	0.97	0.96	0.93	1.00
24	medium	0.98	0.98	0.96	1.00
24	low	1.00	1.00	1.00	1.00
32	high	0.90	0.88	0.79	1.00
32	medium	0.99	0.99	1.00	0.98
32	low	0.97	0.96	1.00	0.95
64	high	0.95	0.93	0.87	1.00
64	medium	0.99	0.98	0.98	1.00
64	low	1.00	1.00	1.00	1.00

is observed in medium and low similarities by adjusting k and increasing the number of patches. Using patch size of 32, the model shows high sensitivity for medium similarity but lower accuracy and sensitivity for high similarity. This behaviour may be attributed to a lower sensitivity score of 0.79 but a high specificity score of 1.00 which indicates a potential bias favouring one class.

4   Conclusion

This study presents a convolutional model trained using multiple-instance learning, which shows promising results for lymph node detection in synthetic data. By evaluation of different combinations of key parameters, including patch size and number of patches, the model excels in discriminating between varying shapes, showing robust performance even in the presence of challenges like darker colour and noise. In this simplified scenario, optimal k and amount of patches were identified, ensuring F1-scores of 0.86 up to 1.00 despite shape similarities. Future research involves adjustment of the model to CT data, using the knowledge about the behaviour of key parameters for efficient modification.

Acknowledgement

The work has been carried out and supervised by the Institute of Medical Informatics, Universität zu Lübeck.

Authors' Statement

Conflict of interest: Authors state no conflict of interest.

5   References

[1] M. Y. Lu, D. F. Williamson, T. Y. Chen, R. J. Chen, M. Barbieri, and F. Mahmood, "Data-efficient and weakly supervised computational pathology on whole-slide images," *Nature biomedical engineering*, vol. 5, no. 6, pp. 555–570, 2021.

[2] S. H. Feng and S.-T. Yang, "The new 8th tnm staging system of lung cancer and its potential imaging interpretation pitfalls and limitations with ct image demonstrations," *Diagnostic and Interventional Radiology*, vol. 25, no. 4, p. 270, 2019.

[3] C. M. Walker, J. H. Chung, G. F. Abbott, B. P. Little, A. H. El-Sherief, J.-A. O. Shepard, and M. Lanuti, "Mediastinal lymph node staging: from noninvasive to surgical," *American Journal of Roentgenology*, vol. 199, no. 1, pp. W54–W64, 2012.

[4] A.-I. Iuga, T. Lossau, L. L. Caldeira, M. Rinneburger, S. Lennartz, N. G. Hokamp, M. Püsken, H. Carolus, D. Maintz, T. Klinder *et al.*, "Automated mapping and n-staging of thoracic lymph nodes in contrast-enhanced ct scans of the chest using a fully convolutional neural network," *European Journal of Radiology*, vol. 139, p. 109718, 2021.

[5] Y. Manjunatha, V. Sharma, Y. Iwahori, M. Bhuyan, A. Wang, A. Ouchi, and Y. Shimizu, "Lymph node detection in ct scans using modified u-net with residual learning and 3d deep network," *International Journal of Computer Assisted Radiology and Surgery*, vol. 18, no. 4, pp. 723–732, 2023.

[6] C. Li, C. Shi, H. Zhang, Y. Chen, and S. Zhang, "Multiple instance learning for computer aided detection and diagnosis of gastric cancer with dual-energy ct imaging," *Journal of biomedical informatics*, vol. 57, pp. 358–368, 2015.

[7] M. López-Pérez, A. Schmidt, Y. Wu, R. Molina, and A. K. Katsaggelos, "Deep gaussian processes for multiple instance learning: Application to ct intracranial hemorrhage detection," *Computer Methods And Programs In Biomedicine*, vol. 219, p. 106783, 2022.

[8] J. Yang, R. Shi, D. Wei, Z. Liu, L. Zhao, B. Ke, H. Pfister, and B. Ni, "Medmnist v2-a large-scale lightweight benchmark for 2d and 3d biomedical image classification," *Scientific Data*, vol. 10, no. 1, p. 41, 2023.

Automatic prostate MRI sequence identifier based on image features

Rebecca Köhler [1] and Leonard Wägele [2]

[1] Medical Informatics, Universität zu Lübeck, rebecca.koehler@student.uni-luebeck.de
[2] FUSE-AI GmbH, Hamburg, leonard.waegele@fuse-ai.de

Abstract

Multiparametric MRI is used to diagnose prostate cancer and comprises several types of MRI sequences. The metadata that identifies the sequence of these images is often unavailable due to anonymisation or has a high variability in its representation. This makes automatic, rule-based identification for machine learning-assisted diagnosis unreliable for unknown data sources. To address this, we developed a gradient tree boosting classifier based on image features to distinguish between five sequence classes: T2W transversal images, low, middle, and high b-values in DWI, and in this use-case unwanted sequences containing DCE, ADC, T2W sagittal and coronal images. The best model, trained with cross-validation, consisted of 1000 estimators and depth 6, achieving a mean accuracy of nearly 99.45% with a standard deviation of 0.03%. Although these results are promising, it is important to note that they only involve certain use-case specific MRI sequences and additional datasets are required to increase data variability.

1 Introduction

Prostate cancer is one of the most commonly diagnosed cancers in men [1]. Traditional methods of diagnosing prostate cancer include tests of the level of prostate-specific-antigen (PSA) and random transrectal ultrasonography (TRUS)-guided prostate biopsy [2][3]. However, elevated PSA can occur in both benign and malignant prostate disease, and TRUS-guided biopsies are associated with several risks [2]. These methods may lead to overdiagnosis of low-risk cancers and underdiagnosis of clinically significant cancers [3]. Multiparametric Magnetic Resonance Imaging (mpMRI) can improve the diagnosis of clinically significant prostate cancer and reduce the number of unnecessary biopsies [2]. The current standard for prostate imaging and reporting is the Prostate Imaging Reporting and Data System (PI-RADS), which aims to standardise the acquisition, interpretation and reporting of prostate mpMRI examinations. Multiparametric MRI consists of different MRI sequences and the PI-RADS guideline lists the following: T1-weighted images (T1W) are recommended for detecting bleeding and seminal vesicles in the prostate and outlining the gland. T2-weighted images (T2W) are used, for example, to see the zonal anatomy of the prostate. The latter should be acquired in the transversal plane and in at least one additional orthogonal plane. Diffusion-weighted imaging (DWI) should also be used, including high b-value images and an apparent diffusion coefficient (ADC) map. The ADC map is usually obtained by using two or more b-values. Finally, as part of the mpMRI, dynamic contrast-enhanced (DCE) MRI should be employed which is based on the administration of gadolinium-based contrast agent (GBCA) [4]. However, automatic identification of MRI sequences remains a current challenge. When MRI data is available in the form of the international standard Digital Imaging and Communications in Medicine (DICOM), it is not guaranteed that the metadata contains easily usable information about the MRI sequence of an image. For example, DICOM includes the *Series Description* attribute, which is optional and which lacks standardisation. This leads to significant variability, depending on factors such as the manufacturer, the radiology professionals or specific in-house protocols [6]. In addition, DICOM defines a so called *Basic Attribute Confidentiality Profile* that outlines how to anonymize data for those attributes containing identifiable patient information. During this process, *Series Description* is completely removed, resulting in the loss of all its information [5]. Several works have been conducted on the automatic classification of MRI sequence types using machine learning: Some of them use DICOM metadata attributes as features [6][8] while others use Deep Learning methods with the MRI images directly as input [7]-[9]. The primary objective of this work is to construct a classifier capable of distinguishing between various MRI image types in prostate MRI images. This is achieved through the usage of image features and a non-deep learning method with the motivation of achieving fast inference on limited hardware. Specifically, the classifier should be capable of identifying transversal T2W images and DWI images, including the classification into high, middle and low b-value images. All other MRI sequence types should be classified as unwanted.

2 Material and Methods

2.1 Dataset

For the experiments, prostate data from ten distinct sources was used. Each MRI slice was considered individually, resulting in a total of 335930 slices selected randomly from the sources. The dataset consists of six different MRI image types, including T2W transversal, sagittal and coronal images, ADC and DCE images, as well as DWI images with b-values ranging from 50 to 2000. Fig. 1 shows an overview of the number of available MRI slices for each sequence type in each data source.

Image sequence type	A	B	C	D	E	F	G	H	I	J
ADC	1553	916	1113	4490	27711		1088	370	328	1224
T2W-tra	1552	916	1112	4517	29862	4166	1366	422	416	1534
T2W-sag					32817					
T2W-cor					32586					
DCE					30000					
DWI: $b \leq 200$	1553	916	1113	4470	27933	3970	1088	370	328	1224
DWI: $200 < b \leq 800$	1553	916	1113	4470	28013	7940	1088	370	328	1224
DWI: $b > 800$	1553	916	1113	4470	55775		1088	370	102	504

Data source

Figure 1: Overview of the dataset. The figure shows the number of available MRI slices for each sequence type in each data source.

2.2 Feature Engineering

The dataset is divided into five classes. The first class consists of T2W transversal images. As the classifier should distinguish between low, middle and high b-values in DWI, the DWI images were grouped into three classes as follows:

1. low b-value: $b \leq 200$

2. middle b-value: $200 < b \leq 800$

3. high b-value: $b > 800$

The remaining sequences DCE, ADC, T2W coronal and sagittal form the last class *unwanted*. The MRI images were resized to 128×128 pixels to ensure consistent feature extraction despite their varying shapes. The features were extracted as follows: We computed the $[0, 1]$-normalized p-percentile values of all pixel values in the images with $p = \{70, 80, 90, 99, 99.5\}$. Additionally, the original shapes of the MRI images were utilized to obtain the x-dimension and the y-dimension as features. The Histograms of Oriented Gradients, which are explained in the following section, were also used as features. A total of 3207 features have been extracted for each image.

2.2.1 Histograms of Oriented Gradients descriptors

The Histogram of Oriented Gradient (HOG) descriptors were first introduced by Dalal and Triggs for human detection [10]. They provide information about the frequency of different intensity gradient orientations in image blocks. We used *scikit-image* which offers a method for extracting the HOG features. Each image is devided into cells of size 16×16 pixels and blocks that contain 4×4 cells. For each cell, a histogram of gradients is created with 8 bins. The final feature vector is created by normalising and concatenating the cell histograms in each block. The blocks are obtained by shifting them one cell at a time. In total, this results in 3200 features. The HOG features are visualized in Fig. 2 using the parameters described above.

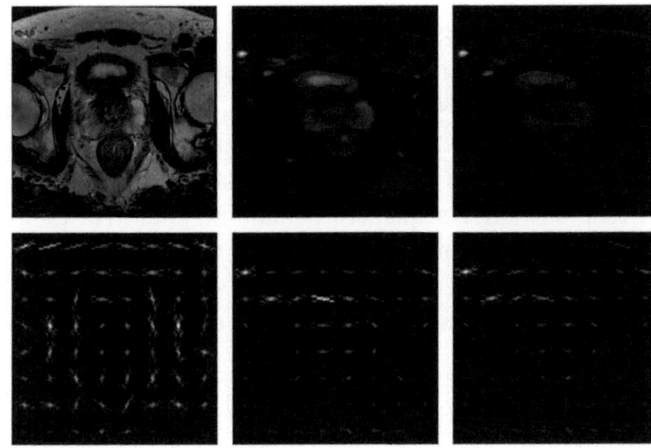

Figure 2: The upper row displays a T2W image (left), a DWI image with a b-value of 50 (middle) and a DWI image with a b-value of 400 (right). The lower row shows a visualization of the corresponding HOG features. These features were calculated using a cell size of 16×16 pixels and a block size of 4×4 cells.

2.3 Classifier

Gradient tree boosting is the applied classification method. It was introduced by Friedman [11] as an ensemble method, in which multiple weak learners are combined to obtain a strong classifier. With boosting, the individual models are trained sequentially with respect to the error of the currently existing ensemble. Given a dataset $(x, y)_{i=1}^{N}$ with $x = (x_1, ..., x_d)$ input variables and the corresponding labels y, the goal is to minimize a loss function $\psi(y, f)$, with f as the true functional dependence of x and y, to obtain $\hat{f}(x) = y$, the estimate function. Starting with a constant as the inital estimate $\hat{f}_0$, for every following iteration the negative gradient of the loss function is calculated with respect to the previous prediction. The new learner is fitted to approximate this negative gradient and the ensemble is updated with the new learner. As base-learner models in the ensemble we used decision trees, which is a common choice in gradient boosting [12]. Therefore, the open source package system *Extreme Gradient Boosting (XGBoost)* has been applied.

2.4 Training and Hyperparameter Tuning

The entire dataset was divided into 80% training data and 20% test data with an equal percentage of slices taken from each sequence type. This resulted in 268743 slices for training and 67187 slices for testing. To tune the *XG-Boost* model, we examined the maximum depth of each tree (*numDepths*) and the number of decision trees (*numTrees*), while leaving the other hyperparameters at their default values and setting the learning rate to 0.1. To determine the optimal hyperparameters for the classifier, we utilized cross-validation grid search (*GridSearchCV*) from *scikit-learn*. The hyperparameters were tuned with *numTrees* = {300, 350, 400, 450, 500, 700, 1000} and *numDepths* = {4, 6, 8, 10}. After cross validation of all possible models, the best model was trained using the entire training dataset and tested with the test dataset.

3 Results and Discussion

3.1 Results of the grid search process

Fig. 3 shows the mean accuracy scores for the validation data, visualized for the results of the cross validations performed in the grid search process. All variations of classifiers perform well, achieving mean accuracy scores of over 98.6%. It can be observed that among classifiers using 300 estimators, the one with a depth of eight performs the best. However, as the number of estimators increases, classifiers with a depth of six show better results. Using 1000 estima-

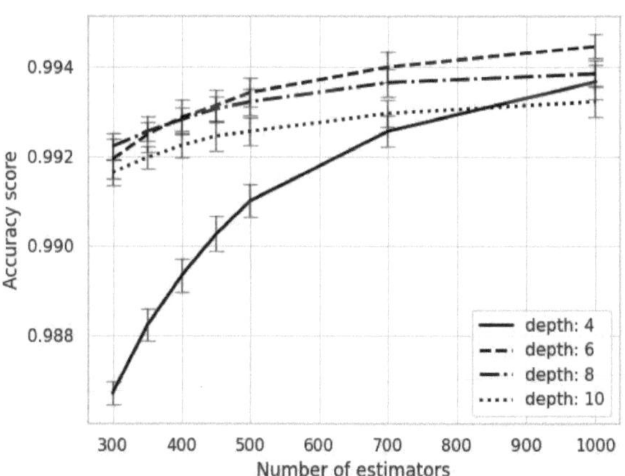

Figure 3: Mean accuracy score and standard deviation for all classifiers trained with cross validation in the grid search process, varying in depth and number of estimators.

tors and a depth of six produces the highest score, achieving a mean accuracy score of nearly 99.45% in cross-validation with a standard deviation of 0.03%. It should be noted that the accuracy curves continue to increase slowly, particularly for a depth of four. However, it is expected that overfitting will occur at some point. The plot also displays the standard deviations of the accuracy scores for each model, all

of which are very small. This indicates a strong robustness of the classifiers.

3.2 Detailed results of the best classifier

The best classifier, here with *numTrees* = 1000 and *numDepths* = 6, was trained on the whole training dataset and tested afterwards with the unseen test dataset. It achieved a perfect accuracy of 100% during training and 99.55% during testing. In Fig. 4, the confusion matrix is displayed for the test data. The diagonal shows the number of correctly classified slices. The T2W-transversal im-

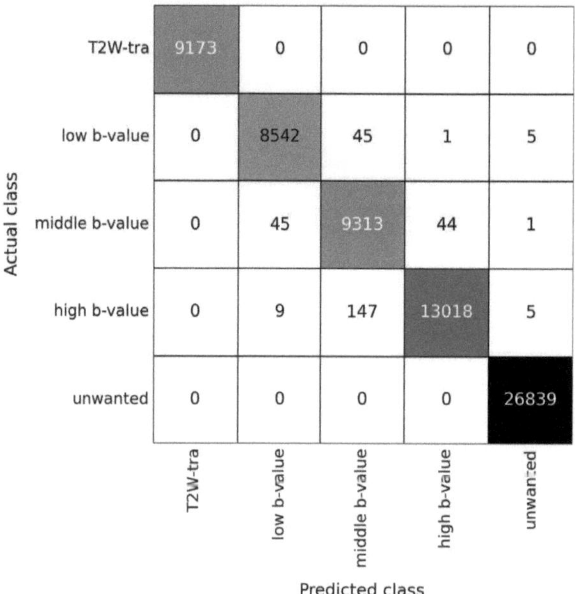

Figure 4: Confusion matrix of the test dataset for the gradient boosting model with 1000 estimators and a depth of 6.

ages were classified accurately, while none of the other sequences were classified as T2W. No unwanted images were classified as wanted. However, there were several misclassifications within the DWI class, and some images were predicted to belong to the *unwanted* class. Fig. 5 shows two plots of feature importance. The top plot shows the top eight features based on the gain, which measures the improvement in accuracy provided by a feature. The most important features are all HOG features, except for the x-shape, which is also listed. Although the percentile features do not appear in the gain feature importance, this does not mean that they have no influence: The lower plot shows the weight feature importance, which is the number of times a feature appears in a classifier tree. Here, unlike the gain plot, four of the five percentiles are listed, as well as the x- and y-shape. The significance of the image shape, which is evident in both plots, may be attributed to the higher resolution of the T2W images.

As one of the goals is to achieve fast inference, the time was measured for about 1000 slices to go through the whole pipeline including image loading, preprocessing, feature extraction and classification. The average time for a slice is

0.2 seconds, which is fast but may be improved by changing more hyperparameters or features and accepting a slightly decreased accuracy as a trade-off for faster inference, depending on the requirements for the model.

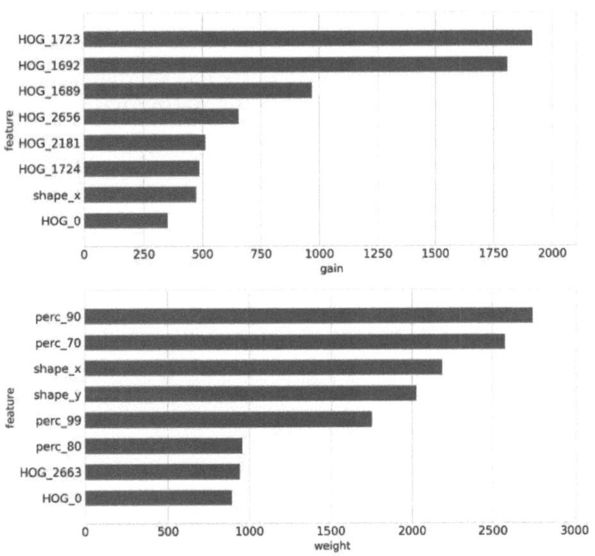

Figure 5: Feature importance of the gradient boosting model with 1000 estimators and a depth of 6, based on the gain (top) and on the weight (bottom).

4   Conclusion

It was shown that the approach of using an image-based gradient tree boosting classifier to distinguish between different MRI sequence classes was successful, achieving over 99% accuracy in testing. The classification seems to be most influenced by the HOG features and the original image shapes, with the percentiles being the features that appear most often in the trees. A next step could be to investigate whether the number of HOG-features could be reduced by tuning the HOG-hyperparameters to obtain a less complex classifier which takes even less time in inference. Furthermore, a simple neural network could also be implemented to compare the inference time. In addition, it is important to note that the MRI sequences used in training and testing represent only a subset of all existing sequence types. Further testing is required to determine the continued success of this approach for a more general classifier with additional sequences. Furthermore, to increase the variability of the data, it may be beneficial to include additional sources beyond the ten that were used in this work.

Acknowledgement

The work has been carried out at FUSE-AI GmbH, Hamburg and supervised by Prof. Mattias Heinrich, Institute of Medical Informatics, Universität zu Lübeck.

Authors' Statement

Conflict of interest: Authors state no conflict of interest.

5   References

[1] B. S. Chhikara and K. Parang, *Global Cancer Statistics 2022: the trends projection analysis.* Chemical Biology Letters, vol. 10, no. 1, pp. 451–451, 2023.

[2] A. O'Shea and M. Harisinghani, *PI-RADS: multiparametric MRI in prostate cancer.* Magnetic Resonance Materials in Physics, Biology and Medicine, vol. 35, no. 4, pp. 523–532, 2022.

[3] A. Stabile, et al., *Multiparametric MRI for prostate cancer diagnosis: current status and future directions.* Nature reviews urology, vol. 17, no. 1, pp. 41–61, 2020.

[4] American College of Radiology Committee on PI-RADS®, *Prostate Imaging-Reporting and Data System 2019 v2.1.* Available: https://www.acr.org/-/media/ACR/Files/RADS/PI-RADS/PIRADS-V2-1.pdf [last accessed on 2024-01-10].

[5] NEMA PS3 / ISO 12052, Digital Imaging and Communications in Medicine (DICOM) Standard, National Electrical Manufacturers Association, Rosslyn, VA, USA (available free at http://medical.nema.org/)

[6] R. Gauriau, et al., *Using DICOM metadata for radiological image series categorization: a feasibility study on large clinical brain MRI datasets.* Journal of digital imaging, vol. 33, pp. 747-762, 2020.

[7] J. P. V. de Mello, et al., *Deep Learning-based Type Identification of Volumetric MRI Sequences.* 2020 25th International Conference on Pattern Recognition (ICPR), IEEE, pp. 1-8, 2021.

[8] N. Kasmanoff, M. D. Lee, N. Razavian and Y. W. Lui, *Deep multi-task learning and random forest for series classification by pulse sequence type and orientation.* Neuroradiology, vol. 65, no. 1, pp. 77-78, 2023.

[9] G. L. Baumgärtner, et al., *Metadata-independent classification of MRI sequences using convolutional neural networks: Successful application to prostate MRI.* European Journal of Radiology, vol. 166, p. 110964, 2023.

[10] N. Dalal and B. Triggs, *Histograms of oriented gradients for human detection.* 2005 IEEE computer society conference on computer vision and pattern recognition (CVPR'05), Ieee, vol. 1, pp. 886-893, 2005.

[11] J. H. Friedman, *Greedy function approximation: a gradient boosting machine.* Annals of statistics, pp. 1189-1232, 2001.

[12] A. Nateki and A. Knoll, *Gradient boosting machines, a tutorial.* Frontiers in neurorobotics, vol. 7, p. 21, 2013.

13

Biomedical optics

Laser induced microbubbles in blood vessel phantoms

Andra-Karina Poetschki [1], Eric Seifert [2], Selma Bolterauer [3], and Ralf Brinkmann [2,4]

[1] Medical Engineering, Universität zu Lübeck, andrakarina.poetschki@student.uni-luebeck.de

[2] Medizinisches Laserzentrum Lübeck GmbH, Lübeck, e.seifert@uni-luebeck.de, brinkmann@mll.uni-luebeck.de

[3] Medical Engineering, Universität zu Lübeck, selma.bolterauer@student.uni-luebeck.de

[4] Institute of Biomedical Optics, Universität zu Lübeck, ralf.brinkmann@uni-luebeck.de

Abstract

The blood-retina barrier (BRB) prevents systemically administered drugs from reaching the retina. Therefore intraocular injections are commonly used for treatment of retinopathies, e.g. age-related macular degeneration (AMD) or diabetic macula edema (DME). This project investigates increasing the BRB permeability by laser-induced microbubbles in retinal vessels, by inducing temporal vessel dilation and potentially allowing a spatial confined drug delivery by bypassing the BRB. It examines the effects of laser irradiation on blood as absorber in a silicone vessel model, visualized by a high-speed camera and optoacoustically recorded by a hydrophone. Next to microbubble formation, coagulation was observed. The probabilities of both effects as a function of the pulse energy were investigated. The pulse energy is an important parameter to guarantee safe treatment. Threshold values are proposed in section 5. Research on phantom dilation, permeability, pulse duration and repitition rate will be the next step.

1 Introduction

The sense of sight is essential for human perception. In order to protect photoreceptors and other neurons of the retina from harmful particles, there is the blood-retina-barrier (BRB). Although it is an important mechanism for ocular health, it makes the systemic administration of medication difficult. The standard procedure for this purpose is intraocular injections, but there is the risk of infections and inflammation [1]. For example, 0.93% (95% CI) of patients with neovascular age-related macular degeneration developed an inflammation inside the eye due to an infection with microorganisms (Endophthalmitis) after injection [2][3]. Therefore other ways to administer drugs to the retina are of interest.

There is an approach to treat retinal neovascularization with photo-mediated ultrasound therapy, which is a combination of low-intensity ultrasound and nanosecond-pulse-duration laser [4]. Recently a group of researchers has shown in rodents that an increased BRB permeability by focused ultrasound is feasible [5][6]. Another potentially even more precise approach might be laser irradiation. Laser-induced microbubbles possibly cause dilation of the vessels and thus increase permeability.

This work examines the possibility to induce microbubbles in a vessel model, by focusing a laser beam on blood as absorber (pig's blood from a local butcher) flowing through a silicone tube. The effects are photographed by a high-speed camera and recorded by a hydrophone.

2 Material and Methods

2.1 Experimental Setup

The described setup is visualized in Fig.1.

A cuboid *cuvette* filled with water is located in the center of the setup. It is open at the top and there are round windows on three sides. The silicone *vessel model* (inner/outer diameter: 0,28mm/0,61mm) is installed vertically in the cuvette as a loop. Because of transparency, reproducibility and easier handling in terms of stability and size, a silicon tube instead of retinal vessels are used. A small *peristaltic pump* is connected to pump a highly absorbent liquid (e.g. pig's blood) through the tube to simulate the physiological blood flow. The tip of a *hydrophone* for optoacoustic recording is also located in the cuvette. It is aligned with the laser spot on the vascular model, to increase its sensitivity to perceive occuring microbubbles.

The *slit lamp* is aligned to a window of the cuvette with an integrated contact lens. Such lenses are used in ophthalmology to examine the fundus of the eye during retinal laser treatments and are adapted to the diameter and refractive power of the human eye. When the slit lamp is correctly adjusted and positioned, it provides a sharp and magnified image of the plane of the vessel model. The red pilot laser of the *SRT II Laser* (section 2.1.1) coupled in an optical fiber (core diameter 50μm, NA=0.11) is continuously visible. This way, the laser can be aimed at the right spot in the vessel.

There is one window each for the high-speed camera and the illumination. A *Photron FASTCAM SA3* is used as the

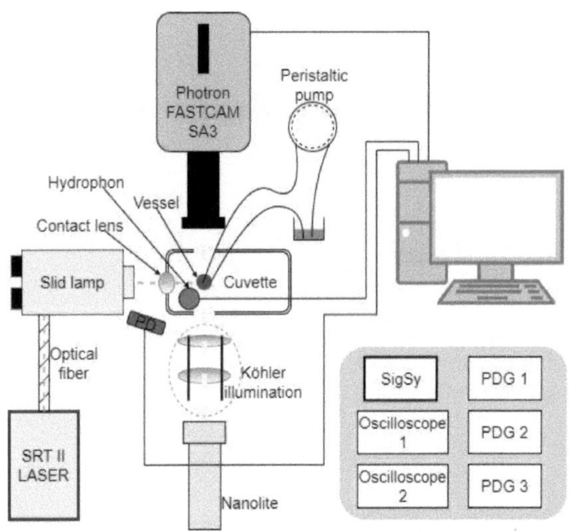

Figure 1: Schematic representation of the experimental setup. All essential components are shown, with the exception of the trigger setup (see Fig.3). PD: Photodiode, PDG: Pulse Delay Generator, SigSy: Signal Synchronizer.

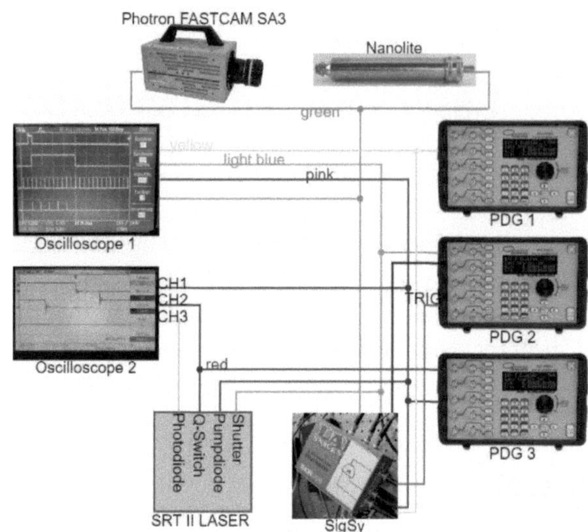

Figure 3: Schematic representation of the trigger setup. Colors of the BNC connections are based on the display of oscilloscope 1.

camera. A *Nanolite* flash lamp, operated at 49,999875Hz to 49,999995Hz, is used for the illumination. Two lenses, *collector* (f=75mm) and *condenser* (f=100mm), as well as an aperture diaphragm are installed in the beam path between the lamp and cuvette. This Köhler structure is used for homogeneous illumination of the object plane with the vessel model.

In addition, a *photodiode* is used as an energy sensor and trigger for the recording process of the hydrophone signal. A *computer* with a Peripheral Component Interconnect Express (PCIe)-oscilloscope digitizes the amplified signals from the photodiode and the acoustic sensor. The live image from the camera is also included in the experimental setup. The *pulse-delay generators*, the *oscilloscopes*, and the self-built signal synchronization (*SigSy*) device are described in more detail in section 2.1.2.

2.1.1   The SRT II Laser

The aforementioned SRT II Laser was built at the Medizinisches Laserzentrum Lübeck GmbH. The diode-pumped solid-state crystal laser uses neodymium-doped yttrium lithium fluoride (Nd:YLF) as active laser medium. The resonator also contains an acousto-optic modulator (AOM) as Q-switch, as well as a non-linear, frequency-doubling potassium titanyl phosphate (KTP-) crystal. It halves the wavelength of the emitted photons to 526,5nm in the green spectral range.

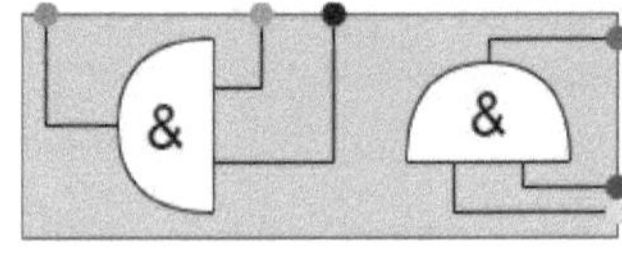

Figure 2: Logical circuit of SigSy. Colors of connections based on Fig.3 Oscilloscope 1.

The light is coupled into an optical fiber (the same one as the pilot laser) leading to the slit lamp. The spot diameter, set on the slit lamp, is $50\mu m$. The pulse energy is variable between ca. $5\mu J$ and $260\mu J$. The laser is operated with a frequency of 100Hz and a pulse duration of $1\mu s$. This means one pulse every 10ms. The number of pulses in a series varies from 24 to 120.

2.1.2   The Trigger Setup

It is not trivial to trigger laser, illumination, camera, and hydrophone, each at the right time, with a fixed temporal reference to each other. Therefore, and due to shortcomings of the available frequency generators, a device (SigSy, Fig. 2) was developed that logically processes multiple transistor-transistor-logic (TTL) signals. The complete setup is shown in Fig. 3. For example, the logical signal that controls the opening of the laser shutter (light blue) and a PDG-generated ca. 50Hz (49,999875Hz to 49,999995Hz) signal (black) are processed in the SigSy by a logical AND operation. The output-TTL (green) is input to camera and flash lamp, to switch them on for the duration of the irradiation, and with a defined delay respectively to every laser pulse.

The flash lamp should not be operated at more than 50Hz for a longer time, to prevent damage. It generates broadband light for a pulse duration of 18ns. Camera and flashlight are triggered with a frequency slightly less than 50Hz, which is every 20,000002ms to 20,000050ms. This way there is an increasing temporal delay of $k\cdot2$ns to $k\cdot50$ns between laser (one pulse every 10ms, 100Hz) and photograph so that a process (e.g. bubble formation and collapse) can be depicted. The minimal value of parameter k is 1 and the maximum value is the number of laser pulses in one irradiation sequence (e.g. 24 pulses).

2.2 Methodology of the measurement series

A measurement is done by irradiating the vessel model filled with a (non-)absorbent liquid with a laser pulse series of 24 to 120 pulses. Imaging and optoacoustic recording is performed simultaneously with every pulse. For a test series, one parameter (e.g. pulse energy, vessel material) is varied, and repetitive measurements are carried out.

Measurement series A: In this case the pulse energy of the laser beam was varied from $5\mu J$ to $200\mu J$ and the effects were documented. Optoacoustic data were screened for indications of microbubble formation (sudden increase in pressure amplitude and/or phase shifts). The thermoelastic expansion and collapse of the bubble surface is a source of sound, leading to a dephased hydrophone signal. The camera visualizes macroscopic alterations.

Measurement series B: Furthermore, a test series was done with different vessel materials (silicone, glass) combined with multiple (non)absorbers (blood, inc, water). This way the interaction of the silicone vessel wall with the blood can be investigated.

Analysis: For examination of the results, the probit-plot method was used. Therefore binary data of the occurrence of an effect are collected and plotted over a metric variable (e.g. pulse energy). The probability of the effect as a function of the metric quantity is determined by fitting in a normal distribution.

3 Results

It was observed, that the irradiation produces dark accumulations remaining on the vessel wall where the laser was focused on, as visible in Fig. 4(a) and (c). The arrow indicates the location of current irradiation on the vessel model. In subsequent images, these accumulations still exist.
In addition, in the case of deactivated blood flow, the accumulations grew in size as a consequence of repetitive irradiation. In the case of activated blood flow, the emerging accumulation was swept away, which is also visible in Fig. 4(a) as a small flag floating downwards.
Other photographs document bubble formation, as visible in Fig. 4(b). The initially unshaped object developed a spherical shape and grew in size, before floating upwards. Notice, that the pulse energy of this measurement was $256\mu J$. Fig. 5 shows the corresponding hydrophone signal, strongly dephased and with a high amplitude.
In the lower energy range (below $100\mu J$), the hydrophone data provides a more distinct indication of bubble formation, than the camera images.

Measurement series A: It was observed, that optoacoustic microbubble detection occurred at lower laser pulse energies than accumulation. For this reason, the energy thresholds for microbubble formation and accumulation were investigated.

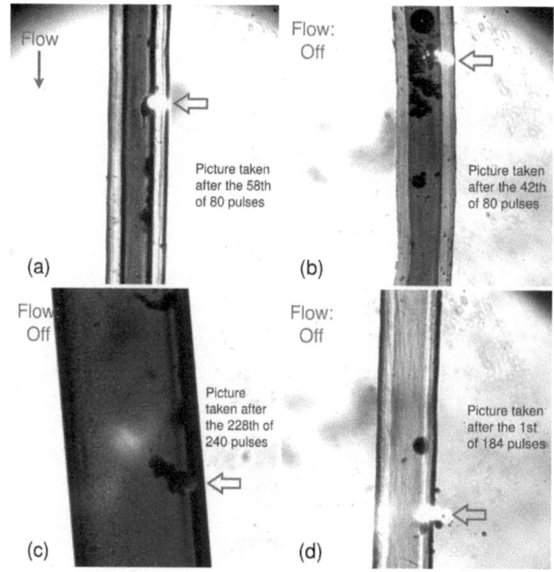

Figure 4: Various photographs acquired by the set up shown in Fig.1. Laser pulse energy ca. $256\mu J$. (a), (b): Silicone tube with pig's blood. (c): Glass cannula with pig's blood. (d): Silicone tube with water.

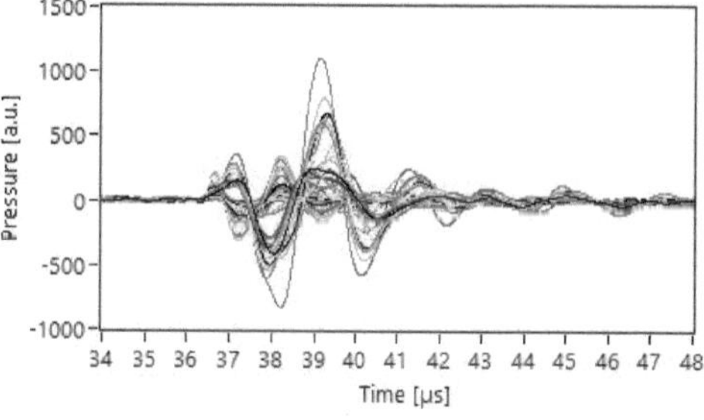

Figure 5: Hydrophone signal of the effect pictured in Fig. 4(b) indicates microbubble formation. Each curve represents one pulse in the sequence.

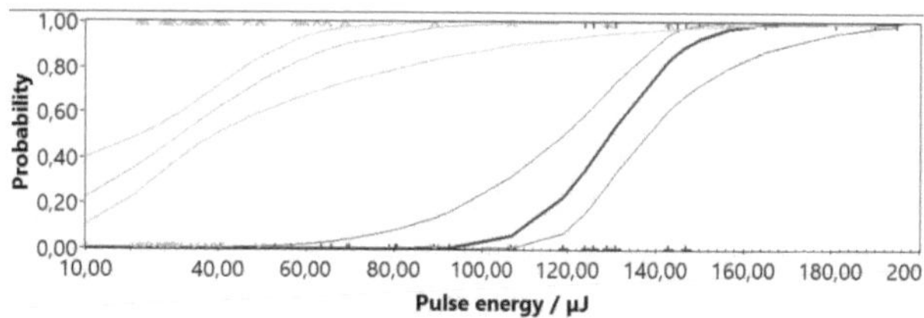

Figure 6: Probability of microbubble formation (green/left) and accumulations (red/right) as a function of pulse energy. CI corresponds to 95%.

The probit curve resulting from the measurement series is visualized in Fig. 6. It supports the hypothesis that the energy threshold for microbubble formation is far below the threshold for accumulation of particles. For example, with a pulse energy of $68,4\mu J$ (95% CI = $55,5\mu J$-$107,2\mu J$), the probability of bubble formation is 90%, and the probability of accumulations is almost 0% (95% CI = 0-3.8%).
This allows the suggestion of a "safe" energy range to induce microbubbles but no accumulation.

Measurement series B: Because of the unexpected observations, it was analyzed whether the accumulations were caused by an interaction of the silicone vessel wall with the absorber. The occurrence of bubbles and accumulated particles depending on the combination of vessel material and absorber (same pulse energy, $256\mu J$) is documented in table 1, and in Fig. 4(c) and (d).

It is noticeable that in the silicone tube even with water as nonabsorbent liquid, it is possible to produce bubbles (see Fig. 4(d)). Microbubbles were also observed in case of the glass cannula. Accumulations in the blood can be observed independent of the vessel material.

Table 1: Effects dependent on vessel material and absorber

Material	Blood	Inc	Water
Silicone	○✳	○-	○-
Glass	○✳	- -	n.a.

○: Bubble formation, ✳: Accumulations

4 Discussion

In addition to microbubbles, dark, irreversible accumulations were detected, which can be interpreted as coagulated blood. In clinical application, a thrombus of clotted blood that is washed away can cause a vascular occlusion in the heart or the brain, which can lead to a stroke, or an embolism.

The accumulation of blood particles appears independent of the vessel material (see *measurement series B*). This observation suggests, that there is a thermal effect, due to repetitive irradiation. Possibly this can be solved by a shorter pulse sequence or shorter pulse duration, with calculations based on the Arrhenius damage integral. Bubble formation seems to depend on the vessel material, possibly also caused by an interaction of the laser photons with the silicone wall. It is essential to define a "safe" range of laser pulse energy or to establish reliable microbubble detection to avoid coagulation at all costs. A critical value of ca. $70\mu J$ is proposed (see *measurement series A*).

Tube dilation due to laser-induced microbubble formation, to increase permeability, has not been observed so far. However, this requires phantoms with vessel-like dimensions and elasticity. Next steps will be to investigate more adequate vessel models, and to analyze the effect of pulse duration and single-pulse irradiation.

5 Conclusion

This study examined the effects of pulsed laser irradiation on pig's blood as absorber in a silicone vessel model. It is possible to induce microbubbles in the blood vessel phantom using laser irradiation.

Due to coagulation, detected in images made by a high-speed camera, the threshold energies for microbubble formation and coagulation were investigated. With a pulse energy of $68,4\mu J$ (95% CI = $55,5\mu J$-$107,2\mu J$) on a set spot diameter of $50\mu m$, the probability of bubble formation is 90%, and the probability of coagulation is almost 0% (95% CI = 0-3.8%). Coagulation is possibly caused by repetitive irradiation and thus thermal effects, independent of the vessel material. Further research is required, to approach the goal of increased BRB permeability by laser-induced microbubbles in retinal blood vessels.

Acknowledgement

The work has been carried out at Medizinisches Laserzentrum Lübeck GmbH and supervised by the Institute of Biomedical Optics, Universität zu Lübeck. It is funded by the ministry of education and research (BMBF).

Authors' Statement

Conflict of interest: Authors state no conflict of interest.

6 References

[1] J. E. Knickelbein, E. Y. Chew, and H. N. Sen, "Intraocular inflammation following intravitreal injection of anti-vegf medications for neovascular age-related macular degeneration," *Ophthalmic epidemiology*, vol. 23, no. 2, pp. 69–70, 2016.

[2] T. A. Meredith, C. A. McCannel, C. Barr, B. H. Doft, E. Peskin, M. G. Maguire, D. F. Martin, J. L. Prenner, C. of Age-Related Macular Degeneration Treatments Trials Research Group *et al.*, "Postinjection endophthalmitis in the comparison of age-related macular degeneration treatments trials (catt)," *Ophthalmology*, vol. 122, no. 4, pp. 817–821, 2015.

[3] F. Antwerpes, M. Schuster, F. Walter, and B. Abels. (2023) Endophthalmitis. Zugriff am 12.01.2024. [Online]. Available: https://flexikon.doccheck.com/de/Endophthalmitis

[4] Y. Qin, Y. Yu, J. Fu, M. Wang, X. Yang, X. Wang, and Y. M. Paulus, "Photo-mediated ultrasound therapy for the treatment of retinal neovascularization in rabbit eyes," *Lasers in Surgery and Medicine*, vol. 54, no. 5, pp. 747–757, 2022.

[5] S. Bleker, Y. Zhang, J. S. Allen, E. Ranaei Pirmardan, A. Hafezi-Moghadam, and C. Russmann, "Modulation of the blood-retina-barrier permeability by focused ultrasound: Computational and experimental approaches," *The Journal of the Acoustical Society of America*, vol. 153, no. 3, pp. A68–A68, 2023.

[6] S. Bleker, Y. Zhang, J. Allen, E. R. Pirmaradan, M. Brettmacher, A. Hafezi-Moghadam, and C. Russmann, "Real-time monitoring of the focused ultrasound-induced permeabilization of the blood-retinal barrier," *Investigative Ophthalmology & Visual Science*, vol. 64, no. 8, pp. 1147–1147, 2023.

Optimization of the triggering sequence in Phase Sensitive Full Field Swept Source OCT

Baris Bargu [1], Sarvesh Thakur [2], Pepijn Klooster [2], Gereon Hüttmann [3], and Dierck Hillmann [2]

[1] Medical Engineering Science, Universität zu Lübeck, baris.bargu@student.uni-luebeck.de

[2] Department of Physics,Vrije Universiteit Amsterdam, {s.a.thakur,d.w.a.hillmann}@vu.nl, n.p.klooster@student.vu.nl

[3] Institute of Biomedical Optics, Universität zu Lübeck, gereon.huettmann@uni-luebeck.de

Abstract

Functional retinal imaging using full field swept source optical coherence tomography (FF-SS-OCT) techniques holds substantial promise for clinical diagnostics and research in the area of signal transduction of photoreceptors. The contrast mechanism, based on minimal phase differences after optical stimulation, enables the visualization of functionality in these images. To ensure precise measurement reproduction, we have developed a meticulous timing sequence for triggering two lasers in succession. Using a conventional and cost-effective Arduino board, we can trigger the system, rapidly acquiring multiple volumes with equidistant timing intervals with nanosecond-level time resolution for the trigger signal. Within this time-sensitive context, we implement a highly adjustable stimulation unit without compromising timing accuracy. This work is an important contribution to retinal functional imaging.

1 Introduction

Optical coherence tomography (OCT) is a non-invasive tomographic imaging technique that relies on the interference pattern of backscattered light from the target tissue and light from a reference arm. Full field swept source OCT (FF-SS-OCT) is an extension of this technology that illuminates and uses an area camera to record the entire sample surface at once, avoiding the need to scan the sample, as in conventional OCT imaging. This improves the imaging speed tremendously, up to $100\,\text{MHz}$ in the system that is being used in this work. Utilizing the high imaging speed ensures phase stability across various imaging depths. Phases across a single en face image in FF-SS-OCT are even inherently stable, because all data points at a specific depth are simultaneously measured. The phase information can be used as a contrast mechanism to enable functional imaging, in particular in the retina. In order to create phase differences, the retina is being optically stimulated. As the photon is absorbed by photopigments in the outer segment of the photoreceptor, a cascade of chemical events is initialized within the inner segment, resulting in the transmission of an electrical signal through ganglion cells [5]. This process results in optical path length changes in the cells that can be made visible by using the phase differences of the acquired OCT data [1], [2].

A high axial resolution is very important when tracking the signal throughout different layers in the retina. Axial resolution in FF-SS-OCT is described by

$$\Delta z = \frac{1.21\pi}{\Delta k}, \tag{1}$$

with k describing the wave number. The axial resolution is increased by combining two different lasers into a single light source, increasing the bandwidth of the light source (Δk) from one laser (805-880 nm) with a bandwidth of 75 nm to two lasers (805-880 nm and 875-950 nm) with a combined bandwidth of 145 nm. This increases the axial resolution from $5.71\,\mu\text{m}$ to $3.19\,\mu\text{m}$. But because the lasers have different sweep speeds and are not perfectly synchronized with the camera sensor, artefacts result in the acquired volumes.

Here a stimulation unit was designed and integrated into an already established FF-SS-OCT system. The stimulation unit is an optical relay system that projects a light source onto the retina to stimulate the photoreceptors.

A conventional Arduino board is used to perfectly synchronize the activation time of all components with nanosecond precision, including the stimulation unit, of the FF-SS-OCT system. This is done by coding a hardware trigger sequence that uses the Arduino pins to externally triggers the components. A reliable and precise triggering sequence ensures repeatability in all measurements under the same conditions and eliminates any major artefacts, that result from varying activation times of the lasers, in the image reconstruction process. Additionally, a stimulation unit is designed to be able to stimulate a designated area of the retina without causing damage to the photoreceptors.

2 Material and Methods

The system layout, including the newly designed stimulation unit, is illustrated in Fig. 1. The laser beams are introduced into a multimode fiber, creating uniform illumination and suppressing multiple scattered light. The light beam is then directed through a collimator and two cylindrical lenses, which widen and shape the beam into a rectangular form matching the camera sensor's area. Subsequently, the beam is split into a reference arm and a sample arm via a beam splitter. Prior to the beam splitter, a biconvex lens is positioned to act as a telescope together with a lens in the sample arm (the second lens is replaced by the human lens in retinal measurements), ensuring even illumination of the retina in the sample arm. Given the full-field nature of this OCT system, the laser light is not focused onto a single spot at the sample, avoiding the need for lateral scanning. Because the entire sample is illuminated, an area camera sensor can also acquire the entire data in parallel. An additional lens in the reference arm and the lens before the beam splitter also function as a telescope.

A mirror at the end of the reference arm reflects the light back to the beam splitter, where it interferes with the backscattered light from the sample, and the resulting interference pattern is directed onto the sensor. The sensor, a high-speed area camera (Phantom TMX 6410), operates at a sampling rate exceeding 66 000 frames per second for an image resolution of 1280 x 800 pixels.

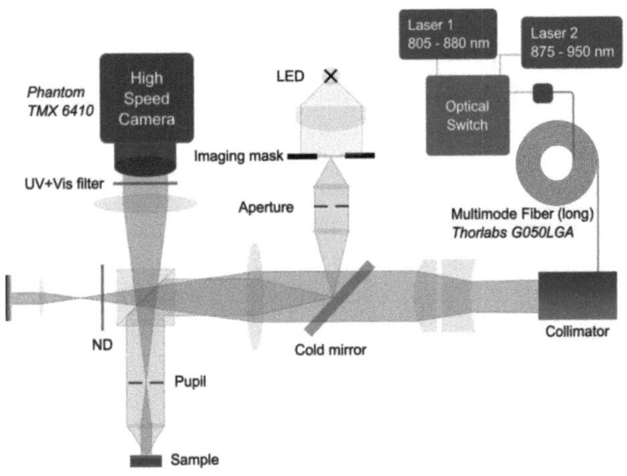

Figure 1: Layout of the FF-SS-OCT system. Light from the stimulation is collected by a condensor and projected onto the eye with the use of a cold mirror.

Using two lasers requires the integration of an optical switch into the system to sequentially couple the lasers into the multimode fiber. Furthermore, an algorithm is implemented to determine the different sweep speeds and slopes of the two lasers and to seamlessly stitch together the resulting images.

At the camera, each pixel simultaneously captures one A-scan by acquiring the spectrum of the interference pattern over time. Notably, scanning occurs solely in the axial direction, and each frame captured by the camera corresponds to a different wavelength from the light source. As mentioned in the introduction, combining two lasers and reconstructing volumes from raw data is challenging.

Moreover, the unintentional fluctuations in the activation time of lasers across different measurements poses a significant obstacle to the implementation of a reconstruction algorithm and results in artefacts in the volumes. Due to the specific order in which various components in the system must be activated, a precise trigger sequence has been established in the previous setup. Fig. 2 illustrates the configuration of the triggers, with an Arduino Mega board managing the triggering process by setting TTL signals at the appropriate times, that triggers the corresponding component.

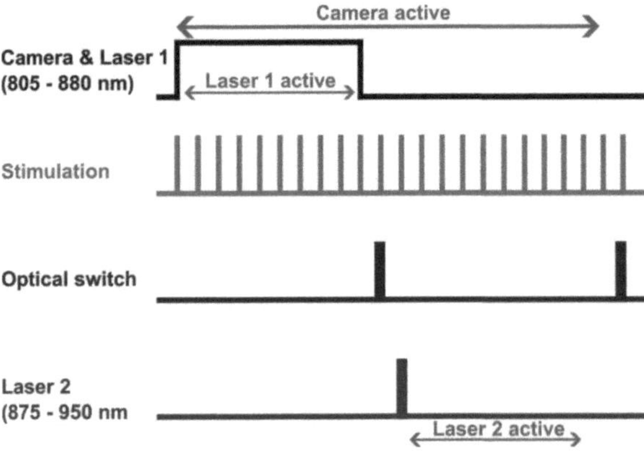

Figure 2: Overview of used triggers, as well as their timing and duration.

2.1 Timing code

The previous code for the triggering sequence in Fig. 3 utilized internal time functions provided by the Arduino board to time trigger signals, primarily relying on the `micros()` function, which returns the number of microseconds since the Arduino board began running [3].

However, this timing approach doesn't consistently enter the if-statement at the same time due to the `micros()` function's reliance on an internal Arduino register, also known as a timer register. The timer register increments every set number of clock cycles and overflows when its value exceeds the maximum stored value, triggering an interrupt service routine (ISR) about once every second. The 3.5 µs the routine needs to execute are not included in the return value of the `micros()` function, which makes it become less accurate over time. For more precise time measurement, a new timer register was manually configured, and a custom ISR was written, bypassing internal timing functions. Other timers were disabled to ensure consistency and reproducibility. Setting up this new ISR required specific register values [4], to enable the timer with the correct parameters. The resolution depends on the Arduino clock, with the Arduino Mega board in the old setup having a clock speed of 16 MHz, providing a resolution of 62.5 ns. Higher resolutions could be achieved with boards like the Arduino Giga board (480 MHz), yielding about 2.08 ns resolution. The Arduino Mega board

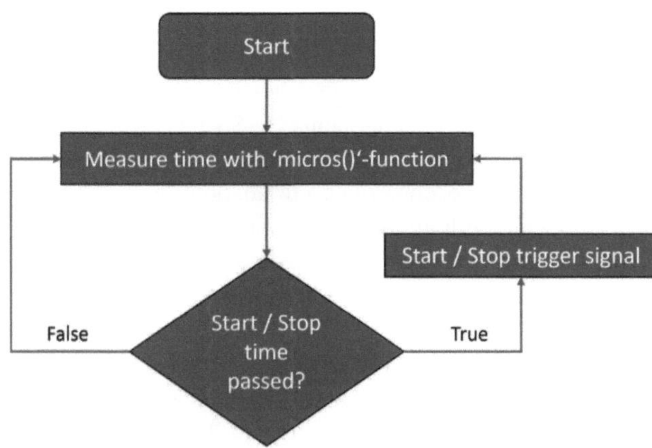

Figure 3: Previous way of timing triggers. The current time is updated every few clock cycles and then compared with a start time in a loop.

was also used for the new code. The timing reproducibility is not dependent on the clock and is more important than the accuracy for image reconstruction.

While running the new ISR, the output pins from the Arduino board need to be raised to a logical high (5 V) to trigger the appliances in the system. To achieve this, pins were set using the `digitalWrite()` function, and the code was then delayed by a certain amount of time with the `delayMicroseconds()` function until it was time for the next trigger signal to be set. This function counts clock cycles independently of timers, ensuring precision, when not interrupted by an ISR. Accurate and consistent time delays were achieved by using the function only inside the new ISR, because ISRs cannot be interrupted. The custom ISR just sets the pins and then delays the code by the amount of time these pins need to stay high, for the trigger signal to successfully trigger the component.

The new ISR is a more robust method for setting pins because this function is executed periodically, every few clock cycles. To elaborate, the timer register is set to execute the ISR every 20 ms, corresponding to 320,000 clock cycles on a 16 MHz board. The timing is now only dependant on the accuracy and speed of the clock of the Arduino board. The period can be changed in the code and can be set in a range of [1/clock speed, timer register max value]. For the Mega board that range is between 62.5 ns and 4.19 s.

2.2 Stimulation unit

The upcoming implementation of the stimulation unit must meet various criteria to ensure its effectiveness. It must be capable of stimulating a designated area of the retina without causing damage to the photoreceptors. The stimulation should be delivered in the form of a series of light flashes, and the frequency and duration of these flashes should be adjustable. Precise control over the onset of stimulation is necessary to synchronize with the recording of a measurement. Importantly, the stimulation must not interfere with

or exert any influence on the FF-SS-OCT system.

The stimulation unit in Fig. 1 consists of several components, including the light source, which is a warm white LED, a condenser to efficiently capture light from the LED, a 4f optic, an imaging mask positioned within the 4f optic, and an aperture for regulating the intensity of the stimulation. The imaging mask right now is an aperture, but future plans include designing thin sheets of varying shapes to stimulate the photoreceptors in a distinct pattern.

With the current setting each functional measurement consists of 70 volumes, with stimulation starting after acquiring 5 volumes, forming the baseline of the measurement. Any alterations in the light path within the cells can be identified by comparing the phases of stimulated volumes to the baseline phases. These settings can be adjusted in the code. To ensure that the stimulation is not harmful for the human retina measurements with a power meter were taken.

3 Results and Discussion

3.1 Time resolution of system

The temporal precision of the modified code was assessed using an oscilloscope and compared to the precision of the old code. Fig. 4 presents a direct comparison of the Arduino pin output. The output signal is supposed to give an 8 ms long pulse starting every 20 ms. The spectra of the measured data is displayed using a discrete Fourier transformation (DFT) to reveal the frequencies contributing to the final output. By using the DFT all frequencies that are in the signal can be made visible. Since the code set a period of 20 ms, a single peak and its harmonics at 50 Hz should remain in the spectrum after the DFT, if the output has perfect timing and is perfectly square shaped. The narrower peaks in the top plot in Fig. 4 clearly demonstrate a significant enhancement in timing accuracies with the implementation of the new code, thereby improving the repeatability of measurements.

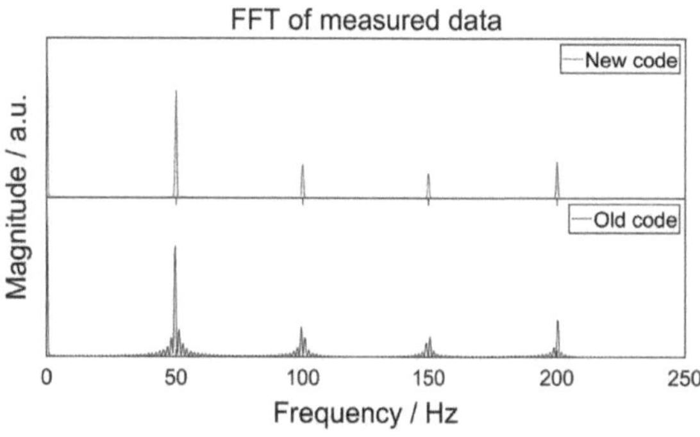

Figure 4: Fourier transform of the original signals measured with an oscilloscope. The data output was measured for 2.4 s with a preset period of 20 ms in both codes. The narrow peaks (top) display a high accuracy and show no fluctuations in the timing. The broader peaks (bottom) show large deviations from the set frequency.

Fig. 5 shows a direct comparison of the measured periods. The timing, produced by the new code (top), shows fewer bins, which are close to the 20 ms mark. This means that the timing is more precise, with a standard deviation of 107.49 ns and more accurate, with the peak centred around 19.99625 ms. Most importantly there is only one peak in the top graph, meaning that the timing does not fluctuate as it does in the previous timing sequence (bottom). The previous timing sequence has two peaks centred around 20.1173 ms and 20.1235 ms, with standard deviations of 171.56 ns and 192.40 ns, meaning that the activation time of the components fluctuates, which creates artefacts in the reconstructed images. The measurements additionally re-

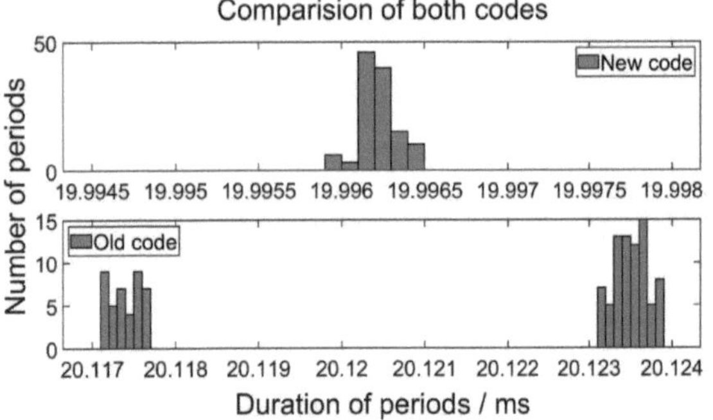

Figure 5: Measurement of the duration of 120 periods created by the new code (top) and the old code (bottom). The new code shows improved accuracy, precision and reproducibility compared to the old code.

vealed an unintentional constant delay of approximately 3 μs, induced by the `digitalWrite()` function. Although this delay could be mitigated by directly addressing the pins without using the function, if necessary. Since the delay remains constant across all measurements and does not introduce timing fluctuations, it was decided to keep the existing implementation. Another unintentional but also constant delay results from the `delayMicroseconds()` function, which runs 1.25 μs too fast. This also explains why the peak in Fig. 5 in the top graph is not centred around exactly 20 ms

3.2   Functional imaging

The power meter measurements were taken at the sample level of the system to measure the actual power arriving at the retina, with an iris placed just before (8.0 mm) to represent the human pupil. The LED was supplied by 3.3 V with a 470 Ω resistor. The measurement showed a range from 13.5 μW to 110 μW with the aperture in the stimulation arm being completely closed (1.2 mm) or open (20.0 mm). Although the power is below any danger level to the human retina, the power range could be further adjusted by choosing higher or lower resistors. No image mask was used during the power measurements.
The spectrum of the LED is well within the range of visible light, matching the sensitivity of photoreceptors.

4   Conclusion

This paper introduces an enhanced timing sequence designed for triggering components FF-SS-OCT system. The novel code enables the optimal utilization of two lasers as a unified light source, ensuring accurate, precise reproducible measurements with an improved axial resolution. Furthermore, the system and its associated code were expanded to incorporate a new stimulation unit, thereby enabling functional imaging capabilities.
With a robust timing sequence, the system can now be extended by further timing sensitive components, such as a third laser or a second camera, in the future. The implemented interrupt service routines can also be applied to any time sensitive project outside the range of FF-SS-OCT.

Acknowledgement

The work has been carried out at Vrije Universiteit Amsterdam, Departement of Physics and supervised by the Biophotonics and Medical Imaging section and the Institute of Biomedical Optics, Universität zu Lübeck.

Authors' Statement

Conflict of interest: Authors state no conflict of interest.

5   References

[1] D. Hillmann et al., *In vivo optical imaging of physiological responses to photostimulation in human photoreceptors.* In: Proceedings of the National Academy of Sciences, vol. 113, no 46, pp. 13138–13143, 2016.

[2] K. Bizheva et al., *Optophysiology: Depth-resolved probing of retinal physiology with functional ultrahigh-resolution optical coherence tomography.* In: Proceedings of the National Academy of Sciences, vol. 103, no 13, pp 5066–5071, 2006.

[3] Arduino 2023, *Arduino reference.* Available: https://www.arduino.cc/reference/en/ [last accessed on 2023-12-18].

[4] Autodesk Instructables, *Arduino Timer Interrupts.* Available: https://www.instructables.com/Arduino-Timer-Interrupts/ [last accessed on 2023-12-21].

[5] D. Baylor, *How photons start vision.* In: Proceedings of the National Academy of Sciences, vol. 93, no 2, pp. 560–565, 1996.

Towards Microcontact Printing and Nanoimprint Lithography of Polymer-based Photonic Biosensors

Julian Lennard Völter [1], Maik Rahlves [2]

[1] Medical Engineering Science, Universität zu Lübeck, julian.voelter@student.uni-luebeck.de
[2] Institute of Biomedical Optics, Universität zu Lübeck, maik.rahlves@uni-luebeck.de

Abstract

Fabrication of chip-based photonic biosensors require surface structures with feature sizes down to a few hundred nanometers. Microcontact printing (μCP) and nanoimprint lithography (NIL) are promising candidates for structuring poylmer material surfaces meeting such demands. In this work, we present a prototype of a combined μCP and NIL device for sensor surface structuring consisting of an xyz-positioning stage with a UV light source, force sensor, stamp holder, sample table and an Arduino controller. Device characterization and NIL replications results are presented using stamps made from polydimethylsiloxane (PDMS), which is commonly used in soft lithography. The capability of our prototype is demonstrated by successful replication of optical gratings in optical adhesive. Replication results are analyzed by confocal microscopy.

1 Introduction

In recent years, chip-based photonic biosensors for quantitative detection of medically relevant biomarkers such as thyroid-stimulating hormone (TSH) have gained increasing interest [1, 2]. Established sensor principles such as photonic crystal based sensors often rely on micro- or nanostructured surfaces [2] which require, depending on the sensor's materials, extensive microfabrication techniques. To reduce sensor costs, polymer based sensors offer the possibility for low-cost and large scale fabrication by, e.g., nanoimprint lithography (NIL) [2]. Instead of modifying the sensor's surface geometry, other sensor principles are based on a microstructured functionalization by antibodies which act as diffractive optical elements (DOE) such as gratings or even Fresnel lenses [1] in the case of Molography. Here, binding of biomarkers change the diffraction efficiency of the DOE, leading to an intensity modulation of the sensors's signal. However, reliable fabrication of such devices remain challenging and are in the focus of current research.

To address these challenges, combining NIL and microcontact printing (μCP) lends itself for both, creating nanostructured surfaces as well as achieving micro-patterned functionalization. μCP is a widely utilized technique in nanoscience and nanotechnology for creating nanostructures of bioreceptors such as proteins and nucleic acids. This versatile method involves inking probes onto an elastomeric stamp, enabling precise and large-scale patterning with high resolution, making it popular for biosensing applications due to its simplicity, multiplexing capability, label-free features, and cost-effectiveness. Microcontact

printing generally consists of two processes. The first process consists of the inking of the stamp and the stamping on a solid substrate. The second process is divided into direct μCP and indirect μCP. Direct μCP is the standard process in which the probes are patterned by stamping. In indirect μCP, the backfilling agents are stamped first and the probes are then physisorbed on the gaps by incubation [3].

An alternative technique for high-resolution surface structuring is NIL. Nanoimprint lithography (NIL) has established itself as a technology for the high-resolution structuring of material surfaces in the micro- and nanometer range. In contrast to conventional lithography methods, NIL offers a cost-efficient alternative with high flexibility and resolution. The embossing process is based on the use of a transparent stamp, typically made of elastic material such as polydimethylsiloxane (PDMS), which has a predefined surface structure [4]. An outstanding feature of NIL is its ability to create almost any sample with extraordinary precision. This is made possible by the ability to produce individual stamps, making the technology extremely flexible for different applications. The surface structure of the stamp is transferred directly to the material to be structured. High-resolution details down to the sub-10 nm range can be achieved [5].

Another decisive advantage of nanoimprint lithography is its high throughput. As the embossing process essentially functions as a "stamping" process, large areas can be structured quickly and efficiently. This leads to a considerable increase in production speed compared to direct writing processes such as electron beam lithography. The high reproducibility and scalability make NIL particularly attractive for the industrial production of micro- and nanostructures. An important area of application for NIL is in the manufacture of micro- and nanoelectromechanical systems (MEMS and NEMS) and in the production of optoelectronic components. Additionally, NIL has found applications in biotechnology, for example in the production of biochips and microfluidic systems. The precise structuring of surfaces on the nanoscale opens up new perspectives for the development of sensors, optical components and advanced materials [6].

In this work, a prototype of a combined µCP and NIL systems for the development of biophotonic sensor structures is presented. UV nanoimprint lithography is used as the basic setup which was combined with an additional lateral xy-translation stage for later µCP pattern transfer. In addition, a force sensor was added to the embossing vertical z-axis, which allows for a controlled touch force which is essential for both, µCP and NIL.

2 Material and Methods

2.1 Experimental Setup

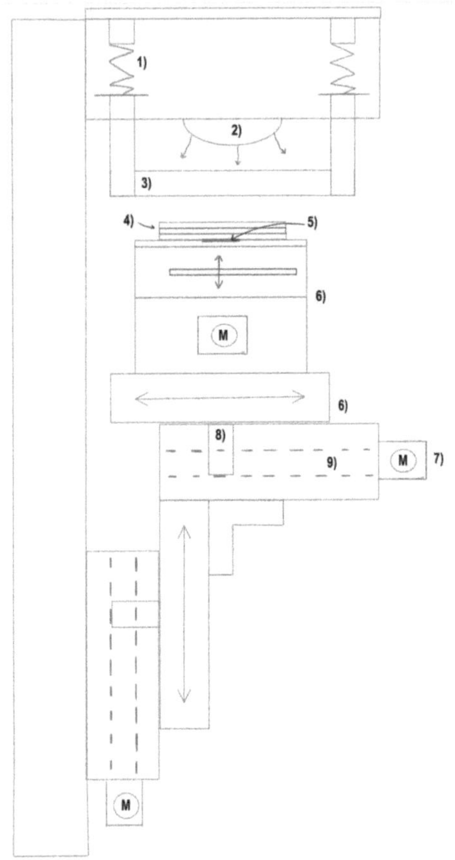

Figure 1: Sketch of the UV-NIL/µCP system: (1) spring; (2) UV-LED; (3) UV-transparent glass; (4) PDMS mold, UV-curing optical adhesives, polymersubstrate; (5) pressure sensor; (6) positioning units; (7) stepper motor; (8) threaded nut; (9) threaded rod.

A sketch of the developed µCP/NIL prototype is shown in Fig. 1. For lateral substrate translation and embossing, a stepper motor-based xyz positioning unit is manufactured with a travel distance of 27 mm per axis. High-torque stepper motors (ST2818S1006-KFE12 by Nanotec) [7] with a step angle of 1.8° (full step) are used for the drive and are controlled via an Arduino Uno and two Adafruit Motor Shields v2.3.[8] Two stepper motors are connected per motor shield, which are supplied with 5 volts and a current of 1.2 amps. A 3D-printed sample table equipped with a pressure sensor (thin film pressure sensor by Az-Delivery) [9] is attached to the positioning unit. The pressure sensor regulates the contact pressure of the stamp in the range of 0 - 5 N. The pressure unit is mounted above the positioning unit with a distance of 5 mm between the UV-transparent glass and the sample table. The 3D-printed parts are produced with an Anycubic Photon Mono 4k printer utilizing TR300 resin (phrozen, China). The stamping unit is 3D-printed and consists of a holder equipped with a UV-LED (Nichia NCSU276C, 1050 mW, 365 nm) [10], which is mounted in the holder's center with the optical active area facing downwards. The UV-LED is connected to the Arduino Uno via a CN5711 LED driver [11]. The driver can be set as a constant current source between 30 mA and 1500 mA. To apply the print to the stamp, a holder with a UV-transparent glass (WG41050, Thorlabs, USA) is attached to the stamp unit. The glass is held in place by four pins that can be pressed 7 mm into the holder. The counter pressure is regulated by the weight of the glass and four springs with a spring constant of 75.117 Nm^{-1} each. The springs could be exchanged to regulate the maximum pressure. The UV-LED supplies 30 mW of power to the stamp for curing the optical adhesive. The Arduino controls the stepper motors, regulates the contact pressure via a pressure sensor and supplies the UV-LED with voltage.

2.2 Replication Process and Characterisation

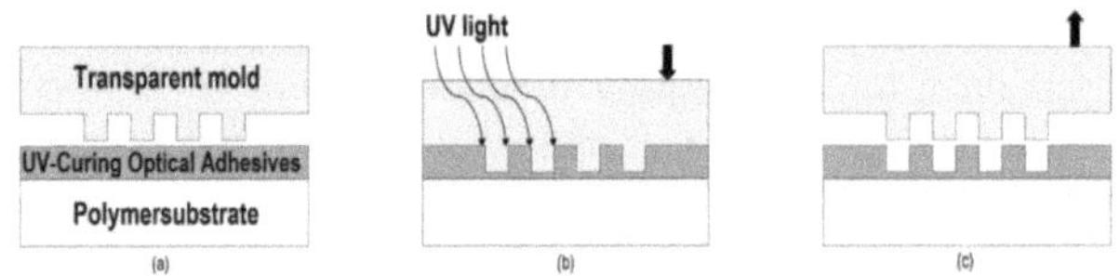

Figure 2: Standard NIL fabrication sequence: a) initial state; b) UV-curing of optical adhesives; c) demolding of stamp and substrate.

The UV-NIL process essentially consists of a transparent mold, UV-curing optical adhesives and a polymersubstrate as shown schematically in Fig. 2. The transparent stamp is made from PDMS (SYLGARD 184, Sigma-Aldrich, GER) using replica molding. A pre-structured mold (master) is used for this process. PDMS is poured onto the master for molding and then cured at room temperature. As UV-curing optical adhesive, NOA 61 (Norland, USA) is used. This has a low viscosity at room temperature and can be cured at wavelengths ranging from 350 nm - 380 nm. Once fully cured, the material is heat resistant up to 125 °C, non-sticky and solvent resistant. This is achieved by complete cross-linking when exposed to UV light [12]. The material used as polymer substrate is polymethyl methacrylate (PMMA; 99524 GT, Röhm GmbH, GER). The thickness of the PMMA is 50 μm. For the printing process, the optical adhesive is applied to the polymer substrate in a thin layer. The stamp is positioned on the adhesive. Suitable pressure must then be applied to the stamp to ensure that it is fully seated and the adhesive is distributed as evenly as possible. When UV-radiation hits the adhesive, it solidifies and remains in the pressed form. After complete curing, the stamp is removed and the pattern adheres to the polymer substrate.

3 Results and Discussion

The developed and constructed device is shown in Fig. 1. The xyz-positioning unit was assembled from three individual positioners so that adjustments are ultimately possible in all of the three axes. The force sensor of the sample table is controlled with an Arduino, so a contact pressure of 5 N is always guaranteed.

Fig. 3 shows the progression of the contact pressure. It can be seen that there is no linear increase, as the stamp has increased friction during pressing due to the 3D-printed parts. The surface warped slightly during curing, resulting in a curvature. As a result, the overlying pressure sensor transmits incorrect values to the Arduino. Pressing the curvature flat results in an inconstant deviation from the real position of the sample. The shape illustrates that the pressure only increases slightly at the start of the embossing process due to the curved plate. Once the curvature has been overcome, the pressure rises sharply. The own weight of the spring pressure and the stamping unit influence the shape. These irregularities are caused by an increased resistance of the individual pins of the stamp holder. This could be avoided by mounting the stamping unit with less friction.

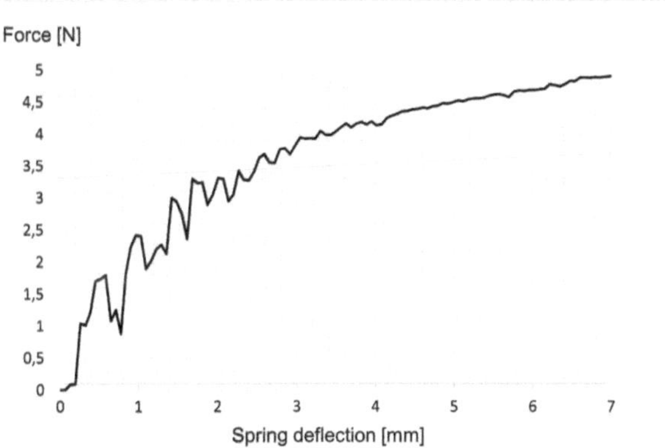

Figure 3: Force measured by the device's sensor as function of the stamp holder's spring defection.

Fig. 4 shows an optical grating with a line spacing of 10 μm on a PMMA substrate on the left side and the original grating on the right side in BK7 glass. The image was taken by a confocal microscope (VK-9710, Keyence, Japan). The print shows a clear representation of the grid structure. Artifacts such as dust inclusions, detachment of the optical adhesive when the stamp is removed and irregular distribution of the adhesive are visible. The formation of dust inclusions is explained by the manual application of the optical adhesive and the intermediate time for applying the stamp. The optical adhesive can lose its adhesion to the PMMA sheet, if the curing time was set for too long. Optimal curing depends on the layer thickness of the optical adhesive and the exposure time, so this problem occurs mainly if the adhesive is applied unevenly. The adhesive should be applied thinly and dripped on by hand from the tube. The thickness of the adhesive layer was not regulated.

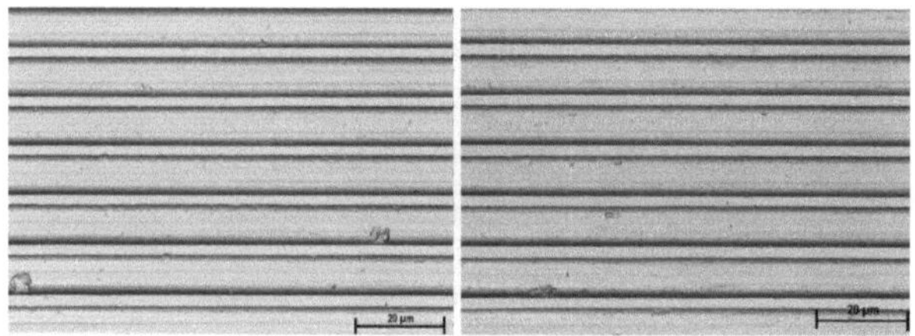

Figure 4: Imprinted grating in NOA 61 optical adhesive (left) and BK7 glass master grating (right).

Fig. 5 shows the 3D-topography of replica and master measured by confocal microscopy. Fig. 6 shows corresponding profiles obtained from the same measurements of the BK7 glass grating with 10 μm pitch (light gray) and its NOA 61 replica (black). In comparison, the master was accurately replicated in NOA 61, which is readily observed when comparing smaller features, e.g., in the grating's valleys. Deviations between both profiles, which are observed at the flanks of the gratings may be attributed to measurement artifacts of the microscope due to a loss of signal intensity.

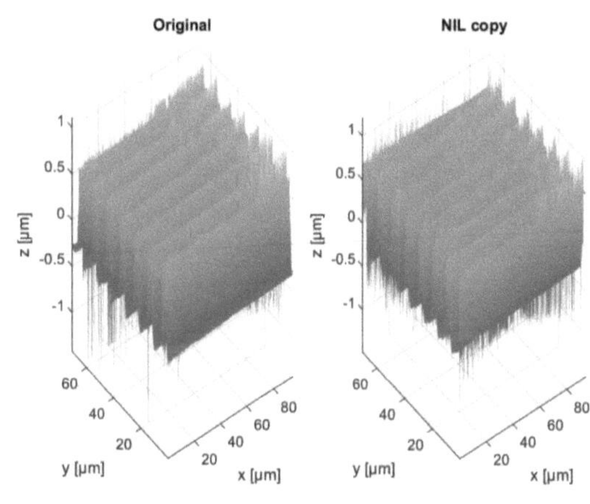

Figure 5: 3D topography of NOA 61 replica with 10 μm grating pitch measured by confocal microscopy.

For future work, the 3D-printed parts could be made of metal to enable more precise measurement. The optical adhesive can be applied by lightly wetting the surface to ensure even distribution. This makes it possible to deter-

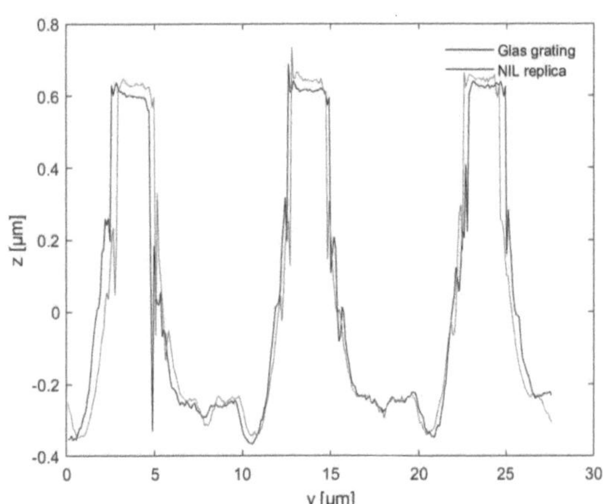

Figure 6: Measured side profile of the original glass grating with a line distance of 10 µm on BK7 glass and the NIL replica recorded by a Keyence VK-9710 confocal microscope.

mine the exact exposure time. Another approach would be a camera-controlled alignment of the sample table. This would ensure that the sample to be stamped is always positioned exactly in the center under the stamping unit and would enable uniform contact pressure. The system could also be converted into a micro contact printing method by attaching the PDMS stamp to the stamping unit and then immersing it in a bath of liquid.

4   Conclusion

This work presents a prototype of a UV-NIL and µCP device. The aim is to reproduce low-cost nanostructures. The prototype essentially consists of an xyz-positioning unit, a stamping unit including a force sensor and a sample table. The sample consists of a PMMA plate and optical adhesive, which is cured using UV radiation. Its functionality was tested by replicating a PDMS grating with a pitch of 10 µm into NOA 61 UV-curing optical adhesive. Confocal surface topography measurements reveal that the grating was accurately replicated and printing forces can be measured by the force sensor integrated in the vertical translation stage used for embossing. Remaining force measurement errors can be attributed to a slip-stick effect of the stamp holder during embossing, which needs to be improved in future work. The presented prototype also features a lateral translation stage which enables µCP by controlled wetting and pattern-transfer to a substrate, which will also be analyzed in future work. This will allow us to functionalize surfaces with specially designed micro-patterned, e.g., antibodies areas enabling future application in photonic sensing.

Acknowledgement

The work has been carried out and supervised by the Institute of Biomedical Optics, Universität zu Lübeck.

Author's Statement

Conflict of interest: Authors state no conflict of interest.

5   References

[1] V. Gatterdam, A. Frutiger, K. P. Stengele, D. Heindl, T. Lübbers, J. Vörös, and C. Fattinger, "Focal molography is a new method for the in situ analysis of molecular interactions in biological samples," *Nature Nanotechnology*, vol. 12, pp. 1089–1095, 11 2017.

[2] M. Shahbaz, M. A. Butt, and R. Piramidowicz, "A concise review of the progress in photonic sensing devices," vol. 10, no. 6, p. 698, 2023.

[3] S. Alom, R. Ab, and C. S. Chen, "Microcontact printing: A tool to pattern," 1999. [Online]. Available: www.rsc.org/softmatter

[4] H. Lan and H. Liu, "Uv-nanoimprint lithography: Structure, materials and fabrication of flexible molds," *Journal of Nanoscience and Nanotechnology*, vol. 13, pp. 3145–3172, 5 2013.

[5] W. Zhou, "Nanoimprint lithography: An enabling process for nanofabrication," *Nanoimprint Lithography: An Enabling Process for Nanofabrication*, vol. 9783642344282, pp. 1–256, 8 2013.

[6] K. Mohamed, "Nanoimprint lithography for nanomanufacturing," *Comprehensive Nanoscience and Nanotechnology*, vol. 1-5, pp. 357–386, 1 2019.

[7] "St2818 - schrittmotor – nema 11 | nanotec," Access: 07.01.2024. [Online]. Available: https://de.nanotec.com/produkte/173-st2818-schrittmotor-nema-11

[8] "Adafruit motor/stepper/servo shield for arduino v2 kit [v2.3] : Adafruit industries, unique & fun diy electronics and kits," Access: 09.01.2024. [Online]. Available: https://www.adafruit.com/product/1438

[9] "Flexibler dünnfilm drucksensor | präziser kraftsensor 3,3 / 5v analog," Access: 10.01.2024. [Online]. Available: https://www.az-delivery.de/products/flexibler-dunnfilm-drucksensor

[10] "Nichia corporation specifications for uv led part no. ncsu276ct," Access: 12.01.2024. [Online]. Available: https://365nm.ru/NCSU276C-E.pdf

[11] "Led-treiber cn5711," Access: 04.01.2024. [Online]. Available: https://elektro.turanis.de/html/prj464/index.html

[12] "Uv-curing optical adhesives," Access: 03.01.2024. [Online]. Available: https://www.thorlabs.com/newgrouppage9.cfm?objectgroup_id=196

Setup for comparative studies to find optimal measurement parameters for specific tissue using dmOCT

Jana Hadjichrysanthou [1], Noah Heldt [2], Martin Ahrens[2], Svea Höhl[3], Tim Eixmann [3] and Gereon Hüttmann [2]

[1] Medical Engineering Science, Universität zu Lübeck, jana.hadjichrysanthou@student.uni-luebeck.de

[2] Institute of Biomedical Optics, Universität zu Lübeck, {noa.heldt, m.ahrens, gereon.huettmann}@uni-luebeck.de

[3] Medizinisches Laserzentrum Lübeck GmbH, Peter-Monnik-Weg 4, 23562 Lübeck, Germany, {s.hoehl, tim1.eixmann}@uni-luebeck.de

Abstract

Cholesteatoma is a form of chronic otitis media that spreads as an aggressive growth in the middle ear. The only treatment for cholesteatoma to date is surgical removal of the affected tissue. A reliable intraoperative diagnosis of cholesteatoma still requires the trained eye of the surgeon. In an earlier study by Leichtle et al., it was shown that cholesteatoma tissue can be visualised ex vivo using dynamic microscopic optical coherence tomography (dmOCT). The aim of this study is to investigate the minimum requirements that a dmOCT system must fulfil in order to be able to differentiate specific tissue types endoscopically in vivo. For this purpose, a setup was developed that allows comparison of two OCT systems by measuring simultaneously the same sample section at the same position. The functionality of the system was demonstrated by the measurement of a mouse trachea sample.

1 Introduction

One form of chronic otitis media is cholesteatoma, a benign growth that spreads aggressively in the middle ear and can affect surrounding structures such as the auditory ossicles [1]. It is a progressive disease that, if left untreated, can lead to serious complications such as hearing loss, balance problems and even life-threatening brain abscesses or further infections [1, 2, 3]. It is characterized by the whitish, shimmering cells of the resulting epithelium [4].

The current treatment option involves microsurgical ear surgery to remove all inflammatory tissue. The challenge is to differentiate the destructive inflammatory process from the sensitive surrounding tissue. The anatomical structures in this area, including blood vessels, nerves, bone tissue and the underlying brain, are in proximity to each other. As described above, treatment options have been limited to invasive surgery. Moreover, the differentiation between tissue and surrounding structures is entirely dependent on the surgeon's judgement. This has led to the development of the concept of a theragnostic probe. This probe aims to provide marker-free differentiation of tissue using dynamic microscopic optical coherence tomography (dmOCT) and simultaneous treatment of the tissue by laser ablation. dmOCT describes a method of microscopic resolution and contrast enhancement of regular OCT imaging. Fluctuations in the signal are used to visualize cellular and subcellular structures, whereby a histology-like contrast is achieved [5]. Fluctuations of the absolute values of the OCT signal are analyzed and Fourier transformed over time [5]. Then the integral values are divided into three frequency ranges and color-coded. The three different colors represent the different speeds of the detected cellular movements. For example, Leichtle et al. [6] displays high signal fluctuations (4.9-25 Hz) in red, medium ones (0.4-4.9 Hz) in green and slow ones (<0.4 Hz) in blue. The pixels of the dmOCT image are allocated to the mentioned color channels depending on their signal fluctuations and displayed in an RGB image.

In a previous study by Leichtle et al. [6], it was shown that cholesteatoma tissue can be visualized ex vivo using dmOCT. Fig. 1 shows OCT images of cholesteatoma and chronic otitis media samples with dynamic contrast. Individual regions and structures of the samples can be distinguished. Both samples were analyzed with a microscopic OCT system (mOCT), which has a higher axial and lateral resolution of nearly 1 μm which is considerably higher compared to regular OCT systems [5]. For implementation in a probe, the dmOCT system must be adapted to its requirements. A probe limits the size and complexity of the system. This plays an important role in the development of the probe. In order to determine the minimum requirements that a dmOCT must fulfil in order to be able to differentiate specific tissue types endoscopically in vivo, a setup was developed that makes it possible to compare two OCT systems and different imaging optics with each other. For a solid comparison of the systems, it is necessary that tissue samples are measured simultaneously and at the same position. If the measurements were carried out one after the other, a comparison would be difficult, as it cannot be guaranteed that the sample is measured in the same position. In addi-

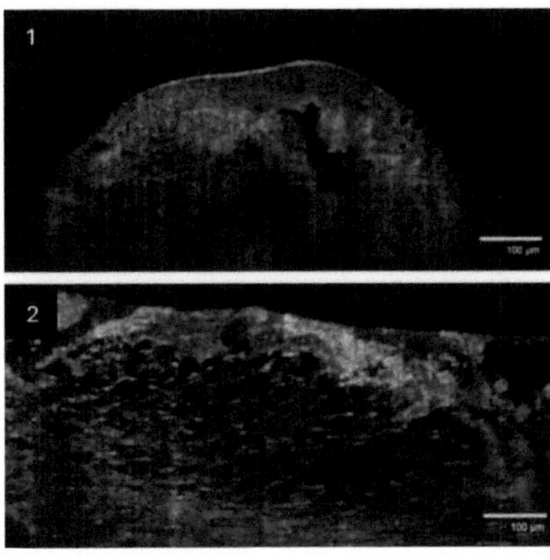

Figure 1: dmOCT images of tissue samples extracted from a patient suffering from chronic otitis media (1) and a patient suffering from cholesteatoma (2). Adapted from [6]

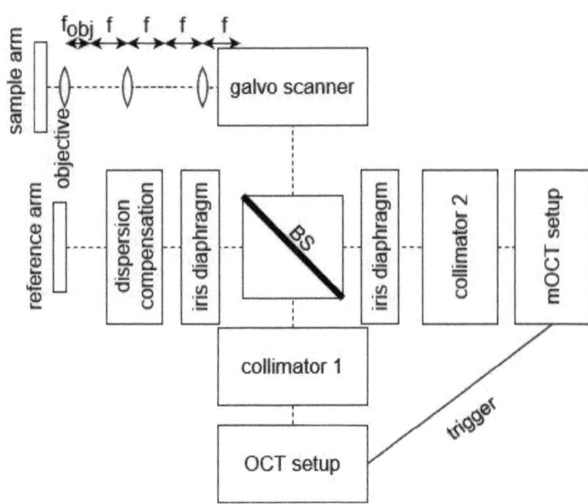

Figure 2: Schematic drawing of the setup. It contains an OCT and a mOCT setup. Each setup consists of a light source, a spectrometer and evaluation electronics. They are connected by a master/slave trigger for simultaneous data acquisition. The light sources are each connected to a collimator via an optical fiber. A beam splitter divides the light 50/50 into a sample arm and a reference arm. In the sample arm, a two-axis galvo scanner images the beam through a 4f optical system into the rear focal plane of the interchangeable objective.

tion, the metabolism of the sample changes over time. The aim is to determine the minimal axial, lateral and temporal resolution of the OCT system. For optimal orientation in the surgical field and quick tissue assessment, a large field of view (FOV) is advantageous. However, this leads to a trade-off between lateral resolution and the FOV. The functionality of the system was demonstrated by measuring of a mouse trachea sample.

2 Material and Methods

2.1 OCT imaging setup

The setup for simultaneous measurement using two OCT systems is shown in Fig. 2. One system is a spectral domain OCT (Ganymede 621, Thorlabs Inc., USA) with a center wavelength of 900 nm and a spectral range from 820-980 nm and 3 µm axial resolution. This is connected to a built-in spectrometer and a dedicated PC for recording and processing the data. The second system is a microscopic OCT setup. It consists of a super continuum light source (SuperK FIANIUM FIU-4, NKT Photonics, Birkerod, Denmark) with a spectral range from ~400-2400 nm of which ~550-900 nm is used resulting in 1.3 µm axial resolution. The super continuum laser has a repetition rate of 312 MHz. Furthermore, it is also connected to a spectrometer (OctoPlus CL, Teledyne, e2v, Canada) and its own PC. The possible use of other light sources/OCT systems in other spectral ranges, e.g., center wavelength 1300 nm, would be desirable. In order to record the incoming light signals on both systems simultaneously, the cameras of both systems are triggered via a master/slave connection. The two OCT setups are each connected to a collimator (60FC 4-A15-02, Schäfter + Kirchhoff GmbH, Germany and F260APC-B, Thorlabs Inc., USA) via an optical fiber to generate a collimated beam. In order to adjust the beam intensity of the supercontinuum light source, an iris diaphragm is located behind the collimator. A beam splitter is used to direct 50% of the light beam into the reference arm and 50%

into the sample arm. The choice of beam splitter posed a challenge in this setup and is discussed in more detail in section 3.1. The reference arm consists of a lens with a mirror behind it, which reflects the light back, a 5 mm thick SF57 glass for dispersion correction and an iris diaphragm to regulate the intensity of the beam. In the sample arm, the light beam is directed into a two-axis galvanometric scanner (CT 6210KB, Cambridge Technology, Germany). This ensures that the beam is deflected vertically and horizontally by moving the two mirrors installed in it, so that the sample is scanned over its entire surface. After the galvanometric scanner, a 4f optic ensures that the beam is imaged into the rear focal plane of the interchangeable objective lens (LSM03 or LSM02, Thorlabs Inc., USA), which focuses the beam onto the sample.

2.2 Optical spectrum analysis

For the setup described in section 2.1 to work, the selected beam splitter must be compatible with the two light sources. This means that after transmission or reflection of the beam, the intensity of the two light sources should be evenly distributed across their spectrum. In order to select a suitable beam splitter, it is essential to have accurate knowledge of the spectrum of the light sources. To determine the spectrum of both light sources, an optical spectrum analyzer (OSA, AQ6374, Yogogawa, Germany) was used (Fig. 3). Measurments showed that the supercontinuum light source of the mOCT setup (FIU-4) has a relatively uniform intensity distribution in the spectral range of 550-900 nm, while the light source of the OCT setup (Ganymede) has a relatively uniform behavior in the spectral range of 820-980 nm.

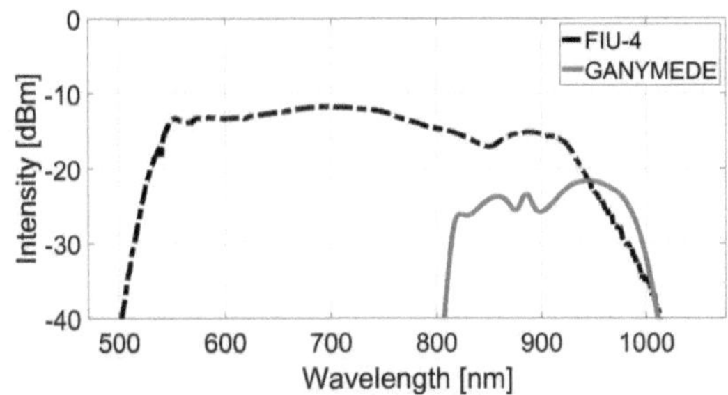

Figure 3: Used spectrum of the two light sources FIU-4 and Ganymede.

3 Results

3.1 Beamsplitter

Using the OSA measurement of the two light sources described above in section 2.2, it was possible to formulate the requirements for a beam splitter. The transmission and reflection ratio must be uniform over the broadband spectrum of 500-980 nm. A uniform R/T ratio up to 1300 nm is desirable for the utilization of other light sources. This ensures that the spectrum is not cut off or that individual wavelengths are attenuated. In addition, the transmission on average (Tav) and the reflection on average (Rav) should be 50% each. Various beam splitters were compared with each other with regard to the above criteria, including fiber-based beam splitters, beam splitter cubes and plates. As many of the considered beamsplitters were rejected due to their specifications in the wavelength range of the comparison system, only a selection is shown in Fig. 4. Two fiber-based beamsplitters (TW630 and TW670-50:50, Thorlabs Inc., USA), each with a center wavelength of 630 nm and 670 nm, a beamsplitter cube (custom-made, VM-Tim GmbH, Germany) and a beamsplitter plate (BSW26R, Thorlabs Inc., USA) are illustrated. The data for the cube and the plate are taken from the raw data in the data sheet. The two fiber beam splitters were measured using the OSA. It is evident that the cube's transmission sharply decreases below 700 nm, resulting in a significant portion of the spectrum being cut off. Therefore, it does not meet our criteria. The two fiber beam splitters exhibit a ratio of approximately 50% at wavelengths above 600 nm. However, the TW630 is unsuitable for the desired spectral range of 500-980 nm as Tav and Rav drift apart significantly at around 750 nm. The spectrum was measured up to a wavelength of 900 nm using the OSA, which is sufficient to evaluate the criteria. The TW670-50:50 exhibits comparable behavior, but the ratio is almost 50% up to approximately 850 nm before diverging significantly. In contrast, the beam splitter plate has a ratio of around 50% across the entire relevant spectral range and up to 1200 nm. Therefore, we chose this component.

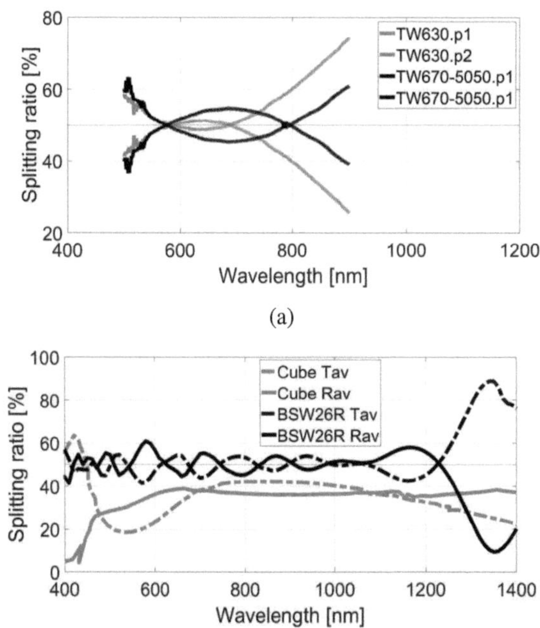

(a)

(b)

Figure 4: Splitting ratio of different beamsplitters. (a) TW630 and TW670-5050 are fiber based beamsplitters and therefore they are measured in part 1 (p1) and part 2 (p2). (b) The ratios of the beamsplitter cube and plate (BSW26R) are depicted as transmission (Tav) and reflection (Rav).

3.2 Calibration

The scan position was calibrated by measuring a grid target (G2300C, 300 square mesh, Copper 3.05 mm, Agar Scientific, UK). The grid has a fixed line spacing of 83 μm. To establish the relationship between the actuation voltage of the galvos and their deflection, the number of pixels between the grid lines can be determined from the en-face representation of the measurement in Fig. 5a. To do this, an amplitude plot was made by drawing a line orthogonal to the grid through the grid target image and plotting the gray values as shown in Fig. 5b. The scanners can be calibrated by determining the average number of pixels between the peaks of the amplitude plot and the known peak spacing. After cal-

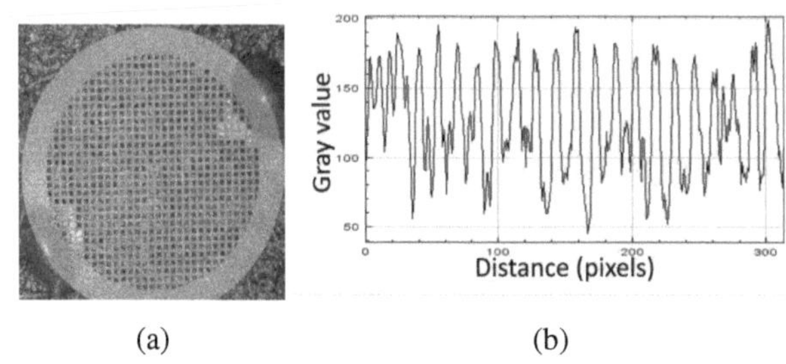

(a)   (b)

Figure 5: (a) OCT grid image. (b) Gray values of a line drawn orthogonally to the grid in the OCT grid image.

ibration the lateral resolution was determined using a scattering sample (silver particles 200-400 nm in epoxy resin (25 μL/ml)). The LSM02 objective achieved a lateral resolution of 8.5 µm for the mOCT and 9.6 µm for the OCT, while the LSM03 objective achieved a lateral resolution of 14.8 µm for the mOCT and 16.8 µm for the OCT.

3.3 OCT images

Simultaneous measurements of a mouse trachea sample were carried out using OCT and mOCT. Examples of these simultaneously achieved images are shown in Fig. 6. The LSM03 objective was used in one instance and the LSM02 objective in another. Additionally, the dynamic contrast was calculated and is visible as a colored signal on the images. The cilia layer is clearly visible in yellow, indicating active movements in this area. The uppermost tissue layer consists of cilia and efficiently transports mucus and particles.

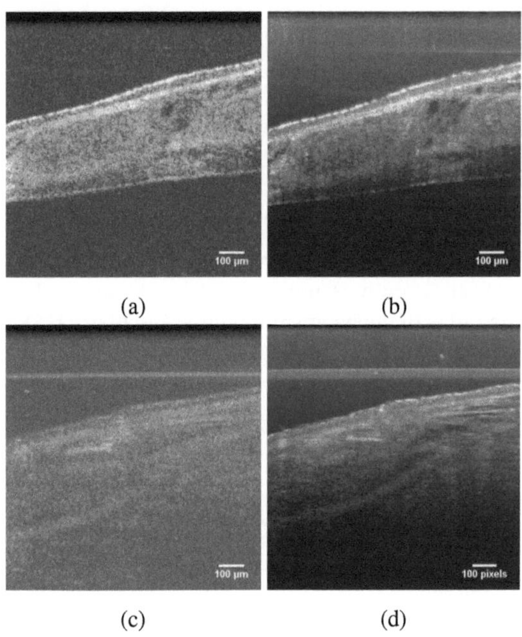

Figure 6: dmOCT images of a mouse trachea. Image (a) and (c) were captured using the OCT setup, whereas image (b) and (d) were achieved by the mOCT setup. The LSM03 objective was employed for (a) and (b), while the LSM02 objective was utilized for (c) and (d).

As anticipated, the mOCT images exhibit significantly superior contrasts compared to the OCT images due to the system's higher resolution.

4 Discussion

The OSA measurements indicated that the fiber-based beam splitters are highly batch-dependent, exhibiting varying transmission and reflection spectra. This variability is likely due to the mechanical production method of the fiber beam splitters. However, the use of a beam splitter plate presents the challenge of a beam offset. To enable an ideal comparison of different objectives, it would have been advantageous to include an objective changer in the setup to ensure measurements are taken at the same point on a sample. Although it was expected that the OCT setup could produce

higher quality images, this may also depend on the specific sample. It is possible that the image quality is better for other tissue types and localizations.

5 Conclusion

A measuring system was developed to simultaneously measure a sample section at the same position using two OCT systems. The setup proved to be challenging. In the future, it must be clarified whether the used beam splitter generates artifacts due to its broadband nature, and if this is the case, these artifacts must be quantified. Future applications of this setup open up the possibility of carrying out additional measurements to determine the parameters required to differentiate specific tissue as initially intended.

Acknowledgement

The work has been carried out at Institute of Biomedical Optics, University of Lübeck. The authors thank Thorlabs for providing galvanometric scanner mirrors. And Mario Pieper for providing the mouse trachea.

Authors' Statement

Conflict of interest: Authors state no conflict of interest.

6 References

[1] C.-L. Kuo, A.-S. Shiao, M. Yung, M. Sakagami, H. Sudhoff et al., *Updates and knowledge gaps in cholesteatoma research*. BioMed Res. Int. 2015, 854024, 2015.

[2] L.M. Haile, K. Kamenov, P.S. Briant, A.U. Orji, J.D. Steinmetz et al., *Hearing loss prevalence and years lived with disability*. 1990–2019: Findings from the Global Burden of Disease Study 2019. Lancet, 397, pp. 996–1009, 2021.

[3] R. Mittal, J.M. Parrish, M. Soni, J. Mittal, K. Mathee, *Microbial otitis media: Recent advancements in treatment, current challenges and opportunities*. J. Med. Microbiol., 67, pp. 1417–1425, 2018.

[4] Gelbe Liste, *Cholesteatom*. Available: https://www.gelbe-liste.de/krankheiten/cholesteatom [last accessed on 2024-01-21].

[5] M. Münter, M. vom Endt, M. Pieper, M. Casper, M. Ahrens et al., *Dynamic contrast in scanning microscopic OCT*. Optics Letters, 45(17), pp.4766–4769, 2020.

[6] A. Leichtle, Z. Penxová, T. Kempin, D. Leffers, M. Ahrens et al., *Dynamic Microscopic Optical Coherence Tomography as a New Diagnostic Tool for Otitis Media*. Photonics, 10(6), 685, 2023.

Robotically-enabled angular compounding for speckle reduction in optical coherence tomography

Tjalfe Laedtke [1], Sazgar Burhan [2], Simon Lotz [2], and Robert Huber [2]

[1] Medical Engineering, Universität zu Lübeck, tjalfe.laedtke@student.uni-luebeck.de

[2] Institute of Biomedical Optics, Universität zu Lübeck, {sa.burhan, si.lotz, robert.huber}@uni-luebeck.de

Abstract

Speckle noise in optical coherence tomography (OCT) arises due to the high coherence of the utilized light sources. This high contrast noise degrades image quality and limits the diagnostic potential. Angular compounding has been used as a method to decrease speckle noise without sacrificing resolution. It involves the acquisition of images from different angular viewpoints and combining them to obtain a speckle-reduced image. Our proposed method incorporates a robot to position an OCT probe at different angles, makes use of an optimization-based image registration technique to align images and constructs a fused image as an average of the registered angular images. The result is a speckle-reduced composite image with enhanced details. The robot allows images to be captured at many various angles, artificially increasing the numerical aperture. We present initial outcomes of our novel approach and outline the obstacles and potential advancements.

1 Introduction

Optical coherence tomography (OCT) produces high-resolution cross-sectional images with many applications in clinical diagnostics, particularly in ophthalmology [1]. The coherent nature of this imaging modality makes it susceptible to image degrading interference effects called speckle. Interference occurs and speckles emerge when backscattered waves from the sample reach the detector out of phase at intervals smaller than the coherence time of the light source [2]. The result is a speckled OCT image presenting with a granular appearance due to localized regions of constructive and destructive interference producing very bright and very dark spots that have no obvious relationship to the sample. This high contrast noise makes it difficult or impossible to distinguish fine structures, making it necessary to invoke methods of speckle reduction.

A possible way to obtain a speckle-reduced image is to construct an average image based on images with uncorrelated speckle patterns [2]. Spatial compounding achieves this by averaging images that correspond to adjacent regions. The inherent spatial mismatching of this method induces a loss of resolution and concurrent blurring of the speckle-reduced image.

An alternative is angular compounding, where the resultant average image originates from images containing the same spatial information but decorrelated speckles. It relies on modulation of speckle patterns by changing the phase difference between scatterers achieved by altering the incident angle of illumination [2], [3]. The intensities of corresponding voxels in images acquired from different viewpoints are modulated according to the angle-dependent phase shift and converted into a more uniform one by subsequent averaging.

Before averaging, image registration must be performed to find the correspondence between voxels in images acquired from different viewpoints as they are rotated with respect to a reference. The quality of image registration is decisive regarding preservation of resolution, as mismatches produce spatial compounding and blurring.

Commonly, angular images are acquired by offsetting a collimated incident beam from the optical axis of the objective lens to manipulate the incident angle [3], [4]. In this study, we used a robot for image acquisition as it offers maximum flexibility in terms of axis of rotation and the ability to produce very large angles.

2 Material and Methods

2.1 Image acquisition

Image acquisition was performed using a system (Fig. 1) that has been developed in context of large area robotically-assisted OCT (LARA-OCT) for skin imaging [5].

The system includes a 7-axis collaborative robot (LBR iiwa 14 R820, KUKA-AG, Germany) that permits precise movement of the scan unit, also referred to here as the OCT probe, along 7 degrees of freedom (DOF) as it is attached to the robots' flange. The associated MHz-OCT imaging

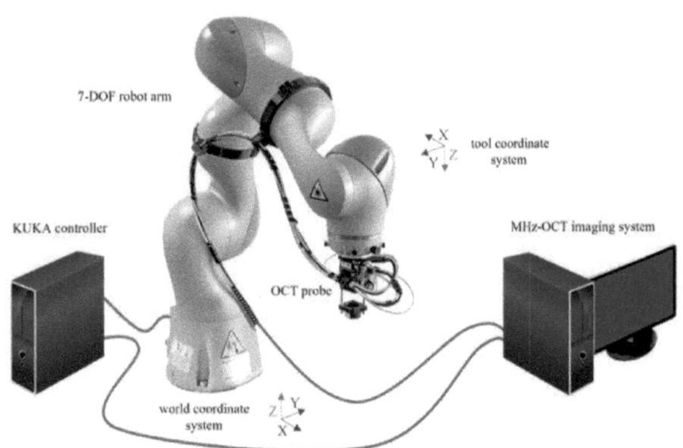

Figure 1: Setup of the LARA-OCT system.

system is based on a Fourier domain mode locked (FDML) laser which allows for A-scan rates of 3.2 MHz [6]. LabVIEW-based software allows images to be acquired at well-defined orientations using robotic positioning.

The system was used to obtain the image data including angular volumes, which are called moving volumes in the context of image registration. These 3D-images are captured at different angles by alternating rotation of the scan unit and image acquisition. The pivot point of rotation is defined by the tool center point (TCP) for which the robots' movement is planned. The TCP can be programmatically set to any location. A desirable position would be aligned with the optical focus of the OCT imaging system to have the field of view cover approximately the same area in angular volumes and a non-rotated reference. Consequently, a z-translation of 160 mm from the flange was chosen which is the theoretical position of the focal point of the scan lens.

2.2 Image registration

Image registration describes the process of finding the transformation that maps a moving volume V_M onto a reference volume V_R. A transformed moving volume $V_M(T_\mu(x))$ is spatially aligned with the reference volume $V_R(x)$ with respect to the voxels x in the reference domain, which is necessary for subsequent image compounding.

The transformation T_μ that describes such an alignment is parameterized by the vector μ. In our case it models a three-dimensional rigid transformation that accounts for rotational and translational displacement between the angular volumes and the reference volume. This is reflected in the parameter vector μ which consists of three Euler angles $\theta_x, \theta_y, \theta_z$ and the translation vector $(t_x, t_y, t_z)^T$:

$$\mu = (\theta_x, \theta_y, \theta_z, t_x, t_y, t_z)^T \quad (1)$$

The sources of these displacements are either controlled and known with reliable accuracy as it is the case for the programmed rotation of the robot or unpredictable as a manifestation of robot inaccuracy and movement of the sample.

The quality of image registration is defined as optimal when the cost function C is minimized with respect to the parameter vector μ. Finding the minimizing parameter vector $\hat{\mu}$ poses a multi-dimensional optimization problem, which can be written as:

$$\hat{\mu} = \arg \min_{\mu} C(\mu; V_R, V_M) \quad (2)$$

The cost function C was modeled using the definition of the mutual information criterion MI given in [7] as a metric for measuring similarity between the reference volume and the transformed moving volume:

$$MI(\mu; V_R, V_M) =$$
$$\sum_{m \in L_M} \sum_{r \in L_R} p(r, m; \mu) \log_2 \left(\frac{p(r, m; \mu)}{p_R(r) p_M(m; \mu)} \right), \quad (3)$$

where m, r are intensity bins associated with the discrete sets of intensities L_M, L_R that make up the histograms of V_M and V_R. The discrete joint probabilities p are estimated using B-spline Parzen windows, whereas the marginal discrete probabilities p_R, p_M are obtained by summing p over m and r, respectively.

The chosen optimizer is called adaptive stochastic gradient descent (ASGD) as proposed in [8] and defined as:

$$\mu_{k+1} = \mu_k - \gamma(t_k)\tilde{g}_k, \quad k = 0, 1, \dots, K, \quad (4)$$

$$t_{k+1} = [t_k + f(-\tilde{g}_k^T \tilde{g}_{k-1})]^+, \quad (5)$$

where $\tilde{g}_k$ denotes a stochastic approximation of the derivative $g = \partial C / \partial \mu$ at μ_k based on a randomly selected small subset of voxels. It can be computed with reduced cost compared to deterministic gradient descent while maintaining similar convergence properties. An adaptive step size mechanism is implemented using an image-driven method of determining the parameters of optimization. The step size is given by evaluating the decaying function γ at time t_k. A time t_{k+1} is determined by evaluating the inner product of the approximate gradients $\tilde{g}_k$ and $\tilde{g}_{k-1}$ which is normalized by the sigmoid function f according to (5). The monotone decreasing function γ evaluates to a larger gain $\gamma(t_{k+1})$ for a decreased time which results from a positive inner product of two consecutive gradients, illustrating that the gradients point in the same direction, justifying larger steps, whereas smaller steps shall be taken when the gradients point in different directions.

The optimization process aims to minimize the cost function by iteratively updating μ according to (4) starting with a zero initiation of μ_0 and ending after a predefined number of iterations.

To further increase the chances of a successful registration, a multi-resolution approach was used, which consists of building an image pyramid and performing the optimization at each level of the pyramid. A Gaussian pyramid was chosen, where the images are not reduced in resolution but are increasingly blurred at higher levels of the pyramid. Starting at the top of the pyramid, the registration of the most smoothed input images leads to a coarser alignment that focuses on larger, dominant structures. This alignment is then refined by moving down the pyramid and ends with a final registration of the original versions of the images.

The methods described are implemented in the open-source image registration framework `elastix` [9], [10], which was used to align the angular volumes to a reference volume.

2.3 Image compounding

Image compounding was performed after registration of every angular volume to the same reference. A mean volume was constructed as an arithmetic average of N registered angular volumes V_{M_i}:

$$\overline{V}(\boldsymbol{x}) = \frac{1}{N} \sum_{i=1}^{N} V_{M_i}(\boldsymbol{T_\mu}(\boldsymbol{x})), \qquad (6)$$

where $\boldsymbol{T_\mu}(\boldsymbol{x}) = \boldsymbol{y}$ denotes the position at which the angular volume is interpolated and mapped to the reference domain $\boldsymbol{x}$.

MelastiX [11] was used to call `elastix` from within MATLAB to allow for manipulation of image data as MATLAB matrices.

3 Results and Discussion

3.1 Image quality

The proposed methods were applied using a polyvinyl alcohol phantom as an imaging sample and produced speckle-reduced composite volumes based on an arbitrary number of registered angular volumes. Exemplary, one slice of a composite volume is compared with the underlying reference in Fig. 2. The speckle reduction achieved by angular compounding is clearly visible, resulting in decreased granularity and more sharply defined structures.

The preservation of resolution can be appreciated in Fig. 3. Here, a region of interest in the original reference is cropped and compared to corresponding regions in an angular-compounded image and a spatially-compounded image. Both techniques show similar speckle reduction, but only angular compounding preserves sharp edges, whereas blurring and loss of information are introduced by spatial averaging.

3.2 Speed considerations

The computational cost for registering an angular volume with a reference volume is approximately 2 minutes for images of a size of 600 x 600 x 1000 voxels ($\approx$ 2.5 GB) using the ASGD optimizer and the mutual information

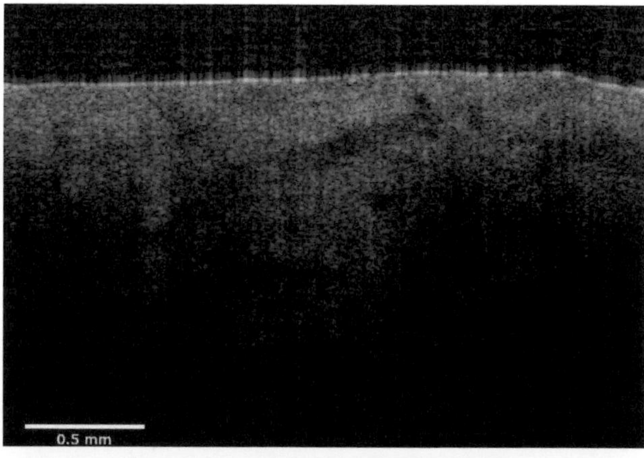

(a) Reference image

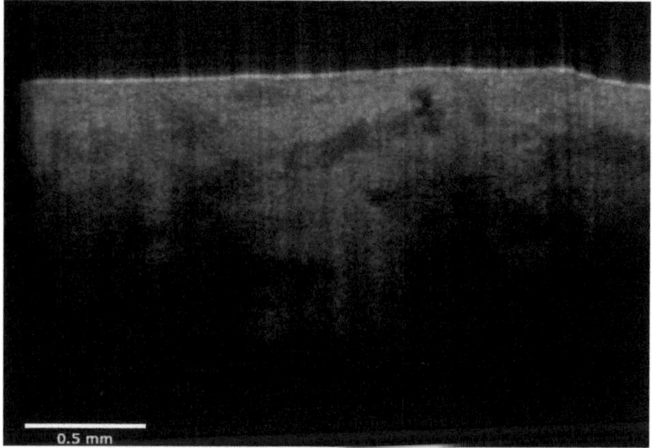

(b) Composite image

Figure 2: An illustration of speckle reduction by angular compounding. (**a**) A speckled slice of a non-rotated reference volume. (**b**) The corresponding despeckled slice of a composite volume that has been constructed as an arithmetic average of ten registered angular volumes, [0.4°, 0.8°, 1.2°, 1.6°, 2.0°, 2.4°, 2.8°, 3.2°, 3.6°, 4.0°], acquired by rotating around an axis perpendicular to the image plane. The left outer area shows a loss of information as the angular volumes only approximately depict the reference volume, with non-overlapping parts of their fields of view.

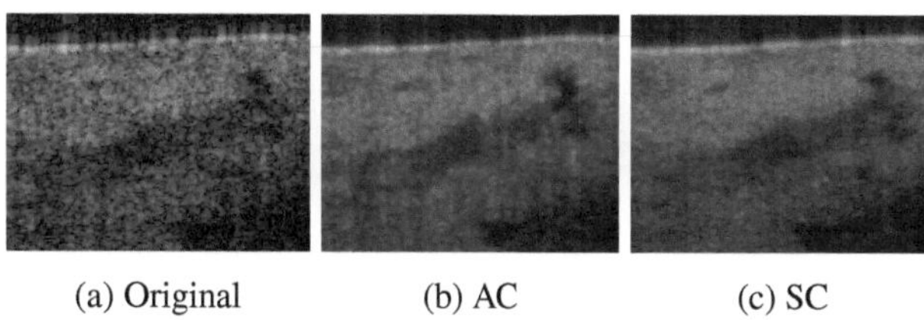

(a) Original   (b) AC   (c) SC

Figure 3: An illustration of the preservation of spatial resolution by angular compounding (AC) compared to spatial compounding (SC) based on Fig. 2. (**a**) A speckled region of interest in one slice of the non-rotated reference volume. (**b**) The corresponding despeckled region in the composite volume obtained by angular compounding. (**c**) The same region as an average of 21 adjacent slices of the reference volume presenting despeckled but blurred.

metric, including preparation of the image pyramids (performed on an Intel Xeon W-2245). However, the whole process of registering 10 angular volumes to one reference and compounding of the individual registered images takes about 45 minutes. The reason for this is that the conversion between MATLAB matrices and image files, which can be handled by `elastix` is computationally expensive. An alternative option would be to use SimpleITK [12] which comes with different interfaces for `elastix` allowing for faster data conversion.

Another approach to improve speed is by tuning the parameters of the entire registration process. Currently, the optimizer performs 1000 iterations at four pyramid levels, which may be excessive for the image data we provided as convergence is reached much earlier.

4 Conclusion

A robotically-enabled implementation of angular compounding is proposed as a viable method of speckle reduction, leading to an improvement in image quality. A denoised image is constructed by fusing rotated OCT images with uncorrelated speckles. These are obtained by altering robotic positioning and aligned with well established methods of image registration. The composite images produced do not exhibit any visible blurring, indicating that the spatial resolution has been maintained.

As this work only qualitatively examined speckle reduction and preservation of spatial resolution, subsequent studies will need to include quantitative assessment using appropriate metrics. This would also allow for examination of the relationship between speckle reduction in composite images and the number of underlying angular images, as well as the range from which they have been acquired.

The use of robotic positioning enables the evaluation of the boundaries of angular compounding for speckle reduction in upcoming studies. The range of usable positions and orientations of viewpoints to produce images with uncorrelated speckle is vast and has not been fully explored in this work.

Acknowledgement

The work has been carried out and supervised by the Institute of Biomedical Optics, Universität zu Lübeck.

Authors' Statement

Conflict of interest: Authors state no conflict of interest.

5 References

[1] J. G. Fujimoto, C. Pitris, S. A. Boppart, and M. E. Brezinski, "Optical coherence tomography: An emerging technology for biomedical imaging and optical biopsy," *Neoplasia*, vol. 2, no. 1–2, pp. 9–25, Jan. 2000.

[2] J. M. Schmitt, S. H. Xiang, and K. M. Yung, "Speckle in optical coherence tomography," *Journal of Biomedical Optics*, vol. 4, no. 1, p. 95, 1999.

[3] J. Zhao, Y. Winetraub, E. Yuan, W. H. Chan, S. Z. Aasi, K. Y. Sarin, O. Zohar, and A. de la Zerda, "Angular compounding for speckle reduction in optical coherence tomography using geometric image registration algorithm and digital focusing," *Scientific Reports*, vol. 10, no. 1, Feb. 2020.

[4] Y. Winetraub, C. Wu, G. P. Collins, S. Chu, and A. de la Zerda, "Upper limit for angular compounding speckle reduction," *Applied Physics Letters*, vol. 114, no. 21, May 2019.

[5] M. Göb, S. Lotz, L. Ha-Wissel, S. Burhan, S. Böttger, F. Ernst, J. Hundt, and R. A. Huber, "Large area robotically assisted optical coherence tomography (lara-oct) for skin imaging with mhz-oct surface tracking," in *Optical Coherence Tomography and Coherence Domain Optical Methods in Biomedicine XXVII*, J. A. Izatt and J. G. Fujimoto, Eds. SPIE, Mar. 2023.

[6] T. Pfeiffer, M. Petermann, W. Draxinger, C. Jirauschek, and R. Huber, "Ultra low noise fourier domain mode locked laser for high quality megahertz optical coherence tomography," *Biomedical Optics Express*, vol. 9, no. 9, p. 4130, Aug. 2018.

[7] M. Unser and P. Thevenaz, "Optimization of mutual information for multiresolution image registration," *IEEE Transactions on Image Processing*, vol. 9, no. 12, pp. 2083–2099, 2000.

[8] S. Klein, J. P. W. Pluim, M. Staring, and M. A. Viergever, "Adaptive stochastic gradient descent optimisation for image registration," *International Journal of Computer Vision*, vol. 81, no. 3, pp. 227–239, Aug. 2008.

[9] S. Klein, M. Staring, K. Murphy, M. Viergever, and J. Pluim, "elastix: A toolbox for intensity-based medical image registration," *IEEE Transactions on Medical Imaging*, vol. 29, no. 1, pp. 196–205, Jan. 2010.

[10] D. Shamonin, "Fast parallel image registration on cpu and gpu for diagnostic classification of alzheimer's disease," *Frontiers in Neuroinformatics*, vol. 7, 2013.

[11] R. Campbell, "Elastix for matlab," GitHub. [Online]. Available: https://github.com/raacampbell/matlab_elastix

[12] B. C. Lowekamp, D. T. Chen, L. Ibáñez, and D. Blezek, "The design of simpleitk," *Frontiers in Neuroinformatics*, vol. 7, 2013.

Motion artifact correction for in-vivo corneal OCT in rabbits

Lukas Pohl [1,2,3], Lara Buhl [3,4], Maron Dolling [5], Michael Wang-Evers [2], Felix Hilge [2,5], Dieter Manstein [2], and Reginald Birngruber [3,4,5]

[1] Medical Engineering Science, Universität zu Lübeck, lukas.pohl@student.uni-luebeck.de

[2] Department of Dermatology, Cutaneous Biology Research Center, Massachusetts General Hospital, Charlestown, MA, USA, {mevers, fhilge, dmanstein, lpohl}@mgh.harvard.edu

[3] Wellman Center for Photomedicine, Massachusetts General Hospital, Harvard Medical School, Boston, MA, USA, lbuhl@mgh.harvard.edu

[4] Ophthalology Department, Ludwig-Maximilians-Universität München, München, Bayern, Germany

[5] Institute for Biomedical Optics, Universität zu Lübeck, {m.dolling, reginald.birngruber}@uni-luebeck.de

Abstract

This study addresses movement artifacts during OCT imaging to look for corneal refraction change in a New Zealand White Rabbit's eye using an injected filler. To correct these artifacts, the OCT device's scan sequence was programmed, capturing three additional reference scans. A Matlab code determines corneal surface positions, enabling the adjustment of volume B-scans based on height differences. The surface roughness parameter is used to make a quantitative assessment. The procedure significantly reduces movement-induced corneal roughness, enhancing both visual and quantitative aspects of imaging. The calculated surface roughness parameter shows a 27.77% reduction, decreasing from an initial ΔR_a of 1.8159 to a corrected value of ΔR_a of 1.3116.

1   Introduction

The cornea is a crucial component in the eye's refraction process, providing two thirds of its total refractive power. As such, it is a vital instrument for human vision [2, 3]. It is important to ensure accurate imaging of the cornea to enable precise diagnoses and post-operative assessments.

The two most important imaging techniques for this purpose are Scheimpflug imaging and optical coherence tomography (OCT). Scheimpflug imaging illuminates different sections of the eye with slit lamps, allowing for cross-sectional imaging through a camera [4]. OCT utilizes a light source that emits light towards the cornea. Interferometry is used to reconstruct an image based on the reflected light signals received, resulting in a high-resolution image of the cornea [6].

Imaging methods are commonly used for quantitative evaluation after eye surgery. However, movement artifacts pose a challenge in this context. The eye is subject to natural movements such as tremor, drift, and microsaccades [7], even when the person is looking at a stable image. Additionally, movements can be triggered by changes in blood pressure [9]. All of these factors affect the resulting image, as devices require a specific amount of time to capture the picture.

Previous methods for correcting eye movement include predictive approaches, such as eye trackers, which incorporate the movements into the image composition when enough specific features are present [8]. Another option is retrospective correction, which involves correcting the image using additional features recorded after the image has been captured, such as cross-correlation between different scans [1]. This paper is part of a project that examines the refraction of a rabbit's eye after surgery using OCT images. During the procedure, a filler material is injected into a pocket within the cornea that was previously created by a laser. This causes a change in the radius of the corneal surface, which automatically leads to a change in the refraction [12]. This technique may be used in the future to improve defective vision. The in-vivo experiment faces the challenge of natural, uncontrolled movements of the rabbit's eye despite anesthesia or sedation. Unlike humans, rabbits cannot fixate on a fixed point in an OCT image, resulting in visible movement artifacts on the volume images along the slow axis as visible in Fig. 1.

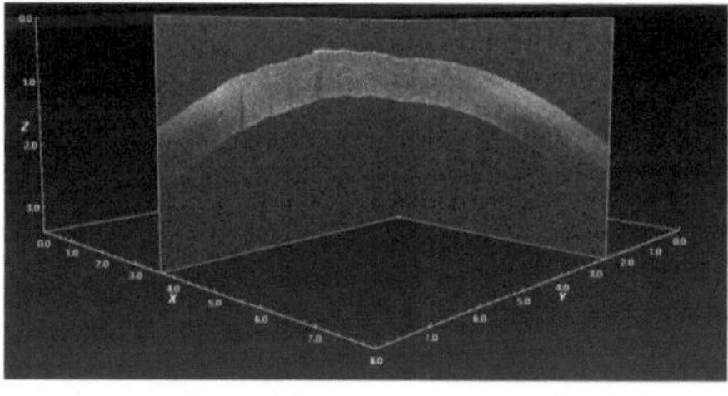

Figure 1: From the right: B-Scan 173 along the x-axis (fast axis) without motion artefacts, from the left: B-Scan 236 in along the y-axis (slow axis) with motion artifacts.

These artifacts make it difficult to determine the radius of the cornea. Similar to the method described in [1], surface detection and displacement can be determined using the intensity values of the images. It is important to correct the image before calculating refractivity. Initially, movement corrections are attempted only vertically since there are too few features in the cornea for horizontal correction.

2   Material and Methods

2.1   Animal Experiments

These in-vivo experiments examine the cornea of White New Zealand Rabbits after laser exposure and filler injection. The animals are fully anesthetized during the post-operation OCT acquisition (day 0), and sedated during the subsequent control examinations (day 1, day 4, day 7, day 13).

2.2   Imaging

The Telesto 320 Spectral Domain Optical Coherence Tomography (OCT) device from Thorlabs is utilized to capture images of the cornea. It operates at a nominal center wavelength of $1310\ nm$ and has a scan rate of $76\ kHz$, allowing for the imaging of a 3D volume. The 3D volume is composed of 512 B-scans, each containing 512 A-scans with a depth of $3.58\ mm$ on $1024\ pixels$. The total volume is 3.58 mm $\times$ 8 mm $\times$ 8 mm. The axial resolution is $5.5\ \mu m$ and the lateral resolution is $13\ \mu m$.

2.3   Software

The ThorImageOCT software allows for the initiation of standard scan sequences for the OCT device. This includes the option for individual B-scans or volume scans. However, only one scan can be performed at a time and the necessary adjustments for a new scan must be set manually.

2.3.1   Programming the Scan Sequence

To perform the motion correction, a specific scan sequence is required, which is programmed using a software development kit (SDK) based on the ThorImageOCT software and written in C++. This allows for running several scan sequences in succession without a large time offset. The scan includes 3 individual B-scans along the slow axis (Y-axis) of the volume, located at $-4\ mm$, $0\ mm$, and $4\ mm$ of the x-axis of the volume. The code provides for a standard B-scan (rotated by $90°$ to the fast axis of the volume) at the beginning, in the middle, and at the end of the volume scan as illustrated in Fig. 2.

The reference scans intersect with the volume scans shown as examples. At these positions, the reference scans should show the actual position of the cornea, while it may be displaced by movement in the volume scans. To view the intersection points, select the respective A-scan at this position, as described in 2.3.2.

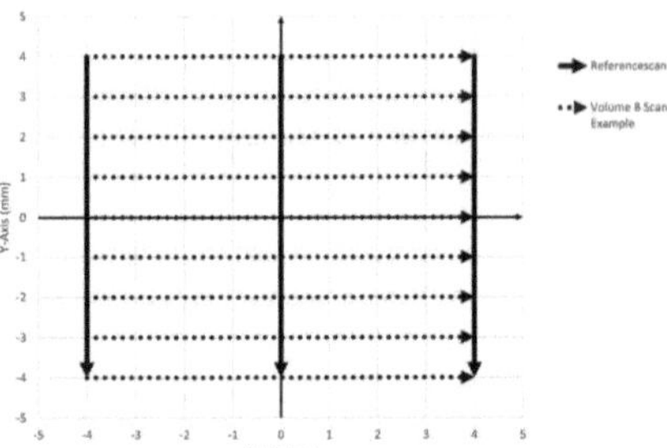

Figure 2: Scan locations, view of the volume from above, 3 reference scans (continuous lines), example volume scans (dashed lines).

2.3.2   Motion Correction

Matlab is utilized to perform motion correction on the volume scans. The code accesses the intensity values of the three reference scans and the composite B-scans (motion artifacts) along the Y-axis of the volume. The gradient values are then calculated in these scans using the Sobel operator, and the position of the maximum gradient is determined to detect the corneal surface. The positions of each B-scan within the volume are compared to those of the reference scans, and the difference values are calculated. The difference value of each individual B-scan to the three reference points determined by the mean value results in the height shift of the B-scan in the volume. The B-scans are then shifted along the Z-axis to create a new, corrected volume image. No changes in content have been made, because the B-scan is only adjusted vertically within the volume, and no further modifications are made to the image.

2.4   Surface Roughness

For evaluation, the material roughness parameter R_a is determined to quantitatively evaluate the corrected images. The formula used is:

$$R_a = \frac{1}{M} \int_1^M |z(x_M)|\, dx, \qquad (1)$$

which describes the average deviation value of the surface from the mean value of the surface in number of pixels [10]. The number of A-scans, denoted by $M = 512$, and the pixel position of the surface within the respective A-scan, x_M, are related. The curvature of the cornea is not regarded as roughness. First, the mean value of the surface is determined. This represents a layer within the image at the height of this value. Now the deviations of the surface from this value are considered and averaged. This is now the surface roughness value.

To assess the surface after motion correction, the difference between the moving surface R_{a2} and the reference scan R_{a1} and the difference between the corrected surface R_{a3} and the reference scan R_{a1} are compared. The following calculation is used for this:

$$\Delta R_a = |R_{a1} - R_{a2}|\ and\ \Delta R_a = |R_{a1} - R_{a3}|. \qquad (2)$$

The respective difference is described by ΔR_a.

3 Results and Discussion

3.1 Results

The following results are listed as examples and do not describe a complete series of measurements. The complete surface can only be accurately determined in the second scan (3(c)), as the two outer scans have a low corneal height. The surface in the middle of the image is recognized, but the accuracy decreases significantly towards the outside (3(a),3(b)). B-scan 256 of the volume along the Y-axis correctly identifies the surface, with clear visibility of eye movements during the scan. At the edges of the second reference scan, the cornea is partially unrecognized, causing deviations upwards. It is important to note that this is a subjective evaluation and should be clearly marked as such.

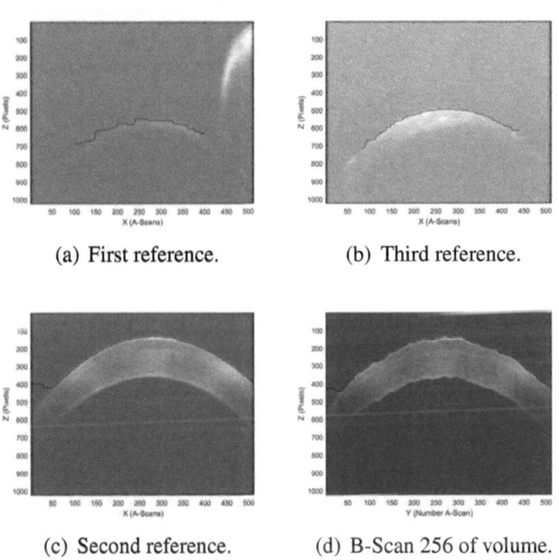

(a) First reference.

(b) Third reference.

(c) Second reference.

(d) B-Scan 256 of volume.

Figure 3: Surface detection for the reference scans (a-c) and the B-Scan 256 of the volume. No correct detection for the first (a) and third (b) reference scan.

Fig. 4 shows the images of the cornea before and after movement correction. Fig. 4(a) and 4(c) show the cornea before the program was applied, while Fig. 4(b) and 4(d) show the results.

The results of the material roughness parameter are shown in Fig. 5. The reference shows the value for the reference B-scan, while the original volume value is assigned to the moving image and the corrected volume value was evaluated after the motion correction. Since the values of the second volume, which has significantly more motion artifacts, are closer together, the differences between the roughness value of the respective reference and the corresponding roughness values of the volume B scans are now calculated. The results of (2) are shown in Fig. 6. Here, the value of the difference amount of the corrected image is smaller than that of the moving image. The difference between Volume 1 and Volume 2 is significant. For the volume

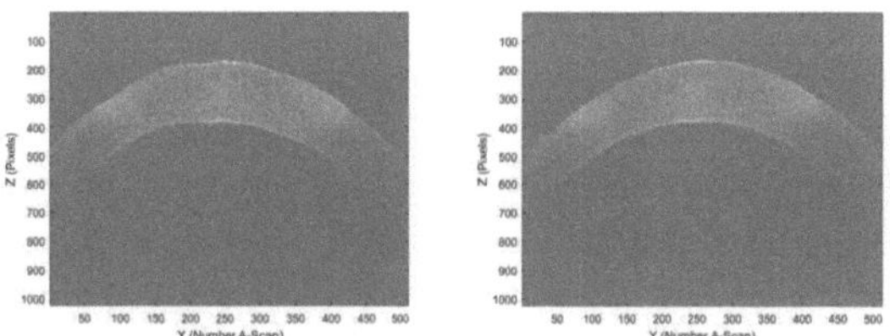

(a) Volume 1 B-Scan with moderate movement.

(b) Volume 1 B-Scan corrected.

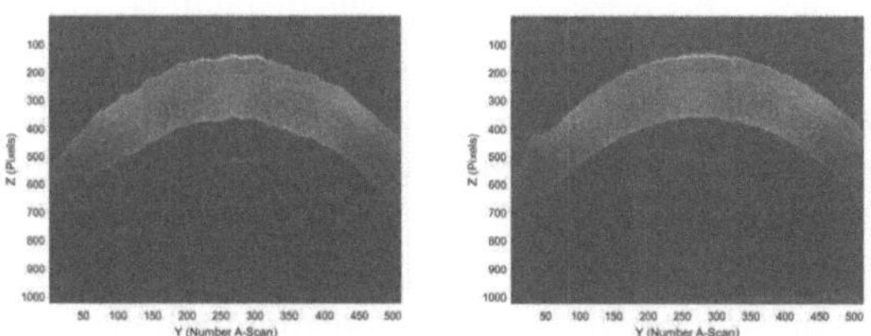

(c) Volume 2 B-Scan with increased movement.

(d) Volume 2 B-Scan corrected.

Figure 4: B-Scans 256 of volumes before (a,c) and after (b,d) moving correction.

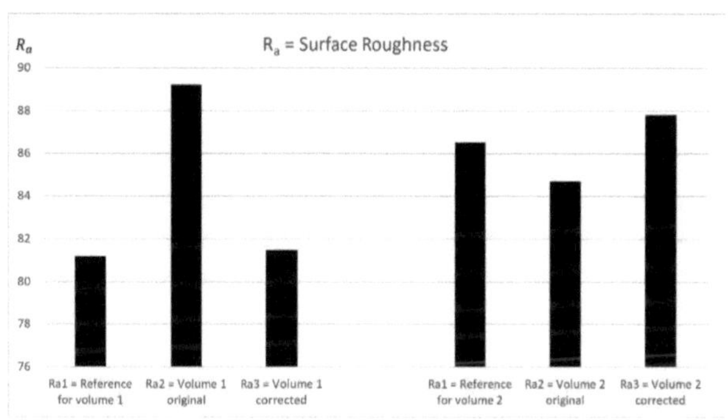

Figure 5: Surface roughness R_a for each scans (R_{a1} = reference scan, R_{a2} = moved Volume B-Scan 256, R_{a3} = corrected Volume B-Scan 256).

with less motion artefacts a decrease of $96,35\%$ is achieved, from $\Delta R_{a2} = 8,0338$ to $\Delta R_{a3} = 0,293$. Also the difference for the volume with more artifacts is decreased: $\Delta R_{a2} = 1,8159$ to $\Delta R_{a3} = 1,3116$, corresponding to a reduction of $27,77\%$.

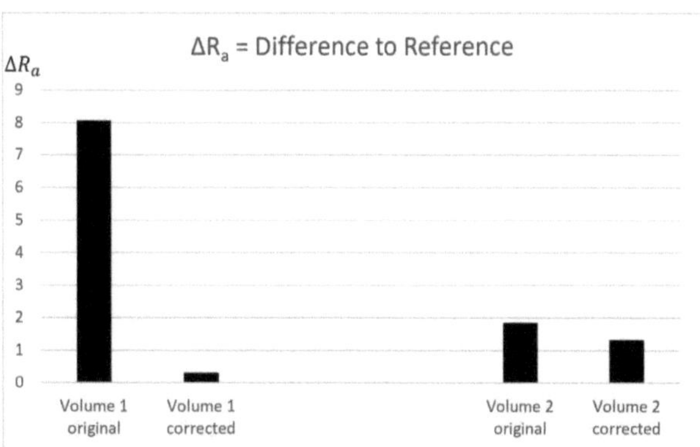

Figure 6: ΔR_a for Volume B-Scan 256 before (original) and after (corrected) moving correction

3.2 Discussion

The two outer reference scans were unable to determine the surface with sufficient precision, and therefore were not utilized for motion correction. However, the resulting motion correction demonstrates a noticeable improvement in motion artifacts. The calculated material roughness parameter further emphasizes the success of the motion correction. It would be beneficial to have at least two additional reference scans to compensate for inaccuracies at the edges of the images. This is a significant issue that will be addressed in future tests by creating four more reference scans before volume acquisition. However, measurements will now be taken at positions $-2\ mm$, $-1\ mm$, $0\ mm$, $1\ mm$, and $2\ mm$ along the X-axis as shown in Fig. 2. These positions were chosen because the cornea's curvature has not yet significantly affected its height, allowing for better surface recognition. In conclusion, the procedure has improved the images in terms of motion correction. However, further improvement may be possible with additional reference scans, particularly regarding the unwanted deviations at the edge of the image. However, the current results are acceptable for the time being and demonstrate that the approach is appropriate. A further benefit of the data collected could be that the radii of the cornea could be determined to calculate astigmatism.

To achieve more precise quantification of the results, it is possible to take a Scheimpflug image of the eye and compare the astigmatisms in both images. This method was previously explored in [11] where the scans were arranged in a star shape, recorded multiple times, and the movement through a meridian was observed. The advantage of the approach presented here is that movement correction can be performed with only three to five additional scans.

4 Conclusion

In summary, the procedure described has successfully reduced motion artifacts, as evidenced by the corrected images and surface roughness parameter. However, there is still room for improvement, as using a single reference scan may result in errors in movement correction if the cornea surface is not properly recognized in the peripheral area. However, it appears that the chosen path is the correct one.

Acknowledgement

The work has been carried out at the Wellman Center for Photomedicine, Mass General Hospital, Boston, USA and Cutaneous Biology Research Center, Mass General Hospital, Charlestown, USA and supervised by the Institute for Biomedical Optics, Universität zu Lübeck.

Authors' Statement

Conflict of interest: Authors state no conflict of interest.

5 References

[1] B. Potsaid, I. Gorczynska, V. J. Srinivasan, Y. Chen, J. Jiang, A. Cable, and J. G. Fujimoto , *Ultrahigh speed Spectral / Fourier domain OCT ophthalmic imaging at 70,000 to 312,500 axial scans per second* . In: Opt. Express, pp. 15149-15169, 2008.

[2] H. Davson, *Physiology of the Eye*. The MacMillan Press Ltd, 1990.

[3] Y. Le Grand and S. G. El Hage, *Physiological Optics*. Springer-Verlag, 1980.

[4] H. Hashemi and S. Mehravaran, *Day to day clinically relevant corneal elevation, thickness, and curvature parameters using the orbscan II scanning slit topographer and the pentacam scheimpflug imaging device*. In: Middle East African Journal of Ophthalmology, 17: pp. 44–55, 2010.

[5] E. A. Swanson, J. A. Izatt, M. R. Hee, D. Huang, C. P. Lin, J. S. Schuman, C. A. Puliafito, and J. G. Fujimoto, *In vivo retinal imaging by optical coherence tomography*. In : Opt Lett. Nov 1;18(21): pp. 1864-1866, 1993.

[6] F. Abtahian, and I.-K. Jang, *Optical coherence tomography: basics, current application and future potential*. In: Current Opinion in Pharmacology, Volume 12, Issue 5, pp. 583-591, 2012.

[7] S. Martinez-Conde, S. Macknik, and D. Hubel, *The role of fixational eye movements in visual perception*. In: Nat Rev Neurosci 5, pp. 229–240, 2004.

[8] L. Sánchez Brea, D. Andrade De Jesus, M. F. Shirazi, M. Pircher, T. van Walsum, and S. Klein, *Review on Retrospective Procedures to Correct Retinal Motion Artefacts in OCT Imaging*. In: Applied Sciences, 2019.

[9] M. D. Robinson, S. J. Chiu, J. Lo, C. Toth, J. Izatt, and S. Farsiu, *New applications of super-resolution in medical imaging*. In: Super-Resolution Imaging 2010, pp. 384-412, 2010.

[10] E. Macherauch and H. W. Zoch, *Topographie von Werkstoffoberflächen*. In: Praktikum in Werkstoffkunde, Springer Vieweg, Wiesbaden, pp. 181–187, 2019.

[11] E. Pavlatos, D. Huang, and Y Li, *Eye motion correction algorithm for OCT-based corneal topography*. In: Biomed. Opt, 2020.

[12] C. M. Wertheimer, K. Brandt, S. Kaminsky, C. Elhardt, S. A. Kassumeh, L. Pham, H. Schulz-Hildebrandt, S. Priglinger, R. Anderson, and R. Birngruber, *Refractive Changes After Corneal Stromal Filler Injection for the Correction of Hyperopia*. In: Journal of Refractive Surgery, 2020.

Estimation of Absorption Using Spatially Resolved Spectroscopy Method

Rafat Saeed [1], Maik Rahlves [2], and Dirk Grosenick [3]

[1] Biomedical Engineering, University of Applied Sciences Lübeck, rafat.saeed@stud.th-luebeck.de
[2] Institute of Biomedical Optics , Universität zu Lübeck, maik.rahlves @uni-luebeck.de
[3] Institute of Biomedical Optics, Physikalisch -Technische Bundesanstalt, dirk.grosenick@ptb.de

Abstract

This research project focuses on the evaluation of the spatially resolved spectroscopy (SRS) method to estimate the absorption of human brain tissues. Measurements were conducted on several phantoms with optical properties known from a reference method. Intensity profiles measured with high spatial resolution by a fiber-based scanner showed too much light for source-detector distances beyond about 2.5 cm compared to diffusion theory. Accordingly, the depth sensitivity would not be sufficient to reach the human brain. When the surface of the phantom was covered by black rubber, the intensity profiles could be measured correctly up to a maximum distance of 4 cm, which should enable sensitivity to the optical properties of the brain. Effective attenuation and absorption coefficients obtained at these long distances by the SRS method were consistent with the known coefficients within an uncertainty of 5 to 10% recommending SRS as a simple method to characterize brain oxygenation.

1   Introduction

Understanding light distribution in biological tissues is crucial for the safe and effective application of optical diagnostic methods and laser therapies [1]. In this study, we focus on continuous-wave near-infrared (NIR) spectroscopy as a method to investigate hemodynamics and oxygenation in the human brain. Oxygenation measurements require the estimation of brain absorption at selected wavelengths in the NIR [2]. A simple but effective method is spatially resolved spectroscopy (SRS) which comprises the measurement of reflected intensities at two source-detector distances only [3]. To sample the human brain of adults rather than superficial layers, source-detector distances of 3 and more centimeters should be applied. Using diffusely scattering phantoms with known optical properties we evaluate the SRS method for selected source-detector configurations.

2   Material and Methods

2.1   Experimental Setup

Measurements were done with a continuous-wave setup designed to record spatially resolved diffuse reflectance of phantoms at selected NIR wavelengths. Two different fiber interfaces were used to either scan the uncovered phantom with high spatial resolution or to measure spatially resolved reflectance (SRR) on a coarse grid with a black absorbing layer on the phantom surface. Fig.1 provides a basic sketch

illustrating the components and the operation of the measuring system. The radiation of six diode lasers emitting at 685 nm, 785 nm, 808 nm, 852 nm, 905 nm, and 980 nm, (LuxX with LightHUB, Omicron Laserage, Germany) is combined into a multimode optical fiber. The output of this fiber is collimated, and a beam splitter is employed to direct about 10% of the light to an avalanche photodiode used to monitor the stability of the lasers. Neutral density filters with transmittance values of T= 0.144 and T= 0.380, positioned in front of the reference diode and free beam output, counteract over modulation and reduce room-interfering light.

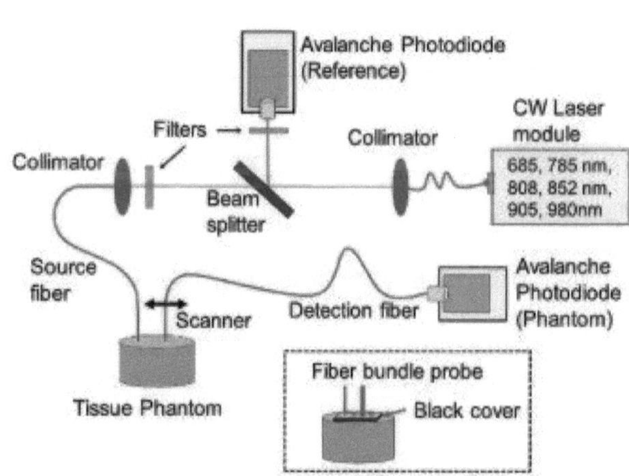

Figure 1: Sketch for the measurement setup of the spatially resolved diffuse reflection.

To record spatially resolved reflectance for source-detector distances from 0.5 mm to 50 mm with a step size of 0.5 mm, the main beam is coupled into a multimode fiber with a core diameter of 365 µm, directed at the phantom. Light detection is done with a second multimode fiber with 365 µm core fixed on a motorized translation stage to scan this fiber over the surface of the phantom. The output of the detection fiber is connected to a second avalanche photodiode used to measure the diffusely reflected light. A long-pass filter in front of the photodiode suppresses ambient light. To partially cover the surface of the phantom under investigation with a black absorber, a 100 mm × 40 mm rectangular fiber probe with a 2 mm thick layer of black rubber was employed. Drill holes were used to position the detection fiber either 10 mm, 20 mm, 30 mm, 35 mm, or 40 mm away from the source fiber. A second probe with a black cover was made for a detection fiber bundle (4 mm diameter) to perform measurements on the same coarse grid with increased sensitivity.

The APD voltages were acquired by two separate data acquisition boards (NI USB-6353 and NI USB-6363, National Instruments) to avoid any crosstalk from the high voltages in the reference channel to the partially very low voltages in the phantom channel. The NI-6353 board was used to sequentially switch the six lasers on for 100 ms to separate the wavelengths at the detectors by time multiplexing. Laser powers at the surface of the phantom were set to about 1 mW. Data acquisition and control of the experiment were carried out using a LabView program. Sampled voltages were averaged over the 100 ms intervals and plotted as a function of the source-detector distance using MATLAB.

Solid phantoms made from polydimethylsiloxane (PDMS) or epoxy resin with different amounts of titanium oxide for light scattering and black absorber were used to measure spatially resolved reflectance over a large range of absorption and reduced scattering coefficients [4]. The absorption and reduced scattering coefficients of these phantoms were known from independent measurements of time-resolved reflectance, analyzed by a Monte Carlo model of light transport.

2.2 Spatially resolved spectroscopy

Illustration of the partial volumes sampled by light detection at different distances ρ from the source position for a simplified layer structure of the human head. The gray matter of the brain will be reached by photons only for sufficiently large values of ρ. Accordingly, absorption coefficients should preferably be estimated from measurements at longer distances to reduce the effect of the top layers.

We use the spatially resolved spectroscopy (SRS) method following the approach outlined by Suzuki [3]. SRS involves analyzing the slope of light attenuation versus distance. We approximate the slope from measurements

at two distances ρ_S (shorter distance) and ρ_L (longer distance) as illustrated in Fig.3

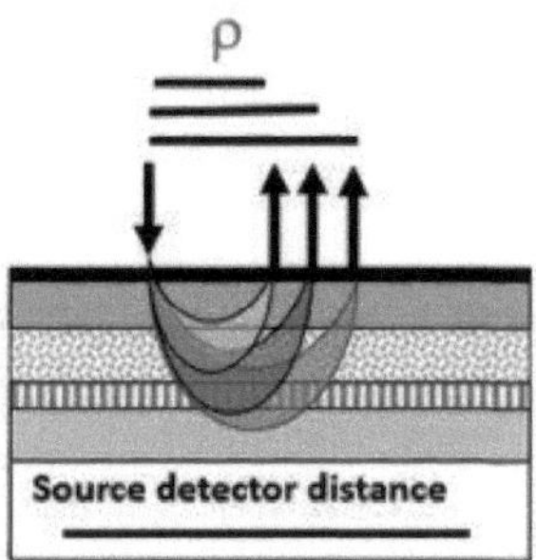

Figure 2: Sampled volume (photon bananas) as a function of the source-detector distance ρ for a layered structure (simplified head model). From top to bottom: hair, skin, skull, cerebrospinal fluid, gray matter, white matter.

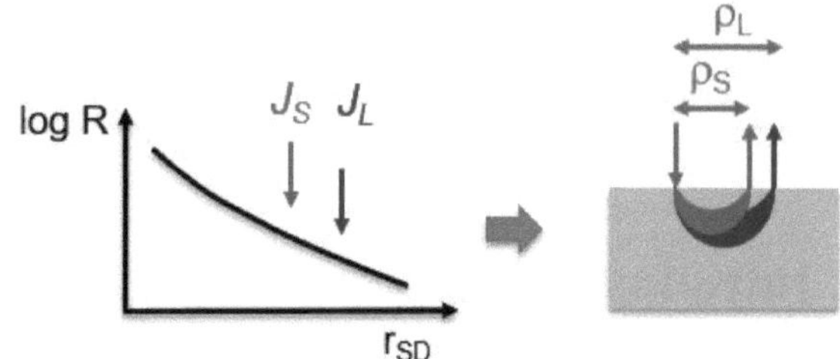

Figure 3: Measurement scheme for the spatially resolved spectroscopy (SRS) method involves measuring the diffuse reflectance at two selected distances ρ_S (shorter distance) and ρ_L (longer distance) from the source position.

With J_S and J_L denoting the detector voltages measured at ρ_S and ρ_L respectively, the effective attenuation coefficient μ_{eff} is given by the following equation:

$$\mu_{\text{eff}} = \frac{\ln\left(\frac{J_L}{J_S}\right) + 2\ln\left(\frac{\rho_L}{\rho_S}\right)}{\rho_S - \rho_L} \tag{1}$$

This equation follows from the solution of the continuous wave (CW) diffusion equation for the semi-infinite medium under the assumption of long distances ρ_S and ρ_L. It is based on the zero-boundary condition. The related absorption coefficient at wavelength λ is then given by:

$$\mu_a(\lambda) = \frac{\mu_{\text{eff}}(\lambda)^2}{3\mu_s'(\lambda)} \tag{2}$$

The reduced scattering coefficient μ_s' cannot be derived from the continuous wave (CW) measurement. Here, we use the true value of μ_s' for each phantom obtained from the time-domain reference measurement. In the case of in vivo investigations, an average value for the tissue taken from the literature could be used as an approximation.

3   Results and Discussion

Measured Spatial Profiles

The scanning approach enables the spatially resolved reflectance profile to be scanned with high spatial resolution. Fig. 4 provides an example for a phantom with reference optical properties $\mu_a = 0.115\,\mathrm{cm}^{-1}$ and $\mu'_s = 12.0\,\mathrm{cm}^{-1}$ (at 785 nm). However, the measured profile follows the reference profile calculated from the true optical properties by a Monte Carlo model up to distances of typically 2.5 cm only. At longer distances, the measured intensity becomes much stronger than expected from the theory. In the example of Fig.4 Cover the measured intensity at 5 cm source-detector distance is about one order of magnitude larger. A possible explanation could be that a small amount of light is traveling directly below the surface of the phantom. At longer distances, this light becomes stronger than the diffusely reflected light coming from the deeper parts of the phantom. This explanation is supported by time-domain measurements showing a small peak at the leading edge of the distribution of times of flight of photons measured at distances beyond 3 cm.

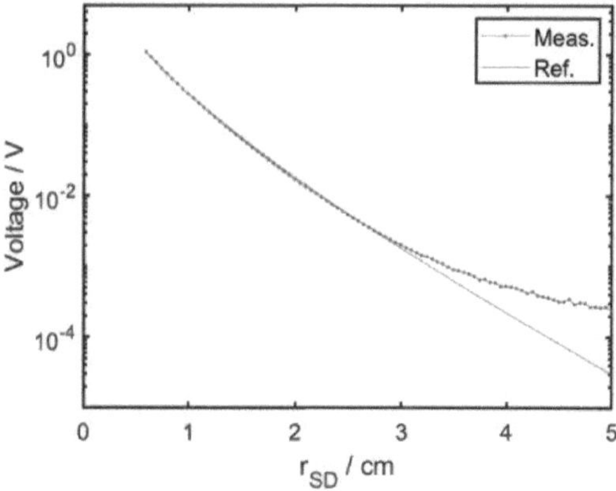

Figure 4: SRR profile for phantom D2 at 785 nm measured with the scanner (no black cover) and the theoretical reference profile.

The SRS method requires fully diffusive light transport. Considering the discrepancy between measurement and diffusion theory, the method could be applied here only for distances up to about 2.5 cm. The depth sensitivity at such small distances is limited. Accordingly, the analysis of reflectance profiles as given in Fig.4 is useful for applications on, for example, small animal organs [5]. The application of the SRS method to study the human brain requires suppression of the photons traveling along the surface to the detector.

This suppression could be successfully realized with the black-covered fiber probes. Fig. 5 shows the spatial profile for the same phantom as in Fig. 4 obtained with the fiber bundle and black cover on top of the phantom surface. Now, the measured profile follows the theoretical one also at longer distances. The high sensitivity obtained with

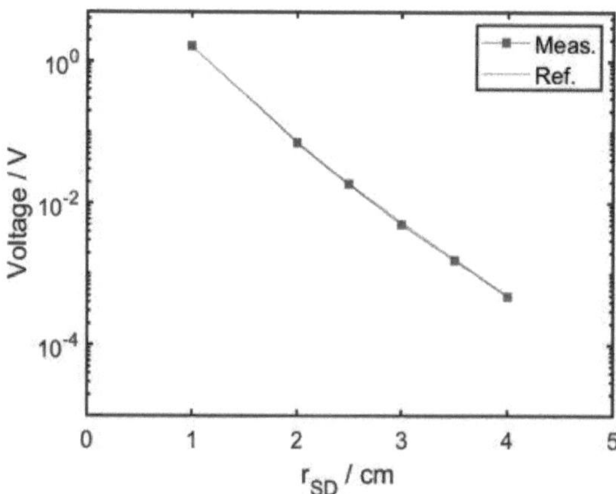

Figure 5: SRR profile measured with black cover (fiber bundle probe) compared to the theoretical curve, for phantom D2 at 785 nm.

the fiber bundle allows measurements up to a maximum source-detector distance of 4 cm with a high signal-to-noise ratio. When the black-covered fiber probe with $365\,\mu\mathrm{m}$ diameter fiber is used instead, the maximum possible distance for the various phantoms is between 3 and 3.5 cm only due to the smaller sensitivity. The black cover disturbs the propagation of light along the surface of the phantom since it changes the conditions of reflection at the phantom surface. Compared to the scanner, the black cover limits the number of possible source-detector distances. However, this is not a drawback for the SRS method, since only two distances are required here, preferably at 3 to 4 cm.

In the following, the measurements with the fiber bundle in the black-covered probe are used to estimate the optical properties by the SRS method. The effective attenuation and the related absorption coefficients were calculated for all available pairs of source-detector distances between 1 cm and 4 cm at the six laser wavelengths.

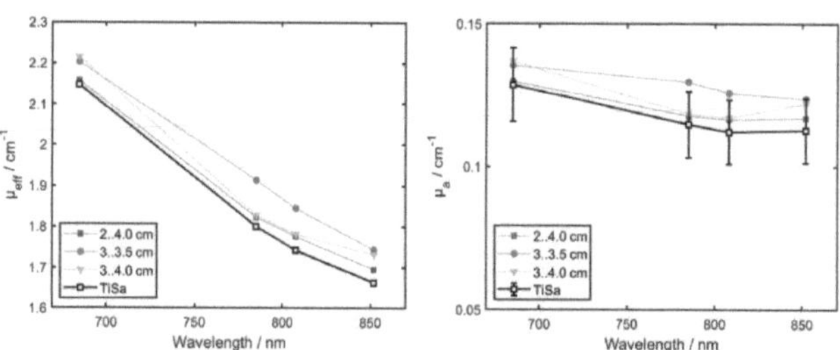

Figure 6: Effective attenuation and absorption coefficients of phantom D2 estimated by the SRS method for selected source-detector pairs and reference optical properties for comparison. Error bars amount to 10%

As an example, Fig.6 illustrates results for phantom D2 obtained with the longer distance set to 3.5 cmcm or 4 cm to ensure a large depth sensitivity. The coefficients are compared to the true optical properties from the time-domain reference measurement. Results at 905 nm and 980 nm are not plotted since reference values at these two wavelengths are not available due to an upper spectral limit of 890 nm of the time-domain setup. The derivation of the absorption coefficients using Eq. 2 involves μ_s' taken from the known reference data. The results are closest to the reference values for the combination $\rho_S = 2$ cm, $\rho_L = 4$ cm. This combination is very stable due to the large difference between ρ_S and ρ_L. For the distance pair (3 cmcm or 4 cm), deviations from the reference go up to +10%. Results become worse for the pair (3 cmcm or 3.5 cm) possibly due to the smaller difference between the two measured detector signals. Overall, despite a tendency for SRS to overestimate μ_a, it consistently yields μ_a within an uncertainty of 10% with respect to the true values. The 3 cmcm to 4 cm range proves favorable for brain characterization due to the large depth sensitivity combined with sufficient stability of the result.

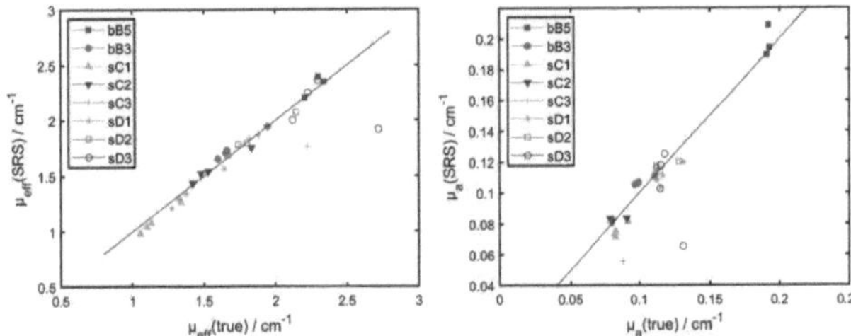

Figure 7: Effective attenuation and absorption coefficients for all investigated phantoms (PDMS phantoms: bB3, bB5, epoxy phantoms: sC1 to sD3) plotted versus reference values. Data were obtained from $(\rho_S, \rho_L) = (2$ cm or 4 cm$)$.

In Fig.7 the effective attenuation and absorption coefficients are compared to the reference values for all 8 phantoms investigated at the four wavelengths 685 nm, 785 nm, 808 nm, 852 nm. 30 of the 32 data points follow the straight line, confirming that SRS yields optical properties close to the true values. The two outliers refer to epoxy phantoms with high reduced scattering values ($\mu_s' \approx 14.7$ cm at 785 nm). Phantoms with either low scattering (≈ 5 cm^{-1} for sC1, sD1) or moderate scattering (≈ 10 cm^{-1} for remaining phantoms) are well described by the SRS method, making this method well-suited for measurements of absorption on the human brain. Brain oxygenation could then be estimated from absorption at selected wavelengths.

4   Conclusion

In conclusion, we evaluated spatially resolved spectroscopy (SRS) as a potential method to estimate absorption of the human brain. To look sufficiently deep into the tissue, measurements of diffuse reflectance at source-detector distances of three or more centimeters are required. Our experiments on phantoms have shown that a black cover on the phantom surface is required to suppress light guiding along the surface of the phantom that distorts diffuse reflectance measurements at distances larger than 2.5 cm.

Notably, SRS performed well in estimating the absorption of various phantoms when measurements from 2 cmcm and 4 cm were combined. When increasing the short distance to 3 cm or even 3.5 cm, absorption coefficients from SRS show deviations from the true optical properties of the phantoms in the order of 10%. Hence, the gain of in-depth sensitivity arising from the larger distance is associated with reduced accuracy.

This observation prompts further consideration and optimization of the SRS method for enhanced precision, especially in the critical range of 3 cmcm to 4 cm. Despite this limitation, our findings contribute valuable insights into brain tissue optical properties, laying the foundation for future refinements and applications in clinical settings.

Acknowledgement

This work was carried out at the Physikalisch Technische Bundesanstalt (PTB), Berlin. Many thanks to PTB for supervising me.

Authors' Statement

The authors state no conflict of interest. Informed consent

5   References

[1] P. Pinti et al., *The present and future use of functional near-infrared spectroscopy (fNIRS) for cognitive neuroscience*. Ann. N. Y. Acad. Sci. vol. 1464, pp. 5–29 (2020).

[2] W. Zhong et al., ., *A Review of Monitoring Methods for Cerebral Blood Oxygen Saturation. Healthc. (Basel), vol. 9, no. 9: 1104, 2021.*.

[3] S. Suzuki, S. Takasaki, T. Ozaki, Y. Kobayashi, *A tissue oxygenation monitor using NIR spatially resolved spectroscopy*. Proc SPIE vol. 3597, pp. 582–592, 1999

[4] S. Schieck, *Herstellung und Charakterisierung strukturierter, gewebeähnlicher Phantome für die biomedizinische Optik.*. Diploma thesis, Technische Fachhochschule Wildau, 2005.

[5] D. Grosenick, et al., *Detailing renal hemodynamics and oxygenation in rats by a combined near-infrared spectroscopy and invasive probe approach*. Biomedical Optics Express, vol. 6, no. 2, pp. 309-323, 2015.

Development of an optical distance measurement system for fiber Fabry-Pérot tunable filters

Allegra Behr [1], Philipp Lamminger [2], and Robert Huber [2]

[1] Biomedical Engineering, Luebeck University of Applied Sciences, allegra.behr@stud.th-luebeck.de

[2] Institute of Biomedical Optics, Universität zu Lübeck, {phi.lamminger, robert.huber}@uni-luebeck.de

Abstract

Fiber Fabry-Pérot tunable filters (FFP-TFs) are used as wavelength selective elements in wavelength agile lasers. Exact positioning of the filter elements directly affects the center wavelength of the FFP-TF. However, FFP-TFs are susceptible to creep, non-linearity, aging, temperature and hysteresis due to the piezo ring in the filter. To accurately measure the distance between the filter elements, an optical distance measurement system was developed. The motivation was to verify, if building a system based on an optical light barrier with sufficient precision in the nanometer (nm) range is possible. Here we present a system that successfully determined the free spectral range and the center wavelength of the FFP-TF with an accuracy of 7 nm for a single measurement. This enables developing a wavelength control system in the future.

1 Introduction

Wavelength agile lasers can be used in various biomedical applications such as anti-Stokes Raman scattering microscopy [1], spectro-temporal laser imaging by diffracted excitation (SLIDE) two-photon microscopy [2] or swept-source optical coherence tomography (SS-OCT) [3]. In SS-OCT for example, a frequency-tunable light source such as a Fourier domain mode locked (FDML) laser for faster OCT imaging is used. These FDML lasers consist of a semiconductor optical amplifier (SOA), a long fiber ring cavity and a Fiber Fabry-Pérot tunable filter (FFP-TF). The FFP-TF is used as the wavelength selective element in the ring cavity [3].

The deployed FFP-TF consists of a dimpled ferrule and a glass plate, both coated with a highly reflective coating. The distance between the aforementioned elements is varied by a voltage controlled piezo ring mounted on the back of the glass plate. By applying voltage to the piezo ring, the material is being deformed thus moving the glass plate. Owing to the high level of stiffness, rapid frequency response and high resolution, piezoelectric actuators are often used in positioning tasks requiring high precision in the nanometer (nm) range [4].

One issue that has been observed with piezo rings is that over time the displacement of the glass plate by the piezo ring changes. This is because the piezo is prone to creep, non-linearity, aging, temperature dependence and hysteresis. These problems are well known from other piezoelectric actuator applications [4]. An additional problem is that not only the piezo but also other components in the FFP-TF are temperature-sensitive. This leads to a drift of the center wavelength of the FFP-TF, as the resonator length of the FFP-TF changes. To address these problems, a prototype optical distance measurement system for the gap between the filter elements in the FFP-TF has been designed and built. The goal of this work is to verify whether it is possible to measure the gap in the FFP-TF with sufficient accuracy in the nm range. With this the free spectral range (FSR) in which the FFP-TF is operating and the center wavelength of the FFP-TF shall be determined. Based on this system a control loop for precise control of the resonator length in the FFP-TF can be implemented later.

2 Material and Methods

2.1 Principles of the FFP-TF

The FFP-TF consists mainly of a dimpled ferrule and a glass plate, both coated with a highly reflective coating. The light is guided within a 125 micrometer (µm) single mode (SM) fiber. One side of the filter allows a small amount of light to pass through with each reflection. This light interferes, allowing only the wavelength that exhibits constructive interference to pass through the filter. The remaining wavelengths destructively interfere and therefore do not pass through. The sequentially reflected beams have a path difference of twice the length of the resonator $2l$. Consequently, constructive interference occurs when the round trip path is covered by an integer number N of the wavelengths λ [5]:

$$N \cdot \lambda = 2l. \qquad (1)$$

Behind the glass plate, the light is recoupled into a fiber. The cavity length l is determined by the FSR $\Delta\lambda_{FSR}$ and the wavelength λ_0 [5]:

$$l = \frac{\lambda_0{}^2}{2 \cdot \Delta\lambda_{FSR}}. \tag{2}$$

2.2 Measurement setup

The main challenge is the small distance of about 3 to 5 μm between the ferrule and the glass plate in the FFP-TF. Therefore, the idea was to measure the distance optically from the side with light emitted from a laser diode (LD). As the voltage applied to the piezo ring increases, the gap size between the filter elements decreases and therefore the intensity detected on the photodiode (PD) decreases as well. This measurement principle is similar to that of an optical light barrier. Because of the small distance of the filter elements it is challenging to get enough light through the gap of the FFP-TF that results in a signal on the PD. Because of that the light has to be focused to a small spot in order to detect the intensity of the light passing the gap on the PD. To focus the light emitted from the LD a fiber has been manufactured (see subsection 2.3). The following Figure 1 shows the design of the FFP-TF and the measurement setup of the optical distance measurement system.

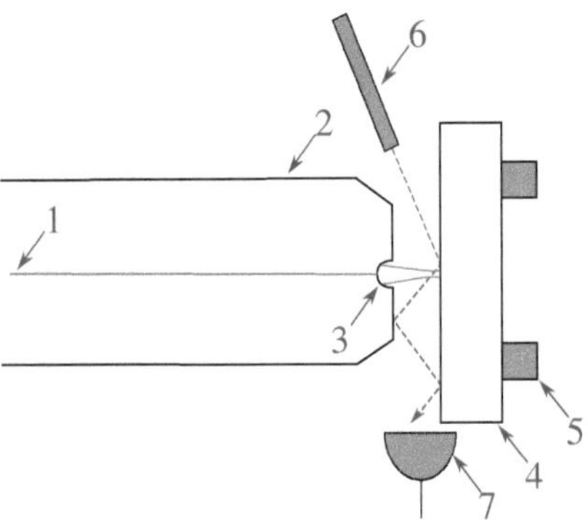

Figure 1: Design of the Fiber Fabry-Pérot tunable filter (components with continuous lines) and the optical distance measurement system (components with dashed lines). 1: Optical fiber, 2: Ferrule, 3: Dimple, 4: Glass plate, 5: Piezo ring, 6: End of optical fiber carrying the light from the laser diode (LD), 7: Photodiode (PD).

The FFP-TF utilized has a FSR of 108 nm at a center wavelength of $\lambda = 1080$ nm. Therefore the cavity length l calculated with equation (2) is 5.4 μm, when operating the FFP-TF at 1080 nm. For this we decided to use a LD (SPL520-1-3-PD, Roithner LaserTechnik GmbH, Austria) with $\lambda = 520$ nm for the measurement setup. This is about half of the wavelength selected in the FFP-TF and will therefore be likely reflected by the coating in the filter as well. A silicon PD (S6931-01, Hamamatsu Photonics, Japan) with high sensitivity at $\lambda = 520$ nm is employed for the detection of the light emitted by the LD. By using $\lambda = 520$ nm and therefore a silicon PD, the PD is less sensitive to the $\lambda = 1080$ nm light running in the FFP-TF, which could interfere with the distance measurement. The LD is controlled by an external voltage supply (RND 320-

KA3305P, Distrelec Group AG, Switzerland) and a voltage divider circuit (see Figure 2), resulting in a maximum power output of $P = 3.1$ mW at a temperature of 25 °C. To ensure a constant power output, the LD is temperature controlled with a temperature controller (MTD1020T, Thorlabs). The controller is set to 25 °C with a maximal allowable temperature difference of 10 mK. The circuit for the PD includes an external voltage supply (RND 320-KA3305P, Distrelec Group AG, Switzerland) and a series resistor of 11 $k\Omega$. The PD is used in reverse bias. The voltage drop over the series resistor is measured differentially by an Analog-to-Digital-Converter (ADC) (ADS1115, AZDelivery, Germany) as visible in Figure 2. The ADC sends the measurement values (ADC A0 and ADC A1 in Figure 2) via I^2C-communication to an Arduino Mega 2560 R3. The ADC has a 16-bit resolution and sends measurement values at a sampling frequency of 860 samples per second. The values are read into the PC by a self-programmed Python module for further analysis.

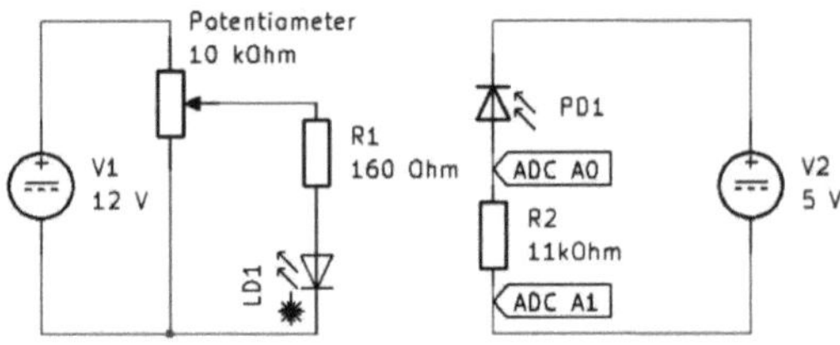

Figure 2: Schematic of the laser diode circuit and the photodiode circuit.

2.3 Fiber selection and manufacturing

To increase the power going through the gap of the FFP-TF, the light emitted has to be focused to a small spot with a similar diameter as the gap of the FFP-TF. To be able to create that spot the light of the employed LD is coupled into a SM fiber suitable for the wavelength of the LD (SM450, Thorlabs, USA). In order to broaden the light and then focus it accordingly afterwards (see Figure 3 b)), a simulation tool developed at our institute was used to simulate the spot sizes and working distances for different fiber types and lengths. A 250 μm piece of coreless fiber (CF) (FG125LA, Thorlabs) and a 125 μm long piece of a graded index (GRIN) multimode fiber (GIF625, Thorlabs) were found to be suitable and were spliced onto the end of the SM450 fiber as shown in Figure 3.
The lengths of the GRIN and the Coreless fiber part were analyzed using a transmitted light microscope (Orthoplan, Leica, Germany) and found to be 127 μm and 253 μm, respectively. A mode field analysis was conducted using a fiber focusing stage and the minimum spot size was 5.9 μm at a distance of 255 μm from the fiber tip (see Figure 4).
The fiber was mounted onto a five axis linear stage to be able to align the position and the angle of the fiber in the FFP-TF. The fiber was aligned such that the maximum intensity on the PD of the optical distance measurement system was achieved.

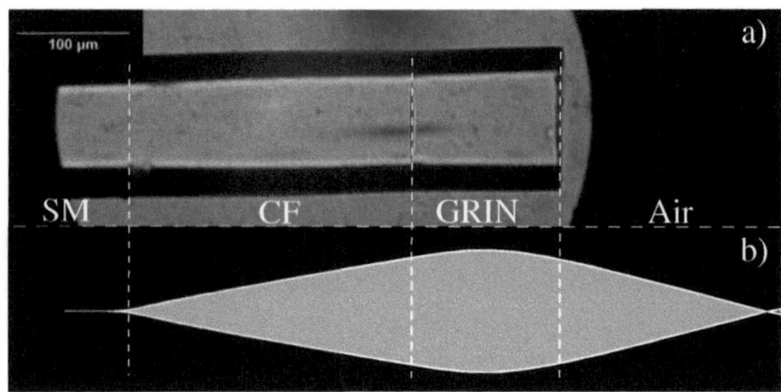

Figure 3: a) Manufactured fiber under the microscope. (From left to right: SM450 fiber, CF, GRIN fiber.) b) Schematic beam path inside and outside the fiber.

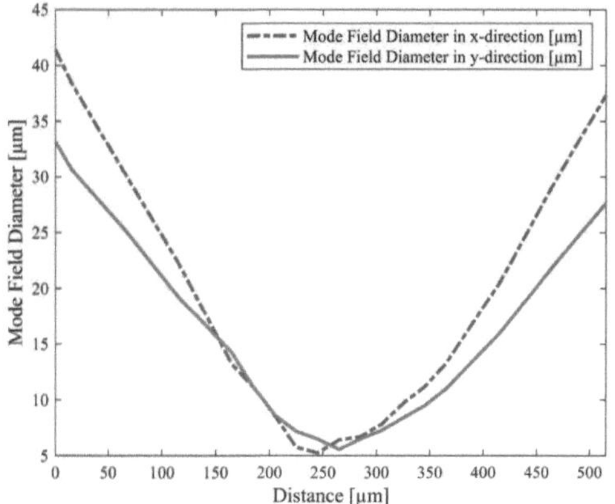

Figure 4: Mode field analysis of the manufactured fiber

2.4 Experimental setup

Figure 5 displays the measurement setup for the experiments conducted. The SOA is used as broadband light source (SOA-1070-70-HI-24dB, innolume, Germany). The light is fed to the circulator (used as an isolator) and subsequently to the FFP-TF. Behind the glass plate the light is coupled in using a lens, which is connected to the Optical Spectrum Analyzer (OSA) (AQ6370, Yokogawa, Japan). The optical distance measurement system is detecting the light beam coming from the LD as intensity values on the PD. The ADC measures the voltage drop over the PD, converts the voltage values to digital counts and sends the measurement values to the Arduino. The Arduino measurement values are read out on a PC.

All measurements were recorded by the OSA and by the optical distance measurement system. The system was covered to avoid noise caused by ambient light. The first measurement was a dark measurement where both, the LD and the SOA, were turned off. For the second measurement, the SOA was turned on, to see if the light coming from the OSA has an influence on the measurement. For the third measurement additionally the LD was turned on. For the fourth and fifth measurement -20 V and 160 V were applied to the piezo ring, respectively. These two settings represent the minimum and maximum applied voltage to the piezo ring. An additional experiment was conducted, tuning the wavelength to 1080 nm in the first FSR and afterwards to

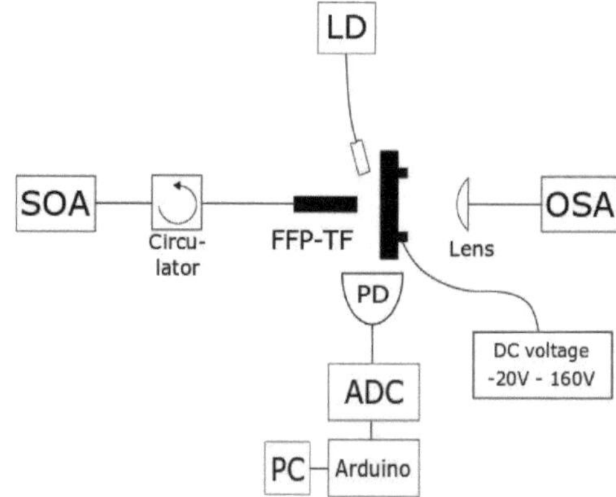

Figure 5: Design of the measurement setup including a SOA, a circulator, the FFP-TF, a lens, an OSA and the components of the optical distance measurement system (LD, PD, ADC, Arduino and PC). Direct current (DC) voltage is applied to the piezoring of the FFP-TF.

1100 nm in the first FSR. The aim of this experiment is to find out how many digital counts correspond to a shift of 20 nm and therefore how many counts correspond to one nm.

3 Results and Discussion

The dark measurement resulted in a mean of 18.7 digital counts (Standard deviation (SD) = 1.6 counts) when the SOA was turned off and a mean of 35.8 digital counts (SD = 1.9 counts) when the SOA was turned on. Thus, the influence of light with a wavelength of 1080 nm is low on the measurement values detected by the silicon PD. When the optical distance measurement system with the LD is turned on, the mean digital value increases to 10782.9 (SD = 27.6 counts). Applying -20 V to the piezo ring leads to a further increase of the mean digital value to 11153.1 digital counts (SD = 25.4 counts). When the piezo ring is tuned to 160 V the mean digital counts decrease to 8781.2 counts (SD = 25.6 counts). The digital counts over time are displayed for the three measurements when the optical distance measurement is turned on (see Figure 6).

The range of digital counts from the minimal to the maximum distance between the ferrule and the glass plate in the FFP-TF therefore is $11153 - 8781$ counts $= 2372$ counts.

In the next experiment the mean digital count was 10304.3 digital counts (SD = 49.8 counts), when the position of the peak on the OSA in the first FSR was 1080 nm. When shifting the wavelength by 20 nm to 1100 nm the mean was 10144.2 digital counts (SD = 55.8 counts). This can also be seen in the boxplot diagram in Figure 7. The SD is by around 30 counts higher than in the first experiment. Several outliers are visible for the first measurement. The difference of the two positions detected is 160 counts which corresponds to 8 counts per nm. The maximum SD of 55.8 digital counts corresponds to 7 nm for this experiment and is twice as high as in the previous measurements. This can be due to more noise in the surrounding.

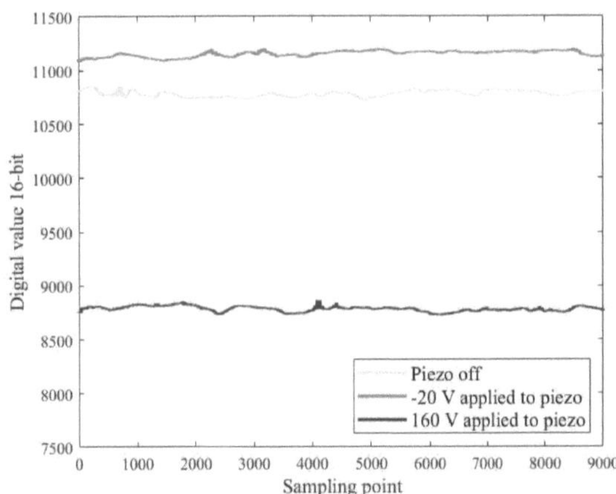

Figure 6: Measurement of the digital counts over 9000 sampling points during different states of the piezo ring (Piezo off, Piezo tuned to -20 V and 160 V).

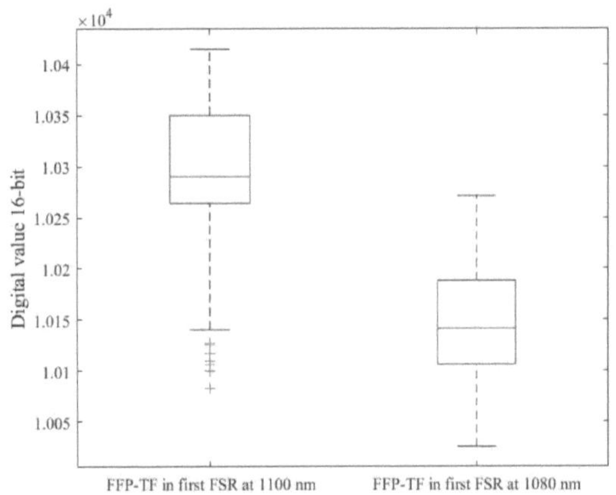

Figure 7: Boxplot diagram of the digital counts over 9000 sampling points during different states of the piezo ring (Shift of the wavelength peak of 20 nm on the OSA).

There are several sources for noise in the system. Firstly the fiber mounted on the five axis stage is only fixed at one position several centimeters away from the filter so the end is hanging free, thus making the system sensible to noise in the surrounding. With suitable bandpass filtering the electrical noise can be reduced. Also, it is possible that noise is coming from the LD, as the power output is not perfectly stable. Reading out the internal PD of the LD could help to reduce this problem. Another option would be using a second PD mounted on the side of the LD to offset the two signals in order to remove the noise from the signal. Furthermore, remaining noise from ambient light, as the surrounding is not perfectly dark, and from the external power supply are possible.

4   Conclusion

An optical distance measurement for FFP-TFs was successfully developed. A fiber with a spot size of 5.9 μm was manufactured and positioned within the filter using a five axis linear stage. The system determined the center wavelength of the FFP-TF as a direct result of the cavity length within a range of 7 nm precision. Therefore, it is able to determine the FSR of 108 nm. To make the measurement even more precise, gluing the fiber to the glass plate is planned in the future. Also, suitable filtering of electronic noise and averaging measurement values in the software is possible. In addition, the noise coming from the LD can be measured with a second PD and can be removed from the signal with suitable calculation. By implementing these measures, a higher precision in the range of one nm or below can be achieved. Building a control system to accurately control the distance between the filter elements in the future is feasible. With the current speed of several measurements per second, the system would be helpful in determining the center wavelength of the FFP-TF. If in the future it is desired to control the sweep amplitude of the FDML laser, a faster ADC operation may be required.

Acknowledgement

The work has been carried out at the Institute of Biomedical Optics, Universität zu Lübeck.

Authors' Statement

Conflict of interest: Authors state no conflict of interest.

5   References

[1] T. Gottschall, T. Meyer-Zedler, M. Schmitt, R. Huber, J. Popp, A. Tünnermann, and J. Limpert, "Ultra–compact tunable fiber laser for coherent anti–stokes raman imaging," *Journal of Raman Spectroscopy*, vol. 52, no. 9, pp. 1561–1568, 2021.

[2] S. Karpf, C. T. Riche, D. Di Carlo, A. Goel, W. A. Zeiger, A. Suresh, C. Portera-Cailliau, and B. Jalali, "Spectro-temporal encoded multiphoton microscopy and fluorescence lifetime imaging at kilohertz frame-rates," *Nature Communications*, vol. 11, no. 1, p. 2062, 2020.

[3] R. Huber, M. Wojtkowski, and J. G. Fujimoto, "Fourier domain mode locking (fdml): A new laser operating regime and applications for optical coherence tomography," *Optics Express*, vol. 14, no. 8, pp. 3225–3237, 2006.

[4] M. Kanchan, M. Santhya, R. Bhat, and N. Naik, "Application of modeling and control approaches of piezo-electric actuators: A review," *Technologies*, vol. 11, no. 6, p. 155, 2023.

[5] Frank L. Pedrotti, Leno S. Pedrotti, Werner Bausch, Hartmut Schmidt, *Optik für Ingenieure: Grundlagen ; mit 28 Tabellen*, 3rd ed.   Berlin and Heidelberg: Springer, 2005.

Fiber-Based Fabry-Pérot Filters: Design, Manufacturing and Characterization

Marlin Kasdorf [1], Marie Klufts [2], and Robert Huber [2]

[1] Medical Engineering Science, University of Lübeck, marlin.kasdorf@student.uni-luebeck.de
[2] Institute of Biomedical Optics, University of Lübeck, {marie.klufts , robert.huber} @uni-luebeck.de

Abstract

This paper presents two fiber Fabry-Pérot (FFP) filter designs with different properties: a fixed-FFP filter and a temperature-tunable-FFP filter. Combining both filters in a laser would provide the narrow linewidth of the fixed-FFP filter while avoiding mode-hoping by temperature-controlling the second filter with a control loop. A stable, tunable, continuous wave fiber-based laser would then be obtained. The fixed-FFP filter built with a spacer of $2\ mm$ in length has a calculated linewidth of $0.1\ pm$ and a measured free spectral range (FSR) of $0.2\ nm$. Despite a major alignment effort to reduce losses, it has been measured and calculated as $\sim 24\ dB$. With a few micrometers gap, the tunable filter presents an FSR above $143\ nm$ with a linewidth of $0.21\ nm$. The stability and step response of the temperature-tunable-FFP are studied and detailed in this paper.

1 Introduction

Nowadays, two kinds of solid-state lasers can be found on the market: fiber and bulk free-space lasers. In the past few years, the worldwide sales of fiber lasers have shown a significantly higher growth rate than that of traditional solid-state lasers [1]. Fiber lasers rely on fiber waveguides to guide the beam and provide optical amplification. Bulk lasers, on the other hand, use bulk crystals for optical amplification. Inside the resonator of a bulk laser, the beam propagates in free space. This difference in design brings some advantages to fiber lasers over bulk lasers. Fiber lasers have a compact footprint and have higher reliability [2]. In addition, they are almost maintenance-free and do not require complex alignment [1, 2]. Furthermore, fiber lasers can be built from relatively inexpensive components and need fewer mechanical components [2]. All these advantages motivate building a stable, continuous wave, narrow linewidth fiber-based laser. The difference between this fiber-based laser and a fiber laser is the different types of gain medium. The planned fiber-based laser uses a semiconductor instead of a doped fiber core as a gain medium. Since one of the main components of a laser is the resonator, two fiber Fabry-Pérot (FFP) filters are built and characterized in this paper. One FFP filter should have a small free spectral range (FSR), a narrow linewidth, and a fixed length, called the fixed-FFP filter in this paper. The other FFP filter, called tunable-FFP filter, should have a large FSR, a large linewidth, and be temperature-tunable. Combining these two FFP filters would provide a new stable and temperature-tunable laser usable for biomedical applications, such as stimulated Raman microscopy [3]. It will provide the narrow linewidth of the fixed-FFP filter and the large FSR of the tunable-FFP filter. The large FSR ensures that only one cavity mode is within the gain profile; thus, only one mode lases at a time. Therefore, mode hopping, the phenomenon in which a laser exhibits sudden jumps in optical frequency associated with transitions between different modes of its cavity, cannot occur. The small line width has the advantage of a long coherence length and is also desired in biomedical applications such as Raman microscopy [3]. The complete development of both filters and their characterization is detailed in the following.

2 Material and Methods

2.1 Design of both FFP filters

The tunable-FFP filter, designed to have a large FSR and a large linewidth, is composed of two fiber-ferrules and a bronze phosphate mating sleeve. The bronze phosphate material has been specifically selected because of its larger thermal expansion coefficient than zirconia, which is usually used in standard mating sleeves. This ensures that the thermal linear expansion of the ferrules behaves differently from the thermal linear expansion of the mating sleeve. In this work, only HI1060 single-mode fibers are used. Both fiber ferrules are positioned in the bronze-phosphate mating sleeve, with a micrometer scale air gap between them. Both ferrules have a concave indent and are coated at their tip. The coating has a high reflectivity of 99.8 % at a wavelength of $1060\ nm$ ensuring a high filter finesse. The same coating is used in the second filter presented below. The dimples are used to refocus the beam, propagating between the two reflective-coated surfaces. The tunable FFP filter is placed in a custom designed holder for temperature control. This holder contains a $10\ k\Omega$ NTC (Vishay, NTCLE413E2103F102L) and a Peltier element (reichelt elektronik GmbH, TEC1-017 03) connected to a temperature controller (Thorlabs, MTD1020T). The fixed-FFP filter, designed to achieve a narrow FSR and a narrow linewidth, contains two coated fiber-ferrules separated by a spacer. The spacer is a self-polished fiber-ferrule of about $2\ mm$ in length. It guides the light propagating between both fiber ferrules. A zirconia mating sleeve is used in this case due to its more precise manufacture. After assembling both filters, as shown in Fig. 1, their alignment capabilities and performances are studied following the detailed experiments in the next part.

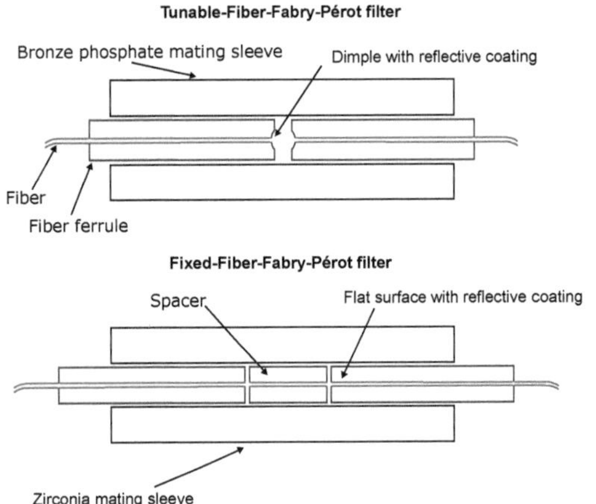

Figure 1: Schematic of both FFP filters as cross-section. Dimensions: fixed-FFP filter: $23\ mm$ length, tunable-FFP filter: $21\ mm$ length. The schematic is not to scale.

2.2 Experimental setups and conducted experiments with the fixed-FFP filter

To ensure good coupling efficiency between the three fibers, the mode field of the beam directly behind the first fiber and after the spacer is analyzed first. For this, a semiconductor optical amplifier (SOA) (Innolume GmbH, SOA-1070-HI-24dB) centered around $1070\ nm$ is used as a light source. It is connected to a circulator, used in this work as an isolator, to suppress all back reflection from the filter. The output of the circulator is then connected to an uncoated fiber ferrule, replacing here the coated fiber ferrules. The uncoated ferrule is used here because it does not reflect light back and makes it easier to find the beam on the camera. The mating sleeve with one fiber ferrule and the spacer is inserted into a custom mode field analyzer setup. This custom-mode field analyzer comprises a CMOS camera (IDS Imaging Development Systems GmbH, UI-1490LE) connected to an x10 objective (Nikon, MSB50100) via a lens barrel. The camera is mounted on a linear stage, which allows finely adjusting the focus on the fiber tip. The losses must be reduced to achieve the best performance of an FFP filter. In this view, it was tried to actively align the filter by using a custom-made fiber-focuser stage. A mating sleeve containing the spacer and the fiber ferrule is positioned at the bottom in a holder. Two stepper motors enable precise movement in both the x- and y-directions. Above, the second fiber fer-

rule is held in another adjustable holder, capable of vertical motion via a stepper motor. Additionally, two piezo elements allow slight rotations in two directions for alignment. One fiber ferrule is connected to a circulator, which in turn is connected to an SOA (Innolume GmbH, SOA-1060-20-PM-32dB) centered around $1060\ nm$. The other fiber ferrule is connected to an optical spectrum analyzer (OSA) (Yokogawa, AQ6370C). The alignment is then performed by gradually reducing the distance between the mating sleeve and the top fiber ferrule while changing the x and y position of the mating sleeve until the transmitted power is at its maximum. Because of the reflective coating, the intensity of the transmitted light is very low. Thus, a booster optical amplifier (Covega, BOA-6165) with a center wavelength of $1300\,nm$ is used for a first coarse alignment. After that, the $1060\ nm$ SOA is used for the final alignment. After the final alignment, the filter spectrum can be measured, and the FSR and the linewidth can be determined. Once aligned, a distributed feedback (DFB) laser diode (AeroDIODE, 1064-LD-2-0-0) is used instead of the SOA to determine the loss. This DFB laser diode is used because of its small linewidth, 200 kHz. This ensures that the emitted spectrum is narrow enough to fit inside one filter transmission peak. Thus, in an ideal filter, the light should be transmitted completely. To ensure achieving the best alignment the top holder´s piezo elements in the fiber focuser stage are driven by a sawtooth waveform. This dynamic adjustment periodically changes the distance between the spacer and the top fiber ferrule, ensuring periodic transmission maxima. This enables the alignment to be less sensitive to vibrations. For this reason, an oscilloscope (Rigol Technologies EU GmbH, DS1104Z) with a photodiode (Thorlabs, Inc., PDA20CS2) is used for this measurement instead of an OSA, as only this can record a time trace of the transmitted maxima.

2.3 Experimental setups and conducted experiments with the tunable-FFP filter

For the characterization of the tunable-FFP filter and the evaluation of its temperature behavior, the same SOA, circulator, and OSA as mentioned above are used. Important characteristics of the tunable-FFP filter are determined with this setup, for instance, the losses, the FSR, and the linewidth. To evaluate the impact of the temperature on the transmission peak, the temperature is varied from $23\ ^\circ C$ to $24\ ^\circ C$ in $0.1\ ^\circ C$ steps. Since we assume that the linear thermal expansion coefficient is linear, this narrow range is sufficient to determine the wavelength shift per $^\circ C$. The step response is also an important characteristic of the temperature control loop, since the step response must be faster than the drift of the fixed-FFP filter. If so, the temperature control loop of the tunable-FFP filter can compensate for this. The step response is measured by changing the target temperature by $0.1\ ^\circ C$, $0.5\ ^\circ C$, and $1\ ^\circ C$ while periodically measuring the filtered transmission spectrum with the OSA. The time needed until the filter spectrum is stable again is measured. To determine the stability of the temperature control loop the temperature is set to $23\,°C$, and the filter transmission spectrum is measured every 5 minutes.

3 Results and Discussion

3.1 Fixed-FFP Filter

The mode field behind the first fiber ferrule shows a Gaussian shape. It has a mode field diameter of $6.6\,\mu m$ in the x-direction and $7.3\,\mu m$ in the y-direction. These results agree with the mode field diameter given in the datasheet of the fiber. However, the mode field after the spacer reveals a multimode intensity profile as shown in Fig. 2.

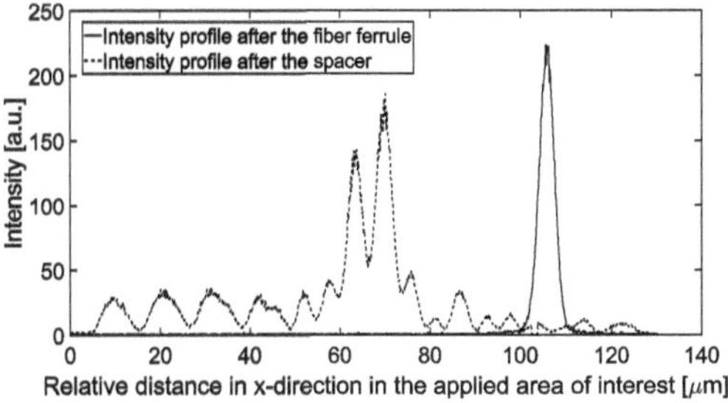

Figure 2: Cross-section of the measured mode fields after the fiber ferrule (solid line) and after the spacer (dashed line).

This indicates that the light is not coupling completely in the fiber's core but also in the cladding. This implies a misalignment of the fiber cores within the mating sleeve. The spacer's dual-sided polishing suggests that the ferrules may only be highly precise at the tip. Usually, the light coupling only happens at the tip of the ferrule and not on the backside, explaining a possible higher precision in the tip area. The company Kyocera advertises that its ferrules have high precision at both ends, unlike competitors [5]. To support this assumption, a spacer is only polished from the backside, and the mode field behind the spacer is reevaluated. It now shows a Gaussian shape, and the mode field diameter is $7\,\mu m$ in x- and in y-direction. Once again, this corresponds well with the value in the datasheet. The filtered spectrum of the actively aligned fixed-FFP filter can be seen in Fig. 3.

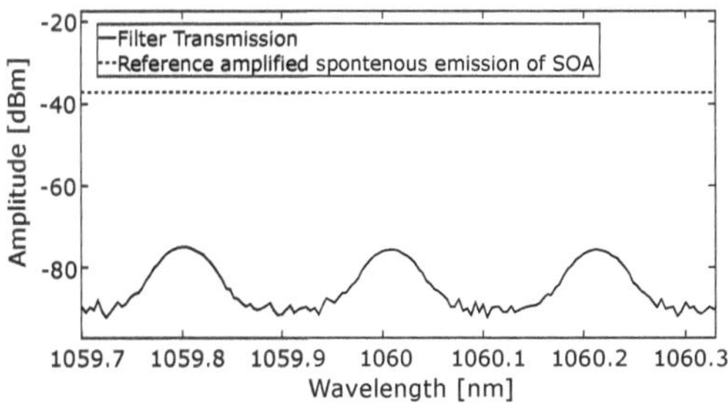

Figure 3: Filter Transmission of fixed-FFP filter.

A desired narrow FSR of the fixed-FFP filter of $0.2\,nm$ is obtained. The linewidth of one filter transmission peak can not be determined because it is assumed to be smaller

than the resolution of the OSA $(20\,pm)$. If the peak is smaller than the resolution of the OSA, the OSA will integrate the power in this $20\,pm$ interval, which makes the peak less high and wider. This fact is also the reason why the losses of the filter can not be determined with this measurement setup. However, with an assumed finesse of 2000, the linewidth of one filter peak can be calculated and equals $0.1\,pm$. The result of the loss measurement of the fixed-FFP filter with the DFB diode as described in the methods section can be seen in Fig. 4.

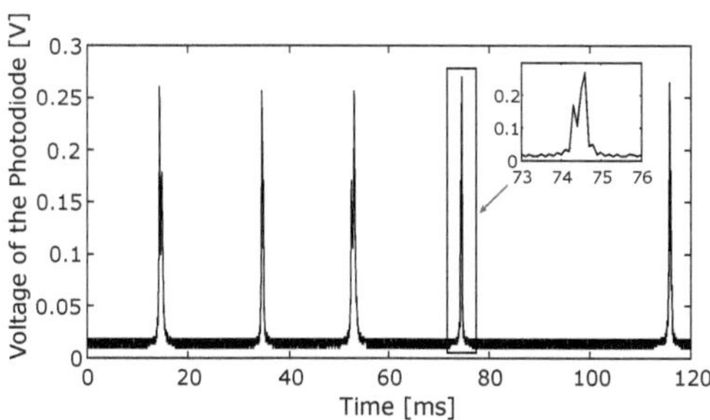

Figure 4: Transmitted power through the aligned fixed-FFP filter.

After optimization, the transmitted power through the filter leads to a max peak voltage of $U_T = 904\,mV$ with an internal photodiode gain set to $30\,dB$. The output power of the DFB diode leads to an output voltage of $U_0 = 4,24\,V$, with a gain set to $0\,dB$. To calculate the losses of the filter, the following equation can be used:

$$L_P = 10\,dB \cdot \log_{10}\left(\frac{P_T}{P_0}\right) \quad (1)$$

, where P_0 is the output power of the DFB diode, P_T is the non-amplified transmitted power through the filter, and L_P is the loss of the filter. Since the incident optical power on a photodiode is proportional to the output voltage, the voltages can be inserted into equation 1 instead of the optical powers. The loss of the fixed-FFP filter is $21.7\,dB$, which is high compared to other FFP filters [4]. A possible explanation for the measured high loss could be an angled end surface of the polished spacer, which leads to misalignment. Another explanation can be misalignment due to poor precision at the backside of the ferrule used, as mentioned earlier [5]. The laser diode driver could also influence the measured loss of the filter. If it supplies a noisy voltage to the DFB diode, its linewidth can be wider than specified in the datasheet. Thus, the linewidth of the laser diode may be wider than the linewidth of the filter, resulting in a mismatch between the measured and actual filter loss. Another way to estimate the loss of the filter is based on the measured transmitted power of one peak of the filter transmission spectrum and the calculated linewidth. As already mentioned, the limited OSA resolution leads to a wider and lower peak because of the integration of the power in $20\,pm$ intervals. However, this only affects the optical power distribution but not the transmitted power of one peak. The transmitted power of one filter peak equals the area under it and must be constant. With this in mind, a Gaussian curve with a linewidth of $0.2\ pm$ and an equal area below compared to one of the measured peaks in the transmission spectrum is calculated. The peak height of this calculated Gaussian curve can then be used to calculate the loss. This calculation results in a loss of $24\ dB$, which corresponds well with the measured loss. The difference of $2.3\ dB$ between the calculated and the measured loss can be caused by errors in the Gaussian fit of the peak or the guessed finesse. In the zoomed figure in Fig. 4, it can be seen that all peaks have a side peak. An explanation can be the occurrence of higher-order longitudinal modes in the filter due to poor alignment.

3.2 Tunable-FFP Filter

The measured optical spectra at the input and the output of the filter can be seen in Fig. 5.

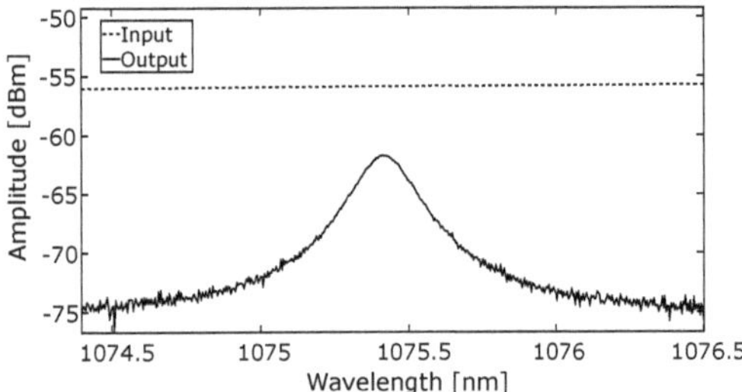

Figure 5: Input and output of the tunable-FFP filter.

The measured loss of the tunable-FFP filter is $5.8\ dB$. A normal connection between two fiber ferrules should have a maximal loss of $1\ dB$. An explanation of why the loss of the tunable-FFP filter is that high could be the use of the bronze-phosphate mating sleeve instead of a zirconia one. Usually, zirconia components are used in optics for their more precise manufacturing. The FSR is measured to be above $143\ nm$. An exact value cannot be given due to the too-narrow spectrum of the SOA. A super-continuum light source could be used to measure it. However, the exact FSR is not relevant here since the FSR is larger than the gain profile of the SOA, only one filter peak is within the gain profile, and therefore, only one wavelength can lase, and mode hopping is impossible. The calculated linewidth of the transmission peak in Fig. 5 is $212\ pm$. The wavelength of the filtered peak can be linearly temperature-tuned by $13.7 \frac{nm}{^\circ C}$. This means that only a small change in temperature is required to sweep over the full-width half maximum of the gain profile of the SOA. Furthermore, the step response of the control loop led to the following results: a stabilization time of $20\ s$ is required after a temperature step of $0.1\ ^\circ C$, after a temperature step of $0.5\ ^\circ C$ a stabilization time of approximately $22\ s$ is required and after a $1\ ^\circ C$ a stabilization time of $26\ s$ is required. For a possible series connection of both FFP filters, the step response must be faster than the fixed-FFP filter's thermal drift. However, the stability measurement of the tunable-FFP filter shows that the peak wavelength varies over a span of $257.7\ pm$ with a standard deviation of $72.9\ pm$. This means that the transmission spectrum varies in the Sub-nanometer regime, but

it is above $0.2\ nm$, the FSR of the fixed-FFP filter. For a series connection of both filters, this would mean, that the filter peak of the tunable-FFP filter would drift between two filter peaks of the fixed-FFP filter. This makes the transmitted power through the series connection fluctuate. A better thermal isolation and optimized PID controller could improve the stability of the filter and reduce the wavelength variation.

4 Conclusion

The experimental results in this paper show that both filter designs can be used to build FFP filters with the desired properties. However, the results also show that there is room for improvement. The high loss of the fixed-FFP filter prevents the build of a laser with it since the gain must be higher than the losses. Investigating higher-quality ferrules might lead to better alignment and, thus, lower loss. The loss of the tunable-FFP filter can also be improved with another, more precisely manufactured mating sleeve with a similar linear thermal expansion coefficient. Regarding the temperature control loop of the tunable-FFP filter, the stability must be improved with better temperature isolation and fine-tuning of the PID parameters. Also, the use of a more precise temperature controller would be the conceivable solution to that problem.

Acknowledgement

The work has been carried out at the Institute of Biomedical Optics, University of Lübeck.

Authors' Statement

Conflict of interest: Authors state no conflict of interest.

5 References

[1] RP Photonics, *Fiber Lasers Versus Bulk Lasers*, https://www.rp-photonics.com/ fiber_lasers_versus_ lbulk _lasers.html [last accessed on 2024-01-12].

[2] FindLight Blog, *Fiber Laser: 7 Advantages and Differences*, https://www.findlight.net/blog/7-advantages-of-fiber-laser/ [last accessed on 2024-01-12].

[3] T. Gottschall, T. Meyer-Zedler, M. Schmitt, R. Huber, J. Popp, A. Tünnermann and J. Limpert, *Ultracompact tunable fiber laser for coherent anti-Stokes Raman imaging*, Journal of Raman Spectroscopy, 2021, 52(9), 1561, https://doi.org/10.1002/jrs.6171.

[4] R. Huber, M. Wojtkowski, and J. G. Fujimoto, *Fourier Domain Mode Locking (FDML): A new laser operating regime and applications for optical coherence tomography*, Opt. Express **14**, 3225-3237 (2006), https://doi.org/10.1364/OE.14.003225.

[5] KYOCERA Corporation, *Ferrules*. Available: https://global.kyocera.com/prdct/semicon/semi/fiber /ferrule.html [last accessed on 2024-01-05].

Intracellular localization of light-sensitive liposomes

Nina Kleineberg[1], Astrid Link[2], Paula Enzian[3] and Ramtin Rahmanzadeh[2]

[1] Medical Engineering Science, Universität zu Lübeck, nina.kleineberg@student.uni-luebeck.de
[2] Institute of Biomedical Optics, Universität zu Lübeck, {astrid.link, ramtin.rahmanzadeh}@uni-luebeck.de
[3] Medical Laser Center Lübeck, Lübeck GmbH, p.enzian@uni-luebeck.de

Abstract

Light-sensitive liposomes can release drugs intracellularly at a specific target site in the body, triggered by externally controlled light irradiation. To investigate the intracellular localization of liposomes before and after irradiation in more detail, light-sensitive liposomes were applied to HeLa cells in this study. A confocal microscope and various fluorescent markers were used to examine the localization. It was demonstrated that the liposomes accumulated in the cytoplasma after 1 hour. In addition, an accumulation of liposomes in the endosomes and lysosomes of the cells was suggested.

1 Introduction

The development of systems for the targeted release of drugs in the human body is attracting great interest in medicine. These drug delivery systems include liposomes, which are particularly suitable due to their biocompatibility, biodegradability and low toxicity [1]. The structure of liposomes consists of an inner cavity and a double lipid layer, which gives the liposome an amphiphilic character [1]. The targeted release of active substances encapsulated in liposomes can be triggered by various mechanisms, one possibility being the incorporation of photosensitizers into the liposome membrane [2]. When excited by light of a certain wavelength, a photosensitizer generates reactive oxygen species (ROS), which decomposes the lipids in the liposome membrane and releases the encapsulated drug [2].

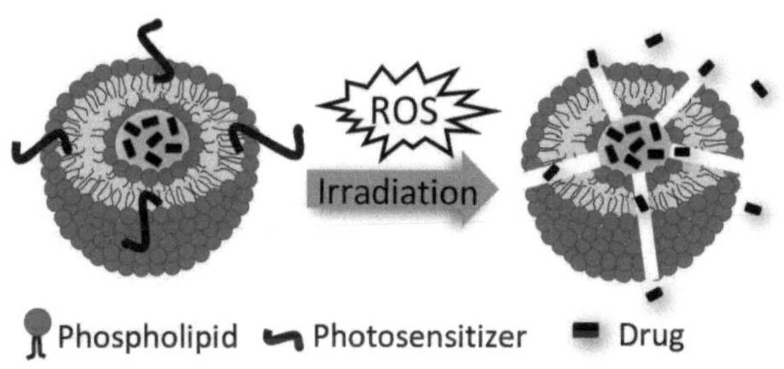

Figure 1: Schematic representation of a light-sensitive liposome. After irradiation with light, the liposome membrane is permeabilized due to the ROS produced, allowing the contents to leak out.

In this work, liposomes including the photosensitizer 5,10-di-(4-hydroxyphenyl)-15,20-di-phenyl-21,23H-porphyrins (5,10-DiOH) were incubated with cervical carcinoma cells and examined microscopically. Fluorescent dyes that stain the mitochondria, endoplasmatic reticu-

lum, endosomes and lysosomes were used to investigate where the liposomes accumulate intracellularly in more detail. The last two fluorescent dyes were employed under the assumption that the liposomes enter the cell via endocytic signaling pathways [3]. To gain a more comprehensive understanding of the liposome uptake process into the cell, microscopy was performed at different time points. In addition, the endocytotic uptake of the liposomes was analysed by functional imaging using photon arrival times to generate fluorescence lifetime-based information (TauSense technology).

2 Material and Methods

2.1 Cell culture

The human cervical cancer cell line (HeLa) was obtained from American Type Culture Collection (ATCC, No. ACC 57). The adherent cells were cultivated in T-75 cell culture flasks (Greiner Bio-One, Germany) with Dulbecco's Modified Eagle's Medium (DMEM high glucose, Sigma-Aldrich, USA). The culture medium was supplemented with 10 % fetal bovine serum (FBS, Sigma-Aldrich, USA) and 1 % penicillin-streptomycin (10,000 units/ml penicillin and 10 mg/ml streptomycin, Sigma-Aldrich, USA). All cells were incubated in a humid incubator with 5 % CO_2 concentration at 37 °C before and between the experiments. For each series of experiments, 2 ml Hela cells at a density of 40,000 cells/ml were seeded either in glass-bottom Petri dishes (Greiner Bio-One, Austria) or on cover slips (Superior Marienfeld, Germany) and grown for 24 hours.

2.2 Preparation of L-calcein-5,10-DiOH

To prepare the liposomes, the lipids (table 1) were dissolved individually in chloroform and mixed in a molar ratio of 1 : 4.8 : 0.25 : 2.3 (DOTAP : DPPC : DSPE-PEG(2000) : cholesterol). All lipids were purchased from Avanti Polar Lipids (Alabama, USA). Then, 50 μl of the stock solution (1 mM) of photosensitizer 5,10-DiOH was added to the lipid solution. The chloroform was evaporated under constant rotation in a nitrogen stream to achieve a homogeneous lipid film. To prepare a liposomal release test system, 500 μl of a calcein solution was added to the lipid mixture instead of a drug. The entire solution was incubated for 40 min at 50 °C, a temperature greater than the transition temperature of the lipids. This was followed by six 10 min freezing and thawing cycles. This process led to the formation of multilamellar vesicles, which were pushed through an extruder system (Avanti Polar Lipids, USA) heated to 50 °C with a 100 nm diameter membrane. Non-encapsulated calcein was removed using gel filtration (CL-4B, Sigma-Aldrich,USA). The encapsulated 5,10-DiOH concentration was determined by absorption spectroscopy.

Table 1: Lipid composition of the L-calcein-5,10-DiOH

lipid	concentration	molar mass
DOTAP	25 mg/ml	698.5 M
DPPC	25 mg/ml	734.0 M
DSPE-PEG(2000)	10 mg/ml	2790.5 M
Cholesterol	50 mg/ml	386.7 M

2.3 Imaging experiments to investigate intracellular localisation

Liposomes with a concentration of 400 nM 5,10-DiOH were added to the cell. Afterwards, the cells were incubated in the humid incubator and analyzed at different time points (15 min, 1 h, 2.5 h). After incubation cells were washed with Phosphate-Buffered Saline with calcium chloride and magnesium chloride (PBS$^+$, Sigma-Aldrich, USA) and viewed in 2 ml PBS$^+$ under a confocal microscope (Stellaris 5, Leica, Germany) using the life science microscope software Leica Application Suite X (LAS X). The images were taken with a plan apochromat oil immersion objective (numerical aperture 1.4). To capture the images, calcein was excited according to its absorption spectrum at a wavelength of 520 nm.

2.4 Investigation of intracellular localization using cell markers

Mitochondrial tracker (MitoTracker® Orange CMTMRos), lysosome tracker (LysoTracker® Deep Red) and endoplasmic reticulum tracker (ERTracker™ Blue-White DPX), were employed to determine the specific cellular organelles where the liposomes were localized. The concentrations at which the trackers were applied to the cells and the subsequent incubation times were chosen according to the manufacturer's instructions. After the incubation times, cells were washed with PBS$^+$ and liposomes were added

($c_{5,10-DiOH}$ = 400 μM). The assessment of liposome accumulate within specific cell organelles was conducted at various time points (1 h, 2 h, 4 h, 24 h) in PBS$^+$ using a confocal microscope. The excitation wavelength for each tracker was chosen according to the manufacturer's specifications (MitoTracker = 554 nm, LysoTracker = 647 nm and ERTracker = 374 nm).

2.5 Antibody staining with Rab7

To determine the presence of liposome accumulation in endosomes, HeLa cells were cultivated on coverslips and subsequently labeled with the primary antibody Rab7 and secondary antibody AlexaFluor 568. For this purpose, the cells were fixed with 4 % Histofix (Sigma-Aldrich, USA). The Rab7 antibody was diluted 1:100 in PBS$^+$ with 10 % FBS and incubated for 30 min on HeLa cells. The cells were then washed three times and the secondary antibody AlexaFluor 568 was added. The antibody was diluted 1:500 in PBS$^+$ with 10 % FBS and incubated for 30 min. Subsequently, the cells were washed three times with PBS$^+$ and attached to a glass slide (Thermo Fisher, USA) using a mounting medium (ProLong™ Glass, Thermo Fisher, USA).

2.6 Functional imaging based on fluorescence lifetime information (TauSense)

TauSense technology calculates the photon arrival time (AT_i) for each image pixel [4]. For this purpose, the excitation time (t_{exc}) of the laser used is subtracted from the arrival time of the photons (t_i) at the scan head (1) and subsequently the average photon arrival time value (AAT) is calculated [4].

$$AT_i = t_i - t_{exc} \qquad (1)$$

The photon arrival time depends on the fluorescence lifetime of the molecule, which can change with the pH value for certain fluorophores [4]. For this reason, the technology was used to investigate the endocytic uptake of liposomes, which is associated with a change in pH.

3 Results and Discussion

To investigate the intracellular localization of L-calcein-5,10-DiOH, fluorescence images were supplemented by an image captured in the Trans PMT (Transmission Photomultiplier) setting (Fig. 2). After 15 min, it can be seen that the liposomes have accumulated primarily in the cell membrane. At 1 h, the first liposomes begin to accumulate in the cytoplasm. After 2.5 h an accumulation of liposomes around the cell nucleus is visible. The individual lipids in the liposome have an influence on cellular uptake [5]. In this work, the lipids DOTAP, DPPC, DSPE-PEG(2000) and cholesterol were used to prepare the liposomes. The use of the lipid DOTAP gives the liposome a positive charge, which according to Botet-Carreras et al. improves intracellular uptake [5]. The reason for this is the electrostatic interaction with the negatively charged plasma membrane of the cell [5]. The cholesterol according to Yang et al. should lead to increased fusion of the liposomes with the cell membrane and thus promote subsequent internalization [6]. Overall, the lipid composition used appears to be well suited to deliver drugs quickly and efficiently into the cells.

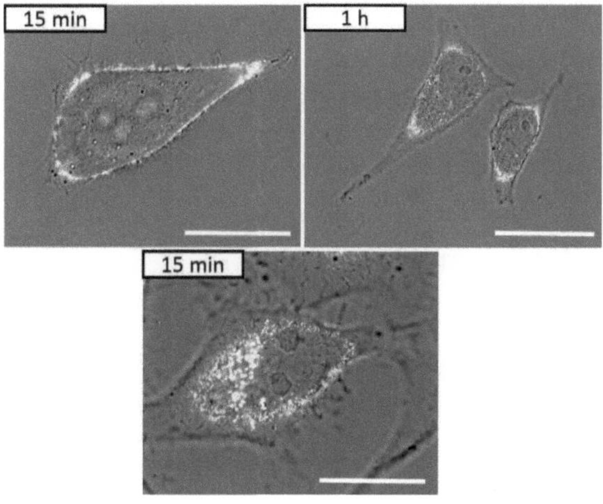

Figure 2: Intracellular localization of L-calcein-5,10-DiOH (white) in HeLa cells at different time points. After 2.5 h liposomes are localized in the cytoplasm around the cell nucleus. Scale bar: 25 μm.

Fluorescent markers were used to distinguish cellular organelles to investigate the liposomes localization in more detail. Given that numerous research studies have indicated that liposomes are predominantly taken up via endocytosis, Rab7 was used to label endosomes and a lysoTracker was utilized to label lysosomes. Two of the four time points investigated (1 h, 2 h, 4 h, 24 h) are shown in a representative way (Fig. 3). In individual cases, it can be assumed that the liposomes are colocalised with the endosomes and lysosomes (white arrows in Fig. 3). Further series of experiments are necessary to determine the exact intracellular localisation. However, the colocalisation suggested in the images is confirmed by other studies, for example Takikawa et al. were able to show a pronounced colocalisation of positively charged liposomes by labelling the lysosomes with a LysoTracker [3].

One reason why the localisation of liposomes in endosomes and lysosomes could not be clearly confirmed in the images of this study could be the presence of the photosensitizer. During the capture of fluorescence images, the photosensitizer is stimulated, leading to the generation of reactive oxygen species. This chemical reaction damages the endolysosomal membranes, causing the liposomes to escape from the endolysosomes [2]. While this process adds complexity to achieving clear localization results under the microscop, it remains crucial for prospective medical applications. This is due to the fact that liposomes can leave the endolysosomal compartments before undergoing enzymatic digested [7].

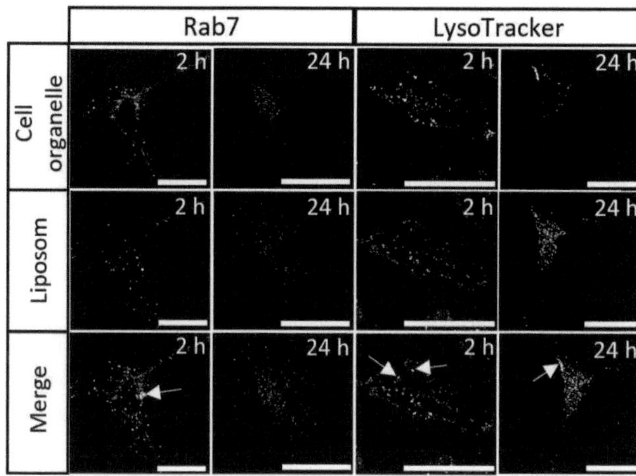

Figure 3: HeLa cells with Rab7 (left), LysoTracker (right) and liposomes (L-calcein-5,10-DiOH) after different incubation times. The white arrows indicate colocalization of the liposomes with the fluorescent markers. Scale bar : 25 μm.

The images taken with the MitoTracker show no colocalisation of the liposomes with the mitochondria (Fig. 4). The HeLa cells stained with the ERTracker show partial accumulation, especially after 24 hours of incubation. Further studies are necessary to determine exactly whether the liposomes accumulate in the endoplasmic reticulum.

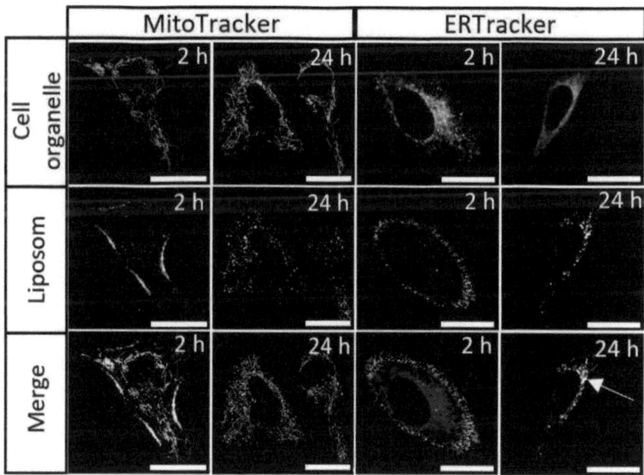

Figure 4: HeLa cells with MitoTracker (left), ERTracker (right) and liposomes (L-calcein-5,10-DiOH) after different incubation times. No colocalization of the liposomes in the mitochondria is visible. In the area of the endoplasmic reticulum an accumulation is recognizable especially after 24 h (white arrow). Scale bar: 25 μm.

Due to the aforementioned challenges in detecting liposomes in endosomes and lysosomes, a fluorescence lifetime-based imaging technique (TauSense) was also used. This imaging technique measures the arrival time of photons at the detector, which depends on the fluorescence lifetime of a molecule, and displays it as a colour contrast in the image [4]. For certain molecules, such as calcein, the fluorescence lifetime and thus the photon arrival time changes with the pH-value [7]. When the liposomes with encapsulated calcein are taken up by endocytosis, the pH-value changes depending on whether the liposomes are located in the plasma membrane, in endosomes or in lysosomes. PH-values in the range of 6.0 to 5.5 typically indicate early endosomes, while pH-values of 5.0 to 4.5 tend to indicate lysosomes [7]. To test if measuring photon arrival times could provide information on whether liposomes with encapsulated calcein were located in endosomes or lysosomes, HeLa cells were incubated with liposomes for 1 hour (Fig. 5). To verify that the dots visible in the image are indeed liposomes, an additional image was taken showing a HeLa cell without liposomes. The white dots in the image represent photons that were detected after 0.3-0.5 ns and the lighter grey dots represent photons that were detected after 0.8-1.1 ns. In another study, fluorescence lifetime-based imaging and a pH-sensitive near-infrared membrane dye were used to detect the uptake of versicles by endocytosis [4]. In this work, longer photon arrival times of 0.8-1.1 ns correspond to pH-values in early endosomes and shorter photon arrival times of 0.3-0.5 ns correspond to pH-values in lysosomes [8]. These times are in good agreement with those measured in our study and support the suggestion that the use of fluorescence lifetime-based imaging, such as TauSense technology, can indeed detect liposome uptake by endocytosis.

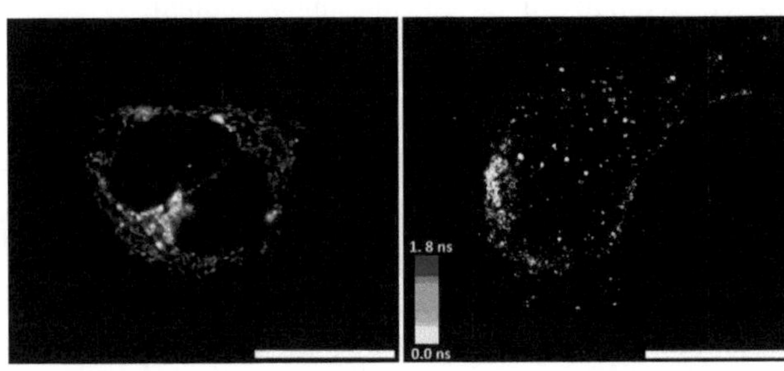

Figure 5: HeLa cells with L-calcein-5,10-DiOH after 1 h incubation (right) and HeLa cells without L-calcein-5,10-DiOH (left) in combination with TauSense. The dots visible in the right image represent the liposomes. The different colours indicate the average arrival time of the measured photons at the detector. Scale bar: 25 μm.

4 Conclusion

The aim of this work was to investigate the interaction of light-sensitive liposomes with cell membranes and to identify the cell organelles where accumulation oocurs. Liposomes containing calcein and 5,10-DiOH were incubated with HeLa cells for this purpose. Results demonstrated that the liposomes had already accumulated in the cytoplasm after 1 hour and in the vicinity of the cell nucleus after 2.5 hours. This could be helpful for later medical applications, as the intracellular availability of encapsulated drugs can be achieved quickly. In addition, the hypothesis that the liposomes enter the cell via endocytic signaling pathways could be substantiated with the help of fluorescence dyes and a fluorescence lifetime-based imaging.

Acknowledgement

The work was carried out at the Institute of Biomedical Optics at the University of Lübeck.

Authors' Statement

The authors declare no conflict of interest.

5 References

[1] Akbarzadeh, A., Rezaei-Sadabady, R., Davaran, S., Joo, S., Zarghami, N., Hanifehpour, Y., Samiei, M., Kouhi, M. & Nejati-Koshki, K. Liposome: classification, preparation, and applications. *Nanoscale Research Letters*. **8** pp. 102 (2013)

[2] Chen, W., Goldys, E. & Deng, W. Light-induced liposomes for cancer therapeutics. *Progress In Lipid Research*. **79** pp. 101052 (2020)

[3] Takikawa, M., Fujisawa, M., Yoshino, K. & Takeoka, S. Intracellular Distribution of Lipids and Encapsulated Model Drugs from Cationic Liposomes with Different Uptake Pathways. *International Journal Of Nanomedicine*. **15** pp. 8401-8409 (2020)

[4] Microsystems, L. TauSense: the potential of STELLARIS. *Nature Methods*. **17**, i-iv (2020)

[5] Botet-Carreras, A., Marimon, M., Millan-Solsona, R., Aubets, E., Ciudad, C., Noé, V., Montero, M., Domènech, Ò. & Borrell, J. On the uptake of cationic liposomes by cells: From changes in elasticity to internalization. *Colloids And Surfaces B: Biointerfaces*. **221** pp. 112968 (2023)

[6] Yang, S., Kreutzberger, A., Lee, J., Kiessling, V. & Tamm, L. The role of cholesterol in membrane fusion. *Chemistry And Physics Of Lipids*. **199** pp. 136-143 (2016)

[7] Hausig-Punke, F., Richter, F., Hoernke, M., Brendel, J. & Traeger, A. Tracking the Endosomal Escape: A Closer Look at Calcein and Related Reporters. *Macromolecular Bioscience*. **22**, e2200167 (2022)

[8] Microsystems, L. Understanding cellular systems using life-time based information. (2023), https://www.youtube.com/watch?v=g44W0s8ZFXo

Influence of blue light irradiation on keratinocyte viability

Anna-Lena Urban [1], Zuzana Penxová [2], Birgit Lange [3], Paula Enzian [3], Ramtin Rahmanzadeh [4], Ralf Brinkmann [3,4], Karl-Ludwig Bruchhage [2], Anke Leichtle [2]

[1] Biophysics, Universität zu Lübeck, annalena.urban@student.uni-luebeck.de

[2] Clinic of Otorhinolaryngology, Head and Neck Surgery, University Medical Center Schleswig-Holstein, Lübeck,{Zuzana.Penxova, Karl-Ludwig.Bruchhage, Anke.Leichtle}@uksh.de

[3] Medical Laser Center Lübeck, {birgit.lange, p.enzian, ralf.brinkmann}@uni-luebeck.de

[4] Institute of Biomedical Optics, Universität zu Lübeck, {ramtin.rahmanzadeh, ralf.brinkmann}@uni-luebeck.de

Abstract

Cholesteatoma, a growing aggressive lesion in the middle ear, can cause severe health issues (hearing loss, etc.). The only effective treatment so far is surgery, which aims to remove the pathogenic tissue and reduce the bacterial load locally. As lasers become more popular in otology, expanding the blue laser application, e.g., for microbial reduction or wound healing, seems promising. To better understand the interactions between blue laser and cholesteatoma, keratinocytes were irradiated as an in vitro model. The focus was on the laser parameter modulation, especially the influence of different irradiation intensities and, thus, exposure times by the same irradiation dose on cell viability. For the irradiation dose of 256 J/cm^2 (applied with 445 nm), there seems to be a threshold around 0.85 W/cm^2 above which cell vitality drops significantly, indicating heat-induced cell death. By applying the dose of 20 J/cm^2 with the aim of photostimulated proliferation using different irradiation intensities, the intensity variation does not play a crucial role.

1 Introduction

The acute microbial-caused inflammation of the middle ear can persist and may lead to chronic inflammation, sometimes manifested as cholesteatoma. Cholesteatoma is an ingrowth of keratinizing squamous epithelium in the middle ear. Standard therapy consists of surgical removal of this hyperplastic tissue while keeping the healthy vicinity of the middle ear intact [1]-[3].

Laser applications in otology are becoming increasingly important. There is particular interest in blue light, as it can be used for cholesteatoma ablation and has an antibacterial effect. Through this, the influence of blue light irradiation on surrounding tissue remains subject to further investigation [4].

Hence, we evaluated the effects of blue light on keratinocyte viability in vitro. For this purpose, we tested two wavelengths: 445 nm and 405 nm. The question of the influence of blue light and whether induced cell death is of photothermal or photochemical origin will be investigated in this paper. Therefore, the same dose was applied at different intensities and, in consequence, different exposure times.

Additional considerations include blue light-stimulated cell proliferation, similar to low-level laser therapy [5]. We aimed to identify the dose and irradiation parameter range for blue light-stimulated proliferation.

2 Material and Methods

2.1 Culturing of the HaCaT-Cells

Due to the crucial role of keratinocytes in the pathogenesis of acquired cholesteatoma, HaCaTs, a spontaneously transformed immortal keratinocyte cell line from adult human skin, was used as an in vitro model.

HaCaTs were cultured in Dulbecco's Modified Eagle Medium (DMEM) supplemented with 1% Penicillin-Streptomycin (P/S) and 10% Fetal Bovine Serum (FBS) and incubated at standard conditions (at 37°C in a 5% CO_2 atmosphere and 90% humidity). On the 3rd day after seeding (starting seeding density of 5×10^5 cells / mL), the cells were irradiated; the next day, the medium was replaced by a fresh one.

2.2 Cell viability using MTT-assay

The MTT-assay was executed three days after the irradiation. MTT (3-(4,5-dimethylthiazol-2-yl)-2,5-diphenyltetrazolium bromide) readily penetrates viable eukaryotic cells and is converted into a purple-colored formazan, while dying cells lose this capability.

The supplemented medium with 10% MTT working solution (a half-saturated MTT solution prepared in PBS) was added to each technical triplicate/quadruplicate well. Cells were incubated with MTT under standard incubation conditions for two hours. After removing the media, Dimethyl Sulfoxide (DMSO) was added to lyse the cells. The absorbance measurement was executed at the wavelength near the formazan absorption maximum of 550 nm and at 655 nm as the reference wavelength in each well, using the ELISA plate reading spectrophotometer (Microplate Manager, Bio-Rad Laboratories, USA). The cell viability expressed in percent was determined as a quotient between the mean of treated cells and the mean of negative control (non-irradiated cells) multiplied by 100%.

2.3 Irradiation set-up for 445 nm

The 445 nm irradiation set-up involved two lasers. For the laser powers 0.5 - 4 W, diode laser FOX IV (A.R.C. Laser GmbH, Germany) was used. Diode laser WOLF (A.R.C. Laser GmbH, Germany) was chosen for applied laser powers 4 W - 7.5 W. The laser beam was transmitted through an optical fiber (Ø 400 µm) using a vibrating motor and fixed in an optical lens system for improved laser beam expansion and homogenization (±11%). The irradiation intensities of $0.28 - 2.26$ W/cm^2 or $2.26 - 4.24$ W/cm^2 could be reached in the irradiation plane (cells). Cell irradiation was performed on the sterile bench at room temperature.

2.4 Irradiation set-up for 405 nm

The 405 nm irradiation was performed using Roboterarm Rotrics with a 405 nm laser module (Rotrics Dexarm Luxury Kit, Rotrics, China). The diode laser provided stable irradiation by the set laser power of $10 - 100$ %, corresponding to irradiation intensities of $34 - 414$ mW/cm^2, respectively.

3 Results and Discussion

3.1 Keratinocyte viability after 445 nm irradiation

The viability (in %) is plotted in Fig. 1 against the irradiation intensities (W/cm^2) corresponding to the following applied range of laser power (500 mW -7.5 W) while keeping the applied irradiation dose of 256 J/cm^2 constant.

The viability of the cells is calculated and normalized using the control (untreated cells). The control non-irradiated cells were treated the same way, despite the irradiation, as the irradiated cells (e.g., time outside the incubator during the irradiation) to exclude other external influences.

The irradiance of 0.28, 0.40, and 0.56 W/cm^2 caused no significant changes in the cell viability decrease; on the contrary, the 0.56 W/cm^2 led to a slight increase in keratinocyte viability.

Starting at 0.85 W/cm^2, corresponding to exposure times lower than 301 s up to 34 s (4.24 W/cm^2), the plotted data demonstrate an abrupt decrease in cell vitality. In the measured range from 0.85 W/cm^2 to 4.24 W/cm^2, the vitality of keratinocytes remains under 10 % compared to those of control (untreated cells). The instantaneous reduction in keratinocyte viability at the same applied dose of 256 J/cm^2 occurs between irradiance of 0.56 W/cm^2 and 0.85 W/cm^2

(laser power 1 W and 1.5 W). This is a strong indication of heat shock-induced cell death [6].

Since the irradiation dose is the same, the way of applying the dose in the manner of modifying the irradiation intensity and exposure time must play a decisive role. The spectral proximity of blue light to ultraviolet (UV) wavelengths raises a question of possible photochemical harm. The photochemical mechanism is dose-dependent; the higher the dose, the more cellular chromophores absorb the energy and the higher the probability of causing cell damage [7]. Presumed, the same dose should lead to almost the same cellular response independently of irradiation intensities and exposure times due to nearly the same amount of absorbed energy. The modulation of intensity and exposure time at the same dose means higher energy deposits in a shorter time and vice versa, causing thermal accumulation. That supposed the photothermal induced mechanism of cell death, either heat shock-induced apoptosis by accumulation of misfolded proteins or heat-induced (primary) necrosis [8].

Further investigations of cell death mechanism and temperature accumulation in three-dimensional in vitro models after 445 nm irradiation will rule out critical points and guarantee a more precise control during treatment in a compressed specimen of a middle ear full of delicate structures (i.e., nerves).

3.2 Keratinocyte viability after 405 nm irradiation

Besides the well-established 445 nm used in diode lasers for otological surgeries, further investigations were realized using the shorter wavelength of 405 nm. The 405 nm borders the UV-A (315 – 400 nm) and is of a shorter wavelength, i.e., greater frequency and higher energy than 445 nm. Hence, tendentially lower doses are required for pathogens inactivation. The expected dose-dependent response regarding the cell viability, namely, the higher the dose, the less viable the cells, was confirmed in the pre-experiments. Moreover, we encountered an effect similar to that of low-level laser therapy (LLLT). LLLT is usually observed in a spectral range of 300 - 10,600 nm, applying the irradiation doses of 10^{-2} - 10^2J/cm^2 by irradiation intensities 10^2 up to 1 W/cm^2 [5]. The HaCaTs were exposed to 20 J/cm^2 at 405 nm by varying the irradiation intensities to investigate this cell growth-stimulating effect in more detail. The measured data are graphically depicted in Fig. 2 as a function of different irradiances and cell viability by a constant irradiation dose of 20 J/cm^2. The HaCaTs' viability fluctuates around 100%, meaning cell vitality is comparable to the viability of non-irradiated cells. Thus, a direct relation is not recognizable as to whether the different irradiance intensities influence cell proliferation. A proliferative stimulation occurs, e.g., at irradiance intensities of 82 mW/cm^2 and 128 mW/cm^2, increasing the HaCaTs' vitality by approximately 45% and 35%, respectively. Comparing the mean values of the biological replicates, the tendency of cell proliferative stimulation diminishes with higher irra-

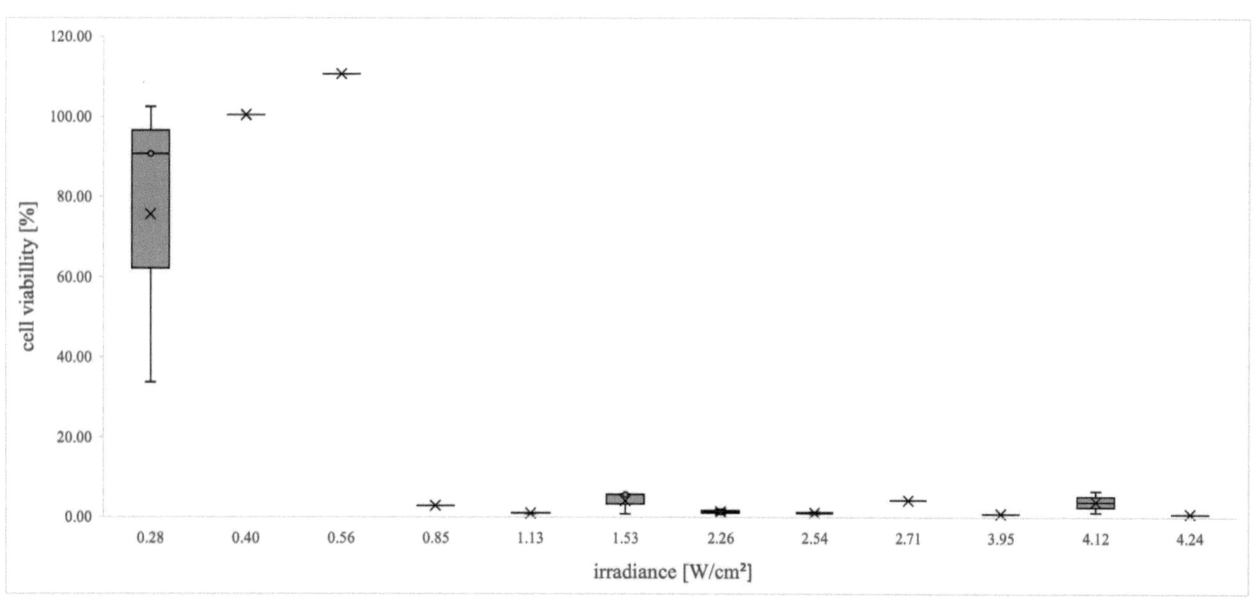

Figure 1: Cell viability after irradiation with 445 nm is plotted against the irradiance [W/cm^2] while keeping the irradiation dose of 256 J/cm^2 constant. Cell viability is expressed in percentage [%] and normalized using a control (untreated cells set as 100% viable); values lower than 100% mean a decrease in cell viability and vice versa. Median values and standard deviations throughout all biological replicates (n=3/4) are indicated in the diagram through vertical lines. Mean values are displayed through a cross.

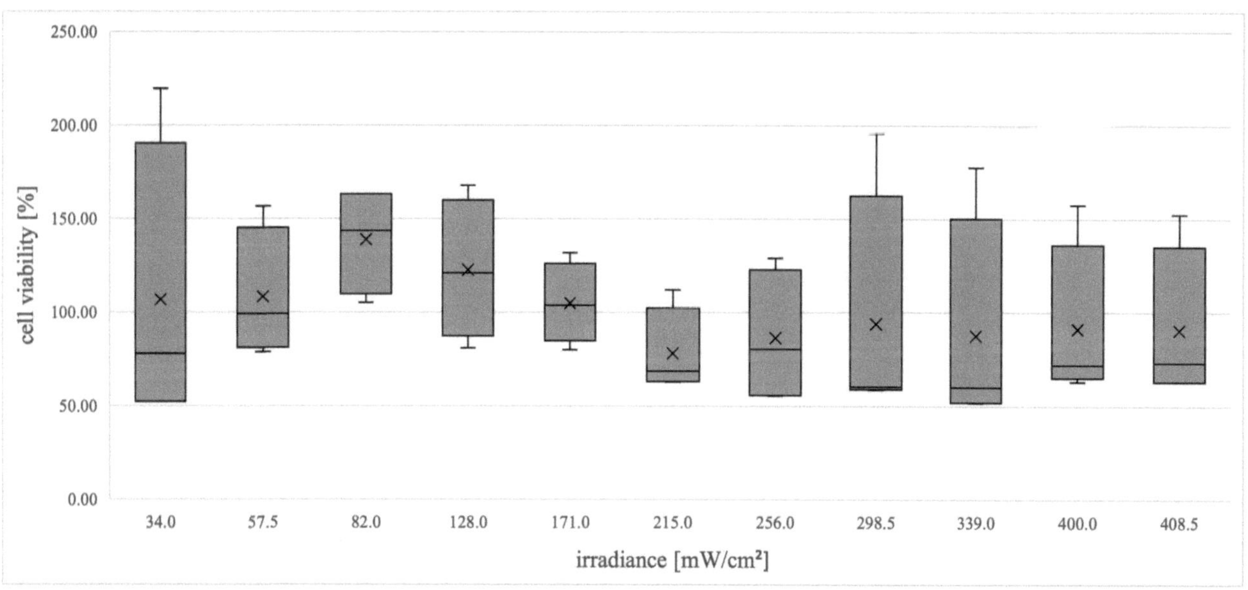

Figure 2: After the irradiation with 405 nm, cell viability is plotted against the irradiance [mW/cm^2] for the same 20 J/cm^2 irradiation dose. Cell viability is expressed in percentage [%] and normalized using a control (untreated cells set as 100% viable); values above the 100% mean increase in cell viability and vice versa (see Fig. 1). Median values and standard deviations throughout all biological replicates (n=3) are indicated in the diagram through vertical lines, mean values through a cross.

diance intensities. The exact mechanism behind the proliferation stimulation or low-level laser therapy has yet to be fully understood. However, an assumption is that mitochondria absorb photons, which leads to increased ATP production. Since no direct cell counting accompanied the MTT-assay, and the MTT-assay is directly proportional to mitochondrial activity; it is not possible to conclude either the cell number or cellular metabolic activity was stimulated. The MTT-assay is generally used as a cell viability and proliferation assay. Regarding the assumed mechanism behind the photomodulation, it is a lack of our study, and in upcoming investigations, the experimental design should be extended relating to that. This stimulating phenomenon has a helpful effect on wound healing [9]. However, in the case of cholesteatoma as a proliferative lesion consisting of a complex matrix comprised of keratinocytes, the blue light-stimulated proliferation is not desirable, f.e, during the irradiation of the middle ear cavity with the aim of pathogenic load reduction.

4 Conclusion

The dose-response behavior, the higher the dose, the lower the viability, is well-known and prominent. However, the presented experiments targeted the effect of laser power modulation and the irradiation intensities at the same dose in a blue wavelength range. On the one hand, at higher doses (256 J/cm^2 at 445 nm), the changes in irradiation intensities are decisive for cell well-being or cell death. On the other hand, at lower doses (20 J/cm^2 at 405 nm), the varying intensities do not seem pivotal for proliferation stimulation. Furthermore, these results support photothermal tissue-laser interaction and, thus, most likely, heat shock-induced cell death mechanism (cf. photochemical effects of UV light). Further investigations in more complex models (3D cell culture, tissue ex vivo) in concordance with microbial experiments are required to bring blue light irradiation to clinical application in middle ear surgery.

Acknowledgement

The work has been carried out and supervised at the Clinic of Otorhinolaryngology, Head and Neck Surgery, University Medical Center Schleswig-Holstein. Research funding: The internship was realized in the frame of the research project "Endoscopic OCT-laser-theragnostic of microbial inflammation in the middle ear (OLE)" funded by the Federal Ministry of Education and Research (BMBF) (Grants No.: 13N15846 and 13N15847).

Authors' Statement

Conflict of interest: Authors state no conflict of interest.

5 References

[1] Chin-Lung Kuo, Wen-Huei Liao, An-Suey Shiao, *A review of current progress in acquired cholesteatoma management*. Eur Arch Otorhinolaryngol, vol. 272, no. 12, pp: 3601-3609, 2015.

[2] Alma Maniu, Oana Harabagiu, Maria Perde Schrepler, Carmen Aure Mogoant, *Molecular biology of cholesteatoma*. Rom J Morphol Embryol, vol. 55, no. 1, pp: 7-13, 2014.

[3] L. Louw, *Acquired cholesteatoma pathogenesis: stepwise explanations*. The Journal of Laryngology & Otology, vol. 124, pp: 587-593, 2010.

[4] Kimberley Lau, Marios Stavrakas, Mark Yardley, Jaydip Ray, *Lasers in Cholesteatoma Surgery: A Systematic Review*. Annals of the New York Academy of Sciences, vol. 100, pp. 94-99, 2021.

[5] Wiliam Posten, David A. Wrone, Jeffrey S. Dover, Kenneth A. Arndt, Sirunya Silapunt, Murad Alam, *Low-Level Laser Therapy for Wound Healing: Mechanism and Efficacy*. American Society for Dermatologic Surgery, Inc., 2005.

[6] Kerstin Bellmann, Steve J. Charette, Philippe J. Nadeau, Dominic J. Poirier, Anne Loranger, Jacques Landry, *The mechanism whereby heat shock induces apoptosis depends on the innate sensitivity of cells to stress*. Cell Stress and Chaperones, vol 15., pp: 101-113, 2010.

[7] Rajesh P. Rastogi, Richa, Ashok Kumar, Madhu B. Tyagi, Rajeshwar P. Sinha, *Molecular Mechanisms of Ultraviolet Radiation-Induced DNA Damage and Repair*. Journal of Nucleic Acids, vol. 2010, 2010.

[8] Christina Janko, Luis Munoz, Ricardo Chaurio, Christian Maueröder, Christian Berens, Kirsten Lauber, Martin Herrmann, *Navigation to the Graveyard-Induction of Various Pathways of Necrosis and Their Classification by Flow Cytometry*. Methods in Molecular Biology, vol. 1004, 2013.

[9] D. Hawkins, N. Houreld, H. Abrahamse, *Low Level Laser Therapy (LLLT) as an Effective Therapeutic Modality for Delayed Wound Healing*. Annals of the New York Academy of Sciences, vol. 1056, no. 1, pp. 486-493, 2005.

14

Auditory technology

Transmission behaviour of the Samba 2 Hi audio processor for different programs

Lisa Marie Schneider [1], Daniela Hollfelder [2]
[1] Auditory Technology, Universität zu Lübeck, li.schneider@student.uni-luebeck.de
[2] Department of Otorhinolaryngology, Head and Neck Surgery, University Hospital Schleswig-Holstein, Lübeck, Daniela.Hollfelder@uksh.de

Abstract

The aim of this project is to be able to compare the transmission of different preset programs for specific situations of the "SAMBA 2 Hi" audio processor (AP) of the Vibrant Sound Bridge (VSB). Using an audio adapter, the associated implant can be simulated for the purpose to test the function of the AP and the signal output can be emitted acoustically. This allows to connect the adapter to a test box and determine the transmission curve for different input signals (International Speech Test Signal (ISTS) and stepped-sine). The stimuli were each presented with an input sound pressure level (SPL) of 65 dB SPL and 90 dB SPL. Because such an adapter was not designed for the purpose of describing transmission curves, there are difficulties in interpreting results above 2 kHz, such as the decreasing SPL. Nevertheless, the results suggest differences in processing within the programs, which have to be reevaluated by further measurements.

1 Introduction

There are different types of hearing losses that are not treatable with conventional hearing aids for various reasons. Due to their origin and severity, some hearing losses can only be compensated with an implant. The Vibrant Soundbridge (VSB) is the only active transcutaneous semi-implantable middle ear implant (MED-EL, Innsbruck, Austria) for rehabilitation of such indication. The latest outer worn Samba 2 Hi audio processor (AP) was introduced in 2020 and unites an advanced signal processing strategy, multi-microphone technology and 11 preset programs for different hearing situations. Reverberation is particularly difficult for hearing impaired people. In this project, transfer curves of the AP for three pre-configured programs (*Universal*, *Privacy* and *Reverberant Room*) were recorded taking the slighty varied setup for conventional hearing aids (HA) in an isolated and controlled setting of a test box in order to compare the processing method.

2 Material and Methods

2.1 Indication

Medical indication criteria for the rehabilitation with VSB are chronic otitis media, atresia and otorrhoe for patients with at least five years of age and a constant hearing threshold within 24 months [1], [2], [3]. Audiological indication citeria (Fig. 1) are an intelligibility of monosyllables (Freiburg monosyllables) of min. 50 % and a frequency dependant hearing loss for air conduction (AC, black) up to a

maximum of 85 dB HL at 2 kHz and bone conduction (BC, grey) 65 dB HL at 2 kHz [1], [3].

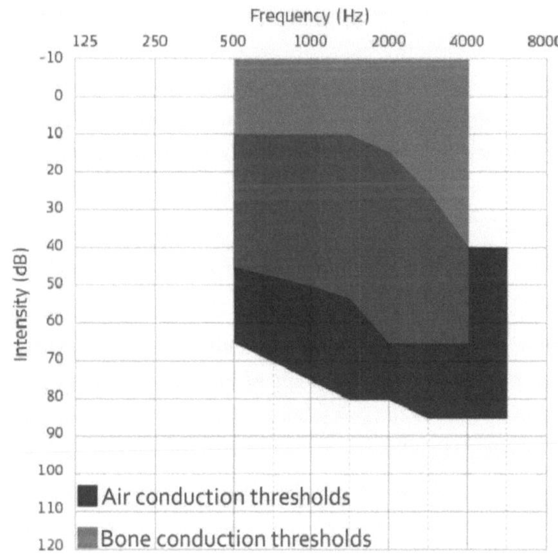

Figure 1: Audiological criteria for the Vibrant Soundbridge (VSB) for air (grey) and bone conduction (black) hearing losses as a function of intensity in relation to frequency [1].

2.2 Audio Processor fitting

The AP is configured with the "SYMFIT 8.0" software from MED-EL using the interface HI-PRO® 2. For the fitting process the hearing with AP is evaluated with the minimum SPL for the frequencies of 500 Hz up to 6 kHz with the implant-internal hearing test (vibrogram). These results undergo a "DSL v5" adjustment strategy and syllable compression and covered in the *Universal* program. The gain setting is adjusted using the "first-fit" setting with a defined broadband sensorineural hearing loss of 60 dB HL. In addition to the *Universal* automatic program, *Privacy* and *Reverberant Room* programs is also implemented with the first fit. These programs differ minimally in their feature settings. *Privacy* is the only program with a maximum sound smoothing setting, while *Reverberant Room* is the only program with a minimum directionality setting for the microphones. The directionality may have a minimal influence on the result due to the construction of the test box. The test box is intended to simulate a free-field situation without reflections. The directionality of the microphones reduces the sound incidence coming from the side or rear, which cannot appropriately tested in a test box.

2.3 Audio processor setting

Each program setting of the AP was measured with all stimuli, using an input level of 65 dB SPL as well as 90 dB SPL. This provided six transmission curves for each programs. To switch between the programs, a remote control device *Samba 2 Go* was used, which was placed around the measurement setup without touching anything.

2.4 Stimuli

For stimuli, the ISTS and a stepped-sine signal were used. The ISTS is a construct of six different languages: Arabic, English, Mandarin, Spanish, French and German. It covers all the sounds of language. It was spoken by a female speaker and mixed in such a way that it resembles normal speech but remains unintelligible. The ISTS was presented at an overall level of 65 dB SPL determined with the root mean square. It was developed for the evaluation of speech processing of hearing systems with a bandwidth of 100 Hz to 16 kHz [4].

The norm IEC 60118-0 specifies to record transmission curves of hearing systems with pure tone signals at different frequencies. The stepped sine-sweep, also known as "sweep", was used as the second measurement signal. This signal represents a series of sine frequencies which are played one after the other. In this case, it was generated in 1/12 octave steps from 100 Hz to 8 kHz [5].

2.5 Measurement Setup and Procedure

The measurements were taken in a anechoic chamber type "BK 4232" of the manufacturer "Brüel & Kjaer". As the measurement and reference microphone a "BK 4192" is used. The reference microphone is positioned 12 mm ($\pm$ 3 mm tolerance) apart from the AP microphones in accordance with IEC 60118-0 [5]. The reference microphone and the measurement microphone on the coupler (2cc coupler "BK 4964" with in-ear adapter) were calibrated to a level of 123.97 dB SPL using the pistonphone. Both microphones each use a "BK 2669-L" microphone amplifier

and both are connected to a "BK 2829" power supply. The loudspeaker inside the measuring box is connected to the power amplifier "BK 2735" from the output of the sound card. The sound card used is an RME "Fireface 802". The loudspeaker was equalised immediately before the measurement, after positioning the measurement objects in the test box. The Samba 2 Hi AP's audio adapter (Fig. 2) was closed airtight to the coupler with modelling clay. The AP was positioned with the front microphone perpendicular to the loudspeaker (Fig. 3).

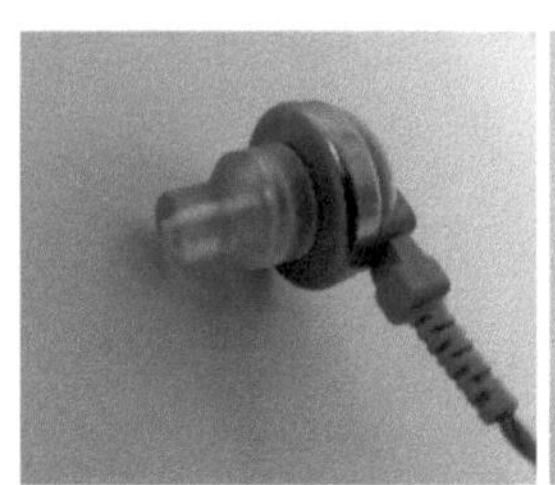

Figure 2: Used adapter for measuring the transfer function of the Samba 2 Hi audio processor (AP).

The playback of the measurement signals, as well as the recording and further processing of the data, was controlled by a Matlab script. All three testing programs were measured with both stimuli at an input level of 65 dB SPL and 90 dB SPL. This results in twelve transmission curves of the AP. The level of 65 dB SPL is selected because the normal speech level corresponds at an average of this value [4]. For the second input level, the gain of the system was set high enough to reach the limit of the Maximum Output Pressure (MPO) setting.

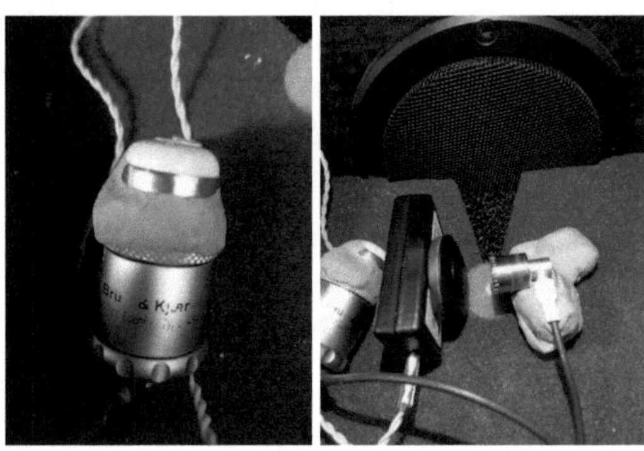

Figure 3: Positioning of the audio processor (AP) connected to the adapter (closed airtight to the coupler) and reference microphone in the test box.

3 Results and Discussion

3.1 Input-signal: ISTS

Fig. 4 compares the recorded output signal of all three programs with the ISTS input signal at 65 dB SPL as a reference. It should be noted that the output signal is filtered using a third-octave band. The ordinate shows the sound power level as a function of the corresponding frequency. Adding all the frequency levels, results in total signal level. The whole ISTS signal has a length of 60 sec, whereby only the time frame of 30 to 59 sec gets analysed. This ensures that the automatic program has detected the sound situation and can fully adapt. All results consist of two measurement repetitions of averaged transmission curves to be sure that there are no unwanted errors during the measurement to influence the results.

The ISTS reference signal is presented with a level of 65 dB SPL (Fig. 4). Because this is a constructed speech signal, the frequencies are not evenly balanced in the signal. Summed up, the level corresponds to the level of the input signal. According to the fundamental frequency of female voices [4], the highest level (80 dB SPL) can be found at around 200 Hz.

The gain oft the *Privacy* program is lower than the other two programs. Due to this the output level is 8 dB SPL lower across all frequencies (Fig. 4). The level of the program *Reverberant Room* is only slightly lower (78 dB SPL) than the automatic program *Universal*. Up to around 2 kHz, the behaviour at an input level of 65 dB SPL is similar for all curves, which indicates a frequency-linear amplification. This also corresponds to the preset broadband hearing loss (section 2.2). Above this frequency, the level decreases rapidly for all programs. This could be a result of the audio adapter, which was not developed for technical measurements and is therefore limited in its components and design. Considering the software setting, the amplification of the AP should remain roughly constant up to approx. 6 kHz.

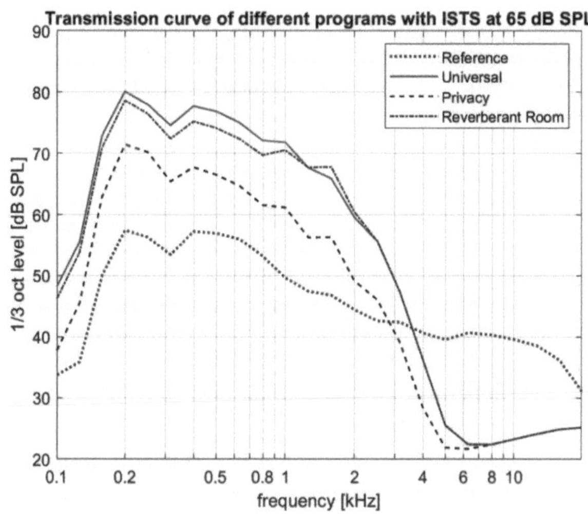

Figure 4: Comparison of transfer curves for different programs (*Universal*, *Privacy* and *Reverberant Room*) with an ISTS at an input level of 65 dB SPL.

At an input level of 90 dB SPL, the curve for *Universal* and *Reverberant Room* are more or less at the same level ($\pm$ 1 dB SPL) and broadband, which indicates that the AP has reached the MPO limit of the application. The limit for *Privacy* is only reached in the low-frequency range up to around 550 Hz before decreasing. These reference response

curve is basically similar to Fig. 4 and is only shifted by the level. The difference between the programs in the processing method is less noticeable in this setting. This suggests that the gain setting is not the fundamental factor in the setting, but that the focus is much more on the directionality of the feature setting.

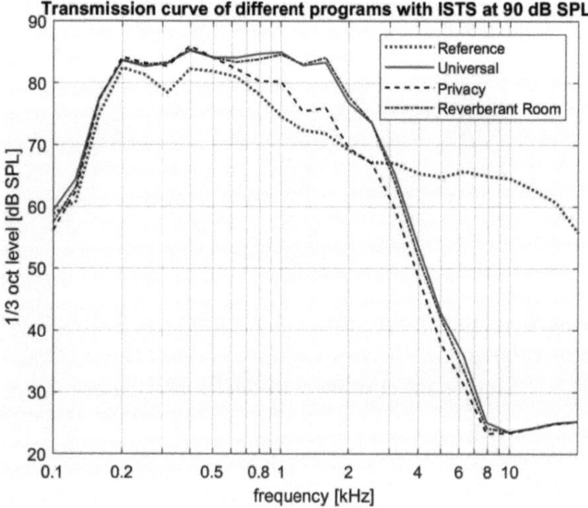

Figure 5: Comparison of transfer curves for different programs (*Universal*, *Privacy* and *Reverberant Room*) with an ISTS at an input level of 90 dB SPL.

The measurements with the ISTS speech signal result in frequency-independent processing of the single programs. Only a broadband reduction in amplification can be seen, where the *Privacy* program is amplified least. However, the directionality setting of the microphones is not taken into account in the test box measurement as in the *Reverberant Room* program.

3.2   Input-signal: Sweep

The transfer function recorded with the sweep signal does not require any pre-processing of the results like the ISTS. Because of that, the level of the reference curve corresponds to the level of the input. They are also averaged from two measurement repetitions. The frequency-dependent amplification in Fig. 6 is not linear as expected for the hearing loss. In this case, a higher gain can be detected at about 100 to 150 Hz and between 1 and 2 kHz. This signal also drops in gain above 2 kHz.

This signal allows larger differences between the *Universal* and *Reverberation Room* programs to be analysed. The gain is reduced, except the frequency around 150 Hz, in the program *Reverberant Room* by 3 to 11 dB SPL up to around 1.5 kHz. In this view, the output level of the *Privacy* program is also reduced for 18 dB SPL over all frequencies. This confirms that the programs are handled differently in the AP at a normal speech volume of 65 dB SPL. Compared to ISTS, frequencies below 200 Hz are slightly more amplified in the *Universal* program, which may indicate speech recognition and noise reduction.

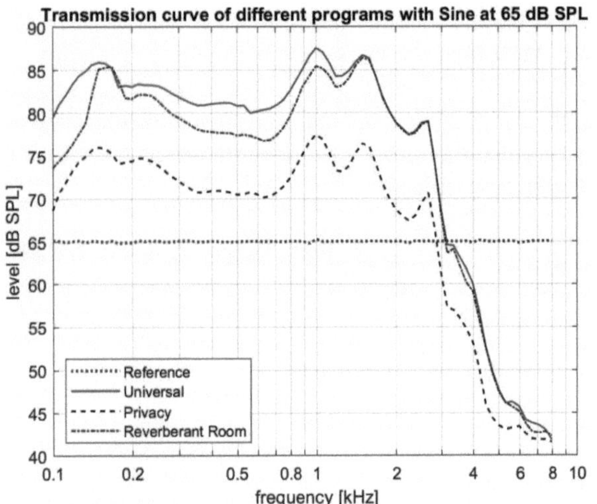

Figure 6: Comparison of transfer curves for different programs (*Universal*, *Privacy* and *Reverberant Room*) with an Stepped-Sine at an input level of 65 dB SPL.

At 90 dB SPL, the curve of the *Universal* program in Fig. 7 is almost completely within the MPO limit. The *Reverberant Room* curve is 1 to 2 dB SPL below the *Universal* curve between 150 Hz and 1 kHz. Apart from that, it is also within the MPO limit. Between 4 and 8 kHz there are distortion products in both programs, probably generated by the adapter or a product of internal resonance behaviour. Again, the *Privacy* program has a lower level of approx. 6 dB SPL compared to the other programs and therefore a lower gain setting and minimal distortion products at the same frequencies.

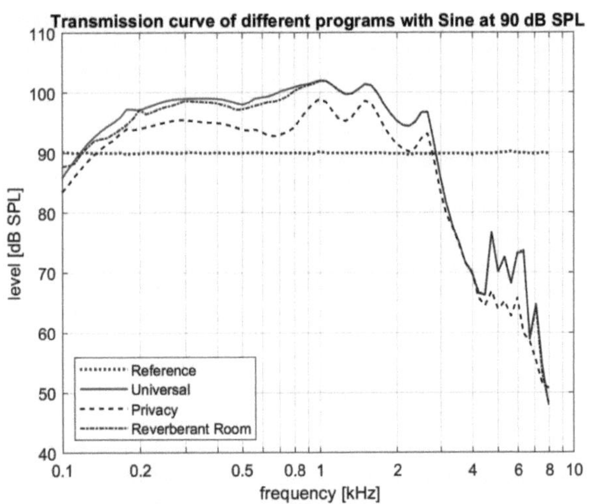

Figure 7: Comparison of transfer curves for different programs (*Universal*, *Privacy* and *Reverberant Room*) with an Stepped-Sine at an input level of 90 dB SPL.

4   Conclusion

In speech processing (ISTS) as well as in the processing of sine tones, the programs show a different transmission curve larger at low input levels (65 dB SPL) than in higher input levels (90 dB SPL). The conclusions offer potential for further studies and various measurement set-ups for example, by directly measuring the vibrations of the implant. The audio adapter limits these measurement results due to its components and design. This project can therefore only consider a limited frequency range. All in all, the test box measurements provide a approach to the different frequency-specific processing methods of the programs *Universal*, *Privacy* and *Reverberant Room*.

Acknowledgement

This work has being carried out and supervised by Dr. med. Daniela Hollfelder at the University Hospital Schleswig-Holstein, Department of Otorhinolaryngology, Head and Neck Surgery. In addition, special thanks are due to Dr. Hendrik Husstedt, Dr. Florian Denk and Markus Kemper of the "Deutsches Hörgeräte Institut GmbH" for providing the measuring instruments and expert advice on carrying out and converting the measurements.

Authors' Statement

Authors declare no conflict of interest.

5   References

[1] MED-EL, BU VIBRANT, "Systematic review and meta-analysis of audiological and patient-reported outcomes with the vibrant soundbridge active middle ear implant.," May 2022. Available: `https:// go.medel.pro/OutcomesVSB`, last accessed on 01.01.2024.

[2] Medizinischer Dienst der Spitzenverbände, *Begutachtungsanleitung Schwerhörigkeit. Begutachtungsanleitung zur apparativen Versorgung bei Funktionsstörungen des Ohres*. Medizinischer Dienst der Spitzenverbände Bund der Krankenkassen e.V. (MDS), 2020.

[3] L. K. Schemke, *Evaluation der audiologischen Ergebnisse bei unterschiedlicher Ankopplung des aktiven transkutanen Mittelohrimplantats Vibrant Soundbridge bei Patienten mit chronischer Otitis media*. PhD thesis, Universität zu Lübeck, 2022.

[4] M. V. Inga Holube, Stefan Fredelake and B. Kollmeier, "Development and analysis of an international speech test signal (ists)," *International Journal of Audiology*, vol. 49, no. 12, 2010.

[5] International Electrotechnical Commission, "IEC 60118-0," *Electroacoustics–Hearing aids–Part 0: Measurement of the performance characteristics of hearing aids. Geneva: IEC*, 2015.

Validation of the spatial reproduction of realistic acoustic scenes with regard to the listener's head orientation.

Peter Schumacher [1,3], Tim Jürgens [2], Juliane Raether [3], Stefan Klockgether [3]

[1] Auditory Technology, Universität zu Lübeck, peter.schumacher@student.uni-luebeck.de
[2] Institute of Acoustics, Technische Hochschule Lübeck, tim.juergens@th-luebeck.de
[3] Sonova AG, Stäfa, {juliane.raether, stefan.klockgether, peter.schumacher}@sonova.com

Abstract

The performance of hearing devices or speech enhancement algorithms is typically evaluated using methods that require listeners to remain in a fixed position and minimize head movement. While this allows for controlled testing conditions, it does not reflect realistic human behavior in real-world acoustic environments.
The Sonova Real Life Lab (RLL) addresses this issue by utilizing a large speaker array and motion-capturing setup to reproduce acoustic scenes. Within a $25\,\mathrm{m}^2$ stage, objects can be tracked and listeners are free to move around and interact with the sound sources by turning their heads or adjusting their distance. To ensure accurate spatial reproduction and understand the actual sounds reaching the listener's ears, acoustic measurements were conducted using multiple microphones. This paper specifically focuses on the impact of head orientation, which was simulated by recordings with an artificial head positioned in the center of the stage.

1   Introduction

Hearing aid algorithm developments and hearing impaired benefit measurements are often conducted in laboratory environments because they are more comparable, reproducible and controllable than measurements in real life. One limitation is that human behavior is much more variable in changing natural environments. In recent years, many different approaches with virtual acoustics in combination with visual representations were made to represent a realistic environment. Examples for that are the Toolbox for acoustic scene creation and rendering (TASCAR) at the implant center in Freiburg or the virtual reality lab in Oldenburg. Also the loudspeaker-based room auralization-Toolbox (LoRa) for measurements of virtual sound environments at Oticon has been implemented. All of them can create realistic virtual acoustic and visual scenes at the centre of a loudspeaker arrangement without any additional headphones. That means the participant can wear hearing aids or other devices. One remaining disadvantage of these setups is that the participant is required to sit in the setup's center at the acoustic sweet spot. When communication is becoming difficult by an increase of background noise or additional visual or acoustic distracter, listeners instinctively adapt their behavior to improve the situation [1], [2], [3]. One examples is altering the distance to the source, which was observable with people move closer together as the acoustic environment gets more difficult [2]. People also raise the voice and altering the pitch they are talking with in noisy environments [1], [3]. The latter behavior is

called the Lombard Effect and was first described in 1911 by the French otolaryngologist (ENT) Étienne Lombard in the article "Le signe de l'élévation de la voix" [1]. As listeners behave differently in different sound environments, this also influences the evaluation of hearing aid features. To observe this natural behavior behind the Lombard Effect, Sonova has setup a lab that is called Real Life Laboratory (RLL). It combines the benefits of realistic acoustic and visual reproduction with the ability for the subjects to move freely around in the reproduced environment. The realistic reproduction of acoustic environments is made possible by the Coding and Multidirectional Parameterization of Ambisonic Sound Scenes (COMPASS) audio reproduction method used in the RLL. To evaluate the audio reproduction method in the RLL, sound level recordings at ear level were made to quantify the acoustic behavior at different points in the room representing different head orientations. In this paper, a first comparison of the transfer functions of two talkers recording from different directions is presented. It serves as a starting point for ongoing technical measurements.

2   Material and Methods

2.1   Material

2.1.1   Real Life Lab

The RLL of Sonova AG, shown in Fig. 1, is a cuboid arrangement combining video screens on all four sides,

105 *KEF E301* passive coaxial loudspeakers, four *Sonible* custom-made passive subwoofers and up to 31 optional *KEF E301* loudspeakers powered by six *Sonible d24 mio* amplifiers. The dimensions of the reproduction room are 8.37 m × 6.60 m × 3.49 m and its reverberation time has been reduced by acoustic treatments to about 0.2 s to 0.3 s. The usable space for experiments is 5 m × 5 m × 2.5 m, with a stage size of 25 m²25 m² to move within the reproduced scene. To track the movement of listeners inside the laboratory an *Optitrack* motion capturing system with 12 *Prime 13W* cameras is installed.

Figure 1: Real Life Lab of Sonova AG with KEMAR for the used measurement setup.

2.1.2 Microphones

Seven microphones are used for the recordings. For measuring the sound at the eardrum an artificial head and torso (KEMAR) *GRAS 45BB* with anthropometric ears is used (Fig. 1, 3). It is equipped with an IEC 711 Ear Simulator (*GRAS RA0045*) with build-in ½" microphone on each side at the position of the eardrum. The KEMAR is a reproduction of a human upper body and head for acoustic measurements. Four of the microphones are the front and rear microphones of two receiver-in-canal hearing aid shells (Phonak *Audéo*) positioned on the left and right ear of the KEMAR as shown in Fig. 3. These are subsequently referred to as satellites (e.g. $Sat_{front,right}$). Their signal is directly captured with no further signal processing of the hearing aid. As the reference microphone, an omnidirectional microphone (*GRAS 26AK*) is positioned directly above the head of the KEMAR.

2.1.3 Stimuli

The used sound files are provided by the Sound Scene Repository (HOA-SSR) [4]. The HOA-SSR dataset consists of audio-visual scenes of 360° video with 4th-order ambisonics in B-format as a usable format for sound processing toolboxes to create everyday life scenes. As a first approach a simple dining scene is recorded and analyzed. It represents a two female talker conversation in a living room without any other background noise and the $L_{A,Eq}$ of 60 dB as shown in Table 1. The left talker is located approx. 30° to the left of the center of the video screen and the right talker is located 15° to the right of the center of the video screen, as shown in Fig. 2.

2.2 Method

2.2.1 Higher Order Ambisonics

The concept of constructive and destructive superimposition of loudspeaker signals which is achieved by Higher Order Ambisonics (HOA) is the fundamental of the reproduction of sound in the RLL. With this technique, it is possible to create a desired sound image at a defined position. The sound field is decomposed using spherical harmonics, which are truncated at a specific order M. The size of the order M and thus also the limit of how narrow the lobe can become, is determined by the number of loudspeakers N in the measurement setup. With the given 4th-order and the equation for a three-dimensional setup ($N \geq (M+1)^2$ [5]) at least 25 loudspeakers has to used to represent the scenes.

2.2.2 Preparation of the sound files

In order to make use of the higher order ambisonics, they need to be decomposed by an appropriate tool. To account for the special arrangement of the RLL and the need for a larger artifact-free sweet spot, [6] described various methods for decoding HOA impulse responses. COMPASS also meets the requirement for a wider artifact-free sweet spot and is therefore applied to the sound files. This is necessary because it has been shown that once the hearing aids move away from the sweet spot in the center of the simulation, acoustic inconsistencies arise. These inconsistencies are that the interaural cues and the performance of the hearing aid beamformers deteriorate compared to the corresponding real-life scenario [7], [8].

2.2.3 Calibration

Sound files and loudspeaker For the representation of plausible loudness and perception of the scenes, the level and the frequency spectrum of the original sound scene from the SSR database has to be adjusted to the loudspeaker setup in the RLL. To get the right perception of the scene in later subjective measurements the impulse response (IR) of the center position at a specific height has to be applied to the sound files. This is done by measuring the IR at the height of 118 cm (sitting) and at 160 cm (standing) with an omnidirectional microphone (*GRAS 26AK*). This height refers to the ear position in a sitting or standing position and these IRs are applied to the sound files. The level of the calibrated sound files will be adjusted to the desired level of the scene according to the level in Table 1. The $L_{A,Eq}$ is measured for 60 s at the position of the KEMAR to obtain the correct level for the scene.

Microphone The reference microphone is calibrated with a *Norsonic AS Sound Calibrator type 1251* with 1 kHz at 94 dB SPL. which corresponds to −40.2 dB FS. All sound level are coded with 24 Bit. In order to apply the calibration of the reference microphone to the other microphones, a warble tone is played into the room, varying up to 10 % around 1 kHz, and picked up by all the microphones at their positions for subsequent recordings. The level offset of the microphones to the reference microphone is corrected by applying the difference between the reference microphone and the other microphones to the measured levels of the microphones.

2.2.4 Measurement setup

As shown in Fig. 2, measurements are taken at the center (E5) of the stage. Four KEMAR orientations (**E**ast, **N**orth, **W**est, **S**outh) are measured at the average height of sitting (SI) and standing (ST) listeners. This corresponds to the height of KEMAR's ears at 118 cm and 160 cm. According to that definition, the name of the condition shown in Fig. 2 in the center, facing east, while the KEMAR is at a standing height is E5E_ST. The video screens shown are not used during the measurement, since visual cues are irrelevant for measurements with the KEMAR. However, they would be important for a realistic audio-visual reproduction.

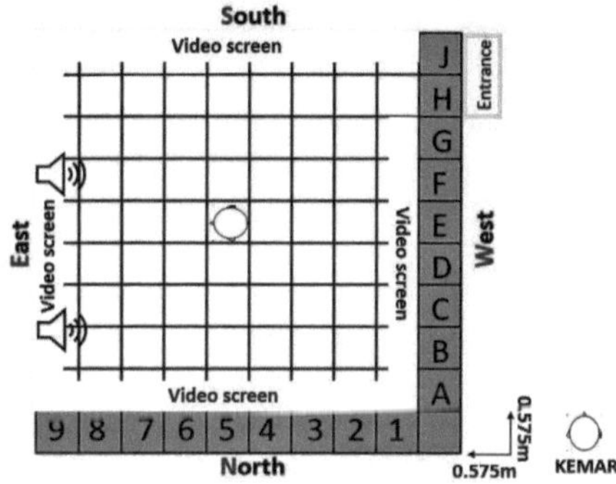

Figure 2: Schematic layout of the RLL for the measurement with the artificial head at the centre position E5 facing towards the east and the speakers represented as loudspeakers on the east video screen (view from the top).

7-channel recordings are made to investigate the variations in transfer functions between different microphone positions on the KEMAR. Additionally, the interaural level difference (ILD) at the KEMAR's position is considered when it is rotated to different orientations within the presented scene. Four channels are the satellite microphones, which are connected directly by cables. Two channels are the ears of the KEMAR. One channel is the reference microphone above the head of the KEMAR as shown in Fig. 3.

To get a realistic variety of the scenes, the four listed condition in Table 1 with their corresponding directions and height are measured.

Table 1: List of conditions.

Scene	level	conditions	
Dinner 2 people	60 dB (A)	E5E_ST	E5E_SI
		E5S_ST	E5S_SI
		E5N_ST	E5N_SI
		E5W_ST	E5W_SI

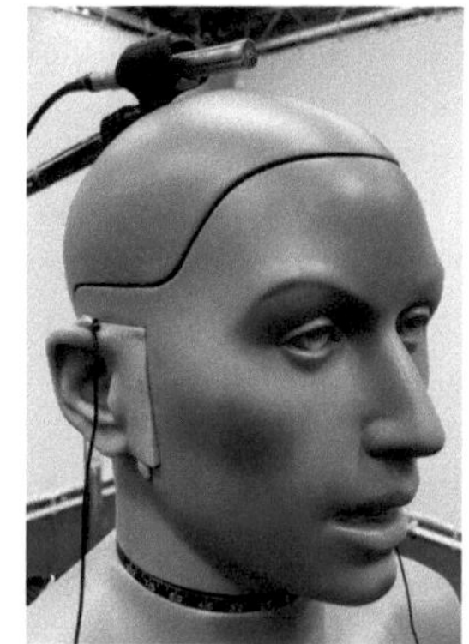

Figure 3: Measurement setup of the KEMAR with the hearing aid shells and reference microphone.

2.2.5 Measurements

Separate pink noise measurements are taken from all four sides (0°, 90°, 180° and 270°) of a single loudspeaker. This is done to confirm that the calibration has been applied correctly and that the ILDs from the literature [9] are comparable with the measurements in the RLL. Afterwards, the dinner scene *LivingDinner2-ppl* is recorded in the four directions listed in Table 1.

3 Results and Discussion

No ILDs are visible when pink noise is presented from 0° or 180°. A small bias of 1 dB to 2 dB is measured for all the channels of the left microphones. In the 90° and 270° directions, there is an increasing ILD starting at 300 Hz. The ILDs increase up to 15 dB at 10 kHz, which fits with the literature for this frequency range [9]. The subsequent measurements of the scene *LivingDining2-ppl* show an ILD of 1 dB to 3 dB to the left for the E5E_ST condition. This ILD is plausible due to the position of the two talkers, since the left talker is located further to the side than the right one. Both data confirm a correct spacial reproduction of the acoustic scene in the center position.

Fig. 4 shows the third-octave spectrums of the KEMAR facing to the north in the upper graph and the south in the lower graph. The x-axis shows the measured frequency in Hz with the corresponding level in dB SPL on the y-axis. All microphone channels show the characteristic frequency curve of a female voice with its fundamental frequency between 200 Hz and 300 Hz. The real ear unaided gain (REUG) of the KEMAR also shows an additional gain in the frequency range of 2 kHz to 4 kHz. The transfer function of the satellites is different from the one of the reference microphone, resulting in an additional level drop above 2 kHz for the reference microphone. In contrast to the measurements with pink noise, no ILD can be recognized with the selected scene for the condition E5N_ST. In this condition the right side of the KEMAR faces the east screen where the talkers are located. Turning the KEMAR towards the video screen to the south (E5S_ST) with the left side facing the talkers, a similarly large ILD is visible as in the pink noise measurements. This difference could be a hint, that the spatial reproduction of the scene is not completely correct. This deviation could also be caused by the usage of the static microphone calibration, which was applied to all the measurements, even though the recordings were not made at the same day or with a perfectly reproduced measurement setup. The asymmetric position of the two speakers could also contribute to the measured difference.

One learning is to always calibrate the different microphones at the beginning of each measurement. To gain more insights on the spatial reproduction the scene could be analyzed in smaller time frames to get a separate ILD for each talker.

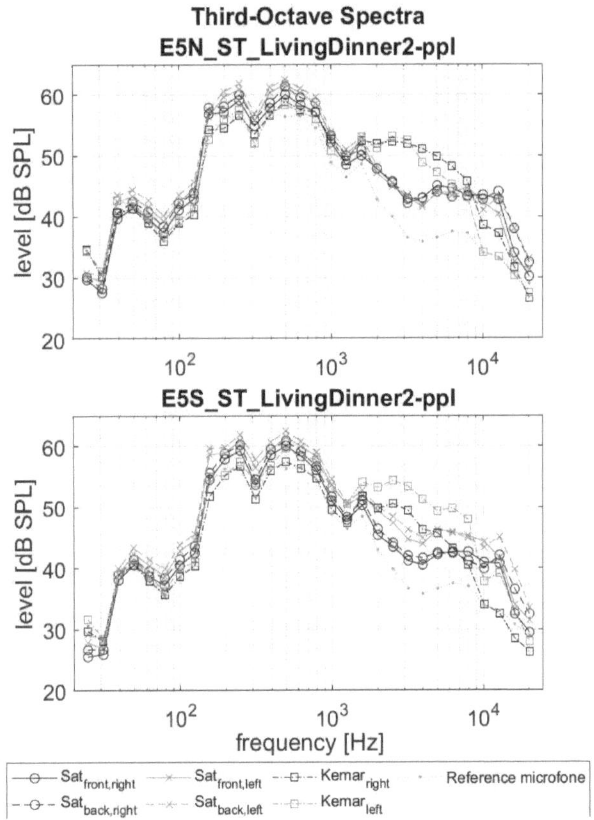

Figure 4: Comparison of the third-octave spectrums of the microphone channels with the KEMAR facing north (upper graph) and south (lower graph) at the height of a standing person with levels in dB SPL.

4   Conclusion

The results show that the setup is capable of generating and recording virtual acoustic scenes from different directions. The frequency differences caused by the orientation of the KEMAR and the associated ILD show that it is in principle possible to visualize the effect of the direction of view with this setup. The next steps are to correct the calibration and to evaluate the sound reproduction in other positions on the stage. Once it is clear how to interpret the ILDs of this relatively simple scene, more complex scenes with background noise and several competing talkers at different positions on the stage will be recorded and analyzed. This will be the starting point for measurements with subjects moving freely on the stage of the RLL.

Acknowledgement

The work has been carried out at Sonova AG and supervised by the Institute of Acoustics, Technische Hochschule Lübeck.

Authors' Statement

Conflict of interest: Authors state no conflict of interest.

5   References

[1] S. A. Zollinger and H. Brumm, "The lombard effect," *Current Biology*, vol. 21, no. 16, pp. R614–R615, 2011.

[2] S. Klockgether, C. Kleinstein, M. Schneeberger, and R. P. Derleth, "Communication behavior observation," *The Journal of the Acoustical Society of America*, vol. 151, no. 4, pp. A134–A134, 2022.

[3] K. Miles, A. Weisser, R. W. Kallen, M. Varlet, M. J. Richardson, and J. M. Buchholz, "Behavioral dynamics of conversation,(mis) communication and coordination in noisy environments," *Scientific Reports*, vol. 13, no. 1, p. 20271, 2023.

[4] FORCE Technology, "Higher-order ambisonic sound scene repository (hoa-ssr) dataset." https://forcetechnology.com/en/articles/download-hoa-ssr-dataset. last accessed on 2024-01-08.

[5] D. B. Ward and T. D. Abhayapala, "Reproduction of a plane-wave sound field using an array of loudspeakers," *IEEE Transactions on speech and audio processing*, vol. 9, no. 6, pp. 697–707, 2001.

[6] L. S. Simon and S. Klockgether, "Near-field source reproduction in lage loudspeaker arrays," in *10th Convention of the European Acoustics Association*, 2023.

[7] G. Grimm, S. Ewert, and V. Hohmann, "Evaluation of spatial audio reproduction schemes for application in hearing aid research," *Acta Acustica United with Acustica*, vol. 101, no. 4, pp. 842–854, 2015.

[8] L. S. Simon, N. Dillier, and H. Wüthrich, "Comparison of 3d audio reproduction methods using hearing devices," *Journal of the Audio Engineering Society*, vol. 68, no. 12, pp. 899–909, 2021.

[9] W. A. Yost and R. R. Fay, *Human psychophysics*, vol. 3, pp. 162–163. Springer Science & Business Media, 2012.

Reproducibility of responses in the Nijmegen Cochlear Implant Questionnaire (NCIQ): Independent vs. supervised performance and its relation to patient satisfaction and speech intelligibility

Jennifer Schmidt [1], Daniela Hollfelder [2]

[1] Auditory Technology, Universität zu Lübeck, je.schmidt@student.uni-luebeck.de
[2] Universitätsklinikum Schleswig-Holstein, Abteilung HNO, Daniela.Hollfelder@uksh.de

Abstract

The Nijmegen Cochlear Implant Questionnaire (NCIQ) is employed to assess patient satisfaction during the cochlear implant (CI) fitting process. This paper explores the reproducibility of patient responses in the NCIQ and investigates the correlation between patient satisfaction and speech intelligibility. 27 CI patients took part who filled out the NCIQ first alone and then with the help of the study director. Afterwards, the Freiburg monosyllable test was conducted to be able to establish the connection. The results show that on average it makes no difference whether patients fill out the questionnaire independently or with help, but there were also patients whose answers differed greatly. The results also showed that there are significant connections between patient satisfaction and the determined speech intelligibility in percent. Categories "speech & music perception", "self-esteem" and "social contacts" show significant values. However, age and length of CI experience (after first activation) have no considerable influence on speech understanding.

1 Introduction

When conventional hearing aids no longer suffice for hearing loss, and individuals experience a decline in quality of life, cochlear implants (CIs) serve as electronic inner ear prostheses, enabling comprehensive hearing and speech rehabilitation. However, first, the indications for CI treatment must be clarified so that later CI care can be successful. In addition to clarifying the indication and anatomical conditions, subsequent quality assurance also plays a key role. Both immediately after the CI fitting, but also in the months and years afterwards, a structured and lifelong monitoring of the CI patient takes place to maintain the optimal quality within care. In addition to many audiological tests, a main measurement approach is the the Freiburg monosyllable test, which is only used in German-speaking countries. This test determines speech intelligibility in percent. However, optimal care is not only based on audiological measurements. Questionnaires are also used to assess and evaluate the patients within their living circumstances. In this study the Nijmegen Cochlear Implant Questionnaire (NCIQ) was used [6]. This questionnaire can be used to ensure patient satisfaction during care. The CI-supplying institutions and CI centers make their quality reports publicly available. They are also part of the national CI register, which requires structured internal documentation. The CI register collects data related to implants in accordance with applicable privacy laws. This is important because this is the only way "possible systematic implant errors can be

identified at an early stage or the influence of process or treatment changes on the quality of the care result can be recognized" [1]. This study focuses on the interaction between two assessment criteria: the Freiburg monosyllable test and NCIQ. The following section will provide a detailed examination of this interaction. The study first assessed the reproducibility of answering the NCIQ twice, once individually filled out by the subject and once in collaboration with the study director. The second part determined the extent to which patient satisfaction in quality of life was correlated to the speech intelligibility.

2 Material and Methods

A total of 27 patients, 11 men and 16 women, participated in the study. The patients differed in terms of the date of initial activation, their age and the type of care they received. The subjects' initial activation was between 2 and 101 months ago, and their age ranged from 41 to 83 years. The two substudies included patients with hearing loss in one ear only (unilateral deafness, SSD), with CI in both ears, or with a cochlear implant in one ear and a hearing aid in the other ear (bimodal).

2.1 Cochlear Implants

Unlike hearing aids, CI function as hearing prostheses. While hearing aids amplify sound to support the ear, CI bypass the ear's acoustic function and directly stimulate the auditory nerve through electrical signals. In a healthy ear, electrical impulses called action potentials are triggered via the hair cells in the inner ear that are connected to the nerve fibers. For CI users, the cochlear implant takes on this function and attaches directly to the nerve fibers. The subsequent frequency resolution within the cochlea is also replicated by the cochlear implant by stimulating the auditory nerve along the cochlea with up to twenty-two electrodes (depending on the manufacturer). This makes it possible to recreate the tonotopy within the cochlea, with the representation of high frequencies at the base of the cochlea and low frequencies at the apex [2], [4].

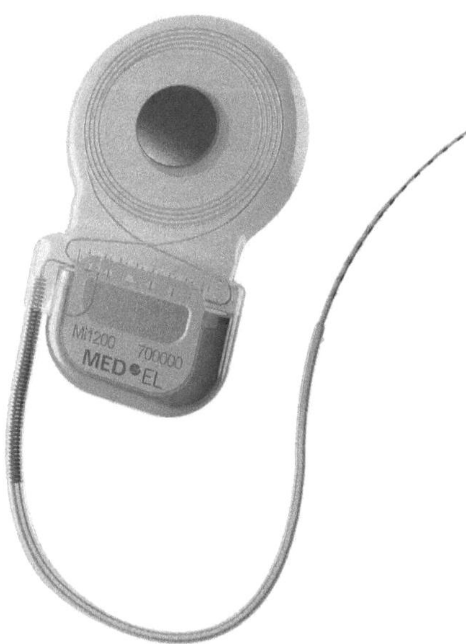

Figure 1: Cochlea Implant.

Fig. 1 shows a cochlear implant. The components include a signal/speech processor, a transmitter coil, an implant body (receiver/stimulator), stimulation electrodes and a reference electrode. The speech or signal processor is usually worn behind the ear or directly on the head. It records the incoming signal using microphones and analyzes it. The recorded signals are encoded using advanced microchips using various stages of signal pre-processing. The transmitter coil, which is held on the head by magnets, is connected to the speech processor. The receiver sits under the skin just below the transmitter coil. It is surgically placed under the skin. This receiver contains the electronics for controlling the electrode carrier. This means that both the electrical impulses and the required energy are passed on from the transmitting coil to the receiving coil. The reference electrode is also placed under the skin and is used for electrical stimulation. The electrode carrier inserted within the cochlea and stimulates the auditory nerve, as described above, and thus enables the signal to be passed on to the brain [2], [5].

2.2 Reproducibility of the answers in the NCIQ

The NCIQ is a questionnaire specifically developed for cochlear implant users to record their health-related quality of life throughout their care process. This questionnaire is used both before and after the implantation of CI. The NCIQ consists of a total of sixty questions, of which fifty-five can be answered as follows: "The answer categories of the questionnaire for fifty-five questions are: "never" (1), "rarely" (2), "sometimes" (3), "often" (4) or "always" (5). For the remaining five questions, "never" (1), "severely" (2), "somewhat" (3), "good" (4) and "very good" (5) are available as answer options. There is also the option of not answering the question and ticking "no answer" [3]. The remaining five questions allow for a personal evaluation and feedback. For this study, twenty-seven patients were tasked to complete the sixty questions of the NCIQ during the adaptation period, similar to their routine appointments, without explicit awareness of their participation in a study. The objective was for patients to complete the questionnaire in a manner consistent with our usual approach, minimizing the likelihood of exerting significantly more effort or providing more detailed responses than in previous appointments. Subsequently, the sixty questions were revisited under the supervision of the study director. In this session, all questions were read aloud, allowing patients to seek clarification or explanations in case of any uncertainties. To facilitate better evaluation, the questions were categorized into six groups, each addressing related aspects of the patients' daily lives. These categories include elementary sound perception (1), speech & music perception (2), advanced sound perception (3), self-esteem (4), activity behavior (5), and social contacts (6). Additionally, an overall score (7) encompassing all questions was calculated. All six categories belong to the main topic "health-related quality of life". The domains "basic sound perception," "advanced sound perception," and "speech production" pertain to the physical functional aspect, while "self-esteem" corresponds to the psychological functional aspect. The areas of "activity" and "social interaction" encompass the social functional dimension [5]. This first part of our study aimed to establish whether NCIQ questions are accurately repeated by patients. It is important to ensure that all questions are answered and not skipped due to a lack of understanding. By collaborating with the study management, we were able to resolve any questions of understanding. The differences between the answers given in the first run (subject alone) and the second run (supervised) were then compared.

2.3 Relationship between patient satisfaction and the assessed speech comprehension

The second part of the study aimed to find out whether patients who achieved a higher percentage of speech intelligibility in the Freiburg monosyllable test also automatically reported higher values in the NCIQ. First, each patient's speech intelligibility percentage was determined using the

Freiburg monosyllabic test in quiet at 65 db. The basis was once again the division of the NCIQ questions into 6 categories, as well as the overall score. However, only the questionnaires that were filled out together with the course director were used for this evaluation to ensure that all questions were understood correctly. For each of the individual categories, it was checked whether there was a linear relationship between the individual categories and the percentage of speech intelligibility. In addition, an analysis was done whether the influence of age and the length of the implantation period had an impact on speech understanding.

3 Results and Discussion

3.1 Results of sub-study 1

Fig. 2 illustrates the discrepancies in patients' responses between the initial round, where they independently completed the questionnaire, and the subsequent round conducted with the assistance of the study leader.

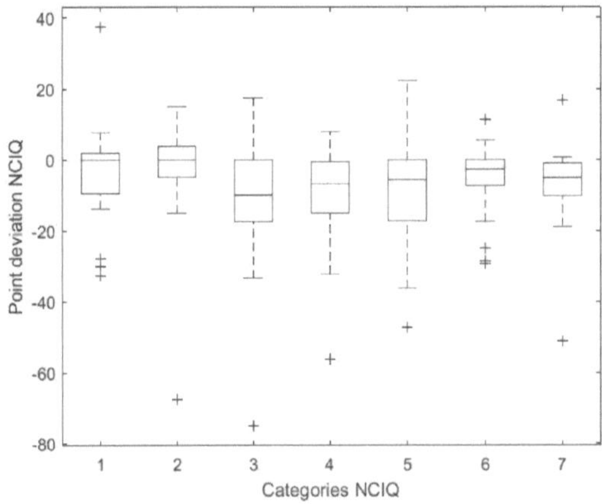

Figure 2: Point deviation in NCIQ responses between the independent and supervised runs.

Boxplots 1-6 depict the categories elaborated in Section 2. For each category, both the median and the upper and lower quartiles were calculated. Considering only the median, it is evident that there were no differences between the first and second run in the initial two categories, as the median for these is 0. However, categories 3, 4 & 5, in particular, stand out, where a median deviation of up to 10 question points was observed compared to when the test subjects completed them independently. Nevertheless, there are between 7 and 12 questions in each of the categories, so the average overall deviation is very small. What is striking, however, is the high number of outliers in categories 1 and 6, which suggests that some patients provided significantly different information within the two runs. In summary, the results indicate that, on average, the NCIQ can be reliably completed independently across all test subjects. However, there are instances where some subjects may misinterpret the meaning of specific questions, leading to varied responses.

3.2 Results of sub-study 2

Sub-study 2 investigates the linear relationships between speech intelligibility in percent and the respective category of the NCIQ. For all categories, the regression line increases, i.e. with a higher speech intelligibility value, a higher score within the category was given. The individual categories can be checked for significance using the associated p- and R-squared values. In linear regression it is assumed that there is no linear relationship between the pairs of values to be examined as null hypothesis. A significance level of 0.05 (5%) was set for this study. Categories 3,4,6 & the overall result in category 7 show a significant result. This means that in these categories there is a significant relationship between the patients' subjective satisfaction and the percentage of recorded speech intelligibility. Fig. 3 shows an example of the regression line of the linear relationship between speech intelligibility and self-esteem.

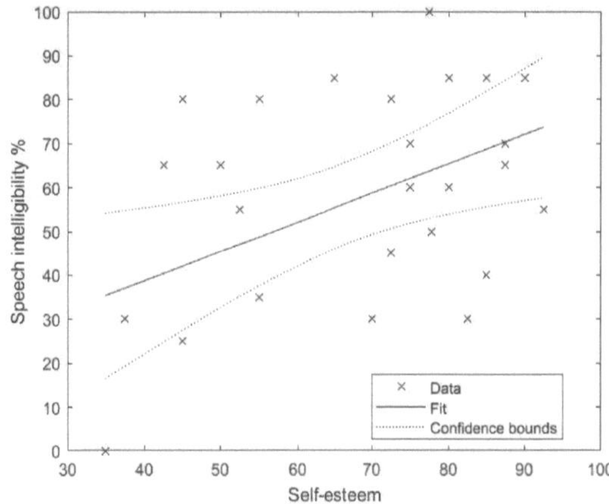

Figure 3: Linear relationships between speech intelligibility in percent for the category self-esteem.

In categories 1 & 7, the value is only slightly above the significance threshold, so we hypothesize that a significant result can also be expected here with a larger number of subjects. Only category 2, which deals with speech and music perception, clearly shows no significant connection to speech understanding, as the value is well above the assigned significance level of 5%. In addition to these relationships, we also looked at whether age had an influence on the percentage of speech intelligibility and whether how long ago the first activation occurred, had an impact on the relationship to speech intelligibility. When it comes to the connection between age and the associated percentage speech intelligibility, the regression line has a negative trend. This means that with increasing age, the measured speech intelligibility decreased. However, the p-value of 0.3 at a significance level of 0.05 shows that this relationship is not significant. The regression line of the connection between CI experience and speech comprehension only increases slightly and therefore suggests that there is no significant connection between the duration of CI experience and the percentage of speech comprehension in the Freiburg monosyllable test. The corresponding p-value with a value of 0.83 also shows that there is no significant relationship between experience and speech intelligibility.

4   Conclusion

The aim of the study was to examine the NCIQ based on two research questions. Firstly, it investigated whether patients could correctly understand and meaningfully answer the NCIQ questions, some of which are formulated negatively and are lengthy and intricate. Secondly, the study aimed to determine if patients with better speech understanding automatically reported higher satisfaction with their CI fitting. The measurement results from Chapter 3 indicate that responses in the NCIQ are generally reproducible, with little or no significant difference on average between self-conducted and supervised runs. However, it was observed that certain patients faced more difficulty than others, highlighting the importance of recognizing the questionnaire as a challenge for specific individuals. To ensure accurate results, supervision may be crucial for such cases. This is the only way to ensure adequate adaptation for these patients. Moreover, the high number of questions reduces motivation and leads to unanswered questions. The study's second part yields significant insights into the NCIQ and its relation to patient satisfaction. Measurement results reveal significant correlations in almost all categories between the respective category and speech intelligibility. This indicates that patients who performed well in the Freiburg monosyllable test also reported higher satisfaction with their CI fitting and daily situations. However, it's crucial to consider that these results stem from a small and non-homogeneous group of test subjects. The type of care, particularly concerning satisfaction, plays a significant role and might have influenced the results of the study's second part. Given that the NCIQ doesn't explicitly inquire about the CI side in everyday questions but assesses satisfaction in everyday life concerning the overall hearing situation, it's plausible that patients with single-sided deafness (SSD) and those with bimodal care may exhibit higher satisfaction due to having a second ear that contributes to acoustic perception in daily life. Thus, the significant result may not only be determined by the level of speech intelligibility but also influenced by the type of care. In summary, it's essential to recognize that CI signify a crucial advancement in the field of hearing prostheses. They provide individuals with missing ear functions, an opportunity to restore their quality of life and and help them to actively participate in daily activities.

Acknowledgement

The study was conducted and supervised at the Universitätsklinikum Schleswig Holstein, Abteilung HNO.

Authors' Statement

Conflict of interest: Authors state no conflict of interest.

5   References

[1] Presidium of the DGHNO-KHC, *White Paper Cochlear Implant (CI) Care*. Revised, 2nd Edition, 2021.

[2] R. Mühler, *Technical Guide Cochlear Implant University ENT Clinic Magdeburg*. 2010.

[3] H. Wizemann, *Effects of structured, intensive listening training on the communicative competence of cochlear implant users*. 2017.

[4] A. Ernst, *Cochlear Implant today*. Springer Berlin, no. 1, pp. 45-53, 2009.

[5] W. Rötz and B. Bertram, *Cochlear implant in adults: Care and rehabilitation in speech and language therapy*. Springer Berlin, no. 1, pp. 25-38, 2022

[6] H. J. Krabbe. *Development and application of a health-related quality-of-life instrument for adults with cochlear implant: The Nijmegen Cochlear Implant Questionnaire. Otolaryngology-Head and Neck Surgery*. no. 123, 756-765, 2000.

Optimizing Illuminance for Eye-Tracking in Auditory Experiments
Ensuring Robustness Through Controlled Illuminance

Luca Wiederschein [1 2], Markus Kemper [2], Florian Denk [2], and Hendrik Husstedt [2]

[1] Auditory Technology, Universität zu Lübeck, luca.wiederschein@student.uni-luebeck.de
[2] Deutsches Hörgeräte Institut GmbH, Lübeck, {m.kemper, f.denk, h.husstedt}@dhi-online.de

Abstract

This research investigates the correlation between illumination levels and their effect on eye-tracking data, with a specific focus on pupillometric responses and the degree of eye openness during auditory experiments. To investigate this, we conducted a speech intelligibility test in the presence of background noise while simultaneously measuring pupil diameter and eyelid openness under various illumination conditions. Eye-tracking is utilized in this field to gain further insights into cognitive processing, such as listening effort. The study demonstrates that illuminance in the examined range has minimal impact on eye-tracking metrics and only becomes slightly noticeable in highly illuminated conditions. Therefore, these results demonstrate the robustness of current eye-tracking technology against varying levels of illumination. No significant correlations were found between different levels of illumination and factors such as blink rate, drowsiness, or overall performance in the auditory task. These findings suggest that eye-tracking techniques can be reliably used in hearing research.

1 Introduction

Eye-tracking technology has become an increasingly valuable tool in auditory experiments, providing unique insights into how we process and respond to auditory stimuli. Researchers can investigate various cognitive processes by tracking eye-reactions [1, 3]. One method to use eye-tracking in auditory experiments is pupillometry, which measures changes in pupil diameter and provides information about various cognitive and physiological processes. Non-illuminance-mediated pupillary dynamics refer to changes in pupil diameter that are not due to illuminance intensity but to cognitive workload or emotional arousal. Researchers hypothesize that changes in cognitive processing performance have a positive correlation between effort and pupil diameter. This relationship has been observed in various contexts, such as hearing, including speech understanding, pattern recognition and generall listening effort. Therefore, pupillometry can be a valuable tool for analyzing cognitive processing performance and understanding the underlying neural mechanisms [1]. Other measurments, especially spontaneous eye-blink parameters have been employed as objective, non-invasive measures for assessing drowsiness or fatigue. These parameters, especially blink duration and percentage of eye-closure, have demonstrated consistency with drowsiness states [4]. A particularly reliable method of detecting drowsiness is the percentage of eyelid closure over time (PERCLOS). For this technique, the maximum eye-openness value is calculated

for each subject and percentage of samples during which the eyes is at least 80 % closed is measured [5]. Previous studies have shown that environmental factors, such as illuminance, can significantly affect pupillometric indices, including pupil diameter, pupil diameter variation, constriction velocity, and dilation velocity [2]. To enhance the precision and reliability of experimental outcomes in auditory research, it is crucial to conduct a comprehensive and systematic investigation of the influence of illuminance on pupillometric assessments. We want to examine the potential impact of such variables on the results and thus improve to the robustness and validity of findings. Additionally, we wanted to investigate the effects of illuminance on drowsiness and fatigue during auditory experiments. This knowledge will offer a more comprehensive understanding of how environmental factors affect auditory testing, ultimately improving the research's overall quality and applicability.

2 Material and Methods

2.1 Subjects

The study was conducted with 10 adults who had normal hearing, as determined by pure tone audiometry and self-reported normal or corrected vision (2 with normal vision, 6 with glasses not worn during the experiment, 1 with contact lenses, and 1 with laser eye surgery). The test subjects were recruited through private contacts and most were employees of the German Institute of Hearing Aids. They consisted of 6 males and 4 females between the ages of 24 and 36 (mean age 29.7, standard deviation 3.8). All subjects were familiar with the stimuli used for the study.

2.2 Study Setup

In this study, we collected the pupil diameter and eye-openness data using a Tobii Pro Spectrum eye-tracker in the 300 Hz version with a EIZO FlexScan EV2451 monitor. Matlab R2023a in combination with the Psychophysics Toolbox Version 3 was used to record both the eye-tracking data and the data from the auditory task. The auditory task was performed using a Genelec 8351 coaxial loudspeaker. The "Oldenburger Measurement Application for Research and Development" (version 2.2.2.2) was used to determine the individual speech reception thresholds (SRT) with the the OLSA (Oldenburg Sentence Test). To measure the different illuminance levels used in the study, we utilized a Volcraft LX-1108 Luxmeter.

2.3 Stimuli

A common speech test material (German version of the OLSA sentence test) was used for the listening effort test. The OLSA is a widely used sentence test in German-speaking countries. During the test, participants are asked to repeat sentences played to them while being exposed to background noise specifically designed as "OLSA noise". This noise is created by overlaying the entire sentence material of the test and simulates speech. The noise level remains constant while the speech level is adaptively varied. Mostly, the SRT50 (Speech Reception Threshold, i.e. Signal-to-noise ratio at 50 % intelligibility) is determined wiht the OLSA. OLSA sentences follow a consistent structure: name-verb-number-adjective-object [1, 6].

2.4 Procedure

First, the standard method to measure the SRT50 with the OLSA described in 2.3 was used. Participants were asked to repeat short sentences presented in background noise. Second, we used the signal-to-noise ratio (SNR) of the SRT50 as a fixed SNR for the main study. In the standard OLSA procedure, there is a break in the noise in the response period after each sentence, whereas in the main study the noise was constant.

The eye-tracker monitored the participant's pupil diameter and eye-openness. The core experiment consisted of five segments, each with 30 trails (randomly selected OLSA sentences at the SNR of the SRT50). Each trial began with a short pause lasting between 1.9 and 3.0 s, followed by the presentation of an OLSA sentence. In two-thirds of the trials, participants were instructed by a speech bubble on the eye-tracker screen to repeat the sentence after a pause lasting between 2.6 and 3.7 s. Both pauses were determined by a right-skewed function.

Each of the five segments was conducted under different illuminance conditions. The illuminance was measured by placing the luxmeter as close to the participant's eye position as possible. The intensities were set logarithmically from the lowest to the highest illuminance setting available in the measurement room (9, 21, 48, 111, and 255 Lux). The order of the condions was determined by a randomized Latin square. Setting up the different illuminance conditions resulted in a small brake of around 3-5 minutes between each condition. The eye-tracker was calibrated with a five-point grid at the beginning of each condition.

2.5 Data Processing and Statistical Method

The pupil data was linearly interpolated to compensate for blinks and potential gaps in pupil tracking, thereby reducing the creation of outliers associated with such interruptions. Additionally, the data was smoothed using a low-pass filter and downsampled from 300 to 50 Hz. We used only the left eye for pupil analyses. This eye had more interpretable data on average across all subjects [7]. We used the PERCLOS method for further evaluating the eye-openness data, looking at the ratio of collected data in which the eye was closed 80 % compared to the maximum [5]. Statistical analysis was carried out using multiple one-way analysis of variance (ANOVA) for the different data sets (pupil, eye-openness and task performance). A t-test was also used to analyze the significance of the pupil data compared to a hypothetical mean of zero and unspecified variance.

3 Results and Discussion

3.1 Auditory Task

We analyzed speech intelligibility during the experiment to verify the measurement setup and to ensure the consistency of the auditory task. Upon examination of the data, no significant differences were found between the illumination levels ($F(1,4)=0.11$, $p=0.77$). When examining the conditions in presentation order, the differences are also not significant. ($F(1,4)=0.92$, $p=0.46$).

3.2 Eye-Tracking Data

3.2.1 Pupil Data

The baseline pupil diameter, which was measured immediately before the onset of the OLSA sentence, serves as the reference point for evaluating changes in pupil diameter seen in Fig. 1 and Fig. 2. In Fig. 1, we observe the average change of pupil diameter in relation to the baseline as a mean over all trials, and subjects for the 5 different illuminance levels. There is an increase after the onset of the OLSA, reaching its first maximum after around 2.5 s. During the repetition process of the OLSA sentence, a second maximum is reached due to speech production. However, for further analysis, we will focus on the first maximum as it relates to the auditory task. A t-test performed on the pupil data within the specified time window around the first peak (2-3 s) showed statistical significance for all illuminance levels ($p<0.001$).

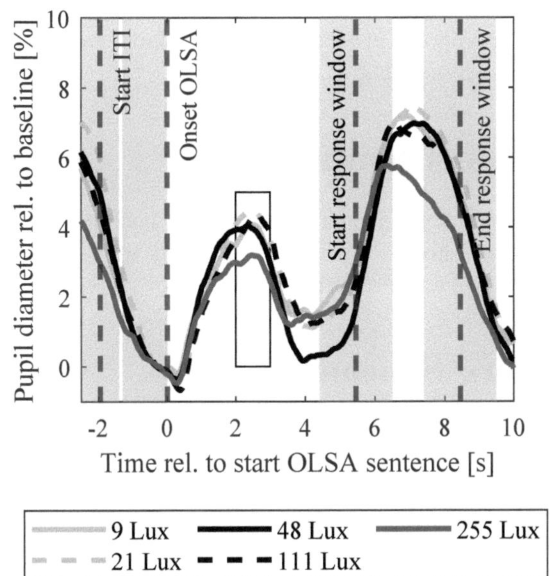

Figure 1: Mean change in pupil diameter across all subjects for the conditions of the main task, with a box around the first maximum in pupil diamater (2-3 s after OLSA onset).

To look further at the window between 2 and 3 s we split up the data into the different conditions. When examining various illumination levels, the biggest effect is observed at 48 Lux, with a median 4 % increase in diameter. The smallest change is observed at 255 Lux, with a median 2 % increase. Median changes at other illumination levels range between 3.5 % and 3.8 %. The standard diviation gets progressivly lower with increaced illumination (1.4 % at 9 Lux to 0.9 % at 255 Lux). Despite the differences observed in Fig. 2, an analysis of the data shows no significant effects between the different illumination levels ($F(1,4)=2$, $p=0.092$). Although there may be a trend towards a lower range of interpretable change at very high illumination levels.

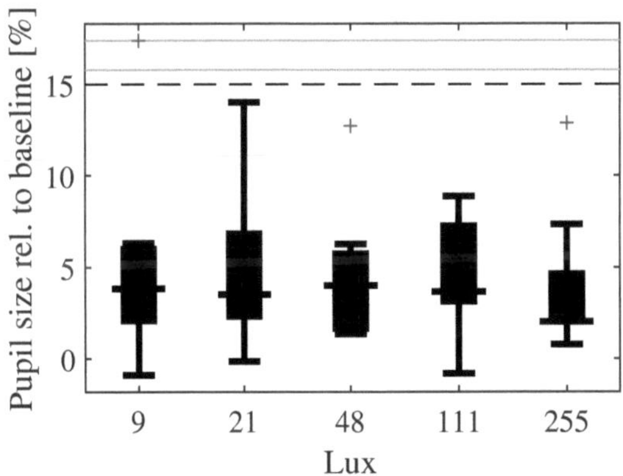

Figure 2: Analysis of change in pupil diameter from baseline from 2 to 3 s after OLSA onset, with at different illumination levels. Average values for each subject over the time period and trials under varied illuminance conditions.

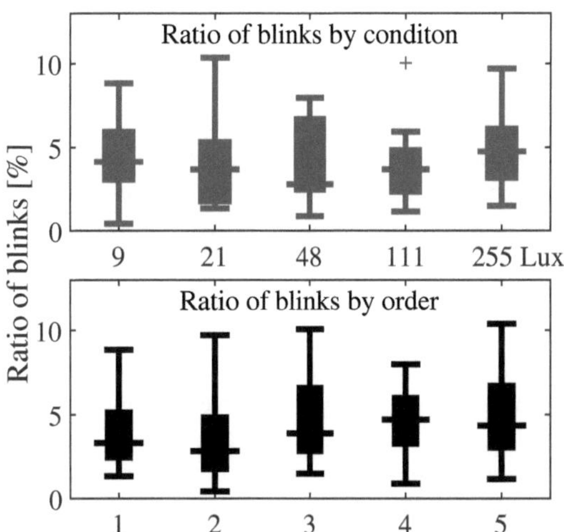

Figure 3: Analysis of blink duration under varied illuminance conditions and the sequential order, optained using not tracked pupil data.

3.2.2 Blink Data

There were no significant differences between the illumination levels in terms of the combined duration of blinks during each trial ($F(1,4)=0.16$, $p=0.96$) seen in Fig. 3. The median combined blink duration ranged from 2.8 % at 48 Lux to 4.8 % at 255 Lux. When reviewing conditions in the order in which they were presented, no significant effects were found either ($F(1,4)=0.46$, $p=0.76$).

3.2.3 Eye-Openness Data

In our analysis of drowsiness using the PERCLOS method (Fig. 4), we found no significant diffcrences in varying levels of illumination ($F(1,4)=0.05$, $p=0.99$). The results were not statistically significant in the order of the presented condition either ($F(1,4)=0.73$, $p=0.58$).

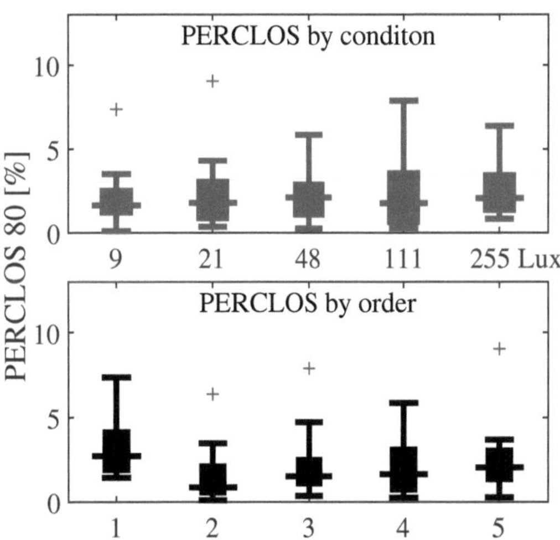

Figure 4: Percentage were the eyelid closure over time (PERCLOS) is over 80 % for varied illuminance conditions and the sequential order

3.3 Discussion

Upon examining the pupil diameter data in relation to the illumination levels, our results did not show any significant effects. Based on our small sample size (n=10) and the trend towards a smaller effect at the highest illumination level, it is suggested that a larger subject group may give significant results at the highest measured levels and that even higher illumination levels may influence the results significantly. It should be noted that all subjects reported a strong influence of the light emitted by the eye-tracker screen for the two lowest illumination levels, even though the luxmeter was placed at the subject's eye position. Most subjects reported feeling more comfortable during mid-level (48 and 111 Lux) illumination conditions because of this effect. Based on our findings, we recommend using lower to mid illumination levels (between 9 and 111 Lux) for auditory studies. If possible, a medium illumination level is preferable to ensure a higher comfort level. This suggestion is based on observing more pronounced effects within the illuminance range of 9 to 111 Lux accompanied by a stable window to operate in that is not sensitive to illuminance changes. Lower illumination levels seem to amplify measurable impacts and may provide an strategic advantage in identifying smaller yet significant auditory effects.

The method used was an effective way to gather pupil responses for an auditory task. We found a significant increase in response due to listening compared to the pupil baseline. This outcome was expected since we used a well-established method to evaluate listening effort through pupilometry [1].

The frequency of blinks appeared to be relatively low, particularly with one subject only blinking around 1 % of the time during one condition. After thorough analysis, testing, and consultation with the eye-tracker manufacturer, we found no issues with the data collection process or any preprocessing that could have caused this result in the measurement setup used. No significant effects were observed regarding blink ratio (i.e. the amount of interpretable data) and drowsiness using the PERCLOS method.

The average speech intelligibility was 51.5 %, aligning with expectations. The auditory data confirmed the stability and reliability of our measurement setup, despite differences between the standard OLSA operating method and the one used for the main study. This demonstrates the effectiveness of our measurement setup. Despite the non-significant results in terms of presentaion order, we still recommend the use of a randomized experimental design in this type of study.

4 Conclusion

In conclusion, this study demonstrates the reliability of modern eye-tracking technology and suggests that comparable relative changes in pupil diameter can be measured, despite different illuminance conditions. It is still recommended to use lower and consistent illuminance levels to ensure optimal results.

We now feel confident that the illuminance level in our test room is appropriate to conduct further auditory experiments including eye-tracking.

Acknowledgement

This study was conducted at the German Institute of Hearing Aids in Lübeck. I would like to thank my supervisors for their great support during this work.

Authors' Statement

Conflict of interest: Authors state no conflict of interest.

5 References

[1] Winn MB, Wendt D, Koelewijn T, Kuchinsky SE. Best Practices and Advice for Using Pupillometry to Measure Listening Effort: An Introduction for Those Who Want to Get Started. Trends in Hearing. 2018;22. doi:10.1177/2331216518800869

[2] Ong C, Hutch M, Smirnakis S. The Effect of Ambient Light Conditions on Quantitative Pupillometry. Neurocrit Care. 2019 Apr;30(2):316-321. doi: 10.1007/s12028-018-0607-8. PMID: 30218349.

[3] Chin-An Wang, Douglas P Munoz, A circuit for pupil orienting responses: implications for cognitive modulation of pupil size, Current Opinion in Neurobiology, Volume 33, 2015, Pages 134-140, ISSN 0959-4388, https://doi.org/10.1016/j.conb.2015.03.018.

[4] Jennifer M. Cori, Clare Anderson, Shamsi Shekari Soleimanloo, Melinda L. Jackson, Mark E. Howard, Narrative review: Do spontaneous eye blink parameters provide a useful assessment of state drowsiness?, Sleep Medicine Reviews, Volume 45, 2019, Pages 95-104, ISSN 1087-0792, https://doi.org/10.1016/j.smrv.2019.03.004.

[5] Takashi Abe, PERCLOS-based technologies for detecting drowsiness: current evidence and future directions, SLEEP Advances, Volume 4, Issue 1, 2023, zpad006, https://doi.org/10.1093/sleepadvances/zpad006

[6] Kuehnel, V., Kollmeier, B., Wagener, K. Entwicklung und Evaluation eines Satztests für die deutsche Sprache I: Design des Oldenburger Satztests. Zeitschrift für Audiologie. 38. 4–15. (1999)

[7] Geller, J., Winn, M.B., Mahr, T. et al. GazeR: A Package for Processing Gaze Position and Pupil Size Data. Behav Res 52, 2232–2255 (2020). https://doi.org/10.3758/s13428-020-01374-8

Technical Characterisation of a Spherical Loudspeaker Array

Cleo Noak [1], Markus Kallinger [2], Jürgen Tchorz [2], Fabian Hettler [2], Florian Denk [3], Hendrik Husstedt [3], Tim Jürgens [2]

[1] Auditory Technology, Universität zu Lübeck, cleo.noak@student.uni-luebeck.de

[2] Auditory Aoustics, Technische Hochschule Lübeck, {markus.kallinger, juergen.tchorz, fabian.hettler, tim.juergens} @th-luebeck.de

[3] Deutsches Hörgeräte Institut GmbH, Lübeck, {f.denk, h.husstedt}@dhi-online.de

Abstract

The evaluation of hearing aids and their functionalities mainly takes place in laboratories, with one- or two-loudspeaker systems. However, these simple arrangements cannot capture the complexity of acoustic situations that hearing aid users face in everyday life. To address this issue, a spherical loudspeaker array with processing hardware has recently been installed in the anechoic chamber of the TH Lübeck. The underlying research programme aims to develop new evaluation methods based on Higher Order Ambisonics (HOA), that are expected to replicate everyday acoustic scenes in the laboratory.

In this work, three key measurements were carried out to assess the performance of each loudspeaker and the overall system. The results showed low levels of background noise, sufficiently flat frequency responses after equalisation and negligible directivity within the central sweet spot. For a 7th order HOA reproduction up to $4\,\mathrm{kHz}$ with a more than head sized sweet spot the quality was satisfactory.

1   Introduction

Modern hearing aids are intelligent devices that can improve the daily lives of many people with a hearing impairment. However, laboratory evaluations alone are not sufficient for assessing their functionality in real-life situations. Standardised evaluation methods involve technical measurements of the hearing aids placed in a closed measuring box or test scenarios that utilise one or two loudspeakers to play back speech and noise. Conducting evaluations of hearing aids directly in real everyday environments is rarely a suitable option as these situations are neither controllable nor comparable. Therefore, simulating everyday situations in a laboratory setting using virtual acoustics has become a popular topic in hearing aid research.

A new research programme at the Lübeck University of Applied Sciences (TH Lübeck) in cooperation with the German Institute of Hearing Aids (DHI) aims to develop reproducible everyday life scenes to enable a more realistic evaluation of hearing aid performance. To this end, a spherical loudspeaker array has recently been installed in the institute's anechoic chamber.

A first step of the project and the focus of this paper implies a technical characterisation of the spherical loudspeaker system to ensure its sufficient performance for an HOA of order 7 up to $4\,\mathrm{kHz}$ and a sweet spot of at least head size. The theoretical background for this evaluation was taken from [2] and [3], the structure of this paper was inspired by [1].

The technical characteristics measured here included the Background Noise Level (BNL) in different settings, the loudspeaker transfer functions as in [5], and the directivity pattern of a single loudspeaker. The actual measuring methods are described in Section 2. The results are then evaluated in Section 3 and analysed for their sufficiency for the 7th order HOA application, using theoretical calculations as guidelines.

2   Material and Methods

This section describes the speaker array, including all its major system components, and provides an overview of the measurements taken to characterise it.

2.1   The Spherical Loudspeaker System

The spherical array consists of 65 passive KEF Ci130QRfl coaxial mid- to high-frequency loudspeakers and four active KEF KF92 subwoofers. The mid-high speakers are arranged in five horizontal rings, with an additional ceiling loudspeaker (Voice of God). The array was installed in the anechoic chamber of the TH Lübeck by the company sonible GmbH. This anechoic chamber absorbs reflections down to a lower frequency limit of $100\,\mathrm{Hz}$.

Fig. 1 shows a photograph of the system inside the anechoic chamber with a head and torso simulator as a representative listener. Its head is positioned in the centre of the sphere and its line of sight is directed towards the $0°$ azimuth loudspeaker of the central ring. Each speaker is encased in PET

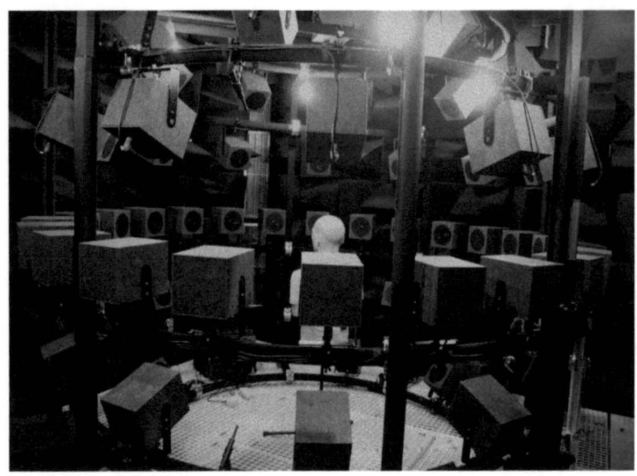

Figure 1: Spherical speaker array inside the anechoic chamber of the TH Lübeck.

felt and the metal frame is covered in Velcro® to reduce reflections from the system components themselves.

The arrangement of the individual speakers is shown in Fig. 2 in the form of a two-dimensional grid. Each dot represents a coaxial mid- to high-frequency loudspeaker while the four diamonds indicate the positions of the subwoofers. The speakers shaded in light grey are all mounted within the saggital plane at $0°$ azimuth. The darker grey represents the movable door area of the framework.

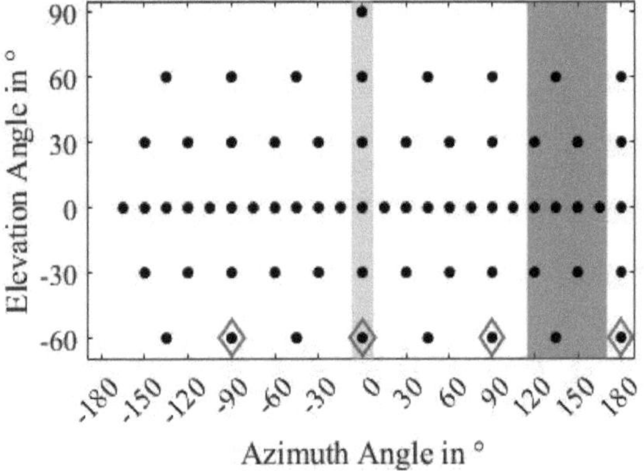

Figure 2: Two-dimensional layout of the speaker positions.

The technical backbone of the system is a rendering computer connected via an internal network (switch) to an RME HDSPe MADI FX audio interface and 3 sonible d:24 amplifiers. The processing system in use is a MADI interface. There are eight free-field microphones evenly distributed throughout the array and an additional talkback microphone in the listener's line of sight.

A frequency of 100 Hz was set as the crossover frequency between the subwoofers and mid-high speakers. The subwoofers always play a sum signal of all channels without contributing to the spatial sound. The mid- to high-frequency loudspeakers were equalised by adjusting five 2nd order parametric IIR filters in the three amplifiers. Each mid-high channel was equalised with identical filters to achieve a preferably flat transfer function. However, indi-

vidual filters could be applied if necessary. Fig. 3 shows the frequency responses of the applied filters in darker and lighter grey and the resulting overall transfer function H in black. The dark grey plot depicts a shelving filter with a cutoff frequency of 500 Hz, a gain of 1.5 dB and a quality of 1. All light grey lines represent peak filters.

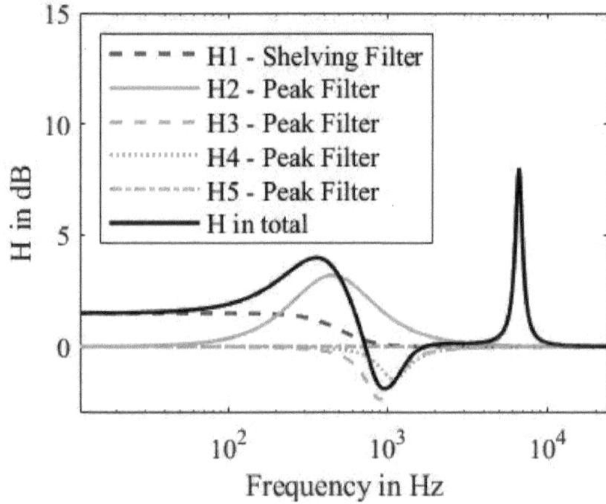

Figure 3: Frequency responses of the IIR-filters for the equalisation of mid- to high-frequency speakers.

In the same order as seen in Fig. 3 the peak filters have centre frequencies of 450 Hz, 900 Hz, 1.1 kHz and 6.6 kHz, gains of 3.2 dB, -2.4 dB, -1.5 dB and 8 dB and qualities of 0.7, 1.7, 1.8 and 9.5.

The filter coefficients a_1, a_2, b_0, b_1 and b_2 for shelving and peak filters were calculated using the formulae given in tables 6.3 and 6.4 of [4] with the following 2nd-order transfer function:

$$H(z) = \frac{b_0 + b_1 z^{-1} + b_2 z^{-2}}{1 + a_1 z^{-1} + a_2 z^{-2}}. \tag{1}$$

The array with its 65 mid- to high-frequency speakers was designed to meet the prerequisites of a $M = 7^{th}$ order HOA reproduction. The required number of loudspeakers L for three dimensions was calculated as shown in [3]:

$$L = (M + 1)^2. \tag{2}$$

The minimum number of speakers was therefore $L = 64$. The listening area in the centre of the spherical array where the sound can be reproduced correctly is called the sweet spot. The radius r of this sweet spot was estimated by the following rule of thumb which indicates a physically correct soundfield reproduction of less than 4 % error [3]:

$$r = \left\lceil \frac{cM}{2\pi f} \right\rceil. \tag{3}$$

The desired upper frequency limit was $f = 4$ kHz and the speed of sound was assumed to be $c = 340\,\mathrm{m\,s^{-1}}$. This resulted in a sweet spot radius of $r = 9.47$ cm.

2.2   Background Noise Level

The first measurement was concerned with the total noise level and should describe both the anechoic chamber and the loudspeaker array. The broadband background noise level (BNL) was therefore determined in three settings, as listed below in Table 1.

Table 1: Measurement settings in the anechoic chamber

Setting	Description
1	Lights and system turned off
2	Lights turned on and system turned off
3	Lights and system turned on

The BNL of each setting was measured using a low-noise Brüel & Kjær 2250 class 1 sound level meter with a type 4189 $1/2$" microphone. The average A-weighted equivalent level (LAeq) was calculated over five rounds of $60\,\mathrm{s}$ for each scenario.

2.3   Transfer Function of the Speakers

The second measurement focused on the transfer functions of the individual mid- to high-frequency speakers and was carried out with the equalisation filters both on and off. To determine these transfer functions, an exponential sweep of $3\,\mathrm{s}$ with a sampling rate of $48\,\mathrm{kHz}$ and frequencies between $20\,\mathrm{Hz}$ and $24\,\mathrm{kHz}$ was played through each loudspeaker and recorded by a class 1 NTi M2211 $1/2$" microphone positioned in the centre of the sphere and aimed at the ceiling. The measuring microphone was connected via a MADI channel, replacing one of the mounted microphones. This allowed synchronous playback and recording of the sweeps. In order to analyse the playback signals of the non-equalised as well as the equalised loudspeakers, the respective recordings were filtered in $1/6$-octave bands and compared by their effective values in each band.

2.4   Directivity Pattern of a single Speaker

The final measurement for this paper was concerned with the directivity pattern of a single speaker. One of the reasons for this measurement was to see if the loudspeakers could meet the requirements for a sufficient sweet spot in 7th-order Higher Order Ambisonics (HOA) reproduction. The same NTi M2211 $1/2$" class 1 microphone as in the second measurement was placed in front of a particular speaker on the $0°$ elevation axis at a radius of $1.2\,\mathrm{m}$ to the speakers centre. The measuring microphone was moved around the loudspeaker from $0°$ to $90°$ azimuth angles always facing the speaker while maintaining the radius. As with the second measurement, the microphone was connected to the system via one of the MADI channels to record a sweep represented by the same speaker for each of the 19 azimuth angles.

The analysis involved calculating the directivity pattern or the sensitivity of the loudspeaker in relation to its $0°$ azimuth transfer function. Rotationally symmetric behaviour was assumed due to the coaxial construction of the speakers to give a directivity pattern from $-90°$ to $90°$ azimuth. In addition the speaker's radius inside a tolerated sensitivity range of $\pm 3\,\mathrm{dB}$ was compared to the radius r of the sweet spot.

3   Results and Discussion

The three measurements carried out were analysed with the aim to confirm the sufficient dimensions and quality of the loudspeaker array to achieve a 7th order HOA sound reproduction with a sweet spot larger than a human head.

3.1   Backgrond Noise Level

The results of the first measurement focusing on the BNL in the anechoic chamber are listed in the following Table 2.

Table 2: Results of the BNL-Measurements

Setting	Average BNL
1	17.4 dB (A)
2	17.3 dB (A)
3	19.5 dB (A)

Table 2 shows that the light only resulted in an average level difference of $-0.1\,\mathrm{dB}$ (A), while the noise emitted by the loudspeaker array increases the total broadband BNL by just over $2\,\mathrm{dB}$ (A). This small difference in level is due to the passive nature of the mid- to high-frequency loudspeakers. The level increase is similar to the results reported in [1], where the total BNL was $28\,\mathrm{dB}$ and the increase was also $2\,\mathrm{dB}$. This comparison should be treated with caution, however, as the A-rating of the broadband level may have inadvertently reduced the influence of the active subwoofers on the BNL.

3.2   Transfer Functions of the Speakers

The comparison between the non-equalised and equalised mid-high loudspeakers is shown in Fig. 4.
Although the overall variation of the individual transfer functions is similar, the average transfer functions of non-equalised and equalised speakers differ significantly. A dip in the lower frequencies up to $500\,\mathrm{Hz}$, a slight rise between $800\,\mathrm{Hz}$ and $1\,\mathrm{kHz}$ and a notch at $7\,\mathrm{kHz}$ have been equalised by the IIR filters shown in Fig. 3. The limited number of low order IIR filters is probably the reason why the equalised average transfer function is not perfectly flat. At this stage, faster and more precise FIR filters were not applicable as the equalisation took place within the amplifiers, which only allowed for five parametric IIR filters. It is questionable whether the benefit of higher order FIR filters outweighs the effort of their implementation for each individual speaker.

3.3   Directivity Pattern of a single Speaker

The directivity pattern of a single equalised mid- to high-frequency speaker can be seen in Fig. 5.

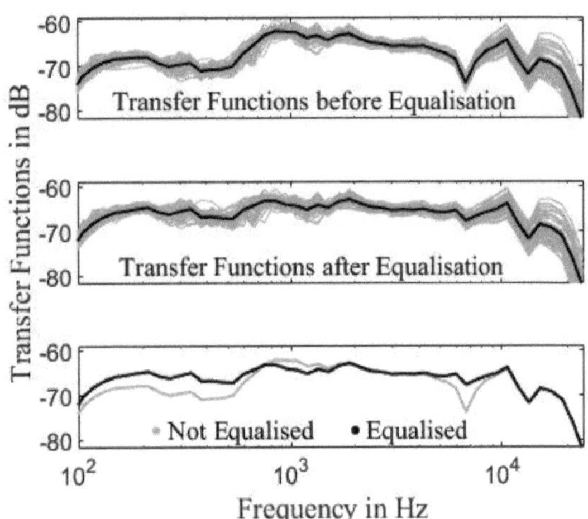

Figure 4: Transfer functions of non-equalised and equalised loudspeakers.

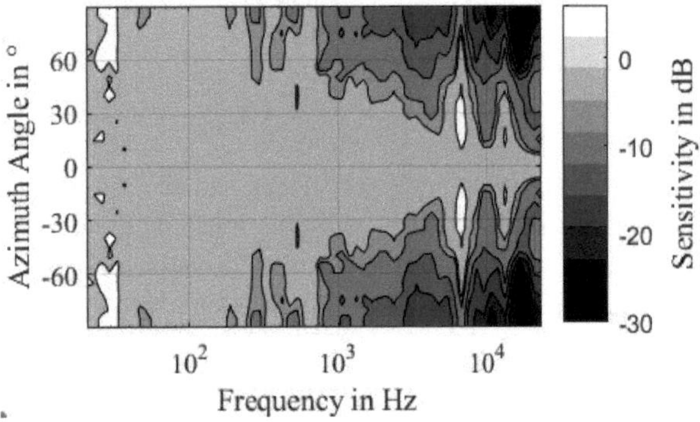

Figure 5: Directivity pattern of a single mid- to high-frequency loudspeaker.

A loudspeaker sensitivity range between $-3\,\text{dB}$ and $3\,\text{dB}$ is represented by the light grey area surrounding the $0°$ azimuth axis. The speakers are capable of maintaining this sensitivity tolerance within an azimuth angle range of $\pm 20°$ in the critical frequency range of $100\,\text{Hz}$ and $4\,\text{kHz}$. This results in a cone of total angle $40°$ around the loudspeaker axis within which the speakers will operate accurately and have a flat transfer function.

With an average distance of $d = 1.2\,\text{m}$ between the centre of one of the mounted loudspeakers to the centre of the sphere and the azimuth angle of $20°$, the corresponding horizontal area within which there is less than $3\,\text{dB}$ difference has a radius of $\tan(20°) \cdot d = 43.68\,\text{cm}$. The sweet spot radius $r = 9.47\,\text{cm}$ that was calculated by (3) is considerably smaller, so the sweet spot fits well into the loudspeaker's working area. Also, the average human head has a radius of about $8.7\,\text{cm}$, so the theoretical r of this system is slightly larger than the size of the listener's head and therefore includes a tolerance for slight head movement. As mentioned in [2], the perceived sweet spot is usually larger than the theoretical one. Therefore, the latter can be considered as a minimum.

4   Conclusion

The spherical loudspeaker array has been evaluated to have a sufficient dimension and technical quality for the purpose of a 7th order HOA sound reproduction up to a frequency limit of $4\,\text{kHz}$. The theoretical sweet spot is at least the size of an average human head. When all amplifiers and speakers are activated, the BNL is only increased by $2.3\,\text{dB}$ due to the passive nature of the mid- to high-frequency loudspeakers and remains below $20\,\text{dB}$ (A). This value should be validated by repeating the BNL measurement with an unweighted measurement in octave bands to justify the comparison with [1] and to consider the active subwoofer's influence. The amplifiers' equalisation filters significantly flattens the transfer function of each speaker in the system, although a perfectly flat transfer function is not achieved. Furthermore, the horizontal working area of a single speaker is significantly larger than the sweet spot. To gain further insight, it may be worth considering an elevation measurement and the analysis of interferences between speakers. The directivity pattern could therefore be measured outside the array to eliminate reflections from other loudspeakers' cases or the array frame.

Acknowledgement

The work has been carried out at the Lübeck University of Applied Sciences (TH Lübeck).

Authors' Statement

Conflict of interest: Authors state no conflict of interest.

5   References

[1] F. Pausch, G. Behler and J. Fels, *SCaLAr – A surrounding spherical cap loudspeaker array for flexible generation and evaluation of virtual acoustic environments.* Acta Acustica, vol. 4, no. 5, art. no. 19, 2020.

[2] G. Grimm, S. Ewert and V. Hohmann, *Evaluation of Spatial Audio Reproduction Schemes for Application in Hearing Aid Research.* Acta Acustica United with Acustica, vol. 101, no. 4, pp. 842–854, 2015.

[3] M. Marschall, T. Dau, E. MacDonald and J. Buchholz, *Capturing and Reproducing Realistic Acoustic Scenes for Hearing Research.* Technical University of Denmark, Department of Electrical Engineering. Contributions to Hearing Research, vol. 18, 2014.

[4] U. Zölzer, *Digital Audio Signal Processing.* 3rd edition, Hoboken: Wiley, 2022.

[5] S. Müller and P. Messarani, *Transfer-Function Measurement with Sweeps.* Journal of the Audio Engineering Society, vol. 49, no. 6, pp. 443-471, 2001.

Measurement and Evaluation of Auditory Evoked Potentials using Deconvolution and Conventional Methods in Normal Hearing Subjects

Sonja Taupadel [1,2], Nina Aldag [2], Philip Gouverneur [1], Marcin Grzegorzek [1] and Waldo Nogueira [2]

[1] Medical Informatics, Universität zu Lübeck, taupadel.sonja@mh-hannover.de
{philip.gouverneur, marcin.grzegorzek}@uni-luebeck.de
[2] German Hearing Center, Medical School of Hannover

Abstract

Auditory evoked potential (AEP), such as auditory brainstem responses (ABRs) and cortical auditory evoked potentials (CAEPs), can be obtained for both clinical and research settings using electroencephalography (EEG). A simultaneous measurement of AEPs, which differ in latency and selected inter-stimulus interval (ISI) depending on the type, would be more time-saving than a separate measurement. Furthermore, a combined recording of these AEPs could provide a comprehensive overview of the auditory pathway and how acoustic stimuli are transmitted. This work evaluates iterative randomised stimulation and averaging as a deconvolution algorithm with an efficient matrix formulation based on latency-dependent filtering and down-sampling subspace-constrained-least-squares for separating overlapping responses of simultaneously measured potentials. Deconvolution showed smaller or even absent amplitudes in ABRs and CAEPs. In conclusion, the comparison of the conventionally recorded potentials with the deconvolved potentials reveals that the separately measured AEPs have greater amplitudes than the deconvolved estimated response.

1 Introduction

The human auditory system can be measured through auditory evoked potentials (AEPs) using electroencephalography (EEG). AEPs have distinct latencies and are named according to their originating area. The interval between stimuli is restricted to ensure that acoustic trigger stimuli are presented with an inter-stimulus interval (ISI) that is longer than the response duration and latency of the recorded signal. With this method, each stimulus evokes an individual response and there is no overlap between the responses. The latency for auditory brainstem responses (ABRs), which originate in the brain-stem, is 1-15ms. The waves of an ABR can be numbered by Roman numbers from I-VI, whereby V is one of the most important waves [1]. ABRs differ if they are measured with an ipsi- or contra-lateral laying electrode compared to the probe tone [1]. The latency for cortical auditory evoked potentials (CAEPs) is 50-500ms [2] and the waves are named with the 'P' and 'N' as an abbreviation for positive and negative waves [2]. ABRs and CAEPs are measured separately due to their latency differences, which can be time-consuming. The iterative randomised stimulation and averaging (IRSA) based on latency-dependent filtering and down-sampling (LDFDS) was developed to circumvent the limitation that AEP have to be measured separately with different ISI [3, 4]. With this time efficient algorithm, AEPs can be measured simultaneously as it deconvolves the overlapping responses. The research question in this study is whether the decon-

volved AEPs are equivalent to separately measured ABRs and CAEPs obtained by the Compumedics Neuroscan EEG system. This study suggests that the system can be adapted for normal hearing listeners, as well as for individuals with a cochlear implant (CI).

2 Material and Methods

2.1 Setup

A 64 electrodes cap (Quick-CapTM, Compumedics Neuroscan, USA Inc., NC, USA 2826) was placed on the subject's head measured from nasion to inion beforehand to use the adequate cap size. The electrodes consist of silver with a silver chloride coating. Two electrodes (M1, M2) placed on the cleaned mastoid behind the ear were connected to the EEG cap. The montage consisted of the middle sagittal plane electrodes FPZ, ground, FCZ, CZ and REF electrodes from the 10-20 system. The mastoid electrodes M1 and M2 were the 'active' electrodes in this study, although mastoid electrodes on the ear are commonly used as reference electrodes. This would allow better recording of ABR. If REF does not work, one of the mid-sagittal electrodes on the head (FPZ, FCZ, CZ) were used as a reference. Despite the possibility that ABR could be more apparent at the mastoid of the same ear where the probe tone is applied [1], the contralateral electrode M1 was chosen. The amplitudes and latencies of M1 and M2 generally did not differ substantially, and the contralateral side is more suit-

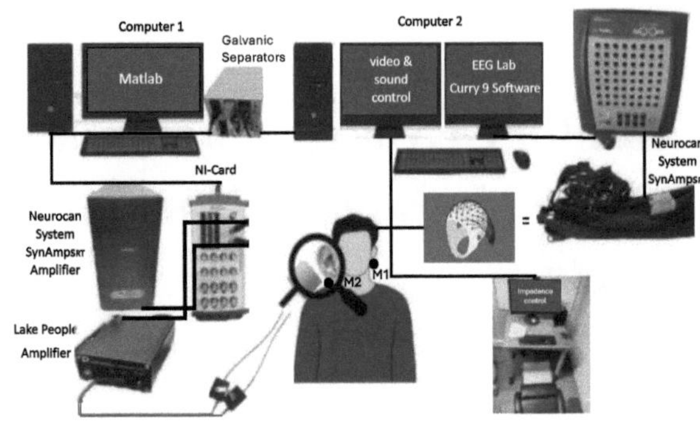

Figure 1: The measurement setup can be divided into three parts: 1. The subject's headphones are inserted into foam ear tips and the EEG-cap connected to the Neuroscan-EEG-system. 2. Computer 1 to run the Matlab-script and the hardware to apply the probe and masker to the earphones. 3. Computer 2: For the EEG Lab: Curry 9 Software to record the EEGsignals. The study control: video and sound of the subject inside the cabinet of the laboratory.

Table 1: The 10 parameter configurations. Measurement numbers 1-5 have the same parameters as measurement numbers 6-10 besides the most comfortable level (MCL) and the MCL-40dB.

#	AEP	probe stimulus	ISI	num. of stimuli	level
1, 2	deconv.	click	120-240ms randomised	1254	MCL
3	ABR	click	15ms	1254	MCL
4	CAEP	click	1000ms	150	MCL
5	CAEP	white Gaussian noise for 60 ms	1000ms	150	MCL
6, 7	deconv.	click	120-240ms randomised	1254	MCL-40dB
8	ABR	click	15ms	1254	MCL-40dB
9	CAEP	click	1000ms	150	MCL-40dB
10	CAEP	white Gaussian noise for 60 ms	1000ms	150	MCL-40dB

able with regard to further measurements on CI users who might experience electrical artefacts due to the implant on the ipsi-lateral ear. The electrodes were filled with conductive electrogel (eci: NET: 32 fl. oz./946 ml, Electro-Cap International Inc, Ohio, USA) using circular movements of the applicator. This was done to decrease the impedance between the electrode and the scalp. The impedances were below 10 kΩ and the impedance differences in between the electrodes were below 2 kΩ [1]. The EEG-cap was connected to the Compumedics Neuroscan SynAmpsRT amplifier headbox, as can be seen in Fig. 1. For recording the EEG-software Curry 9 (Compumedics) was used. The neuroscan-amplifier is connected to an adapter, which connects the trigger BNC-cable (TTL) to the NI-card (NI USB 6216, National Instruments, Debrecen, Hungary). The headphones (Compumedics) were plugged into the Lake People (phone Amp G103) amplifier. These devices of the hardware belong to a standard setup and standard procedures at the auditory prosthetic group (APG) of the Hannover Medical School (MHH). For the probe and masker application, Matlab 2020b (The MathWorks, Inc., Natick, Massachusetts) was used. Afterwards, EEGlab (Version 2023.1) and Matlab R2023b (The MathWorks) were used for evaluation.

2.2 Subjects

Four normal hearing (NH) subjects from MHH participated in this study (three male, one female, mean age: 28.5 $\pm$ std: 3.3). The young adults were asked to sit in an electromagnetically and acoustically soundproof cabinet of the EEG laboratory without moving or chewing as this would lead to movement artefacts in the EEG signal. All subjects were stimulated on the right ear to have the M1 electrode at the contra-lateral side, as can be seen in Fig. 1.

2.3 Measurements

A ten-point loudness-rating scale was used to determine the individual MCLs for different types of stimuli and ISIs. The ten parameter configurations, which were selected with regard to the total duration of the measurement, were performed on each of the four subjects. Table 1 shows the measurements at MCL (#1-#5) and 40dB below the MCL (#6-#10), five measurements each. The measurement at MCL-40dB was used to evaluate if there is a valid auditory response compared to the responses at MCL. The order of acquisition was randomised. For the deconvolution measurement (Table 1, #1, #2, #6, #7), two measurements of the same type were averaged [2]. For a randomised ISI between 120-240ms, the average stimulation rate correlates to 5.5Hz, so that the number of stimuli can be calculated including the duration of each deconvolution measurement(228s) as follows: 228s $\times$ 5.5Hz = 1254 [3]. To compare the deconvolved signals with conventionally measured signals, ABRs and CAEPs were recorded at parameters, such as ISI, that were typically used in clinic (Table 1, #3 and #5). The CAEP signal takes time to arrive to the cortex. Therefore, the ISI of 1000ms is a lot larger than for the signals arriving at the brain stem (ABR). CAEPs were evoked through linearly-ramped white Gaussian noise stimuli. While the plateau time was set to 40ms, the rise and fall time of the ramp was 10ms. Since CAEPs are usually not recorded using clicks as probe tones, but e.g. white Gaussian noise instead, CAEPs were also measured with a click as probe tone to make the measurements more comparable (Table 1, #4).

2.4 Data processing

ABRs and CAEPs were processed differently. High-pass filters were employed to remove resting and myogenic ac-

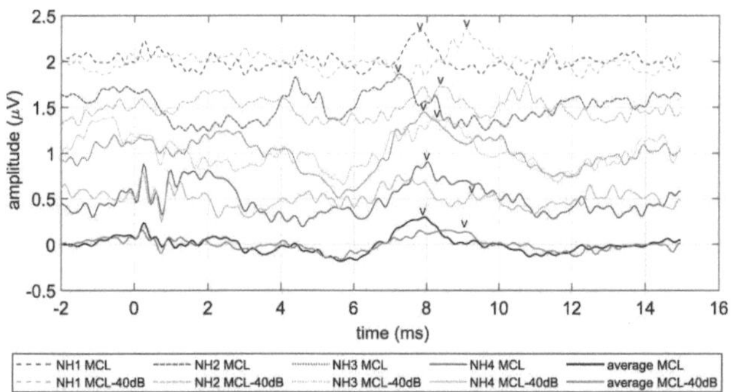

Figure 2: Band pass (100-3000Hz) and RMS filtered ABR (Table 1, #3, #8) for subject 1-4 and mean of all 4 for MCL and MCL-40.

tivity below 20Hz/100Hz from the brain's EEG, while the main purpose of low-pass filters was to avoid aliasing at higher frequencies [1]. Low-pass filters were used to filter ABRs between 100-3000Hz and CAEPs between 1 and 30Hz [2]. Both ABRs and CAEPs can be assigned to epochs by a trigger. The epochs can then be filtered using a root-mean-square (RMS) algorithm to reject any epochs with an RMS value more than 2 standard deviations away from the mean RMS value of all epochs. Finally, the responses can be displayed as an average of all responses over the period of a single epoch. The simultaneously measured responses were filtered using a band pass filter ranging from 1 to 3000Hz.

2.5 Deconvolution

IRSA deconvolves overlapping AEPs in a computational cost effective way [3] to overcome the ISI-restriction. The initialisation is the expected EEG, estimated as the convolution of a EEG-signal $x(t)$ out of the randomised stimulation algorithm (RSA) with the sequence of indices for probe tone stimulation $s(t)$. In each step of the iterative convergent algorithm the expected EEG can be corrected by averaging the result of the subtraction of the recorded EEG from the expected EEG-response. Each correction of the expected EEG-response accompanies with computational load, which can be reduced by the matrix-based formulation [3]. If the IRSA is performed after an LDFDS, an subspace-constrained (SC) is defined which reduces the execution time. The LDFDS leads to an optimal least-squares (LS) estimation, a reduced computational load and a better signal-to-noise-ratio (SNR) than LS-based IRSA only [4, 5]. Following previous studies, LDFDS-based SC-LS deconvolution implemented with IRSA fulfilled the requirements of a time-efficient deconvolution algorithm and was chosen for this study.

3   Results and Discussion

Fig. 2 shows the measured ABR at MCL and at MCL-40dB. In the two curves, belonging to subject 1, the ABR

can be seen clearly, since the peak at 7.9ms with an amplitude of 0.33μV (curve measured at MCL) occurs with a smaller amplitude and a larger latency at 9.1ms in the response corresponding with MCL-40dB. A similar effect can be seen in subject 2 at 7.3ms (0.36μV) for the ABR at MCL with a smaller peak (0.23μV) at 8.4ms in the response at the MCL-40dB loudness condition. In the average of all measurements of the normal hearing subjects at both loudness levels for the probe tone, the wave V peak move is visible. This means that ABRs can be measured with the Neuroscan system for these subjects. The peaks that are most likely to be one of the important peaks of the AEP (V in ABRs and N1, P2 in CAEPs) were marked in all plots. To make the plots more structured and the curves of the subjects more distinguishable, they are arranged with an artificial offset.

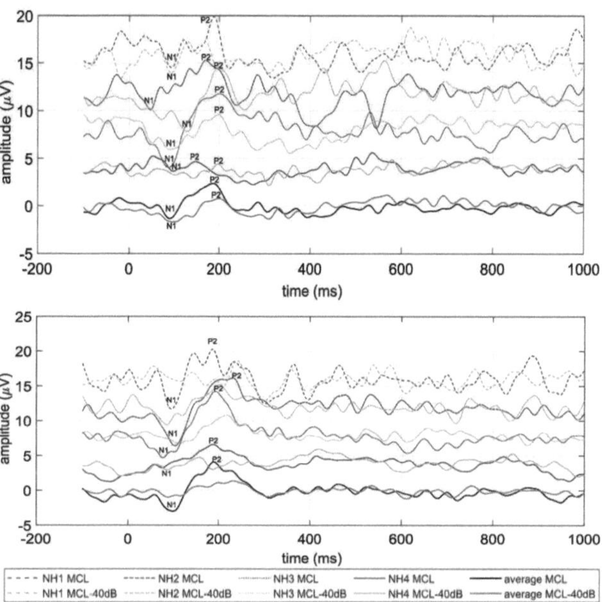

Figure 3: Band pass (1-30Hz) and RMS filtered CAEP with Click (upper graphic, Table 1, #4, #9) vs. white Gaussian noise (lower graphic, Table 1, #5, #10) as the probe for subjects 1-4 and mean of all 4 at MCL and at MCL-40dB.

Fig. 3 shows the amplitude (μV) of the potential recorded in a time period of 1100 ms. Typically, CAEPs occur in a period of 50-500ms [2]. The lower graphic of Fig. 3, which presents CAEPs elicited by white Gaussian noise, shows a delayed N1: 115ms at MCL-40dB compared to 98.3ms at MCL, (P2: 232.2ms at MCL-40dB vs. 189.8ms at MCL) and smaller (P2:1.35μV at MCL-40dB compared to 4.1μV at MCL) peak in the average of the subjects, when they listened to a Gaussian noise at their MCL-40dB (Table 1, #10) compared to the peak at the their individual chosen MCL (Table 1, #5). In the response of NH subjects 1-3, the characteristic negative peak at approximately 100ms (-5.58 til -3.29μV) and the positive peak (4.1-6.23μV) at approximately 200ms is more evident than for subject 4 (0.54μV at 120ms, 2.61μV at 188.3ms). When recording a CAEP, if a click is used instead of white Gaussian noise, the amplitude of the N1 (-1.34μV) and P2 (2.38μV) peaks are less pronounced. The average of all four subjects presented in the

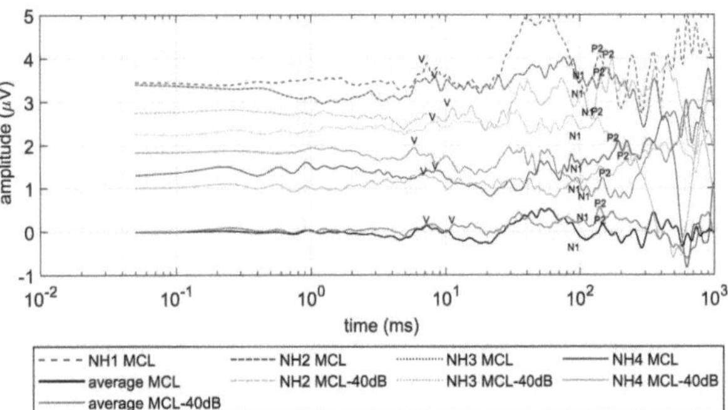

Figure 4: With IRSA deconvolved signal (Table 1, average of #1, #2 vs. average of #9, #10) for subject 1-4 at MCL and MCL-40dB.

upper graphic of Fig. 3 show a substantial amplitude change at MCL and a smaller and shifted peak at the MCL-40dB. These findings can also be detected in each subject separately. However, it has to be taken into account, that only four NH subjects were included in this study and a single subject outlier can affect the whole response average. The responses of the simultaneously recorded responses of the auditory pathway which were deconvolved with the time efficient IRSA are displayed in Fig. 4. The graphic displays all estimated responses on a logarithmic scale. The average for measurements #1 and #2 from Table 1 were taken for each subject, as well as the average of measurement #9 and #10, except for subject 1 due to a not usable signal because of the presence of too many artefacts. For all four subjects, the amplitude peak of the ABR wave V was detectable at 7ms on average. An increase in amplitude can be observed between 300ms and 500ms. Except for the negative peak N1 at 100ms, the AEPs are not very well detectable due to reduced amplitudes compared to separately measured AEPs. For randomised stimuli within the range of 120-240ms, typically auditory evoked potentials like the middle latency response (MLR) are measured, with its latency falling between the latencies of the ABR and the CAEP. The ISI may still elicit both ABR and CAEP despite their differing latencies, but may not be optimal for evoking both simultaneously. Another possible explanation would be that the brain does not produce a response for each stimulus individually and adaptation effects arise. This hypothesis is based on a review of temporal resolution and inter-stimulus gaps. [6]

4   Conclusion

This work compares the measured ABRs and CAEPs with deconvolved estimated responses based on recorded EEG signals. Typical ABRs and CAEPs could be obtained by the Neuroscan system with clinical parameters and even for CAEPs with a click as acoustic probe tone. In contrast, the estimated responses of each subject and the averages of the deconvolved signals for MCL and MCL-40dB show decreased response amplitudes compared to the wave V for

the ABR and N1 or P2 waves for the CAEP, making them more difficult to detect and distinguish from background noise. To address this issue, the system can be adapted to remove background noise and adjust algorithms so that LDFDS parameters become even more relevant and filtering becomes more adequate for the AEP. In future studies, it may be beneficial to include a weighting of CAEP evoked by stimuli that are followed by a greater ISI in the range of 120-240ms. Further research could also adapt the deconvolution algorithm to data from CI users.

Acknowledgement

The work has been carried out at Auditory Prosthetic Group at Deutsches Hörzentrum Hannover at the Medizinische Hochschule Hannover and the work has administered by the Institut für Medizinische Informatik, Universität zu Lübeck.

Authors' Statement

Conflict of interest: Authors state no conflict of interest. Informed consent: Informed consent has been obtained from all individuals included in this study.

5   References

[1] J. Durrant, C. Fowler, J. Ferraro, and S. Purdy, *Basic Concepts of Clinical Electrophysiology in Audiology.* Plural Publishing, Incorporated, 2022. [Online]. Available: https://books.google.de/books?id=MDJYEAAAQBAJ

[2] A. de la Torre, J. T. Valderrama, J. C. Segura, and I. M. Alvarez, "Latency-dependent filtering and compact representation of the complete auditory pathway response," *The Journal of the Acoustical Society of America*, vol. 148, no. 2, pp. 599–613, 2020.

[3] ——, "Matrix-based formulation of the iterative randomized stimulation and averaging method for recording evoked potentials," *The Journal of the Acoustical Society of America*, vol. 146, no. 6, pp. 4545–4556, 2019.

[4] A. de la Torre, J. T. Valderrama, J. C. Segura, I. M. Alvarez, and J. Garcia-Miranda, "Subspace-constrained deconvolution of auditory evoked potentials," *The Journal of the Acoustical Society of America*, vol. 151, no. 6, pp. 3745–3757, 2022.

[5] F. Bardy, B. Van Dun, H. Dillon, and R. Cowan, "Least-squares (ls) deconvolution of a series of overlapping cortical auditory evoked potentials: A simulation and experimental study," *Journal of neural engineering*, vol. 11, p. 046016, 06 2014.

[6] T. Picton, "Hearing in time: evoked potential studies of temporal processing," *Ear and hearing*, vol. 34, no. 4, pp. 385–401, 2013.

15

Image Processing

Are Transformers Better KITTI Detectors Than YOLO?: A Comparative Study of Object Detectors for Mobile Applications

Vincent Raphael Jansen[1], and Ahmed Mahmoudi[2]

[1] Robotics and Autonomous Systems, student of Robotics and Autonomous Systems, vincent.jansen@student.uni-luebeck.de

[2] Institute of Computer Engineering, Universität zu Lübeck, ahmed.mahmoudi@uni-luebeck.de

Abstract

Object detectors are crucial for the future of self-driving cars. They play a big role in preventing accidents and ensuring road safety by spotting and tracking objects. Since speed and power consumption are important factors in this field, there's a push for more compact object detection models that use less power and can detect objects at enhanced speeds.

In this paper, we chose three up-to-date object detectors: MobileNetV2-SSD, YOLOv7, and DETR. The training routine involved the utilization of the KITTI dataset, a dataset tailored to emulate real-world self-driving scenarios. After the training phase, a comprehensive evaluation and comparative analysis were conducted using diverse scenarios to validate their performance and suitability for mobile applications.

1 Introduction

Object detection is a fundamental challenge in computer vision and a popular use-case of AI. An object detector searches for features in images to assign a certain object category based on these features. These methods are already in use in autonomous driving [1] and other fields. Often object detectors are also used in mobile applications [2]. In mobile applications, detection models have a very limited amount of computation resources to solve the detection problem. Detectors are often designed as large complex AI models to achieve high detection accuracy. On the contrary, in mobile applications, it can be more important to have a more compact or faster model, even at the expense of some precision. In this research, we picked three modern object detectors and trained them on autonomous driving images since it is one of the most popular use cases for detection.

As the first two models, we choose MobileNetV2-SSD [3] and YOLOv7 [4]. Both models are popular mobile object detectors. The third model we picked is DETR [5]. DETR architecture is based on a transformer, a model designed for sequence prediction. Transformers are more complex models than MobileNet and YOLO, but they show good detection results in modern research.

For the data, we used the KITTI [6] standard dataset. KITTI contains different street scenes and is rather popular, which makes the results better for comparison.

Our goal is to investigate how well the chosen models perform on the detection task, how robust the detections are, and which model may be most suitable to use in mobile applications in general.

2 Material and Methods

2.1 Models

There are two categories of modern object detectors. One-stage and two-stage detectors [7]. The stages here refer to the number of iterations each image is processed for the detection of the objects. While one-stage detectors usually are less accurate than two-stage detectors, the detection speed is much higher. All models we used during this research belong to the one-stage family.

Classic-designed object detectors often consist of two parts. The first is the backbone. The backbone is a CNN (Convolutional Neural Network) which extracts key points of the input image and creates a low-dimensional feature map. Then a detection head predicts the location and class of one or multiple objects in the image, based on the features detected in the map.

MobileNetV2-SSD: MobileNetV2 is a classic model when it comes to object detection for mobile devices and is popular in the object detection community. The MobileNetV2-SSD model is a combination of the feature extraction model MobileNetV2 [3] and the detection head SSD [8]. The MobileNet architecture makes use of depthwise separable convolution. While normal convolution operates over all three image dimensions at once (*height* $\times$ *width* $\times$ *filters*), the depthwise convolution computes one feature map for each color filter and combines all three feature maps in the end. This method is highly efficient for devices with little computational power.

The prediction head SSD is a network of standard convolutions where each layer influences the output of the last layer to improve the final prediction. SSD, like many one-stage detectors, uses anchor boxes to generate precise predictions on one iteration. The anchor boxes are predefined detection boxes and are located in every part of the image. Instead of creating a prediction for the object's location on a blank space, one object can have multiple anchor box candidates that describe its location. In the end, the candidate which describes the object location best is chosen. **YOLOv7:** The YOLO (You Only Look Once) family is a group of object detectors designed for real-time applications. YOLOv7 [4] is one of its newest members and is state-of-the-art when it comes to object detection in images and videos. On many datasets, it scores a high accuracy with fast inference time.

The architecture of YOLOv7 contains an FPN (feature pyramid network) between the backbone and the prediction head. The FPN creates a feature pyramid out of the map created by the backbone. The advantage of these pyramids is that each CNN layer influences every prediction in every layer of the detection head and improves the final detection.

YOLOv7 is based on the YOLOv5 architecture but includes new features. These features are called freebies. These freebies improve the model's accuracy without losing detection speed.

DETR: The DETR (**DE**tection **TR**ansformer) [5] is different from a normal AI-based object detector. For detecting objects the DETR uses a transformer, a network originally designed for sequence prediction. The advantage of transformers for object detection is that they model pairwise interactions between elements in the network. This can be used to constrain outputs to detect different objects. Other models, need pre- or postprocessing to filter out the duplicate detections.

The transformer input is generated by a simple CNN backbone. The generated feature map is mapped to a sequence-like form which is used as an input "sequence" for the DETR. The output of the transformer is then processed by a simple Feed-Forward Network to generate the final detections.

2.2 Dataset

We trained our models on detecting objects in street scenes since our target application is autonomous driving. There are many datasets that contain street scene images. We trained our models on the KITTI [6] dataset since it is a popular dataset which makes comparing our results easier to other research. KITTI contains street scenes from the inner city, rural areas, and highways. There are nine different object classes to detect: car, pedestrian, cyclist, truck, van, tram, person sitting, misc, and don't care. The most occurrent classes are car and pedestrian.

To test the robustness of our trained models we tested them on the DeepDrive [9] dataset as well. The DeepDrive data is more diverse than KITTI. It contains different weather conditions, various day- and nighttimes, light reflections, and in general more variation in its street scenes.

Model	mAP (KITTI)	mAP (COCO)	Inf. Time	Model Size
MobileNetV2-SSD	37.9%	20.2%	34.8ms	33.8MB
YOLOv7	74.13%	51.4%	8.01ms	74.8MB
DETR	72.4%	42.0%	71.34ms	158.9MB

Table 1: Results of the model's performance after training them on the KITTI dataset. The inference time was measured on NVIDIA DGX-2.

2.3 Training

Detecting objects is a combination of two problems: Regression and Classification. The regression problem describes finding an object and classification to identify the corresponding class. Therefore training an object detector is mostly based on optimizing these features.

For training the models we used transfer learning. In transfer learning the weights of a trained model are fine-tuned on another dataset. All models we trained can be found as fully trained models. They are optimized to detect objects of the COCO [10] dataset. By fine-tuning the models instead of training them from scratch, we saved a lot of time and resources.

As training parameters, we used the recommended training settings for each model. All models used different loss functions for optimization, with the same underlying minimization problems (classification and regression). We trained MobileNetV2-SSD with TensorFlow and used Pytorch to train YOLOv7 and DETR.

3 Results and Discussion

3.1 Training Results

After training MobileNetV2-SSD, YOLOv7, and DETR the models performed quite well in the evaluation process. All models show a higher accuracy when trained on KITTI rather than COCO. The results are shown in table 1. This can be due to a less diverse dataset and fewer classes. MobileNetV2-SSD scored an mAP (mean average Precision) of nearly 38% and YOLOv7 and DETR reached over 70%. We tested the model inference time by computing the average computation time the models need for processing one image. The fastest model by far is YOLOv7 which takes 8ms seconds for a prediction. MobileNetV2-SSD takes around 35ms and in last place is DETR with 71ms
The most compact model is the MobileNetV2-SSD which is 34MB small and is quite small compared with YOLOv7 which is more than double the size and DETR which is nearly five times as big.

3.2 Testing - KITTI

Analyzing our models on the test data we can see some patterns and properties of the quality of the predictions. As expected from the evaluation, the overall prediction results

(a) MobileNetV2-SSD

(b) YOLOv7

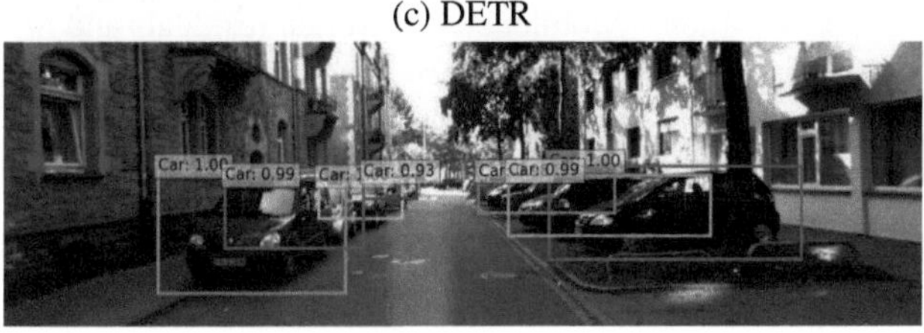

(c) DETR

Figure 1: Test detection for the KITTI test data.

(a) MobileNetV2-SSD

(b) YOLOv7

Figure 2: Test detection of MobileNetV2-SSD and YOLOv7 for the DeepDrive dataset.

look better in YOLOv7 and DETR than in MobileNetV2-SSD. MobileNetV2-SSD often doesn't detect all objects no matter the distance or light conditions. Also, the predicted bounding boxes often don't fit the object. This can be seen in Fig.1a. The class that MobileNets predicts the best is car. It struggles to detect pedestrians, which is after all the second most occurring class.

YOLOv7 performs quite well in detecting most objects and makes good predictions about the class and location of objects that are close as well as objects that are far. When analyzing the test data there is also no clear preference for better detection of one object class seen (besides maybe a higher confidence score).

The detection results of the DETR are nearly as good as the YOLO, but there are a few patterns that stay consistent throughout the test data. DETR also has good predictions of big and small objects in general but sometimes misses smaller ones that YOLOv7 recognizes. An example can be seen in Fig.1c, where objects in the back of the street are not detected. Even if the DETR catches most of the objects, the quality of the location description is often not as accurate as it could be. Mostly the object boxes include the objects but are bigger than required. Sometimes, even if not frequently, the DETR also groups multiple small objects (mostly pedestrians) in one box.

In terms of confidence, the DETR shows on average higher confidence for its predictions than YOLOv7. The confidence of most DETR predictions is often between 90% and 100% while the confidence of YOLOv7 is more of-

ten around 75% to 95%. Both compared we can say that YOLOv7 is better at detecting the objects, while DETR is better at identifying the objects.

3.3   Testing - DeepDrive

After testing the models on KITTI we also tested them on the DeepDrive dataset [9]. This dataset also contains street scenes but with more variation.

In general, we can observe several properties of the prediction we saw before in the KITTI test data.

Similar to the test with the KITTI data, the MobileNetV2-SSD performs worse than YOLOv7 and DETR. Often the model doesn't even detect any object, even on a busy street. Additionally, MobileNetV2 predicts objects where none are present and on reflecting surfaces (also when no learned object is reflected). Mostly these non-existing objects are labeled as cars. Fig.2a shows a good example of this effect.

YOLOv7 still shows good results on street scenes that are similar to those in KITTI but the model is strongly influenced by lighting effects and reflections and it significantly worsens in dark images like images captured during the night. These effects cause the model to often overlook many objects and when it detects them, worsen the quality of the bounding boxes. Fig.2b shows a good example, where YOLOv7 detects the cars left and right, but misses the car with the reflecting headlights.

The DETR seems to be more robust against reflections and variations in the data. Even though KITTI doesn't contain nighttime images, the DETR can detect objects in those DeepDrive images. It also seems not to be influenced as much by reflections and other light effects like YOLOv7. The quality of the bounding box also doesn't seem to worsen as badly compared to the results from KITTI.

In the end, when we compare the better-performing models, we can say that even though YOLOv7 performed better during the training and testing on the KITTI data the DETR is much better at detecting objects of unknown datasets. It is more robust against disturbance factors and has a higher confidence in the predictions.

4 Conclusion

In our work, we selected three different kinds of object detectors that are suited more or less for mobile applications. We trained these models on object detection for autonomous driving. Next, we tested how these models perform on test data. The test data consisted on the one hand of data from the same dataset and on the other we took data from a different dataset to test the flexibility and robustness of the trained models.

In our results, we show that MobileNetV2-SSD is the smallest of all three models but is also the model with the worst predictions. It performs especially badly when used on data that is too different from the training data.

Using a transformer for object detection is quite an innovation since they were initially designed for sequence prediction. In our results, we show that the DETR produced very accurate results and can compete with YOLOv7, which is one of the best object detectors today. DETR also surpasses the performance of YOLOv7 when confronted with more difficult unknown data. However, transformers are quite big and complex models that take a lot of time for computation. The DETR might be a good choice for tasks where these requirements can be neglected but since we are comparing the models for mobile applications, the DETR is a little bit too expensive in resources.

YOLOv7 is a great compromise in detection quality and size. It is (much) smaller than the DETR and performs (much) better than MobileNetV2-SSD. Even if the Deep-Drive test showed that it is vulnerable to light disturbance, YOLOv7 showed better performance in training and testing on KITTI than DETR. Another big plus for YOLOv7 is the detection speed. When used for live detection like moving robot navigation (kind of *mobile* mobile application) a fast computation is a necessity.

In the end, YOLOv7 proves again that it is a great model, suitable for many applications.

Acknowledgement

The work has been carried out at Vincent Jansen, and supervised by Ahmed Mahmoudi at the Institute of Control Engineering, Universität zu Lübeck.

Authors' Statement

Conflict of interest: Authors state no conflict of interest.

5 References

[1] X. Chen, K. Kundu, Z. Zhang, H. Ma, S. Fidler, and R. Urtasun, "Monocular 3d object detection for autonomous driving," in *Proceedings of the IEEE conference on computer vision and pattern recognition*, 2016.

[2] F. Zhang, Q. Li, Y. Ren, H. Xu, Y. Song, and S. Liu, "An expression recognition method on robots based on mobilenet v2-ssd," in *2019 6th International Conference on Systems and Informatics (ICSAI)*, 2019.

[3] M. Sandler, A. Howard, M. Zhu, A. Zhmoginov, and L.-C. Chen, "Mobilenetv2: Inverted residuals and linear bottlenecks," in *Proceedings of the IEEE conference on computer vision and pattern recognition*, 2018.

[4] C.-Y. Wang, A. Bochkovskiy, and H.-Y. M. Liao, "Yolov7: Trainable bag-of-freebies sets new state-of-the-art for real-time object detectors," in *Proceedings of the IEEE/CVF Conference on Computer Vision and Pattern Recognition*, 2023.

[5] N. Carion, F. Massa, G. Synnaeve, N. Usunier, A. Kirillov, and S. Zagoruyko, "End-to-end object detection with transformers," in *European conference on computer vision*, 2020.

[6] A. Geiger, P. Lenz, and R. Urtasun, "Are we ready for autonomous driving? the kitti vision benchmark suite," in *Conference on Computer Vision and Pattern Recognition (CVPR)*, 2012.

[7] Z. Zou, K. Chen, Z. Shi, Y. Guo, and J. Ye, "Object detection in 20 years: A survey," *Proceedings of the IEEE*, 2023.

[8] W. Liu, D. Anguelov, D. Erhan, C. Szegedy, S. Reed, C.-Y. Fu, and A. C. Berg, "Ssd: Single shot multibox detector," in *Computer Vision–ECCV 2016: 14th European Conference, Amsterdam, The Netherlands, October 11–14, 2016, Proceedings, Part I 14*, 2016.

[9] F. Yu, H. Chen, X. Wang, W. Xian, Y. Chen, F. Liu, V. Madhavan, and T. Darrell, "Bdd100k: A diverse driving dataset for heterogeneous multitask learning," in *The IEEE Conference on Computer Vision and Pattern Recognition (CVPR)*, 2020.

[10] T.-Y. Lin, M. Maire, S. Belongie, J. Hays, P. Perona, D. Ramanan, P. Dollár, and C. L. Zitnick, "Microsoft coco: Common objects in context," in *Computer Vision–ECCV 2014: 13th European Conference, Zurich, Switzerland, September 6-12, 2014, Proceedings, Part V 13*, 2014.

Magnetic Particle Imaging, Quantification and Reduction of Artefacts for non-ideal FFL Positions

Maria Angela Qato [1]
[1] Biomedical Engineering, Luebeck University of Applied Sciences, maria.angela.qato@stud.th-luebeck.de

Abstract

In Magnetic Particle Imaging (MPI), artifacts related to the Field-Free Line (FFL) significantly compromise image quality, posing accuracy challenges in biomedical imaging. Precise interpretations rely on improved image quality. This research aims to mitigate FFL-associated artifacts in MPI, enhancing imaging reliability for accurate biomedical diagnoses and research. Our research focuses on mitigating inconsistencies in field-free line positioning. Through simulations of the 1D System Matrix image reconstruction pipeline—encompassing sinogram generation and filtered back projection techniques—we aimed to identify a suitable phantom to rectify these inconsistencies. A phantom featuring a central dot was utilized. Observing shifts in sinogram patterns prompted an investigation into quantifying these shifts and suggesting corrective measures. Addressing these challenges is crucial for producing more reliable images in biomedical contexts. Ultimately, this endeavor seeks to fortify MPI efficiency, ensuring clearer and more reliable images essential for precise medical diagnostics and advancement in biomedical fields.

1 Introduction

Magnetic Particle Imaging represents a groundbreaking leap in medical imaging, departing significantly from traditional Magnetic Resonance Imaging (MRI) and Computer Tomography (CT). It relies on detecting the unique behavior of superparamagnetic iron oxide nanoparticles (SPIO), offering outstanding 3D imaging with impressive spatial and temporal resolutions, and promising real-time in vivo imaging applications [1].

Spatial encoding constitutes the core of MPI, involving the creation and manipulation of an FFL within the Field of View (FOV) through specific magnetic fields, specifically the drive field and the selection field [2].

The selection field is a stable magnetic force responsible for creating a Field-Free Line. Its primary function is precisely establishing the region where particles contribute to the detected signal. In contrast, the drive field initiates the stimulation of superparamagnetic nanoparticles within the sample, creating a dynamic magnetic field, that prompts the nanoparticles to align and realign in response to changes in the magnetic field.

Thus, the FFL is generated at the center of the area being imaged. As the FFL traverses the SPIO, it induces the flipping of these particles. The flipping process emits signals that are subsequently detected by the receiver coils. These signals are then processed to derive their position and intensity information. By associating each signal with a specific position along the FFL, the MPI system encodes this spatial information. The image represents the corresponding spatial distribution of superparamagnetic nanoparticles in the imaging space. This not only facilitates accurate imaging and reconstruction of particle spatial arrangements but also enables spatial encoding [3]-[4]. This encoding technique shares similarities with CT scans using X-rays, allowing for analogous image reconstruction methodologies [1]-[4].

While MPI shows promise in biomedical diagnostics due to its high sensitivity and non-invasive nature, its accuracy and reliability can be compromised by artifacts resulting from deviations in the FFL position. The FFL is expected to be precisely centered and acts as a reference line for imaging accuracy. Deviations from this centered position create artifacts that impact image quality and diagnostic precision. Non-ideal FFL positions pose a challenge, influenced by factors such as system calibration errors, external interferences, or mechanical inaccuracies. This paper aims to quantify the artifacts caused by non-centered FFLs, identifying a suitable phantom that can effectively mitigate and correct the artifacts, thereby contributing to enhanced precision in medical diagnoses.

2 Material and Methods

The project was executed with the Julia Programming language and it concentrates on simulating the MPI process and processing data within MPI environments. It is built upon a framework directly derived from the contributions made by Fabian Mohn [5].

The simulations were carried out by performing the following steps: the 1D system matrix was measured and 1D image reconstruction was then performed. Then the sinogram was generated and the Filtered Backprojection (FBP) algorithm was executed to reconstruct the image.

The different steps are explained in more detail below.

The system matrix in MPI represents the fundamental relationship between the acquired signals and the distribution of magnetic particles within the imaged object, as shown in (1)

$$S \times c = u. \tag{1}$$

This equation demonstrates how the positions of tracer particles, denoted as c, relate to the measured signal u. The computation for each pixel involves evaluating how tracer particles at different positions contribute to the resulting signal [3]-[4].

The process of measuring the system matrix involves conducting controlled experiments using a well-defined phantom with known magnetic properties. These experiments consisted of applying controlled magnetic fields or sequences to the object, capturing the signals received by the MPI system, and assembling a comprehensive dataset.

The fundamental objective of the system matrix is to facilitate the reconstruction of images from the gathered measured data.

Utilizing the system matrix S, the reconstruction of 1D MPI images for signals within a specified angular range is performed.

When reconstructing 1D MPI images, an inverse problem is addressed. The inverse problem involves estimating the distribution of magnetic particles within the imaged object based on the signals received by the MPI system.

Given the measured signals U_Θ for a specific angle Θ, the corresponding 1D image I_Θ is reconstructed using the equation (2)

$$I_\Theta = S^{-1} \cdot U_\Theta, \tag{2}$$

where I_Θ represents the reconstructed 1D image for a particular angle Θ, S^{-1} denotes the inverse of the system matrix S, and U_Θ signifies the measured signals acquired at angle Θ.

The primary purpose of this reconstruction procedure is to deduce and visualize the distribution and arrangement of magnetic particles within the imaged object, focusing on the angle Θ. This mathematical process aims to generate a 1D image that illustrates the distribution of magnetic particles along a specific direction or angle, utilizing both the acquired signals and the system matrix.

Upon obtaining the reconstructed 1D images I_Θ, for various angles Θ, these images are systematically arranged into a sinogram $P(x,\Theta)$. The latter is structured such that each row represents a different position x along the object and each column represents a different angle Θ.

Filtered Backprojection is employed for image reconstruction.

Similar to the procedure in CT, Fig.1 showcases the collection of a lone projection $g(l,\Theta)$, at a specific rotation angle

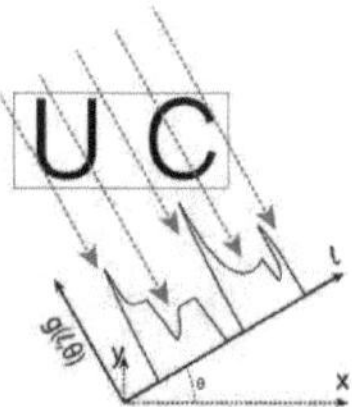

Figure 1: Diagram of an image projection reconstruction system. The magnetic field FFL undergoes rotation at an angle Θ, after which a single projection, denoted as $g(l,\Theta)$, is obtained. Multiple projections are acquired by altering the value of Θ [6].

Θ. Variation in Θ allows the acquisition of multiple projections.

As implied by its name, this technique initially filters projections through a ramp filter. Then, it proceeds to project these filtered projections across the resulting image space corresponding to the acquisition angle Θ. Subsequently, the accumulation of multiple back-projected images is conducted to generate the output image [6].

Simulations were performed by moving the FFL up to 3 mm from the center; to move the FFL, the focus field offset was increased from 0 mT to 15 mT.

The displacement of the FFL caused a discernible shift in the sinogram data. An attempt was then made to estimate the focus field offset directly by analyzing the sinogram and the image reconstruction, without prior knowledge of the applied shift.

In this effort, an analysis centered on identifying the positions of the peak maxima intensity in the ideal sinogram, obtained at a 0 mT focus field offset, so with the FFL centered, was conducted. These positions were also measured for the peak maxima intensity of the sinograms that had been shifted using various offsets. Afterward, the difference between the mean positions identified in the ideal sinogram and those found in the shifted sinograms served as a quantitative measure of the shift magnitude.

The subsequent line presents the mathematical expression. Let P_{Ideal} represent the set of positions of the peak of maxima intensity of the ideal sinogram, $P_{Shifted}$ denote those in the shifted sinogram, and n the total number of peak of maxima intensity, then the shift was given by (3)

$$S_{\text{shift}} = \frac{1}{n} \sum_{i=1}^{n} (P_{\text{Ideal}}[i] - P_{\text{Shifted}}[i]). \tag{3}$$

This mathematical formula provides a way to quantitatively measure the shift of the FFL.

The same procedure was conducted analyzing the image reconstruction obtained from the ideal sinogram and the image reconstruction achieved from the sinogram varying the focus field offset.

Two different phantoms were evaluated to determine the shift caused by varying the focus field offset. Through multiple simulations, it was clear that having phantoms with particles positioned in the middle is crucial. This observation stemmed from noticing a loss of information in the middle at larger focus field offsets.

The phantom selected for this project is shown in Fig.2 . It consists of a single dot centered within the field of view.

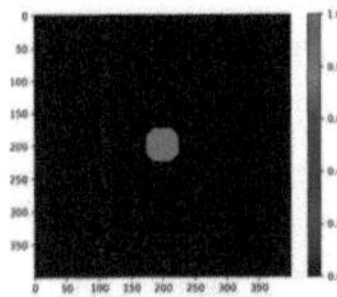

Figure 2: It shows the phantom used for the experiment; the phantom is a dot positioned in the center of the Field of View.

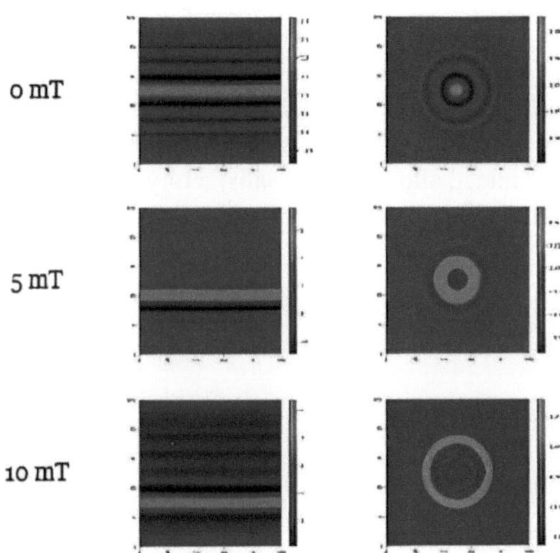

Figure 3: It describes, in the first column, the sinogram evaluated with focus field offset values of 0, 5, and 10 mT, from top to bottom. The second column illustrates the image reconstruction after the Filtered Back Back Projection.

2.1 Experiments

After generating the phantom, several simulations were conducted in *Julia* using custom-authored code. This code facilitated continuous rotation of the FFL, capturing projection angles. The process included measuring the sinogram and ultimately applying the FBP technique.

The simulations involved systematically adjusting the focus field offset, varying in increments of 0.5 mT from 0 mT to 15 mT. This deliberate variation aimed to discern and anticipate any potential artifacts that might emerge.

When the artifact was found to be a displacement inside the sinogram, a code was implemented to rectify the issue. This procedure entailed shifting the entire sinogram upwards by a specific number of pixels corresponding to the measured shift applied in mT. Following this adjustment, an image reconstruction process was carried out.

Efforts were undertaken to address the shift by employing custom-designed code. This code was intended to rectify the observed displacement in the image domain, attributed to the aforementioned artifact. The primary objective was to precisely understand and mitigate the detected shift, enhancing the accuracy and fidelity of the resulting images, and aimed to assess the shift's magnitude without prior knowledge of the initially applied shift, but solely relying on image observation.

The ultimate goal was to refine the software program by understanding and correcting this undetermined shift inferred from the images themselves.

offset, indicating the centering of the field free line. In the second row, the representations depict a 5 mT offset, signifying a 1 mm shift of the field free line. Moving to the third row, the images correspond to a 10 mT focus field offset, with the field free line shifted by 2 mm.

As the focus field offset increases, signifying a greater distance of the field free line from the center, the sinogram descends within the image. As a result, the reconstructed phantom, following the image reconstruction, displays diminished quality and resolution, marked by the presence of ring artifacts and information loss at the center. Despite these challenges, the phantom remains reconstructible, yielding results of reasonable quality.

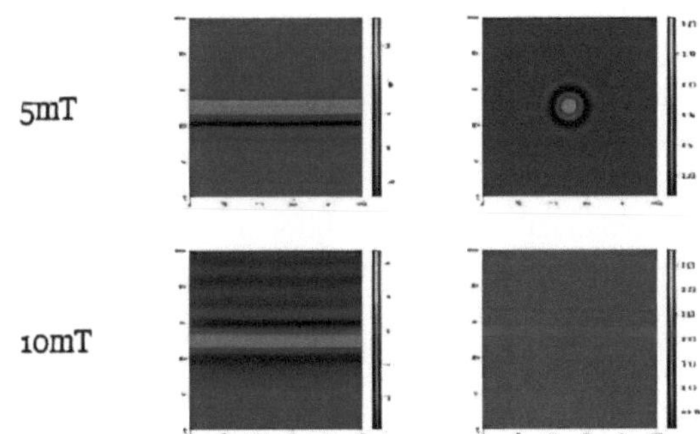

Figure 4: The top images illustrate the adjusted sinogram and its corresponding image reconstruction when applying a 5 mT focus field offset. In the two images below, the same sequence is presented with a 10 mT focus field offset.

3 Results and Discussion

The focus field offset denotes the shift of the FFL and implementing this offset results in artifacts evident in both the sinogram and the subsequent image reconstruction.
In the sinogram, the artifact becomes apparent as a noticeable shift. As illustrated in Fig.3 , the first row displays the sinogram and image reconstruction with a 0 mT focus field

To address the shift shown in Fig.3 , a specialized code was developed to precisely reposition the sinogram, aligning it ideally as if the focus field offset were zero, ensuring the

FFL is centered. As depicted in Fig.4, the sinogram with a 5 mT offset underwent a 1 mm upward shift, while the one with a 10 mT offset experienced a 2 mm upward adjustment. This realignment centers the sinogram, prompting a subsequent re-execution of the Filtered Back Projection. It is demonstrated that with a moderate shift, the sinogram remains intact, allowing for a satisfactory image reconstruction, albeit with some information loss. In our attempts

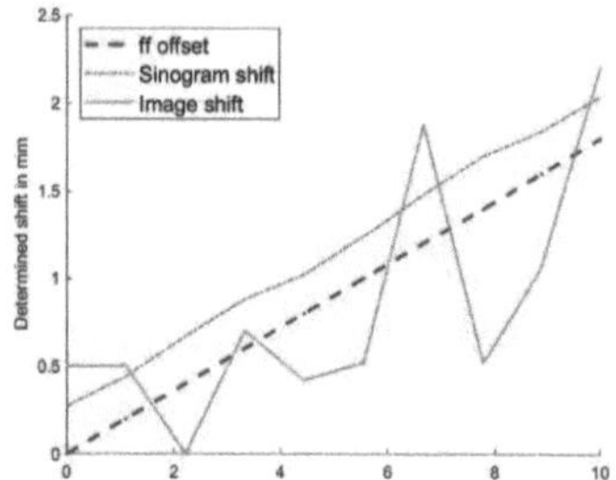

Figure 5: The plot illustrates shift values in mm obtained from sinogram analysis (Sinogram shift) and image reconstruction analysis (Image shift). The line for ff offset represents the expected values. Divergence was observed between anticipated shifts and image-derived shifts. Sinogram analysis aligns closely with expectations.

to refine the process, we explored image processing techniques on both the sinogram and the reconstructed image.

The plotted comparison in Fig.5 reveals that the shift values derived from sinogram analysis (Sinogram shift), represented by the dash-dot line, exhibit a closer alignment with the expected shift values. Parallel patterns suggest that sinogram analysis might yield more accurate results closely resembling anticipated shifts.

Conversely, a substantial disparity between expected values (ff offset) and those derived from the image reconstruction process (Image shift) is evident. In particular, the smearing phenomenon inherent in Filtered Back Projection is a primary factor behind the increased variability seen in the image domain. This inherent smearing effect introduces distortions, broadening the spectrum of features in the reconstructed images and, as a result, contributing to greater errors in the final output. Therefore, while directly identifying and manually correcting shifts in the sinogram proves advantageous, this approach may not be optimally suited for image reconstruction.

Thus, to enhance the alignment between determined and expected shifts, further investigation is recommended and other image manipulation methods must be used.

4   Conclusion

The study evaluated the implications of the Field-Free Line being mispositioned away from its ideal centered location. Significant outcomes were observed in simulations and an analysis of the 1D System Matrix image reconstruction pipeline involving sinogram generation and fil-

tered back projection techniques. A downward shift in the sinogram was identified when the FFL was not centered. Consequently, corrective actions were employed to realign the sinogram, mirroring the expected shift at a focus field offset of zero. This refinement maintains the integrity of the sinogram, enabling a successful image reconstruction, albeit with a slight loss of information. Additionally, the focus field offset quantity was assessed directly from both the sinogram and the reconstructed image. Sinogram analysis suggests accurate results, but a notable disparity with expected values arises during image reconstruction. Directly correcting sinogram shifts is beneficial but may not optimize image reconstruction. Further investigation and other image manipulation methods are recommended for improved alignment between determined and expected shifts.

Acknowledgement

The work has been carried out at Fraunhofer IMTE/ Universität zu Lübeck and supervised by Dr.-Ing. Mandy Ahlborg.

Authors' Statement

Conflict of interest: Authors state no conflict of interest.

5   References

[1] T. Knopp, N. Gdaniec, M. Möddel, *Magnetic particle imaging: from proof of principle to preclinical applications*. Phys Med Biol.;62(14): R124-R178. doi: 10.1088/1361-6560/aa6c99. PMID: 28398219., 2017.

[2] H. Medimagh, P. Weissert, G. Bringout, K. Bente, M. Weber,K. Gräfe, A. Cordes, T.M. Buzug, *Artifacts in field free line magnetic particle imaging in the presence of inhomogeneous and nonlinear magnetic fields*. Current Directions in Biomedical Engineering, vol. 1, no. 1, pp. 245-248, 2015.

[3] G. Jiaa, L. Huangb, Z. Wanga, X. Lianga,Y. Zhanga, Y. Zhanga, *High-resolution MPI with spatially resolved measurement on field free lines*. International Journal on Magnetic Particle Imaging IJMPI, Vol. 8 No. 1, Suppl 1, 2022.

[4] M. Erbe, *Field Free Line Magnetic Particle Imaging*. Springer Vieweg Wiesbaden, 2014.

[5] F.Mohn, T. Knopp, M. Boberg, F. Thieben, P. Szwargulski, M. Graeser, *System Matrix Based Reconstruction for Pulsed Sequences in Magnetic Particle Imaging*. IEEE transactions on medical imaging, 41(7), 1862–1873.

[6] J. J. Konkle, P. W. Goodwill, O. M. Carrasco-Zevallos, S. M. Conolly, *Projection Reconstruction Magnetic Particle Imaging*. IEEE Transactions on Medical Imaging, vol. 32, no. 2, 2013.

Quantifying Spatial Resolution in Magnetic Particle Imaging Through Analysis of FFL Sinogram Data

Bansari Shah [1]
[1] Biomedical Engineering, Luebeck University of Applied Sciences, bansari.shah@stud.th-luebeck.de

Abstract

Magnetic Particle Imaging (MPI) is a cutting-edge medical imaging technology that utilizes superparamagnetic nanoparticles to produce detailed and rapid images of biological structures. By directly detecting the magnetic signals emitted by these nanoparticles, MPI offers a unique approach to imaging, promising enhanced sensitivity and potential advancements in medical diagnostics. Within the rapidly advancing field of medical imaging, MPI is a novel technique that has the potential to completely transform diagnosis. This work explores the complexities of spatial resolution determination with a focus on Sinogram Resolution Quantification for Field Free Line Imaging in MPI. Using a two-dot idealized phantom, the research uses image processing algorithms to achieve effective resolution quantification in sinograms.

1 Introduction

Magnetic Particle Imaging (MPI) is a medical imaging technique that utilizes the magnetic properties of superparamagnetic nanoparticles to create high-resolution, real-time images of biological tissues. Promising as a novel medical imaging technique, MPI is distinguished by its fast acquisition times and high sensitivity via the field-free line (FFL) selection field technique [1]. FFL imaging technique selects a particular line in the magnetic field where particles create strong signals, to produce crisp and detailed images [2].

This paper investigates Sinogram Resolution Quantification for FFL Imaging in MPI. The data used for experimentation is an image phantom with two unique dots. By using the Julia programming language, a thorough validation of the resulting spatial resolution expression through simulations and tests is performed. By delving into sinogram resolution quantification, we aim to uncover whether this specific approach can offer practical advantages [3].

2 Material and Methods

The primary methodology utilizes the FFL selection field method. The Julia programming language is used to implement, analyze, and visualize results efficiently.

2.1 Ideal Phantom

An idealized phantom featuring two distinct dots is utilized to have controlled experimental conditions. This serves as a reproducible representation for spatial resolution analysis, allowing for a focused examination of the imaging system's ability to discern features under optimized conditions [4].

The test image is in 2D with a single pixel set to a high value, while all other pixels are set to zero.

The ideal phantom has a field of view (FOV) of 40 x 40 mm. The image size of the data is set to 400 x 400 pixels (0.1 x 0.1 mm). The phantom consists of two points. The size of points parameter specifies the size of each point in the particle distribution which is 50 x 50 pixels (0.8 x 0.8 mm). However, the size of 10 x 10 pixels (4 x 4 mm) was used throughout to test the robustness of the algorithm. The distance between the two points in the particle distribution was initially set to 100 pixels (0.4 mm). This distance was varied for some tests as described later in the paper. The pivot point of the particle distribution is (175,125). It represents the center of rotation for the phantom.

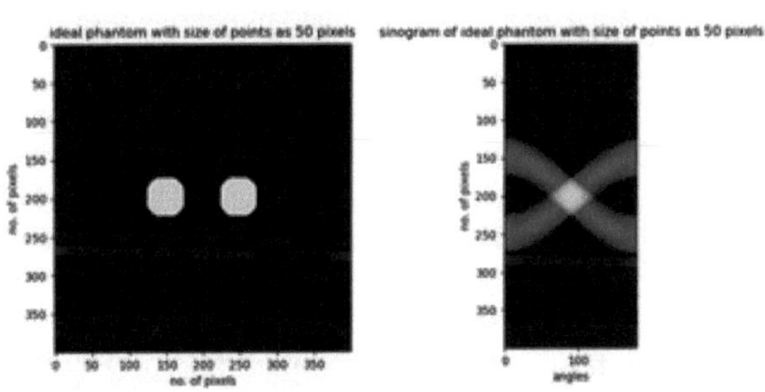

Figure 1: Ideal phantom with two dots and its sinogram

As shown in Fig. 1 Ideal phantom with two dots and its sinogram, was used to determine the spatial resolution of the imaging process. The sinogram output was obtained by applying the radon transform to the phantom.

2.2 Peak Determinations

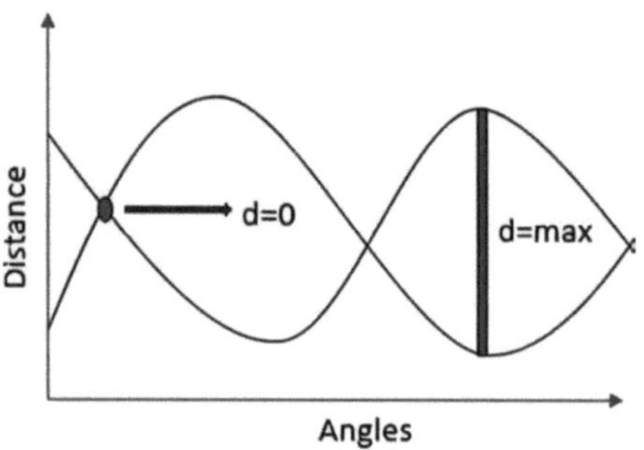

Figure 2: Distance calculation for identifying distances between peaks

The Peaks package [5] is integrated to identify and analyze the two highest local maxima in each column of the sinogram. These peaks represent spatial features within the image. Subsequently, the distances between these peaks are calculated as shown in Figure 2 Distance calculation for identifying distances between peaks providing insights into the system's ability to resolve features at different spatial frequencies. Let y1 and y2 be the positions of the two highest local maxima in a particular column of the sinogram. The distance between these two peaks (d) can be calculated as in (1):

$$d = |y_2 - y_1|. \tag{1}$$

2.3 Experiment with Simulated MPI Data

Simulated MPI data is introduced to augment the analysis. The analysis of peaks and distances is performed similarly as with the ideal phantom, on this simulated data enabling a comprehensive evaluation of the methodology's robustness and applicability under controlled and simulated conditions. The MPI simulation involves a sequence with a duration of 1 second, rotation frequency of 1 Hz, and rotation phase of 0 radians, utilizing the Kolibri Field-Free Line Center Plane system [6] with a measured base system matrix file. The simulation employs 100 threads and a particle distribution created from an image with a point size of 10x10 pixels (0.6 x 0.6 mm), a distance between points ranging from 0 to 200 pixels (0.03 mm), a circular shape, and a field of view of 6x6 mm.

The 1D system matrix is a fundamental and essential element. This matrix is instrumental in depicting the relationship between the excitation field and the resulting particle signal, particularly along offsets that are orthogonal to the Field-Free Line (FFL). In the 1D system matrix, each column signifies a unique location along the FFL or imaging dimension.

3 Results and Discussion

The results obtained from the experiments are discussed here.

3.1 Robustness of the algorithm

The phantom was varied in different ways. Ideally, we had the phantom with a distance of 100 pixels (0.06 mm), with the size of points to be (10,10) pixels. The corresponding unit of size of points in millimeters is (0.6 x 0.6 mm). Then, the phantom was rotated to an angle of 45 degrees. Lastly, the phantom was introduced with an asymmetric distance of 200 mm to observe the results.

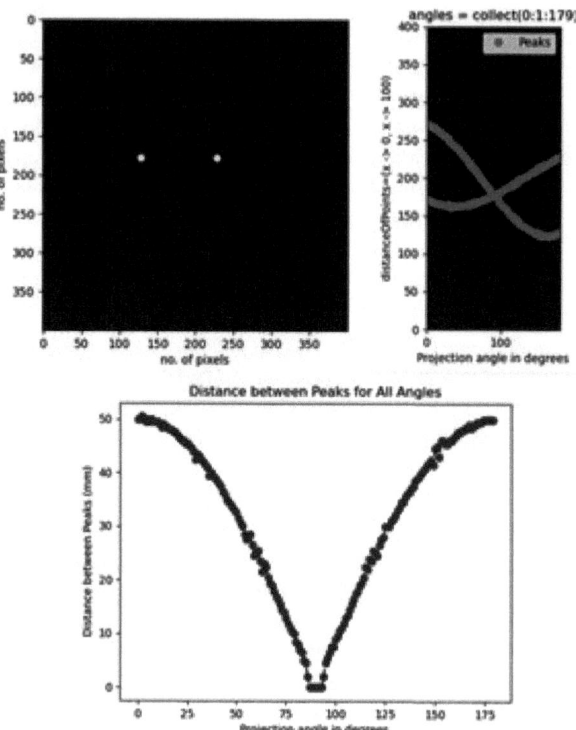

Figure 3: a) Phantom with size (10.,10) pixels. b) Sinogram of the phantom. c) The maximum distance plot between points to verify that it corresponds to the number of pixels between the points in the image space.

Then, the maximum distance between the peaks was calculated using (2), for all the angles and then verified, to correspond to the number of pixels between the points in the image space.

$$d_{\max} = \max_{i} \left(|y_{2i} - y_{1i}| \right) \tag{2}$$

The maximum distance was verified to correspond to the number of pixels between the points in the image space. The known distance between the points was 100 pixels (50 millimeters), and the maximum distance was 101 (50.5 millimeters) pixels at an angle of 2 degrees. Therefore, the maximum distance corresponds to the number of pixels between the points in the image space. The mathematical formula for verification is typically based on checking if the calculated maximum distance is approximately equal to the known distance with a certain tolerance. The formula can be expressed as (3).

$$\text{verification} = |d_{\max} - d| \leq \text{tol.} \tag{3}$$

Where d_{max} is the calculated maximum distance between peaks. d is the known or expected distance. tol is a tolerance value that defines how close the calculated distance should be to the known distance to be considered verified. It was set to 1 mm in this case. A defined angle of 45 degrees then rotated the phantom, and previous steps were repeated to verify that the method does not depend on the rotation angle but just on the distance of the points. The maximum distance was calculated and verified to correspond to the number of pixels between the points in the rotated sinogram.

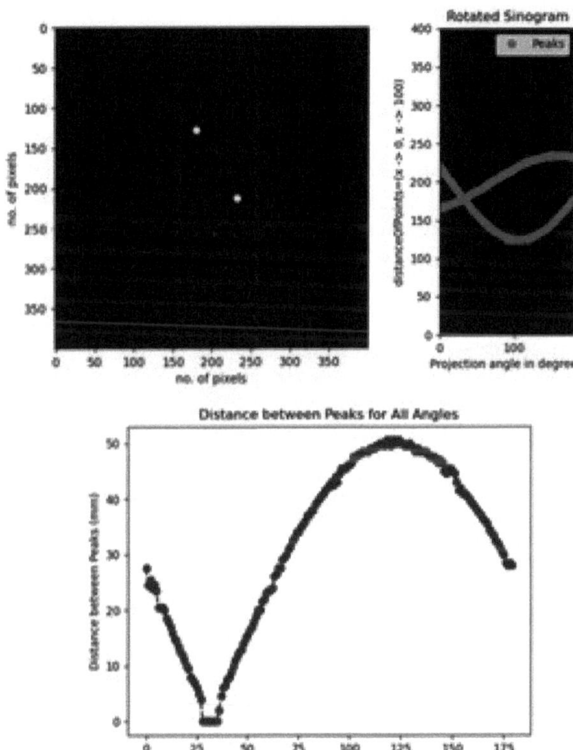

Figure 4: a) Rotated Phantom b) Sinogram of the phantom. c) distance plot to verify that the method does not depend on the rotation angle but just on the distance of the points

Furthermore, the distance between the two points in the particle distribution was varied asymmetrically to 200 mm so that one point was further away from the center than the other, and again, all the previous steps were repeated. The maximum distance (asymmetric) was calculated and verified to correspond to the number of pixels between the points in the asymmetric case.

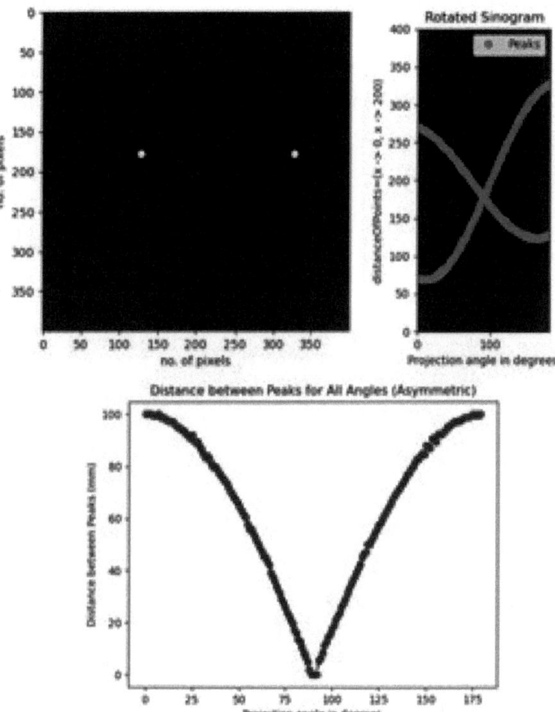

Figure 5: a)Phantom with points far from each other (here, 200 mm). b) Sinogram of the phantom c) Distance plot for asymmetric data

3.2   Real MPI Data

In the simulation data, we are performing a MPI simulation mimicking the Kolibri Field-Free Line (FFL) Center Plane system and a specific sequence. The system matrix is constructed from a measured base system matrix file. The simulation parameters include the sequence duration T of 1 second, rotation frequency ω_{rot} of $2\pi \times 1$ Hertz, and rotation phase ϕ_{rot} of 0 radians. The particle signal is generated based on the defined distribution and sequence. The Fourier domain of the signal is modified to clean frequencies below the excitation frequency in the particle signal.

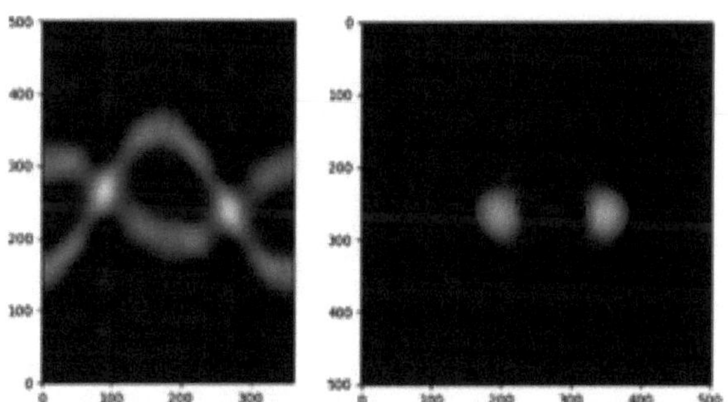

Figure 6: Sinogram and reconstruction of simulated MPI data

The reconstruction aims to produce a clear and detailed representation of the distribution of magnetic particles within the imaged object [7].

3.3 Evaluation of resolution quantification

In Fig. 7 The distance plot for distances from 0 to 100 mm for the ideal phantom, the sinogram's minimum distance between peaks is displayed.

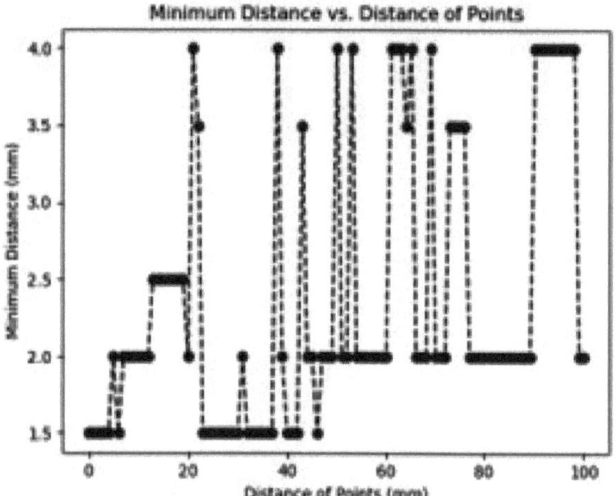

Figure 7: The distance plot for distances from 0 to 100 mm for the ideal phantom. It is to determine the minimum distance between two peaks for different distances of points. The minimum distance obtained from the sinogram peaks is then plotted as a measure of resolution

The variation in the minimum distance values indicates how much the sinusoidal patterns in the sinogram change as the distance between points in the original image is altered. In the context of MPI, resolution often refers to the ability of the imaging system to distinguish two closely spaced magnetic particles. Full Width at Half Maximum (FWHM) is a common metric for spatial resolution, representing the width of the peak at half of its maximum height. It indicates how well the system can separate two adjacent features.

4 Conclusion

The research proposes an expression for the spatial resolution, which includes the distance between the two highest local maxima for all angles, and validates it through simulations and experiments. The method used to determine the spatial resolution is independent of the rotation angle of the phantom. The results show that resolution quantification is feasible directly from sinogram data of MPI devices featuring an FFL. This could lead to more comparable resolution quantifications between different FFL imaging systems.

In the case of real MPI data, the particle properties at 5 T/m depict that the resolution in MPI is influenced by various factors, including the magnetic field gradient, particle properties, and the system matrix. The gradient strength of 5 T/m is specified in my algorithm (gradient = 5). Higher gradient strengths generally lead to better spatial resolution, but this is also influenced by the particle size, the tracer distribution, and the imaging sequence.

Acknowledgement

The work has been carried out at the Fraunhofer IMTE, and supervised by the Universität zu Lübeck.

Authors' Statement

Conflict of interest: The authors state no conflict of interest.

5 References

[1] N. Panagiotopoulos, R.L.Duschka, M. Ahlborg, G. Bringout, C. Debbeler, M. Graeser, C. Kaethner, K. Lüdtke-Buzug, H. Medimagh, J. Stelzner, T.M. Buzug, J. Barkhausen, F.M. Vogt, J. Haegele, *Magnetic particle imaging: current developments and future directions*. Int J Nanomedicine. 2015 Apr 22;10:3097-114. doi: 10.2147/IJN.S70488. PMID: 25960650; PMCID: PMC4411024.

[2] Weizenecker, Jürgen Gleich, Bernhard Borgert, Joern. (2008), *Magnetic particle imaging using a field free line.*. Journal of Physics D: Applied Physics. 41. 105009-. 10.1088/0022 3727/41/10/105009.

[3] Yin, L., Li, W., Du, Y. et al., *Recent developments of the reconstruction in magnetic particle imaging.*. Vis. Comput. Ind. Biomed. Art 5, 24 (2022). https://doi.org/10.1186/s42492-022 00120-5

[4] AnselmvonGladiss, MatthiasGraeser, AileenCordes, AnnaC.Bakenecker, AndréBehrends, XinChen, ThorstenM.Buzug, *International Journal on Magnetic Particle Imaging* . Vol6,No1,ArticleID2003004,11Pages

[5] Allen Hill, MrHenning [Online] Gitlab GitHub - halleysfifthinc/Peaks.jl: Find peaks (local extrema) of signals

[6] [Online] Fraunhofer Research Institution for Individualized and Cell-Based Medical Engineering. [Cited: 02 04, 2024.] https://www.imte.fraunhofer.de/en/Kompetenzfelder/Medizintechnik/Medizinische-Bildgebung-und Elektronik.html

[7] X. Chen, Z. Jiang, X. Han, X. Wang, X. Tang. *The Reconstruction of Magnetic Particle Imaging: Current Approaches Based on the System Matrix*. Diagnostics (Basel). 2021 Apr 26;11(5):773. doi: 10.3390/diagnostics11050773.

Detection of Repetitive Phases in Surgical Training Videos

Felicia Bader [1], Georg Wolf [2], Alexandra Eberenz [2], Georg Maennel [2], Mattias Heinrich [3], Freschta Malekzada [4], Michael Thomaschewski[4], and Jannis Hagenah [2]

[1] Medical Informatics, Universität zu Lübeck, felicia.bader@student.uni-luebeck.de

[2] Fraunhofer Research Institution for Individualized and Cell-Based Medical Engineering IMTE, {georg.wolf, alexandra.eberenz, georg.maennel, jannis.hagenah}@imte.fraunhofer.de

[3] Medical Informatics, Universität zu Lübeck, mattias.heinrich@uni-luebeck.de

[4] Department of Surgery, University Medical Center Schleswig-Holstein Campus Luebeck, {Freschta.Malekzada, Michael.Thomaschewski}@uksh.de

Abstract

Automatic surgical skill assessment received an increasing amount of attention over the past few years. Research so far has focused on generating scores corresponding to whole videos of conducted trials. A score with higher temporal resolution, i.e. assessing different stages of a training video, would enable trainees to understand the cause and effect of given scores. We propose an approach to detect repetitive actions in laparoscopic training exercises as a first step towards higher temporal resolution in surgical skill assessment. Our frame-wise classification approach achieves a decent performance even without temporal information. This is tested on a self-recorded dataset, including three different training exercises out of the Lübecker Toolbox Curriculum. The results show the feasibility of detecting different phases in endoscopic videos and build a foundation for future skill assessments with a higher temporal resolution.

1 Introduction

Over the past years, more robotic interventions are carried out in several medical fields [1]. However, as this development goes on, aspiring surgeons are in need of training to use these new and different technologies in a controlled setting, before practising on a patient [2]. It was shown, that more laparoscopic training reduces the probability of complications in real surgeries [3]. For the most part, this training is assessed manually by experts, which is not only time-consuming and expensive but also subject to bias. Due to this, there are several approaches, to automate this process with the help of deep learning. So far, the manual assessment as well as automatic approaches evaluate a whole training video [4], [5]. However, additional feedback on specific video-phases could enable a better understanding of score-causality and facilitate faster learning curves. For this purpose, we propose an approach to identify repetitive actions as different phases in laparoscopic training videos. This is done through a frame-wise classification without the additional use of temporal information. Our approach is a first step towards surgical skill assessment systems with a higher temporal resolution.

2 Experimental Setup

The following section illustrates the dataset creation, labeling and pre-processing process as well as the used network architecture, training and evaluation routine.

2.1 Dataset Creation

For the purpose of phase detection, existing endoscopic videos out of an ongoing study for laparoscopic training at the Fraunhofer IMTE were used. There, nine probands trained three exercises out of the Lübecker Toolbox Curriculum [6] with a specific time goal. Per participant and exercise, 21-35 training trials were conducted. The exercises carried out were "Pack Your Luggage", "Weaving" and "Chinese Jump Rope", depicted in Fig. 1. All three task are meant to train instrument handling, hand-eye-coordination and bimanual work. In the first task, 18 pegs with two different colors have to be transferred into two wells closed by a lid. The task of the instruments, i.e. grabbing the peg or opening and closing the lid, alternates for every peg. For the second exercise, a 24cm cord has to be placed below and above eight rubber bands in an alternating manner. In the third exercise, five pegs have to be transferred from their position on one side to the opposite one and then back again. These pegs need to be grabbed and placed while holding a rubber string aside with the other instrument. Images to illustrate the three different exercises are given in Fig 1.

For these three exercises, different phases were defined, based on repetitive actions. The first task consists of 18 repetitive actions of placing the peg into the compartment. This resulted in 18 phases, one for each peg. The second exercise consisted of nine phases, one for each time, the cord was transferred below and then above a rubber band.

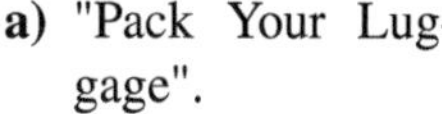

a) "Pack Your Luggage".

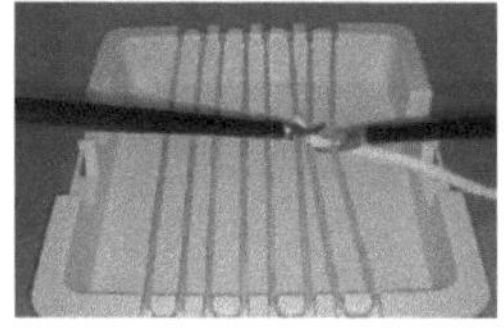

b) "Weaving".

c) "Chinese Jump Rope".

Figure 1: An illustration of the three different training exercises used for the phase detection.

Table 1: The four configurations that were tested.

Configuration	Frozen denseblocks	learning rate decay
1.	1,2 and 3	0
2.	1,2 and 3	0.95
3.	1	0
4.	1	0.95

Table 2: F1-score results for exercise 1

Configuration	Training	Validation	Test
E1.1	0.8835	0.8672	0.7650
E1.2	0.9992	0.9200	0.9464
E1.3	0.9144	0.8983	0.9270
E1.4	1.0	0.9382	**0.9511**

Ten phases were defined for the third exercise, one for each time a peg is transferred to the opposite side. The frames given in Fig. 1 each illustrate the start of a new phase for all three exercises.

For the annotation, a subset of data was selected. Due to the fact, that there should be no significant difference of the phases in videos from various probands, we only used videos from the first four probands. We decided to use 14 videos per exercise for the training, two for validation and four for testing. Per participant and exercise, five videos were chosen, which were sampled uniformly over all trials of this participant. This ensured a high variability considering the skills of the participants in the chosen videos. The Computer Vision Annotation Tool (CVAT) [7] was used to annotate our dataset. Since we needed an efficient annotation of the phases, we looked for the first frame of a phase and set a tag to mark this frame. In a post-processing pipeline, the phase of a tag was transferred to all consecutive frames, until a new tag with a higher phase was found. This way, every frame was annotated with the corresponding phase number.

2.2 Training and Evaluation

A Densenet161 [8], which consists of four denseblocks and was pre-trained on ImageNet [9], was fine-tuned for the proposed classification task.
The densenet was trained task-wise. To fine-tune the pre-trained network, the linear layer was replaced by one that matched the desired output dimension for each task, i.e. the number of phases. During the training, four configurations were tested which can be seen in Table 1. For each configuration the learning rate was set to 0.0001 and the number of epochs to 80. Furthermore, only every 6th frame out of the video was used for training, to avoid very similar images. For all exercises, only the frames from the left side of the endoscope were used. As a loss function,

the Cross-Entropy-Loss was chosen and as an optimizer the Adam Optimizer [10]. Every tested configuration was trained five times. As augmentation, a color jitter, random rotation and random resized crop were used. The color jitter compromised a brightness and contrast change between $[max(0, 0.5), 1.5]$ and a saturation change between $[max(0, 0.7), 1.3]$. The rotation degree was uniformly chosen between $0°$ and $20°$ and the scale for the crop between 0.8 and 1.0. After all frames of a video were classified by the network, the predictions were filtered to improve the accuracy. This was done with a sliding-window technique, where each prediction is set to the majority value of its neighborhood.

3 Results and Discussion

In the following section the results of each exercise are compared and analyzed.

3.1 Exercise 1

As explained in Sec. 2.2, the Densenet161 was fine-tuned in four different configurations. The F1-score for each configuration, averaged over five repetitions, on the training, validation and test dataset can be seen in Table 2. The F1-score on the test dataset was achieved, after the post-processing described in Sec. 2.2. The best results over all where achieved for the last configuration, i.e. when the weights in the first denseblock were frozen and a learning rate decay of 0.95 was utilized. It seems that the value of the learning rate decay had a bigger impact on the score than the amount of frozen denseblocks. Considering the degree of frozen denseblocks, freezing the first, second and third was probably too much, which can be seen in the lower F1-score. In Fig. 2, the confusion matrix on the test dataset after the post-processing is depicted. The filtering of classification outputs yielded an improvement from 0.931 to 0.9511 for the F1-score of the best configuration.

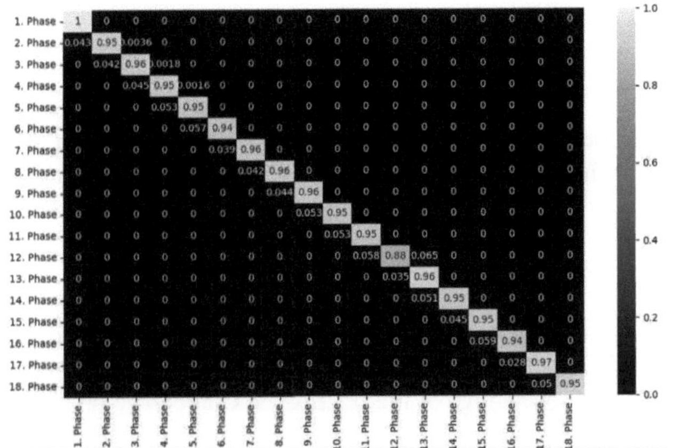

Figure 2: Distribution of prediction and target labels for exercise 1.

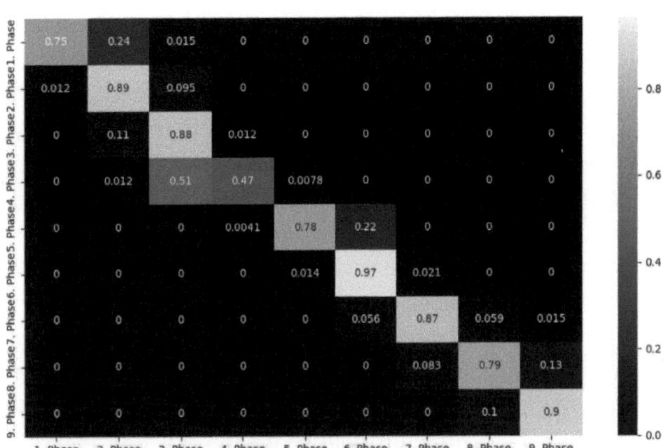

Figure 3: Distribution of prediction and target labels for exercise 2.

Table 3: F1-score results for exercise 2

Configuration	Training	Validation	Test
E2.1	0.8295	0.7754	0.6226
E2.2	0.9958	0.8380	0.7134
E2.3	0.8739	0.7909	0.7076
E2.4	0.9995	0.8667	**0.7989**

Table 4: F1-score results for exercise 3

Configuration	Training	Validation	Test
E3.1	0.9507	0.9201	0.9203
E3.2	0.9999	0.9706	0.9710
E3.3	0.9672	0.9086	0.9049
E3.4	1.0	0.9713	**0.9763**

3.2 Exercise 2

The results for exercise 2 with the four configurations can be seen in Table 3. Again, the specified F1-score is averaged over all five trained networks for each configuration and the F1-score on the test dataset is achieved after the post-processing. The filtering of classification outputs yielded an improvement from 0.7915 to 0.7989 for the F1-score of the best configuration. Just like for exercise 1, the scores are increased when introducing a learning rate decay of 0.95 and with frozen weights in the first denseblock. Fig. 3 illustrates the confusion matrix of the test dataset after post-processing. Overall, the F1-score is not as high as for exercise 1. Especially the miss-classification between phase 3 and phase 4 pulls down the score. One assumption for the lower F1-score is the encountered ambiguity in the labeling, concerning the start and end of each phase. Still, due to the high degree of miss-classifications, one can conclude that not only frames in the border area of a phase are miss-classified, but also during a phase. This means, that so far, this network cannot be used for the phase detection, until the results are substantially improved.

3.3 Exercise 3

For exercise 3, again the four different configurations were tested. All F1-scores can be seen in Table 4. The filtering of classification outputs yielded an improvement from 0.9512 to 0.9763 for the F1-score of the best configuration.
As it was the case for the two previous exercises, the results are improved when a learning rate decay of 0.95 is used and only the weights in the first three denseblocks are frozen. Again, the distribution of prediction and target labels after post-processing can be seen in Fig. 4. The results are the highest for this exercise compared to the others. This was probably due to the fact that the labeling was the easiest and least ambiguous of all the tasks, since the first frame where the rubber band is not touched by an instrument anymore was simple to detect. Thereby, the end of a phase and the start of the next was very well defined.

4 Conclusion

This work assessed the ability of a pre-trained neural network to detect different phases of surgical training exercises, solely based on a frame-wise classification. For exercise 1 and 3 of the Lübecker Toolbox curriculum, the networks achieved decent results and should be capable to detect phases frame-wise in a video with high confidence. Due

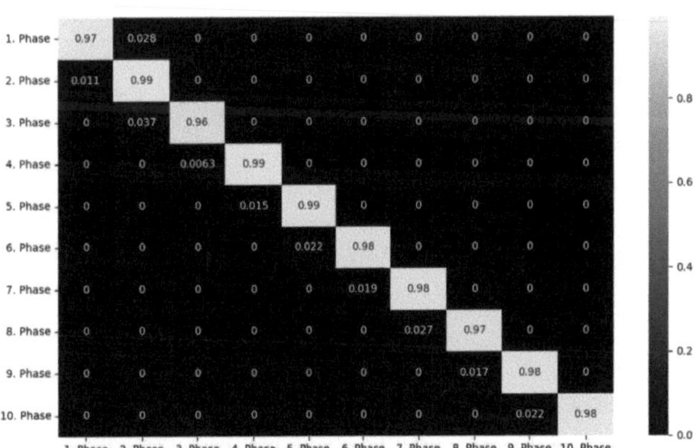

Figure 4: Distribution of prediction and target labels for exercise 3.

to this, these networks can be the foundation for a high temporal resolution feedback system for prospective surgeons, by assessing each phase of an exercise separately. On the other hand, there is reason for improvement of the results of exercise 2. The phase detection there is not reliable enough to be used for an accurate phase detection. A possible approach to increase the accuracy could be, to use a network pre-trained for segmentation or object detection, since these networks could detect more useful features, based on the position of the instruments. It is also possible, that a temporal component is needed for this exercise, to confidently detect the different phases. Another idea would be to use a two-stream architecture with both the left and the right frames of the endoscopic video. It might be, that the results can be further improved for all exercises, when the weights in the denseblocks are fine-tuned together over all networks and only the last linear layer for the classification is trained separately for each task.

In summary, this work showed the possibility of using only a frame-wise classification system without any temporal information to accurately predict phases in an endoscopic training video, thereby enabling a more refined skill assessment in the future.

Acknowledgement

The work has been carried out at the Fraunhofer Research Institution for Individualized and Cell-Based Medical Engineering IMTE and was supervised by the Institute of Medical Informatics, Universität zu Lübeck.

Authors' Statement

Research funding: This work was partially funded by the Euro- pean Union - European Regional Development Fund (ERDF), the Federal Government and Land Schleswig Holstein, Project No. 12420002 and No. 12422005. Conflict of interest: Authors state no conflict of interest. Informed consent and ethical approval: Approved by ethics committee university of Luebeck, ref. No. 2023-448.

5   References

[1] Anderson, J. E., Chang, D. C., Parsons, J. K., Talamini, M. A. (2012). The first national examination of outcomes and trends in robotic surgery in the United States. Journal of the American College of Surgeons, 215(1), 107-114.

[2] Ahmed, K., Khan, R., Mottrie, A., Lovegrove, C., Abaza, R., Ahlawat, R., ... Dasgupta, P. (2015). Development of a standardised training curriculum for robotic surgery: a consensus statement from an international multidisciplinary group of experts. BJU international, 116(1), 93-101.

[3] See, W. A., Cooper, C. S., Fisher, R. J. (1993). Predictors of laparoscopic complications after formal training in laparoscopic surgery. Jama, 270(22), 2689-2692.

[4] Tang, Y., Ni, Z., Zhou, J., Zhang, D., Lu, J., Wu, Y., Zhou, J. (2020). Uncertainty-aware score distribution learning for action quality assessment. In Proceedings of the IEEE/CVF conference on computer vision and pattern recognition (pp. 9839-9848).

[5] Liu, D., Li, Q., Jiang, T., Wang, Y., Miao, R., Shan, F., Li, Z. (2021). Towards unified surgical skill assessment. In Proceedings of the IEEE/CVF Conference on Computer Vision and Pattern Recognition (pp. 9522-9531).

[6] Laubert, T., Esnaashari, H., Auerswald, P., Höfer, A., Thomaschewski, M., Bruch, H. P., ... Benecke, C. (2018). Conception of the Lübeck Toolbox curriculum for basic minimally invasive surgery skills. Langenbeck's Archives of Surgery, 403, 271-278.

[7] CVAT.ai Corporation. (2023). Computer Vision Annotation Tool (CVAT) (Version 2.8.2) [Computer software]. https://github.com/opencv/cvat

[8] Huang, Gao, et al. "Densely connected convolutional networks." Proceedings of the IEEE conference on computer vision and pattern recognition. 2017.

[9] J. Deng, W. Dong, R. Socher, L. -J. Li, Kai Li and Li Fei-Fei, "ImageNet: A large-scale hierarchical image database," 2009 IEEE Conference on Computer Vision and Pattern Recognition, Miami, FL, USA, 2009, pp. 248-255, doi: 10.1109/CVPR.2009.5206848 .

[10] Kingma, D. P., Ba, J. (2014). Adam: A method for stochastic optimization. arXiv preprint arXiv:1412.6980.

Detection and Recognition of Racing Bib Numbers

Leander Busch [1]

[1] Robotics and Autonomous Systems, Universität zu Lübeck, leander.busch@student.uni-luebeck.de

Abstract

For this project a system for the detection and recognition of racing bib numbers from images of marathon runners is implemented. The system consists of two parts: a detection part and a text recognition part. First, a detection network locates all the bib numbers in an input image and predicts bounding boxes around each number. Then, small cut out images of just the bib numbers are created by cropping around the bounding boxes. These small images are forwarded to a recognition network, which outputs the bib number present in each image. The detection network is trained on a custom dataset for bib number detection of 2000 images that is created and labeled for this project. The combined system achieves an F1-score of 0.64 on the Racing Bib Number Recognition (RBNR) [1] dataset and outperforms other methods tested on this dataset.

1 Introduction

The goal of this project is to build a computer vision system that aids in an image management software, where users can search for images containing specific text instances from a big collection of images by an input search query. This project is focused on a subset of text instances: bib numbers, or starting numbers, that are worn by athletes participating in a race. In this scenario a photographer that has taken many pictures of athletes participating in a marathon might want to find all images of a specific participant, identifying them through their bib number. For this, a system that detects and recognizes bib numbers from images needs to be implemented. There are different approaches to bib number detection and recognition. In [1] a face detector and the stroke width transformation are used to generate bib number proposal regions and then detect text in them. The Tesseract OCR engine [2] is used to recognize the detected text. In [3] bib number regions are extracted using a histogram of oriented gradients and a support vector machine. Tesseract OCR engine is applied as well to recognize bib numbers. For this project a two-part system using deep learning is implemented. First, all bib number instances in an image are detected and localized. Then, based on these detections each bib number is recognized and its transcription is output.

2 Material and Methods

2.1 System Architecture

In order to detect and recognize the bib numbers of athletes participating in a race inside an image, the following system architecture is proposed: First, the image is fed into a bib number detection network. The detection network predicts bounding boxes around each detected bib number in the image. Then, for every detected number a small image is created by cropping around the respective bounding boxes. Finally, the cropped images tightly containing a bib number are used as inputs for a text recognition network. The text recognition network receives the cropped images as inputs one by one and outputs the identified number for each input image. The system architecture is illustrated in Fig. 1.

2.2 Dataset for Bib Number Detection

For the detection part of the system, a dedicated bib number detection network is trained. Since availability of such datasets is sparse, a new dataset for training a bib number detection network is created. The basis of the dataset is a collection of images taken of participants from several runs in Germany. From this collection, images are selected in which at least one bib number was in focus and clearly visible. In total the dataset consists of 2000 images that are labeled by hand. For every image bounding boxes are defined around each bib number that is recognizable by the human eye. An example image from the dataset can be seen in Fig. 2.

2.3 Bib Number Detection with YOLOv8

For bib number detection the YOLOv8m network with 26 M parameters is trained on the dataset mentioned in the previous section. YOLOv8 is the latest version of the you only look once (YOLO) models released by ultralytics [4] and is based on YOLOv5. The YOLO series of models are one-stage object detection models, meaning they are models which skip the region proposal stage of two-stage models and run detection directly using learned features from the whole image. YOLOv8 is chosen as a network for this task because it achieves high accuracy for object detection tasks

Figure 1: Illustration of the proposed system architecture for bib number detection and recognition. First, a bib number detector predicts bounding boxes around every number instance. Then, small images containing the numbers are obtained by cropping around these bounding boxes. A text recognition network outputs the numbers in these small images.

Figure 2: An example image of the bib number detection dataset that was created for this project.

while maintaining a lightweight model size. The YOLOv8 is an anchor-free detection model, meaning it directly predicts the bounding box parameters instead of an offset from known anchor boxes. The YOLOv8 network is trained on the custom bib number detection dataset with a batch size of 16. The input images are resized to a dimension of 640×640. For training an early stopping strategy was employed so that the training would stop if the performance on the test set would not improve for 50 epochs. The network is initialized with pre-trained weights on the Common Objects in Context (COCO) [5] dataset to take advantage of transfer learning. For class prediction Binary Cross-Entropy loss is used. For the prediction of the bounding box parameters the Complete IoU loss [6] is used. As the optimizer the Adam algorithm [7] is picked with an initial learning rate of 0.001 and weight decay of 0.0005. The effects of different image augmentation techniques are explored when training the YOLOv8 net. First, a network is trained without additional image augmentation. Then, a version of the YOLOv8 network is trained where all images were converted to grayscale. The intention behind this was to force the network to detect bib numbers based on patterns independently of colours. Additionally, the brightness, contrast and saturation of the grayscale images are randomly altered to introduce as much variation as possible to the grayscale images.

2.4   Text Recognition with SATRN

In order to recognize the bib numbers detected by the first part of the system, the Self-Attention Text Recognition Network (SATRN) [8] is used. SATRN is chosen because of its great performance on popular Scene Text Recogni-

tion datasets. The network is based on the Transformer [9] with some modifications to the Transformer's architecture to make it more suitable for text recognition in scene images. SATRN consists of an encoding and a decoding part. In the encoding part a shallow convolutional network extracts two dimensional features from the input image. The image features are passed to a stack of six encoders where the self-attention mechanism is used to learn how much positions on the feature representation of the image relate to each other. The information learned about the image is then passed to a stack of six decoders. In the decoders self-attention is used to predict the next character of the sequence found on the image based on the previous predicted characters and the visual information provided by the encoders. The SATRN network is trained on a combination of two popular Scene Text Recognition datasets: MJSynth [10] and SynthText [11]. Both are synthetically generated datasets of images containing one English word each. The combined dataset consists of 17 million images. For training the Adam optimizer [7] is used with an initial learning rate of 0.0003 for five epochs with a batch size of 64. Cross-Entropy is used as the loss function. The input images are resized to 32×100 for training and testing. During training a random rotation transformation is applied to the input images, causing a rotation of up to 35 degrees in both directions.

3   Results and Discussion

To assess whether the proposed system and its individual parts perform well for their required tasks they are validated on an external dataset. The RBNR [1] dataset is used as the validation dataset. This dataset consists of 217 images of runners participating in races with labeled bounding boxes around their individual bib numbers. First, the networks used for the detection and recognition parts are evaluated in isolation, then the performance of the combined system is evaluated.

3.1   Bib Number Detection Evaluation

The YOLOv8 networks trained for bib number detection are evaluated on the validation dataset. For this, the average precision and average recall metrics are used. First, the YOLOv8 detector trained without any additional image augmentation is evaluated. This version of the YOLOv8

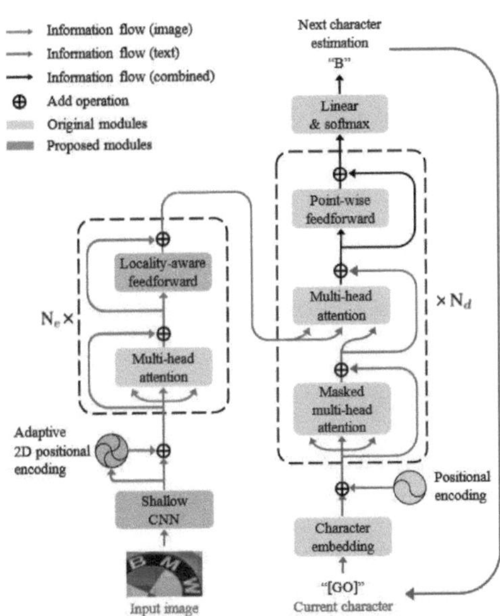

Figure 3: SATRN architecture overview, from [8]. On the left column the encoding part of the network is shown, on the right is the decoding part. The encoders process 2D visual features from a convolutional neural network and forward them to the decoders. The decoders uses the visual information from the encoding part and the embeddings of the previously predicted characters in order to predict the next character of the sequence present in the image.

detector achieves an average precision of 0.6535 and an average recall of 0.6866. The rather low score for precision and recall indicate that the network failed to generalize well on the external dataset. From the detection results it can be inferred that the network struggles in particular with detecting bib numbers with lighter fonts and numbers on darker background. An example of such images is shown in Fig. 4. Then, the YOLOv8 detector trained on grayscale images with additional image augmentation is evaluated. This version of the YOLOv8 detector achieves an average precision of 0.8710 and an average recall of 0.8856. The great increase in performance shows that the image augmentation strategies help the network to reliably detect bib numbers, even those the previous version of the network struggles with. An example of the detection results of this network is shown in Fig. 5.

3.2 Bib Number Recognition Evaluation

To ensure that the SATRN network performs well for the recognition of bib numbers, it is also evaluated on the external validation dataset. For this small images tightly containing bib numbers where created by cropping around every ground truth bounding box of the dataset. Then, for every image created this way the transcription of the bib number present in the image is added as a label. The performance of the SATRN network is evaluated for the recognition of complete bib numbers and for the recognition of individual digits. For complete bib number recognition the metric

Figure 4: Left: The YOLOv8 detector without additional image augmentation does not detect certain bib numbers well, such as numbers with a light red font. Right: When trained on grayscale images and with additional image augmentation, the network is able to detect those numbers.

Figure 5: An example of the detection results for the YOLOv8 detector trained on grayscale images with additional image augmentation. The image shows that this version of the YOLOv8 detector is capable of detecting bib numbers even on cluttered images.

used is accuracy. On the digit level precision and recall are used. The SATRN network achieves a complete bib number accuracy of 0.9379 and a character precision of 0.9689 and a character recall of 0.9779. The results for the different metrics indicate that the SATRN network can recognize bib numbers well provided that the input images tightly contain the bib number and show the whole number.

3.3 Evaluation of Combined System

Next, the performance of the combined system is evaluated on the validation dataset. The system is evaluated based on complete matches between the predicted bib numbers and the ground truth bib numbers for each image. For this the metrics used are precision, recall and F-score. In order to compare the performance of the proposed system with other methods that were evaluated on this dataset, F-score is defined as in [1], [3] with α set to 0.5:

$$F = \frac{\text{precision} \cdot \text{recall}}{\alpha \cdot \text{recall} + (1 - \alpha) \cdot \text{precision}} \qquad (1)$$

Table 1 shows that the proposed system outperforms other methods that were tested for bib number detection and recognition on the RBNR dataset. The lower performance of the complete system compared to the performance of its individual parts is to be expected. Since it is a two-part system, any inaccuracy made in the detection part of the system would likely lead to an error in the following recognition part. An example of such an imperfect detection causing a recognition error is shown in Fig. 6.

Table 1: Performance comparison of bib number detection and recognition systems

Method	Precision	Recall	F-Score
Proposed	0.68	0.60	0.64
RBNR [1]	0.39	0.69	0.50
MMM [3]	0.41	0.72	0.52

Figure 6: An example of an inaccurate detection that causes an incorrect recognition. Since parts of the bib number are cut off by the bounding box, the recognition network is unable to correctly recognize the complete bib number.

4   Conclusion

The proposed system with YOLOv8 trained as a dedicated bib number detector and SATRN as the text recognition network achieves decent performance on an external validation dataset. Considering the possible use case where a photographer wants to find some (but not necessarily all) images of a specific athlete by their bib number from a large collection of images of a race, the proposed system would help find a sufficient amount of those desired images. The bottleneck for the performance of the whole system is clearly the first part, the detection module. In this project, the performance of the YOLOv8 detector is improved significantly through image augmentation. In order to improve the performance of the combined system, improving the detection part even further would be the most efficient.

Acknowledgement

The work has been carried out at Pattern Recognition Company, Lübeck and supervised by Thomas Käster, Institute for Neuro- and Bioinformatics, University of Lübeck.

Authors' Statement

Conflict of interest: Author states no conflict of interest.

5   References

[1] I. Ben-Ami, T. Basha, and S. Avidan, "Racing bib number recognition," 01 2012.

[2] R. Smith, "An overview of the tesseract ocr engine," in *Ninth International Conference on Document Analysis and Recognition (ICDAR 2007)*, vol. 2, 2007, pp. 629–633.

[3] P. Shivakumara, R. Ramachandra, L. Qin, K. Raja, T. Lu, and U. Pal, "A new multi-modal approach to bib number/text detection and recognition in marathon images," *Pattern Recognition*, vol. 67, 08 2017.

[4] G. Jocher, A. Chaurasia, and J. Qiu, "YOLO by Ultralytics," Jan. 2023. [Online]. Available: https://github.com/ultralytics/ultralytics

[5] T. Lin, M. Maire, S. J. Belongie, L. D. Bourdev, R. B. Girshick, J. Hays, P. Perona, D. Ramanan, P. Doll'a r, and C. L. Zitnick, "Microsoft COCO: common objects in context," *CoRR*, vol. abs/1405.0312, 2014. [Online]. Available: http://arxiv.org/abs/1405.0312

[6] Z. Zheng, P. Wang, W. Liu, J. Li, R. Ye, and D. Ren, "Distance-iou loss: Faster and better learning for bounding box regression," 2019.

[7] D. P. Kingma and J. Ba, "Adam: A method for stochastic optimization," 2017.

[8] J. Lee, S. Park, J. Baek, S. J. Oh, S. Kim, and H. Lee, "On recognizing texts of arbitrary shapes with 2d self-attention," 2019.

[9] A. Vaswani, N. Shazeer, N. Parmar, J. Uszkoreit, L. Jones, A. N. Gomez, L. Kaiser, and I. Polosukhin, "Attention is all you need," 2023.

[10] M. Jaderberg, K. Simonyan, A. Vedaldi, and A. Zisserman, "Reading text in the wild with convolutional neural networks," *International Journal of Computer Vision*, vol. 116, no. 1, pp. 1–20, jan 2016.

[11] A. Gupta, A. Vedaldi, and A. Zisserman, "Synthetic data for text localisation in natural images," in *IEEE Conference on Computer Vision and Pattern Recognition*, 2016.

Infinite Science
Publishing